ROTORCRAFT AEROMECHANICS

Rotorcraft is a class of aircraft that uses large-diameter rotating wings to accomplish efficient vertical takeoff and landing. The class encompasses helicopters of numerous configurations (single main rotor and tail rotor, tandem rotors, coaxial rotors), tilting proprotor aircraft, compound helicopters, and many other innovative concepts. Aeromechanics includes much of what the rotorcraft engineer needs: performance, loads, vibration, stability, flight dynamics, and noise. These topics cover many of the key performance attributes and many of the often encountered problems in rotorcraft designs. This comprehensive book presents, in depth, what engineers need to know about modeling rotorcraft aeromechanics. The focus is on analysis, and calculated results are presented to illustrate analysis characteristics and rotor behavior. The first third of the book is an introduction to rotorcraft aerodynamics, blade motion, and performance. The remainder of the book covers advanced topics in rotary-wing aerodynamics and dynamics.

Wayne Johnson worked at the U.S. Army Aeromechanics Laboratory from 1970 to 1981, at the NASA Ames Research Center. He was with NASA from 1981 to 1986, including several years as Assistant Branch Chief. Dr. Johnson founded Johnson Aeronautics in 1986, where he developed rotorcraft software. Since 1998, he has worked at the Aeromechanics Branch of NASA Ames Research Center. Dr. Johnson is the author of the comprehensive analysis CAMRADII and the rotorcraft design code NDARC and of the book *Helicopter Theory* (1980). He is a Fellow of the American Institute of Aeronautics and Astronautics (AIAA) and the American Helicopter Society (AHS) and received a U.S. Army Commander's Award for Civilian Service, NASA Medals for Exceptional Engineering Achievement and Exceptional Technology Achievement, the AHS Grover E. Bell Award, the Ames H. Julian Allen Award, the AIAA Pendray Aerospace Literature Award, and the 2010 AHS Alexander Nikolsky Honorary Lectureship.

Cambridge Aerospace Series

Editors
Wei Shyy
and
Vigor Yang

1. J. M. Rolfe and K. J. Staples (eds.): *Flight Simulation*
2. P. Berlin: *The Geostationary Applications Satellite*
3. M. J. T. Smith: *Aircraft Noise*
4. N. X. Vinh: *Flight Mechanics of High-Performance Aircraft*
5. W. A. Mair and D. L. Birdsall: *Aircraft Performance*
6. M. J. Abzug and E. E. Larrabee: *Airplane Stability and Control*
7. M. J. Sidi: *Spacecraft Dynamics and Control*
8. J. D. Anderson: *A History of Aerodynamics*
9. A. M. Cruise, J. A. Bowles, C. V. Goodall, and T. J. Patrick: *Principles of Space Instrument Design*
10. G. A. Khoury (ed.): *Airship Technology*, Second Edition
11. J. P. Fielding: *Introduction to Aircraft Design*
12. J. G. Leishman: *Principles of Helicopter Aerodynamics*, Second Edition
13. J. Katz and A. Plotkin: *Low-Speed Aerodynamics*, Second Edition
14. M. J. Abzug and E. E. Larrabee: *Airplane Stability and Control: A History of the Technologies that Made Aviation Possible*, Second Edition
15. D. H. Hodges and G. A. Pierce: *Introduction to Structural Dynamics and Aeroelasticity*, Second Edition
16. W. Fehse: *Automatic Rendezvous and Docking of Spacecraft*
17. R. D. Flack: *Fundamentals of Jet Propulsion with Applications*
18. E. A. Baskharone: *Principles of Turbomachinery in Air-Breathing Engines*
19. D. D. Knight: *Numerical Methods for High-Speed Flows*
20. C. A. Wagner, T. Hüttl, and P. Sagaut (eds.): *Large-Eddy Simulation for Acoustics*
21. D. D. Joseph, T. Funada, and J. Wang: *Potential Flows of Viscous and Viscoelastic Fluids*
22. W. Shyy, Y. Lian, H. Liu, J. Tang, and D. Viieru: *Aerodynamics of Low Reynolds Number Flyers*
23. J. H. Saleh: *Analyses for Durability and System Design Lifetime*
24. B. K. Donaldson: *Analysis of Aircraft Structures*, Second Edition
25. C. Segal: *The Scramjet Engine: Processes and Characteristics*
26. J. F. Doyle: *Guided Explorations of the Mechanics of Solids and Structures*
27. A. K. Kundu: *Aircraft Design*
28. M. I. Friswell, J. E. T. Penny, S. D. Garvey, and A. W. Lees: *Dynamics of Rotating Machines*
29. B. A. Conway (ed): *Spacecraft Trajectory Optimization*
30. R. J. Adrian and J. Westerweel: *Particle Image Velocimetry*
31. G. A. Flandro, H. M. McMahon, and R. L. Roach: *Basic Aerodynamics*
32. H. Babinsky and J. K. Harvey: *Shock Wave–Boundary-Layer Interactions*
33. C. K. W. Tam: *Computational Aeroacoustics: A Wave Number Approach*
34. A. Filippone: *Advanced Aircraft Flight Performance*
35. I. Chopra and J. Sirohi: *Smart Structures Theory*
36. W. Johnson: *Rotorcraft Aeromechanics*
37. W. Shyy, H. Aono, C. K. Kang, and H. Liu: *An Introduction to Flapping Wing Aerodynamics*
38. T. C. Lieuwen and V. Yang: *Gas Turbine Engines*

Rotorcraft Aeromechanics

Wayne Johnson
NASA Ames Research Center

CAMBRIDGE UNIVERSITY PRESS

CAMBRIDGE UNIVERSITY PRESS
Cambridge, New York, Melbourne, Madrid, Cape Town,
Singapore, São Paulo, Delhi, Mexico City

Cambridge University Press
32 Avenue of the Americas, New York, NY 10013-2473, USA

www.cambridge.org
Information on this title: www.cambridge.org/9781107028074

© Wayne Johnson 2013

This publication is in copyright. Subject to statutory exception
and to the provisions of relevant collective licensing agreements,
no reproduction of any part may take place without the written
permission of Cambridge University Press.

First published 2013

Printed in the United States of America

A catalog record for this publication is available from the British Library.

ISBN 978-1-107-02807-4 Hardback

Cambridge University Press has no responsibility for the persistence or accuracy of
URLs for external or third-party Internet Web sites referred to in this publication
and does not guarantee that any content on such Web sites is, or will remain, accurate
or appropriate.

Contents

Preface *page* xvii

1 Introduction 1
   1.1 The Helicopter   1
      1.1.1 The Helicopter Rotor   3
      1.1.2 Helicopter Configuration   6
      1.1.3 Helicopter Operation   7
   1.2 Design Trends   8
   1.3 History   14
   1.4 Books   25

2 Notation 27
   2.1 Dimensions   27
   2.2 Nomenclature   27
      2.2.1 Physical Description of the Blade   27
      2.2.2 Blade Aerodynamics   29
      2.2.3 Blade Motion   29
      2.2.4 Rotor Angle-of-Attack and Velocity   30
      2.2.5 Rotor Forces and Power   30
      2.2.6 Rotor Disk Planes   31
   2.3 Other Notation Conventions   31
   2.4 Geometry and Rotations   32
   2.5 Symbols, Subscripts, and Superscripts   33
   2.6 References   38

3 Hover 39
   3.1 Momentum Theory   39
      3.1.1 Actuator Disk   40
      3.1.2 Momentum Theory in Hover   40
      3.1.3 Momentum Theory in Climb   42
   3.2 Hover Power   43
   3.3 Figure of Merit   44
   3.4 Extended Momentum Theory   45

	3.4.1 Rotor in Hover or Climb	46
	3.4.2 Swirl in the Wake	48
3.5	Blade Element Theory	52
	3.5.1 History of Blade Element Theory	52
	3.5.2 Blade Element Theory for Vertical Flight	54
	3.5.3 Combined Blade Element and Momentum Theory	59
3.6	Hover Performance	60
	3.6.1 Scaling with Solidity	61
	3.6.2 Tip Losses	61
	3.6.3 Induced Power due to Nonuniform Inflow	63
	3.6.4 Root Cutout	64
	3.6.5 Blade Mean Lift Coefficient	64
	3.6.6 Equivalent Solidity	65
	3.6.7 The Ideal Rotor	66
	3.6.8 The Optimum Hovering Rotor	67
	3.6.9 Elementary Hover Performance Results	68
3.7	Vortex Theory	70
	3.7.1 Vortex Representation of the Rotor and Wake	73
	3.7.2 Actuator Disk Vortex Theory	74
	3.7.3 Finite Number of Blades	77
3.8	Nonuniform Inflow	83
	3.8.1 Hover Wake Geometry	84
	3.8.2 Hover Performance Results from Free Wake Analysis	86
3.9	Influence of Blade Geometry	87
	3.9.1 Twist and Taper	87
	3.9.2 Blade Tip Shape	90
3.10	References	91

4 Vertical Flight 92

4.1	Induced Power in Vertical Flight	92
	4.1.1 Momentum Theory for Vertical Flight	93
	4.1.2 Flow States of the Rotor in Axial Flight	96
	4.1.3 Induced Velocity Curve	98
4.2	Vortex Ring State	102
4.3	Autorotation in Vertical Descent	105
4.4	Climb in Vertical Flight	109
4.5	Optimum Windmill	111
4.6	Twin Rotor Interference in Hover	111
	4.6.1 Coaxial Rotors	112
	4.6.2 Tandem Rotors	115
4.7	Vertical Drag and Download	117
4.8	Ground Effect	119
4.9	References	121

5 Forward Flight Wake 123

5.1	Momentum Theory in Forward Flight	123
	5.1.1 Rotor Induced Power	123
	5.1.2 Climb, Descent, and Autorotation in Forward Flight	128
	5.1.3 Rotor Loading Distribution	129

5.2	Vortex Theory in Forward Flight		133
	5.2.1	Actuator Disk Results	134
	5.2.2	Induced Velocity Variation in Forward Flight	136
5.3	Twin Rotor Interference in Forward Flight		137
	5.3.1	Tandem and Coaxial Configurations	138
	5.3.2	Side-by-Side Configuration	140
5.4	Ducted Fan		141
5.5	Influence of Ground in Forward Flight		144
	5.5.1	Ground Effect	144
	5.5.2	Ground Vortex	144
5.6	Interference		146
	5.6.1	Rotor-Airframe Interference	146
	5.6.2	Tail Design	147
	5.6.3	Rotor Interference on Horizontal Tail	147
	5.6.4	Pylon and Hub Interference on Tail	148
	5.6.5	Tail Rotor	149
5.7	References		150

6 Forward Flight 152

6.1	The Helicopter Rotor in Forward Flight	152
	6.1.1 Velocity	153
	6.1.2 Blade Motion	154
	6.1.3 Reference Planes	161
6.2	Aerodynamics of Forward Flight	165
6.3	Rotor Aerodynamic Forces	168
6.4	Power in Forward Flight	173
6.5	Rotor Flapping Motion	178
6.6	Linear Inflow Variation	184
6.7	Higher Harmonic Flapping Motion	186
6.8	Reverse Flow	188
6.9	Blade Weight Moment	192
6.10	Compressibility	193
6.11	Reynolds Number	194
6.12	Tip Loss and Root Cutout	195
6.13	Assumptions and Examples	195
6.14	Flap Motion with a Hinge Spring	201
6.15	Flap-Hinge Offset	208
6.16	Hingeless Rotor	212
6.17	Gimballed or Teetering Rotor	213
6.18	Pitch-Flap Coupling	215
6.19	Tail Rotor	219
6.20	Lag Motion	220
6.21	Helicopter Force and Moment Equilibrium	223
6.22	Yawed Flow and Radial Drag	226
6.23	Profile Power	230
6.24	History	237
	6.24.1 The Beginning of Aeromechanics	237
	6.24.2 After Glauert	239
6.25	References	241

7 Performance ... 243

- 7.1 Rotor Performance Estimation — 245
  - 7.1.1 Hover and Vertical Flight Performance — 245
  - 7.1.2 Forward Flight Performance — 246
  - 7.1.3 D/L Formulation — 248
  - 7.1.4 Rotor Lift and Drag — 249
  - 7.1.5 P/T Formulation — 250
  - 7.1.6 Rotorcraft Performance — 251
  - 7.1.7 Performance Charts — 253
- 7.2 Rotorcraft Performance Characteristics — 254
  - 7.2.1 Hover Performance — 254
  - 7.2.2 Power Required in Level Flight — 259
  - 7.2.3 Climb and Descent — 261
  - 7.2.4 Maximum Speed — 262
  - 7.2.5 Ceiling — 263
  - 7.2.6 Range and Endurance — 264
  - 7.2.7 Referred Performance — 266
- 7.3 Performance Metrics — 266
- 7.4 References — 270

8 Design ... 271

- 8.1 Rotor Configuration — 271
- 8.2 Rotorcraft Configuration — 275
- 8.3 Anti-Torque and Tail Rotor — 281
- 8.4 Helicopter Speed Limitations — 283
- 8.5 Autorotation, Landing, and Takeoff — 285
- 8.6 Helicopter Drag — 291
- 8.7 Rotor Blade Airfoils — 294
- 8.8 Rotor Blade Profile Drag — 298
- 8.9 References — 301

9 Wings and Wakes ... 303

- 9.1 Rotor Vortex Wake — 303
- 9.2 Lifting-Line Theory — 306
- 9.3 Perturbation Solution for Lifting-Line Theory — 312
- 9.4 Nonuniform Inflow — 318
- 9.5 Wake Geometry — 326
- 9.6 Examples — 335
- 9.7 Vortex Core — 339
- 9.8 Blade-Vortex Interaction — 345
- 9.9 Vortex Elements — 350
  - 9.9.1 Vortex Line Segment — 351
  - 9.9.2 Vortex Sheet Element — 353
  - 9.9.3 Circular-Arc Vortex Element — 356
- 9.10 History — 359
- 9.11 References — 363

10 Unsteady Aerodynamics 366
10.1 Two-Dimensional Unsteady Airfoil Theory 366
10.2 Lifting-Line Theory and Near Shed Wake 376
10.3 Reverse Flow 380
10.4 Trailing-Edge Flap 381
10.5 Unsteady Airfoil Theory with a Time-Varying Free Stream 382
10.6 Unsteady Airfoil Theory for the Rotary Wing 387
10.7 Two-Dimensional Model for Hovering Rotor 391
10.8 Blade-Vortex Interaction 403
10.9 References 412

11 Actuator Disk 414
11.1 Vortex Theory 414
11.2 Potential Theory 424
11.3 Dynamic Inflow 432
11.4 History 436
11.5 References 439

12 Stall 442
12.1 Dynamic Stall 443
12.2 Rotary-Wing Stall Characteristics 448
12.3 Elementary Stall Criteria 451
12.4 Empirical Dynamic Stall Models 457
12.5 References 460

13 Computational Aerodynamics 462
13.1 Potential Theory 462
13.2 Rotating Coordinate System 464
13.3 Lifting-Surface Theory 466
    13.3.1 Moving Singularity 466
    13.3.2 Fixed Wing 468
    13.3.3 Rotary Wing 469
13.4 Boundary Element Methods 471
    13.4.1 Surface Singularity Representations 471
    13.4.2 Integral Equation 473
    13.4.3 Compressible Flow 474
13.5 Transonic Theory 477
    13.5.1 Small-Disturbance Potential 477
    13.5.2 History 480
13.6 Navier-Stokes Equations 481
    13.6.1 Hover Boundary Conditions 482
    13.6.2 CFD/CSD Coupling 483
13.7 Boundary Layer Equations 485
13.8 Static Stall Delay 487
13.9 References 488

14 Noise 493
14.1 Helicopter Rotor Noise 493

14.2	Rotor Sound Spectrum	496
14.3	Broadband Noise	499
14.4	Rotational Noise	502
	14.4.1 Rotor Pressure Distribution	503
	14.4.2 Hovering Rotor with Steady Loading	505
	14.4.3 Vertical Flight and Steady Loading	510
	14.4.4 Stationary Rotor with Unsteady Loading	511
	14.4.5 Forward Flight and Steady Loading	512
	14.4.6 Forward Flight and Unsteady Loading	514
	14.4.7 Doppler Shift	516
	14.4.8 Thickness Noise	516
14.5	Sound Generated Aerodynamically	518
	14.5.1 Lighthill's Acoustic Analogy	518
	14.5.2 Ffowcs Williams-Hawkings Equation	519
	14.5.3 Kirchhoff Equation	522
	14.5.4 Integral Formulations	522
	14.5.5 Far Field Thickness and Loading Noise	527
	14.5.6 Broadband Noise	532
14.6	Impulsive Noise	535
14.7	Noise Certification	540
14.8	References	542

15 Mathematics of Rotating Systems 545

15.1	Fourier Series	545
15.2	Sum of Harmonics	547
15.3	Harmonic Analysis	548
15.4	Multiblade Coordinates	549
	15.4.1 Transformation of the Degrees of Freedom	549
	15.4.2 Matrix Form	552
	15.4.3 Conversion of the Equations of Motion	553
	15.4.4 Reactionless Mode and Two-Bladed Rotors	557
	15.4.5 History	560
15.5	Eigenvalues and Eigenvectors of the Rotor Motion	562
15.6	Analysis of Linear, Periodic Systems	564
	15.6.1 Linear, Constant Coefficient Equations	566
	15.6.2 Linear, Periodic Coefficient Equations	568
15.7	Solution of the Equations of Motion	573
	15.7.1 Early Methods	573
	15.7.2 Harmonic Analysis	575
	15.7.3 Time Finite Element	576
	15.7.4 Periodic Shooting	577
	15.7.5 Algebraic Equations	578
	15.7.6 Successive Substitution	579
	15.7.7 Newton-Raphson	579
15.8	References	580

16 Blade Motion 582

16.1	Sturm-Liouville Theory	582

16.2	Derivation of Equations of Motion	584
	16.2.1 Integral Newtonian Method	585
	16.2.2 Differential Newtonian Method	585
	16.2.3 Lagrangian Method	586
	16.2.4 Normal Mode Method	586
	16.2.5 Galerkin Method	588
	16.2.6 Rayleigh-Ritz Method	589
	16.2.7 Lumped Parameter and Finite Element Methods	590
16.3	Out-of-Plane Motion	590
	16.3.1 Rigid Flapping	590
	16.3.2 Out-of-Plane Bending	592
	16.3.3 Non-Rotating Frame	596
	16.3.4 Bending Moments	597
16.4	In-Plane Motion	599
	16.4.1 Rigid Flap and Lag	599
	16.4.2 Structural Coupling	602
	16.4.3 In-Plane Bending	603
	16.4.4 In-Plane and Out-of-Plane Bending	604
16.5	Torsional Motion	606
	16.5.1 Rigid Pitch and Flap	606
	16.5.2 Structural Pitch-Flap and Pitch-Lag Coupling	610
	16.5.3 Torsion and Out-of-Plane Bending	613
	16.5.4 Non-Rotating Frame	618
16.6	Hub Reactions	619
	16.6.1 Rotating Loads	619
	16.6.2 Non-Rotating Loads	624
16.7	Shaft Motion	628
16.8	Aerodynamic Loads	633
	16.8.1 Section Aerodynamics	633
	16.8.2 Flap Motion	639
	16.8.3 Flap and Lag Motion	641
	16.8.4 Non-Rotating Frame	644
	16.8.5 Hub Reactions in Rotating Frame	649
	16.8.6 Hub Reactions in Non-Rotating Frame	653
	16.8.7 Shaft Motion	655
	16.8.8 Summary	660
	16.8.9 Large Angles and High Inflow	664
	16.8.10 Pitch and Flap Motion	666
16.9	References	670

17 Beam Theory 671

17.1	Beams and Rotor Blades	671
17.2	Engineering Beam Theory for a Twisted Rotor Blade	672
17.3	Nonlinear Beam Theory	680
	17.3.1 Beam Cross-Section Motion	681
	17.3.2 Extension and Torsion Produced by Bending	685
	17.3.3 Elastic Variables and Shape Functions	685
	17.3.4 Hamilton's Principle	687
	17.3.5 Strain Energy	688

		17.3.6	Extension-Torsion Coupling	694
		17.3.7	Kinetic Energy	694
		17.3.8	Equations of Motion	696
		17.3.9	Structural Loads	697
	17.4	Equations of Motion for Elastic Rotor Blade		699
	17.5	History		703
	17.6	References		706
18	**Dynamics**			710
	18.1	Blade Modal Frequencies		710
	18.2	Rotor Structural Loads		715
	18.3	Vibration		717
	18.4	Vibration Requirements and Vibration Reduction		722
	18.5	Higher Harmonic Control		729
		18.5.1	Control Algorithm	730
		18.5.2	Helicopter Model	731
		18.5.3	Identification	732
		18.5.4	Control	736
		18.5.5	Time-Domain Controllers	737
		18.5.6	Effectiveness of HHC and IBC	740
	18.6	Lag Damper		740
	18.7	References		746
19	**Flap Motion**			749
	19.1	Rotating Frame		749
		19.1.1	Hover Roots	750
		19.1.2	Forward Flight Roots	752
		19.1.3	Hover Transfer Function	758
	19.2	Non-Rotating Frame		758
		19.2.1	Hover Roots and Modes	760
		19.2.2	Hover Transfer Functions	761
	19.3	Low-Frequency Response		766
	19.4	Hub Reactions		769
	19.5	Wake Influence		773
	19.6	Pitch-Flap Coupling and Feedback		782
	19.7	Complex Variable Representation of Motion		783
	19.8	Two-Bladed Rotor		784
	19.9	References		787
20	**Stability**			788
	20.1	Pitch-Flap Flutter		788
		20.1.1	Pitch-Flap Equations	788
		20.1.2	Divergence Instability	790
		20.1.3	Flutter Instability	791
		20.1.4	Shed Wake Influence	795
		20.1.5	Forward Flight	796
		20.1.6	Coupled Blades	796
	20.2	Flap-Lag Dynamics		797
		20.2.1	Flap-Lag Equations	798

		20.2.2	Articulated Rotors	800
		20.2.3	Stability Boundary	802
		20.2.4	Hingeless Rotors	802
		20.2.5	Pitch-Flap and Pitch-Lag Coupling	805
		20.2.6	Blade Stall	808
		20.2.7	Elastic Blade and Flap-Lag-Torsion Stability	808
	20.3	Ground Resonance		810
		20.3.1	Ground Resonance Equations	811
		20.3.2	No-Damping Case	813
		20.3.3	Damping Required for Ground Resonance Stability	818
		20.3.4	Complex Variable Representation of Motion	821
		20.3.5	Two-Bladed Rotor	822
		20.3.6	Air Resonance	827
		20.3.7	Dynamic Inflow	827
		20.3.8	History	828
	20.4	Whirl Flutter		832
		20.4.1	Whirl Flutter Equations	832
		20.4.2	Propeller Whirl Flutter	834
		20.4.3	Tiltrotor Whirl Flutter	836
	20.5	References		841
21	**Flight Dynamics**			844
	21.1	Control		844
	21.2	Aircraft Motion		846
	21.3	Motion and Loads		849
	21.4	Hover Flight Dynamics		853
		21.4.1	Rotor Forces and Moments	853
		21.4.2	Hover Stability Derivatives	856
		21.4.3	Vertical Dynamics	859
		21.4.4	Directional Dynamics	860
		21.4.5	Longitudinal Dynamics	861
		21.4.6	Response to Control and Loop Closures	867
		21.4.7	Lateral Dynamics	872
		21.4.8	Coupled Longitudinal and Lateral Dynamics	874
	21.5	Forward Flight		878
		21.5.1	Forward Flight Stability Derivatives	879
		21.5.2	Longitudinal Dynamics	880
		21.5.3	Short Period Approximation	883
		21.5.4	Lateral-Directional Dynamics	886
	21.6	Static Stability		888
	21.7	Twin Main Rotor Configurations		889
		21.7.1	Tandem Helicopter	890
		21.7.2	Side-by-Side Helicopter or Tiltrotor	894
	21.8	Hingeless Rotor Helicopters		895
	21.9	Control Gyros and Stability Augmentation		895
	21.10	Flying Qualities Specifications		901
		21.10.1	MIL-H-8501A	902
		21.10.2	Handling Qualities Rating	906

	21.10.3	Bandwidth Requirements	907
	21.10.4	ADS-33	909
21.11	References		913

22 Comprehensive Analysis 915
    22.1   References   919

Index   921

Preface

Rotorcraft is a class of aircraft that uses large-diameter rotating wings to accomplish efficient vertical takeoff and landing. The class thus encompasses helicopters of numerous configurations, tilting proprotor aircraft, compound helicopters, and many other innovative concepts.

Defining "aeromechanics" is more difficult. Today's dictionaries do not capture what the term means for the rotorcraft community. The definitions are not broad enough, and they do not reflect the multidisciplinary facet of the word as applied to rotorcraft. In my 2010 Nikolsky Lecture for the American Helicopter Society, I proposed the following definition:

> Aeromechanics: The branch of aeronautical engineering and science dealing with equilibrium, motion, and control of elastic rotorcraft in air.

Aeromechanics covers much of what the rotorcraft engineer needs: performance, loads, vibration, stability, flight dynamics, and noise. These topics cover many of the key performance attributes and many of the often encountered problems in rotorcraft designs.

As with my previous book *Helicopter Theory* (written in 1976, published in 1980 by Princeton University Press, republished in 1994 by Dover Publications), this text is focused on analysis, with only occasional reference to test data to develop arguments or support results, and with nothing at all regarding the techniques of testing in wind tunnels or flight. Calculated results are presented to illustrate analysis characteristics and rotor behavior. Generally these results were obtained using computer programs that I have developed: the rotorcraft comprehensive analysis CAMRAD II and the sizing code NDARC.

I aim to be comprehensive in coverage, presenting in as much depth as possible what engineers need to know about modeling rotorcraft aeromechanics. Although connections to *Helicopter Theory* are apparent throughout this text, many significant advances in the theory have occurred since 1976. I assume the reader has a general knowledge of aeromechanics fields, such as classical dynamics, beam theory, lifting-line theory, and two-dimensional airfoils. I do provide introductory material where needed as a foundation for rotary-wing developments. Unlike *Helicopter Theory*, no attempt is made to be comprehensive in the bibliography. Sources are cited where considered historically important or the development is associated with a specific work and also to direct the reader to expanded coverage of a subject. Several topics conclude with an outline of the history of their theoretical development.

The scope of this text is still a subset of rotorcraft aeromechanics. A shaft-driven helicopter is the primary focus, but other rotorcraft configurations are discussed, and most of the analysis is relevant to all configurations. Based mainly on my experience and interests, important areas such as structures, materials, and propulsion are not covered. The topic of flight dynamics here encompasses the aircraft behavior, but not handling qualities, which would require treatment of the control system and the pilot. Computational fluid dynamics is covered with an emphasis on identifying the unique aspects of the methods as applied to rotary wings. The equations are developed with an emphasis on rotating wings, but nothing is presented regarding solution procedures.

Chapter 1 describes the helicopter rotor and helicopter configurations. Design trends are shown to illustrate the aircraft characteristics. A brief history of helicopter invention is presented. Chapter 2 summarizes the notation used in this book, providing an important overview of the description of a rotor for analysis.

Chapters 3 to 8 constitute an introduction to rotorcraft aerodynamics and performance. Chapters 9 to 14 cover advanced aerodynamic topics, and Chapters 15 to 20 cover advanced topics in dynamics.

Chapter 3 begins the analysis of aeromechanics by considering hover, which is the key to helicopter effectiveness; it presents the first description of momentum theory, blade element theory, and vortex theory. Chapter 4 extends the analysis to vertical flight, both climb and descent. Chapter 5 examines the wake in forward flight in terms of momentum and vortex theories. Aerodynamic interference is covered, including rotor-to-airframe and rotor-to-tail effects. Chapter 6 on the edgewise flight of a rotor is the longest chapter in the book, dealing with blade element theory calculation of rotor forces and power and beginning the analysis of blade motion, particularly flapping. Chapter 7 summarizes performance analysis for the isolated rotor and for the complete aircraft. Chapter 8 discusses rotor and rotorcraft configurations further, as well as special topics related to design.

Chapter 9 deals with wings and wakes, as the start of advanced aerodynamics. Lifting-line theory and nonuniform inflow from a vortex wake are covered, including free wake geometry. Chapter 10 presents unsteady aerodynamic theory, beginning with the classical two-dimensional analysis, and covers special models and problems for the rotary wing. Chapter 11 is on actuator disk models, concluding with dynamic inflow theory. Chapter 12 describes dynamic stall of airfoils, stall of rotor blades, and stall effects on rotor performance and loads. Chapter 13 on computational aerodynamics focuses on the unique aspects of applications to rotorcraft. Chapter 14 deals with the theory of rotor-generated noise.

Chapter 15 introduces advanced dynamics by describing the mathematics of rotating wings, including multiblade coordinates and Floquet theory, and the solution of equations of motion. Chapter 16 on blade motion is a long chapter that derives the equations of motion for blade flap, lag, and pitch degrees of freedom, including hub reactions and shaft motion, and the aerodynamic loads. Solutions of these equations are found in subsequent chapters. Chapter 17 derives linear and nonlinear beam theory for rotor blades. Chapter 18 on rotor dynamics covers blade frequencies, structural loads, vibration, and higher harmonic control. Chapter 19 is on the stability and response of the blade flap motion, which is fundamental to the behavior of helicopter rotors. Chapter 20 solves the equations of motion for several stability problems, including pitch-flap flutter, flap-lag dynamics, ground resonance, and whirl flutter.

Chapter 21 analyzes rotorcraft flight dynamics, including hover and forward flight operation, single main rotor and multi-rotor configurations, control gyros, and flying qualities specifications.

Chapter 22 concludes the book with a discussion of rotorcraft comprehensive analyses.

My thanks go to my wife Juliet for her support. The time I spent on this book was as much hers as mine. I am indebted to Michael P. Scully, William Warmbrodt, Gloria K. Yamauchi, Franklin D. Harris, Anubhav Datta, Christopher Silva, and Gareth D. Padfield for reviewing the draft manuscript. Their numerous suggestions have resulted in a much improved work.

I got into helicopter research through some interesting thesis topics at the Massachusetts Institute of Technology. I stay in the field because I like the multidisciplinary part of aeromechanics and because we have not run out of problems to solve. Since graduating from MIT, I have been associated with NASA and the U.S. Army at Ames Research Center, even during the 12 years I spent working alone as Johnson Aeronautics. My first assignment at Ames was with the 40- by 80-Foot Wind Tunnel branch and the latest is with the Aeromechanics Branch, always doing rotorcraft research. I have enjoyed collaboration with many capable people, both at Ames and around the world. I am fortunate that they are good friends as well as good engineers.

Wayne Johnson
Palo Alto, California
August 2012

1 Introduction

1.1 The Helicopter

The helicopter is an aircraft that uses rotating wings to provide lift, propulsion, and control. Figure 1.1 shows the principal helicopter configurations. The rotor blades rotate about a vertical axis, describing a disk in a horizontal or nearly horizontal plane. Aerodynamic forces are generated by the relative motion of a wing surface with respect to the air. The helicopter with its rotary wings can generate these forces even when the velocity of the vehicle is zero, in contrast to fixed-wing aircraft, which require a translational velocity to sustain flight. The helicopter therefore has the capability of vertical flight, including vertical take-off and landing. The efficient accomplishment of heavier-than-air hover and vertical flight is the fundamental characteristic of the helicopter rotor.

The rotor must supply a thrust force to support the helicopter weight. Efficient vertical flight means a high power loading (ratio of rotor thrust to rotor power required, T/P), because the installed power and fuel consumption of the aircraft are proportional to the power required. For a rotary wing, low disk loading (the ratio of rotor thrust to rotor disk area, T/A) is the key to a high power loading. Conservation of momentum requires that the rotor lift be obtained by accelerating air downward, because corresponding to the lift is an equal and opposite reaction of the rotating wings against the air. Thus the air left in the wake of the rotor possesses kinetic energy that must be supplied by a power source in the aircraft if level flight is to be sustained. This is the induced power, a property of both fixed and rotating wings that constitutes the absolute minimum power required for equilibrium flight. For the rotary wing in hover, the induced power loading is inversely proportional to the square root of the rotor disk loading ($P/T \propto \sqrt{T/A}$). Hence the efficiency of rotor thrust generation increases as the disk loading decreases.

For a given gross weight the induced power is inversely proportional to the rotor radius, and therefore the helicopter is characterized by large diameter rotors. The disk loading characteristic of helicopters is in the range of 5 to 15 lb/ft^2. The small diameter rotating wings found in aeronautics, including propellers and turbofan engines, are used mainly for aircraft propulsion. For such applications a high disk loading is appropriate, since the rotor is operating at high axial velocity (which reduces the induced power) and at a thrust equal to only a fraction of the gross weight. However, the use of high disk loading rotors for direct lift severely compromises the

single main rotor helicopter (UH-1)

single main rotor helicopter (SH-60)

single main rotor helicopter (Bo-105)

tandem helicopter (CH-47D)

compound helicopter (EC X³)

coaxial helicopter (Ka-32)

tiltrotor (XV-15)

tiltrotor (V-22)

Figure 1.1. Rotorcraft configurations; drawings by Eduardo Solis.

vertical flight capability in terms of greater installed power and much reduced hover endurance. The helicopter uses the lowest disk loading of all VTOL (vertical takeoff and landing) aircraft designs and hence has the most efficient vertical flight capability. The helicopter can be defined as an aircraft using large diameter, low disk loading rotary wings to provide the lift for flight.

Because the helicopter must also be capable of translational flight, a means is required to produce a propulsive force to oppose the aircraft drag and rotor drag in forward flight. For low speeds at least, this propulsive force is obtained from the rotor, by tilting the thrust vector forward. The rotor is also the source of the forces and moments on the aircraft that control its position, attitude, and velocity. In a fixed-wing aircraft, the lift, propulsion, and control forces are provided by largely separate aerodynamic surfaces. In the helicopter, all three are provided by the rotor.

Vertical flight capability is not achieved without a cost, which must be weighed against the value of VTOL capability in the desired applications of the aircraft. The task of the engineer is to design an aircraft that accomplishes the required operations in the most effective manner. The price of vertical flight includes a higher power requirement than for fixed-wing aircraft, a factor that influences the purchase price and operating cost. For most configurations, a large transmission is required to deliver the power to the rotor at low speed and high torque. The fact that the rotor is a mechanically complex system increases purchase price and maintenance costs. The rotor is a source of vibration, requiring a vibration alleviation system to avoid increased maintenance costs, passenger discomfort, and pilot fatigue. There are high alternating loads on the rotor, reducing the structural component life and in general resulting in increased maintenance cost. Aircraft noise is an important factor in air transportation, as the primary form of interaction of the aircraft with a large part of society. The helicopter is among the quietest of aircraft (or at least can be), but utilization of VTOL capability often involves operation close to urban areas, leading to stricter noise requirements. All these factors can be overcome to design a highly successful aircraft. The engineering analysis required for that task is the subject of this book.

1.1.1 The Helicopter Rotor

The conventional helicopter rotor consists of two or more identical, equally spaced blades attached to a central hub. The blades are maintained in uniform rotational motion, usually by a shaft torque from the engine. The lift and drag forces on these rotating wings produce the torque, thrust, and other forces and moments of the rotor. The large diameter rotor required for efficient vertical flight and the high aspect ratio blades required for good aerodynamic efficiency of the rotating wing result in blades that are considerably more flexible than high disk loading rotors such as propellers. Consequently, there is substantial elastic motion of the rotor blades in response to the aerodynamic forces in the rotary-wing environment. This motion can produce high stresses in the blades or large moments at the root, which are transmitted through the hub to the helicopter. Attention must therefore be given in the design of the helicopter rotor blades and hub to keeping these loads small. The centrifugal stiffening of the rotating blade results in the motion being predominantly about the blade root. Hence the design task focuses on the configuration of the rotor hub. Figure 1.2 shows some typical helicopter rotor hubs.

teetering rotor (UH-1D)

articulated rotor (AS 355)

articulated rotor (UH-60A)

articulated rotor (AH-64)

hingeless rotor (Bo-105)

hingeless rotor (Lynx)

bearingless rotor (MD 900)

coaxial rotor (Ka-29)

Figure 1.2. Rotor hub configurations; photos courtesy Burkhard Domke.

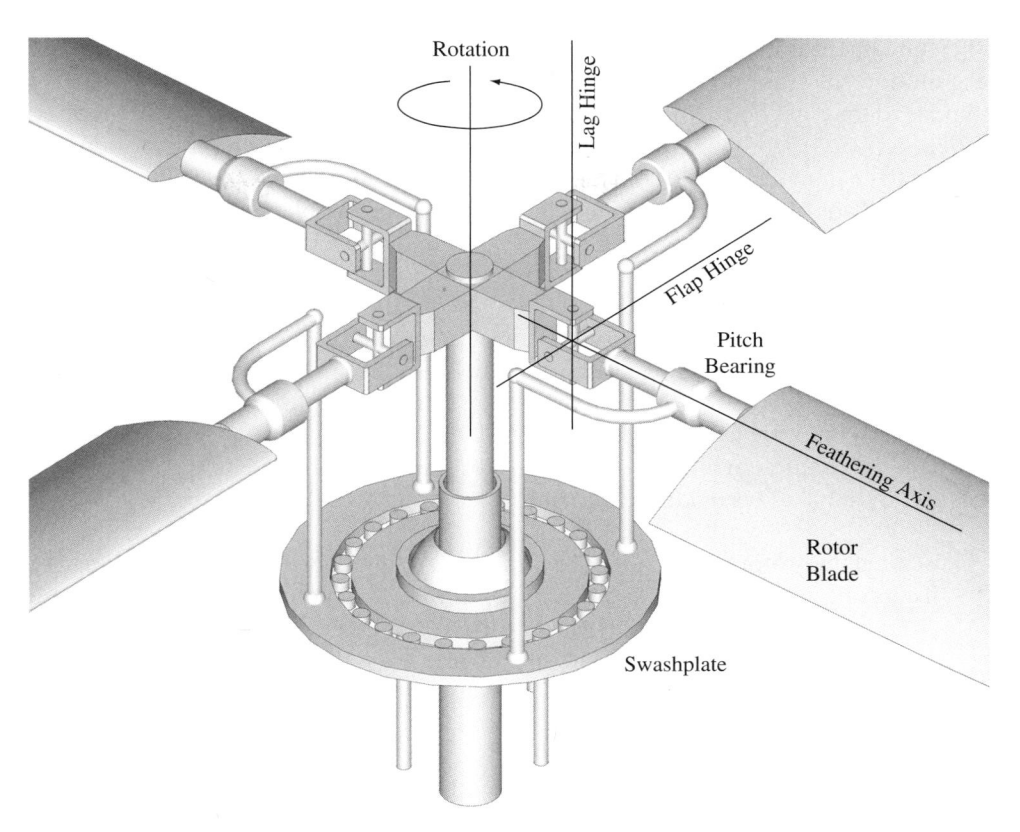

Figure 1.3. Schematic of an articulated rotor hub and root.

A common design solution that was adopted early in the development of the helicopter is to use hinges at the blade root that allow free motion of the blade normal to and in the plane of the disk. A schematic of the root hinge arrangement is given in Figure 1.3. The bending moment is zero at the blade hinge and small throughout the root area, and no hub moment is transmitted through the blade hinges to the helicopter. This configuration makes use of the blade motion to relieve the bending moments that would otherwise arise at the root of the blade. The motion of the blade allowed by these hinges has an important role in the behavior of the rotor and in the analysis of that behavior. Some current rotor designs eliminate the hinges at the root, so that the blade motion involves structural bending. The hub and blade loads are necessarily higher for a hingeless design. The design solution is basically the same, because the blade must be provided with enough flexibility to allow substantial motion or the loads would be unacceptably high even with advanced materials and design technology. Hence blade motion remains a dominant factor in rotor behavior, although the root load and hub moment capability of a hingeless blade has a significant influence on helicopter design and operating characteristics.

The motion of a hinged blade consists basically of rigid-body rotation about each hinge, with restoring moments due to the centrifugal forces acting on the rotating blade. Motion about the hinge lying in the rotor disk plane (and perpendicular to the blade radial direction) produces out-of-plane deflection of the blade and is called flap motion. Motion about the vertical hinge produces deflection of the blade in the plane of the disk and is called lag motion (or lead-lag). For a blade without hinges the fundamental modes of out-of-plane and in-plane bending define the flap and lag motion. Because of the high centrifugal stiffening of the blade these modes are similar to the rigid body rotations of hinged blades, except in the vicinity of the root, where most of the bending takes place. In addition to the flap and lag motion, the ability to change the pitch of the blade is required to control the rotor. Pitch motion allows control of the aerodynamic angle-of-attack of the blade, and hence control of the aerodynamic forces on the rotor. This blade pitch change, called feathering motion, is usually accomplished by movement about a hinge or bearing. The pitch bearing on a hinged or articulated blade is typically outboard of the flap and lag hinges; on a hingeless blade the pitch bearing can be either inboard or outboard of the major flap and lag bending at the root. There are also rotor designs that eliminate the pitch bearings as well as the flap and lag hinges; the pitch motion then occurs about a region of torsional flexibility at the blade root.

The mechanical arrangement of the rotor hub to accommodate the flap and lag motion of the blade provides a fundamental classification of rotor types:

Articulated rotor: The blades are attached to the hub with flap and lag hinges.
Teetering rotor: Two blades forming a continuous structure through the hub are attached to the rotor shaft with a single flap hinge in a teetering or seesaw arrangement. The rotor has no lag hinges.
Gimballed rotor: Three or more blades are attached to the hub without hinges, and the hub is attached to the rotor shaft by a gimbal or universal joint arrangement.
Hingeless rotor: The blades are attached to the hub without flap or lag hinges, although still with a feathering bearing or hinge. The blade is attached to the hub with cantilever root restraint, so that blade motion occurs through bending at the root. This rotor configuration is also called a rigid rotor. The limit of a truly rigid blade, which is so stiff that there is no significant bending motion, is applicable only to high disk loading rotors (such as a propeller).
Bearingless rotor: The blades are attached to the hub without flap or lag hinges and without load-carrying pitch bearings. Flap, lag, and torsion motion occur through deflection of a flexbeam at the root. Pitch control is accomplished using a torque rod or torque tube.

Chapter 8 discusses rotor configurations in more detail.

1.1.2 Helicopter Configuration

The arrangement of the rotors is the most distinctive external feature of a rotorcraft, and is an important factor in the aircraft behavior, particularly the stability and control characteristics. Usually the power is delivered to the rotor through the shaft, accompanied by a torque. The aircraft in steady flight has no net force or moment acting on it, so the rotor torque must be balanced in some manner. The method chosen to accomplish this torque balance is the primary determinant of the helicopter configuration. Two methods are in general use: a configuration with a single main rotor and a tail rotor, and configurations with twin contra-rotating rotors. Chapter 8 has a further discussion of rotorcraft configurations.

The single main rotor and tail rotor configuration uses a small auxiliary rotor to provide the torque balance and yaw control. This rotor is on the tail boom, typically slightly beyond the edge of the main rotor disk. The tail rotor is normally vertical, with the shaft horizontal and parallel to the helicopter lateral axis. The torque balance is produced by the tail rotor thrust acting on an arm about the main rotor shaft. The main rotor provides lift, propulsive force, and roll, pitch, and vertical control for this configuration. The tail rotor provides yaw control.

A twin main rotor configuration uses two contra-rotating rotors, of equal size and loading, so that the torques of the rotors are equal and opposing. There is then no net yaw moment on the helicopter due to the main rotors. This configuration automatically balances the main rotor torque without requiring a power-absorbing auxiliary rotor. The rotor-rotor and rotor-airframe aerodynamic interference typically absorbs about the same amount of power as a tail rotor. The most common twin rotor arrangement is the tandem helicopter configuration: fore and aft placement of the main rotors on the fuselage, usually with significant overlap of the rotor disks and the rear rotor raised vertically above the front rotor. Coaxial and side-by-side twin rotor arrangements have also found some application.

1.1.3 Helicopter Operation

Operation in vertical flight, with no translational velocity, is the particular role for which the helicopter is designed. Operation with no velocity at all relative to the air, either vertical or translational, is called hover. Lift and control in hovering flight are maintained by rotation of the wings to provide aerodynamic forces on the rotor blades. General vertical flight involves climb or descent with the rotor horizontal, and hence with purely axial flow through the rotor disk. A useful aircraft must be capable of translational flight as well. The helicopter accomplishes forward flight by keeping the rotor nearly horizontal, so that the rotor disk sees a relative velocity in its own plane in addition to the rotational velocity of the blades. The rotor continues to provide lift and control for the aircraft. It also provides the propulsive force to sustain forward flight, by means of a small forward tilt of the rotor thrust.

Safe operation after a loss of power is required of any successful aircraft. The fixed-wing aircraft can maintain lift and control in power-off flight, descending in a glide at a shallow angle. Rotary-wing aircraft also have the capability of sustaining lift and control after a loss of power. Power-off descent of the helicopter is called autorotation. The rotor continues to turn and provide lift and control. The power required by the rotor is taken from the air flow provided by the aircraft descent. The procedure on recognition of a loss of power is to set the controls as required for autorotative descent and to establish equilibrium flight at the minimum descent rate. Then near the ground the helicopter is flared, using the rotor stored kinetic energy of rotation to eliminate the vertical and translational velocity just before touchdown. The helicopter rotor in vertical power-off descent is nearly as effective as a parachute of the same diameter as the rotor disk; about half that descent rate is achievable in forward flight.

A rotary-wing aircraft called the autogyro uses autorotation as the normal working state of the rotor. In the helicopter, power is supplied directly to the rotor, and the rotor provides propulsive force as well as lift. In the autogyro, no power or shaft torque is supplied to the rotor. The power and propulsive force required to sustain

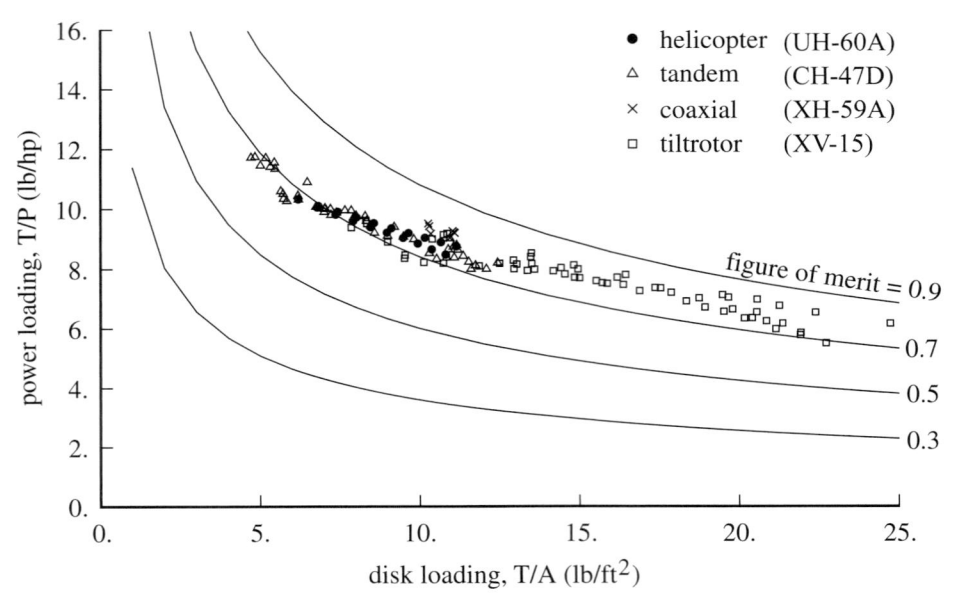

Figure 1.4. Rotor hover performance: rotor power loading as a function of rotor disk loading (based on projected disk area) for four rotors.

level forward flight are supplied by a propeller or other propulsion device. Hence the autogyro is like a fixed-wing aircraft, because the rotor takes the role of the wing in providing only lift for the vehicle, not propulsion. Sometimes the aircraft control forces and moments are supplied by fixed aerodynamic surfaces as in the airplane, but obtaining the control from the rotor is better. The rotor performs much like a wing and has a fairly good lift-to-drag ratio. Although rotor performance is not as good as that of a fixed wing, the rotor is capable of providing lift and control at much lower speeds. Hence the autogyro is capable of flight speeds much slower than fixed-wing aircraft. Without power to the rotor, it is not capable of actual hover or vertical climb. Because autogyro performance is not that much better than the performance of an airplane with a low wing loading, usually the requirement of actual VTOL capability is necessary to justify a rotor on an aircraft.

1.2 Design Trends

The power of a hovering rotor is dominated by the power resulting from the wing drag due to lift, which is called induced power. The induced power is the product of the thrust T and induced velocity v_i: $P_i = Tv_i$ (see Chapter 2 for a complete description of notation). Dimensional analysis shows that the induced velocity scales with the disk loading: $v_i = \kappa v_h$, where $v_h^2 = T/2\rho A$. With the factor of 2, v_h is the ideal (momentum theory) induced velocity for hover. In hover the induced power is typically 10–15% larger than ideal, so $\kappa = 1.10$–1.15. Hence $P_i/T = \kappa\sqrt{T/2\rho A}$, and low power requires low disk loading, thus large disk area and a large rotor diameter. The rotor also has a power loss due to the wing viscous drag, called profile power P_o. A figure of merit for a hovering rotor is defined as the ratio of the ideal induced power to the total rotor power: $M = T\sqrt{T/2\rho A}/P$. Because of profile and non-ideal induced losses, this figure of merit must be less than 1. Figure 1.4 plots for

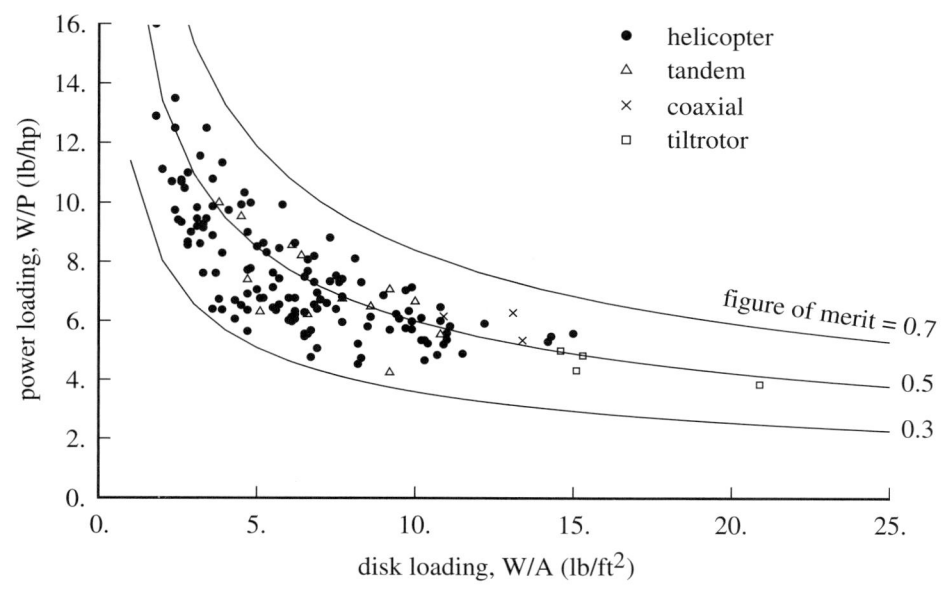

Figure 1.5. Rotorcraft design: power loading of various aircraft as a function of disk loading; from takeoff power, maximum takeoff weight, and projected disk area.

several rotors the measured power loading as a function of disk loading (based on projected disk area), as the rotor thrust is varied. The peak figure of merit is typically $M = 0.74$–0.77 for helicopters. Tiltrotors achieve a somewhat larger figure of merit, $M = 0.78$–0.81, partly due to higher twist but primarily the result of operating at larger disk loading (and thus lower power loading). Mutual interference between coaxial rotors increases the figure of merit (for M based on the projected disk area).

The hover power required dominates the determination of installed power for a rotorcraft, as illustrated in Figure 1.5. Each point in Figure 1.5 is for a single aircraft, calculated from the takeoff power and the maximum takeoff weight (and sea-level-standard air density). The aircraft figure of merit in these terms ranges from 0.40 to 0.55, smaller than the figure of merit of the rotor alone because of other power losses (such as accessory, tail rotor, and interference losses) and the need to operate over a range of flight conditions. As an aircraft metric, the large range of figure of merit values reflects design choices more than rotor performance. In addition to low disk loading, efficient hover operation depends on low propulsion system weight, including rotor, transmission, engine, and fuel weight. At constant disk loading, the propulsion system weight would increase faster than the gross weight. Thus for good empty weight fraction, the design choice for disk loading tends to increase with rotorcraft size, as shown in Figure 1.6. The power loading tends to decrease with size (Figure 1.7). The trend lines shown are $W/A = 0.15 W^{0.4}$ and $W/P = 18 W^{-0.1}$, so with an increase in rotorcraft size the rotor diameter increases less and the installed power increases more than would be implied by constant disk loading.

The induced power of the rotor decreases with forward speed. Viewed as a circular wing, the induced power of the edgewise-moving rotor at speed V is $P_i/T = \kappa T/2\rho A V$. For a circular wing with uniform disk loading, hence elliptical span loading, $\kappa = 1$. For a helicopter rotor the loading is far from uniform, because of the asymmetry of the rotor blade aerodynamic environment, and at high speed κ

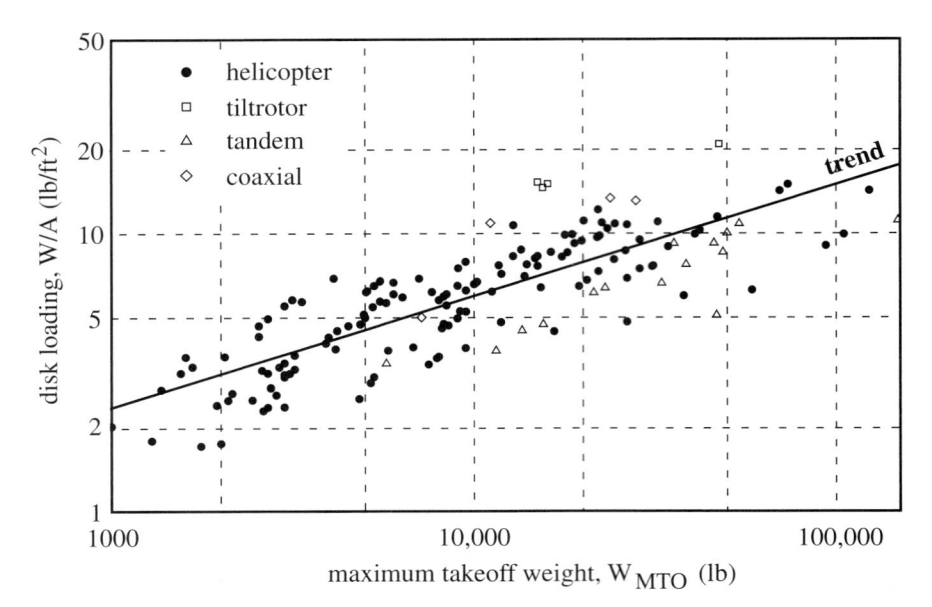

Figure 1.6. Disk loading as a function of maximum takeoff weight.

is much greater than 1, and the induced power can even increase with speed. Profile power P_o increases with speed, especially when that asymmetry leads to significant areas of high Mach number or stall on the rotor disk. The helicopter in forward flight has a parasite drag $D = \frac{1}{2}\rho V^2 f$ that must be balanced by a propulsive force from the rotor. The rotor power required to provide the propulsive force, $P_p = DV$, is called parasite power. Eventually the parasite power dominates the power required. If the installed power is determined by hover, then the maximum flight speed is fallout and is largely determined by the aircraft drag. Figure 1.8 shows the drag area

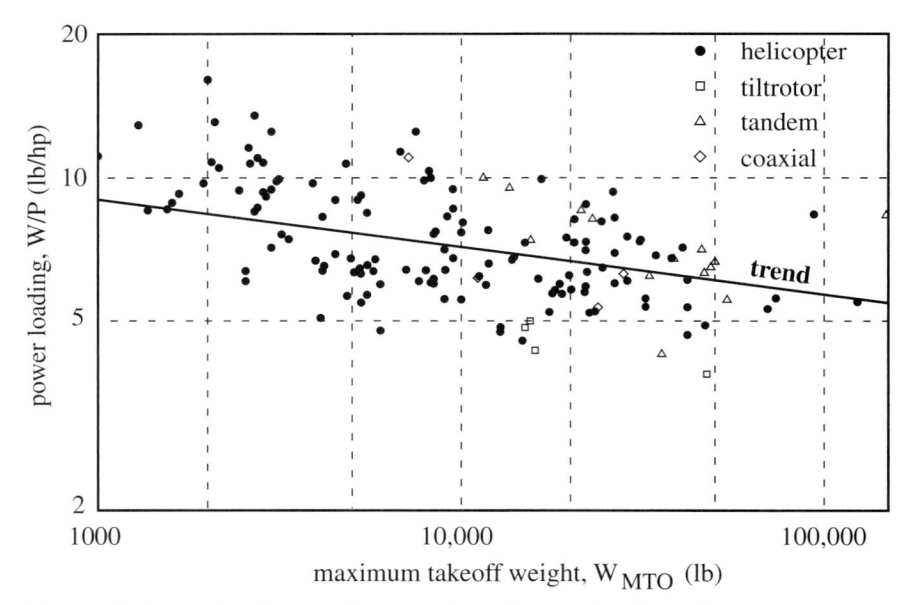

Figure 1.7. Power loading as a function of maximum takeoff weight.

1.2 Design Trends

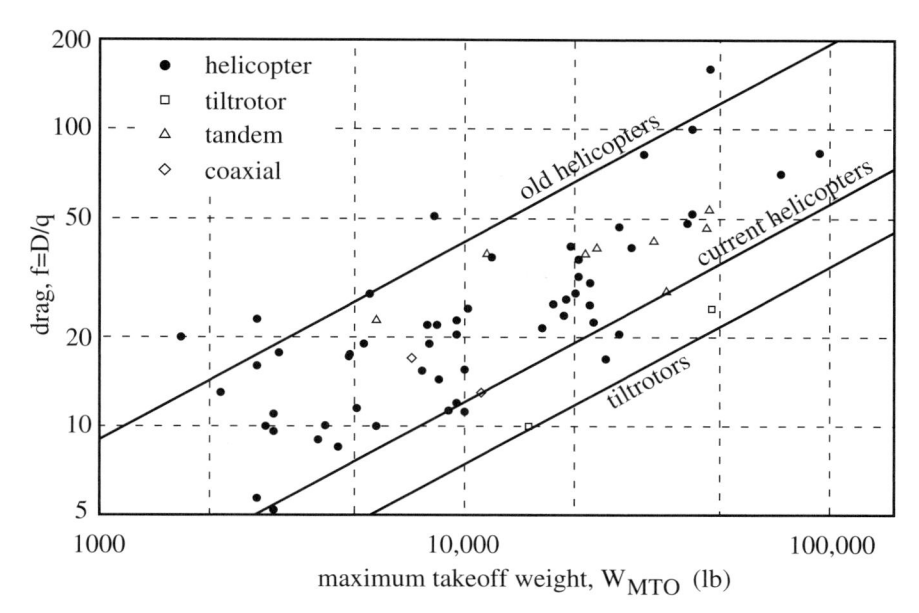

Figure 1.8. Aircraft drag as a function of maximum takeoff weight.

$f = D/q$ as a function of takeoff weight for various helicopters and tiltrotors. The drag area tends to scale with the $2/3$ power of the weight. With operations dominated by low speed conditions, or maximum speed limited by rotor behavior, low aircraft drag has often not been a priority in helicopter design. Thus the drag of current helicopters is twice that of typical turboprop aircraft.

Considered as a circular wing, the lifting rotor in edgewise flight has an effective drag of $D_e = (P - P_p)/V = (P_i + P_o)/V$. The effective lift-to-drag ratio L/D_e of the rotor is a measure of the rotor efficiency in forward flight. A measure of aircraft efficiency is the lift-to-drag ratio $L/D = WV/P$. The range is obtained from this lift-to-drag ratio, together with the specific fuel consumption and the fuel weight fraction. Figure 1.9 plots for several aircraft the lift-to-drag ratio $L/D = WV/P$ as a function of flight speed. The data are from flight test measurements of rotor power (plus propeller power for the compound, but not tail rotor power). Also shown is the rotor effective lift-to-drag ratio L/D_e for a typical helicopter rotor (measured in a wind tunnel). Figure 1.10 shows $L/D = WV/P$ for various aircraft, calculated from the takeoff weight, cruise speed, and installed power. Each point in Figure 1.10 is for a single aircraft. As an aircraft metric, $L/D = WV/P$ reflects design requirements as well as aerodynamic, structural, and propulsive efficiency.

Evident in Figure 1.10 is the fact that a helicopter is not a high-speed machine, largely because of the asymmetric aerodynamic environment of the edgewise-flying rotor. Integrating the blade section lift gives the rotor thrust, and the section lift in hover can be written in terms of the section lift coefficient: $T = \int NL\,dr$, $L = \frac{1}{2}\rho(\Omega r)^2 cc_\ell$. So the mean lift coefficient of the blade is $\bar{c}_\ell = 6C_T/\sigma$ for the hovering rotor, where $C_T = T/\rho A(\Omega R)^2$ is the thrust coefficient and the solidity σ is the ratio of total blade area to rotor disk area. The asymmetry of the rotor aerodynamics is governed by the advance ratio μ, which is the ratio of the flight speed and the blade tip speed. In forward flight the velocity of the blade relative to the air is increased on the advancing side and reduced on the retreating side. Consequently

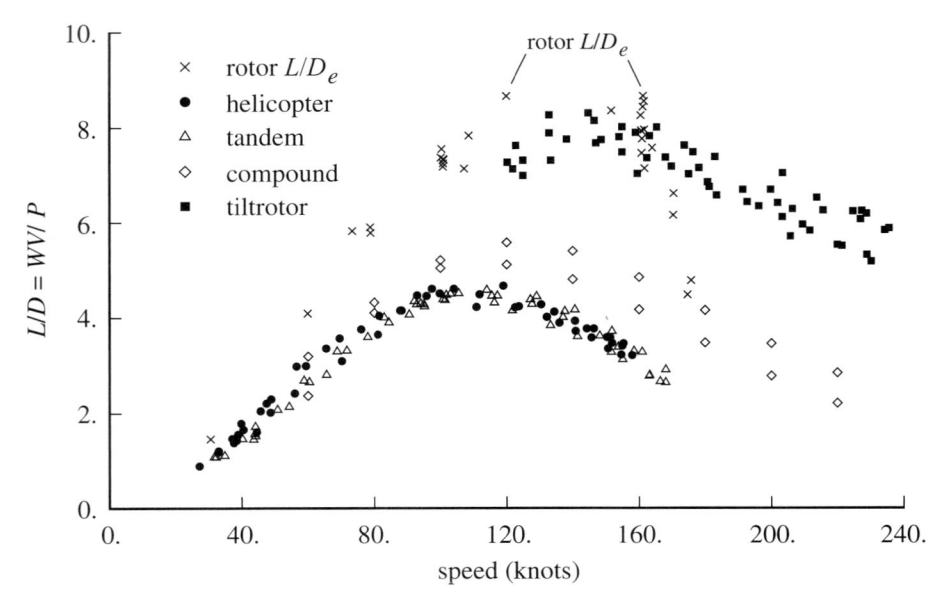

Figure 1.9. Rotor forward flight performance: lift-to-drag ratio $L/D = WV/P$ (from power required and thrust) as a function of flight speed for four rotorcraft: UH-60A, CH-47D, AH-56A, and XV-15.

with increasing μ, the Mach number on the advancing tip (M_{at}) increases, while the lift coefficient increases on the retreating side (especially if the rotor must maintain roll moment balance). Stall limits the maximum blade section lift coefficient. Figure 1.11 shows for various rotorcraft the design choices of rotor blade loading C_T/σ (a measure of blade lift coefficient) and maximum advance ratio μ (a measure of

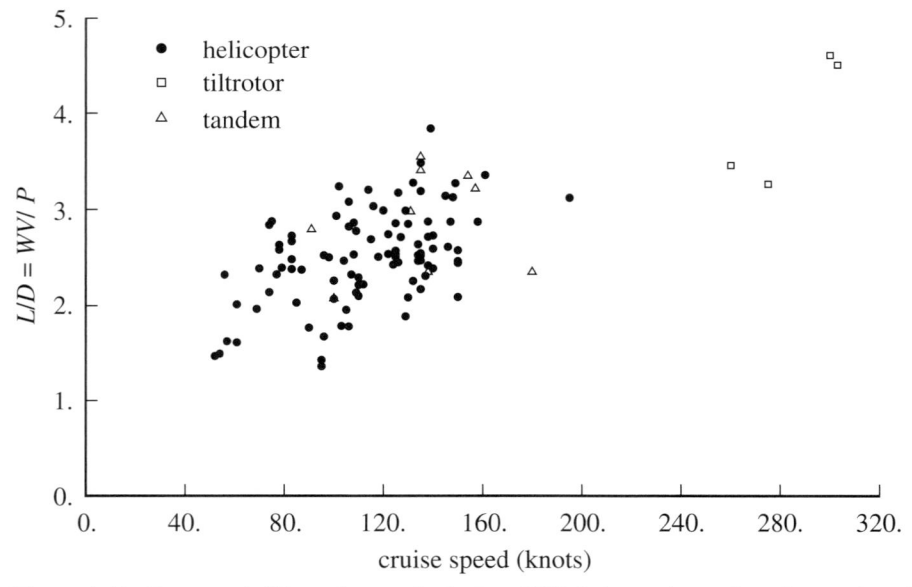

Figure 1.10. Rotorcraft lift-to-drag ratio $L/D = WV/P$ for various aircraft as a function of cruise speed; from installed power, takeoff weight, and cruise speed.

1.2 Design Trends

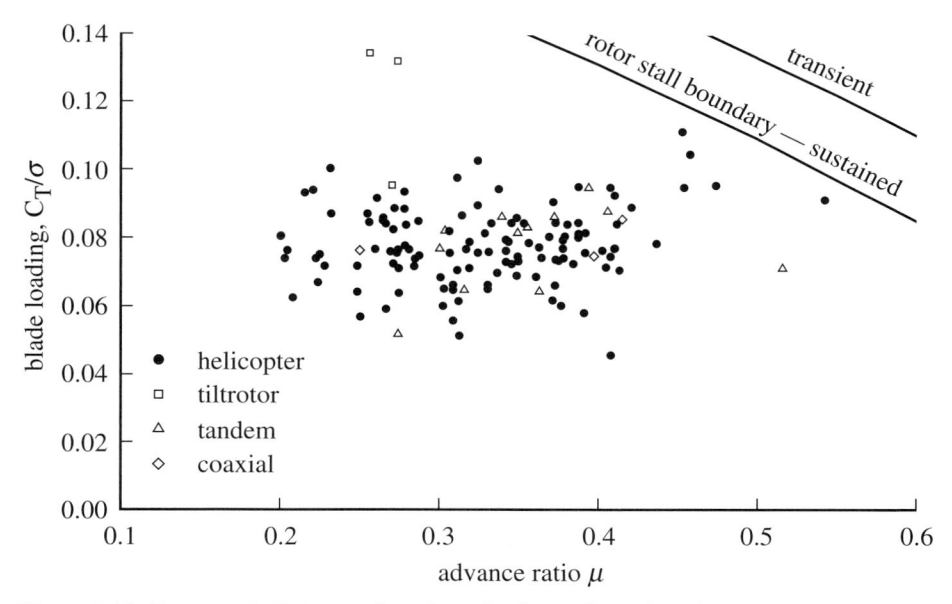

Figure 1.11. Rotorcraft C_T/σ as a function of μ for various aircraft.

aerodynamic asymmetry), relative to rotor stall boundaries for sustained and transient flight conditions (see Chapter 12). Design C_T/σ (for maximum takeoff weight and sea-level-standard conditions) is in the range 0.07 to 0.09 for helicopters and is higher for tiltrotors. Compressible aerodynamics of the blade tip limit the advancing tip Mach number M_{at}. Limitations on μ and M_{at} combine to restrict the design choices for rotor tip speed and aircraft flight speed, as shown in Figure 1.12. With design values of C_T/σ and rotor tip speed constrained to relatively small ranges, the

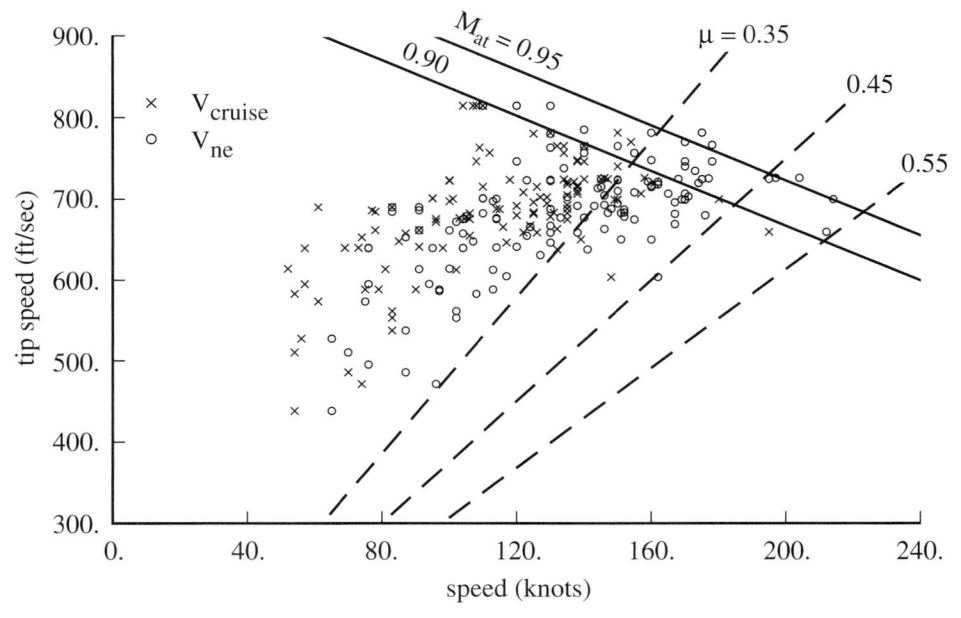

Figure 1.12. Rotor tip speed and rotorcraft flight speed.

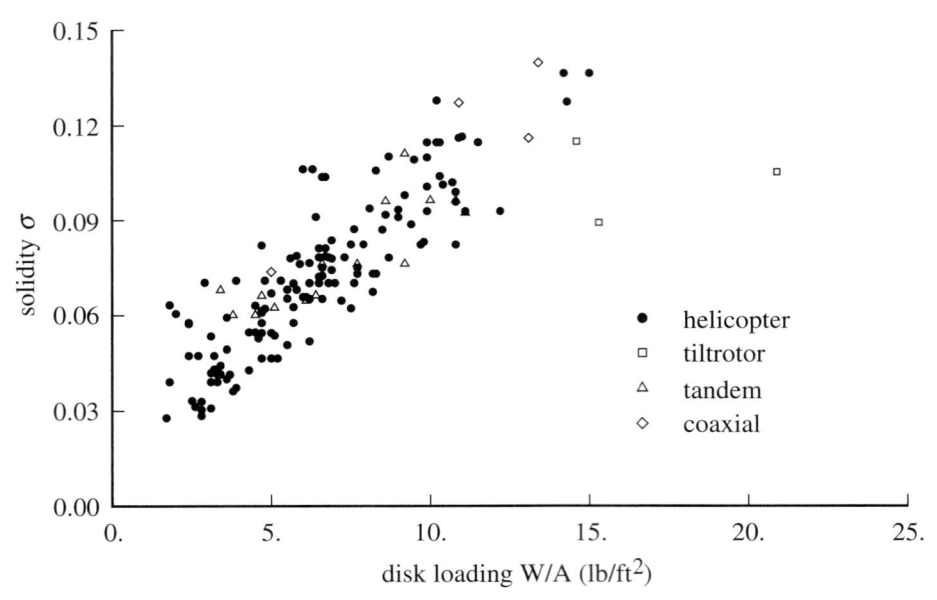

Figure 1.13. Rotor solidity as a function of disk loading.

rotor solidity varies roughly linearly with disk loading (Figure 1.13). The low solidity values for helicopter disk loadings imply high blade aspect ratios, typically 10 to 20.

Figure 1.14 shows the trends over time of several rotorcraft design parameters. Disk loading has increased and power loading has decreased, largely due to improvements in propulsion capability. Drag has decreased. Empty weight fraction decreased until the mid-1960s, particularly with the introduction of the turboshaft engine. Rotor tip speed and advancing tip Mach number have increased to the limits imposed by compressible tip aerodynamics. Aircraft flight speed and advance ratio have increased within the limits imposed by rotor blade stall.

1.3 History

The initial development of rotary-wing aircraft faced three major problems that had to be overcome to achieve a successful vehicle. The first problem was to find a light and reliable engine. The reciprocating internal combustion engine was the first to fulfill the requirements, and the later adoption of the turboshaft engine for the helicopter was a significant advance. The second problem was to develop a light and strong structure for the rotor, hub, and blades while maintaining good aerodynamic efficiency. The final problem was to understand and develop means of controlling the helicopter, including balancing the rotor torque. These problems were essentially the same as those that faced the development of the airplane and were solved eventually by the Wright brothers. The development of the helicopter in many ways paralleled that of the airplane. That helicopter development took longer can be attributed to the challenge of vertical flight, which required a higher development of aeronautical technology before the problems could be satisfactorily overcome.

The history of helicopters begins with toys and imagination. Among Leonardo da Vinci's work (1483) were sketches of a machine for vertical flight utilizing a screw-type propeller (Figure 1.15). The Chinese flying top was a stick with a propeller

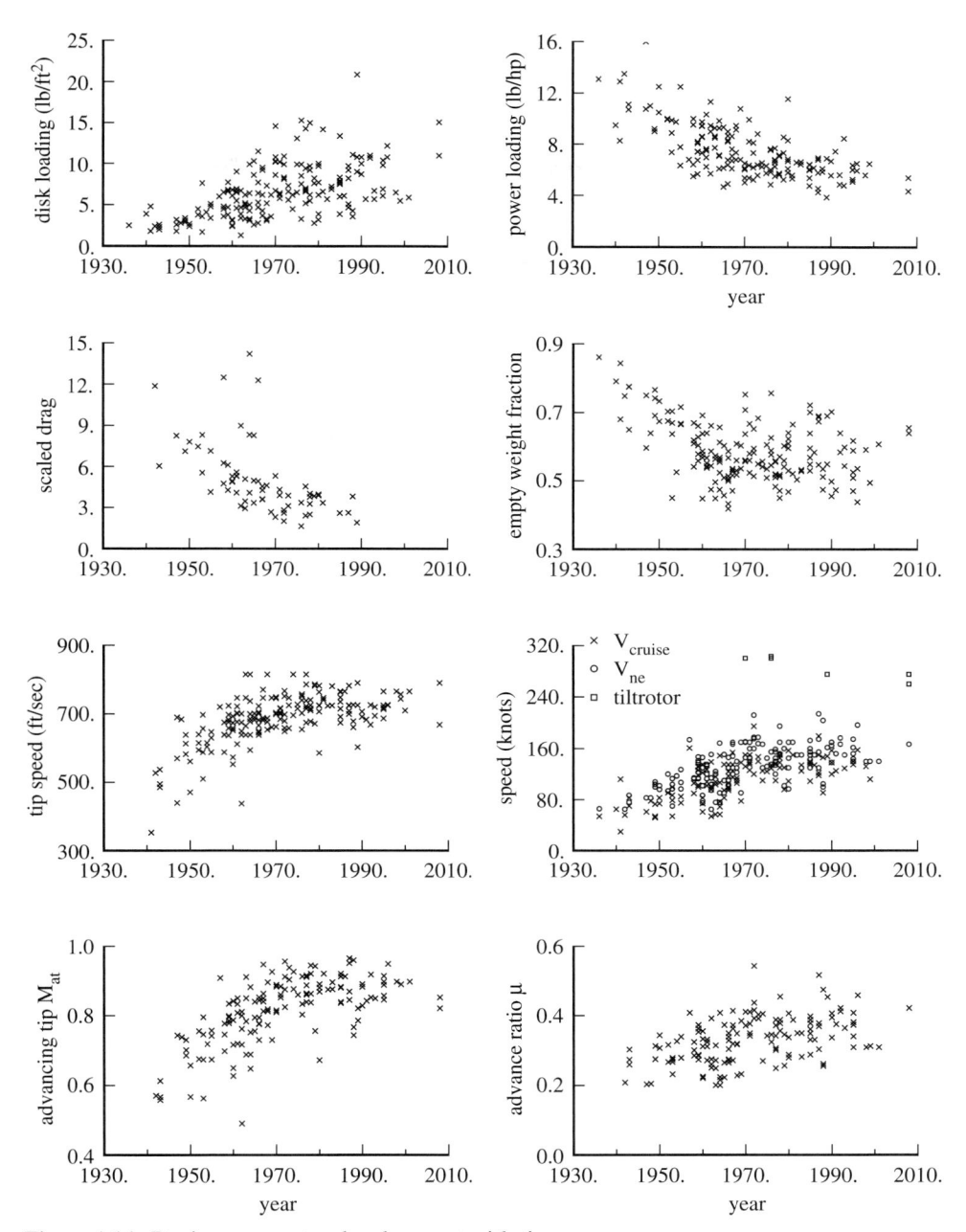

Figure 1.14. Design parameter development with time.

attached, which was spun by the hands and released. In the 18th century there was some work with models. Mikhail V. Lomonosov (Russia, 1754) demonstrated a spring-powered model to the Russian Academy of Sciences. Launoy and Bienvenu (France, 1784) demonstrated a spring-powered model to the French Academy of Sciences. This model had two contra-rotating rotors of four blades each, powered by a flexed bow. Sir George Cayley (England) constructed models powered by elastic elements and made sketches of helicopters.

Figure 1.15. Imagination of Leonardo da Vinci.

In the last half of the 19th century many inventors were working on developing a helicopter. There was some practical progress, but no successful vehicle. The problem was the lack of a light and reliable engine. A number of attempts used a steam engine. W.H. Phillips (England, 1842) constructed a 20-lb steam-powered model. Gustave de Ponton d'Amecourt (France, 1863) built a small steam-driven model; he also coined the word "helicopter" – helicoptere in French, from the Greek heliko (spiral) and pteron (wing).

Alphonse Penaud (France, 1870s) experimented with models. Enrico Forlanini (Italy, 1878) built a 3.5-kg flying steam-driven model. Thomas Edison (United States, 1880s) experimented with models. He recognized that the problem was the lack of an adequate engine. Edison concluded that no helicopter would be able to fly until engines were available with a weight-to-power ratio below 3 to 4 lb/hp. These were still only models, but they were beginning to address the problem of an adequate power source for sustained flight. The steam engine was not effective for aircraft, especially the helicopter, because of the low power-to-weight ratio of the system.

Around 1900 the internal combustion reciprocating gasoline engine became available, making airplane flight possible, and eventually helicopter flight as well. Renard (France, 1904) built a helicopter with two side-by-side rotors, using a two-cylinder engine; he introduced the flapping hinge for the helicopter rotor. G.A. Crocco (Italy, 1906) patented a cyclic pitch design, as a means to counter the asymmetry in aerodynamic loads on the advancing and retreating sides of the disk in forward flight.

Early experimentation with vertical flight was limited by the lack of sufficient power to achieve hover. Figure 1.16 shows the helicopters of this period. The Breguet-Richet Gyroplane No.1 (France, 1907) had four rotors with four biplane blades each (8-m diameter rotors, gross weight 580 kg, 45-hp Antoinette engine). This aircraft made a tethered flight with a passenger at an altitude of about 2 ft for about 1 minute. Paul Cornu (France 1907) constructed a machine that made the first flight with a pilot. Cornu's aircraft had two contra-rotating rotors in tandem configuration with two fabric-covered blades each (6-m diameter rotors, gross weight

1.3 History

Breguet-Richet (France, 1907)
Gyroplane No. 1

Cornu (France, 1907)

Sikorsky (Russia, 1910)

Petroczy, von Karman, Zurovec
PKZ-2 (Austria 1916)

Figure 1.16. Experimentation: search for power. Sikorsky image courtesy Sikorsky Historical Archives; other images courtesy American Helicopter Society.

260 kg, 24-hp Antoinette engine connected to the rotors by belts). Control was provided by vanes in the rotor slipstream, but was not very effective. This helicopter achieved an altitude of about 1 ft for about 20 sec, but had problems with mechanical design and with lack of stability. Although often cited as the first helicopters to achieve flight, these machines lacked sufficient power to hover, even in ground effect. These aircraft did not have effective means of control and were quite unstable.

Emile Berliner and John Newton Williams (United States, 1909) built a two-engine coaxial helicopter that lifted a pilot untethered. Igor Sikorsky (Russia, 1910) built a helicopter with two coaxial three-bladed rotors (5.8-m diameter rotors, 25-hp Anzani engine) that could lift 400 lb but not its own weight plus the pilot. Sikorsky would return to the development of the helicopter (with considerably more success) after building airplanes in Russia and in the United States. Boris N. Yuriev (Russia, 1912) built a machine with a two-bladed main rotor and a small anti-torque tail rotor (8-m diameter main rotor, gross weight 200 kg, 25-hp Anzani engine). This helicopter made no successful flight, but Yuriev went on to supervise helicopter development in the Soviet Union. Stefan von Petroczy and Theodor von Kármán, with the assistance of Wilhelm Zurovec (Austria, 1916), built the PKZ-2, a tethered observation helicopter that achieved an altitude of 50 m with payload.

The development of better engines during and after World War I solved the problem of an adequate power source, at least enough to allow experimenters to tackle the task of finding a satisfactory solution for helicopter control (Figure 1.17).

George de Bothezat (United States, 1922) built a helicopter with four six-bladed rotors at the ends of intersecting beams (26.5-ft diameter rotors, gross weight 1700 kg, 180-hp engine at the center). This aircraft had good control capability, utilizing

de Bothezat (USA, 1922)    Oemichen (France, 1924)

Pescara (Spain, 1924)    Berliner (USA, 1920–1925)

von Baumhauer (Holland, 1924-1929)    d'Ascenio (Italy, 1930)

Figure 1.17. Experimentation: search for control. Images courtesy American Helicopter Society.

differential collective of the four rotors, and made many flights with passengers up to an altitude of 15 ft. Collective pitch is a change made in the mean blade pitch angle to control the rotor thrust magnitude. This was the first rotorcraft ordered by the U.S. Army, but after the expenditure of $200,000 the project was finally abandoned as being too complex mechanically.

Etienne Oemichen (France, 1924) built a machine with four two-bladed rotors (two 7.6 m in diameter and two 6.4 m in diameter) to provide lift, five horizontal propellers for attitude control, two propellers for propulsion, and one propeller in front for yaw control – all powered by a single 120-hp Le Rhone engine. Oemichen's aircraft set the first helicopter distance record, 360 m. Raul Pateras Pescara (an Argentine working in Spain, 1924) constructed a helicopter with two coaxial rotors of four biplane blades each (7.2-m diameter rotors, gross weight 850 kg, 180-hp Hispano-Suiza engine; a 1920 craft of similar design that used 6.4-m diameter rotors and a 45-hp Hispano engine had inadequate lift). For control, he warped the biplane blades to change their pitch angle. Pescara was the first to demonstrate effective cyclic pitch for control of the main rotors. Cyclic pitch is a sinusoidal, once-per-revolution change made in the blade pitch to tilt the rotor disk. Pescara's helicopter set a distance record (736 m), but had stability problems.

Emile and Henry Berliner (United States, 1920–1925) built a helicopter using two rotors positioned on the tips of a biplane wing in a side-by-side configuration. They used rigid wooden propellers for the rotors and obtained control by tilting the entire rotor. Louis Brennan (England, 1920s) built a helicopter with a rotor turned by propellers on the blades, to eliminate the torque problem (60-ft diameter two-bladed rotor, gross weight 3300 lb, 230-hp Bentley engine); the machine was mechanically too complex.

A.G. von Baumhauer (Holland, 1924–1929) developed a helicopter with a single main rotor and a vertical tail rotor for torque balance (15.2-m diameter two-bladed main rotor, gross weight 1300 kg, 200-hp Bentley rotary engine). A separate engine was used for the tail rotor (80-hp Thulin rotary engine mounted directly to the tail rotor). The main rotor blades were free to flap, but were connected by cables to form a teetering rotor. Control was by cyclic pitch of the main rotor, produced using a swashplate. Flights were made, but never at more than 2 m altitude. There were difficulties with directional control because of the separate engines for the main rotor and tail rotor, and the project was abandoned after a bad crash.

Corradino d'Ascanio (Italy, 1930) constructed a helicopter with two coaxial rotors (13-m diameter rotors, gross weight 800 kg, 95-hp Fiat engine). The two-bladed rotors had flap hinges and free-feathering hinges. Servo tabs on the blades produced cyclic and collective pitch changes. For several years this machine held records for altitude (17 m), endurance (8 min 45 sec), and distance (1078 m). The stability and control characteristics were marginal, however.

M.B. Bleecker (United States, 1930) built a helicopter with four wing-like blades. Power was delivered to a propeller on each blade from an engine in the fuselage. Control was by aerodynamic surfaces on the blades and by a tail on the aircraft. The Central Aero-Hydrodynamic Institute of the Soviet Union developed a series of single rotor helicopters under the direction of Yuriev. The TsAGI I-EA (1931) had a four-bladed main rotor (11-m diameter rotor, gross weight 1100 kg, 120-hp engine) with cyclic and collective control, and two small contra-rotating anti-torque rotors.

The development of the helicopter was fairly well advanced at this point, but the stability and control characteristics were still marginal, as were the forward flight and power-off (autorotation) capabilities of the designs. During this period, in the 1920s and 1930s, the autogyro was developed (Figure 1.18). The autogyro was the first practical use of the direct-lift rotary wing. It was developed largely by Juan de la Cierva (Spanish), who coined (and trademarked) the word "Autogiro." In this aircraft, a windmilling rotor replaces the wing of the airplane. Essentially, it uses the fixed-wing aircraft configuration, with a propeller supplying the propulsive force. The initial designs even used conventional airplane-type aerodynamic surfaces for control (ailerons, rudder, and elevator). With no power directly to the rotor, hover and vertical flight are not possible, but the autogyro is capable of very slow flight and in cruise behaves much like an airplane.

Juan de la Cierva had designed an airplane that crashed in 1919 due to stall near the ground. He then became interested in designing an aircraft with a low takeoff and landing speed that would not stall if the pilot dropped the speed excessively. He

Cierva C.6A (Spain, 1925)   Pitcairn PCA-2 (USA, 1932–1935)
used for NACA research

Figure 1.18. Autogyros. Cierva image courtesy American Helicopter Society; Pitcairn image from NACA.

determined from wind-tunnel tests of model rotors that, with no power to the shaft but with a rearward tilt of the rotor, good lift-to-drag ratio could be obtained even at low speed. The best results were at low, positive collective pitch of the rotor. In 1922, Cierva built the C.3 autogyro with a five-bladed rigid rotor and "a tendency to fall over sideways." He had a model with blades of flexible palm wood that flew properly. The flexible rotor blades accounted for the successful flight of the model, suggesting the use of articulated rotor blades on the autogyro. Cierva consequently incorporated flapping blades in his design. The flap hinge eliminated the rolling moment on the aircraft in forward flight due to the asymmetry of the flow over the rotor. Cierva was the first to use the flap hinge in a successful rotary-wing aircraft. In 1923, the C.4 autogyro was built and flown. It had a four-bladed rotor with flap hinges on the blades (9.8-m diameter rotor, 110-hp Le Rhone engine). Control was by conventional airplane aerodynamic surfaces. In 1924, the C.6A autogyro with flapping rotor blades was built (11-m diameter four-bladed rotor, 100-hp Le Rhone rotary engine). An Avro 504K aircraft fuselage and ailerons on outrigger spars were used. The demonstration of this autogyro in 1925 at the Royal Aircraft Establishment was the stimulus for the early analysis of the rotary wing in England by Glauert and Lock. The C.6A is generally regarded as Cierva's first successful autogyro (1926).

In 1925, Cierva founded the Cierva Autogiro Company in England, which was his base thereafter. In the next decade about 500 of his autogyros were produced, many by licensees of the Cierva Company, including A.V. Roe, de Havilland, Weir, Westland, Parnell, and Comper in Britain; Pitcairn, Kellett, and Buhl in the United States; Focke-Wulf in Germany; Loire and Olivier in France; and the TsAGI in Russia. A crash in 1927 led to an appreciation of the high in-plane blade loads due to flapping, and a lag hinge was added to the rotor blades. This completed the development of the fully articulated rotor hub for the autogyro. In 1932, Cierva added rotor control to replace the airplane control surfaces, which were not very effective at low speeds. He used direct tilt of the rotor hub for longitudinal and lateral control. Raoul Hafner (England, 1935) developed an autogyro incorporating cyclic pitch control by means of a "spider" control mechanism to replace the direct tilt of the rotor hub. E. Burke Wilford (United States, 1930s) developed a hingeless rotor autogyro that also used cyclic control of the rotors.

By 1935, the autogyro was well developed in both Europe and America. Its success preceded that of the helicopter because of the lower power required without

1.3 History

Breguet-Dorand (France, 1935)

Focke Fa-61 (Germany, 1936)

Flettner Fl-282 (Germany, 1938-1940)

Focke Fa-223 (Germany, 1941)

Sikorsky VS-300 (USA, 1939-1941)

Sikorsky R-4 (USA, 1942)

Figure 1.19. First successful helicopters. R-4 image from U.S. Army; other images courtesy American Helicopter Society.

actual vertical flight capability and because the unpowered rotor is mechanically simpler. The autogyro was able to use a great deal of airplane technology, for example in the propulsion system, and initially even the control system. Lacking true vertical flight capability, however, the autogyro was never able to compete effectively with fixed-wing aircraft. However, autogyro developments, including experimental and practical experience, did influence helicopter development and design. The autogyro had a substantial impact on the development of rotary-wing analysis. Much of the work of the 1920s and 1930s, which forms the foundation of helicopter analysis, was originally developed for the autogyro.

Meanwhile, the development of the helicopter continued, with the first successful helicopters flying in Europe in the mid-1930s (Figure 1.19). Louis Breguet and Rene

Dorand (France, 1935) built the Gyroplane Laboratoire, a helicopter with coaxial two-bladed rotors (16.4-m diameter rotors, gross weight 2000 kg, 350-hp Hispano engine). The rotors had an articulated hub (flap and lag hinges); control was by cyclic for pitch and roll, and differential torque for directional control. The aircraft had satisfactory control characteristics and held records for speed (44.7 kph), altitude (158 m), duration (62 min), and closed-circuit distance (44 km).

Henrich Focke (Germany, 1936) constructed the Focke-Achgelis Fa-61, a helicopter with two three-bladed rotors mounted on trusses in a side-by-side configuration (7-m diameter rotors, gross weight 950 kg, 160-hp Bramo engine). Focke had licensed the C.19 and C.30 autogyros, and the rotor components were supplied by the Weir-Cierva company. The rotor had an articulated hub and tapered blades. Directional and longitudinal control were by cyclic, and roll control by differential collective. Height control, however, was obtained by varying rotor speed. Vertical and horizontal tail surfaces were used for stability and trim in forward flight, and the rotor shafts were inclined inward for stability. This helicopter set records for speed (122.5 kph), altitude (3427 m), endurance (81 min), and distance (230 km). The Fa-61 was a well-developed machine, with good control, performance, and reliability. This aircraft was demonstrated by Hanna Reitsch inside the Berlin Deutschlandhalle in February 1938.

C.G. Pullin (Britain, 1938) built for G. & J. Weir Ltd. helicopters with a side-by-side configuration: in 1938 the W-5 (15-ft diameter two-bladed rotors, gross weight 860 lb, 50-hp Weir engine), and in 1939 the W-6 (25-ft diameter three-bladed rotors, gross weight 2360 lb, 205-hp de Havilland engine). Ivan P. Bratukhin (TsAGI in the USSR, 1939–1940) constructed the Omega I helicopter with two three-bladed rotors in a side-by-side configuration (7-m diameter rotors, gross weight 2050 kg, two 220-hp engines).

There was considerable effort in rotary-wing development in Germany during World War II. Anton Flettner (Germany, 1938–1940) developed a synchropter design, with two rotors in a side-by-side configuration but highly intermeshed (hub separation 0.6 m). The FL-282 had two-bladed articulated rotors (12-m diameter rotors, gross weight 1000 kg, 140-hp Siemens Halske engine). Orders were placed for one thousand aircraft, but only about 30 were completed. Focke (Germany 1941) developed the Focke-Achgelis Fa-223 with two three-bladed rotors in the side-by-side configuration (rotors 12 m in diameter, gross weight 4300 kg, 1000-hp Bramo engine). This helicopter had an absolute ceiling of 5000 m, a range of 320 km, a cruise speed of 176 kph with six passengers, and a useful load of 900 kg. Orders were placed for 100, with 10 delivered. These helicopter developments reached a dead end because of World War II.

Igor Sikorsky (United States, 1939–1941) returned to helicopter development in 1938 after designing and building airplanes in Russia and the United States. In 1941, Sikorsky built the VS-300, a helicopter with a single three-bladed main rotor and a small anti-torque tail rotor (28-ft diameter rotor, gross weight 1150 lb, 100-hp Franklin engine). Lateral and longitudinal control was by main rotor cyclic, and directional control was by means of the tail rotor. The tail rotor was driven by a shaft from the main rotor. The pilot's controls were like the present standard (cyclic stick, pedals, and a collective stick with a twist grip throttle). Considerable experimentation was required to develop a configuration with suitable control characteristics. After tethered flights in late 1939 of an aircraft with main rotor cyclic control, a more stable and controllable configuration was flown in 1940 with three auxiliary rotors (one vertical and two horizontal) on the tail. In 1941, the number of auxiliary rotors was reduced to two: a vertical tail rotor for yaw and a horizontal tail rotor for pitch control. Finally in December 1941, the horizontal propeller was removed and main rotor cyclic used for longitudinal control. This was Sikorsky's 18th version: the single main rotor and tail rotor configuration that has become the most common helicopter type. Sikorsky also tried a two-bladed main rotor, which had comparable performance and was simpler, but was not pursued because the vibration was considered excessive.

In 1942 the R-4 (VS-316), was constructed (38-ft diameter three-bladed rotor, gross weight 2540 lb, 185-hp Warner engine). This helicopter model went into production, and with the R-5 and R-6 about 600 aircraft were built and delivered by September 1945. Igor Sikorsky's R-4 was the first widely used, practical helicopter. The R-4 was controllable and mechanically simple (relative to other helicopter designs of the time at least), was produced in significant quantities, and saw operational use. In 1946 the R-5 set records for speed (178 kph), altitude (5842 m), and distance (1132 km). The R-5 evolved to the Sikorsky S-51, which was the second helicopter to receive commercial certification (April 1947).

The 1940s saw the development of helicopters that formed the basis for the new industry (Figure 1.20). Arthur Young (United States, 1943) developed for Bell Aircraft the Model 30, a helicopter with a two-bladed teetering main rotor and a tail rotor (32-ft diameter rotor, gross weight 2300 lb, 160-hp Franklin engine), using the gyro stabilizer bar that he invented in the 1930s. In March 1946, the Bell Model 47 (35-ft diameter rotor, gross weight 2100 lb, 178-hp Franklin engine) received the first American certificate of airworthiness for helicopters. Frank N. Piasecki (United States, 1943) for the P-V Engineering Forum developed the second successful helicopter prototype in the United States: the PV-2 single main rotor and tail rotor helicopter (22-ft diameter rotor, gross weight 1000 lb, 90-hp Franklin engine). In 1945 Piasecki developed the PV-3, a tandem rotor helicopter (41-ft diameter three-bladed rotors, gross weight 6900 lb, 600-hp Pratt and Whitney engine); the Navy prototypes were designated XHRP-1. Piasecki's company eventually became the Boeing Vertol Company, with the tandem configuration remaining its basic production type. Stanley Hiller (United States, 1944–1948) experimented with several types of helicopters, including the XH-44 coaxial helicopter, eventually settling on the single main rotor and tail rotor configuration. Hiller developed the control rotor, which is a gyro stabilizer bar with aerodynamic surfaces that the pilot controlled to adjust the rotor orientation. He built the Model 360 helicopter in 1947 (35.4-ft diameter two-bladed rotor, gross weight 2100 lb, 178-hp Franklin engine). Charles Kaman (United States, 1946–1948) developed the servotab control method of rotor pitch control, in which the rotor blade is twisted rather than rotated about a pitch bearing at the root. Kaman also developed a helicopter of the synchropter configuration. The Bristol Aircraft Company (British, 1947) developed the Bristol-171 Sycamore. Mikhail Mil' (USSR, 1949) developed a series of helicopters of the single main rotor and tail rotor configuration, including in 1949 the Mi-1 (14-m diameter three-bladed rotor, gross weight 2250 kg, 570 hp engine). Nikolai Kamov (USSR, 1952) developed helicopters with the coaxial configuration, including the Ka-15 helicopter (10-m diameter three-bladed rotors, gross weight 1370 kg, 225-hp engine).

An important development was the introduction of the turboshaft engine to helicopters, replacing the reciprocating engine (Figure 1.21). A substantial improvement in performance and reliability was realized because of the lower specific weight (lb/hp) of the turboshaft engine. Kaman Aircraft Company (United States, 1951)

Young, Bell Model 30 (USA, 1943)    Young, Bell Model 47 (USA, 1946)

Piasecki PV-2 (USA, 1943)    Piasecki HRP-1 (USA, 1945)

Hiller XH-44 (USA, 1944)    Hiller Model 360 (USA, 1947)

Figure 1.20. Start of the helicopter industry. Model 30 image courtesy Bell Helicopter; Model 47 image from U.S. Army; other images courtesy American Helicopter Society.

constructed the first helicopter with turbine power, installing a single turboshaft engine (175-hp Boeing engine) in its K-225 helicopter. In 1954, Kaman also developed the first twin-engine turbine-powered helicopter, an HTK-1 synchropter with two Boeing engines (total 350 hp) replacing a single 240-hp piston engine of the same weight in the same position. The Sud Aviation Alouette II (France, 1955) was the first helicopter to be produced with a turboshaft engine. Since that time the turboshaft engine has become the standard powerplant for all but the smallest helicopters.

After the 1950s, the history of rotorcraft development involved programs and companies, rather than inventors. So this exposition is concluded at a point where the helicopter was successful if not mature, and the foundations of the industry were established. In the years that followed, several helicopter designs achieved extremely

Kaman, K-225 (USA, 1951)   Sud Aviation, Alouette II (France, 1955)

Figure 1.21. Power from the turboshaft engine.

successful production runs, and some very large helicopters were constructed. Rotorcraft configurations such as compound helicopters and tiltrotors have been explored, particularly in the search for higher speed. The operational use of the helicopter has grown to be a major factor in the air transportation system.

1.4 Books

A large number of books on helicopter aeromechanics have been published since the 1940s. The early texts included Nikolsky (1951), Shapiro (1955), and Stepniewski (1955). Gessow and Myers (1952) is a concise introduction, based on the excellent NACA research program on autogyros and helicopters. Important recent books are Bramwell (1976), Stepniewski and Keys (1979), Johnson (1980), Hodges (2006), and Leishman (2006). Padfield (2007) covers rotorcraft flight dynamics and handling qualities in depth. A bibliography of books on helicopter engineering, listed in chronological order, follows.

Glauert, H. "Airplane Propellers." In *Aerodynamic Theory*, Durand, W.F. (Editor). New York: Julius Springer, 1935.
Nikolsky, A.A. *Notes on Helicopter Design Theory*. Princeton, NJ: Princeton University Press, 1944.
Glauert, H. *The Elements of Aerofoil and Airscrew Theory*. Cambridge: Cambridge University Press, 1947.
Young, R.A. *Helicopter Engineering*. New York: The Ronald Press Company, 1949.
Nikolsky, A.A. *Helicopter Analysis*. New York: John Wiley & Sons, Inc., 1951.
Gessow, A., and Myers, G.C., Jr. *Aerodynamics of the Helicopter*. New York: The Macmillan Company, 1952.
Dommasch, D.O. *Elements of Propeller and Helicopter Aerodynamics*. London: Sir Isaac Pitman and Sons, Ltd., 1953.
Shapiro, J. *Principles of Helicopter Engineering*. London: Temple Press Limited, 1955.
Stepniewski, W.Z. *Introduction to Helicopter Aerodynamics*. Morton, PA: Rotorcraft Publishing Committee, 1955 (first edition 1950).
Payne, P.R. *Helicopter Dynamics and Aerodynamics*. London: Sir Isaac Pitman and Sons, Ltd., 1959.
Shapiro, J. *The Helicopter*. New York: The Macmillan Company, 1959.
Legrand, F. *Rotorcraft*. France: Ecole Nationale Superieure de l'Aeronautique, 1964 (translation NASA TT F-11530, April 1968).
Seckel, E. *Stability and Control of Airplanes and Helicopters*. New York: Academic Press, Inc., 1964.

Mil, M.L., Nekrasov, A.V., Braverman, A.S., Grodko, L.N., and Leykand, M.A. *Helicopter, Calculation and Design*. Moscow: Izdatel'stvo Mashinostroyeniye, 1966 (translation – Volume I: *Aerodynamics*, NASA TT F-494, September 1967, Volume II: *Vibrations and Dynamic Stability*, TT F-519, May 1968).

McCormick, B.W., Jr. *Aerodynamics of V/STOL Flight*. New York: Academic Press, Inc., 1967.

Bramwell, A.R.S. *Helicopter Dynamics*. London: Edward Arnold (Publishers) Ltd., 1976.

Tischenko, M.N., Nekrasov, A.V., and Radin, A.S. *Helicopters. Selection of Design Parameters*. Moscow: Mashinostroyeniye Press, 1976 (translation April 1979).

Stepniewski, W.Z., and Keys, C.N. *Rotary-Wing Aerodynamics*. New York: Dover Publications, Inc., 1979.

Johnson, W. *Helicopter Theory*. Princeton, NJ: Princeton University Press, 1980.

Layton, D.M. *Helicopter Performance*. Beaverton, OR: Matrix Publishers, Inc., 1984.

Prouty, R.W. *Helicopter Performance, Stability, and Control*. Malabar, Florida: Robert E. Krieger Publishing Company, Inc., 1986.

Prouty, R.W. *Military Helicopter Design Technology*. Malabar, Florida: Krieger Publishing Company, 1989.

Seddon, J. *Basic Helicopter Aerodynamics*. Oxford: BSP Professional Books, 1990.

Bielawa, R.L. *Rotary Wing Structural Dynamics and Aeroelasticity*. Washington, DC: American Institute of Aeronautics and Astronautics, Inc., 1992.

Bramwell, A.R.S., Done, G., and Balmford, D. *Bramwell's Helicopter Dynamics*. Reston, VA: American Institute of Aeronautics and Astronautics, Inc., 2001.

Franklin, J.A. *Dynamics, Control, and Flying Qualities of V/STOL Aircraft*. Reston, VA: American Institute of Aeronautics and Astronautics, Inc., 2002.

Hodges, D.H. *Nonlinear Composite Beam Theory*. Reston, VA: American Institute of Aeronautics and Astronautics, 2006.

Leishman, J.G. *Principles of Helicopter Aerodynamics*. Second Edition. Cambridge: Cambridge University Press, 2006.

Tischler, M.B., and Remple, R.K. *Aircraft and Rotorcraft System Identification*. Reston, VA: American Institute of Aeronautics and Astronautics, 2006.

Dreier, M.E. *Introduction to Helicopter and Tiltrotor Flight Simulation*. Reston, VA: American Institute of Aeronautics and Astronautics, 2007.

Padfield, G.D. *Helicopter Flight Dynamics*. Second Edition. Oxford: Blackwell Science Ltd, 2007.

Seddon, J., and Newman, S. *Basic Helicopter Aerodynamics*. Third Edition. Reston, VA: American Institute of Aeronautics and Astronautics, Inc., 2011.

Harris, F.D. "Introduction to Autogyros, Helicopters, and Other V/STOL Aircraft, Volume I: Overview and Autogyros." NASA SP 2011-215959, Vol I, May 2011. "Volume II: Helicopters." NASA SP 2012-215959, Vol II, October 2012.

2 Notation

This chapter summarizes the principal nomenclature to be used in the text. The intention is to provide a reference for the later chapters and also to introduce the basic elements of the rotor and its analysis. Only the most fundamental parameters are included here; the definitions of the other quantities required are presented as the analysis is developed. A number of the basic dimensionless parameters of helicopter analysis are also introduced. An alphabetical listing of symbols is provided at the end of the chapter.

2.1 Dimensions

Generally the analyses in this text use dimensionless quantities. The natural reference length scale for the rotor is the blade radius R, and the natural reference time scale is the rotor rotational speed Ω (rad/sec). For a reference mass the air density ρ is chosen.

For typographical simplicity, no distinction is made between the symbols for the dimensional and dimensionless forms of a quantity when the latter are based on ρ, Ω, and R. New symbols are introduced for those dimensionless parameters normalized using other quantities.

2.2 Nomenclature

2.2.1 Physical Description of the Blade

$R =$ the rotor radius; the length of the blade, measured from center of rotation to tip.
$\Omega =$ the rotor rotational speed or angular velocity (rad/sec).
$\rho =$ air density.
$\psi =$ azimuth angle of the blade (Figure 2.1), defined as zero in the downstream direction. This is the angle measured from downstream to the blade span axis, in the direction of rotation of the blade. Hence for constant rotational speed, $\psi = \Omega t$.
$r =$ radial location on the blade (Figure 2.1), measured from the center of rotation ($r = 0$) to the blade tip ($r = R$, or when dimensionless $r = 1$).

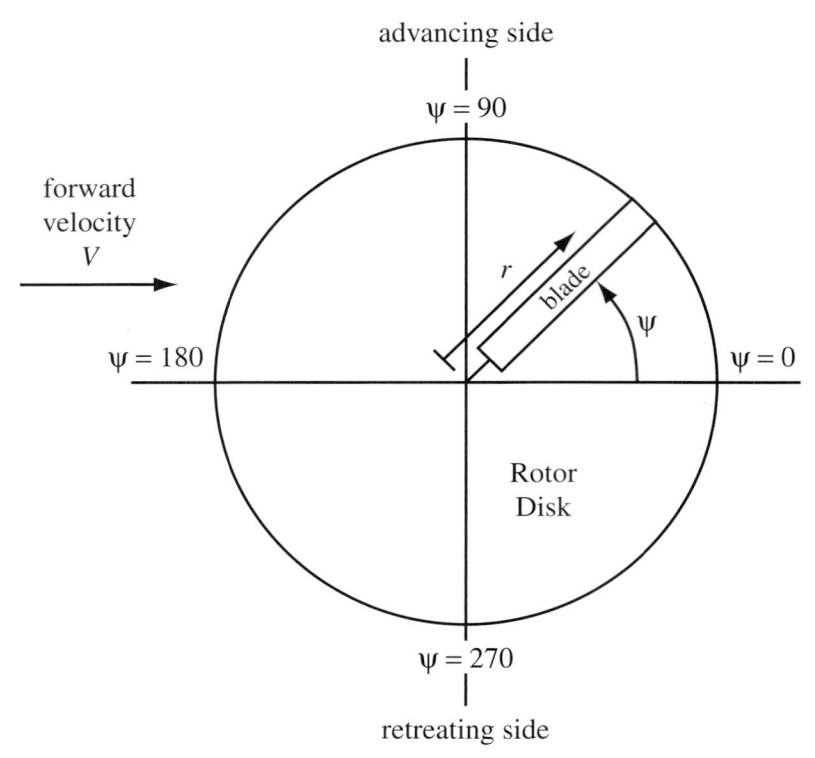

Figure 2.1. Rotor disk in edgewise flight, showing definition of ψ and r.

In the United States the convention is to assume that the rotor rotation direction is counterclockwise (viewed from above). Then the right side of the rotor disk is called the advancing side, and the left side is called the retreating side. The variables r and ψ usually refer to the radial and azimuthal position of the blade, but they can also be used as polar coordinates for the rotor disk.

c = blade chord, which for tapered blades is a function of r.
N = number of blades.
m = blade mass per unit length, as a function of r.
$I_b = \int_0^R r^2 m \, dr$ = moment of inertia of the blade about the center of rotation.

The rotor blade normally is twisted along its length. The analysis often considers linear twist, for which the built-in variation of the blade pitch with respect to the root is $\Delta\theta = \theta_{tw} r$. The linear twist rate θ_{tw} (equal to the tip pitch minus the root pitch) is normally negative for the helicopter rotor. The following derived quantities are important:

$A = \pi R^2$ = rotor disk area.
$\sigma = Nc/\pi R$ = rotor solidity.
$\gamma = \rho a c R^4 / I_b$ = blade Lock number.

The solidity σ is the ratio of the total blade area (NcR for constant chord) to the rotor disk area (πR^2). The Lock number γ represents the ratio of the aerodynamic and inertial forces on the blade.

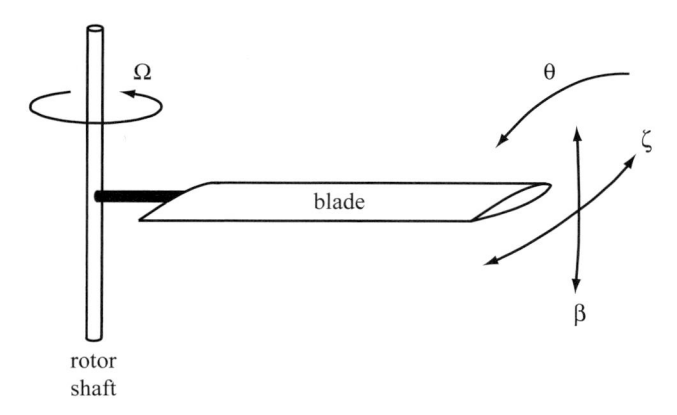

Figure 2.2. Fundamental blade motion.

2.2.2 Blade Aerodynamics

    a = blade section two-dimensional lift curve slope.
    α = blade section angle-of-attack.
    M = blade section Mach number.

The subscript (r, ψ) on α or M is used to indicate the point on the rotor disk being considered; for example, the retreating-tip angle-of-attack $\alpha_{1,270}$ or the advancing-tip Mach number $M_{1,90}$ (also written M_{at}).

2.2.3 Blade Motion

The basic motion of the blade is essentially rigid body rotation about the root, which is attached to the hub (Figure 2.2).

    β = blade flap angle. This degree of freedom produces blade motion out of the disk plane, about either an actual flap hinge or a region of structural flexibility at the root. Flapping is positive for upward motion of the blade (as produced by the thrust force on the blade).
    ζ = blade lag angle. This degree of freedom produces blade motion in the disk plane. Lagging is positive when opposite the direction of rotation of the rotor (as produced by the blade drag forces).
    θ = blade pitch angle, or feathering motion, produced by rotation of the blade about a hinge or bearing at the root with the pitch axis parallel to the blade spar. Pitching is positive for nose-up rotation of the blade.

The degrees of freedom β, ζ, and θ can also be viewed as rotations of the blade about hinges at the root, with axes of rotation as follows: β is the angle of rotation about an axis in the disk plane, perpendicular to the blade spar; ζ is the angle of rotation about an axis normal to the disk plane, parallel to the rotor shaft; and θ is the angle of rotation about an axis in the disk plane, parallel to the blade spar. The description of more complex blade motion – for example, motion that includes blade flexibility – is introduced as required in later chapters.

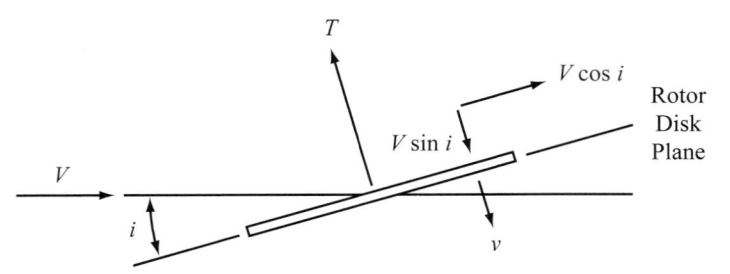

Figure 2.3. Rotor disk velocity and orientation.

In steady-state operation of the rotor, blade motion is periodic around the azimuth and hence can be expanded as a Fourier series in ψ:

$$\beta = \beta_0 + \beta_{1c}\cos\psi + \beta_{1s}\sin\psi + \beta_{2c}\cos 2\psi + \beta_{2s}\sin 2\psi + \ldots$$

$$\zeta = \zeta_0 + \zeta_{1c}\cos\psi + \zeta_{1s}\sin\psi + \zeta_{2c}\cos 2\psi + \zeta_{2s}\sin 2\psi + \ldots$$

$$\theta = \theta_0 + \theta_{1c}\cos\psi + \theta_{1s}\sin\psi + \theta_{2c}\cos 2\psi + \theta_{2s}\sin 2\psi + \ldots$$

The mean and first harmonics of the blade motion (the 0, 1c, and 1s Fourier coefficients) are the harmonics most important to rotor performance and control. The rotor coning angle is β_0; β_{1c} and β_{1s} are, respectively, the pitch and roll angles of the tip-path plane relative to the hub plane. The rotor collective pitch is θ_0, and θ_{1c} and θ_{1s} are the cyclic pitch angles.

2.2.4 Rotor Angle-of-Attack and Velocity

> i = rotor disk plane incidence angle or angle-of-attack, positive for forward tilt (as required if a component of the rotor thrust is to provide the propulsive force for the helicopter).
> V = rotor or helicopter velocity with respect to the air.
> v = rotor induced velocity, normal to the disk plane and positive when downward through the disk (as produced by a positive rotor thrust).

The resultant velocity seen by the rotor, resolved into components parallel and normal to the disk plane and made dimensionless with the rotor tip speed ΩR, gives the following velocity ratios (Figure 2.3):

> $\mu = V\cos i/\Omega R =$ rotor advance ratio.
> $\mu_z = V\sin i/\Omega R =$ normal velocity ratio.
> $\lambda = (V\sin i + v)/\Omega R =$ rotor inflow ratio (positive for flow downward through the disk).
> $\lambda_i = v/\Omega R =$ induced inflow ratio.

The advance ratio μ is the ratio of the forward velocity to the rotor tip speed. The inflow ratio λ is the ratio of the total inflow velocity to the rotor tip speed.

2.2.5 Rotor Forces and Power

Rotor forces, relative to an appropriate axis system:

> T = rotor thrust, normal to the disk plane and positive when directed upward.
> H = rotor drag force in the disk plane; positive when directed rearward, opposing the forward velocity of the helicopter.

Y = rotor side force in the disk plane; positive when directed toward the advancing side of the rotor.
Q = rotor shaft torque, positive when an external torque is required to turn the rotor (helicopter operation).
P = rotor shaft power, positive when power is supplied to the rotor.

In coefficient form based on air density, rotor disk area, and tip speed these quantities are as follows:

C_T = thrust coefficient = $T/\rho A(\Omega R)^2$.
C_H = H force coefficient = $H/\rho A(\Omega R)^2$.
C_Y = Y force coefficient = $Y/\rho A(\Omega R)^2$.
C_Q = torque coefficient = $Q/\rho A(\Omega R)^2 R$.
C_P = power coefficient = $P/\rho A(\Omega R)^3$.

The rotor shaft power and torque are related by $P = \Omega Q$, so the coefficients are equal, $C_P = C_Q$. The rotor disk loading is the ratio of the thrust to the rotor area, T/A, and the power loading is the ratio of the thrust to the power, T/P. The rotor blade loading is the ratio of the thrust to the blade area, $T/A_b = T/(\sigma A)$, or in coefficient form the ratio of the thrust coefficient to solidity, C_T/σ.

2.2.6 Rotor Disk Planes

The rotor disk planes (defined in section 6.1.3) are denoted by:

TPP   tip-path plane
NFP   no-feathering plane
HP   hub plane
CP   control plane

2.3 Other Notation Conventions

No true standard nomenclature is used throughout the helicopter literature, so one must always take care to determine the definitions of the quantities used in any particular work, including the present text.

Helicopter and rotor force, moment, and power coefficients can be based on the quantity $\frac{1}{2}\rho A(\Omega R)^2$, instead of $\rho A(\Omega R)^2$ as in this text, often using the same symbols. See for example, Shapiro (1955).

One system of notation that is common enough in the literature to deserve attention is that used by the National Advisory Committee for Aeronautics (NACA). The primary deviations of NACA notation from the practice in this text are as follows:

$b =$ number of blades.
$x = r/R =$ dimensionless span variable.
$\theta_1 =$ linear twist rate (from the expansion $\theta = \theta_0 + \theta_1 r$).
$I_1 =$ rotor blade flapping inertia.
$\lambda = (V \sin\alpha - v)/\Omega R =$ rotor inflow ratio, positive when upward through the disk.
$\alpha =$ rotor disk incidence angle or angle-of-attack, positive for rearward tilt.

In addition, λ and α are assumed to refer to the no-feathering plane if there are no subscripts or other indication that another reference plane is being used. The blade motion is represented by Fourier series with the following definitions for the harmonics:

$$\beta = a_0 - a_1 \cos\psi - b_1 \sin\psi - a_2 \cos 2\psi - b_2 \sin 2\psi \ldots$$

$$\theta = A_0 - A_1 \cos\psi - B_1 \sin\psi - A_2 \cos 2\psi - B_2 \sin 2\psi \ldots$$

$$\zeta = E_0 + E_1 \cos\psi + F_1 \sin\psi + E_2 \cos 2\psi + F_2 \sin 2\psi \ldots$$

A subscript s is used for quantities measured with respect to the shaft or hub plane, for example A_{1s} and B_{1s}. The differences in sign from the present notation arise because the NACA notation was designed for autogyro analysis, and quantities were defined such that the parameters would usually have a positive value. The complete NACA notation system for helicopter analysis is given by Gessow (1948) and by Gessow and Myers (1952).

2.4 Geometry and Rotations

The nomenclature for geometry and rotations employs the following conventions. A vector x is a column matrix of three elements, measuring the vector relative to a particular basis (or axes, or frame). The basis is indicated as follows:

a) x^A is a vector measured in axes A.
b) $x^{EF/A}$ is a vector from point F to point E, measured in axes A.

A rotation matrix C is a three-by-three matrix that transforms vectors from one basis to another:

c) C^{BA} transforms vectors from basis A to basis B, so $x^B = C^{BA} x^A$.

The matrix C^{BA} defines the orientation of basis B relative to basis A and can also be viewed as rotating the axes from A to B. For a vector u, a cross-product matrix $\tilde{u}$ is defined as follows:

$$\tilde{u} = \begin{bmatrix} 0 & -u_3 & u_2 \\ u_3 & 0 & -u_1 \\ -u_2 & u_1 & 0 \end{bmatrix}$$

such that $\tilde{u}v$ is equivalent to the vector cross-product $\mathbf{u} \times \mathbf{v}$. The cross-product matrix enters the relationship between angular velocity and the time derivative of a rotation matrix:

$$\dot{C}^{AB} = -\tilde{\omega}^{AB/A} C^{AB} = C^{AB} \tilde{\omega}^{BA/B}$$

(the Poisson equations). For rotation by an angle α about the x, y, or z axis (1, 2, or 3 axis), the following notation is used:

$$X_\alpha = \begin{bmatrix} 1 & 0 & 0 \\ 0 & \cos\alpha & \sin\alpha \\ 0 & -\sin\alpha & \cos\alpha \end{bmatrix}$$

$$Y_\alpha = \begin{bmatrix} \cos\alpha & 0 & -\sin\alpha \\ 0 & 1 & 0 \\ \sin\alpha & 0 & \cos\alpha \end{bmatrix}$$

$$Z_\alpha = \begin{bmatrix} \cos\alpha & \sin\alpha & 0 \\ -\sin\alpha & \cos\alpha & 0 \\ 0 & 0 & 1 \end{bmatrix}$$

Thus for example, $C^{BA} = X_\phi Y_\theta Z_\psi$ means that the axes B are located relative to the axes A by first rotating by angle ψ about the z-axis, then by angle θ about the y-axis, and finally by angle ϕ about the x-axis.

2.5 Symbols, Subscripts, and Superscripts

Listed next alphabetically are the principal symbols used in this text. Symbols appearing only within one chapter are not included.

a	blade section two-dimensional lift-curve slope
A	rotor disk area, πR^2; total aircraft disk area
A_b	rotor blade area, $NcR = \sigma A$
B	tip loss factor
c	blade chord
C	Theodorsen's lift deficiency function
C'	Loewy's lift deficiency function
c_d	section drag coefficient, $D/(\tfrac{1}{2}\rho U^2 c)$
C_H	H-force coefficient, $H/\rho A(\Omega R)^2$
c_ℓ	section lift coefficient, $L/(\tfrac{1}{2}\rho U^2 c)$
c_m	section pitch moment coefficient, $M_a/(\tfrac{1}{2}\rho U^2 c^2)$
C_{Mx}	roll moment coefficient, $M_x/\rho A(\Omega R)^2 R$
C_{My}	pitch moment coefficient, $M_y/\rho A(\Omega R)^2 R$
C_P	power coefficient, $P/\rho A(\Omega R)^3$
C_{Pc}	climb power
C_{Pi}	induced power
C_{Po}	profile power
C_{Pp}	parasite power
C_Q	torque coefficient, $Q/\rho A(\Omega R)^2 R$
c_s	speed of sound
C_T	thrust coefficient, $T/\rho A(\Omega R)^2$
C_T/σ	ratio of thrust coefficient to solidity, $T/\rho A_b(\Omega R)^2$
C_Y	Y-force coefficient, $Y/\rho A(\Omega R)^2$
D	section aerodynamic drag force; rotorcraft drag; rotor diameter
$d(C_T/\sigma)/dr$	section blade loading, $F_z/(\rho(\Omega R)^2 c)$
D_e	rotor equivalent drag, $P/V + X$
e	flap or lag hinge offset from center of rotation
EI, EI_{zz}	flapwise bending stiffness
EI_{xx}	chordwise bending stiffness
f	equivalent drag area of rotorcraft airframe, including rotor hubs, $D/(\tfrac{1}{2}\rho V^2)$
F_r	section radial aerodynamic force
F_x	section aerodynamic force component parallel to disk plane
F_z	section aerodynamic force normal to disk plane
g	acceleration due to gravity
GJ	torsion stiffness

h	rotor mast height, distance of hub above helicopter center of gravity
H	rotor drag force, positive rearward; blade aerodynamic in-plane shear force coefficient (with subscript)
i	rotor disk plane incidence angle or angle-of-attack, positive for forward tilt
I_b	characteristic inertia of the rotor blade, normally $\int_0^R r^2 m\,dr$ or the flapping moment of inertia
I_f	blade pitch inertia, $\int_0^R I_\theta\,dr$
I_{pk}	generalized mass of k-th torsion mode, $\int_0^R \xi_k^2 I_\theta\,dr$
$I_{qk}, I_{\beta k}$	generalized mass of k-th out-of-plane bending mode, $\int_0^R \eta_{zk}^2 m\,dr$
I_x	rotorcraft roll moment of inertia; inertial flap-pitch coupling, $\int_0^R x_I r m\,dr$
I_y	rotorcraft pitch moment of inertia
I_z	rotorcraft yaw moment of inertia
I_β	generalized mass of fundamental flap mode, $\int_0^R \eta_\beta^2 m\,dr$
$I_{\beta\alpha}$	inertial coupling of flap and hub motion, $\int_0^R \eta_\beta r m\,dr$
$I_{\beta\zeta}$	Coriolis flap-lag coupling, $\int_0^R \eta_\beta \eta_\zeta m\,dr/(1-e)$
I_ζ	generalized mass of fundamental lag mode, $\int_0^R \eta_\zeta^2 m\,dr$
$I_{\zeta k}$	generalized mass of k-th in-plane bending mode, $\int_0^R \eta_{xk}^2 m\,dr$
$I_{\zeta\alpha}$	inertial coupling of lag and hub motion, $\int_0^R \eta_\zeta r m\,dr$
I_θ	section moment of inertia about feathering axis
I_0	blade rotational inertia, $\int_0^R r^2 m\,dr$
k	reduced frequency, $\omega b/U$ (ω is frequency, b airfoil semichord, and U free stream velocity)
k_x	rotorcraft roll radius of gyration, $I_x = Mk_x^2$
k_y	rotorcraft pitch radius of gyration, $I_y = Mk_y^2$
k_z	rotorcraft yaw radius of gyration, $I_z = Mk_z^2$
$K_P, K_{P\beta}$	pitch-flap coupling, $\Delta\theta = -K_P\beta$ ($K_P = \tan\delta_3$), positive for flap-up/pitch-down
$K_{P\zeta}$	pitch-lag coupling, $\Delta\theta = -K_{P\zeta}\zeta$, positive for lag-back/pitch-down
K_β	flap hinge spring constant
K_ζ	lag hinge spring constant
K_θ	control system spring constant
L	section aerodynamic lift force; rotorcraft roll moment stability derivative (with subscript); rotor wind-axis lift force
L/D_e	rotor equivalent lift-to-drag ratio, $L/(P/V + X)$
L/D	aircraft equivalent lift-to-drag ratio, WV/P
ℓ_{tr}	tail rotor distance behind main rotor shaft
m	blade index, $m = 1$ to N; aerodynamic pitch moment coefficient (with subscript); blade mass per unit length
M	hover figure of merit, $P_{\text{ideal}}/P = Tv/P = T\sqrt{T/2\rho A}/P$; blade section Mach number; rotorcraft mass, including rotor; rotorcraft pitch moment stability derivative (with subscript); blade aerodynamic flap moment coefficient (with subscript)
$\dot{m}$	mass flux through the rotor disk (momentum theory)
M_a	section aerodynamic pitch moment
M_b	blade mass, $\int_0^R m\,dr$

M_f	aerodynamic pitch moment
M_F	aerodynamic flap moment
M_L	aerodynamic lag moment
M_{tip}	blade tip Mach number, $\Omega R/c_s$
M_x	rotor hub roll moment, positive toward retreating side
M_y	rotor hub pitch moment, positive rearward
$M_{at}, M_{1,90}$	blade advancing-tip Mach number
$M^2 c_n$	blade section normal force, $N/(\frac{1}{2}\rho c_s^2 c)$
N	number of blades; rotorcraft yaw force stability derivative (with subscript); blade section normal force
$N_\star$	coupling parameter of flap dynamics, $N_\star = \frac{8}{\gamma}(\nu^2 - 1) + K_P$
N_F	blade root flapwise moment
N_L	blade root lagwise moment
p	sound pressure
P	power; rotor shaft power; rotorcraft power
p_k	generalized coordinate of k-th torsion mode (p_0 is the rigid pitch degree of freedom)
Q	rotor shaft torque, positive when external torque is required to turn rotor; blade aerodynamic torque or lag moment coefficient (with subscript)
q_k, q_{zk}	generalized coordinate of k-th out-of-plane bending mode
q_{xk}	generalized coordinate of k-th in-plane bending mode
r	blade or rotor disk radial coordinate
R	rotor radius; blade aerodynamic radial shear force coefficient (with subscript)
Re	Reynolds number
s	eigenvalue or Laplace variable
S_b	blade first moment of inertia, $\int_0^R rm\,dr$
S_r	blade root radial shear force
S_x	blade root in-plane shear force
S_z	blade root vertical shear force
S_β	first moment of flap mode, $\int_0^R \eta_\beta m\,dr$
S_ζ	first moment of lag mode, $\int_0^R \eta_\zeta m\,dr$
t	time
T	rotor thrust, positive upward for lifting rotor or forward for propelling rotor; blade aerodynamic thrust force coefficient (with subscript)
T/A	rotor disk loading; rotorcraft disk loading
T/A_b	rotor blade loading
U	section resultant velocity, $(u_T^2 + u_P^2)^{1/2}$
u_G	longitudinal gust velocity component
u_P	air velocity of blade section, perpendicular to the disk plane
u_R	air velocity of blade section, radial
u_T	air velocity of blade section, tangent to the disk plane
v	rotor induced velocity (positive down through the disk)
V	rotor or rotorcraft velocity with respect to the air
v_G	lateral gust velocity component
v_h	ideal hover induced velocity, $\sqrt{T/2\rho A}$
w	rotor induced velocity in the far wake

W	rotorcraft gross weight
w_G	vertical gust velocity component
x	rotor non-rotating coordinate axis, positive aft; blade in-plane deflection; blade chordwise coordinate
X	rotorcraft longitudinal force derivative (with subscript); rotor wind-axis drag force
x_A	chordwise offset of blade aerodynamic center behind pitch axis
x_B	rotorcraft rigid-body longitudinal degree of freedom
x_h	hub longitudinal displacement
x_I	chordwise offset of blade center of gravity behind pitch axis
y	rotor non-rotating coordinate axis, positive to right (advancing side)
Y	rotor side force, positive toward advancing side; rotorcraft side force stability derivative (with subscript)
y_B	rotorcraft rigid body lateral degree of freedom
y_h	hub lateral displacement
z	rotor non-rotating coordinate axis, positive upward; blade out-of-plane deflection
Z	rotorcraft vertical force stability derivative (with subscript)
z_B	rotorcraft rigid-body vertical degree of freedom
z_h	hub vertical displacement
α	blade section angle-of-attack
α_x	hub roll perturbation
α_y	hub pitch perturbation
α_z	hub yaw perturbation
$\alpha_{1,270}$	blade retreating tip angle-of-attack
β	blade flap angle (positive upward)
β_p	precone angle
β_0	coning angle
β_{1c}	longitudinal tip-path-plane tilt angle, positive forward
β_{1s}	lateral tip-path-plane tilt angle, positive toward retreating side
γ	blade Lock number, $\rho a c R^4 / I_b$
Γ	blade bound circulation
$\delta_0, \delta_1, \delta_2$	coefficients in expansion for section drag, $c_d = \delta_0 + \delta_1 \alpha + \delta_2 \alpha^2$
δ_3	pitch-flap coupling ($K_P = \tan \delta_3$), positive for flap-up/pitch-down
ζ	blade lag angle, positive opposite the direction of rotation of the rotor
η, η_β	mode shape of fundamental flap mode
η_ζ	mode shape of fundamental lag mode
η_k, η_{zk}	mode shape of k-th out-of-plane bending mode
η_{xk}	mode shape of k-th in-plane bending mode
θ	blade pitch or feathering angle, positive nose upward
θ_B	rotorcraft rigid body pitch degree of freedom
θ_{con}	pitch control input (collective and cyclic)
θ_e	elastic torsion deflection
θ_{FP}	flight path angle, climb velocity $= V \sin \theta_{FP}$
θ_{tw}	linear twist rate (negative for tip pitch smaller than root pitch)
θ_0	collective pitch angle
θ_{1c}	lateral cyclic pitch angle
θ_{1s}	longitudinal cyclic pitch angle

$\theta_{.75}$	collective pitch angle at 75% radius
κ	induced power factor, ratio induced power to ideal power
λ	rotor inflow ratio, $(V\sin i + v)/\Omega R$, positive down through disk
λ_c	climb inflow ratio
λ_i	induced inflow ratio, $v/\Omega R$
λ_x	coefficient of longitudinal variation of induced velocity
λ_y	coefficient of lateral variation of induced velocity
λ_0	rotor mean induced velocity
μ	rotor advance ratio, $V\cos i/\Omega R$
μ_z	rotor axial velocity ratio, $V\sin i/\Omega R = \mu\tan i$
ν, ν_β	rotating natural frequency of blade fundamental flap mode
$\nu_e, \nu_{\beta e}$	effective flap frequency including pitch-flap coupling, $\nu_e^2 = \nu^2 + (\gamma/8)K_P$
ν_k, ν_{zk}	natural frequency of k-th out-of-plane bending mode
ν_{xk}	natural frequency of k-th in-plane bending mode
ν_ζ	rotating natural frequency of blade fundamental lag mode
ξ_k	mode shape of k-th elastic torsion mode
ρ	air density; blade radial coordinate in spanwise integration
σ	rotor solidity, $A_b/A = Nc/\pi R$
ϕ	section inflow angle, $\tan^{-1} u_P/u_T$
ϕ_B	rotorcraft rigid body roll degree of freedom
ψ	azimuth angle of the blade or rotor disk; dimensionless time, Ωt
ψ_B	rotorcraft rigid body yaw degree of freedom
ψ_m	azimuth position of m-th blade ($m = 1$ to N)
$\omega, \omega_0, \omega_\theta$	natural frequency of rigid pitch motion (control system stiffness)
ω_k	natural frequency of k-th elastic torsion mode
Ω	rotor rotational speed (rad/sec)

Subscripts and Superscripts

$0, 1c, 1s, \ldots nc, ns, \ldots \infty$	harmonics of a Fourier series representation of a periodic function
$0, 1c, 1s, \ldots nc, ns, \ldots N/2$	multiblade coordinate degrees of freedom (total number N)
c	climb
CP	control plane
h	hover
HP	hub plane
i	induced
m	blade index, $m = 1$ to N
mr	main rotor
NFP	no-feathering plane
o	profile
p	parasite
p	rotorcraft stability derivative due to roll rate
q	rotorcraft stability derivative due to pitch rate
r	rotorcraft stability derivative due to yaw rate
TPP	tip-path plane

tr	tail rotor
u	rotorcraft stability derivative due to longitudinal velocity
v	rotorcraft stability derivative due to lateral velocity
w	rotorcraft stability derivative due to vertical velocity
β	rotor aerodynamic force due to blade flap displacement
$\dot\beta$	rotor aerodynamic force due to blade flapping velocity or hub angular motion
ζ	rotor aerodynamic force due to blade lag displacement
$\dot\zeta$	rotor aerodynamic force due to blade lagging velocity or hub yawing motion
θ	rotor aerodynamic force due to blade pitch motion
$\dot\theta$	rotor aerodynamic force due to blade pitch rate
λ	rotor aerodynamic force due to hub vertical velocity or induced velocity perturbation
μ	rotor aerodynamic force due to hub in-plane velocity
$(\dot{\ })$	$d(\ )/dt$ or $d(\ )/d\psi$
$(\ )'$	$d(\ )/dr$
$\widetilde{(\ )}$	normalized: rotor blade inertias divided by I_b; rotorcraft inertias divided by $\tfrac{1}{2}NI_b$; dimensionless section aerodynamic forces divided by c

Abbreviations

BEM	combined blade element and momentum theory
BVI	blade-vortex interaction
CFD	computational fluid dynamics
CP	control plane
HHC	higher harmonic control
HP	hub plane
HQR	handling qualities rating
HSI	high-speed impulsive
IBC	individual blade control
IGE	in ground effect
MBC	multi-blade coordinates
NFP	no-feathering plane
OGE	out of ground effect
TPP	tip-path plane
VRS	vortex ring state
VTOL	vertical takeoff and landing

2.6 REFERENCES

Gessow, A. "Standard Symbols for Helicopters." NACA TN 1604, June 1948.

Gessow, A., and Myers, G.C., Jr. *Aerodynamics of the Helicopter*. New York: The Macmillan Company, 1952.

Shapiro, J. *Principles of Helicopter Engineering*. London: Temple Press Limited, 1955.

3 Hover

Hover is the operating state in which the lifting rotor has no velocity relative to the air, either vertical or horizontal. General vertical flight involves axial flow with respect to the rotor. Vertical flight implies axial symmetry of the rotor flow field, so the velocities and loads on the rotor blades are independent of the azimuth position. Axial symmetry greatly simplifies the dynamics and aerodynamics of the helicopter rotor, as is evident when forward flight is considered. The basic analyses of a rotor in axial flow originated in the 19th century with the design of marine propellers and were later applied to airplane propellers. The principal objectives of the analysis of the hovering rotor are to predict the forces generated and power required by the rotating blades and to design the most efficient rotor.

3.1 Momentum Theory

Momentum theory applies the basic conservation laws of fluid mechanics (conservation of mass, momentum, and energy) to the rotor and flow as a whole to estimate the rotor performance. The theory is a global analysis, relating the overall flow velocities to the total rotor thrust and power. Momentum theory was developed for marine propellers by W.J.M. Rankine in 1865 and R.E. Froude in 1885, and extended in 1920 by A. Betz to include the rotation of the slipstream; see Glauert (1935) for the history.

The rotor disk supports a thrust created by the action of the air on the blades. By Newton's law there must be an equal and opposite reaction of the rotor on the air. As a result, the air in the rotor wake acquires a velocity increment directed opposite to the thrust. There is kinetic energy in the wake flow field, energy that must be supplied by the rotor. This energy constitutes the induced power of a rotary wing, corresponding to the induced drag of a fixed wing.

Momentum conservation relates the rotor thrust per unit mass flow through the disk, $T/\dot{m}$, to the induced velocity in the far wake, w. Energy conservation relates $T/\dot{m}$, w, and the induced velocity at the rotor disk, v. Finally, mass conservation gives $\dot{m}$ in terms of the induced velocity v. Eliminating w then gives a relation between the induced power and the rotor thrust, which is the principal result of momentum theory. Momentum theory is not concerned with the details of the rotor airloads or flow, and hence is not sufficient for designing the blades. What momentum theory provides is an estimate of the induced power requirement of the rotor and of the ideal performance limit.

3.1.1 Actuator Disk

In momentum theory the rotor is modeled as an actuator disk, which is a circular surface of zero thickness that can support a pressure difference (but no velocity change) and thus accelerate the air through the disk. The loading is assumed to be steady, but in general can vary over the surface of the disk. The actuator disk can also support a torque, which imparts angular momentum to the fluid that passes through the disk. The task of the analysis is to determine the influence of the actuator disk on the flow and, in particular, to find the induced velocity and power for a given thrust. Momentum theory solves this problem using the basic conservation laws of fluid motion; vortex theory uses the Biot-Savart law for the velocity induced by the wake vorticity; and potential theory solves the fluid dynamic equations for the velocity potential or stream function. For the same model, all three methods must give identical results.

The actuator disk model is only an approximation to the actual rotor. Distributing the rotor blade loading over a disk is equivalent to considering an infinite number of blades. The detailed flow of the actuator disk is thus very different from that of a real rotor with a small number of blades. The real flow field is actually unsteady, with a wake of discrete vorticity corresponding to the discrete loading. The actual induced power is therefore larger than the momentum theory result because of the nonuniform and unsteady induced velocity. The approximate nature of the actuator disk model imposes a fundamental limit on the applicability of extended momentum or vortex theories. The principal use of the actuator disk model is to obtain a first estimate of the wake-induced flow, and hence of the ideal induced power.

3.1.2 Momentum Theory in Hover

Consider an actuator disk of area A and total thrust T (Figure 3.1). The loading is assumed to be distributed uniformly over the disk. Let v be the induced velocity at the rotor disk and w the wake-induced velocity infinitely far downstream. A well-defined, smooth slipstream is assumed, with v and w uniform over the slipstream cross-section. The rotational energy in the wake due to the rotor torque is neglected for now. The fluid is incompressible and inviscid. The mass flux through the disk is $\dot{m} = \rho A v$; by conservation of mass, the mass flux is constant all along the wake. Momentum conservation equates the rotor force to the rate of change of momentum, which is the momentum flowing out at station 1 less the momentum flowing in at station 0 (Figure 3.1). The flow far upstream is at rest for the hovering rotor, so $T = \dot{m}v$. Energy conservation equates the work done by the rotor to the rate of change of energy in the fluid, which is the kinetic energy flowing out at station 1 less the kinetic energy flowing in at station 0; hence $Tv = \frac{1}{2}\dot{m}w^2$. Eliminating $T/\dot{m}$ from the momentum and energy conservation relations gives $w = 2v$; the induced velocity in the far wake is twice that at the rotor disk. This is the same result as for an elliptically loaded fixed wing. Since the mass flux and density are constant, the area of the slipstream in the far wake (station 1) is $\frac{1}{2}A$.

Alternatively, this result can be obtained using Bernoulli's equation, which is an integrated form of the energy equation for the fluid. The pressure in the far wake

3.1 Momentum Theory

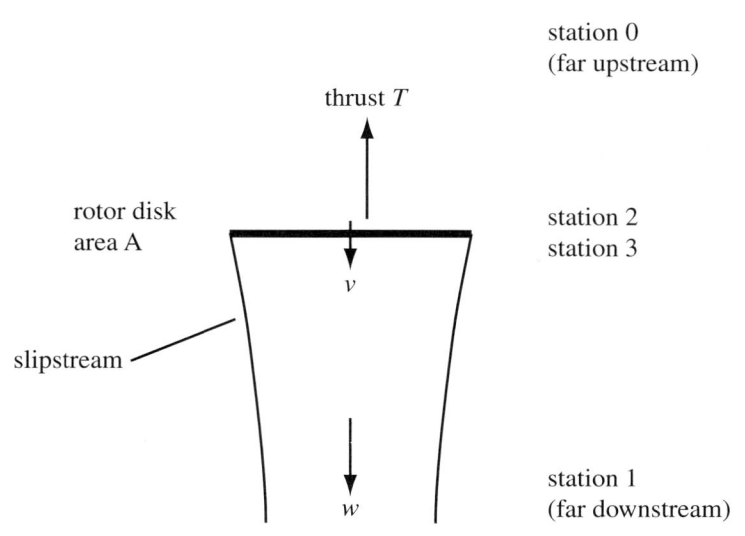

Figure 3.1. Momentum theory flow model for hover.

(station 1) is assumed to be at the ambient level p_0; this is equivalent to neglecting the swirl in the wake. Applying Bernoulli's equation between stations 0 and 2 gives $p_0 = p_2 + \frac{1}{2}\rho v^2$; applying it between stations 3 and 1 gives $p_3 + \frac{1}{2}\rho v^2 = p_0 + \frac{1}{2}\rho w^2$. Combining these equations, we obtain

$$T/A = p_3 - p_2 = \frac{1}{2}\rho w^2 \qquad (3.1)$$

With $\dot{m} = \rho A v$, this becomes

$$Tv = \frac{1}{2}\dot{m}w^2 \qquad (3.2)$$

as before. The total pressure in the fully developed wake is $p_0 + \frac{1}{2}\rho w^2 = p_0 + T/A$. The increase in total head due to the actuator disk is equal to the disk loading T/A, which for helicopters is very small compared to p_0. Therefore, the over-pressure in the helicopter wake is small, although the wake velocities can still be fairly high. The pressure in the slipstream falls from p_0 to $p_2 = p_0 - \frac{1}{2}\rho v^2 = p_0 - \frac{1}{4}T/A$ just above the disk, and from $p_3 = p_0 + \frac{3}{2}\rho v^2 = p_0 + \frac{3}{4}T/A$ just below the disk to p_0 in the far wake. So there is always a falling pressure except across the rotor disk, where the pressure increase accelerates the flow.

Momentum theory thus relates the rotor thrust to the induced velocity at the rotor disk by $T = \dot{m}v = 2\rho A v^2$. The induced velocity in hover is

$$v = \sqrt{T/2\rho A} \qquad (3.3)$$

The induced power for hover is

$$P = Tv = T\sqrt{T/2\rho A} \qquad (3.4)$$

In coefficient form, based on the rotor tip speed ΩR, these results become $\lambda = \sqrt{C_T/2}$ and $C_P = C_T \lambda = C_T^{3/2}/\sqrt{2}$. The subscript "$h$" designates the ideal induced velocity of the hovering rotor: $v_h = \sqrt{T/2\rho A}$ or $\lambda_h = \sqrt{C_T/2}$.

Momentum theory gives the induced power per unit thrust for a hovering rotor:

$$P/T = v = \sqrt{T/2\rho A} \qquad (3.5)$$

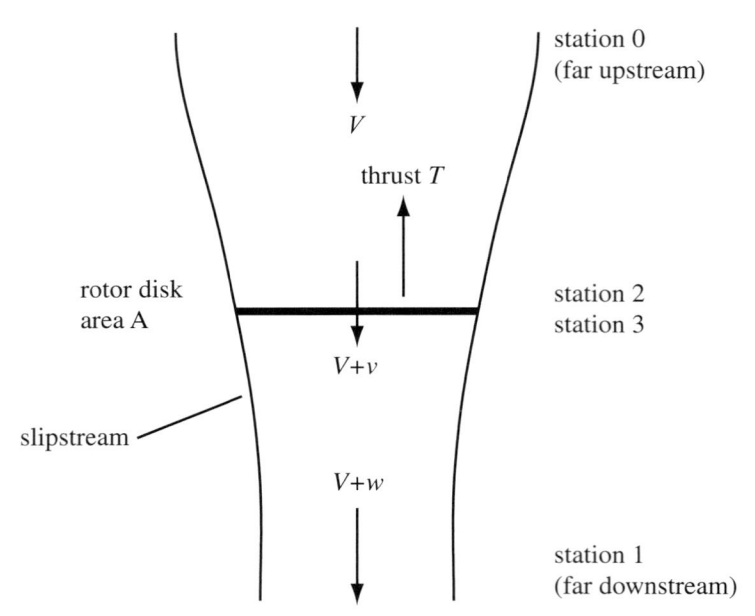

Figure 3.2. Momentum theory flow model for climb.

This relation determines the basic characteristics of the helicopter. Equation 3.5 is based on the fundamental physics of fluid flow, which imply that for a low inflow velocity and hence low induced power the air must be accelerated through the disk by a small pressure differential. To hover efficiently requires a small value of P/T (for low fuel and engine weight), which demands that the disk loading T/A be low. With $T/A = 5$ to 15 lb/ft², the helicopter has the lowest disk loading and therefore the best hover performance of all VTOL aircraft. The parameter determining the induced power is really $T/\rho A$, so the induced velocity increases with altitude and temperature, as the air density decreases. The dynamic pressure increase in the far wake remains T/A, independent of density.

As for fixed wings, uniform induced velocity gives the minimum induced power for a given thrust. This can be proved using the calculus of variations, as follows. The problem is to minimize the kinetic energy of the wake $KE \propto \int v^2 dA$ for a given thrust or wake momentum $\int v \, dA$. Write the induced velocity, $v = \bar{v} + \delta v$, as a mean or uniform value $\bar{v}$ plus a perturbation δv, for which $\int \delta v \, dA = 0$. Then $\int v^2 dA = \bar{v}^2 A + \int (\delta v)^2 dA$, and minimum kinetic energy requires that $\delta v = 0$ over the entire disk; hence that the inflow be uniform. With nonuniform inflow the areas on the disk with high local loading cost more power than is gained from the areas with low loading.

3.1.3 Momentum Theory in Climb

Now consider momentum theory for a rotor in a vertical climb at velocity V (Figure 3.2). The basic assumptions are the same as for the hover analysis: actuator disk model, uniform loading, a well-defined and smooth slipstream, uniform induced velocity, slipstream circumferential velocities neglected, and ideal fluid. The mass flux is now $\dot{m} = \rho A (V + v)$. Momentum and energy conservation give $T = \dot{m}(V + w) - \dot{m}V = \dot{m}w$ and $T(V + v) = \frac{1}{2}\dot{m}(V + w)^2 - \frac{1}{2}\dot{m}V^2 = \frac{1}{2}\dot{m}w(w + 2V)$, respectively. The momentum conservation equation is independent of V. Eliminating $T/\dot{m}$ gives again $w = 2v$, as for hover; the induced velocity in the far wake is twice that at the rotor disk. The total pressure in the far wake is now $p_0 + \frac{1}{2}\rho(V + w)^2 = p_0 + \frac{1}{2}\rho V^2 + T/A$.

For the climbing rotor, the relation between the thrust and induced velocity becomes $T = \dot{m}w = 2\rho A(V + v)v$. Again define v_h as

$$v_h = \sqrt{T/2\rho A} \tag{3.6}$$

so that $v(V + v) = v_h^2$, which has the solution

$$v = -\frac{V}{2} + \sqrt{\left(\frac{V}{2}\right)^2 + v_h^2} \tag{3.7}$$

Thus climb reduces the induced velocity v. The induced and climb power is

$$P = T(V + v) = T\left(\frac{V}{2} + \sqrt{\left(\frac{V}{2}\right)^2 + v_h^2}\right) \tag{3.8}$$

Finally, the velocity in the far wake is $V + w = V + 2v = \sqrt{V^2 + 4v_h^2}$. For very large climb rates the induced velocity v is approximately v_h^2/V, and the power approaches only the climb power TV. For small rates of climb ($V \ll v_h$, which is generally true for helicopter rotors), the induced power P is approximately equal to $T(V/2 + v_h) = P_h + \frac{1}{2}TV$. The power required increases with V, but the climb power increment is reduced by the induced power decrease.

3.2 Hover Power

Momentum theory gives the induced power of an ideal rotor in hover, $C_{Pi} = C_T^{3/2}/\sqrt{2}$. A real rotor has other power losses as well, in particular the profile power due to the drag of the blades in a viscous fluid. There is also an induced power due to the nonuniform inflow of a non-optimum rotor design. The swirl in the wake due to the rotor shaft torque is another loss, although it is usually small for helicopter rotors. Finally, the hovering rotor has tip losses as a result of the discreteness and periodicity in the wake, when the number of blades is finite. The distribution of the power losses of the rotor in hover is approximately as follows:

Power component	At peak efficiency	Off peak
Ideal induced power	74% to 78%	65%
Profile power	10% to 19%	25%
Nonuniform inflow	5% to 7%	6%
Swirl in the wake	less than 1%	less than 1%
Tip losses	2% to 4%	3%

The main rotor absorbs most of the helicopter power, but there are other losses as well. The engine and transmission absorb 3% to 5% of the total power with turbine engines, or 5% to 9% with reciprocating engines. The turbine engine has larger transmission losses since its high rotational speed requires more reduction, whereas the piston engine has significant losses for cooling. The tail rotor absorbs about 7% to 9% of the total helicopter power, and there is an additional loss of about

2% due to aerodynamic interference (rotor-fuselage and rotor-rotor). A tandem rotor helicopter has about the same total loss of 9% to 11%, which is primarily due to aerodynamic interference but also includes some additional drive train losses. The tail rotor and aerodynamic interference power losses are much smaller for the helicopter in forward flight. Because of the fuselage download produced by the rotor wake, the rotor thrust is 4% to 7% (or more) larger than the gross weight, producing a corresponding increase in required power.

3.3 Figure of Merit

The figure of merit is a measure of rotor hovering efficiency, defined as the ratio of the minimum possible power required to hover to the actual power required to hover. Thus the figure of merit compares the actual rotor performance with the performance of an ideal rotor, which has only the inescapable induced power:

$$M = \frac{P_{\text{ideal}}}{P} \tag{3.9}$$

Momentum theory gives the optimum induced power as $P_{\text{ideal}} = Tv = T\sqrt{T/2\rho A}$. Hence the figure of merit is

$$M = \frac{T\sqrt{T/2\rho A}}{P} = \frac{C_T^{3/2}/\sqrt{2}}{C_P} \tag{3.10}$$

The figure of merit $M = Tv/P$ is similar to the propulsive efficiency $\eta =$ propulsive power/input power $= TV/P$. The latter is appropriate for a propulsive device but not for a hovering rotor, where the useful power is that required to produce static thrust. The generalized efficiency factor $\eta = T(V+v)/P$ can be used over the entire range of axial flow.

In terms of the induced and profile power contributions to the rotor power, the figure of merit can be written as $M = C_{P\text{ideal}}/(C_{Pi} + C_{Po})$. Usually the profile power C_{Po} is at least 10% to 20% of the total power, and the induced power C_{Pi} is 10% to 15% higher than the ideal power. Thus the figure of merit is a measure of the ratio of the profile power to the induced power. However, the figure of merit can be misleading, since M is not directly concerned with the total hover power. By increasing the disk loading T/A, the induced power is increased relative to the profile power, resulting in a higher figure of merit. However, the total power required also then increases, which is unlikely to be considered an improvement in the rotor efficiency. The use of the figure of merit to compare rotors is thus best restricted to constant disk loading. Within this limitation M is a valuable measure of the rotor aerodynamic efficiency. The figure of merit is particularly useful for comparing rotors with different airfoil sections and for examining the influence of other design parameters such as twist or planform.

The ideal figure of merit is $M = 1$. M is lower for a real rotor because of profile and non-optimum induced power losses. The figure of merit for a given rotor is typically presented as a function of the blade loading C_T/σ (the ratio of the rotor thrust coefficient to solidity), which is a measure of the mean angle-of-attack of the blade. For current well-designed rotors the maximum figure of merit is typically $M = 0.74$ to 0.78. An inefficient rotor could have a maximum figure of merit as low as $M = 0.50$. The figure of merit decreases at low C_T/σ because of the reduced disk loading, and at high C_T/σ because of stall (which increases the profile power). At the design loading of the rotor, a figure of merit of $M = 0.70$ is typical. For sea level density, the definition of the figure of merit gives $T/P = 37.9M/\sqrt{T/A}$ when the power loading T/P is in lb/hp and the disk loading T/A is in lb/ft^2. Thus a helicopter disk loading of 5 to 15 lb/ft^2 implies a corresponding power loading of 12 to 7 lb/hp.

3.4 Extended Momentum Theory

The most important and useful results of momentum theory can be obtained by quite simple analyses, as in section 3.1. A more detailed analysis is not easily justified because of the basic limitations of the actuator disk model, yet there are some useful things to be learned from an extended momentum theory for axial flow. Momentum theory was extensively developed in the early part of the 20th century for airplane propellers. Here we examine momentum theory more rigorously for the rotor in climb or hover, including the effects of swirl in the rotor wake.

The integral conservation laws of fluid dynamics are as follows:

$$\rho \int \mathbf{q} \cdot \mathbf{n} \, dS = 0 \tag{3.11}$$

$$\rho \int \mathbf{q}\mathbf{q} \cdot \mathbf{n} \, dS + \int p\mathbf{n} \, dS = \mathbf{F}_{\text{body}} \tag{3.12}$$

$$\rho \int \mathbf{r} \times \mathbf{q}\mathbf{q} \cdot \mathbf{n} \, dS + \int p\mathbf{r} \times \mathbf{n} \, dS = \mathbf{M}_{\text{body}} \tag{3.13}$$

$$\int (p + \frac{1}{2}\rho q^2) \mathbf{q} \cdot \mathbf{n} \, dS = \frac{dE}{dt} \tag{3.14}$$

for mass, momentum, angular momentum, and energy conservation, respectively (in vector form). The flow is assumed to be steady (in a frame moving with the rotor) and incompressible, and there are no viscous losses on the surface of a body in the fluid. Here dS is the differential area at $\mathbf{r}$ of a surface enclosing the fluid, $\mathbf{n}$ is the outward normal to the surface, and $\mathbf{q}$ is the velocity of the fluid. The force and moment on the body in the fluid (the rotor in this case) are $\mathbf{F}_{\text{body}}$ and $\mathbf{M}_{\text{body}}$, and dE/dt is the power being added to the flow. Application of the mass and energy conservation laws to a stream tube gives Bernoulli's equation: $p + \frac{1}{2}\rho q^2 = $ constant, if no energy is added. The momentum theory extensions to be developed here are based on the application of these conservation relations to a rotor in axial flow.

Momentum theory is a calculus of variations problem. A function of the rotor radius r must be found, such as the induced velocity $v(r)$, that minimizes the power for a given thrust. Consider expressions for the power and thrust as integrals over the rotor disk: $P = \int F(r, v) dA$ and $T = \int G(r, v) dA$, where $dA = 2\pi r \, dr$ for axisymmetric conditions. Using a Lagrange multiplier λ, let $I = P - \lambda T$. The solution $v(r)$ for minimum P subject to the constraint T is given by the stationary values of the first variation of I, namely

$$\delta I = \delta \int (F - \lambda G) dA = \int \left(\frac{\partial F}{\partial v} - \lambda \frac{\partial G}{\partial v} \right) \delta v \, dA = 0 \tag{3.15}$$

Thus the optimum $v(r)$ is given by the solution of the Euler equation,

$$\frac{\partial}{\partial v}\left(F - \lambda G\right) = 0 \tag{3.16}$$

Hover

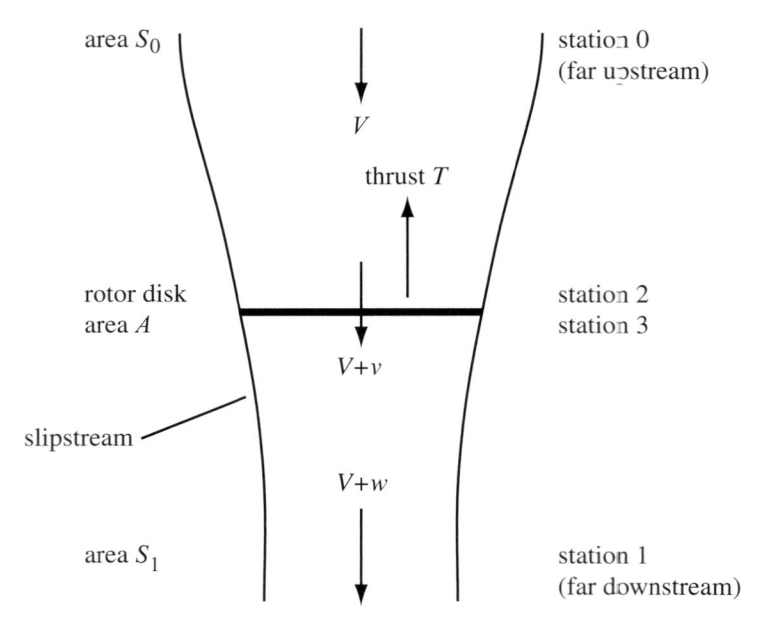

[Figure: Flow model diagram showing rotor with area S₀ far upstream, velocity V; rotor disk area A with thrust T, velocity V+v at disk (stations 2 and 3); slipstream; area S₁ far downstream with velocity V+w (station 1).]

Figure 3.3. Flow model for rotor in hover or climb.

If the integrands F and G are independent of r, the Euler equation is of the form

$$\text{function}(v) = \text{constant} \tag{3.17}$$

which has the solution $v = $ constant.

3.4.1 Rotor in Hover or Climb

Consider a rotor with thrust T operating in hover or in vertical climb at speed V (Figure 3.3). The rotor is represented by an actuator disk, which can support a pressure jump but not an axial velocity discontinuity. A well-defined, smooth slipstream is assumed, and for now the energy losses due to the angular momentum in the wake are neglected. Consider a control volume bounded by the slipstream and the disks of area S_0 and S_1 far upstream and downstream, respectively. The pressure at station 0 is p_0, and the velocity is V. At station 1, far downstream, the pressure is again p_0 since the wake swirl is neglected. Energy conservation then shows that outside the slipstream at station 1 the velocity is everywhere equal to V. Mass and momentum conservation give

$$V S_0 = \int (V + v) dA = \int (V + w) dS_1 \tag{3.18}$$

$$T = \int \Delta p \, dA = \int \rho (V + w) w \, dS_1 \tag{3.19}$$

where Δp is the pressure difference across the rotor disk. There is a net pressure reaction on the ends of the control volume equal to $p_0(S_0 - S_1)$, which is exactly canceled by the pressure on the slipstream, as can be established by considering momentum conservation for the fluid outside the slipstream. Energy conservation gives

$$P = \int \Delta p (V + v) dA = \int \frac{1}{2} \rho (V + w)^3 dS_1 - \frac{1}{2} \rho V^3 S_0 \tag{3.20}$$

3.4 Extended Momentum Theory

or, using mass conservation,

$$P = \int \Delta p (V + v) dA = \int \frac{1}{2}\rho(V + w)(2Vw + w^2) dS_1 \qquad (3.21)$$

The first expression is the work done in moving the air through the disk, and the second is the kinetic energy added to the slipstream. Subtracting TV from P gives $\int \Delta p\, v\, dA = \int \frac{1}{2}\rho(V + w)w^2 dS_1$, which can be interpreted as $\int v\, dT = \int \frac{1}{2}w\, dT$. Thus the rotor thrust and power have been expressed in terms of the induced velocity in the far wake, $w(r)$, which in general can vary over the wake section. Consider now the following optimization problem: find the function $w(r)$ that minimizes the power $P = \int \frac{1}{2}\rho(V + w)(2Vw + w^2) dS_1$ for a given thrust $T = \int \rho(V + w)w\, dS_1$. This is a calculus of variations problem with a constraint. Since the integrands of P and T are independent of r, the solution of the Euler equation is simply $w = $ constant.

Bernoulli's equation applied along streamlines above and below the rotor gives

$$p_0 + \frac{1}{2}\rho V^2 = p_2 + \frac{1}{2}\rho(V + v)^2 \qquad (3.22)$$

$$p_2 + \Delta p + \frac{1}{2}\rho(V + v)^2 = p_0 + \frac{1}{2}\rho(V + w)^2 \qquad (3.23)$$

Combining these equations, we obtain $\Delta p = \frac{1}{2}\rho(2Vw + w^2)$. Since the wake-induced velocity w is uniform, the disk loading Δp is also uniform. At the rotor disk, $p_2 + \frac{1}{2}\rho(V + v)^2 = $ constant, so the pressure and induced velocity there are not necessarily uniform. The assumptions of uniform Δp and w made in section 3.1 are thus validated, but momentum theory does not provide the distribution of the induced velocity v at the rotor disk. This result is analogous to the Trefftz plane analysis of a fixed wing, which shows that the minimum induced drag is obtained with uniform downwash in the far wake and elliptical loading, but tells nothing about the induced angle-of-attack at the wing. Lifting-line theory or lifting-surface theory is required to find the induced angle-of-attack, which is needed to design the wing to achieve the optimum loading. From lifting-line theory the downwash at the wing is one-half the downwash at the far wake, and therefore the optimal solution can be used directly in the wing design. Similarly with rotors the inflow is assumed to be also uniform at the disk, and $v = \frac{1}{2}w$. The assumption of uniform inflow for the rotor in axial flight is consistent with the accuracy of the actuator disk model.

For uniform loading Δp and far wake inflow w, the conservation relations become

$$(V + \bar{v})A = (V + w)S_1 \qquad (3.24)$$

$$T = \Delta p A = \rho(V + w)w S_1 \qquad (3.25)$$

$$P = (V + \bar{v})T = \left(V + \frac{1}{2}w\right)T \qquad (3.26)$$

where $\bar{v} = \int v\, dA/A$ is the mean induced velocity at the rotor disk. The energy balance then gives $\bar{v} = \frac{1}{2}w$, so that, although the distribution of v is not known, the mean $\bar{v}$ has the same value as that obtained earlier assuming uniform inflow. When the far wake parameters S_1 and w are eliminated, the rotor thrust and power are given by

$$T = 2\rho A(V + \bar{v})\bar{v} \tag{3.27}$$

$$P = T(V + \bar{v}) \tag{3.28}$$

This is the same result as in section 3.1.3, here given in terms of the mean induced velocity.

Replacing the integral conservation relations by their differential forms is customary:

$$(V + v)dA = (V + w)dS_1 \tag{3.29}$$

$$dT = \Delta p\, dA = \rho(V + w)w\, dS_1 \tag{3.30}$$

$$dP = \Delta p(V + v)dA = \frac{1}{2}\rho(V + w)(2Vw + w^2)dS_1 \tag{3.31}$$

The energy equation then gives $v = \frac{1}{2}w$, and eliminating dS_1 and w gives

$$dT = 2\rho\, dA(V + v)v \tag{3.32}$$

$$dP = dT(V + v) \tag{3.33}$$

There is no strict justification for this differential form of momentum theory, which assumes that there is no mutual interference of the disk elements. The key to the result is the assumption that $v = \frac{1}{2}w$ is valid for individual streamlines, which allows the thrust and power to be expressed entirely in terms of v. The differential form of momentum theory can be applied to rotors with nonuniform loading and inflow.

3.4.2 Swirl in the Wake

Consider next the effect of the swirl velocities in the rotor wake, which are due to the rotor induced torque. For shaft-driven rotors, the power and torque are related by $P = \Omega Q$, where Ω is the rotor rotational speed. The rotor must therefore add rotational kinetic energy to the wake corresponding to the torque. For helicopter rotors the swirl energy is small compared to the axial downwash energy, so only a small correction to the induced power is sought. Figure 3.4 shows the flow model considered. There are circumferential velocities $u(r)$ just below the rotor disk and $u_1(r_1)$ in the far wake. Angular momentum conservation inside the slipstream shows that there can be no swirl velocities above the disk; that is, the flow remains irrotational until it passes through the rotor. When there is a rotational velocity in the far wake, the pressure no longer has the static value p_0; instead $dp_1/dr_1 = \rho u_1^2/r_1$, and $p_1 = p_0$ at the boundary of the slipstream (r_1 is the radial coordinate at station 1). This pressure gradient provides the centripetal force required to support the rotational velocity of the fluid inside the wake.

Requirements for the conservation of mass, axial momentum, angular momentum, and energy give the following relations:

$$VS_0 = \int (V + v)dA = \int (V + w)dS_1 \tag{3.34}$$

$$T = \int \Delta p\, dA = \int \rho(V + w)w\, dS_1 + \int (p_1 - p_0)dS_1 \tag{3.35}$$

3.4 Extended Momentum Theory

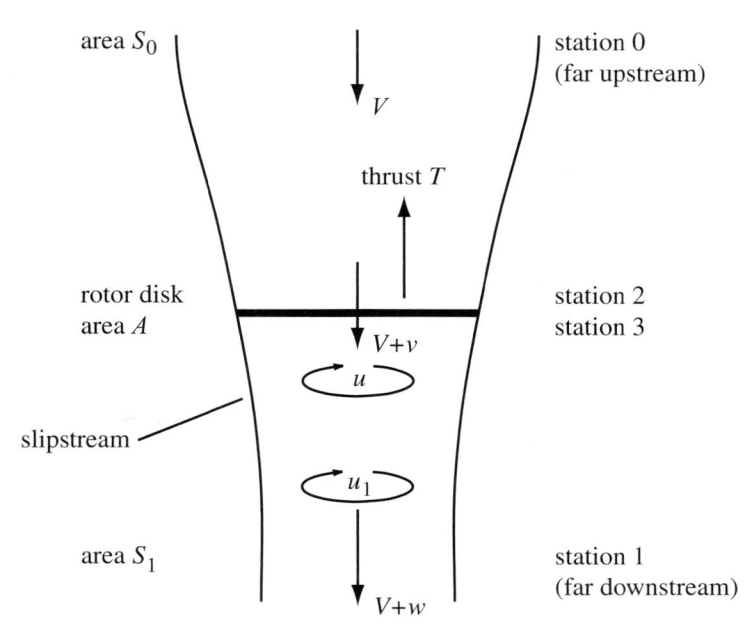

Figure 3.4. Flow model including swirl velocities in the wake.

$$Q = \int \rho(V+v)ur\,dA = \int \rho(V+w)u_1 r_1 dS_1 \tag{3.36}$$

$$P = \int \Delta p(V+v)dA + \int \frac{1}{2}\rho u^2(V+v)dA \tag{3.37}$$

$$= \int \frac{1}{2}\rho(2Vw + w^2 + u_1^2)(V+w)dS_1 + \int (p_1 - p_0)(V+w)dS_1 \tag{3.38}$$

The thrust, torque, and power can be expressed as functionals of the far wake velocities w and u_1 alone, by using

$$p_1 - p_0 = -\int_{r_1}^{R_1} \rho \frac{u_1^2}{r_1} dr_1 \tag{3.39}$$

The resulting optimization problem – finding w and u_1 to minimize P subject to the constraints of the given T and of $Q = P/\Omega$ – is more complex than is needed to estimate the power losses due to swirl.

To formulate a simpler optimization problem, note that from $P = \Omega Q$ there follows $P = \int \rho(V+v)u\Omega r\,dA$. Then equating the expressions for power gives

$$\int \Delta p(V+v)dA = \int \rho(V+v)\left(\Omega r - \frac{1}{2}u\right)u\,dA \tag{3.40}$$

This relation can be interpreted as equating alternative expressions for the work performed, $\int (V+v)dT = \int (\Omega - \frac{1}{2}u/r)dQ$. Based on the results of section 3.4.1, the approximation $dT = \Delta p\,dA \cong \rho(V+v)2v\,dA$ is used. The thrust can then be written $T = \int 2\rho(V+v)v\,dA$, and the differential form of $P = \Omega Q$ becomes simply

$$2(V+v)v = (\Omega r - \frac{1}{2}u)u \tag{3.41}$$

Thus the momentum theory for the rotor with swirl in the wake is formulated as follows: minimize the power P for a given thrust T:

$$P = \int \rho(V+v)u\Omega r\,dA \tag{3.42}$$

$$T = \int 2\rho(V+v)v\,dA \tag{3.43}$$

subject to the constraint $P = \Omega Q$. The Euler equation for this calculus of variations problem,

$$\frac{(V+v)\Omega r}{\Omega r - u} + \frac{u\Omega r}{2V+4v} = \text{constant} = (V+v_0)\left(\frac{V+3v_0}{V+2v_0}\right) \tag{3.44}$$

together with equation 3.41 determine the inflow and swirl velocities at the rotor disk. The constant in the Euler equation is written such that $v \to v_0$ at large r (small u).

Equations 3.41 and 3.44 can be written in terms of v/v_0 and $\hat{u} = u/\Omega r$ as functions of $\Omega r/v_0$ and V/v_0. For hover, the implicit solution is

$$\frac{\Omega r}{v_0} = \frac{3(1-\hat{u})\sqrt{(2-\hat{u})\hat{u}}}{\hat{u}(3-2\hat{u})} \tag{3.45}$$

$$\frac{v}{v_0} = \frac{3(1-\hat{u})(2-\hat{u})}{2(3-2\hat{u})} \tag{3.46}$$

This scaled solution is plotted in Figure 3.5 in the form v/v_0 and u/v_0 as a function of r/R, which requires specifying $\lambda_0 = v_0/\Omega R = 0.05$. An approximate solution

$$\frac{v}{v_0} = \frac{(\Omega r)^2}{(\Omega r)^2 + \sqrt{2}v_0^2} \tag{3.47}$$

$$\frac{u}{v_0} = \frac{2\Omega r\,v_0}{(\Omega r)^2 + \sqrt{2}v_0^2} \tag{3.48}$$

is accurate for large r (see Figure 3.5). Since this approximation gives $u\Omega r = 2v_0 v$, the hover power is $P = \int \rho vu\Omega r\,dA = Tv_0$. This approximate solution for v can be used to evaluate the thrust:

$$T = \int 2\rho v^2\,dA = 2\rho v_0^2 \pi \int_0^R \left[\frac{(\Omega r)^2}{(\Omega r)^2 + \sqrt{2}v_0^2}\right]^2 2r\,dr \tag{3.49}$$

$$= 2\rho A v_0^2 \left[1 - 2\sqrt{2}\lambda_0^2 \ln\left(1 + \frac{1}{\sqrt{2}\lambda_0^2}\right) + \frac{\sqrt{2}\lambda_0^2}{1+\sqrt{2}\lambda_0^2}\right] \tag{3.50}$$

where $\lambda_0 = v_0/\Omega R$. Then v_0 can be obtained from $v_h = \sqrt{T/2\rho A}$:

$$v_h^2 \cong v_0^2 \left[1 + 2\sqrt{2}\lambda_h^2 \ln(\sqrt{2}\lambda_h^2)\right] = v_0^2 \left[1 + 2\frac{C_T}{\sqrt{2}} \ln\frac{C_T}{\sqrt{2}}\right] \tag{3.51}$$

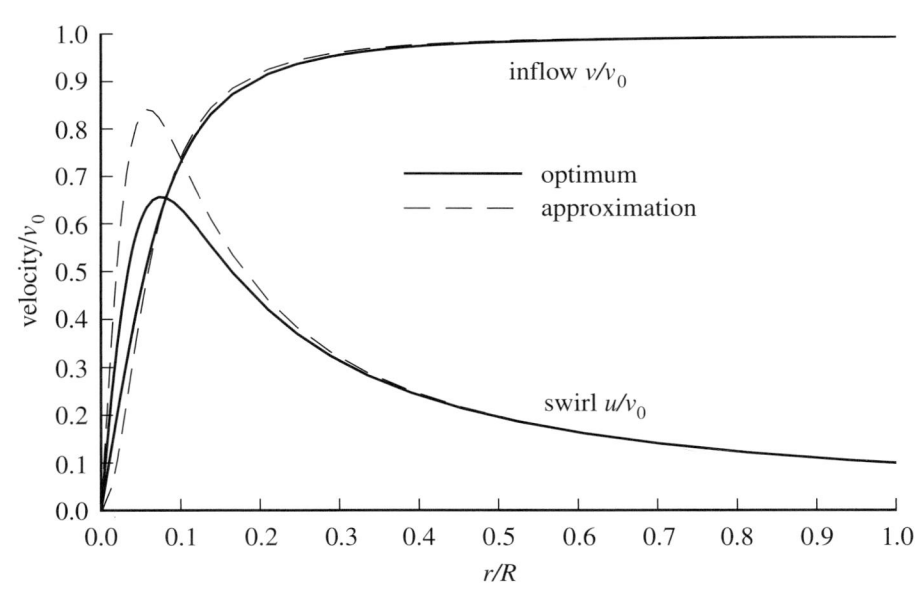

Figure 3.5. Radial distribution of inflow v and swirl u in hover (for $\lambda_0 = 0.05$).

The factor in brackets is negative, so $v_0 > v_h$. Thus including swirl, the hover power is $P = \kappa_s T \sqrt{T/2\rho A}$, where

$$\kappa_s = \left[1 + 2\frac{C_T}{\sqrt{2}} \ln \frac{C_T}{\sqrt{2}}\right]^{-1/2} \tag{3.52}$$

For typical C_T, $\kappa_s \cong 1.02$, so there is about a 2% increase in the induced velocity and induced power because of swirl in the wake, or about a 1% increase in the total rotor power. Using $v_0 \cong v_h$, the velocities are

$$v \cong v_h \frac{(\Omega r)^2}{(\Omega r)^2 + \sqrt{2}v_h^2} \tag{3.53}$$

$$u \cong v_h \frac{2\Omega r\, v_h}{(\Omega r)^2 + \sqrt{2}v_h^2} \tag{3.54}$$

and the loading $\Delta p = dT/dA = 2\rho v^2$ is

$$\Delta p \cong \frac{T}{A} \left[\frac{(\Omega r)^2}{(\Omega r)^2 + \sqrt{2}v_h^2}\right]^2 \tag{3.55}$$

The inflow and loading distributions are nearly uniform except at the root, inboard of say 20%R. Equation 3.48 for the swirl has a peak value at $\Omega r = \sqrt[4]{2}v_h$, or about $r = 6\%R$; the exact solution has a lower peak, which is more outboard (Figure 3.5). Vortex theory (see section 3.7) shows that for a uniformly loaded rotor the wake vorticity is distributed on the slipstream boundary and in a line vortex along the axis with circulation $\gamma = 2\pi T/\rho A\Omega$. The present solution for the swirl velocity at large r is $u \cong 2v_h^2/\Omega r = T/\rho A\Omega r = \gamma/2\pi r$, consistent with vortex theory.

The previous paragraphs derived the wake swirl velocity due to the rotor-induced torque. The rotor also has a profile power loss, caused by the viscous drag of the blades, and hence a profile torque that adds more rotational kinetic energy to the wake. In terms of the blade section drag-to-lift ratio c_d/c_ℓ, the profile power can be written as

$$P_o = \int \Omega r \, dD = \int \Omega r \frac{c_d}{c_\ell} dT = \int \Omega r \frac{c_d}{c_\ell} 2\rho(V+v)v \, dA \qquad (3.56)$$

Then the differential form of $P = \Omega Q$ becomes

$$2(V+v)v + 2\frac{c_d}{c_\ell} v\Omega r = (\Omega r - \frac{1}{2}u)u \qquad (3.57)$$

replacing equation 3.41.

In summary, except for very near the rotor axis, the momentum theory solution consists of uniform inflow v, uniform loading Δp, and swirl u due to a line vortex on the axis. The influence of the swirl velocities in the wake is small outboard of about 20% radius, and so can generally be neglected for helicopter rotors.

3.5 Blade Element Theory

Blade element theory calculates the forces on the blade caused by its motion through the air, and hence the forces and performance of the entire rotor. Blade element theory is lifting-line theory applied to the rotating wing. Each blade section is assumed to act as a two-dimensional airfoil to produce aerodynamic forces, with the influence of the wake and the rest of the rotor contained entirely in an induced angle-of-attack at the section. The solution thus requires an estimate of the wake-induced velocity at the rotor disk, which is provided by momentum theory, vortex theory, or nonuniform inflow calculations. Lifting-line theory is based on the assumption that the wing has a high aspect ratio. For a rotor, the aspect ratio of a single blade is related to the solidity and number of blades by $AR = R/c = (N/\pi)\sigma$. For low disk loading helicopter rotors, the assumption of high aspect ratio is usually valid. However, although the geometric aspect ratio can be large, in areas where the loading or induced velocity has high gradients the effective aerodynamic aspect ratio can still be small. Examples of such high gradients for the rotating wing include blade sections near the tip or near an encounter with a vortex from a preceding blade.

Blade element theory is the foundation of most analyses of helicopter aerodynamics because it deals with the detailed flow and loading of the blade and hence relates the rotor performance and other characteristics to the rotor design parameters. In contrast, momentum theory (or any actuator disk analysis) is a global analysis, which provides useful results but cannot alone be used to design the rotor.

3.5.1 History of Blade Element Theory

The early development of rotary-wing theory followed two separate lines, momentum theory and blade element theory, which were finally brought together in the 1920s. The names "momentum theory" and "blade element theory" in fact had somewhat different meanings from the current usage, referring in the early work to separate and seemingly independent approaches to airscrew analysis. The key factor was the concept of induced drag, which fluid dynamicists were still working to understand for both fixed and rotating wings in the early decades of the 20th century. The concept of an induced power (the power required to produce lift on a three-dimensional wing) and its association with the velocity induced at the wing by the wake vorticity had to be fully developed before an accurate calculation of the rotor loading was possible.

The origins of blade element theory can be traced to the work of William Froude in 1878, but the first major treatment was developed by Stefan Drzewiecki between 1892 and 1920; see Glauert (1935). Drzewiecki considered the blade sections to act independently, but he was uncertain of the aerodynamic characteristics that should be used for the airfoils. Thus he proposed to obtain the required airfoil characteristics from measurements on a series of propellers. This was typical of the early approaches to blade element theory. These theories used only the velocities Ωr and V at the blade section, which are due to the rotation and axial velocity of the rotor, respectively, and then considered what airfoil characteristics to use. Momentum theory describes the velocity at the rotor disk as $V + v$, which is greater than the free stream velocity V because of the rotor lift (and also a circumferential velocity at the disk due to the rotor torque). However, Drzewiecki maintained that there was no logical connection between the momentum theory axial velocity and the velocity actually experienced by the blade section. The former is a mean velocity, whereas the latter is the local value, and a rigorous momentum theory analysis does not in fact give information about the induced velocity at the rotor disk (momentum theory is really concerned with the velocities in the far wake). Lacking a sound theoretical treatment of the velocities at the rotor disk, Drzewiecki considered only the terms Ωr and V. When two-dimensional airfoil characteristics were used in such an analysis, the calculated performance exhibited a significant error that was therefore attributed to the airfoil characteristics. For fixed wings the effective aerodynamic characteristics clearly varied with aspect ratio, so Drzewiecki proposed that three-dimensional wing characteristics (for the appropriate aspect ratio) be used in the rotor blade element theory, with any remaining discrepancies to be established from tests on a series of propellers. The results of this theory had the right general behavior, but were quantitatively inaccurate.

There were several attempts from 1915 to 1919 to use the increased axial velocity from momentum theory in a blade element analysis; none developed to the point of using the two-dimensional airfoil characteristics, however, so all resorted at some stage to experiments to establish what characteristics to use. A. Betz in 1915 used the $V + v$ result of momentum theory and remarked that the appropriate aspect ratio to use was higher than that of the actual blade. While recognizing that the aspect ratio was tending toward infinity, he still considered the correct value to depend on the blade planform. G. de Bothezat in 1918 also used the $V + v$ result of momentum theory (and the corresponding circumferential velocity at the disk), but he adopted Drzewiecki's plan of a series of special propeller tests to determine the airfoil characteristics. A. Fage and H. E. Collins in 1917 used an empirical fraction of $V + v$; they retained the airfoil characteristics of a wing with aspect ratio 6, and hence a correction to the induced velocity was required to handle aspect ratio variations. Thus blade element theory remained on an empirical basis with regard to both the magnitude of the interference flow and the appropriate airfoil characteristics.

A correct accounting for the influence of the propeller wake on the aerodynamic environment at the blade section followed the development of Prandtl's wing theory, which gave a clear explanation of the role of the wake-induced velocity at the wing. Prandtl, Lanchester, and others developed the concept that the lift on an airfoil is due to a bound circulation, resulting in trailed vorticity in the wake that induces a velocity at the wing. Lifting-line theory for fixed wings involved a calculation of the induced

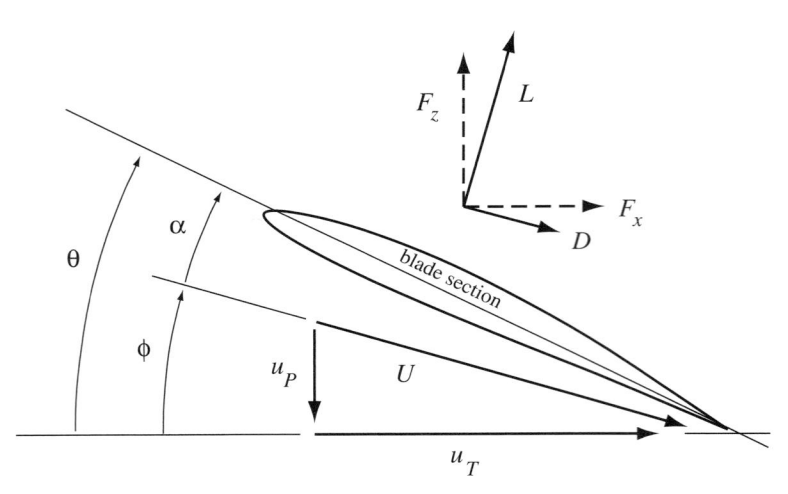

Figure 3.6. Blade section aerodynamics.

velocity from the vortex wake properties. Thus rotary-wing theory also turned to consideration of the vortex wake to define the velocities seen by a blade section. The resulting analysis is called vortex theory, and through this approach rather than momentum theory the induced velocity was finally incorporated correctly into blade element theory. For a rotor or propeller, the vortices in the wake are trailed in helical paths rather than straight back as for fixed wings. This transcendental geometry makes the mathematical task of calculating the induced velocity much more difficult than for fixed wings. Consequently vortex theory, like momentum theory, frequently used the actuator disk model of the rotor, for which analytical solutions were possible.

A general airscrew theory was developed in the early 1920s on the basis of vortex theory and Prandtl's wing theory. By applying the concept of induced velocity, the aerodynamic environment at the rotor disk was established from the vortex theory results. The appropriate airfoil characteristics for this analysis are those of the two-dimensional wing. Later work established that for the same model the momentum theory and vortex theory results are indeed identical, so blade element theory is now usually derived using momentum theory results for the induced velocity. In the early development of rotary-wing analysis, however, the vortex concepts of Prandtl had so great an impact that vortex theory completely superseded momentum theory. Momentum theory lacked the basis for understanding the induced velocity at the rotor disk, which was required to complete the development of blade element theory. As a result, vortex theory became regarded as the more reliable and logical foundation for both fixed- and rotary-wing analyses.

3.5.2 Blade Element Theory for Vertical Flight

Blade element theory is based on the lifting-line assumption; for the present derivation we also assume low disk loading and neglect stall and compressibility effects in order to obtain an analytical solution. Figure 3.6 defines the geometry, velocities, and forces of the blade section. The blade section has a pitch angle θ, measured from the plane of rotation to the airfoil zero-lift line. The air velocity seen by the blade has components u_T and u_P, which are tangent to and perpendicular to the disk plane, respectively. The resultant velocity magnitude and inflow angle are then given by

$$U = \sqrt{u_T^2 + u_P^2} \qquad (3.58)$$

$$\phi = \tan^{-1} u_P/u_T \qquad (3.59)$$

The aerodynamic angle-of-attack of the blade is $\alpha = \theta - \phi$. The air flow at the blade section produces lift and drag forces, L and D, which are normal to and parallel to the resultant velocity, respectively. The components of the total aerodynamic force normal to and parallel to the disk plane are F_z and F_x. Writing the section forces in terms of the lift and drag coefficients gives

$$L = \frac{1}{2}\rho U^2 c c_\ell \qquad (3.60)$$

$$D = \frac{1}{2}\rho U^2 c c_d \qquad (3.61)$$

where ρ is the air density and c is the blade chord. In general, the section coefficients c_ℓ and c_d are complicated functions of the angle-of-attack, Mach number, and other parameters, but quite simple forms are used here. Resolving the aerodynamic forces normal and parallel to the disk plane gives

$$F_z = L\cos\phi - D\sin\phi \qquad (3.62)$$

$$F_x = L\sin\phi + D\cos\phi \qquad (3.63)$$

Finally, the elemental thrust, torque, and power on the rotor blade are

$$dT = NF_z dr \qquad (3.64)$$

$$dQ = NF_x r\, dr \qquad (3.65)$$

$$dP = \Omega\, dQ = NF_x \Omega r\, dr \qquad (3.66)$$

where N is the number of blades. The total forces on the rotor are obtained by integrating over the blade span from root to tip.

For the rotor in hover or vertical flight, the normal velocity u_P consists of the climb velocity V (zero for hover) and the induced velocity v; the in-plane velocity u_T is due only to the rotation of the blades at rate Ω. Therefore $u_P = V + v$ and $u_T = \Omega r$. Now from the assumption of low disk loading for the helicopter rotor, the inflow ratio $\lambda = (V + v)/\Omega R$ is small. The momentum theory result for hover typically gives $\lambda_h = 0.05$ to 0.07. Then $u_P/u_T = (V + v)/\Omega r = \lambda(R/r)$ is also small, except near the blade root, where the dynamic pressure is low and thus the loads are negligible anyway. Therefore the small angle assumption is appropriate for helicopter rotors, namely $\phi, \theta, \alpha \ll 1$, from which $\phi \cong u_P/u_T$, $\cos\phi \cong 1$, $\sin\phi \cong \phi$, and $U \cong u_T$. The next assumption is that stall and compressibility effects are negligible, so that the lift coefficient is linearly related to the angle-of-attack: $c_\ell = a\alpha$. Here a is the slope of the blade two-dimensional lift curve; typically $a = 5.7$, including real flow effects. Then the blade section forces reduce to

$$L \cong \frac{1}{2}\rho u_T^2 ca(\theta - u_P/u_T) \qquad (3.67)$$

$$D \cong \frac{1}{2}\rho u_T^2 c c_d \qquad (3.68)$$

and
$$dT \cong NL\,dr \tag{3.69}$$
$$dQ \cong N(L\phi + D)r\,dr \tag{3.70}$$

Next, all quantities are made dimensionless, normalized with respect to the air density, rotor speed, and rotor radius (ρ, Ω, and R). In coefficient form, the results for the contribution of a blade section to the rotor thrust and power are

$$dC_T = \frac{\sigma a}{2}\left(\theta u_T^2 - u_T u_P\right) dr = \frac{\sigma a}{2}\left(\theta r^2 - \lambda r\right) dr \tag{3.71}$$

$$dC_P = dC_Q = \left[\frac{\sigma a}{2}\left(\theta u_T u_P - u_P^2\right) + \frac{\sigma c_d}{2} u_T^2\right] r\,dr \tag{3.72}$$

$$= \left[\frac{\sigma a}{2}\left(\theta r\lambda - \lambda^2\right) + \frac{\sigma c_d}{2} r^2\right] r\,dr \tag{3.73}$$

where $\lambda = (V + v)/\Omega R$ is the inflow ratio and $\sigma = Nc/\pi R$ is the solidity ratio. In general this σ is a function of radius, except for constant-chord blades. Given the blade geometry, inflow, and section drag, these expressions can be integrated numerically over the blade span. With certain additional assumptions the integration can be performed analytically – for example, with uniform inflow, constant chord, linear twist, and constant drag coefficient.

3.5.2.1 Rotor Thrust

Blade element theory gives the rotor thrust coefficient as

$$C_T = \int_0^1 \frac{\sigma a}{2}\left(\theta r^2 - \lambda r\right) dr \tag{3.74}$$

For a blade with constant chord and linear twist ($\theta = \theta_0 + r\theta_{tw} = \theta_{.75} + (r - 0.75)\theta_{tw}$), and assuming uniform inflow ($\lambda =$ constant), we obtain

$$C_T = \frac{\sigma a}{2}\left(\frac{\theta_{.75}}{3} - \frac{\lambda}{2}\right) \tag{3.75}$$

where $\theta_{.75}$ is the pitch of the blade at 75% radius.

For uniform inflow, constant chord, and a twist distribution given by $\theta = \theta_t/r$, the thrust coefficient is

$$C_T = \frac{\sigma a}{4}\left(\theta_t - \lambda\right) \tag{3.76}$$

or with $\phi = \lambda/r = \phi_t/r$

$$C_T = \frac{\sigma a}{4}\left(\theta_t - \phi_t\right) = \frac{\sigma a}{4}\alpha_t \tag{3.77}$$

where the subscript "t" refers to the value at the blade tip. This twist distribution, although not physically realizable at the root, gives uniform inflow with the constant-chord blades. It is called the ideal twist distribution, since momentum theory shows that the minimum induced power is obtained with uniform inflow.

3.5.2.2 Induced Velocity

Blade element theory gives the rotor thrust as a function of the pitch angle and inflow ratio. The induced velocity is required if C_T is to be expressed as a function of θ alone. Momentum theory gives the following induced velocity for the rotor in hover or vertical climb:

$$\lambda = \frac{\lambda_c}{2} + \sqrt{\left(\frac{\lambda_c}{2}\right)^2 + \frac{C_T}{2}} \tag{3.78}$$

where $\lambda_c = V/\Omega R$. In hover $\lambda = \sqrt{C_T/2}$, so for a constant chord, linearly twisted blade equation 3.75 can be solved for the induced velocity:

$$\lambda = \sqrt{\frac{C_T}{2}} = \frac{\sigma a}{16}\left[\sqrt{1 + \frac{64}{3\sigma a}\theta_{.75}} - 1\right] \tag{3.79}$$

or

$$\theta_{.75} = \frac{6C_T}{\sigma a} + \frac{3}{2}\sqrt{\frac{C_T}{2}} \tag{3.80}$$

The first term in equation 3.80 corresponds to the mean angle-of-attack of the rotor blade, whereas the second term is the additional pitch required because of the induced inflow angle ϕ. These relations allow λ and C_T to be obtained for a given collective pitch $\theta_{.75}$, or alternatively λ and $\theta_{.75}$ for a given thrust.

For a constant chord, ideally twisted blade the momentum theory value for the inflow ratio and equation 3.76 give

$$\lambda = \frac{\sigma a}{16}\left[\sqrt{1 + \frac{32}{\sigma a}\theta_t} - 1\right] \tag{3.81}$$

or

$$\theta_t = \frac{4C_T}{\sigma a} + \sqrt{\frac{C_T}{2}} \tag{3.82}$$

3.5.2.3 Power or Torque

The differential power coefficient can be written as

$$dC_P = \left[\lambda\frac{\sigma a}{2}\left(\theta r^2 - \lambda r\right) + \frac{\sigma c_d}{2}r^3\right]dr = \lambda\,dC_T + \frac{\sigma c_d}{2}r^3\,dr \tag{3.83}$$

hence

$$C_P = \int \lambda\,dC_T + \int_0^1 \frac{\sigma c_d}{2}r^3\,dr \tag{3.84}$$

The first term in C_P is the climb plus induced power, $C_{Pc} + C_{Pi} = \int \lambda\,dC_T$, which arises from the in-plane component of the lift due to the induced angle-of-attack ($dP_c + dP_i = (V + v)dT$). The second term is the profile power C_{Po}, which is due to the viscous drag forces on the rotor blade.

For uniform inflow the climb plus induced power is simply $C_{Pc} + C_{Pi} = \lambda C_T$, which agrees with the momentum theory result. For vertical flight, λ includes the inflow due to the climb velocity, $\lambda_c = V/\Omega R$, so that P includes the climb power $P = TV$. In hover, the momentum theory result for λ gives $C_{Pi} = C_T^{3/2}/\sqrt{2}$. This induced velocity value is for an ideal rotor. A real rotor with a practical twist and planform and a finite number of blades has a higher induced power than the minimum given by momentum theory. One way to calculate the true induced power is to integrate $\int \lambda \, dC_T$ using the actual induced velocity distribution, which in general is nonuniform as well as larger than the ideal value. An alternative approach is to use the momentum theory expression for the induced power, but with an empirical factor to account for the additional losses of a real rotor:

$$C_{Pi} = \kappa \lambda_h C_T = \kappa C_T^{3/2}/\sqrt{2} \tag{3.85}$$

Typically the factor κ has a value around 1.15 (see section 4.1.3.1).

For a constant-chord blade, and assuming a constant drag coefficient $c_d = c_{do}$, the profile power coefficient can be evaluated as

$$C_{Po} = \frac{\sigma c_{do}}{8} \tag{3.86}$$

For an accurate calculation of the profile power, the variation of the drag coefficient with angle-of-attack and Mach number should be included, which probably requires a numerical integration. Consider a profile drag polar of the form

$$c_d = \delta_0 + \delta_1 \alpha + \delta_2 \alpha^2 \tag{3.87}$$

By properly choosing the constants δ_0, δ_1, and δ_2 the variation of drag with lift for a given airfoil can be well represented for angles of attack below stall. This representation for c_d was introduced by Sissingh (1939) and Bailey (1941). Bailey's numerical example $c_d = 0.0087 - 0.0216\alpha + 0.400\alpha^2$ is frequently found in early helicopter research. See section 8.8 for a further discussion. Then the profile power coefficient is

$$C_{Po} = \int_0^1 \frac{\sigma}{2} \left[\delta_0 + \delta_1 (\theta - \lambda/r) + \delta_2 (\theta - \lambda/r)^2 \right] r^3 \, dr \tag{3.88}$$

For a constant chord, ideally twisted rotor with uniform inflow this integrates to

$$C_{Po} = \frac{\sigma \delta_0}{8} + \frac{\sigma \delta_1}{6}(\theta_t - \lambda) + \frac{\sigma \delta_2}{4}(\theta_t - \lambda)^2 = \frac{\sigma \delta_0}{8} + \frac{2\delta_1}{3a} C_T + \frac{4\delta_2}{\sigma a^2} C_T^2 \tag{3.89}$$

using $\theta_t - \lambda = 4C_T/\sigma a$. Similarly, for a constant chord, linearly twisted blade with uniform inflow, the profile power is

$$C_{Po} = \frac{\sigma \delta_0}{8} + \frac{\sigma \delta_1}{8}\left(\theta_{.75} + \frac{1}{20}\theta_{tw} - \frac{4}{3}\lambda\right)$$
$$+ \frac{\sigma \delta_2}{8}\left(\theta_{.75}^2 + \frac{1}{10}\theta_{.75}\theta_{tw} + \frac{7}{240}\theta_{tw}^2 + 2\lambda^2 - \frac{8}{3}\theta_{.75}\lambda\right) \tag{3.90}$$

The simplest relation for the total hover power of a real rotor is

$$C_P = \frac{\kappa C_T^{3/2}}{\sqrt{2}} + \frac{\sigma c_{do}}{8} \tag{3.91}$$

This result gives the basic features of the hover performance and is reasonably accurate when the appropriate empirical factor κ is used for the induced power and an appropriate mean drag coefficient c_{do} is used for the profile power. A plot of the power coefficient as a function of thrust coefficient (or C_P/σ as a function of C_T/σ) is called the rotor polar. For an ideal rotor (no profile power and minimum induced power, hence a figure of merit of $M = 1$), the polar is given by $C_P = \kappa C_T^{3/2}/\sqrt{2}$. The polar for a real rotor has an offset compared to the ideal polar because of the profile power, and the power increases faster with C_T because of the larger induced power. The figure of merit corresponding to the above expression for the rotor power is

$$M = \frac{C_{Pideal}}{C_{Pi} + C_{Po}} = \frac{C_T^{3/2}/\sqrt{2}}{\kappa C_T^{3/2}/\sqrt{2} + \sigma c_{do}/8} \tag{3.92}$$

Even such a simple result leads to some conclusions about the rotor blade design. Recall that the proper use of the figure of merit is for comparisons of rotors at constant disk loading. For a given C_T, high M requires a low value of σc_{do}. If the rotor solidity is too low, however, high angles-of-attack are needed to achieve the required lift, and as a result the profile drag increases. Therefore the rotor should have as small a solidity as possible (small chord) with an adequate stall margin. The blade loading (hence twist and chord) distribution influences both the induced and profile power, but a more detailed calculation is required to examine such effects.

3.5.3 Combined Blade Element and Momentum Theory

The rotor performance calculations in the preceding section used the momentum theory result for the induced velocity, which was assumed to be uniform over the rotor disk. A nonuniform inflow distribution can be obtained by considering the differential form of momentum theory for hover or vertical flight. The resulting analysis is called combined blade element and momentum theory, commonly abbreviated "BEM." Blade element theory describes the differential thrust on an annulus of the disk (on all N blades) of width dr at radial station r as

$$dC_T = \frac{\sigma a}{2}(\theta - \lambda/r)r^2 dr \tag{3.93}$$

From section 3.4.1, the differential form of momentum theory is $dT = 2\rho\, dA(V + v)v$ or

$$dC_T = 4\lambda\lambda_i r\, dr \tag{3.94}$$

where $\lambda_i = v/\Omega R$ is the induced inflow ratio, $\lambda_c = V/\Omega R$ is the climb inflow ratio, and $\lambda = \lambda_c + \lambda_i$. By using the differential form of momentum theory, the induced velocity at radial station r is assumed to be due only to the thrust dT at that station. Equating the blade element and momentum theory expressions for dC_T then gives

$$\lambda^2 + \left(\frac{\sigma a}{8} - \lambda_c\right)\lambda - \frac{\sigma a}{8}\theta r = 0 \tag{3.95}$$

which has the solution

$$\lambda = \sqrt{\left(\frac{\sigma a}{16} - \frac{\lambda_c}{2}\right)^2 + \frac{\sigma a}{8}\theta r} - \left(\frac{\sigma a}{16} - \frac{\lambda_c}{2}\right) \tag{3.96}$$

For hover, $\lambda_c = 0$, the solution for the induced velocity is

$$\lambda = \frac{\sigma a}{16}\left[\sqrt{1 + \frac{32}{\sigma a}\theta r} - 1\right] \tag{3.97}$$

This is the nonuniform inflow distribution; compare with the uniform inflow results in section 3.5.2.2. For a given pitch, twist, and chord, the inflow can be calculated as a function of r, and then the rotor thrust and power can be evaluated. Although the resulting rotor performance is more accurate than that obtained with uniform inflow, differential momentum theory is still only an approximate model of the rotor. A further refinement of the inflow calculation requires a consideration of the details of the rotor vortex wake. Observe that for a constant-chord blade, uniform inflow is obtained if $\theta r = $ constant; that is, if the blade has the ideal twist distribution $\theta = \theta_t/r$. From this uniform inflow, the ideally twisted rotor also has uniform disk loading and the minimum possible induced power.

3.6 Hover Performance

To summarize the equations involved in the calculation of rotor hover performance, blade element theory gives the thrust and power as

$$C_T = \int_0^1 \frac{\sigma}{2} r^2 c_\ell \, dr \qquad (3.98)$$

$$C_P = \int \lambda \, dC_T + \int_0^1 \frac{\sigma}{2} r^3 c_d \, dr \qquad (3.99)$$

where the section lift and drag coefficients are functions of the angle-of-attack $\alpha = \theta - \lambda/r$ and Mach number $M = r M_{\text{tip}}$. In general, the chord and pitch can be functions of the radial station r. The most frequent case encountered is a constant chord, linearly twisted blade: $\sigma = $ constant, $\theta = \theta_0 + \theta_{tw} r$. If actual airfoil characteristics are not available, the simple relations $c_\ell = a\alpha$ and $c_d = $ constant can be used. From combined blade element and momentum theory, the inflow distribution is

$$\lambda = \frac{\sigma a}{16}\left[\sqrt{1 + \frac{32}{\sigma a}\theta r} - 1\right] \qquad (3.100)$$

Alternatively, uniform inflow with an empirical factor can be used: $\lambda = \kappa\sqrt{C_T/2}$. In general the rotor loads must be integrated numerically over the span. Then stall and compressibility can be included by use of the appropriate section airfoil data. The limitations in the performance calculated according to these expressions arise principally from the neglect of three-dimensional flow effects at the tip and the use of the differential momentum theory for the induced velocity.

By using combined blade element and momentum theory the rotor thrust and induced power can equivalently be written in terms of the induced velocity as $dC_T = 4\lambda^2 r \, dr$ and $dC_{P_i} = 4\lambda^3 r \, dr$.

The actuator disk analysis, particularly with vortex theory, requires a relation among the rotor disk loading, blade span loading, and blade bound circulation. The section loading L (lift per unit span) and circulation Γ are related by $L = \rho \Omega r \Gamma$. Hence

$$\frac{dT}{dA} = \frac{N}{2\pi r} L = \frac{\rho \Omega N}{2\pi} \Gamma \qquad (3.101)$$

So uniform disk loading corresponds to triangular blade loading and constant bound circulation. In dimensionless form, for uniform disk loading and constant chord, $\Gamma/\Omega R^2 = (2\pi/N)C_T = 2(c/R)C_T/\sigma$.

3.6.1 Scaling with Solidity

Blade element theory results can be scaled with the solidity ratio by writing in terms of C_T/σ^2, C_P/σ^3, θ/σ, and λ/σ, as observed by Knight and Hefner (1937). For constant chord, equations 3.98, 3.99, and 3.100 become

$$\frac{C_T}{\sigma^2} = \int_0^1 \frac{a}{2} \left(\frac{\theta}{\sigma} r^2 - \frac{\lambda}{\sigma} r \right) dr \qquad (3.102)$$

$$\frac{C_P}{\sigma^3} = \int \frac{\lambda}{\sigma} d\frac{C_T}{\sigma^2} + \frac{1}{\sigma^3} \int_0^1 \frac{\sigma}{2} r^3 c_d dr \qquad (3.103)$$

$$\frac{\lambda}{\sigma} = \frac{a}{16} \left[\sqrt{1 + \frac{32}{a}\frac{\theta}{\sigma} r} - 1 \right] \qquad (3.104)$$

For example, the inflow solution for linear twist (equation 3.79) becomes

$$\frac{\lambda}{\sigma} = \sqrt{\frac{1}{2}\frac{C_T}{\sigma^2}} = \frac{a}{16} \left[\sqrt{1 + \frac{64}{3a}\frac{\theta_{.75}}{\sigma}} - 1 \right] \qquad (3.105)$$

The simple power formula (equation 3.91) is

$$\frac{C_P}{\sigma^3} = \frac{\kappa}{\sqrt{2}} \left(\frac{C_T}{\sigma^2} \right)^{3/2} + \frac{1}{\sigma^2} \frac{c_{do}}{8} \qquad (3.106)$$

and

$$M = \frac{(C_T/\sigma^2)^{3/2}/\sqrt{2}}{C_P/\sigma^3} \qquad (3.107)$$

is the figure of merit. These relations define the influence of solidity on hover performance. With the induced velocity obtained from a wake model, a separate influence of the blade aspect ratio and number of blades is found, not simply the dependence on $\sigma = Nc/\pi R$.

3.6.2 Tip Losses

By using momentum theory (or differential momentum theory) instead of a vortex wake model to calculate the induced velocity at the rotor disk, blade element theory is an approximation for lifting-line theory, but an approximation that breaks down near the blade tips. When the chord at the tip is finite, blade element theory gives a nonzero lift all the way out to the end of the blade. In fact, the blade loading drops to zero at the tip over a finite distance because of three-dimensional flow effects (Figure 3.7). Since the dynamic pressure is proportional to r^2, the loading for a rotary wing is concentrated at the tip and drops off even faster than the loading on fixed wings. The loss of lift at the tip is an important factor in calculating the rotor performance. If this loss is neglected, the rotor thrust for a given power or collective is significantly overestimated. A rigorous treatment of the tip loading requires a vortex wake model or a lifting-surface analysis. Here we consider an approximate representation of the tip loss effects.

The tip loss can alternatively be considered in terms of the influence of the rotor wake. With an actuator disk model, a nonzero loading extending to the edge of the disk is perfectly acceptable. Thus the tip loss can be viewed as the influence

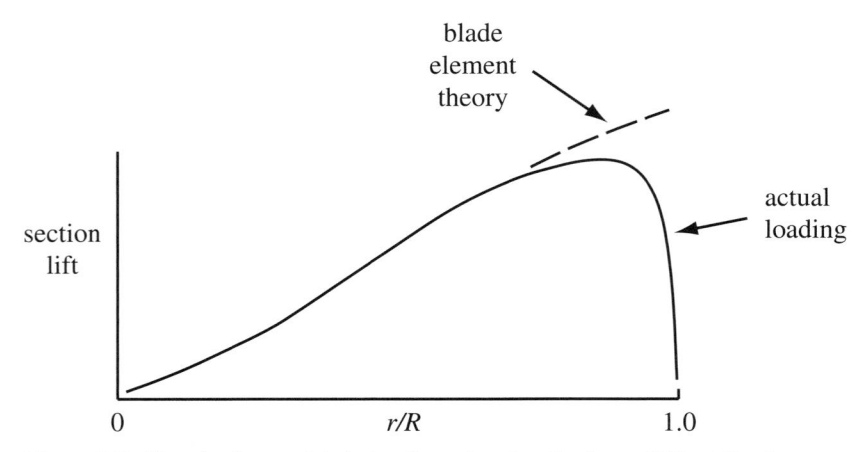

Figure 3.7. Sketch of rotor blade loading, showing the loss of lift at the tip.

of the finite number of blades. The loading being concentrated on a finite number of blades, rather than distributed around the disk, introduces the three-dimensional flow effects. Figure 3.8 sketches the influence of the discrete wake on the flow through the rotor. With a finite number of blades, the discrete vortices in the wake constrain the flow to a volume smaller than the nominal wake boundary. The tip loss in this sense is like having a smaller effective area in the wake or equivalently a higher effective disk loading, which implies a higher induced power loss.

An approximate method to account for tip losses is to assume that the blade elements outboard of the radial station $r = BR$ have profile drag but produce no lift. The parameter B is called the tip loss factor. A number of methods are available for calculating the appropriate value of B. Prandtl gave an expression based on a two-dimensional model of the rotor wake; for a low inflow rotor

$$B = 1 - \frac{\sqrt{2C_T}}{N} \tag{3.108}$$

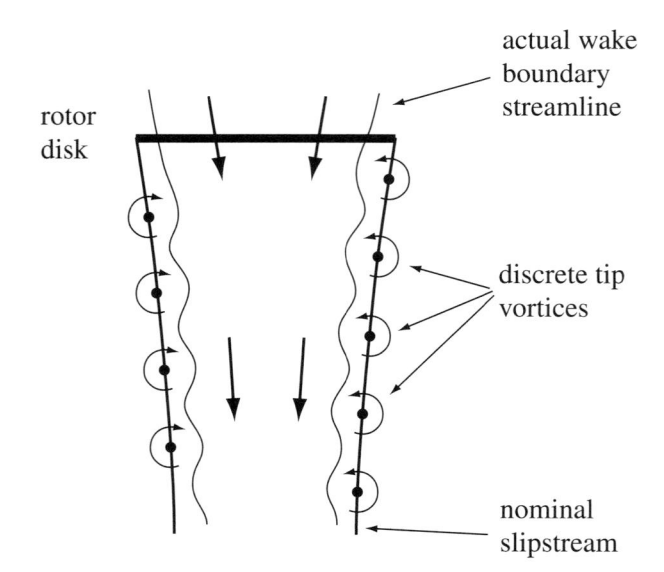

Figure 3.8. Influence of the discrete wake on the flow through the rotor.

where N is the number of blades; this result is derived in section 3.7.3.2. The tip loss depends on the spacing of vorticity sheets in the wake, which is proportional to λ/N. Prandtl's result typically gives $B = 0.96$ to 0.98. Wheatley (1934) suggested

$$B = 1 - \frac{1}{2}\frac{\text{tipchord}}{R} = 1 - \frac{c(r=1)}{2R} \qquad (3.109)$$

The outer half-chord length of the blade therefore develops no lift. Similarly, Sissingh (1939) suggested

$$B = 1 - \frac{2}{3}\frac{c(r=0.7)}{R} \qquad (3.110)$$

Often the tip loss factor is simply set to $B = 0.97$, which generally gives a good correlation with experimental data.

When the tip loss factor is included in the rotor thrust, the result from blade element theory becomes

$$C_T = \int_0^B \frac{\sigma}{2} r^2 c_\ell \, dr = \int_0^B \frac{\sigma a}{2} \left(\theta r^2 - \lambda r\right) dr \qquad (3.111)$$

Then for the constant chord, linearly twisted blade with uniform inflow, it is

$$C_T = \frac{\sigma a}{2}\left(\theta_0 \frac{B^3}{3} + \theta_{tw}\frac{B^4}{4} - \lambda \frac{B^2}{2}\right) \qquad (3.112)$$

and, for the ideally twisted blade,

$$C_T = \frac{\sigma a}{4} B^2 \left(\theta_t - \lambda\right) \qquad (3.113)$$

There is about a 6% to 9% reduction in the rotor thrust for a given collective pitch due to tip losses. The tip loss affects the required rotor power by increasing the induced velocity. The effective disk area of the rotor is reduced by a factor of B^2, and since the induced velocity is proportional to the square root of the disk loading, the induced velocity is higher than the momentum theory result by a factor of B^{-1}. Thus the rotor induced power becomes

$$C_{Pi} = \frac{1}{B}\lambda_h C_T = \frac{1}{B}C_T^{3/2}/\sqrt{2} \qquad (3.114)$$

There is then about a 3% induced power increase due to the tip loss ($\kappa = B^{-1} \cong 1.03$). Other effects, particularly nonuniform inflow, also increase the induced power.

There are more rigorous approaches to the calculation of rotor performance including tip losses, such as vortex theory for a finite number of blades or lifting-line theory with a discrete vortex wake model. The tip loss factor is a crude representation of the three-dimensional flow effects, but is widely used because of its simplicity and reasonable accuracy.

3.6.3 Induced Power due to Nonuniform Inflow

The hover induced power is written as $C_{Pi} = \kappa C_T^{3/2}/\sqrt{2}$, where κ is an empirical factor accounting for the additional losses of a real rotor. The losses due to nonuniform inflow can be estimated using the momentum theory results:

$$C_{Pi} = \int_0^B 4\lambda^3 r \, dr \qquad C_T = \int_0^B 4\lambda^2 r \, dr \qquad (3.115)$$

For uniform inflow, these relations give $C_{Pi} = 2\lambda^3 B^2$ and $C_T = 2\lambda^2 B^2$, or $C_{Pi} = (1/B)C_T^{3/2}/\sqrt{2}$. Hence tip loss alone gives $\kappa = B^{-1} \cong 1.03$.

Considering a linear inflow distribution, $\lambda = \lambda_t r$, we obtain $C_{Pi} = (4/5)B^5\lambda_t^3$ and $C_T = B^4\lambda_t^2$, or $C_{Pi} = (4/5B)C_T^{3/2}$. Hence

$$\kappa = \frac{4\sqrt{2}}{5B} = \frac{1.13}{B} \cong 1.17 \tag{3.116}$$

Other simple nonuniform inflow distributions give similar results. Thus the hovering rotor is expected to have around an 8% to 12% increase in induced power due to nonuniform inflow, and a 2% to 4% increase due to tip loss. The parameter κ is best obtained by correlation with measured rotor performance.

3.6.4 Root Cutout

Performance losses also arise from the root cutout. The lifting portion of the blade starts at radial station $r = r_R$, which is typically 10% to 30% of the blade radius. The area inboard of this station, called the root cutout, is taken up with the rotor hub, flap and lag hinges, pitch bearing, and blade shank. Since the root cutout is aerodynamically an area of high drag coefficient and low lift, the blade element theory evaluation of the thrust should use integration from $r = r_R$ to $r = B$: $C_T = \int_{r_R}^{B} dC_T$. The dynamic pressure is low in the root cutout area, so generally the correction to the performance calculation is minor.

In hover, an effect of the root cutout is to reduce the effective rotor disk area and hence to increase the disk loading and induced velocity. With both root cutout and the tip loss factor, the effective disk area gives the induced power parameter κ

$$\kappa = \sqrt{\frac{A}{A_{\text{eff}}}} = \frac{1}{\sqrt{B^2 - r_R^2}} \tag{3.117}$$

For usual values of r_R, the root cutout effect on the induced power is small compared to the tip loss effect. The root cutout is an area of high drag coefficient, even with fairings, so its effect on profile power can be significant.

3.6.5 Blade Mean Lift Coefficient

A useful measure of the aerodynamic operating state of the rotor is a mean lift coefficient for the blades. The mean lift coefficient $\bar{c}_\ell$ is defined to give a thrust coefficient with the value $C_T = \int \frac{\sigma}{2} r^2 c_\ell \, dr$ when the entire blade is assumed to be working at $\bar{c}_\ell$. So

$$C_T = \int_0^1 \frac{\sigma}{2} r^2 \bar{c}_\ell \, dr = \frac{1}{2}\bar{c}_\ell \int_0^1 \sigma r^2 \, dr = \frac{1}{6}\sigma \bar{c}_\ell \tag{3.118}$$

and then

$$\bar{c}_\ell = 6\frac{C_T}{\sigma} \tag{3.119}$$

Thus C_T/σ, the ratio of the rotor thrust coefficient to the solidity, is a measure of the blade lift coefficient. Correspondingly, $6C_T/\sigma a$ can be interpreted as the mean angle-of-attack of the blade. Now

$$C_T/\sigma = \frac{T/(\rho A_{\text{rotor}}(\Omega r)^2)}{A_{\text{blade}}/A_{\text{rotor}}} = \frac{T}{\rho A_{\text{blade}}(\Omega R)^2}$$

is the dimensionless blade loading, whereas C_T is the dimensionless disk loading. The parameter C_T/σ has an important role in rotor aerodynamics, since many of the characteristics of the rotor and helicopter depend on the blade lift coefficient. Using the simple power expression (equation 3.91), the rotor figure of merit can be written as

$$M = \frac{\lambda_h C_T}{\kappa \lambda_h C_T + \dfrac{\sigma c_{do}}{8}} = \frac{1}{\kappa + \dfrac{3}{4}\dfrac{c_{do}/\bar{c}_\ell}{\lambda_h}} \tag{3.120}$$

showing that a high section lift-to-drag ratio is required for a good hover figure of merit.

3.6.6 Equivalent Solidity

In the expressions for the rotor hover performance, the rotor chord and number of blades have been accounted for by using a local solidity, $\sigma = Nc/\pi R$, which varies along the blade span if the chord is not constant. The rotor solidity then is

$$\sigma_{\text{rotor}} = \frac{\text{blade area}}{\text{rotor area}} = \int_0^1 \sigma \, dr \tag{3.121}$$

For constant-chord blades, the local solidity and rotor solidity are identical. When comparing the performance of two rotors with different blade planforms, using an equivalent solidity that accounts for the major effects of the varying chord is desirable.

Rotors with tapered blades are conventionally compared to a rotor with rectangular blades and an equivalent solidity ratio, operating at the same thrust coefficient. The equivalent solidity σ_e is defined by $C_T = \int \frac{1}{2}\sigma r^2 c_\ell dr = \frac{1}{2}\sigma_e \int r^2 c_\ell dr$, or (assuming constant lift coefficient)

$$\sigma_e = 3 \int_0^1 \sigma r^2 dr \tag{3.122}$$

Equation 3.122 defines the thrust-weighted solidity, whereas equation 3.121 is the geometric solidity. Similarly, for rotors compared on the basis of the same power or torque, the equivalent solidity is

$$\sigma_e = 4 \int_0^1 \sigma r^3 dr \tag{3.123}$$

For linearly tapered blades ($\sigma = \sigma_0 + \sigma_1 r$), the equivalent solidity is

$$\sigma_e = \begin{cases} \sigma(r = 0.75) & \text{thrust basis} \\ \sigma(r = 0.80) & \text{power basis} \end{cases} \tag{3.124}$$

For a blade with tip taper ratio $t = c_{\text{tip}}/c_{\text{root}}$, the taper starting at $r = r_0$ (constant chord c_{root} from $r = 0$ to $r = r_0$, linear variation to c_{tip} at $r = 1$), the thrust-weighted and geometry solidity are

$$\sigma_e = \frac{\sigma_{\text{root}}}{4}\left(1 + 3t + (1-t)(r_0 + r_0^2 + r_0^3)\right) \tag{3.125}$$

$$\sigma_g = \frac{\sigma_{\text{root}}}{2}\left(1 + t + (1-t)r_0\right) \tag{3.126}$$

in terms of the root chord $\sigma_{\text{root}} = Nc_{\text{root}}/\pi R$. The thrust-weighted ($r^2$-weighted) equivalent solidity is generally used for comparisons of rotor power at a given thrust.

3.6.7 The Ideal Rotor

Consider a rotor with constant chord and the ideal twist distribution $\theta = \theta_t/r$. Section 3.5.3 showed that this twist results in uniform induced velocity over the rotor disk and hence corresponds to the minimum induced power. With the ideal twist, the blade loading is triangular:

$$dC_T = \frac{\sigma a}{2}\alpha r^2 dr = \frac{\sigma a}{2}(\theta_t - \lambda)r\,dr \tag{3.127}$$

The corresponding bound circulation and disk loading are

$$\frac{N}{\pi}\Gamma = \frac{1}{r}\frac{dC_T}{dr} = \frac{\sigma a}{2}(\theta_t - \lambda) \tag{3.128}$$

$$\frac{dT}{dA} = \frac{\pi\,dC_T}{2\pi r\,dr} = \frac{\sigma a}{4}(\theta_t - \lambda) \tag{3.129}$$

Thus the ideal twist gives constant bound circulation and uniform disk loading, which is indeed the loading required by momentum theory to produce uniform induced velocity.

From sections 3.5.2 and 3.6.2, the performance of the ideal rotor (constant chord, ideal twist, uniform inflow) is given by

$$C_T = \frac{\sigma a B^2}{4}(\theta_t - \lambda) = \frac{\sigma a B^2}{4}\alpha_t \tag{3.130}$$

$$C_P = \lambda C_T + \frac{\sigma}{8}\left(\delta_0 + \frac{4}{3}\delta_1\alpha_t + 2\delta_2\alpha_t^2\right) \tag{3.131}$$

Momentum theory gives the induced velocity $\lambda = \sqrt{C_T/2B^2}$ for hover; the pitch is $\theta_t = \alpha_t + \lambda$. The local angle-of-attack of the blade section is thus

$$\alpha = \frac{\alpha_t}{r} = \frac{4C_T/\sigma}{B^2 a}\frac{1}{r} \tag{3.132}$$

and the section lift coefficient is $c_\ell = a\alpha$. The section lift coefficient is limited by stall at the blade root and the ideal twist distribution is not realizable at the root, but the inboard sections of the blade have a minor role in the rotor performance. The real practical difficulty is that a different twist distribution is required for every operating condition of the rotor. From $\alpha = (\theta_t - \lambda)/r$:

$$\theta_t = \alpha_t + \lambda = \frac{4C_T/\sigma}{B^2 a} + \frac{1}{B}\sqrt{C_T/2} \tag{3.133}$$

3.6 Hover Performance

The ideal rotor is useful as a limiting case, if not as a practical design, indicating the form the twist distribution must approach to achieve the best rotor hover performance.

3.6.8 The Optimum Hovering Rotor

The ideal rotor is designed to have minimum induced power. The angle-of-attack is $\alpha = \alpha_t/r$, so only one blade section can be operating at the best lift-to-drag ratio and the ideal rotor does not have the least profile power possible. Consider now a rotor optimized for both induced and profile power. Minimum induced power requires uniform inflow. Minimum profile power requires that each blade section operate at its optimum condition, $\alpha = \alpha_{\text{opt}}$, where the best c_ℓ/c_d is achieved. These two criteria determine the twist and taper for the optimum rotor, which has the best hover performance.

Combined blade element and momentum theory gives

$$dC_T = \frac{\sigma a}{2} \alpha_{\text{opt}} r^2 dr = 4\lambda^2 r \, dr \tag{3.134}$$

or

$$\lambda^2 = \frac{\sigma a}{8} r \alpha_{\text{opt}} \tag{3.135}$$

Assuming that α_{opt} is the same for all blade sections, uniform inflow requires $\sigma r = $ constant; hence a blade taper distribution given by $\sigma = \sigma_t/r$ ($c = c_t/r$). Then the blade twist required is

$$\theta = \alpha_{\text{opt}} + \lambda/r = \alpha_{\text{opt}} + \sqrt{\frac{\sigma_t a \alpha_{\text{opt}}}{8}} \frac{1}{r} \tag{3.136}$$

The rotor thrust and profile power are

$$C_T = \int_0^B \frac{\sigma}{2} r^2 a \alpha_{\text{opt}} dr = \frac{\sigma_t a B^2}{4} \alpha_{\text{opt}} \tag{3.137}$$

$$C_{Po} = \int_0^1 \frac{\sigma}{2} r^3 c_d \, dr = \frac{\sigma_t c_{do}}{2} \int_0^1 r^2 dr = \frac{\sigma_t c_{do}}{6} \tag{3.138}$$

because the blade drag coefficient is constant over the span for the optimum rotor. The total rotor power is then

$$C_P = \frac{C_T^{2/3}}{B\sqrt{2}} + \frac{\sigma_t c_{do}}{6} \tag{3.139}$$

The thrust-weighted equivalent solidity for the optimum rotor is $\sigma_e = 3 \int \sigma r^2 dr = (3/2)\sigma_t$, so the profile power is

$$C_{Po} = \frac{\sigma_e c_{do}}{9} \tag{3.140}$$

Compared to the profile power with rectangular blades ($C_{Po} = \sigma c_{do}/8$), there is at least an 11% reduction in the profile power for the optimum rotor. The difference is even greater because of the higher value of the mean drag coefficient for the constant-chord blade. The figure of merit becomes

$$M = \frac{C_T^{3/2}/\sqrt{2}}{\frac{1}{B}C_T^{3/2}/\sqrt{2} + \frac{\sigma_e c_{do}}{9}} = \frac{1}{\frac{1}{B} + \frac{2}{3}\frac{c_{do}/c_\ell}{\lambda_h}} \quad (3.141)$$

To summarize the optimum rotor design, for a given airfoil section (which defines α_{opt}) the taper and twist required are

$$\sigma = \frac{4C_T}{B^2 a \alpha_{opt}} \frac{1}{r} \quad (3.142)$$

$$\theta = \alpha_{opt} + \sqrt{\frac{C_T}{2B^2}} \frac{1}{r} \quad (3.143)$$

As for the ideal rotor, the design depends on the operating state, in addition to having both the chord and twist singular at the blade root. The optimum rotor solution shows the maximum benefits that are attainable with twist and taper of the blade and indicates the design trends required for real rotors. In general, a washout of the blade pitch at the tip (negative twist) is required, and the blade should be tapered, although the performance gains with taper often cannot justify the added manufacturing cost. In the past, most blades were designed with linear twist and constant chord, and occasionally with linear taper as well. With modern materials and manufacturing techniques, nonlinear twist and non-constant-chord designs are being produced. The optimum rotor solution implies that the design of a real rotor must always be a compromise, because a fixed chord and twist distribution cannot be optimal for all operating conditions.

3.6.9 Elementary Hover Performance Results

Examples of hover performance results are presented in this section, based on the expressions that have been derived for ideal and real rotors. These calculations are for a rotor with solidity $\sigma = 0.08$ (thrust-weighted solidity for tapered blades), tip loss factor $B = 0.97$, and lift curve slope $a = 5.7$. Where relevant, the induced power factor is $\kappa = 1.12$, and the blade drag coefficient is $c_d = 0.0120$. For the section drag as a function of angle-of-attack, $c_d = 0.0087 - 0.0216\alpha + 0.400\alpha^2$ is used, representing an NACA 23012 airfoil (Bailey (1941)).

Figure 3.9 shows the hover power C_P as a function of thrust C_T for three cases of limiting rotor behavior: the rotor with a figure of merit $M = 1$, which has no profile power and minimum induced power, so $C_P = C_T^{3/2}/\sqrt{2}$; the optimum rotor, which has twist for uniform inflow and taper for constant section angle-of-attack, and therefore minimum profile power and induced power; and the ideal rotor, which has twist for uniform inflow, hence minimum induced power, and constant chord. The power of the optimum rotor and of the ideal rotor are given by equations 3.139 and 3.131, respectively. For these cases the induced power is increased by the tip loss, $\kappa = 1/B$, and the mean drag coefficient is $c_{do} = 0.0120$. The real rotor performance in Figure 3.9 is obtained from equation 3.91 with $\kappa = 1.12$ and $c_{do} = 0.0120$. Finally, Figure 3.9 shows the real rotor performance obtained from blade element theory (equations 3.85 and 3.90) for constant chord and linear twist $\theta_{tw} = -8°$, using $\kappa = 1.12$ and $c_d(\alpha)$. The accuracy of the simple performance estimate (equation 3.91) depends on the proper choice of the empirical parameters κ and c_{do}.

3.6 Hover Performance

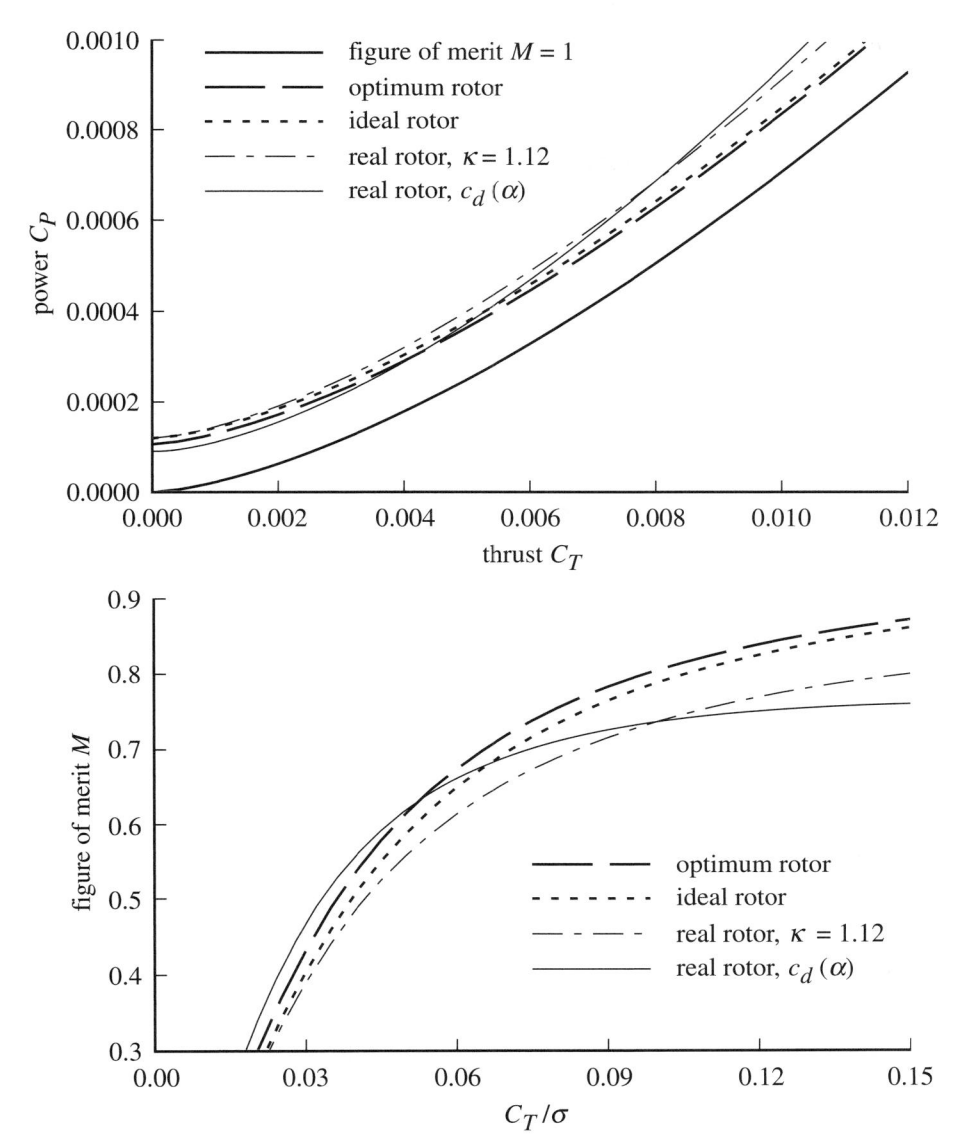

Figure 3.9. Hover performance for ideal and optimum rotors.

This performance comparison is clearer in terms of figure of merit M as a function of blade loading C_T/σ, which is also presented in Figure 3.9. The figure of merit is small at low loading because then the induced power is small relative to the profile power. None of the solutions account for stall, however, so the figure of merit continues to increase for large C_T/σ. The function $c_d = \delta_0 + \delta_1\alpha + \delta_2\alpha^2$ is intended to model the airfoil viscous drag increase with lift coefficient, but does not model stall. The $M = 1$ case has only the minimum induced power; the optimum rotor adds the minimum profile power; the ideal rotor increases the profile power slightly because of the constant chord; and for the real rotor the induced power has been further increased by the factor $\kappa = 1.12$.

Figure 3.10 shows the spanwise distribution of the induced velocity obtained for $C_T/\sigma = 0.08$ by the various methods. The momentum theory value is $\lambda = \sqrt{C_T/2}$.

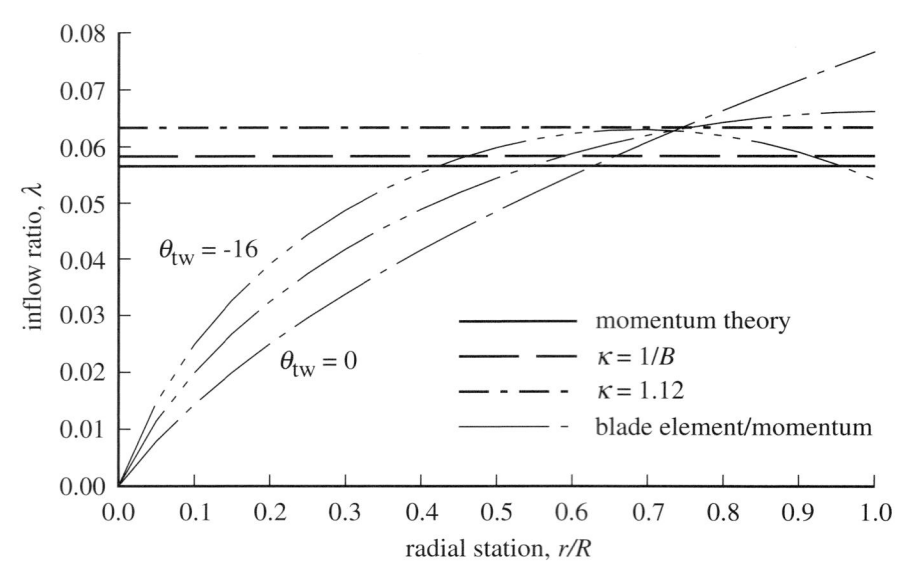

Figure 3.10. Calculated inflow distribution for a hovering rotor at $C_T/\sigma = 0.08$.

For the optimum rotor and the ideal rotor, the induced velocity is increased by the tip loss, $\kappa = 1/B$. The blade element theory results of Figure 3.9 used $\kappa = 1.12$ (equation 3.85). Figure 3.10 also shows the inflow calculated using combined blade element and momentum theory (equation 3.100) for a constant-chord blade with linear twist of $\theta_{tw} = 0$, $-8°$, and $-16°$. Increasing the twist rate produces a more uniform radial distribution of inflow. Tapering the blade would further improve the inflow distribution.

The hover performance calculated using combined blade element and momentum theory (equations 3.100, 3.98, and 3.99) is shown in Figure 3.11 in terms of three metrics: hover figure of merit M, induced power factor $\kappa = C_{Pi}/C_{Pideal}$, and mean drag coefficient $c_{dmean} = 8C_{Po}/\sigma$. Here κ and c_{do} are not specified; rather the induced velocity is calculated by equation 3.100, from which the angle-of-attack is obtained and then the drag coefficient $c_d(\alpha)$. The ratio of the induced power $C_{Pi} = \int \lambda\, dC_T$ to the ideal power $C_{Pideal} = C_T^{3/2}/\sqrt{2}$ (momentum theory) gives κ. The profile power is interpreted as a mean drag coefficient c_{dmean} (in general the influence of flight speed must be included in the definition of c_{dmean}). Figure 3.11 give results for linear blade twist of $\theta_{tw} = 0$, $-8°$, and $-16°$; and for constant chord and taper ratio $c_{tip}/c_{root} = 0.5$. Increasing twist and tapering the blade improve the calculated hover performance by reducing both induced power and profile power. By accounting for the radial variation of the inflow, combined blade element and momentum theory allows the impact of blade twist and planform on hover performance to be considered in the design. The model still does not account for stall, and the representation of tip loss effects is simplistic. As a result, the power is underestimated at high thrust, and the induced power calculation is optimistic.

3.7 Vortex Theory

The lift on a wing is associated with bound circulation; consequently vorticity is trailed into the wake from a three-dimensional wing. On the rotating wing, the

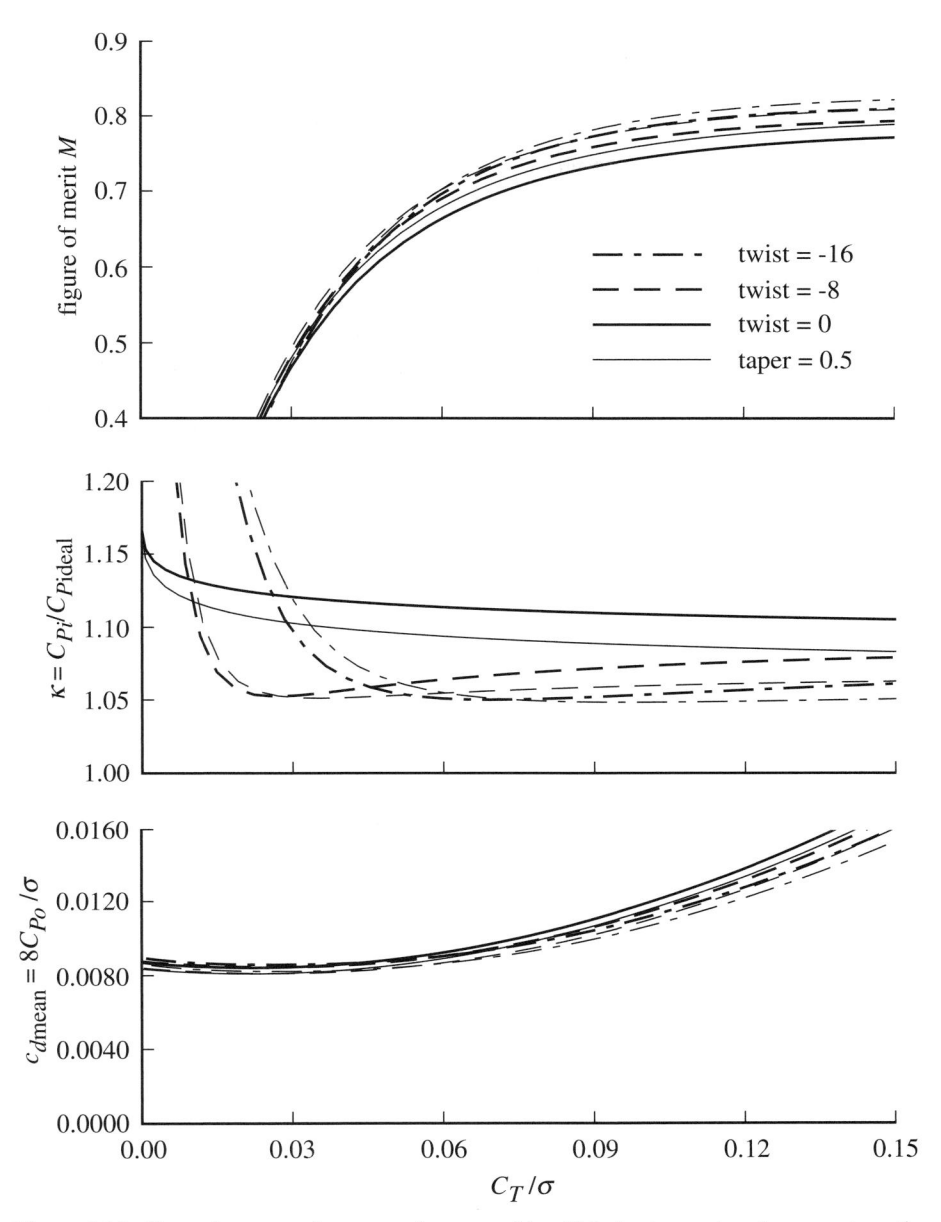

Figure 3.11. Rotor hover performance from combined blade element and momentum theory.

change in the blade loading occurs mostly at the tip, and the rotor wake vorticity is therefore concentrated in tip vortices that lie in helices below the rotor disk. Unlike the fixed wing, the rotary wing has close encounters with its own wake and the wake from preceding blades. These encounters have a significant impact on the induced velocity and blade loads. Vortex theory is a rotor analysis that calculates the flow field of the rotor wake, in particular the induced velocity at the rotor disk, by using the fluid dynamic laws governing the action and influence of vorticity (the Biot-Savart law, Kelvin's theorem, and Helmholtz's laws). The simplest version of vortex theory uses an actuator disk model. The actuator disk neglects the discreteness in the rotor and wake associated with a finite number of blades, and distributes the vorticity throughout the wake volume. The actuator disk model produces a tractable mathematical problem, at least for vertical flight. When considering the same model as momentum theory, vortex theory must give identical results. Vortex theory is better suited than momentum theory to extensions of the model (such as to a nonuniform disk loading), since it is based on a consideration of the local flow characteristics rather than global properties.

If the discreteness in the wake is retained for the vortex theory model, the wake consists of lines and sheets of vorticity trailed behind each blade. Because of the fundamentally transcendental geometry of the rotor wake, integration to evaluate the induced velocity for such a model must be performed numerically. The result is a large numerical problem, which became practical to solve only with the availability of high-speed digital computers for helicopter engineering. With the current availability of computers, use of a discrete vortex model to represent the rotor and wake has become nearly universal when detailed information about the flow field and loading is required. The name "vortex theory" is generally restricted now to the classical work, which primarily used the actuator disk model. The use of a vortex wake model in numerical calculations of the induced velocity is discussed in Chapter 9.

N.E. Joukowski laid the foundations for vortex theory from 1912 to 1929. He investigated the induced velocity due to the helical wake system of a propeller, but had to use the infinite-blade model because of the mathematical complexities. The results of momentum theory were duplicated using this vortex theory and actuator disk analysis. In 1918, Joukowski proposed the use of airfoil characteristics for a cascade of two-dimensional airfoils with the induced velocity taken from vortex theory. This approach gave the elements of modern blade element theory since the cascade effect is negligible for helicopter rotors.

In 1919, A. Betz analyzed the vortex system of the propeller wake in detail, determining the minimum power and best thrust distribution by vortex theory. Around 1920, investigations furthering vortex theory were made by R. Wood and H. Glauert, and by E. Pistolesi. In 1929 S. Goldstein considered more accurately the vortex wake of a propeller with a finite number of blades. For more on the development of vortex theory, see Glauert (1935).

The velocity vector $\mathbf{u}(\mathbf{x})$ induced by a line vortex of strength κ is given by the Biot-Savart law:

$$\mathbf{u}(\mathbf{x}) = -\frac{\kappa}{4\pi} \int \frac{(\mathbf{x} - \mathbf{y}) \times d\ell(\mathbf{y})}{|\mathbf{x} - \mathbf{y}|^3} \qquad (3.144)$$

where the integration is along the entire length of the vortex and $d\ell$ is the tangent to the vortex at position $\mathbf{y}$. This result can also be written as

$$\mathbf{u}(\mathbf{x}) = -\frac{\kappa}{4\pi} \nabla \Sigma \qquad (3.145)$$

where Σ is the solid angle subtended at $\mathbf{x}$ by the line vortex. For an infinite straight line vortex, the induced velocity is entirely circumferential, with magnitude $|\mathbf{u}| = \kappa/2\pi h$, where h is the perpendicular distance to the vortex. In a real fluid, viscosity eliminates the infinite velocity at the vortex line by diffusing the vorticity into a tube of small but finite cross-section radius called the vortex core. Stokes' theorem equates the flux of vorticity through a surface S with the circulation about the boundary of that surface. Kelvin's circulation theorem states that for an inviscid, incompressible fluid of uniform density the circulation $\Gamma = \oint \mathbf{u} \cdot d\ell$ is constant, moving with the fluid.

3.7 Vortex Theory

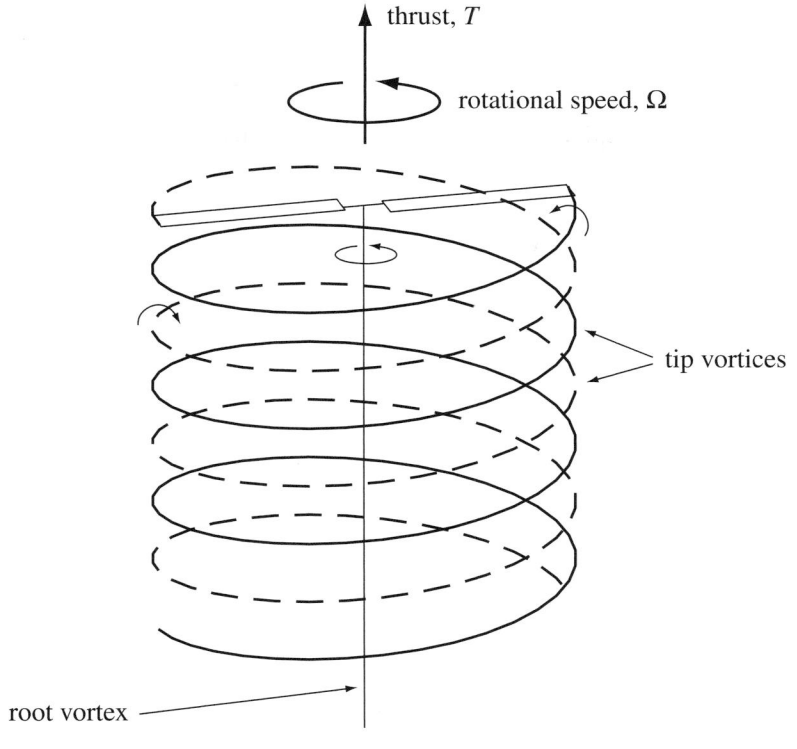

Figure 3.12. Rotor vortex wake in vertical flight.

Helmholtz's laws of vorticity then follow: a fluid initially irrotational remains so; a vortex tube (in particular, a line vortex) moves with constant strength with the fluid; and vortex lines must either be closed or end at solid surfaces. By means of these laws, vortex theory determines the flow of the helicopter rotor.

3.7.1 Vortex Representation of the Rotor and Wake

Associated with the lift L at a wing section is a circulation Γ about the section, such that $L = \rho U \Gamma$ (where U is the free stream velocity and ρ the air density). Thus the rotor blade can be modeled by bound vorticity with strength determined by the rotor lift distribution. Since vortex lines cannot end, this bound vorticity must be trailed into the rotor wake from the blade tips and trailing edges.

With constant blade circulation (which corresponds to uniform loading), vorticity is trailed into the wake only from the blade root and tip. The tip vortex is trailed in a helix because of the combination of the rotational motion of the blade and the axial velocity of the flow through the rotor disk (Figure 3.12). In hover, this axial velocity is entirely due to the wake-induced inflow. There is a tip vortex from each blade, trailed in interlocking helices. The root vortices are trailed along the axis of the rotor in a straight line (ignoring any root cutout). With positive thrust on the rotor, the signs of the vorticity are such that the root vortex and the axial components of the tip spirals induce a swirl in the wake in the same direction as the rotor rotation, and the circumferential components of the tip vortex spirals (ring vortices) induce an axial velocity inside the wake in the opposite direction to the thrust. Thus the wake vortex system produces the velocities that are required by conservation of axial and angular momentum.

More generally, the blade bound circulation varies along the span, requiring that vorticity be trailed all along the wing trailing edge. The wake then consists of helical vortex sheets behind each blade. For the real rotor, the edges of the vortex sheet quickly roll up into concentrated tip vortices, which are well represented by line vortices. There also is considerable self-induced distortion of the wake geometry from the nominal helical form. Classical vortex theory usually ignores the rollup of the vortex sheets, which is a successful approach for propellers, where the high axial velocity sweeps the wake downstream. For low inflow helicopter rotors a more detailed model for the wake is preferable. In forward flight the blade loading varies with azimuth as well as radially, so radial vorticity in addition to the axial and circumferential vorticity is shed into the wake. Radial vorticity can be present in vertical flight if the blade motion is unsteady.

3.7.2 Actuator Disk Vortex Theory

Consider now vortex theory for the actuator disk model of the rotor in hover. The bound vorticity of the blades is distributed in a sheet over the rotor disk in this infinite-blade approximation. So the wake vorticity is distributed throughout the volume of the wake rather than being concentrated in helical sheets or lines. This model greatly reduces the difficulties in calculating the velocity induced by the wake. We have considered this model already in the momentum theory analysis of the rotor. Although the results are not new, vortex theory gives more information about their source, which is valuable background for more sophisticated analyses.

Consider first a uniformly loaded actuator disk, for which $dT/dA = $ constant. The blades then have triangular loading and constant bound circulation:

$$\Gamma = \frac{1}{\rho \Omega r} \frac{dT}{dr} = \frac{2\pi}{\rho \Omega} \frac{dT}{dA} = \frac{2\pi}{\rho \Omega} \frac{T}{A} \tag{3.146}$$

(here Γ is the bound circulation of all the blades). Therefore, the wake consists of only a vortex sheet at the boundary of the slipstream and a line vortex on the axis (Figure 3.13). The line vortex on the axis is the root vortex and is the sum of all the bound vorticity, with strength Γ. The rotor disk is a sheet of radial vorticity. Because the bound circulation of the rotor is spread over the entire disk, the radial vorticity has strength $\gamma_b = \Gamma/2\pi r = T/\rho A \Omega r$. With uniform bound circulation the wake consists of only tip and root vortices, and in the actuator disk limit of infinite blades, the interlocking tip vortex spirals become a vortex sheet on the boundary of the wake, with axial and circumferential components. The axial component of the tip vortex sheet has strength $\gamma = \Gamma/2\pi R_1$, where R_1 is the radius of the wake. The vortex lines form a continuous path (as required by Helmholtz's law) consisting of the root vortex, the radial bound circulation of the disk, and the axial vorticity components of the tip vortex sheet. Because of the helical geometry of the tip vortices, in the infinite-blade limit the wake also contains a circumferential component of vorticity, which can be viewed as consisting of ring vortices. The ring vortex strength is $\gamma = \Gamma/h$, where h is the distance the wake moves down during one rotor revolution. Relating h to the axial velocity at the wake boundary gives $h = 2\pi v/\Omega$, and so $\gamma = T/\rho A v$.

3.7 Vortex Theory

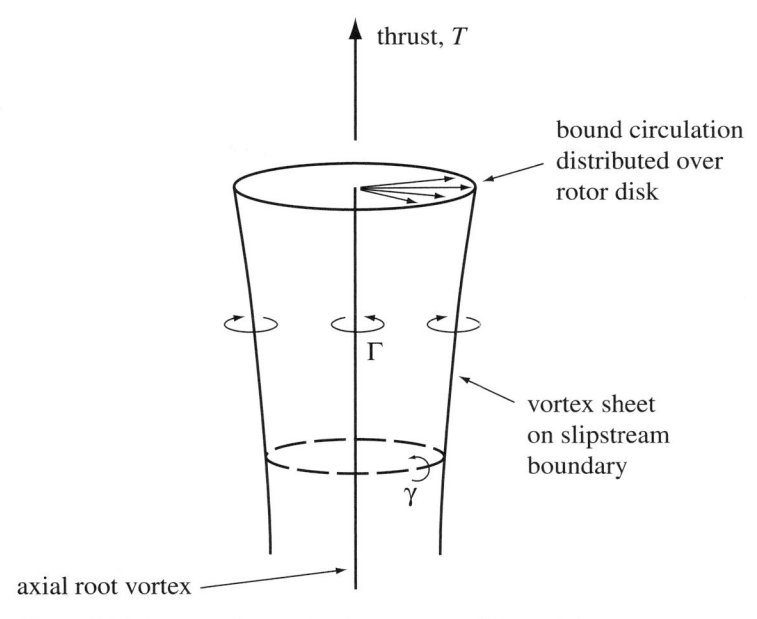

Figure 3.13. Vortex theory for the actuator disk model.

The ring vorticity in the wake produces the axial velocity inside the slipstream. The axial velocity at the rotor disk and in the far wake is due to the wake vorticity alone, with no contributions from the bound circulation. If the wake contraction and swirl velocity are ignored, the induced velocity at the disk is due to a semi-infinite vortex cylinder, and the velocity in the far wake is due to an infinite cylinder. Hence the induced velocity at the disk is one-half the velocity in the far wake, $v = \frac{1}{2}w$. Since the fluid is irrotational far upstream of the rotor, the flow must always be irrotational unless it passes through the rotor disk. Thus there can be no circulation about any path lying entirely outside the wake, and in particular a circumferential velocity can only exist inside the rotor wake. Just above the rotor disk there is no circumferential velocity, whereas just below the disk there is a circumferential velocity u due to the rotor torque. The root vortex induces a circumferential velocity component $u_1 = \Gamma/4\pi r$ both above and below the rotor disk; there is no contribution to the swirl inside the wake from the vorticity on the slipstream boundary (Stokes' theorem). The bound vorticity induces a circumferential velocity u_b just below the disk and $-u_b$ just above the disk. Satisfying the requirement of no swirl outside the wake then requires $u_b = u_1$, and the total swirl just below the disk is $u = 2u_1$. Indeed, since the jump in velocity across the vortex sheet of the rotor disk equals the vortex strength, we have $2u_b = \gamma_b = \Gamma/2\pi r$ again. The velocity seen by the blade due to its own rotation, and the wake-induced swirl is then $\Omega r - u_1 = \Omega r - \frac{1}{2}u$, which explains the appearance of this factor in the expression for the rotor torque in section 3.4.2.

To examine the axial induced velocity further, consider the relation

$$\mathbf{u}(\mathbf{x}) = -\frac{\kappa}{4\pi} \nabla \Sigma \qquad (3.147)$$

where $\mathbf{u}$ is the velocity induced by a line vortex of strength κ that subtends at the point $\mathbf{x}$ a solid angle Σ. The rotor axial velocity is due to a semi-infinite cylinder of ring vortices with strength $\kappa = \gamma\, dz_1$. The axial component of the induced velocity is thus

$$v(z) = -\int_0^\infty \frac{\gamma}{4\pi} \frac{\partial}{\partial z} \Sigma \, dz_1 \tag{3.148}$$

where Σ is the angle of the ring vortex at z_1 as seen at z; the rotor disk is at $z = 0$. Now if the wake contraction rate is slow, the change in Σ as the observer at z moves is primarily due to the distance change $(z - z_1)$ and only secondarily to the change in the ring size. Motion of the observer and of the ring are then equivalent, $\partial \Sigma/\partial z = -\partial \Sigma/\partial z_1$, or

$$v = \int_{z=0}^{z=\infty} \frac{\gamma}{4\pi} d\Sigma \tag{3.149}$$

Next neglect any change in the spacing of the wake spirals, so that the ring vortex strength is constant. With these assumptions, the induced velocity is given by

$$v = \frac{\gamma}{4\pi} \Delta \Sigma \tag{3.150}$$

where $\Delta \Sigma$ is the total solid angle covered by the wake surface, as seen at the location of v. We use this result to evaluate the induced velocity at several points in the flow. For any point on the rotor disk, $\Delta \Sigma = 2\pi$, so $v = \gamma/2$. Recalling that the ring strength is $\gamma = T/\rho A v$, we obtain again for the induced velocity at the rotor disk

$$v = \sqrt{T/2\rho A} \tag{3.151}$$

Moreover, the induced velocity is constant over the disk for this uniformly loaded rotor. Consider the points still in the disk plane but now outside the rotor disk; then $\Delta \Sigma = 0$ and $v = 0$, so there is no axial induced velocity except at the disk. For a point inside the far wake, $\Delta \Sigma = 4\pi$, so $w = \gamma$; the induced velocity is uniform in the far wake and $w = 2v$, as in momentum theory. Finally, for an arbitrary point on the central axis of the wake and at a distance z below the rotor, the induced velocity is

$$v = \frac{\gamma}{4\pi} \left[4\pi - \Sigma_0 \right] \tag{3.152}$$

where Σ_0 is the angle subtended by the rotor disk,

$$\Sigma_0 = 2\pi \left[1 + \frac{z/R}{\sqrt{1 + (z/R)^2}} \right] \tag{3.153}$$

On the wake axis, the axial velocity is therefore

$$v = v(0) \left[1 + \frac{z/R}{\sqrt{1 + (z/R)^2}} \right] \tag{3.154}$$

which has the proper limits far above and far below the rotor (at $z = -\infty$ and $z = \infty$, respectively).

Now consider an actuator disk with nonuniform loading. With varying bound circulation on the blade, the trailed vorticity is distributed throughout the wake cylinder rather than concentrated on the boundary. The wake can be viewed as constructed from shells consisting of the cylindrical sheet at radius r plus the corresponding inboard bound vorticity and root vortex required for conservation of vortex lines. The bound vorticity then is built up from the contributions of all shells outboard of r, and the change in bound vorticity at r is due to the trailed wake there. From the previous paragraph, only the shells outboard of r contribute to the induced velocity $v(r)$, since only for these is the point inside the disk. Hence the axial induced velocity is

$$v = \int_r^R \frac{1}{2}\gamma\, dr \qquad (3.155)$$

where γ is the strength of the trailed vorticity, which is related to the change in the bound circulation Γ by

$$\gamma = -\frac{d\Gamma}{dr}\frac{1}{h} = -\frac{d\Gamma}{dr}\frac{\Omega}{2\pi(V+v)} \qquad (3.156)$$

Then

$$v = -\int_r^R \frac{\Omega}{4\pi(V+v)}\frac{d\Gamma}{dr}\,dr = \frac{\Omega}{4\pi(V+v)}\Gamma + \int_r^R \frac{\Omega}{4\pi}\Gamma\frac{d}{dr}\left(\frac{1}{V+v}\right)dr \qquad (3.157)$$

In terms of the loading distribution $dT/dA = \rho\Omega\Gamma/2\pi$, this vortex theory result for the induced velocity becomes

$$2\rho(V+v)v = \frac{dT}{dA} + (V+v)\int_r^R \frac{dT}{dA}\frac{d}{dr}\left(\frac{1}{V+v}\right)dr \qquad (3.158)$$

Compare this result with the differential form of momentum theory, $dT = 2\rho(V+v)v\,dA$ (equation 3.32), which was obtained (without proof) by application of the conservation laws to the annulus of the rotor disk at r. The induced velocity obtained from differential momentum theory (such as in combined blade element and momentum theory), although not exact, appears to be reasonably accurate as long as the inflow is reasonably uniform. Similarly, recall that the relation $w = 2v$ between the induced velocities at the disk and in the far wake is not an exact result of momentum theory. The assumptions required in vortex theory to duplicate these momentum theory results give a better idea of the approximations involved in their application.

3.7.3 Finite Number of Blades

Vortex theory for vertical flight is elementary with the actuator disk model, especially for uniform loading. With a finite number of blades, vortex theory models the wake by vortex lines and sheets trailed in helices behind each blade. This problem is mathematically much more difficult than the case of distributed wake vorticity, but in axial flow some analytical solutions are still possible. Finite-blade vortex theory is analogous to the Trefftz plane analysis of fixed wings. The wake is studied far downstream, where the wing has negligible influence on the flow. The solution for the wake vorticity also determines the loading on the wing. By solving the simpler flow problem in the far wake (where there is no axial dependence), an exact loading distribution that includes tip effects is obtained. The accuracy of the solution depends on the wake model used. The classical analyses use approximate models, with vortex sheets rather than concentrated tip vortices and no self-induced wake distortion. Moreover, a far wake analysis does not provide information about the blade design needed to obtain the desired loading; for that the induced velocity at the rotor disk is needed.

The bound circulation Γ varies along the blade span and must be zero at the root and at the tip. Trailed vorticity of strength $-\partial\Gamma/\partial r$ springs from each radial station. This vortex system is initially formed as a helical or screw surface behind each blade.

The edges of the vortex sheet roll up, so at a sufficient distance downstream the vorticity is concentrated in contracted, helical lines emanating from the blade tips, of strength nearly equal to the maximum bound circulation, plus a straight root vortex on the axis (see Figure 3.12). For an idealized model of the wake, the rollup and contraction are ignored. For a lightly loaded propeller, the distortion of the vortex sheets is neglected, and the helical geometry is produced by the blade rotation and the flight speed. Assuming the flow is inviscid and irrotational, the bound circulation gives the jump in velocity potential at the corresponding point on the screw surface. Once the wake geometry has been established, the boundary condition of no flow through the sheets determines the flow field, from which the vorticity strength in the wake and hence the loading on the blade are obtained.

Betz showed that the loading distribution for a lightly loaded propeller with minimum energy loss is such that the vorticity forms a helicoidal sheet moving backward undeformed. Prandtl developed an approximate solution for the flow around a helicoidal sheet, from the two-dimensional flow around a cascade of semi-infinite sheets. Goldstein solved for the potential field and distribution of circulation for a helicoidal vortex system for the case of a lightly loaded propeller. Theodorsen removed the limitation to lightly loaded propellers by considering the wake far downstream of the propeller.

3.7.3.1 Wake Structure for Optimum Performance

Let us examine the wake geometry implied for optimum performance of a propeller. This discussion follows Betz, as described by Glauert (1935). Consider the optimum loading in terms of the section aerodynamic lift, $L = \rho U \Gamma$. The inflow angle $\phi = \tan^{-1}(V+v)/(\Omega r - \frac{1}{2}u)$ gives the normal and in-plane components of L, producing the thrust and torque of the rotor:

$$T = \rho N \int \Gamma U \cos\phi \, dr = \rho N \int \Gamma \left(\Omega r - \frac{1}{2}u\right) dr \qquad (3.159)$$

$$Q = \rho N \int \Gamma U \sin\phi \, r \, dr = \rho N \int \Gamma (V+v) r \, dr \qquad (3.160)$$

As in section 3.4.2, v is the inflow velocity and u is the swirl velocity induced by the wake. These equations are consistent with momentum theory results (equations 3.42 and 3.43),

$$T = \int 2\rho(V+v)v \, dA = \int \rho\left(\Omega r - \frac{1}{2}u\right) u \, dA \qquad (3.161)$$

$$Q = \int \rho(V+v)u\Omega r \, dA \qquad (3.162)$$

since $u \, dA = u 2\pi r \, dr = N\Gamma \, dr$. The variational problem is to find the loading Γ for minimum torque given the thrust. Using a Lagrange multiplier λ,

$$\delta Q - \lambda \delta T = \rho N \int \left[r(V+v) - \lambda\left(\Omega r - \frac{1}{2}u\right)\right] \delta\Gamma \, dr = 0 \qquad (3.163)$$

The optimum solution is $r(V+v)/(\Omega r - \frac{1}{2}u) = \lambda =$ constant, or

$$r \tan\phi = \text{constant} \qquad (3.164)$$

Now the surface generated by rotating the blade at Ω and convecting without distortion the wake at constant velocity W is a helicoid or screw, described in cylindrical coordinates (r, θ, z) by $\theta = \Omega r + \theta_0$ and $z = Wt$. This rigid surface has a helix angle at r of

$$\tan\phi = \frac{1}{r}\frac{dz}{d\theta} = \frac{W}{\Omega r} \qquad (3.165)$$

The corresponding pitch of the helix is $p = 2\pi r \tan\phi = 2\pi W/\Omega = $ constant. For a constant pitch (screw-like) helix, wake elements shed from the trailing edge at a given instant always remain in the same horizontal radial line. The pitch angle is the angle of the wake surface with respect to the horizontal.

Hence the optimum solution $r\tan\phi = $ constant corresponds to a helicoid in uniform axial translation. Thus Betz demonstrated that the ideal efficiency of a propeller (minimum energy loss for a given thrust) is obtained with a blade loading that produces a rigidly moving helicoidal wake. The flow behind the propeller is the same as if the screw surfaces formed by the trailing vortices are rigid and moving axially at constant velocity.

Although the wake sheets move rigidly with axial velocity W, the fluid about the sheets has axial velocity u_z, swirl velocity u_θ, and radial velocity u_r. The trailed vorticity in the wake surface corresponds to a jump in the radial velocity across the surface. No flow through the wake surface means that at the surface the normal velocity of the air must equal the velocity of the sheet: $u_n = W\cos\phi$. So the components of the air velocity at the sheet must be $u_z = u_n\cos\phi = W\cos^2\phi$ and $u_\theta = u_n\sin\phi = W\cos\phi\sin\phi$. These conditions are sufficient to determine the circulation distribution Γ. With a finite number of blades, the normal velocity decreases between the sheets, and there is radial flow as well, which decreases the lift at the blade tip. In the limit of an infinite number of blades, the sheets are very close, and as a consequence all the fluid is carried with the wake. There are then no losses due to flow around the edges.

A complication of this argument is that the inflow angle ϕ in fact varies with distance along the axis, from $\tan\phi = (V+v)/(\Omega r - \frac{1}{2}u)$ at the disk to $\tan\phi = (V+w)/(\Omega r_1 - u_1)$ in the far wake. This complication is avoided by assuming the propeller is lightly loaded, so $\tan\phi \cong V/\Omega r$.

3.7.3.2 Prandtl's Tip Loading Solution

In 1919, A. Betz analyzed the vortex system of the propeller wake in detail, determining the minimum power and best thrust distribution by vortex theory. In an appendix to Betz's paper, L. Prandtl gave an approximate correction for the tip effect on the thrust distribution of a rotor with a finite number of blades. This solution is described by Glauert (1935).

The rotor induced velocity and loading can be obtained by considering the wake far downstream from the rotor disk, with the solution depending on the model used for the wake. Distributed wake vorticity implies distributed loading on the rotor disk, hence an actuator disk model. In fact, the rotor consists of discrete lifting surfaces. The simplest wake model including the effects of a finite number of blades consists of helical vortex sheets trailed from each blade. The major effect of the finite number of blades is a reduction of the loading at the blade tip. In terms of the wake flow, there is a flow around the edges of the vortex sheets from the lower surface to the

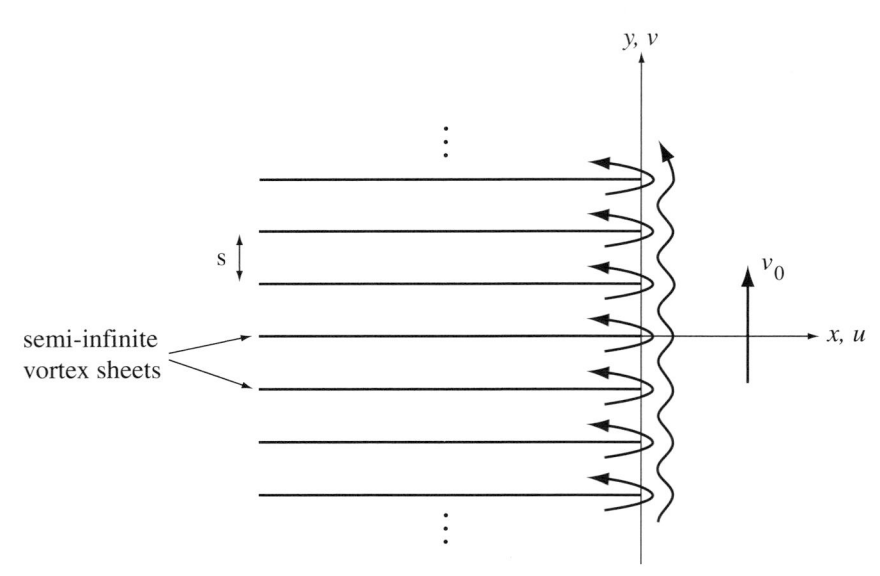

Figure 3.14. Two-dimensional model for the rotor wake.

upper surface that reduces the total downward momentum. Prandtl derived a tip loss correction for a finite number of blades by using a two-dimensional model for the vorticity in the far wake.

The system of helical vortex sheets are modeled as a series of semi-infinite, parallel vortex lines (Figure 3.14), thereby replacing the axisymmetric wake by a two-dimensional flow model that can be solved using complex potential methods. Since this model is equivalent to considering the flow only near the edges of the wake helices, for low inflow (small spacing of the sheets) the model should be fairly accurate. A coordinate system moving downward with the wake at velocity v_0 is used, such that the wake sheets are stationary and the external flow is moving upward at v_0. The fluid velocities are u and v (see Figure 3.14), and dimensionless quantities are used (based on ρ, Ω, and R). For a lightly loaded rotor, the wake spacing is

$$s = \frac{2\pi}{N} \frac{\lambda}{\sqrt{1+\lambda^2}} \cong \frac{2\pi\lambda}{N} \quad (3.166)$$

where λ is the inflow ratio and N is the number of blades.

The complex potential satisfying the condition of no flow through the vortex sheets and the requirement that v and u approach v_0 and zero, respectively, as x approaches infinity is

$$w = -v_0 \frac{s}{\pi} \cos^{-1} e^{\pi z/s}$$

where $z = x + iy$. Then the velocity is

$$u - iv = \frac{dw}{dt} = v_0 \frac{e^{\pi z/s}}{\sqrt{1 - e^{2\pi z/s}}} \quad (3.167)$$

For example, at $y = 0$ (one of the sheets)

$$u - iv = v_0 \frac{e^{\pi x/s}}{\sqrt{1 - e^{2\pi x/s}}} \quad (3.168)$$

3.7 Vortex Theory

In the fixed frame, the sheets are moving downward at v_0, and the fluid far away from the wake is at rest. There is air flowing up around the edges of the sheets, however, which reduces the mean downward velocity of the air between the sheets. Momentum conservation implies that there must be a reduction of the lift near each blade tip. In the fixed frame, the average vertical velocity between the sheets is

$$\frac{1}{s}\int_0^s (v_0 - v)dy = v_0 \frac{2}{\pi} \cos^{-1} e^{\pi x/s} \tag{3.169}$$

or $\bar{v}(x) = v_0 F$, where for the helicopter rotor with $\pi x/s = (r-1)N/2\lambda$

$$F = \frac{2}{\pi} \cos^{-1} e^{(r-1)N/2\lambda} \tag{3.170}$$

The function F is the principal result of this analysis. The vorticity γ in the wake sheet (which is related to the rotor bound circulation distribution) is

$$\gamma = v_{y=0} - v_{y=s} = 2v_0 \frac{e^{\pi x/s}}{\sqrt{1 - e^{2\pi x/s}}} \tag{3.171}$$

Then the blade bound circulation is

$$\Gamma(x) = \int_x^0 \gamma \, dx = v_0 s \frac{2}{\pi} \cos^{-1} e^{\pi x/s} = v_0 s F \tag{3.172}$$

Substituting for $s = 2\pi\lambda/N = 2\pi(\lambda_c + \lambda_i)/N$ and using $v_0 = 2\lambda_i$ gives $\Gamma = (4\pi/N)(\lambda_c + \lambda_i)\lambda_i F$, or

$$\frac{dC_T}{dr} = 4(\lambda_c + \lambda_i)\lambda_i r F \tag{3.173}$$

which is simply the momentum theory result (equation 3.94), corrected for the effect of the blade tip by the factor $F(r)$. The function F is significantly less than unity only over the outer 5% to 10% of the blade.

In combined blade element and momentum theory, the hover induced velocity now becomes

$$\lambda = \frac{\sigma a}{16F}\left[\sqrt{1 + \frac{32F}{\sigma a}\theta r} - 1\right] \tag{3.174}$$

The effect of the tip, expressed through the factor F, is to increase the induced velocity and thus reduce the loading at the blade tip and increase the induced power. The factor F also affects the planform for the optimum rotor, necessitating the introduction of rounded tips. Figure 3.15 shows the spanwise distribution of the induced velocity λ_i and the section thrust dC_T/dr for a constant-chord blade at $C_T/\sigma = 0.08$, as calculated by combined blade element and momentum theory using Prandtl's function. For comparison, the results obtained using a simple tip loss factor are also shown (see Figure 3.10). With Prandtl's function, the inflow is increased at the tip, as required to reduce the section lift to zero. The effect on integrated performance is approximately a 2% increase in the induced power factor κ and a 1% reduction in the peak figure of merit.

Rather than using the factor F to correct the span loading near the tips, we can instead use this model to obtain an equivalent tip loss factor B for the rotor loading and performance calculations. An equivalent infinite-blade model (with a smaller effective disk area) is found that produces the same thrust for a given power as the finite-blade rotor. If the vortex sheets were infinitely close, the fluid between the

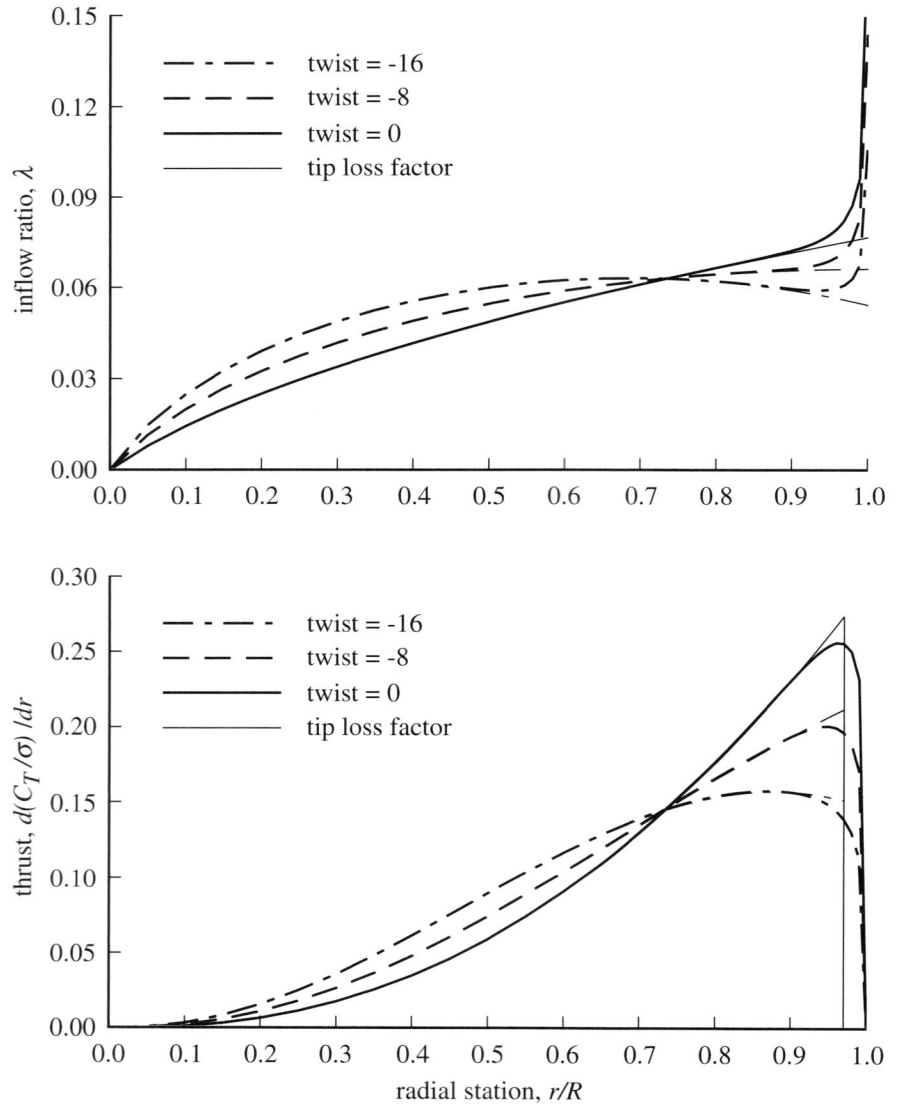

Figure 3.15. Inflow and loading distribution for a hovering rotor, calculated by combined blade element and momentum theory using Prandtl's function.

sheets would all be carried downward at velocity v_0 and the fluid outside would be at rest. With a finite distance between the sheets some fluid flows up around the edges, reducing the downward momentum. By equating the momentum reduction $v_0(1-B)$ of an infinite-blade model with reduced wake area to the momentum reduction due to the finite number of blades, the tip loss factor B can be evaluated:

$$1 - B = \frac{1}{v_0} \int_0^\infty (v - v_0) dx = \int_0^\infty \left[\frac{e^{\pi x/s}}{\sqrt{e^{2\pi x/s} - 1}} - 1 \right] dx$$
$$= \frac{s}{\pi} \ln 2 = \frac{\lambda}{N} 2 \ln 2 = 1.39 \frac{\lambda}{N} \qquad (3.175)$$

Here the quantity λ is the inflow velocity, which determines the wake spiral spacing. For hover with linear inflow, $\lambda = r\lambda_t = r\sqrt{C_T}$ (section 3.6.3), so

$$B = 1 - 1.39 \frac{\sqrt{C_T}}{N} \cong 1 - \frac{\sqrt{2C_T}}{N} \qquad (3.176)$$

which is the result generally cited (see equation 3.108).

3.7.3.3 Propeller Analyses of Goldstein and Theodorsen

Goldstein (1929) developed a vortex theory for propellers with a finite number of blades in axial flow. The wake was modeled as helical trailed vortex sheets moving axially at a constant velocity like rigid surfaces. The boundary condition of no flow through the sheets completely defines the vortex strength in the wake, which can be related to the bound circulation distribution on the blade. Goldstein solved the potential flow problem of N intermeshed helical surfaces of infinite extent axially (i.e. in the far wake) but finite radius moving with axial velocity v_0. The solution takes the form of a tip loading factor F, which is a function of the inflow ratio, the number of blades, and the blade radial station. Assuming the propeller is lightly loaded, the wake contraction is ignored, and the helix angle is $\tan\phi = V/\Omega r$.

Prandtl's function is a good approximation to Goldstein's more complete result for low inflow, specifically when $\lambda/N < 0.1$ or so. Thus Prandtl's solution is good for helicopter rotors, but Goldstein's solution is more appropriate for propellers.

Theodorsen (1944) generalized the vortex theory solution to heavily loaded propellers, by making the distinction between the conditions at the rotor disk and in the wake far downstream. The analysis was based on the inflow angle in the far wake, $\tan\phi = (V + w)/\Omega r$. Neglecting contraction, the potential jump in the far wake can be mapped to the bound circulation on the blade.

A vortex sheet has a singular induced velocity at its edge. Thus the trailed vorticity rolls up about the outer edge to form a tip vortex, and the sheets roll up at the center to form a root vortex. In models that neglect the rollup, assuming rigid vortex surfaces, there must exist edge forces to suppress the rollup. In the contraction region near the rotor disk, these edge forces are tilted. As a result, the models produce a thrust that is too small and a torque that is too large. For a propeller with small contraction, this effect is small. Moreover, such vortex theory models neglect the interactions between blades and rolled-up tip vortices that are important for low inflow rotors.

3.8 Nonuniform Inflow

For low inflow helicopter rotors, the trailed vorticity quickly rolls up into concentrated tip vortices that remain close to the disk and strongly influence the loading near the tips of both the generating and following blades. The self-induced distortion in the wake and the resulting interference between the rotor blade and the tip vortices have a significant influence on the rotor performance. Such effects must be included in the vortex theory if an accurate calculation of the blade loading and power is to be achieved. Such an analysis has to be numerical because of the complexities of the vortex wake structure and geometry of a real rotor. The modern variant of vortex theory is a numerical solution for the rotor induced velocity, loads, and performance that uses a detailed model of the vortex wake.

Chapter 9 presents a further exposition of nonuniform inflow and free wake geometry calculation.

3.8.1 Hover Wake Geometry

The fundamental character of the wake geometry of a hovering rotor was recognized by Gray (1955, 1956). He conducted flow visualization studies of hovering rotor wakes to guide the development of improved hover performance prediction. The flow visualization experiment involved high-speed photographs of the wake geometry of a model rotor, the wake made visible by smoke emanating from the blade tips. Initially a two-bladed rotor was used, but he observed the overtaking of the tip vortex from one blade by that from the other. Gray attributed this behavior to differences in tip vortex strength caused by differences in blade thrust. To eliminate this interaction, a one-bladed rotor was used. The rotor had an untwisted, constant-chord blade, with a radius of 4 ft. Based on this work, Gray developed a sketch of the hovering rotor wake geometry (Figure 3.16). The flow visualization led to a quantitative description of the geometry that was unexpectedly simple: exponential contraction plus a two-stage vertical convention velocity (Figure 3.16).

Gray's work was followed by more extensive flow visualization tests, which were used to establish prescribed wake geometry methods for hovering rotor performance calculation. Analyses based on this realistic wake geometry provided significantly improved prediction of rotor performance attributes.

Landgrebe (1971, 1972) used externally generated smoke filaments to visualize the wake of a 2.2-ft radius model rotor (Figure 3.16). The test matrix included a range of blade number (2 to 8), twist, blade aspect-ratio (13.6 to 18.2), and collective pitch. Landgrebe (1971) also noted the "discovery of a reduction in wake stability with increasing distance from the rotor." Landgrebe (1972) described the structure of the wake:

> The wake contains two primary components. The first, and most prominent, is the strong tip vortex which arises from the rapid rolling up of the portion of the vortex sheet shed from the tip region of the blade. The second feature is the vortex sheet shed from the inboard section of the blade. The vertical or axial transport velocity near the outer end of the inboard vortex sheet is much greater than that of the tip vortex. The vertical velocity of the vortex sheet also increases with radial position, resulting in a substantially linear cross section of the vortex sheet at any specific azimuth position. These characteristics largely result from the velocities induced by the strong tip vortex.

From the flow visualization data, a generalized representation of the near wake geometry was constructed. The form of the equations for contraction and convection followed Gray.

Kocurek and Tangler (1977) conducted a flow visualization test of a model rotor in hover. The wake of a 1.0-ft radius rotor was visualized with Schlieren techniques (Figure 3.16). The test matrix included a range of blade number (1 to 4) and blade aspect ratios (7.1 to 18.2) beyond that of Landgrebe, specifically covering low aspect-ratio two-bladed rotors. The form of the equations for contraction and convection were the same as used by Landgrebe, but with different dependence on thrust and blade geometry.

The hover wake geometry is described by two-stage vertical convection and exponential spanwise contraction. In terms of the dimensionless wake age ϕ (azimuth

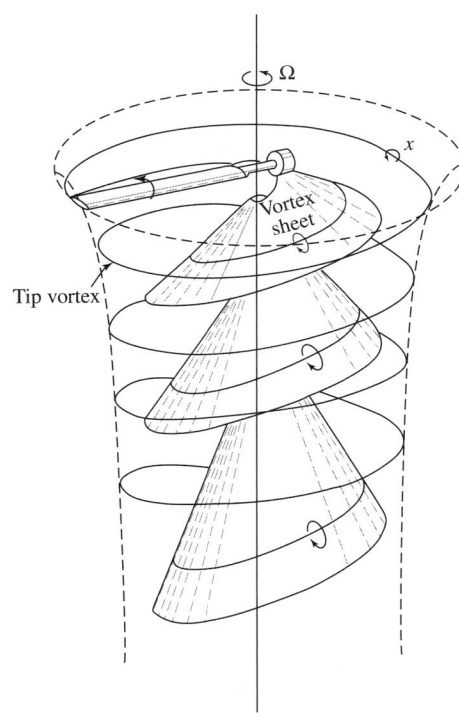

3.8 Nonuniform Inflow

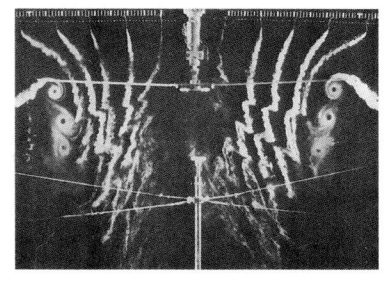

Sample flow visualization photograph; Landgrebe (1971).

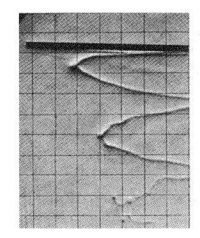

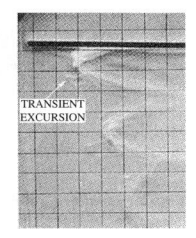

The vortex pattern in the wake of a single-bladed hovering helicopter rotor as obtained from smoke studies; Gray (1956).

Single and multiple exposure schlieren photgraphs; Kocurek and Tangler (1977).

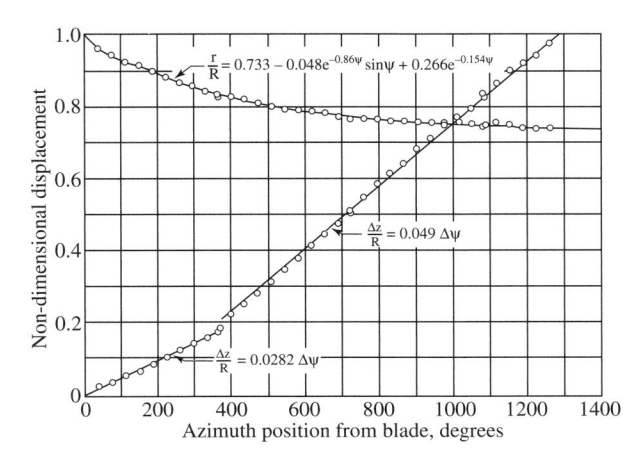

$$\frac{r}{R} = 0.733 - 0.048 e^{-0.86\psi} \sin\psi + 0.266 e^{-0.154\psi}$$

$$\frac{\Delta z}{R} = 0.049 \, \Delta\psi$$

$$\frac{\Delta z}{R} = 0.0282 \, \Delta\psi$$

Non-dimensional radial and vertical displacement of tip vortex as a function of azimuth position from blade; Gray (1955).

Figure 3.16. Hover wake geometry from flow visualization tests.

angle in the wake, measured from the blade in the direction opposite to rotation), the vertical and spanwise position of the wake (scaled with radius R) are

$$z = \begin{cases} K_1 \phi & \phi < \phi_1 \\ K_1 \phi_1 + K_2(\phi - \phi_1) & \phi > \phi_1 \end{cases} \quad (3.177)$$

$$r = K_4 + e^{-K_3 \phi}(1 - K_4) \quad (3.178)$$

where $\phi_1 = 2\pi/N$ is the age at the encounter with the following blade (N is the number of blades). For the inside edge of the inboard sheet, r is multiplied by the root cutout ratio. This geometry is steady relative to the moving wing, so z and r are functions of only wake age ϕ, not time. The vertical convection is defined by K_1 and K_2, the rates before and after encountering the first following blade. These are dimensionless velocities, scaled with ΩR. The spanwise contraction is defined by K_3 and K_4, the rate of contraction and the maximum contraction ratio, respectively. The contraction time constant is $1/K_3 \Omega$, and K_4 is the dimensionless distance, scaled with R.

Landgrebe (1971) developed a prescribed wake model of the above form. The vertical convection constants are

	K_1	K_2
tip vortex	$0.25(C_T/\sigma + 0.001\theta_{tw})$	$(1 + 0.01\theta_{tw})\sqrt{C_T}$
outside sheet edge	$1.55\sqrt{C_T}$	$1.90\sqrt{C_T}$
inside sheet edge	0	$-(0.0025\theta_{tw}^2 + 0.099\theta_{tw})\sqrt{C_T}$

for the twist rate θ_{tw} in degrees. The spanwise contraction parameters for the tip vortex and the inboard sheet are

$$K_3 = 0.145 + 27 C_T \quad (3.179)$$

$$K_4 = 0.78 \quad (3.180)$$

Kocurek and Tangler (1977) revised the tip vortex geometry based on experimental data including low aspect-ratio, two-bladed rotors, obtaining

$$K_1 = B + C \left(\frac{C_T}{N^n}\right)^m \quad (3.181)$$

$$K_2 = \left[C_T - N^n \left(-\frac{B}{C}\right)^{1/m}\right]^{1/2} \quad (3.182)$$

$$K_3 = 4.0\sqrt{C_T} \quad (3.183)$$

where $B = 0.000729 \theta_{tw}$, $C = 2.3 - 0.206 \theta_{tw}$, $m = 1.0 - 0.25 e^{0.04 \theta_{tw}}$, and $n = 0.5 - 0.0172 \theta_{tw}$. The other constants are the same as for the Landgrebe model. These two prescribed models are based on flow visualization results, in which typically only four spirals of the wake are observed.

3.8.2 Hover Performance Results from Free Wake Analysis

The blade loading of a hovering rotor is strongly influenced by the tip vortex from the preceding blade and hence is very sensitive to the radial and vertical position of that vortex relative to the blade. As established by the flow visualization tests described in section 3.8.1, the tip vortex is initially convected downward (at a rate smaller than the momentum-theory value for the mean induced velocity) and inward (the beginning of the wake contraction). Consequently the tip vortex encounters the following blade inboard of the tip, with a small vertical separation. The effect of this vortex is to increase the angle-of-attack outboard of its position and to decrease the angle-of-attack inboard. Depending on the twist and planform, the rotor performance can be adversely affected, especially with thick tip airfoil sections at high Mach number. If the wake contraction is neglected, then all blade stations are inboard of the vortex and nowhere does the vortex increase the angle-of-attack. With a distributed vorticity wake model or even simpler theories, the effects of rolled-up tip vortices cannot be observed at all. Thus classical methods for rotor hover performance, or analyses with an uncontracted wake, tend to underestimate the hover power required. An accurate wake geometry is crucial to refinements in the calculation of hover performance.

Figure 3.17 shows the hover performance calculated using a free wake analysis. As in section 3.6.9, these calculations are for a four-bladed rotor with solidity $\sigma = 0.08$, and constant-chord blade with linear twist of $\theta_{tw} = 0$, $-8°$, and $-16°$. The airfoil tables used (lift and drag coefficient as a function of angle-of-attack and Mach number) corresponded to current technology sections, with a thin section at the tip. The blade is modeled as rigid with a flap hinge; hence the blade cones in response to the thrust. The hover performance is shown in terms of three metrics: hover figure of merit M, induced power factor $\kappa = C_{Pi}/C_{Pideal}$, and mean drag coefficient $c_{dmean} = 8C_{Po}/\sigma$. The ratio of the induced power to the ideal power $C_{Pideal} = C_T^{3/2}/\sqrt{2}$ (momentum theory) gives κ. The profile power is interpreted as a mean drag coefficient c_{dmean} (in forward flight the influence of flight speed in the definition of c_{dmean} must also be considered). The corresponding results from combined blade element and momentum theory were presented in Figure 3.11. Increasing twist improves the calculated hover performance by reducing both induced power and profile power. Nonuniform inflow calculation with free wake geometry changes the magnitude of the induced power and the variation of κ with C_T/σ. Because the airfoil tables include the effect of stall on section drag, the profile power increases substantially at high thrust, resulting in a decrease in the figure of merit.

Figure 3.18 shows for $C_T/\sigma = 0.08$ the spanwise distribution of the induced velocity, angle-of-attack, tip vortex vertical position, section thrust, and section power. At its first encounter with the following blade, the tip vortex has moved about $7\% R$ inboard of the tip, and is about $3\% R$ below the blade. There is a strong influence of this first blade-vortex encounter on the inflow and angle-of-attack at the tip, as well as the expected influence of the twist inboard.

3.9 Influence of Blade Geometry

3.9.1 Twist and Taper

For a given rotor thrust, radius, and tip speed, both the induced and profile power losses can be minimized by the proper choice of the blade twist and taper. A linear variation of chord or pitch can closely approximate the optimum distributions over the outer portions of the blade, where the loading is most important. In fact, $-8°$ to $-12°$ of linear twist produces most of the induced power gains of blades with ideal twist, compared to untwisted blades. Linear twist is easily built into the blades,

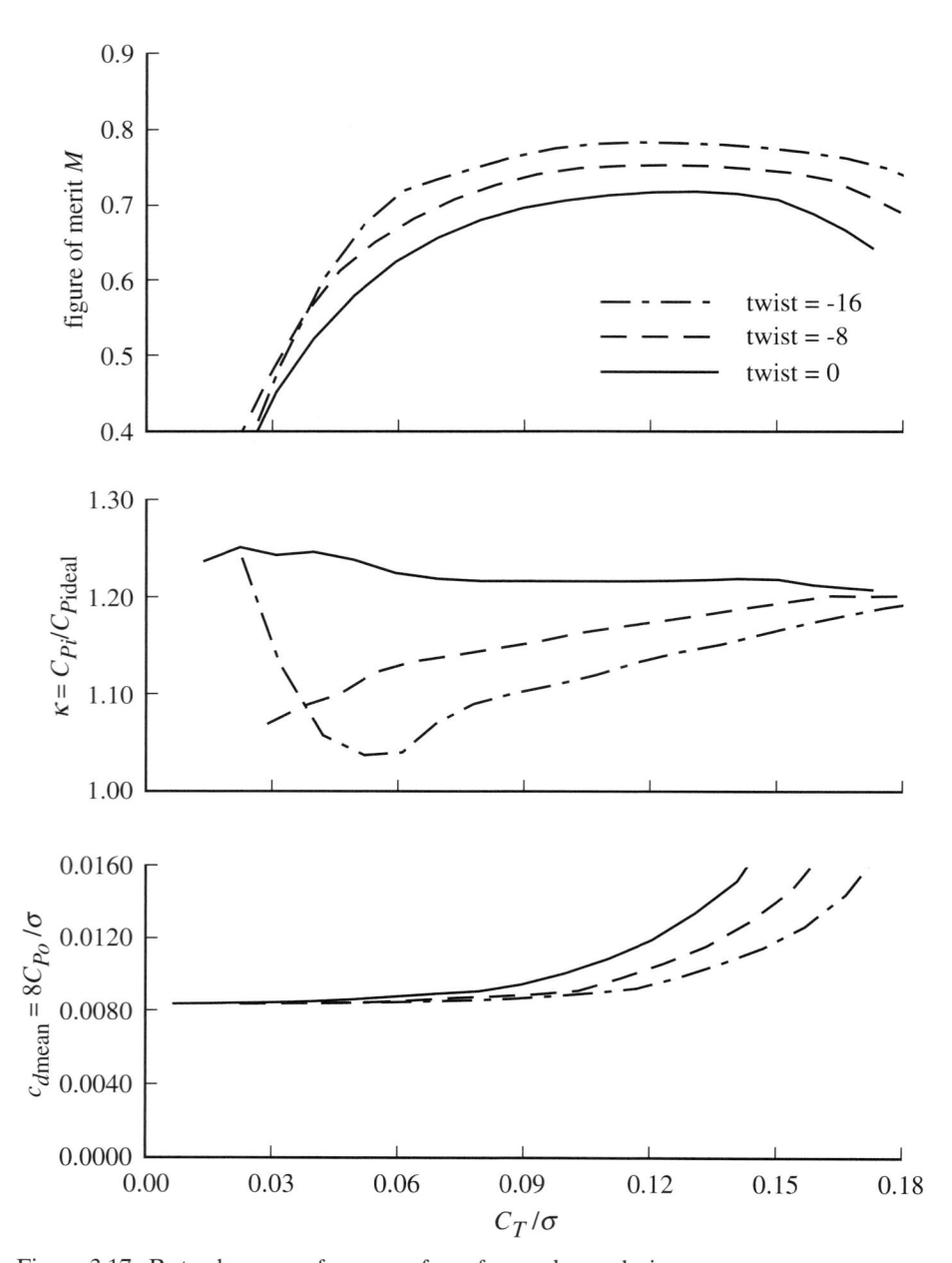

Figure 3.17. Rotor hover performance from free wake analysis.

with little production cost increase for a significant performance benefit. Taper improves the rotor performance also, but because of production costs was in the past justified only for very large rotors, and in recent designs the focus is more often on tip planform. Table 3.1, based on the calculations presented in this chapter, gives the percentage reduction in rotor hover power that is obtained with various combinations of twist and taper. The reduction (positive for better performance, lower power) is measured relative to the result for −8° twist and no taper. Also shown are results for the ideal rotor (ideal twist, section 3.6.7) and the optimum rotor

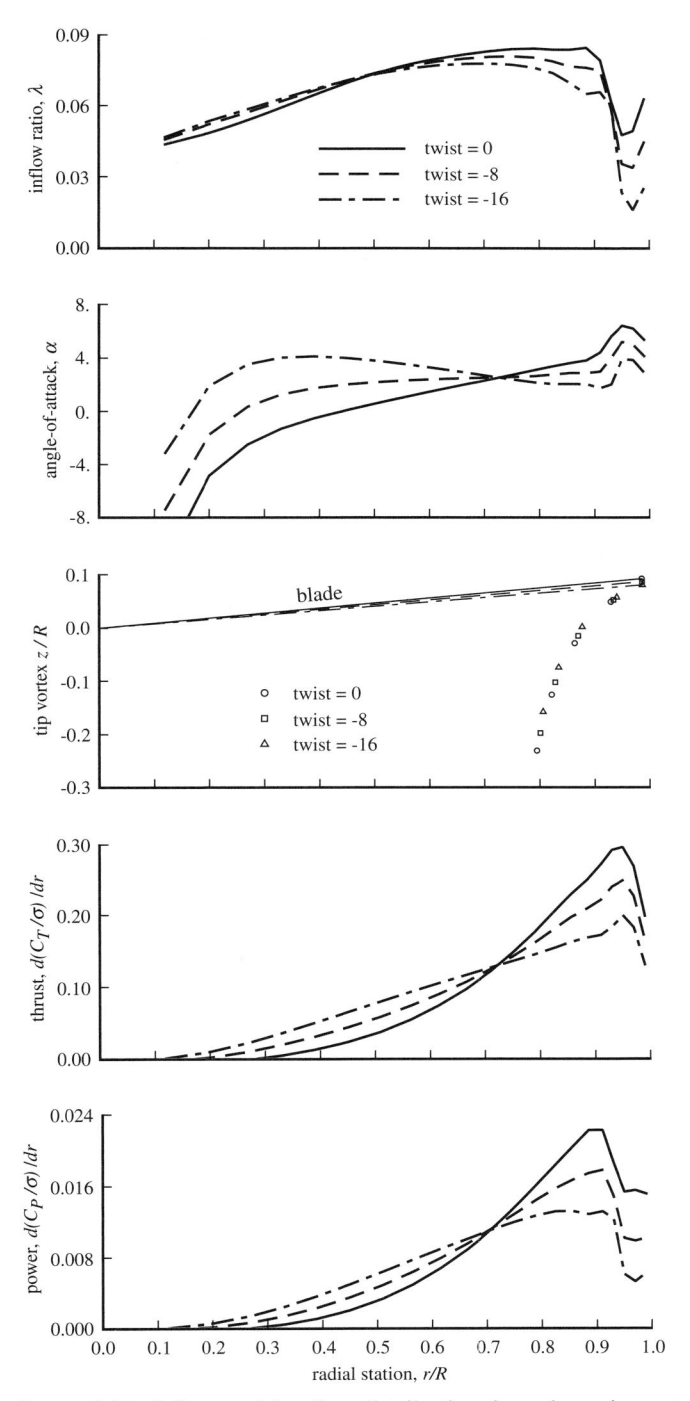

Figure 3.18. Inflow and loading distribution for a hovering rotor, calculated by free wake analysis.

Table 3.1. *Reduction of rotor hover power due to twist and taper*

Twist (degrees)	Taper ratio	Blade element & momentum	BEM with Prandtl's function	Nonuniform inflow and free wake
0	1	−3.0%	−3.1%	−6.0%
−8°	1	—	—	—
−16°	1	1.3%	1.5%	4.2%
0	0.5	−0.5%	−0.6%	−6.2%
−8°	0.5	1.8%	1.9%	0.3%
−16°	0.5	2.4%	2.6%	6.1%
ideal	1	2.8%	3.4%	1.5%
optimum	vary	5.4%	6.0%	4.2%

(twist and taper, section 3.6.8). All the calculations for Table 3.1 are for a blade loading $C_T/\sigma = 0.08$ and thrust-weighted solidity of $\sigma = 0.08$. The combined blade element and momentum theory (BEM) results correspond to section 3.6.9, but with a constant blade drag coefficient ($c_d = 0.0120$). The results for BEM with Prandtl's tip loss function correspond to section 3.7.3.2. The calculated power is increased by using the tip loss function (note the values for the ideal rotor and the optimum rotor, compared to BEM), but the relative effects of linear twist and linear taper are about the same as BEM (which here used $B = 0.97$). The free wake results correspond to section 3.8.2 and reflect the use of airfoil tables for better section aerodynamic load estimation, as well as the influence of the discrete tip vortices in the wake.

Negative twist improves rotor hover performance, but also tends to increase hover download and reduce hover in-ground-effect thrust augmentation. Moderate values of negative twist improve rotor performance in forward flight, since unloading the tips delays stall on the retreating blade and compressibility effects on the advancing blade. Large values of negative twist, although beneficial for hover performance, contribute to the appearance of negative loading on the advancing side in forward flight. Thus twist reduces forward flight maximum lift-to-drag ratio and increases hub vibratory loads and blade alternating loads at maximum speed. Twist also has some effect on autorotation performance.

Thus selection of the rotor twist and taper is a complex task requiring consideration of all the operating conditions of the helicopter.

3.9.2 Blade Tip Shape

The angle-of-attack change produced by the tip vortex suggests modifications to the blade tip geometry. Increasing the blade twist rate (more negative pitch) over the outer 5–7% of the blade can counter the upwash induced by the tip vortex. Some blade designs also use zero or positive twist rate for the last part of the blade, intended to reduce the peak bound circulation and resulting tip vortex strength by increasing the loading near the tip.

Anhedral or droop of the blade tip can increase the hover performance by 2–3%. With such droop, the tip vortex forms at a lower position. Because of the wake contraction, this tip vortex encounters the following blade under the inboard (not drooped) part of the blade, hence with increased vertical separation. Larger vertical separation reduces the blade-vortex interaction and improves the rotor hover performance.

Tip taper and sweep improve rotor performance by reducing compressibility effects near the tip. The hover figure of merit can be increased at high tip Mach numbers. With a square tip, nose-down pitching moments at high Mach number lead to high control loads in high-speed or high load-factor flight. With a swept tip, the effective Mach number is reduced and the center of lift shifted aft. A swept tip can improve the loading in the second quadrant during forward flight, since the aft loading tends to twist the blade to counter occurrence of negative lift.

3.10 REFERENCES

Bailey, F.J., Jr. "A Simplified Theoretical Method of Determining the Characteristics of a Lifting Rotor in Forward Flight." NACA Report 716, 1941.

Glauert, H. "Airplane Propellers." In *Aerodynamic Theory*, Durand, W.F. (Editor). New York: Julius Springer, 1935.

Goldstein, S. "On the Vortex Theory of Screw Propellers." Proceedings of the Royal Society of London, Series A, *123*:792 (April 1929).

Gray, R.B. "On the Motion of the Helical Vortex Shed from a Single-Bladed Hovering Model Helicopter Rotor and Its Application to the Calculation of the Spanwise Aerodynamic Loading." Princeton University, Aeronautical Engineering Department Report No. 313, September 1955.

Gray, R.B. "An Aerodynamic Analysis of a Single-Bladed Rotor in Hovering and Low-Speed Forward Flight as Determined from Smoke Studies on the Vorticity Distribution in the Wake." Princeton University, Aeronautical Engineering Department Report No. 356, September 1956.

Knight, M., and Hefner, R.A. "Static Thrust Analysis of the Lifting Airscrew." NACA TN 626, December 1937.

Kocurek, J.D., and Tangler, J.L. "A Prescribed Wake Lifting Surface Hover Performance Analysis." Journal of the American Helicopter Society, *22*:1 (January 1977).

Landgrebe, A.J. "An Analytical and Experimental Investigation of Helicopter Rotor Hover Performance and Wake Geometry Characteristics." USAAMRDL TR 71-24, June 1971.

Landgrebe, A.J. "The Wake Geometry of a Hovering Helicopter Rotor and Its Influence on Rotor Performance." Journal of the American Helicopter Society, *17*:4 (October 1972).

Sissingh, G. "Contribution to the Aerodynamics of Rotating-Wing Aircraft." NACA TM 921, December 1939.

Theodorsen, T. "The Theory of Propellers." NACA Reports 775, 776, 777, and 778, 1944.

Wheatley, J.B. "An Aerodynamic Analysis of the Autogiro Rotor with a Comparison Between Calculated and Experimental Results." NACA Report 487, 1934.

4 Vertical Flight

Vertical flight of the helicopter rotor at speed V includes the operating states of hover ($V = 0$), climb ($V > 0$), and descent ($V < 0$) and the special case of vertical autorotation (power-off descent). Between the hover and autorotation states, the helicopter is descending at reduced power. Beyond autorotation, the rotor is producing power for the helicopter. The principal subject of this chapter is the induced power of the rotor in vertical flight, including descent. The key physics are associated with the flow states of the rotor in axial flight. Axial flight of a rotor also encompasses the propeller in cruise ($V > 0$) and static ($V = 0$) operation, and a horizontal axis wind turbine ($V < 0$).

4.1 Induced Power in Vertical Flight

In Chapter 3, momentum theory was used to estimate the rotor induced power P_i for hover and vertical climb. Momentum theory gives a good power estimate if an empirical factor is included to account for additional induced losses, particularly tip losses and losses due to nonuniform inflow. In the present chapter these results are extended to include vertical descent. Momentum theory is not applicable for a range of descent rates because the assumed wake model is not correct. Indeed, the rotor wake in that range is so complex that no simple model is adequate. In autorotation, the operating state for power-off descent, the rotor is producing thrust with no net power absorption. The energy to produce the thrust (the induced power P_i) and turn the rotor (the profile power P_o) comes from the change in gravitational potential energy as the helicopter descends. The range of descent rates where momentum theory is not applicable includes autorotation.

Momentum theory gives the rotor power as $P = T(V + v)$ (not including the profile power). The rotor thrust T is positive. TV is the power input to the rotor for climb at vertical speed V or for descent at speed $|V|$, in which case the airflow supplies the power $T|V|$ to the rotor. The induced power is $P_i = Tv$, where v is the induced velocity at the rotor disk. The induced power is always positive, so $v > 0$ for positive thrust. Since the induced velocity is seldom uniform, especially in vertical descent, v is best viewed as being equivalent to the induced power by the definition $v = P_i/T$. This view is consistent with the way v is obtained from measured rotor performance. The induced velocity or power is a function of the speed, thrust, rotor

4.1 Induced Power in Vertical Flight

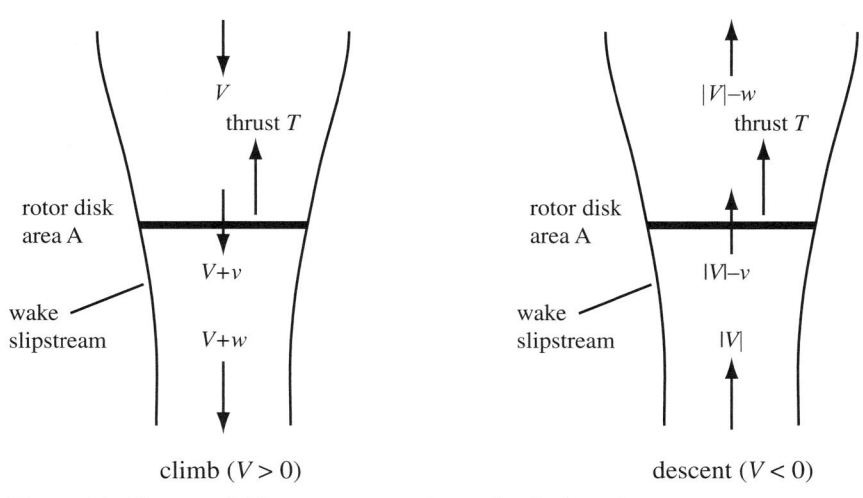

Figure 4.1. Flow model for momentum theory in climb or descent.

disk area, and air density:

$$v = f(V, T, A, \rho) \tag{4.1}$$

For forward flight, an additional parameter is the disk incidence angle i (see Chapter 5). There are other parameters influencing the induced velocity that are not considered here, such as the rotor tip speed and the distribution of the loading over the rotor disk. From dimensional analysis, the functional form for v must be

$$\frac{v}{v_h} = f\left(\frac{V}{v_h}, i\right) \tag{4.2}$$

where $v_h^2 = T/2\rho A$ (the momentum theory result for the hover induced velocity). The induced power and the momentum theory hover power are $P_i = Tv$ and $P_h = Tv_h$, so $v/v_h = P_i/P_h$. The function $f(V/v_h, i)$ can be obtained by analysis (such as momentum theory) or by experiment. A measurement or calculation of P_i and T for a given V is plotted in the form of v/v_h as a function of V/v_h. Any discrepancies in the empirical correlation of measured performance by this function are due to other factors that influence the induced power, such as the twist distribution, number of blades, planform and airfoil shape, and tip Mach number. For the purposes of obtaining a first estimate of the induced power in vertical flight, $v/v_h = f(V/v_h)$ covers the primary functional dependence.

4.1.1 Momentum Theory for Vertical Flight

As in section 3.1, consider momentum theory for an actuator disk model of a uniformly loaded rotor. The rotor is climbing at velocity V, and therefore the flow is downward through the rotor disk (Figure 4.1). The induced velocities v and w at the rotor disk and in the far wake, respectively, are assumed to be uniform. The sign convention (important when the descent case is considered) is that the thrust is positive upward and the velocities positive downward. The mass flux is $\dot{m} = \rho A(V + v)$. Momentum conservation gives $T = \dot{m}(V + w) - \dot{m}V = \dot{m}w$, and energy conservation gives $P = T(V + v) = \frac{1}{2}\dot{m}(V + w)^2 - \frac{1}{2}\dot{m}V^2 = \frac{1}{2}\dot{m}(2Vw + w^2)$. Eliminating $T/\dot{m}$ gives $w = 2v$, and hence $T = 2\rho A(V + v)v$. On writing $v_h^2 = T/2\rho A$, the momentum theory result for the rotor in climb becomes

$$\frac{v}{v_h}\left(\frac{V}{v_h}+\frac{v}{v_h}\right)=1 \tag{4.3}$$

with solution

$$v=-\frac{V}{2}+\sqrt{\left(\frac{V}{2}\right)^2+v_h^2} \tag{4.4}$$

since v must be positive. The net velocities at the disk and far downstream are then

$$V+v=\frac{V}{2}+\sqrt{\left(\frac{V}{2}\right)^2+v_h^2} \tag{4.5}$$

$$V+w=V+2v=\sqrt{V^2+4v_h^2} \tag{4.6}$$

The key to the momentum theory analysis is to use the correct model for the flow. The climb model cannot be used with $V<0$, because in descent the free stream velocity is directed upward and therefore the far downstream wake is above the rotor disk. The flow model for descent is also shown in Figure 4.1. The mass flux is still $\dot{m}=\rho A(V+v)$. Now momentum and energy conservation give $T=\dot{m}V-\dot{m}(V+w)=-\dot{m}w$ and $P=T(V+v)=\frac{1}{2}\dot{m}V^2-\frac{1}{2}\dot{m}(V+w)^2=-\frac{1}{2}\dot{m}(2Vw+w^2)$. V is negative now, whereas T, v, and w are still positive. Since $V+v$ is negative (upward flow through the disk), $P=T(V+v)$ is negative, and the rotor is extracting power from the airstream in excess of the induced power. This flow condition is called the windmill brake state. Eliminating $T/\dot{m}$ gives $w=2v$ again. The momentum theory result for the induced velocity in descent is $T=-2\rho A(V+v)v$, or

$$\frac{v}{v_h}\left(\frac{V}{v_h}+\frac{v}{v_h}\right)=-1 \tag{4.7}$$

with solution

$$v=-\frac{V}{2}-\sqrt{\left(\frac{V}{2}\right)^2-v_h^2} \tag{4.8}$$

The net velocities at the disk and far downstream are

$$V+v=\frac{V}{2}-\sqrt{\left(\frac{V}{2}\right)^2-v_h^2} \tag{4.9}$$

$$V+w=V+2v=-\sqrt{V^2-4v_h^2} \tag{4.10}$$

The other solution of the quadratic for v gives $v>0$ and $V+v<0$ as required, but has $V+w>0$. Thus the flow in the far wake would be downward, contrary to the assumed flow model.

Figure 4.2 shows the momentum theory solution for the rotor in vertical climb or descent. The dashed portions of the curves are branches of the solution that do not correspond to the assumed flow state. The line $V+v=0$ is where the flow through the rotor disk and the total power $P=T(V+v)$ change sign. At the line $V+2v=0$ the flow in the far wake changes sign. The lines $V=0$, $V+v=0$, and $V+2v=0$ divide the plane into four regions, where the rotor operating condition is named the normal working state (climb and hover), vortex ring state, turbulent

4.1 Induced Power in Vertical Flight

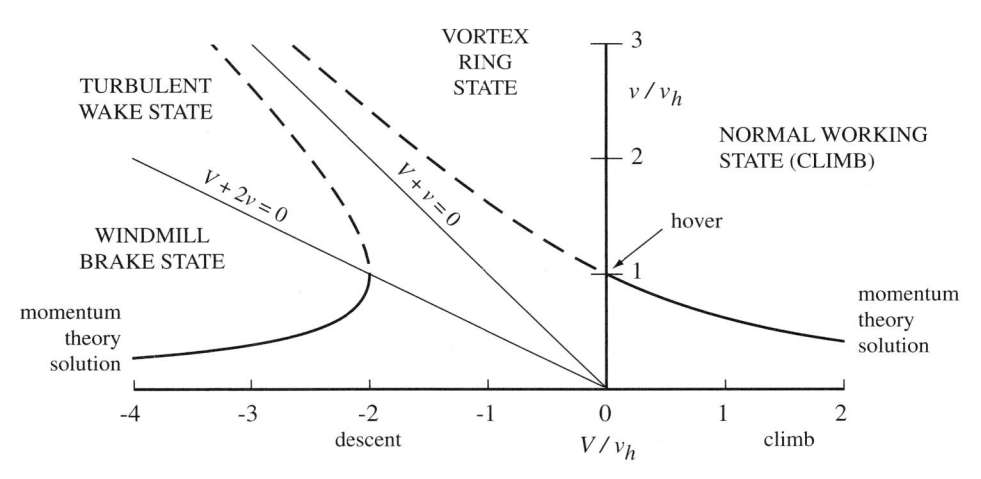

Figure 4.2. Momentum theory results for the induced velocity in vertical flight.

wake state, and windmill brake state (see Figure 4.2). For climb, the air is assumed to be moving downward throughout the flow field (V, $V + v$, and $V + 2v$ all positive). For the branch of the solution given by $V < 0$, however, the flow through the disk and in the wake is downward while the flow outside the slipstream is upward; this is not a physically realizable condition for the entire flow field. The climb solution can be expected to be valid for small rates of descent, however, where at least near the rotor the flow is all downward. Thus the region of validity for the momentum theory solution does include hover. For the rotor in descent, the air is assumed to be moving upward throughout the flow field (V, $V + v$, and $V + 2v$ all negative). For the upper branch of the descent solution, however, $V + 2v > 0$, so the flow is downward in the far wake while upward everywhere else, including outside the wake slipstream. Again this is not a physically realizable condition. Thus, in the vortex ring and turbulent wake states the flow outside the slipstream is upward while the flow inside the far wake is nominally downward. Because such a flow state is not possible, there is no valid momentum theory solution for the moderate rates of descent between $V = 0$ and $V = -2v_h$. The line $V + v = 0$ corresponds to ideal autorotation, $P = 0$, and is in the center of the range where momentum theory is not valid. The momentum theory results become infinite at $V + v = 0$ because the theory implies that thrust is produced without mass flow through the rotor disk ($\dot{m} = 0$).

In summary, momentum theory is based on a wake model consisting of a definite slipstream and a well-defined wake downstream, with the air moving in the same direction throughout the flow field. Since this a good model for the rotor in climb or a high rate of descent, in the normal working state and windmill brake state, momentum theory gives a good estimate of the induced power. The momentum solution for climb is actually good for small rates of descent as well and hence in a range including hover; the flow model is really incorrect, but near the rotor there is no drastic change in the flow until perhaps $V/v_h < -\frac{1}{2}$. For moderate rates of descent, $-2v_h < V < 0$, there is no valid wake model for momentum theory. The flow would like to be upward everywhere except in the far wake, where it wants to be downward. The result is an unsteady, turbulent flow with no definite slipstream. Thus the induced velocity law for the vortex ring and turbulent wake states must be determined empirically from a correlation of measured rotor performance.

4.1.2 Flow States of the Rotor in Axial Flight

Figure 4.3 shows sketches, more impressionistic than representational, of the flow states of rotor in axial flight. The four states are shown: normal working state, vortex ring state, turbulent wake state, and windmill brake state.

4.1.2.1 Normal Working State

The normal working state includes climb and hover (see Figures 4.3a and 4.3b). For climb, the velocity throughout the flow field is downward with both V and v positive. From mass conservation, the wake contracts downstream of the rotor. A wake model with a definite slipstream is valid for this flow state (although the wake really consists of discrete vorticity), and momentum theory gives a good estimate of the performance. There is also entrainment of air into the slipstream below the rotor and some recirculation near the disk, particularly for hover. Although such phenomena are not included in the momentum theory model, their effect on the induced power is secondary.

Hover ($V = 0$) is the limit of the normal working state. By mass conservation, the area of the slipstream becomes infinite upstream of the rotor. Still, momentum theory models the flow well in the vicinity of the rotor disk and hence gives a good performance estimate, even though hover is nominally a limiting case.

4.1.2.2 Vortex Ring State

When the rotor starts to descend, a definite slipstream ceases to exist because the flows inside and outside the slipstream in the far wake want to be in opposite directions. Therefore, from hover to the windmill brake state the flow has large recirculation and high turbulence. Sometimes this entire region is called the vortex ring state, but the convention here is that the vortex ring state is defined by $P = T(V + v) > 0$, so that the power extracted from the airstream is less than the induced power. The region with $P = T(V + v) < 0$ is called the turbulent wake state. Partial power descents occur in the vortex ring state, and equilibrium autorotation usually occurs in the turbulent wake state.

Figures 4.3c and 4.3d sketch the flow about the rotor in the vortex ring state. At small rates of descent, recirculation near the disk and unsteady, turbulent flow above the disk begin to develop. The flow in the vicinity of the disk is still reasonably well represented by the momentum theory model, however. Because the change in flow state for small rates of climb or descent is gradual, the momentum theory solution remains valid for some way into the vortex ring state. Eventually, at descent rates beyond about $V = -\frac{1}{2}v_h$, the flow even near the rotor disk becomes highly unsteady and turbulent. The rotor in this state experiences a high vibration level, and aircraft motion can develop that is difficult to control. In particular, in the vortex ring state the power required is not very sensitive to vertical velocity, and hence controlling the descent rate is difficult in this region.

The flow pattern in the vortex ring state is like that of a vortex ring in the plane of the rotor disk or just below it (hence the name given the state; the flow is highly turbulent as well). The upward free-stream velocity in descent keeps the blade tip

4.1 Induced Power in Vertical Flight

(a) climb

(b) hover, $V = 0$

normal working state

(c) low descent rate

(d) higher descent rate

vortex ring state

(e) ideal autorotation ($V+v = 0$)

(f) turbulent wake state

turbulent wake state

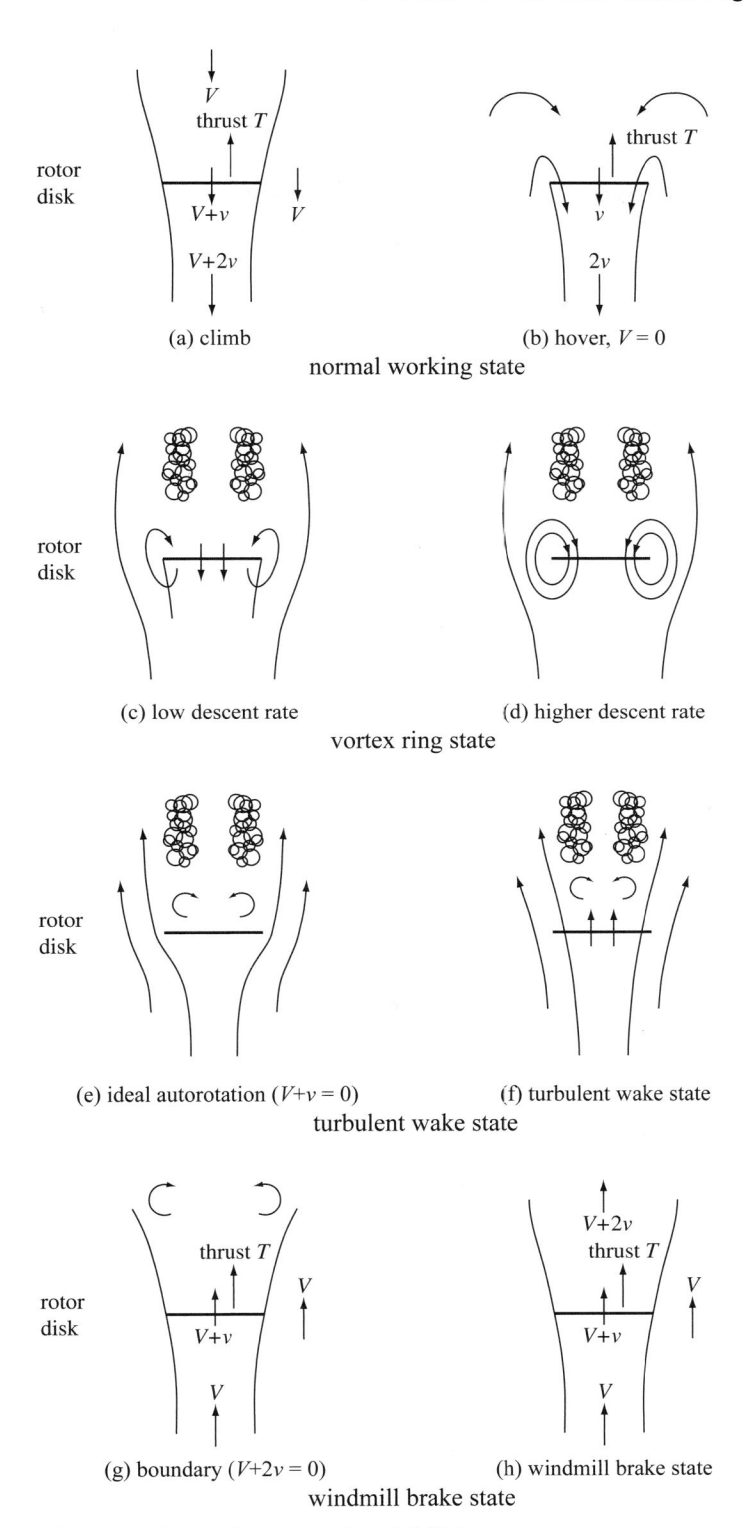

(g) boundary ($V+2v = 0$)

(h) windmill brake state

windmill brake state

Figure 4.3. Rotor flow states in axial flight.

vortex spirals piled up under the disk, forming the ring. With each revolution of the rotor the ring vortex builds up strength until it breaks away from the disk plane in a sudden breakdown of the flow. The flow field is thus unsteady, the vortex ring periodically being allowed to escape and rise into the flow above the rotor. This behavior is a source of disturbing low-frequency vibration. In the turbulent wake state, $V + v < 0$ and so the flow is nominally upward through the rotor disk. The tip vortices are then carried upward, away from the disk again.

4.1.2.3 Turbulent Wake State

Figure 4.3e shows the flow state for ideal autorotation, $V + v = 0$. If the rotor had no profile power losses, power-off descent would be at this condition, since there $P = T(V + v) = 0$. Although nominally there is no flow through the disk, actually there is considerable recirculation and turbulence. The flow state is similar to that of a circular plate of the same area: no flow through the disk, with a turbulent wake above.

Figure 4.3f sketches the flow for the turbulent wake state. The flow still has a high level of turbulence, but since the velocity at the disk is upward there is much less recirculation through the rotor. The flow pattern above the rotor disk in the turbulent wake state is similar to the turbulent wake of a bluff body (hence the name given the state). The rotor in this state experiences some roughness due to the turbulence, but nothing like the high vibration in the vortex ring state.

4.1.2.4 Windmill Brake State

At large rates of descent ($V < -2v_h$) the flow is again smooth, with a definite slipstream. Figures 4.3g and 4.3h show this flow condition in the windmill brake state. The velocity is upward throughout the flow field, the slipstream expanding in the wake above the rotor. In the windmill brake state the rotor is producing a net power $P = T(V + v) < 0$ for the helicopter by the action of the airstream on the rotor. The simple wake model of momentum theory is again applicable, and a good performance estimate is obtained.

At the windmill brake state boundary ($V + 2v = 0$ at $V = -2v_h$), the velocity in the far wake above the rotor is nominally zero. Thus the slipstream area approaches infinity above the disk as the flow tries to stagnate. The flow outside the slipstream is still upward, however, so in contrast to the hover case this limit is an unstable condition. At the boundary between the windmill brake and turbulent wake states, the flow changes rather abruptly from a state with a smooth slipstream to one with recirculation and turbulence as the nominal velocity in the far wake changes direction. Thus the validity of the momentum theory solution ceases abruptly at the windmill brake boundary.

4.1.3 Induced Velocity Curve

Figure 4.4 presents the universal law for the induced power in vertical flight in terms of v/v_h as a function of V/v_h, a form originated by Hafner (1947). The induced velocity v is not measured directly, but rather the law is a correlation of measured rotor power and thrust at various axial speeds. The ordinate is therefore best interpreted as P_i/P_h. The measured rotor power also includes profile power ($P = T(V + v) + P_o$),

4.1 Induced Power in Vertical Flight

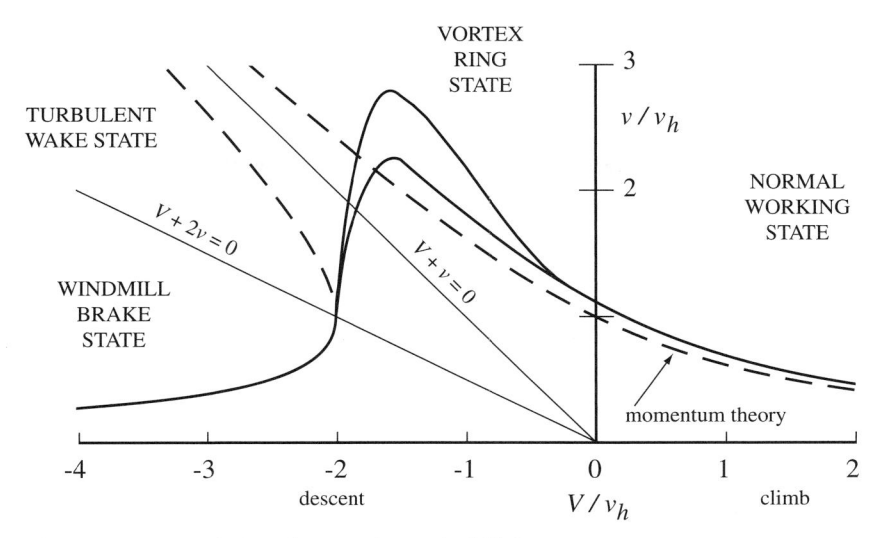

Figure 4.4. Rotor induced power in vertical flight.

which must be accounted for to obtain the induced power:

$$\frac{V+v}{v_h} = \frac{P-P_o}{T\sqrt{T/2\rho A}} = \frac{C_P - C_{Po}}{C_T^{3/2}/\sqrt{2}} \qquad (4.11)$$

Obtaining the induced velocity thus requires an estimate of the profile power coefficient. The simple result $C_{Po} = \sigma c_d/8$ might be used, but a more detailed calculation of C_{Po} is desirable since any errors in C_{Po} result in corresponding scatter in the induced power correlation. By this means the universal induced velocity curve can be constructed, as in Figure 4.4.

Momentum theory indeed gives a good performance estimate in the normal working and windmill brake states. In hover and climb, the measured induced power is higher than the momentum theory result by a small, relatively constant factor. This power increase is due to the additional induced losses of the real rotor, particularly nonuniform inflow and tip losses. The induced velocity correlation always shows some scatter, due to errors in the profile power calculation, variations in the nonoptimum losses, and the influence of other design parameters, such as tip Mach number and blade twist. At hover, for example, the result could be 5% or 10% different from that shown in Figure 4.4. The scatter is greatest in the vortex ring state. Because of the highly turbulent and unsteady flow condition, the induced velocity is not well represented by a single line in this range of descent rates. Moreover, the vortex ring state is basically an unstable flow condition and hence is sensitive to factors such as ground proximity and wind or ground speed, making good performance measurements in this region difficult to obtain.

An alternative presentation of the induced velocity law, developed by Lock (1947), is in terms of $(V+v)/v_h$ as a function of V/v_h (Figure 4.5). In this case the total power $P/P_h = (V+v)/v_h$ is given, rather than just the induced power. Such a presentation is more consistent with the way the curve is obtained and used, since the total power is of interest in the rotor performance calculation. Figure 4.5 also shows the lines $V+v = 0$ (the abscissa) and $V+2v = 0$, which define the four flow states of the rotor in axial flow. The line $v = 0$ goes through the origin at 45°; the

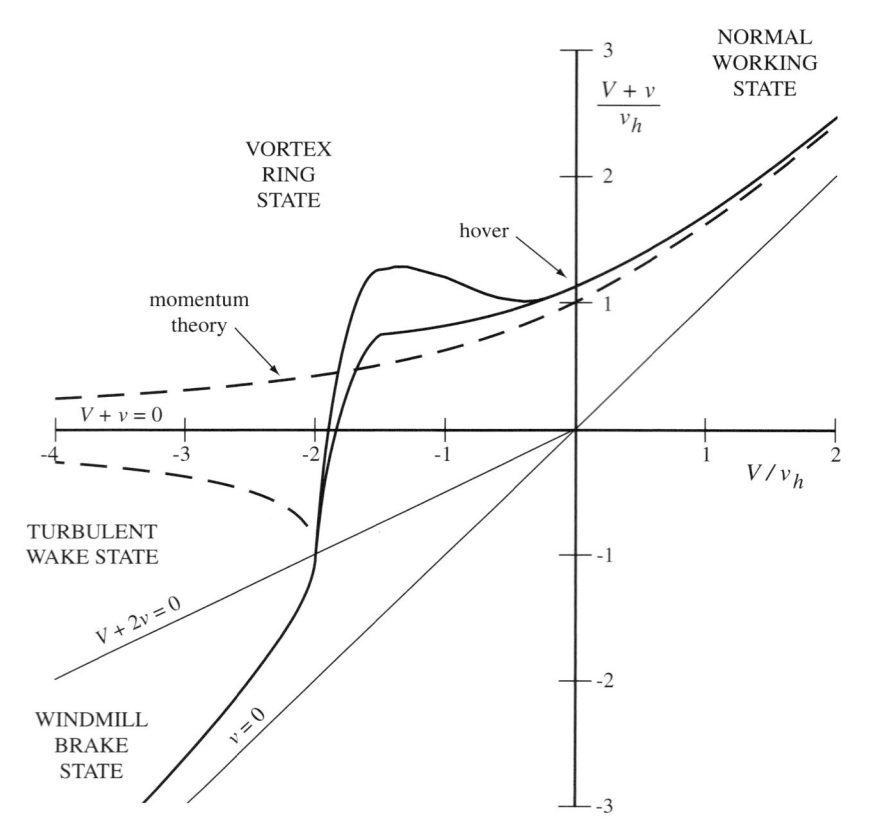

Figure 4.5. Total rotor induced and climb power in vertical flight.

induced velocity v is given by the vertical distance between the inflow curve and the $v = 0$ line. The abscissa, $V + v = 0$, is the ideal autorotation case here; for points above ideal autorotation the rotor is absorbing power, and for points below the rotor is producing power for the helicopter.

In the early British literature, the induced velocity curve was often plotted in terms of $1/f = (V/v_h)^2$ as a function of $1/F = ((V + v)/v_h)^2$, which followed from writing the loading distribution as $dT/dr = 4\pi r \rho (V + v)^2 F = 4\pi r \rho V^2 f$; see Lock, Bateman, and Townend (1925). This form is not very useful, because the sign of V and $(V + v)$ is lost, and by squaring the velocities their behavior near zero (including hover) is obscured.

To interpret the scale of these inflow curves, note that for sea level density $v_h = \sqrt{T/2\rho A} = 14.5\sqrt{T/A}$ ft/sec $= 870\sqrt{T/A}$ ft/min (disk loading in lb/ft²), or $v_h = 0.64\sqrt{T/A}$ m/sec (disk loading in N/m²). For the disk loading range typical of helicopter rotors, $T/A = 5$ to 15 lb/ft², the velocity $v_h = 2000$ to 3400 ft/min.

4.1.3.1 Hover Performance

The measured rotor hover performance indicates that the induced power is consistently higher than the momentum theory result by about 10% to 15%. The momentum theory power estimate is the best possible performance. The additional induced power is due to nonuniform inflow, tip losses, swirl, and other factors. Thus, in hovering performance calculations (such as in section 3.5.2.3) the induced power can be obtained using the momentum theory result with an empirical correction factor, $C_{Pi} = \kappa C_T^{3/2}/\sqrt{2}$. A value of $\kappa = 1.15$ is typical for hover.

4.1.3.2 Vortex Ring State

Although no theoretical inflow curve is available in the vortex ring and turbulent wake states, a fairly accurate approximation is given by the cubic relation

$$\frac{V+v}{v_h} = a\left(\frac{V}{v_h}\right)^3 - b\left(\frac{V}{v_h}\right) \tag{4.12}$$

Matching to the momentum theory results at the windmill brake state boundary $V/v_h = -2$ (where $(V+v)/v_h = -1$) and in the vortex ring state at $V/v_h = -1$ (where $(V+v)/v_h = (\sqrt{5}-1)/2$) gives these constants: $a = \sqrt{5}/6 = 0.373$ and $b = (4\sqrt{5}-3)/6 = 0.991$. If the empirical factor κ is included,

$$\frac{v}{v_h} = \kappa \frac{V}{v_h}\left[0.373\left(\frac{V}{v_h}\right)^2 - 1.991\right] \tag{4.13}$$

which describes the inflow curve in the range $-2 < V/v_h < -1$. For climb, hover, and low rates of descent ($V/v_h > -1$) and for high rates of descent in the windmill brake region ($V/v_h < -2$) the momentum theory results with an appropriate empirical correction are valid.

However, measured data suggest that using momentum theory down to $V/v_h = -1.5$ is appropriate, and the extension of equation 4.13 to edgewise flow has small jumps at the transitions from this curve to momentum theory. An alternative result is obtained using a third-order polynomial to match momentum theory v and dv/dV at $V/v_h = -1.5$ and v at $V/v_h = -2.1$. This is the basis of the lower curve for vortex ring state that is shown in Figures 4.4 and 4.5; see Johnson (2005) for details.

4.1.3.3 Autorotation and Turbulent Wake State

The universal inflow curve crosses the ideal autorotation line $V+v = 0$ at about $V/v_h = -1.8$ (the scatter extends over roughly $V/v_h = -1.6$ to -1.9; see Figure 4.5). Real autorotation occurs at a higher rate of descent, in the turbulent wake state. In the turbulent wake state the induced velocity curve can be approximated fairly well by a straight line on the $(V+v)/v_h$ vs. V/v_h plane. Joining the ideal autorotation intercept $(V+v = 0$ at $V/v_h = -x)$ and the windmill brake state boundary $((V+v)/v_h = -1$ at $V/v_h = -2)$ gives

$$\frac{V+v}{v_h} = \frac{x}{2-x} + \frac{1}{2-x}\frac{V}{v_h} \tag{4.14}$$

So for $V/v_h = -1.8$ at ideal autorotation, we obtain

$$\frac{V+v}{v_h} = 9 + 5\frac{V}{v_h} \tag{4.15}$$

in the turbulent state. This relation is useful in estimating the descent rate in real autorotation (see section 4.3).

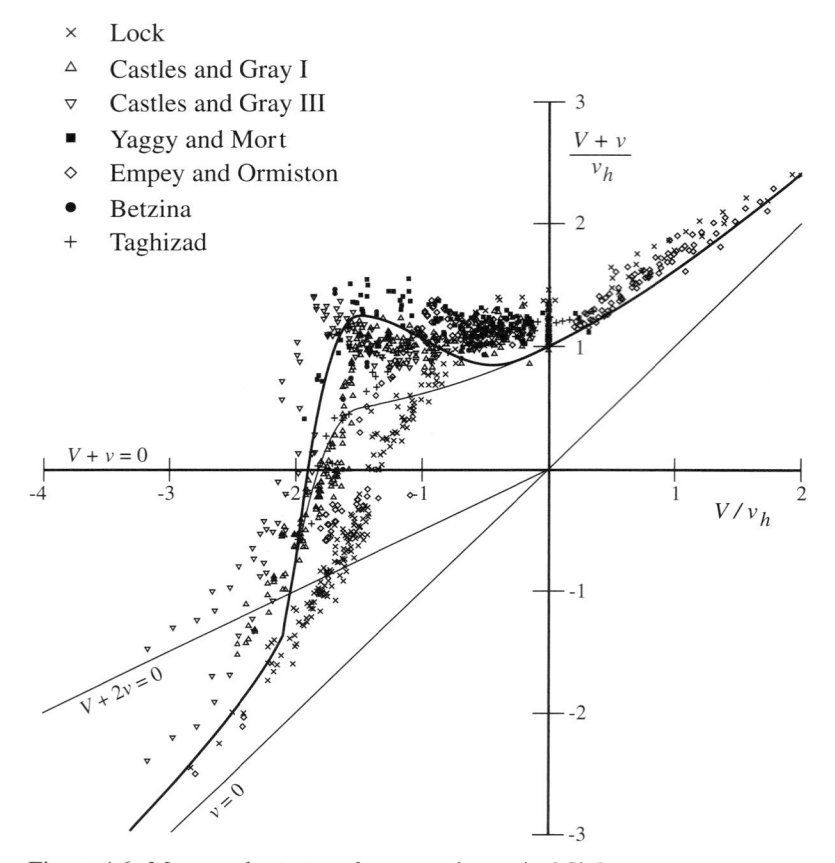

Figure 4.6. Measured rotor performance in vertical flight.

4.2 Vortex Ring State

The experimental data that form the basis for the empirical correction of the inflow curve are shown in Figure 4.6. The data are from Lock, Bateman, and Townend (1925); Glauert (1926); Lock (1947); Castles and Gray (1951); Yaggy and Mort (1963); Empey and Ormiston (1974); Betzina (2001); and Taghizad, Jimenez, Binet, and Heuze (2002).

A rotor is operating in vortex ring state (VRS) when descending at zero or low forward speed with a vertical velocity that approaches the value of the wake-induced velocity at the rotor disk. In this condition the rotor tip vortices are not convected away from the disk rapidly enough, and the wake builds up and periodically breaks away (Figure 4.7). The tip vortices collect in a vortex ring, producing a circulating flow down through the rotor disk, then outward and upward outside the disk. The resulting flow is unsteady and hence is a source of considerable low-frequency vibration and possible control problems. For descent at forward speeds sufficiently high enough that the wake is convected away from the rotor, vortex ring state does not develop.

Vortex ring state encounter can produce a significant increase in the descent rate of a helicopter or a rolloff of a tiltrotor. Figure 4.8 shows flight test data for the vertical velocity increase (and stabilization) of a helicopter, from Taghizad, Jimenez, Binet, and Heuze (2002) and rolloff of a tiltrotor in helicopter configuration, from Kisor, Blyth, Brand, and MacDonald, (2004). In Figure 4.8, V_z is the vertical velocity

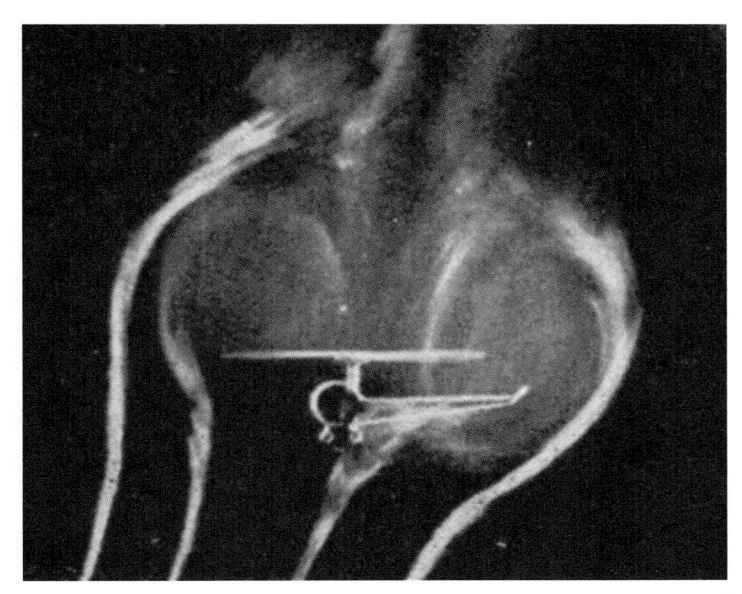

Figure 4.7. Rotor flow visualization in vortex ring state; Drees and Hendal (1951).

and V_x the horizontal velocity; also shown is a stability boundary developed by Johnson (2005). This motion is an instability of the helicopter vertical or tiltrotor roll dynamics. If the aircraft rate becomes sufficiently large as a result of the instability, recovering is not possible using collective control for the helicopter or lateral cyclic control (differential collective) for the tiltrotor. Although the response to control is still a positive acceleration increment, the control authority is not sufficient to reverse the motion. Hence recovery from VRS encounter requires a drop in collective and forward cyclic for a helicopter or a forward nacelle tilt for a tiltrotor. Basically it is necessary to fly out of the instability region. The flight test data define essentially the same VRS boundary for a helicopter and a tiltrotor, in spite of the different manifestation of the instability (vertical drop or rolloff), and large differences in twist and solidity between the rotors of the two aircraft. Figure 4.8 also compares the VRS boundaries from a number of investigations (Reeder and Gustafson (1949), Drees and Hendal (1951), Yeates (1958), Washizu, Azuma, Koo, and Oka (1966), Taghizad, Jimenez, Binet, and Heuze (2002), and Johnson (2005)). The ONERA and Johnson boundaries are based on the control and flight dynamics behavior; the other boundaries are based on the vibration and roughness that a helicopter encounters in VRS. Of particular note are the boundaries that Washizu constructed for thrust fluctuation levels of $\Delta T/T = 0.15$ and 0.30, which are found in numerous documents on VRS.

The instability of the aircraft in vortex ring state is a consequence of the form of the rotor inflow as a function of the descent rate. The measured data (Figure 4.6) show that at moderate descent rates (in VRS), the total velocity $V + v$ increases as the descent rate increases. As the rotor descends into VRS, the energy losses resulting from the recirculating flow increase; hence the power (total inflow $V + v$) can increase. Where $d(V + v)/dV$ is negative (roughly $V/v_h = -0.5$ to -1.5 in Figures 4.5 and 4.6), the vertical motion (and roll motion of a tiltrotor) is unstable, because an increase in descent rate at constant collective produces an increase in total inflow and hence a reduction in thrust – which is negative heave damping of the rotor.

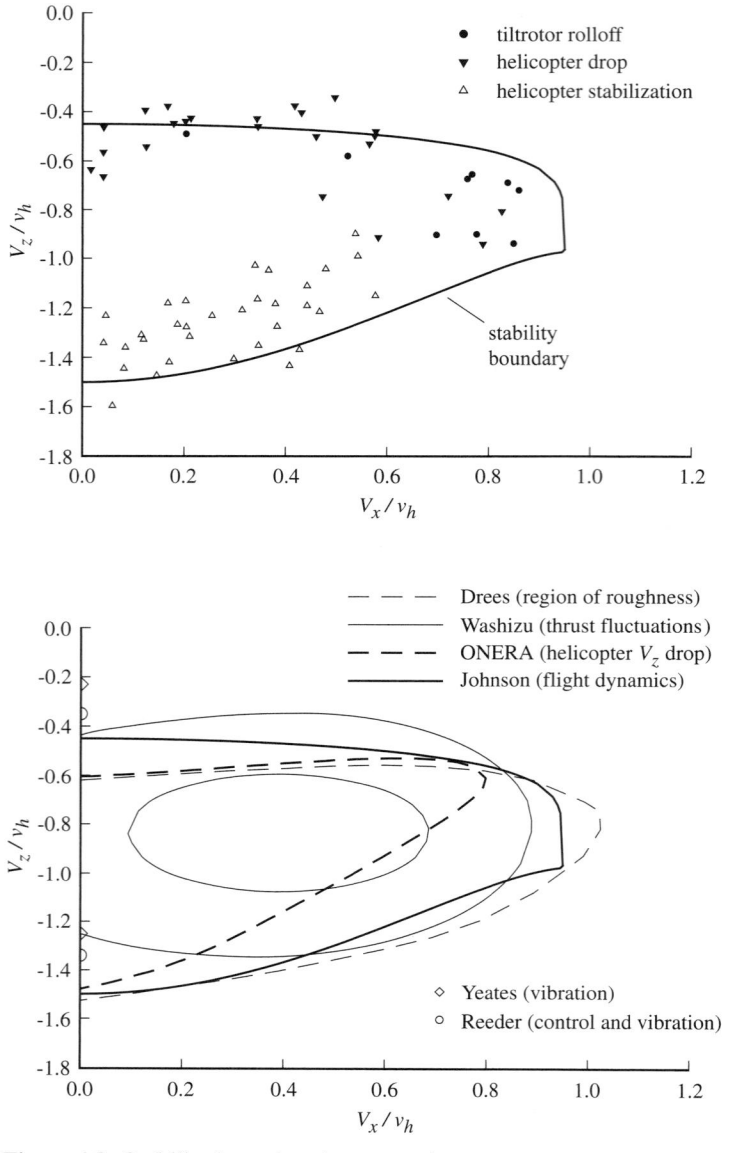

Figure 4.8. Stability boundary in vortex ring state.

The experimental data exhibit considerable scatter due to the essential unsteadiness of the flow and resulting loads, and error due to the influence of test facilities on the flow field. Some of the data, notably that of Lock, do not show the region of negative $d(V + v)/dV$. Flight test data (Figure 4.8) show where $d(V + v)/dV = 0$ occurs, which defines the points of entry into and recovery from VRS. This is the basis of the upper curve for vortex ring state that is shown in Figures 4.4 to 4.6; see Johnson (2005) for details.

This instability mechanism has been recognized in several efforts. More investigations have been focused on the unsteady nature of VRS aerodynamics. In the range $V/v_h = -0.5$ to -1.5, the flow is characterized by a high level of roughness. There are large periodic variations in the velocity at the disk and hence in the rotor loads as the vortex ring alternately builds up and then escapes the rotor disk. The low-frequency thrust variations produce a disturbing vibration of the helicopter that is a dominant feature of the vortex ring state. Such roughness is a symptom of VRS and can be a cue for the pilot, but it is the negative heave damping that is important for the aircraft behavior. Moreover, the unsteady motion of the aircraft must be considered and modeled, not just the quasistatic performance defined by Figure 4.5; see Brand, Dreier, Kisor, and Wood (2007).

4.3 Autorotation in Vertical Descent

Autorotation is the state of rotor operation with no net power requirement. The power to produce the thrust and turn the rotor is supplied either by auxiliary propulsion (the autogyro) or by descent of the helicopter. In an autogyro the rotor is functioning as a wing. A component of the aircraft forward velocity directed upward through the rotor disk supplies the power to the rotor, so the autogyro requires a forward speed to maintain level flight. In autorotative descent of the helicopter, the source of power is the decrease of the gravitational potential energy. More directly, the descent velocity upward through the disk supplies the power to the rotor. Although the lowest descent rate is achieved in forward flight, the helicopter rotor is also capable of power-off autorotative descent in vertical flight.

The net rotor power is zero for vertical descent in autorotation: $P = T(V + v) + P_o = 0$. The decrease in potential energy (TV) balances the induced (Tv) and profile (P_o) losses of the rotor. Neglecting the profile power gives ideal autorotation, $P = T(V + v) = 0$. When the profile power is included, autorotation occurs at $(V + v) = -P_o/T$. Thus the descent rate can be obtained from the universal inflow curve in the form $(V + v)/v_h$ vs. V/v_h by finding the intercept of the curve with $-P_o/P_h$. In coefficient form,

$$\frac{V+v}{v_h} = -\frac{C_{Po}}{C_T^{3/2}/\sqrt{2}} \qquad (4.16)$$

This intercept typically is at $(V + v)/v_h \cong -0.3$, which is in the turbulent wake state at a descent rate slightly higher than ideal autorotation. Because the slope of the inflow curve is large in this region, the increase in descent rate required to supply the profile power is small. Tail rotor and aerodynamic interference losses should also be included in finding the power $(V + v)/v_h$ of a real helicopter; such losses are only 15% to 20% of the profile power and therefore contribute only a small increase to the descent rate. The limit of the descent rate in vertical autorotation can be obtained from the boundary of the turbulent wake state, at roughly $V/v_h = -1.8$ to -2.

For a more quantitative estimate of the autorotative performance of real rotors, recall the definition of the figure of merit for hover:

$$M = \frac{C_T^{3/2}/\sqrt{2}}{\kappa C_T^{3/2}/\sqrt{2} + C_{Po}} \qquad (4.17)$$

so that

$$\frac{C_{Po}}{C_T^{3/2}/\sqrt{2}} = \frac{1}{M} - \kappa \qquad (4.18)$$

Assuming now that C_{P_o} and C_T do not change from hover to autorotation (hence assuming that the blade drag coefficient and tip speed are the same), this is the quantity required to define the autorotation point on the inflow curve. So

$$\frac{V+v}{v_h} = -\left(\frac{1}{M} - \kappa\right) \tag{4.19}$$

which typically gives $(V+v)/v_h = -0.2$ to -0.4. Low profile power gives both good hover performance (high figure of merit) and good autorotation performance (low descent rate). Now we use equation 4.15 for the inflow curve in the turbulent wake state: $(V+v)/v_h = 9 + 5V/v_h$. Combining the two relations for $(V+v)/v_h$ gives the descent rate

$$\frac{V}{v_h} = -\left[1.8 + 0.2\left(\frac{1}{M} - \kappa\right)\right] \tag{4.20}$$

Typically, then, vertical autorotation takes place at $V/v_h = -1.85$, or $V = 26.8\sqrt{T/A}$ ft/sec $= 1610\sqrt{T/A}$ ft/min (disk loading in lb/ft^2), or $V = 1.18\sqrt{T/A}$ m/sec (disk loading in N/m^2), which gives $V = 3600$ to 6200 ft/min for the range of disk loadings typical of helicopters.

The autorotation performance can be considered in terms of a drag coefficient based on the rotor disk area and the descent velocity:

$$C_D = \frac{T}{\frac{1}{2}\rho V^2 A} = \frac{T/2\rho A}{V^2/4} = \left(\frac{2}{V/v_h}\right)^2 \tag{4.21}$$

Hence, a low rate of descent corresponds to a high drag coefficient. This parameter is a useful description of the performance since it is independent of the helicopter disk loading. At the descent rates typical of real helicopters, the drag coefficient has a value in the range $C_D = 1.1$ to 1.2. For comparison, a circular flat plate of area A has a drag coefficient of about $C_D = 1.28$, and a parachute of frontal area A has $C_D = 1.40$. The helicopter rotor in power-off vertical descent is thus quite efficient in producing the thrust to support the helicopter. The rotor is nearly as good as a parachute of the same diameter. The descent rate in vertical autorotation is high because the rotor is a rather small parachute for such a weight. A much lower descent rate is possible in forward flight, however. The rotor flow state in autorotation is similar to that of a bluff body of the same size, so producing comparable drag forces is reasonable. Because the rotor efficiency is about as high as possible, a low descent rate can be achieved only with very low disk loadings. Usually, the disk loading is selected primarily on the basis of the rotor performance; the design of the helicopter for good autorotation characteristics is usually concerned with the ability to maintain rotor speed and to flare at the ground (see section 8.5).

Consider now power-off descent in terms of the blade aerodynamic loading. The inflow ratio $\lambda = (V+v)/\Omega R$ is directed upward through the disk, so there is a forward tilt of the lift vector (Figure 4.9). For power equilibrium at the blade section, the inflow angle must be such that there is no net in-plane force and hence no contribution to the rotor torque: $dQ = (D - \phi L)r\,dr = 0$. Because autorotation involves induced and profile torques of the entire rotor, generally only one section is in equilibrium, while the others are either producing or absorbing power. Since $\phi = \tan^{-1}|V+v|/\Omega r$, the inflow angle is large inboard and decreases toward the tip. Then $dQ < 0$ on the inboard sections, which produce an accelerating torque on the rotor and absorb power from the air; and $dQ > 0$ on the outboard sections, which

4.3 Autorotation in Vertical Descent

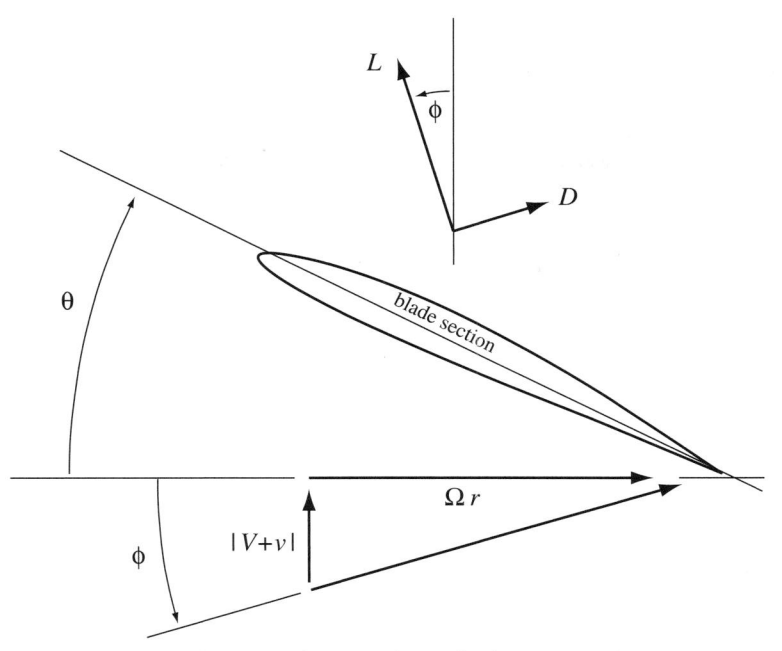

Figure 4.9. Rotor blade section aerodynamics in autorotation.

produce a decelerating torque and deliver power to the airstream. Since there is no net power to the rotor, the accelerating and decelerating torques must balance. For a given descent rate, the rotor tip speed ΩR adjusts until this equilibrium is achieved. Figure 4.10 illustrates the section aerodynamic environment on the rotor in autorotation. If the equilibrium rotor speed is decreased slightly, the inflow angle increases. Then the accelerating region moves outboard, increasing in size, and there is a net accelerating torque on the rotor that acts to increase the rotor speed back to the equilibrium value. Thus the rotor speed in autorotation is stable. The angle-of-attack $\alpha = \theta + \phi$ increases inboard because of the inflow angle increase. At the blade root, then, the sections are stalled. The negative twist that rotors generally have to improve hover and forward flight performance further increases the angle-of-attack of the inboard sections. Although negative twist is undesirable for autorotation, most of the work of the rotor is done at the blade tips, where the velocity is high, so the stall at the root does not usually have a particularly adverse effect on autorotation performance.

In hover the inflow is downward through the disk, whereas in autorotation $(V + v)$ is upward. Hence between hover and autorotation, there is a net increase in angle-of-attack due to the inflow change if the collective pitch is not changed after the loss of power in hover. The excess decelerating torque attributable to this angle-of-attack change decreases the rotor speed. In addition, the stall region increases in extent, limiting the blade lift, which is required for the accelerating torque, and increasing the drag, which produces the decelerating torque. With a stalled rotor, autorotation may therefore not be possible. To avoid excessive blade stall and rotor speed decrease, the blade pitch must be reduced as soon as possible after power failure. The best collective pitch for autorotation is usually found to be a slightly positive angle; the rotor speed can then be held near the normal value. The rate of descent does not actually vary much with collective or rotor speed as long as large

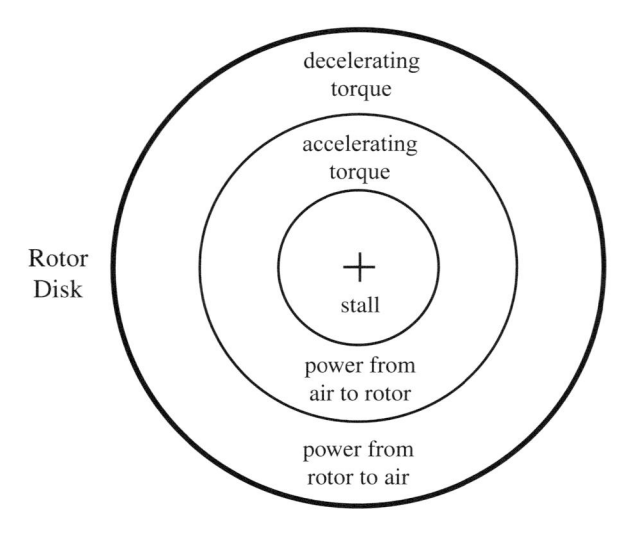

Figure 4.10. Aerodynamic environment of the rotor blade in autorotation.

regions of stall are avoided, because the profile power does not change much and the inflow curve is steep in the turbulent wake state.

For section equilibrium, recall that $D - \phi L = 0$, or $D/L = c_d/c_\ell = \phi$. Consider a plot of the blade airfoil characteristics, drawn to show c_d/c_ℓ as a function of α (Figure 4.11). Section equilibrium requires that $c_d/c_\ell = \phi = \alpha - \theta$, which for a given θ is a line on the c_d/c_ℓ vs. α plane. The intersection of this line with the curve of the airfoil characteristics determines the angle-of-attack for which equilibrium is achieved at this section. Although only one blade section is in true equilibrium, the inboard sections working at higher angle-of-attack and the outboard sections at lower, the autorotation diagram (Figure 4.11) does give a good indication of the characteristics of the entire rotor. Minimum descent rate means minimum ϕ, which therefore requires that the blade operate at the angle-of-attack with lowest c_d/c_ℓ, resulting in minimum profile power. The collective pitch for this optimum operation is easily determined from the autorotation diagram. At both higher and lower collective, the blade drag-to-lift ratio is higher and therefore the descent rate is higher. At low angles of attack, c_d/c_ℓ increases because c_ℓ is low, and it increases at high angles because of stall. However, for many airfoils the drag-to-lift ratio tends to be fairly flat around the minimum, so the descent rate is not too sensitive to θ near the optimum value. Although the entire blade cannot be working at the optimum angle-of-attack for autorotation, most of the blade is still at a low value of c_d/c_ℓ. The rotor tip speed is more sensitive than the descent rate to collective pitch changes. The relation $c_d/c_\ell = \phi = |V + v|/\Omega r$ indicates that the maximum rotor speed is obtained at the minimum c_d/c_ℓ, and that the rotor slows down at higher or lower collective pitch values. The autorotation diagram further shows that there is a maximum collective pitch value $\theta_{\max}$, above which equilibrium is not possible (Figure 4.11). When the angle-of-attack is high because of the high collective, the rotor stalls and not enough lift becomes available to balance the decelerating torque created by the high drag. The importance of reducing the collective pitch soon after power loss derives from the necessity of avoiding this collective limit, where the rotor speed decreases and the descent rate increases with no possibility of achieving equilibrium.

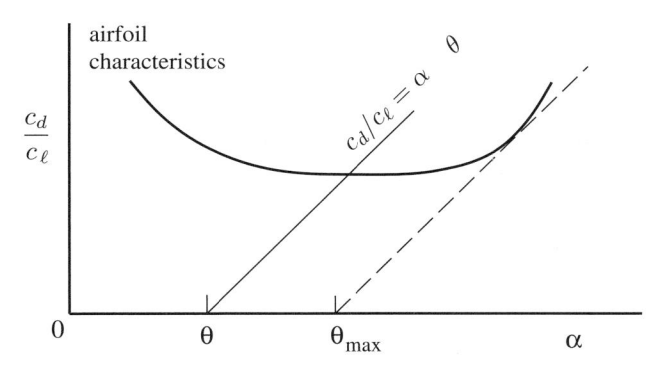

Figure 4.11. Section autorotation diagram.

Blade element theory gives for autorotative descent

$$C_P = \lambda C_T + \frac{\sigma c_{do}}{8} = 0 \qquad (4.22)$$

$$C_T = \frac{\sigma a}{2}\left(\frac{\theta_{.75}}{3} - \frac{\lambda}{2}\right) \qquad (4.23)$$

Solving for the inflow ratio then gives

$$\lambda = \frac{\theta_{.75}}{3} - \sqrt{\left(\frac{\theta_{.75}}{3}\right)^2 + \frac{c_{do}}{2a}} \qquad (4.24)$$

For a given collective pitch then, λ and C_T can be calculated. The disk loading gives the rotor speed from C_T, and the inflow curve gives the descent rate from λ. Thus the autorotation descent rate as a function of collective can be plotted, and the optimum collective pitch value determined. A more detailed numerical analysis is desirable, however, because of the importance of blade stall in the autorotation behavior of the rotor. Blade element theory can at least be used to estimate the collective pitch reduction required between hover and autorotation. From equation 4.23, and assuming the tip speed ΩR is not changed,

$$\Delta\theta_0 = \frac{3}{2}(\lambda - \lambda_h) = \frac{3}{2}\left(\frac{V+v}{v_h} - 1\right)\lambda_h = -\frac{3}{2}\left(\frac{C_{Po}}{C_T^{3/2}/\sqrt{2}} + 1\right)\lambda_h$$

$$= -\frac{3}{2}\left(\frac{1}{M} + 1 - \kappa\right)\sqrt{C_T/2} \qquad (4.25)$$

which is typically about 5 degrees.

4.4 Climb in Vertical Flight

The momentum theory result for the power required in vertical climb is

$$V + v = \frac{V}{2} + \sqrt{\left(\frac{V}{2}\right)^2 + v_h^2} \cong \frac{V}{2} + v_h \qquad (4.26)$$

where the last approximation is valid for small climb rates (roughly $V/v_h < 1$; see Figure 4.5). Then the induced velocity $v \cong v_h - V/2$ is reduced by the climb velocity

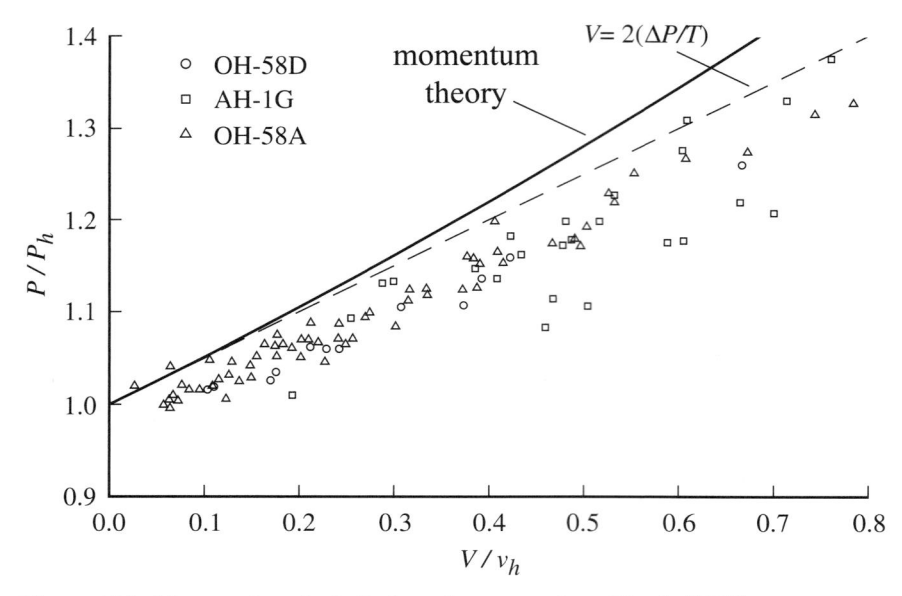

Figure 4.12. Measured vertical climb performance; from Harris (1987).

because of the increased mass flow through the rotor disk. The power required in climb is $P = T(V + v) + P_o$. Assuming that the profile power is unchanged by the climb velocity, the power increment between climb and hover is

$$\Delta P = P - P_h = T(V + v - v_h) \qquad (4.27)$$

Using the small climb rate result for $V + v$, the excess power for climb is given by $\Delta P/T \cong V/2$, and the climb rate for a given power increase is

$$V \cong 2\frac{\Delta P}{T} \qquad (4.28)$$

The power required just to increase the potential energy of the helicopter is $\Delta P/T = V$. Hence the reduction in the induced power required doubles the climb rate possible with a given power increase. Flight test data show that the vertical rate-of-climb achieved for a given power increment is greater than this momentum theory result (see Figure 4.12), an effect attributed to the influence of climb on the wake geometry.

For an exact formulation, the excess power (equation 4.27) gives $V = (\Delta P/T) + v_h - v$. Now the momentum theory result for climb, $(V + v)v = v_h^2$, can be written:

$$v = \frac{v_h^2}{V + v} = \frac{v_h^2}{\Delta P/T + v_h} \qquad (4.29)$$

Eliminating v then gives

$$V = \frac{\Delta P}{T}\frac{2v_h + \Delta P/T}{v_h + \Delta P/T} \qquad (4.30)$$

from which the climb rate can be obtained if the excess power and rotor thrust are specified. For small V this reduces to $V \cong 2\Delta P/T$ again.

Blade element theory can be used to estimate the collective pitch increase required in climb. From $2C_T/\sigma a = (\theta_{.75}/3 - \lambda/2)$,

$$\Delta\theta = \frac{3}{2}(\lambda - \lambda_h) \cong \frac{3}{4}\lambda_c \qquad (4.31)$$

for small climb rates where $\lambda \cong \lambda_h + \frac{1}{2}\lambda_c$ ($\lambda_c = V/\Omega R$). Alternatively, without the assumption of small climb velocity,

$$\Delta\theta = \frac{3}{2}\frac{V + v - v_h}{\Omega R} = \frac{3}{2}\frac{\Delta P/T}{\Omega R} = \frac{3}{2}\frac{\Delta C_P}{C_T} \qquad (4.32)$$

4.5 Optimum Windmill

The momentum theory solution in the windmill brake state (see section 4.1.1) can be written as

$$P = -\frac{1}{2}\dot{m}(2Vw + w^2) = -\frac{1}{2}\rho A\left(V + \frac{w}{2}\right)(2Vw + w^2) \qquad (4.33)$$

Here the axial velocity V is negative, and the rotor is generating power as a windmill ($P < 0$). Since $w = 2v$ is determined by the wind speed V and the rotor thrust T (really $v_h^2 = T/2\rho A$), equation 4.33 gives the power as a function of thrust and speed. For fixed wind speed, the maximum power is determined by $dP/dw = 0$, which has the solution $w = -\frac{2}{3}V$. The maximum power is $P_{\max} = \frac{8}{27}\rho A V^3$. The energy available from the wind flowing through the rotor disk is $E = \frac{1}{2}\rho A V^3$. Hence the efficiency of the optimum windmill is

$$\eta = \frac{P_{\max}}{E} = \frac{16}{27}$$

At most, 59% of the wind energy can be extracted by the windmill. This result is called the Betz limit (see Betz (1928)), but was also obtained by Lanchester (Bergey (1979)).

4.6 Twin Rotor Interference in Hover

The operation of two or more rotors in close proximity modifies the flow field at each, and hence the performance of the rotor system is not the same as for the isolated rotors. Examples of such configurations are the coaxial helicopter, the tandem rotor helicopter (typically with 30% to 50% overlap), and the side-by-side configuration. For most designs, all the main rotors are identical except for direction of rotation. In particular all the rotors have the same diameter and the same disk area A. To assess the influence of interference on the rotor performance, the power of the twin rotors can be compared with that of isolated rotors operating at the same thrust. A limiting case is the coaxial rotor system with no vertical separation, which has just one-half the disk area of the isolated rotors and hence twice the disk loading. So by operating the rotors coaxially, the induced power required is increased by a factor of $\sqrt{2}$, a 41% induced power increase.

Consider the case of two rotors operating close together but with no overlap. According to vortex theory, in the disk plane but outside the rotor disk circle there is no normal induced velocity component and hence no interference power. With some vertical separation there can be an interference, favorable or unfavorable, even with no overlap of the rotors.

For rotors with some overlap and small vertical separation, there is a common rate of flow through the overlapped portions of the disk. For the same total thrust, the overlap area has a higher disk loading than the isolated rotors, which increases the local induced power. As the separation decreases, the increase in power approaches the 41% of coaxial rotors. As the vertical separation of coaxial rotors is increased, the wake of the top rotor contracts and thus affects less area of the lower rotor, reducing the interference power.

The efficiency of the twin rotor system can be assessed by comparing it to the power of an ideal system, which is taken as the momentum theory (uniform loading) power of a single rotor with the same total thrust and an area equal to the projected area of the twin rotor system.

4.6.1 Coaxial Rotors

Coaxial main rotors generally have better hover efficiency than the equivalent single rotor (that is, no vertical separation), a consequence of the contraction of the wake of the upper rotor, as well as a reduction in wake swirl losses. For vertical separation of 10% of the rotor diameter, the upper rotor wake in hover has contracted to about $0.85R$ when it reaches the lower rotor (Akimov, Butov, Bourtsev, and Selemenev (1994), based on tip vortex visualization). Far below the disk, the upper rotor wake has contracted to $0.82R$ and the lower rotor wake to $0.91R$, compared to the far wake contraction of $0.78R$ for a single rotor.

Consider coaxial rotors with area A of each rotor and total thrust $T = T_u + T_\ell$. Define the reference velocity as $v_h^2 = T/2\rho A$ (based on the area of a single rotor). For coaxial rotors with zero vertical spacing (that is, a single rotor with the same total solidity), the momentum theory solution for ideal induced power is $P = Tv_h$. For two separate isolated rotors, the solution is $P = 2(T/2)\sqrt{(T/2)/(2\rho A)} = 2^{-1/2}Tv_h = 0.7071Tv_h$.

A simple approach to estimate coaxial rotor efficiency is to consider the area of the lower rotor that is outside the upper rotor slipstream as an extra active area of the rotor system (Bourtsev, Kvokov, Vainstein, and Petrosian (1997)). For large separation the effective area is $\frac{3}{2}A$ (since by momentum theory the upper rotor wake contracts to area $\frac{1}{2}A$), and $P = T\sqrt{T/(2\rho(3/2)A)} = (2/3)^{1/2}Tv_h = 0.8165Tv_h$. For finite spacing with contraction ratio x, the effective area is $A_e = (2 - x^2)A$, and $P = T\sqrt{T/(2\rho A_e)} = (2 - x^2)^{-1/2}Tv_h$. So for 85% contraction, $P = 0.8847Tv_h$, a 13% decrease in induced power. Measured rotor performance typically shows an 8–11% increase in the figure of merit for coaxial rotors.

Momentum theory can be used to determine the induced power for coaxial rotors with very large vertical separation. Figure 4.13 illustrates the flow model at the lower rotor, showing the velocities and areas in the flow field and the pressure on the rotor disk. For large vertical separation, the lower rotor has no effect on the upper rotor, and the momentum theory solution for the upper rotor is $v_u^2 = T_u/2\rho A$ and $P_u = T_u v_u$. The far wake velocity of the lower rotor $w_u = 2v_u$ is uniform over the cross-section area $A/2$. This far wake velocity acts on the lower rotor.

Momentum theory for the lower rotor follows the derivation of section 3.4, with the addition of the interference velocity $w_u = 2v_u$ above the rotor. Mass, momentum, and energy conservation become

$$\dot{m} = \int v_\ell dA = \int w_\ell dS \qquad (4.34)$$

4.6 Twin Rotor Interference in Hover

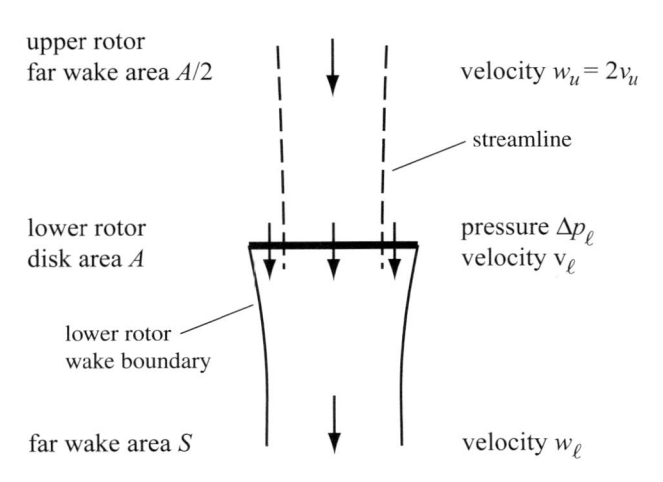

Figure 4.13. Momentum theory model of coaxial rotors in hover.

$$T_\ell = \int \Delta p_\ell dA = \int \rho w_\ell^2 dS - \rho(2v_u)^2(A/2) \quad (4.35)$$

$$P_\ell = \int \Delta p_\ell v_\ell dA = \int \frac{1}{2}\rho w_\ell^3 dS - \frac{1}{2}\rho(2v_u)^3(A/2) \quad (4.36)$$

where S is the area in the far wake of the lower rotor. Calculus of variations shows that the solution for minimum power with constrained thrust is w_ℓ uniform over the wake. Thus

$$\dot{m} = \int v_\ell dA = w_\ell S \quad (4.37)$$

$$T_\ell = \int \Delta p_\ell dA = \rho w_\ell^2 S - 2\rho A v_u^2 \quad (4.38)$$

$$P_\ell = \int \Delta p_\ell v_\ell dA = \frac{1}{2}\rho S w_\ell^3 - 2\rho A v_u^3 \quad (4.39)$$

Momentum theory does not give information about the distribution of the induced velocity over the rotor disk. Bernoulli's equation can be used to relate the loading on the rotor disk, $\Delta p_\ell = dT_\ell/dA$, to the far wake velocity w_ℓ. Bernoulli's equation is applied from far above the rotor (where the pressure equals ambient) to just above the rotor disk, and from just below the rotor disk to far below (where the pressure again equals ambient), for streamlines starting from within and without the upper rotor wake (subscripts I and O, for inboard and outboard, respectively), giving

$$\Delta p_{\ell O} = \frac{1}{2}\rho w_\ell^2 \quad (4.40)$$

$$\Delta p_{\ell I} = \frac{1}{2}\rho w_\ell^2 - 2\rho v_u^2 \quad (4.41)$$

For an isolated rotor (without the effect of the upper rotor wake), uniform far wake velocity w_ℓ implies uniform disk loading Δp_ℓ. For coaxial rotors, the loading is significantly different in the inboard and outboard regions, although uniform in each. Roughly the inboard loading is $1/3$ the outboard loading for this optimum power solution. Let α be the ratio of the inboard or outboard local loading Δp_ℓ to the lower rotor disk loading: $\Delta p_\ell = \alpha(T_\ell/A)$. So $\int \Delta p_\ell dA = T_\ell$ gives $\alpha_I A_I + \alpha_O A_O = A$

(A_I and A_O are the inboard and outboard areas at the rotor disk, which can be determined from mass conservation if v_ℓ is known). From $\Delta p_{\ell I} = \Delta p_{\ell O} - 2\rho v_u^2$, $\alpha_I = \alpha_O - v_u^2/(T_\ell/2\rho A)$. Define the mean induced velocity $\bar{v}_\ell = \int v_\ell dA/A$, and a nonuniform loading parameter $\bar{\alpha} = \int \Delta p_\ell v_\ell dA/(T_\ell \bar{v}_\ell) = \int \alpha v_\ell dA/(\bar{v}_\ell A)$, so the power can be written $P_\ell = \int \Delta p_\ell v_\ell dA = \bar{\alpha} T_\ell \bar{v}_\ell$. The loading distribution is characterized by the single parameter $\bar{\alpha}$; in the absence of a detailed analysis, a range of $\bar{\alpha}$ values are considered. For an isolated rotor, the optimum solution is uniform disk loading, hence $\bar{\alpha} = 1$; in general $\bar{\alpha}$ is the average of the disk loading weighted by the induced velocity, giving $\bar{\alpha} > 1$. With these definitions, the conservation equations are

$$\dot{m} = \bar{v}_\ell A = w_\ell S \tag{4.42}$$

$$T_\ell = \rho A \bar{v}_\ell w_\ell - 2\rho A v_u^2 \tag{4.43}$$

$$P_\ell = \bar{\alpha} T_\ell \bar{v}_\ell = \frac{1}{2}\rho A \bar{v}_\ell w_\ell^2 - 2\rho A v_u^3 \tag{4.44}$$

using the mass flux relation in the momentum and energy equations. For an isolated rotor ($v_u = 0$ and $\bar{\alpha} = 1$), w_ℓ is easily eliminated and the usual solution for the mean induced velocity obtained. Define the lower rotor reference velocity $v_r^2 = T_\ell/2\rho A$; recall $v_u^2 = T_u/2\rho A$, so $v_h^2 = T/2\rho A = v_u^2 + v_r^2$. Substituting $T_\ell = 2\rho A v_r^2$ in equations 4.43 and 4.44, and eliminating w_ℓ, gives the relation

$$\bar{\alpha} v_r^2 \bar{v}_\ell^2 + v_u^3 \bar{v}_\ell = (v_u^2 + v_r^2)^2 \tag{4.45}$$

Write $v_r = r v_u$, $T_\ell = r^2 T_u = \tau T_u$, and $\bar{v}_\ell = s v_u$. Then equation 4.45 becomes

$$\bar{\alpha} \tau s^2 + s = (1 + \tau)^2 \tag{4.46}$$

and the solution gives the total power $P = T_u v_u + \bar{\alpha} T_\ell \bar{v}_\ell = (1 + \bar{\alpha} \tau s) T_u v_u$ or

$$P = (1 + \tau)^{-3/2}(1 + \bar{\alpha}\tau s) T v_h \tag{4.47}$$

Given τ, the ratio of the lower and upper induced velocities is

$$s = \frac{1}{2\bar{\alpha}\tau}\left(\sqrt{1 + 4(1+\tau)^2 \bar{\alpha}\tau} - 1\right) \tag{4.48}$$

The thrust and power ratios are then $T_u/T = 1/(1+\tau)$ and $P_u/P = 1/(1+\bar{\alpha}\tau s)$. Also, since $\alpha_I = \alpha_O - 1/\tau$, the inboard and outboard loading ratios are $\alpha_I = 1 - A_O/(A\tau)$ and $\alpha_O = 1 + A_I/(A\tau)$.

From the thrust distribution (hence τ) and weighted loading ($\bar{\alpha}$), equation 4.48 gives the mean inflow (s) and then the power. The solution for equal thrust of the two rotors follows from $\tau = 1$:

$$s = \frac{1}{2\bar{\alpha}}\left(\sqrt{1 + 16\bar{\alpha}} - 1\right) \tag{4.49}$$

$$P = 2^{-3/2}(1 + \bar{\alpha}s) T v_h \tag{4.50}$$

For uniform loading as well ($\bar{\alpha} = 1$), $s = \frac{1}{2}(\sqrt{17} - 1)$ and $P = 2^{-5/2}(\sqrt{17} + 1) T v_h = 0.9056 T v_h$. The solution for equal power of the two rotors follows from $\bar{\alpha}\tau s = 1$:

$$\frac{2}{\bar{\alpha}\tau} = (1 + \tau)^2 \tag{4.51}$$

$$P = 2(1+\tau)^{-3/2} T v_h = 2\left(\frac{2}{\bar{\alpha}\tau}\right)^{-3/4} T v_h \tag{4.52}$$

Table 4.1. *Coaxial rotor induced power*

	Thrust share T_u/T	Power share P_u/P	Power relative to No separation $P_i/(Tv_h)$	Power relative to Independent rotors $P_i/(Tv_h/\sqrt{2})$
No separation	0.5	0.5	1.0	1.4142
Large vertical separation				
$\bar{\alpha} = 1.10$				
equal T	0.5	0.3765	0.9382	1.3282
equal P	0.6024	0.5	0.9352	1.3226
$\bar{\alpha} = 1.05$				
equal T	0.5	0.3832	0.9226	1.3048
equal P	0.5962	0.5	0.9208	1.3022
$\bar{\alpha} = 1.00$				
equal T	0.5	0.3904	0.9056	1.2808
equal P	0.5898	0.5	0.9058	1.2810
Independent rotors	0.5	0.5	0.7071	1.0

For uniform loading as well ($\bar{\alpha} = 1$), $\tau(1+\tau)^2 = 2$ has the solution $\tau = 0.6956$, so $P = 0.9058 Tv_h$. The usual requirement for torque balance between the two main rotors makes the equal power solution more relevant.

Table 4.1 gives the results for $\bar{\alpha} = 1.00$, 1.05, and 1.10 for both equal thrust and equal power cases. Although the difference between upper and lower rotor power (for equal thrust) or thrust (for equal power) is substantial, the momentum theory solution is a weak function of the thrust ratio T_u/T. The equal thrust and equal power results are remarkably close, so the total power depends primarily on the nonuniform loading parameter $\bar{\alpha}$. Although the equal loading case $\bar{\alpha} = 1.00$ is interesting, the loading on the lower rotor is far from uniform. The results for large separation and no separation bound the performance for practical coaxial rotors, with vertical separation typically around 10% of the rotor diameter.

In Table 4.1 the power is presented relative to no vertical separation (reference power Tv_h) and relative to independent rotors (reference power $Tv_h/\sqrt{2}$). The coaxial rotor system with large vertical separation has 28–32% more induced power than independent rotors, increasing to 41% for no separation. The hovering coaxial rotor has at most (for large separation) 6–9% less induced power than for no separation. So defining the hover figure of merit for the coaxial rotor in terms of the total thrust T, the total power P, and the area A of one rotor is appropriate: $M = T\sqrt{T/2\rho A}/P$.

4.6.2 Tandem Rotors

For a momentum theory analysis of overlapped rotors, consider two rotors of the same radius but perhaps with different thrusts. Let mA be the overlap area; T_1 and T_2 the thrusts on the two rotors, with $T_1 + T_2 = T$ fixed; P_1 and P_2 the induced power outside the overlap area and P_m the induced power of the overlap area; and v_1, v_2, and v_m the corresponding induced velocities. With uniform loading, $T_1(1-m)$ and $T_2(1-m)$ are the thrusts of the areas outside the overlap, and $m(T_1 + T_2)$ the thrust of the overlap area. Negligible vertical separation is assumed, so that in the overlap area both rotors have the common induced velocity v_m. Based on the differential momentum theory results $dT = 2\rho v^2 dA$ and $dP = v\,dT$:

$$v_1 = \sqrt{T_1/2\rho A} \qquad v_2 = \sqrt{T_2/2\rho A} \qquad v_m = \sqrt{(T_1+T_2)/2\rho A} \qquad (4.53)$$

$$P_1 = T_1(1-m)v_1 \qquad P_2 = T_2(1-m)v_2 \qquad P_m = m(T_1+T_2)v_m \qquad (4.54)$$

Then the total power is $P = P_1 + P_2 + P_m$. For the isolated rotors the total power is

$$P_{\text{isolated}} = (P_1 + P_2)|_{m=0} = (T_1^{3/2} + T_2^{3/2})/\sqrt{2\rho A} \qquad (4.55)$$

The interference power is therefore

$$\begin{aligned}\Delta P &= (P_1 + P_2 + P_m) - (P_1 + P_2)|_{m=0} \\ &= m\left[(T_1+T_2)^{3/2} - (T_1^{3/2} + T_2^{3/2})\right]/\sqrt{2\rho A}\end{aligned} \qquad (4.56)$$

or, as a fraction of the power for the isolated rotors,

$$\frac{\Delta P}{P_{\text{isolated}}} = m\left[\frac{1}{\tau_1^{3/2} + \tau_2^{3/2}} - 1\right] \qquad (4.57)$$

where $\tau_1 = T_1/T$ and $\tau_2 = T_2/T$ (so $\tau_1 + \tau_2 = 1$) give the distribution of thrust between the two rotors. When the thrust on the two rotors is the same, $\tau_1 = \tau_2 = \frac{1}{2}$, the interference power becomes $\Delta P/P = 0.41m$, which gives 41% for coaxial rotors (100% overlap, hence $m = 1$). In general, the interference power is directly proportional to the fraction of area overlapped.

Alternatively, for hovering twin rotors with overlap area mA, the performance estimate can be based on the effective disk loading of the system as a whole: $P = T\sqrt{T/2\rho A_{\text{sys}}}$, where the total rotor area is $A_{\text{sys}} = A(2-m)$. The ratio of the total power to that of the isolated rotors then is

$$\frac{P}{P_{\text{isolated}}} = \left(\frac{T}{T_{\text{isolated}}}\right)^{3/2}\left(\frac{2}{2-m}\right)^{1/2} \qquad (4.58)$$

and for the same thrust the interference power is

$$\frac{\Delta P}{P_{\text{isolated}}} = \left(\frac{2}{2-m}\right)^{1/2} - 1 \qquad (4.59)$$

In the coaxial limit ($m = 1$) this gives $\Delta P/P = 0.41$ as required. For small overlap, however, $\Delta P/P_{\text{isolated}} \cong 0.25m$, and initially the power does not increase with overlap as quickly as in the previous result. The difference is that the present model has a lower disk loading in the overlap region than the previous model, and hence a greater efficiency for small overlap. The larger estimate of the interference power is probably more representative of tandem rotor behavior. Finally, with a shaft separation ℓ the overlap fraction is

$$m = \frac{2}{\pi}\left[\cos^{-1}\frac{\ell}{2R} - \frac{\ell}{2R}\sqrt{1 - \left(\frac{\ell}{2R}\right)^2}\right] \qquad (4.60)$$

For small overlap ($\ell = 2R - \Delta\ell$, with $\Delta\ell/R \ll 1$), $m \cong 1.20(\Delta\ell/2R)^{3/2}$.

Stepniewski (1955a, 1955b) developed a combined blade element and momentum theory for twin rotors in hover. The theory assumes that the vertical separation of the rotors is small, so that the overlapped area has a common rate of flow through both rotors. Outside the overlap region, the induced velocities v_1 and v_2 are given by the usual combined blade element and momentum theory expression (see section 3.5.3). Inside the overlap region, consider an area dA located at radial stations r_1 and r_2 on the two rotors; let $\lambda_m = v_m/\Omega R$ be the inflow ratio in the overlap region. Momentum theory gives $dT = 2\rho v_m^2 dA$ or $dC_T = (2/\pi)\lambda_m^2 dA$. Blade element theory gives $dC_{T1} = (\sigma_1 a/4\pi)(\theta_1 r_1 - \lambda_m)dA$ and $dC_{T2} = (\sigma_2 a/4\pi)(\theta_2 r_2 - \lambda_m)dA$, where θ_1 and θ_2 are the pitch of the two blades at r_1 and r_2, respectively. Equating $dC_T = dC_{T1} + dC_{T2}$ gives a quadratic equation for λ_m with the solution

$$\lambda_m = -\left(\frac{\sigma_1 a}{16} + \frac{\sigma_2 a}{16}\right) + \sqrt{\left(\frac{\sigma_1 a}{16} + \frac{\sigma_2 a}{16}\right)^2 + \frac{\sigma_1 a}{8}\theta_1 r_1 + \frac{\sigma_2 a}{8}\theta_r r_2} \quad (4.61)$$

Using v_1, v_2, and v_m, the thrust and power can be evaluated from

$$T = \int 2\rho v_1^2 dA + \int 2\rho v_2^2 dA + \int 2\rho v_m^2 dA \quad (4.62)$$

$$P = \int 2\rho v_1^3 dA + \int 2\rho v_2^3 dA + \int 2\rho v_m^3 dA \quad (4.63)$$

where the three integrals cover, respectively, the first and second rotors outside the overlap area, and the overlap area. Alternatively, the blade element theory expressions for the two rotors can be used if integration is performed azimuthally as well as radially along the blade. Stepniewski (1955a) obtained good results from this analysis, based on a comparison of the downwash and power prediction with test data. He found no aerodynamic interference of practical importance in hover with no overlap, and for overlap in the range $\Delta\ell/rR = 0$ to $0.4R$, the thrust and power were $T/T_{\text{isolated}} \cong 1.0$ to 0.94 and $P/P_{\text{isolated}} \cong 1.1$ to 1.2. Here P is just the induced power, and P_{isolated} is the induced power for the isolated rotors with uniform inflow; hence the interference power also includes the nonuniform inflow losses of the isolated rotors.

4.7 Vertical Drag and Download

The rotor downwash acting on the fuselage produces a vertical drag force or download on the helicopter in hover and vertical flight. This drag force requires an increase in the rotor thrust for a given gross weight and hence degrades the helicopter performance. To estimate the vertical drag force, consider the downwash velocity in the fully developed rotor wake. In hover $w_h = 2v_h$, and in vertical flight

$$V + w = \sqrt{V^2 + 4v_h^2} \cong 2v_h \quad (4.64)$$

So $V + w \cong w_h$ independent of the climb velocity, as long as $V^2/v_h^2 \ll 1$. The vertical drag characteristics of the fuselage can be described by either an equivalent drag area f or by a drag coefficient C_D based on some relevant area S, related by $f = SC_D$. Then the vertical drag requires a rotor thrust increase

$$\Delta T = \tfrac{1}{2}\rho w_h^2 f = \frac{T}{A} f \quad (4.65)$$

or

$$\frac{\Delta T}{T} = \frac{f}{A} = \frac{S}{A} C_D \quad (4.66)$$

The fuselage is very near the rotor and hence may not really be in the far wake; moreover, the downwash field is highly nonuniform and unsteady. Such effects can be included in an empirical factor. Assume that the downwash velocity at the fuselage is nv_h, where the parameter n theoretically varies from 1 at the disk to 2 in the far wake. Then

$$\frac{\Delta T}{T} = \frac{n^2}{4}\frac{f}{A} = \frac{S}{A}\left(\frac{n^2 C_D}{4}\right) \qquad (4.67)$$

The parameter $(n^2 C_D/4)$ can then be obtained from measurements of the force on bodies in the rotor wake. Typically $(n^2 C_D/4) \cong 0.7$, but the value depends highly on the position of the body in the wake, its size relative to the rotor, and its shape.

Glauert (1935) suggested using the following expression for the vertical drag:

$$\frac{\Delta T}{T} = \frac{S}{A}C_D\left(1.22 + 0.254/C_D\right) \qquad (4.68)$$

The last factor is the effect of the pressure gradient in the wake on the forces acting on the body. Makofski and Menkick (1956) suggested

$$\frac{\Delta T}{T} = 0.66\,\frac{S}{A}\frac{b}{2R} \qquad (4.69)$$

based on measurements with rectangular panels $0.2R$ to $0.64R$ below the disk. Here b is the panel span, so the factor $b/2R$ accounts for the radial variation of the downwash. Another approach is to estimate n and C_D separately for the components of the fuselage in the wake. From vortex theory for the velocity on the wake axis, at a distance z below the disk,

$$n = 1 + \frac{z/R}{\sqrt{1 + (z/R)^2}} \qquad (4.70)$$

(equation 3.154), and the appropriate drag coefficient can be found from the standard literature.

These approaches are crude, but fairly large errors can be tolerated since $\Delta T/T$ is small. A good analysis of the problem is difficult, since an accurate model for the helicopter wake is required, including the interference between the body and wake, and there is not much experimental data. Acceptable results are obtained from a properly calibrated calculation based on the rotor induced velocity and a drag coefficient of sections of the fuselage below the rotor disk. Computational fluid dynamics methods are useful now in calculating download.

There is a significant radial variation of the downwash in the wake, which must be accounted for when estimating the download. There are also large periodic variations in the drag, a possible source of helicopter vibration. In fact, the drag is largest when the body is closest to the rotor disk, diminishing rapidly as the body moves from the disk plane. This behavior is due to the periodic variation of the wake downwash. Although the mean downwash does increase from the rotor disk to the far wake about as expected from the vortex theory results, the mean dynamic pressure is significantly increased near the disk because of the periodic components in the flow. If the object in the wake is large enough, wake blockage must also be considered. A reduction of the effective disk area, particularly near the tips, decreases the rotor efficiency. Forward speed of the helicopter sweeps the wake rearward, so there is little vertical drag above transition speeds.

Download in hover can depend on the height above the ground, an effect depending greatly on the shape of the fuselage. For example, Flemming and Erickson (1983) in direct measurements of the download found about $\Delta T/T = 4\%$ at large height z above the ground, reducing to zero at about $z = 0.65R$. On some aircraft the download is negative (an upward force) very close to the ground.

4.8 Ground Effect

The proximity of the ground to the hovering rotor disk constrains the rotor wake and reduces the induced velocity at the rotor, which means a reduction in the power required for a given thrust; this behavior is called ground effect. Equivalently, ground proximity increases the rotor thrust at a given power. Because of this phenomenon, a helicopter can hover in ground effect (IGE) at a higher gross weight or altitude than is possible out of ground effect (OGE). The thrust increase near the ground also helps flare the helicopter when landing. Ground effect must be considered in testing helicopter rotors in hover, since the rotor must either be far enough above the ground for its influence to be neglected or the data must be corrected for the influence of the ground. Ground effect has been examined analytically using the method of images, where a mirror-image rotor is placed below the ground plane so that the boundary condition of no flow through the ground is automatically satisfied. Most of the useful information about the phenomenon has come from rotor performance measurements.

The influence of the ground can be viewed as a reduction of the rotor induced velocity by a factor $\kappa_g \leq 1$. Ground effect in hover can be described in terms of the figure of merit $M = (T^{3/2}/\sqrt{2\rho A})/P$. Usually the test data are given as the ratio of the thrust to OGE thrust, for constant power: $T/T_\infty = (M/M_\infty)^{2/3} = f_g \geq 1$. The effect on power at constant thrust is $P/P_\infty = (M/M_\infty)^{-1} = (T/T_\infty)^{-3/2} = f_g^{-3/2}$. Constant thrust implies $P/P_\infty = v/v_\infty = \kappa_g = f_g^{-3/2}$. The basic parameter is the height above the ground z, expressed as a fraction of the rotor radius or diameter, $z/D = z/2R$. There is a secondary dependence of ground effect on the rotor blade loading, C_T/σ. Ground effect is generally negligible when the rotor is more than one diameter above the ground, $z/R > 2$. Ground effect decreases rapidly with forward speed, since the wake is swept backward rather than being directed at the ground. It is also sensitive to winds, which displace the wake from under the rotor.

Figure 4.14 shows experimental data for the influence of height above the ground on T/T_∞. The data are from Zbrozek (1947), Cheeseman and Bennett (1955), Rabbott (1969), Hayden (1976), Cerbe, Reichert, and Curtiss (1988), and Schmaus, Berry, Gross, and Koliais (2012).

Zbrozek (1947) employed model and flight test data to express the influence of the ground in terms of the thrust increase T/T_∞ at constant power as a function of rotor height and C_T/σ. A curve fit of Zbrozek's interpolation has been developed. Betz (1937) analyzed the performance of a rotor in ground effect. For small distances above the ground ($z/R \ll 1$), he obtained the power at constant thrust: $P/P_\infty = 2z/R$. Knight and Hefner (1941) conducted an experimental and theoretical investigation of ground effect. They modified vortex theory to account for the ground by including image vortices below the ground plane. For a uniformly loaded actuator disk, the wake consisted of a cylindrical vortex sheet from the rotor to the ground and the corresponding image vortex cylinder below the ground. They obtained good correlation with measurements of the effect of the ground on the rotor performance. Cheeseman and Bennett (1955) made a simple analysis based on the method of images, representing the rotor by a source, and developed a model using blade element theory (BE) to incorporate the influence of thrust. Law (1972)

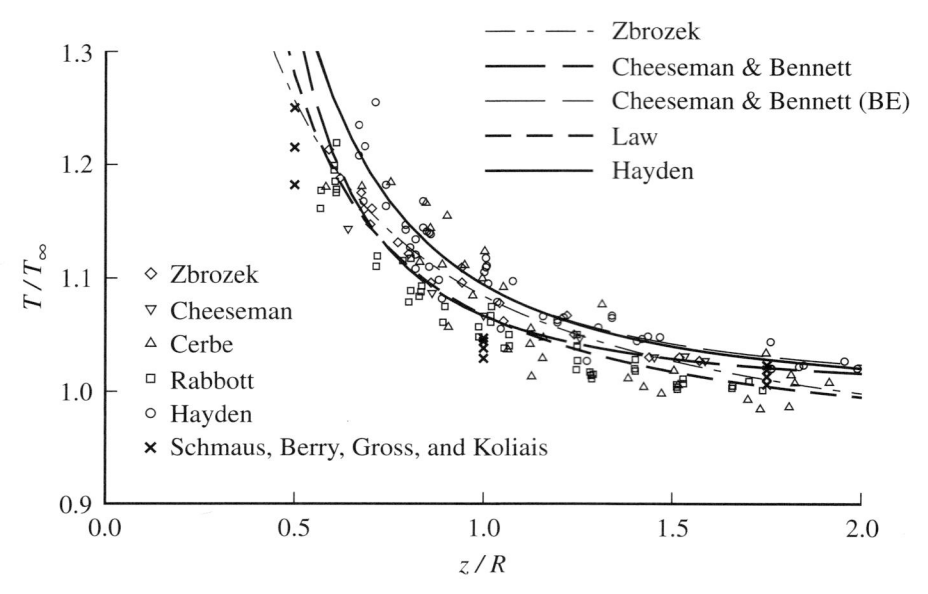

Figure 4.14. Ground effect: a thrust increase at constant power.

developed a method to account for the influence of the ground, based on flight test data from seven helicopters. Hayden (1976) correlated flight test data to obtain the influence of the ground on hovering performance. Schmaus, Berry, Gross, and Koliais (2012) measured ground effect on rotor performance down to $z/R = 0.1$. The following empirical ground effect models have been developed for $T/T_\infty = f_g$:

$$f_g = \left[1 - \frac{1}{(4z/R)^2}\right]^{-1} \quad \text{Cheeseman and Bennett} \tag{4.71}$$

$$f_g = \left[1 + 1.5 \frac{\sigma a \lambda_i}{4 C_T} \frac{1}{(4z/R)^2}\right] \quad \text{Cheeseman and Bennett (BE)} \tag{4.72}$$

$$f_g = \left[\frac{1.0991 - 0.1042/(z/D)}{1 + (C_T/\sigma)(0.2894 - 0.3913/(z/D))}\right]^{-1} \quad \text{Law} \tag{4.73}$$

$$f_g = \left[0.9926 + \frac{0.03794}{(z/2R)^2}\right]^{2/3} \quad \text{Hayden} \tag{4.74}$$

$$f_g = \left[0.9122 + \frac{0.0544}{(z/R)\sqrt{C_T/\sigma}}\right] \quad \text{Zbrozek} \tag{4.75}$$

$$f_g = \left[0.146 + 2.090\left(\frac{z}{R}\right) - 2.068\left(\frac{z}{R}\right)^2 + 0.932\left(\frac{z}{R}\right)^3 - 0.157\left(\frac{z}{R}\right)^4\right]^{-2/3}$$
$$\text{Schmaus, Berry, Gross, and Koliais} \tag{4.76}$$

The data base for ground effect extends down to about $z/R = 0.6$, and the equations are generally restricted to $z/R \geq 0.3$. The exception is the results from Schmaus, Berry, Gross, and Koliais (2012), which cover $z/R \geq 0.1$. Figure 4.14 shows T/T_∞

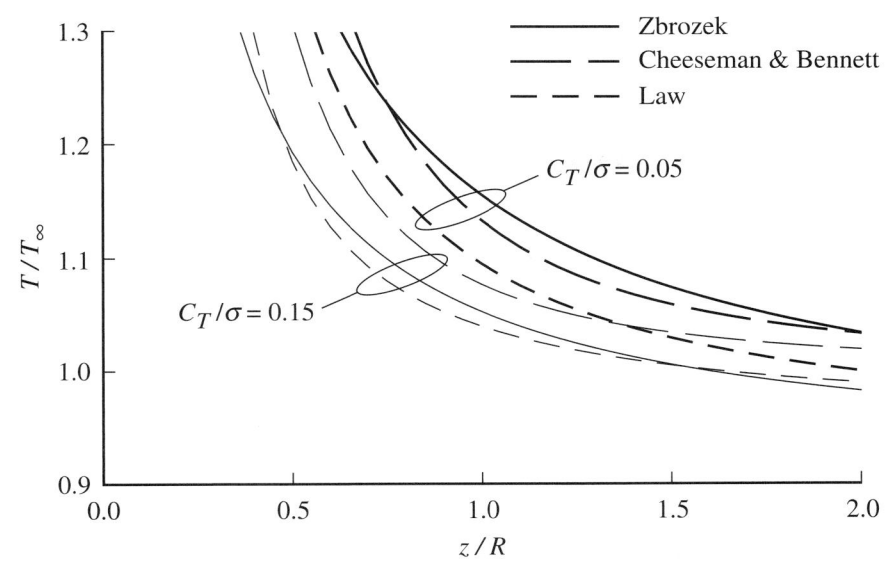

Figure 4.15. Ground effect: influence of blade loading.

as a function of z/R for these models (at $C_T/\sigma = 0.05$). Figure 4.15 illustrates the variation of the models with blade loading C_T/σ.

4.9 REFERENCES

Akimov, A.I., Butov, V.P., Bourtsev, B.N., and Selemenev, S.V. "Flight Investigation of Coaxial Rotor Tip Vortex Structure." American Helicopter Society 50th Annual Forum, Washington, DC, May 1994.

Bergey, K.H. "The Lanchester-Betz Limit." Journal of Energy, 3:6 (November-December 1979).

Betz, A. "Windmills in the Light of Modern Research." NACA TM 474, August 1928.

Betz, A. "The Ground Effect on Lifting Propellers." NACA TM 836, August 1937.

Betzina, M.D. "Tiltrotor Descent Aerodynamics: A Small-Scale Experimental Investigation of Vortex Ring State." American Helicopter Society 57th Annual Forum, Washington, DC, May 2001.

Bourtsev, B.N., Kvokov, V.N., Vainstein, I.M., and Petrosian, E.A. "Phenomenon of a Coaxial Helicopter High Figure of Merit at Hover." Twenty-Third European Rotorcraft Forum, Dresden, Germany, September 1997.

Brand, A., Dreier, M., Kisor, R., and Wood, T. "The Nature of Vortex Ring State." American Helicopter Society 63th Annual Forum, Virginia Beach, VA, May 2007.

Castles, W., Jr., and Gray, R.B. "Empirical Relation Between Induced Velocity, Thrust, and Rate of Descent of a Helicopter rotor as Determined by Wind-Tunnel Tests on Four Model Rotors." NACA TN 2474, October 1951.

Cerbe, T., Reichert, G., and Curtiss, H.C., Jr. "Influence of Ground Effect on Helicopter Takeoff and Landing Performance." Fourteenth European Rotorcraft Forum, Milan, Italy, September 1988.

Cheeseman, I.C., and Bennett, W.E. "The Effect of the Ground on a Helicopter Rotor in Forward Flight." ARC R&M 3021, September 1955.

Drees, J.M., and Hendal, W.P. "Airflow Patterns in the Neighbourhood of Helicopter Rotors." Aircraft Engineering, 23:266 (April 1951).

Empey, R.W., and Ormiston, R.A. "Tail-Rotor Thrust on a 5.5-Foot Helicopter Model in Ground Effect." American Helicopter Society 30th Annual National Forum, Washington, DC, May 1974.

Flemming, R.J., and Erickson, R.E. "An Evaluation of Vertical Drag and Ground Effect Using the RSRA Rotor Balance System." Journal of the American Helicopter Society, 28:4 (October 1983).

Glauert, H. "The Analysis of Experimental Results in the Windmill Brake and Vortex Ring States of an Airscrew." ARC R&M 1026, February 1926.

Hafner, R. "Rotor Systems and Control Problems in the Helicopter." Royal Aeronautical Society Anglo-American Aeronautical Conference, September 1947.

Harris, F.D. "Rotary Wing Aerodynamics – Historical Perspective and Important Issues." American Helicopter Society National Specialists' Meeting on Aerodynamics and Aeroacoustics, Arlington, TX, February 1987.

Hayden, J.S. "The Effect of the Ground on Helicopter Hovering Power Required." American Helicopter Society 32nd Annual National V/STOL Forum, Washington, DC, May 1976.

Johnson, W. "Model for Vortex Ring State Influence on Rotorcraft Flight Dynamics." NASA TP 2005-213477, December 2005.

Kisor, R., Blyth, R., Brand, A., and MacDonald, T. "V-22 Low-Speed/High Rate of Descent (HROD) Test Results." American Helicopter Society 60th Annual Forum, Baltimore, MD, June 2004.

Knight, M., and Hefner, R.A. "Analysis of Ground Effect on the Lifting Airscrew." NACA TN 835, December 1941.

Law, H.Y.H. "Two Methods of Prediction of Hovering Performance." USAAVSCOM TR 72-4, February 1972.

Lock, C.N.H., Bateman, H., and Townend, H.C.H. "An Extension of the Vortex theory of Airscrews with Applications to Airscrews of Small Pitch, Including Experimental Results." ARC R&M 1014, September 1925.

Lock, C.N.H. "Note on the Characteristic Curve for an Airscrew or Helicopter." ARC R&M 2673, June 1947.

Makofski, R.A., and Menkick, G.F. "Investigation of Vertical Drag and Periodic Airloads Acting on Flat Panels in a Rotor Slipstream." NACA TN 3900, December 1956.

Rabbott, J.P., Jr. "Model vs. Full Scale Rotor Testing." CAL/AVLABS Symposium on Aerodynamics of Rotary Wing and V/STOL Aircraft, Buffalo, NY, June 1969.

Reeder, J.P., and Gustafson, F.B., "On the Flying Qualities of Helicopters," NACA TN 1799, January 1949.

Schmaus, J., Berry, B., Gross, W., and Koliais, P. "Experimental Study of Rotor Performance in Deep Ground Effect with Application to a Human-Powered Helicopter." American Helicopter Society 68th Annual Forum, Fort Worth, TX, May 2012.

Stepniewski, W.Z. *Introduction to Helicopter Aerodynamics*. Morton, PA: Rotorcraft Publishing Committee, 1955a.

Stepniewski, W.Z. "A Simplified Approach to the Aerodynamic Rotor Interference of Tandem Helicopters." American Helicopter Society 11th Annual Forum, Washington, DC, April 1955b.

Taghizad, A., Jimenez, J., Binet, L., and Heuze, D. "Experimental and Theoretical Investigation to Develop a Model of Rotor Aerodynamics Adapted to Steep Descents." American Helicopter Society 58th Annual Forum, Montreal, Canada, June 2002.

Washizu, K., Azuma, A., Koo, J., and Oka, T. "Experiments on a Model Helicopter Rotor Operating in the Vortex Ring State," Journal of Aircraft, 3:3, May-June 1966.

Yaggy, P.F., and Mort, K.W. "Wind-Tunnel Tests of Two VTOL Propellers in Descent." NASA TN D-1766, March 1963.

Yeates, J.E. "Flight Measurements of the Vibration Experienced by a Tandem Helicopter in Transition, Vortex-Ring State, Landing Approach, and Yawed Flight," NACA TN 4409, September 1958.

Zbrozek, J. "Ground Effect on the Lifting Rotor." ARC R&M 2347, July 1947.

5 Forward Flight Wake

During translational motion of the helicopter, when the rotor is nearly horizontal, the rotor blades see a component of the forward velocity as well as the velocity due to their own rotation (Figure 5.1). In forward flight the rotor does not have axisymmetry as in hover and vertical flight; rather, the aerodynamic environment varies periodically as the blade rotates with respect to the direction of flight. The advancing blade has a velocity relative to the air higher than the rotational velocity, whereas the retreating blade has a lower velocity relative to the air. This lateral asymmetry has a major influence on the rotor and its analysis in forward flight. Thus the rotor blade loading and motion are periodic with a fundamental frequency equal to the rotor speed Ω. The analysis is more complicated than for hover because of the dependence of the loads and motion on the azimuth angle.

As a consequence of the axisymmetry, the analysis of the hovering rotor primarily involves a consideration of the aerodynamics. In forward flight, however, the lateral asymmetry in the basic aerodynamic environment produces a periodic motion of the blade, which in turn influences the aerodynamic forces. The analysis in forward flight must therefore consider the blade dynamics as well as the aerodynamics. This chapter covers a number of aerodynamic topics that are familiar from the analysis of the rotor in vertical flight. In particular, we are concerned with the momentum theory treatment of the induced velocity and power in forward flight. Then the rotor blade motion and its behavior in forward flight are considered in Chapter 6.

5.1 Momentum Theory in Forward Flight

5.1.1 Rotor Induced Power

The momentum theory analysis of rotor induced power in forward flight was introduced by Glauert (1926). As in hover, the power is represented by an induced velocity $v = P_i/T$. When used in blade element theory, the induced velocity is assumed to be uniform over the disk; although that is not as good as an assumption in forward flight as in hover, at high forward speed the induced velocity is small compared to the other velocity components at the rotor blade. At low forward speeds the variation of the inflow over the disk is important, particularly for vibration and blade loads. A uniformly loaded actuator disk is again used to represent the rotor. In forward flight such an actuator disk can be viewed as a circular wing.

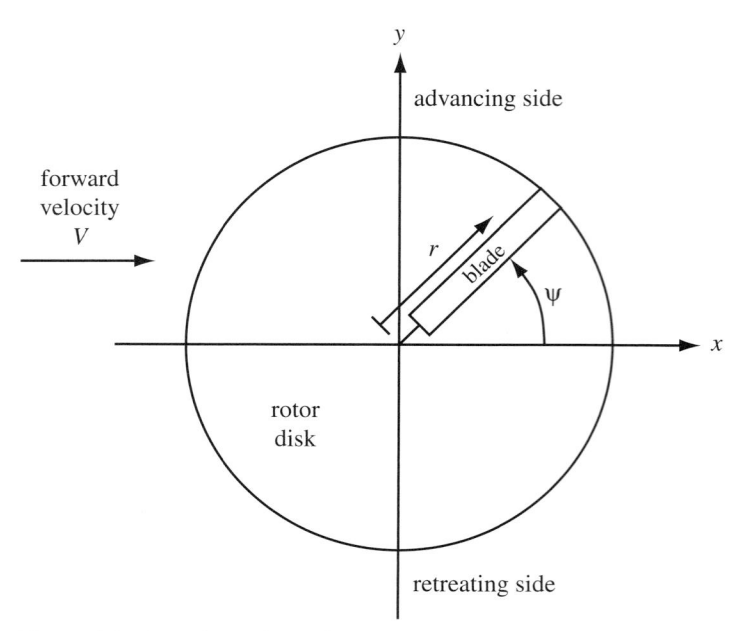

Figure 5.1. Aerodynamic environment of the rotor in forward flight.

Fixed-wing theory gives the minimum induced drag for a thin, planar wing of span b, operating at velocity V and lift T:

$$D_i = \frac{T^2}{2\rho A V^2} \tag{5.1}$$

where $A = \pi(b/2)^2$ is the area of a circle with diameter b (a more familiar form perhaps is $C_{Di} = C_L^2/\pi AR$, where the aspect ratio AR equals the span-squared divided by the wing area). In terms of the induced velocity,

$$v = \frac{P_i}{T} = \frac{VD_i}{T} = \frac{T}{2\rho AV} \tag{5.2}$$

This minimum drag is achieved with elliptical loading of the wing. The uniformly loaded rotor has a circular span loading, which is a special case of elliptical loading; at high forward speeds the rotor wake vorticity is swept back in the plane of the disk, like the fixed-wing wake. Moreover, the induced drag solution is based on a Trefftz plane analysis in the far wake, so is valid for wings of arbitrary aspect ratio. Therefore, $v = T/2\rho AV$ is an appropriate solution for the induced velocity of the helicopter rotor in high-speed forward flight. For the rotor, the wing span is the rotor diameter, so A is simply the rotor disk area. Lifting-line theory interprets v as the actual induced velocity at the wing, uniform over the span for high aspect ratios. For the circular wing, which has the aspect ratio $AR = 4/\pi = 1.27$, considerable variation of the induced velocity over the disk can be expected.

Expressions for the rotor induced power are now available for the rotor in vertical flight and in high-speed forward flight. A connection between the two regions is required, so the inflow can be obtained for all operating conditions of the rotor. The forward flight result can be written as $T = \dot{m}2v$, where $\dot{m} = \rho AV$ is the mass flux through an area equal to the rotor disk area. This is exactly the form of the momentum theory results for vertical flight. In hover and climb, for example, $T = \dot{m}2v$ and

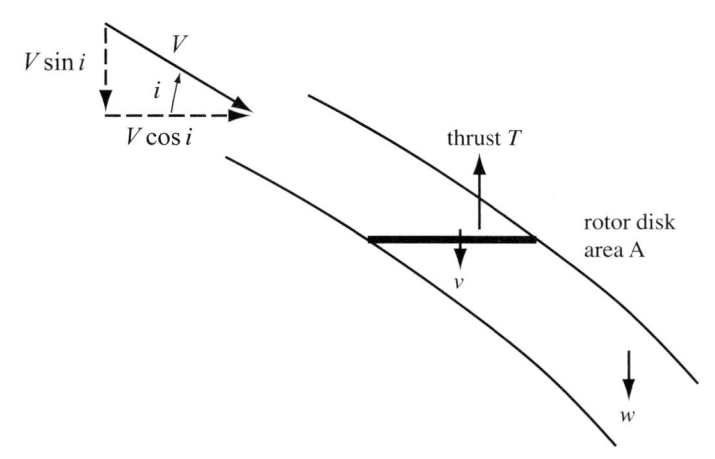

Figure 5.2. Flow model for momentum theory analysis of rotor in forward flight.

$\dot{m} = \rho A(V + v)$. Thus a uniformly valid expression for induced velocity can be obtained by considering the mass flux through the area A for all operating conditions. This observation was first made by Glauert (1926).

Consider a rotor operating at velocity V, with angle of incidence i between the free stream velocity and the rotor disk (Figure 5.2). The induced velocity at the disk is v. In the far wake, the velocity $w = 2v$ and is assumed to be parallel to the rotor thrust vector. Momentum conservation gives the rotor thrust $T = \dot{m}2v$, where the mass flux is $\dot{m} = \rho A U$. The resultant velocity U is given by

$$U^2 = (V \cos i)^2 + (V \sin i + v)^2 = V^2 + 2Vv \sin i + v^2 \qquad (5.3)$$

Hence

$$T = 2\rho A v \sqrt{V^2 + 2Vv \sin i + v^2} \qquad (5.4)$$

Energy conservation gives the rotor power

$$P = \dot{m}\left(\frac{1}{2}\left[(V \sin i + 2v)^2 + (V \cos i)^2\right] - \frac{1}{2}V^2\right) = T(V \sin i + v) \qquad (5.5)$$

For high forward speeds ($V \gg v$) we have $T = 2\rho A V v$, and in hover ($V = 0$) $T = 2\rho A v^2$, so this expression does have the proper limits. Although there is no strict theoretical justification for this approach at intermediate forward speeds, there is good agreement with measured rotor performance and with vortex theory results; thus the model can be accepted over the entire range of rotor speeds. In the expression for the rotor power (equation 5.5), the term Tv is the induced power, and the term $TV \sin i$ is the power required to climb and to propel the helicopter forward (the parasite power). As for vertical flight, we can write $(V \sin i + v)/v_h = P/Tv_h = P/P_h$.

The solution for the induced velocity is

$$v = \frac{v_h^2}{\sqrt{(V \cos i)^2 + (V \sin i + v)^2}} = \frac{v_h^2}{\sqrt{V_x^2 + (V_z + v)^2}} \qquad (5.6)$$

where $v_h^2 = T/2\rho A$ as usual. The edgewise and axial components of the rotor velocity V are $V_x = V \cos i$ and $V_z = V \sin i$, respectively (which are not the horizontal and vertical components unless the rotor disk is horizontal). Define now the dimensionless components of the velocity parallel to and normal to the rotor disk, the advance ratio μ, and the inflow ratio λ:

$$\mu = \frac{V \cos i}{\Omega R} \tag{5.7}$$

$$\lambda = \frac{V \sin i + v}{\Omega R} = \mu \tan i + \lambda_i \tag{5.8}$$

Use is also made of $\mu_z = V_z/\Omega R = V \sin i/\Omega R = \mu \tan i$. Then in coefficient form the induced inflow ratio is

$$\lambda_i = \frac{C_T}{2\sqrt{\mu^2 + \lambda^2}} \tag{5.9}$$

This is the Glauert inflow formula. In general it is a quartic equation for λ_i (although it can be solved directly for μ_z given μ and λ). An iterative Newton–Raphson solution for λ is

$$\widehat{\lambda}_{in} = \frac{\lambda_h^2}{\sqrt{\lambda_n^2 + \mu^2}} \tag{5.10}$$

$$\lambda_{n+1} = \lambda_n - \frac{\lambda_n - \mu_z - \widehat{\lambda}_{in}}{1 + \widehat{\lambda}_{in}\lambda_n/(\lambda_n^2 + \mu^2)} f \tag{5.11}$$

where $\lambda_h^2 = C_T/2$. A relaxation factor of $f = 0.5$ is used to improve convergence. Three or four iterations are usually sufficient, using

$$\lambda \cong \frac{\lambda_h^2}{\sqrt{(\lambda_h + \mu_z)^2 + \mu^2}} + \mu_z \tag{5.12}$$

to start the solution. To eliminate the singularity of the momentum theory result at ideal autorotation ($\lambda = \lambda_i + \mu_z = 0$), an extension of equation 4.13 is

$$\lambda = \mu_z \left[\frac{0.373\mu_z^2 + 0.598\mu^2}{\lambda_h^2} - 0.991 \right] \tag{5.13}$$

The equation $\lambda = \mu_z(a\mu_z^2 - b\lambda_h^2 + c\mu^2)/\lambda_h^2$ is an approximation for the induced power measured in the turbulent-wake and vortex-ring states. Matching this equation to the axial-flow momentum theory result at $\mu_z = -2\lambda_h$ and $\mu_z = -\lambda_h$ gives $a = \sqrt{5}/6 = 0.373$ and $b = (4\sqrt{5} - 3)/6 = 0.991$. Then matching to the forward-flight momentum theory result at $\mu = \lambda_h$ and $\mu_z = -1.5\lambda_h$ gives $c = 0.598$. Equation 5.13 should be used when

$$1.5\mu^2 + (2\mu_z + 3\lambda_h)^2 < \lambda_h^2 \tag{5.14}$$

which covers the axial domain ($-2\lambda_h < \mu_z < -\lambda_h$) exactly and gives good results for $\mu > 0$ (only small jumps to the momentum theory results). A good representation of the inflow in vortex ring state requires a more complicated approach; see Johnson (2005).

For high forward speeds ($\mu \gg \lambda$), the momentum theory solution becomes $\lambda_i \cong C_T/2\mu$ (or $v = T/2\rho AV \cos i$, which is just the circular wing result). The usefulness of this approximation is that iteration is not necessary to obtain λ_i. Figure 5.3 shows the induced velocity in forward flight for the case $i = 0$, for which an exact analytical

5.1 Momentum Theory in Forward Flight

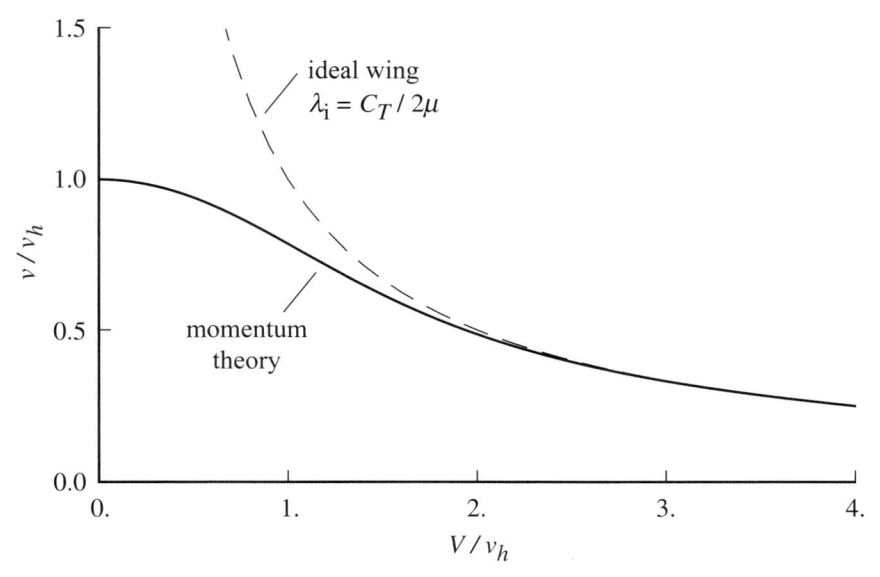

Figure 5.3. Rotor induced velocity in forward flight with $i = 0$.

solution is possible: $v^2(V^2 + v^2) = v_h^4$, so

$$v^2 = -\frac{V^2}{2} + \sqrt{\left(\frac{V^2}{2}\right)^2 + v_h^4} \tag{5.15}$$

Forward speed reduces the induced power as a result of the increased mass flux. Figure 5.3 also shows the approximation $\lambda_i \cong C_T/2\mu$, which is good when $\mu/\lambda_h > 1.5$ or so.

In Figure 4.5 the momentum theory solution was plotted in the form of $P/P_h = (V + v)/v_h$ as a function of the vertical velocity V/v_h. For forward flight this is generalized to $P/P_h = (V_z + v)/v_h$ as a function of the normal velocity component V_z/v_h, for a fixed value of the in-plane velocity component V_x/v_h (or λ/λ_h as a function of μ_z/λ_h, given μ/λ_h). The result is shown in Figure 5.4. The effect of forward flight in this figure is to always reduce the induced power. The results in Figure 5.4 have been corrected in two respects on the basis of performance measurements; the corresponding momentum theory results are shown also. First, the measured performance data show that the actual induced power in hover and low speed is 10% to 15% higher than the momentum theory estimate. Thus an empirical correction factor κ should be included in the induced power calculation, $P_i = \kappa T v$. Second, for vertical flight and for low forward speed, measured performance is the only means to define the induced velocity curve in the vortex ring state (see section 4.1.2.2). The boundary for the negative heave damping in VRS extends to about $V_x/v_h = 1$ (Figure 4.8). With sufficient forward velocity a moderate descent rate of the helicopter presents no problems, since the wake of the rotor is swept back instead of being allowed to build up under the disk. Indeed, the momentum theory result is also well behaved above $V_x/v_h = 1$. Figure 5.4 shows the region of VRS instability and roughness.

The forward speed scale is defined by $v_h = 8.59\sqrt{T/A}$ knots for sea-level-standard density and the disk loading in lb/ft^2 ($v_h = 1.25\sqrt{T/A}$ knots with disk loading in N/m^2), or typically $v_h = 19$ to 33 knots.

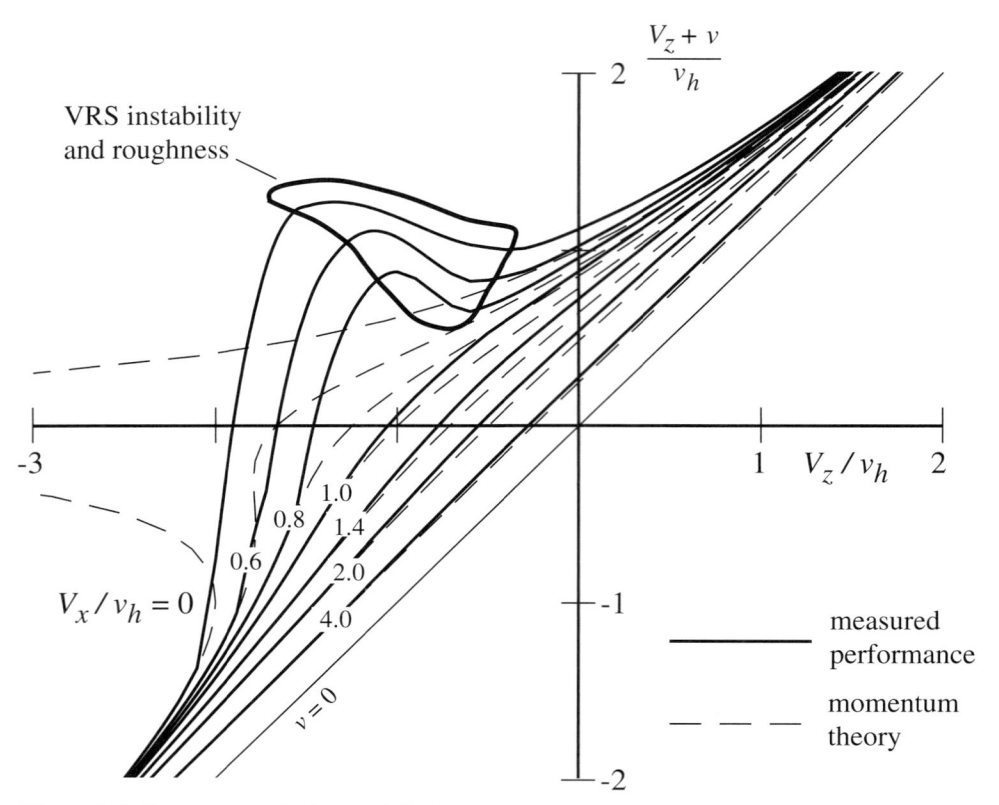

Figure 5.4. Rotor power in forward flight.

The high-speed approximation $\lambda_i \cong C_T/2\mu$ can be written $v \cong v_h^2/V_x$, which in Figure 5.4 is a straight line parallel to the $v = 0$ line. This approximation is good for $V_x/v_h > 1.5$ or so, which corresponds to forward speeds above 30 to 50 knots for the disk loadings typical of helicopters. In terms of the rotor advance ratio, $\mu/\lambda_h > 1.5$, which typically gives $\mu > 0.1$. Thus, according to momentum theory, the rotor wake system is like a circular wing except at very low speeds. The speed range in which the rotor wake is no longer directly under the rotor but still has a significant vertical extent, roughly $0 < \mu < 0.1$, is called the transition region of the helicopter. The transition region has a number of special characteristics in addition to the requirement for the general induced velocity expression, notably a high level of blade loads and vibration and noise due to interactions of the blades with the rotor vortex wake.

5.1.2 Climb, Descent, and Autorotation in Forward Flight

The power required in forward flight, including now the profile power P_o, is

$$P = P_o + TV \sin i + \kappa T v \tag{5.16}$$

The term $(TV \sin i)$ combines the rotor parasite power and climb power, which require the thrust component $(T \sin i)$ in the direction of V. To determine the rotor disk incidence angle i, consider the equilibrium of forces on the helicopter, as shown in Figure 5.5. The forces acting on the helicopter are the rotor thrust T, the helicopter

5.1 Momentum Theory in Forward Flight

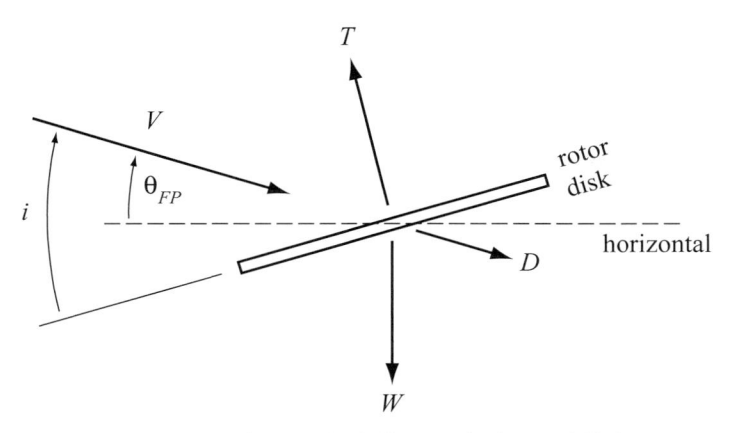

Figure 5.5. Forces acting on the helicopter in forward flight.

weight W, and the helicopter drag D. Here θ_{FP} is the flight path angle, so the climb speed is $V_c = V\theta_{FP}$. For small angles, vertical and horizontal force equilibrium gives $i = \theta_{FP} + D/T$ and $T = W$. Hence

$$TV \sin i = TV_c + DV \qquad (5.17)$$

where the first term is the climb power and the second is the parasite power. A more detailed derivation of helicopter force equilibrium and performance is given in section 6.4. For high enough forward speed, the rotor induced velocity is $v \cong T/2\rho AV \cos i \cong T/2\rho AV$. Thus the power equation can be solved for the climb velocity:

$$V_c = \frac{P - (P_o + VD + \kappa T^2/2\rho AV)}{T} \qquad (5.18)$$

Because the induced power in forward flight is independent of the climb or descent rate, a simple and direct expression for V_c has been obtained. Assuming that the rotor profile power and the helicopter drag are not influenced by the climb or descent velocity, we have

$$V_c = \frac{P - P_{\text{level}}}{T} = \frac{\Delta P}{T} \qquad (5.19)$$

where P_{level} is the power required for level flight at the same forward speed. The helicopter climb or descent rate is determined simply by the excess power ΔP. The climb and autorotation characteristics of the helicopter in forward flight can then be obtained from the power available and the power required for level flight. In particular, the maximum climb rate is achieved with maximum available power at the speed for minimum power in level flight, and the minimum power-off descent rate is achieved at the same forward speed. The helicopter flight performance characteristics are considered in more detail in Chapter 7.

5.1.3 Rotor Loading Distribution

The basis for the Glauert inflow formula (equation 5.9) is the result for minimum drag of a circular wing. Let us examine this solution further, following Ashley and Landahl (1965). The induced drag of a system of wings can be calculated from the energy in the far wake. For the ideal case, there is no rollup or distortion of the wake vorticity, so the wake far downstream is represented by potential jumps on lines that are projections of the wing geometry.

Consider incompressible, irrotational flow with velocity potential ϕ (velocity $\mathbf{Q} = \nabla\phi$) satisfying $\nabla^2\phi = 0$. Start with Gauss's theorem for an arbitrary vector field $\mathbf{A}$:

$$\int \mathbf{A} \cdot \mathbf{n}\, dS = -\int \nabla \cdot \mathbf{A}\, dV \tag{5.20}$$

where $\mathbf{n}$ is the normal vector into the fluid. Setting $\mathbf{A} = \phi\nabla\phi'$ gives Green's theorem:

$$\int \phi \frac{\partial \phi'}{\partial n}\, dS = -\int \left(\nabla\phi \cdot \nabla\phi' + \phi\nabla^2\phi'\right) dV \tag{5.21}$$

Now let $\phi = \phi'$, and ϕ be a velocity potential. Then

$$\int \phi \frac{\partial \phi}{\partial n}\, dS = -\int |\nabla\phi|^2 dV = -\int Q^2 dV \tag{5.22}$$

and the total kinetic energy in the fluid is

$$KE = -\frac{\rho}{2}\int \phi \frac{\partial \phi}{\partial n}\, dS \tag{5.23}$$

For a wing, the surface of integration includes the wake. In steady flow at velocity U, the flow far downstream (the Trefetz plane) gives the induced drag as the change in kinetic energy per unit distance:

$$D_i = -\frac{\rho}{2}\int_{\text{wake}} \phi \frac{\partial \phi}{\partial n}\, dy \tag{5.24}$$

where y is the spanwise coordinate. A thin planar wing with bound circulation distribution $\Gamma(y)$ generates a wake sheet of trailed vorticity strength $\delta(y) = \partial(\Delta\phi_{\text{wake}})/\partial y = d\Gamma/dy$. Hence

$$\Delta\phi_{\text{wake}}(y) = \Gamma(y) \tag{5.25}$$

$$\frac{\partial \phi}{\partial n} = \pm\phi_z = \mp\frac{1}{2\pi}\int \frac{d\Gamma}{d\eta}\frac{d\eta}{y-\eta} \tag{5.26}$$

where equation 5.26 is the wake-induced velocity w from the Biot-Savart law, on the upper and lower surfaces of the wake. Then the induced drag is

$$D_i = -\frac{\rho}{2}\int \Delta\phi_{\text{wake}}\phi_z(y)dy = \frac{\rho}{4\pi}\int\int \Gamma(y)\frac{d\Gamma}{d\eta}\frac{d\eta\, dy}{y-\eta} \tag{5.27}$$

The induced drag is minimum for elliptical loading: $\Gamma = \Gamma_0\sqrt{1-(2y/b)^2}$ (for wing span b) gives $D_{i\min} = L^2/(\frac{\pi}{2}\rho U^2 b^2)$. In general the wing spanwise loading can be written as a series: $\Gamma = Ub\sum A_n \sin n\theta$, $y = \frac{b}{2}\cos\theta$. Evaluating the integrals gives the wing lift and induced drag:

$$L = \frac{\pi}{4}\rho U^2 b^2 A_1 \tag{5.28}$$

$$D_i = \frac{\pi}{8}\rho U^2 b^2 \sum n A_n^2 = D_{i\min}\sum n(A_n/A_1)^2 \tag{5.29}$$

These results apply to wings with arbitrary planform, and (for appropriate interpretation of the integrations) to non-planar and multiple-surface wings. In terms of the

5.1 Momentum Theory in Forward Flight

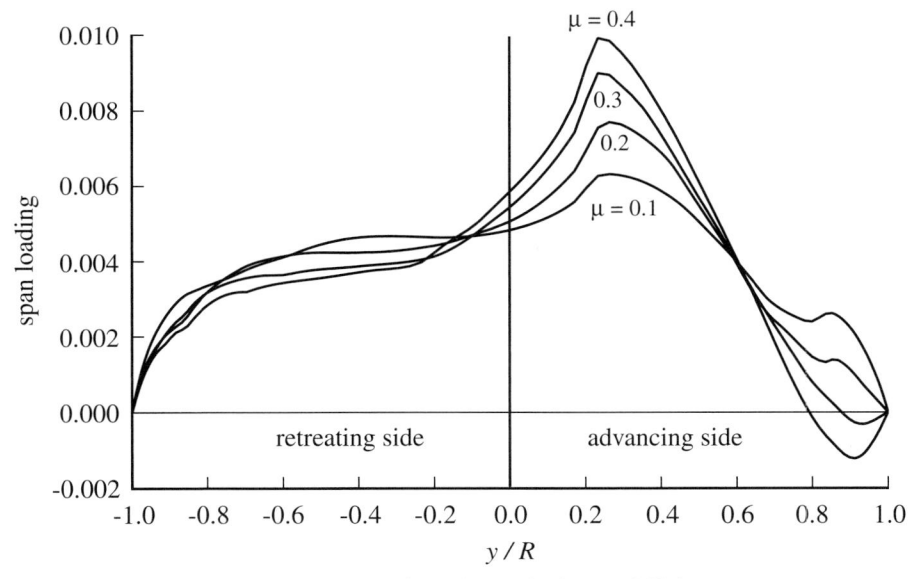

Figure 5.6. Spanwise loading distribution of rotor in forward flight.

section lift $\ell = \rho U \Gamma$ and the wake induced velocity w (equation 5.26), the induced power is

$$P_i = U D_i = \frac{1}{2} \int \ell w \, dy = \int \ell v \, dy \qquad (5.30)$$

The lifting-line concept collapses a high aspect-ratio wing to a bound vortex line and evaluates the induced velocity v at the bound vortex. From symmetry $v = \frac{1}{2}w$, hence the last result in equation 5.30.

For the circular wing as a model for the rotor in forward flight, elliptical span loading corresponds to uniform loading over the disk. In fact, the loading on a rotor in high-speed flight is far from uniform, and the span load distribution is far from elliptical.

Figure 5.6 shows the span loading distribution on a rotor for several advance ratios, calculated by integrating the blade loading from the front edge to the rear edge of the rotor disk (and averaging over time) at a fixed lateral distance y from the hub. A modern four-bladed rotor was used for this example. See also Prouty (1976). The lateral asymmetry of the loading is a consequence of the asymmetry of the rotor aerodynamic environment in forward flight. The blade on the advancing side of the disk sees an increase in the relative air velocity, while the blade on the retreating side sees a decrease. Because of reduced dynamic pressure, the lifting capability of the retreating blade is limited by stall. Then the necessity for roll moment balance means the lifting capability of the advancing blade is also constrained. As the advance ratio increases, the loading becomes concentrated on the front and rear of the disk, effectively lowering the span of the lifting system. Moreover, the differences in dynamic pressure distribution mean that the loading distribution is different on the advancing and retreating sides, a source of asymmetric span loading that persists even if the requirement for roll moment balance is relaxed. Observe also in Figure 5.6 the loading on the advancing tip as speed changes. At low speed, blade-vortex interaction

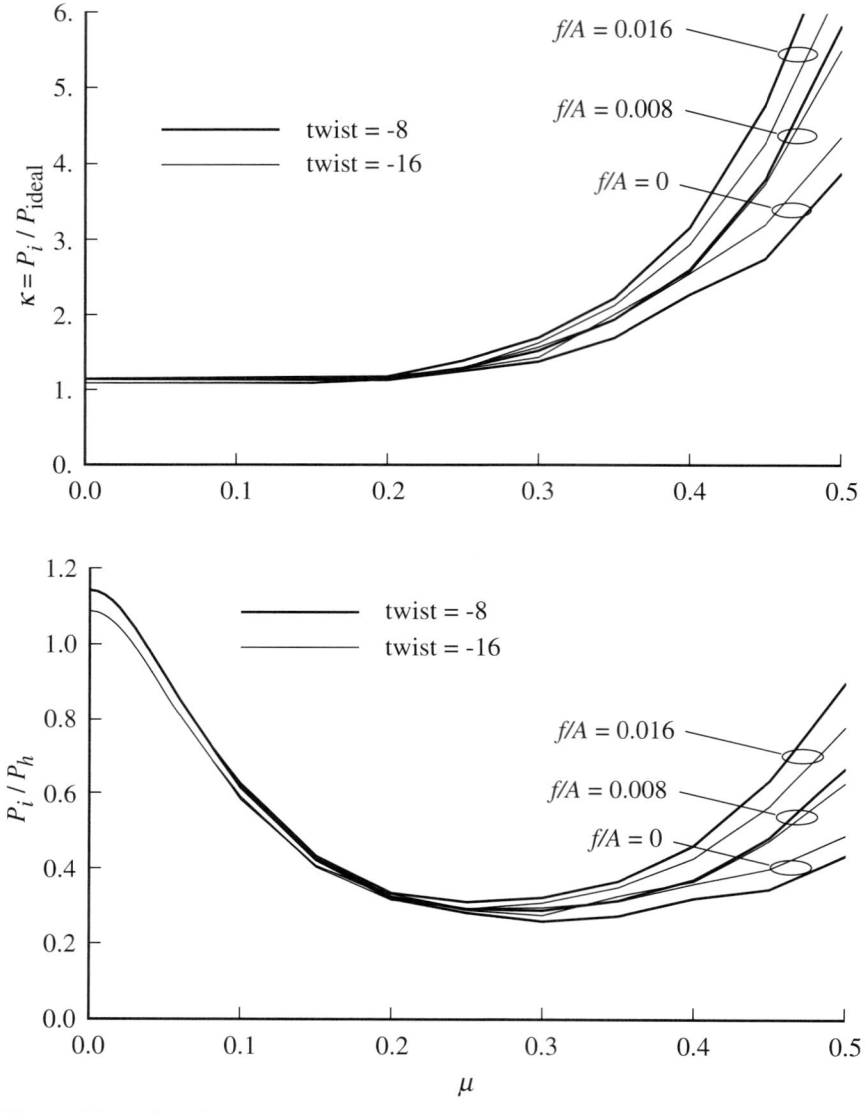

Figure 5.7. Induced power factor in forward flight.

produces an increase in the loading at the tip. At a higher advance ratio, the loading on the advancing tip can be negative, particularly with highly twisted blades.

The loading distribution that follows from the aerodynamic environment of the rotating wing in edgewise flight produces a significant increase in the rotor induced power. Figure 5.7 shows the induced power factor $\kappa = P_i/P_{ideal}$ as a function of the advance ratio, calculated using a free wake analysis. The ideal power here is from momentum theory (equation 5.9). Above an advance ratio of about $\mu = 0.2$, κ increases substantially. A value $\kappa = 4.0$ corresponds to an effective wing span of one-half the rotor diameter. The induced power factor is presented for $C_T/\sigma = 0.08$, two values of blade twist, and propulsive force corresponding to aircraft drag $f = D/q = 0$, $0.008A$, and $0.016A$. Increasing rotor propulsive force increases the induced power. Figure 5.7 also shows the ratio of the induced power to the hover

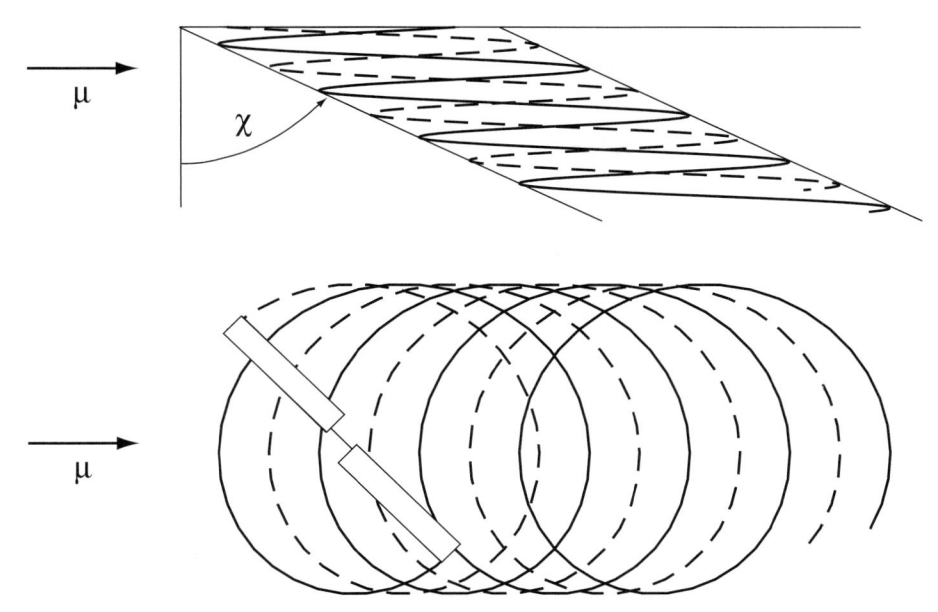

Figure 5.8. Tip vortex geometry of the rotor wake in forward flight (without self-induced distortion).

momentum theory value, P_i/P_h. Initially the induced power rapidly decreases with forward speed because of the increase in mass flux through the disk. Eventually the asymmetry of the loading becomes significant, and the induced power actually increases with speed.

5.2 Vortex Theory in Forward Flight

In forward flight the helical vortices trailed from the blade tips are carried rearward by the free stream velocity component parallel to the disk (μ) as well as downward by the component normal to the disk (λ). Thus the wake geometry consists of concentrated vortices from each blade trailed in skewed, interlocking helices (Figure 5.8). The wake skew angle $\chi = \tan^{-1} \mu/\lambda$ can be estimated fairly well using momentum theory. The helicopter transition operating region $0 < \mu/\lambda_h < 1.5$ corresponds approximately to wake angles from $\chi = 0$ to $\chi = 60°$. The relative positions of the rotor blade and the individual wake vortices vary periodically as the blade rotates, producing a strong variation in the wake-induced velocity encountered by the blade and hence in the blade loading. The induced velocity is thus highly nonuniform in forward flight. The interaction between the blades and the wake is particularly strong on the advancing and retreating sides of the disk, where the tip vortex from the preceding blade sweeps radially along the blade. Under certain flight conditions where the wake is close to the rotor disk, the vortex-induced loads are very high.

The vortex wake of the rotor in forward flight rolls up in a two-stage process. The individual tip vortices quickly roll up into concentrated lines as they are trailed from the blades. Then the interlocking, overlapping spirals in the far wake interact, and they roll up to form two vortices like those behind a circular wing. Such behavior has been observed experimentally; the two tip vortices from the edges of the disk are seen forming several rotor radii downstream from the disk. This behavior downstream of the rotor is of little consequence as far as the downwash and loading at the disk are concerned, but can be significant for interference effects of the rotor far wake. The wake behavior also demonstrates the validity of viewing the rotor as a circular wing in high-speed flight.

Classical vortex theory for forward flight is based on the actuator disk model, so the vorticity is distributed throughout the wake rather than being concentrated in discrete lines. Often uniform loading is also assumed, so the vorticity is only on the surface of the wake cylinder and in a root vortex. These assumptions yield the simplest wake model, but in contrast to hover, the mathematical problem is still not trivial, because of the skewed cylindrical geometry. With the exception of a few special locations, numerical calculations are required to obtain the induced velocity at or near the rotor disk. For uniform loading the results are the same as from momentum theory; in particular, the high-speed results must approach the wing theory solution. Actuator disk models in forward flight are the subject of Chapter 11; some results are presented in the next section. Detailed calculations of the induced velocity are best obtained from a nonuniform inflow analysis, including a representation of the discrete vorticity in the wake (see Chapter 9).

5.2.1 Actuator Disk Results

Coleman, Feingold, and Stempin (1945) conducted a vortex theory analysis of the induced velocity along the fore-aft diameter of the rotor disk. They considered an actuator disk with uniform loading and decomposed the vorticity into rings and axial lines (neglecting the latter) to calculate the induced velocity. Along the fore-aft diameter of the disk the normal component of the induced velocity can be obtained in closed form, but the solution involves elliptic integrals even there. A good approximation to the numerical results was $v = v_0(1 + \kappa_x r \cos \psi)$, where v_0 is the usual momentum theory result and

$$\kappa_x = \tan \frac{\chi}{2} \tag{5.31}$$

based on the slope of the downwash at the center of the disk. Using $\tan \chi = \mu/\lambda$, this result becomes

$$\kappa_x = \sqrt{1 + (\lambda/\mu)^2} - |\lambda/\mu| \tag{5.32}$$

For high speeds ($\mu \gg \lambda$), $\kappa_x = 1$.

Drees (1949) calculated the rotor induced velocity using vortex theory. He considered an actuator disk with radially constant bound circulation, but allowed an azimuthal variation of the form $\Gamma = \Gamma_0 - \Gamma_1 \sin \psi$. The trailed vorticity is still only on the surface of the wake cylinder, but now the cylinder is filled with shed vorticity as well. The velocity seen by the blade is $U = \Omega r + \Omega R \mu \sin \psi$, so the total blade lift and flap moment are

$$L = \int_0^R \rho U \Gamma \, dr$$

$$= \frac{1}{2} \rho \Omega R^2 \Gamma_0 \left[1 + \left(2\mu - \frac{\Gamma_1}{\Gamma_0} \right) \sin \psi - 2\mu \frac{\Gamma_1}{\Gamma_0} \sin^2 \psi \right] \tag{5.33}$$

$$M = \int_0^R \rho U \Gamma r \, dr$$

$$= \frac{1}{3}\rho\Omega R^3 \Gamma_0 \left[1 + \left(\frac{3}{2}\mu - \frac{\Gamma_1}{\Gamma_0}\right)\sin\psi - \frac{3}{2}\mu\frac{\Gamma_1}{\Gamma_0}\sin^2\psi\right] \quad (5.34)$$

Requiring that the mean blade lift equal the rotor thrust per blade ($L = T/N$) and that the first harmonic of the flap moment be zero (for moment equilibrium on the articulated blade) gives the distribution of the blade bound circulation:

$$\rho\Omega R^2 N\Gamma = \frac{2T}{1 - \frac{3}{2}\mu^2}\left(1 - \frac{3}{2}\mu\sin\psi\right) \quad (5.35)$$

Drees found the induced velocity due to the bound, trailed, and shed vorticity associated with this circulation distribution. The induced velocities at $r = 0$ and $r = 0.75$ were

$$\lambda(0) = \frac{C_T}{2\mu}\sin\chi \quad (5.36)$$

$$\lambda(0.75) = \frac{C_T}{2\mu}\left[\sin\chi + (1 - \cos\chi - 1.8\mu^2)\cos\psi - \frac{3}{2}\mu\sin\chi\sin\psi\right] \quad (5.37)$$

where $\chi = \tan^{-1}\mu/\lambda$ is the wake skew angle. Here a factor of $(1 - (3/2)\mu^2)$ in the denominator has been omitted; although often found in vortex theory results, Heyson (1960) showed that this factor was present only because the axial wake vorticity was ignored. Since $\sin\chi = \mu/\sqrt{\mu^2 + \lambda^2}$, the mean induced velocity is

$$\lambda_i = \frac{C_T}{2\sqrt{\mu^2 + \lambda^2}} \quad (5.38)$$

as in equation 5.9. Assuming a linear variation of the velocity over the rotor disk, this result can be generalized to

$$\lambda_i = \frac{C_T}{2\sqrt{\mu^2 + \lambda^2}}\left(1 + \frac{4}{3}\frac{1 - \cos\chi - 1.8\mu^2}{\sin\chi}r\cos\psi - 2\mu r\sin\psi\right) \quad (5.39)$$

Drees also suggested an empirical correction for the momentum theory results in order to remove the singularity at ideal autorotation in vertical flight:

$$\lambda_i = 1.2\frac{1}{\sqrt{\mu^2 + \lambda^2}}\left(\lambda_h^2 - \frac{\mu_z^2 C_{W0}}{4(1 + 8\lambda^2/\lambda_h^2)(1 + 8\mu^2/\lambda_h^2)}\right) \quad (5.40)$$

where $\lambda_h^2 = C_T/2$, and C_{W0} is the drag coefficient of the rotor in ideal autorotation. Drees suggested $C_{W0} = 1.38$, which gives $V/v_h = -1.70$ for ideal autorotation.

Castles and De Leeuw (1954) presented tables and graphs for the normal component of the induced velocity in the longitudinal plane of symmetry of the flow (the vertical plane through the center of the disk and the wake axis) and on the lateral axis in the disk plane. The velocities were calculated numerically using vortex theory for a uniformly loaded actuator disk. They concluded that the downwash reaches its maximum far wake value about one rotor radius downstream of the center of the disk for high-speed flight, that is, about at the trailing edge of the rotor disk. For hover and low-speed flight the far wake value is achieved about $2R$ downstream of the center of the disk.

5.2.2 Induced Velocity Variation in Forward Flight

For a first (and very rough) approximation to the nonuniform inflow distribution at the rotor in forward flight, consider a linear variation over the disk:

$$v = v_0(1 + \kappa_x x + \kappa_y y) = v_0(1 + \kappa_x r \cos\psi + \kappa_y r \sin\psi) \qquad (5.41)$$

(the x and y coordinates on the rotor disk are defined in Figure 5.1). Here v_0 is the mean value of the induced velocity, which can be obtained from momentum theory. The form $v = v_0(1 + \kappa_x r \cos\psi)$ was first suggested by Glauert (1926). Typically κ_x is positive and κ_y is negative, so that the induced velocity is larger at the rear of the disk and on the retreating side. At high speeds κ_x is approximately 1, which gives a velocity near zero at the leading edge of the disk and about twice the mean value at the trailing edge; κ_y is generally smaller in magnitude. Both κ_x and κ_y must be zero with axisymmetric loading in hover. An induced velocity variation of this form is easily incorporated in the analysis of the rotor behavior in forward flight. At best, though, a linear variation over the disk can only be expected to improve the estimate of the mean and first harmonic quantities, assuming that good values for κ_x and κ_y are available. The actual nonuniform induced velocity distribution in forward flight is more complicated, and the higher harmonics of the inflow can be important.

The classical vortex theory results described in section 5.2.1 give estimates of the factors κ_x and κ_y. Coleman, Feingold, and Stempin (1945) suggested

$$\kappa_x = \tan\frac{\chi}{2} = \sqrt{1 + (\lambda/\mu)^2} - |\lambda/\mu| \qquad (5.42)$$

where χ the skew angle of the wake at the disk. Here κ_x indeed approaches unity for high speed. Drees (1949) gave

$$\kappa_x = \frac{4}{3}\frac{1 - \cos\chi - 1.8\mu^2}{\sin\chi} = \frac{4}{3}\left[(1 - 1.8\mu^2)\sqrt{1 + (\lambda/\mu)^2} - \lambda/\mu\right] \qquad (5.43)$$

$$\kappa_y = -2\mu \qquad (5.44)$$

from which κ_x is zero at $\mu = 0$, has a maximum of about 1.1 at $\mu = 0.2$, and is approximately 1 at $\mu = 0.35$.

White and Blake (1979) developed the inflow distribution based on a horseshoe vortex model of the rotor. Consider a bound vortex of constant strength Γ on the y-axis of the disk, extending from $y = -R$ to $y = R$, together with the trailed vortices from the ends of the bound vortex. Then velocity along the x-axis induced by this horseshoe vortex is

$$w = w_0\left[1 + \frac{x}{\sqrt{x^2 + R^2}} + \frac{R^2}{x\sqrt{x^2 + R^2}}\right] \qquad (5.45)$$

with $w_0 = \Gamma/2\pi R$. The first two terms are from the trailed vortices and the third term from the bound vortex. By including a small core in the bound vortex, the velocity becomes zero at $x = 0$, so w_0 is the induced velocity at the center of the disk. The change of the velocity from the leading edge of the disk ($x = -R$) to the trailing edge ($x = R$) is $(w_{te} - w_{le})/2 = \sqrt{2}w_0$. Extending this result to low speed and hover by introducing a factor $\sin\chi$ gives

$$\kappa_x = \sqrt{2}\sin\chi = \sqrt{2}\frac{\mu}{\sqrt{\mu^2 + \lambda^2}} \qquad (5.46)$$

White and Blake found good agreement with measured flapping data using this expression.

A net aerodynamic moment on the rotor disk also produces an inflow variation. To estimate this inflow, consider the differential form of momentum theory in high speed, $dT = 2\rho V v\, dA$. The local disk loading dT/dA is assumed to have a linear variation over the disk due to the pitch and roll moments: hence the inflow is

$$v = \frac{1}{2\rho V}\left[\frac{T}{A} - 4\frac{M_y}{RA}r\cos\psi + 4\frac{M_x}{RA}r\sin\psi\right] \tag{5.47}$$

or

$$\lambda = \frac{C_T}{2\mu} - \frac{2C_{My}}{\mu}r\cos\psi + \frac{2C_{Mx}}{\mu}r\sin\psi \tag{5.48}$$

where C_{My} and C_{Mx} are the pitch and roll moment coefficients. So $\kappa_x = -4C_{My}/C_T$ and $\kappa_y = 4C_{Mx}/C_T$. Thus the inflow variation is proportional to the offset of the thrust vector from the center of rotation, which can be significant for hingeless rotors.

5.3 Twin Rotor Interference in Forward Flight

The mutual interference of a multi-rotor system can be accounted for by writing the induced velocity at the m-th rotor as

$$v_m = \kappa_m v_{im} + \sum_{n\neq m} x_{mn} v_{in} \tag{5.49}$$

Here v_{in} is the ideal induced velocity for the isolated n-th rotor; κ_m is the correction for the additional induced losses of a real rotor; and x_{mn} is the interference factor at the m-th rotor due to the thrust of the n-th rotor. For a power loss the interference factor x_{mn} is positive, whereas x_{mn} is negative for favorable interference. This expression is applicable to all speeds, including hover, although the interference factors x_{mn} depend on speed. In high-speed forward flight, wing theory gives the induced velocity $v_{in} = T_n/2\rho AV$. The total induced power in forward flight is therefore

$$P = \sum_m T_m v_m = \frac{1}{2\rho AV}\left(\sum_m \kappa_m T_m^2 + \sum_m \sum_{n\neq m} x_{mn} T_m T_n\right) \tag{5.50}$$

(assuming that all the rotors have the same area A, which is usually the case). Since the isolated rotor power is just $P_{\text{isolated}} = \sum \kappa_m T_m^2/2\rho AV$,

$$\frac{P}{P_{\text{isolated}}} = 1 + \frac{\sum_m \sum_{n\neq m} x_{mn} T_m T_n}{\sum \kappa_m T_m^2} \tag{5.51}$$

The second term is the interference power, which is usually a significant positive increment. For some configurations a modest favorable interference is possible. For twin main rotors of equal area, the induced power of the individual rotors is

$$P_1 = \frac{1}{2\rho AV}\left(\kappa_1 T_1^2 + x_{12} T_1 T_2\right) \tag{5.52}$$

$$P_2 = \frac{1}{2\rho AV}\left(\kappa_2 T_2^2 + x_{21} T_2 T_1\right) \tag{5.53}$$

and the total induced power is

$$\frac{P}{P_{\text{isolated}}} = 1 + \frac{x_{12} + x_{21}}{\kappa_1 T_1/T_2 + \kappa_2 T_2/T_1} = 1 + X \tag{5.54}$$

where X is the interference factor for the entire rotor system.

The wing theory of section 5.1.3 was not derived with any assumption about the geometry of the wake surfaces far downstream. For multiple wings, equation 5.30 becomes the sum of the power on the m-th wing due to the wake of the n-th wing:

$$P_i = \sum_m \sum_n P_{mn} = \sum_m \sum_n \int \ell_m v_{mn} dy_m \tag{5.55}$$

from which the induced power factors κ_m and interference factors x_{mn} can be evaluated, given the section loading ℓ_m on all the wings. With the idealization of a rotor as an actuator disk (circular wing), this result applies to a system of rotors in forward flight, with arbitrary loading distributions. The assumptions are high speed (wake angle χ nearly $90°$) and an infinite number of blades.

Wing theory for a single lifting surface shows that the induced power is proportional to the thrust squared divided by the span squared, $P \propto (T/\text{span})^2$. So the total induced power of a multi-rotor system depends on the span of the effective lifting surface. For twin isolated rotors of thrust T and span $2R$, $P = 2(T^2/2\rho AV)$. The same two rotors in coaxial configuration act like a single rotor with twice the span loading; hence the induced power is doubled, or $X \cong 1$.

Wing theory shows that the total induced drag is independent of the longitudinal separation of lifting elements (Munk's stagger theorem). As a result, tandem rotors with no vertical separation have about the same loss as coaxial rotors, $X \cong 1$. The distribution of the loss between the two rotors is the property that changes with longitudinal separation. For the coaxial configuration, the two rotors are identical, so $x_{12} = x_{21} = 1$. For large longitudinal separation, however, the front rotor is not influenced by the rear rotor, while the rear rotor sees the fully developed wake of the front one. Hence for the tandem configuration, $x_{FR} = 0$ and $x_{RF} = 2$ is expected as a limit. As the vertical separation of the tandem or coaxial rotors increases, they approach isolated rotors in forward flight; hence $X < 1$, decreasing to $X = 0$ for a vertical spacing of about one rotor radius. The vertical spacing of the rotor wakes, not the spacing of the disks, determines the interference.

A useful reference power is the ideal power of a single rotor (area A) with the total thrust T of both rotors: $P_{\text{ref}} = T^2/2\rho AV$. Then the total induced power can be written $P_i = CP_{\text{ref}}$. For uniform loading and no vertical or lateral separation of the rotors, $C = 1$; for large separation, $C = \frac{1}{2}$. A different perspective of the interference is given by $P = (1 + X)P_{\text{isolated}}$, where P_{isolated} is the sum of power with the same distribution of thrust between the rotors. For no separation of the rotors, $X = 1$; for large separation $X = 0$.

5.3.1 Tandem and Coaxial Configurations

In forward flight, the biplane effect reduces the induced power of twin rotors at moderate speed, compared to the induced power for no vertical separation. From Munk's stagger theorem, this is true for both tandem and coaxial configurations, as long as the vertical separation is measured in the wake.

5.3 Twin Rotor Interference in Forward Flight

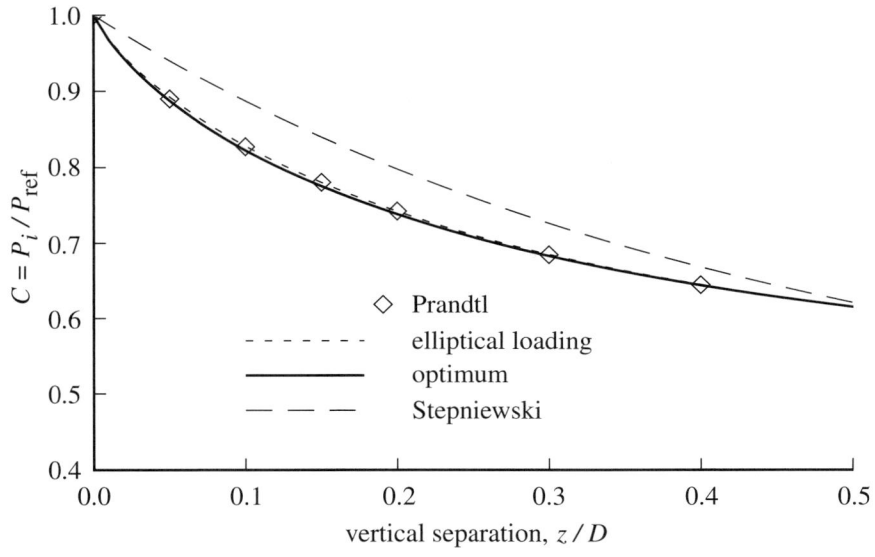

Figure 5.9. Ideal induced power in forward flight for twin rotors.

For the coaxial or tandem configuration, let z/D be the vertical spacing of the two rotors. The wake spacing far downstream is assumed to equal z/D, although the aircraft pitch angle affects the wake spacing with tandem rotors. Figure 5.9 shows the ideal power P_i/P_{ref} as a function of vertical separation, calculated using equation 5.55. The optimum span loading was found numerically by varying the span loading in terms of the series in $\theta = \cos^{-1} 2y/b$ (from symmetry, the loading is the same on the two wings). For comparison, the induced power obtained assuming elliptical loading on each wing is shown (this is the optimum solution for zero and infinite spacing). Also shown in Figure 5.9 is Prandtl's biplane result for elliptical loading (quoted by Glauert (1947)) and Stepniewski's result (derived at the end of this section). The optimum solution is nearly elliptical, with at most $A_3/A_1 = 0.052$ at $z/D = 0.12$. For typical coaxial rotors, the benefit of the vertical spacing is 8 to 20% reduction in induced power (compared to zero spacing), which is a significant effect at low speed, but is overwhelmed by the effect of nonelliptical span loading at high speed.

Consider a momentum theory analysis of the tandem rotor helicopter in forward flight. Assuming that rear rotor wake has no influence on the front rotor, and that the rear rotor is operating in the fully developed wake of the front rotor, the total induced velocities at the front and rear rotors are $v_{\text{front}} = v_F$ and $v_{\text{rear}} = v_R + 2v_F$, respectively, where $v_F = T_F/2\rho AV$ and $v_R = T_R/2\rho AV$. The total induced power is then

$$P = T_F v_F + T_R(v_R + 2v_F) = \frac{T_F^2 + T_R^2 + 2T_F T_R}{2\rho AV} = \frac{(T_F + T_R)^2}{2\rho AV} \tag{5.56}$$

The interference factor is

$$X = \frac{\Delta P}{P_{\text{isolated}}} = \frac{2T_F T_R}{T_F^2 + T_R^2} \tag{5.57}$$

which is $X = 1$ for equal thrust on the two rotors. Unequal thrust changes the interference, but not the total power. This result is for zero vertical separation.

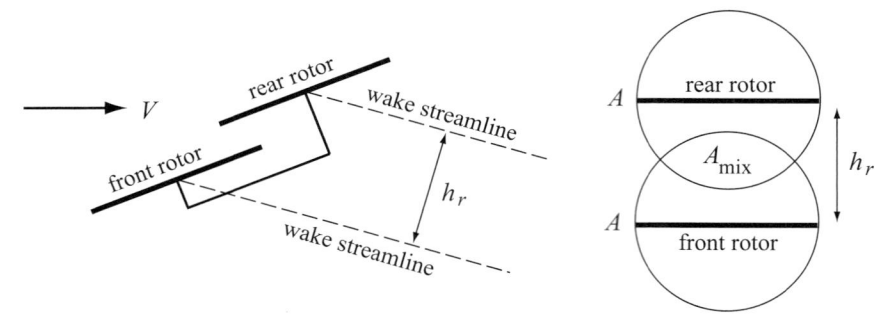

Figure 5.10. Momentum theory analysis of tandem rotors in forward flight.

Stepniewski (1955) developed a momentum theory analysis for tandem rotors in forward flight that includes the effect of vertical separation. Figure 5.10 shows the configuration. Because of the elevation of the rear rotor on a pylon and the forward tilt of the helicopter, the rear rotor wake is a distance h_r above the front rotor wake; typically $h_r \cong 0.3R$ to $0.5R$. The rear rotor sees an interference velocity less than $2v_F$ because of the separation of the wakes, and as a consequence the efficiency of the rotor system is improved. Based on the idea that the wing influences a volume of air contained in a cylinder circumscribing the wing tips (Figure 5.10), Stepniewski proposed that the tandem rotor interference be estimated from the overlap area $A_{\text{mix}} = \tilde{m}A$ of the cylinders about the two wings. The overlap fraction depends on the separation:

$$\tilde{m} = \frac{2}{\pi}\left[\cos^{-1}\frac{h_r}{2R} - \frac{h_r}{2R}\sqrt{1 - \left(\frac{h_r}{2R}\right)^2}\right] \tag{5.58}$$

so $\tilde{m} = 1$ at $h_r = 0$ (corresponding to full interference, $2v_F$) and $\tilde{m} = 0$ at $h_r \geq 2R$ (no interference). Using $v_{\text{rear}} = v_R + 2v_F\tilde{m}$, the total induced power is

$$P = T_F v_F + T_R(v_R + 2v_F\tilde{m}) = \frac{T_F^2 + T_R^2 + 2T_F T_R \tilde{m}}{2\rho AV} \tag{5.59}$$

and the interference factor is

$$X = \frac{\Delta P}{P_{\text{isolated}}} = \frac{2T_F T_R}{T_F^2 + T_R^2}\tilde{m} \tag{5.60}$$

Although this is a simple result, the approach does not give as large an effect as does biplane theory, as shown in Figure 5.9. Stepniewski found that this theory compares well with the measured losses of tandem rotor helicopters in forward flight. From the effective area $A_e = A(2 - \tilde{m})$ follows $C = P_i/P_{\text{ref}} = 1/(2 - \tilde{m})$.

5.3.2 Side-by-Side Configuration

Consider the side-by-side configuration. For zero lateral spacing (coaxial rotors), $X \cong 1$ again. When the shaft spacing is $2R$ (rotor disks just touching), the system is like a single rotor with the same span loading as the two isolated rotors. Hence the total induced power should be reduced by up to a factor of two, or $X \cong -\frac{1}{2}$. This favorable interference is due to each rotor operating in the upwash flow field of the other. The span loading of the side-by-side configuration is far from elliptical for the

5.4 Ducted Fan

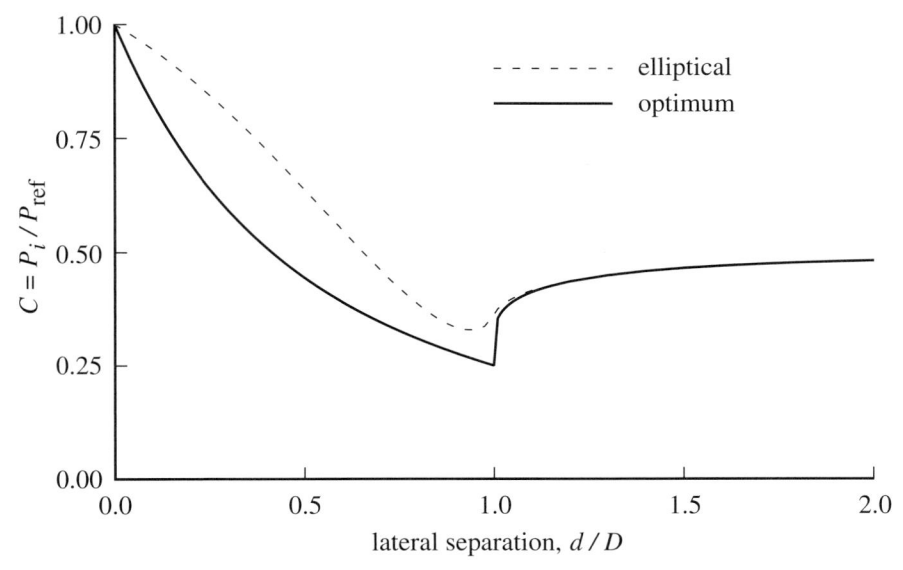

Figure 5.11. Ideal induced power in forward flight for twin rotors.

entire system, however. Thus the actual interference, although still favorable, is not as great as indicated by the span loading. As the lateral separation increases further, X approaches zero again.

Let d/D be the lateral separation of the two rotors; the vertical separation is zero. Figure 5.11 shows the ideal power P_i/P_{ref} as a function of lateral separation, calculated using equation 5.55. The interference factor is $X = 2C - 1$ for this ideal power. The optimum span loading was found numerically for $d/D > 1$ (from symmetry, the loading is the same on the two wings). For $d/D < 1$, the optimum loading is elliptical for the two rotors combined; hence $C = 1/(1 + d/D)^2$; this is not, however, a practical loading for d/D near 1. The result in terms of the isolated rotors is

$$X = \frac{P}{P_{\text{isolated}}} - 1 = \frac{2}{(1 + d/D)^2} - 1 \quad (5.61)$$

For comparison, Figure 5.11 shows the induced power obtained assuming elliptical loading on each wing (this is the optimum solution for zero and infinite spacing). Measured performance data give $X \cong -0.2$ to -0.3 ($C \cong 0.4$ to 0.35) with the disks just touching, and the most favorable interference (at $d/R \cong 1.75$) has $X \cong -0.25$ to -0.45 ($C \cong 0.375$ to 0.275). Thus measured results are close to the solution for elliptical loading in Figure 5.11. The increased effective span of the side-by-side configuration significantly reduces the induced power.

5.4 Ducted Fan

Ducted fans can be used in rotorcraft for primary lift, for auxiliary propulsion, or as an alternative to the tail rotor in the single main-rotor configuration. As an antitorque device, a ducted fan is referred to as a Fenestron™ or fan-in-fin and, being part of the vertical tail, has a rather short duct. The effect of the duct in hover is to keep the wake from contracting and to generate aerodynamic thrust independent of the rotor blades. The aerodynamic force is generated by the duct acting as an airfoil in the entrained flow of the rotor, which guides shaping the duct cross-section. The duct influence on the rotor wake is highly dependent on the gap between the blade tips and the duct surface, a small gap being required for good performance. In edgewise flight of a ducted fan, the flow usually separates on the upstream edge of the duct, leading to complex aerodynamic interactions with the blades. In all operating conditions of hover and forward flight, calculating the aerodynamic loads on the duct is challenging.

Rotor momentum theory can be extended to the case of a ducted fan. Consider a rotor system with disk area A, operating at speed V, with an angle i between V and the disk plane. The induced velocity at the rotor disk is v, and in the far wake it is $w = f_W v$. The far wake area is $A_\infty = A/f_A$. The axial velocity at the fan is $f_{Vz}V_z$, with f_{Vz} accounting for acceleration or deceleration through the duct. The edgewise velocity at the fan is $f_{Vx}V_x$, with $f_{Vx} = 1.0$ for wing-like behavior, or $f_{Vx} = 0$ for tube-like behavior of the flow. The total thrust (rotor plus duct) is T, and the rotor thrust is $T_{\text{rotor}} = f_T T$. For this model, the duct aerodynamics are defined by the thrust ratio f_T or far wake area ratio f_A, plus the fan velocity ratios f_{Vx} and f_{Vz}. The mass flux through the rotor disk is $\dot{m} = \rho A U = \rho A_\infty U_\infty$, where U and U_∞ are, respectively, the total velocity magnitudes at the fan and in the far wake:

$$U^2 = (f_{Vx}V\cos i)^2 + (f_{Vz}V\sin i + v)^2 \qquad (5.62)$$

$$U_\infty^2 = (V\cos i)^2 + (V\sin i + w)^2 \qquad (5.63)$$

Mass conservation ($f_A = A/A_\infty = U_\infty/U$) relates f_A and f_W. Momentum and energy conservation give

$$T = \dot{m}w = \rho A U_\infty w/f_A = \rho A U f_W v \qquad (5.64)$$

$$P = \frac{1}{2}\dot{m}w(2V\sin i + w) = T\left(V\sin i + \frac{w}{2}\right) \qquad (5.65)$$

With these expressions, the span of the lifting system in forward flight is assumed to be equal to the rotor diameter $2R$. Next, the power must equal the rotor induced and parasite power:

$$P = T_{\text{rotor}}(f_{Vz}V\sin i + v) = T f_T(f_{Vz}V\sin i + v) \qquad (5.66)$$

In axial flow, this result can be derived from Bernoulli's equation for the pressure in the wake. In forward flight, any induced drag on the duct is being neglected. From these two expressions for power, $V_z + f_W v/2 = f_T(f_{Vz}V_z + v)$ is obtained, relating f_T and f_W.

For a ducted fan, the thrust C_T is calculated from the total load (rotor plus duct). To define the duct effectiveness, either the thrust ratio $f_T = T_{\text{rotor}}/T$ or the far wake area ratio $f_A = A/A_\infty$ is specified (and the fan velocity ratio f_V). The wake-induced velocity is obtained from the momentum theory result for a ducted fan:

$$\frac{C_T}{2} = \lambda_h^2 = \frac{f_W \lambda_i}{2}\sqrt{(f_{Vx}\mu)^2 + (f_{Vz}\mu_z + \lambda_i)^2} \qquad (5.67)$$

If the thrust ratio f_T is specified, this can be written as

$$f_{Vz}\mu_z + \lambda_i = \frac{\lambda_h^2/f_T}{\sqrt{(f_{Vz}\mu_z + \lambda_i)^2 + (f_{Vx}\mu)^2}} + \frac{\mu_z}{f_T} \qquad (5.68)$$

In this form, λ_i can be determined following the solution of equation 5.9. Then from λ_i the velocity and area ratios are obtained:

$$f_W = 2\left(f_T - (1 - f_T f_{Vz})\frac{\mu_z}{\lambda_i}\right) \tag{5.69}$$

$$f_A = \sqrt{\frac{\mu^2 + (\mu_z + f_W \lambda_i)^2}{(f_{Vx}\mu)^2 + (f_{Vz}\mu_z + \lambda_i)^2}} \tag{5.70}$$

If instead the area ratio f_A is specified, it is simplest to first solve for the far wake velocity $f_W \lambda_i$:

$$\mu_z + f_W \lambda_i = \frac{\lambda_h^2 2 f_A}{\sqrt{(\mu_z + f_W \lambda_i)^2 + \mu^2}} + \mu_z \tag{5.71}$$

In this form, $f_W \lambda_i$ can be determined following the solution of equation 5.9. The induced velocity is

$$(f_{Vz}\mu_z + \lambda_i)^2 = \frac{1}{f_A^2}\left[\mu^2 + (\mu_z + f_W \lambda_i)^2\right] - (f_{Vx}\mu)^2 \tag{5.72}$$

The velocity ratio is $f_W = (f_W \lambda_i)/\lambda_i$, and

$$f_T = \frac{\mu_z + f_W \lambda_i/2}{f_{Vz}\mu_z + \lambda_i} \tag{5.73}$$

is the thrust ratio.

With no duct ($f_T = f_{Vx} = f_{Vz} = 1$), the far wake velocity is always $w = 2v$ so $f_W = 2$, and the far wake area ratio is

$$f_A = \sqrt{\frac{\mu^2 + (\mu_z + 2\lambda_i)^2}{(f_{Vx}\mu)^2 + (f_{Vz}\mu_z + \lambda_i)^2}} \tag{5.74}$$

With an ideal duct ($f_A = f_{Vx} = f_{Vz} = 1$), the far wake velocity is $w = v$ so $f_W = 1$, and the thrust ratio is

$$f_T = \frac{\mu_z + \lambda_i/2}{\mu_z + \lambda_i} \tag{5.75}$$

In hover (with or without a duct), $f_W = f_A = 2f_T$, and the induced velocity is $v = \sqrt{2/f_W}\, v_h$. The rotor ideal induced power is

$$P_{\text{ideal}} = T\frac{w}{2} = \frac{f_W}{2}Tv = \sqrt{\frac{f_W}{2}}Tv_h$$

For a long, constant diameter duct in hover, there is no wake contraction so $f_A = f_W = 1$; hence half the thrust is from the rotor and half from the duct ($f_T = \frac{1}{2}$). Compared to an unducted rotor of the same area and same thrust, the induced velocity at the rotor disk is 41% larger, $v = \sqrt{2}v_h$; but since the rotor has half the thrust, the ideal induced power is 29% smaller, $P = P_h/\sqrt{2}$.

The factors required to use this momentum theory must be obtained from tests or more sophisticated analysis. A higher fidelity analysis is needed to calculate the loads on the duct, especially in the complex flow field of a helicopter in forward flight.

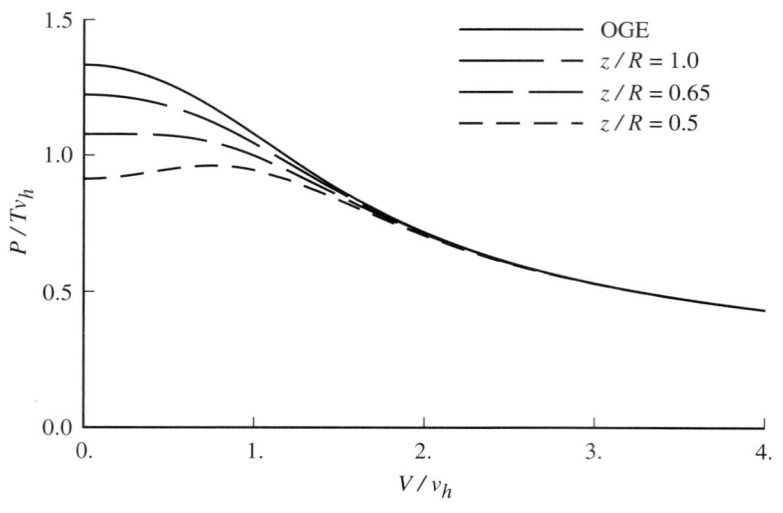

Figure 5.12. Influence of forward speed on ground effect.

5.5 Influence of Ground in Forward Flight

5.5.1 Ground Effect

As discussed in section 4.8, the rotor induced power is decreased by the proximity of the ground, or equivalently the thrust is increased for a given power. In forward flight, where the wake is swept behind the rotor, the effect of the ground diminishes rapidly with forward speed. Ground effect is negligible for speeds above about $V/v_h = 1.5$ to 2, or roughly $\mu = 0.15$. Figure 5.12 illustrates the influence of forward speed on ground effect. In hover the ground proximity significantly reduces the helicopter power required. The effect continues at low speeds but decreases rapidly beyond transition until ground effect is negligible at around 40 knots. A net effect of the ground is to reduce the sensitivity of the power required to changes in speed near hover. Very close to the ground, the power can even increase with speed, if ground effect decreases more rapidly than translational lift develops. The sensitivity of ground effect to the helicopter velocity or, equivalently, to winds can be of considerable importance to helicopter operations.

Cheeseman and Bennett (1955) extended their result for $T/T_\infty = f_g$ in hover to forward flight by measuring the distance to the ground along the rotor wake axis. Hence it is only necessary to use

$$\frac{z}{\cos \chi} = z\sqrt{1 + (\mu/\lambda)^2} \tag{5.76}$$

in place of z in the expressions for f_g of section 4.8. Figure 5.12 was constructed from $P = P_o + \kappa_g \kappa T v$, using equation 4.71 for $\kappa_g = f_g^{-3/2}$, induced velocity v/v_h for disk incidence $i = 0$ and $\kappa = 1.2$, and a profile power estimate from hover figure of merit $M = 0.75$.

5.5.2 Ground Vortex

The rotor downwash is turned outward by the ground. In low-speed forward flight (or hovering in low winds), the outwash in the direction of flight tends to stagnate

5.5 Influence of Ground in Forward Flight

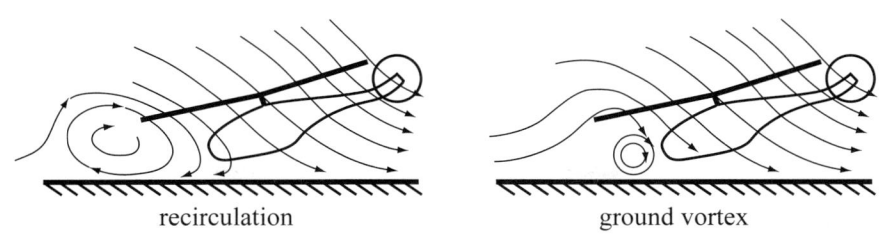

Figure 5.13. Flow regimes in flight near the ground.

at some distance from the rotor, the distance decreasing as forward speed increases. Thus a ground vortex forms in the direction of the flight, the flow swirling with the velocity away from the helicopter at the ground and toward the helicopter above. The ground vortex is a three-dimensional flow feature, a horseshoe vortex oriented relative to the direction of flight or wind. This description of ground vortex and its effects is based on Sheridan and Wiesner (1977), Sheridan (1978), Curtiss, Sun, Putman, and Hanker (1984), Curtiss, Erdman, and Sun (1987), and Cerbe, Reichert, and Curtiss (1988).

Two regimes of flow are observed in visualization studies: recirculation and ground vortex. These flows are illustrated in Figure 5.13 (based on Sheridan (1978), Curtiss, Sun, Putman, and Hanker (1984), and Cerbe, Reichert, and Curtiss (1988)). Recirculation develops at very low speed, as part of the rotor wake flows forward and upward and recirculates through the rotor disk. This is a large-scale flow feature, with some unsteadiness.

As speed increases, the diameter of the recirculation pattern reduces, until the diameter is the order of the height of the rotor above the ground, and the flow structure passes under the leading edge of the rotor disk. At this point a well-defined, concentrated horseshoe vortex has formed on the ground under the rotor. The downwash through the rotor is maximum with the structure just forward of the disk. With the ground vortex under the disk, high upwash is produced at the disk leading edge. Consequently the ground vortex is a source of irregular changes in the rotor hub moment, reflecting variation of the longitudinal inflow gradient with speed and height above the ground.

As speed increases further, the vortex size decreases. Above about $V/v_h = 1.0$ the ground vortex disappears and all of the wake flows downstream. The flow changes from hover-like to wing-like, and super-vortices begin to form from the rotor disk tips.

Experimental evidence for the boundaries of the recirculation and ground vortex regimes is summarized in Figure 5.14 (based on data from Curtiss, Sun, Putman, and Hanker (1984), Sheridan (1978), Empey and Ormiston (1974), and Nathan and Green (2010)). The boundary depends on whether the helicopter is in forward flight (air and ground moving relative to the rotor and its wake) or hovering in a wind (air moving relative to the rotor, wake, and ground), with the ground vortex persisting to higher speeds for the latter condition.

The ground vortex changes the longitudinal gradient of downwash over the disk, so its presence is seen in the lateral cyclic required to trim the rotor (for articulated rotors), or equivalently in hub moments. The increased mean downwash can increase the rotor power, countering ground effect. The flow associated with the ground vortex can also counter the reduction of download near the ground.

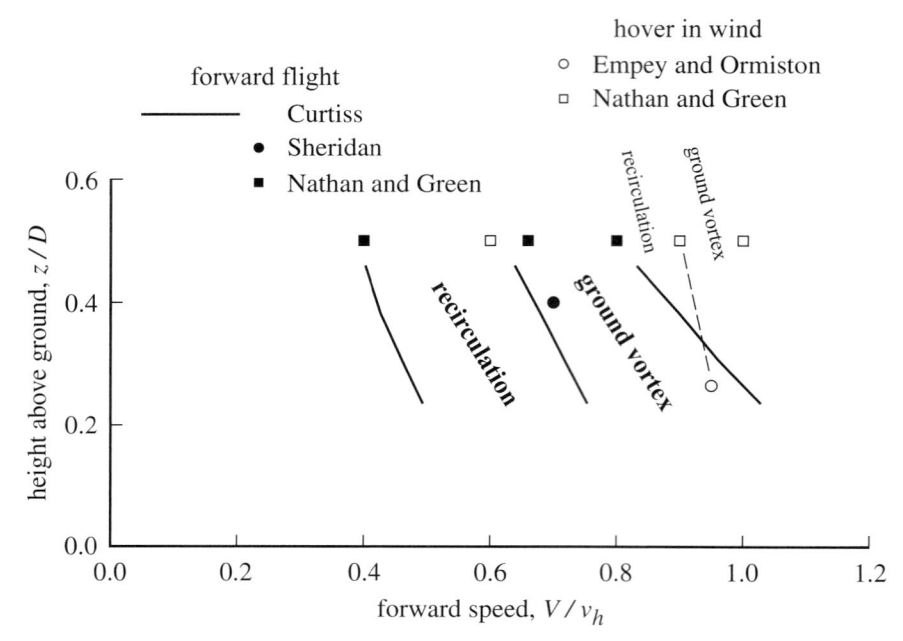

Figure 5.14. Domains of recirculation and ground vortex regimes.

The ground vortex can influence tail rotor power and effectiveness in rearward or sideward flight and also influence engine exhaust reingestion.

5.6 Interference

Mutual aerodynamic interference between the components of a rotorcraft can have a substantial influence on flight dynamics and performance. Often adverse effects of interference are encountered in flight tests, requiring significant resources and time to identify the phenomena and develop corrected designs. A helicopter tail configuration rarely survives unchanged through the developmental flight tests. The interference problems and solutions have been documented for a number of helicopters, including the BK 117 (Huber and Masue (1981)); EH 101 (Mazzucchelli and Wilson (1981)); SA 365N (Roesch and Vuillet (1982)); CH-53E, UH-60A, S-76 (Hansen (1988)); Tiger (Cassier, Weneckers, and Pouradier (1994)); and EC 135 (Hamel and Humpert (1994)).

5.6.1 Rotor-Airframe Interference

The mean static pressure on the airframe is increased because of the energy in the main rotor wake. The rotor wake produces unsteady pressures on the airframe surface, largely periodic at the blade passage frequency N/rev, with fluctuations much larger than the mean interaction. There is a large pressure pulse as the blade passes over the airframe (from the bound circulation). The rotor tip vortices produce low pressure peaks as the vortex impinges on the surface, with viscous effects on the vortex as it passes around the airframe. The blade-passage effect is greatly reduced by increasing the rotor-airframe separation, although the tip vortex effect is strong at long distances in the wake.

The lateral asymmetry of main rotor wake produces a side load on the tail boom, requiring a larger anti-torque force. A strake can be used to reduce the side force, at the cost of increased drag. On the EH 101, the tail boom side force in sideward flight is reduced by a strake on the top, to force separation of the flow. The EH 101 reduces tail boom download by using skirts on the lower edge of the tail boom.

The rotor wake geometry is strongly affected by the airframe, particularly at low speed. The airframe's influence on the main rotor in forward flight can be significant, either favorable or adverse.

5.6.2 Tail Design

The horizontal tail of a helicopter contributes to static and dynamic longitudinal stability and to aircraft trim in cruise. The horizontal tail can be positioned to guard the tail rotor, thereby improving safety. Typically trim requires downward lift on the tail. An inverted cambered airfoil or a Gurney flap is used to improve maximum tail lift. A large tail produces significant hover download. Interference of the main rotor wake on the horizontal tail can produce a pitch attitude change in transition from hover to forward flight.

The vertical tail of a helicopter contributes to static stability in yaw and to lateral-directional dynamic stability. The tail typically cancels the unstable yaw moment from the fuselage, and positive static stability comes from the tail rotor. The vertical tail can provide the anti-torque force in cruise, unloading the tail rotor and thereby reducing tail rotor power. The vertical tail could provide anti-torque force at low speed in the event of tail rotor damage by operating with a large sideslip angle (on the order of 20°). A requirement to fly home after the loss of tail rotor thrust is not often achieved, due to competing requirements. The fin area is limited by tail rotor blockage in hover and by the side force on the tail in sideward flight. Accommodating a tail rotor drive shaft within the vertical tail requires a thick section.

The fin yaw moment is typically nonlinear for small sideslip, from partial shielding of the fin by the fuselage and hub. The dynamic pressure loss inside the wake reduces the tail effectiveness. Also, airfoils with large trailing-edge angles can have an unstable lift-curve slope at small angle-of-attack. Changes to the trailing edge, such as a Gurney flap or double trailing-edge strip, can improve the vertical tail separation characteristics and reduce the nonlinear behavior of the lift with angle-of-attack. Vertical tail area can be increased using a ventral fin or end plates on the horizontal tail. A ventral fin (below the tail boom) is clear of the fuselage wake in descent.

Static and dynamic stability problems are often encountered because of the poor aerodynamic environment of the tail surfaces, which operate in the wake created by the fuselage and hub.

5.6.3 Rotor Interference on Horizontal Tail

Impingement of the main rotor wake on the horizontal tail can produce a pitch up of several degrees from hover to about 40 knots. A rapid forward displacement of the cyclic stick is required to maintain trim as speed increases. A number of design features have been used to counter this pitch-up phenomenon. A variable-incidence stabilizer can be used, although it increases weight and complexity. The horizontal tail can be located on the tail boom, so it is inside the rotor wake even in hover;

but then the tail has a short moment arm and contributes download in hover. The horizontal tail can be located high (on the vertical tail), as high as the main rotor plane to keep the tail out of the wake; yet a high tail is structurally inefficient and introduces additional vibration modes.

On the CH-53E, UH-60A, and S-76 the initial location of the horizontal tail was at the base of the vertical fin. This location reduced wake-induced tail shake associated with high tails and improved safety by guarding the tail rotor. Low-speed, nose-up pitch was experienced due to impingement of the rotor wake on the tail. On the CH-53E, the horizontal tail was moved to the top of the vertical fin, out of the main rotor wake. On the UH-60A a stabilator (moving stabilizer) was designed, so the incidence could be changed with airspeed to align the surface with the rotor flow. Typically the incidence was $40°$ at low speed, and $\pm 8°$ (depending on collective) at high speed; the incidence increased with collective in high speed to decouple pitch response to collective. Pitch rate feedback to the tail was used to improve handling qualities. On the S-76 a fixed-incidence tail was acceptable, because of less tail area and no tail rotor cant.

The Tiger helicopter started with a large tail with end plates, mounted aft and low. Significant nose-down variation of the pitch attitude during transition was encountered, due to the main rotor downwash on the tail, with a corresponding large longitudinal cyclic stick displacement. A forward tail position put the tail always in the main rotor wake during hover and transition, but the helicopter was unstable in IGE hover, download was increased, and the tail rotor was unguarded. The tail area was reduced to 51% of the original, with spoilers and end plates. Then the pitch and longitudinal cyclic variations were acceptable, and the aircraft had sufficient static stability.

The EH 101 encountered the pitch-up phenomenon: an unacceptably high pitch increase as speed increased from hover ($5°$ over 15 to 35 knots), from the influence of the main rotor wake on the horizontal tail. The initial tail configuration was symmetrical and mounted low. A high location was best for pitch up and dynamic stability, but was not good for folding. Most of the interference in low speed (75% of the download) was on the port side. With a single-sided, high aspect-ratio horizontal tail, mounted low, the pitch up was acceptable (less than $2°$).

5.6.4 Pylon and Hub Interference on Tail

The wake of the main rotor pylon and hub reduces the tail effectiveness. The dynamic pressure at the tail is reduced (by 40–50%), and the flow is unsteady. The change in the flow direction at the tail is reduced for aircraft angle-of-attack and sideslip angle variations. The oscillatory flow results in structural vibration. These effects are present in most helicopters, with varying intensity.

Reducing the drag of the hub and pylon generally reduces the dynamic pressure deficit and the unsteadiness. A fairing between the top of the fuselage and rotor hub or one on top of the hub (a hub cap) can help. Such fairings generate vortices that reduce the dynamic pressure deficit and unsteadiness and move the interfering wake downward.

The SA 365N encountered tail shake in high-speed descent: a structural vibration due to intermittent tail loads, caused by unsteady wake from the main rotor hub. In descent this wake struck high on the vertical tail. A pylon fairing was designed (in wind tunnel tests) to depress the wake downward and attenuate the turbulence by creating tip vortices at the edge of the fairing top and by contouring the sides. A hollow cap on the rotor head was used to depress the hub wake. The hub cap forms a cambered flow vane and creates vortices that disperse before reaching the tail. The hub cap depresses the center of the wake, and pylon reduces the dynamic pressure loss. The fairing reduced the aircraft drag by 0.15 m^2. There was an adverse effect on directional stability, countered by increasing the vertical tail area.

The EC 135 encountered tail shake. There was a significant reduction in dynamic pressure at the tail. The tail shake depended on speed and climb rate, reaching a maximum in descent at moderate speed. A hub cap reduced the intensity of the shake. An improved pylon and hub cap were designed through wind-tunnel tests. Flight test found the tail shake to be almost eliminated.

The Tiger helicopter had upper engine cowlings shaped by drag, weight, and access considerations. A low-frequency lateral vibration was encountered, from erratic aerodynamic excitation of lateral bending modes by vortices from the upper cowling. A hub cap deflected the wake down, but was not sufficient. Through wind-tunnel tests the upper cowling was reshaped.

The BK 117 encountered tail shake (manifested as random lateral vibrations in the cabin), which was worst at high speed and a moderate descent rate. There was a 50% reduction in dynamic pressure at the tail, during descent centered on the vertical tail. A hub cap was added, and the aft fuselage spoiler was removed. The hub cap increased the aircraft drag.

The EH 101 encountered a shuffle: random vibration resulting from aerodynamic excitation of the lateral fuselage bending by the wake shed from the main rotor head and cowlings, and the wake striking the vertical fin and tail rotor. Drag reduction of the hub and cowlings reduced the flow unsteadiness at the tail. The initial design solution consisted of a large beanie (to deflect the flow downward), a fairing of the blade tension link and arm, a horse collar and forward pylon fairing, and a nose-up tail plane setting. This solution was not good in terms of production, ship stowage, and blade fold. The final design solution consisted of a small beanie, tension link fairings, horse collar, and extended engine cowling.

5.6.5 Tail Rotor

Tail rotor design and operation are strongly influenced by aerodynamic interactions, particularly with the main rotor wake and the vertical tail. See Lynn, Robinson, Batra, and Duhon (1970) and Cook (1978).

Blockage of the tail rotor flow by the vertical fin reduces the tail rotor thrust. For the tractor configuration, with the fin in the wake, the thrust loss is approximately $T_{\text{loss}}/T = -0.75(S/A)$, typically 10 to 20%. Here S/A is the ratio of the blocked tail rotor disk area to the total area. There is only a small dependence on the separation between the fin and tail rotor. For the pusher configuration, with the fin on the inflow side, the thrust loss is about the same as for the tractor at small separations, but reduces as the separation increases, down to about 2% for separation $d = 0.6R$. There is some reduction in rotor efficiency due to the fin blockage. Hence pusher is the preferred configuration. The fin can be canted to increase the separation between the fin and tail rotor. This interaction is not a factor for the fan-in-fin.

On the CH-53E, the sideward flight requirement was not met because of vertical tail interference on the tail rotor. Removing the fin trailing edge was necessary, although the resulting reduction in camber reduced the vertical tail effectiveness.

On the UH-60A, the fin trailing edge was removed to reduce tail rotor power in hover and climb and to reduce drag. On the EH 101, truncation of the trailing edge was used to reduce the tail rotor power, but the loss of vertical tail area was not acceptable, so a modified trailing edge and better tail rotor fairing were used instead. The fin was also canted to increase clearance between the fin and tail rotor for reduced blockage in hover.

Top-aft is the preferred rotation direction for the tail rotor to minimize adverse interaction with the main rotor wake. The tail rotor effectiveness is reduced if it rotates in the same direction as swirl in the flow field. In very low speed flight, the main rotor downwash at the edge of the disk is downward at the front of the tail rotor, producing a local reduction in tail rotor loading. During rearward or sideward IGE flight, typically at 15 to 25 knots and heights less than $z/D = 0.3$, the tail rotor can be operating in the ground vortex. The flow in the ground vortex swirls in the top-forward direction. During sideward flight, the tail rotor can be immersed in the super-vortices (rolled-up tip vortices) that emanate from the edges of the main rotor disk in forward flight. This is the most significant interaction of the main rotor and tail rotor, typically encountered in quartering winds (40° to 90° from left or right) at 20 to 35 knots. The swirl in these super-vortices is in the top-forward direction. Thus with the tail rotor rotating top-forward, the helicopter has poor directional control characteristics, particularly in sideward flight (to the left for counter-clockwise rotation of the main rotor), either OGE or IGE. There is a loss of pedal effectiveness, requiring up to 50% more tail rotor thrust capability to compensate. With the tail rotor rotating top-aft, the directional control characteristics are acceptable. There is also for top-aft rotation a beneficial influence of the main rotor on tail rotor VRS encounter in sideward flight.

The main rotor tip vortices encountering the tail rotor produce non-harmonic loading on the tail rotor, which is a source of structural loads, vibration, and noise. The tip vortex encounters with tail rotor blades at the top of the tail rotor disk occur at increased relative velocity if the rotation is top-forward. Thus top-aft rotation of the tail rotor minimizes the interaction noise; see Leverton (1982).

The tail rotor and fin can adversely affect the main rotor, increasing the main rotor hover OGE power by up to about 3%. The loss is more for the tractor configuration than for the pusher configuration. The tail rotor can also increase main rotor noise.

5.7 REFERENCES

Ashley, H., and Landahl, M. *Aerodynamics of Wings and Bodies.* Reading, MA: Addison-Wesley Publishing Company, Inc., 1965.

Cassier, A., Weneckers, R., and Pouradier, J.-M. "Aerodynamic Development of the Tiger Helicopter." American Helicopter Society 50th Annual Forum, Washington, DC, May 1994.

Castles, W., Jr., and De Leeuw, J.H. "The Normal Component of the Induced Velocity in the Vicinity of a Lifting Rotor and Some Examples of Its Application." NACA Report 1184, 1954.

Cerbe, T., Reichert, G., and Curtiss, H.C., Jr. "Influence of Ground Effect on Helicopter Takeoff and Landing Performance." Fourteenth European Rotorcraft Forum, Milan, Italy, September 1988.

Cheeseman, I.C., and Bennett, W.E. "The Effect of the Ground on a Helicopter Rotor in Forward Flight." ARC R&M 3021, September 1955.

5.7 References

Coleman, R.P., Feingold, A.M., and Stempin, C.W. "Evaluation of the Induced-Velocity Field of an Idealized Helicopter Rotor." NACA ARR L5E10, June 1945.

Cook, C.V. "A Review of Tail Rotor Design and Performance." Vertica, *2*:3/4 (1978).

Curtiss, H.C., Jr., Erdman, W., and Sun, M. "Ground Effect Aerodynamics." Vertica, *11*:1/2 (1987).

Curtiss, H.C., Jr., Sun, M., Putman, W.F., and Hanker, E.J., Jr. "Rotor Aerodynamics in Ground Effect at Low Advance Ratios." Journal of the American Helicopter Society, *29*:1 (January 1984).

Drees, J.M. "A Theory of Airflow Through Rotors and Its Application to Some Helicopter Problems." Journal of the Helicopter Association of Great Britain, *3*:2 (July-August-September 1949).

Empey, R.W., and Ormiston, R.A. "Tail-Rotor Thrust on a 5.5-Foot Helicopter Model in Ground Effect." American Helicopter Society 30th Annual National Forum, Washington, DC, May 1974.

Glauert, H. "A General Theory of the Autogyro." ARC R&M 1111, November 1926.

Glauert, H. *Elements of Aerofoil and Airscrew Theory*. Cambridge: Cambridge University Press, 1947.

Hamel, D., and Humpert, A. "Eurocopter EC135 Initial Flight Test Results." Twentieth European Rotorcraft Forum, Amsterdam, Netherlands, October 1994.

Hansen, K.C. "Handling Qualities Design and Development of the CH-53E, UH-60A, and S-76." Royal Aeronautical Society International Conference on Helicopter Handling Qualities and Control, London, UK, November 1988.

Heyson, H.H. "A Note on the Mean Value of Induced Velocity for a Helicopter Rotor." NASA TN D-240, May 1960.

Huber, H., and Masue, T. "Flight Characteristics Design and Development of the MBB/KHI BK 117 Helicopter." Seventh European Rotorcraft and Powered Lift Aircraft Forum, Garmisch-Partenkirchen, Germany, 1981.

Johnson, W. "Model for Vortex Ring State Influence on Rotorcraft Flight Dynamics." NASA TP 2005-213477, December 2005.

Leverton, J.W. "Reduction of Helicopter Noise by Use of a Quiet Tail Rotor." Vertica, *6*:1 (1982).

Lynn, R.R., Robinson, F.D., Batra, N.N., and Duhon, J.M. "Tail Rotor Design. Part I: Aerodynamics." Journal of the American Helicopter Society, *15*:4 (October 1970).

Mazzucchelli, C., and Wilson, F.T. "The Achievement of Aerodynamic Goals on the EH101 Project Through the "Single Site" Concept." Seventeenth European Rotorcraft Forum, Berlin, Germany, September 1991.

Nathan, N.D., and Green, R.B. "Wind Tunnel Investigation of Flow around a Rotor in Ground Effect." American Helicopter Society Specialists' Conference on Aeromechanics, San Francisco, CA, January 2010.

Prouty, R.W. "A Second Approximation to the Induced Drag of a Rotor in Forward Flight." Journal of the American Helicopter Society, *21*:3 (July 1976).

Roesch, P., and Vuillet, A. "New Designs for Improved Aerodynamic Stability on Recent Aerospatiale Helicopters." Vertica, *6*:3 (1982).

Sheridan, P.F. "Interactional Aerodynamics of the Single Rotor Helicopter Configuration." USARTL TR 78-23A, September 1978.

Sheridan, P.F., and Wiesner, W. "Aerodynamics of Helicopter Flight Near the Ground." American Helicopter Society 33rd Annual National Forum, Washington, DC, May 1977.

Stepniewski, W.Z. *Introduction to Helicopter Aerodynamics*. Morton, PA: Rotorcraft Publishing Committee, 1955.

White, F., and Blake, B.B. "Improved Method of Predicting Helicopter Control Response and Gust Sensitivity." American Helicopter Society 35th Annual Forum, Washington, DC., May 1979.

6 Forward Flight

6.1 The Helicopter Rotor in Forward Flight

Efficient hover capability is the fundamental characteristic of the helicopter, but without good forward flight performance the ability to hover has little value. During translational flight of the helicopter, the rotor disk is moving edgewise through the air, remaining nearly horizontal, generally with a small forward tilt to provide the propulsive force for the aircraft. A tiltrotor cruises with the rotors tilted to operate as propellers. A compound helicopter reduces the lift and propulsive force required of the rotor. Yet all rotorcraft configurations execute low-speed forward flight with the flapping rotor in edgewise flow, which is the subject of this chapter.

Thus in forward flight the rotor blade sees both a component of the helicopter forward velocity and the velocity due to its own rotation. On the advancing side of the disk the velocity of the blade is increased by the forward speed, whereas on the retreating side the velocity is decreased. For a constant angle-of-attack of the blade, the varying dynamic pressure of the rotor aerodynamic environment in forward flight would tend to produce more lift on the advancing side than on the retreating side; that is, a rolling moment on the rotor. If nothing were done to counter this moment, the helicopter would respond by rolling toward the retreating side of the rotor until equilibrium was achieved, with the rotor moment balanced by the gravitational force acting at the helicopter center-of-gravity. The rotor moment could possibly be so large that an equilibrium roll angle would not be achieved. A number of crashes of early helicopter and autogyro designs as they attempted forward flight were due to this phenomenon. In addition, the rolling moment on the rotor disk corresponds to a large bending moment at the blade root that oscillates once per revolution, from maximum positive on the advancing side to maximum negative on the retreating side.

Since the rotor blade loading (T/A_blade) is limited by stall of the airfoil sections, for a given thrust (and tip speed) the rotary wing tends to have about the same blade area regardless of the rotor diameter. The low disk loading (T/A_rotor) helicopter rotor has low solidity $\sigma = A_\text{blade}/A_\text{rotor}$ and thus high aspect-ratio blades. The high aspect-ratio, thin blades required for aerodynamic efficiency limit the structural load-bearing capability at the root, and as a consequence the 1/rev loads due to forward flight are a severe problem. Some means is required to alleviate the root bending-moments and reduce the blade stresses to an acceptable level. With stiff

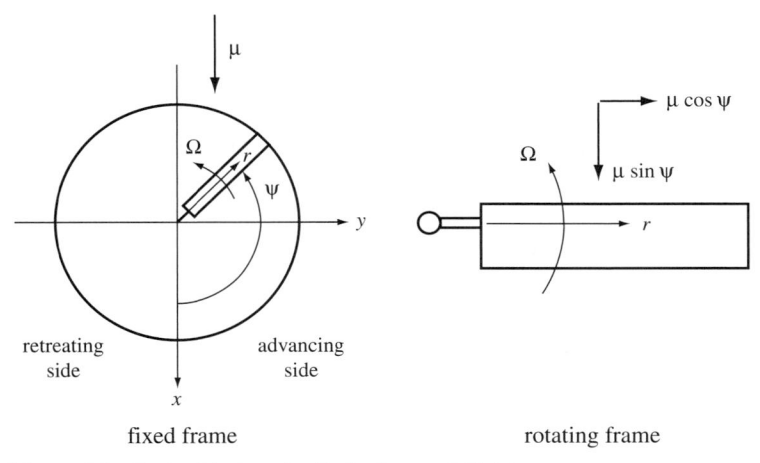

Figure 6.1. Rotor blade velocity in forward flight.

blades, such as on propellers, the structure must absorb all of the aerodynamic loads. In contrast, flexible blades respond to the aerodynamic forces with considerable bending motion, so the blade loads can be countered by the aerodynamic forces due to blade motion rather than by structural forces. Hence in response to the lateral aerodynamic moment in forward flight there is a 1/rev motion of the blades out of the plane of the disk, called flapping motion. When the inertial and aerodynamic forces due to this flapping motion are accounted for, the net blade loads at the root and the rolling moment on the helicopter are small.

The conventional approach has been to use a flap hinge at the blade root, about which the blade can rotate as a rigid body to produce the flap motion (see Figure 1.3). Since the moment at the flap hinge must be zero, no hub moment at all can be transmitted to the helicopter (unless the hinge is offset from the center of rotation), and the bending moment throughout the blade root must be low. A rotor with mechanical flap hinges is called an articulated rotor. Since the 1960s, there have been successful helicopter designs without flap hinges, which are called hingeless rotors. With modern materials, the blade root can be strong while still flexible enough to provide the flapping motion necessary to eliminate most of the root loads. Because of the large centrifugal forces on the blade, the flap motion of hingeless rotors is in fact similar to that of articulated rotors. The root loads of a hingeless rotor are naturally higher than those of articulated rotors, and the increased hub moments have a significant effect on the helicopter handling qualities. Regardless of the blade root design, the flap motion of the helicopter blades has the effect of reducing the asymmetry of the rotor lift distribution in forward flight. Thus the flap motion is a principal concern of the analysis of the forward flight performance of the rotor.

6.1.1 Velocity

Let us examine the velocity components seen by the rotating blades in forward flight (Figure 6.1). The helicopter has forward velocity V and disk incidence angle i (positive for forward tilt of the rotor). The rotor rotates with speed Ω. In this book the rotation direction is counter-clockwise when viewed from above (as in Figure 6.1), so the advancing side of the disk is to the right (starboard). The fixed-frame coordinates

x, y, and z are aft, right, and up, respectively, with origin at the center of rotation of the rotor. The component of the helicopter velocity in the plane of the rotor disk is $V \cos i$. Define the rotor advance ratio as the in-plane forward velocity component normalized by the rotor tip speed:

$$\mu = \frac{V \cos i}{\Omega R} \quad (6.1)$$

Thus μ is the dimensionless forward speed of the rotor. The blade position is given by the azimuth angle $\psi = \Omega t$, measured from downstream. In a frame rotating with the rotor blade, the tangential component of the velocity seen by the blade is $(\Omega r + V \cos i \sin \psi)$, and the radial component is $(V \cos i \cos \psi)$. The dimensionless velocity components in the rotating frame are thus

$$u_T = r + \mu \sin \psi \quad (6.2)$$

$$u_R = \mu \cos \psi \quad (6.3)$$

The 1/rev variation of the tangential velocity u_T has a major influence on the aerodynamics of the rotor in forward flight. The advance ratio μ is small for typical helicopter cruise speeds. Early designs had a maximum speed corresponding to $\mu_{\max} \cong 0.25$, whereas current helicopter designs have perhaps $\mu_{\max} = 0.35$ to 0.40. For a tip speed of $\Omega R = 675$ ft/sec, an advance ratio of $\mu = 0.5$ corresponds to about $V = 200$ knots.

A phenomenon introduced by forward flight is the reverse flow region, an area on the retreating side of the rotor disk where the velocity relative to the blade is directed from the trailing edge to the leading edge. The forward velocity component ($\Omega R \mu \sin \psi$) is negative on the retreating side of the rotor ($\psi = 180°$ to $360°$), whereas the rotational velocity Ωr is positive and linearly increasing along the blade. Consequently, there is always a region at the blade root where the rotational velocity is smaller in magnitude than the forward speed component, so that the flow is reversed. Specifically, for $\psi = 270°$ the total velocity is $\Omega R(r - \mu)$, and the flow is reversed for blade stations inboard of $r = \mu$. In general, the reverse flow region is defined as the area of the disk where $u_T < 0$, which has the boundary $(r + \mu \sin \psi) = 0$. The reverse flow boundary is a circle of diameter μ, centered at $r = \mu/2$ on the $\psi = 270°$ radial on the retreating side (Figure 6.2). When $\mu \geq 1$, the reverse flow region includes the entire blade at $\psi = 270°$ and has a significant impact on the rotor aerodynamics. An advance ratio of $\mu = 0.35$ to 0.4 is more typical of current helicopter forward speeds. For low advance ratio, the reverse flow region occupies only a small portion of the disk (the ratio of the reverse flow area to the total disk area is $\mu^2/4$). Moreover, since by definition $u_T = 0$ at the boundary, the entire reverse flow region is characterized by low dynamic pressure until the advance ratio gets large. The root cutout, extending to typically 10% to 25% of the rotor radius, covers much of the reverse flow region. Thus the effects of the reverse flow region are negligible up to an advance ratio of about $\mu = 0.5$.

6.1.2 Blade Motion

The asymmetry of the aerodynamic environment in forward flight, which is due to the combination of the forward velocity and rotor rotation, means that the blade loads and motion depend on the azimuthal position ψ. For steady-state conditions, the behavior of the blade as it revolves must always be the same at a given azimuth,

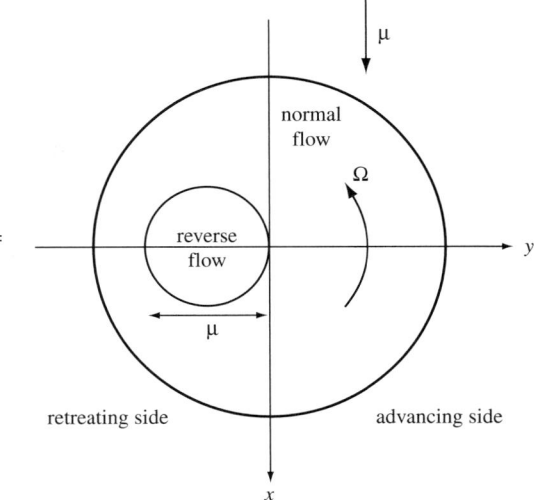

Figure 6.2. Reverse flow region (shown for $\mu = 0.7$).

which implies that the blade loads and motion are periodic around the azimuth, with period 2π. In dimensional terms, the rotor blade behavior is periodic with a fundamental frequency equal to the rotor speed Ω and a period $T = 2\pi/\Omega$. Periodic functions can be represented by a Fourier series. For example, the flap angle β can be written

$$\beta(\psi) = \beta_0 + \beta_{1c} \cos\psi + \beta_{1s} \sin\psi + \beta_{2c} \cos 2\psi + \beta_{2s} \sin 2\psi + \ldots$$

$$= \beta_0 + \sum_{n=1}^{\infty} \left(\beta_{nc} \cos n\psi + \beta_{ns} \sin n\psi \right) \tag{6.4}$$

The periodic function $\beta(\psi)$ is then defined by the harmonics β_0, β_{1c}, β_{1s}, and so on. Generally only the lowest few harmonics are required to adequately describe the rotor motion, so the complete time behavior is described by a small number of parameters. The Fourier coefficients, or harmonics, are obtained from $\beta(\psi)$ as follows:

$$\beta_0 = \frac{1}{2\pi} \int_0^{2\pi} \beta \, d\psi \tag{6.5}$$

$$\beta_{nc} = \frac{1}{\pi} \int_0^{2\pi} \beta \cos n\psi \, d\psi \tag{6.6}$$

$$\beta_{ns} = \frac{1}{\pi} \int_0^{2\pi} \beta \sin n\psi \, d\psi \tag{6.7}$$

The motion of the blade degrees of freedom is described by differential equations, which must be solved for the periodic motion in the rotating frame. One method of solution is the substitutional method, which consists of the following steps. The Fourier series representations of the degrees of freedom and their time derivatives are substituted into the equations of motion. Products of harmonics are reduced to sums of harmonics by means of trigonometric relations. All the terms in the equation for a given harmonic are then collected, and the coefficient of each harmonic (1, $\cos\psi$, $\sin\psi$, $\cos 2\psi$, $\sin 2\psi$, etc.) is set to zero. The result is a set of algebraic

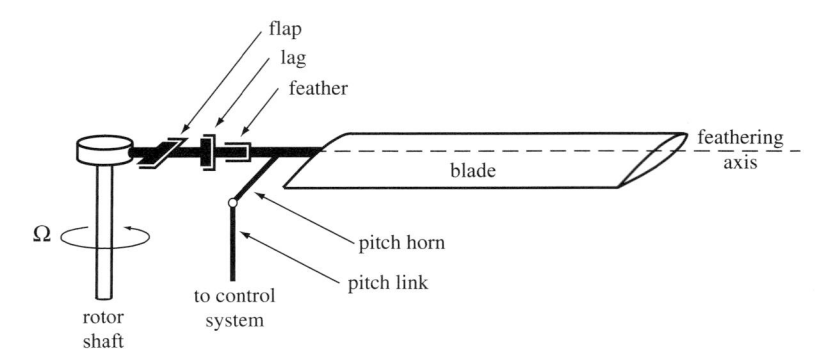

Figure 6.3. Schematic of the flap hinge, lag hinge, and pitch bearing at the hub of an articulated blade.

equations for the harmonics of the blade motion. An alternative approach is the operational method, in which each of the operators

$$\frac{1}{2\pi}\int_0^{2\pi}(\ldots)\,d\psi \tag{6.8}$$

$$\frac{1}{\pi}\int_0^{2\pi}(\ldots)\cos\psi\,d\psi \qquad \frac{1}{\pi}\int_0^{2\pi}(\ldots)\sin\psi\,d\psi \tag{6.9}$$

$$\frac{1}{\pi}\int_0^{2\pi}(\ldots)\cos 2\psi\,d\psi \qquad \frac{1}{\pi}\int_0^{2\pi}(\ldots)\sin 2\psi\,d\psi \tag{6.10}$$

and so on, is applied to the differential equations of motion. Then the definitions of the harmonics are used to replace the integrals of the blade motion by the Fourier coefficients. The result is the same set of algebraic equations as obtained by the substitutional method, although the operational method obtains the equations one at a time. Linear differential equations reduce to linear algebraic equations for the harmonics. The solution for the blade motion is necessarily approximate, because the Fourier series must be truncated to obtain a finite set of equations.

The flapping hinge is needed to alleviate the root stresses and hub moments by allowing out-of-plane motion of the blade. The flap motion also introduces aerodynamic and inertial forces, particularly Coriolis forces, in the plane of the rotor disk. Therefore, a lag hinge is also frequently used to alleviate the chordwise root loads by allowing in-plane motion of the blade. The lag hinge increases the mechanical complexity of the hub and introduces the possibility of a mechanical instability called ground resonance, which requires a mechanical lag damper to stabilize the motion. Ground resonance involves the coupled motion of the blades about the lag hinges and the in-plane motion of the rotor hub, which is usually due to the flexibility of the landing gear when the helicopter is on the ground; see section 20.3. The alternative is to make the blade root strong and heavy enough to take the in-plane loads without a lag hinge. A pitch bearing or hinge is also required at the blade root to allow the blade pitch angle to be changed in response to control inputs. Thus the blade of a fully articulated rotor has three hinges at the root: flap, lag, and feather (sketched in Figure 6.3). The motion about the flap and lag hinges is restrained by centrifugal forces when the blade is rotating, whereas the motion about the feathering hinge is restrained by the control system. The phrase "rotor blade hinges" usually means

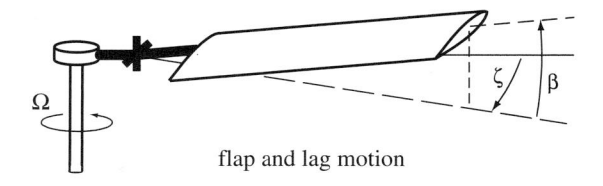

Figure 6.4. Rotor blade motion.

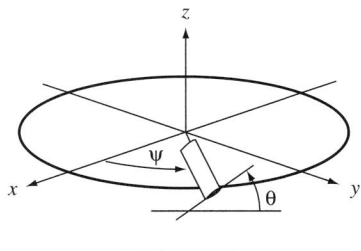

the flap and lag hinges; specifically, a hingeless rotor does have a pitch bearing. Mechanical considerations for fully articulated rotors require that the flap and lag hinges be offset slightly from the center of rotation. An offset of the lag hinge is necessary in any case, or transmitting torque from the shaft to the blades in order to turn the rotor would not be possible. A flap-hinge offset improves the helicopter handling qualities by allowing some pitch and roll moments to be transmitted to the helicopter. In teetering or gimballed rotors the flap hinge is at the center of rotation, and these rotors are designed without lag hinges. With hingeless rotors the flap and lag motion is primarily due to bending at the blade root; such blades can be considered as roughly equivalent to articulated rotors with large hinge offsets.

The basic blade motion is represented by the flap, lag, and pitch degrees of freedom (Figure 6.4). The out-of-plane or flap motion is due to rigid-body rotation about the flap hinge by the angle β (positive upward). The in-plane or lag motion is due to rotation about the lag hinge by the angle ζ (positive aft, opposite the direction of rotor rotation). Finally, the blade pitch or feather motion is due to rotation about the feathering axis by the angle θ (positive when nose up). The flap and pitch angles are measured from a disk reference plane. The various reference planes used in rotor analyses are discussed in section 6.1.3. The steady-state flap motion is described by a Fourier series:

$$\beta(\psi) = \beta_0 + \beta_{1c} \cos\psi + \beta_{1s} \sin\psi + \beta_{2c} \cos 2\psi + \beta_{2s} \sin 2\psi + \dots \quad (6.11)$$

Let us examine what these harmonics imply in terms of the rotor motion as viewed in the fixed frame (Figure 6.5). The zero-th harmonic or mean value β_0 is the coning angle. When $\beta = \beta_0$, the flap motion is independent of ψ, and therefore the blades describe a cone as they rotate. The blade tips describe a circle that lies in a plane parallel to the reference plane. The first harmonic β_{1c} generates a once-per-revolution variation of the flap angle, $\beta = \beta_{1c} \cos\psi$. The blade out-of-plane deflection is then $z = r\beta = r\beta_{1c} \cos\psi = x\beta_{1c}$. Thus as they rotate the blades describe a plane tilted forward about the lateral (y) axis by the angle β_{1c} relative to the reference plane. Similarly, the first harmonic β_{1s} generates an out-of-plane deflection of $z = r\beta = r\beta_{1s} \sin\psi = y\beta_{1s}$, which corresponds to a plane tilted to the left (toward the retreating side) about the longitudinal axis by the angle β_{1s} relative to the reference

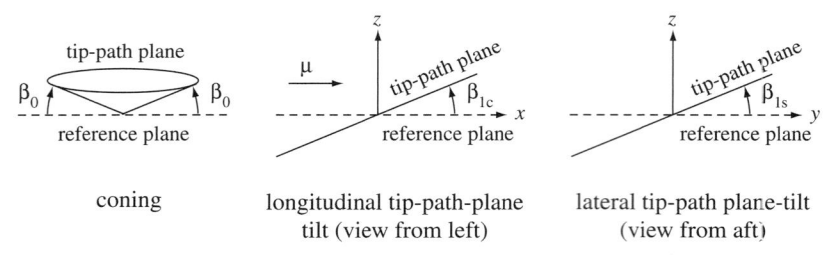

Figure 6.5. Interpretation of the blade flap harmonics.

plane. The combination of the harmonics β_0, β_{1c}, and β_{1s} forms a cone that has been tilted laterally and longitudinally. The circular path described by the blade tips still lies in a plane, which is called the tip-path plane. The orientation of the tip-path plane relative to the reference plane is given by β_{1c} and β_{1s}. The higher harmonics of the flap motion (β_{2c}, β_{2s}, etc.) produce a distortion of the tip-path plane. These harmonics are usually small, so the rotor flap motion is described primarily by β_0, β_{1c}, and β_{1s} for the helicopter in forward flight.

The lag motion can also be written as a Fourier series:

$$\zeta(\psi) = \zeta_0 + \zeta_{1c} \cos\psi + \zeta_{1s} \sin\psi + \ldots \tag{6.12}$$

The zero-th harmonic ζ_0 is the mean lag angle of the blades relative to the rotor hub and shaft (Figure 6.6). The first harmonic cyclic lag ζ_{1c} produces a lateral shift of the blades, to the left when $\zeta_{1c} > 0$ (see Figure 6.6). Neglecting the hinge offset, the center-of-gravity of the blade is at $x_{CG} = r_{CG}\cos(\psi - \zeta) \cong r_{CG}(\cos\psi + \zeta\sin\psi)$ and $y_{CG} = r_{CG}\sin(\psi - \zeta) \cong r_{CG}(\sin\psi - \zeta\cos\psi)$, where r_{CG} is the radial location of the center-of-gravity. The mean center-of-gravity location, which is also the center-of-gravity for the entire rotor, is obtained by averaging over the rotor azimuth and multiplying by the number of blades. Using the definition of the lag harmonics, we obtain

$$(x_{CG})_{\text{rotor}} = \frac{N}{2\pi}\int_0^{2\pi} x_{CG}\,d\psi = \frac{N}{2}r_{CG}\zeta_{1s} \tag{6.13}$$

$$(y_{CG})_{\text{rotor}} = \frac{N}{2\pi}\int_0^{2\pi} y_{CG}\,d\psi = -\frac{N}{2}r_{CG}\zeta_{1c} \tag{6.14}$$

Thus the cyclic lag ζ_{1c} produces a lateral shift of the rotor center-of-gravity. Similarly, the cyclic lag ζ_{1s} produces a longitudinal shift of the blades in the plane of rotation (aft when $\zeta_{1s} > 0$) and a longitudinal shift in the rotor center-of-gravity.

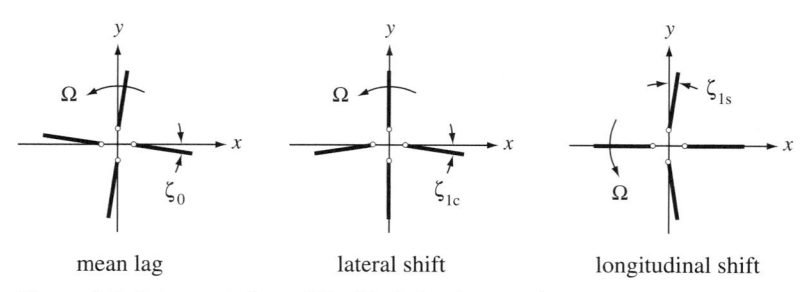

Figure 6.6. Interpretation of the blade lag harmonics.

From the character of the rotor motion associated with the lowest harmonics of flap and lag, the coning is the reaction to the mean blade lift, ζ_0 is the reaction to the mean rotor torque, the cyclic flap β_{1c} and β_{1s} are the response to moments on the rotor disk, and the cyclic lag ζ_{1c} and ζ_{1s} are the response to in-plane motion of the rotor hub.

The Fourier series representation for the blade pitch motion is

$$\theta(\psi) = \theta_0 + \theta_{1c} \cos \psi + \theta_{1s} \sin \psi + \ldots \quad (6.15)$$

The zero-th harmonic θ_0 is the average blade pitch, whereas the first harmonics give a once-per-revolution variation of the pitch angle. Blade pitch or feathering motion has two sources. First, there is the elastic deformation of the control system and blade, described by dynamic degrees of freedom. Such motion is determined by equilibrium of feathering moments on the blade. The second source of blade pitch is the commanded input from the helicopter control system. The pilot controls the helicopter by commanding the rotor blade pitch. The feathering moments on the blade are low, and the lift changes due to pitch are large because the angle-of-attack is directly changed. Controlling the blade pitch is therefore a very effective means of controlling the forces on the rotor. The present chapter is concerned only with the blade pitch as a control variable. The control inputs usually consist of just the mean and first harmonics: $\theta(\psi) = \theta_0 + \theta_{1c} \cos \psi + \theta_{1s} \sin \psi$. The mean angle θ_0 is called the collective pitch, and the 1/rev harmonics θ_{1c} and θ_{1s} are called the cyclic pitch angles. Basically, collective pitch controls the average blade force, and hence the rotor thrust magnitude, whereas cyclic pitch controls the tip-path-plane tilt (that is, the 1/rev flapping) and hence the thrust vector orientation (θ_{1c} controls the lateral orientation, and θ_{1s} controls the longitudinal orientation).

The rotor must have a mechanical means of producing collective and cyclic pitch changes on the rotor blades. The blade pitch motion takes place about a pitch bearing or hinge (Figure 6.7). A pitch horn is rigidly attached to the blade outboard of the pitch bearing, and a pitch link is attached to the pitch horn in such a way that vertical motion of the link produces blade pitch motion. The top of the pitch link is shown aligned with the flap hinge axis, so flap motion is not accompanied by pitch motion. Then what is required is a way to produce a steady and 1/rev sinusoidal vertical motion of the pitch link. This arrangement or its mechanical equivalent is fairly standard in rotor designs. There are other means of producing the blade lift control, such as the Kaman servoflap, and there are many variations in the mechanical implementation of this arrangement. All means of controlling the rotor can be viewed along these general lines, however.

A widely used method of providing the blade pitch control is by means of a swashplate. A swashplate is a mechanical device that transmits the pilot's control motion in the nonrotating frame to the blade cyclic pitch motion in the rotating frame. Figure 6.7 is a schematic of the swashplate arrangement. The actual mechanical arrangement varies widely, but this figure defines the principal components that must be present in some form. The swashplate has rotating and non-rotating rings concentric with the shaft, with bearings between the two rings. The rotating ring is gimballed to the shaft in an arrangement that allows an arbitrary orientation of the plane of the swashplate relative to the rotor shaft while one ring is stationary and the other rotates. The blade pitch links attach to the rotating ring, and links from the pilot's controls attach to the stationary ring. Vertical displacement of the swashplate provides a vertical motion of the pitch links that is independent of azimuth, thereby

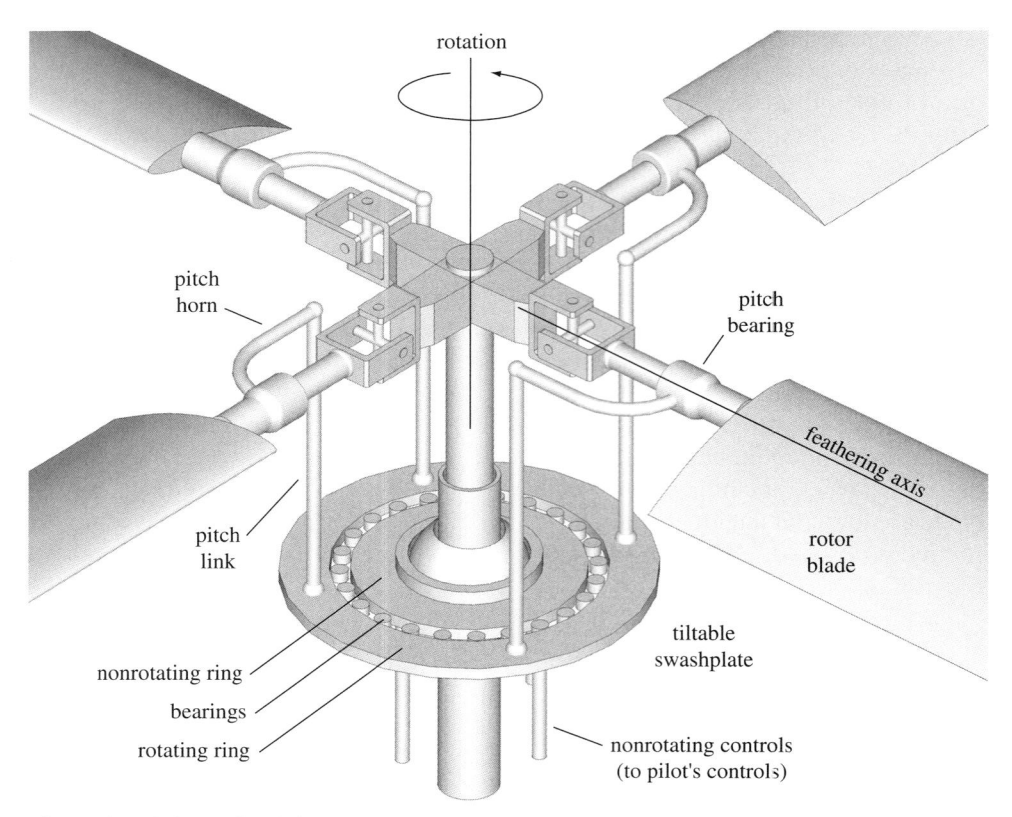

Figure 6.7. Schematic of the rotor swashplate.

changing the rotor collective pitch angle θ_0. If the swashplate is given a longitudinal tilt ϕ_{SP}, the vertical position at the pitch link exhibits a 1/rev sinusoidal variation: $z_{PL} = \phi_{SP} x_{PL} = \phi_{SP} r_{PL} \cos \psi$. Similarly, lateral tilt by ϕ_{SP} gives $z_{PL} = \phi_{SP} r_{PL} \sin \psi$. The swashplate tilt in response to the pilot's stick motion thus produces the cyclic blade pitch control, and vertical motion of the swashplate (or its equivalent, perhaps in an entirely separate mechanism) produces the collective pitch control. In general, the control system can be represented by a control plane, its tilt corresponding to cyclic control and its vertical position corresponding to collective control. Since there can be other sources of blade pitch motion, such as kinematic pitch-flap coupling, the control plane alone does not necessarily represent the entire blade pitch motion.

There always exists a reference plane relative to which the blade pitch has no 1/rev variation. Since the pitch angle θ as measured from this plane is constant, it is called the no-feathering plane. To locate the no-feathering plane, consider an arbitrary reference plane relative to which the cyclic pitch, θ_{1c} and θ_{1s}, is nonzero. The no-feathering plane then is obtained by rotating rearward about the lateral (y) axis by θ_{1s}, and to the left about the longitudinal (x) axis by θ_{1c}. The component of these rotations about the feathering axis of the blade at azimuth ψ is $(\theta_{1c} \cos \psi + \theta_{1s} \sin \psi)$, which indeed cancels the cyclic pitch of the original reference plane. Hence the longitudinal tilt of the no-feathering plane represents the sine cyclic θ_{1s}, whereas lateral tilt represents the cosine cyclic θ_{1c}. In response to control, the tip-path plane (and with it the thrust vector) tilts parallel to the no-feathering plane. Thus θ_{1s} provides longitudinal control of the helicopter and is called longitudinal cyclic, whereas θ_{1c}

provides lateral control and is called lateral cyclic. The no-feathering plane was often used in early rotor analyses, because the absence of cyclic pitch simplifies the calculations. The no-feathering plane and control plane are not necessarily equivalent; the former represents the total blade pitch, whereas the latter represents the control system, hence just the commanded pitch.

6.1.3 Reference Planes

Next, consider the geometry of the blade motion relative to the tip-path plane and the no-feathering plane. The flap and pitch angles (specifically, the 1/rev harmonics of β and θ) define the orientation of the plane of the rotor blade relative to the reference disk plane. We shall examine how β and θ transform as we change from one reference plane to another while maintaining the same orientation of the blade with respect to space. The orientation of the blade relative to space and the air has physical meaning. The choice of a reference plane is arbitrary, although a particular reference plane can be more useful than others for certain theories and calculations. There are invariants of the transformation between frames, which represent the orientation of the blade with respect to space and so must be independent of the reference plane chosen.

Consider two reference planes, the second tilted forward by the angle ϕ_y relative to the first. This tilt decreases β by ϕ_y at $\psi = 0$, and increases it by ϕ_y at $\psi = 180°$, while at $\psi = 90°$ the pitch is increased by ϕ_y and at $\psi = 270°$ is decreased by ϕ_y. This suggests that the cyclic flap β_{1c} has been decreased by ϕ_y and the cyclic pitch θ_{1s} increased by ϕ_y as a result of the reference plane tilt. The flap and pitch angles transform in such a way that the 1/rev harmonics of β and θ as measured relative to the reference plane change by the same magnitude, but with a 90° shift in phase. Similarly, with a lateral tilt of the reference plane by ϕ_x, β_{1s} and θ_{1c} are decreased by the angle ϕ_x. The quantities $(\beta_{1c} + \theta_{1s})$ and $(\beta_{1s} - \theta_{1c})$ should therefore be independent of the reference plane. Now let us derive the reference plane transformation more rigorously.

The angles β and θ in the rotating frame define the orientation of the plane of the blade with respect to a particular reference plane. The components of the blade plane tilt in the nonrotating frame are $(\theta \cos \psi + \beta \sin \psi)$ laterally and $(\theta \sin \psi - \beta \cos \psi)$ longitudinally. Now tilt the reference plane by the angles ϕ_x laterally and ϕ_y longitudinally. Since the position of the blade in space is unchanged, the orientation of the blade relative to the first and second reference planes must be as follows:

$$(\theta)_2 \cos \psi + (\beta)_2 \sin \psi = (\theta)_1 \cos \psi + (\beta)_1 \sin \psi - \phi_x \tag{6.16}$$

$$(\theta)_2 \sin \psi - (\beta)_2 \cos \psi = (\theta)_1 \sin \psi - (\beta)_1 \cos \psi - \phi_y \tag{6.17}$$

or in the rotating frame:

$$(\theta)_2 = (\theta)_1 - \phi_x \cos \psi - \phi_y \sin \psi \tag{6.18}$$

$$(\beta)_2 = (\beta)_1 - \phi_x \sin \psi + \phi_y \cos \psi \tag{6.19}$$

which defines the transformation of the flap and pitch angles (only the first harmonics are affected). Writing β and θ as Fourier series gives

$$(\theta_{1c})_2 = (\theta_{1c})_1 - \phi_x \qquad (\theta_{1s})_2 = (\theta_{1s})_1 - \phi_y \tag{6.20}$$

$$(\beta_{1c})_2 = (\beta_{1c})_1 + \phi_y \qquad (\beta_{1s})_2 = (\beta_{1s})_1 - \phi_x \tag{6.21}$$

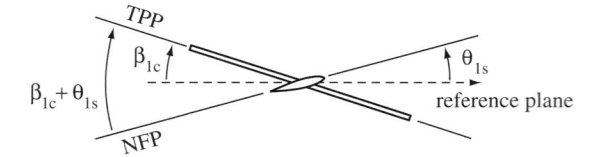

longitudinal tilt of tip-path plane relative to no-feathering plane
(view from advancing side)

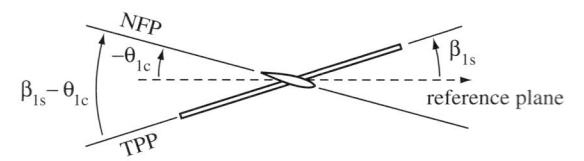

lateral tilt of tip-path plane relative to no-feathering plane
(view from aft)

Figure 6.8. Equivalence of flapping and feathering.

Eliminating ϕ_x and ϕ_y shows that in the transformation from one reference plane to another the quantities $(\beta_{1c} + \theta_{1s})$ and $(\beta_{1s} - \theta_{1c})$ are constant. In terms of the no-feathering plane (NFP) and tip-path plane (TPP) variables,

$$\beta_{1s} - \theta_{1c} = (\beta_{1s})_{\text{NFP}} = -(\theta_{1c})_{\text{TPP}} \tag{6.22}$$

$$\beta_{1c} + \theta_{1s} = (\beta_{1c})_{\text{NFP}} = (\theta_{1s})_{\text{TPP}} \tag{6.23}$$

Relative to a general reference plane, θ_{1c} and θ_{1s} define the orientation of the no-feathering plane, whereas β_{1c} and β_{1s} define the orientation of the tip-path plane. Figure 6.8 shows that the quantities $(\beta_{1c} + \theta_{1s})$ and $(\beta_{1s} - \theta_{1c})$ are simply the longitudinal and lateral angles between the tip-path plane and the no-feathering plane, which indeed must be independent of the choice of reference plane. The two planes of no flap motion and no pitch motion are physically relevant; hence their association with the invariants of flap and pitch in a reference plane transformation. The fact that, by the reference plane transformation, cyclic flapping can be exchanged for cyclic feathering, and vice versa, is referred to as the equivalence of flapping and feathering motion.

Consider a blade with no offset of the flapping hinge or pitch bearing. Although not mechanically practical, such a configuration is simple and well represents the basic behavior of an articulated rotor. In this case the flap and pitch hinges form a gimbal connecting the blade root to the hub, allowing an arbitrary orientation of the rotor shaft while the blade remains fixed with respect to space. So the shaft orientation has no influence on the blade aerodynamics or dynamics; only the relative orientation of the no-feathering plane and tip-path plane is significant. The analysis can therefore be conducted in the no-feathering plane or tip-path plane, ignoring the shaft orientation except to determine the actual cyclic pitch control required. Flapping-feathering equivalence simply expresses the change in β and θ for the various possible shaft orientations. For a hingeless rotor or an articulated rotor with offset hinges, the shaft orientation relative to the no-feathering plane and tip-path

6.1 The Helicopter Rotor in Forward Flight

Figure 6.9. Rotor reference planes: tip-path plane (TPP), no-feathering plane (NFP), hub plane (HP), and control plane (CP).

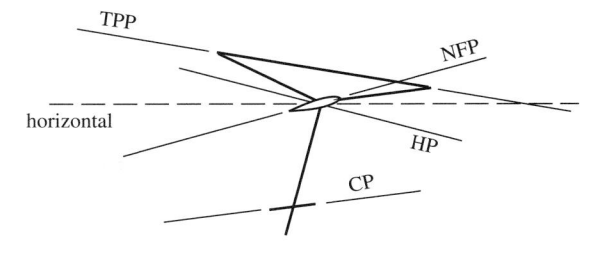

plane has physical importance. The reference plane perpendicular to the rotor shaft is called the hub plane.

Figure 6.9 summarizes the various reference planes used for the helicopter rotor in forward flight. In vertical flight, the natural reference disk plane is the horizontal. With axial symmetry the tip-path plane and no-feathering plane are horizontal. The hub plane is not necessarily horizontal in vertical flight unless the helicopter center-of-gravity is on the rotor shaft axis. In Chapters 3 and 4 the hub plane was not considered, because the hover analysis is primarily concerned with rotor aerodynamics. In forward flight, however, a number of reference planes have physical meaning, and due to the asymmetry of the aerodynamics in forward flight, these planes do not in general coincide with the horizontal plane or with each other. The tip-path plane (TPP) is parallel to the plane described by the blade tips, so there is no 1/rev flapping motion. The orientation of the tip-path plane defines the cyclic flapping β_{1c} and β_{1s} relative to any other plane. The no-feathering plane (NFP) has no 1/rev pitch variation; its orientation thus defines the cyclic pitch θ_{1c} and θ_{1s} relative to any other plane. The control plane (CP) represents the commanded cyclic pitch from the rotor control system and can be considered the swashplate plane. The hub plane (HP) is normal to the rotor shaft. The hub plane is the natural reference frame when there are important physical effects of the blade orientation relative to the hub, such as in the case of offset hinges or a hingeless rotor. Both cyclic flapping and cyclic feathering motion occur in the hub plane. Although in general no two of these planes coincide, there are special cases. For a flapping rotor with no cyclic pitch control (such as the tail rotor and some autogyros), the hub plane and control plane are equivalent; if there is no pitch-flap coupling or other pitch sources, then the control plane and no-feathering plane coincide as well. For a feathering rotor with no flapping (such as a propeller with cyclic pitch) the hub plane and tip-path plane are equivalent.

Figure 6.10 summarizes the quantities defining the rotor motion, velocity, and forces relative to a reference plane. In the non-rotating axis system, x and y lie in the reference plane and z is normal to it. The flap and pitch angles are measured relative to the reference plane. The forward velocity has magnitude V and lies in the x-z plane at an incidence angle i (positive for forward tilt of the disk). The rotor induced velocity v is assumed to be normal to the reference plane. The advance ratio μ and inflow ratio λ are the respective dimensionless velocity components parallel to and normal to the reference plane:

$$\mu = \frac{V \cos i}{\Omega R} \qquad (6.24)$$

$$\lambda = \frac{V \sin i + v}{\Omega R} = \mu \tan i + \lambda_i \qquad (6.25)$$

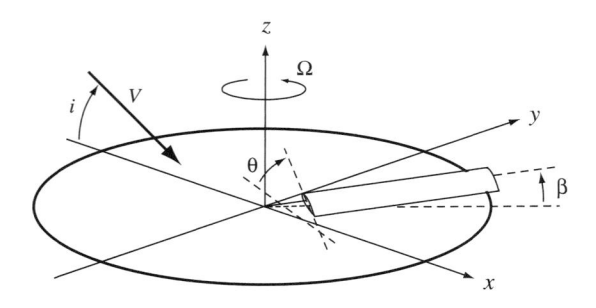

Figure 6.10. Definition of rotor motion, velocity, and forces relative to a reference plane.

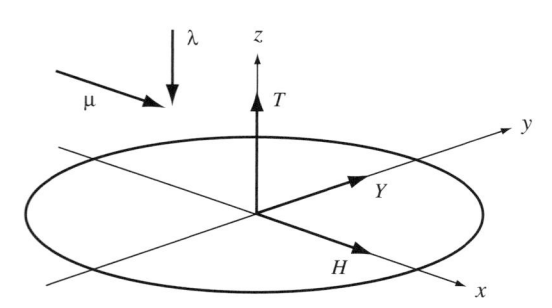

For small disk inclination, $\mu \cong V/\Omega R$ and $\lambda = \mu i + \lambda_i$, so that although i (hence also λ) depends on the reference plane orientation, the advance ratio μ is approximately independent of the reference plane used. The rotor force components are also defined relative to the reference plane chosen: the thrust T is normal to the disk, whereas the rotor drag force H and side force Y are in the reference plane. The coefficients are defined as

$$C_T = T/\rho A(\Omega R)^2 \qquad (6.26)$$

$$C_H = H/\rho A(\Omega R)^2 \qquad (6.27)$$

$$C_Y = Y/\rho A(\Omega R)^2 \qquad (6.28)$$

Similarly, the hub moments C_{Mx} and C_{My} and the torque C_Q are defined relative to this reference plane. The resultant force of the rotor must be independent of the reference plane. Since the thrust is normally much greater than the drag or side force, the rotor thrust is approximately independent of the reference plane. Invariants of the velocity and forces under a transformation of reference planes can be obtained by extending the derivation of the invariants of the blade motion, $(\beta_{1c} + \theta_{1s})$ and $(\beta_{1s} - \theta_{1c})$ (equivalence of flapping and feathering). Thus in terms of quantities in the no-feathering plane or tip-path plane,

$$\lambda = \lambda_{\text{NFP}} + \mu\theta_{1s} = \lambda_{\text{TPP}} - \mu\beta_{1c} \qquad (6.29)$$

$$i = i_{\text{NFP}} + \theta_{1s} = i_{\text{TPP}} - \beta_{1c} \qquad (6.30)$$

$$H = H_{\text{NFP}} + T\theta_{1s} = H_{\text{TPP}} - T\beta_{1c} \qquad (6.31)$$

$$Y = Y_{\text{NFP}} - T\theta_{1c} = Y_{\text{TPP}} - T\beta_{1s} \qquad (6.32)$$

and
$$\lambda = \mu i + \lambda_i = \mu(i_{\text{NFP}} + \theta_{1s}) + \lambda_i = \mu(i_{\text{TPP}} - \beta_{1c}) + \lambda_i \qquad (6.33)$$
for the inflow ratio.

6.2 Aerodynamics of Forward Flight

Now we begin the analysis of the aerodynamics and dynamics of the helicopter rotor in forward flight. At first only the simplest possible case is considered: a fully articulated rotor with constant chord, no hinge offset, no hinge spring, and no pitch-flap coupling; rigid flapping is the only blade motion, with rigid pitch control; and the effects of reverse flow, tip loss, and root cutout are neglected. The aerodynamics of the blade in forward flight are derived, and the forces on the rotor obtained. Then the dynamics of the blade flapping motion are investigated. The remaining sections of the chapter consider some of the factors neglected with this simple model.

This section derives the aerodynamic forces on the rotor blade in forward flight. A fully articulated rotor is considered, with no hinge offset. The blade motion consists of rigid flapping β and rigid pitch due to collective and cyclic control inputs. Elastic bending and torsion of the blade are neglected. Such a model is sufficient to determine the performance and control characteristics of an articulated rotor. Blade element theory is used to find the section aerodynamic forces. The effects of the reverse flow region are neglected for now. A general reference plane is considered.

Blade element theory assumes that each blade section acts as a two-dimensional airfoil for which the influence of the rotor wake consists entirely of an induced velocity at the section. Two-dimensional airfoil characteristics can then be used to evaluate the section loads in terms of the blade motion and aerodynamic environment at that section alone. The induced velocity can be obtained by various means: momentum theory, vortex theory, or nonuniform inflow calculations. Blade element theory requires that the aspect ratio be high, which is normally true for rotary wings. However, near the blade tip or in the large induced velocity gradients of a vortex-blade interaction, advanced aerodynamic theories should be used for best results.

Consider the velocities seen by the blade section (Figure 6.11). The blade section pitch θ is measured from the reference plane to the zero-lift line; θ includes the collective and cyclic pitch control and the built-in twist of the blade. The components of the velocity of the air relative to the blade are u_T (tangential to the disk plane, positive toward the trailing edge), u_P (perpendicular to the disk plane, positive downward), and u_R (radial, positive outward). The resultant velocity and inflow angle of the section are $U = \sqrt{u_T^2 + u_P^2}$ and $\phi = \tan^{-1} u_P/u_T$. The section angle-of-attack then is $\alpha = \theta - \phi$. The velocity seen by the blade is due to the rotor rotation, the helicopter forward speed and induced velocity, and the blade flap motion. To lowest order the tangential and radial components u_T and u_R are due solely to the rotor rotation and advance ratio (see Figure 6.12):

$$u_T = r + \mu \sin \psi \qquad (6.34)$$
$$u_R = \mu \cos \psi \qquad (6.35)$$

These components are then also independent of the reference plane used. The normal velocity u_P has three terms: $\Omega R \lambda$, which is the induced velocity plus the component of the free stream velocity normal to the rotor disk (recall $\lambda = \mu i + \lambda_i$);

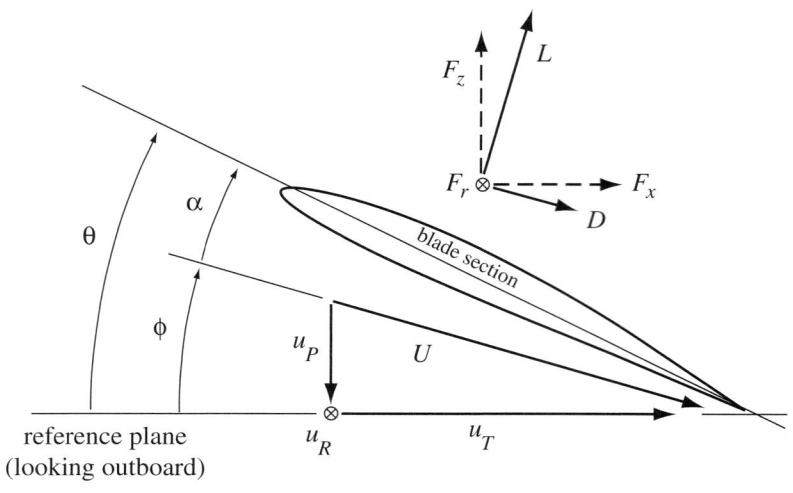

Figure 6.11. Aerodynamics of the rotor blade section.

$rd\beta/dt$, which is the angular velocity of the blade about the flap hinge; and $\Omega R\beta\mu \cos\psi$, which is a component of the radial velocity u_R normal to the blade when the blade is flapped up by the angle β (see Figure 6.12). Thus the dimensionless normal velocity is

$$u_P = \lambda + r\dot{\beta} + \beta\mu \cos\psi \qquad (6.36)$$

Each term in u_P depends on the reference plane. In deriving these expressions for the velocity components, the flap angle β was assumed to be small. Finally, although the pitch angle and velocity components depend on the reference plane, the section aerodynamic environment defined by the resultant velocity and angle-of-attack must not. For small angles $U \cong u_T$, and $\alpha \cong \theta - u_P/u_T$ is easily shown to be invariant during a reference plane transformation.

Figure 6.11 also shows the aerodynamic forces on the blade section. The aerodynamic lift and drag (L and D) are, respectively, normal to and parallel to the

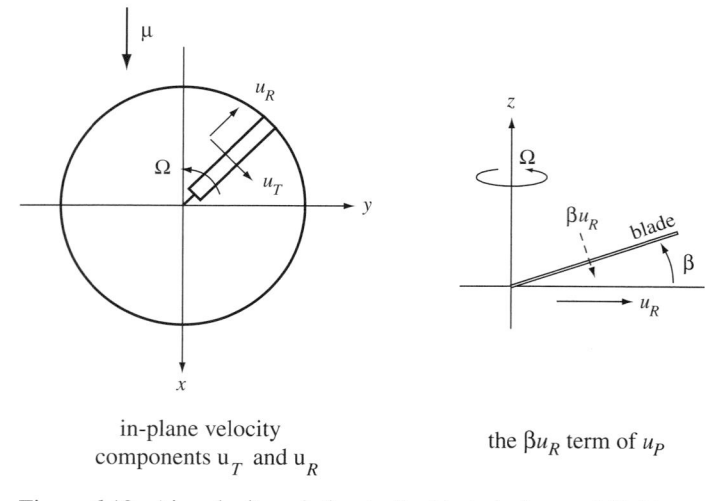

Figure 6.12. Air velocity relative to the blade in forward flight.

resultant velocity U. The components of the section lift and drag resolved in the reference plane are F_z and F_x (normal and in-plane, respectively). The radial force on the section is F_r, defined to be positive when outward. The section forces can be expressed in terms of the lift and drag coefficients:

$$L = \frac{1}{2}\rho U^2 c\, c_\ell \tag{6.37}$$

$$D = \frac{1}{2}\rho U^2 c\, c_d \tag{6.38}$$

where c_ℓ and c_d are in general functions of the section angle-of-attack α and the Mach number $M = M_{\text{tip}} U$. Here ρ is the air density (which is omitted when dimensionless quantities are used), c is the blade chord, and $M_{\text{tip}} = \Omega R/c_s$ is the tip Mach number. The normal and in-plane forces are

$$F_z = L\cos\phi - D\sin\phi \tag{6.39}$$

$$F_x = L\sin\phi + D\cos\phi \tag{6.40}$$

The first term in F_x is the induced drag, and the second term is the profile drag. The blade radial force is

$$F_r = -\beta F_z + D_{\text{radial}} \tag{6.41}$$

The first term in F_r is the radial component of the normal force when the blade flaps up. The second term is a radial drag force due to radial flow along the blade, which is neglected until sections 6.22 and 6.23. Now substitute for L and D in terms of the section coefficients, divide by the chord c and the section two-dimensional lift-curve slope a, and use dimensionless quantities. The resulting section forces in the reference axis system are

$$\frac{F_z}{ac} = U^2\left(\frac{c_\ell}{2a}\cos\phi - \frac{c_d}{2a}\sin\phi\right) \tag{6.42}$$

$$\frac{F_x}{ac} = U^2\left(\frac{c_\ell}{2a}\sin\phi + \frac{c_d}{2a}\cos\phi\right) \tag{6.43}$$

$$\frac{F_r}{ac} = -\beta\frac{F_z}{ac} \tag{6.44}$$

Next we make the small angle assumption and neglect stall and compressibility effects. Assuming λ, β, ϕ, and θ are all small angles, it follows that u_P/u_T and α are small; that $\phi \cong u_P/u_T$, $\sin\phi \cong \phi$, and $\cos\phi \cong 1$; and that $U^2 \cong u_T^2$ and $\alpha \cong \theta - u_P/u_T$. Assuming a constant lift-curve slope and absorbing the zero-lift angle in the definition of blade pitch, the lift coefficient is $c_\ell = a\alpha$. Neglecting stall gives $c_d/c_\ell \ll 1$, so with the small angle assumption $F_z \cong L$ and $F_x \cong L\phi + D$. Hence the section aerodynamic forces become

$$\frac{F_z}{ac} = \frac{1}{2}u_T^2\alpha = \frac{1}{2}\left(u_T^2\theta - u_P u_T\right) \tag{6.45}$$

$$\frac{F_x}{ac} = u_T^2\left(\frac{\alpha}{2}\phi + \frac{c_d}{2a}\right) = \frac{1}{2}\left(u_P u_T\theta - u_P^2\right) + \frac{c_d}{2a}u_T^2 \tag{6.46}$$

$$\frac{F_r}{ac} = -\beta\frac{F_z}{ac} \tag{6.47}$$

6.3 Rotor Aerodynamic Forces

Now we derive the aerodynamic forces acting on the rotor. A general reference plane is used, although some of the results are examined in the no-feathering plane and tip-path plane. The thrust T is normal to the rotor disk; the rotor drag force H is in the disk plane, positive aft; and the rotor side force Y is in the disk plane, positive toward the advancing side (see Figure 6.10). The rotor drag and side forces are usually small in the tip-path plane, so in general H/T and Y/T are of the order of the tip-path-plane tilt angles. In addition, there is a torque moment Q on the rotor, positive for a rotor absorbing power. For an articulated rotor with no flap hinge offset, there can be no net pitch or roll moment transmitted to the rotor hub. The rotor forces are obtained by integrating the blade section forces along the span. The rotor thrust is due to the normal force F_z, the drag and side forces are due to the in-plane forces F_x and F_r resolved in the non-rotating frame, and the torque is due to the in-plane force F_x. Multiplying by the number of blades N to obtain the forces on the entire rotor, the aerodynamic forces are

$$T = N \int_0^R F_z \, dr \tag{6.48}$$

$$H = N \int_0^R (F_x \sin\psi + F_r \cos\psi) \, dr \tag{6.49}$$

$$Y = N \int_0^R (-F_x \cos\psi + F_r \sin\psi) \, dr \tag{6.50}$$

$$Q = N \int_0^R r F_x \, dr \tag{6.51}$$

It is also necessary to average these expressions over the azimuth to obtain the steady forces, by operating with $\frac{1}{2\pi}\int_0^{2\pi}(\ldots)d\psi$. The rotor thrust coefficient is

$$C_T = \frac{T}{\rho A (\Omega R)^2} = \frac{N}{\pi}\int_0^R \frac{F_z}{\rho R(\Omega R)^2}\frac{dr}{R} = \int_0^R \frac{Nc}{\pi R}\frac{F_z}{\rho c(\Omega R)^2}\frac{dr}{R} \tag{6.52}$$

or, using dimensionless quantities, $C_T = \int_0^1 \sigma(F_z/c)\,dr$. In general, the blade chord can be a function of r, but here only a constant chord is considered. Then the solidity σ is a constant. Hence the rotor coefficients are

$$\frac{C_T}{\sigma a} = \int_0^1 \frac{F_z}{ac}\,dr \tag{6.53}$$

$$\frac{C_H}{\sigma a} = \int_0^1 \left(\frac{F_x}{ac}\sin\psi + \frac{F_r}{ac}\cos\psi\right)dr \tag{6.54}$$

$$\frac{C_Y}{\sigma a} = \int_0^1 \left(-\frac{F_x}{ac}\cos\psi + \frac{F_r}{ac}\sin\psi\right)dr \tag{6.55}$$

$$\frac{C_Q}{\sigma a} = \int_0^1 r\frac{F_x}{ac}\,dr \tag{6.56}$$

6.3 Rotor Aerodynamic Forces

Assuming small angles, and neglecting the tip loss and root cutout, substitute for the section forces to obtain

$$\frac{C_T}{\sigma a} = \int_0^1 \frac{1}{2}(u_T^2 \theta - u_P u_T) dr \tag{6.57}$$

$$\frac{C_H}{\sigma a} = \int_0^1 \left\{ \sin\psi \left[\frac{1}{2}(u_P u_T \theta - u_P^2) + \frac{c_d}{2a} u_T^2\right] \right.$$
$$\left. - \beta \cos\psi \left[\frac{1}{2}(u_T^2 \theta - u_P u_T)\right] \right\} dr \tag{6.58}$$

$$\frac{C_Y}{\sigma a} = \int_0^1 \left\{ -\cos\psi \left[\frac{1}{2}(u_P u_T \theta - u_P^2) + \frac{c_d}{2a} u_T^2\right] \right.$$
$$\left. - \beta \sin\psi \left[\frac{1}{2}(u_T^2 \theta - u_P u_T)\right] \right\} dr \tag{6.59}$$

$$\frac{C_Q}{\sigma a} = \int_0^1 r \left[\frac{1}{2}(u_P u_T \theta - u_P^2) + \frac{c_d}{2a} u_T^2\right] dr \tag{6.60}$$

where

$$u_T = r + \mu \sin\psi \tag{6.61}$$
$$u_P = \lambda + r\dot\beta + \beta\mu \cos\psi \tag{6.62}$$
$$\beta = \beta_0 + \beta_{1c}\cos\psi + \beta_{1s}\sin\psi + \beta_{2c}\cos 2\psi + \beta_{2s}\sin 2\psi + \ldots \tag{6.63}$$
$$\theta = \theta_0 + \theta_{tw} r + \theta_{1c}\cos\psi + \theta_{1s}\sin\psi \tag{6.64}$$

Linear twist has been assumed, and usually uniform inflow is used in this chapter. The flap motion has been written as a complete Fourier series, but in fact only the mean and first harmonics are considered for most of this chapter.

The moments on the rotor hub can be obtained in a similar fashion. The pitch moment M_y and roll moment M_x (positive rearward and toward the retreating side, respectively) are

$$M_y = -N \int_0^R \cos\psi \, r F_z dr \tag{6.65}$$

$$M_x = N \int_0^R \sin\psi \, r F_z dr \tag{6.66}$$

Then in coefficient form

$$\frac{C_{My}}{\sigma a} = -\int_0^1 \cos\psi \, \frac{F_z}{ac} r \, dr \tag{6.67}$$

$$\frac{C_{Mx}}{\sigma a} = \int_0^1 \sin\psi \, \frac{F_z}{ac} r \, dr \tag{6.68}$$

The root flapping moment on a single blade in the rotating frame is $M_F = \int_0^R r F_z dr$. Writing M_F as a Fourier series and remembering that the rotor forces and moments must be averaged over the azimuth, the pitch and roll moments are

$$M_y = -\frac{N}{2}M_{F1c} \tag{6.69}$$

$$M_x = \frac{N}{2}M_{F1s} \tag{6.70}$$

Hence the 1/rev flap moments at the center of rotation lead to the steady pitch and roll moments on the helicopter. In the case of an articulated rotor with the flap hinge at the center of rotation there is no moment on the hinge, and for that reason there can be no hub moment acting on the helicopter. In general, the pitch and roll moment can be related to the rotor tip-path-plane tilt, which is a measure of the 1/rev flapping moments.

It is convenient to separate the drag and side forces and the torque into two terms: a profile term due to the drag coefficient c_d, and an induced term due to the lift coefficient c_ℓ. The former is denoted by the subscript "o" and the latter by "i". Although such a separation is suggested by the division of induced and profile power, it is not quite consistent here because the induced terms include the inflow ratio λ, part of which is due to the disk tilt necessary to counter the rotor profile drag C_{Ho}. The division here is strictly formal, based on whether the source of the section force is the drag coefficient or the lift coefficient. In section 6.4 the rotor profile power and induced power are obtained. Thus $C_H = C_{Hi} + C_{Ho}$, $C_Y = C_{Yi} + C_{Yo}$, and $C_Q = C_{Qi} + C_{Qo}$; the rotor thrust has no drag terms (with the small angle assumption). The profile terms are

$$C_{Ho} = \int_0^1 \frac{\sigma c_d}{2} \sin\psi \, u_T^2 \, dr \tag{6.71}$$

$$C_{Yo} = \int_0^1 \frac{\sigma c_d}{2}(-\cos\psi) u_T^2 \, dr \tag{6.72}$$

$$C_{Qo} = \int_0^1 \frac{\sigma c_d}{2} r u_T^2 \, dr \tag{6.73}$$

and the induced terms are

$$C_{Hi} = \frac{\sigma a}{2}\int_0^1 (u_T\theta - u_P)(u_P \sin\psi - u_T\beta\cos\psi)\,dr \tag{6.74}$$

$$C_{Yi} = \frac{\sigma a}{2}\int_0^1 (u_T\theta - u_P)(-u_P \cos\psi - u_T\beta\sin\psi)\,dr \tag{6.75}$$

$$C_{Qi} = \frac{\sigma a}{2}\int_0^1 r(u_P u_T\theta - u_P^2)\,dr \tag{6.76}$$

Furthermore,

$$u_P \sin\psi - u_T\beta\cos\psi = \lambda\sin\psi + r\dot\beta\sin\psi - r\beta\cos\psi \tag{6.77}$$

$$u_P \cos\psi + u_T\beta\sin\psi = \lambda\cos\psi + r\dot\beta\cos\psi + r\beta\sin\psi + \mu\beta \tag{6.78}$$

6.3 Rotor Aerodynamic Forces

The next step is to average over the rotor azimuth. By using the definitions of the Fourier coefficients, integrals of β and θ can be replaced by the appropriate harmonics. For example, the rotor thrust requires the term

$$\frac{1}{2\pi}\int_0^{2\pi} \theta u_T^2 d\psi = \frac{1}{2\pi}\int_0^{2\pi} \theta\left[r^2 + 2r\mu\sin\psi + \frac{1}{2}\mu^2(1-\cos 2\psi)\right]d\psi$$
$$= (\theta_0 + r\theta_{tw})\left(r^2 + \frac{\mu^2}{2}\right) + \theta_{1s}r\mu - \theta_{2c}\frac{\mu^2}{4} \quad (6.79)$$

where the definitions

$$\theta_0 = \frac{1}{2\pi}\int_0^{2\pi}\theta\, d\psi, \quad \theta_{1s} = \frac{1}{\pi}\int_0^{2\pi}\theta\sin\psi\, d\psi, \quad \theta_{2c} = \frac{1}{\pi}\int_0^{2\pi}\theta\cos 2\psi\, d\psi \quad (6.80)$$

have been used. Also required is the term

$$\frac{1}{2\pi}\int_0^{2\pi} u_P u_T d\psi = \frac{1}{2\pi}\int_0^{2\pi}(\lambda + r\dot\beta + \mu\beta\cos\psi)(r + \mu\sin\psi)d\psi = \lambda r + \frac{\mu^2}{4}\beta_{2s} \quad (6.81)$$

using the definition of β_{2s} and noting that

$$\int_0^{2\pi}(r^2\ddot\beta + r\mu\dot\beta\sin\psi + r\mu\beta\cos\psi)d\psi = \int_0^{2\pi}\frac{d}{d\psi}(r^2\dot\beta + r\mu\beta\sin\psi)d\psi = 0 \quad (6.82)$$

since the quantity $(r^2\dot\beta + r\mu\beta\sin\psi)$ is periodic. There is no higher harmonic control, so $\theta_{2c} = 0$; and the higher harmonics of flapping are small, so β_{2s} is neglected. Thus the rotor thrust coefficient is

$$C_T = \frac{\sigma a}{2}\int_0^1\left[(\theta_0 + r\theta_{tw})\left(r^2 + \frac{\mu^2}{2}\right) + \theta_{1s}r\mu - \lambda r\right]dr$$
$$= \frac{\sigma a}{2}\left[\frac{\theta_0}{3}\left(1 + \frac{3}{2}\mu^2\right) + \frac{\theta_{tw}}{4}(1 + \mu^2) + \frac{\mu}{2}\theta_{1s} - \frac{\lambda}{2}\right] \quad (6.83)$$

Similarly, the induced drag and side force terms are

$$C_{Hi} = \frac{\sigma a}{2}\left[\frac{\theta_0}{3}\left(-\beta_{1c} + \frac{3}{2}\mu\lambda\right) + \frac{\theta_{tw}}{4}(-\beta_{1c} + \mu\lambda)\right.$$
$$-\frac{1}{6}\theta_{1c}\beta_0 + \frac{1}{4}\theta_{1s}(-\mu\beta_{1c} + \lambda)$$
$$\left.+\frac{3}{4}\lambda\beta_{1c} + \frac{1}{6}\beta_0\beta_{1s} + \frac{1}{4}\mu\left(\beta_0^2 + \beta_{1c}^2\right)\right] \quad (6.84)$$

$$C_{Yi} = -\frac{\sigma a}{2}\left[\frac{\theta_0}{3}\left(\beta_{1s}\left(1 + \frac{3}{2}\mu^2\right) + \frac{9}{4}\mu\beta_0\right) + \frac{\theta_{tw}}{4}(\beta_{1s}(1 + \mu^2) + 2\mu\beta_0)\right.$$
$$+\frac{1}{4}\theta_{1c}(\lambda + \mu\beta_{1c}) + \frac{1}{6}\theta_{1s}(\beta_0(1 + 3\mu^2) + 3\mu\beta_{1s})$$
$$\left.-\frac{3}{4}\lambda\beta_{1s} + \frac{1}{6}\beta_0\beta_{1c}(1 - 6\mu^2) - \frac{3}{2}\mu\lambda\beta_0 - \frac{1}{4}\mu\beta_{1c}\beta_{1s}\right] \quad (6.85)$$

The induced torque is considered in section 6.4. For the profile terms, the section drag coefficient is assumed to be constant over the entire rotor disk and has an appropriate mean value c_{do}. Then averaging over the azimuth gives

$$C_{Ho} = \int_0^1 \frac{\sigma c_d}{2} \sin\psi\, u_T^2 dr = \int_0^1 \frac{\sigma c_d}{2} r\mu\, dr = \frac{\sigma c_{do}}{4}\mu \qquad (6.86)$$

$$C_{Yo} = \int_0^1 \frac{\sigma c_d}{2}(-\cos\psi) u_T^2 dr = 0 \qquad (6.87)$$

$$C_{Qo} = \int_0^1 \frac{\sigma c_d}{2} r u_T^2 dr = \int_0^1 \frac{\sigma c_d}{2} r \left(r^2 + \frac{\mu^2}{2}\right) dr = \frac{\sigma c_{do}}{8}(1+\mu^2) \qquad (6.88)$$

The profile side force is always zero because of the longitudinal symmetry of the flow, as long as the variation of the drag coefficient is neglected. These results have been obtained neglecting reverse flow and radial flow effects. In section 6.23 the profile drag force, torque, and power are extended to include reverse flow, radial flow, and the radial drag force. Since the radial drag cannot produce a torque on the rotor, and C_{Yo} remains zero because of symmetry, the only influence of radial drag is on C_{Ho}.

In terms of the blade pitch at 75% radius ($\theta_{.75} = \theta_0 + \frac{3}{4}\theta_{tw}$) and the inflow in the no-feathering plane ($\lambda_{\text{NFP}} = \lambda - \mu\theta_{1s}$), the rotor thrust in forward flight is

$$C_T = \frac{\sigma a}{2}\left[\frac{\theta_{.75}}{3}\left(1 + \frac{3}{2}\mu^2\right) - \frac{\theta_{tw}}{8}\mu^2 - \frac{\lambda_{\text{NFP}}}{2}\right] \qquad (6.89)$$

Although this is the most compact form, the inflow relative to the tip-path plane ($\lambda - \mu\theta_{1s} = \lambda_{\text{TPP}} - \mu(\beta_{1c} + \theta_{1s})$) has the most physical significance, since the incidence angle of the tip-path plane is determined directly by the drag of the helicopter and rotor. Thus the angle of the tip-path plane relative to the no-feathering plane ($\beta_{1c} + \theta_{1s}$) is needed to complete the evaluation of the thrust.

Relative to the tip-path plane, the rotor drag force is

$$C_{H\text{TPP}} = \frac{\sigma c_{do}}{4}\mu + \frac{\sigma a}{2}\left[\frac{1}{2}\mu\lambda_{\text{TPP}}\left(\theta_0 + \frac{1}{2}\theta_{tw}\right)\right.$$
$$\left. -\frac{1}{6}\theta_{1c}\beta_0 + \frac{1}{4}\theta_{1s}\lambda_{\text{TPP}} + \frac{1}{4}\mu\beta_0^2\right] \qquad (6.90)$$

Then for a general reference plane the rotor drag is found by adding the term due to tilt of the thrust vector, $C_H = C_{H\text{TPP}} - \beta_{1c}C_T$. Similarly, C_H in the no-feathering plane can be found by dropping the θ_{1c} and θ_{1s} terms from the general result, and then $C_H = C_{H\text{NFP}} + \theta_{1s}C_T$. The rotor side force in the tip-path plane is

$$C_{Y\text{TPP}} = -\frac{\sigma a}{2}\left[\frac{3}{4}\mu\beta_0\left(\theta_0 + \frac{2}{3}\theta_{tw}\right)\right.$$
$$\left. + \frac{1}{4}\theta_{1c}\lambda_{\text{TPP}} + \frac{1}{6}\theta_{1s}\beta_0(1+3\mu^2) - \frac{3}{2}\mu\beta_0\lambda_{\text{TPP}}\right] \qquad (6.91)$$

and then $C_Y = C_{Y\text{TPP}} - \beta_{1s}C_T$ or $C_Y = C_{Y\text{NFP}} - \theta_{1c}C_T$. Since $C_{H\text{TPP}}/C_T$ and $C_{Y\text{TPP}}/C_T$ are usually small, the rotor thrust vector is tilted from the normal to the tip-path plane by only 0.5° to 1.0° or less in forward flight (the tilt is proportional to μ).

6.4 Power in Forward Flight

The expression for the rotor torque obtained in section 6.3 is

$$\frac{C_Q}{\sigma a} = \int_0^1 r \frac{F_x}{ac} dr = \int_0^1 r \left[\frac{1}{2}(u_P u_T \theta - u_P^2) + \frac{c_d}{2a} u_T^2 \right] dr \qquad (6.92)$$

By integrating this expression over the disk, as for the rotor force coefficients, the torque can be evaluated. With all power transmitted to the rotor through the shaft, $P = \Omega Q$ or $C_P = C_Q$. An alternative procedure, which yields a simpler result, is the energy balance formulation of the rotor power. The energy expression is also more general, since many of the assumptions required in the force-balance method are not necessary.

Consider the general expressions for the rotor thrust, drag force, and torque in terms of the section forces normalized by blade chord:

$$C_T = \int_0^1 \sigma \widehat{F}_z dr = C_{Ti} + C_{To} \qquad (6.93)$$

$$C_H = \int_0^1 \sigma \left(\widehat{F}_x \sin \psi + \widehat{F}_r \cos \psi \right) dr = C_{Hi} + C_{Ho} \qquad (6.94)$$

$$C_Q = \int_0^1 \sigma \widehat{F}_x r \, dr = C_{Qi} + C_{Qo} \qquad (6.95)$$

As usual, an average over the azimuth is also required. The subscripts "i" and "o" refer to the contributions from c_ℓ and c_d. In terms of the section lift and drag,

$$\widehat{F}_z = \frac{F_z}{c} = U^2 \left(\frac{c_\ell}{2} \cos \phi - \frac{c_d}{2} \sin \phi \right) = \frac{U}{2}(c_\ell u_T - c_d u_P) \qquad (6.96)$$

$$\widehat{F}_x = \frac{F_x}{c} = U^2 \left(\frac{c_\ell}{2} \sin \phi + \frac{c_d}{2} \cos \phi \right) = \frac{U}{2}(c_\ell u_P + c_d u_T)$$

$$= \tan \phi \, \widehat{F}_z + U^2 \frac{c_d}{2} \frac{1}{\cos \phi}$$

$$= \frac{u_P}{u_T} \widehat{F}_z + \frac{U^3}{u_T} \frac{c_d}{2} \qquad (6.97)$$

$$\widehat{F}_r = \frac{F_r}{c} = -\eta' \beta \, \widehat{F}_z + \widehat{F}_{ro} \qquad (6.98)$$

where $\tan \phi = u_P/u_T$ and $U^2 = u_T^2 + u_P^2$. No small angle assumptions have been made here. The radial drag term $\widehat{F}_{ro}$ is dealt with in sections 6.22 and 6.23. An important generalization has been introduced. The out-of-plane blade deflection is $z = \beta \eta$, where $\eta(r)$ is an arbitrary radial mode shape. For rigid flapping $\eta = r$, but with hinge offset or a hingeless rotor a more complicated mode shape is required. Then the normal velocity becomes $u_P = \lambda + \eta \dot{\beta} + \eta' \beta \mu \cos \psi$, and the radial force due to tilt of F_z is $F_r = -\eta'\beta F_z$. Substituting for $\widehat{F}_x$ and $\widehat{F}_r$ in terms of $\widehat{F}_z$ gives

$$C_H = \int_0^1 \sigma \widehat{F}_z \left(\frac{u_P}{u_T}\sin\psi - \eta'\beta\cos\psi\right) dr$$

$$+ \int_0^1 \frac{\sigma c_d}{2}\frac{U^3}{u_T}\sin\psi\, dr + \int_0^1 \sigma \widehat{F}_{ro}\cos\psi\, dr$$

$$= C_{Hz} + C_{Hx} \tag{6.99}$$

$$C_Q = \int_0^1 \sigma \widehat{F}_z \frac{u_P}{u_T} r\, dr + \int_0^1 \frac{\sigma c_d}{2}\frac{U^3}{u_T} r\, dr$$

$$= C_{Qz} + C_{Qx} \tag{6.100}$$

The subscript "z" designates the terms produced by the section normal force $\widehat{F}_z$. The remainder (subscript "x") depends on the section drag (c_{do} and $\widehat{F}_{ro}$). Now consider the quantity $C_{Qz} + \mu C_{Hz}$:

$$C_{Qz} + \mu C_{Hz} = \int_0^1 \sigma \left(r\frac{u_P}{u_T} + \mu\sin\psi\frac{u_P}{u_T} - \eta'\beta\mu\cos\psi\right)\widehat{F}_z dr$$

$$= \int_0^1 \sigma (u_P - \eta'\beta\mu\cos\psi)\widehat{F}_z dr$$

$$= \int_0^1 \sigma (\lambda + \eta\dot\beta)\widehat{F}_z dr \tag{6.101}$$

The second term is zero, as is now shown. Since $\eta\dot\beta = \dot z$ is the velocity of the out-of-plane motion of the blade,

$$\int_0^{2\pi} \eta\dot\beta F_z d\psi = \int_0^{2\pi} \dot z F_z d\psi = \oint F_z\, dz \tag{6.102}$$

is the work done on the blade section by the periodic aerodynamic force F_z during one revolution of the rotor. The total energy of the blade from one revolution to the next must be unchanged, since the steady-state rotor motion is periodic. Then the net work done on the blade during one revolution must be zero. This result can also be obtained by considering the dynamics of the blade flap motion. The differential equation for the blade flapping motion (derived in section 6.5) is

$$\ddot\beta + \nu^2\beta = \gamma \int_0^1 \eta\frac{F_z}{ac} dr \tag{6.103}$$

where ν is the natural frequency. Averaging over the azimuth gives

$$\frac{1}{2\pi}\int_0^{2\pi}\left(\int_0^1 \eta\dot\beta\frac{F_z}{ac}dr\right)d\psi = \frac{1}{2\pi}\int_0^{2\pi}\dot\beta\left(\int_0^1 \eta\frac{F_z}{ac}dr\right)d\psi$$

$$= \frac{1}{2\pi}\int_0^{2\pi}\frac{1}{\gamma}\dot\beta(\ddot\beta + \nu^2\beta)d\psi$$

$$= \frac{1}{2\pi}\int_0^{2\pi}\frac{1}{2\gamma}\frac{d}{d\psi}(\dot\beta^2 + \nu^2\beta^2)d\psi$$

$$= 0 \tag{6.104}$$

6.4 Power in Forward Flight

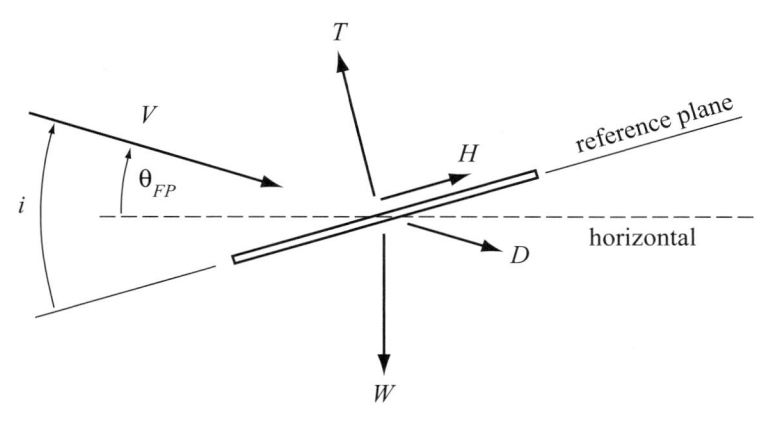

Figure 6.13. Longitudinal forces acting on the helicopter.

since the energy of the out-of-plane motion $(\dot{\beta}^2 + \nu^2\beta^2)$ is periodic. This result is for an arbitrary flapping mode shape and frequency and hence applies to all rotors. The same result is obtained when more than one mode is used to represent the out-of-plane blade deflection. Thus

$$C_{Qz} + \mu C_{Hz} = \int_0^1 \sigma \lambda \widehat{F}_z dr = \int \lambda \, dC_T \tag{6.105}$$

The rotor power can be written

$$\begin{aligned} C_P = C_Q &= (C_{Qz} + \mu C_{Hz}) - \mu C_H + (C_{Qx} + \mu C_{Hx}) \\ &= \int \lambda \, dC_T - \mu C_H + (C_{Qx} + \mu C_{Hx}) \\ &= \int \lambda_i \, dC_{Ti} + \left(C_{Qx} + \mu C_{Hx} + \int \lambda_i \, dC_{To} \right) + \mu_z C_T - \mu C_H \end{aligned} \tag{6.106}$$

where the inflow ratio $\lambda = \mu \tan i + \lambda_i = \mu_z + \lambda_i$ has been substituted in the last step. From equation 6.96, $dC_{Ti} = \frac{1}{2}\sigma U u_T c_\ell dr$ and $dC_{To} = -\frac{1}{2}\sigma U u_P c_d dr$. Figure 6.13 shows the rotor velocity V, disk incidence angle i, and the rotor forces T and H relative to the reference plane. The rotor wind-axis lift L (perpendicular to V) and drag X (parallel to V, positive aft) are

$$L = T \cos i + H \sin i = (T\mu + H\mu_z)/\sqrt{\mu^2 + \mu_z^2} \tag{6.107}$$

$$X = H \cos i - T \sin i = (H\mu - T\mu_z)/\sqrt{\mu^2 + \mu_z^2} \tag{6.108}$$

which implies $i = \tan^{-1}(H/T) - \tan^{-1}(X/L)$. From equation 6.108,

$$\mu_z C_T - \mu C_H = -\sqrt{\mu^2 + \mu_z^2}\, C_X = -(V/\Omega R)C_X \tag{6.109}$$

The product of the propulsive force $-X$ and the rotor speed V is the propulsive power, which is the sum of parasite and climb power: $P_p + P_c = -XV$. Thus

$$\begin{aligned} C_P &= \int \lambda_i \, dC_{Ti} + \left(C_{Qx} + \mu C_{Hx} + \int \lambda_i \, dC_{To} \right) - (V/\Omega R)C_X \\ &= C_{Pi} + C_{Po} - (V/\Omega R)C_X \end{aligned} \tag{6.110}$$

or
$$P = P_i + P_o - VX \tag{6.111}$$

introducing the induced power $C_{Pi} = \int \lambda_i dC_{Ti}$ and profile power C_{Po}. This result is applicable to all rotors, lifting or propelling, in hover or axial or edgewise flight. In terms of isolated rotor performance, separating parasite power and climb power is not possible.

To identify the contributions to the rotor drag force X, consider force equilibrium for the helicopter in steady flight (Figure 6.13). The rotor thrust T and drag H are defined relative to the reference plane used. The helicopter drag D is directed opposite to the free stream velocity V. The remaining force on the helicopter is the weight W, which is vertical. Auxiliary propulsion or lifting devices on the helicopter can be included by subtracting their forces from W and D. The helicopter flight path angle θ_{FP} gives a climb speed $V_c = V \sin \theta_{FP}$ (and $\lambda_c = V_c/\Omega R$). For small angles $W = L$, and longitudinal force equilibrium gives $D + H = T(i - \theta_{FP})$. The disk incidence is then

$$i = \theta_{FP} + \frac{D}{W} + \frac{C_H}{C_T} \tag{6.112}$$

and

$$\mu_z = \mu \tan i \cong \mu i = \lambda_c + \mu \frac{D}{W} + \mu \frac{C_H}{C_T} \tag{6.113}$$

Since the in-plane forces on the rotor are small when measured in tip-path-plane axes, the tip-path-plane incidence i_{TPP} is nearly the sum of the climb angle θ_{FP} and the tilt of the rotor thrust vector needed to provide a propulsive force equal to the helicopter drag. A derivation valid for large angles is given in section 6.21. Longitudinal force equilibrium in the direction of the velocity gives

$$-X = D + W \sin \theta_{FP} = D + W \frac{V_c}{V} \tag{6.114}$$

(without small angle assumptions). So $P_p + P_c = -XV = DV + WV_c$ defines the parasite power and climb power.

Substituting for the rotor drag force, the power is

$$C_P = \int \lambda_i dC_{Ti} + \left(C_{Qx} + \mu C_{Hx} + \int \lambda_i dC_{To} \right) + \frac{DV}{\rho A(\Omega R)^3} + \frac{V_c W}{\rho A(\Omega R)^3}$$
$$= C_{Pi} + C_{Po} + C_{Pp} + C_{Pc} \tag{6.115}$$

C_{Pi} is the rotor induced power, which is required to produce the thrust; C_{Po} is the rotor profile power, required to turn the rotor in a viscous fluid; $P_p = DV$ is the rotor parasite power, required to overcome the drag of the helicopter; and $P_c = WV_c$ is the rotor climb power, required to increase the gravitational potential energy. This is the energy balance expression for the helicopter performance in forward flight, relating the power required to all the sources of energy loss. The energy balance expression is independent of the reference plane used.

The rotor induced power is $C_{Pi} = \int \lambda_i dC_{Ti}$, where dC_{Ti} is the c_ℓ terms in $dC_T = \sigma \widehat{F}_z dr$ (and an average over the azimuth is required as well). With uniform inflow this is simply $C_{Pi} = \lambda_i C_T$. For edgewise forward flight above about $\mu = 0.1$, $\lambda_i \cong \kappa C_T/2\mu$ is a good approximation; hence $C_{Pi} =\cong \kappa C_T^2/2\mu$. The empirical factor κ accounts for tip loss, nonuniform inflow, and other losses.

6.4 Power in Forward Flight

The profile power is

$$C_{Po} = C_{Qx} + \mu C_{Hx} + \int \lambda_i dC_{To}$$

$$= \int_0^1 \frac{\sigma c_d}{2}\left(r\frac{U^3}{u_T} + \mu \sin\psi \frac{U^3}{u_T} - \lambda_i U u_P\right) dr + \int_0^1 \sigma \widehat{F}_{ro}\mu \cos\psi \, dr$$

$$\cong \int_0^1 \frac{\sigma c_d}{2} U^3 dr \tag{6.116}$$

where $U^2 = u_T^2 + u_P^2$. This result neglects reverse flow, yawed flow, and radial drag (see section 6.23). The $\lambda_i U u_P$ term is small compared to U^3 and is absent if $C_{Pi} = \int \lambda_i dC_T$ is used for the induced power. Equation 6.116 can be written $C_{Po} = \int_0^1 (\sigma/c)DU \, dr$, where DU is the power absorbed by the blade section. In terms of the profile (c_d) contributions,

$$C_{Po} = C_{Qo} + \mu C_{Ho} - \mu_z C_{To} + \int \frac{\sigma c_d}{2} U u_P(\eta \dot{\beta}) dr \tag{6.117}$$

Assuming constant chord, a mean value for the drag coefficient, and low inflow edgewise flight so $U^3 \cong u_T^3$, we obtain

$$C_{Po} = \frac{\sigma c_{do}}{2}\int_0^1 u_T^3 dr = \frac{\sigma c_{do}}{2}\int_0^1\left(r^3 + \frac{3}{2}r\mu^2\right) dr = \frac{\sigma c_{do}}{8}(1 + 3\mu^2) \tag{6.118}$$

which could also have been obtained using $C_{Qo} = (\sigma c_{do}/8)(1 + \mu^2)$ and $C_{Ho} = (\sigma c_{do}/4)\mu$ from section 6.3. Two-thirds of the increase of the profile power with speed is thus due to C_{Ho}. When reverse flow and radial drag effects are included, a further increase of C_{Po} with speed is found. The simple expression

$$C_{Po} = \frac{\sigma c_{do}}{8}(1 + 4.65\mu^2) \tag{6.119}$$

can be used for low advance ratio. Section 6.23 gives better approximations for the speed dependence of the profile power.

The parasite power is $C_{Pp} = DV/\rho A(\Omega R)^3 \cong \mu(D/W)C_T$. If the helicopter drag is written in terms of an equivalent area f, so $D = \frac{1}{2}\rho V^2 f$, then

$$C_{Pp} = \frac{1}{2}\frac{V^3 f}{(\Omega R)^3 A} \cong \frac{1}{2}\mu^3 \frac{f}{A} \tag{6.120}$$

or $P = DV = \frac{1}{2}\rho V^3 f$. The climb power is $P_c = V_c W$, or $C_{Pc} \cong \lambda_c C_T$.

In summary, the helicopter rotor power in edgewise forward flight can be estimated from the individual sources of energy loss:

$$C_P = C_{Pi} + C_{Po} + C_{Pp} + C_{Pc}$$

$$\cong \frac{\kappa C_T^2}{2\mu} + \frac{\sigma c_{do}}{8}(1 + 4.65\mu^2) + \frac{1}{2}\mu^3\frac{f}{A} + \lambda_c C_T \tag{6.121}$$

which gives the power required as a function of gross weight and speed. The performance estimate can be improved by using a nonuniform induced velocity distribution; by considering the actual drag coefficient of the rotor blade, which requires the angle-of-attack distribution over the disk; and by refining the helicopter drag representation.

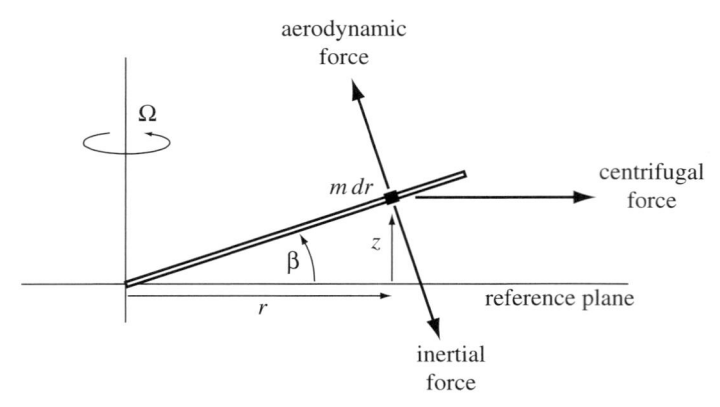

Figure 6.14. Rotor blade flapping moments.

Numerical calculations of the helicopter performance generally use the force balance method, obtaining the power from the rotor torque by $C_P = \int_0^1 \sigma \widehat{F}_x r\, dr$. The energy breakdown of the power follows from the definitions of induced power, $C_{Pi} = \int \lambda_i dC_T$ and the propulsive power (sum of parasite and climb power), $P_p + P_c = -XV$; then the profile power is $C_{Po} = C_P - (C_{Pi} + C_{Pp} + C_{Pc})$.

6.5 Rotor Flapping Motion

To complete the solution for the rotor behavior in forward flight, the harmonics of the blade flapping motion are required, particularly the coning and tip-path-plane tilt angles (β_0, β_{1c}, and β_{1s}). The angle of the tip-path plane relative to the no-feathering plane is derived in this section. From the tip-path-plane orientation (established by equilibrium of forces on the helicopter) the no-feathering-plane orientation can be found, and hence the cyclic pitch control required to fly the helicopter in the given operating condition. The blade flapping motion is determined by equilibrium of inertial and aerodynamic moments about the flap hinge. To introduce the analysis of the flapping motion, the simplest model is used: a rigid articulated rotor blade with no flap-hinge offset or spring restraint.

Consider equilibrium of the inertial and aerodynamic moments about the flapping hinge (Figure 6.14). The out-of-plane deflection is $z = \beta r$ for rigid motion with no hinge offset. Acting on a mass element $m\,dr$ (m is the blade mass per unit length) at radial station r are the following section forces:

i) an inertial force $m\ddot{z} = mr\ddot{\beta}$ opposing the flap motion, with moment arm r about the flap hinge
ii) a centrifugal force $m\Omega^2 r$ directed radially outward, with moment arm $z = r\beta$
iii) an aerodynamic force F_z normal to the blade, with moment arm r

Recall that for small angles F_z is just the section lift L. The centrifugal force is the influence of the blade rotation. Since the centrifugal force always acts radially outward in a plane normal to the rotation axis, it acts as a spring force opposing the blade flap motion.

The moments about the flap hinge are given by integrals over the span of the section forces times their corresponding moment arms. Since there is no flap-hinge spring, the sum of the moments must be zero. Thus the equation of

6.5 Rotor Flapping Motion

motion for the flap motion is

$$\int_0^R mr\ddot{\beta}r\,dr + \int_0^R m\Omega^2 r(r\beta)\,dr - \int_0^R F_z r\,dr = 0 \tag{6.122}$$

or

$$\left(\int_0^R r^2 m\,dr\right)(\ddot{\beta} + \Omega^2 \beta) = \int_0^R rF_z\,dr \tag{6.123}$$

Now define the moment of inertia about the flap hinge, $I_b = \int_0^R r^2 m\,dr$, and use dimensionless quantities based on ρ, Ω, and R. Then

$$\ddot{\beta} + \beta = \frac{1}{I_b}\int_0^1 rF_z\,dr \tag{6.124}$$

The dimensionless time variable is the rotor azimuth, $\psi = \Omega t$. Next define the blade Lock number γ:

$$\gamma = \frac{\rho a c R^4}{I_b} \tag{6.125}$$

The Lock number is a dimensionless parameter representing the ratio of aerodynamic forces to inertial forces on the blade. Typically $\gamma = 8$ to 10 for articulated rotors and $\gamma = 5$ to 7 for hingeless rotors. Note that γ contains the sole influence of the air density on the flap motion. Assuming a constant chord and introducing the Lock number, the flapping equation becomes

$$\ddot{\beta} + \beta = \gamma \int_0^1 r\frac{F_z}{ac}\,dr = \gamma M_F \tag{6.126}$$

The left-hand side is a mass-spring system with a natural frequency of 1/rev (Ω dimensionally) due to the centrifugal spring. The right-hand side is the aerodynamic forcing moment. So 1/rev aerodynamic forces excite the blade flap motion at its resonant frequency. The amplitude of a system forced at resonance is determined by its damping alone, which in this case comes from the aerodynamic forces. The phase of the 1/rev response is exactly a 90° lag, regardless of the magnitude of the damping.

The aerodynamic force normal to the blade is $F_z/ac = L/ac = \frac{1}{2}u_T^2\alpha = \frac{1}{2}(u_T^2\theta - u_P u_T)$. The aerodynamic flap moment is therefore

$$M_F = \int_0^1 r\frac{F_z}{ac}\,dr = \int_0^1 r\frac{1}{2}\left[(r + \mu\sin\psi)^2\theta - (\lambda + r\dot\beta + \mu\beta\cos\psi)(r + \mu\sin\psi)\right]dr \tag{6.127}$$

Assuming uniform inflow and linear twist, the integration over the span can be performed:

$$M_F = M_\theta \theta_{\text{con}} + M_{tw}\theta_{tw} + M_\lambda \lambda + M_{\dot\beta}\dot\beta + M_\beta \beta \tag{6.128}$$

$$= \theta_{\text{con}}\left(\frac{1}{8} + \frac{\mu}{3}\sin\psi + \frac{\mu^2}{4}\sin^2\psi\right)$$

$$+ \theta_{tw}\left(\frac{1}{10} + \frac{\mu}{4}\sin\psi + \frac{\mu^2}{6}\sin^2\psi\right) - \lambda\left(\frac{1}{6} + \frac{\mu}{4}\sin\psi\right)$$

$$- \dot\beta\left(\frac{1}{8} + \frac{\mu}{6}\sin\psi\right) - \beta\mu\cos\psi\left(\frac{1}{6} + \frac{\mu}{4}\sin\psi\right) \tag{6.129}$$

where $\theta_{\text{con}} = \theta_0 + \theta_{1c}\cos\psi + \theta_{1s}\sin\psi$ is the collective and cyclic pitch control. The flapping equation of motion is thus

$$\ddot{\beta} + \beta = \gamma\left(M_\theta\theta_{\text{con}} + M_{tw}\theta_{tw} + M_\lambda\lambda + M_{\dot{\beta}}\dot{\beta} + M_\beta\beta\right) \tag{6.130}$$

The aerodynamic coefficients are the flap moments due to angle-of-attack changes produced by the blade pitch, twist, inflow, flapping velocity, and flapping displacements, respectively. A flapping velocity produces an angle-of-attack perturbation that changes the blade lift to oppose the motion; hence the blade has aerodynamic damping given by the coefficient $M_{\dot{\beta}}$.

The steady-state solution for the blade flapping is required, namely the harmonics of the periodic motion. Only the mean and first harmonics are derived for now. Without higher harmonic control (2/rev and above), the higher harmonics of flapping are small. The solution procedure involves operating on the flapping equations with

$$\frac{1}{2\pi}\int_0^{2\pi}(\ldots)d\psi, \quad \frac{1}{\pi}\int_0^{2\pi}(\ldots)\cos\psi\,d\psi, \quad \frac{1}{\pi}\int_0^{2\pi}(\ldots)\sin\psi\,d\psi \tag{6.131}$$

By using the definitions of the Fourier coefficients expressed in terms of the integrals of $\beta(\psi)$ and $\theta(\psi)$, linear algebraic equations are obtained for the harmonics of β. This operational method evaluates the mean and 1/rev flap moments; the latter correspond to pitch and roll moments on the rotor disk (see section 6.3). An alternative approach is the substitutional method discussed in section 6.1.2. For the inertial and centrifugal terms we obtain

$$\frac{1}{2\pi}\int_0^{2\pi}(\ddot{\beta} + \beta)d\psi = \frac{1}{2\pi}\int_0^{2\pi}\beta\,d\psi = \beta_0 \tag{6.132}$$

$$\frac{1}{\pi}\int_0^{2\pi}(\ddot{\beta} + \beta)\cos\psi\,d\psi = \frac{1}{\pi}\int_0^{2\pi}(-\beta\cos\psi + \beta\cos\psi)d\psi = 0 \tag{6.133}$$

$$\frac{1}{\pi}\int_0^{2\pi}(\ddot{\beta} + \beta)\sin\psi\,d\psi = \frac{1}{\pi}\int_0^{2\pi}(-\beta\sin\psi + \beta\sin\psi)d\psi = 0 \tag{6.134}$$

where the $\ddot{\beta}\cos\psi$ and $\ddot{\beta}\sin\psi$ terms have been integrated by parts twice. The centrifugal force gives an average flap moment when the rotor is coned by β_0. The 1/rev components of the inertial and centrifugal forces exactly cancel. So with no flap spring or hinge offset, the 1/rev components of the aerodynamic flap moments must also be zero. The requirement of zero aerodynamic pitch and roll moment on the rotor disk determines the tip-path-plane tilt angles β_{1c} and β_{1s}. The inertial and spring terms exactly cancel because the 1/rev aerodynamic forces are forcing the flap motion at its resonant frequency. Thus with no aerodynamic forces, there would be no means of controlling the rotor, because the tip-path plane would be in equilibrium for any orientation.

Applying the operators to the flapping equation of motion, using the aerodynamic coefficients given in equation 6.129, and neglecting second and higher harmonics of flap and pitch, we obtain

$$\beta_0 = \gamma\left[\frac{\theta_0}{8}(1 + \mu^2) + \frac{\theta_{tw}}{10}\left(1 + \frac{5}{6}\mu^2\right) + \frac{\mu}{6}\theta_{1s} - \frac{\lambda}{6}\right] \tag{6.135}$$

6.5 Rotor Flapping Motion

$$0 = \frac{1}{8}\theta_{1c}\left(1 + \frac{1}{2}\mu^2\right) - \frac{1}{8}\beta_{1s} - \frac{\mu}{6}\beta_0 - \frac{\mu^2}{16}\beta_{1s} \tag{6.136}$$

$$0 = \frac{1}{8}\theta_{1s}\left(1 + \frac{3}{2}\mu^2\right) + \frac{\mu}{3}\theta_0 + \frac{\mu}{4}\theta_{tw} - \frac{\mu}{4}\lambda + \frac{1}{8}\beta_{1c} - \frac{\mu^2}{16}\beta_{1c} \tag{6.137}$$

With $\lambda - \mu\theta_{1s} = \lambda_{\text{NFP}}$, the solution for the rotor flapping motion is

$$\beta_0 = \gamma\left[\frac{\theta_{.8}}{8}(1 + \mu^2) - \frac{\mu^2}{60}\theta_{tw} - \frac{\lambda_{\text{NFP}}}{6}\right] \tag{6.138}$$

$$\beta_{1s} - \theta_{1c} = \frac{-\frac{4}{3}\mu}{1 + \frac{1}{2}\mu^2}\beta_0 \tag{6.139}$$

$$\beta_{1c} + \theta_{1s} = \frac{-\frac{8}{3}\mu}{1 - \frac{1}{2}\mu^2}\left[\theta_{.75} - \frac{3}{4}\lambda_{\text{NFP}}\right] \tag{6.140}$$

Alternatively, in terms of $\lambda_{\text{TPP}} = \lambda_{\text{NFP}} + \mu(\beta_{1c} + \theta_{1s})$,

$$\beta_0 = \gamma\left[\frac{\theta_{.8}}{8}(1 + \mu^2) - \frac{\mu^2}{60}\theta_{tw} - \frac{\lambda_{\text{TPP}}}{6} + \frac{\mu}{6}(\beta_{1c} + \theta_{1s})\right] \tag{6.141}$$

$$\beta_{1s} - \theta_{1c} = \frac{-\frac{4}{3}\mu}{1 + \frac{1}{2}\mu^2}\beta_0 \tag{6.142}$$

$$\beta_{1c} + \theta_{1s} = \frac{-\frac{8}{3}\mu}{1 + \frac{3}{2}\mu^2}\left[\theta_{.75} - \frac{3}{4}\lambda_{\text{TPP}}\right] \tag{6.143}$$

Although the expression for β_0 is simpler using λ_{NFP}, the expressions using λ_{TPP} are more appropriate since the tip-path-plane orientation has a direct physical meaning: because rotor in-plane forces are small relative to the tip-path plane, the tip-path-plane orientation specifies the thrust vector orientation, which is determined by helicopter longitudinal force equilibrium. The singularity of the $(\beta_{1c} + \theta_{1s})$ solution at $\mu = \sqrt{2}$ also disappears when the tip-path-plane inflow is used. In any case, $\mu = \sqrt{2}$ is beyond the range of validity of these expressions; including reverse flow removes the singularity even for the expression in terms of λ_{NFP}. The coning angle, roughly $\beta_0 \cong \frac{3}{4}\gamma(C_T/\sigma a)$, is proportional to the blade loading. The first harmonics β_{1c} and β_{1s} are proportional to the advance ratio μ and to C_T/σ. Typically β_0 and β_{1c} have values of a few degrees, whereas the lateral flapping β_{1s} is somewhat smaller.

For hover the flapping solution reduces to

$$\beta_0 = \gamma\left(\frac{\theta_{.8}}{8} - \frac{\lambda}{6}\right) \tag{6.144}$$

$$\beta_{1s} - \theta_{1c} = 0 \tag{6.145}$$

$$\beta_{1c} + \theta_{1s} = 0 \tag{6.146}$$

Recall that $(\beta_{1c} + \theta_{1s})$ and $(\beta_{1s} - \theta_{1c})$ give the orientation of the tip-path plane relative to the no-feathering plane. The hover solution for the flap motion is thus that the tip-path plane and no-feathering plane are always parallel. The flap motion in hover can be obtained as follows. The aerodynamic coefficients are constant when $\mu = 0$, and $M_\beta = 0$. There are no net 1/rev flap moments due to the inertial forces, the inflow, or the twist. Consequently, the equation of motion reduces to

$$M_\theta \theta + M_{\dot\beta} \dot\beta = 0 \qquad (6.147)$$

which has components $M_\theta \theta_{1c} + M_{\dot\beta} \beta_{1s} = 0$ and $M_\theta \theta_{1s} - M_{\dot\beta} \beta_{1c} = 0$. Hence the solution is $\beta_{1s}/\theta_{1c} = -\beta_{1c}/\theta_{1s} = -M_\theta/M_{\dot\beta}$. As expected, the phase of the response of flapping to cyclic pitch is exactly 90°, with the magnitude determined by the ratio of the control moment to the damping. For the present simple model, $M_\theta = -M_{\dot\beta} = \frac{1}{8}$, so $-M_\theta/M_{\dot\beta} = 1$, and the tip-path plane is always parallel to the no-feathering plane. In forward flight, the rotor operating state uniquely determines the relative orientation of the tip-path plane and no-feathering plane, because $(\beta_{1c} + \theta_{1s})$ and $(\beta_{1s} - \theta_{1c})$ are obtained as functions of the helicopter speed and loading alone. As the no-feathering plane is tilted in response to pilot control inputs, the rotor tip-path plane and hence the thrust vector are tilted. By this means the pilot can control the attitude of the helicopter, using cyclic pitch (swashplate tilt) to produce moments about the helicopter center-of-gravity by tilting the thrust vector.

Let us examine further the role of inertial and aerodynamic forces in the rotor flap response. For the case of no aerodynamic forces, the rotor without flap hinge offset or restraint has the equation of motion $\ddot\beta + \beta = 0$. The solution of this equation is $\beta = \beta_{1c} \cos\psi + \beta_{1s} \sin\psi$, where β_{1c} and β_{1s} are arbitrary constants. The orientation of the rotor is thus arbitrary, but fixed in space since in the absence of aerodynamic forces or a hinge offset there is no means by which blade pitch or shaft tilt can produce a moment on the disk. The rotor behaves as a gyro, maintaining its orientation relative to inertial space in the absence of external moments. The rotor in air has the capability of producing an aerodynamic flap moment due to blade pitch (M_θ), which can be used to precess the rotor and hence control its orientation. If M_θ were the only moment, the rotor would respond to cyclic with a constant rate of tilt. However, the rotor also has aerodynamic flap damping moments ($M_{\dot\beta}$). A tilt of the tip-path plane by β_{1c} or β_{1s} produces a flapping velocity in the rotating frame. Consequently, a moment due to control-plane tilt precesses the rotor, tilting the tip-path plane until the flapping produces through $M_{\dot\beta}$ a moment just sufficient to counter the control moment on the disk. Because the moments due to θ and $\dot\beta$ balance, the rotor has achieved a new equilibrium position.

Thus there are two ways to view the rotor flap dynamics. The rotor blade can be considered a system with natural frequency of 1/rev, so that aerodynamic moments due to cyclic pitch excite the system at resonance. The response has a phase lag of exactly 90° (one quarter of a cycle, which at 1/rev means an azimuth angle of 90° also) and a magnitude determined by the damping. Alternatively, the rotor can be considered a gyro, with the flap hinges at the center of rotation forming the gyro gimbal. A control moment on the disk due to cyclic pitch precesses the rotor with a 90° phase lag characteristic of a gyro, until the flap damping produces a moment to stop the precession.

The rotor coning is proportional to the Lock number γ because the coning is determined by the balance of the centrifugal and aerodynamic flap moments. The coning is essentially proportional to the rotor thrust coefficient, the difference arising because of the extra factor r in the integrand to obtain the flap moment rather than the total blade lift force. Because the rotor thrust produces a flap moment, the rotor

6.5 Rotor Flapping Motion

cones upward until a centrifugal flap moment sufficient to cancel the aerodynamic moment is generated.

Since the longitudinal flap motion ($\beta_{1c} + \theta_{1s}$) is negative, in forward flight the tip-path plane tilts backward relative to the no-feathering plane. The lateral asymmetry of the blade velocity u_T in forward flight means that for constant pitch (i.e., relative to the no-feathering plane) the blade has a higher lift on the advancing side than on the retreating side of the disk. The result is a lateral flap moment on the rotor disk. In the rotating frame, where this flap moment is at the resonant frequency 1/rev, the blade responds with a 90° phase lag, the maximum flap displacement occurring on the front of the disk. The tip-path plane therefore tilts longitudinally (rearward) in response to the lateral moment. Now β_{1c} gives a flapping velocity $\dot\beta = -\beta_{1c}\sin\psi$, which has maximum amplitude at the sides of the disk. Through the flap damping, the tip-path-plane tilt then produces a lateral flap moment. The rotor flaps back until this lateral moment due to the flap damping is just enough to counter the lateral moment due to the aerodynamic asymmetry. With this balance of the aerodynamic forces the new equilibrium orientation is achieved.

Since the lateral flap motion ($\beta_{1s} - \theta_{1c}$) is negative, in forward flight the tip-path plane tilts toward the advancing side relative to the no-feathering plane. When the blade is at the coning angle β_0 there is a component of the forward velocity normal to the blade surface: $\beta_0\mu\cos\psi$ (see Figure 6.12). The angle-of-attack due to this normal velocity term is maximum positive at the front of the disk and maximum negative at the rear of the disk, hence producing a longitudinal aerodynamic moment on the rotor. The blade responds to this 1/rev aerodynamic forcing in the rotating frame with a maximum flap amplitude 90° after the maximum moment; that is, with a lateral (to the advancing side) tilt of the tip-path plane. Now β_{1s} gives a flapping velocity $\dot\beta = \beta_{1s}\cos\psi$, which through the flap damping also produces a longitudinal moment. The rotor flaps to the advancing side until the longitudinal moment due to the flap damping balances the longitudinal moment due to the coning, and the rotor is in equilibrium with this orientation of the tip-path plane.

The tip-path-plane tilt is roughly proportional to the advance ratio μ. To keep the thrust vector orientation fixed as the speed increases, the no-feathering plane must be tilted forward and toward the retreating side to counter the increased tip-path-plane tilt. Thus as speed increases, a forward cyclic stick displacement is required in addition to the forward displacement to increase the propulsive force. An increasing cyclic stick displacement to the retreating side is also required. The control required to trim the helicopter is determined by helicopter force and moment equilibrium. As found in section 6.4, longitudinal force equilibrium determines the orientation of the tip-path plane relative to the horizontal (i_{TPP}, and also λ_{TPP}). The equilibrium of pitching moments on the helicopter determines the orientation of the hub plane relative to the horizontal (i_{HP}) as a function of the helicopter longitudinal center-of-gravity position and the aerodynamic forces on the aircraft (see section 6.21). The combination then determines the longitudinal flapping in the hub plane: $\beta_{1cHP} = i_{TPP} - i_{HP}$. Rotor flapping equilibrium determines the orientation of the tip-path plane relative to the no-feathering plane, and from this the longitudinal cyclic control θ_{1sHP} can be obtained. Similarly, side force and roll moment equilibrium on the helicopter give the lateral flapping β_{1sHP} and therefore the lateral cyclic control θ_{1cHP}.

The flapping solution has not been obtained in a form that can be directly used to calculate the cyclic required from the tip-path-plane tilt. The C_T and β_{1c} equations must be solved simultaneously for the collective pitch and longitudinal cyclic. Then the coning angle can be evaluated and the lateral cyclic obtained from β_0. The result is

$$\theta_{.75} = \frac{1}{\Delta}\left[\left(1+\frac{3}{2}\mu^2\right)\left(\frac{6C_T}{\sigma a}+\frac{3}{8}\mu^2\theta_{tw}\right)+\frac{3}{2}\lambda_{TPP}\left(1-\frac{1}{2}\mu^2\right)\right] \quad (6.148)$$

$$\theta_{1s} = -\beta_{1c} - \frac{1}{\Delta}\left[\frac{8}{3}\mu\left(\frac{6C_T}{\sigma a}+\frac{3}{8}\mu^2\theta_{tw}\right)+2\mu\lambda_{TPP}\left(1-\frac{3}{2}\mu^2\right)\right] \quad (6.149)$$

$$\beta_0 = \frac{\gamma/8}{\Delta}\left[\left(1-\frac{19}{18}\mu^2+\frac{3}{2}\mu^4\right)\frac{6C_T}{\sigma a}+\left(\frac{1}{20}+\frac{29}{120}\mu^2-\frac{1}{5}\mu^4+\frac{3}{8}\mu^6\right)\theta_{tw}\right.$$
$$\left.+\lambda_{TPP}\left(\frac{1}{6}-\frac{7}{12}\mu^2+\frac{1}{4}\mu^4\right)\right] \quad (6.150)$$

$$\theta_{1c} = \beta_{1s} + \frac{\frac{4}{3}\mu}{1+\frac{1}{2}\mu^2}\beta_0 \quad (6.151)$$

where $\Delta = 1 - \mu^2 + \frac{9}{4}\mu^4$. Using these equations, the collective and cyclic control can be obtained from the thrust, tip-path-plane incidence, and flapping relative to the hub plane.

6.6 Linear Inflow Variation

As a first approximation to the effects of nonuniform inflow in forward flight, consider an induced velocity of the form

$$\lambda_i = \lambda_0(1 + \kappa_x r \cos\psi + \kappa_y r \sin\psi) \quad (6.152)$$

This is a linear variation over the rotor disk, with λ_0 being the mean induced velocity. The coefficients κ_x and κ_y are functions of μ, since they must be zero in hover. For high speed, κ_x is around 1 and κ_y is somewhat smaller in magnitude and negative. Section 5.2.2 discussed a number of estimates for κ_x and κ_y. This linear inflow variation can be considered the first term in an expansion of the general nonuniform induced velocity $\lambda_i(r, \psi)$. The lowest order terms are important for the rotor performance and flapping, whereas the higher order terms (which can be large in certain flight conditions) are important for the blade loads and vibration. So far in this chapter a uniform inflow distribution has been used. Now additional contributions to the rotor forces and blade motion are found, produced by the inflow increment

$$\Delta\lambda = \lambda_0(\kappa_x r \cos\psi + \kappa_y r \sin\psi) = \lambda_x r \cos\psi + \lambda_y r \sin\psi \quad (6.153)$$

Here λ_x gives the longitudinal induced velocity variation and λ_x the lateral variation. The rotor thrust increment is

$$\Delta C_T = \frac{\sigma a}{2}\int_0^1 (-\Delta\lambda\, u_T)dr = \frac{\sigma a}{2}\int_0^1 \left(-\frac{1}{2}\lambda_y\right)r\mu\, dr = \frac{\sigma a}{2}\left(-\frac{1}{4}\lambda_y\mu\right) \quad (6.154)$$

6.6 Linear Inflow Variation

Hence

$$C_T = \frac{\sigma a}{2}\left[\frac{\theta_{.75}}{3}\left(1+\frac{3}{2}\mu^2\right) - \frac{\theta_{tw}}{8}\mu^2 - \frac{1}{2}\left(\bar{\lambda}_{\text{NFP}} + \frac{\mu}{2}\lambda_y\right)\right] \qquad (6.155)$$

where $\bar{\lambda}_{\text{NFP}}$ is the mean inflow. Thus there is a change of order μ^2 in the thrust for a given collective. The increments in the rotor drag and side forces are

$$\Delta C_{H\text{TPP}} = \frac{\sigma a}{2}\left[\lambda_x\left(\frac{\mu}{16}\theta_{1c} + \frac{1}{6}\beta_0\right)\right.$$
$$\left. + \lambda_y\left(\frac{1}{6}\theta_{.75} + \frac{3\mu}{16}\theta_{1s} - \frac{1}{2}\bar{\lambda}_{\text{TPP}}\right)\right] \qquad (6.156)$$

$$\Delta C_{Y\text{TPP}} = \frac{\sigma a}{2}\left[-\lambda_x\left(\frac{1}{6}\theta_{.75} + \frac{\mu}{16}\theta_{1s} - \frac{1}{2}\bar{\lambda}_{\text{TPP}}\right)\right.$$
$$\left. + \lambda_y\left(\frac{\mu}{16}\theta_{1c} - \frac{1}{6}\beta_0\right)\right] \qquad (6.157)$$

The flap moment increment is

$$\Delta M_F = \int_0^1 (-\Delta\lambda)\frac{1}{2}(r^2 + r\mu\sin\psi)dr = -(\lambda_x\cos\psi + \lambda_y\sin\psi)\left(\frac{1}{8} + \frac{\mu}{6}\sin\psi\right) \qquad (6.158)$$

The flap equations become

$$\beta_0 = \gamma\left[\frac{\theta_{.8}}{8}(1+\mu^2) - \frac{\mu^2}{60}\theta_{tw} - \frac{1}{6}\left(\bar{\lambda}_{\text{NFP}} + \frac{\mu}{2}\lambda_y\right)\right] \qquad (6.159)$$

$$0 = \frac{1}{8}(\theta_{1c} - \beta_{1s})\left(1 + \frac{1}{2}\mu^2\right) - \frac{\mu}{6}\beta_0 - \frac{1}{8}\lambda_x \qquad (6.160)$$

$$0 = \frac{1}{8}(\theta_{1s} + \beta_{1s})\left(1 - \frac{1}{2}\mu^2\right) + \frac{\mu}{3}\theta_{.75} - \frac{\mu}{4}\bar{\lambda}_{\text{NFP}} - \frac{1}{8}\lambda_y \qquad (6.161)$$

with solutions for the tip-path-plane tilt as follows:

$$\beta_{1s} - \theta_{1c} = \frac{1}{1+\frac{1}{2}\mu^2}\left[-\frac{4}{3}\mu\beta_0 - \lambda_x\right] \qquad (6.162)$$

$$\beta_{1c} + \theta_{1s} = \frac{1}{1-\frac{1}{2}\mu^2}\left[-\frac{8}{3}\mu\left(\theta_{.75} - \frac{3}{4}\bar{\lambda}_{\text{NFP}}\right) + \lambda_y\right] \qquad (6.163)$$

There is an order μ^2 change in the coning angle, as for the thrust, because the lateral inflow λ_y decreases (for $\lambda_y < 0$) the mean value of λu_T; this effect is small. The inflow variation has a significant influence on the tip-path-plane tilt. There are longitudinal and lateral angle-of-attack changes on the disk due, respectively, to λ_x and λ_y, which produce lateral and longitudinal flapping. The longitudinal flapping (and hence the longitudinal cyclic trim) change is small but not negligible, whereas the change in the lateral flapping and cyclic is large. Thus the rotor cyclic flapping and cyclic pitch trim are sensitive to nonuniform inflow.

Nonuniform inflow gives an increment in the rotor induced power:

$$\Delta C_{Pi} = \int \Delta\lambda \, dC_T = \sigma a \int_0^1 \Delta\lambda \frac{F_z}{ac} \, dr \qquad (6.164)$$

Here we shall consider an arbitrary induced velocity distribution $\lambda(r, \psi)$. To evaluate ΔC_{Pi}, expand the inflow as a Fourier series in azimuth and radially in terms of the orthogonal blade bending modes:

$$\Delta\lambda = \sum_{n=1}^{\infty}\sum_{i=1}^{\infty} \left(\lambda_{nc}^i \cos n\psi + \lambda_{ns}^i \sin n\psi\right) \eta_i(r) \qquad (6.165)$$

The functions η_i are the mode shapes of the out-of-plane bending of the blade with corresponding natural frequencies ν_i (a general blade is considered, but for an articulated rotor with no hinge offset $\eta_1 = r$ and $\nu_1 = 1$). In section 16.3.2 the differential equation for the blade bending modes is derived:

$$\widehat{I}_i(\ddot{q}_i + \nu_i^2 q_i) = \gamma \int_0^1 \eta_i \frac{F_z}{ac} \, dr \qquad (6.166)$$

where $\widehat{I}_i$ is the generalized mass of the i-th mode. Then the integration over the span and azimuth in ΔC_{Pi} can be performed, giving

$$\Delta C_{Pi} = \sigma a \sum_{n=1}^{\infty}\sum_{i=1}^{\infty} \frac{1}{2\pi} \int_0^{2\pi} \left(\lambda_{nc}^i \cos n\psi + \lambda_{ns}^i \sin n\psi\right) \int_0^1 \eta_i \frac{F_z}{ac} \, dr \, d\psi$$

$$= \sigma a \sum_{n=1}^{\infty}\sum_{i=1}^{\infty} \frac{1}{2\pi} \int_0^{2\pi} \left(\lambda_{nc}^i \cos n\psi + \lambda_{ns}^i \sin n\psi\right) \frac{\widehat{I}_i}{\gamma}(\ddot{q}_i + \nu_i^2 q_i) \, d\psi$$

$$= \frac{\sigma a}{2} \sum_{n=1}^{\infty}\sum_{i=1}^{\infty} \frac{\widehat{I}_i(\nu_i^2 - n^2)}{\gamma} \left(\lambda_{nc}^i q_{nc}^i + \lambda_{ns}^i q_{ns}^i\right) \qquad (6.167)$$

where q_{nc}^i and q_{ns}^i are the harmonics of the steady-state response of the i-th bending mode. Integrating $C_{Pi} = \int \lambda_i dC_T$ numerically is simpler. For a linear inflow variation, only the $n = 1$ terms are present. Considering in addition just the first mode of an articulated blade with no hinge offset (so $\nu_1 = 1$) gives $\Delta C_{Pi} = 0$.

6.7 Higher Harmonic Flapping Motion

Consider next the solution of the flapping equation of motion for the second harmonics β_{2c} and β_{2s}. The higher harmonics of the blade motion are strongly influenced by nonuniform inflow and elastic blade bending, but the present solution serves as a guide to the basic behavior of the higher harmonics. Assuming that β_{2c} and β_{2s} are much smaller than β_{1c} and β_{1s}, the previous results for the first harmonic flapping are not changed. The algebraic equations for β_{2c} and β_{2s} are obtained by operating on the flap differential equation with

$$\frac{1}{\pi}\int_0^{2\pi}(\ldots)\cos 2\psi \, d\psi, \quad \frac{1}{\pi}\int_0^{2\pi}(\ldots)\sin 2\psi \, d\psi \qquad (6.168)$$

By neglecting the influence of the second harmonics on the mean and first harmonics, it is only necessary to solve these two additional equations for β_{2c} and β_{2s} rather than five simultaneous equations for all five coefficients. The inertial and centrifugal terms give

$$\frac{1}{\pi}\int_0^{2\pi}(\ddot{\beta}+\beta)\cos 2\psi\,d\psi = -3\beta_{2c} \qquad (6.169)$$

$$\frac{1}{\pi}\int_0^{2\pi}(\ddot{\beta}+\beta)\sin 2\psi\,d\psi = -3\beta_{2s} \qquad (6.170)$$

Since the 2/rev flap moments are acting above the system resonant frequency, the response is dominated by the blade inertia. In general the equations for β_{nc} and β_{ns} have the terms $(1-n^2)\beta_{nc}$ and $(1-n^2)\beta_{ns}$ from the inertial and centrifugal forces. As a result, the higher harmonic flapping in response to the aerodynamic flap moments decreases rapidly with harmonic order, roughly as n^{-2}. When the blade bending modes with natural frequencies above 1/rev are considered, there is again a possibility of large-amplitude motion due to excitation near resonance.

With the aerodynamic terms as well, the equations of motion for β_{2c} and β_{2s} are

$$-3\beta_{2c} = \gamma\left(-\frac{\mu^2}{8}\theta_0 - \frac{\mu^2}{12}\theta_{tw} - \frac{\mu}{6}\theta_{1s} - \frac{1}{4}\beta_{2s} - \frac{\mu}{6}\beta_{1c} + \frac{\mu}{12}\lambda_y\right) \qquad (6.171)$$

$$-3\beta_{2s} = \gamma\left(\frac{\mu}{6}\theta_{1c} + \frac{1}{4}\beta_{2c} - \frac{\mu}{6}\beta_{1s} - \frac{\mu^2}{8}\beta_0 - \frac{\mu}{12}\lambda_x\right) \qquad (6.172)$$

A linear inflow variation has been included. The solution for the second harmonic flapping is

$$\begin{pmatrix}\beta_{2c}\\ \beta_{2s}\end{pmatrix} = \frac{6\gamma}{144+\gamma^2}\begin{bmatrix}1 & \frac{\gamma}{12}\\ -\frac{\gamma}{12} & 1\end{bmatrix}\begin{pmatrix}\mu^2\theta_{.67} + \frac{4}{3}\mu(\beta_{1c}+\theta_{1s}) - \frac{2}{3}\mu\lambda_y\\ \mu^2\beta_0 + \frac{4}{3}\mu(\beta_{1s}-\theta_{1c}) + \frac{2}{3}\mu\lambda_x\end{pmatrix} \qquad (6.173)$$

Note that β_{2c} and β_{2s} are smaller than the first harmonic flapping by at least order μ. Typically the second harmonics have values of a few tenths of a degree. In general the solution for β_{nc} and β_{ns} is of order μ^n/n^2 (with uniform inflow).

The primary excitation of the higher harmonics of blade motion is provided by nonuniform inflow, which has not been considered except for the simple linear variation. With nonuniform inflow the higher harmonic blade motion has a significantly larger amplitude than that found here. In addition, the blade bending modes must also be included for a consistent and accurate calculation of the blade response at higher frequencies. The higher harmonic motion usually has little influence on the rotor performance and control, but is important for the helicopter vibration and blade loads.

Let us examine the response to higher harmonic pitch control. Consider a hovering rotor only, so that there is no inter-harmonic coupling of the pitch control and flap response as occurs in forward flight because of the periodic aerodynamics of the blade. Then n/rev blade pitch gives just n/rev flapping. The flapping equation of motion in hover is

$$\ddot{\beta}+\beta = \frac{\gamma}{8}(-\dot{\beta}+\theta) \qquad (6.174)$$

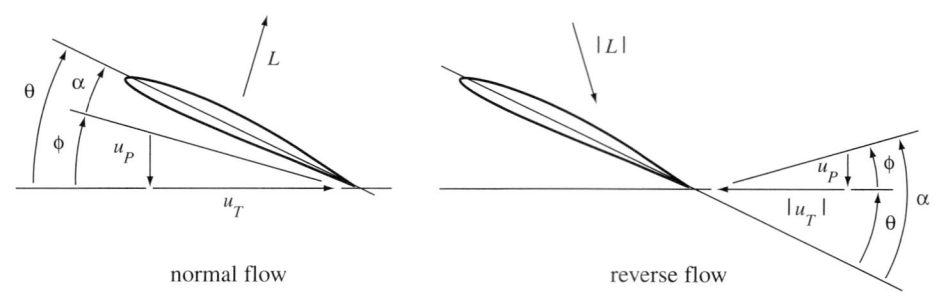

Figure 6.15. Rotor blade section aerodynamics in normal and reverse flow.

For an input of $\theta = \overline{\theta}\cos[n(\psi + \psi_0)]$, the flap response is $\beta = \overline{\beta}\cos[n(\psi + \psi_0) - \Delta\psi]$. The equation of motion gives the magnitude and phase of the response:

$$\overline{\beta}/\overline{\theta} = \frac{\gamma/8}{\sqrt{(n\gamma/8)^2 + (n^2 - 1)^2}} \qquad (6.175)$$

$$\Delta\psi = 90° + \tan^{-1}\frac{n^2 - 1}{n\gamma/8} \qquad (6.176)$$

For the first harmonic, $\overline{\beta}/\overline{\theta} = 1$ and $\Delta\psi = 90°$ as expected. For large harmonic numbers, the amplitude decreases as $\overline{\beta}/\overline{\theta} \cong \gamma/8n^2$, since the blade inertia dominates the response, and the phase lag approaches $\Delta\psi = 180°$. The effectiveness of 1/rev cyclic pitch in controlling the rotor lies in the fact that the flap motion is being excited at resonance.

6.8 Reverse Flow

The reverse flow region is a circle of diameter μ on the retreating side of the rotor disc (Figure 6.2). For low advance ratio the influence of reverse flow is small, since reverse flow is confined to a small area where the dynamic pressure is low. Generally, reverse flow effects can be neglected up to about $\mu = 0.5$. At higher advance ratios, the reverse flow region occupies a large portion of the disk and must be accounted for in calculating the aerodynamic forces on the blade. An elementary model for the blade aerodynamics in the reverse flow region is developed here. Near the reverse flow boundary at least, there is significant separated and radial flow, which requires a better model.

Figure 6.15 compares the section aerodynamics in the normal and reverse flow regions. Section 6.2 gives the normal aerodynamic force for small angles and neglecting stall:

$$\frac{F_z}{ac} \cong \frac{L}{ac} \cong \frac{1}{2}u_T^2\alpha = \frac{1}{2}u_T(u_T\theta - u_P) \qquad (6.177)$$

This result also neglects reverse flow. The positive directions of the various quantities are as follows: F_z and L upward, θ nose up, u_P downward, and u_T from the leading edge to the trailing edge. Figure 6.15 shows that in the reverse flow region the angle-of-attack is

$$\alpha = \theta + \phi = \theta + \frac{u_P}{|u_T|} = \theta - \frac{u_P}{u_T} \qquad (6.178)$$

6.8 Reverse Flow

just as in normal flow. However, in reverse flow a positive α gives a negative (downward) lift:

$$\frac{L}{ac} = -\frac{1}{2}u_T^2\alpha = \frac{1}{2}|u_T|u_T\alpha \tag{6.179}$$

Thus an expression valid in both reverse and normal flow is

$$\frac{F_z}{ac} \cong \frac{L}{ac} \cong \frac{1}{2}|u_T|u_T\alpha = \frac{1}{2}|u_T|(u_T\theta - u_P) \tag{6.180}$$

This convention is appropriate with the small angle assumption, specifically the angle-of-attack α small in both normal and reverse flow. A different convention is required when airfoil tables are used, because then reverse flow is associated with an angle-of-attack near 180°.

The inertial and centrifugal flap moments are unaffected by reverse flow, so the only change to the flap dynamics is in the aerodynamic moment:

$$M_F = \int_0^1 r\frac{F_z}{ac}dr = \int_0^1 \frac{1}{2}|u_T|(u_T\theta - u_P)r\,dr$$

$$= M_\theta\theta_{\mathrm{con}} + M_{tw}\theta_{tw} + M_\lambda\lambda + M_{\dot\beta}\dot\beta + M_\beta\beta \tag{6.181}$$

The reverse flow introduces a change in sign, so that the aerodynamic coefficients now require integrations of the form $\int_0^1 \mathrm{sign}(u_T)f(r,\psi)dr$. The evaluation of this integral depends on the azimuth angle. Assuming $\mu \leq 1$, it is only necessary to distinguish between the advancing and retreating sides of the disk:

$$\int_0^1 \mathrm{sign}(u_T)f(r,\psi)dr = \begin{cases} \int_0^1 f\,dr & 0° < \psi < 180° \\ \int_0^1 f\,dr - 2\int_0^{-\mu\sin\psi} f\,dr & 180° < \psi < 360° \end{cases} \tag{6.182}$$

Thus on the advancing side the aerodynamic coefficients are identical to the results already obtained, whereas on the retreating side a correction for the changed sign in the reverse flow region is required. If $\mu > 1$, there is also a region extending from $\psi = 270° - \cos^{-1}(1/\mu)$ to $\psi = 270° + \cos^{-1}(1/\mu)$, where the blade is entirely in the reverse flow region, so that

$$\int_0^1 \mathrm{sign}(u_T)f(r,\psi)dr = -\int_0^1 f\,dr \tag{6.183}$$

which is just the negative of the expression for the advancing side. Evaluating the flapping aerodynamic coefficients for $\mu \leq 1$ gives

$$M_\theta = \frac{1}{8} + \frac{\mu}{3}\sin\psi + \frac{\mu^2}{4}\sin^2\psi \tag{6.184}$$

$$M_{tw} = \frac{1}{10} + \frac{\mu}{4}\sin\psi + \frac{\mu^2}{6}\sin^2\psi \tag{6.185}$$

$$M_\lambda = -\left(\frac{1}{6} + \frac{\mu}{4}\sin\psi\right) \tag{6.186}$$

$$M_{\dot\beta} = -\left(\frac{1}{8} + \frac{\mu}{6}\sin\psi\right) \tag{6.187}$$

$$M_\beta = -\mu\cos\psi\left(\frac{1}{6} + \frac{\mu}{4}\sin\psi\right) \tag{6.188}$$

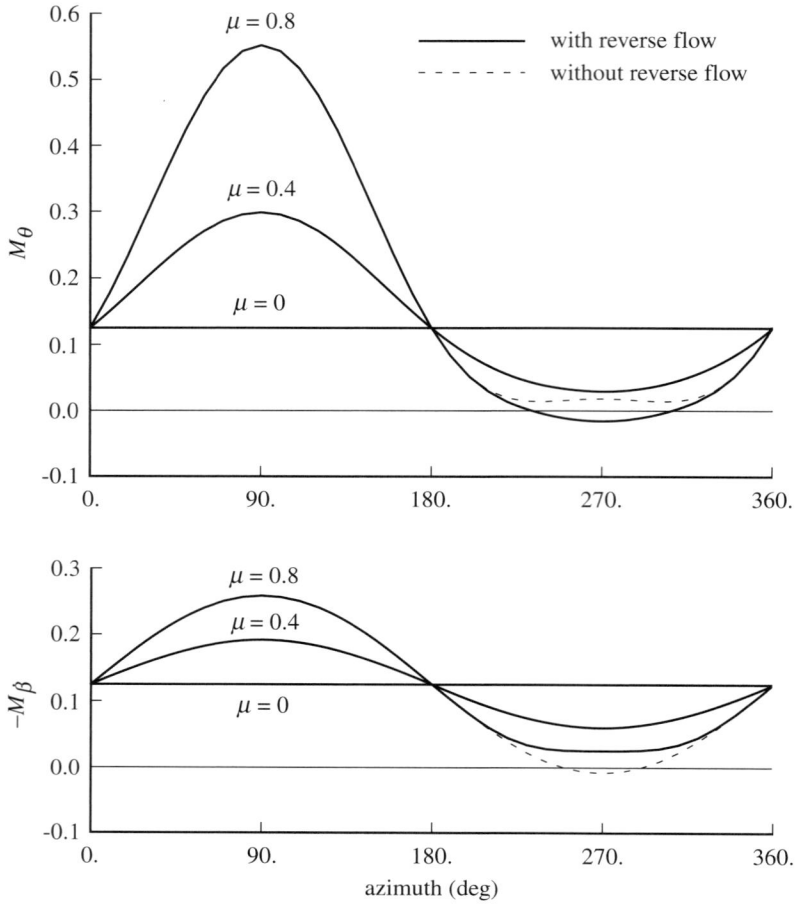

Figure 6.16. Flap damping and pitch control coefficients.

on the advancing side (equation 6.129) and

$$M_\theta = \frac{1}{8} + \frac{\mu}{3} \sin\psi + \frac{\mu^2}{4} \sin^2\psi - \frac{\mu^4}{12} \sin^4\psi \tag{6.189}$$

$$M_{tw} = \frac{1}{10} + \frac{\mu}{4} \sin\psi + \frac{\mu^2}{6} \sin^2\psi + \frac{\mu^5}{30} \sin^5\psi \tag{6.190}$$

$$M_\lambda = -\left(\frac{1}{6} + \frac{\mu}{4} \sin\psi - \frac{\mu^3}{6} \sin^3\psi\right) \tag{6.191}$$

$$M_{\dot\beta} = -\left(\frac{1}{8} + \frac{\mu}{6} \sin\psi + \frac{\mu^4}{12} \sin^4\psi\right) \tag{6.192}$$

$$M_\beta = -\mu \cos\psi \left(\frac{1}{6} + \frac{\mu}{4} \sin\psi - \frac{\mu^3}{6} \sin^3\psi\right) \tag{6.193}$$

on the retreating side.

Figure 6.16 shows the flap damping coefficient $M_{\dot\beta}$ and pitch control coefficient M_θ for several advance ratios. The flap damping is always positive ($M_{\dot\beta} < 0$). In hover, the damping is constant at $M_{\dot\beta} = -0.125$; when $\mu > 0$, the damping is higher on the advancing side and lower on the retreating side. When $\mu > 0.794$, the damping reaches a minimum value of $M_{\dot\beta} = -0.0258$ on the retreating side, with a local maximum at $\psi = 270°$. The pitch control has a value $M_\theta = 0.125$ for hover, and it is higher on the advancing side and lower on the retreating side in forward flight. When $\mu > 0.641$, M_θ is negative on the retreating side. The twist coefficient M_{tw} behaves like M_θ, whereas the inflow coefficient M_λ and flap spring $M_\beta/(\mu \cos \psi)$ are similar to $M_{\dot\beta}$. The flap damping is always positive, even when $\mu > 1$, but the aerodynamic flap spring M_β is negative on the front of the disk because of the $\mu \cos \psi$ factor.

To solve the flapping equations, the aerodynamic coefficients are expressed as Fourier series. Because of the symmetry of the flow, half of the harmonics are found to be zero:

$$M_\theta = M_\theta^0 + M_\theta^{1s} \sin\psi + M_\theta^{2c} \cos 2\psi + M_\theta^{3s} \sin 3\psi + \ldots \tag{6.194}$$

and similarly for M_{tw}, $M_{\dot\beta}$, M_λ; and

$$M_\beta = M_\theta^{1c} \cos\psi + M_\theta^{2s} \sin 2\psi + M_\theta^{3c} \cos 3\psi + \ldots \tag{6.195}$$

For an articulated rotor, the equations for the flap harmonics β_0, β_{1c}, and β_{1s} then become

$$\beta_0 = \gamma \left[\theta_0 M_\theta^0 + \theta_{tw} M_{tw}^0 + \theta_{1s} \frac{1}{2} M_\theta^{1s} \right.$$

$$\left. + \beta_{1c} \left(\frac{1}{2} M_\beta^{1c} - \frac{1}{2} M_{\dot\beta}^{1s} \right) + \lambda M_\lambda^0 \right] \tag{6.196}$$

$$0 = \theta_{1c} \left(M_\theta^0 + \frac{1}{2} M_\theta^{2c} \right) + \beta_{1s} \left(M_{\dot\beta}^0 + \frac{1}{2} M_{\dot\beta}^{2c} + \frac{1}{2} M_{\dot\beta}^{2s} \right) + \beta_0 M_\beta^{1c} \tag{6.197}$$

$$0 = \theta_0 M_\theta^{1s} + \theta_{tw} M_{tw}^{1s} + \theta_{1s} \left(M_\theta^0 - \frac{1}{2} M_\theta^{2c} \right)$$

$$+ \beta_{1c} \left(-M_{\dot\beta}^0 + \frac{1}{2} M_{\dot\beta}^{2c} + \frac{1}{2} M_{\dot\beta}^{2s} \right) + \lambda M_\lambda^{1s} \tag{6.198}$$

To be consistent, the second harmonics of the flap motion should be considered when the advance ratio is high enough to require modeling reverse flow. Such a calculation is best performed numerically. Using $\lambda = \lambda_{\text{NFP}} + \mu\theta_{1s} = \lambda_{\text{TPP}} - \mu\beta_{1c}$, the solution is

$$\beta_0 = \gamma \left[\theta_0 M_\theta^0 + \theta_{tw} M_{tw}^0 + \theta_{1s} \left(\frac{1}{2} M_\theta^{1s} + \mu M_\lambda^0 \right) \right.$$

$$\left. + \beta_{1c} \left(\frac{1}{2} M_\beta^{1c} - \frac{1}{2} M_{\dot\beta}^{1s} \right) + \lambda_{\text{NFP}} M_\lambda^0 \right] \tag{6.199}$$

$$\beta_{1s} = \frac{-\beta_0 M_\beta^{1c} - \theta_{1c} \left(M_\theta^0 + \frac{1}{2} M_\theta^{2c} \right)}{M_{\dot\beta}^0 + \frac{1}{2} M_{\dot\beta}^{2c} + \frac{1}{2} M_{\dot\beta}^{2s}} \tag{6.200}$$

$$\beta_{1c} = \frac{\theta_0 M_\theta^{1s} + \theta_{tw} M_{tw}^{1s} + \theta_{1s} \left(M_\theta^0 - \frac{1}{2} M_\theta^{2c} + \mu M_\lambda^{1s} \right) + \lambda_{\text{NFP}} M_\lambda^{1s}}{M_{\dot\beta}^0 - \frac{1}{2} M_{\dot\beta}^{2c} - \frac{1}{2} M_{\dot\beta}^{2s}} \tag{6.201}$$

$$= \frac{\theta_0 M_\theta^{1s} + \theta_{tw} M_{tw}^{1s} + \theta_{1s} \left(M_\theta^0 - \frac{1}{2} M_\theta^{2c} \right) + \lambda_{\text{TPP}} M_\lambda^{1s}}{M_{\dot\beta}^0 - \frac{1}{2} M_{\dot\beta}^{2c} - \frac{1}{2} M_{\dot\beta}^{2s} + \mu M_\lambda^{1s}} \tag{6.202}$$

Evaluating the required harmonics then gives

$$\beta_0 = \gamma \left[\frac{\theta_{.8}}{8}\left(1+\mu^2 - \frac{\mu^4}{8}\right) - \frac{\theta_{tw}}{60}\left(\mu^2 - \frac{9}{8}\mu^4\right) \right.$$
$$\left. - \frac{\lambda_{\mathrm{NFP}}}{6}\left(1 + \frac{2}{3\pi}\mu^3\right) - \frac{\mu^4}{15\pi}(\beta_{1c} + \theta_{1s}) \right] \tag{6.203}$$

$$\beta_{1s} - \theta_{1c} = \frac{-\frac{4}{3}\mu\left(1 + \frac{4}{15\pi}\mu^3\right)}{1 + \frac{1}{2}\mu^2 - \frac{1}{24}\mu^4}\beta_0 \tag{6.204}$$

$$\beta_{1c} + \theta_{1s} = \frac{-\frac{8}{3}\mu}{1 - \frac{1}{2}\mu^2 + \frac{7}{24}\mu^4}\left[\theta_{.75}\left(1 + \frac{4}{15\pi}\mu^3\right) - \frac{23}{75\pi}\theta_{tw}\mu^3 \right.$$
$$\left. - \frac{3}{4}\lambda_{\mathrm{NFP}}\left(1 - \frac{1}{4}\mu^2\right)\right] \tag{6.205}$$

$$= \frac{-\frac{8}{3}\mu}{1 + \frac{3}{2}\mu^2 - \frac{5}{24}\mu^4}\left[\theta_{.75}\left(1 + \frac{4}{15\pi}\mu^3\right) - \frac{23}{75\pi}\theta_{tw}\mu^3 \right.$$
$$\left. - \frac{3}{4}\lambda_{\mathrm{TPP}}\left(1 - \frac{1}{4}\mu^2\right)\right] \tag{6.206}$$

Thus reverse flow introduces flap motion of order μ^4. When $\mu > 0.5$, stall and compressibility must be considered, as well as reverse flow, and other blade degrees of freedom also become important. For example, the lift acting at the three-quarter chord in the reverse flow region produces a significant response in the blade elastic pitch above $\mu \cong 0.7$, which alters the blade loading. Generally a numerical calculation of the blade loading and motion is required at high advance ratio in order to use a consistent model.

The rotor thrust coefficient including the effects of reverse flow is

$$C_T = \sigma a \int_0^1 \frac{1}{2}|u_T|(u_T\theta - u_P)dr$$
$$= \frac{\sigma a}{2}\left[\frac{\theta_{.75}}{3}\left(1 + \frac{3}{2}\mu^2 - \frac{4}{3\pi}\mu^3\right) - \frac{\theta_{tw}}{8}\left(\mu^2 - \frac{8}{3\pi}\mu^3 + \frac{1}{4}\mu^4\right)\right.$$
$$\left. - \frac{\lambda_{\mathrm{NFP}}}{2}\left(1 + \frac{1}{2}\mu^2\right) - \frac{1}{8}\mu^3(\beta_{1c} + \theta_{1s})\right] \tag{6.207}$$

Reverse flow primarily introduces terms of higher order in μ. The $\lambda\mu^2$ term is the one effect of reverse flow aerodynamics that is significant even at fairly low advance ratio.

6.9 Blade Weight Moment

The force of gravity, approximately normal to the helicopter rotor disk, acts on the blade to produce a weight moment about the flap hinge. The blade weight opposes the lift force and thereby reduces the coning angle. The gravitational force on a blade element is mg, directed downward, with a moment arm r about the flap hinge.

The additional flap moment is thus

$$\int_0^R mgr\, dr = g\int_0^R rm\, dr = gS_b \qquad (6.208)$$

where $S_b = \int_0^R rm\, dr = M_b r_{CG}$ is the first moment about the flap hinge (M_b is the blade mass, and r_{CG} is the radial location of the center-of-gravity). This flap moment is added to the equation of motion, dimensionless quantities are used, and the equation is normalized by dividing by I_b. The result is

$$\ddot{\beta} + \beta = \gamma \int_0^1 r\frac{F_z}{ac}\, dr - \widehat{S}_b \frac{g}{\Omega^2 R} \qquad (6.209)$$

where

$$\widehat{S}_b = \frac{RS_b}{I_b} = \frac{\int_0^1 rm\, dr}{\int_0^1 r^2 m\, dr} \cong \frac{3}{2} \qquad (6.210)$$

The approximation $\widehat{S}_b \cong \frac{3}{2}$ is for a uniform mass distribution. The value of the dimensionless gravitational constant $g/\Omega^2 R$ is small, typically around 0.002. Assuming a constant tip speed, it scales with R, whereas for a given radius it is inversely proportional to the square of the tip speed.

The term $\widehat{S}_b g$ is a constant, so it only affects the solution for the coning angle. The coning is decreased by $\Delta\beta_0 = -\widehat{S}_b(g/\Omega^2 R)$, typically $-0.1°$ to $-0.2°$, which is small enough to be neglected for most purposes. The deflection due to the blade weight can be significant for very large rotors or very low tip speeds.

The dimensionless gravitational constant $g/\Omega^2 R$ can be viewed as the ratio of gravitational forces to centrifugal forces on the blade. Its small value implies that the rotor behavior is dominated by the centrifugal forces, and the blade weight generally has only a small influence.

6.10 Compressibility

Compressibility of the air influences the rotor performance and motion by its effects on the blade forces. Of particular importance are the increase in lift-curve slope with Mach number and the sharp increase in drag and pitching moment above a certain critical Mach number. When the blade is operating at high, unsteady angle-of-attack, as on the retreating side with high rotor loading, compressibility effects are important even at a low Mach number. The primary effect of compressibility on the rotor performance is a rapid increase in the profile power C_{Po} when the tip Mach number exceeds the critical Mach number for drag divergence. The critical Mach number depends on the angle-of-attack and is increased by the three-dimensional flow at the tip. The larger compressible lift-curve slope has little effect on β_{1c} or β_{1s}/β_0 (which involve only a balance of aerodynamic moments), but significantly increases the rotor thrust and coning angle at high tip speeds. The only practical means of accounting for the compressibility effects in detail is to use data for the airfoil aerodynamic characteristics expressed as a function of Mach number and angle-of-attack in a numerical calculation of the rotor loads and motion.

The blade normal Mach number in forward flight is

$$M = \frac{u_T}{c_s} = M_{\text{tip}}(r + \mu \sin\psi) \qquad (6.211)$$

where $M_{\text{tip}} = \Omega R/c_s$ is the Mach number based on the tip speed and c_s is the speed of sound. Write $M_{r,\psi}$ for the Mach number at radial station r and azimuth angle ψ. The highest Mach number occurs on the tip of the advancing blade:

$$M_{at} = M_{1,90} = M_{\text{tip}}(1 + \mu) \tag{6.212}$$

for edgewise flight. M_{tip} is a good parameter to represent the average effects of compressibility, whereas the advancing tip Mach number $M_{at} = M_{1,90}$ is a measure of the extreme effects. Compressibility limits the maximum speed of the helicopter. Since $M_{at} = (\Omega R + V)/c_s$, the critical Mach number of the blade constrains the forward speed of the helicopter, because the tip speed cannot be decreased too much without encountering other limits (see section 8.4).

The effect of the increased lift-curve slope on the helicopter loads and flapping can be estimated by using the Prandtl-Glauert relation for the lift-curve slope a:

$$a = \frac{a_{\text{incomp}}}{\sqrt{1 - M^2}} \tag{6.213}$$

Because the Mach number varies over the disk, the compressible lift-curve slope does as well. Thus the Prandtl-Glauert factor $1/\sqrt{1 - M^2}$ should be included in the integrands, which then cannot be evaluated analytically. Alternatively, an averaged lift-curve slope can be used for the entire rotor; for example, a slope based on the mean Mach number at effective radius r_e:

$$a = \frac{a_{\text{incomp}}}{\sqrt{1 - (r_e M_{\text{tip}})^2}} \tag{6.214}$$

Peters and Ormiston (1975) found that when the advancing tip Mach number is less than 0.9, this simple correction based on the 75% radius is sufficient. When M_{at} is above about 0.9, the radial and azimuthal variations of the compressibility effects must be included.

Gessow and Crim (1956) calculated the effects of high tip Mach number on the flapping, thrust, and power of a helicopter rotor in forward flight, using models similar to those developed in this chapter. They found a minor increase in the rotor flapping and thrust due to compressibility. The largest effect was an increase in profile power on the advancing side of the disk when the drag-divergence Mach number was exceeded. The increase in profile power correlated well with ΔM_d, the amount by which the advancing tip Mach number M_{at} exceeded the section drag-divergence Mach number, roughly according to the expression

$$\Delta C_{Po}/\sigma = 0.007 \, \Delta M_d + 0.052 (\Delta M_d)^2 \tag{6.215}$$

Such an estimate of compressibility effects on power can be useful, particularly for conceptual design studies, but the constants depend on the rotor technology considered (the results of Gessow and Crim (1956) are for a 12% thick airfoil).

6.11 Reynolds Number

The effect of an increase in Reynolds number on two-dimensional airfoil characteristics is to reduce the drag and increase the maximum lift. In terms of rotor tip speed and chord, the Reynolds number is $Re = \Omega Rc/\nu$, with $\nu = \mu/\rho$ the kinematic viscosity. For fixed velocities and air properties, the Reynolds number increases with the size of the rotorcraft. Hence a small-scale model has higher profile power and an earlier onset of stall compared to the full-scale aircraft. The Reynolds number for an aircraft decreases as the operating altitude increases, due to the increase in kinematic viscosity ν for the standard atmosphere.

Static airfoil table data consist of lift, drag, and moment coefficients as a function of angle-of-attack and Mach number. The Reynolds number $Re = Vc/\nu$ and Mach number $M = V/c_s$ both depend linearly on the velocity, so the Reynolds number of the airfoil table is proportional to the Mach number ($Re = M(c_s c/\nu)$), if the airfoil test is conducted at constant temperature and pressure. Typically the wind-tunnel temperature increases with speed for airfoil tests, so the relation is not quite linear.

If the Reynolds number Re of the rotor section does not equal the Reynolds number Re_t of the airfoil table, the drag and lift coefficients can be corrected:

$$c_\ell = K c_{\ell_{2D}}((\alpha - \alpha_z)/K + \alpha_z) \tag{6.216}$$

$$c_d = \frac{1}{K} c_{d_{2D}} \tag{6.217}$$

where $K = (Re/Re_t)^n$. Experimental data for turbulent flow suggest $n = 0.125$ to 0.2 for the exponent; see Yamauchi and Johnson (1983). The 1/5-th power law for a turbulent flat plate boundary layer implies $n = 0.2$. If the Reynolds number is larger for the airfoil than for the table, then $K > 1$, and these corrections reduce the drag coefficient and increase the maximum lift coefficient.

6.12 Tip Loss and Root Cutout

The decrease of the lift to zero over a finite distance at the blade tip can be accounted for by using a tip loss factor B such that the blade has drag but no lift when $r > BR$. In addition, the rotor has a root cutout, so that the blade airfoil starts at radial station r_R rather than at $r = 0$. When tip loss and root cutout are taken into account, the expression for the rotor thrust becomes

$$\begin{aligned}C_T &= \frac{\sigma a}{2} \int_{r_R}^{B} \left[(\theta_0 + r\theta_{tw})\left(r^2 + \frac{\mu^2}{2}\right) + \lambda_{\text{NFP}} r \right] dr \\ &= \frac{\sigma a}{2} \left[\frac{\theta_0}{3}\left(B^3 - r_R^3 + \frac{3}{2}(B - r_R)\mu^2\right) \right. \\ &\quad \left. + \frac{\theta_{tw}}{4}\left(B^4 - r_R^4 + (B^2 - r_R^2)\mu^2\right) - \frac{\lambda_{\text{NFP}}}{2}(B^2 - r_R^2) \right]\end{aligned} \tag{6.218}$$

The principal effect of the tip loss is to reduce the thrust for a given collective pitch, roughly by a factor B^3. The root cutout has only a minor influence on C_T. The tip loss has a similar effect on the coning, reducing it for fixed collective, but has less of an effect on β_{1c} or β_{1s}/β_0.

6.13 Assumptions and Examples

Simplifying the model was necessary to obtain an analytical solution for the rotor forces and flap motion. In particular, this analysis has neglected stall and compressibility effects in the rotor aerodynamics; nonuniformities in the induced velocity distribution, beyond the simplest linear variation; the higher harmonics of blade flap motion; and all degrees of freedom except for the fundamental flap mode. In general, small angles have been assumed. The section aerodynamic characteristics have been described by a constant lift-curve slope and a mean profile drag coefficient. The effects of stall, compressibility, and radial flow have been neglected. The blade has constant chord and linear twist.

So far in this chapter, only rigid flap motion and collective and cyclic pitch control have been considered. In this model there is no elastic flap motion and no lag or pitch degree of freedom. The rotor is articulated, with no flap-hinge offset, spring restraint, or pitch-flap coupling.

The solutions for thrust and flapping of the helicopter rotor in edgewise flight are thus

$$C_T = \frac{\sigma a}{2} \left[\frac{\theta_0}{3} \left(1 + \frac{3}{2}\mu^2 \right) B^3 + \frac{\theta_{tw}}{4} \left(1 + \mu^2 \right) B^4 \right.$$
$$\left. - \frac{1}{2} \left(\bar{\lambda}_{\text{NFP}} + \frac{\mu}{2}\lambda_y \right) \left(1 + \frac{1}{2}\mu^2 \right) B^2 \right] \quad (6.219)$$

$$\beta_0 = \gamma \left[\frac{\theta_0}{8} \left(1 + \mu^2 \right) B^4 - \frac{\theta_{tw}}{10} \left(1 + \frac{5}{6}\mu^2 \right) B^5 \right.$$
$$\left. - \frac{1}{6} \left(\bar{\lambda}_{\text{NFP}} + \frac{\mu}{2}\lambda_y \right) B^3 \right] - \widehat{S}_b \frac{g}{\Omega^2 R} \quad (6.220)$$

$$\beta_{1s} - \theta_{1c} = \frac{1}{1 + \frac{1}{2}\mu^2} \left[-\frac{4}{3}\mu\beta_0 - \lambda_x \right] \quad (6.221)$$

$$\beta_{1c} + \theta_{1s} = \frac{1}{1 - \frac{1}{2}\mu^2} \left[-\frac{8}{3}\mu \left(\theta_{.75} - \frac{3}{4}\bar{\lambda}_{\text{NFP}} \right) + \lambda_y \right] \quad (6.222)$$

including the principal effects of tip loss factor B, reverse flow (just the $\mu^2\lambda$ term in C_T), linear inflow gradient (λ_x and λ_y), and the weight moment ($\widehat{S}_b g$). These expressions are nearly as accurate as the results of Bailey (1941) for speeds up to $\mu = 0.5$.

Peters and Ormiston (1975) extended the calculation of the rotor steady-state flapping to hingeless rotors, considering the influence of various elements in the analytical model on the solution. Their work produced the following conclusions regarding the modeling requirements for the rotor flapping and loads analysis. For an accurate calculation of the n/rev flapping harmonics (β_{nc} and β_{ns}), the analysis must include all the harmonics up to m/rev, where $m = n$ for $0 < \mu < 0.4$, and $m = n + 1$ for $0.4 < \mu < 1.0$. Reverse flow is required only for $\mu > 0.6$. The tip loss factor is always important, but the root cutout must be accounted for only above $\mu = 1.0$. The effects of compressibility are important, but when the advancing tip Mach number $M_{at} < 0.9$, a simple correction based on the 75% radius is sufficient. The equivalent hinge spring and offset model is not really a very good representation of the hingeless rotor mode shape; the use of the actual elastic cantilever modes is preferred. To calculate the rotor motion, it is necessary to include only a single flapwise mode for $0 < \mu < 0.6$, two modes for $0.6 < \mu < 1.2$, and three modes for $1.2 < \mu < 1.6$.

6.13 Assumptions and Examples

The solution obtained under these approximations provides information about the rotor behavior and is reasonably accurate over a wide range of helicopter operating conditions. For a helicopter operating in more extreme flight conditions, including high speed, high tip Mach number, and high gross weight, one or more of the assumptions is no longer valid and a better model is required. Moreover, even at operating conditions for which a simple model gives a good estimate of performance and flapping, a better model is required to calculate blade loads and vibration.

Lifting-line theory is a fundamental assumption of rotor aerodynamics and is valid except near the blade tip or in the vicinity of a tip vortex from a preceding blade. The lag and pitch degrees of freedom and blade bending are important for vibration, blade loads, and aeroelastic stability, but can often be neglected for helicopter performance and control. Similarly, the higher harmonic blade motion, important for vibration and loads, can be neglected. Reverse flow can be neglected up to about $\mu = 0.5$, which covers the speed range of most helicopters. Neglect of stall and compressibility limits the validity of the theory at extreme operating conditions (high μ or high C_T/σ). Uniform inflow can be satisfactory for performance calculations at high speed (with the appropriate empirical factor κ), but leads to significant errors in the calculation of flap motion, particularly β_{1s}. Nonuniform inflow is also important for rotor loads and vibration. The constant-chord, linearly twisted blade is a typical rotor design, but more complex planform and twist are now common.

As an example of the solutions developed in this chapter, consider an articulated rotor with solidity $\sigma = 0.08$, Lock number $\gamma = 8$, and twist $\theta_{tw} = -8°$. Ideal aerodynamics are used, ignoring stall and compressibility: linear lift $c_\ell = a\alpha$ with lift-curve slope $a = 5.7$ and constant drag $c_d = 0.0080$. The tip loss factor is $B = 0.98$, and the blade has no root cutout. The induced velocity is obtained from momentum theory (uniform inflow), using $\kappa = 1.2$. The only blade motion is rigid flap and pitch, with no hinge offsets from the center of rotation. The influence of twist is examined by considering $\theta_{tw} = 0$ and $\theta_{tw} = -16°$.

The baseline operating condition is thrust $C_T/\sigma = 0.08$ and propulsive force corresponding to fuselage drag $f/A = 0.008$. Thrust variations of $C_T/\sigma = 0.12$ and $C_T/\sigma = 0.16$ are examined, as are propulsive force variations of $f/A = 0$ and $f/A = 0.016$. At high advance ratio, this rotor can attain $C_T/\sigma = 0.16$ only because stall has been neglected. The calculations were performed by adjusting collective pitch and shaft incidence to match the required thrust and propulsive force, with cyclic pitch for zero flapping relative to the shaft. Hence the shaft incidence angle is also the tip-path-plane incidence.

Figure 6.17 shows the inflow, disk incidence, coning, and collective and cyclic pitch control as a function of advance ratio. The inflow is

$$\lambda_{\text{TPP}} = \lambda_i + \mu \tan i_{\text{TPP}} = \frac{\kappa C_T}{2\sqrt{\mu^2 + \lambda^2}} + \frac{1}{2}\mu^3 \frac{f/A}{C_T} \qquad (6.223)$$

using $i_{\text{TPP}} = \theta_{FP} + D/W + H_{\text{TPP}}/T \cong D/W$ for level flight (equation 6.112). At low speed the inflow relative to the tip-path plane is due to the induced velocity, whereas at high speed it is due to the rotor propulsive force. The mean wake-induced velocity actually increases at high speed because of the asymmetric loading on the rotor disk (see Figure 5.7), an effect not shown here because $\kappa = 1.2$ is fixed. The inflow

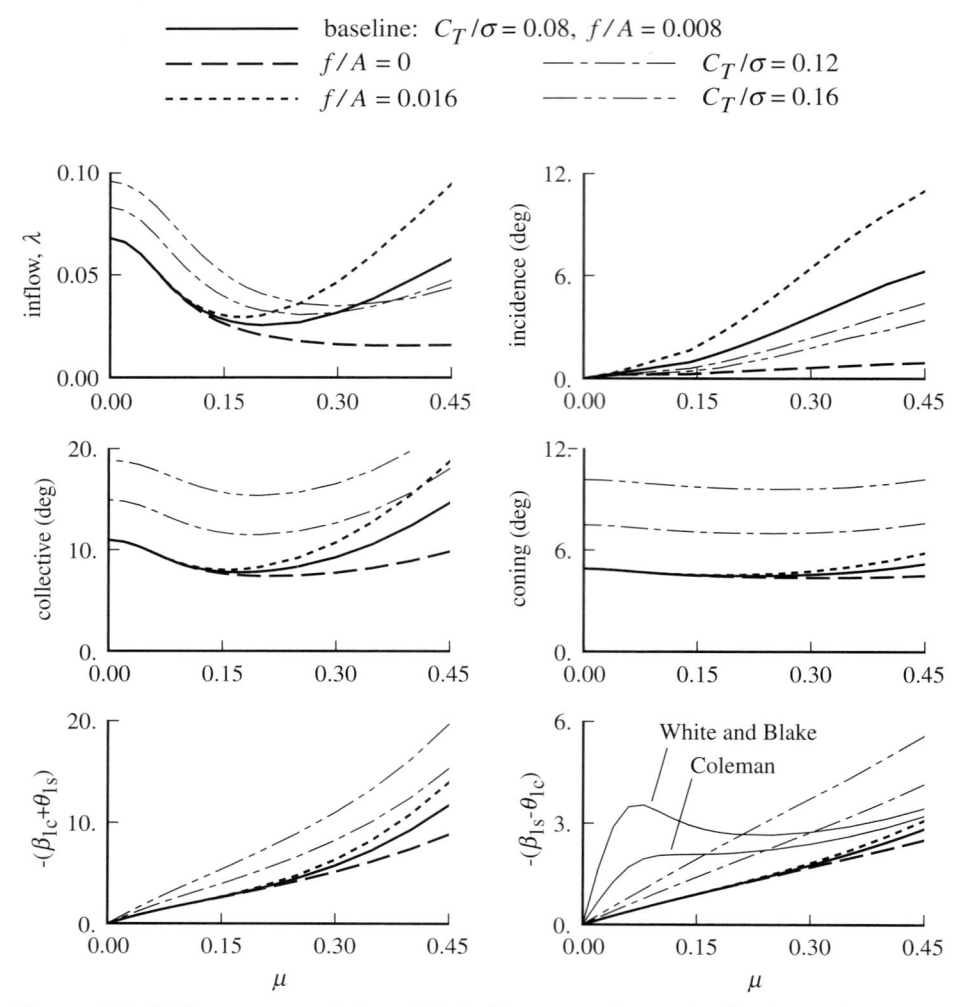

Figure 6.17. Helicopter rotor inflow, disk incidence, coning, and collective and cyclic pitch control variation with edgewise forward speed (uniform inflow).

variation is essentially the power variation as well, since with this simple blade aerodynamic model the profile power is nearly the same for all conditions. The disk incidence reflects the propulsive force; increasing the thrust requires less i_{TPP} to produce the required propulsive force. The collective pitch control tends to follow the variation of the inflow, since fixed rotor C_T/σ implies fixed mean blade angle-of-attack, which equates to the difference between pitch and inflow. The coning angle depends primarily on the rotor thrust. The flapping relative to the no-feathering plane is basically proportional to thrust and to speed. Figure 6.17 also shows the lateral flapping β_{1s} obtained using the White and Blake (1979) and Coleman, Feingold, and Stempin (1945) formulas for longitudinal inflow gradient λ_x (described in section 5.2.2). There is little effect of twist on the parameters shown in Figure 6.17.

Figure 6.18 shows the variation of the section angle-of-attack as the blade moves around the rotor disk, for the baseline condition ($C_T/\sigma = 0.08$, $f/A = 0.008$,

6.13 Assumptions and Examples

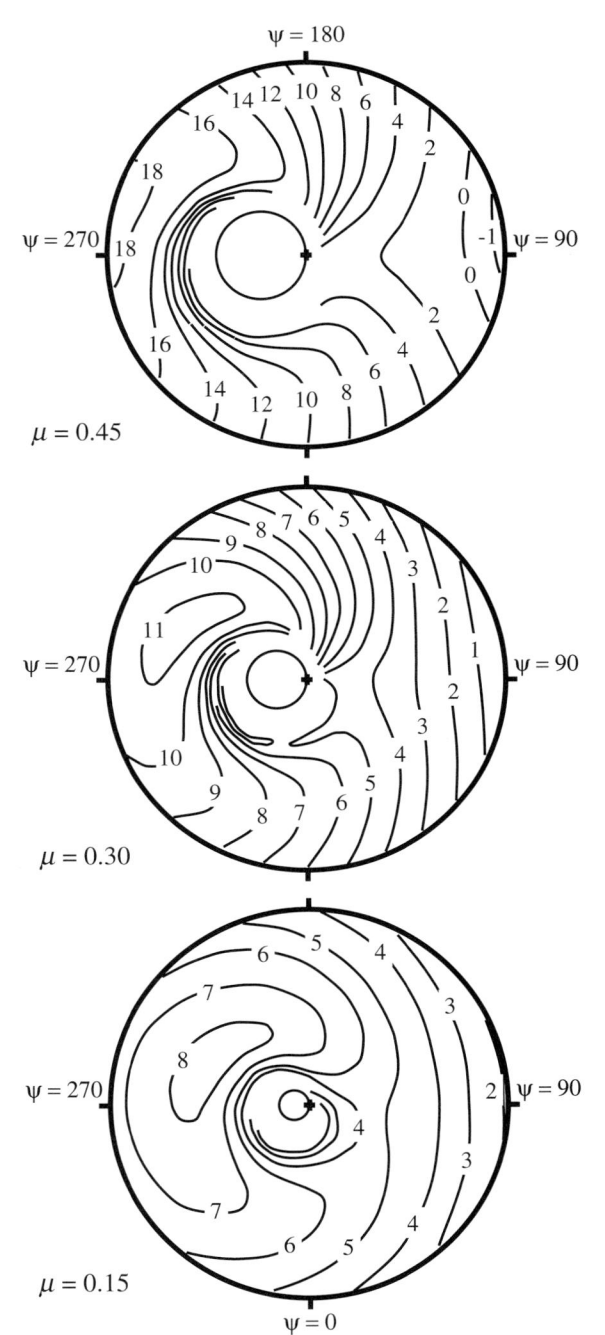

Figure 6.18. Blade angle-of-attack distribution (in degrees) for $C_T/\sigma = 0.08$, $f/A = 0.008$, $\theta_{tw} = -8°$, $\sigma = 0.08$ (uniform inflow); $\mu = 0.15, 0.30, 0.45$.

$\theta_{tw} = -8°$) at advance ratios of $\mu = 0.15, 0.30$, and 0.45. With the linear aerodynamic model and small angle assumption, the angle-of-attack is given by

$$\alpha = \theta - u_P/u_T = \theta_0 + r\theta_{tw} + (\beta_{1c} + \theta_{1s})\sin\psi$$
$$- (\beta_{1s} - \theta_{1c})\cos\psi - \frac{\lambda_{TPP} + \mu\beta_0\cos\psi}{r + \mu\sin\psi} \quad (6.224)$$

Forward speed reduces the angle-of-attack on the advancing side and increases the angle-of-attack on the retreating side as the blade flaps to maintain the same lift on both sides of the disk in the asymmetric aerodynamic environment of forward flight (really, to maintain zero 1/rev flapping moment). In forward flight the working angles-of-attack are concentrated on the front and rear of the disk. The maximum angle-of-attack occurs in the third quadrant of the disk (around $\psi = 240°$) for this twist, and it increases with advance ratio. At high speed the angle-of-attack on the advancing tip is negative.

The blade angle-of-attack distribution depends on the operating condition (thrust and propulsive force) and twist, as shown in Figure 6.19. The blade is well into stall angles-of-attack on the retreating side for $C_T/\sigma = 0.12$ (here there are no negative consequences of high angle-of-attack, such as high drag and power). Increasing propulsive force increases the areas of high angle-of-attack on the retreating side, which implies a degradation of rotor efficiency. Note the significant change in angle-of-attack distribution for the rotor with zero twist. In forward flight, negative twist reduces the maximum angle-of-attack and shifts the stall region inboard from the blade tip. Although a twist of $\theta_{tw} = -16°$ is good for hover performance, it leads to negative angle-of-attack on the advancing tip even for $\mu = 0.30$. All these results are for uniform inflow, however. The interference from discrete tip vortices in the rotor wake, captured by a nonuniform inflow calculation, can significantly alter the angle-of-attack distribution (see section 9.6).

The blade bound circulation ($\Gamma/\Omega^2 R$) and section lift ($d(C_T/\sigma)/dr = F_z/(\rho(\Omega R)^2 c)$) corresponding to the angle-of-attack distribution are shown in Figure 6.20 for the baseline condition. The change in pattern is produced by the additional factors of $u_T = r + \mu \sin\psi$. Figure 6.21 shows the radial and azimuthal variation of the blade section lift at advance ratios of $\mu = 0.15, 0.30$, and 0.45, for the baseline condition ($C_T/\sigma = 0.08$, $f/A = 0.008$, $\theta_{tw} = -8°$). Observe that at high speed, the requirement for roll moment balance between the advancing and retreating sides is achieved with very different radial distribution of the loading, a consequence of rigid flap and rigid pitch motion in the aerodynamic environment of forward flight. The result is negative lift on the advancing blade tip at high advance ratio. The small variation of loading with azimuth at low advance ratio follows from the assumption of uniform inflow. Blade-vortex interaction in fact produces substantial higher-harmonic loading variation at low speed.

At high advance ratio, the rotor behavior changes significantly. Figure 6.22 shows the derivatives of rotor thrust and flapping as a function of advance ratio. Simple calculations (uniform inflow and rigid blade) and complex calculations (nonuniform inflow and elastic blade) are compared with measurements. The baseline calculations are for 2° collective, zero shaft angle, and 1/rev flapping trimmed to zero. The measurements were derived by Harris (1987) from wind-tunnel tests of teetering and articulated rotors by McCloud, Biggers, and Stroub (1968) and Charles and Tanner (1969); and by Datta, Yeo, and Norman (2011). As advance ratio increases, the thrust change with collective pitch ($\partial(C_T/\sigma)/\partial\theta$) decreases, becoming negative around $\mu \cong 1$. The thrust change with shaft angle ($\partial(C_T/\sigma)/\partial i$) continues to increase with μ, with a positive thrust increment for aft tilt of the shaft. So at high speed, collective is not effective in controlling the rotor thrust, but rotor shaft or aircraft incidence remains effective. Rotor cyclic can still control rotor flapping ($\partial\beta_{1c}/\partial\theta_{\text{cyc}}$ and $\partial\beta_{1s}/\partial\theta_{\text{cyc}}$) at high advance ratio.

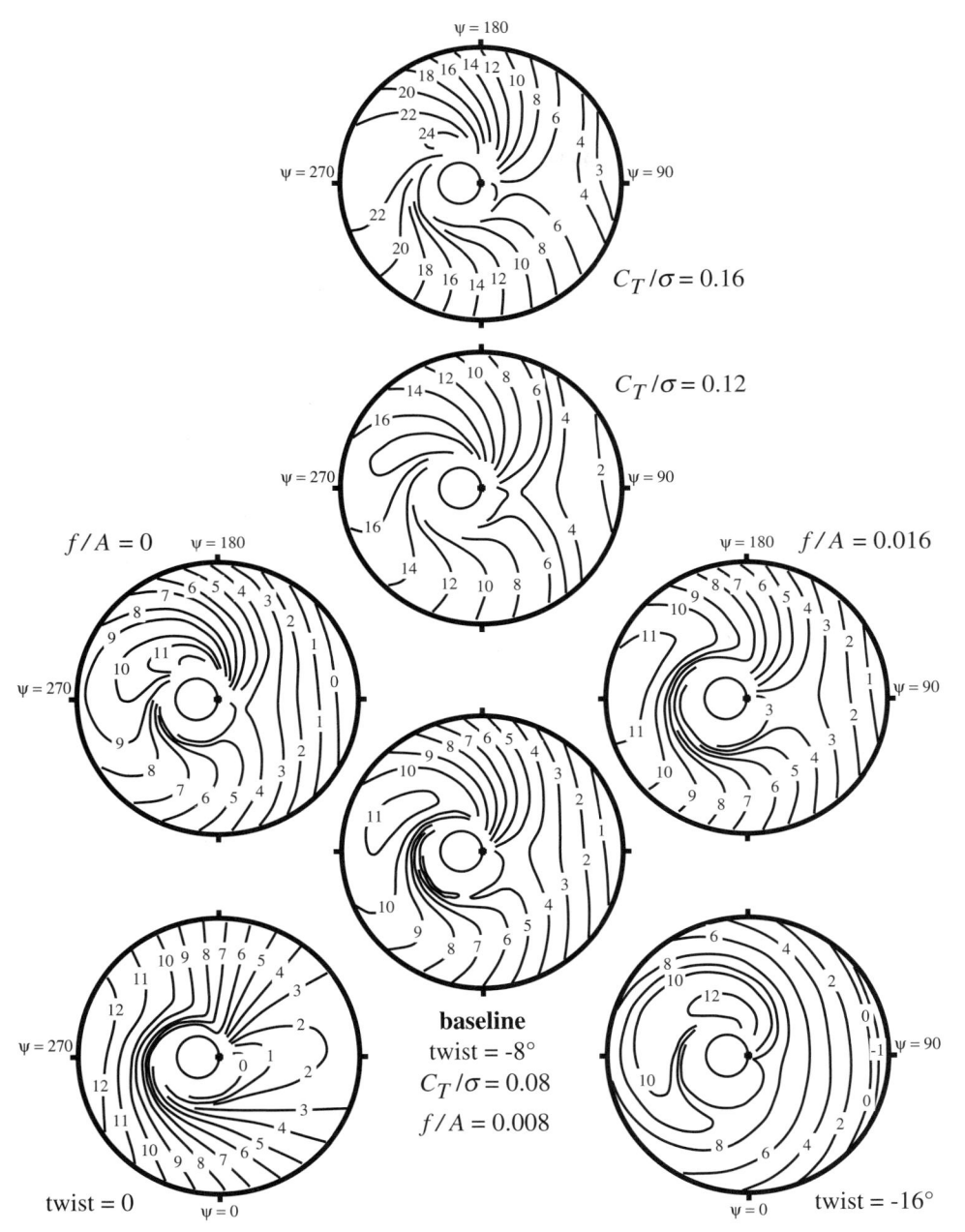

Figure 6.19. Variation of blade angle-of-attack distribution at $\mu = 0.30$ (in degrees) with thrust C_T/σ, propulsive force f/A, and twist θ_{tw} (uniform inflow).

6.14 Flap Motion with a Hinge Spring

Consider an articulated rotor blade with no hinge offset from the center of rotation, but now with a spring about the flap hinge that produces a restoring moment on the blade (Figure 6.23). Such a spring might be used to augment the rotor control power, because with a spring the flap motion not only tilts the rotor thrust vector but also directly produces a moment on the hub. Since a hingeless rotor has a structural

202　　　　　　　　　　　　　　　Forward Flight

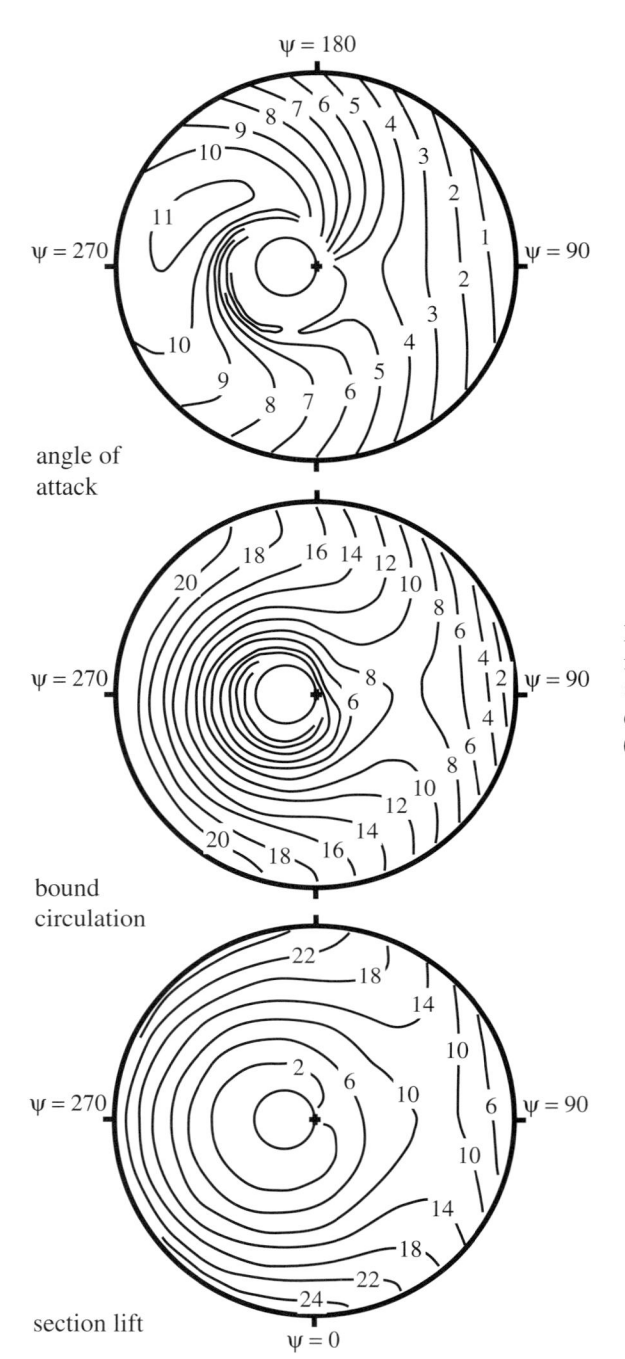

Figure 6.20. Blade angle-of-attack (degrees), blade bound circulation ($1000\,\Gamma/\Omega^2 R$) and section lift ($100d(C_T/\sigma)/dr$) for $C_T/\sigma = 0.08$, $f/A = 0.008$, $\theta_{tw} = -8°$, $\sigma = 0.08$, $\mu = 0.30$ (uniform inflow).

spring at the blade root, consideration of the blade with a flap-hinge spring serves as a guide to hingeless rotor behavior as well. The blade motion is assumed to still consist of only rigid rotation about the flap hinge, so the out-of-plane deflection is $z = r\beta$. For a very stiff spring the blade root restraint would approach that of a cantilevered blade, introducing considerable bending into the fundamental flapping mode shape. The spring stiffness that might be used on a rotor blade would be small

6.14 Flap Motion with a Hinge Spring

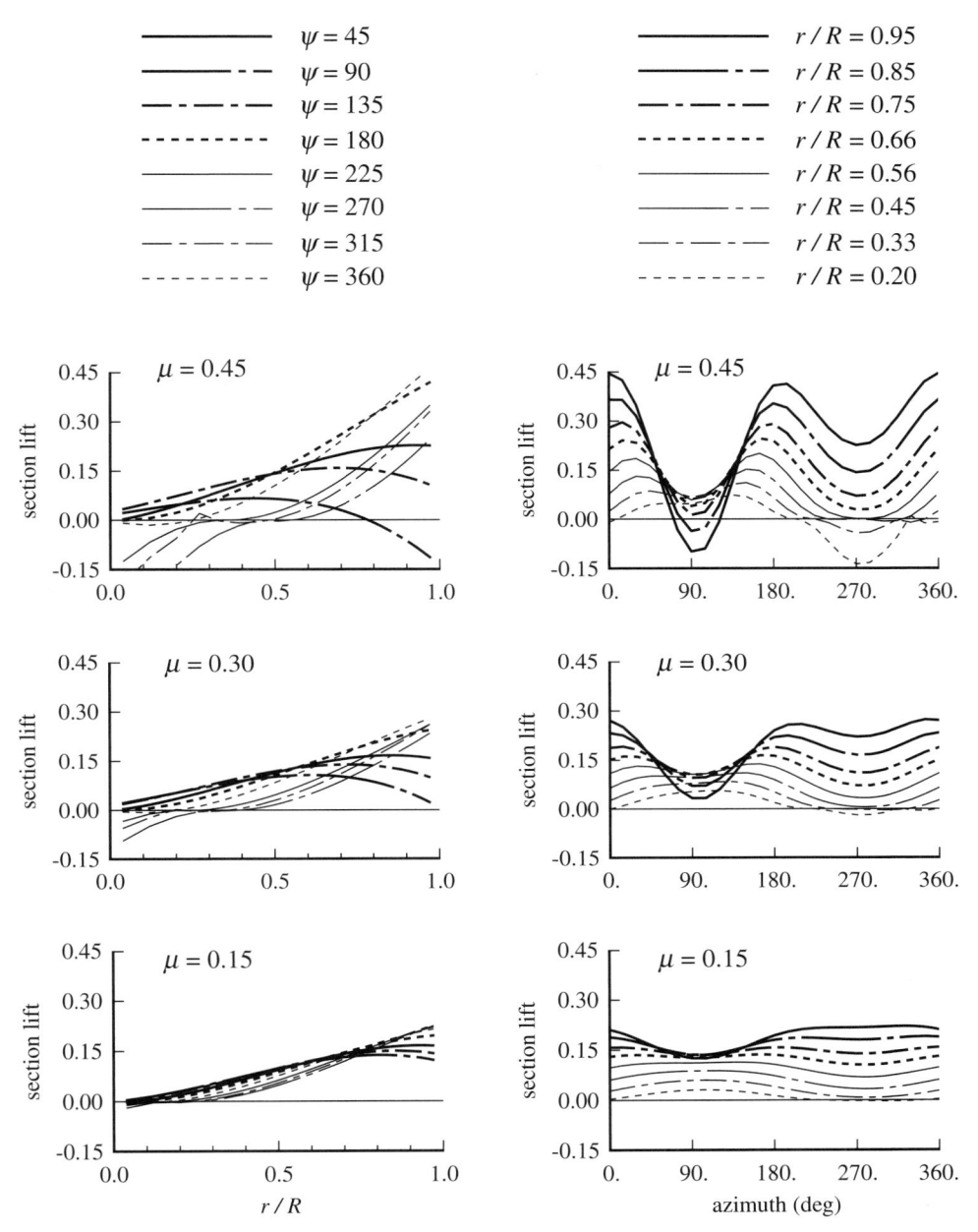

Figure 6.21. Blade section lift $(d(C_T/\sigma)/dr)$ for $C_T/\sigma = 0.08$, $f/A = 0.008$, $\theta_{tw} = -8°$, $\sigma = 0.08$ (uniform inflow); $\mu = 0.15, 0.30, 0.45$.

compared to the centrifugal stiffening, however, so the rigid flapping assumption is reasonable. With rigid flapping motion, the equations for the rotor forces and power are unchanged. The hinge spring does change the rotor flapping equation of motion, introducing an additional flap moment. Because the spring moment is proportional to the flapping displacement relative to the shaft, the hub plane is the appropriate reference plane in this case.

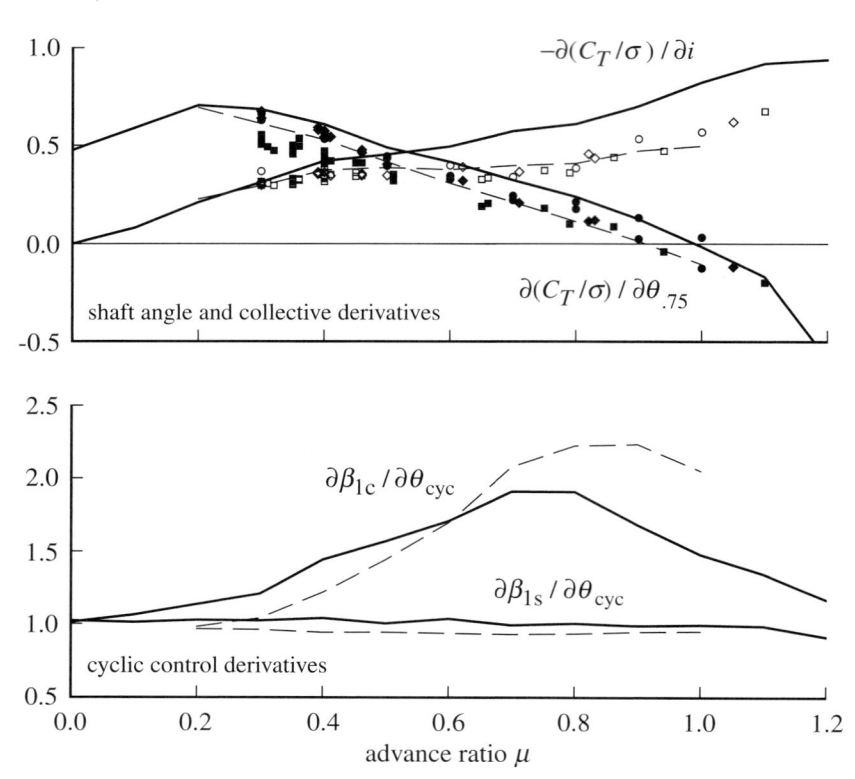

Figure 6.22. Rotor behavior at high advance ratio.

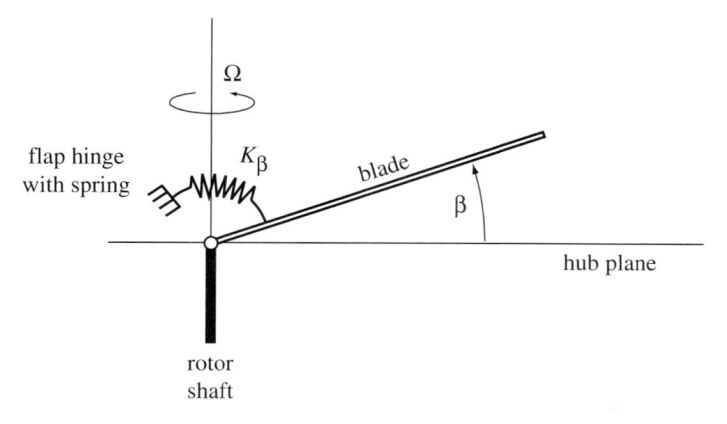

Figure 6.23. Blade flap motion, with a hinge spring.

6.14 Flap Motion with a Hinge Spring

In the derivation of the flapping equation of motion, the only addition is the flap moment due to the hinge spring: $K_\beta (\beta - \beta_p)$, where K_β is the spring rate and β_p is the precone angle. With a spring at the flap hinge, the blade coning would produce a steady root moment except for the precone angle, which biases the hinge moment to zero for $\beta - \beta_p$. Then the flapping equation becomes

$$I_b \left(\ddot{\beta} + \Omega^2 \beta\right) + K_\beta (\beta - \beta_p) = \int_0^R r F_z \, dr \tag{6.225}$$

or

$$\ddot{\beta} + \nu^2 \beta = (\nu^2 - 1)\beta_p + \gamma \int_0^1 r \frac{F_z}{ac} \, dr \tag{6.226}$$

where

$$\nu^2 = 1 + \frac{K_\beta}{I_b \Omega^2} \tag{6.227}$$

is the dimensionless natural frequency of the flap motion in the rotating frame. For practical flap springs, ν is just slightly greater than 1. When $\nu > 1$, the aerodynamic forces acting at 1/rev are no longer forcing the flap motion exactly at resonance. Thus the rotor responds to this excitation with a reduced magnitude, and the lag is somewhat less than 90° in azimuth due to the spring quickening the response. Flap-hinge offset or cantilever root restraint also increases the natural frequency of the flapping. By considering the articulated blade with a hinge spring, the fundamental influence of the flap frequency is isolated, since the hinge spring changes nothing else. If the present problem is considered a model for an arbitrary rotor with flap frequency ν, the approximation lies in using rigid flapping for the blade mode shape.

The aerodynamic flap moments are unchanged by the hinge spring, but the inertial, centrifugal, and spring terms of the flapping equation now give

$$\frac{1}{2\pi}\int_0^{2\pi} \left(\ddot{\beta} + \nu^2 \beta - (\nu^2 - 1)\beta_p\right) d\psi = \nu^2 \beta_0 - (\nu^2 - 1)\beta_p \tag{6.228}$$

$$\frac{1}{2\pi}\int_0^{2\pi} \left(\ddot{\beta} + \nu^2 \beta - (\nu^2 - 1)\beta_p\right) \cos\psi \, d\psi = (\nu^2 - 1)\beta_{1c} \tag{6.229}$$

$$\frac{1}{2\pi}\int_0^{2\pi} \left(\ddot{\beta} + \nu^2 \beta - (\nu^2 - 1)\beta_p\right) \sin\psi \, d\psi = (\nu^2 - 1)\beta_{1c} \tag{6.230}$$

The flapping equations thus become

$$\nu^2 \beta_0 = (\nu^2 - 1)\beta_p + \gamma \left[\frac{\theta_{.8}}{8}(1 + \mu^2) - \frac{\mu^2}{60}\theta_{tw} - \frac{\lambda_{\text{NFP}}}{6}\right] \tag{6.231}$$

$$(\nu^2 - 1)\beta_{1c} = \gamma \left[\frac{1}{8}(\theta_{1c} - \beta_{1s})\left(1 + \frac{1}{2}\mu^2\right) - \frac{\mu}{6}\beta_0\right] \tag{6.232}$$

$$(\nu^2 - 1)\beta_{1s} = \gamma \left[\frac{1}{8}(\theta_{1s} + \beta_{1c})\left(1 - \frac{1}{2}\mu^2\right) + \frac{\mu}{3}\theta_{.75} - \frac{\mu}{4}\lambda_{\text{NFP}}\right] \tag{6.233}$$

The solution for the coning is

$$\beta_0 = \frac{\nu^2 - 1}{\nu^2}\beta_p + \frac{\gamma}{\nu^2}\left[\frac{\theta_{.8}}{8}(1 + \mu^2) - \frac{\mu^2}{60}\theta_{tw} - \frac{\lambda_{\text{NFP}}}{6}\right] \tag{6.234}$$

The flap spring reduces the coning angle. The solution can be written in terms of the coning with no spring:

$$\beta_0 = \frac{1}{\nu^2}\beta_{\text{ideal}} + \frac{\nu^2 - 1}{\nu^2}\beta_p \tag{6.235}$$

where β_{ideal} is the coning angle for $\nu = 1$. The role of the hub precone angle is to reduce the steady moments in the blade root. The mean hinge spring moment is

$$(\nu^2 - 1)(\beta_0 - \beta_p) = \frac{\nu^2 - 1}{\nu^2}(\beta_{\text{ideal}} - \beta_p) \tag{6.236}$$

The mean moment is thus nonzero when $\nu > 1$, unless the precone angle is selected to be $\beta_p = \beta_{\text{ideal}}$. With ideal precone, the coning angle $\beta_0 = \beta_{\text{ideal}}$ is independent of the flap frequency. Although the proper choice of precone reduces the root loads, the ideal value is a function of the rotor loading, and therefore the precone must be selected for a particular design operating condition.

Now consider the tip-path-plane tilt. In hover the equations become

$$\beta_{1s} + \frac{\nu^2 - 1}{\gamma/8}\beta_{1c} = \theta_{1c} \tag{6.237}$$

$$\beta_{1c} - \frac{\nu^2 - 1}{\gamma/8}\beta_{1s} = -\theta_{1s} \tag{6.238}$$

which have the solution

$$\beta_{1s} = \frac{1}{1 + N_\star^2}(\theta_{1c} + N_\star\theta_{1s}) \tag{6.239}$$

$$\beta_{1c} = \frac{1}{1 + N_\star^2}(-\theta_{1s} + N_\star\theta_{1c}) \tag{6.240}$$

The parameter $N_\star = \frac{8}{\gamma}(\nu^2 - 1)$ is the ratio of the hinge spring to the aerodynamic flap damping. The inertia and centrifugal forces still exactly cancel, so the spring and damping forces determine the response. The effect of a flap frequency ν greater than 1 is to introduce lateral flapping due to θ_{1s} and longitudinal flapping due to θ_{1c}. Write the cyclic flapping as $\bar{\beta}\cos(\psi + \psi_0 - \Delta\psi)$ and the cyclic pitch as $\bar{\theta}\cos(\psi + \psi_0)$. Then the magnitude and phase of the response are

$$\bar{\beta}/\bar{\theta} = \frac{1}{\sqrt{1 + N_\star^2}} \tag{6.241}$$

$$\Delta\psi = 90° - \tan^{-1} N_\star \tag{6.242}$$

Increasing the flap frequency so that the system is excited below resonance slightly reduces the amplitude of the flap response to cyclic and, most importantly, reduces the lag in the response. For example, with $\nu = 1.15$ and $\gamma = 8$, the amplitude is reduced only about 5%, but the lag becomes 72° instead of the 90° of an articulated rotor. This phase change constitutes a coupling of the lateral and longitudinal response of the tip-path plane to the no-feathering-plane control inputs. As far as control of the helicopter is concerned, such coupling can be eliminated by introducing a compensating phase shift between the control plane and no-feathering plane. That is, the control system geometry is changed so that the rotor still responds with purely longitudinal tip-path-plane tilt due to longitudinal cyclic stick displacements.

6.14 Flap Motion with a Hinge Spring

In forward flight, the solution for the cyclic required to trim the helicopter is

$$\theta_{1c} = \beta_{1s} + \frac{1}{1+\frac{1}{2}\mu^2}\left[N_\star\beta_{1c} + \frac{4}{3}\mu\beta_0\right] \tag{6.243}$$

$$\theta_{1s} = -\beta_{1c} + \frac{1}{1+\frac{3}{2}\mu^2}\left[N_\star\beta_{1s} - \frac{8}{3}\mu\left(\theta_{.75} - \frac{3}{4}\lambda_{\mathrm{TPP}}\right)\right] \tag{6.244}$$

The tip-path-plane tilt relative to the hub plane (β_{1c} and β_{1s}) is determined by the equilibrium of helicopter forces and moments. The second term in these expressions gives the phase shift arising when $\nu > 1$. Forward speed has an influence on the phase shift, and moreover that influence is not the same for both axes of cyclic. The ideal control rigging to compensate for the lateral-longitudinal coupling varies with speed (by typically 5% to 15% between hover and maximum speed) and is not the same for lateral and longitudinal cyclic. The influence of forward flight is only of order μ^2, so a single value for the control system phase can be chosen that is satisfactory over the entire speed range of the helicopter.

The helicopter is controlled by using the rotor to produce moments about the aircraft center of gravity. An articulated rotor (with no hinge offset) has no moment at the blade root and thus can produce moments on the helicopter only by tilting the rotor thrust vector. With a hinge spring, tilt of the rotor tip-path plane also produces a moment on the rotor hub. In the rotating frame, the hub moment due to the flap deflection of a single blade is

$$M = K_\beta(\beta - \beta_p) = (\nu^2 - 1)I_b\Omega^2(\beta - \beta_p) \tag{6.245}$$

The pitch and roll moments on the hub are obtained by resolving the flap moment in the non-rotating frame, multiplying by the number of blades, and averaging over the azimuth:

$$M_y = -\frac{N}{2\pi}\int_0^{2\pi}\cos\psi\, M\, d\psi, \quad M_x = \frac{N}{2\pi}\int_0^{2\pi}\sin\psi\, M\, d\psi, \tag{6.246}$$

(see also section 6.3). In coefficient form, the hub pitch moment C_{My} and roll moment C_{Mx} are

$$\frac{2C_{My}}{\sigma a} = -\frac{\nu^2 - 1}{\gamma}\beta_{1c} \tag{6.247}$$

$$\frac{2C_{Mx}}{\sigma a} = \frac{\nu^2 - 1}{\gamma}\beta_{1s} \tag{6.248}$$

The in-plane forces on the rotor can be written $H_{\mathrm{HP}} = H_{\mathrm{TPP}} - T\beta_{1c}$ and $Y_{\mathrm{HP}} = Y_{\mathrm{TPP}} - T\beta_{1s}$. Neglecting the forces relative to the tip-path-plane, the pitch and roll moments about the helicopter center-of-gravity a distance h below the hub are $M_y = hH_{\mathrm{HP}} = -hT\beta_{1c}$ and $M_x = -hY_{\mathrm{HP}} = hT\beta_{1s}$. Combining the moments due to the thrust tilt and the hinge spring, the total moments about the helicopter center-of-gravity due to the rotor tip-path-plane tilt are

$$\begin{pmatrix} -\dfrac{2C_{My}}{\sigma a} \\ \dfrac{2C_{Mx}}{\sigma a} \end{pmatrix} = \left(\frac{\nu^2 - 1}{\gamma} + h\frac{2C_T}{\sigma a}\right)\begin{pmatrix}\beta_{1c} \\ \beta_{1s}\end{pmatrix} \tag{6.249}$$

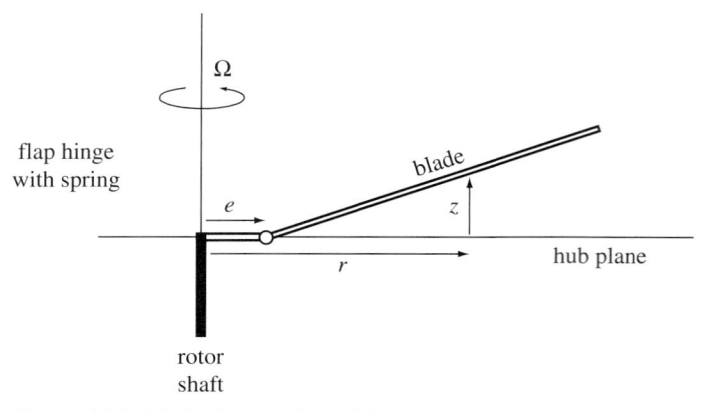

Figure 6.24. Blade flap motion with hinge offset.

The moment generating capability of the helicopter is increased greatly when $\nu > 1$. An articulated rotor normally obtains about half its moment from hinge offset and half from the thrust tilt. For a hingeless rotor the direct hub moment can be two to four times the thrust tilt term. Moreover, the direct hub moment term is independent of the helicopter load factor.

Including the effect of the flap frequency ν, the solution for the collective and cyclic control from the thrust, tip-path-plane incidence, and flapping relative to the hub plane is

$$\theta_{.75} = \frac{1}{\Delta}\left[\left(1 + \frac{3}{2}\mu^2\right)\left(\frac{6C_T}{\sigma a} + \frac{3}{8}\mu^2\theta_{tw}\right) + \frac{3}{2}\lambda_{TPP}\left(1 - \frac{1}{2}\mu^2\right) - \frac{3}{2}\mu N_\star\beta_{1s}\right] \tag{6.250}$$

$$\theta_{1s} = -\beta_{1c} - \frac{1}{\Delta}\left[\frac{8}{3}\mu\left(\frac{6C_T}{\sigma a} + \frac{3}{8}\mu^2\theta_{tw}\right) + 2\mu\lambda_{TPP}\left(1 - \frac{3}{2}\mu^2\right) - \left(1 + \frac{3}{2}\mu^2\right)N_\star\beta_{1s}\right] \tag{6.251}$$

$$\beta_0 = \frac{\gamma/8\nu^2}{\Delta}\left[\left(1 - \frac{19}{18}\mu^2 + \frac{3}{2}\mu^4\right)\frac{6C_T}{\sigma a} + \left(\frac{1}{20} + \frac{29}{120}\mu^2 - \frac{1}{5}\mu^4 + \frac{3}{8}\mu^6\right)\theta_{tw}\right.$$
$$\left. + \lambda_{TPP}\left(\frac{1}{6} - \frac{7}{12}\mu^2 + \frac{1}{4}\mu^4\right) - \frac{\mu}{6}(1 - 3\mu^2)N_\star\beta_{1s}\right] + \frac{\nu^2 - 1}{\nu^2}\beta_p \tag{6.252}$$

$$\theta_{1c} = \beta_{1s} + \frac{1}{1 + \frac{1}{2}\mu^2}\left[\frac{4}{3}\mu\beta_0 + N_\star\beta_{1c}\right] \tag{6.253}$$

where $\Delta = 1 - \mu^2 + \frac{9}{4}\mu^4$ (see section 6.5). This result neglects tip losses and reverse flow, assumes uniform inflow, and uses $\eta = r$ for the flap mode shape.

6.15 Flap-Hinge Offset

Consider next an articulated rotor with the flap-hinge offset from the center of rotation by a distance eR (Figure 6.24). Such an arrangement is usually mechanically simpler than one with no offset and in addition has a favorable influence on the helicopter handling qualities, because the offest produces a flap frequency above

6.15 Flap-Hinge Offset

1/rev. Articulated rotors typically have an offset of $e = 0.03$ to 0.05. The analysis in this section also considers a hinge spring. The blade radial coordinate r is still measured from the center of rotation. The blade motion is rigid rotation about the flap hinge, with degree of freedom β and mode shape $\eta(r)$, such that the out-of-plane deflection is $\beta\eta$.

Rigid rotation about a flap-hinge offset by e corresponds to a mode shape

$$\eta = \begin{cases} k(r - e) & r \geq e \\ 0 & r < e \end{cases} \qquad (6.254)$$

where k is a constant determined by the mode shape normalization. The normalization used here requires that the mode shape be equal to unity at the blade tip: $\eta(1) = 1$. Thus $k = 1/(1 - e)$, and the mode shape is $\eta = (r - e)/(1 - e)$. This reduces to $\eta = r$ for the case of no offset. Normalizing the mode shape to unity at the tip means that the degree of freedom β can be interpreted as the angle between the disk plane and a line extending from the center of rotation to the blade tip. This normalization is easily extended to the higher bending modes. An alternative mode shape is $\eta = (r - e)$, which makes β the actual angle of rotation about the flap hinge. The physically relevant quantities of the solution, such as the out-of-plane deflection $z = \beta\eta$, must of course be independent of the normalization chosen for the mode shape.

The normal velocity of the blade with an arbitrary flapping mode shape becomes

$$u_P = \lambda + \dot{z} + u_R \frac{dz}{dr} = \lambda + \eta\dot{\beta} + \eta'\beta\mu\cos\psi \qquad (6.255)$$

There are no other changes to the blade aerodynamics. The rotor thrust is then

$$C_T = \sigma a \int_0^1 \frac{1}{2}(u_T^2\theta - u_P u_T)dr$$

$$= \frac{\sigma a}{2}\left[\frac{\theta_{.75}}{3}\left(1 + \frac{3}{2}\mu^2\right) - \frac{\theta_{tw}}{8}\mu^2 - \frac{1}{2}(\lambda - \mu\theta_{1s}) - \frac{\mu}{2}\beta_{1c}\frac{e}{1-e}\right] \qquad (6.256)$$

So the effect of the hinge offset on C_T is small. Similar results can be obtained for C_H and C_Y. Recall that the rotor power was derived for a general mode shape. The principal influence of the hinge offset is on the rotor flapping motion.

Consider again equilibrium of moments about the flap hinge. The forces acting on the blade section are as follows:

i) an inertial force $m\ddot{z} = m\eta\ddot{\beta}$ opposing the flap motion, with moment arm $(r - e)$ about the flap hinge
ii) a centrifugal force $m\Omega^2 r$ directed radially outward, with moment arm $z = \eta\beta$
iii) an aerodynamic force F_z normal to the blade, with moment arm $(r - e)$

There is also a spring moment at the flap hinge, $K_\beta(\beta - \beta_p)$, as in section 6.14. For now a general mode shape $\eta = k(r - e)$ is allowed. On integrating over the blade span, equilibrium of the flap moments gives

$$\int_e^R \eta(r - e)m\,dr\,\ddot{\beta} + \int_e^R \eta r m\,dr\,\Omega^2\beta + K_\beta(\beta - \beta_p) = \int_e^R (r - e)F_z dr \qquad (6.257)$$

Now use dimensionless quantities and multiply by $k = \eta(1)/(1 - e)$:

$$\int_e^1 \eta^2 m\,dr\,\ddot{\beta} + k\int_e^1 \eta r m\,dr\,\beta + \frac{K_\beta k}{\Omega^2}(\beta - \beta_p) = \int_e^1 \eta F_z dr \qquad (6.258)$$

Let $I_b = \int_e^1 \eta^2 m \, dr$, and

$$k \int_e^1 \eta r m \, dr = \int_e^1 \eta^2 m \, dr + ke \int_e^1 \eta m \, dr = I_b + \frac{e\eta(1)}{1-e} \int_e^1 \eta m \, dr \qquad (6.259)$$

Then the flapping equation of motion is

$$\ddot{\beta} + \nu^2 \beta = \frac{K_\beta \eta(1)}{I_b \Omega^2 (1-e)} \beta_p + \gamma \int_e^1 \eta \frac{F_z}{ac} \, dr \qquad (6.260)$$

The Lock number is defined as $\gamma = \rho a c R^4 / I_b$ again, but here the definition of the characteristic moment of inertia I_b depends on the mode shape. If the definition $I_b = \int_0^1 r^2 m \, dr$ were retained, the normalized flapping inertia $\widehat{I}_\beta = \int_e^1 \eta^2 m \, dr / I_b$ would be introduced on the left-hand side of the flap equation. Such an approach is best when more degrees of freedom are considered, but here using the flapping generalized mass for I_b and γ is simplest.

The natural frequency of the flap motion for the blade with hinge offset and spring is

$$\nu^2 = 1 + \frac{e}{1-e} \frac{\eta(1) \int_e^1 \eta m \, dr}{\int_e^1 \eta^2 m \, dr} + \frac{K_\beta}{I_b \Omega^2 (1-e)} \qquad (6.261)$$

The first term is the centrifugal spring, the second term is the hinge offset effect (also due to the centrifugal forces), and the third term is the hinge spring. For a uniform mass distribution and no hinge spring, the result is

$$\nu^2 = 1 + \frac{3}{2} \frac{e}{1-e} \qquad (6.262)$$

In general the flap frequency can be written $\nu^2 = 1 + e r_{CG} M / I$, where M is the blade mass, I the moment of inertia about the flap hinge, and r_{CG} the radial center-of-gravity location relative to the hinge. Hinge offset thus raises the flap frequency above 1/rev. For the offsets of articulated rotors the increase is small, typically giving $\nu = 1.02$ to 1.04. This increase in flap frequency is the primary influence of the hinge offset. The flap dynamics of a rotor with $\nu > 1$ were examined in section 6.14. With a hinge offset there are also small changes in the aerodynamic flapping moments, due to the mode shape change.

Now consider the aerodynamic forces. Again defining the aerodynamic coefficients by

$$M_F = \int_0^1 \eta \frac{F_z}{ac} \, dr = M_\theta \theta_{\text{con}} + M_{tw} \theta_{tw} + M_\lambda \lambda + M_{\dot{\beta}} \dot{\beta} + M_\beta \beta \qquad (6.263)$$

we obtain

$$M_\theta = \frac{1}{8} c_2 + \frac{\mu}{3} c_1 \sin \psi + \frac{\mu^2}{4} c_0 \sin^2 \psi \qquad (6.264)$$

$$M_{tw} = \frac{1}{10} c_3 + \frac{\mu}{4} c_2 \sin \psi + \frac{\mu^2}{6} c_1 \sin^2 \psi \qquad (6.265)$$

$$M_\lambda = -\left(\frac{1}{6} c_1 + \frac{\mu}{4} c_0 \sin \psi \right) \qquad (6.266)$$

$$M_{\dot{\beta}} = -\left(\frac{1}{8} d_1 + \frac{\mu}{6} d_0 \sin \psi \right) \qquad (6.267)$$

$$M_\beta = -\mu \cos \psi \left(\frac{1}{6} f_1 + \frac{\mu}{4} f_0 \sin \psi \right) \qquad (6.268)$$

6.15 Flap-Hinge Offset

where $c_n = (n+2)\int_e^1 \eta r^n dr$, $d_n = (n+3)\int_e^1 \eta^2 r^n dr$, and $f_n = (n+2)\int_e^1 \eta\eta' r^n dr$. With the mode shape $\eta = (r-e)/(1-e)$, the required constants are

$$c_0 = 1 - e \tag{6.269}$$

$$c_1 = 1 - (e + e^2)/2 \tag{6.270}$$

$$c_2 = 1 - (e + e^2 + e^3)/3 \tag{6.271}$$

$$c_3 = 1 - (e + e^2 + e^3 + e^4)/4 \tag{6.272}$$

$$d_0 = 1 - e \tag{6.273}$$

$$d_1 = 1 - (2e + e^2)/3 \tag{6.274}$$

$$f_0 = 1 \tag{6.275}$$

$$f_1 = 1 + e/2 \tag{6.276}$$

These constants should actually be evaluated by integrating from r_R (root cutout) to B. The solution of the flapping equations is now

$$\beta_0 = \frac{K_\beta}{v^2 I_b \Omega^2 (1-e)} \beta_p + \frac{\gamma}{v^2}\left[\frac{\theta_{.8}}{8}(c_2 + \mu^2 c_0)\right.$$
$$+ \frac{\theta_{tw}}{10}\left(c_3 - c_2 + \mu^2\left(\frac{5}{6}c_1 - c_0\right)\right)$$
$$\left. - \frac{\lambda_{NFP}}{6}c_1 + \beta_{1c}\frac{\mu}{12}(d_0 - f_1)\right] \tag{6.277}$$

$$\left(c_2 + \frac{1}{2}\mu^2 c_0\right)\theta_{1c} = \left(d_1 + \frac{1}{2}\mu^2 f_0\right)\beta_{1s} + \frac{v^2 - 1}{\gamma/8}\beta_{1c} + \frac{4}{3}\mu f_1 \beta_0 \tag{6.278}$$

$$\left(c_2 - \frac{1}{2}\mu^2 c_0\right)\theta_{1s} = -\left(d_1 - \frac{1}{2}\mu^2 f_0\right)\beta_{1c} + \frac{v^2 - 1}{\gamma/8}\beta_{1s}$$
$$- \frac{8}{3}\mu\left(\theta_{.75}c_1 + \frac{3}{4}\theta_{tw}(c_2 - c_1) - \frac{3}{4}\lambda_{NFP}\right) \tag{6.279}$$

Thus the hinge offset produces small changes in the constants arising from the aerodynamic forces. The primary effect of the hinge offset on the flap response is the coupling of the lateral and longitudinal control that arises because $v > 1$. For hover, the phase lag between the flapping response and cyclic pitch input is reduced by

$$\Delta\psi = -\tan^{-1}\frac{v^2 - 1}{\gamma/8} \cong -\frac{12}{\gamma}e \tag{6.280}$$

which is small for articulated rotors (the hinge spring term has been omitted here).

Finally, consider the hub moment for a rotor with offset flapping hinges. The contributions to the moment about the hub ($r = 0$) are

i) the inertial force $m\ddot{z} = m\eta\ddot{\beta}$, with moment arm r
ii) the centrifugal force $m\Omega^2 r$, with moment arm $z = \eta\beta$
iii) the aerodynamic force F_z, with moment arm r

Then the flapwise moment on the hub produced by one blade is

$$M = -(\ddot{\beta} + \beta)\int_e^1 m\eta r\, dr + \int_e^1 rF_z\, dr \tag{6.281}$$

Substituting for $\ddot\beta$ from the flapping equation gives

$$M = -\left[\frac{K_\beta \eta(1)}{I_b\Omega^2(1-e)}\beta_p + \frac{1}{I_b}\int_e^1 \eta F_z dr + (1-\nu^2)\beta\right]\int_e^1 \eta rm\, dr + \int_e^1 rF_z dr \qquad (6.282)$$

The precone term is constant, so does not contribute to the pitch or roll moments on the hub. Using $r = (1-e)\eta + e$,

$$-\int_e^1 \eta F_z dr \int_e^1 \eta rm\, dr + \int_e^1 \eta^2 m\, dr \int_e^1 rF_z dr$$

$$= e\left[-\int_e^1 \eta F_z dr \int_e^1 \eta m\, dr + \int_e^1 \eta^2 m\, dr \int_e^1 F_z dr\right] \qquad (6.283)$$

The factor in brackets is zero if the lift distribution is proportional to the mode shape, $F_z \propto (r-e)$. In general, this sum is second-order small and can be neglected. The hub moment thus reduces to

$$M = I_b(\nu^2 - 1)\beta \qquad (6.284)$$

and the pitch and roll components from the N blades give

$$\begin{pmatrix}-\dfrac{2C_{My}}{\sigma a}\\[4pt] \dfrac{2C_{Mx}}{\sigma a}\end{pmatrix} = \frac{\nu^2 - 1}{\gamma}\begin{pmatrix}\beta_{1c}\\ \beta_{1s}\end{pmatrix} \qquad (6.285)$$

This is the same result as that obtained for the hinge spring alone. A more general derivation of the result is given in Chapter 16. Although all the other effects of hinge offset examined have been only small refinements of the basic rotor behavior, the hub moment capability with offset hinges is important. Articulated rotor helicopters generate about half the moment about the center of gravity by the thrust tilt and about half by the direct hub moment.

6.16 Hingeless Rotor

Hingeless rotors have no flap or lag hinges; instead the blades are attached to the hub with a cantilever root restraint. Such a rotor has the advantages of a mechanically simple hub and generally improved handling qualities. The fundamental out-of-plane bending mode of a hingeless rotor blade is similar to the rigid flapping mode of an articulated blade, because of the dominance of the centrifugal stiffening relative to the structural stiffening. The fundamental natural flap frequency of a hingeless blade is thus not far above 1/rev, although significantly greater than the frequency achieved with offset-hinged blades. Typically the flap frequency $\nu = 1.10$ to 1.15 for hingeless rotors.

In section 6.15, the flapping equation was obtained for an arbitrary mode shape:

$$\ddot\beta + \nu^2\beta = \gamma\int_0^1 \eta \frac{F_z}{ac}\, dr \qquad (6.286)$$

With the proper value for the flap frequency ν, this equation can be used for the hingeless rotor blade as well. We have seen that the influence of the mode shape is secondary to that of the flap frequency. Thus a hingeless rotor can be modeled by using the correct flap frequency, but with a simple approximate mode shape. The flap frequency is either specified in the investigation or must be obtained from a free vibration analysis of the blade. An appropriate mode shape is that of rigid rotation about an offset hinge, $\eta = (r-e)/(1-e)$. The offset e can be chosen by matching the slope of the actual mode shape at an appropriate station such as 75% radius: $e = 1 - 1/\eta'(0.75)$. Although such an approximate model must be used with care, it does give the fundamental behavior of the hingeless rotor. When other degrees of freedom (such as the lag or torsional motion) are involved, a more accurate model of the rotor motion that includes the correct mode shapes must be used.

The hingeless rotor flap frequency is often described in terms of an equivalent hinge offset. From equation 6.262 for a uniform offset-hinge blade, follows

$$e_{\text{equiv}} = \frac{\frac{2}{3}(\nu^2 - 1)}{1 + \frac{2}{3}(\nu^2 - 1)} \tag{6.287}$$

Flap frequencies of $\nu = 1.10$ to 1.15 thus are characterized as offsets $e_{\text{equiv}} = 0.12$ to 0.18.

6.17 Gimballed or Teetering Rotor

A gimballed rotor has three or more blades attached to the hub without flap or lag hinges (cantilever root restraint), and the hub is attached to the rotor shaft by a universal joint or gimbal. The motion of the gimballed hub relative to the shaft is described by two degrees of freedom: the longitudinal and lateral tilt angles β_{1c} and β_{1s}, which correspond to the tip-path-plane tilt of an articulated rotor by cyclic flapping. The hub can include spring restraint of the gimbal motion. During the coning motion of the blades the hub does not tilt, because there is no net pitch or roll moment on the rotor. Hence for the coning motion the blades behave as on a hingeless rotor. For the higher harmonics of the flap motions (β_{2c}, β_{2s}, etc.) the hub also remains fixed.

The flapwise moment on the m-th blade of a gimballed rotor is

$$\frac{M^{(m)}}{I_b \Omega^2} = -(\ddot{\beta} + \nu^2 \beta) + \gamma \int_0^1 r \frac{F_z}{ac} \, dr \tag{6.288}$$

(see section 6.15). The mode shape $\eta = r$ that has been used corresponds to rigid body motion of the rotor about the gimbal. The equations of motion for longitudinal and lateral tilt of the gimbal are obtained from equilibrium of moments on the entire rotor. Summing the pitch moments from all N blades, adding a hub spring moment, and averaging over the azimuth gives

$$\frac{M_{\text{spring}}}{I_b \Omega^2} + \frac{1}{2\pi} \int_0^{2\pi} \left[\sum_{m=1}^N \cos \psi_m \frac{M^{(m)}}{I_b \Omega^2} \right] d\psi = 0 \tag{6.289}$$

where $\psi_m = \psi + m(2\pi/N)$ is the azimuth position of the m-th blade. For the steady-state solution, since all the blades have the same periodic motion, the sum over N blades followed by the average over ψ is equivalent to N times the average for one blade:

$$\frac{M_{\text{spring}}}{I_b \Omega^2} + \frac{1}{2\pi} \int_0^{2\pi} \left[N \cos \psi \frac{M^{(m)}}{I_b \Omega^2} \right] d\psi = 0 \tag{6.290}$$

Now the longitudinal hub spring moment is

$$M_{\text{spring}} = -K_\beta \beta_{1c} = -K_\beta \frac{1}{\pi} \int_0^{2\pi} \beta \cos\psi \, d\psi \qquad (6.291)$$

Thus the equation of motion is

$$\frac{1}{\pi} \int_0^{2\pi} \cos\psi \left[-\frac{K_\beta \beta}{\frac{1}{2}NI_b\Omega^2} + \frac{M}{I_b\Omega^2} \right] d\psi = 0 \qquad (6.292)$$

Similarly, for roll moments on the rotor we obtain

$$\frac{1}{\pi} \int_0^{2\pi} \sin\psi \left[-\frac{K_\beta \beta}{\frac{1}{2}NI_b\Omega^2} + \frac{M}{I_b\Omega^2} \right] d\psi = 0 \qquad (6.293)$$

The operators $\frac{1}{\pi}\int_0^{2\pi}(\ldots)\cos\psi\, d\psi$ and $\frac{1}{\pi}\int_0^{2\pi}(\ldots)\sin\psi\, d\psi$ are those used to obtain the equations for β_{1c} and β_{1s} of the articulated rotor. The equations of motion for the gimbal tilt are therefore the same as for the tip-path-plane tilt of an equivalent single blade with differential equation

$$\ddot\beta + \nu^2 \beta = \gamma \int_0^1 \eta \frac{F_z}{ac} dr \qquad (6.294)$$

and the solution is then the same as for the articulated rotor. Here the flap natural frequency is

$$\nu^2 = 1 + \frac{K_\beta}{\frac{1}{2}NI_b\Omega^2} \qquad (6.295)$$

Unless there is a hub spring, the frequency is $\nu = 1$ as for an articulated blade with no hinge offset. With a gimbal, the hub spring can be put in the non-rotating system, so it does not have to operate with a continual 1/rev motion. Moreover, different spring rates can then be used for longitudinal and lateral motions. For the coning and the second and higher harmonics of the flap motion the blade acts as a hingeless rotor. Again the solution can be obtained by considering an equivalent single blade and using the flap frequency corresponding to the cantilevered blade.

A teetering rotor has two blades attached to the hub without flap or lag hinges. The hub is attached to the shaft by a single flapping hinge, the two blades forming a single structure. The flapping motion is like that of a seesaw or teeter board, hence the name given this rotor. Such a hub configuration has the advantage of being mechanically very simple. As for the gimballed rotor, the coning motion gives no net moment about the teeter hinge and in effect the blades have cantilever root restraint. In general, the steady-state motion of the teetering rotor must be obtained by considering equilibrium of moments on the entire rotor. Since both blades must be executing the same periodic motion, the root flapping moment of the m-th blade is a periodic function of ψ_m:

$$M^{(m)} = M_0 + \sum_{n=1}^\infty M_{nc} \cos n\psi_m + M_{ns} \sin n\psi_m \qquad (6.296)$$

where $\psi_1 = \psi + \pi$ and $\psi_2 = \psi$. This can be written as

$$M^{(m)} = M_0 + \sum_{n=1}^\infty (-1)^{mn}\left(M_{nc}\cos n\psi + M_{ns}\sin n\psi\right) \qquad (6.297)$$

The total flap moment about the teeter hinge is then

$$M = M^{(2)} - M^{(1)} = \sum_{n=1}^{\infty}[1-(-1)^n]\left(M_{nc}\cos n\psi + M_{ns}\sin n\psi\right)$$

$$= 2\sum_{n\text{ odd}}\left(M_{nc}\cos n\psi + M_{ns}\sin n\psi\right) \qquad (6.298)$$

So for all even harmonics (including the coning motion) the flap moments from the two blades cancel each other. Only the odd harmonics, in particular the tip-path-plane tilt degrees of freedom β_{1c} and β_{1s}, produce a net moment about the hinge and hence teetering motion of the blade.

For the odd harmonics of the teetering rotor flap motion, the hinge moment consists of the root flapping moments from the two blades (which is equivalent to twice the moment of one of the blades) and a possible hub spring moment:

$$-\frac{K_\beta \beta}{I_b\Omega^2} + 2\left[-(\ddot{\beta}+\nu^2\beta) + \gamma\int_0^1 \eta\frac{F_z}{ac}dr\right] = 0 \qquad (6.299)$$

The equation of motion is therefore

$$\ddot{\beta} + \nu^2\beta = \gamma\int_0^1 \eta\frac{F_z}{ac}dr \qquad (6.300)$$

where the natural frequency of the flapping is

$$\nu^2 = 1 + \frac{K_\beta}{2I_b\Omega^2} \qquad (6.301)$$

Usually a teetering rotor does not have a hub spring, so $\nu = 1$. The tip-path-plane tilt motion of the teetering rotor is thus the same as that of an articulated rotor with no hinge offset.

To summarize the behavior of gimballed and teetering rotors, for those harmonics of the flap motion that give a net moment on the hub, including the tip-path-plane tilt, the blade acts as an articulated rotor with no hinge offset ($\eta = r$ and $\nu = 1$). For those harmonics (including the coning motion) where the flap moments are reacted internally in the hub, the blade acts as a hingeless rotor of high stiffness. With these considerations, the solutions obtained for an articulated rotor are also applicable to gimballed and teetering rotors.

6.18 Pitch-Flap Coupling

Pitch-flap coupling is a kinematic feedback of the flapping displacement to the blade pitch motion, which can be described by $\Delta\theta = -K_P\beta$. For positive pitch-flap coupling ($K_P > 0$), flap up decreases the blade pitch and hence the blade angle-of-attack. The resulting lift reduction produces a change in flap moment that opposes the original flap motion. Thus positive pitch-flap coupling acts as an aerodynamic spring on the flap motion.

Pitch-flap coupling can be obtained entirely by mechanical means. The simplest approach is to skew the flap hinge by an angle δ_3, so the hinge is no longer perpendicular to the radial axis of the blade (Figure 6.25). Then a rotation about the hinge with a flap angle β must also produce a pitch change of $-\beta\tan\delta_3$. The feedback gain for this arrangement is $K_P = \tan\delta_3$. Pitch-flap coupling is usually defined in terms of the

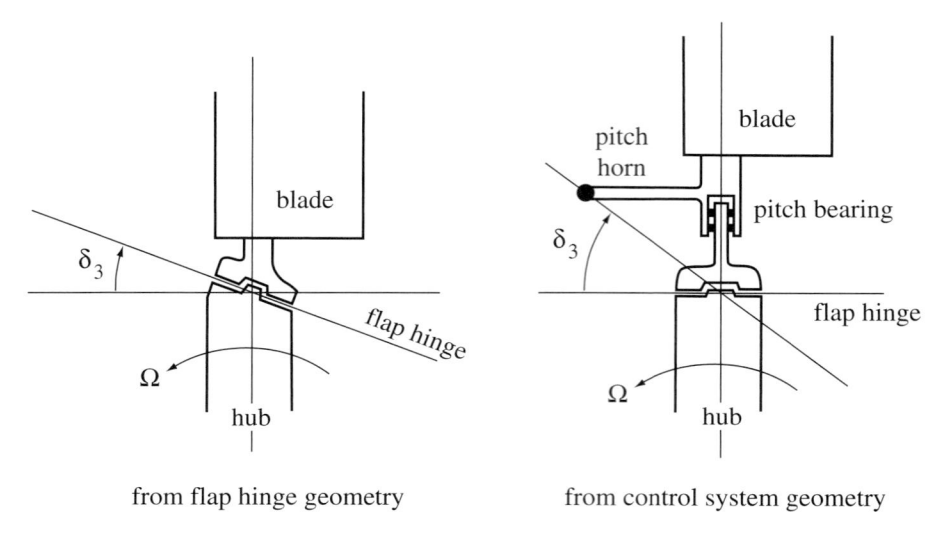

Figure 6.25. Pitch-flap coupling of a rotor blade.

delta-three angle. Positive coupling $\delta_3 > 0$ represents negative feedback, decreasing the blade pitch for a flap increase. Pitch-flap coupling can also be introduced by the control system geometry (Figure 6.25). When the pitch bearing is outboard of the flap hinge (the usual arrangement), the blade experiences a pitch change due to flapping if the pitch link is not in line with the axis of the flap hinge. For a fixed swashplate position, the flap motion can be viewed as occurring about a virtual hinge axis joining the end of the pitch horn and the actual flap hinge. The δ_3 angle then is the angle between this virtual hinge axis and the real flap hinge axis. Another source of pitch-flap coupling is the mean lag angle due to the rotor torque. If the flap hinge is outboard of the lag hinge, the mean lag angle is equivalent to a skew of the flap hinge, so $\delta_3 = \zeta_0$. There are similar coupling effects on hingeless rotors. Although pitch-flap and other couplings are determined for an articulated rotor by the hub, root, and control system geometry, for hingeless rotors the structural and inertial characteristics of the blade must also be considered. Often the δ_3 angle depends on the blade pitch because of changes in the control system geometry with collective, so in general $K_P = -\partial\theta/\partial\beta$ must be evaluated for a given collective, coning, and mean lag angle of the blade.

The equation of motion for the blade flapping was derived in section 6.5 considering only the pitch due to the control system input, θ_{con}. The solution relates the flapping to the actual blade pitch. That solution remains valid with pitch-flap coupling, but the root pitch and the control input are no longer identical. The difference can be accounted for by noting that the root pitch is now $(\theta - K_P\beta)$ if θ retains its meaning as the control input. Pitch-flap coupling thus changes the relative orientation of the control plane and no-feathering plane, whereas the solution for the orientation of the no-feathering plane relative to the tip-path plane is unchanged. Since pitch-flap coupling acts on the flapping with respect to the hub plane, $\theta_{\text{HP}} = \theta_{\text{CP}} - K_P\beta_{\text{HP}}$ is the actual blade root pitch. The flapping solution of section 6.5 determines θ_{HP} in terms of β_{HP}. There are two possible approaches to analyze the effects of pitch-flap coupling. The quantity $\theta_{\text{CP}} - K_P\beta_{\text{HP}}$ can be substituted for θ_{HP} in the differential equation of motion for the flapping, the solution of which then gives the control

6.18 Pitch-Flap Coupling

required θ_{CP} and shows the other effects of K_P. Alternatively, the previous solutions can be used directly, with $\theta_{CP} = \theta_{HP} + K_P \beta_{HP}$ determining the control required.

Consider the differential equation obtained for the flap motion of a rotor with flap frequency ν. Substitute $(\theta_{\text{con}} - K_P\beta)$ for θ_{con}, giving

$$\ddot{\beta} + \nu^2 \beta = \gamma \left(M_\theta \left(\theta_{\text{con}} - K_P \beta \right) + M_{tw}\theta_{tw} + M_\lambda \lambda + M_{\dot{\beta}} \dot{\beta} + M_\beta \beta \right) \quad (6.302)$$

For hover this becomes

$$\ddot{\beta} + \frac{\gamma}{8}\dot{\beta} + \left(\nu^2 + \frac{\gamma}{8}K_P\right)\beta = \frac{\gamma}{8}\theta_{\text{con}} + \frac{\gamma}{10}\theta_{tw} - \frac{\gamma}{6}\lambda \quad (6.303)$$

The mode shape $\eta = r$ has been used to evaluate the aerodynamic coefficients. Thus pitch-flap coupling introduces an aerodynamic spring that increases the effective natural frequency of the flap motion to

$$\nu_e^2 = \nu^2 + \frac{\gamma}{8} K_P \quad (6.304)$$

The flapping response to cyclic depends on the effective spring ν_e. However, pitch-flap coupling does not produce a hub moment, which is still determined by $(\nu^2 - 1)$. The solution for the cyclic pitch control is

$$\theta_{1c} = \beta_{1s} + K_P \beta_{1c} + \frac{1}{1+\frac{1}{2}\mu^2}\left[\frac{8}{\gamma}(\nu^2-1)\beta_{1c} + \frac{4}{3}\mu\beta_0 \right] \quad (6.305)$$

$$\theta_{1s} = -\beta_{1c} + K_P \beta_{1s} + \frac{1}{1+\frac{3}{2}\mu^2}\left[\frac{8}{\gamma}(\nu^2-1)\beta_{1s} \right.$$

$$\left. - \frac{8}{3}\mu\left(\theta_{.75} - \frac{3}{4}\lambda_{\text{TPP}}\right) + \frac{8}{3}\mu K_P \beta_0 \right] \quad (6.306)$$

or in hover

$$\theta_{1c} = \beta_{1s} + \left(\frac{\nu^2-1}{\gamma/8} + K_P\right)\beta_{1c} = \beta_{1s} + N_\star \beta_{1c} \quad (6.307)$$

$$\theta_{1s} = -\beta_{1c} + \left(\frac{\nu^2-1}{\gamma/8} + K_P\right)\beta_{1s} = -\beta_{1c} + N_\star \beta_{1s} \quad (6.308)$$

with $N_\star = \frac{8}{\gamma}(\nu^2 - 1) + K_P = \frac{8}{\gamma}(\nu_e^2 - 1)$. The magnitude and phase of the tip-path-plane response to cyclic are

$$\overline{\beta}/\overline{\theta} = \frac{1}{\sqrt{1+N_\star^2}} \quad (6.309)$$

$$\Delta\psi = 90° - \tan^{-1} N_\star \quad (6.310)$$

For an articulated rotor ($\nu = 1$), the result is

$$\overline{\beta}/\overline{\theta} = \frac{1}{\sqrt{1+K_P^2}} = \cos\delta_3 \quad (6.311)$$

$$\Delta\psi = 90° - \tan^{-1} K_P = 90° - \delta_3 \quad (6.312)$$

Thus the swashplate phasing required is equal to the δ_3 angle.

Now consider the effect of pitch-flap coupling in terms of the change of the control-plane orientation relative to the no-feathering plane. The relation

$\theta_{CP} = \theta_{HP} + K_P \beta_{HP}$ gives the collective and cyclic pitch required:

$$\begin{pmatrix} \theta_0 \\ \theta_{1c} \\ \theta_{1s} \end{pmatrix}_{CP} = \begin{pmatrix} \theta_0 \\ \theta_{1c} \\ \theta_{1s} \end{pmatrix}_{HP} + K_P \begin{pmatrix} \beta_0 \\ \beta_{1c} \\ \beta_{1s} \end{pmatrix}_{HP} \tag{6.313}$$

For a given thrust and positive pitch-flap coupling, the collective input must thus be increased to counter the feedback of the coning angle and keep the actual root collective at θ_{0HP}. Similarly, the cyclic pitch required can be determined from these relations. A special case is that of a rotor with no cyclic pitch control, an important example of which is the tail rotor. In that case the helicopter operating condition fixes the orientation of the control plane instead of the tip-path plane. With no cyclic in the control plane, $\theta_{CP} = \theta_{HP} + K_P \beta_{HP}$ gives

$$\theta_{1cHP} + K_P \beta_{1cHP} = 0 \tag{6.314}$$

$$\theta_{1sHP} + K_P \beta_{1sHP} = 0 \tag{6.315}$$

The orientation of the tip-path plane relative to the no-feathering plane,

$$\beta_{1cNFP} = \beta_{1cHP} + \theta_{1sHP} \tag{6.316}$$

$$\beta_{1sNFP} = \beta_{1sHP} - \theta_{1cHP} \tag{6.317}$$

is fixed by flap moment equilibrium. Eliminating θ_{HP} gives

$$\beta_{1cHP} = \frac{1}{1+K_P^2} (\beta_{1cNFP} + K_P \beta_{1sNFP}) \tag{6.318}$$

$$\beta_{1sHP} = \frac{1}{1+K_P^2} (\beta_{1sNFP} - K_P \beta_{1cNFP}) \tag{6.319}$$

or

$$|\beta|_{HP} = \frac{1}{\sqrt{1+K_P^2}} |\beta|_{NFP}$$

Thus pitch-flap coupling reduces the flapping magnitude relative to the rotor shaft. Negative coupling is as effective as positive coupling, because the effect of K_P is to move flap motion from resonant excitation. The sign of the feedback influences the phase of the response, and large negative pitch-flap coupling does have an adverse effect on the flapping stability. Using 45° of delta-three on tail rotors ($K_P = 1$) is common, to reduce the transient and steady-state flapping relative to the shaft.

The "delta-three" notation for pitch-flap coupling has its origins in a convention for describing flap and lag hinge orientation that was developed at the Cierva Autogiro Company; see Bennett (1961). When flap and lag hinge inclination was introduced on the autogyro rotors, the inboard (flap) hinge was designated the "delta hinge," and the outboard (lag) hinge designated the "alpha-hinge." A general orientation of each hinge was described in terms of the angles made by the hinge axis as projected on each of the three planes of the axis system (Figure 6.26). With this definition, a flap hinge has δ_2 about 90°, and a lag hinge has α_1 about 90°. A "delta-three hinge" is in the x-y plane, at an angle δ_3 from the x-axis. A lag hinge tilted from the z-axis in the y-z plane produced an automatic change of the blade pitch for a jump takeoff of the autogyro; it was called an "alpha-one hinge" and

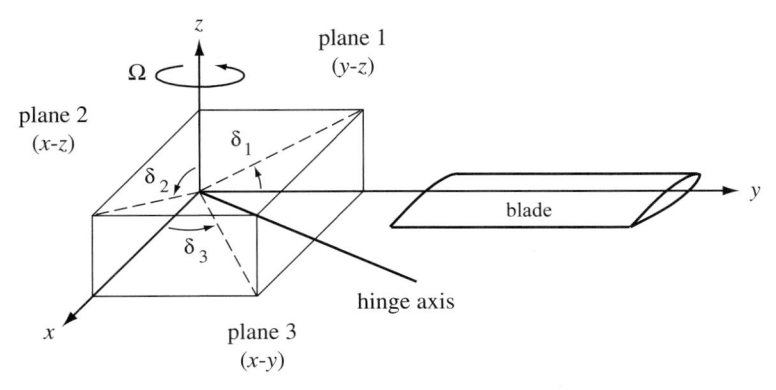

Figure 6.26. Geometry of the delta hinge.

introduced lag-pitch coupling. An "alpha-two hinge" introduced lag-flap coupling and was used to improve ground resonance stability. The angle δ_3 is still used to characterize pitch-flap coupling of a rotor blade. Less common is the characterization of pitch-lag coupling by an α_1 angle, and likely the convention is that α_1 is the tilt of the lag hinge from the z-axis, or $\tan\alpha_1 = -\partial\theta/\partial\zeta$.

6.19 Tail Rotor

The tail rotor of a single main rotor helicopter is a small diameter rotary wing with the functions of balancing the main rotor torque and providing yaw control, which are achieved through the action of the tail rotor thrust on a longitudinal arm (usually somewhat longer than the main rotor radius) about the main rotor shaft. The tail rotor is usually a flapping rotor with a low disk loading, to which the analysis developed in this chapter is applicable. The special features of the tail rotor configuration make the use of the analysis somewhat different than for a main rotor. First, the tail rotor has no cyclic pitch control, just collective to control the thrust magnitude. Second, the tail rotor shaft angle is fixed by the geometry of the tail rotor installation and the helicopter yaw angle, instead of being determined by force equilibrium of the rotor. The tail rotor drag or propulsive force adds to the airframe drag and is balanced by the main rotor.

When there is no cyclic pitch control, the expressions for blade flap moment equilibrium give the flapping produced, rather than the cyclic required for the rotor operating state. The tail rotor usually has pitch-flap coupling, which gives the actual pitch in the hub plane in terms of the flapping: $\theta_{1cHP} = -K_P\beta_{1cHP}$ and $\theta_{1sHP} = -K_P\beta_{1sHP}$. A delta-three value of 45° is typical; hence $K_P = 1$. With the shaft orientation fixed, the hub-plane angle-of-attack i_{HP} is known, and then the tip-path-plane incidence depends on the longitudinal flapping. Hence the inflow ratio is

$$\lambda_{TPP} = \lambda_{HP} + \mu\beta_{1cHP} = \lambda_i + \mu(i_{HP} + \beta_{1cHP}) \tag{6.320}$$

and the tail rotor drag force, which must be reacted by the main rotor, is

$$D_{tr} = H_{TPP} - Ti_{TPP} = H_{TPP} - T(i_{HP} + \beta_{1cHP}) \tag{6.321}$$

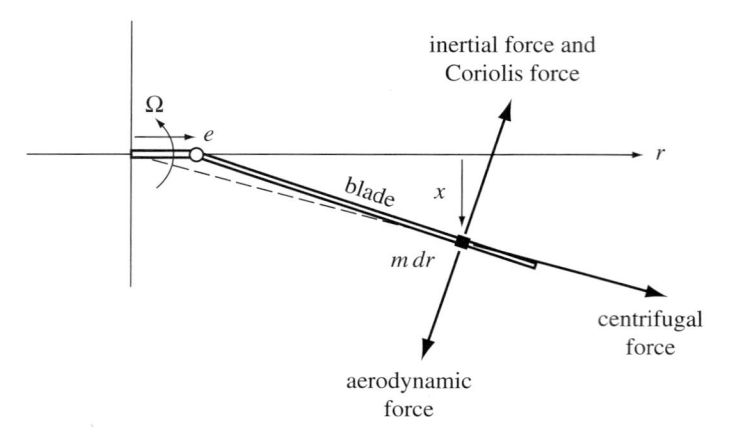

Figure 6.27. Rotor blade lag moments.

6.20 Lag Motion

In addition to flap motion, the helicopter rotor blade has motion in the plane of the disk, called lag or lead-lag. An articulated rotor has a lag or drag hinge, so the lag motion consists of rigid-body rotation about a vertical axis near the center of rotation. Generally the lag motion requires a more complicated analysis than does the flap motion. The flapping motion produces in-plane inertial forces that couple the flap and lag degrees of freedom of the blade. Also, for low inflow rotors the in-plane forces on the blade are small compared to the out-of-plane forces, and consequently more care is required in analyzing the motion resulting from lag moment balance. The present section is only an introduction to the topic; the rotor lag dynamics are covered in more detail in Chapter 16.

Consider the in-plane motion of a blade with a lag-hinge offset by a distance eR from the center of rotation (Figure 6.27). If there is no lag-hinge spring, the offset cannot be zero, or there would be no way to deliver torque to the rotor. Rigid-body rotation about the lag hinge is represented by the lag degree of freedom ζ, defined to be positive for motion opposing the rotor rotation direction. With a mode shape $\eta = (r - e)/(1 - e)$, the in-plane deflection is $x = \eta\zeta$. A lag-hinge spring with constant K_ζ is included. The in-plane forces acting on the blade section at r, and their moment arms about the lag hinge at $r = e$, are as follows:

i) an inertial force $m\ddot{z} = mr\ddot{\zeta}$ opposing the lag motion, with moment arm $(r - e)$ about the lag hinge
ii) a centrifugal force $m\Omega^2 r$ directed radially outward from the center of rotation, hence with moment arm $x(e/r) = \eta\zeta(e/r)$ about the lag hinge
iii) an aerodynamic force F_x in the drag direction, with moment arm $(r - e)$
ii) a Coriolis force $2\Omega \dot{z} z' m = 2\Omega \dot{\beta}\beta r m$ in the same direction as the inertial force, with moment arm $(r - e)$

If the lag hinge were at the center of rotation the centrifugal force would produce no lag moment. The Coriolis force is due to the product of the rotor angular velocity Ω and the radially inward section velocity $\dot{z}z'$. This radial velocity is the in-plane component of the flap velocity $\dot{z} = r\dot{\beta}$, produced when the blade is coned upward at angle $z' = \beta$. The Coriolis force is in the blade lead direction when $\beta\dot{\beta} > 0$.

6.20 Lag Motion

Equilibrium of moments about the lag hinge, including a spring moment $K_\zeta \zeta$, gives the equation of motion:

$$\int_e^R \left[(m\eta\ddot{\zeta})(r-e) + m\Omega^2 r \left(\frac{e}{r}\eta\zeta\right) + 2\Omega\beta\dot{\beta}rm(r-e) \right] dr + K_\zeta \zeta = \int_e^R F_x(r-e)dr \qquad (6.322)$$

Expressed in terms of dimensionless quantities and divided by $(1-e)$, the equation becomes

$$I_b \ddot{\zeta} + \left[\frac{e}{1-e} \int_e^1 \eta m \, dr + \frac{K_\zeta}{\Omega^2(1-e)} \right] \zeta + 2I_b \beta \dot{\beta} = \int_e^1 \eta F_x dr \qquad (6.323)$$

with $I_b = \int_e^1 \eta^2 m \, dr$. Introducing the blade Lock number $\gamma = \rho a c R^4 / I_b$ (here in terms of the lag moment of inertia), the differential equation for the blade lag motion is

$$\ddot{\zeta} + v_\zeta^2 \zeta + 2\beta\dot{\beta} = \gamma \int_e^1 \eta \frac{F_x}{ac} dr \qquad (6.324)$$

The lag dynamics are described by a mass and spring system excited by the in-plane aerodynamic forces (profile and induced drag) and a Coriolis force due to the blade flapping. The aerodynamic forces damp the lag motion, but much less effectively than out-of-plane motion. The natural frequency of the lag motion is

$$v_\zeta^2 = \frac{e}{1-e} \frac{\int_e^1 \eta m \, dr}{\int_e^1 \eta^2 m \, dr} + \frac{K_\zeta}{I_b \Omega^2 (1-e)} \qquad (6.325)$$

The first term, the centrifugal spring on the lag motion, is zero if there is no hinge offset. For uniform mass distribution and no hinge spring, the result is simply

$$v_\zeta^2 = \frac{3}{2} \frac{e}{1-e} \qquad (6.326)$$

Articulated rotors typically have a lag frequency of $v_\zeta = 0.2$ to 0.3/rev. With hingeless rotors (or with a lag-hinge spring) a higher lag frequency can be attained. To avoid excessive blade loads, the lag frequency must not be too near 1/rev. Thus hingeless rotors naturally fall into two classes: soft in-plane rotors, for which the lag frequency is below 1/rev (typically $v = 0.65$ to 0.75/rev); and stiff in-plane rotors, for which the lag frequency is above 1/rev (typically $v = 1.4$ to 1.6/rev). Articulated rotors are soft in-plane. Gimballed and teetering rotors usually are in the stiff in-plane class. Soft in-plane rotors exhibit a mechanical instability called ground resonance (see section 20.3) if the lag frequency or the lag damping is too low. For this reason an articulated rotor and even some soft in-plane hingeless rotors must have mechanical dampers.

The Coriolis force is a second-order term, but because all the in-plane forces on the blade are small it is an important factor in the blade behavior. The purpose of the lag hinge on an articulated rotor is alleviation of the in-plane loads generated by Coriolis forces due to blade flapping. Linearizing the Coriolis term about the blade position (appropriate for aeroelastic stability analysis) gives

$$\beta\dot{\beta} \cong \beta_{\text{trim}} \delta\dot{\beta} + \dot{\beta}_{\text{trim}} \delta\beta \qquad (6.327)$$

For hover, or when averaged trim values are used in forward flight, this becomes $\beta\dot{\beta} \cong \beta_0 \delta\dot{\beta}$. The Coriolis force is therefore due primarily to the radial component of the flapping velocity of the blade coned at a trim angle β_0. For the steady-state solution, the Coriolis term acts as a forcing function and can be evaluated by considering the coning and first harmonics of the flap response:

$$\beta\dot{\beta} = (\beta_0 + \beta_{1c}\cos\psi + \beta_{1s}\sin\psi)(-\beta_{1c}\sin\psi + \beta_{1s}\cos\psi)$$

$$= \beta_0\beta_{1s}\cos\psi - \beta_0\beta_{1c}\sin\psi + \beta_{1c}\beta_{1s}\cos 2\psi + \frac{1}{2}(\beta_{1s}^2 - \beta_{1c}^2)\sin 2\psi \quad (6.328)$$

Consider the steady-state lag motion, which is periodic and therefore can be written as a Fourier series. The inertial and Coriolis forces have zero mean values, so the mean lag angle is

$$\zeta_0 = \frac{\gamma}{\nu_\zeta^2}\frac{C_Q}{\sigma a} \quad (6.329)$$

since the mean value of $\int_0^1 r(F_x/ac)dr$ is the rotor torque coefficient $C_Q/\sigma a$. The mean lag angle is typically a few degrees, varying from slightly negative in autorotation to perhaps 10° at maximum power.

The solution for the first harmonic lag motion due to the aerodynamic and Coriolis forces is

$$\zeta_{1c} = \frac{1}{1-\nu_\zeta^2}\left(-\gamma(C_Q/\sigma a)_{1c} + 2\beta_0\beta_{1s}\right) \quad (6.330)$$

$$\zeta_{1s} = \frac{1}{1-\nu_\zeta^2}\left(-\gamma(C_Q/\sigma a)_{1s} - 2\beta_0\beta_{1c}\right) \quad (6.331)$$

A lag frequency near 1/rev gives large 1/rev lag motion and hence high in-plane blade loads. The damping, which determines the response amplitude for $\nu_\zeta = 1$, is low for the blade lag motion and therefore does not alter this conclusion. (An articulated blade with high damping from a mechanical damper also has a small lag frequency.) Thus the lag frequency of a soft in-plane rotor is generally a compromise between the requirements of low blade loads (low lag frequency) and ground resonance stability (high lag frequency). These expressions for ζ_{1c} and ζ_{1s} are somewhat misleading, because there are actually flap terms in the 1/rev aerodynamic lag moments that cancel part of the Coriolis excitation.

The solution for the 2/rev lag motion due to the Coriolis forces alone is

$$\Delta\zeta_{2c} = \frac{2\beta_{1c}\beta_{1s}}{4-\nu_\zeta^2} \quad (6.332)$$

$$\Delta\zeta_{2s} = \frac{\beta_{1s}^2 - \beta_{1c}^2}{4-\nu_\zeta^2} \quad (6.333)$$

or

$$|\Delta\zeta|_{2/\text{rev}} = \sqrt{\zeta_{2c}^2 + \zeta_{2s}^2} = \frac{\beta_{1c}^2 + \beta_{1s}^2}{4-\nu_\zeta^2} = \frac{|\beta|^2_{1/\text{rev}}}{4-\nu_\zeta^2} \quad (6.334)$$

The Coriolis forces thus produce a 2/rev lag motion proportional to the square of the 1/rev flap amplitude.

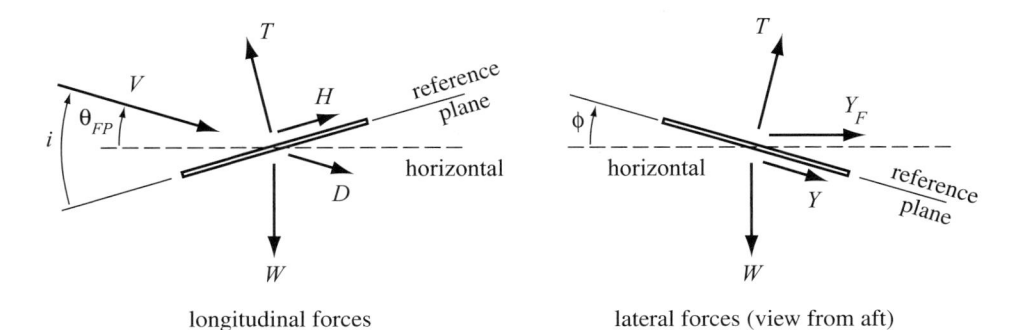

Figure 6.28. Helicopter force equilibrium.

6.21 Helicopter Force and Moment Equilibrium

The operating condition of the rotor is determined by force and moment equilibrium on the entire helicopter. In this section the longitudinal and lateral equilibrium for a helicopter in steady unaccelerated flight is examined. In the case of longitudinal force equilibrium the result for large angles is derived. This result is then used to determine the rotor power required. Although in numerical calculations the simultaneous equilibrium of all six components of force and moment on the helicopter can be found, the basic behavior is illustrated by considering lateral and longitudinal equilibrium separately.

Longitudinal force equilibrium considers the forces in the vertical, longitudinal plane of the helicopter (Figure 6.28; see also section 6.4). The helicopter has speed V and a flight path angle θ_{FP}, so the climb or descent velocity is $V_c = V \sin\theta_{FP}$. The forces on the rotor are the thrust T and rotor drag H, defined relative to the reference plane used. The reference plane has angle-of-attack i with respect to the forward speed (i is positive for forward tilt of the rotor). The forces acting on the helicopter are the weight W (vertical) and the aerodynamic drag D (in the same direction as V). Auxiliary propulsion or lifting devices can be accounted for by including their forces in W and D. From vertical and horizontal force equilibrium,

$$W = T\cos(i - \theta_{FP}) - D\sin\theta_{FP} + H\sin(i - \theta_{FP}) \qquad (6.335)$$

$$D\cos\theta_{FP} + H\cos(i - \theta_{FP}) = T\sin(i - \theta_{FP}) \qquad (6.336)$$

which for small angles is $W = T$ and $D + H = T(i - \theta_{FP})$. So the rotor thrust equals the helicopter weight, and horizontal force equilibrium gives the angle-of-attack:

$$i = \theta_{FP} + \frac{D}{W} + \frac{H}{T} = \frac{\lambda_c}{\mu} + \frac{D}{W} + \frac{C_H}{C_T} \qquad (6.337)$$

where $\lambda_c = V_c/\Omega R \cong \mu\theta_{FP}$. Then with $H = H_{TPP} - \beta_{1c}T$,

$$i = \frac{\lambda_c}{\mu} + \frac{D}{W} + \frac{C_{H\text{TPP}}}{C_T} - \beta_{1c} \qquad (6.338)$$

and the inflow ratio is

$$\lambda = \lambda_i + \mu i \cong \lambda_i + \lambda_c + \mu\frac{D}{W} + \mu\frac{C_H}{C_T} \qquad (6.339)$$

This is the same result as in section 6.4. If H_{TPP} is neglected, the tip-path-plane inclination is determined by the helicopter drag and climb velocity alone: $i_{TPP} \cong \theta_{FP} + D/W$.

For large angles, using the horizontal force equation to eliminate the drag force allows the vertical force equation to be written:

$$W = \frac{T \cos i}{\cos \theta_{FP}} \left(1 + \frac{H}{T} \tan i\right) \quad (6.340)$$

Then the horizontal force equation can be written as

$$\frac{D}{T \cos i} + \frac{H}{T}(1 + \tan i \tan \theta_{FP}) = \tan i - \tan \theta_{FP} \quad (6.341)$$

or

$$\frac{D}{W \cos \theta_{FP}}\left(1 + \frac{H}{T}\tan i\right) + \frac{H}{T}(1 + \tan i \tan \theta_{FP}) = \tan i - \tan \theta_{FP} \quad (6.342)$$

Solving for $\tan i$ gives

$$\tan i = \frac{\tan \theta_{FP} + \frac{D}{W \cos \theta_{FP}} + \frac{H}{T}}{1 - \frac{H}{T}\left(\tan \theta_{FP} + \frac{D}{W \cos \theta_{FP}}\right)} \quad (6.343)$$

from which the inflow ratio $\lambda = \mu \tan i + \lambda_i$ can be obtained. This result can be written:

$$i = \tan^{-1}\left(\tan \theta_{FP} + \frac{D}{W \cos \theta_{FP}}\right) + \tan^{-1}\frac{H}{T} = i|_{H=0} + \tan^{-1}\frac{H}{T} \quad (6.344)$$

For small angles this reduces to the previous result. In summary, a forward tilt of the disk is required to produce the propulsive force opposing the helicopter and rotor drag, and to provide the climb velocity.

Lateral force equilibrium (Figure 6.28) determines the roll angle ϕ of the reference plane relative to the horizontal. The rotor thrust T and side force Y are defined relative to the reference plane used. The forces on the helicopter are the weight W and a side force Y_F (such as that due to the tail rotor). Horizontal and vertical force equilibrium give

$$Y_F + Y \cos \phi + T \sin \phi = 0 \quad (6.345)$$

$$W = T \cos \phi - Y \sin \phi \quad (6.346)$$

with the solution

$$\tan \phi = \frac{-\frac{Y_F}{W} - \frac{Y}{T}}{1 - \frac{Y_F}{W}\frac{Y}{T}} \quad (6.347)$$

or

$$\phi = -\tan^{-1}\frac{Y_F}{W} - \tan^{-1}\frac{Y}{T} \quad (6.348)$$

The rotor disk must roll to provide a component of thrust to cancel the side forces of the helicopter and rotor. For small angles the result is $\phi = -Y_F/W - C_Y/C_T$, or using $Y = Y_{TPP} - \beta_{1s}T$,

$$\phi = -\frac{Y_F}{W} - \frac{C_{Y_{TPP}}}{C_T} + \beta_{1s} \quad (6.349)$$

6.21 Helicopter Force and Moment Equilibrium

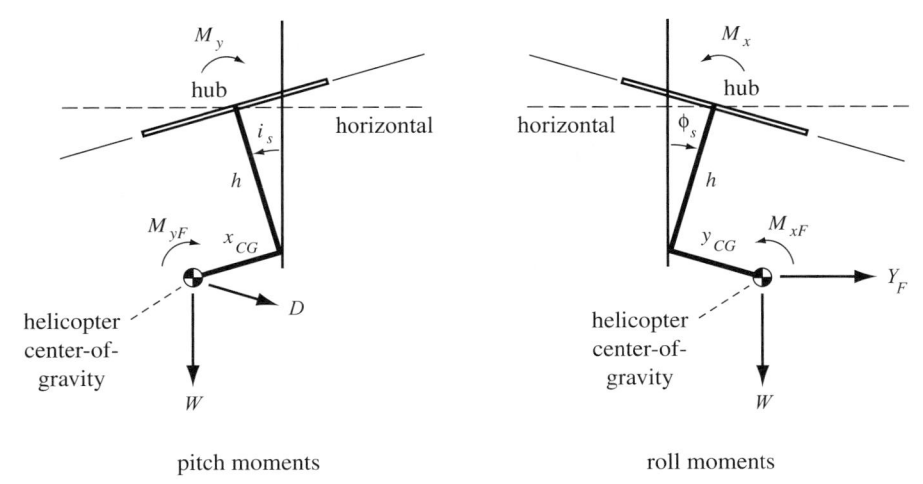

Figure 6.29. Helicopter moment equilibrium.

Next consider the equilibrium of pitch moments on the helicopter (Figure 6.29), which determines the angle-of-attack of the rotor shaft relative to the vertical, i_s. Moments are taken about the rotor hub so that the rotor forces are not involved and the rotor reference plane does not enter the problem. The rotor hub moment M_y must be included, however. Acting at the helicopter center-of-gravity are the weight W, the aerodynamic drag D, and an aerodynamic pitch moment M_{yF}. The position of the helicopter center-of-gravity is defined relative to the rotor shaft (that is, in the hub plane axis system). The center-of-gravity is located a distance h below the hub and a distance x_{CG} forward of the shaft. So x_{CG} is the longitudinal center-of-gravity position. For small angles, the moment equilibrium about the rotor hub gives

$$M_y + M_{yF} + W(hi_s - x_{CG}) - hD = 0 \tag{6.350}$$

which can be solved for the shaft angle-of-attack:

$$i_s = i_{HP} - \theta_{FP} = \frac{x_{CG}}{h} + \frac{D}{W} - \frac{M_{yF}}{Wh} - \frac{M_y}{Wh} \tag{6.351}$$

The shaft angle (the orientation of the hub plane relative to the horizontal) has also been written in terms of the hub-plane tilt (relative to the aircraft velocity) and the flight path angle (between the velocity and the horizontal). Now the rotor hub moment is given by the tilt of the tip-path plane relative to the hub plane:

$$\frac{M_y}{Wh} = \frac{C_{My}}{hC_T} = -\frac{(\nu^2 - 1)/\gamma}{h 2 C_T/\sigma a} \beta_{1cHP} \tag{6.352}$$

Next, recall that longitudinal force equilibrium gave

$$i_{HP} - \theta_{FP} - \frac{D}{W} = \frac{H_{HP}}{T} = \frac{H_{TPP}}{T} - \beta_{1cHP} \tag{6.353}$$

After combining the force and moment equilibrium results to eliminate ($i_{HP} - \theta_{FP} - D/W$), solving for β_{1cHP} gives

$$\beta_{1cHP} = \frac{1}{1 + \frac{(\nu^2-1)/\gamma}{h 2 C_T/\sigma a}} \left(\frac{-x_{CG}}{h} + \frac{M_{yF}}{hW} + \frac{C_{HTPP}}{C_T} \right) \tag{6.354}$$

and then the shaft angle

$$i_{HP} = \frac{D}{W} + \frac{1}{1 + \frac{(\nu^2-1)/\gamma}{h2C_T/\sigma a}} \left(\frac{x_{CG}}{h} - \frac{M_{yF}}{hW} + \frac{(\nu^2-1)/\gamma}{h2C_T/\sigma a} \frac{C_{H\mathrm{TPP}}}{C_T} \right) \quad (6.355)$$

The rotor shaft angle and the tip-path-plane tilt relative to the shaft are determined by helicopter moment equilibrium. From β_{1cHP} the flapping solution then gives the longitudinal cyclic control required, θ_{1sHP}. A forward center-of-gravity position requires a rearward tilt of the rotor and a forward tilt of the helicopter so that the center-of-gravity remains under the hub and the rotor thrust remains vertical. A flap frequency above 1/rev reduces the tilt required for a given center-of-gravity offset and hence reduces the cyclic control travel.

Similarly, roll moment equilibrium gives the shaft roll angle ϕ_s (Figure 6.29). The rotor hub roll moment is M_x, and the forces on the helicopter are the weight W, side force Y_F, and aerodynamic roll moment M_{xF}. The helicopter center-of-gravity is offset to the right of the rotor shaft by the distance y_{CG}. Then for small angles, roll moment equilibrium about the rotor hub gives

$$M_x + M_{xF} + W(h\phi_s - y_{CG}) + hY_F = 0 \quad (6.356)$$

or

$$\phi_s = \phi_{HP} = \frac{y_{CG}}{h} - \frac{Y_F}{W} - \frac{M_{xF}}{Wh} - \frac{M_x}{Wh} \quad (6.357)$$

Now the rotor hub moment is

$$\frac{M_x}{Wh} = \frac{C_{Mx}}{hC_T} = \frac{(\nu^2-1)/\gamma}{h2C_T/\sigma a} \beta_{1sHP} \quad (6.358)$$

and from lateral force equilibrium

$$\phi_{HP} + \frac{Y_F}{W} = -\frac{Y_{HP}}{T} = -\frac{Y_{\mathrm{TPP}}}{T} + \beta_{1sHP} \quad (6.359)$$

Solving for β_{1sHP} yields

$$\beta_{1sHP} = \frac{1}{1 + \frac{(\nu^2-1)/\gamma}{h2C_T/\sigma a}} \left(\frac{y_{CG}}{h} - \frac{M_{xF}}{hW} + \frac{C_{Y\mathrm{TPP}}}{C_T} \right) \quad (6.360)$$

and

$$\phi_s = -\frac{Y_F}{W} + \frac{1}{1 + \frac{(\nu^2-1)/\gamma}{h2C_T/\sigma a}} \left(\frac{y_{CG}}{h} - \frac{M_{xF}}{hW} - \frac{(\nu^2-1)/\gamma}{h2C_T/\sigma a} \frac{C_{Y\mathrm{TPP}}}{C_T} \right) \quad (6.361)$$

From the lateral tip-path-plane tilt β_{1sHP}, the flapping solution gives the lateral cyclic control required, θ_{1cHP}.

6.22 Yawed Flow and Radial Drag

The loading on an infinite wing yawed at an angle Λ to the free stream velocity V is not the same as the loading on an unyawed wing. The spanwise flow on the yawed wing influences the boundary layer and hence the wing drag and stall characteristics. Spanwise flow along the wing generates a spanwise component of the viscous drag force on the wing sections. For rotor blade aerodynamic analysis, the loading on the yawed wing must be estimated using corrections to two-dimensional (unyawed) airfoil data.

6.22 Yawed Flow and Radial Drag

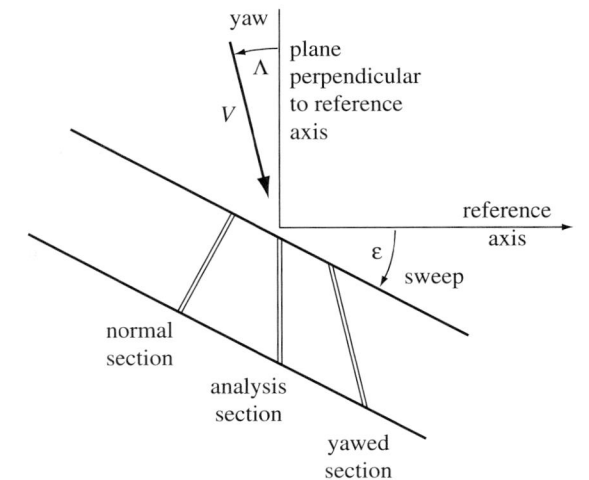

Figure 6.30. Yawed and swept wing aerodynamics.

Yawed flow over the rotor blade section can be accounted for by using the equivalence assumptions for swept wings: that the yawed section drag coefficient is given by two-dimensional airfoil characteristics, and the normal section lift coefficient is not influenced by yawed flow below stall. Since the wing viewed in a frame moving spanwise at a velocity $V \sin \Lambda$ is equivalent to an unyawed wing with free stream velocity $V \cos \Lambda$, except for changes in the boundary layer, there should be no effect of spanwise flow on the loading below stall. Also, the total viscous drag force is assumed to be in the direction of the yawed flow. These assumptions are largely verified by experimental data for yawed wings.

Consider a wing with both yaw Λ of the free stream and sweep ϵ of the quarter chord relative to the analysis section (Figure 6.30). For a rotor blade, the reference axis typically is a straight radial line through the center of rotation. The analysis section is in a plane perpendicular to the wing reference line. Hence the wing sweep angle ϵ is the angle between the tangent to the quarter-chord locus and the tangent to the reference line. The total sweep angle of the free stream relative to the quarter chord is $(\Lambda + \epsilon)$. The normal section is in a plane perpendicular to the quarter-chord locus. The yawed section is in the direction of the free stream V. The yaw angle Λ varies with azimuth in forward flight of the rotor, whereas the sweep angle ϵ is a fixed geometric angle of the blade. Figure 6.30 illustrates a blade section at radial station r, with the normal, yawed, and analysis sections separated for clarity. In the context of lifting-line theory, the section analysis is based on an infinite wing in uniform, yawed flow, so the loading is the same at all spanwise stations.

The section lift, drag, and moment coefficients are based on chord and dynamic pressure of that section, related as $c_n = c \cos \epsilon = c_y \cos(\Lambda + \epsilon)$ and $q_n / \cos^2(\Lambda + \epsilon) = q / \cos^2 \Lambda = q_y$, using the subscripts "$n$" for the normal section and "y" for the yawed section. The pressure on the wing is independent of the spanwise position, so the chordwise integrals are

$$\int p \frac{dx}{c} = q c_\ell = \text{constant} \tag{6.362}$$

$$\int p \frac{x}{c} \frac{dx}{c} = q c_m = \text{constant} \tag{6.363}$$

Thus

$$\cos^2(\Lambda + \epsilon)c_{\ell n} = \cos^2 \Lambda\, c_\ell = c_{\ell y} \qquad (6.364)$$

$$\cos^2(\Lambda + \epsilon)c_{mn} = \cos^2 \Lambda\, c_m = c_{my} \qquad (6.365)$$

The total drag force is in the yawed flow direction. The force acting on a section of width Δr is independent of the section orientation: $c_d q (c\Delta r)/\cos(\text{yaw}) = \text{constant}$. Thus

$$\frac{c_{dn} q_n (c_n \Delta r_n)}{\cos(\Lambda + \epsilon)} = \frac{c_d q (c\Delta r)}{\cos \Lambda} = c_{dy} q_y (c_y \Delta r_y) \qquad (6.366)$$

where the area $(c\Delta r)$ is the same for all sections, or

$$\cos(\Lambda + \epsilon)c_{dn} = \cos \Lambda\, c_d = c_{dy} \qquad (6.367)$$

The radial drag force is $R = D \tan \Lambda$ (analysis section). For a given upwash at the blade section (due to blade motion and wake-induced velocity), the product $V\alpha$ is constant. Hence the angle-of-attack is

$$\cos(\Lambda + \epsilon)\alpha_n = \cos \Lambda\, \alpha = \alpha_y \qquad (6.368)$$

In summary,

$$c_{\ell n} = \frac{1}{K_\epsilon^2} c_\ell = \frac{1}{\cos^2(\Lambda + \epsilon)} c_{\ell y} \qquad (6.369)$$

$$c_{mn} = \frac{1}{K_\epsilon^2} c_m = \frac{1}{\cos^2(\Lambda + \epsilon)} c_{my} \qquad (6.370)$$

$$c_{dn} = \frac{1}{K_\epsilon} c_d = \frac{1}{\cos(\Lambda + \epsilon)} c_{dy} \qquad (6.371)$$

$$\alpha_n = \frac{1}{K_\epsilon} \alpha = \frac{1}{\cos(\Lambda + \epsilon)} \alpha_y \qquad (6.372)$$

where $K_\epsilon = \cos(\Lambda + \epsilon)/\cos \Lambda$. Since $\partial c_{\ell y}/\partial \alpha_y = \cos(\Lambda + \epsilon)\partial c_{\ell n}/\partial \alpha_n$, the lift-curve slope of the yawed section is reduced relative to that of the normal section. These are geometric relations. The only aerodynamic assumption so far is that the total drag force is in the yawed flow direction.

Assume now that in the linear range, the normal section lift and moment coefficients are not affected by the spanwise component of the flow, so

$$c_{\ell n} = c_{\ell_{2D}}(\alpha_n) \qquad (6.373)$$

$$c_{mn} = c_{m_{2D}}(\alpha_n) \qquad (6.374)$$

where $c_{\ell_{2D}} = c_{\ell\alpha}(\alpha - \alpha_z)$ and $c_{m_{2D}} = c_{m\alpha}(\alpha - \alpha_z) + c_{mz}$ for the linear range. The yawed section coefficients are to be evaluated from the two-dimensional characteristics, but at an effective angle-of-attack α_e:

$$c_{\ell y} = c_{\ell_{2D}}(\alpha_e) \qquad (6.375)$$

$$c_{my} = c_{m_{2D}}(\alpha_e) + B \qquad (6.376)$$

with no change in the zero-lift angle:

$$\alpha_e - \alpha_z = A(\alpha_n - \alpha_z) \qquad (6.377)$$

6.22 Yawed Flow and Radial Drag

Then $A = \cos^2(\Lambda + \epsilon)$ and $B = (\cos^2(\Lambda + \epsilon) - 1)c_{mz}$. Compressibility effects depend on the Mach number of the normal section:

$$M_n = K_\epsilon M = \cos(\Lambda + \epsilon) M_y \qquad (6.378)$$

Assume also that the drag is determined by the yawed section aerodynamics: $c_{dy} = c_{d_{2D}}(\alpha_e)$, with $\alpha_e = \alpha_y$.

The equivalence assumptions for swept wings thus lead to the following expressions for the analysis section aerodynamic coefficients in terms of two-dimensional airfoil characteristics:

$$c_\ell = \frac{1}{\cos^2 \Lambda} c_{\ell_{2D}}(\alpha_e) \qquad (6.379)$$

$$c_m = \frac{1}{\cos^2 \Lambda} (c_{m_{2D}}(\alpha_e) - c_{mz}) + K_\epsilon^2 c_{mz} \qquad (6.380)$$

$$\alpha_e - \alpha_z = \cos^2(\Lambda + \epsilon) \left(\frac{1}{K_\epsilon} \alpha - \alpha_z \right) \qquad (6.381)$$

$$c_d = \frac{1}{\cos \Lambda} c_{d_{2D}}(\alpha_e) \qquad (6.382)$$

$$\alpha_e - \alpha_z = \cos(\Lambda + \epsilon) \left(\frac{1}{K_\epsilon} \alpha - \alpha_z \right) \qquad (6.383)$$

$$c_r = c_d \tan \Lambda \qquad (6.384)$$

at the normal Mach number $M_n = K_\epsilon M$. The sweep parameter is

$$K_\epsilon = \frac{\cos(\Lambda + \epsilon)}{\cos \Lambda} = \cos \epsilon - \sin \epsilon \tan \Lambda \qquad (6.385)$$

These yawed flow corrections must be washed out for angles-of-attack approaching $\pm 90°$. Equation 6.384 follows from the assumption that the drag force on the section has the same sweep angle as the local section velocity, but actually only the skin friction drag has a spanwise component, not the pressure drag. The chordwise drag at zero lift can be used as an estimate of the skin friction drag, so $c_r = c_{dz} \tan \Lambda$ instead. The actual spanwise drag is probably between these two estimates. Moreover, the skin friction is in the direction of the velocity at the bottom of the boundary layer, which does not have the same yaw angle as the outer flow.

Without sweep ($\epsilon = 0$), the analysis section is the normal section, and the yawed flow correction is simply

$$c_\ell(\alpha) = c_{\ell_{2D}}(\alpha \cos^2 \Lambda) / \cos^2 \Lambda \qquad (6.386)$$

$$c_d(\alpha) = c_{d_{2D}}(\alpha \cos \Lambda) / \cos \Lambda \qquad (6.387)$$

$$c_m(\alpha) = c_{m_{2D}}(\alpha \cos^2 \Lambda) / \cos^2 \Lambda \qquad (6.388)$$

(for $\alpha_z = 0$ as well). These are the normal section coefficients in terms of the normal section angle-of-attack α and Mach number M, the flow yaw angle Λ, and two-dimensional airfoil characteristics. For angles-of-attack in the linear range, there is no effect of yaw on the lift and moment coefficients. With these transformations, yaw does delay lift and moment stall, since $\alpha_{max} = \alpha_{max_{2D}} / \cos^2 \Lambda$. The longer chord of the yawed section gives the boundary layer more time to grow, increasing the drag coefficient by the factor $(\cos \Lambda)^{-1}$. Yaw delays drag stall as well. The airfoil shape (including thickness and camber ratios and the leading-edge radius) for the two-dimensional characteristics should be that of the yawed section.

For a rotor blade with a straight quarter-chord line, the appropriate reference line for all stations is the blade radial coordinate (the r-axis). The blade section aerodynamic environment has a yaw angle in forward flight due to the radial velocity $u_R = \mu \cos \psi$:

$$\cos \Lambda = \sqrt{\frac{U^2}{U^2 + u_R^2}} = \sqrt{\frac{u_P^2 + u_T^2}{u_P^2 + u_T^2 + u_R^2}} \qquad (6.389)$$

Modern rotors often have swept tips, which could be modeled using either the r-axis as the reference line for the entire blade (so ϵ is the sweep angle of the tip relative to this reference line) or a reference line tangent to the local quarter-chord locus. The choice of the analysis section (which is in the plane perpendicular to the reference line) should not change the results. Using a straight reference line for the entire blade is generally simpler; in particular, the incremental blade area remains $c\,dr$ regardless of the tip planform geometry.

Although the airfoil shape considered for the two-dimensional aerodynamic characteristics should be that of the yawed section, that shape varies with azimuth as the yaw angle varies. In hover, or in forward flight at the mean yaw angle, this yawed section is in planes perpendicular to the r-axis. How airfoil sections are used in building a blade with a swept tip is an entirely separate issue from what airfoil sections should be used in the analysis. Rotor blades have been built with airfoil sections in the swept tip defined relative to the straight blade radial coordinate. Rotor blades have also been built with sections defined relative to the swept quarter-chord line.

6.23 Profile Power

The contributions of blade profile drag to the rotor forces, torque, and power were derived in section 6.4:

$$C_{To} = \int_0^1 \sigma \widehat{F}_{zo}\,dr \qquad (6.390)$$

$$C_{Ho} = \int_0^1 \sigma \left(\widehat{F}_{xo} \sin \psi + \widehat{F}_{ro} \cos \psi \right) dr \qquad (6.391)$$

$$C_{Qo} = \int_0^1 \sigma \widehat{F}_{xo} r\,dr \qquad (6.392)$$

(equations 6.93, 6.94, 6.95), and the profile power

$$C_{Po} = C_{Qo} + \mu C_{Ho} - \mu_z C_{To} \qquad (6.393)$$

(equation 6.117, neglecting the flap motion term). The average over the azimuth is also required. Because of the symmetry of the flow environment, the side force C_{Yo} is zero for all the models considered in this section. The blade section vertical and in-plane forces (divided by chord) due to the drag coefficient are

$$\widehat{F}_{zo} = -\frac{c_d}{2} U u_P \qquad (6.394)$$

$$\widehat{F}_{xo} = \frac{c_d}{2} U u_T \qquad (6.395)$$

6.23 Profile Power

where $U^2 = u_P^2 + u_T^2$. No small angle approximations have been made in these equations, so they are applicable to all operating conditions, including hover, edgewise flight, and high inflow axial flight. From equation 6.384, the radial drag is $D_r = D \tan \Lambda$, so

$$\widehat{F}_{ro} = \frac{1}{2}U^2 c_d \frac{u_R}{U} = \frac{c_d}{2} U u_R \qquad (6.396)$$

The section yaw angle is $\cos \Lambda = U/W$ and the resultant section velocity is W; $W^2 = U^2 + u_R^2 = u_P^2 + u_T^2 + u_R^2$. F_{xo} is positive opposing the rotor rotation, so has the sign of u_T and changes direction in the reverse flow region. The radial drag F_{ro} is in the direction of the radial velocity, so has the sign of u_R. From equation 6.387, the drag coefficient corrected for yawed flow is $c_d = c_{d_{2D}}(\alpha \cos \Lambda)/\cos \Lambda = c_{d_{2D}} W/U$. The section forces are then

$$\widehat{F}_{zo} = -\frac{c_{d_{2D}}}{2} W u_P \qquad (6.397)$$

$$\widehat{F}_{xo} = \frac{c_{d_{2D}}}{2} W u_T \qquad (6.398)$$

$$\widehat{F}_{ro} = \frac{c_{d_{2D}}}{2} W u_R \qquad (6.399)$$

The section velocities are $u_T = r + \mu \sin \psi$, $u_R = \mu \cos \psi$, and $u_P = \mu_z$ (neglecting blade motion terms in u_P). Thus

$$C_{To} = \int_0^1 \frac{\sigma c_{d_{2D}}}{2} W(-\mu_z) dr \qquad (6.400)$$

$$C_{Ho} = \int_0^1 \frac{\sigma c_{d_{2D}}}{2} W (u_T \sin \psi + u_R \cos \psi) \, dr = \int_0^1 \frac{\sigma c_{d_{2D}}}{2} W (r \sin \psi + \mu) \, dr \qquad (6.401)$$

$$C_{Qo} = \int_0^1 \frac{\sigma c_{d_{2D}}}{2} W u_T r \, dr \qquad (6.402)$$

$$C_{Po} = \int_0^1 \frac{\sigma c_{d_{2D}}}{2} W \left(u_T r + \mu(r \sin \psi + \mu) + \mu_z^2 \right) dr = \int_0^1 \frac{\sigma c_{d_{2D}}}{2} W^3 dr \qquad (6.403)$$

with the resultant velocity $W^2 = u_P^2 + u_T^2 + u_R^2 = r^2 + 2r\mu \sin \psi + \mu^2 + \mu_z^2$. Thus the energy form of the rotor power gives the profile power as the product of the total section drag ($c_d W^2$) and the resultant velocity (W). In terms of the section losses,

$$\frac{dC_{Po}}{dr} = \frac{\sigma c_{d_{2D}}}{2} W^3 = \frac{\sigma c_{d_{2D}}}{2} U(U^2 + u_R^2) = \frac{\sigma c_{d_{2D}}}{2} U^2 \left(U + u_R \frac{u_R}{U} \right) \qquad (6.404)$$

or

$$\frac{dP_o}{dr} = UD + u_R D_r \qquad (6.405)$$

which is the sum of the losses due to the normal drag D and the radial drag D_r. Given the angle-of-attack distribution over the rotor disk and the section drag coefficient data (as a function of angle-of-attack and Mach number), these expressions can be numerically integrated.

To proceed analytically, we use a mean drag coefficient $c_{d_{2D}}(\alpha \cos \Lambda) = c_{do}$ and a mean solidity σ (or constant chord). The coefficients are then scaled with the hover profile power:

$$C_{T_o} = \frac{\sigma c_{do}}{8} \int_0^1 4W(-\mu_z) dr = \frac{\sigma c_{do}}{8} F_T(\mu, \mu_z) \qquad (6.406)$$

$$C_{H_o} = \frac{\sigma c_{do}}{8} \int_0^1 4W(r \sin\psi + \mu) dr = \frac{\sigma c_{do}}{8} F_H(\mu, \mu_z) \qquad (6.407)$$

$$C_{Q_o} = \frac{\sigma c_{do}}{8} \int_0^1 4W u_T r \, dr = \frac{\sigma c_{do}}{8} F_Q(\mu, \mu_z) \qquad (6.408)$$

$$C_{P_o} = \frac{\sigma c_{do}}{8} \int_0^1 4W^3 dr = \frac{\sigma c_{do}}{8} F_P(\mu, \mu_z) \qquad (6.409)$$

defining the functions F_T, F_H, F_Q, and F_P that account for the influence of rotor edgewise and axial velocity. Analytical evaluation of the integrals is not possible for $\mu \neq 0$, but numerical integration is straightforward. For high-speed flight ($\mu \gg 1$ or $\mu_z \gg 1$) the profile terms are to highest order

$$C_{T_o} \cong \int_0^1 \frac{\sigma c_d}{2}(\mu^2 + \mu_z^2)^{-1/2}(-\mu_z) dr = -\frac{\sigma c_{do}}{8} 4V\mu_z \qquad (6.410)$$

$$C_{H_o} \cong \int_0^1 \frac{\sigma c_d}{2}(\mu^2 + \mu_z^2)^{-1/2}(\mu) dr = \frac{\sigma c_{do}}{8} 4V\mu \qquad (6.411)$$

$$C_{P_o} \cong \int_0^1 \frac{\sigma c_d}{2}(\mu^2 + \mu_z^2)^{3/2} dr = \frac{\sigma c_{do}}{8} 4V^3 \qquad (6.412)$$

where $V^2 = \mu^2 + \mu_z^2$ is the rotor velocity. The profile power at high speed approaches $P_o = \frac{1}{2}\rho V^3 A_b c_d = (qA_b c_d)V$ (for dimensional total rotor velocity V, $q = \frac{1}{2}\rho V^2$). The total drag area of the stopped rotor is thus $D/q = A_b c_d$.

For axial flight ($\mu = 0$), $W = \sqrt{r^2 + \mu_z^2}$ and the integrals are

$$C_{T_o} = \frac{\sigma c_{do}}{8} \int_0^1 4(r^2 + \mu_z^2)^{1/2}(-\mu_z) dr$$

$$= \frac{\sigma c_{do}}{8} \left[\sqrt{1 + \mu_z^2}(-2\mu_z) - 2\mu_z^3 \ln \frac{1 + \sqrt{1 + \mu_z^2}}{\mu_z} \right] \qquad (6.413)$$

$$C_{H_o} = 0 \qquad (6.414)$$

$$C_{Q_o} = \frac{\sigma c_{do}}{8} \int_0^1 4(r^2 + \mu_z^2)^{1/2} r^2 dr$$

$$= \frac{\sigma c_{do}}{8} \left[\sqrt{1 + \mu_z^2}\left(1 + \frac{1}{2}\mu_z^2\right) - \frac{1}{2}\mu_z^4 \ln \frac{1 + \sqrt{1 + \mu_z^2}}{\mu_z} \right] \qquad (6.415)$$

$$C_{P_o} = \frac{\sigma c_{do}}{8} \int_0^1 4(r^2 + \mu_z^2)^{3/2} dr$$

$$= \frac{\sigma c_{do}}{8} \left[\sqrt{1 + \mu_z^2}\left(1 + \frac{5}{2}\mu_z^2\right) + \frac{3}{2}\mu_z^4 \ln \frac{1 + \sqrt{1 + \mu_z^2}}{\mu_z} \right] \qquad (6.416)$$

6.23 Profile Power

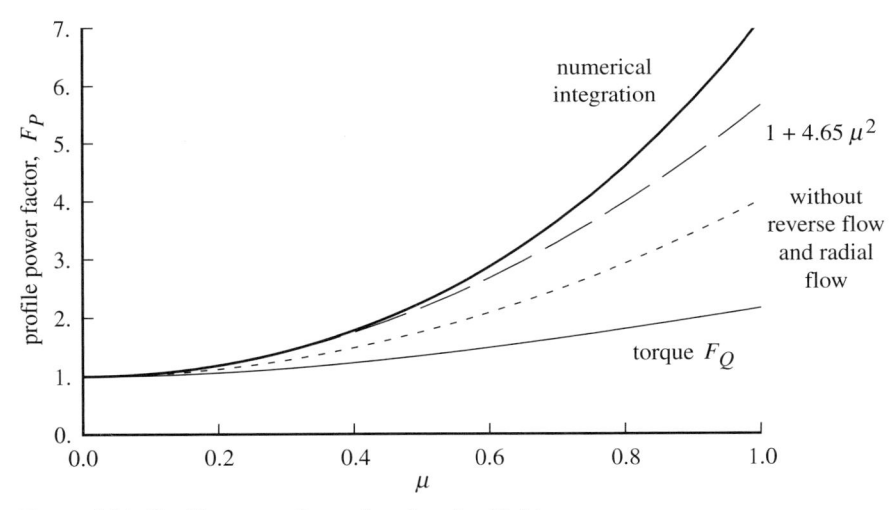

Figure 6.31. Profile power factor in edgewise flight.

For the rotor in edgewise flight ($\mu_z = 0$), $W = \sqrt{u_T^2 + u_R^2}$ and

$$C_{To} = 0 \qquad (6.417)$$

$$C_{Ho} = \frac{\sigma c_{do}}{8} \int_0^1 4 \left(u_T^2 + u_R^2\right)^{1/2} (r \sin \psi + \mu) \, dr \qquad (6.418)$$

$$C_{Qo} = \frac{\sigma c_{do}}{8} \int_0^1 4 \left(u_T^2 + u_R^2\right)^{1/2} u_T r \, dr \qquad (6.419)$$

$$C_{Po} = \frac{\sigma c_{do}}{8} \int_0^1 4 \left(u_T^2 + u_R^2\right)^{3/2} dr \qquad (6.420)$$

The following are approximate expressions for the case $\mu_z = 0$:

$$C_{Ho} = \frac{\sigma c_{do}}{8} \left(3\mu + 1.98\mu^{2.7}\right) \qquad (6.421)$$

$$C_{Qo} = \frac{\sigma c_{do}}{8} \left(1 + 1.5\mu^2 - 0.37\mu^{3.7}\right) \qquad (6.422)$$

$$C_{Po} = \frac{\sigma c_{do}}{8} \left(1 + 4.5\mu^2 + 1.61\mu^{3.7}\right) \qquad (6.423)$$

These expressions are accurate to 1% up to $\mu = 1$, meaning that the error is less than 1% relative to numerical integration. Accurate evaluation of the profile power also depends on the mean drag coefficient. A frequently used approximation is

$$C_{Po} = \frac{\sigma c_{do}}{8} \left(1 + 4.65\mu^2\right) \qquad (6.424)$$

which has an error less than 1% up to $\mu = 0.35$, and less than 4% up to $\mu = 0.5$. The factor $(1 + 4.65\mu^2)$, which gives the profile power increase with speed, has the following contributions: $(1 + \mu^2)$ from the rotor torque and $2\mu^2$ from the rotor drag C_{Ho} due to the normal blade drag force; μ^2 from C_{Ho} due to the radial drag force; $0.45\mu^2$ due to the yawed-flow increase of the drag coefficient; and $0.20\mu^2$ due to reverse flow. Figure 6.31 shows the profile power factor F_P as a function of advance ratio, comparing the exact result (equation 6.409) with the approximation $F_P = 1 + 4.65\mu^2$. Also shown is the power factor with no radial flow or reverse flow, and the profile torque factor F_Q. The profile power increase is significant at moderate μ and is very large at high μ. At high speeds, stall and compressibility effects must also be included in the evaluation of C_{P_o}.

Harris (2008) developed approximations valid for all μ and μ_z by integrating analytically in r (using mathematics software) and approximating the integration in ψ:

$$F_T \cong -\mu_z \sqrt{1+V^2}\left(2 + \frac{1}{2}\mu^2 \frac{1+2V^2}{(1+V^2)^2}\right)$$
$$- \mu_z \left(2\mu_z^2 + \mu^2\right) \ln \frac{1+\sqrt{1+V^2}}{V} \qquad (6.425)$$

$$F_H \cong \mu\sqrt{1+V^2}\left(3 + \frac{1}{4}\mu^2 \frac{V^2-1}{(1+V^2)^2}\right)$$
$$+ \mu \left(\mu_z^2 + \frac{3}{4}\mu^2\right) \ln \frac{1+\sqrt{1+V^2}}{V} \qquad (6.426)$$

$$F_Q \cong \sqrt{1+V^2}\left(1 + \frac{1}{2}V^2 + \frac{1}{8}\mu^2 \frac{4+V^2-4V^4}{(1+V^2)^2} + \frac{3}{16}\frac{\mu^4}{1+V^2}\right)$$
$$- \left(\frac{1}{2}\mu_z^4 + \frac{1}{2}\mu_z^2\mu^2 + \frac{3}{16}\mu^4\right) \ln \frac{1+\sqrt{1+V^2}}{V} \qquad (6.427)$$

$$F_P \cong \sqrt{1+V^2}\left(1 + \frac{5}{2}V^2 + \frac{3}{8}\mu^2 \frac{4+7V^2+4V^4}{(1+V^2)^2} - \frac{9}{16}\frac{\mu^4}{1+V^2}\right)$$
$$+ \left(\frac{3}{2}\mu_z^4 + \frac{3}{2}\mu_z^2\mu^2 + \frac{9}{16}\mu^4\right) \ln \frac{1+\sqrt{1+V^2}}{V} \qquad (6.428)$$

with here $V^2 = \mu^2 + \mu_z^2$. These expressions have an error less than 1% for all speeds. They are exact for axial flow ($\mu = 0$). Note that $F_P \cong 4V^3$ for large V.

It is useful to examine separately the effects of reverse flow, radial drag, and the yawed-flow drag coefficient. The integrals for edgewise flight ($\mu_z = 0$) can be written as

$$C_{Ho} = \frac{\sigma c_{do}}{8} \int_0^1 \frac{4}{\cos \Lambda} (u_T \sin \psi + u_R \cos \psi) |u_T| dr \qquad (6.429)$$

$$C_{Qo} = \frac{\sigma c_{do}}{8} \int_0^1 \frac{4}{\cos \Lambda} u_T |u_T| r\, dr \qquad (6.430)$$

$$C_{Po} = \frac{\sigma c_{do}}{8} \int_0^1 \frac{4}{\cos \Lambda} (u_T^2 + u_R^2) |u_T| dr \qquad (6.431)$$

since $W = |u_T|/\cos \Lambda$. Reverse flow gives the absolute value of u_T (the integrals are handled as in equation 6.182). Radial drag contributes the u_R terms. The $(\cos \Lambda)^{-1}$ factor is the increase of the drag coefficient due to yawed flow. Neglecting all three effects reduces the model to that considered in section 6.4:

$$C_{Ho} = \frac{\sigma c_{do}}{8} \int_0^1 4u_T^2 \sin \psi\, dr = \frac{\sigma c_{do}}{8} 2\mu \qquad (6.432)$$

6.23 Profile Power

$$C_{Qo} = \frac{\sigma c_{do}}{8} \int_0^1 4u_T^2 r\, dr = \frac{\sigma c_{do}}{8}(1+\mu^2) \quad (6.433)$$

$$C_{Po} = \frac{\sigma c_{do}}{8} \int_0^1 4u_T^3\, dr = \frac{\sigma c_{do}}{8}(1+3\mu^2) \quad (6.434)$$

Including the radial drag force gives

$$C_{Ho} = \frac{\sigma c_{do}}{8} \int_0^1 4u_T(r\sin\psi + \mu)\, dr = \frac{\sigma c_{do}}{8} 3\mu \quad (6.435)$$

$$C_{Qo} = \frac{\sigma c_{do}}{8} \int_0^1 4u_T^2 r\, dr = \frac{\sigma c_{do}}{8}(1+\mu^2) \quad (6.436)$$

$$C_{Po} = \frac{\sigma c_{do}}{8} \int_0^1 4(u_T^2 + u_R^2)u_T\, dr = \frac{\sigma c_{do}}{8}(1+4\mu^2) \quad (6.437)$$

The radial drag increases the rotor H-force by 50%, which increases the profile power in forward flight. Including reverse flow only (a form frequently encountered in the literature) gives

$$C_{Ho} = \frac{\sigma c_{do}}{8} \int_0^1 4u_T|u_T|\sin\psi\, dr = \frac{\sigma c_{do}}{8}\left(2\mu + \frac{1}{2}\mu^3\right) \quad (6.438)$$

$$C_{Qo} = \frac{\sigma c_{do}}{8} \int_0^1 4u_T|u_T|r\, dr = \frac{\sigma c_{do}}{8}\left(1 + \mu^2 - \frac{1}{8}\mu^4\right) \quad (6.439)$$

$$C_{Po} = \frac{\sigma c_{do}}{8} \int_0^1 4u_T^2|u_T|\, dr = \frac{\sigma c_{do}}{8}\left(1 + 3\mu^2 + \frac{3}{8}\mu^4\right) \quad (6.440)$$

for $\mu \leq 1$. With both the radial drag and reverse flow included,

$$C_{Ho} = \frac{\sigma c_{do}}{8} \int_0^1 4(r\sin\psi + \mu)|u_T|\, dr = \frac{\sigma c_{do}}{8}\left(3\mu + \frac{3}{4}\mu^3\right) \quad (6.441)$$

$$C_{Qo} = \frac{\sigma c_{do}}{8} \int_0^1 4u_T|u_T|r\, dr = \frac{\sigma c_{do}}{8}\left(1 + \mu^2 - \frac{1}{8}\mu^4\right) \quad (6.442)$$

$$C_{Po} = \frac{\sigma c_{do}}{8} \int_0^1 4\left(u_T^2 + u_R^2\right)|u_T|\, dr = \frac{\sigma c_{do}}{8}\left(1 + 4\mu^2 + \frac{5}{8}\mu^4\right) \quad (6.443)$$

The effect of reverse flow is smaller than the radial drag force effect, because of the low dynamic pressure in the reverse flow region.

Glauert (1926) obtained the energy expression for the profile power, as an extension of the propeller (axial flow) result:

$$C_{Po} = \int_0^1 \frac{\sigma c_d}{2}\left(u_T^2 + u_R^2\right)^{3/2}\, dr \quad (6.444)$$

and by blade element theory,

$$C_{Po} = \int_0^1 \frac{\sigma c_d}{2} u_T^3\, dr = \frac{\sigma c_{do}}{8}(1+3\mu^3) \quad (6.445)$$

(neglecting reverse flow and radial flow). To evaluate the energy expression, he averaged the values at $\psi = 0°$, $90°$, $180°$, and $270°$ (where the radial integration can be performed analytically). Writing this average as $C_{Po} = (\sigma c_{do}/8)(1+n\mu^2)$, he obtained

$$F_P = 1 + n\mu^2 \cong \frac{1}{2}(1 + 6\mu^2 + \mu^4) + \frac{1}{4}(2 + 5\mu^2)\sqrt{1+\mu^2}$$
$$+ \frac{3}{8}\mu^4 \ln\frac{\sqrt{1+\mu^2}+1}{\sqrt{1+\mu^2}-1} \qquad (6.446)$$

To order μ^2 this gives $n = 4.5$. At $\mu = 1$ this approximation gives $n = 6.13$. Equation 6.446 has an error less than 1% up to $\mu = 1.9$. This result was largely ignored, even by Lock and those who built on the work of Glauert and Lock. Subsequent work started with blade element theory rather than the energy method, hence omitting the radial flow effects.

Cierva (1933) developed an approximation for C_{Po} by introducing a change of variables in the integration. Since $U^2 = u_T^2 + u_R^2 = (r + \mu \sin \psi)^2 + (\mu \cos \psi)^2$, it follows that $dr = (U/u_T)dU = dU/(\cos \Lambda)$; so

$$C_{Po} = \frac{\sigma c_{do}}{2}\int_0^1 U^3 dr = \frac{\sigma c_{do}}{2}\int_\mu^{\sqrt{1+\mu^2+2\mu\sin\psi}} \frac{U^3}{\cos \Lambda} dU \qquad (6.447)$$

Setting the lower limit to zero and neglecting the $(\cos \Lambda)^{-1}$ factor (compensating errors), the integrations over U and ψ can be performed:

$$C_{Po} = \frac{\sigma c_{do}}{2}\int_0^{\sqrt{1+\mu^2+2\mu\sin\psi}} U^3 dU = \frac{\sigma c_{do}}{8}\left(1 + 4\mu^2 + \mu^4\right) \qquad (6.448)$$

which has an error less than 1% only up to $\mu = 0.15$, and less than 4% up to $\mu = 0.30$.

Bennett (1940) derived an expression for C_{Po} by expanding the integral for small μ:

$$C_{Po} = \frac{\sigma c_{do}}{8}\left(1 + \frac{9}{2}\mu^2 - \frac{3}{4}\mu^4 \ln \mu + \frac{3}{16}\mu^6 - \frac{3}{128}\mu^8 + \ldots\right) \qquad (6.449)$$

Expressed as $C_{Po} = (\sigma c_{do}/8)(1+n\mu^2)$, equation 6.449 gives $n = 4.5$ at $\mu = 0$ and $n = 4.66$ at $\mu = 1$. Bennett suggested using $n = 4.65$ (the origin of equation 6.424), although the expansion is only valid for small μ. Equation 6.449 has an error less than 1% up to $\mu = 0.3$, and less than 4% up to $\mu = 0.5$.

The best calculation of profile power is obtained by using the force balance method to evaluate the rotor torque, then subtracting induced and propulsive power (section 6.4). A simple calculation of profile power, as for a conceptual design analysis, is obtained using equations 6.409 and 6.428, with an estimate of the mean drag coefficient. The expressions for C_{Po} with various assumptions and approximates illustrate the contributions to the profile power. In particular, equations 6.434, 6.440, and 6.424 are encountered in the literature. The distribution of profile power over the rotor disk is given by equation 6.405. A measure of the rotor profile power is the mean drag coefficient, $c_{dmean} = (8C_{Po}/\sigma)/F_P$, where F_P (from equation 6.428) accounts for the increase of the mean dynamic pressure with rotor edgewise and axial flight speed.

Figure 6.32. C.6A autogyro at Farnborough.

6.24 History

6.24.1 The Beginning of Aeromechanics

Rotorcraft aeromechanics analysis began with Glauert. In a remarkable November 1926 report for the British Aeronautical Research Council and a paper read two months later to the Royal Aeronautical Society, H. Glauert of the Royal Aircraft Establishment presented the foundations of induced and profile power analysis for rotors, and introduced blade element theory for performance and hub loads of flapping rotors in forward flight (Glauert (1926, 1927)). The impetus for this work was given by the demonstration flights in Britain of the Cierva C.6A autogyro, carried out at Farnborough in October 1925 (Figure 6.32), and a lecture by Juan de la Cierva that was read to the Royal Aeronautical Society on October 22 (Cierva (1926)). The Royal Aircraft Establishment (RAE) work was motivated by the need to check the claims that Cierva was making for his aircraft. The first general account of Cierva's invention was received in Britain in February 1925, which led immediately to preliminary wind-tunnel tests and a theoretical investigation. Following the C.6A demonstration, wind-tunnel tests were conducted at the National Physical Laboratory (NPL) and RAE (1926–1927), and model autogyro drop tests were also done (1926). The more detailed theory (Glauert (1926)) provided a satisfactory estimate of maximum lift, an account of flapping, and a qualitative explanation of the side force. The theory was extended by C.N.H. Lock of the National Physical Laboratory in ARC reports of March and May 1927 (Lock (1927)). The history is documented in the original reports, especially Glauert and Lock (1928); see also Johnson (2011).

Glauert proposed an expression (equation 5.9) for the rotor induced velocity, making the connection between propeller and wing induced power. He characterized the result as a "logical generalization of the ordinary aerofoil formula," reducing for small incidence angle to the standard formula for the normal induced velocity of a wing of span $2R$, and for incidence angles nearly $90°$ to the ordinary momentum formula for an airscrew. "It is anticipated therefore that the formula will be valid over a wide range of angle of incidence." The key fact permitting this generalization is that the mass flow affected by a wing is that through a circle around the wing span. Glauert recognized that for a circular wing this circle has the same area as the rotor disk.

In an appendix, Glauert (1926) considered the energy losses of an autogyro. "Two main sources of loss of energy are considered, due respectively to the induced velocity caused by the thrust and to the profile drag of the blades. An additional source of loss of energy is the periodic distribution of thrust over the disc of the windmill but no simple method has been found of estimating its magnitude." First considering just the blade section tangential velocity, the energy loss due to the drag of the blades was written as in equations 6.118 and 6.445, ignoring the axial velocity and the blade section radial velocity. The effect of the former is small for edgewise flight of the helicopter rotor, but the effect of the radial velocity is important. Including the radial velocity gave the profile energy loss as in equations 6.409 and 6.446. This result for the profile power was developed as a check of the blade element theory result, particularly at low disk incidence and high speed. Since the connection between the two approaches is complex, the blade element theory form was viewed as the primary result. That was unfortunate since the energy form was in fact more accurate and predicted better performance for the autogyro.

Glauert introduced blade element theory for edgewise-moving rotors. "The autogyro is essentially a windmill of low pitch working in a sidewind, and it is natural to apply to it the modern methods of strip theory combined with the Prandtl theory of interference, which have been so successful in the case of the monoplane wing and the ordinary airscrew" (Lock (1927)). Expressions were developed for the blade section velocities including flight speed, rotation, flapping, and induced velocity. The section analysis was based on two-dimensional airfoil characteristics with a constant lift-curve slope and a mean drag coefficient. "The aerodynamic characteristics of the aerofoil section must be taken to correspond to two dimensional motion or infinite aspect ratio in accordance with modern aerofoil and airscrew theory" (Glauert (1927)). The flapping equation of motion was derived from equilibrium of inertial, centrifugal, aerodynamic, and weight moments about the flap hinge (with no hinge offset).

Glauert neglected squares and higher powers of advance ratio μ. Considering the breakdown of the small angle assumption near reverse flow, he viewed the limit of validity to be $\mu < 0.5$. The maximum speed of the C.6A gave about $\mu = 0.4$, although later autogyros would operate at much higher advance ratios. However, an order μ blade element analysis led to performance estimates more pessimistic than obtained by energy methods. In a report published the next year, Lock (1927) eliminated the order μ assumption of the blade element analysis and verified that the blade element analysis and energy method gave identical performance results. He also added cyclic pitch and solved for the higher harmonics of flap motion.

The results for lateral flapping and side force in Glauert's model implied a lateral shaft-axis force proportional to thrust and forward speed, the direction being to port (the retreating side) with curved blades and to starboard with straight blades. This did not agree with the observed lateral movement of the shaft on the Cierva autogyro. The discrepancy was attributed to the assumption of uniform inflow over the rotor disk. Considering the wake of a circular wing, the induced velocity is expected to increase from the front to the rear of the disk. So Glauert introduced an increment of the induced velocity proportional to the distance behind the center of the disc; hence $v + v_1(r/R)\cos\psi$ for the total, with an estimate of $v_1 \cong v$. The resulting correction improved the prediction of the lateral shaft angle, but the correction was not sufficiently large.

Between them, Glauert and Lock established many of the key concepts and even the notation of rotor aeromechanics. Glauert (1926) introduced the concept of the tip-path plane, and derived the expression for the mean lift coefficient. He used β for the blade flap angle (but not a Fourier series expansion), and σ for rotor solidity. Lock (1927) described the equivalence of flapping and feathering and used a negative Fourier series for the flap motion (to become NACA notation). He used $\mu = V\cos i/\Omega R$ for the advance ratio. Lock introduced the parameter representing the ratio of aerodynamic and inertial forces on the blade, $\gamma = \rho a c R^4/I_b$, which consequently bears his name (though differing by a factor of 2 from the U.S. definition, because of the different convention for the lift coefficient).

Cierva sent a letter as his contribution to the discussion of Glauert's 1927 RAS lecture, in which he strongly objected to Glauert's results and conclusions. Glauert replied to the discussion: "I believe that [the autogyro] is less economical than an aeroplane, but that it has very considerable advantages as regards safety and ease of landing." There would seem to be common ground with Cierva's remark: "The autogyro will have performance at least as good as, and possibly better than, the aeroplane, since lift/drag is not the sole criterion of performance." However, there was direct disagreement regarding the values of the lift-to-drag ratio and vertical rate of descent, Glauert's estimate of the performance being significantly less favorable than Cierva's statements. From today's perspective, Glauert's analysis is considered optimistic. Cierva was likely most concerned about the possible impact of Glauert's results on his continued development of the autogyro, for he concluded his letter with a remark on "the risk of mistake which is necessarily involved in trying to limit from the beginning the possibilities and improvements of any new system."

6.24.2 After Glauert

In the decades following Glauert's work, development of the basic analysis of autogyro and helicopter rotors progressed in a series of steps built on the foundation provided by Glauert. Glauert considered a flapping rotor with no twist, constant chord, no hinge offset, and no cyclic pitch, as well as small angle aerodynamics with constant lift-curve slope and mean drag coefficient. The rotor loads and flapping (coning and first harmonic) were obtained, with only first-order terms in advance ratio retained. Lock (1927) extended Glauert's analysis by including higher powers of advance ratio, second harmonic flapping, and cyclic pitch. He showed the equivalence of no-feathering plane and tip-path plane analyses, and the equivalence of the blade element theory and energy method results for power (though neglecting radial flow and reverse flow).

Wheatley (1934) extended the theory of Glauert and Lock and evaluated the accuracy of the theory by comparing calculations with test results. He considered a flapping rotor with no hinge offset; linear twisted, constant-chord blades; the tip loss factor; and linear induced velocity variation. Wheatley included reverse flow (accounting for the sign of lift and drag in the reverse flow region), but still with small angle aerodynamics, and he neglected radial flow and radial drag effects in the profile power. The calculations were compared with test data for the Pitcairn autogyro.

The comparison was generally good up to about $\mu = 0.5$. A significant discrepancy was in the calculation of lateral flapping, which was typically 1.5° low in magnitude. Including an estimate of the longitudinal inflow gradient ($\kappa_x = 0.5$) reduced the error to about 1.0°. Wheatley considered the likely source of this discrepancy to be the simple induced velocity variation used.

Sissingh (1939) extended Wheatley's analysis, considering a flap-hinge offset and eliminating the assumption of a constant (mean) drag coefficient by using a general drag polar of the form $c_d = \delta_0 + \delta_1 \alpha + \delta_2 \alpha^2$.

Bailey (1941) put Wheatley's analysis in practical form for routine use by expressing all quantities (thrust, torque, and profile power) as direct functions of the blade collective pitch and no-feathering-plane inflow ratio. The coefficients of these expressions depend on the rotor twist, Lock number, tip loss factor, and advance ratio. Bailey considered a flapping rotor with linearly twisted, constant-chord blades and a drag polar $c_d = \delta_0 + \delta_1 \alpha + \delta_2 \alpha^2$. The aerodynamic model included reverse flow (to order μ^4), but still assumed small angles to evaluate the angle-of-attack and used a constant lift-curve slope, $c_\ell = a\alpha$. The theory used uniform inflow and neglected stall, compressibility, radial flow, and radial drag. Bailey developed a method to find the coefficients δ_0, δ_1, and δ_2 from the aerodynamic characteristics of the airfoil section (see section 8.8). The expression $c_d = 0.0087 - 0.0216\alpha + 0.400\alpha^2$ was obtained for the NACA 23012 airfoil at $Re = 2 \times 10^6$, which gives the drag accurately to about $\alpha = 12°$. This expression has been used often for the blade drag in rotor analyses, even for other airfoil sections. He considered the blade stall limit, in terms of when the quadratic expression for c_d is no longer valid. Bailey presented a solution procedure for the rotor performance. The procedure was iterative for helicopter rotors, but for autogyros involved solving $C_Q = 0$ directly for the inflow ratio. Bailey and Gustafson (1944) and Gustafson (1953) presented helicopter performance charts based on Bailey's theory.

Castles and New (1952) extended the theory of Wheatley and Bailey to large angles of pitch and inflow. They represented the section aerodynamic coefficients as $c_\ell = a \sin \alpha$ and $c_d = \epsilon_0 + \epsilon_1 \sin \alpha + \epsilon_2 \cos \alpha$; substituted trigonometric expansions of

$$\sin \alpha = \sin(\theta - \phi) = \sin \theta \cos \phi - \sin \phi \cos \theta \qquad (6.450)$$

$$\cos \alpha = \cos(\theta - \phi) = \cos \theta \cos \phi + \sin \phi \sin \theta \qquad (6.451)$$

and used large angle expressions $\sin \phi = u_P/(u_T^2 + u_P^2)^{1/2}$ and $\cos \phi = u_T/(u_T^2 + u_P^2)^{1/2}$. They considered an articulated blade with arbitrary twist and chord distributions, and root cutout. A linear inflow distribution was used.

Gessow and Crim (1952) also extended the theory of Wheatley and Bailey to large pitch and inflow angles, for a linearly twisted, constant-chord blade. The angle-of-attack $\alpha = \theta - \phi$ was still assumed to be small. They observed that the flow at high speeds during powered operation is usually stalled in the reverse flow region, so reverse flow aerodynamics were approximated by using stalled values of lift and drag coefficients: constant $c_\ell = 1.2$ and $c_d = 1.1$ for powered flight, $c_\ell = 0.5$ and $c_d = 0.1$ for autorotation. Based on this theory, Tapscott and Gessow (1956) presented charts of the calculated blade flap motion, and Gessow and Tapscott (1956) prepared performance charts.

Gessow and Crim (1955) developed the equations and a solution procedure for the numerical integration of the transient flap motion. They considered an articulated rotor with offset flapping hinge (or a teetering rotor); large angles of flapping, inflow, and pitch; and general airfoil characteristics (lift and drag coefficients as a function of angle-of-attack and Mach number). The solution was obtained by numerical integration using a digital computer. The analysis was developed for investigations of the flap dynamic stability (from the transient motion) and rotor performance (from the converged periodic solution). By using numerical integration, general aerodynamic characteristics could be considered, including stall, compressibility, and reverse flow (assuming that the required airfoil characteristics are available). Gessow (1956) further developed the equations for numerical calculation of the aerodynamic characteristics of rotors in a form intended for digital computer applications. The model included arbitrary blade twist, chord, and mass distribution; general two-dimensional aerodynamic coefficients for the blade airfoil; and large angles of pitch and inflow. The blade flap angle was assumed to be small, and radial flow effects were still neglected. The solution procedure solved the flap equation of motion directly for the harmonics of the blade motion. Gessow and Tapscott (1960) presented tables and charts of calculated rotor performance, including flight conditions well into the stall range, based on the analyses of Gessow and Crim (1955) and Gessow (1956). Tanner (1964a, 1964b) developed a performance calculation method based on the analysis of Gessow and Crim, and prepared performance charts and tables.

This series of analytical developments was accompanied by a shift of the work from Britain (RAE and NPL) to the United States (NACA), and from autogyros to helicopters. The progression from Glauert was clear up to the introduction of the digital computer, which changed the way problems would be formulated and solved.

6.25 REFERENCES

Bailey, F.J., Jr. "A Simplified Theoretical Method of Determining the Characteristics of a Lifting Rotor in Forward Flight." NACA Report 716, 1941.

Bailey, F.J., Jr., and Gustafson, F.B. "Charts for Estimation of the Characteristics of a Helicopter Rotor in Forward Flight; I – Profile Drag-Lift Ratio for Untwisted Rectangular Blades." NACA ACR L4H07, August 1944.

Bennett, J.A.J. "Rotary-Wing Aircraft." Aircraft Engineering, 12:133 (March 1940).

Bennett, J.A.J. "The Era of the Autogiro." The Journal of the Royal Aeronautical Society, 65:610 (October 1961).

Castles, W., Jr., and New, N.C. "A Blade-Element Analysis for Lifting Rotors That Is Applicable for Large Inflow and Blade Angles and Any Reasonable Blade Geometry." NACA TN 2656, July 1952.

Charles, B.D., and Tanner, W.H. "Wind Tunnel Investigation of Semirigid Full-Scale Rotors Operating at High Advance Ratios." USAAVLABS TR 69-2, January 1969.

de la Cierva, J. "The Development of the Autogyro." The Journal of the Royal Aeronautical Society, 30:181 (January 1926).

de la Cierva, J. "Engineering Theory of the Autogiro; Theory of Stresses on Autogiro Rotor Blades." 1933.

Coleman, R.P., Feingold, A.M., and Stempin, C.W. "Evaluation of the Induced-Velocity Field of an Idealized Helicopter Rotor." NACA ARR L5E10, June 1945.

Datta, A., Yeo, H., and Norman, T.R. "Experimental Investigation and Fundamental Understanding of a Slowed UH-60A Rotor at High Advance Ratios." American Helicopter Society 66th Annual Forum, Virginia Beach, VA, May 2011.

Gessow, A. "Equations and Procedures for Numerically Calculating the Aerodynamic Characteristics of Lifting Rotors." NACA TN 3747, October 1956.

Gessow, A., and Crim, A.D. "An Extension of Lifting Rotor Theory to Cover Operation at Large Angles of Attack and High Inflow Conditions." NACA TN 2665, April 1952.

Gessow, A., and Crim, A.D. "A Method for Studying the Transient Blade-Flapping Behavior of Lifting Rotors at Extreme Operating Conditions." NACA TN 3366, January 1955.

Gessow, A., and Crim, A.D. "A Theoretical Estimate of the Effects of Compressibility on the Performance of a Helicopter Rotor in Various Flight Conditions." NACA TN 3798, October 1956.

Gessow, A., and Tapscott, R.J. "Charts for Estimating Performance of High-Performance Helicopters." NACA Report 1266, 1956.

Gessow, A., and Tapscott, R.J. "Tables and Charts for Estimating Stall Effects on Lifting-Rotor Characteristics." NASA TN D-243, May 1960.

Glauert, H. "A General Theory of the Autogyro." ARC R&M 1111, November 1926.

Glauert, H. "The Theory of the Autogyro." The Journal of the Royal Aeronautical Society, *31*:198 (June 1927).

Glauert, H., and Lock, C.N.H. "A Summary of the Experimental and Theoretical Investigations of the Characteristics of an Autogyro." ARC R&M 1162, April 1928.

Gustafson, F.B. "Charts for Estimation of the Profile Drag-Lift Ratio of a Helicopter Rotor Having Rectangular Blades with -8 deg Twist." NACA RM L53G20a, October 1953.

Harris, F.D. "Rotary Wing Aerodynamics – Historical Perspective and Important Issues." American Helicopter Society National Specialists' Meeting on Aerodynamics and Aeroacoustics, Arlington, TX, February 1987.

Harris, F.D. "Rotor Performance at High Advance Ratio; Theory versus Test." NASA CR 2008-215370, October 2008.

Johnson, W. "Milestones in Rotorcraft Aeromechanics." NASA TP 2011-215971, May 2011.

Lock, C.N.H. "Further Development of Autogyro Theory." ARC R&M 1127, March 1927.

McCloud, J.L., III, Biggers, J.C., and Stroub, R.H. "An Investigation of Full-Scale Helicopter rotors at High Advance Ratios and Advancing Tip Mach Numbers." NASA TN D-4632, July 1968.

Peters, D.A., and Ormiston, R.A. "Flapping Response Characteristics of Hingeless Rotor Blades by a Generalized Harmonic Balance Method." NASA TN D-7856, February 1975.

Sissingh, G. "Contribution to the Aerodynamics of Rotating-Wing Aircraft." NACA TM 921, December 1939.

Tanner, W.H. "Charts for Estimating Rotary Wing Performance in Hover and at High Forward Speeds." NASA CR 114, November 1964a.

Tanner, W.H. "Tables for Estimating Rotary Wing Performance at High Forward Speeds." NASA CR 115, November 1964b.

Tapscott, R.J., and Gessow, A. "Charts for Estimating Rotor-Blade Flapping Motion of High-Performance Helicopters." NACA TN 3616, March 1956.

Wheatley, J.B. "An Aerodynamic Analysis of the Autogiro Rotor with a Comparison Between Calculated and Experimental Results." NACA Report 487, 1934.

White, F., and Blake, B.B. "Improved Method of Predicting Helicopter Control Response and Gust Sensitivity." American Helicopter Society 35th Annual Forum, Washington, DC, May 1979.

Yamauchi, G.K., and Johnson, W. "Trends of Reynolds Number Effects on Two Dimensional Airfoil Characteristics for Helicopter Rotor Analyses." NASA TM 84363, April 1983.

7 Performance

The calculation of rotorcraft performance is largely a matter of determining the power required and power available over a range of flight conditions. The power information can then be translated into quantities such as payload, range, ceiling, speed, and climb rate, which define the operational capabilities of the aircraft. The rotor power required is divided into four parts: the induced power, required to produce the rotor thrust; the profile power, required to turn the rotor through the air; the parasite power, required to move the aircraft through the air; and the climb power, required to change the gravitational potential energy. The aircraft has additional contributions to power required, including accessory and transmission losses and perhaps anti-torque power. In hover there is no parasite power, and the induced power is 65% to 75% of the total. As the forward speed increases, the induced power decreases, the profile power increases slightly, and the parasite power increases until it is dominant at high speed. Thus the total power required is high at hover, because of the induced power with a low but reasonable disk loading. At first the total power decreases significantly with increasing speed, as the induced power decreases; then it increases again at high speed, because of the parasite power. Minimum power required occurs roughly in the middle of the helicopter speed range.

The task in rotorcraft performance analysis is the calculation of the rotor forces and power. Procedures to perform these calculations have been developed in the preceding chapters. There are two basic approaches to the calculation of rotor performance: the force-balance method and the energy-balance method. In the force-balance method, the blade section forces are integrated to obtain the net rotor forces and torque. The solution also requires knowledge of the rotor induced velocity and blade motion to define the blade angle-of-attack distribution. The aircraft is trimmed to force and moment equilibrium to determine the rotor forces and attitude required to maintain the specified flight condition. Even in its simplest form, the force-balance method is complicated, so is best suited for numerical calculations, when it can be used with the most advanced models of the rotor motion and aerodynamics.

The second approach to rotorcraft performance analysis is the energy-balance method, in which the power required is expressed in terms of the individually identifiable sources of energy loss in the aircraft. In section 6.4, the energy-balance expression was derived from the force-balance relations, showing that the two methods are equivalent and must give identical results when the same assumptions are used. The energy-balance method is most useful for routine performance calculations for a number of reasons. First, the aircraft longitudinal force equilibrium has already been considered, so the power is obtained directly without the necessity of calculating the rotorcraft trim as well. Second, the parasite and climb power are given in simple yet exact forms, and with separate expressions given for the induced and profile power, using approximate results for these terms is straightforward. With the simplest approximations for the induced and profile power, the energy-balance method is fast and reasonably accurate and hence is well suited for use in conceptual design. For more detailed performance analyses a better estimate of the induced and profile power is needed, requiring again a calculation of the blade angle-of-attack distribution. Thus with numerical methods the force-balance and energy-balance methods are computationally equivalent, although separating the total power into the induced, profile, parasite, and climb components is still useful to interpret the results.

A comprehensive analysis of rotorcraft performance must consider an arbitrary rotor, including general chord, twist, and profile distributions, and it must be applicable to extreme flight conditions, such as high loading or high speed. The climb and parasite power can be obtained exactly, assuming that the aircraft flight path angle and the parasite drag are known. Thus efforts to improve the calculation of rotorcraft performance have been primarily concerned with the induced power and profile power. Improving the estimate of the induced power requires a calculation of the nonuniform induced velocity distribution, as well as an accurate loading distribution. The profile power estimate is improved by considering the actual angle-of-attack and Mach number distribution in calculating the section drag. Obtaining the blade angle-of-attack requires the nonuniform induced velocity calculation and also a solution for the blade motion. At extreme operating conditions, more blade degrees of freedom than the fundamental flapping mode must be considered. Thus an improved performance analysis is a complicated numerical problem requiring more attention to the details of the rotor and its aerodynamics.

Thus rotorcraft performance analysis generally takes one of two forms: the energy-balance method, using fairly simple expressions for the induced and profile power, or a numerical calculation by the force-balance method, using as detailed a model as is possible and appropriate. Before digital computers were widely available, rotor performance charts were commonly prepared. These charts were based on analytical or numerical solutions, usually by the force balance method. Now computational resources permit routine numerical calculations of performance for particular rotor parameters and operating conditions, beginning early in the design process.

The equations for rotor performance in hover and forward flight are summarized in this chapter, and the attributes that describe the rotorcraft performance capabilities are discussed, including maximum speed, rate of climb and descent, ceilings, range, and endurance. In a broader sense, the helicopter performance requirement is the ability to complete a specified mission most efficiently, and the capability to calculate these specific performance quantities is required not only to define the operational limits of the aircraft but also to perform a mission analysis.

7.1 Rotor Performance Estimation

7.1.1 Hover and Vertical Flight Performance

The rotor power required in vertical flight has been obtained in the form $C_P = C_{Pi} + C_{Po} + C_{Pc}$, where the induced, profile, and climb power coefficients are given by

$$C_{Pi} = \int_{r_R}^{B} \lambda_i dC_T \tag{7.1}$$

$$C_{Po} = \int_{r_R}^{1} \frac{\sigma c_d}{2} r^3 dr \tag{7.2}$$

$$C_{Pc} = \lambda_c C_T \tag{7.3}$$

(equation 3.84). In dimensional form

$$P = P_i + P_o + P_c = \int_{r_R}^{B} (V + v)dT + \rho A (\Omega R)^3 C_{Po} \tag{7.4}$$

The induced power C_{Pi} is the energy created in the rotor wake by imparting a downward momentum to the air, from which the lift reaction on the rotor is obtained. Momentum theory gives the simplest estimate of the induced power. Using a correlation of measured rotor performance data to define the law in regions where momentum theory is not applicable, the sum of induced and climb power in vertical flight is given by a universal curve of $(V+v)/v_h = P/P_h$ as a function of V/v_h, where $v_h^2 = T/2\rho A$; see Figure 4.5. For hover, the induced velocity is $v = \kappa v_h = \kappa \sqrt{T/2\rho A}$, where κ is an empirical factor correcting for the additional losses, principally tip losses and losses due to nonuniform inflow. Typically $\kappa = 1.10$ to 1.15 (section 3.3).

The profile power C_{Po} is the energy dissipated by the viscous drag of the blade. A rough estimate of C_{Po} is obtained by using a mean blade drag coefficient: $C_{Po} = \sigma c_{do}/8$ for hover. A more accurate estimate requires an integration of the drag coefficient over the span of the blade, using the actual angle-of-attack distribution and Mach number of the blade.

Hover performance can be expressed in terms of the polar of C_P as a function of C_T. The approximate expression for the hover polar is

$$C_P = \frac{\kappa C_T^{3/2}}{\sqrt{2}} + \frac{\sigma c_{do}}{8} \tag{7.5}$$

(equation 3.91), or

$$P = \frac{\kappa T^{3/2}}{\sqrt{2\rho A}} + \rho A (\Omega R)^3 \frac{\sigma c_{do}}{8} \tag{7.6}$$

In vertical flight the polar is $C_P = (\lambda_i + \lambda_c)C_T + \sigma c_{do}/8$, where for small climb or descent rates $\lambda_i + \lambda_c \cong \kappa \lambda_h + \frac{1}{2}\lambda_c$ (equation 4.26).

The rotor hovering efficiency is expressed in terms of the figure of merit:

$$M = \frac{T\sqrt{T/2\rho A}}{P} = \frac{C_T^{3/2}/\sqrt{2}}{C_P} \tag{7.7}$$

(equation 3.10). The figure of merit is a measure of the relative contributions of the induced and profile power in hover. Typically the profile power is at least 10% to 20% of the total power, and the nonideal induced power is 10% to 15% of the ideal power, giving a maximum figure of merit of $M = 0.74$ to 0.78, and typically $M = 0.70$ at design loading (section 3.3).

The helicopter power available is obtained from the performance data for the engine. The engine power usually decreases as the altitude or temperature increases, and there is some influence of speed. Thus the variations in the power available are important in calculating the helicopter performance. There are power losses in the engine and transmission that must also be accounted for, including gear train losses, any power required to cool the transmission and engine, and the power required to drive accessories such as the generator and oil pump. These losses can be expressed in terms of an overall efficiency factor $\eta < 1$, relating the total power required to the rotor power: $P_{\text{req}} = \frac{1}{\eta} P_{\text{rotor}}$. Typically the engine and drive train losses correspond to $\eta = 0.95$ to 0.96 with turbine engines.

The helicopter has power losses in addition to those of the isolated main rotor. Rotor-rotor and rotor-fuselage aerodynamic interference losses can be a significant fraction of the total power, particularly for a tandem helicopter configuration. For the single main rotor helicopter, the tail rotor power must be included. The tail rotor performance calculation is complicated by the fact that the tail rotor operates in the wake of the main rotor and fuselage. Aerodynamic interference reduces the efficiency of the tail rotor and in particular increases its loads and vibration. During yawing maneuvers, the tail rotor can even be operating in the vortex ring state, which reduces the control power and greatly increases the vibration. The tail rotor thrust is given by the main rotor torque $T_{tr} = Q/\ell_{tr}$, where ℓ_{tr} is the tail rotor moment arm about the main rotor shaft. Given T_{tr}, the tail rotor performance can be calculated. As a first approximation, all power losses except for the isolated main rotor power can be accounted for using an overall helicopter efficiency η. Including tail rotor power and aerodynamic interference in addition to engine and transmission losses, typically $\eta = 0.84$ to 0.87 for hover. The efficiency usually improves in forward flight as the interference and tail rotor power decrease.

7.1.2 Forward Flight Performance

The rotor torque coefficient can be expressed as an integral of the in-plane component of the blade section forces:

$$C_P = C_Q = \int_0^1 \sigma \widehat{F}_x r \, dr = \int_0^1 \sigma \frac{U^2}{2} (c_\ell \sin \phi + c_d \cos \phi) r \, dr \qquad (7.8)$$

where $U^2 = u_T^2 + u_P^2$ and $\tan \phi = u_P/u_T$ (equation 6.92). The c_ℓ term is the accelerating torque, and the c_d term is the decelerating torque. From this, the energy balance relation for the helicopter power required in forward flight was derived:

$$C_P = C_{Pi} + C_{Po} + C_{Pp} + C_{Pc} \qquad (7.9)$$

where the induced, profile, parasite, and climb power terms are

$$C_{Pi} = \int_{r_R}^{B} \lambda_i dC_T \qquad (7.10)$$

$$C_{Po} = C_{Qx} + \mu C_{Hx} + \int \lambda_i dC_{To} \tag{7.11}$$

$$C_{Pp} = \frac{DV}{\rho A (\Omega R)^3} \tag{7.12}$$

$$C_{Pv} = \frac{V_c W}{\rho A (\Omega R)^3} \tag{7.13}$$

(sections 6.4 and 6.21). This result required no small angle assumptions, so is applicable to all operating conditions, including hover, edgewise flight, and high-speed axial flight. Forward flight introduces the helicopter parasite power $P_p = DV$, which is the power required to move the helicopter through the air against the drag force D.

The ideal power (no profile power) of the rotor in forward flight is $P = P_i + P_p + P_c = T(V \sin i + v)$ (equation 5.5). Figure 5.4 presents the solution, based on a combination of momentum theory and experimental results, for $(V \sin i + v)/v_h = P/P_h$ as a function of $(V \cos i)/v_h$ and $(V \sin i)/v_h$. The momentum theory result for the induced velocity in forward flight is

$$\lambda_i = \frac{C_T}{2\sqrt{\mu^2 + \lambda^2}} \tag{7.14}$$

where $\lambda = \lambda_i + \mu \tan i$. For all but the lowest speeds, a good approximation in edgewise forward flight is $\lambda_i \cong C_T/2\mu$ (section 5.1.1). This result is independent of the climb or descent velocity. Including the empirical correction factor κ, the induced power in forward flight is $C_{Pi} = \lambda_i C_T = \kappa C_T^2/2\mu$ or $P_i = \kappa T^2/2\rho AV$.

From section 6.23, the rotor profile power is

$$C_{Po} = C_{Qo} + \mu C_{Ho} - \mu_z C_{To} = \int_0^1 \frac{\sigma c_d}{2}\left(u_P^2 + u_T^2 + u_R^2\right)^{3/2} dr \tag{7.15}$$

which includes the effects of reverse flow, radial flow, and the radial drag force. Using a mean section drag coefficient, $C_{Po} = (\sigma c_{do}/8) F_P(\mu, \mu_z)$ (equation 6.409), with F_P given accurately for all axial and edgewise speeds by equation 6.428. A less accurate but often used approximation for edgewise flight up to speeds of about $\mu = 0.5$ is

$$C_{Po} = \frac{\sigma c_{do}}{8}(1 + 4.65\mu^2) \tag{7.16}$$

(equation 6.424). For helicopter rotors at high speeds or high blade loading, stall and compressibility must be included in calculating the profile power, and this requires a numerical solution including a determination of the blade angle-of-attack distribution in forward flight. A lightly loaded rotor, slowed down to maintain a reasonable advancing tip Mach number, can operate at high edgewise advance ratio μ without significant stall or compressibility losses.

When the helicopter drag is written in terms of an equivalent parasite drag area, $D = \frac{1}{2}\rho V^2 f$, the parasite power is $P_p = DV = \frac{1}{2}\rho V^3 f$, or

$$C_{Pp} = \frac{1}{2}\left(\frac{V}{\Omega R}\right)^3 \frac{f}{A} \cong \frac{1}{2}\mu^3 \frac{f}{A} \tag{7.17}$$

(equation 6.120). Alternatively, in terms of the drag force we have $C_{Pp} \cong \mu(D/W)C_T$. The helicopter climb power is $P_c = V_c W$, where $V_c = V \sin \theta_{FP}$ is the climb velocity and W is the helicopter weight. In terms of $\lambda_c = V_c/\Omega R$ then

$$C_{Pc} = \lambda_c \frac{W}{\rho A (\Omega R)^3} \cong \lambda_c C_T \tag{7.18}$$

Thus the energy-balance method gives the following estimate of the helicopter rotor power required in edgewise forward flight:

$$C_P = \frac{\kappa C_T^2}{2\mu} + \frac{\sigma c_{do}}{8}(1 + 4.65\mu^2) + \frac{1}{2}\mu^3 \frac{f}{A} + \lambda_c C_T \tag{7.19}$$

from which the power as a function of gross weight or speed can be found. At low speed, the induced power must be calculated instead from $C_{Pi} = \kappa C_T^2/2\sqrt{\mu^2 + \lambda^2}$, which is valid down to hover. At high speed, the neglect of stall and compressibility effects in the profile power becomes a significant consideration. In addition, at high speed, the small angle approximations made for the parasite and climb power in this result may not be correct, but the exact results can be easily used instead.

In edgewise forward flight, the induced power is essentially independent of the disk inclination or climb speed: $C_{Pi} \cong \kappa C_T^2/2\mu$. This approximation is valid for $\mu > 0.1$ or so, or for speeds above $V = 30$ to 50 knots. The rotor profile power is also not very sensitive to climb or descent in forward flight, assuming that there is no great change in the blade angle-of-attack distribution; neither is the parasite power if the change in the helicopter drag with angle-of-attack is neglected. With these assumptions, only the climb power $P_c = V_c W$ depends on the climb or descent rate in edgewise flight, and the power required can be written

$$P = P_i + P_o + P_p + P_c \cong (P_i + P_o + P_p)_{V_c=0} + P_c = P_{\text{level}} + P_c \tag{7.20}$$

which gives the climb rate

$$V_c = \frac{P - P_{\text{level}}}{W} = \frac{\Delta P}{W} \tag{7.21}$$

(equation 5.19). Here P_{level} is the power required for level flight at the given thrust and speed, and ΔP is the excess power available. The helicopter climb and descent characteristics in forward flight can be determined from the power required for level flight and the power available. At low forward speeds ($\mu < 0.1$ or so), the change in induced power with climb speed becomes significant. The climb rate approaches $V_c \cong 2\Delta P/W$ for vertical flight (equation 4.28).

7.1.3 D/L Formulation

The rotor power required can be written in terms of an equivalent drag force D by the definition $P = DV$. Hence $D = D_i + D_o + D_p + D_c$, or in terms of the drag-to-lift ratio,

$$\left(\frac{D}{L}\right)_{\text{total}} = \left(\frac{D}{L}\right)_i + \left(\frac{D}{L}\right)_o + \left(\frac{D}{L}\right)_p + \left(\frac{D}{L}\right)_c \tag{7.22}$$

where $L = T \cos i + H \sin i = W \cos \theta_{FP}$ is the rotor lift. The rotor equivalent drag-to-lift ratio is defined as

$$\left(\frac{D}{L}\right)_e = \left(\frac{D}{L}\right)_i + \left(\frac{D}{L}\right)_o \tag{7.23}$$

7.1 Rotor Performance Estimation

The drag-to-lift ratio can also be written as

$$\frac{D}{L} = \frac{P}{VL} \cong \frac{P}{TV\cos i} = \frac{C_P}{\mu C_T} \tag{7.24}$$

For edgewise flight, the induced, profile, parasite, and climb powers give

$$\left(\frac{D}{L}\right)_i = \frac{C_{Pi}}{\mu C_T} \cong \frac{\kappa C_T}{2\mu^2} \tag{7.25}$$

$$\left(\frac{D}{L}\right)_o = \frac{C_{Po}}{\mu C_T} \cong \frac{\sigma c_{do}}{8} \frac{1+4.65\mu^2}{\mu C_T} = \frac{3}{4}\frac{c_{do}}{\bar{c}_\ell}\frac{1+4.65\mu^2}{\mu} \tag{7.26}$$

$$\left(\frac{D}{L}\right)_p = \frac{P_p}{VL} = \frac{D}{W\cos\theta_{FP}} \tag{7.27}$$

$$\left(\frac{D}{L}\right)_c = \frac{P_c}{VL} = \frac{V\sin\theta_{FP}W}{VW\cos\theta_{FP}} = \tan\theta_{FP} \tag{7.28}$$

where $\bar{c}_\ell = 6C_T/\sigma$ has been used in the expression for the parasite power. These results take a simpler form if the induced power and parasite power are written in terms of an aircraft lift coefficient C_L, defined as

$$C_L = \frac{L}{\frac{1}{2}\rho V^2 A} \tag{7.29}$$

Then

$$\left(\frac{D}{L}\right)_i = \frac{T^2/2\rho AV}{LV} \cong \frac{L}{2\rho AV^2} = \frac{C_L}{4} \tag{7.30}$$

$$\left(\frac{D}{L}\right)_p = \frac{DV}{LV} = \frac{\frac{1}{2}\rho V^2 f}{L} = \frac{f/A}{C_L} \tag{7.31}$$

The induced power result is simply the induced drag of a circular wing: for aspect ratio $AR = 4/\pi$, the drag-to-lift ratio is $D_i/L = C_{Di}/C_L = C_L/\pi AR = C_L/4$. Thus, in terms of C_L the aircraft power required is

$$\left(\frac{D}{L}\right)_{total} = \frac{C_L}{4} + \left(\frac{D}{L}\right)_o + \frac{f/A}{C_L} + \tan\theta_{FP} \tag{7.32}$$

The lift coefficient C_L is calculated from the gross weight and speed, and the profile power $(D/L)_o$ can be estimated from these simple expressions, performance charts, or numerical calculations.

The D/L formulation was developed for autogyro performance calculations. The lift coefficient C_L is used because the rotor on an autogyro functions like a fixed wing. Consequently, many early rotor analyses express the results for improved profile power calculations in terms of $(D/L)_o$. For helicopter performance calculations this formulation is only applicable in forward flight, since the drag-to-lift ratio $D/L = P/VL$ is singular at hover.

7.1.4 Rotor Lift and Drag

Theoretical and experimental rotor performance data are often expressed in terms of the rotor lift and drag, defined as the wind-axis components of the total force on

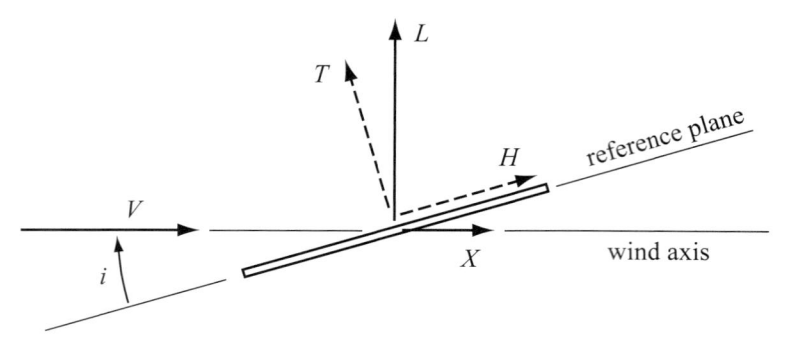

Figure 7.1. Rotor lift and drag forces (wind axes).

the rotor hub (Figure 7.1). In terms of the rotor thrust and H-force, defined relative to some reference plane such as shaft axes, the coefficients C_L and C_X are

$$C_L = C_T \cos i + C_H \sin i \tag{7.33}$$

$$C_X = C_H \cos i - C_T \sin i \tag{7.34}$$

Here $C_L = L/\rho A(\Omega R)^2$ (rotor coefficient form), which is not the same quantity used in the preceding section. The calculated and measured results can be presented in terms of C_L/σ and C_X/σ. The rotor propulsive force is the negative of the X-force. By using a wind-axis presentation of the data, performance charts can be directly interpreted in terms of the helicopter operating condition. The helicopter gross weight determines the rotor lift required, and the helicopter parasite drag determines the rotor propulsive force.

The rotor equivalent drag D_e is defined as the aircraft equivalent drag P/V minus the propulsive force:

$$D_e = \frac{P}{V} + X \tag{7.35}$$

The rotor equivalent lift-to-drag ratio $(L/D)_e$ is useful as a measure of the rotor efficiency in edgewise flight. Moreover, $(L/D)_e$ can be obtained directly from rotor performance measured in a wind tunnel: lift, drag, and power. The rotor propulsive force is determined by the helicopter parasite drag and climb velocity ($-X = D + WV_c/V$), so the rotor equivalent drag-to-lift ratio is

$$\left(\frac{D}{L}\right)_e = \frac{P/V + X}{L} = \left(\frac{D}{L}\right)_{total} - \left(\frac{D}{L}\right)_p - \left(\frac{D}{L}\right)_c = \left(\frac{D}{L}\right)_i + \left(\frac{D}{L}\right)_o \tag{7.36}$$

which is consistent with equation 7.23.

7.1.5 P/T Formulation

It is also useful to express the power required in terms of the power-to-thrust ratio P/T. In contrast with the drag-to-lift formulation $(D/L = P/VL)$, P/T is not singular at hover. In coefficient form

$$\frac{C_P}{C_T} = \frac{P}{\Omega R T} \tag{7.37}$$

7.1 Rotor Performance Estimation

So

$$\left(\frac{C_P}{C_T}\right)_{total} = \left(\frac{C_P}{C_T}\right)_i + \left(\frac{C_P}{C_T}\right)_o + \left(\frac{C_P}{C_T}\right)_p + \left(\frac{C_P}{C_T}\right)_c \quad (7.38)$$

and

$$\left(\frac{C_P}{C_T}\right)_i = \lambda_i \cong \frac{\kappa C_T}{2\mu} \quad (7.39)$$

$$\left(\frac{C_P}{C_T}\right)_o \cong \frac{\sigma c_{do}}{8} \frac{1 + 4.65\mu^2}{C_T} = \frac{3}{4} \frac{c_{do}}{\overline{c}_\ell}(1 + 4.65\mu^2) \quad (7.40)$$

$$\left(\frac{C_P}{C_T}\right)_p = \frac{DV}{\Omega RT} \cong \mu \frac{D}{W} \quad (7.41)$$

$$\left(\frac{C_P}{C_T}\right)_c = \frac{V_c W}{\Omega RT} \cong \lambda_c \quad (7.42)$$

are the induced, profile, parasite, and climb power terms for edgewise forward flight.

7.1.6 Rotorcraft Performance

Consider the level flight performance of rotary-wing aircraft with lifting and propelling rotors. The power required is the sum of the induced, profile, and propulsive power of all rotors:

$$P = \sum_{rotors}(P_i + P_o + P_p) \quad (7.43)$$

plus transmission and accessory losses. The induced power is characterized by the factor κ: $P_i = \kappa T v_i$. Including interference losses, the power for rotor m is

$$P_{im} = T_m\left(\kappa_m v_{im} + \sum_{n \neq m} x_{mn} v_{in}\right) \quad (7.44)$$

(equation 5.50). The profile power is characterized by the mean drag coefficient $c_{d\,mean}$: $P_o = \rho A(\Omega R)^3(\sigma c_{d\,mean}/8)F_P = \rho A_b(\Omega R)^3(c_{d\,mean}/8)F_P$. The propulsive power is given by the rotor drag force X: $P_p = -XV$.

If the shaft power P is specified for a rotor, then the propulsive power is $P_p = P - (P_i + P_o)$, and the rotor drag is

$$X = -\frac{P - (P_i + P_o)}{V} \quad (7.45)$$

In terms of the rotor inflow, $X > 0$ corresponds to a rearward tilt of the disk. Then there is a component of the forward velocity flowing upward through the disk, providing the additional energy required by the rotor when $(P_i + P_o)$ is greater than the shaft power P. If the shaft angle is fixed for a rotor, the performance solution gives the power required and the rotor propulsive or drag force. The forces and moments required to balance this rotor must be supplied by the rest of the aircraft.

For a helicopter with one or more lifting and propelling rotors (no wing or auxiliary propulsion), the rotor lifts sum to the aircraft weight and the rotor propulsive forces sum to the aircraft drag: $\sum(-X) = D_{\text{aircraft}}$. Thus the total aircraft propulsive power is

$$P_p = \sum_{\text{rotors}} P_p = D_{\text{aircraft}} V \tag{7.46}$$

The arrangement and lift distribution of the main rotors influence the induced and interference power.

For the single main rotor and tail rotor configuration, the drag force of the tail rotor must be countered by the main rotor, adding to its propulsive power. The tail rotor drag force is $X_{tr} = H_{\text{HP}} - T_{tr} i_{\text{HP}}$. Since the rotor thrust is nearly perpendicular to the tip-path plane,

$$X_{tr} = H_{\text{TPP}} - T_{tr}(i_{\text{HP}} + \beta_{1c\text{HP}}) \cong -T_{tr}(i_{\text{HP}} + \beta_{1c\text{HP}}) \tag{7.47}$$

Thus finding the rotor drag or propulsive force with the shaft angle fixed requires a solution for the longitudinal flapping relative to the shaft ($\beta_{1c\text{HP}}$), which gives the tip-path-plane angle. In the case of the tail rotor there is no cyclic pitch, and usually there is a large pitch-flap coupling. After the tail rotor drag X_{tr} is obtained, the power absorbed is calculated from $P_{tr} = P_i + P_o + P_p$, where the propulsive power $P_p = -X_{tr} V$.

The tail rotor has two contributions to the power required for the entire helicopter: the power absorbed directly through the tail rotor shaft, including the tail rotor propulsive power, $(P_p)_{tr} = -X_{tr} V$, and the main rotor propulsive power required because of the tail rotor drag force, $(\Delta P_p)_{mr} = X_{tr} V$. The total power attributed to the tail rotor is thus

$$(P_{\text{total}})_{tr} = P_{\text{shaft}} + P_{\text{drag}} = (P_i + P_o + P_p)_{tr} + \Delta(P_i + P_o + P_p)_{mr}$$
$$= (P_i + P_o)_{tr} + \Delta(P_i + P_o)_{mr} \tag{7.48}$$

So the total power loss due to the tail rotor is independent of the tail rotor drag force, which simply determines the distribution of the total loss between the tail rotor and main rotor shaft powers. The helicopter performance can be analyzed by ignoring the tail rotor drag or propulsive force. The result is a small change in the main rotor disk inclination, as determined by horizontal force equilibrium, but the tail rotor flapping solution for its tip-path-plane orientation is not needed. This approach neglects the influence of a main rotor propulsive force increase on the main rotor induced and profile power ($\Delta(P_i + P_o)_{mr}$).

A compound helicopter has a wing to off-load the rotor in cruise and/or auxiliary propulsion systems. Consider a helicopter with just a wing. The aircraft drag includes the wing drag:

$$D_{\text{wing}} = D_0 + \frac{q(L/q - (L/q)_0)^2}{\pi e b^2} \tag{7.49}$$

where b is the wing span, e the Oswald efficiency, and $q = \frac{1}{2}\rho V^2$. The rotor propulsive power is increased by $\Delta P_p = D_{\text{wing}} V$. Off-loading the rotor reduces the ideal induced power $T v_i$, but likely increases κ and reduces the profile power ($c_{d\,\text{mean}}$).

The tiltrotor in cruise generates all the lift by the wing, with the rotors operating as propellers to provide only propulsive force. Tilting the nacelles reduces the aircraft drag substantially. The rotor thrust in cruise is reduced compared to hover by a factor equal to the airframe lift-to-drag ratio. Hence the rotor ideal induced power is reduced, although the hover-cruise compromise in blade twist can increase κ. The axial speed of the rotor increases the profile power through $F_P(\mu_z)$. The profile power is reduced substantially by operating the rotors at lower tip speed, which is possible since the required rotor thrust is small compared to hover. If the aircraft trims with the rotor shaft not horizontal, the rotors produce some lift (in-plane force). The mutual interference between the rotor and wing can be significant and is beneficial with the rotors turning top-blades-outward.

Consider a helicopter or autogyro with propellers for propulsion. The aircraft trim determines the drag X of each main rotor, with the total $X_{mr} = \sum X$. The total rotor propulsive power is then $(P_p)_{mr} = -X_{mr}V$. The aircraft drag is the sum of the airframe drag and rotor drag, $D = D_{af} + X_{mr}$. The propeller power is then

$$P_{\text{prop}} = (P_i + P_o)_{\text{prop}} + DV = \frac{DV}{\eta} \tag{7.50}$$

in terms of the propeller propulsive efficiency η. The rotorcraft power is then

$$P = (P_i + P_o)_{mr} - X_{mr}V + (P_i + P_o)_{\text{prop}} + DV = (P_i + P_o)_{mr} + (P_i + P_o)_{\text{prop}} + D_{af}V \tag{7.51}$$

So if the energy balance sums the induced and profile power of both rotors and propellers, the aircraft propulsive power is

$$P_p = D_{af}V = \eta P_{\text{prop}} - X_{mr}V \tag{7.52}$$

Alternatively, if only the main rotors are considered for the induced and profile power, the aircraft propulsive power is

$$P_p = P_{\text{prop}} - X_{mr}V = \frac{1}{\eta}D_{af}V + \left(\frac{1}{\eta} - 1\right)X_{mr}V \tag{7.53}$$

In both approaches, the propeller shaft power P_{prop} is not a separate component of the energy-balance description of rotorcraft required power, but instead is just part of the propulsive power.

For an autogyro, the rotor shaft power is zero, so the rotor drag is

$$X_{mr} = \frac{P_i + P_o}{V} \tag{7.54}$$

In terms of the drag-to-lift formulation, the result is

$$\frac{X}{W} = \left(\frac{D}{L}\right)_i + \left(\frac{D}{L}\right)_o = \left(\frac{D}{L}\right)_e \tag{7.55}$$

(the flight path angle $\theta_{FP} = 0$ for level flight). The autogyro rotor acts like a wing: aircraft lift is supplied at the cost of induced and profile drag. The rotor drag is balanced by the aircraft propulsion system. The autogyro can also have a wing to off-load the rotor in high-speed flight, but the aircraft always requires a rotor lift sufficient to generate the accelerating torque for rotor equilibrium at zero power.

7.1.7 Performance Charts

Measured or calculated rotor performance characteristics covering a range of operating conditions can be summarized in graphical form. This section presents common and useful formats for rotor performance charts. The examples are from calculations for a single, articulated rotor with solidity $\sigma = 0.08$ and $-8°$ linear twist. The rotor forces and power were obtained at $V/\Omega R = 0.4$ and $M_{\text{tip}} = 0.6$, over a range of shaft incidence i_s (positive forward) and collective pitch $\theta_{.75}$.

For autogyros, a single performance chart can cover a range of flight speeds, showing for example profile power C_{Po}/σ or profile drag-to-lift ratio $(D/L)_o$ as a function of rotor lift C_L/σ, for a range of advance ratio μ and collective pitch. Extending this format to helicopters, a set of charts can be constructed, with each chart showing a specified value of rotor power or rotor drag-to-lift ratio; see Bailey and Gustafson (1944) and Gustafson (1953). Use of such charts is straightforward for autogyros (rotor power $P = 0$), but requires interpolation between charts for helicopters.

Figure 7.2 illustrates performance results as obtained in a wind-tunnel test (for example, McCloud, Biggers, and Stroub (1968)). The data are for the rotor trimmed to zero 1/rev flapping (so the shaft incidence equals the tip-path-plane incidence), over a matrix of shaft incidence angles ($-10°$ to $10°$) and collective angles ($2°$ increments). Only quantities that can be measured in a wind tunnel are shown: rotor wind-axis lift and drag forces (X is positive for drag, negative for a propulsive force), power, and no-feathering-plane incidence ($i_{\text{NFP}} = i_s + \theta_{1s}$). The non-ideal power is obtained by subtracting the propulsive power and ideal induced power: $C_{Pn} = C_P + (V/\Omega R)C_X - C_T/2\mu$.

Measured or calculated performance obtained by trimming to a specified rotor lift C_L/σ and zero 1/rev flapping can be presented in a carpet plot (Figure 7.3; for example, Tanner (1964)). For each C_L/σ value, a sweep of shaft incidence or collective pitch (using the other variable to trim the lift) generates a curve of power C_P/σ as a function of drag C_X/σ, which is plotted with an offset power scale. Connecting points where the constant-lift curves intersect the vertical grid generates constant-power curves. Thus given the required rotor lift and drag (from the helicopter force balance), the rotor power can be found. Other quantities of the operating condition can be plotted (probably by interpolation) on the lift-drag grid (Figure 7.3).

Calculated performance obtained by trimming to a specified rotor power C_P/σ and zero 1/rev flapping can be plotted in terms of rotor lift C_L/σ as a function of rotor drag C_X/σ (Figure 7.4; see Kisielowski, Bumstead, Fissel, and Chinsky (1967)). In a test, trimming to a specified power is more difficult than trimming to a specified lift. Given the required rotor lift and drag (from the helicopter force balance), the rotor power can be found. Using $(C_X/\sigma)/\mu^2 = X/qA_{\text{blade}}$ for the abscissa is appropriate, since then the operating condition for a helicopter at fixed weight W and drag D/q is at the same point on charts for different speeds. For each C_P/σ value, a sweep of shaft incidence (trimming power with collective) generates a curve. Figure 7.4 also shows a boundary for mean drag coefficient $c_{d\,\text{mean}} = 0.0200$ (obtained from the profile power; see section 7.3), as an indication of rotor stall.

7.2 Rotorcraft Performance Characteristics

7.2.1 Hover Performance

The rotor hovering performance can be expressed in terms of C_P as a function of C_T, using collective pitch as the parameter (Figure 7.5). At low thrust, the primary loss is the profile power; at moderate thrust levels C_P increases as $C_T^{3/2}$ because of the induced power rise; and at high thrust there is a steep increase in the profile

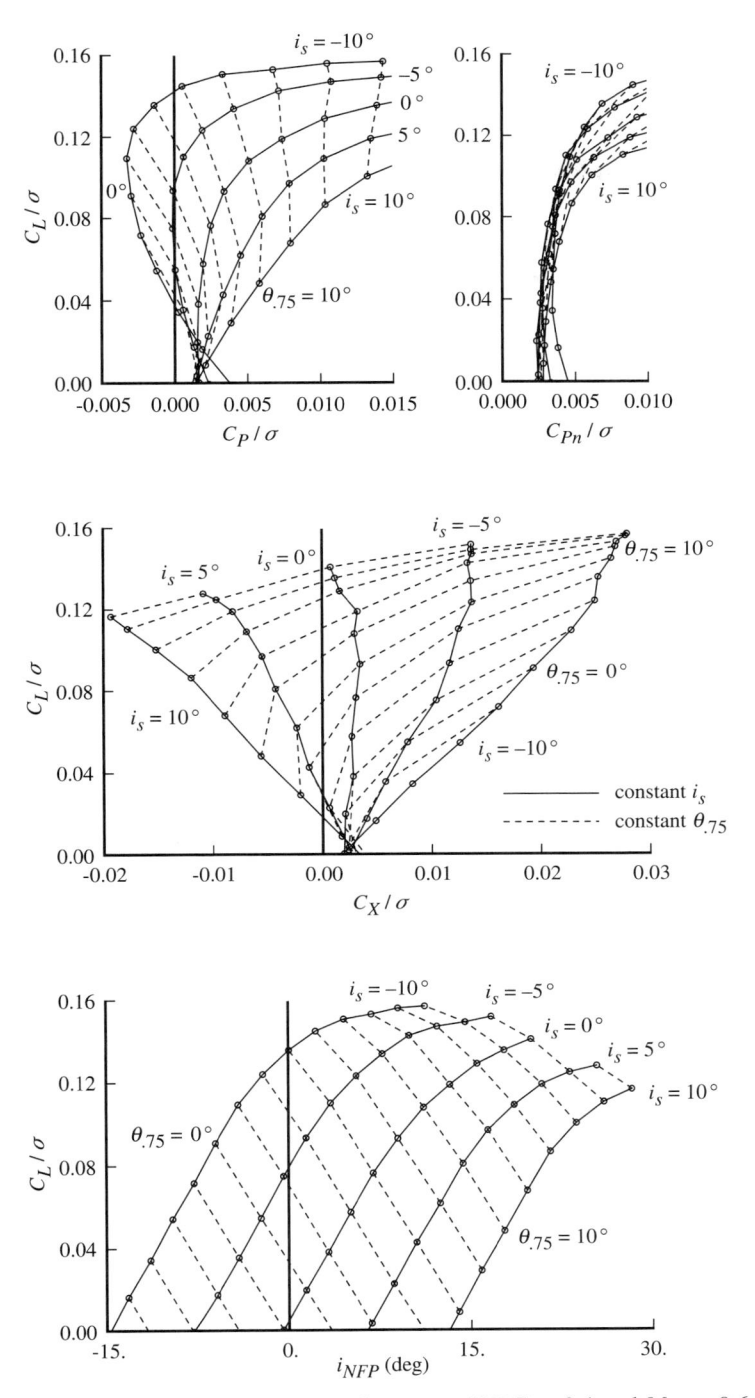

Figure 7.2. Helicopter rotor performance; $V/\Omega R = 0.4$ and $M_{\text{tip}} = 0.6$.

power as a result of stall of the rotor blade. The maximum figure of merit occurs at minimum $C_P/C_T^{3/2}$. Without stall, the maximum figure of merit would be achieved at very high thrust; that is, at very high disk loading where M approaches 1 because of the induced power increase. With stall included in the rotor profile power, the

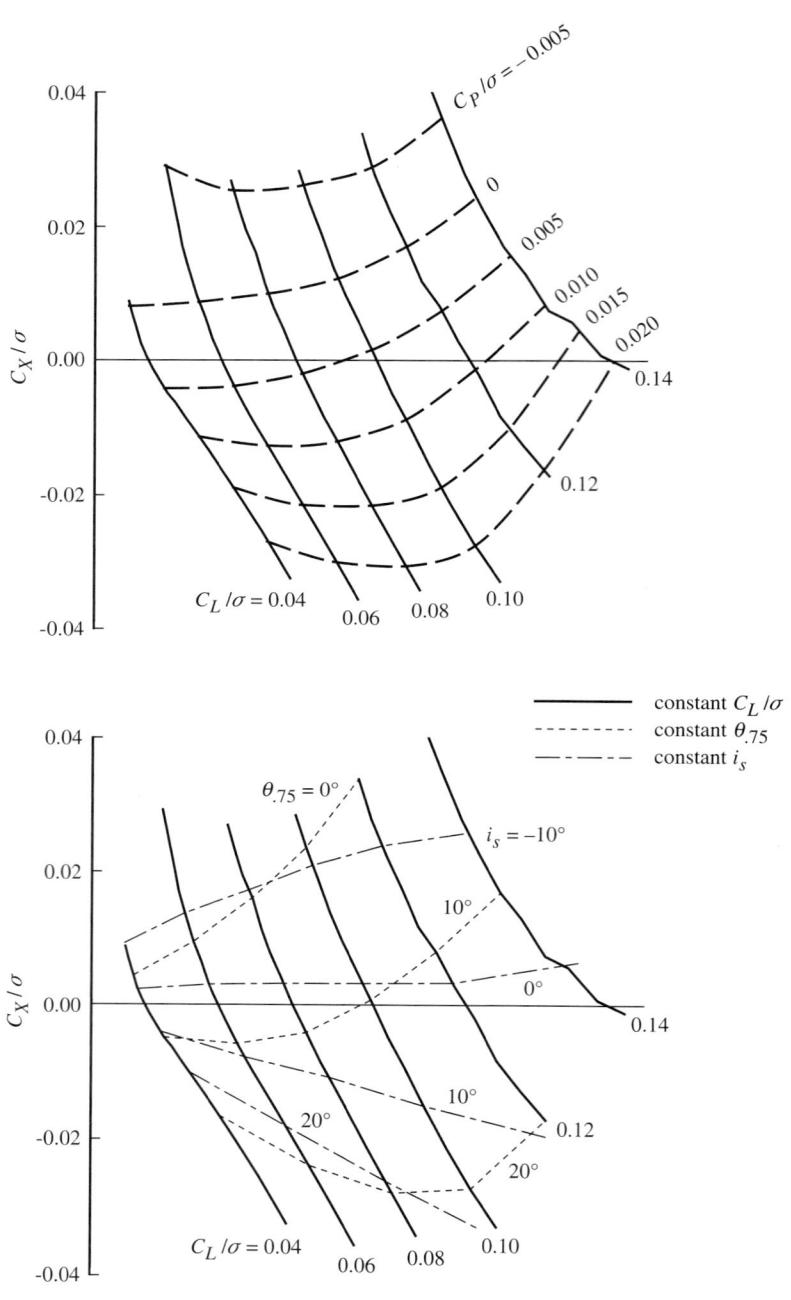

Figure 7.3. Helicopter rotor performance; $V/\Omega R = 0.4$ and $M_{\text{tip}} = 0.6$.

maximum figure of merit is achieved at a value of C_T/σ just above the inception of stall.

The hover power required increases with gross weight, the induced power (which accounts for most of the hover power) varying as $P_i \propto W^{3/2}$. The air density decreases as the altitude or temperature increases, reducing the rotor profile power because of the smaller drag forces on the blades but increasing the induced power because of the higher effective disk loading. Except at very low disk loadings, the induced

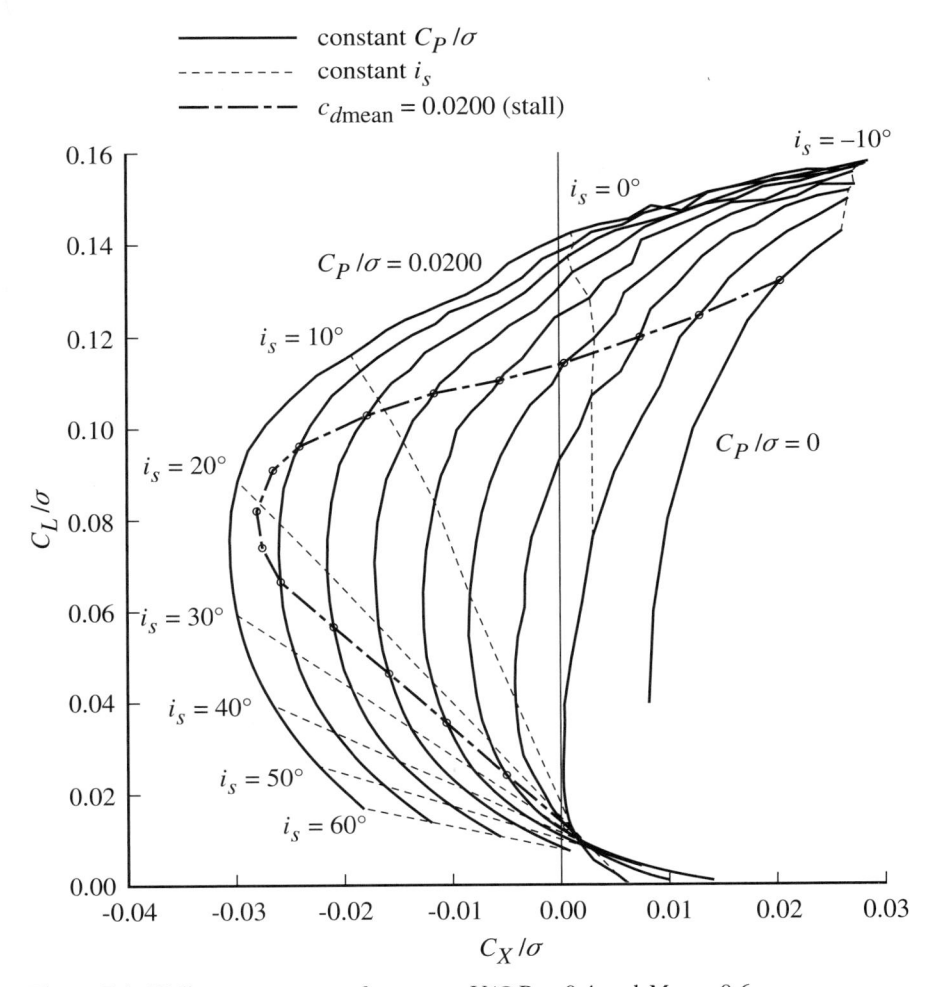

Figure 7.4. Helicopter rotor performance; $V/\Omega R = 0.4$ and $M_{tip} = 0.6$.

power increase dominates, and the total power required increases with altitude and temperature. The hover polar also depends on ground effect, which reduces the power required at a given gross weight for small distances above the ground (Figure 7.5).

Consider now the disk loading for the best power loading of a hovering rotor. Without the profile power, the solution is $T/A = 0$, which implies zero induced power. The design guidance is simply that low disk loading gives good power loading. Including the profile power, the hover power per unit thrust can be written

$$\frac{P}{T} = \kappa\sqrt{T/2\rho A} + \frac{\sigma c_{do}}{8}\frac{(\Omega R)^3}{T/\rho A} \qquad (7.56)$$

or

$$\frac{C_P}{C_T} = \kappa\sqrt{\frac{C_T}{2}} + \frac{\sigma c_{do}}{8C_T} \qquad (7.57)$$

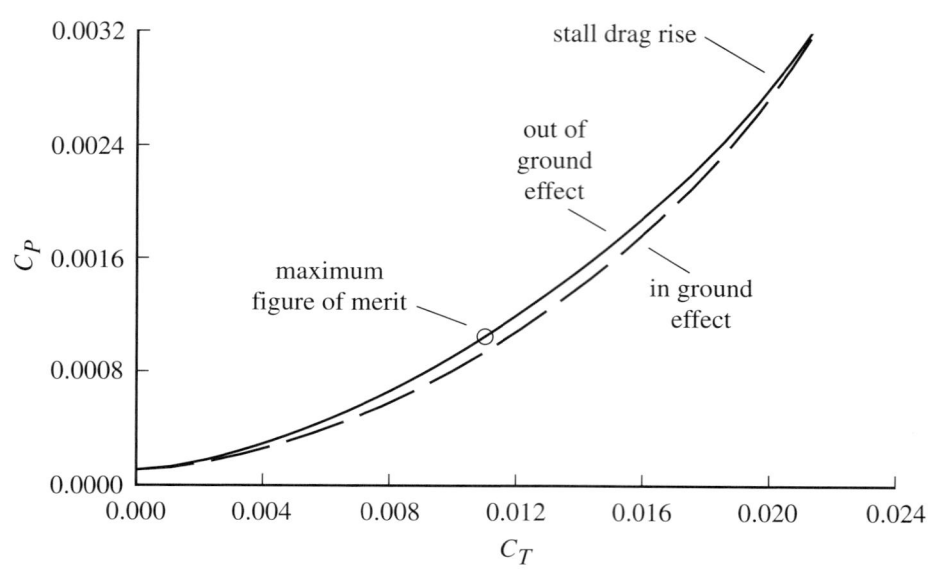

Figure 7.5. Helicopter rotor power in hover.

Minimizing C_P/C_T as a function of C_T (which for fixed tip speed means minimizing P/T as a function of $T/\rho A$) gives the optimum solution

$$C_T = \frac{1}{2}\left(\frac{\sigma c_{do}}{\kappa}\right)^{2/3} \tag{7.58}$$

which occurs at the point where $P_i = 2P_o$, so

$$\frac{C_P}{C_T} = 3\frac{C_{Po}}{C_T} = \frac{3}{4}\left(\kappa^2 \sigma c_{do}\right)^{1/3} \tag{7.59}$$

Dimensionally, the solution is

$$\frac{T}{A} = \frac{1}{2}\rho(\Omega R)^3 \left(\frac{\sigma c_{do}}{\kappa}\right)^{2/3} \tag{7.60}$$

which is the disk loading for best power loading. For a given gross weight, this disk loading determines the optimum radius of the rotor. As the profile power increases, the optimum disk loading increases, and therefore the rotor radius decreases. This solution also gives a figure of merit of

$$M = \frac{T\sqrt{T/2\rho A}}{P} = \frac{C_T^{3/2}/\sqrt{2}}{C_P} = \frac{2}{3\kappa} \tag{7.61}$$

The figure of merit for the rotor hovering at best power loading is a constant, depending only on the induced power factor κ. For $\kappa = 1.10$, this figure of merit is $M = 0.61$. From the fixed value of the figure of merit, at this optimum the power scales with size as $P \propto W^{3/2}$. This optimum solution gives a lower disk loading than is normally used, because considerations in addition to power loading are involved in selecting the disk loading. The weight of the rotor blade and transmission decreases as the radius is reduced, and the aircraft weight depends on the engine specific weight (engine weight per unit power) and specific fuel consumption. The variation of P/T with T/A is fairly small near the optimum value, so the designer has some latitude in choosing the rotor radius. Modern helicopters use significantly higher disk loading than the optimum value from equation 7.60.

Given a constraint on blade loading C_T/σ, the disk loading for best power loading gives a solidity of

$$\sigma = \frac{1}{8} \frac{(c_d/\kappa)^2}{(C_T/\sigma)^3} \tag{7.62}$$

which is rather low. The same solution is obtained if the power loading is optimized as a function of C_T/σ for fixed solidity. Alternatively, consider the optimum power loading for a given disk loading. Then the induced power is fixed, and minimizing the profile power requires a low value of $\sigma(\Omega R)^3$. With a constraint on blade loading,

$$\sigma(\Omega R)^3 = \left(\frac{T/\rho A}{C_T/\sigma}\right)\Omega R = \left(\frac{T/\rho A}{C_T/\sigma}\right)^{3/2} \sigma^{-1/2} \tag{7.63}$$

There is no absolute minimum to this problem, unless system weight is considered. The design guidance therefore is that low tip speed and high solidity are desired to reduce profile power.

7.2.2 Power Required in Level Flight

Figure 7.6 illustrates the power variation with speed for a single main-rotor helicopter in level flight. It shows the power required and power available for design gross weight at sea level and at 10000-ft altitude, and for maximum takeoff weight. The power-required components are main rotor induced, profile, and parasite power (which sum to main rotor power); tail rotor power; and losses (accessory power and transmission losses). The induced power is the largest component in hover, but quickly decreases with speed. The profile power exhibits a slight increase with speed. The parasite power is negligible at low speeds, but increases proportional to V^3 to dominate at high speed. Thus the total power required is high at hover, has a minimum value in the middle of the helicopter speed range, and then increases again at high speed because of the parasite power. At very high speeds, stall and compressibility effects also increase the profile power, and induced power begins to increase due to the low loading on the edges of the rotor disk. Tail rotor power is highest in hover, since the vertical tail contributes to the anti-torque force in forward flight. Ground effect significantly reduces the power required at hover and very low speeds, but has little influence at high speeds.

The effect of gross weight is primarily on the induced power until the loading is high enough to increase profile power because of stall. For different aircraft the parasite drag increases with the gross weight, roughly as $f \propto W^{2/3}$. So the parasite power increases with helicopter size, but P/T decreases with size. The effect of altitude is to increase rotor induced power, decrease profile and parasite power, and reduce the engine power available. In Figure 7.6, at 10000 ft the helicopter can hover only in ground effect.

For a given weight, there is a speed at which the helicopter power required is a minimum. The point at which the power required is a minimum is important, since it determines the best endurance, best climb rate, and minimum descent rate of the aircraft. The speed for minimum power is determined from the power-required curve

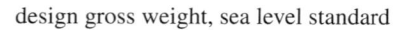

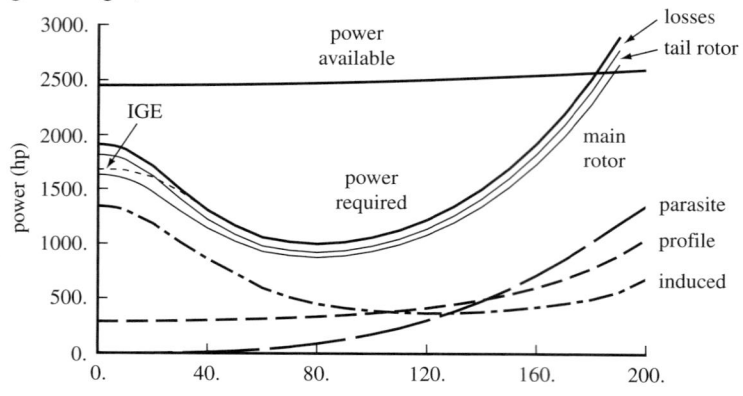

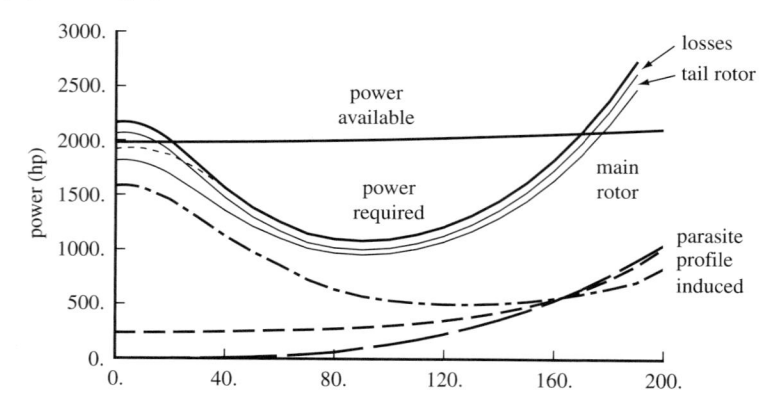

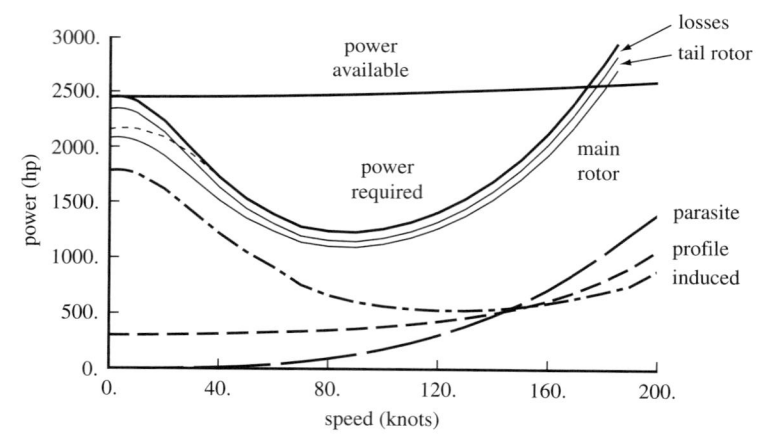

Figure 7.6. Helicopter power in forward flight.

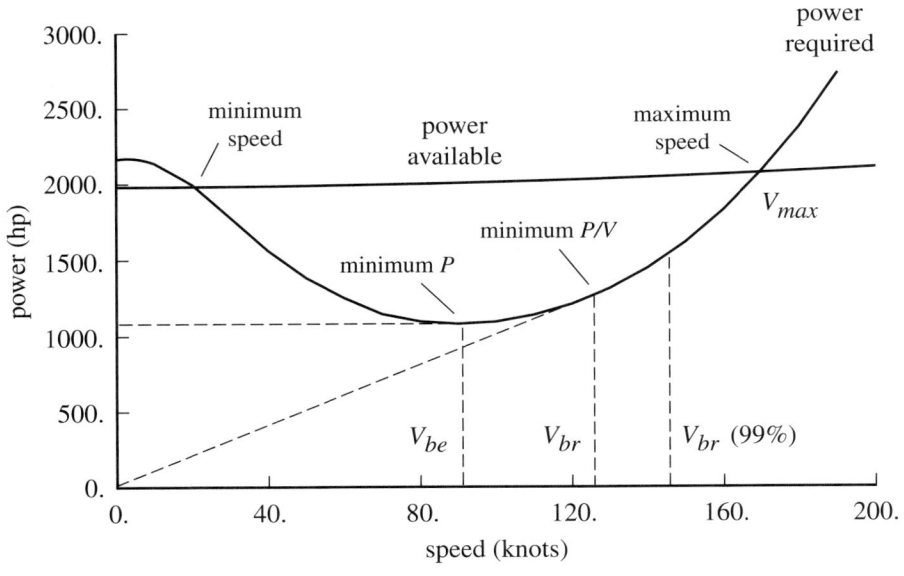

Figure 7.7. Operating speeds from power required and power available.

(Figure 7.7). To estimate this speed, consider the power in edgewise level flight:

$$C_P = \frac{\kappa C_T^2}{2\mu} + \frac{\sigma c_{do}}{8}(1 + 4.65\mu^2) + \frac{1}{2}\mu^3 \frac{f}{A} \qquad (7.64)$$

(equation 7.19). Since the profile power increase relative to hover is initially small, the minimum power point is primarily determined by the changes in the induced power and parasite power. Neglecting the variation of C_{Po}, minimum C_P as a function of μ is at

$$\mu = \left(\frac{\kappa C_T^2}{3f/A}\right)^{1/4} = \lambda_h \left(\frac{4\kappa}{3f/A}\right)^{1/4} \qquad (7.65)$$

or

$$V = v_h \left(\frac{4\kappa}{3f/A}\right)^{1/4} \qquad (7.66)$$

where $v_h^2 = T/2\rho A$ as usual. This solution occurs where $P_i = 3P_p$. The speed for minimum power is typically 60 to 90 knots. Because it is proportional to v_h, the speed for minimum power increases with altitude and gross weight.

We are also interested in the speed for minimum P/V, which is required for the best range and best descent angle. The speed for minimum P/V is found on the power-required curve as the point where a straight line through the origin is tangent to the curve (see Figure 7.7).

7.2.3 Climb and Descent

The vertical climb rate can be calculated for a given excess power. For the low rates typical of helicopters, $V_c \cong 2\Delta P/T$ (equation 4.28). The climb rate at maximum power is reduced by gross weight, because of both the factor T in the denominator and the increase in hover power. The climb rate slows with increasing altitude and temperature because of the hover power increase and the reduction in available engine power. The altitude at which the climb rate is zero defines the absolute hover ceiling. The descent rate in power-off vertical autorotation is proportional to v_h (section 4.3), so increases with gross weight and altitude.

In edgewise forward flight, the climb or descent rate is $V_c = (P - P_{\text{level}})/W = \Delta P/W$ (equation 5.19; the influence of climb rate on the induced power is neglected in this approximation). The maximum climb rate is thus achieved at maximum ΔP, or at the speed for minimum level-flight power required (neglecting the variation of the power available with speed). The best angle of climb is achieved at maximum $V_c/V = \Delta P/WV$. If the helicopter can hover at the given gross weight and altitude, the best angle is vertical. Above the hover ceiling, the speed for the best angle of climb lies between the minimum speed and the speed for minimum power. The minimum power increases and hence the best climb rate decreases with gross weight. The climb rate decreases with altitude. The point where the maximum climb rate reaches zero defines the absolute ceiling of the aircraft.

The descent rate in power-off autorotation in forward flight is $V_d = P_{\text{level}}/W$. The minimum descent rate thus occurs at the speed for minimum level-flight power required. This descent rate is typically about one-half the rate in vertical autorotation. The minimum angle of descent, $V_d/V = P/WV$, is attained at the speed for minimum level-flight P/V. Usually this angle is between 30° and 45° from the horizontal. After power failure at high altitudes above the ground, the pilot establishes equilibrium autorotation at the forward speed giving the minimum descent rate. Near the ground the aircraft is flared to reduce both the vertical and forward speed to zero just before contacting the ground. When power failure occurs near the ground, there is not enough time to achieve a stabilized descent; for a power failure in hover, the optimum descent is purely vertical. Helicopter autorotation characteristics are discussed further in section 8.5.

7.2.4 Maximum Speed

The minimum and maximum velocities of a rotorcraft are determined by the intersection of the power-required and power-available curves for a given gross weight and altitude (Figure 7.7). For $V > V_{\text{max}}$ there is insufficient power available to sustain level flight. If the helicopter can hover, the minimum speed is zero, but at high altitude or high gross weight the power available can be insufficient to hover as well, so V_{min} is positive. The maximum speed of the helicopter may not be power limited, however. The maximum speed might be determined by retreating blade stall and advancing blade compressibility effects, which produce severe vibration and loads at high speed. There can also be drive-train torque limits that define maximum speed. The power-limited maximum speed can be estimated by neglecting the variation of induced power and profile power with speed, compared to the parasite power increase. The result is

$$V_{\text{max}} = \left[\frac{2}{\rho f} (P_{\text{avail}} - P_i - P_o) \right]^{1/3} \quad (7.67)$$

or

$$\mu_{\text{max}} = \left[\frac{2}{f/A} (C_{P\text{avail}} - C_{Pi} - C_{Po}) \right]^{1/3} \quad (7.68)$$

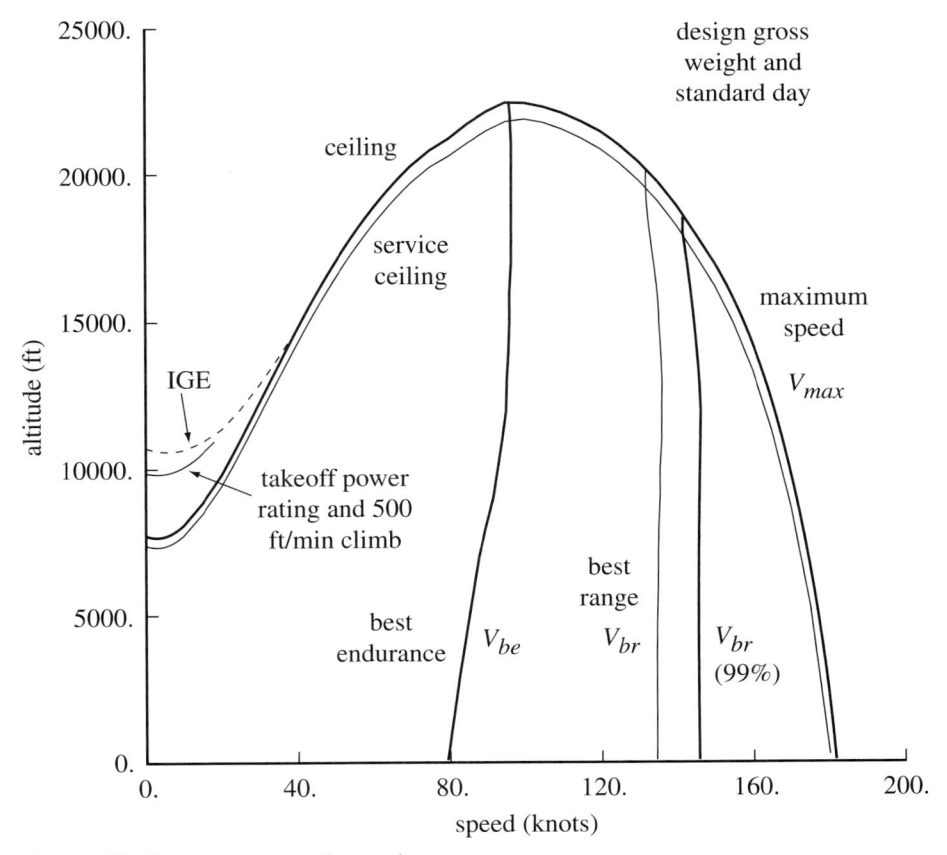

Figure 7.8. Helicopter operating regime.

If the power required at maximum speed is about the same as the power required to hover (a balanced design), then $P_{\text{avail}} - P_i - P_o \cong P_{\text{hover}} - P_o \cong P_{i\text{hover}} = \kappa T \sqrt{T/2\rho A}$, which gives

$$V_{\max} \cong v_h \left(\frac{4\kappa}{f/A}\right)^{1/3} \qquad (7.69)$$

The maximum speed is increased by increasing the installed power or by decreasing the rotorcraft drag. The parasite power rise is proportional to V^3, so a large change in drag or power is needed to achieve a significant maximum speed increment. The parasite power decreases with altitude, so initially the maximum speed may increase. Eventually the reduction in air density reduces the power available, and then the maximum speed decreases with altitude. Above the hover ceiling there is a non-zero minimum speed also. At still higher altitude, the minimum and maximum speeds approach each other until they coincide (together with the speed for minimum power) at the absolute ceiling of the rotorcraft (Figure 7.8).

7.2.5 Ceiling

The rotorcraft ceiling is the altitude at which the maximum power available is just equal to the power required. At a higher altitude, maintaining level flight is not possible (see Figure 7.8). This absolute ceiling is also defined as the altitude at which the climb rate becomes zero. Since the absolute ceiling can be approached from below only asymptotically, working with the service ceiling can be more meaningful. The service ceiling is defined as the altitude where the climb rate is reduced to some small, nonzero value (typically around 100 ft/min). The principal factors defining the ceiling are the reduction of engine power with increasing altitude, the increase in power required with altitude and gross weight, and the variation of the power required with speed.

Several ceilings are of particular interest for the rotorcraft. The hover ceiling out of ground effect (OGE) is where the power available equals the power required to hover at a given gross weight. The hover ceiling in ground effect (IGE) is substantially larger than the OGE hover ceiling, since ground effect reduces the induced power required. The fact that ground effect increases the operational ceiling or weight of the rotorcraft can be used advantageously in operating the aircraft. The maximum ceiling is encountered in forward flight at the speed for minimum power. In both calculations and flight tests these ceilings are obtained by measurements of the rotorcraft climb rate at maximum power. Extrapolating the curves to zero climb rate gives the absolute ceiling.

7.2.6 Range and Endurance

The fuel burn rate is the product of the engine power and the engine specific fuel consumption: $\dot{W}_f = P(\text{sfc})$. The rotorcraft specific range and specific endurance are

$$\frac{dR}{dW_f} = \frac{V}{\dot{W}_f} = \frac{V}{P\,\text{sfc}} \tag{7.70}$$

$$\frac{dE}{dW_f} = \frac{1}{\dot{W}_f} = \frac{1}{P\,\text{sfc}} \tag{7.71}$$

Conventional units (speed in knots, range in nautical miles, specific fuel consumption (sfc) in lb/hp-hr or kg/kW-hr) require introducing a conversion factor in these equations. The rotorcraft range is calculated by integrating the specific range over the total fuel weight, for a given initial gross weight and flight condition:

$$R = \int_{\text{takeoff}}^{\text{landing}} \frac{dR}{dW}\,dW = -\int \frac{dR}{dW_f}\,dW_f = -\int \frac{V}{P\,\text{sfc}}\,dW_f \tag{7.72}$$

Similarly, the endurance is obtained by integrating the specific endurance:

$$E = \int_{\text{takeoff}}^{\text{landing}} \frac{dE}{dW}\,dW = -\int \frac{dE}{dW_f}\,dW_f = -\int \frac{1}{P\,\text{sfc}}\,dW_f \tag{7.73}$$

Generally dR/dW_f and dE/dW_f vary during a flight even if the aircraft is operated at the optimum conditions. Moreover, the power depends on the altitude and gross weight, and the specific fuel consumption depends on the power and altitude. Consequently, these expressions must be numerically integrated for an accurate determination of range and endurance.

The speeds for best range and endurance can be found by examining the specific range and endurance data as a function of velocity. Assuming that the specific fuel consumption is independent of velocity (which is not really true, because of the dependence of the sfc on the engine power), the minimum fuel consumption per unit distance and hence the maximum range are achieved at the speed for minimum P/V (Figure 7.7). This is the best-range speed. The variation of fuel burn rate with speed is not large, so a significant reduction in flight time can be obtained with little loss of range by operating at the speed for 99% maximum specific range (Figures 7.7 and 7.8). Similarly, the maximum endurance is achieved at the speed for minimum P. The speeds for which fuel consumption is a minimum are more accurately obtained from $P(\text{sfc})$ as a function of speed for a given altitude and gross weight. The speed for best endurance is at the minimum of $P(\text{sfc})$, whereas the speed for best range lies at the point where a straight line through the origin is tangent to the curve.

Assuming that P/W, the best-range or best-endurance speed, and the specific fuel consumption are independent of the aircraft weight, the range and endurance can be evaluated analytically. Write

$$\frac{dW}{dR} = -\frac{dW_f}{dR} = -\frac{P\,\text{sfc}}{V} = -\left(\frac{P}{WV}\,\text{sfc}\right)W \tag{7.74}$$

Integrating from the takeoff weight $W = W_{to}$ to the landing weight $W = W_{to} - W_f$ gives the range

$$R = \frac{WV/P}{\text{sfc}} \ln \frac{W_{to}}{W_{to} - W_f} \tag{7.75}$$

where W_f is the total fuel burned. Introducing an equivalent lift-to-drag ratio for the rotorcraft, defined as $L/D = WV/P$, gives the Breguet range equation:

$$R = \frac{L/D}{\text{sfc}} \ln \frac{W_{to}}{W_{to} - W_f} \tag{7.76}$$

for an aircraft with shaft power engines (sfc based on power not on thrust). This expression accounts for the decrease in the gross weight as the fuel is used, a factor that reduces the fuel consumption, since $(P/WV)\,\text{sfc}$ is assumed to be constant. Equation 7.76 is obtained assuming that the aircraft can fly the mission at constant L/D, which is easily achieved with an airplane but not for a helicopter. The corresponding result for endurance is

$$E = \frac{W/P}{\text{sfc}} \ln \frac{W_{to}}{W_{to} - W_f} = \frac{L/D}{V\,\text{sfc}} \ln \frac{W_{to}}{W_{to} - W_f} \tag{7.77}$$

Alternatively, writing $P = WV(D/L) \propto W^{3/2}$ (assuming that the aircraft flies at constant $C_L^{3/2}/C_D$, in order to eliminate V from the expression) leads to a different dependence on $W_{to}/(W_{to} - W_f)$.

Figure 7.9 shows the payload-range diagram for a helicopter, for design gross weight and maximum takeoff weight. Each line is the range for constant takeoff weight W_{to}. The payload is obtained from the operating weight empty W_O and the usable fuel W_f: $W_p = W_{to} - W_O - W_f$. The payload is maximum for zero range, which only requires fuel for mission segments that are independent of range (such as taxi and reserve). The range increases as fuel increases and payload decreases, for constant useful load. The corner point is the range of the rotorcraft for maximum internal fuel, with payload as fallout. The slope dW_p/dR reflects the efficiency of the aircraft (inverse of the specific range). A slightly higher range can be achieved by reducing the payload with maximum fuel on board, since the reduced gross weight improves the fuel consumption. Cruising at higher altitude can increase the

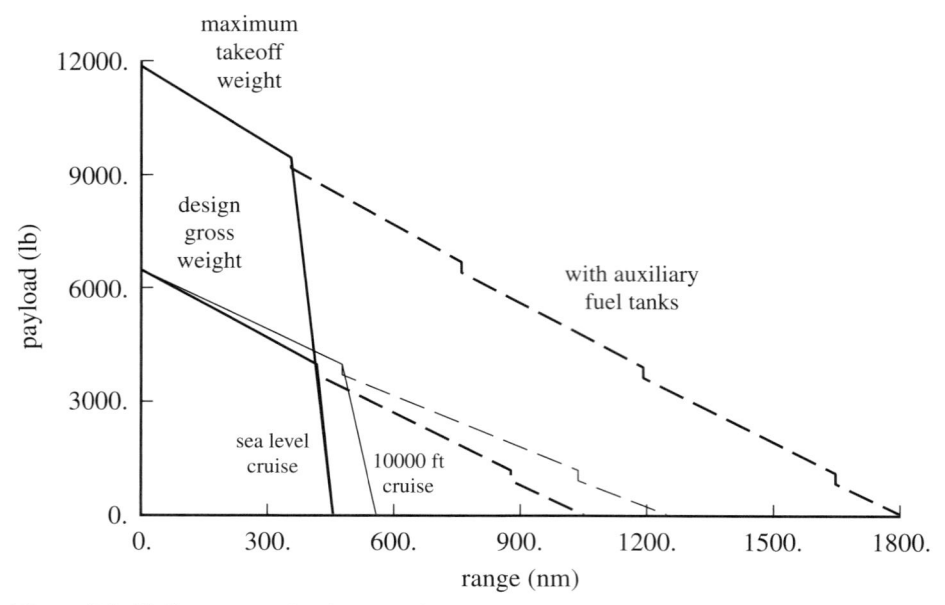

Figure 7.9. Helicopter payload-range diagram.

efficiency, hence increase the range. The vertical jumps in the payload-range curve with auxiliary fuel tanks are due to the additional weight of an auxiliary tank.

7.2.7 Referred Performance

The general analysis of rotor performance is formulated in terms of coefficients, thereby scaling results with air density ρ, rotor disk area $A = \pi R^2$, and tip speed ΩR (Chapter 2). When accounting for scale is not necessary, as for the flight manual of a specific helicopter, referred performance parameters can be used. Referred performance parameters scale results to standard sea level atmospheric conditions, while retaining dimensional form. Define the density, temperature, and pressure ratios: $\sigma = \rho/\rho_0$, $\theta = T/T_0$, and $\delta = p/p_0$ (subscript "0" indicates standard sea level values). From $p = \rho RT$ there follows $\delta = \sigma\theta$. Hence density scales as $\rho \propto \delta/\theta$; speed of sound (and other velocities) scales as $c_s \propto \sqrt{\theta}$; and dynamic pressure scales as $\rho c_s^2 \propto \delta$. We also use the rotor speed ratio $\omega = \Omega/\Omega_0$, based on the normal rotor speed Ω_0. Table 7.1 shows the correspondence between coefficients and referred parameters.

7.3 Performance Metrics

A number of performance indices are useful in assessing the efficiency of the helicopter and its rotor. Let W be the aircraft weight, T the rotor thrust, V the flight speed, and A the rotor disk area. The rotor power $P = P_i + P_o + P_p + P_c$ is a function of the rotor wind-axis lift L and drag X, with the propulsive power (sum of parasite and climb power) $P_p + P_c = -XV$.

The hover figure of merit compares the actual power with the ideal power of the hovering rotor:

$$M = \frac{P_{\text{ideal}}}{P} = \frac{Tv}{P} = \frac{T\sqrt{T/2\rho A}}{P} \tag{7.78}$$

Table 7.1. *Helicopter referred performance*

	Coefficient	Referred parameter
rotor speed	$M_{\text{tip}} = \dfrac{\Omega R}{c_s}$	$\dfrac{\omega}{\sqrt{\theta}}$
flight speed	$M = \dfrac{V}{c_s}$	$\dfrac{V}{\sqrt{\theta}}$
advance ratio	$\mu = \dfrac{V}{\Omega R}$	$\dfrac{V}{\omega}$
weight	$C_W = \dfrac{W}{\rho A (\Omega R)^2}$	$\dfrac{W}{\sigma \omega^2}$
weight	$\dfrac{C_W}{M_{\text{tip}}^2} = \dfrac{W}{\rho A c_s^2}$	$\dfrac{W}{\delta}$
power	$C_P = \dfrac{P}{\rho A (\Omega R)^3}$	$\dfrac{P}{\sigma \omega^3}$
power	$\dfrac{C_P}{M_{\text{tip}}^3} = \dfrac{P}{\rho A c_s^3}$	$\dfrac{P}{\delta \sqrt{\theta}}$

(section 3.3). Momentum theory gives the minimum induced power as $P_{\text{ideal}} = Tv = T\sqrt{T/2\rho A}$. In hover there is no parasite or climb power, so $M = P_{\text{ideal}}/(P_i + P_o)$. The figure of merit is a measure of the relative contributions of the induced and profile power in hover. With twin main rotors, an appropriate expression for the ideal power is $P_{\text{ideal}} = T\sqrt{T/2\rho A_p}$, where T is the total thrust of both rotors and $A_p = (2 - m)A$ is the projected disk area. The overlap ratio is $m = 1$ for the coaxial configuration and $m = 0$ for no overlap. With this convention, the reference power for coaxial rotors is the ideal induced power of a single rotor of area A (the limit of no vertical separation).

The aircraft hover figure of merit is $M = W\sqrt{W/2\rho A_p}/P$, in terms of the aircraft weight and total aircraft power. The aircraft weight is greater than the rotor lift because of download, and the aircraft power includes transmission losses and auxiliary power as well as rotor shaft power, so the aircraft hover figure of merit is significantly less than the isolated rotor figure of merit.

For the rotor operating as a propeller (axial flow at high inflow ratio μ_z), the propulsive efficiency is

$$\eta = \frac{TV}{P} = 1 - \frac{P_i + P_o}{P} \qquad (7.79)$$

where TV is the parasite power. The hover figure of merit and propeller propulsive efficiency can be combined into a single metric

$$M_a = \frac{P_{\text{ideal}}}{P} = \frac{T(V + v)}{P} \qquad (7.80)$$

applicable to axial flow in general.

The rotor equivalent drag is defined as $D_e = P/V + X$, so the rotor equivalent lift-to-drag ratio is

$$\frac{L}{D_e} = \frac{L}{P/V + X} = \frac{LV}{P_i + P_o} \qquad (7.81)$$

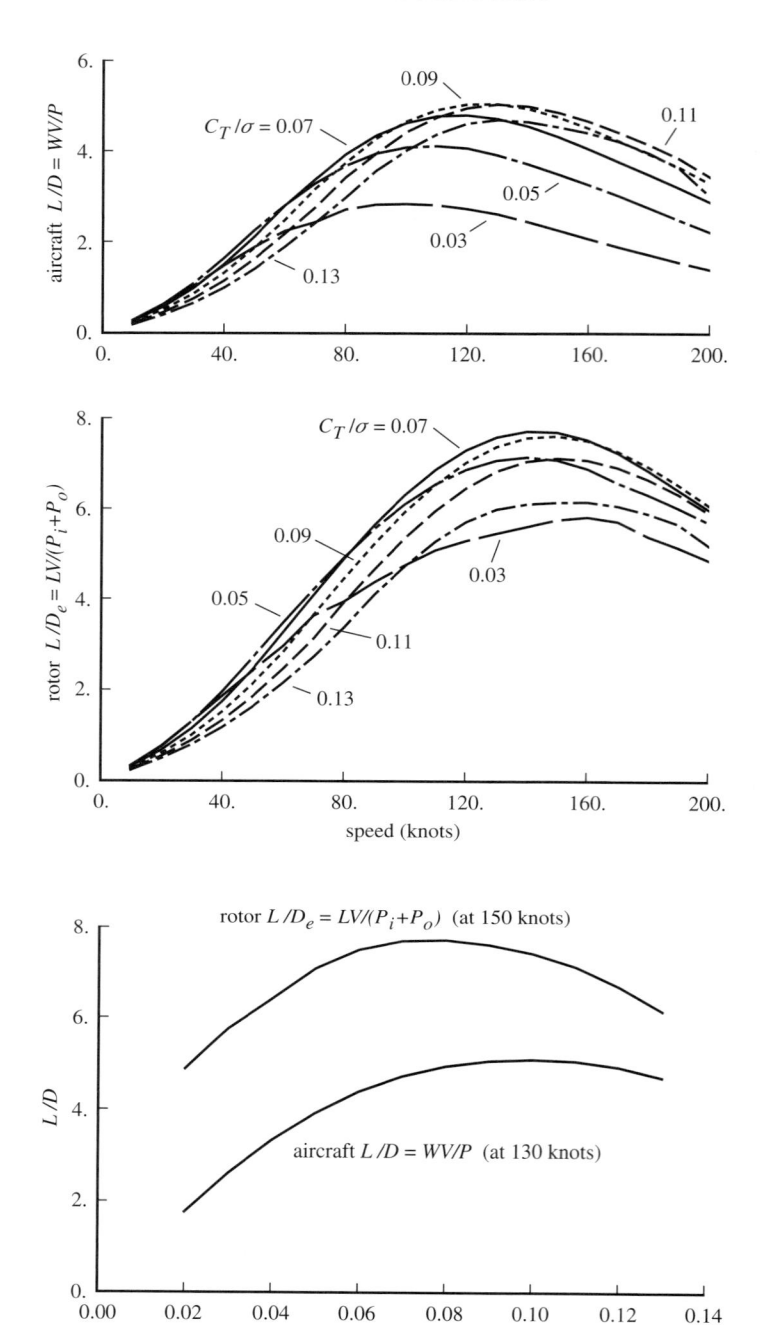

Figure 7.10. Aircraft and rotor lift-to-drag ratios.

(section 7.1.4). Here L and X are the wind-axis rotor lift and drag forces, so L/D_e depends only on quantities that can be measured in a wind tunnel.

The aircraft equivalent drag is defined as $D_e = P/V$; hence the equivalent lift-to-drag ratio of the aircraft is $L/D = WV/P$. Figure 7.10 shows typical aircraft and rotor lift-to-drag ratio variation with speed and rotor thrust, for a conventional helicopter configuration.

For a given operating condition, the specific range is the ratio of the speed to the fuel flow: $V/\dot{W}_f = V/(P\text{sfc}) = (L/D)/(W\text{sfc})$ (nm/lb or nm/kg). From the Breguet range equation, the range for fuel equal 1% of the gross weight is

$$R_{1\%GW} = \frac{L/D}{\text{sfc}} \ln\left(\frac{1}{.99}\right) \tag{7.82}$$

using the aircraft lift-to-drag ratio $L/D = WV/P$ (section 7.2.6). A fuel efficiency measure is the product of the payload and specific range: $e = W_{\text{pay}}(V/\dot{w})$ (ton-nm/lb or ton-nm/kg). A productivity measure is $p = W_{\text{pay}}V/W_O$ (ton-kt/lb or ton-kt/kg), where W_O is the operating weight.

A fuel efficiency measure for a mission is the product of the payload and range divided by the fuel weight: $e = W_{\text{pay}}R/W_{\text{burn}}$ (ton-nm/lb or ton-nm/kg). A productivity measure for a mission is $p = W_{\text{pay}}V/W_O$ (ton-kt/lb or ton-kt/kg), where W_O is the operating weight and V the block speed, or $p = W_{\text{pay}}V/W_{\text{burn}}$ (ton-kt/lb or ton-kt/kg). Metrics that include the payload weight are useful only in the context of a specific mission (including takeoff conditions and range or endurance) to account for fuel burn.

In a wind tunnel it is not possible to measure separately the induced and profile power, only the sum $P_i + P_o = P - XV$. The induced and profile power can be separated in calculations of rotor performance and characterized by an induced power factor κ and a mean drag coefficient $c_{d\,\text{mean}}$. The induced power factor for a single rotor is

$$\kappa = \frac{P_i}{P_{\text{ideal}}} \tag{7.83}$$

where $P_{\text{ideal}} = Tv_{\text{ideal}}$ is the momentum theory solution (sections 5.1.1 and 5.1.3). With twin main rotors, the induced power factor can be defined using a reference power, $\kappa = P/P_{\text{ref}}$. It is best to use as the reference power the ideal momentum theory power of the actual rotor configuration, including the effect of vertical or lateral separation of the two rotors. A simpler approach is to use $P_{\text{ref}} = T\sqrt{T/2\rho A_p}$ in hover (as for the figure of merit) and $P_{\text{ref}} = T^2/(2\rho AV)$ in cruise. This cruise reference is the ideal induced power of a single rotor of area A, carrying the total rotor thrust T (the limit of no separation, vertical or longitudinal or lateral). Using a reference power independent of rotor separation means $\kappa = P/P_{\text{ref}}$ provides an absolute comparison of induced powers.

The mean drag coefficient is a measure of the rotor profile power:

$$c_{d\,\text{mean}} = \frac{8C_{Po}/\sigma}{F_P} \tag{7.84}$$

where $F_P(\mu, \mu_z)$ accounts for the increase of mean dynamic pressure with rotor speed, both axial and edgewise (equation 6.409). The function F_P should be consistent with the assumptions in the calculation of C_{Po} (see section 6.23). For edgewise flight, including reverse flow and radial flow, $F_P = 1 + 4.5\mu^2 + 1.61\mu^{3.7}$ is a good approximation (equation 6.423). For general operating conditions, equation 6.428 should be used.

Although it is not possible experimentally to separate induced and profile powers, or propulsive and lifting efficiencies, local derivatives of the power provide some information. From propeller power in terms of the thrust and propulsive efficiency,

$P = TV/\eta$, a local propulsive efficiency of the rotor can be defined:

$$\frac{1}{\eta} = -\frac{1}{V}\frac{\partial P}{\partial X} = 1 - \frac{1}{V}\frac{\partial (P_i + P_o)}{\partial X} \qquad (7.85)$$

From rotor induced power in terms of lift and induced power factor, $P_i = \kappa L^2/2\rho AV$, the local derivative of the power gives

$$\kappa = 2\rho AV \frac{\partial P}{\partial L^2} \qquad (7.86)$$

which, however, also includes the variation of the profile power with lift. The power required for an aircraft with separate lift and propulsion is

$$P = \frac{TV}{\eta} = \frac{V}{\eta}\left(D_0 + \frac{(L - L_0)^2}{q\pi e b^2}\right) \qquad (7.87)$$

where η is the propeller efficiency and e the Oswald efficiency of the aircraft. Hence the local derivative of the power gives a lift efficiency:

$$\kappa = \frac{1}{e} = 2\eta\rho AV \frac{\partial P}{\partial L^2} = 2\eta \frac{V}{\Omega R}\frac{\partial C_P}{\partial C_L^2} \qquad (7.88)$$

Conventionally e does include the aircraft parasite drag variation with lift.

7.4 REFERENCES

Bailey, F.J., Jr., and Gustafson, F.B. "Charts for Estimation of the Characteristics of a Helicopter Rotor in Forward Flight; I – Profile Drag-Lift Ratio for Untwisted Rectangular Blades." NACA ACR L4H07, August 1944.

Gustafson, F.B. "Charts for Estimation of the Profile Drag-Lift Ratio of a Helicopter Rotor Having Rectangular Blades with -8 deg Twist." NACA RM L53G20a, October 1953.

Kisielowski, E., Bumstead, R., Fissel, P., and Chinsky, I. "Generalized Rotor Performance." USAAVLABS TR 66-83, February 1967.

McCloud, J.L., III, Biggers, J.C., and Stroub, R.H. "An Investigation of Full-Scale Helicopter Rotors at High Advance Ratios and Advancing Tip Mach Numbers." NASA TN D-4632, July 1968.

Tanner, W.H. "Charts for Estimating Rotary Wing Performance in Hover and at High Forward Speeds." NASA CR 114, November 1964.

8 Design

8.1 Rotor Configuration

The helicopter rotor type is largely determined by the construction of the blade root and its attachment to the hub. The blade root configuration has a fundamental influence on the blade flap and lag motion and hence on the helicopter handling qualities, vibration, loads, and aeroelastic stability. The basic distinction between rotor types is the presence or absence of flap and lag hinges, and thus whether the blade motion involves rigid-body rotation or bending at the blade root. A simple classification of rotor hubs has the categories articulated, teetering, hingeless, and bearingless, as sketched in Figures 8.1 to 8.4. With real designs (see Figure 1.2) the distinctions are not as clear as in these drawings.

An articulated rotor has its blades attached to the hub with both flap and lag hinges (Figure 8.1). The flap hinge is usually offset from the center of rotation because of mechanical constraints and to improve the helicopter handling qualities. The lag hinge must be offset for the shaft to transmit torque to the rotor. The purpose of the flap and lag hinges is to reduce the root blade loads (since the moments must be zero at the hinge) by allowing blade motion to relieve the bending moments that would otherwise arise at the blade root. With a lag hinge a mechanical lag damper is also needed to avoid a mechanical instability called ground resonance, involving the coupled motion of the rotor lag and hub in-plane displacement. The articulated rotor is the classical design solution to the problem of the blade root loads and hub moments. The design is conceptually simple, and the analysis of the rigid body motion is straightforward. The articulated rotor is mechanically complex, however, involving three hinges (flap, lag, and feather) and a lag damper for each blade. The flap and lag bearings must transmit both the blade thrust and centrifugal force to the hub and so operate in a high-load environment. The hub also has the swashplate and the rotating and non-rotating links of the control system. The resulting hub requires a high level of maintenance and contributes substantially to the helicopter parasite drag. Recently, the use of elastomeric bearings has been introduced. Replacing the mechanical bearings with elastomeric bearings eliminates a major maintenance problem.

The teetering rotor (also called the semi-articulated or semi-rigid or seesaw rotor) has two blades attached to the hub without flap or lag hinges, and the hub is attached to the rotor shaft with a single flap hinge (Figure 8.2). The two blades form a single structure that flaps as a whole relative to the shaft. The hub usually

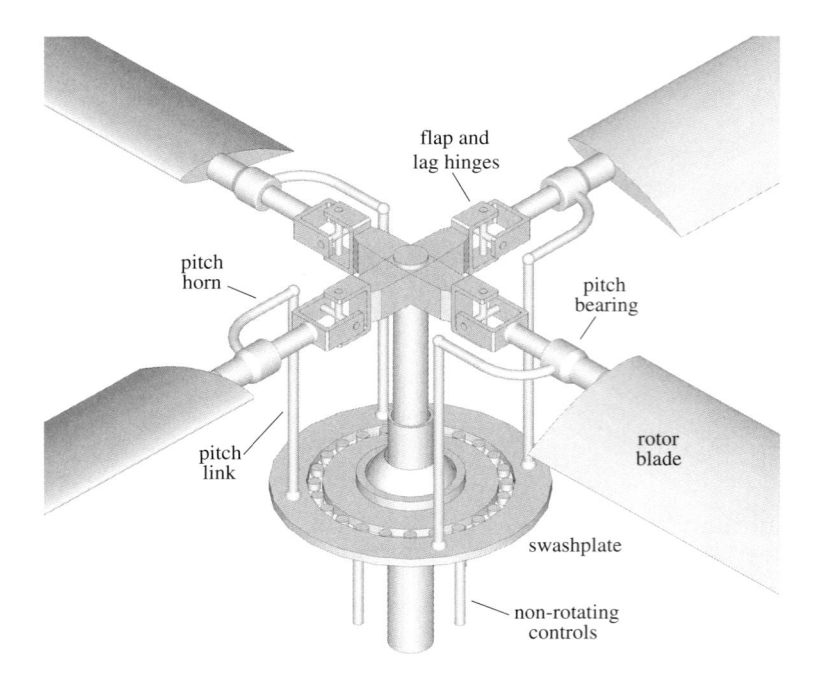

Figure 8.1. Sketch of an articulated rotor hub.

has a built-in precone angle to reduce the steady coning loads, and perhaps also an undersling to reduce Coriolis forces. The blades have feathering bearings. Without lag hinges, the blade in-plane loads must be reacted by the root structure. Torsional flexibility of the rotor shaft and drive train also serves to alleviate lag moments on

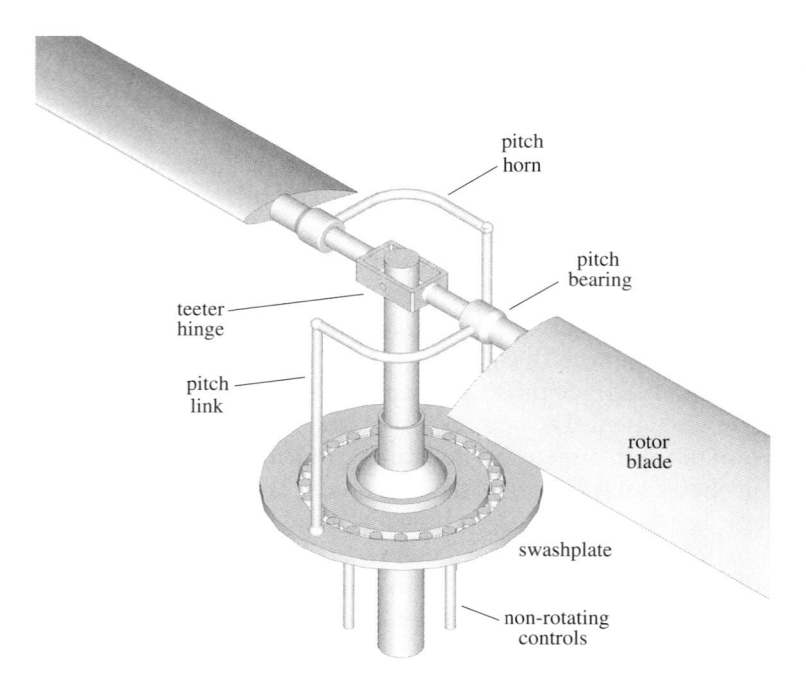

Figure 8.2. Sketch of a teetering rotor hub.

8.1 Rotor Configuration

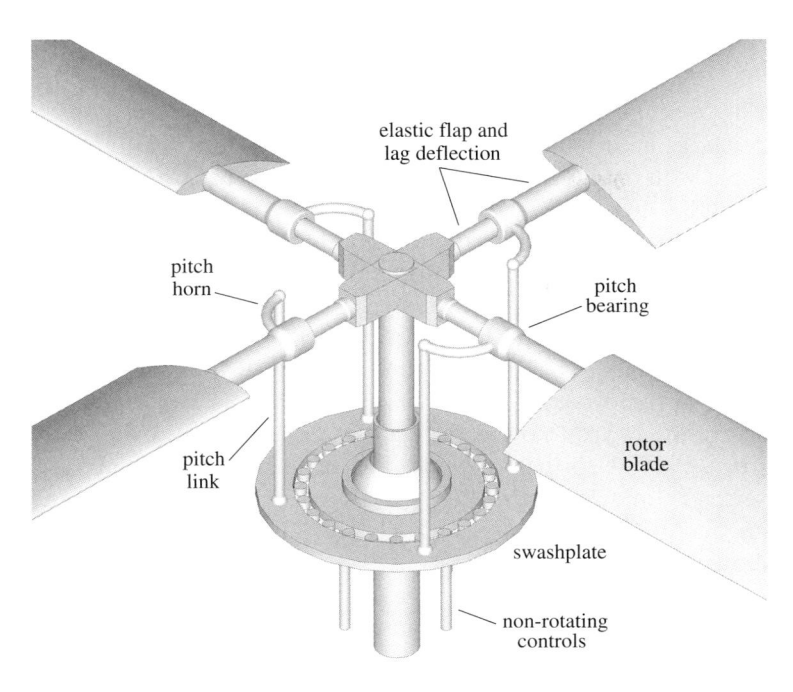

Figure 8.3. Sketch of a hingeless rotor hub.

the blade and hub. The rotor coning produces structural loads, except at the design precone angle. To accommodate these loads the rotor requires additional structure and weight relative to an articulated rotor. This factor is offset by the mechanical simplicity of the teetering configuration, which eliminates all the lag hinges and

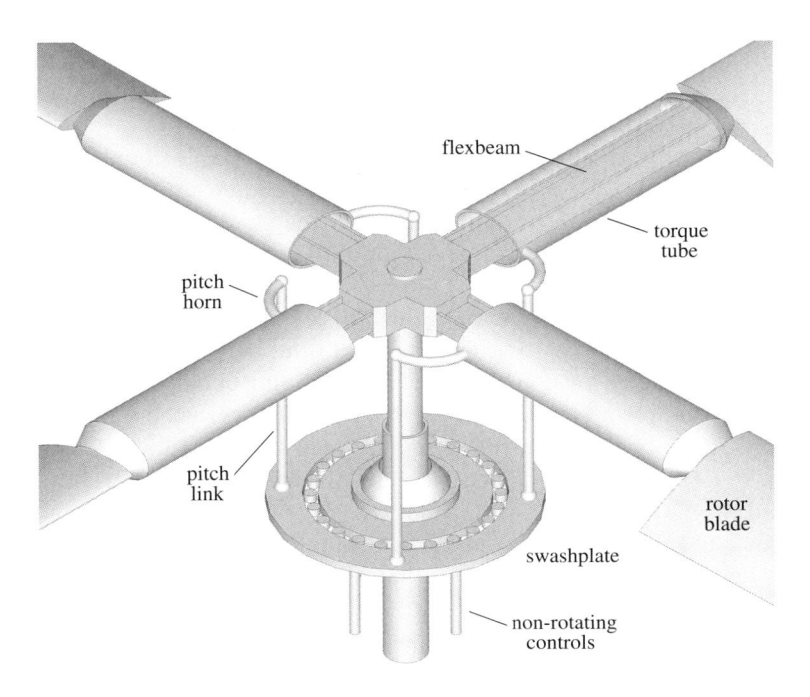

Figure 8.4. Sketch of a bearingless rotor hub.

dampers and all but a single flap hinge. The flap hinge also does not have to carry the centrifugal loads of the blade, only the rotor thrust, since the centrifugal forces are reacted in the hub. The teetering configuration is the simplest for a small helicopter, but is not practical for large helicopters because a large chord is required to obtain the necessary blade area with only two blades. A gimballed rotor has three or more blades attached to the hub without flap or lag hinges (but with feathering bearings); the hub is attached to the shaft by a universal joint or gimbal. The gimballed rotor is the multiblade counterpart of the teetering rotor and similarly has the advantage of a simpler hub than articulated rotors. The teetering and gimballed rotors are characterized by a flap hinge at the center of rotation, giving a flap frequency of exactly 1/rev. The improvements in handling qualities due to offset hinges are not available. For example, flight at low or zero load factor is not possible with a teetering or gimballed rotor, since the control power and damping of the rotor are directly proportional to the thrust. A hub spring can be used to increase the flap frequency, although in the teetering rotor a hub spring leads to large 2/rev loads as well. The lag mode of teetering and gimballed rotors is usually stiff in-plane motion, with a natural frequency above 1/rev.

The hingeless rotor (also called a rigid rotor) has its blades attached to the rotor hub and shaft with cantilever root constraint (Figure 8.3). Although the rotor has no flap or lag hinges, there are hinges or bearings for the feathering motion. The fundamental flap and lag motion involves bending at the blade root. The structural stiffness is still small compared to the centrifugal stiffening of the blade, so the fundamental mode shape is not too different from the rigid body rotation of articulated blades and the flap frequency is not far above 1/rev: typically $\nu = 1.10$ to 1.15 for hingeless rotors. Depending on the structural design of the root, the blade can be either soft in-plane (lag frequency below 1/rev, typically 0.65 to 0.75/rev) or stiff in-plane (lag frequency above 1/rev). The soft in-plane hingeless rotor is susceptible to ground resonance, and although the blade lag damping required can be an order of magnitude less than that required by an articulated rotor, a mechanical lag damper can still be needed. A lag frequency above 0.6/rev is desired for ground resonance and air resonance stability, and the frequency should be below about 0.8/rev for acceptable loads. Without hinges, there can be considerable coupling of the flap, lag, and pitch motions of the blade, which leads to significantly different aeroelastic characteristics than with articulated blades. A matched-stiffness design minimizes couplings, but requires a lag frequency about 0.5/rev. A stiff in-plane rotor has no ground or air resonance problems, but has higher loads. The hingeless rotor is capable of producing a large moment on the hub due to the tip-path-plane tilt. This moment capability has a significant influence on the helicopter handling qualities, including not only increased control power and damping but also increased gust response. The hingeless rotor is a simple design mechanically, with potentially low maintenance requirements and a low hub drag. A stronger hub and blade root are required to take the hub moments, which increases the blade and hub weight. Acceptable loads and strength have been achieved by the use of advanced materials, and most often by the selection of the soft in-plane configuration for main rotors. Acceptable stability can be achieved by designing the rotor for minimum coupling of the blade modes (such as by using a matched-stiffness design), or by designing the rotor for favorable values of pitch-lag and flap-lag couplings. The pitch bearing can be a source of lag damping on a hingeless rotor.

Bearingless rotor designs eliminate the blade pitch bearings as well (Figure 8.4). The flap, lag, and torsion motion occur through deflection of a torsionally soft flexbeam at the root. Pitch control is accomplished using a torque rod or torque tube. Eliminating the highly loaded pitch bearings is another step in simplifying the design, which should reduce cost and maintenance of the rotor hub. However, simplification of the design is accompanied by complication of the analysis. The flap frequency of the bearingless rotor can be lower than that of a hingeless rotor (which is considered too high), since the flap flexure can be put close to the center of rotation. The first applications of bearingless rotor concepts were stiff in-plane tail rotor designs. Bearingless main rotors are generally soft in-plane, with a lag frequency below 1/rev (typically 0.65 to 0.75/rev). Without the pitch bearing, the pitch-bending coupling (from kinematics as well as due to nonlinear structural dynamics) can be quite complicated. Much design flexibility is possible with the bearingless configuration, but production rotors share common characteristics. A stiff torque tube surrounding a soft flexbeam is a good choice for the structure. Since a long flexbeam is needed to accommodate all the motion, the torque tube and flexbeam are connected to the aerodynamic surface of the blade rather far outboard, at say 25% radius. The torque tube does not usually have an airfoil shape, but it can at least be considered an aerodynamic fairing for the large root cutout. A snubber attaches the inboard end of the torque tube to the hub or flexbeam, free in rotation and axial motion, but constraining vertical motion in order to control pitch-flap coupling and to produce pure pitch rotation due to control from vertical motion of the pitch link. The snubber is then also a place where lag damping can be introduced through an elastomeric (or more complicated) damper of chordwise motion. Such snubbers are expensive and do not provide much damping, but are sufficient to deal with ground resonance.

Most rotor designs have a hinge or bearing at the blade root to allow the feathering or pitch motion of the blade for collective and cyclic control. Although it is the most common design solution, the pitch bearing operates under adverse conditions, transmitting the centrifugal and thrust loads of the blade while undergoing a periodic motion due to the rotor cyclic pitch control. Thus there have been other approaches to achieving blade pitch control. To simplify the mechanical design a hinge can be used instead of a bearing, or an elastomeric bearing can be used instead of a mechanical one. Another approach is to allow the pitch motion to take place about torsional flexibility at the root or to use tension-torsion straps between the blade and hub. Kaman developed a rotor that uses a servoflap on the outboard portion of a torsionally flexible blade. Servoflap deflection causes the blade to twist, which can be used for the collective and cyclic control of the rotor in place of root pitch.

Figure 8.5 shows the flap frequency (per-rev) and Lock number for various rotor designs. Recall that a large Lock number implies a lightweight blade. Teetering rotors have a flap frequency of $\nu = 1$, whereas articulated rotors have hinge offsets of 3% to 6%; hence flap frequencies of $\nu = 1.02$ to 1.04 typically. Hingeless rotors have had a flap frequency above $\nu = 1.1$. Modern bearingless rotors have flap frequencies between $\nu = 1.05$ and 1.07. Hingeless rotors tend to have a low Lock number. The tiltrotors shown have gimballed, stiff in-plane hubs, and their coning frequency has been plotted.

8.2 Rotorcraft Configuration

Definition of the rotorcraft configuration involves the number and orientation of the main rotors, the means for torque balance and yaw control, and the fuselage arrangement. Figure 1.1 shows the principal rotorcraft configurations. The basic rotor

276   Design

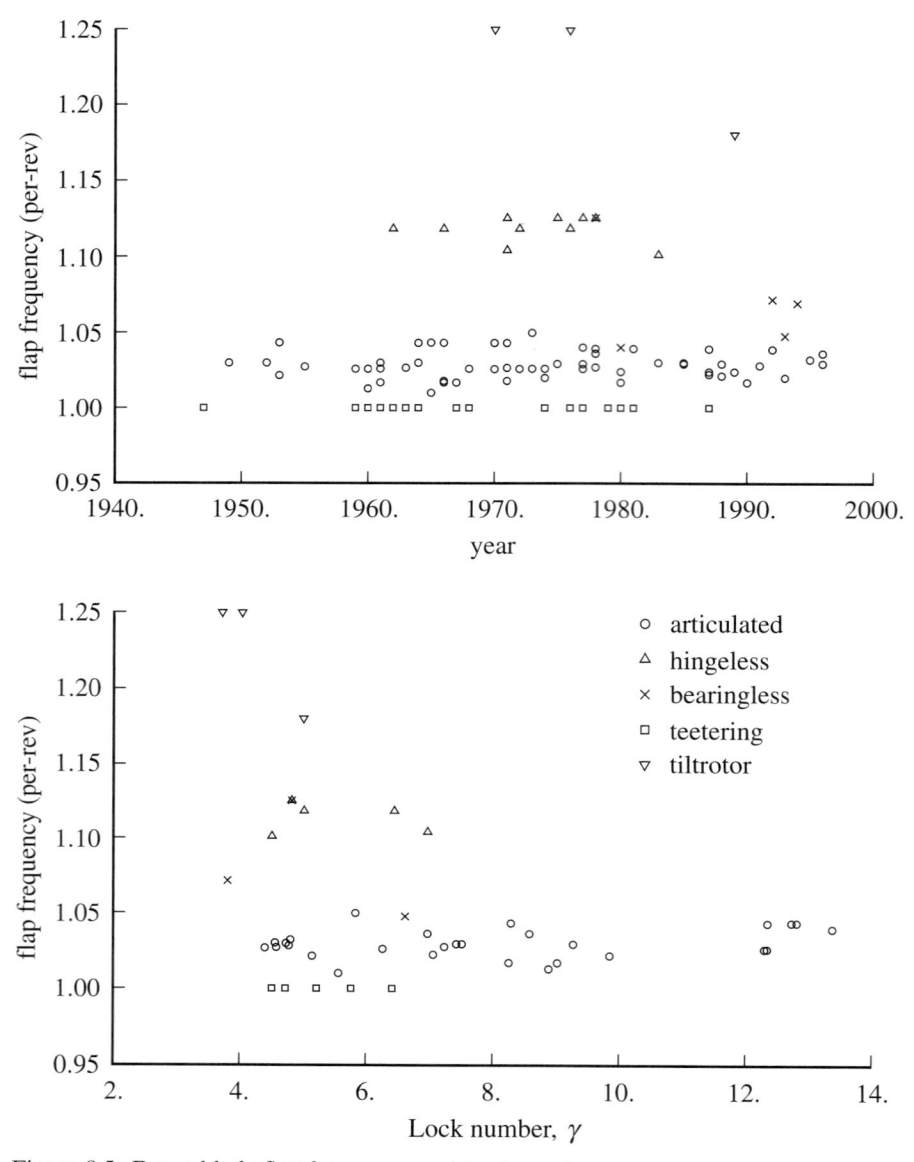

Figure 8.5. Rotor blade flap frequency and Lock number.

analysis is applicable to all helicopter types, but the configuration of the helicopter does have an influence on its behavior, notably on its control characteristics and handling qualities.

A single main rotor and tail rotor is the most common configuration. The tail rotor is a small auxiliary rotor used for torque balance and yaw control. It is mounted vertically on a tail boom, with the thrust acting to the right for a counter-clockwise-rotating main rotor. The moment arm of the tail rotor thrust about the main rotor shaft is usually slightly greater than the main rotor radius. Pitch and roll control of this configuration is achieved by tilting the main rotor thrust using cyclic pitch; height control by changing the main rotor thrust magnitude using collective pitch; and yaw control by changing the tail rotor thrust magnitude using tail rotor collective pitch. The configuration is simple, requiring only a single set of main rotor controls and a single main transmission. The tail rotor gives good yaw control, but absorbs power in balancing the torque, which increases the helicopter power requirement by several percent. The single main rotor configuration typically has a limited center-of-gravity range, although the range is larger with a hingeless rotor. The tail rotor is also a hazard to ground personnel unless it is located very high on the tail, and it can strike the ground during operation of the helicopter. The tail rotor operates in an adverse aerodynamic environment (as do the vertical and horizontal tail surfaces) due to the wake of the main rotor and fuselage, which reduces the aerodynamic efficiency and increases the tail rotor loads and vibration. The single main rotor and tail rotor configuration is the simplest and lightest for small and medium-sized helicopters.

With two (or more) contra-rotating main rotors, torque balance is inherent in the configuration, and no specific anti-torque device with its own power loss is required. However, there are aerodynamic losses from the interference between the main rotors and between the rotors and fuselage; these losses reduce the overall efficiency of twin main rotor configurations to about the same level as for the single main and tail rotor configuration. The mechanical complexity is greater with twin main rotors because of the duplication of control systems and transmissions. Rotor and transmission weight fractions increase with rotor diameter for a fixed disk loading, which favors twin main rotors for large aircraft.

The tandem rotor helicopter has two contra-rotating main rotors with longitudinal separation. The main rotor disks are usually overlapped, typically by 30% to 50% (the shaft separation is around $1.7R$ to $1.5R$). To minimize the aerodynamic interference created by the operation of the rear rotor in the wake of the front rotor, the rear rotor is elevated on a pylon, typically $0.3R$ to $0.5R$ above the front rotor. Longitudinal control is achieved by differential change of the main rotor thrust magnitude, from differential collective; roll control by lateral thrust tilt with cyclic pitch; and height control by main rotor collective. Yaw control is achieved by differential lateral tilt of the thrust on the two main rotors using differential cyclic pitch. A large fuselage is inherent in the design, being required to support the two rotors. The tandem helicopter also has a large longitudinal center-of-gravity range because of the use of differential thrust to balance the helicopter in pitch. Tandem helicopters tend to have lower disk loading than a single-main-rotor helicopter of the same weight, resulting in better hover power loading. The smaller diameter of each rotor means that the effective span of the lifting system in forward flight is lower, resulting in lower cruise efficiency. The operation of the rear rotor in the wake of the front rotor is a significant source of vibration, oscillatory loads, noise, and power loss. The high pitch and roll inertia, unstable fuselage aerodynamic moments, and low yaw control power adversely affect the helicopter handling qualities. There is a structural weight penalty for the rear rotor pylon. Generally the tandem rotor configuration is suitable for medium and large helicopters.

The side-by-side configuration has two contra-rotating main rotors with lateral separation. The rotors are mounted on the tips of wings or pylons, with usually no overlap (so the shaft separation is at least $2R$). Control is as for the tandem helicopter configuration, but with the pitch and roll axes reversed. Roll control is achieved by differential collective pitch, and helicopter pitch control by longitudinal cyclic pitch. The structure to support the rotors is only a source of drag and weight, unless the aircraft has a high enough speed to benefit from the lift of a fixed wing.

The coaxial rotor helicopter has two contra-rotating main rotors with concentric shafts. Some vertical separation of the rotor disks (typically 8% to 10% of the diameter) is required to accommodate lateral flapping. Pitch and roll control is achieved by main rotor cyclic, and height control by collective pitch, as in the single main rotor configuration. Yaw control is achieved by differential torque of the two rotors, and probably rudders in forward flight. The concentric configuration complicates the rotor controls and transmission, but the extensive cross-shafting of other twin rotor configurations is not required. Yaw control by differential torque is typically weak, especially in descent. The symmetry of the configuration (for small vertical separation) results in largely decoupled longitudinal and lateral flight dynamics. The large mast height increases the control power. This helicopter configuration is compact, having small diameter main rotors and requiring no tail rotor. The coaxial rotor has better hover efficiency than the equivalent single rotor (same radius, chord, and number of blades, but no vertical separation), due to the contraction of the upper rotor wake before it encounters the lower rotor, the reduction of swirl in the wake, and the absence of tail rotor power loss. The disk loading of a coaxial rotor (based on projected disk area) tends to be higher than a single main-rotor helicopter of the same weight, resulting in worse hover power loading. In forward flight, the biplane effect of the two rotors operating with small vertical separation reduces the induced power, but the effective span of the lifting system in forward flight is lower than a single main-rotor helicopter, resulting in lower cruise efficiency. The phase of the rotors (whether blades cross at $0°$ or $90°$ azimuth) affects the vibration resulting from oscillatory hub loads.

The synchropter has two contra-rotating main rotors with small lateral separation. The rotors are nearly operating in the coaxial configuration, but the design is simpler mechanically because of the separate shafts.

The manner in which the actions of the main and auxiliary rotors are combined to produce the required control moments and forces depends on the helicopter configuration. Table 8.1 summarizes how control is accomplished for the major helicopter configurations. Rotor cyclic pitch tilts the tip-path plane, and thereby tilts the thrust vector and produces a hub moment. Even with a very stiff hingeless rotor, cyclic pitch can tilt the thrust vector for control, although the tip-path-plane tilt is reduced; see section 19.5. The rotor collective changes the thrust magnitude. For all configurations, vertical control is accomplished by changing the main rotor thrust magnitude using collective pitch. Longitudinal and lateral control is generally obtained using cyclic pitch to tilt the tip-path plane of the main rotor. When there are two main rotors with longitudinal or lateral separation of the shaft, one axis of the helicopter can be controlled by means of differential main rotor collective pitch changes. How a yaw moment is produced for directional control is closely related to the manner in which balance of the main rotor torque is achieved, and as a result varies with the helicopter configuration.

In most helicopter designs the power is delivered to the rotor by a mechanical drive, through the rotor shaft torque. Such designs require a transmission and a means for balancing the main rotor torque. An alternative is to supply the power by a jet reaction drive of the rotor, using cold or hot air ejected out of the blade tips or trailing edges. Helicopters have also been designed with ram jets on the blade tips or with jet flaps on the blade trailing edges that use compressed air generated in the fuselage. Since there is no torque reaction between the helicopter and rotor (except for the small bearing friction), no transmission or anti-torque device is required,

Table 8.1. *Helicopter control*

		longitudinal control	lateral control	height control	directional control
helicopter configuration	torque balance	pitch moment	roll moment	vertical force	yaw moment
single main rotor and tail rotor	tr thrust	mr long cyc	mr lat cyc	mr coll	tr coll
tandem	mr diff torque	mr diff coll	mr lat cyc	mr coll	mr diff cyc
side-by-side	mr diff torque	mr long cyc	mr diff coll	mr coll	mr diff cyc
coaxial	mr diff torque	mr long cyc	mr lat cyc	mr coll	mr diff coll

Notes: mr = main rotor, tr = tail rotor, coll = collective, diff = differential, cyc = cyclic, long = longitudinal, lat = lateral.

resulting in a considerable weight saving. With a jet reaction drive, the propulsion system is potentially lighter and simpler, although the aerodynamic and thermal efficiency are lower. The helicopter must still have a mechanism for yaw control. Fixed aerodynamic surfaces (a rudder) can be used, but at low speeds they are not very effective, depending on the forces generated by the rotor wake velocities.

The autogyro uses autorotation as the normal working state of the rotor. In the helicopter, power is supplied directly to the rotor, and the rotor provides propulsive force as well as lift. In the autogyro, the rotor is free to rotate on the shaft, and no power or shaft torque is supplied to the rotor. The propulsive force required to sustain level flight is supplied by a propeller or other propulsion device. The rotor operates with aft tilt of the tip-path plane, the rotation driven by the resulting air flowing upward through the disk. Sometimes the aircraft control forces and moments are supplied by fixed aerodynamic surfaces as in the airplane, but obtaining the control from the rotor is better. The rotor performs much like a wing (although the aspect ratio $AR = 4/\pi$ is low) and has a fairly good lift-to-drag ratio. Although rotor performance is not as good as that of a fixed wing, the rotor is capable of providing lift and control at much lower speeds than fixed-wing aircraft, because lift on the blades is generated by both the rotation and the forward speed. The power required is less than that of a helicopter, and the weight is less due to the absence of a transmission. Without power being supplied to the rotor, the autogyro is not capable of actual hover or vertical climb. Because autogyro performance is not that much better than the performance of an airplane with a low wing loading, usually the requirement of actual VTOL capability is necessary to justify a rotor on an aircraft. The autogyro enjoyed some success, until helicopter development was completed.

The maximum speed of a rotor in edgewise flight is limited by the asymmetry of the aerodynamic environment. The advancing blade tip encounters shocks and increased drag at high Mach number. The retreating blade encounters high angle-of-attack and stall near the reverse flow region. The asymmetric dynamic pressure seen by the blade, the requirement for roll moment balance on the rotor, and generation of a propulsive force combine to limit helicopter flight speed (see Figure 1.12). There are numerous ideas for modifications to the basic helicopter configuration that are aimed principally at achieving higher speed in level flight. Many of these concepts have been tested in flight.

With auxiliary propulsion (usually a propeller for good efficiency), a propulsive force is not needed from the rotor, resulting in better rotor effective lift-to-drag ratio and delayed rotor stall. If a wing is added to the helicopter, its lift in forward flight allows the rotor loading to be reduced, thus delaying stall effects. Since the rotor lift is also the source of the helicopter propulsive force, significantly reducing the rotor loading requires an auxiliary propulsion device as well. The result is the compound helicopter configuration. Unless a hingeless rotor is used, fixed aerodynamic control surfaces are required to maintain control in forward flight at low rotor thrust. The dual systems for lift and propulsion increase the complexity of the aircraft and increase its empty weight fraction. In forward flight, the rotor and especially the hub are just contributing drag to the aircraft, reducing the cruise efficiency.

With reaction drive of the main rotor, a transmission is required only for the propulsion system. The net effect on the empty weight fraction can be favorable for large aircraft. In cruise the rotor can be operated as an autogyro, with the reaction drive turned off. The noise produced by tip jets in hover and low speed flight is a significant issue. The Fairey Rotodyne demonstrated the compound helicopter using reaction drive (Hislop (1959)).

Avoiding compressibility limits in high-speed cruise flight requires that the rotor be slowed. The slowed and unloaded rotor might then be stopped completely and stowed, to minimize the aircraft drag at high speed. There are also concepts for stopping the rotor and using it as a fixed wing in high-speed forward flight.

The lift-offset rotor maintains lift on the advancing side of the rotor disk in high-speed forward flight by eliminating the requirement for roll moment balance of the rotor. With the rotor lift center-of-action at $0.2R$ to $0.3R$ from the hub on the advancing side, retreating side stall can be avoided. Such operation requires very stiff hingeless rotors (flap frequency around $\nu = 1.45$ has been used) to carry the roll moment. With good blade design and choice of airfoils, the lift offset substantially reduces the growth of induced and profile power with speed. Roll moment balance of the aircraft is achieved using twin main rotors, with counter-acting hub roll moments. The coaxial configuration requires stiff blades also to maintain flapping clearance. The stiff hingeless rotor brings increased control power and damping of the flight dynamics, but also high vibratory hub loads and increased blade and hub weight. As with all high-speed edgewise rotor concepts, reducing hub drag is a major issue. The Sikorsky XH-59A (Advancing Blade Concept) demonstrated the lift-offset rotor in the coaxial configuration (Ruddell (1977)).

In the tilting proprotor configuration, the rotors are tilted forward to act as propellers in cruise, thus avoiding the many problems resulting from the asymmetric aerodynamics of rotors in edgewise flight. The tiltrotor configuration was demonstrated with the Bell XV-15 (Tilt Rotor Research Aircraft; Wernicke and Magee (1979)), and is now in production. The tiltrotor is the simplest of the high-speed rotorcraft configurations, using a single rotor system for helicopter mode lift and control, and propulsion in all flight regimes. This rotorcraft can cruise efficiently at 300 to 350 knots, with about 10% increase in empty weight fraction. Relative to a side-by-side helicopter, the tiltrotor adds the wing for high-speed capability. Relative to a turboprop aircraft, it adds rotor blade area and complexity, and gains VTOL capability. The basic tiltrotor has two rotors mounted in nacelles at the wing tips. The design can tilt the engine with the rotor or not. If the engines tilt, then the engine residual thrust contributes to aircraft lift in hover, but interaction of the engine exhaust with the ground becomes an issue. The typical design couples the wing span and rotor radius (with a requirement for clearance between the rotor tip and the fuselage). A wing extension, outboard of the nacelle, can be used to increase the wing aspect ratio. A cross shaft between the rotors is required for one-engine inoperative conditions. With a rotor spinner and appropriate shaping of the nacelle, the aircraft drag in cruise can be comparable to that of a turboprop.

Fixed-wing aerodynamic surfaces are used for control in airplane mode. The wing brings good cruise efficiency, especially at high altitude, and increased maneuver capability. The wing also is responsible for a large hover download (10% is typical), and avoiding the coupled rotor and wing aeroelastic instability called whirl flutter requires a thick wing section and increased wing stiffness. High lift-to-drag ratio in cruise implies that the rotor thrust requirement as a propeller is an order of magnitude less than the requirement in hover. The tiltrotor design usually has a high disk loading T/A, for lighter rotors and a high hover blade loading C_T/σ, since the wing off-loads the rotor in helicopter mode forward flight. Nonetheless, the blade area required for hover is much more than is needed as a propeller in cruise. To reduce the profile power and increase propulsive efficiency, the rotor is slowed for airplane mode operation. The engine power-turbine speed can be reduced by 15% to 25% before the specific fuel consumption is affected significantly.

The difference in aerodynamic environments between hover and cruise operation leads to compromises in the blade design, particularly twist, which is too much for hover and too little for cruise. For counter-clockwise rotation of the right rotor, there is a favorable influence of the rotor wake on the wing in cruise, reducing the wing induced power. Further reducing the rotor speed, to 40% or 50% of hover, can produce further improvements in propulsive efficiency, both by reducing the profile power and increasing the influence of the rotor on the wing induced drag. Moreover, with large rotor speed reduction a compromise in the rotor blade design is no longer necessary, the twist required for hover being about the same as that required for cruise. Such large changes in rotor speed must be accomplished using a variable-geometry engine or by a variable-speed transmission and is accompanied by an increase in rotor torque.

8.3 Anti-Torque and Tail Rotor

A large number of design solutions have been considered for torque reaction in the single main-rotor configuration. In addition to countering the main rotor torque in hover and low-speed flight, the device must provide directional control and should contribute to yaw damping and directional stability. Factors influencing the choice of device are performance, weight, cost, handling qualities, and noise.

The tail rotor is efficient, provides good directional control, and has inherent yaw damping. However, there are safety issues with a tail rotor, notably hazard to personnel and the possibility of ground contact. On many helicopters the tail rotor is a source of high noise, but there are design choices to alleviate the noise, especially lower tip speed and unequal blade spacing.

The tail rotor must produce thrust, both positive and negative, with the air velocity from all directions. The tail rotor size is typically determined by the requirement to hover in wind (up to 35–45 knots) from any direction, plus maneuver requirements such as yawing at low speed. The tail rotor operates in an adverse aerodynamic environment, with substantial interference from the main rotor, airframe, vertical tail, and ground. It has only collective control, no cyclic. Usually pitch-flap coupling is used to reduce flapping, $\delta_3 = 35°$ to $45°$. The preferred configuration is top-aft and pusher, to minimize adverse interference (see section 5.6.5).

In sideward flight (to the left for counter-clockwise rotation of the main rotor) at 10 to 25 knots (depending on the disk loading), the tail rotor can encounter vortex ring state (VRS). The flow unsteadiness and reversed sign of thrust change with axial speed in vortex ring state (section 4.2) produce difficulties in directional trim, high oscillatory pedal motions, and large oscillatory yaw of the aircraft. Interaction with the main rotor and the ground influences the tail rotor VRS characteristics.

Upward cant of the tail rotor shaft improves the helicopter performance, particularly in hover and vertical climb (see Hansen (1988)). Having the tail rotor share the lift is efficient, even though the required tail rotor thrust is larger. For example, with 20° of cant, 34% of the tail rotor thrust contributes to vertical lift, for a 6% loss in the anti-torque component. With lift from the tail rotor, the aircraft center-of-gravity range is increased, and the aircraft trims with the center-of-gravity aft of the shaft. Cant reduces the tail rotor susceptibility to vortex ring state in sideward flight, especially with bank angle. The canted tail rotor must be a tractor (the fin on the intake side of the disk), but the bottom of the tail rotor is farther from the fin, which reduces the interference. Tail rotor cant introduces control couplings. Pitching moments would be produced by pedal because of the cant and by collective because of the aft center-of-gravity position, requiring pilot's pedal and collective control to produce longitudinal cyclic pitch. Tail rotor cant also introduces couplings in the flight dynamics; in particular, tail rotor thrust changes due to sideslip produce a pitching moment.

A successful alternative to the tail rotor must have satisfactory stability, control power, autorotation capability, weight, and power loss. The tail rotor has satisfactory characteristics in all these areas and excellent characteristics in some. Most candidate replacements are seriously deficient in at least one area.

The most common alternative to the tail rotor is the ducted fan, implemented as a shrouded fan in the vertical tail. The primary deficiencies of the tail rotor are its hazard to personnel, noise, and vibration. The ducted fan offers improvements in safety and vulnerability, particularly regarding personnel hazard. The fan-in-fin tail rotor, or Fenestron™, has been developed and taken to production by Eurocopter. The control and handling qualities with a fan-in-fin are similar to those with a tail rotor. The power required and weight are somewhat increased. The duct is not long, and the flow around the inlet is complex in forward flight.

Also in production is a circulation-control tail boom (operating in the main rotor wake, for anti-torque in hover) with a reaction jet for yaw control. There is a fan inside the boom, driving air out slots in the boom and a jet at the tail. Torque reaction is obtained from the side force on the tail boom, generated by boom circulation control in the wake of the main rotor. Vanes or other devices vary the jet direction for yaw control. Relative to the tail rotor, safety is better and the noise is low, but there is no contribution to directional stability. The flow around the tail boom is complex.

8.4 Helicopter Speed Limitations

As for fixed-wing aircraft, the maximum speed of a helicopter in level flight is limited by the power available, but with a rotary wing there are a number of other speed limitations as well, among them stall, compressibility, and aeroelastic stability effects. The primary limitation with many current designs is retreating blade stall, which at high speed produces an increase in the rotor and control system loads and in helicopter vibration, severe enough to limit the flight speed. The result of these limitations is a cruise speed for a pure helicopter between 150 and 180 knots with current technology. To achieve a higher cruise speed requires either an improvement in rotor and fuselage aerodynamics or a significant change in the rotorcraft configuration.

The absolute maximum level flight speed is the speed at which the power required equals the maximum power available. At high speed, parasite power dominates the power required. To increase the power-limited speed requires an increase in the installed power of the helicopter or a reduction in the hub and body drag. Because the parasite power is proportional to V^3, a substantial change in drag or installed power is required to noticeably influence the helicopter speed. At sufficiently high speed, the rotor profile power also shows a sharp increase as a result of stall and compressibility effects.

A measure of the compressibility effects on the rotor blade in edgewise forward flight is the Mach number of the advancing tip,

$$M_{at} = M_{1,90} = M_{\text{tip}}(1 + \mu) = \frac{\Omega R + V}{c_s} \qquad (8.1)$$

where c_s is the speed of sound and $M_{\text{tip}} = \Omega R/c_s$. The magnitude of compressibility effects on speed and power depends primarily on whether M_{at} is above or below the critical Mach number for the angle-of-attack of the advancing tip. Above the critical Mach number, compressibility increases the rotor profile power due to drag divergence, and the high periodic forces on the blade increase the helicopter vibration and rotor loads. Dynamic stability problems (flapping or flap-pitch flutter) due to compressibility can also be encountered. An increasingly important limit on the rotor Mach number is the rotor noise level. Power and vibration effects do not appear until a significant portion of the rotor disk is above the critical Mach number, so usually a value of M_{at} that is 5% to 10% above the section critical Mach number can be tolerated. If rotor noise is considered, a substantially lower rotor speed may well be required. An alternative to reducing the rotor speed to avoid compressibility effects is to increase the critical Mach number, notably by using thin airfoil sections and sweep at the blade tip. Since the compressibility limitation on the advancing-tip Mach number basically provides a maximum value for $(\Omega R + V)$, the designer must compromise between the rotor speed and flight speed.

A measure of stall effects on the rotor is the ratio of the thrust coefficient to solidity, C_T/σ, which represents the mean lift coefficient of the blade. In hover, high values of C_T/σ can be achieved before the profile power increase due to stall is encountered. In forward flight, however, the angle-of-attack increases on the retreating side of the disk to maintain the same loading as on the advancing side (see section 6.13), so that stall is encountered at significantly lower values of C_T/σ. The rotor profile power increases when a substantial portion of the disk is stalled, and more importantly there is a sharp increase in the rotor loads and vibration, particularly in the control system, as a result of the high transient pitch moments on the periodically stalling blade. Stall of the helicopter rotor is discussed fully in Chapter 12. The stall-limited C_T/σ in forward flight decreases as either forward speed or propulsive force increases, since both increase the nonuniformity of the blade angle-of-attack distribution. Alternatively, for a given C_T/σ severe rotor stall effects are encountered at some critical advance ratio, which increases as the blade loading is reduced. Since the amount the blade area can be increased to reduce C_T/σ is limited by the weight and performance penalties, the advance ratio restriction due to stall is an important helicopter design criterion.

The maximum advance ratio at which the helicopter can be operated depends on several factors. As μ increases, the aeroelastic stability of the blade motion decreases, the blade and control loads increase because of the asymmetry of the flow, and the aerodynamic efficiency and propulsive force capability of the rotor decrease. Retreating blade stall often constitutes the primary restriction on μ. For a specified maximum advance ratio $\mu = V/\Omega R$, the designer must increase the rotor tip speed to obtain a high forward speed of the helicopter. However, compressibility limits the possible tip speed and thus limits the helicopter speed.

Compressibility effects on the advancing blade and stall effects on the retreating blade combine to restrict the maximum forward speed of the helicopter rotor. The advancing-tip Mach number and advance ratio specify the sum and ratio of the tip speed and velocity:

$$M_{at} = (\Omega R + V)/c_s \tag{8.2}$$

$$\mu = V/\Omega R \tag{8.3}$$

Solving for V and ΩR gives

$$V = c_s M_{at} \frac{\mu}{1+\mu} \tag{8.4}$$

$$\Omega R = c_s M_{at} \frac{1}{1+\mu} \tag{8.5}$$

A high helicopter speed thus requires a high tip Mach number and a high advance ratio. This relationship is shown graphically in Figure 8.6, which plots ΩR as a function of V for constant advancing-tip Mach number and advance ratio of the rotor in edgewise forward flight. From this diagram the maximum helicopter speed for given limits on M_{at} and μ can be determined. For example, a critical Mach number of $M_{at} = 0.9$ and a maximum advance ratio of $\mu = 0.5$ produce a tip speed $\Omega R = 670$ ft/sec (200 m/sec) and a maximum velocity $V = 200$ knots.

The rotor tip speed is selected largely as a compromise between the effects of stall and compressibility. A high tip speed increases the advancing-tip Mach number, leading to high profile power, blade loads, vibration, and noise. A low tip speed increases the angle-of-attack on the retreating blade until limiting profile power, control loads, and vibration due to stall are encountered. Thus there is only a limited range of acceptable tip speeds, which becomes smaller as the helicopter velocity increases.

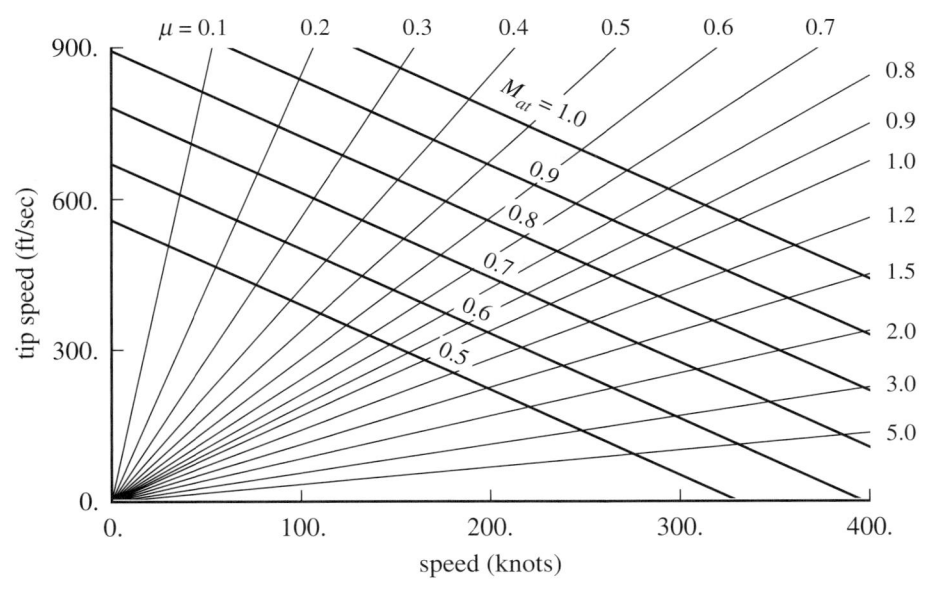

Figure 8.6. Rotor speed and velocity limits for edgewise flight (sea-level-standard speed of sound).

8.5 Autorotation, Landing, and Takeoff

After engine failure, the helicopter has the capability of making an autorotation landing, in which the rotor lift is maintained while the aircraft descends at a steady rate. Because the equilibrium descent rate of the helicopter is fairly high, even in forward flight, autorotational descent is normally an emergency procedure. Moreover, the pilot must take prompt and correct action to establish the optimum flight path both at the beginning and end of the maneuver.

After power failure, the rotor slows down as profile and induced power absorb the rotor kinetic energy, which is the only power source available until the helicopter begins to descend. As the descent rate builds up, the inflow up through the rotor disk increases and therefore the blade angle-of-attack increases. Possibly the helicopter can then achieve an equilibrium descent rate in these conditions, with the angle-of-attack increase countering the rotor speed loss to maintain thrust equal to gross weight. Stall places a limit on the angle-of-attack, and the rotor kinetic energy must be conserved for the end of the maneuver. To keep the angle-of-attack in autorotation low and maintain rotor speed, after a power failure the pilot must reduce collective pitch. The transient lift capability of a rotor is higher than its static capability, which gives the pilot some additional time to react, but still the pilot must recognize the power loss and drop the collective within 2 or 3 seconds to prevent excessive rotor speed decay. The collective pitch required in autorotation is usually a small positive angle. On a single main rotor helicopter, the rotor torque loss also requires a pedal control change to reduce the tail rotor thrust. After the initial control actions, the pilot must establish equilibrium power-off flight at the minimum possible descent rate. The lowest autorotation descent rate is achieved in forward flight at the speed for the minimum power required for level flight (see section 7.2.3); the value is about one-half the descent rate in vertical autorotation.

Near the ground the pilot must flare the helicopter, reducing the vertical and horizontal velocities for touchdown. Ideally, the helicopter has zero velocity just at the instant it contacts the ground. The flare maneuver requires that the collective be raised to increase the thrust and decelerate the helicopter and that aft longitudinal cyclic be used to reduce the forward speed (producing a significant pitch-up motion as well). The power for the rotor during the flare maneuver is supplied by the rotational kinetic energy stored in the rotor. This is a limited power source, so the flare maneuver must be well timed by the pilot. Because the rotor slows down when the collective is increased, blade stall limits the flare capability of the helicopter. The total kinetic energy of the rotor is $\text{KE} = \frac{1}{2}NI_b\Omega^2$ (where NI_b is the rotational moment of inertia of the entire rotor), but the fraction of the energy available before the rotor stalls and the thrust is lost is only $(1 - \Omega_f^2/\Omega_i^2)$. Here Ω_i and Ω_f are the rotor speeds at the beginning and end of the flare. Assuming that the rotor thrust remains fairly constant,

$$\frac{\Omega_f^2}{\Omega_i^2} = \frac{(C_T/\sigma)_i}{(C_T/\sigma)_f} \tag{8.6}$$

The rotor speed and C_T/σ at the beginning of the maneuver are close to the normal operating values of the helicopter, and $(C_T/\sigma)_f$ is determined by the rotor stall limit (taking into account the lift overshoot possible in a transient maneuver).

If a power failure occurs when the helicopter is near the ground, establishing an equilibrium descent condition is not possible. Then the entire power-off landing is a transient maneuver, and the best flight path is somewhat different. If the power loss occurs in hover, the minimum contact velocity at the ground is achieved with a purely vertical flight path. Thus the pilot should not attempt to establish the forward velocity for lowest equilibrium descent rate, but only maintain enough speed to give a view of the landing spot.

Good autorotation characteristics depend on a number of the helicopter primary design parameters. The descent rate in autorotation is proportional to the square root of the rotor disk loading, so the disk loading should be low. Therefore a low autorotation descent rate is associated with low hover power. Helicopter flare capability is even more important for power-off landings than the steady-state descent rate, particularly since the choice of disk loading is influenced primarily by performance considerations. The flare capability depends on the rotor kinetic energy, which requires a high rotor speed and a large blade moment of inertia. The stall margin should be high, both for good flare characteristics and for a minimal loss of rotor speed before the collective is reduced just after the power failure. Thus the helicopter operating C_T/σ should be low. The rotor inertia is the most effective parameter for improving helicopter autorotation characteristics.

The helicopter must have a free-wheeling or overriding clutch so that the engine can drive the rotor but not the other way around. Then when a power failure occurs, the engine automatically disengages from the rotor, and the rotor does not have the drag of the engine during autorotation. The tail rotor of a single main rotor helicopter must be geared directly to the main rotor, so that yaw control can be maintained in the event of power failure.

Figure 8.7 shows the height-velocity diagram, which summarizes the helicopter behavior after engine power failure. If the power loss occurs high above the ground, the pilot has time to establish equilibrium descent. The normal rotor speed can be

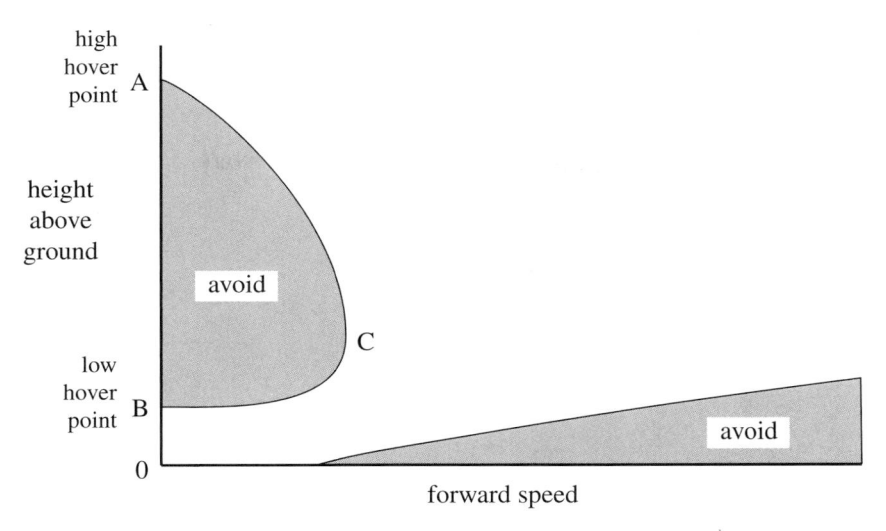

Figure 8.7. Helicopter height-velocity diagram.

recovered by a momentary increase in the descent rate, so that the flare can be initiated with the maximum possible stored energy in the rotor. If the power loss occurs near the ground, it is not possible before the flare is started to make up for the rotor speed drop at the beginning of the maneuver, particularly when the pilot reaction time is accounted for. The result for most helicopters is that there is a range of altitudes for which the flare cannot be initiated with sufficient rotor energy and low enough descent rate to avoid excessive vertical velocity at ground contact. Thus there is a region at low speed on the helicopter height-velocity diagram (Figure 8.7) in which the helicopter should not be operated, because a safe landing after power loss is not possible. The boundary of this region is called the deadman's curve. Above the high hover point (point A) the rotor speed can be recovered sufficiently, and the descent rate kept low enough, to make a safe landing. With enough altitude, equilibrium autorotation descent can be established. For very low heights (below point B in Figure 8.7), the ground is reached before the helicopter has time to accelerate to an excessive velocity. With sufficient forward speed (point C) a safe landing is again possible because of the reduction in autorotative descent rate with forward flight. The high hover point is at 300 to 500 ft altitude for light helicopters, perhaps at 2500 ft for medium helicopters. The low hover point is typically 10 to 15 ft above the ground. The maximum forward speed of the avoid region is at 30 to 60 knots. There is also usually a restriction on high-speed flight near the ground, as shown in Figure 8.7. If a power failure occurs at high speed and low altitude, either there is not time to reduce the horizontal velocity sufficiently to avoid damage to the landing gear (particularly for helicopters with skid-type gear), or cyclic flare to reduce the speed would lead to contact of the tail with the ground.

The two avoid regions on the height-velocity diagram combine to constrain the helicopter takeoff and landing to a specific corridor. The limit on the operational use of the helicopter is not particularly restrictive, however. A purely vertical takeoff or landing is not usually made because of the avoid region. Rather, after a vertical climb to about 15 ft altitude the pilot begins to accelerate the helicopter forward. With two or more engines, the avoid region of the helicopter operation disappears or at least becomes much smaller. The concern with multi-engine helicopters is more with the one-engine-inoperative (OEI) performance capability than with the consequences of a complete power failure.

Consider the initial rate of descent and rotor speed decay following power failure, but before the pilot reacts to the situation, so that the collective pitch is unchanged. The following analysis was introduced by McCormick (1956). The equation of motion for the vertical acceleration of the helicopter is $M\ddot{h} = T - W$, where h is the helicopter height above the ground, W the gross weight, T the rotor thrust, and $M = W/g$ the helicopter mass. The equation of motion for the rotor speed is $NI_b\dot{\Omega} = -Q$, where NI_b is the total rotor moment of inertia and Q is the decelerating torque on the rotor. Before the power failure (at time $t = 0$), the rotor thrust equals the gross weight and the rotor speed is constant. After the power failure, the engine torque no longer balances the rotor decelerating torque, so the rotor slows down; Q then is just due to the rotor power requirement. Since the collective is assumed to be unchanged from the hover value, and at least initially the descent rate has not built up sufficiently to change the inflow ratio, the rotor thrust and torque coefficients (C_T and C_Q, which are functions of $\theta_{.75}$ and λ) must remain fixed at the same values as at the instant of power failure. The rotor thrust and torque are then changed only by variations in the rotor speed: $T = W(\Omega/\Omega_0)^2$ and $Q = Q_0(\Omega/\Omega_0)^2$. Here Ω_0 is the initial rotor speed and Q_0 the rotor torque required in level flight, so $P = \Omega_0 Q_0$ is the helicopter power required for level flight. The equation of motion of the rotor speed for $t > 0$ then becomes $NI_b\dot{\Omega} = -Q_0(\Omega/\Omega_0)^2$, which integrates to

$$\frac{\Omega}{\Omega_0} = \left(1 + \frac{tQ_0}{NI_b\Omega_0}\right)^{-1} = \frac{\tau}{t + \tau} \tag{8.7}$$

and the helicopter vertical velocity (negative for descent) has the solution

$$\dot{h} = -gt^2 \frac{Q_0}{NI_b\Omega_0} \left(1 + \frac{tQ_0}{NI_b\Omega_0}\right)^{-1} = -\frac{gt^2}{t + \tau} \tag{8.8}$$

The time constant in the final expressions is

$$\tau = \frac{NI_b\Omega_0}{Q_0} = 2\frac{\text{KE}}{P} \tag{8.9}$$

where P is the rotor power required for level flight and $\text{KE} = \frac{1}{2}NI_b\Omega_0^2$ is the kinetic energy stored in the rotor. These expressions describe the helicopter behavior fairly well for the first few seconds after power failure.

The flare is a more important part of the power-off landing, but McCormick's analysis introduces the parameter $\tau = 2\,\text{KE}/P$ as a measure of helicopter autorotation characteristics. A small decay of the rotor speed requires a large value of τ; hence, a high rotor kinetic energy and a low required power. The helicopter power enters as a measure of the torque acting to decelerate the rotor after engine power is lost. Typically, $\text{KE}/P = 2$ to 3 seconds, so the time for a significant decay of the rotor speed is around 1 to 2 seconds. The largest permissible reaction time can be estimated by setting the rotor speed decrease equal to the stall limit:

$$\frac{\Omega^2}{\Omega_0^2} = \frac{C_T/\sigma}{(C_T/\sigma)_{\text{stall}}} \tag{8.10}$$

from which

$$t_{max} = 2\frac{KE}{P}\left[\left(\frac{(C_T/\sigma)_{stall}}{C_T/\sigma}\right)^{1/2} - 1\right] \quad (8.11)$$

Typically $t_{max} = 1.5$ to 2.5 sec.

A number of autorotation performance indices have been proposed as measures of overall autorotation characteristics such as represented by the height-velocity diagram:

$$\text{time delay} \quad t/K = \frac{KE}{P}\left(1 - \frac{C_T/\sigma}{0.8(C_T/\sigma)_{max}}\right) \quad (8.12)$$

$$\text{usable kinetic energy} \quad E = \frac{KE}{W}\left(1 - \frac{C_T/\sigma}{(C_T/\sigma)_{max}}\right) \quad (8.13)$$

$$\text{autorotation index} \quad AI = \frac{KE}{P} \quad (8.14)$$

$$\text{energy factor} \quad h = \frac{KE}{W} \quad (8.15)$$

$$\text{flare index} \quad FI = \frac{KE}{W(\rho_0 W/\rho A)} \quad (8.16)$$

Increasing values of an index imply better autorotation characteristics. Here $KE = \frac{1}{2}NI_b\Omega^2$ is the rotor kinetic energy, P the installed power of the helicopter, W the gross weight, $(C_T/\sigma)/(C_T/\sigma)_{max}$ the ratio of rotor blade loading to stall-limited blade loading, and $(\rho_0 W/\rho A)$ the referred disk loading. Wood (1976) and White, Logan, and Graves (1982) discussed the origin of these parameters. Wood (1976) presented a correlation of the first four indices with autorotation characteristics for lightweight, heavy rotor helicopters. The index t/K is based on the decay time of rotor speed after power failure, taken as a measure of how the rotor rotational energy is used during the flare to touchdown or of the time the touchdown can be delayed. The factor K in t/K is included to acknowledge the differences between these events at the start and at the end of the landing. The index E represents the minimum energy level for safe landing, as limited by the stall blade loading. The index $AI = KE/P$ (sec) and energy factor $h = KE/W$ (ft or m) are measures of the relative amount of kinetic energy. Based on hover power required, $P \propto W\sqrt{\rho_0 W/\rho A}$, so the indices t/K and AI have an additional factor of $\sqrt{\rho_0 W/\rho A}$ in the denominator, relative to the indices E and h. Thus t/K and AI have a greater sensitivity to disk loading (smaller values of the indices for high disk loading). The flare index FI was developed by Fradenburgh (1984), based on the ratio of energy available (KE) to energy required ($P\Delta t$) during flare. For a given load factor to produce the deceleration, the time of the flare maneuver is proportional to the rate of descent; hence $\Delta t \propto \sqrt{\rho_0 W/\rho A}$ and $P\Delta t \propto W(\rho_0 W/\rho A)$. So FI is more sensitive to disk loading than t/K and AI. This flare index has not been correlated with helicopter autorotation characteristics. Fradenburgh gave values of FI for a number of Sikorsky helicopters (disk loadings from 4 to 12 lb/ft^2) and noted that all have been landed safely without power.

A simple point-mass model combined with optimal control is effective in calculating the takeoff and landing behavior of rotorcraft, including landing after loss of power. Such an analysis can predict the shape of the height-velocity diagram, although as an optimum solution often predicts a smaller avoid area. See Johnson (1977), Lee, Bryson, and Hindson (1988), Lee (1990), Zhao, Jhemi, and Chen (1996), Carlson, Zhao, and Chen (1998), and Carlson (2001).

Consider a rotorcraft with horizontal motion x (positive forward), height h (positive upward), and rotational speed Ω; these are the degrees of freedom of the motion. The forces acting on the rotorcraft are weight W (downward), drag D (opposite the total velocity), and tip-path-plane thrust T. The decelerating torque is $Q = P/\Omega$, where P is the rotor power required. The tip-path-plane thrust T and angle relative to horizontal i (positive forward) are the control variables. The equations of motion are

$$M\ddot{h} = T\cos i - D\dot{h}/V - W \qquad (8.17)$$

$$M\ddot{x} = T\sin i - D\dot{x}/V \qquad (8.18)$$

$$I_R\dot{\Omega} = Q_{\text{eng}} - Q \qquad (8.19)$$

where $M = W/g$ is the rotorcraft mass, $I_R = NI_b$ the total rotational inertia, and $V = \sqrt{\dot{x}^2 + \dot{h}^2}$ the rotorcraft velocity. The drag is defined by the area $f = D/(\frac{1}{2}\rho V^2)$ and must cover hover download as well as forward flight parasite drag. The torque $Q = P/\Omega = \rho A(\Omega R)^2 R C_P$ can be obtained from energy balance expressions, such as equation 6.121. The rotor induced velocity is obtained from Glauert's formula (equation 5.9), including corrections for the singularity at ideal autorotation (equation 5.13) and ground effect. Using a differential equation for the rotor inflow is best, both to simplify solution of the equations and to introduce a time lag that filters the inflow variation. The model can also include a wind velocity. For powered flight, differential equations are required for the engine torque Q_{eng}, representing the engine dynamics, governor and fuel control, and perhaps throttle command. Constraints are defined for the magnitude and rate of change of the control variables and perhaps to model effects such as C_T/σ limits due to stall.

Inverse simulation methods can find the solution following a takeoff or landing trajectory. Optimal control theory can find the control schedules to minimize a cost function. For the problem of landing after power loss, the goal is to minimize the velocity at touchdown, so the cost function can be a sum of the horizontal and vertical kinetic energies: $J = \frac{1}{2}(\dot{h}_f^2 + \dot{x}_f^2)$, evaluated at the final time t_f. The initial conditions are level flight: altitude h_0, flight speed $\dot{x}_0$, and rotor speed Ω_0 at time $t = 0$. Since the final altitude ($h_f = 0$) is known, but not the final time, the equations can be expressed in terms of altitude h as the independent variable and vertical speed $\dot{h}$ as the dependent variable: $d(\ldots)/dt = \dot{h}\,d(\ldots)/dh$. The cost function can be transformed to an integral over h.

According to Airworthiness Standards for Transport Category Rotorcraft (Federal Aviation Regulation Part 29), transport helicopters can be certified as either Category A or Category B. Rotorcraft with a maximum weight greater than 20000 lb and 10 or more passenger seats must be certified as Category A. An important aspect of the certification is the capability of a multi-engine aircraft in one-engine-inoperative (OEI) conditions. The rotorcraft design and operating procedures must meet the requirements of Category A takeoff:

> If one engine fails at any time during the takeoff, it must be possible to return to and stop safely on the takeoff area, or to continue the takeoff and climb-out.

The takeoff decision point (TDP) is the first point that continued takeoff is assured and the last point that rejected takeoff is assured. The TDP is established as a function of two parameters (such as airspeed and height), including the pilot reaction time following engine failure. If an engine fails before TDP, takeoff must be rejected; if an engine fails after TDP, takeoff must be continued.

The takeoff path must be clear of the avoid regions of the height-velocity diagram. The path cannot descend below 15 ft above the takeoff surface, when the takeoff decision point is above 15 ft. For elevated heliports, the path may descend below the takeoff surface, if the aircraft remains clear of all obstacles. The takeoff safety speed V_{TOSS} is the speed at which the required OEI climb performance can be achieved. After attaining V_{TOSS}, the steady rate of climb must be at least 100 ft/min until 200 ft above the takeoff surface (out of ground effect, at approved engine rating).

The takeoff distance is the horizontal distance from start of the takeoff until the rotorcraft attains and remains at least 35 ft above the takeoff surface, attains and maintains a speed of at least V_{TOSS}, and establishes a positive rate of climb.

Certification is typically obtained for takeoff from runways, ground-level helipads, and elevated helipads. For Category B takeoff, it is only necessary that the landing can be made safely at any point along the flight path if an engine failure occurs; there is no requirement for a capability to continue flight.

8.6 Helicopter Drag

The estimation of helicopter parasite drag is an important aspect of performance calculation because it establishes the propulsive force and power requirement at high speed. The helicopter drag is commonly expressed in terms of the parasite drag area f, such that $D = \frac{1}{2}\rho V^2 f$. Except for compressibility or Reynolds number effects, $f = D/q$ is independent of speed. The variation of the parasite drag with fuselage angle-of-attack is important for accurate performance calculations. The parasite drag area can be calculated from the drag coefficients of the various components of the airframe by

$$f = \sum_n C_{Dn} S_n \qquad (8.20)$$

where S_n is the component wetted area or frontal area, on which C_{Dn} is based.

A major contributor to the helicopter drag is the rotor hub, which typically accounts for 25% to 50% of the total parasite drag area. Blade shank drag can be significant and is usually book-kept with the hub drag. The drag of even a clean helicopter is significantly greater than that of an airplane of similar gross weight, partly because of the large rotor hub drag and partly because of higher fuselage drag. Early helicopter designs in particular tended to have high drag levels. Thus helicopters offer significant potential for drag reduction: main rotor hub fairings, retractable gear, streamlined fuselage, tail rotor hub fairings, and momentum losses. Well-designed hub fairings can reduce the hub drag by 30% to 40%, although they can introduce maintenance issues.

Figure 8.8 shows the parasite drag of various rotorcraft as a function of size (weight or disk area). The helicopter drag can be estimated based on the maximum takeoff weight, $f = D/q = k(W_{MTO}/1000)^{2/3}$. Based on historical data, $k = 9$ for old helicopters, $k = 2.5$ for current low-drag helicopters, $k = 1.6$ for current tiltrotors,

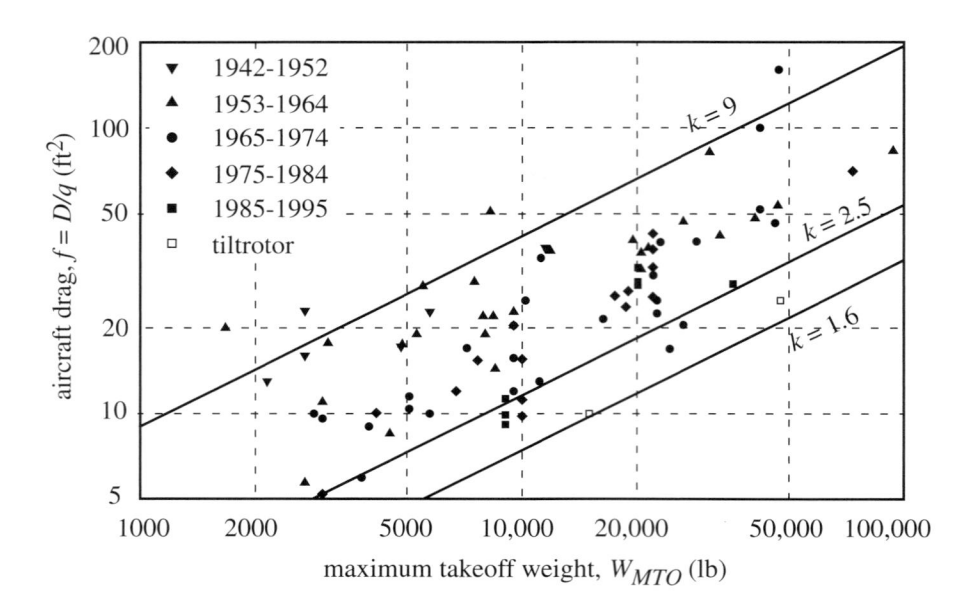

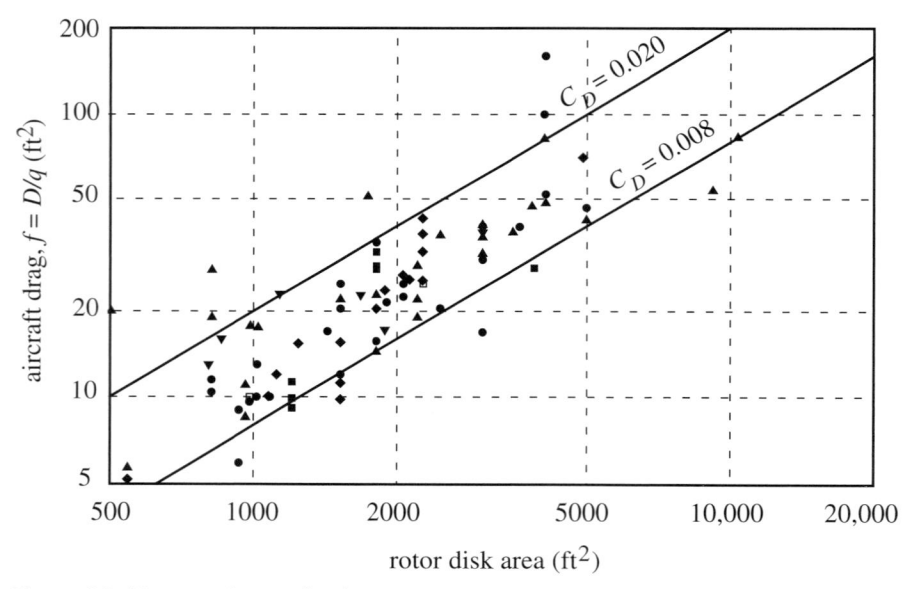

Figure 8.8. Rotorcraft parasite drag.

and $k = 1.4$ for turboprop aircraft (English units, so f in ft^2 and W_{MTO} in lb). Alternatively, the parasite area can scaled with the rotor disk area, $f = A_{\text{rotors}} C_D$, where A_{rotors} is the projected area of the rotors. Based on historical data, the drag coefficient $C_D = 0.02$ for old helicopters and $C_D = 0.008$ for current low-drag helicopters. These trends would imply $A \propto W^{2/3}$, whereas section 1.2 found disk loading scaling as $W/A = 0.15 W^{0.4}$.

Similarly hub drag can be estimated based on the gross weight, using a squared-cubed relationship or a square-root relationship, or it may be scaled in terms of a drag coefficient C_D based on the rotor disk area A (Figure 8.9). Based on historical

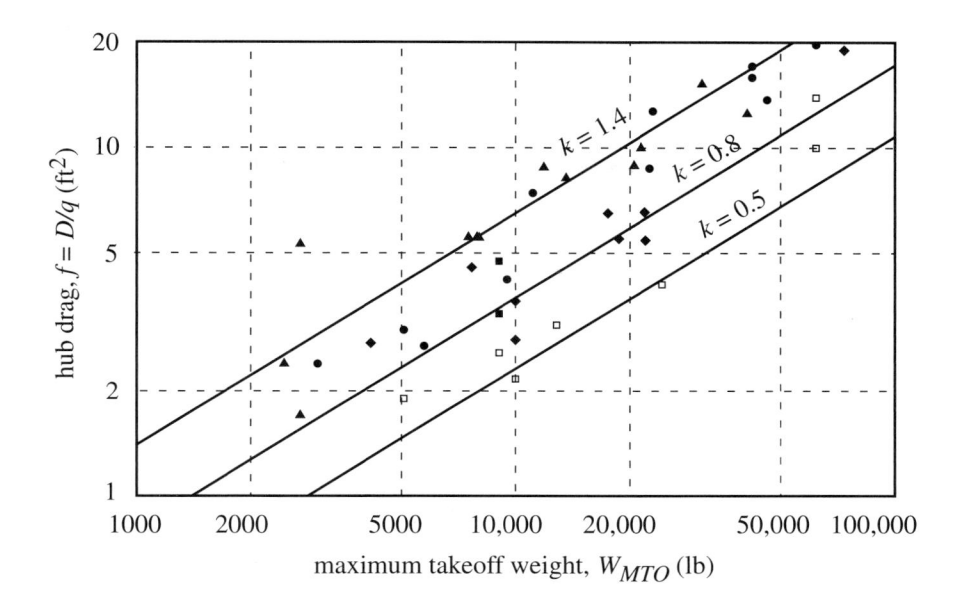

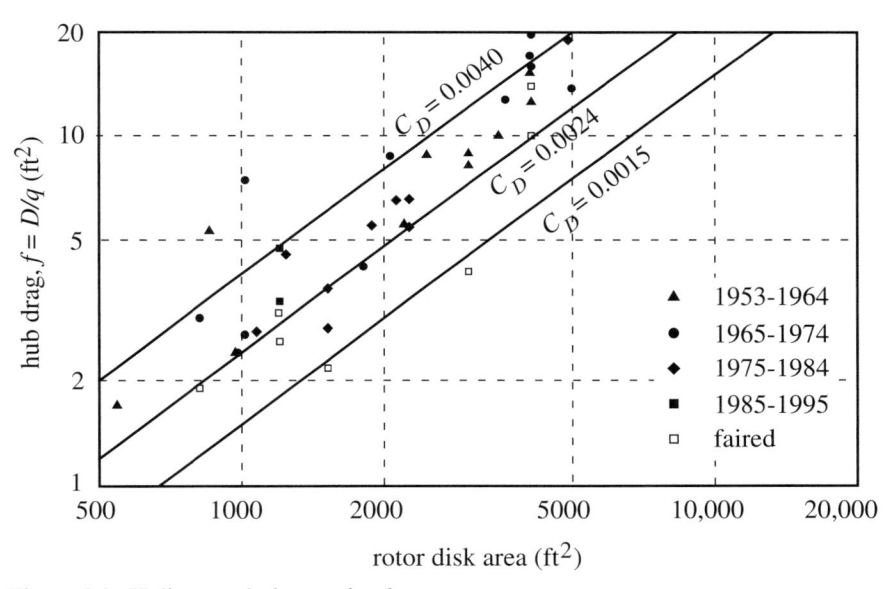

Figure 8.9. Helicopter hub parasite drag.

data, the drag coefficient $C_D = 0.004$ for typical hubs, $C_D = 0.0024$ for current low-drag hubs (30% of the total drag), and $C_D = 0.0015$ for faired hubs (40% reduction). For the squared-cubed relationship, $f_{hub} = k(W_{MTO}/1000)^{2/3}$. Based on historical data, $k = 1.4$ for typical hubs, $k = 0.8$ for current low-drag hubs (30% of the total drag), and $k = 0.5$ for faired hubs (English units). For the square-root relationship, $f_{hub} = k\sqrt{W_{MTO}/N_{rotor}}$ (where W_{MTO}/N_{rotor} is the maximum takeoff gross weight per lifting rotor). Based on historical data (Keys and Rosenstein (1978)), $k = 0.074$ for single rotor helicopters, $k = 0.049$ for tandem rotor helicopters (probably a blade number effect), $k = 0.038$ for hingeless rotors, and $k = 0.027$ for faired hubs (English units).

8.7 Rotor Blade Airfoils

The airfoils for a rotor blade are chosen to give the rotor good aerodynamic efficiency while allowing the structural requirements of the blade to be satisfied. Selecting an airfoil section or designing airfoils specifically for rotor blades is difficult because of the complex aerodynamic environment in which the rotary wing operates. The airfoil used must balance the many constraints imposed by the rotor flow. Although the rotary-wing aerodynamic environment is highly three-dimensional and unsteady, significant improvements in rotor performance and loads can be achieved by considering the two-dimensional, static airfoil characteristics.

The figure of merit is a useful measure of the aerodynamic efficiency of the hovering rotor. The figure of merit can be written

$$M = \frac{\lambda_h C_T}{\kappa \lambda_h C_T + \frac{\sigma c_{do}}{8}} = \frac{1}{\kappa + \frac{3}{4}\frac{c_{do}/\bar{c}_\ell}{\lambda_h}} \qquad (8.21)$$

(equation 3.120 in section 3.6.5). Thus for a fixed disk loading, the key parameter is the blade section drag-to-lift ratio. A high figure of merit requires that the airfoil section have a low drag for moderate to high lift coefficients.

Good stall characteristics are important for any wing, including the rotor blade. The rotor airfoil should have a high maximum lift coefficient, which allows the rotor to be designed to operate at a high C_T/σ value and hence have a low tip speed and blade area. The strictest limitation imposed by stall is on the retreating blade in forward flight, so a high lift coefficient is required at low to moderate Mach numbers. In forward flight, stall occurs periodically as the blade rotates, so really the airfoil must have good unsteady stall characteristics (see Chapter 12). Generally, good static stall characteristics imply good dynamic stall characteristics, so the airfoil selection can reasonably be based on static data.

At high forward speeds, the advancing-tip Mach number is high. The rotor airfoil should therefore have a high critical Mach number for drag divergence and shock formation at the low angles-of-attack characteristic of the advancing side of the disk.

Aerodynamic pitch moments on the blade are transmitted to the control system. A low moment about the airfoil aerodynamic center is required to avoid excessive control system loads, particularly in forward flight, where there is a large periodic variation in angle-of-attack and dynamic pressure. If the control system is entirely mechanical, the aerodynamic pitch moments on the blade are also transmitted to the pilot's cyclic and collective control sticks. A reflexed trailing edge can be added to a cambered airfoil section to reduce the pitch moment.

Figure 8.10 illustrates the principal concerns in selecting or designing an airfoil for a helicopter rotor blade. The rotor blade section operates over a wide range of conditions. Low drag is required at the working conditions of the rotor in hover, namely moderately high angles-of-attack and Mach number. Good stall characteristics, including a high maximum lift coefficient, are required at the low to moderate Mach numbers of the retreating blade in forward flight. A high critical Mach number is required at the low angle-of-attack of the advancing blade in forward flight. The hover criterion is intended to give good lifting performance by the rotor, whereas the forward flight requirements are primarily based on achieving low vibration and loads at high speed. In addition, the airfoil should have a small pitching moment. In general, the stall and compressibility requirements (high maximum lift coefficient at

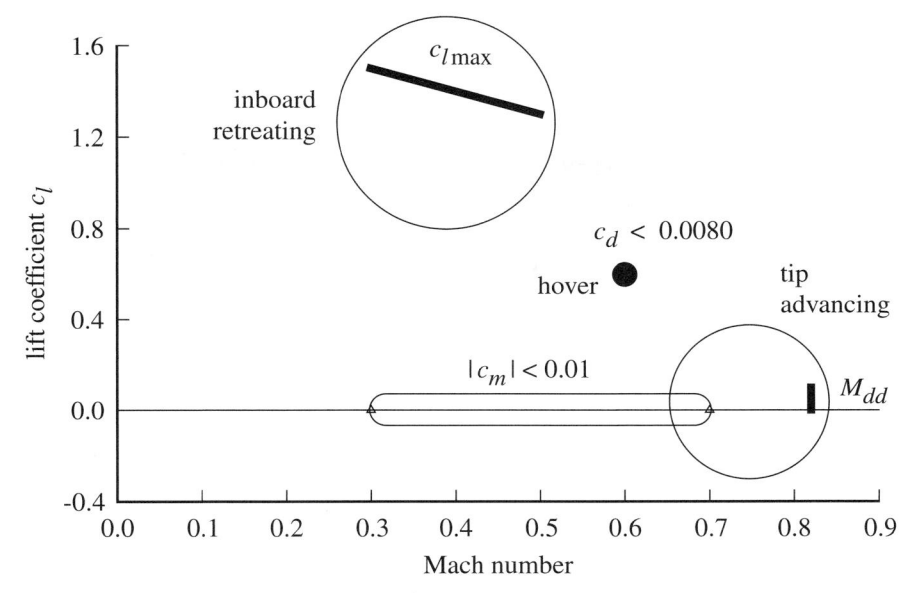

Figure 8.10. Rotor blade airfoil criteria.

moderate Mach numbers and high critical Mach number at low lift) demand significant compromise if met with a single contour. The best approach is to use different airfoils at the tip (where compressibility effects dominate) and at midspan (where stall effects dominate).

A symmetrical, moderately thick airfoil section was frequently the choice for rotor blades in early designs, with the same section over the entire span for simplicity of construction. A symmetrical section assures a zero pitching moment. The thickness ratio (typically 10% to 15%) is a compromise between the thin section desired because of compressibility effects and the thick section desired for structural efficiency. Fortunately, extremely thick sections are required only at the blade root. The NACA 0012 airfoil was a frequent selection for past rotor designs and has come to be considered the standard or reference rotor airfoil. With improved aerodynamic, structural, and manufacturing technology, more sophisticated blade designs are used for current helicopters. A number of airfoils have been developed with characteristics optimized for the rotary-wing environment, and using thinner sections at the blade tip is common.

As a guide in the evaluation and selection of a rotor blade airfoil, both the section operating conditions and the airfoil characteristics can be plotted as a function of angle-of-attack and Mach number (Figure 8.11). The airfoil characteristics shown as a function of Mach number for a hypothetical airfoil are the angles-of-attack for maximum lift coefficient ($\alpha_{\max}$) and for drag divergence or supercritical flow (α_{crit}). Also plotted is the operating condition for several radial stations as the blade moves around the azimuth (a closed curve is generated in forward flight, converging to a single angle-of-attack and Mach number for hover). The tip sections of the blade show the highest Mach numbers, whereas sections somewhat inboard (such as at 75% radius) show the largest angle-of-attack. Thus the requirements dictated by the aerodynamic environment of the blade vary with the radial station. The requirements for a given rotor operating state can be compared with the stall and compressibility characteristics of a particular airfoil by a plot such as shown in Figure 8.11. This

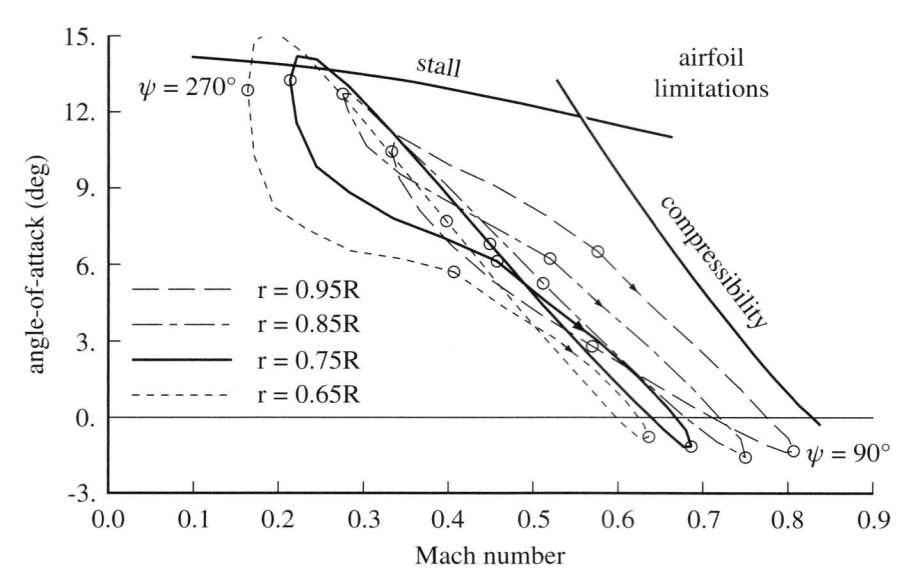

Figure 8.11. Rotor airfoil section operating conditions.

plot can also be used to graphically compare the characteristics of different airfoil sections; an improved airfoil should show increased angle-of-attack limits over the entire Mach number range.

Figure 8.12 shows the maximum lift coefficient ($c_{\ell\max}$ at $M = 0.4$) and drag-divergence Mach number (M_{dd}) for rotor blade airfoil sections. The drag-divergence Mach number is typically defined as where $\partial c_d/\partial M = 0.1$. Families of airfoil sections have been designed specifically for rotor blades, with thin sections for the tip and thicker sections for inboard stations. The symmetric NACA sections (NACA 00xx) found use for their zero pitching moment; the cambered sections (NACA 230xx)

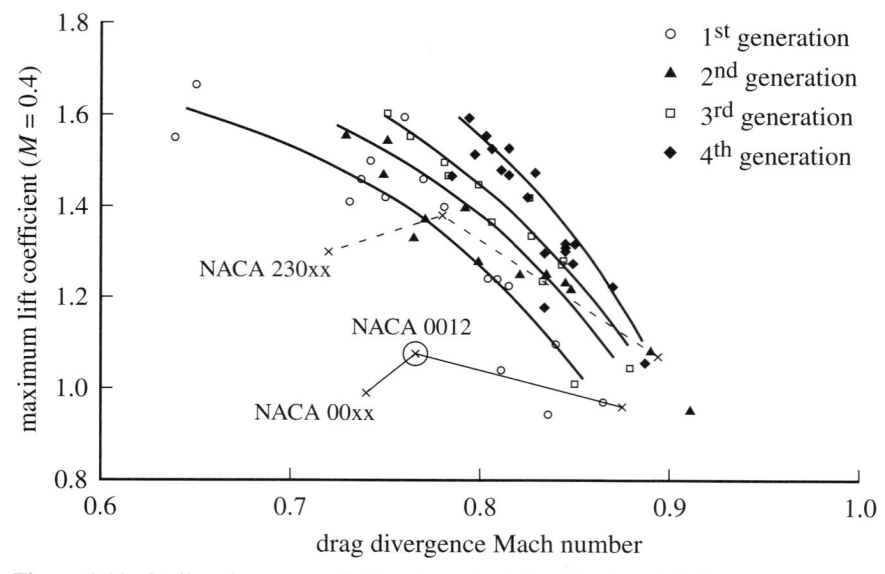

Figure 8.12. Stall and compressibility characteristics of rotor airfoils.

8.7 Rotor Blade Airfoils

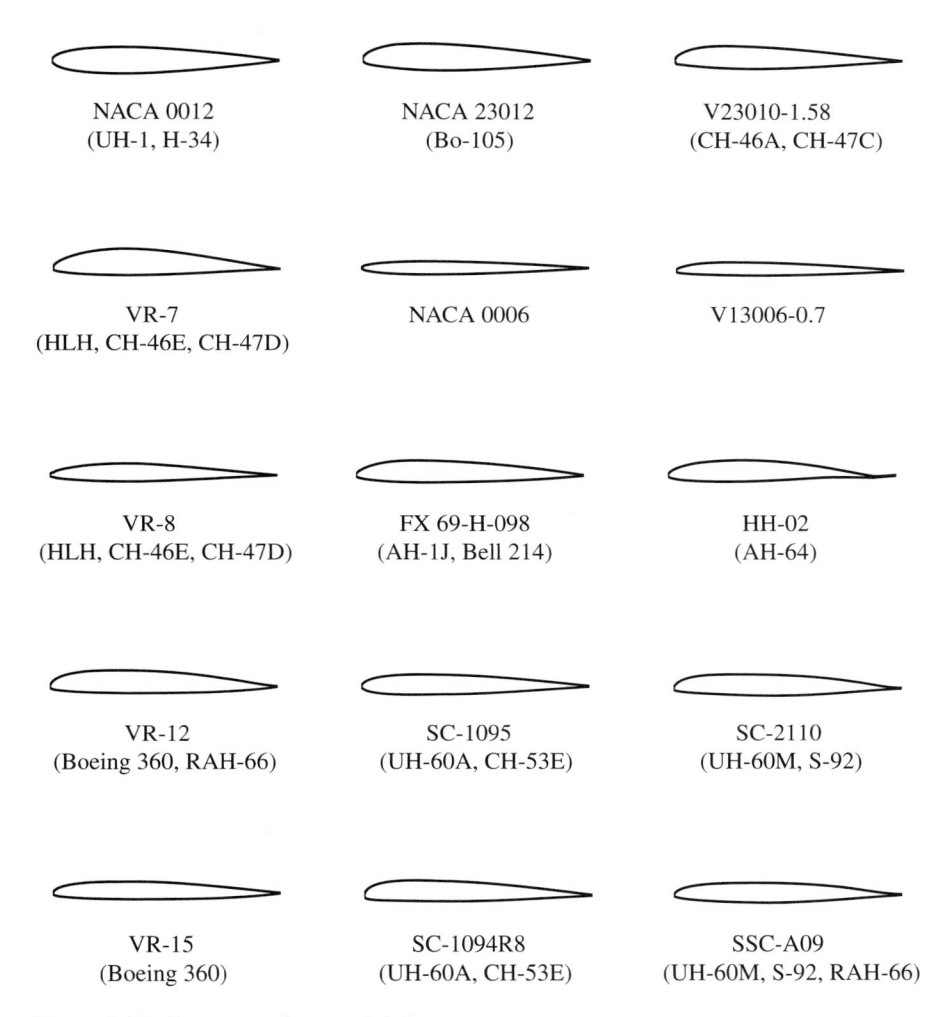

Figure 8.13. Contours of rotor airfoils.

- NACA 0012 (UH-1, H-34)
- NACA 23012 (Bo-105)
- V23010-1.58 (CH-46A, CH-47C)
- VR-7 (HLH, CH-46E, CH-47D)
- NACA 0006
- V13006-0.7
- VR-8 (HLH, CH-46E, CH-47D)
- FX 69-H-098 (AH-1J, Bell 214)
- HH-02 (AH-64)
- VR-12 (Boeing 360, RAH-66)
- SC-1095 (UH-60A, CH-53E)
- SC-2110 (UH-60M, S-92)
- VR-15 (Boeing 360)
- SC-1094R8 (UH-60A, CH-53E)
- SSC-A09 (UH-60M, S-92, RAH-66)

have much higher maximum lift coefficients. The generations of rotor airfoil designs shown cover the 1960s to the 1990s. Figure 8.13 shows the contours of a number of these airfoils.

As airfoil analysis and design methodology improves, more detailed criteria have been developed, tailored to specific rotorcraft and specific design conditions. For helicopter rotor blade airfoil design criteria see Gustafson (1949), Davenport and Front (1966), Benson, Dadone, Gormont, and Kohler (1973), Dadone and Fukushima (1975), Dadone (1978a, 1978b), Thibert and Gallot (1981), Thibert and Philippe (1982), Horstmann, Koster, and Polz (1982, 1984), McCroskey (1987), and Bousman (2002, 2003). Narramore (1987) gives design criteria for tiltrotor airfoils. Typical design criteria for rotor airfoils are as follows:

> For good hover performance, need low drag at the lift coefficient and Mach number of hover: $c_d \leq 0.0080$ at $c_\ell = 0.6$ and $M = 0.6$ ($c_\ell/c_d \geq 75$). Laminar flow should be extended, especially on the lower side, and trailing-edge contours causing premature trailing-edge separation should be avoided.

For stall of the retreating blade, need high maximum lift at moderate Mach number: $c_{\ell\max} > 1.5$ at $M = 0.4$ to 0.5. Good $c_{\ell\max}$ at $M = 0.5$ requires a compressible design method. Maximum lift depends on the leading edge contour and camber. The buildup of velocity at the leading edge must be balanced with the onset of trailing-edge separation, which is complicated by the requirement for low c_m.

For compressibility effects on the advancing tip, need low drag at high Mach number: $M_{dd} \geq 0.8$ at $c_\ell = 0$ to 0.2 inboard, $M_{dd} \geq 0.84$ to 0.9 at the tip. High M_{dd} requires a relatively flat upper and lower surface.

To minimize loads in forward flight, especially control loads, need small pitch moment about the aerodynamic center: $|c_m| \leq 0.010$ at $M = 0.3$ and $c_\ell = 0$. Small c_{m0} with little lift and drag penalties can be obtained by shaping the trailing-edge contour, particularly reflex of the mean line at the trailing-edge.

For constant c_{m0}, camber and thickness changes can improve $c_{\ell\max}$ or M_{dd}, but not both. A camber increase produces an increase in $c_{\ell\max}$ but a decrease in M_{dd}. A thickness decrease improves M_{dd}: M_{dd} increases by about 0.012 for a 1% reduction in thickness ratio. A thin section, $t/c = 6\%$ to 8%, is required for $M_{dd} > 0.85$. For reductions in airfoil thickness below 11% there is a substantial deterioration of $c_{\ell\max}$. Rotor airfoils can have a trailing-edge tab (perhaps for manufacturability), with reflex to reduce c_{m0}. Tab upward deflection reduces $c_{\ell\max}$ as well: approximately $dc_{\ell\max}/d\delta = 0.02$ and $dc_m/d\delta = -0.006$, where δ is the tab angle in degrees (Bousman (2002)). The tab deflection has little effect on drag for $\delta < 3°$.

Airfoil aerodynamic characteristics are used in rotorcraft analyses in the form of tables of lift, drag, and moment coefficients as a function of angle-of-attack and Mach number. An airfoil table format was introduced by Davis, Bennett, and Blankenship (1974) for the Rotorcraft Flight Simulation Program C81. The format consists of a square (angle-of-attack and Mach number) array of data for the coefficients, in a fixed form designed for an IBM card: 10 columns, each 7 characters wide. The C81 airfoil deck remains a useful standard for communicating airfoil table data between organizations and between codes.

8.8 Rotor Blade Profile Drag

The calculation of rotor performance requires a knowledge of the blade section profile drag coefficient, including its dependence on angle-of-attack and Mach number. There are other factors that influence the drag coefficient in the three-dimensional, unsteady aerodynamic environment of the rotor blade in forward flight. In particular, it can be necessary to account for the radial flow, the time-varying angle-of-attack, and three-dimensional flow effects at the tip. Roughness and blade construction quality also influence the section drag, often increasing the drag coefficient by 20% to 50% compared to its value for smooth, ideally shaped airfoils. The general practice in numerical work is to rely on tabular data for c_ℓ, c_d, and c_m as a function of α and M for the particular profile used, with semi-empirical corrections to account for the other factors that are considered important. The measured aerodynamic characteristics can be sensitive to small variations in the airfoil or test facility, leading to different properties for airfoils that are nominally identical. Often obtaining a complete and reliable set of even static, two-dimensional airfoil data is difficult.

The simplest rotor analyses can use a mean profile drag coefficient to represent the overall effects of the blade drag on the rotor. The mean drag coefficient can be evaluated using the mean lift coefficient of the rotor, and the Mach number and Reynolds number at some representative radial station (say 75% radius). The use of a mean drag coefficient greatly simplifies the analysis; it has been frequently used in the preceding chapters to obtain elementary expressions for the rotor profile power. Such an analysis is sufficiently accurate for some purposes, such as preliminary design, or when detailed aerodynamic characteristics for the blade section are not available. A mean drag coefficient is not appropriate when localized aerodynamic phenomena, such as stall and compressibility effects in forward flight, are important. Additional corrections or more detailed analyses are required for rotors at extreme operating conditions.

Helicopter performance analyses can use a drag polar of the form $c_d = \delta_0 + \delta_1\alpha + \delta_2\alpha^2$ (see sections 3.5.2.3 and 6.24.2). This is a better representation than a mean value and is simple enough to allow analytical treatment. The constants δ_0, δ_1, and δ_2 depend on the airfoil section. Hoerner (1965) suggested the following procedure for estimating the profile drag polar. A basic skin friction coefficient is obtained for the appropriate section Reynolds number. As an example, for a turbulent boundary layer in the Reynolds number range $10^6 < Re < 10^8$, Hoerner suggested

$$c_f = 0.44 Re^{-1/6} \tag{8.22}$$

The minimum profile drag coefficient for the airfoil is then twice c_f, multiplied by a factor accounting for the airfoil thickness. For the NACA 4-digit or 5-digit airfoil series the result is

$$c_{d\min} = 2c_f\big(1 + 2(t/c) + 60(t/c)^4\big) \tag{8.23}$$

where t/c is the section thickness ratio. The term $2(t/c)$ accounts for the velocity increase due to thickness, and the term $60(t/c)^4$ is due to the pressure drag. Hoerner then gave the effect of lift on the profile drag as

$$c_d \cong c_{d\min}\big(1 + c_\ell^2\big) \tag{8.24}$$

which completes the construction of the drag polar.

Bailey (1941) developed a procedure for identifying the constants in the drag polar $c_d = \delta_0 + \delta_1\alpha + \delta_2\alpha^2$, given the basic section characteristics. Using this method, the polar $c_d = 0.0087 - 0.0216\alpha + 0.400\alpha^2$ was obtained for an NACA 23012 airfoil at $Re = 2 \times 10^6$. This particular result is quoted and used often in the helicopter literature. Bailey started by writing the profile drag as $c_d = c_{d\min} + \Delta c_d$ and assuming that the minimum drag depends on the Reynolds number and Δc_d depends on the angle-of-attack. For many airfoil sections Δc_d is approximately a unique function of

$$\ell = \frac{c_\ell - c_{\ell\text{opt}}}{c_{\ell\max} - c_{\ell\text{opt}}} \tag{8.25}$$

where $c_{\ell\max}$ is the maximum lift coefficient of the airfoil and $c_{\ell\text{opt}}$ is the lift coefficient at minimum drag. Bailey wrote the profile drag function as

$$c_d = K_0 + K_1\ell + K_2\ell^2 = 0.0003 - 0.0025\ell + 0.0229\ell^2 \tag{8.26}$$

with the constants determined by matching the function to the empirical curve at $\ell = 0.125, 0.4$, and 0.675. This expression is a good approximation to about $\ell = 0.8$. At higher lift the stall effects are large, and the drag is significantly underestimated by this expression. Using $c_\ell = a\alpha$ (the zero-lift angle is ignored), the constants in the drag polar $c_d = \delta_0 + \delta_1\alpha + \delta_2\alpha^2$ can then be evaluated:

$$\delta_0 = c_{d\min} + K_0 - \frac{K_1 c_{\ell\text{opt}}}{c_{\ell\max} - c_{\ell\text{opt}}} + \frac{K_2 c_{\ell\text{opt}}^2}{\left(c_{\ell\max} - c_{\ell\text{opt}}\right)^2} \qquad (8.27)$$

$$\delta_1 = \frac{aK_1}{c_{\ell\max} - c_{\ell\text{opt}}} - \frac{2aK_2 c_{\ell\text{opt}}}{\left(c_{\ell\max} - c_{\ell\text{opt}}\right)^2} \qquad (8.28)$$

$$\delta_2 = \frac{a^2 K_2}{\left(c_{\ell\max} - c_{\ell\text{opt}}\right)^2} \qquad (8.29)$$

Thus, given $c_{d\min}$, $c_{\ell\max}$, $c_{\ell\text{opt}}$, and $a = c_{\ell\alpha}$ at the required Reynolds number, the profile drag polar can be constructed. Bailey's expression for Δc_d does not reduce to zero at $c_{\ell\text{opt}}$; rather $\Delta c_d = 0.0003$ there; the minimum occurs at $\ell = 0.055$, where $\Delta c_d = 0.0002$. Alternative constants that are nearly as accurate and give a minimum $\Delta c_d = 0$ at $c_\ell = c_{\ell\text{opt}}$ are $K_0 = K_1 = 0$ and $K_2 = 0.0200$. Bailey's expression is more accurate in the working range of blade angle-of-attack. The limit $\ell = 0.8$ gives

$$\alpha < \alpha_{\text{limit}} = \frac{1}{a}\left(0.8 c_{\ell\max} + 0.2 c_{\ell\text{opt}}\right) \qquad (8.30)$$

This limit is the drag rise due to stall at high angle-of-attack.

As an example, Bailey considered the NACA 23012 airfoil at $Re = 2 \times 10^6$. For this airfoil, $c_{d\min} = 0.0066$, $c_{\ell\max} = 1.45$, $c_{\ell\text{opt}} = 0.08$, and $a = 5.73$. The minimum drag was increased by 25% to account for roughness, so $c_{d\min} = 0.0082$. The resulting polar is $c_d = 0.0087 - 0.0216\alpha + 0.400\alpha^2$. The limit on the accuracy of this expression is $\alpha < 11.8°$.

Consider the NACA 0012 airfoil at $Re = 2 \times 10^6$. From $c_{d\min} = 0.0065$ (increased to $c_{d\min} = 0.0081$ for roughness), $c_{\ell\max} = 1.08$, $c_{\ell\text{opt}} = 0$, and $a = 5.73$; the drag polar is $c_d = 0.0084 - 0.0133\alpha + 0.645\alpha^2$. The limit on the accuracy of this expression is $\alpha < 8.6°$.

Airfoil aerodynamic characteristics (c_ℓ, c_d, and c_m) depend on the section Reynolds number: $Re = \rho Vc/\mu$, where ρ is the air density, μ the viscosity, V the velocity, and c the chord. The principal effects of Reynolds number are a decrease in drag and an increase in maximum lift as Re increases. Thus a small-scale model has worse performance than the full-scale rotor, with larger profile power and stall at lower thrust. The Reynolds number can be written in terms of Mach number instead of speed: $Re = M\rho(c_s/\mu)c$. So in a wind-tunnel test of an airfoil, the Reynolds number is proportional to the Mach number if the temperature and pressure remain constant during the test. Tabular airfoil characteristics constructed by analysis should also have Re proportional to M. In applications of an airfoil table to a specific rotor, it is often necessary to account for a difference in chord (scale) or a difference in atmospheric conditions (particularly altitude). A simple correction for Reynolds number is

$$c_d(\alpha) = \frac{1}{K} c_{d_{\text{table}}}(\alpha) \qquad (8.31)$$

$$c_\ell(\alpha) = K c_{\ell_{\text{table}}}(\alpha/K) \qquad (8.32)$$

where $K = (Re/Re_{\text{table}})^n$ is a function of the Reynolds number ratio. Experimental data for airfoils in turbulent flow suggest $n = \frac{1}{8}$ to $\frac{1}{5}$ for the exponent; see Yamauchi and Johnson (1983); $n = \frac{1}{5}$ is the 1/5-th power law for a turbulent flat plate boundary layer. For analysis of a small-scale model using full-scale airfoil data ($Re < Re_{\text{table}}$), $K < 1$ and the correction increases the drag and decreases the maximum lift. For example, a 20% scale model has $K = 0.72$ to 0.82.

8.9 REFERENCES

Bailey, F.J., Jr. "A Simplified Theoretical Method of Determining the Characteristics of a Lifting Rotor in Forward Flight." NACA Report 716, 1941.

Benson, G.R., Dadone, L.U., Gormont, R.E., and Kohler, G.R. "Influence of Airfoils on Stall Flutter Boundaries of Articulated Helicopter Rotors." Journal of the American Helicopter Society, *18*:1 (January 1973).

Bousman, W.G. "Airfoil Design and Rotorcraft Performance." American Helicopter Society 58th Annual Forum, Montreal, Canada, June 2002.

Bousman, W.G. "Aerodynamic Characteristics of SC1095 and SC1094 R8 Airfoils." NASA TP 2003-212265, December 2003.

Carlson, E.B. "An Analytical Methodology for Category A Performance Prediction and Extrapolation." American Helicopter Society 57th Annual Forum, Washington, DC, May 2001.

Carlson, E.B., Zhao, Y.J., and Chen, R.T.N. "Optimal Trajectories for Tiltrotor Aircraft in Total Power Failure." American Helicopter Society 54th Annual Forum, Washington, DC, May 1998.

Dadone, L.U. "Design and Analytical Study of a Rotor Airfoil." NASA CR 2988, May 1978a.

Dadone, L. "Rotor Airfoil Optimization: An Understanding of the Physical Limits." American Helicopter Society 34th Annual National Forum, Washington, DC, May 1978b.

Dadone, L.U., and Fukushima, T. "A Review of Design Objectives for Advanced Helicopter Rotor Airfoils." American Helicopter Society 31st Annual National Forum, Washington, DC, May 1975.

Davenport, F.J., and Front, J.V. "Airfoil Sections for Helicopter Rotors – A Reconsideration." American Helicopter Society 22nd Annual National Forum, Washington, DC, May 1966.

Davis, J.M., Bennett, R.L., and Blankenship, B.L. "Rotorcraft Flight Simulation with Aeroelastic Rotor and Improved Aerodynamic Representation." USAAMRDL TR 74-10, June 1974.

Fradenburgh, E.A. "A Simple Autorotative Flare Index." Journal of the American Helicopter Society, *29*:3 (July 1984).

Gustafson, F.B. "The Application of Airfoil Studies to Helicopter Rotor Design." NACA TN 1812, February 1949.

Hansen, K.C. "Handling Qualities Design and Development of the CH-53E, UH-60A, and S-76." Royal Aeronautical Society International Conference on Helicopter Handling Qualities and Control, London, UK, November 1988.

Hislop, G.S. "The Fairey Rotodyne." Journal of the Helicopter Association of Great Britain, 13:1 (February 1959).

Hoerner, S.F. *Fluid-Dynamic Drag*. New Jersey: Published by the Author, 1965.

Horstmann, K.H., Koster, H., and Polz, G. "Development of New Airfoil Sections for Helicopter Rotor Blades." Eighth European Rotorcraft Forum, Aix-en-Provence, France, September 1982.

Horstmann, K.H., Koster, H., and Polz, G. "Improvement of Two Blade Sections for Helicopter Rotors." Tenth European Rotorcraft Forum, The Hague, The Netherlands, August 1984.

Johnson, W. "Helicopter Optimal Descent and Landing After Power Loss." NASA TM 73244, May 1977.

Keys, C.N., and Rosenstein, H.J. "Summary of Rotor Hub Drag Data." NASA CR 152080, March 1978.

Lee, A.Y. "Optimal Autorotational Descent of a Helicopter with Control and State Inequality Constraints." Journal of Guidance, Control, and Dynamics, 13:5 (September-October 1990).

Lee, A.Y., Bryson, A.E., Jr., and Hindson, W.S. "Optimal Landing of a Helicopter in Autorotation." Journal of Guidance, Control, and Dynamics, 11:1 (January-February 1988).

McCormick, B.W., Jr. "On the Initial Vertical Descent of a Helicopter Following Power Failure." Journal of the Aeronautical Sciences, 23:12 (December 1956).

McCroskey, W.J. "A Critical Assessment of Wind Tunnel Results for the NACA 0012 Airfoil." NASA TM 100019, October 1987.

Narramore, J.C. "Airfoil Design, Test, and Evaluation for the V-22 Tilt Rotor Vehicle." American Helicopter Society 43rd Annual Forum, St. Louis, MO, May 1987.

Ruddell, A.J. "Advancing Blade Concept (ABC) Development." Journal of the American Helicopter Society, 22:1 (January 1977).

Thibert, J.-J., and Gallot, J. "Advanced Research on Helicopter Blade Airfoils." Vertica, 5:3 (1981).

Thibert, J.J., and Philippe, J.J. "Studies of Aerofoils and Blade Tips for Helicopters." La Recherche Aerospatiale, 1982:4 (1982).

Wernicke, K.G., and Magee, J.P. "XV-15 Flight Test Results Compared with Design Goals." AIAA Paper No. 79-1839, August 1979.

White, G.T., Logan, A.H., and Graves, J.D. "An Evaluation of Helicopter Autorotation Assist Concepts." American Helicopter Society 38th Annual Forum, Anaheim, CA, May 1982.

Wood, T.L. "High Energy Rotor System." American Helicopter Society 32nd Annual National V/STOL Forum, Washington, DC, May 1976.

Yamauchi, G.K., and Johnson, W. "Trends of Reynolds Number Effects on Two Dimensional Airfoil Characteristics for Helicopter Rotor Analyses." NASA TM 84363, April 1983.

Zhao, Y., Jhemi, A.A., and Chen, R.T.N. "Optimal Vertical Takeoff and Landing Helicopter Operation in One Engine Failure." Journal of Aircraft, 33:2 (March-April 1996).

9 Wings and Wakes

9.1 Rotor Vortex Wake

The behavior of the airloading of a helicopter rotor blade in forward flight is illustrated in Figure 9.1. At low speed, there is impulsive loading on the blade as it flies over the tip vortices from the preceding blade (see Figure 5.8). This phenomenon is called blade-vortex interaction (BVI). There is a down-then-up pulse on the advancing side and an up-then-down pulse on the retreating side, the magnitude of the pulse depending on the extent of the rollup process and the proximity of the distorted tip vortices to the tip-path plane. At high speed, although blade-vortex interaction loading can still be observed, negative loading on the advancing tip is common, a consequence of flap moment balance with stall-limited loads on the retreating side. With a flapping rotor the pitch and roll moments on the hub must be small. In forward flight, the lift capability on the retreating side is limited by the combination of low dynamic pressure and stall of the airfoil sections. The lift on the advancing side must then also be small to maintain roll balance, and at sufficiently high speed the advancing-tip lift can become negative.

Associated with the lift of a rotor blade is a bound circulation. On the three-dimensional blade, conservation of vorticity requires that the bound circulation be trailed into the wake from the blade tip and root (Figure 9.2). Vorticity is also left in the rotor wake as a consequence of radial and azimuthal changes in the bound circulation. The trailed vorticity γ_t, which is generated by the radial variation of the bound circulation, is parallel to the local free stream at the instant the vorticity leaves the blade. The shed vorticity γ_s, generated by the azimuthal variation of the bound circulation, is oriented radially in the wake. The strength of the rotor trailed and shed vorticity is

$$\gamma_t = \left.\frac{\partial \Gamma}{\partial r}\right|_{\psi-\phi} \tag{9.1}$$

$$\gamma_s = -\frac{1}{u_T}\left.\frac{\partial \Gamma}{\partial \psi}\right|_{\psi-\phi} \tag{9.2}$$

where the derivatives of the bound circulation are evaluated at the time the wake element left the blade; namely, the current blade azimuth angle ψ less the wake age ϕ (in dimensionless time). The existence of shed vorticity implies that the trailed

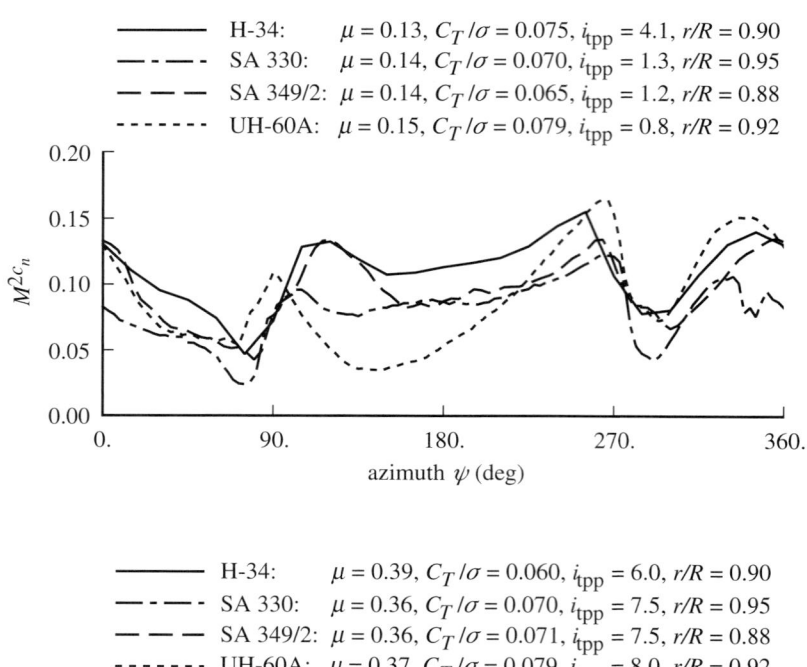

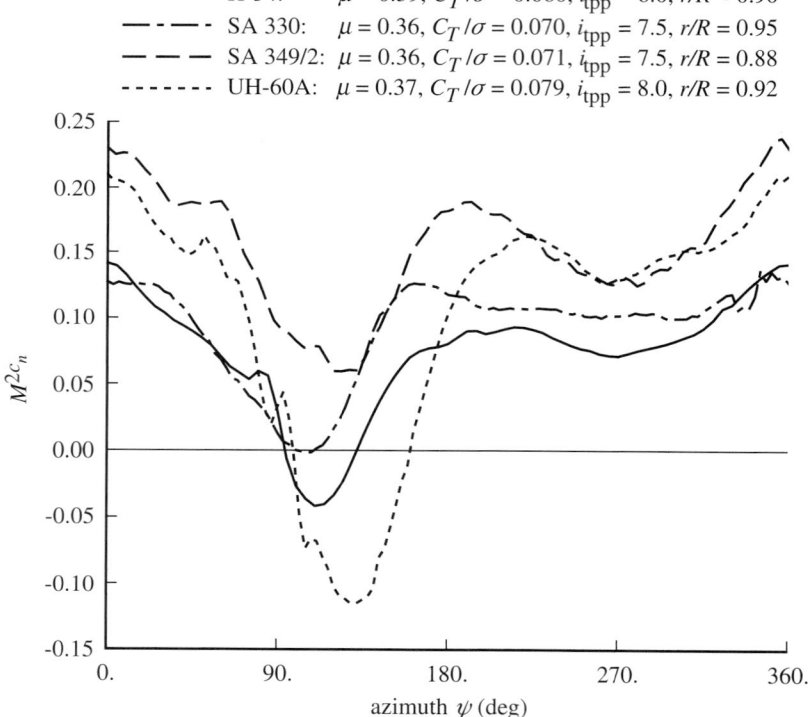

Figure 9.1. Characteristic rotor blade loading in forward flight (from flight and wind-tunnel measurements of blade pressures); $M^2 c_n = \text{section normal force}/(\frac{1}{2}\rho c_s^2 c)$.

vorticity strength varies along the length of the vortex filaments. Similarly, the trailed vorticity implies that the shed vorticity strength varies radially. Specifically,

$$-\frac{\partial}{\partial r} u_T \gamma_s = \frac{\partial}{\partial \psi} \gamma_t \tag{9.3}$$

9.1 Rotor Vortex Wake

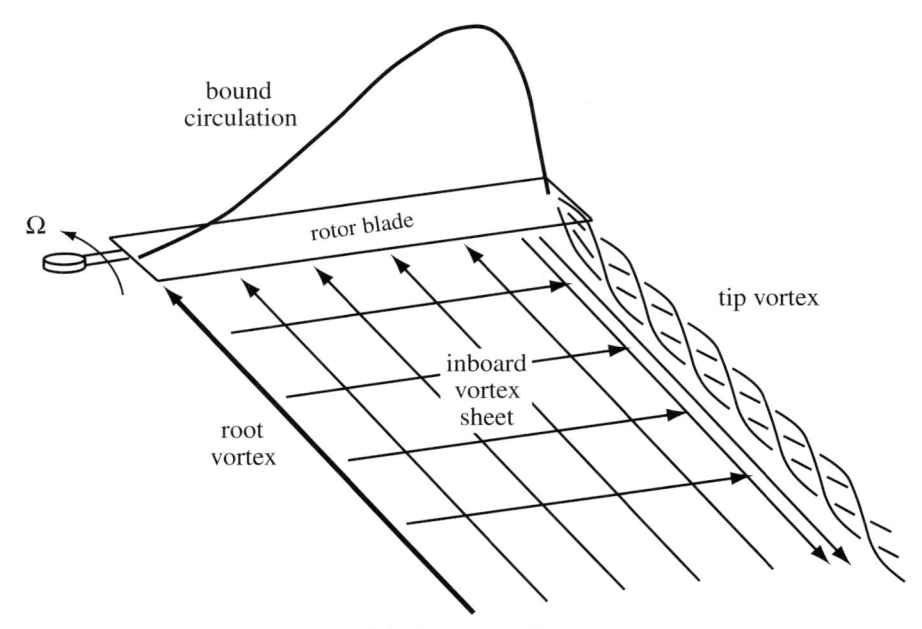

Figure 9.2. Trailed and shed vorticity in rotor wake.

Because of the rotation of the blade, the lift and circulation are concentrated at the tip, as sketched in Figure 9.3. Although the circulation drops to zero at the tip over a finite distance, the rate of decrease is still very high. The result is a large trailing vorticity strength at the outer edge of the wake, causing the vortex sheet to quickly roll up into a concentrated tip vortex. The formation of this tip vortex is also influenced by the blade tip geometry. The rolled-up tip vortex quickly reaches a strength nearly equal to the maximum bound circulation of the blade. The vorticity is distributed over a small but finite region because of the viscosity of the fluid. The vortex core radius is defined at the maximum tangential velocity. The tip vortex has a small core radius that depends on the blade geometry and loading.

On the inboard portion of the blade, the bound circulation drops off gradually to zero at the root. Hence there is an inboard sheet of trailed vorticity in the wake, with opposite sign to the tip vortex. Since the gradient of the bound circulation is low, the root vortex is generally much weaker and more diffuse than the tip vortex. When the bound circulation varies azimuthally, either as a result of the periodic loading in forward flight or because of transient loads, there is also an inboard sheet of shed vorticity in the wake.

The trailed and shed vorticity of the rotor wake is deposited in the flow field as the blades rotate, and then it is convected with the local velocity in the fluid. This

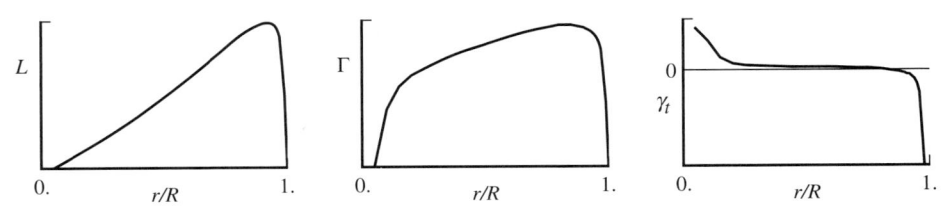

Figure 9.3. Radial distribution of blade lift L, bound circulation Γ, and trailed vorticity γ_t.

Figure 9.4. Tip vortices of hovering helicopters.

local velocity consists of the free stream velocity and the wake self-induced velocity. The wake is transported downward, normal to the disk plane, by a combination of the mean wake-induced velocity and the free stream velocity. The component of the free stream velocity that is normal to the disk is a result of tilt of the disk plane in forward flight or axial velocity in vertical flight. The wake is transported aft of the rotor disk by the in-plane component of the free stream velocity. The self-induced velocity of the wake produces substantial distortion of the vortex filaments as they are convected with the local flow. Thus the wake geometry consists of distorted interlocking helices, one behind each blade, skewed aft in forward flight.

The strong concentrated tip vortices are by far the dominant feature of the rotor wake (Figure 9.4). When the atmospheric conditions are just right and condensation makes the rotor wake visible, the only vortex structures observed are the tip vortices. Because of its rotation, a rotor blade encounters the tip vortex from the preceding blade in both hover and forward flight. A vortex passing close to a blade induces a large velocity and hence a large change in loading on the wing. Such vortex-induced loading is a principal source of the rotor higher harmonic airloading. Thus the rotor wake plays a key role in most problems of helicopters: high vibration, excessive noise, large oscillatory blade and hub loads, poor performance, and adverse aerodynamic interference.

9.2 Lifting-Line Theory

The principal method for calculating rotary-wing airloads has long been lifting-line theory: a wing model to obtain the loading from the section aerodynamic environment and a wake model to evaluate the induced velocity at the wing, from a wake geometry that includes the self-induced distortion. A rotor analysis must treat viscous and compressible flow effects, which are present to some degree in almost all helicopter operating conditions. This is accomplished in lifting-line theory by using experimental data for the two-dimensional airfoil characteristics. Lifting-line theory allows the combination of three-dimensional influences (the wake) with two-dimensional solutions that include stall and compressibility effects. The rotation of the wing introduces a number of features that require special attention, notably the returning vortex wake, the time-varying free stream and radial flow, and a fundamentally transcendental geometry of the wake that necessitates either approximate or numerical solutions.

A three-dimensional wing is characterized by a drag due to lift, called the induced drag, which arises because of the energy convected downstream in the vortex wake of the wing. For high-aspect-ratio wings, the induced drag can be associated with the induced velocity at the wing. The aerodynamic solution for the three-dimensional wing includes the induced drag (the rotor torque for the rotating wing) in the total drag. In addition, the concepts of lifting-line theory allow the induced power to be independently evaluated from the product of the wing loading and induced velocity.

Lifting-line theory assumes that the wing has a high aspect ratio or, more generally, that spanwise variations of the aerodynamic environment are small. This assumption allows the problem to be split into separate wing and wake models, which are solved individually and combined. Classical lifting-line theory was developed by Prandtl (1921). The classical theory treats the case of a high-aspect-ratio, planar, fixed wing in steady flow. In the linearized model both the wing and wake are represented by thin planar sheets of vorticity.

The assumption of a high aspect ratio splits the three-dimensional wing aerodynamic problem (unsteady, compressible, and viscous) into two parts: a wing model and a wake model (Figure 9.5). In the context of perturbation theory, these are called inner and outer problems. Combining the two parts of the lifting-line theory gives the solution for the spanwise loading of the three-dimensional wing. The outer problem is the wake, consisting of trailed and shed vorticity behind the wing, which is just a bound vortex line. The inner problem is a two-dimensional airfoil or, more generally, an infinite wing in a uniform, yawed free stream. The influence of the wake and the rest of the wing is represented entirely by an induced downwash at the section (an induced angle-of-attack change). Two-dimensional airfoil theory or experimental section characteristics are used to obtain the blade section aerodynamic loads (lift, drag, and pitching moment). The wing and wake problems are connected through the wake-induced velocity and the bound circulation. The outer problem calculates the induced velocity at the wing, from a wake with strength determined by the bound circulation. The induced velocity is not needed at an arbitrary point, just at the lifting line. The inner problem calculates the bound circulation from the aerodynamic environment, with the wake-induced velocity included in the free stream. The pressure on the wing is not needed, just the bound circulation (as well as the section lift, drag, and moment in order to calculate performance and couple with the structural dynamics). Compared to the full aerodynamic solution, the assumption of a high aspect ratio in lifting-line theory leads to important simplifications. The inner problem has simpler geometry (two-dimensional) but complex flow (Navier-Stokes equations). In the outer limit the inner solution can be considered irrotational. The outer problem has complex geometry (the vortex wake) but irrotational flow. In the

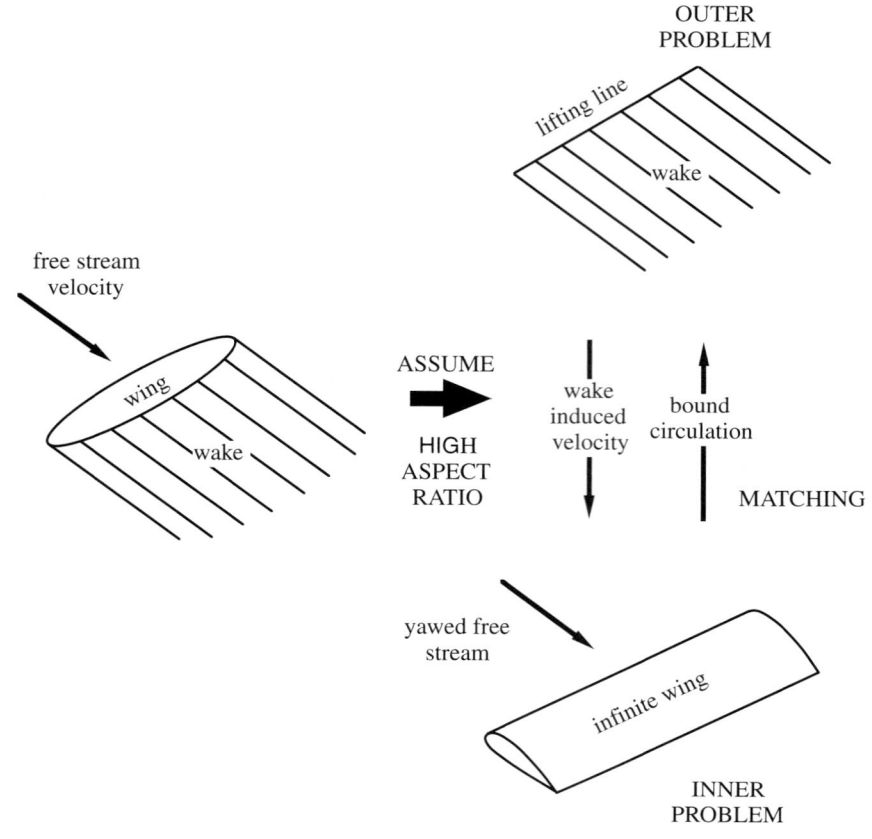

Figure 9.5. Lifting-line theory.

inner limit the outer solution has simple geometry. In the matching domain there are both simple geometry and irrotational flow.

Uniform inflow from ideal-wing (momentum or vortex) theory is an approximation for the solution of the outer problem (the wake). This approximation introduces effects requiring additional treatment, particularly the tip loss factor.

Blade element theory is essentially lifting-line theory for the rotary wing. The linearized wake model consists of helical vortex sheets trailed behind each blade. For a fixed wing the distortion of the wake geometry and the rollup of the tip vortices can generally be neglected, because the wake is convected downstream away from the wing. In contrast, the rotary wing encounters the wake from preceding blades of the rotor. Consequently, a more detailed and more accurate model of the wake is required to obtain an estimate of the induced velocity in lifting-line theory for the rotary wing. The blade vorticity quickly rolls up into concentrated tip vortices, which are best represented by line vortex elements instead of sheets. In many flight conditions the self-induced distortion of the tip vortex helices must also be accounted for to accurately obtain the blade loads. An entirely analytical solution analogous to that of the fixed wing is not possible for the rotary wing because of the helical geometry of the wake, except in the case of the continuous wake of an actuator disk model. To obtain a tractable mathematical problem for calculating the induced velocity, the vorticity in the wake is usually modeled by a series of discrete vortex elements. Even in steady-state forward flight the blade loading is periodic in the rotating frame, and the rotor analysis requires unsteady aerodynamic theory.

The basic assumption of lifting-line theory, that the wing has a high aspect ratio, is almost always satisfied with a helicopter main rotor blade. However, the assumption that there be no rapid spanwise change in the aerodynamic environment of the blade is often not satisfied, notably at the blade tip and near blade-vortex interaction. The loading near a wing tip must drop to zero in a finite distance. The loading on a rotor blade is concentrated at the tip because of the higher velocities there, so the gradient of the lift at the tip is particularly high, and any small distortion of the loading due to three-dimensional flow effects is very important. In certain flight conditions the rotor blade passes quite close to a tip vortex from a preceding blade. The vortex-induced velocity gradients at the blade are large for such close passages, and classical lifting-line theory significantly overestimates the loading produced. Hence lifting-line theory must be extended, modified, or corrected to handle some of the aerodynamic problems of the rotary wing.

Formal lifting-line theory is the solution of the three-dimensional wing loading problem using the method of matched asymptotic expansions. Based on the assumption of a large wing aspect ratio, the problem is split into separate outer (wake) and inner (wing) problems, which are solved individually and then combined through a matching procedure. In general wings with swept and yawed planforms in unsteady, compressible, and viscous flow must be considered. The lowest-order solution is Prandtl's theory (steady and no sweep). Development of higher-order lifting-line theory originated with Weissinger (1947) for intuitive methods and with van Dyke (1963) for singular perturbation methods. The lifting-line theory developments found in the literature, although they include higher orders and unsteady, transonic, and swept flow, are generally analytical methods. They obtain analytical solutions for both the inner and outer problems and are in quadrature rather than integral-equation form. Often the inner solution is inviscid or even a thin airfoil. These theories are not therefore directly applicable to the general case, but do provide a guide and sound mathematical foundation for the development of a rotary-wing analysis. For the rotor blade, stall (high angles-of-attack) must be included in the inner solution, and the distorted, rolled-up wake geometry in the outer solution. Hence the objective is to obtain from lifting-line theory a separate formulation of the inner and outer problems, with numerical not analytical solutions, and a matching procedure that is the basis for an iterative solution. A key consideration is the need to retain the two-dimensional airfoil tables in the inner solution because of the viscous and compressible effects embodied in such tables. Hence whatever approximations that are required to retain the tables are accepted.

Higher-order theory is used to improve the calculation of the airloads without actually resorting to lifting-surface or more advanced methods. Several investigations have shown that second-order lifting-line theory gives nearly the same results as lifting-surface theory, including the lift produced in close blade-vortex interactions. In addition, second-order theory should also improve the load calculations for swept tips, yawed flow, and low-aspect-ratio wings. Second-order lifting-line theory is developed in section 9.3; the results are summarized here. In the second-order outer problem (the wake), the wing is a dipole line plus a quadrupole line, which for a thin airfoil (with no thickness or camber) is equivalent to a dipole at the quarter chord. The dipole solution is a wake of vortex sheets. In the second-order inner problem (the wing), the boundary condition is a wake-induced velocity that varies linearly in space. For a thin airfoil, the same lift is obtained with a uniform induced velocity by using the value at the three-quarter chord. The correct moment is not obtained, however, since a linear induced velocity variation over the chord produces a moment about the quarter chord, but a uniform induced velocity does not. A remainder solution must be defined that is the second-order solution after accounting for the induced velocity at the three-quarter chord. This remainder solution is ignored in practice, so becomes an error estimate. The lift error is small, as is the moment error except for the moment about the quarter chord produced by a linearly varying induced velocity. To retain use of the airfoil tables, the only parts of the second-order theory that can be used are placing the lifting-line at the quarter chord and the collocation point at the three-quarter chord.

The perturbation solution procedure alternates between the inner and outer problems, using only the solution up to the previous order. Combining and then solving all orders simultaneously is equivalent in terms of the perturbation expansions. Combining the inner problems is a natural step. The lowest-order inner problem is the airfoil with just geometry boundary conditions, and the first-order inner problem has the wake-induced velocity boundary conditions. These problems are easily combined because the wake just gives an angle-of-attack change. Combining the outer problems means evaluating the induced velocity using the total bound circulation from the combined inner problem, rather than just from the lowest-order inner problem. This changes the nature of the solution from direct quadrature to an integral equation. Inverting the integral equation is necessary, but with airfoil tables an iterative solution is required anyway. Moreover, the solution of the integral equation is well behaved, whereas the direct solution is singular at the wing tips for normal wing planforms.

An examination of the simplified inner problem shows the differences between the common formulations of lifting-line theory. Let θ be the geometric angle-of-attack of the wing section. The inner problem solution (steady and unswept) can be analytical or numerical:

$$\text{analytical, thin airfoil:} \quad \frac{\Gamma}{\pi c} = U\theta - v \qquad (9.4)$$

$$\text{numerical, airfoil tables:} \quad \frac{\Gamma}{\pi c} = \frac{U}{2\pi} c_\ell(\theta - v/U) \qquad (9.5)$$

where Γ is the bound circulation, c the chord, U the free stream velocity, $c_\ell(\alpha)$ the lift-coefficient as a function of angle-of-attack, and v the wake-induced velocity. The thin airfoil theory result is the lift $L = \rho U \Gamma$ due to the upwash $w = U\theta - v$, with a lift-curve slope of 2π. The outer problem can obtain the induced velocity v at the quarter chord or three-quarter chord by integrating the effects of all wake vorticity (excluding the bound vortex). The following implementations of lifting-line theory are of interest:

a) First-order perturbation theory obtains v from $\Gamma_0 = \pi c U \theta$:

$$\frac{\Gamma}{\pi c} = U\theta - v_{c/4}(\theta) \qquad (9.6)$$

b) Prandtl's integral equation is first order, but obtains v from Γ:

$$\frac{\Gamma}{\pi c} = U\theta - v_{c/4}(\Gamma) \qquad (9.7)$$

c) A common implementation for wings is the first-order theory, but using airfoil tables:
$$\frac{\Gamma}{\pi c} = \frac{U}{2\pi} c_\ell(\theta - v_{c/4}(\Gamma)/U) \qquad (9.8)$$

d) A second-order implementation uses v at the three-quarter chord:
$$\frac{\Gamma}{\pi c} = \frac{U}{2\pi} c_\ell(\theta - v_{3c/4}(\Gamma)/U) \qquad (9.9)$$

e) Weissinger's L-theory includes the contribution of the bound vortex in evaluating v at the three-quarter chord, and it equates the induced angle-of-attack to the geometric angle-of-attack:
$$\left(\frac{\Gamma}{\pi c} + v_{3c/4}(\Gamma)\right) = U\theta \qquad (9.10)$$

which is equivalent to using the thin-airfoil solution of the inner problem in second-order theory.

The second-order theory (d) is a good basis for rotary-wing analysis.

Large sweep angles can be included in the second-order theory, but small curvature must be assumed so the wake-induced velocity effects in the inner problem remain two-dimensional. For many wings the curvature is small, except at kinks in the quarter-chord line. The analysis relies on the integral-equation form and spanwise discretization to keep the loading well behaved at such kinks. With unsteady motion and loading, the inner problem is an unsteady, two-dimensional airfoil with a shed wake, and the outer problem excludes both the bound vortex and the inner shed wake when calculating the induced velocity. The shed wake is retained in the outer rather than the inner problem, so the shed wake and trailed wake can be treated identically, especially since subtracting the inner shed wake from the outer problem is difficult with complex wake geometry. Then the induced velocity from all vorticity (except the bound vortex still) is evaluated at the three-quarter chord and treated as a uniform flow for the inner problem. The shed wake is thus a boundary condition of the inner solution, not a part of the inner model. The assumption that the shed-wake-induced velocity is constant over the chord is a major approximation. With the induced velocity evaluated at a single point, the shed wake model must be modified to obtain the unsteady loads correctly. The Theodorsen and Sears functions (the shed wake effects in two-dimensional airfoil theory) are well approximated for low frequency if the shed wake in the outer problem is created a quarter chord aft of the collocation point, not at the bound vortex as for the trailed wake (see section 10.2).

Guided by the results of perturbation theory, the following is a practical implementation of lifting-line theory. The outer problem is an incompressible vortex wake behind a lifting-line, with distorted geometry and rollup. The lifting-line (bound vortex) is at the quarter chord, as an approximation for the quadrupole line introduced by second-order loading. The trailed wake begins at the bound vortex. The shed wake is created a quarter chord aft of the collocation point on the wing (the lifting-line approximation for unsteady loading). The three components of wake-induced velocity are evaluated at the collocation points, excluding the contributions of the bound vortex. The collocation points are at the three-quarter chord (in the direction of the local flow), as an approximation for a linearly varying induced velocity introduced by the second-order wake. The induced velocity calculated at the three-quarter chord is used only to calculate the angle-of-attack for the loading solution. The local section aerodynamic environment, including the orientation of the lift and drag and hence the magnitude of the induced power, is still obtained from the induced velocity at the quarter chord. The inner problem consists of unsteady, compressible, viscous flow about an infinite wing, in a uniform flow consisting of the yawed free stream (including rigid and elastic motion of wing through the air) and three components of wake-induced velocity. This problem is split into two-dimensional, steady, compressible, viscous flow (airfoil tables), plus corrections. The corrections account for unsteady flow (small angle-of-attack non-circulatory loads, without any shed wake); dynamic stall (an empirical model); swept and yawed flow (the equivalence assumption for a swept wing); tip flow; and Reynolds number.

This formulation is almost second order (in the inverse of the aspect ratio) accurate for lift, including the effects of sweep and yaw, but is less accurate for section moments (basically still first order). In particular, with typical blade-vortex separations, second-order lifting-line theory is as accurate as lifting-surface theory for vortex-induced lift calculations. For a steady, incompressible, non-rotating wing, this theory is equivalent to Weissinger's L-method for tapered and swept wings. Weissinger (1947) found the L-method to be as accurate as a simplified lifting-surface theory. DeYoung and Harper (1948) showed that the L-method gave good predictions of measured span loading for wings with a wide range of aspect ratio, sweep, and taper. Brower (1981) compared vortex lattice lifting-surface theory, Prandtl's (first-order) lifting-line theory, and Weissinger's lifting-line theory for non-rotating blade-vortex interaction. The L-method gave excellent results even for very small blade-vortex separation. Kocurek and Tangler (1977) and Kocurek, Berkowitz, and Harris (1980) compared lifting-surface theory, the L-method (as a one-chordwise panel version of the lifting-surface method), and first-order lifting-line theory for a hovering rotor blade. The one-panel lifting-surface theory gave results nearly as good as those from the theory with five or seven chordwise panels, whereas the first-order lifting-line theory gave significantly different results.

9.3 Perturbation Solution for Lifting-Line Theory

Formal lifting-line theory is the solution for the three-dimensional wing loading using the method of matched asymptotic expansions, based on the small parameter $\epsilon = c/R = 1/AR$. The lowest-order three-dimensional solution for a non-rotating, steady, unswept wing in incompressible flow is Prandtl's theory. Beginning with the work of van Dyke (1963), higher-order theories have been developed by singular perturbation methods, extending lifting-line theory to swept, unsteady, transonic, rotating wings; see Johnson (1986). Ashley and Landahl (1965) provided an introduction to perturbation methods in aerodynamics.

Fixed-wing theories are applicable almost directly to the rotor problem, because the key to the method is the matching process, and in the matching domain, the rotor appears almost the same as the non-rotating wing. For the rotary wing, swept and yawed planforms, unsteady motion, and compressible or transonic flow must be considered. High angle-of-attack (stall) must be included in the inner solution and the helical, distorted, and rolled-up wake geometry in the outer solution. So the rotor theory must use numerical solutions of the inner and outer problems. Simplifications are possible. With undistorted geometry, the wake integrals can be evaluated analytically in the radial direction, although the helical nature requires a

9.3 Perturbation Solution for Lifting-Line Theory

numerical integration over the wake age. Analytical solutions can be obtained for the linearized, inviscid inner problem, just as for fixed wings, although the solutions for higher-order lifting-line theory are not simple. The value of lifting-line theory in rotor problems, however, lies in the complexities that can be retained in the inner and outer problems. In the matching domain, the inner and outer problems can always be consistently simplified without requiring that they be simple in their own domains. Lifting-line theory for the rotor blade provides the formulation of the inner and outer problems and the matching that couples them as the basis for an iterative numerical solution.

In the method of matched asymptotic expansions, the inner and outer solutions are expanded as series, with each term an order ϵ smaller than the previous term. Then the n-term/m-order inner solution is matched to the m-term/n-order outer solution. At each level, matching provides boundary conditions for the next term in the inner or outer expansion, from the solution at previous levels.

With a high aspect ratio, the inner problem has simpler geometry (two-dimensional), but complex flow (Navier-Stokes equations). In the outer limit, the inner solution can be considered irrotational. The outer problem has complex geometry (the vortex wake), but irrotational flow. In the inner limit, the outer solution has simple geometry. In the matching domain, there are both simple geometry and irrotational flow. From irrotational flow in the matching domain, the theory can be developed in terms of the velocity potential ϕ (relative to still air). The wing geometry is defined in moving axes (translating and rotating for the rotor blade), with x chordwise (positive toward the trailing edge) and y spanwise. The wing span is R and the chord is $c(y)$. The wing surface is at $z = g(x, y)$ and is a function of time for the unsteady problem. It is assumed that $\epsilon = c/R \ll 1$.

Consider second-order lifting-line theory for an unswept, steady, fixed wing in incompressible, inviscid flow. The velocity U is in the x-axis direction. The velocity potential satisfies Laplaces's equation $\nabla^2 \phi = 0$, with boundary conditions $\phi = 0$ at infinity and

$$\phi_z = (U + \phi_x)g_x + \phi_y g_y \tag{9.11}$$

on the wing surface. The inner problem is defined by small ϵ (large aspect ratio) for fixed chord, so $x, z = O(\epsilon)$ and $y = O(1)$. Write Laplace's equation as $\nabla^2_{2D} \phi = -\phi_{yy}$, and expand $\phi = \phi_0 + \phi_1 + \phi_2 + \cdots$, such that $\phi_n = O(\epsilon^n)$. Then the inner equation of motion and boundary condition on the wing, for each order, are

$$\text{order 0:} \quad \nabla^2_{2D}\phi_0 = 0 \quad \phi_{0z} = (U + \phi_{0x})g_x \tag{9.12}$$

$$\text{order 1:} \quad \nabla^2_{2D}\phi_1 = 0 \quad \phi_{1z} = \phi_{1x}g_x \tag{9.13}$$

$$\text{order 2:} \quad \nabla^2_{2D}\phi_2 = -\phi_{0yy} \quad \phi_{2z} = \phi_{2x}g_x + \phi_{0y}g_y \tag{9.14}$$

The outer problem is defined by small ϵ (large aspect ratio) for fixed span, so $x, y, z = O(1)$. Expand the outer solution $\phi = \phi_0 + \phi_1 + \phi_2 + \cdots$, such that $\phi_n = O(\epsilon^n)$. The outer equation of motion and boundary condition at each order are $\nabla^2 \phi_n = 0$ and $\phi_n = 0$ at infinity. Matching is required to complete the problem definition in the two domains.

The outer limit of the inner solution is obtained by substituting $x_I = x_O/\epsilon$ and letting $\epsilon \to 0$ with x_O (outer variable) fixed, so $x_I \to \infty$. The inner limit of the outer solution is obtained by substituting $x_O = x_I \epsilon$ and letting $\epsilon \to 0$ with x_I (inner variable) fixed, so $x_O \to 0$. The inner expansion has the form

$$\phi \sim U_i x_i - \frac{\Gamma}{2\pi}\theta + \frac{c_i x_i}{r^2} + \cdots \tag{9.15}$$

using cylindrical coordinates θ and r. In the inner domain, all terms are order ϵ; in the outer domain, the terms are order $1, \epsilon, \epsilon^2$. The outer expansion has the form

$$\phi \sim \phi + \phi_{x_i} x_i + \phi_{x_i x_j} x_i x_j + \cdots \tag{9.16}$$

In the outer domain, all terms are order 1; in the inner domain, the terms are order $1, \epsilon, \epsilon^2$. Matching gives the boundary condition for the next term in the expansion. The terms become increasingly singular. The solution proceeds as follows.

Inner ϕ_0. In the outer problem, the chord goes to zero as $\epsilon \to 0$, so to lowest order there is no wing ($\phi_0 = 0$).

$$\text{inner 1-term/1-order:} \quad \phi = u_{0i} x_i \tag{9.17}$$

$$\text{outer 1-term/1-order:} \quad \phi = \phi_0 = 0 \tag{9.18}$$

Matching gives the inner solution boundary condition $\phi_0 = 0$ at infinity. So the order-0 inner problem is a two-dimensional airfoil (thickness, camber, and geometric angle-of-attack) in a uniform free stream produced by the wing velocity U.

Outer ϕ_1. The outer limit of the inner solution is the solution of $\nabla^2_{2D}\phi = 0$, which for regular boundary conditions can be expanded (for larger inner variables) as a uniform flow plus a bound vortex plus a quadrupole:

$$\phi \sim U_i x_i - \frac{\Gamma}{2\pi}\theta + \frac{c_i x_i}{r^2} + O\left(\frac{1}{r^2}\right) \tag{9.19}$$

For an airfoil with chord c, $\Gamma \sim c$ and $c_i \sim c^2$; so

$$\text{inner 1-term/2-order:} \quad \phi = -\frac{\Gamma_0}{2\pi}\theta \tag{9.20}$$

$$\text{outer 2-term/1-order:} \quad \phi = \phi_1(0, y, 0) \tag{9.21}$$

Matching gives $-(\Gamma_0/2\pi)\theta = \phi_1(0, y, 0)$. The order-1 outer solution is a line of dipoles produced by the inner bound circulation Γ_0 (the bound vortex). So the outer problem for ϕ_1 is a lifting line.

Inner ϕ_1. In the inner limit, the outer solution appears to be an infinite vortex line and the associated trailed vorticity. The solution can be expanded as a regular part and a singular part:

$$\phi \sim -\frac{\Gamma}{2\pi}\theta + \frac{\Gamma''}{4\pi}(z^2\theta - zx \ln r) + Az + Bxz \tag{9.22}$$

(the last two terms represent the regular part of the expansion). So

$$\text{inner 2-term/2-order:} \quad \phi = -\frac{\Gamma_0}{2\pi}\theta + u_{1i} x_i \tag{9.23}$$

$$\text{outer 2-term/2-order:} \quad \phi = -\frac{\Gamma_0}{2\pi}\theta + \frac{\partial \phi_1}{\partial x_i} x_i \tag{9.24}$$

Matching gives $u_{1i} x_i = (\partial \phi_1 / \partial x_i) x_i$, so the inner solution boundary condition is $\phi_1 = v_{1i} x_i$ at infinity. The order-1 inner problem is a two-dimensional airfoil, with

9.3 Perturbation Solution for Lifting-Line Theory

zero boundary conditions at the airfoil surface and the wake-induced velocity at infinity (a uniform free stream). Note v_{1i} is the regular part of the expansion and excludes the bound vortex. In general the cross-flow velocity $\partial \phi_1 / \partial y$ is retained, although it has no effect on the inviscid inner solution. This is Prandtl's first-order theory, obtained by perturbation methods.

Outer ϕ_2.

$$\text{inner 2-term/3-order:} \quad \phi = -\frac{\Gamma_0}{2\pi}\theta + \frac{c_{0i}x_i}{r^2} + u_{1i}x_i - \frac{\Gamma_1}{2\pi}\theta \qquad (9.25)$$

$$\text{outer 3-term/2-order:} \quad \phi = -\frac{\Gamma_0}{2\pi}\theta + \frac{\partial \phi_1}{\partial x_i}x_i + \phi_2(0, y, 0) \qquad (9.26)$$

Matching gives $(c_{0i}x_i/r^2) - (\Gamma_1/2\pi)\theta = \phi_2(0, y, 0)$, so the outer solution for ϕ_2 is a quadrupole line from the inner ϕ_0 and a dipole line from the inner ϕ_1. Note that ϕ_2 (with a quadrupole) is more singular in the inner limit than ϕ_1 (a dipole line); hence the need for singular perturbation methods.

Inner ϕ_2. The inner limit of the outer solution requires an expansion of a quadrupole line:

$$\phi \sim c_x \frac{x}{r^2} - \frac{1}{2}c_x''(2z\theta - x \ln r) + \alpha x$$

$$+ c_z \frac{z}{r^2} - \frac{1}{2}c_z''(z + z \ln r) + \beta z \qquad (9.27)$$

So

$$\text{inner 3-term/3-order:} \quad \phi = -\frac{\Gamma_0}{2\pi}\theta + \frac{c_{0i}x_i}{r^2} + u_{1i}x_i - \frac{\Gamma_1}{2\pi}\theta + \phi_2 \qquad (9.28)$$

$$\text{outer 3-term/3-order:} \quad \phi = -\frac{\Gamma_0}{2\pi}\theta + \frac{\partial \phi_1}{\partial x_i}x_i + \frac{1}{2}\frac{\partial^2 \phi_1}{\partial x_i \partial x_j}x_i x_j$$

$$+ \frac{\Gamma_0''}{4\pi}(z^2\theta - xz \ln r) - \frac{\Gamma_1}{2\pi}\theta + \frac{\partial \phi_2}{\partial x_i}x_i + \frac{c_{0i}x_i}{r^2}$$

$$- \frac{1}{2}c_{0x}''(2z\theta - x \ln r) - \frac{1}{2}c_{0z}''(z + z \ln r) \qquad (9.29)$$

and matching gives

$$\phi_2 = \frac{1}{2}\frac{\partial^2 \phi_1}{\partial x_i \partial x_j}x_i x_j + \frac{\Gamma_0''}{4\pi}(z^2\theta - xz \ln r) + \frac{\partial \phi_2}{\partial x_i}x_i$$

$$- \frac{1}{2}c_{0x}''(2z\theta - x \ln r) - \frac{1}{2}c_{0z}''(z + z \ln r) \qquad (9.30)$$

which is the boundary condition at infinity for the inner solution. The order-2 inner problem is an inhomogeneous equation, with a particular solution produced by the spanwise derivative of the order-0 solution. The boundary condition for ϕ_2 is a regular part (uniform and linearly varying induced velocity) plus a singular part.

As perturbation expansions, equivalent solutions are obtained if the inner problems are combined and all orders solved simultaneously. Up to second order, the equation of motion is $\nabla^2_{2D}\phi = -\phi_{0yy}$, with boundary condition $\phi_z = (U + \phi_x)g_x + \phi_{0y}g_y$ on the surface ($z = g$) plus the Kutta condition. The behavior of the solution at infinity is

$$\phi \sim -\frac{\Gamma}{2\pi}\theta + v_i x_i + \frac{1}{2}\frac{\partial v_i}{\partial x_j}x_i x_j + \frac{\Gamma''}{4\pi}(z^2\theta - xz \ln r) + \frac{c_i x_i}{r^2}$$
$$- \frac{1}{2}c''_x(2z\theta - x \ln r) - \frac{1}{2}c''_z(z + z \ln r) \qquad (9.31)$$

The combined outer problem consists of a bound vortex line due to $\Gamma = \Gamma_0 + \Gamma_1$, plus a quadrupole line due to Γ_0.

The order-2 outer solution has a quadrupole line produced by the order-0 inner solution (Γ_0). The outer limit of the inner ϕ_0 is the same if the quadrupole strength is zero, but the bound vortex is offset. For axes with the origin at the wing midchord, the outer limit of the inner solution is

$$\phi = u_i x_i - \frac{\Gamma}{2\pi}\theta + \frac{c_i x_i}{r^2} + O(r^{-2}) \qquad (9.32)$$

Now consider just a bound vortex, no quadrupole term, located at $x = \xi$ and $z = \zeta$; the same outer limit ($r \gg \xi, \zeta$) is obtained if $(\Gamma/2\pi)\xi = -c_z$ and $(\Gamma/2\pi)\zeta = c_x$. For a thin airfoil, the bound vortex must be at the quarter chord ($\xi = -c/4$ and $\zeta = 0$). Camber and thickness introduce additional chordwise and vertical shifts, respectively. The quadrupole strengths resulting from camber and thickness are not proportional to the bound circulation, so these shifts depend on the lift. Using a bound vortex at the quarter chord means that the outer problem only requires the solution for a line of dipoles; that is, a wake of vortex sheets.

The boundary condition for the order-2 inner problem involves a linear variation of downwash over the chord, $w = w_0 + w_1 x$, from the wake-induced velocity and the wing motion. Thin airfoil theory gives the solution for the lift and moment: $c_\ell = -2\pi w_{3QC}/U$ and $c_{mQC} = \frac{\pi}{8}w_1 c/U$, where w_{3QC} is the downwash w at the three-quarter chord point. So the correct lift is obtained by using the wake-induced velocity evaluated at a single point on the section, the three-quarter chord. However, a linearly varying velocity would give a pitching moment, which the constant velocity does not, so without the gradient w_1 a second-order moment term has been neglected. This version of the second-order theory can be compared with Weissinger's method, in which the induced velocity is evaluated at the three-quarter chord, including the effect of the bound vortex at the quarter chord, and equated to the geometric angle-of-attack. Weissinger's method is equivalent to using thin-airfoil theory for the inner solution (Pistolesi's theorem states that the velocity at one-half chord from the bound vortex gives the correct boundary condition) and combines the inner and outer solutions.

A remainder problem is defined by the order-2 inner problem, after accounting for the use of the induced velocity evaluated at the three-quarter chord. This problem was solved by van Dyke (1963) for a wing at constant angle-of-attack and by van Holten (1975, 1977) for a constant chord wing. In practical implementation of lifting-line theory this remainder problem is usually neglected, so the analytical solutions are useful as error estimates. The lift error is second-order small, and so is the moment error, except probably the neglected moment from the linear variation of induced velocity.

Combining and then solving all orders simultaneously is equivalent in terms of the perturbation expansions. The combined inner problem is a two-dimensional airfoil in a uniform free stream, consisting of the wing motion and the wake-induced velocity. The induced velocity (three components) is evaluated at the three-quarter chord from the wake associated with a bound vortex line at the quarter chord (excluding the bound vortex, which has already been accounted for in the matching). Combining the inner problems is a natural step, since only an angle-of-attack change produced by the induced velocity is required.

The perturbation solution procedure alternates between the inner and outer problems, using only the solution up to the previous order. The two-dimensional airfoil with no wake is solved for Γ_0; the wake-induced velocity produced by Γ_0 is calculated; then the airfoil in the wake flow field is solved for Γ_1. Combining the outer problems means evaluating the induced velocity using the total bound circulation from the combined inner problem rather than just from the lowest-order inner problem. This changes the nature of the solution from direct quadrature to an integral equation (as in Prandtl's theory).

Higher-order lifting-line theories have been developed for swept, unsteady, transonic, rotating wings; see Johnson (1986). Van Holten (1975, 1977) developed a second-order lifting-line theory for a rotary wing using the method of matched asymptotic expansions. For swept wings, the inner limit of the dipole and quadrupole lines has singular terms proportional to $\Gamma' \tan \Lambda$ (where Λ is the local sweep angle). Matching gives an inner problem boundary condition at infinity that consists not only of the induced velocity (regular part of the swept wake solution) but also of these singular terms produced by sweep. If the induced velocity is calculated at the three-quarter chord and these new swept wing boundary conditions are ignored, the wing lift is correct but there is a moment error. Experience with the L-method confirms that this is a good approximation to the second-order lifting-line theory for the wing lift distribution. If the wing curvature is order ϵ^2 small, the differential equations for the inner problem remain two-dimensional. Small curvature is typical of rotor blades, except at kinks in the planform. At kinks, the asymptotic expansion used is invalid to any order, and the integral-equation formulation must be relied on to keep the solution well behaved.

In the unsteady theory, the order-0 inner problem is an unsteady two-dimensional airfoil in a swept free stream produced by the wing motion with a two-dimensional shed wake. The order-1 outer problem is again a lifting line, matching both the bound vorticity and the order-0 inner shed wake. The boundary condition at infinity for the order-1 inner problem again consists of the induced velocity (regular part) and a singular part produced by sweep, from which must be excluded now not only the bound vortex but also the inner shed wake, since both have already been accounted for. The unsteady wake-induced velocity is a function of $(t - x/U)$, like a convected gust. It is also possible in unsteady lifting-line theory to consider the shed wake not in the inner problem, but as part of the boundary conditions for the inner problem. Then only the bound vortex would be excluded from the induced velocity. The entire shed wake would be part of the outer problem and would be excluded from the inner problem. For complex wing motion and wake geometry (such as with a rotor blade), ensuring consistency between the outer and inner wake models is easier with this approach. However, one difficulty is that the inner shed wake begins at the trailing edge, whereas the outer wake emanates from the bound vortex line. Treating the shed wake and trailed wake identically is also a natural step: evaluating the induced velocity at the three-quarter chord and using it as a uniform free stream in the inner problem. Piziali (1966) combined the outer wake and inner shed wake by evaluating the induced velocity along the blade chord from all sources and applying thin-airfoil theory. Miller (1964a) used the induced velocity at the three-quarter chord and modified the shed wake model as required to obtain the correct unsteady loads. For unsteady airfoil motion, the shed wake effect is matched well with the induced velocity evaluated at the three-quarter chord and the shed wake started at the trailing edge.

9.4 Nonuniform Inflow

The wake-induced velocity distribution over the rotor disk is highly nonuniform, because the requirements for uniform inflow (constant bound circulation and a very large number of blades) are far from satisfied with real rotors. The dominant factor in the rotor induced velocity is the discrete tip vortices trailed in helices from each blade. Because of the rotation of the wing, the wake is laid down in spirals close to the rotor disk, to be encountered again by the blades. In particular, a blade passes close to the tip vortex from the preceding blade, both in hover and forward flight. A line vortex induces a tangential velocity that varies inversely with the distance from the vortex and has a maximum value at the vortex core radius. Thus the tip vortices in the wake induce a highly nonuniform flow field through which the blades must pass.

In hover, the tip vortex is convected downward only slightly until after it encounters the next blade, and there is little radial contraction (see Figure 3.16). The vortex-blade encounter takes place near the tip, with the vortex a small distance from the blade. The vortex produces a large variation in the loading at the tip and has a substantial influence on the rotor hover performance (see section 3.8). In forward flight the rotor wake is convected downstream, so the tip vortices are swept past the entire rotor disk instead of remaining in the tip region. The close vortex-blade encounters occur primarily on the advancing and retreating sides of the disk, where the blades sweep over the vortices (see Figure 5.8). Thus in forward flight there is a large azimuthal variation of the induced velocity, which produces a large higher harmonic content of the loading. Nonuniform inflow is an important factor in helicopter vibration, loads, and noise during forward flight. Even the influence on the 1/rev loading, and hence on the cyclic control of the rotor, is large. The effects of nonuniform inflow in forward flight vary greatly with the flight condition, being largest in states such as transition, where the wake is closest to the rotor. In a tandem helicopter, the rear rotor also encounters the wake of the front rotor in forward flight, in particular resulting in large vortex-induced loads on the forward portion of the rear rotor disk. Figure 9.1 shows examples of the rotor blade loading in forward flight. The vortex-induced loading is apparent in the low speed loads.

The accurate calculation of the wake-induced nonuniform velocity, and the resulting airloads and blade motion, is a prerequisite for the satisfactory prediction of rotor blade loads, rotor noise, helicopter vibration, and even the rotor performance and cyclic control. A complicated, detailed model of the rotor aerodynamics and dynamics is clearly required for such a calculation, and only a numerical solution is practical. The wake-induced velocity is obtained by integrating the Biot-Savart law over the vortex wake elements in the rotor wake. The wake strength is determined by the radial and azimuthal variation of the bound circulation. For the wake geometry, a simple assumed model, experimental measurements, or a calculated geometry can be used. Given the vorticity strength and geometry, the induced velocity can be evaluated. For line elements the Biot-Savart law is

$$\mathbf{u}(\mathbf{y}) = -\frac{1}{4\pi} \int \gamma(\mathbf{x}) \frac{\mathbf{s} \times d\mathbf{l}(\mathbf{x})}{s^3} \qquad (9.33)$$

where $\mathbf{s} = \mathbf{y} - \mathbf{x}$ is the distance between the vortex element $\gamma \, d\mathbf{l}$ and the point where the induced velocity $\mathbf{u}$ is required. With the helical geometry of the rotary-wing wake such integrals cannot be evaluated analytically, even if the self-induced distortion of the wake is neglected. Moreover, a direct numerical integration is usually not satisfactory, because the large variations of the induced velocity at close vortex-blade encounters would require a very small step size for accurate results. Calculating the nonuniform inflow with the wake modeled using a set of discrete vortex elements is most accurate and most efficient. The vortex elements used are chosen such that the Biot-Savart law can be integrated efficiently, generally by analytical integration. A finite-length line segment, straight or curved, is the usual choice. Then the induced velocity at any point in the flow field is obtained by summing the contributions from all the elements in the rotor wake. The vortex line segment is particularly important since the helical tip vortices can be modeled very well by a connected series of such elements. The approximations involved in such models of the rotor wake include replacement of the curvilinear geometry by a series of straight-line or planar segments; simplified distribution of vorticity over the discrete wake elements (usually constant or linear); and perhaps physical approximations, such as the use of line elements to represent the vortex sheets. A practical model must balance the accuracy and efficiency of such approximations.

A discretized wake model is developed by assuming that the wing bound circulation is known at discrete points along the span and in the past. Thus the bound circulation $\Gamma(r, t)$ is calculated at the aerodynamic span stations r_i, $i = 1$ to M, and at the wake ages $\phi_k = k \, \Delta \phi$, $k = 0$ to K. A linear variation of Γ between these points means the wing generates a wake of sheet elements behind the wing (Figure 9.6). The strength of the trailed and shed vorticity in an element can be obtained from Γ at the time that the vorticity was created. Thus the strength can be characterized by the bound circulation corresponding to the four corners of the element. The element leading edge corresponds to Γ at $(\psi - \phi_k)$, and the trailing edge to Γ at the earlier time $(\psi - \phi_{k+1})$. A spanwise change in Γ produces trailed vorticity γ_t. A linear variation of Γ means that the strength γ_t is constant spanwise, but linear along its length. A time change in Γ produces shed vorticity γ_s. A linear variation of Γ means that the strength of γ_s is constant in age, but linear along its length. The wake geometry provides the locations of the four corners of the element.

For calculation of performance and structural loads, typically 20 or so radial stations are used, concentrated at the tip where the greatest loading changes occur; the typical azimuthal resolution is 5° to 15°.

An economical approximation for the sheet element is to use line segments, with a large core to avoid velocity singularities near the line. Hence a vortex lattice model is obtained by collapsing all sheet elements to finite-strength line segments (Figure 9.7). The line segments are in the center of the sheet element, so the points at which the induced velocity are calculated (collocation points) are at the midpoints of the vortex lattice grid, both spanwise and in time. Locating the collocation points midway between the trailers is a standard practice, used to avoid the singularities at the lines. Locating the collocation points midway in time is required to correctly

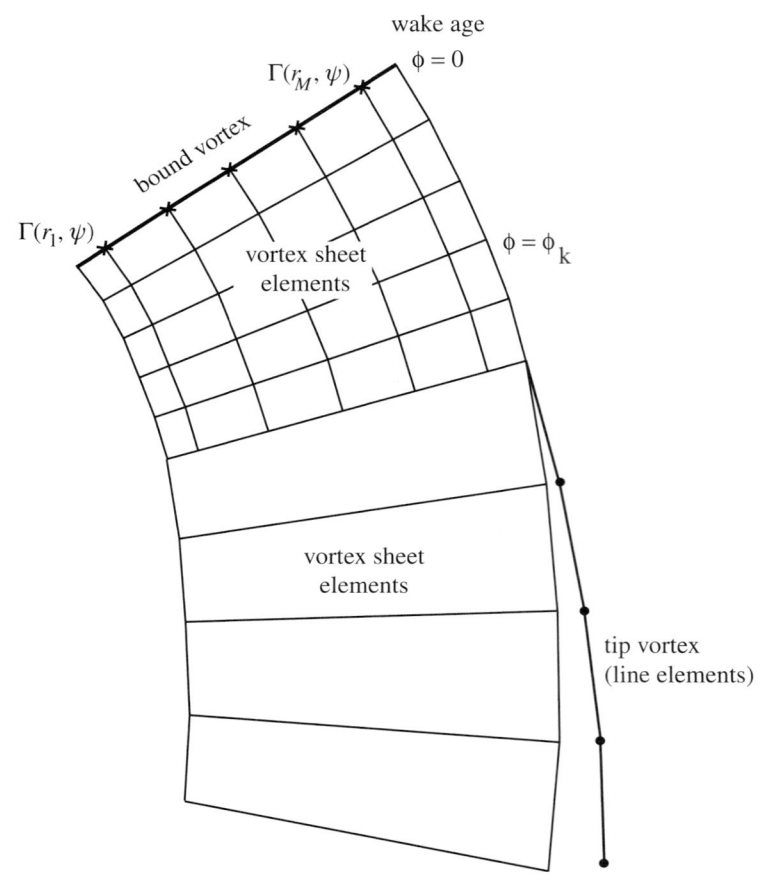

Figure 9.6. Discretized wake model: sheet elements generated by linear variation of bound vorticity, plus line elements for the rolled-up tip vortices.

obtain the unsteady aerodynamic effects of the shed wake. With the sheets collapsed to lines, the strength of the line segments varies linearly along their length. A further approximation is stepped (piecewise constant) variation in strength. Then instead of a linear variation along the line, there is a jump in the strength at the center of the element where the shed and trailed lines cross. This wake model corresponds to a stepped distribution of the bound circulation, both spanwise and in time.

The implementation of lifting-line theory involves primarily the wake directly behind the bound vortex where the induced velocity is calculated. The most important requirement is to model the detailed variation of the wake strength, both spanwise and in time, to accurately obtain the classical three-dimensional (Prandtl) and unsteady (Theodorsen) effects of the wake on the wing loading. A vortex lattice, rather than sheet elements, is best used for the near wake. Discretization of the wake is better behaved with line segments than with sheet elements. With sheet elements, numerical difficulties arise from the edge and corner singularities (particularly when planar-rectangular elements are used, which introduce mismatched edges) and the fact that nonplanar-quadrilateral elements cannot be integrated analytically. Also, the most important case of the downwash from the two sheet elements adjoining the collocation point requires a higher-order element or some other special treatment.

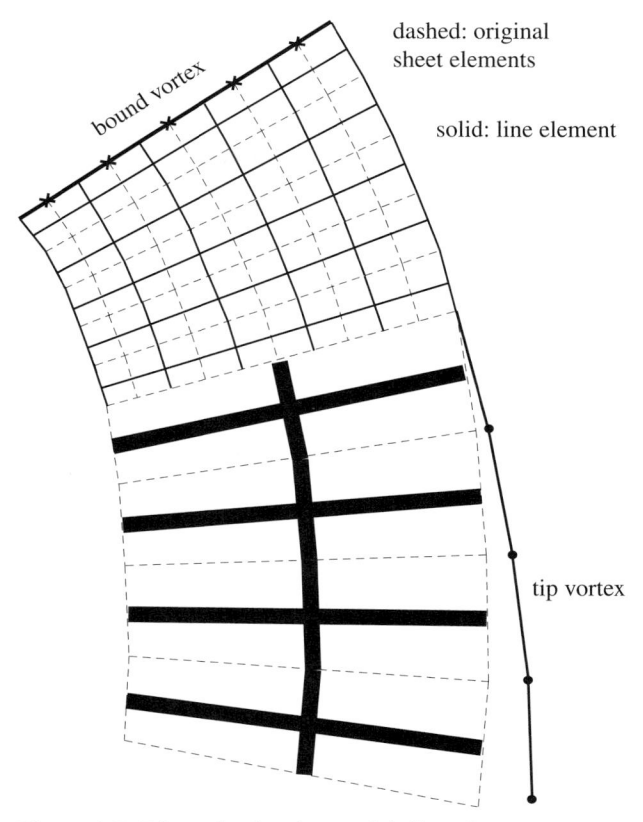

Figure 9.7. Discretized wake model: line elements to approximate vortex sheets, plus line elements for the rolled-up tip vortices.

For rotors, the distortion of the wake can lead to close encounters of the blade with its near wake lattice, requiring that the line elements have a core. By using a small core radius, velocity singularities can be avoided while not changing the induced velocity at the blade for normal wake configurations. To implement second-order lifting-line theory, the geometry of the collocation points involves placing them at the three-quarter chord in the local flow direction. The collocation points are part of the geometry of the wake rather than the wing. The wake geometry is obtained from the position of the lifting line at the current and past time steps. Thus the wake geometry can be used to get the local flow direction for the collocation points.

When the wake reaches a following blade or some other aerodynamic surface (such as the tail rotor, another main rotor, fuselage, or tail), the rolled-up tip vortices are the most important part of the model. The rollup problem involves three-dimensional, unsteady, viscous fluid dynamics. The vortex core is largely formed at the wing trailing edge, so the problem is not the inviscid rollup of a vortex sheet. Moreover, discretization of the wake for a rollup calculation is difficult. Thus with a vortex model of the rotor wake, the rollup process is usually not calculated but rather modeled, meaning that the structure and properties of the rolled-up wake are determined from assumptions and input parameters and from the spanwise distribution of the bound circulation where the wake was created. The model must account for the influence of the strength of the tip vortex when it encounters a following blade, the influence of the core radius when the vortex is fully rolled up, and the effect of the vortex rollup on the tip load of the generating blade.

The inboard vorticity plays a smaller role in any interactions than the concentrated tip vortices, so more approximations are acceptable. The entrainment of vorticity into the tip vortex and the accompanying stretching and distortion of the inboard vortex sheet are details likely not included in the modeling of the tip vortex rollup process. Thus Figures 9.6 and 9.7 show an abrupt change from the vortex lattice behind the blade to the rolled-up structure in the far wake. The use of a single vortex panel reflects the loss of information about the strength and geometry of the inboard vorticity. Options for modeling the inboard wake elements include vortex sheets, either nonplanar-quadrilateral or planar-rectangular, or line segments in the middle of the elements. Line segments are commonly used, producing a vortex lattice model for the entire wake. Line segments that approximate sheet elements require a large core size, not as a representation of a physical effect, but to eliminate any singularities in the velocity for the close passage of following blades.

In many operating conditions, the rotor blade loading is positive all along the blade span (as sketched in Figure 9.2), with peak bound circulation Γ_{max}. The simplest model of the rollup for such a case assumes that in the far wake (where the rollup process is complete) there are concentrated tip vortices with strength equal to the value of Γ_{max} at the time that the wake element was created. If the root vortex is not rolled up, there is corresponding negative trailed vorticity with total strength $-\Gamma_{max}$ in the inboard sheet. The tip vortex model then is a line segment with this strength and a small core radius. The peak bound circulation Γ_{max} is the maximum possible strength of the tip vortex, obtained by entrainment of all the trailed vorticity between the peak and the tip. The strength is smaller if more or less trailed vorticity is entrained. The error in the assumed strength must be compensated for by the value of the core radius. In the absence of a calculation of the stretching and distortion produced by the rollup, the inboard vorticity is modeled as a single sheet element with trailed and shed vorticity. This is an efficient model, consisting of only two elements with a total of three line segments at each age, which minimizes computation, and it depends only on Γ_{max}, which minimizes storage.

Although Figure 9.2 shows a monotonic change in circulation from the peak to the blade tip, blades with more complicated geometry could exhibit more than one local maximum in the loading. A geometric feature such as the edge of a trailing-edge flap or a rapid change in blade chord can produce a large spanwise gradient in the bound circulation. The vortex rollup model can be extended to such a case by assuming that a concentrated trailed vortex (represented by line segments) emanates from each geometric feature. The models can also be made more complex by introducing gradual entrainment of the trailed vorticity into the tip vortex or by prescribing the tip vortex strength as a fraction of the peak bound circulation.

The blade loading at a given time need not be all positive or all negative. For example, a helicopter rotor in high-speed forward flight often has negative lift on the advancing tip, particularly in the second quadrant. With a flapping rotor, the net pitch and roll moment on the hub must be small (zero if there is no flap hinge offset). In forward flight, the lift capability on the retreating side is limited by the combination of low dynamic pressure and stall of the airfoil. Consequently the lift on the advancing side must also be small to maintain roll balance. At sufficiently high speed, the lift on the advancing tip can become negative. Large twist of the blade (built in or elastic) increases the negative loading. A tiltrotor blade can have negative tip loading on

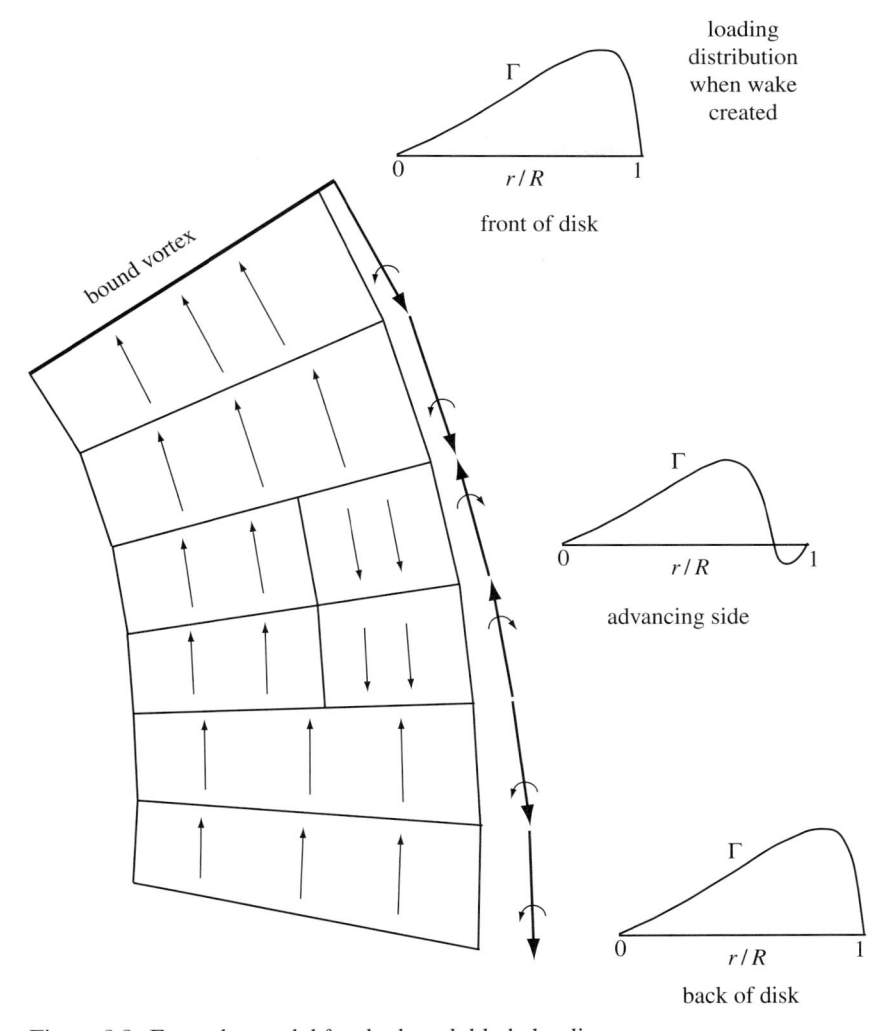

Figure 9.8. Far wake model for dual-peak blade loading.

the entire advancing side at even low-speed edgewise flight. The vortex-induced loading on following blades shows that negative tip loading produces substantial negative trailed vorticity in the wake. Over much of the disk, Γ is still positive all along the span. However, there is a range of azimuth on the advancing side with a negative peak near the tip and a positive peak inboard. Figure 9.8 illustrates the wake structure. At the current blade position (say on the front of the disk) the blade loading is positive. At an earlier azimuth (say on the advancing side) the wake was generated by blade loading that was positive inboard and negative at the tip (called "dual peak"). At a still earlier time (say on the back of the disk), the wake was generated from single-peak loading. For rotors in high-speed flight with negative tip loading, the experimental evidence (Hooper (1984)) implies that the entrainment of trailed vorticity into the tip vortex occurs from the outboard (negative) peak bound circulation, not from the inboard (large positive) peak. Thus the tip vortex that forms from the wake generated by dual-peak loading has the opposite sign and smaller strength than the tip vortices corresponding to single-peak loading (Miller

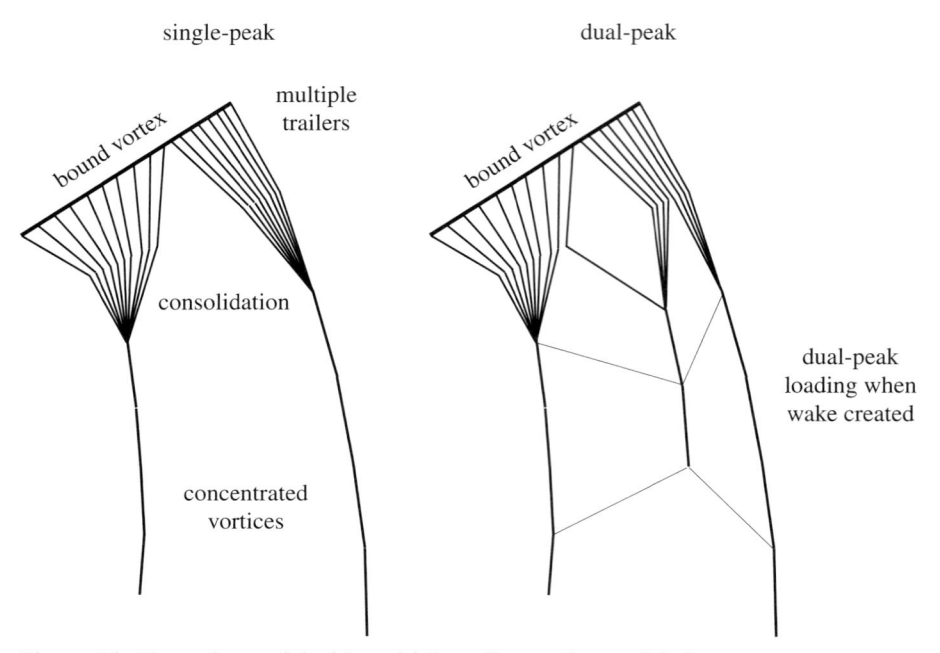

Figure 9.9. Far wake model with multiple trailers and consolidation.

(1985)). The simplest model of the dual-peak rollup assumes that the tip vortex has strength and sign equal to the outboard bound circulation peak. The inboard sheet has then both positive and negative trailed vorticity, divided at a span station corresponding to the inboard peak (Figure 9.8). The trailed vorticity between the two peaks can partially roll up as well, which can be modeled by using a line segment with a physically meaningful core radius. The single-peak model (Figure 9.7) applied to a case of dual-peak loading at least has the appropriate tip vortex strength and sign if the rollup is always driven by the outboard peak, not by the bound circulation maximum magnitude.

For a more detailed representation of the wake, the vortices trailed between collocation points on the blade can be extended into the far wake. The rollup is not well calculated even with many trailed vortex lines, because of the coarse discretization and the neglect of viscosity. Hence it is useful to impose consolidation of individual trailed vortex lines into a small number of rolled-up vortices (illustrated in Figure 9.9). The trailed vorticity at each wake age is partitioned into sets of adjacent lines that have the same sign (bound circulation increasing or decreasing). It is assumed that all the vorticity in a set eventually rolls up into a single vortex, located at the centroid of the original vorticity distribution (Betz (1932) and Rossow (1973)). For each set, the total strength Γ, centroid r_C, and moment (radius of gyration) r_G of the trailed vorticity in the set are calculated. The rate of consolidation scales with the characteristic time r_G^2/Γ (Bilanin and Donaldson (1975) and Quackenbush, Lam, Wachspress, and Bliss (1994)). For a set of trailed lines at the edge of the wake, the rollup is probably dominated by the flow over the wing tip, so the final position of the consolidated lines is at the edge rather than at the centroid. With these assumptions, the structure of the wake adjusts to accommodate dual-peak loading (Figure 9.9) or even more complicated variations of the bound circulation with azimuth and radial station. Note the vortex lines shifting between rolled-up vortices in the dual-peak

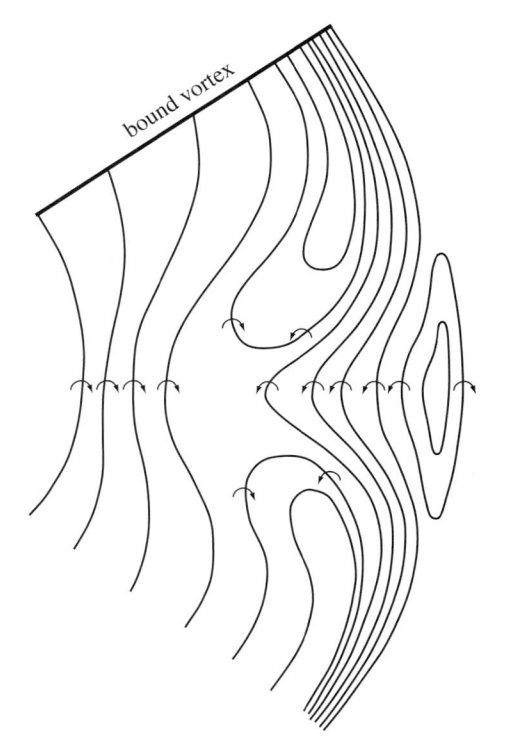

Figure 9.10. Far wake model with constant-strength curved vortex elements; from Quackenbush, Wachspress, and Boschitsch (1995).

case of Figure 9.9. Since each of the trailed vortex lines arise from a fixed radial location on the blade, the movement of a line from one rolled-up structure to another reflects an azimuthal shift in the radial location of a bound circulation peak, resulting in a change in sign of a particular trailed element. An alternative approach is to discretize the rotor wake using curved lines of constant strength instead of a fixed radial location on the blade, as illustrated in Figure 9.10. Azimuthal and radial variation of the bound circulation then produces wake filaments that follow the resultant vorticity in the sheets, instead of a vortex lattice of separate trailed and shed elements. This model has been implemented using curved vortex elements; see Bliss, Dadone, and Wachspress (1987) and Quackenbush, Wachspress, and Boschitsch (1995).

Numerous wake models are found in applications. The strong, concentrated tip vortices are well represented by line elements with an appropriate viscous core radius, and the helical geometry can be reasonably approximated using a connected series of finite-length segments. Model variations for the rolled-up structures in the wake involve the vortex core representation and the choice of straight or curved elements. The influence of the inboard vorticity is much less than that of the tip vortices, allowing greater latitude in developing a model while retaining accuracy. Model variations for the vortex elements include the choice of line or sheet elements, and the vorticity detail captured. Early work in particular involved far wake modeling assumptions intended to minimize computational effort, such as neglecting shed wake elements or neglecting all wake elements except the rolled-up tip vortices.

The induced velocity is evaluated by integrating over all vorticity in the wake. For a given wake geometry and incompressible flow, the Biot-Savart law gives a linear relation between the induced velocity and the wake strength. The wake strength is linearly related to the wing bound circulation at the current and past times. Hence the induced velocity produced by one blade at a collocation point P takes the form:

$$v_P(\psi) = \int_0^\infty \int_0^1 C(\psi, \phi; r)\Gamma(\psi - \phi; r)\, dr\, d\phi \qquad (9.34)$$

where Γ is the blade bound circulation and C the influence coefficients (calculated from the wake geometry and modeling assumptions). C is actually a differential operator, since the wake strength depends on the time and span derivatives of the bound circulation. The blade span variable is r. At dimensionless time ψ (blade azimuth angle), a wake element and its position in the wake are identified by the wake age ϕ, such that $(\psi - \phi)$ is the time that the element was created. Hence the strength of that element depends on the bound circulation at $(\psi - \phi)$, and the induced velocity is evaluated by an integral equation. The age goes from $\phi = 0$ at the current blade position to $\phi = \infty$ far behind the blade.

The bound circulation is discretized spanwise, evaluated at the aerodynamic stations r_i, $i = 1$ to M. The wake must have a consistent spanwise discretization. The strength of the wake directly behind the blade depends on the circulation at all radial stations, but in the rolled-up wake the strength depends only on the peak circulation values. The wake age is discretized by using the geometry and strength only at a set of ages $\phi_k = k\,\Delta\phi$, $k = 0$ to K, for fixed wake age increment $\Delta\phi$. Thus the integral equation for the wake-induced velocity becomes

$$v_P(\psi_\ell) = \sum_{\text{blades}}\left[\sum_{\text{peaks}} \sum_{k=0}^{K} C_{\text{peak}}(\psi_\ell, \phi_k)\Gamma_{\text{peak}}(\psi_\ell - \phi_k) \right.$$
$$\left. + \sum_{k=0}^{K}\sum_{i=1}^{M} C(\psi_\ell, \phi_k; r_i)\Gamma(\psi_\ell - \phi_k; r_i)\right] \qquad (9.35)$$

including a summation over all blades. The summation over "peaks" depends on the structure of the far wake (such as multiple rollup, dual peak, multiple trailer with consolidation). The induced velocity is calculated at a set of times ψ_ℓ, as determined by the solution procedure. The influence coefficients C depend primarily on the gross character of the wake (such as determined by advance ratio μ and thrust C_T) and are usually weak functions of the bound circulation.

9.5 Wake Geometry

In the close encounters of the rotating helicopter rotor blade with the wake from preceding blades, the induced airloads are sensitive to the relative position of the tip vortex and blade. Thus the geometry of the rotor vortex wake is a major factor in determining the nonuniform inflow and aerodynamic loading.

The wake geometry describes the position of the wake vorticity in space. The undistorted geometry is obtained from the motion of the wing: as the blade moves through the air, trailed and shed vorticity is created, and this vorticity is then convected by the wind as the blade continues to move. So the wake geometry reflects the past history of the blade motion. Relative to the rotor disk, the wake is convected downward (normal to the disk plane) by the mean induced velocity and free stream, and aft in forward flight by the in-plane component of the free stream. The self-induced velocity produces substantial distortion of the vortex filaments as they are convected. The wake geometry thus consists of distorted, interlocking helices,

9.5 Wake Geometry

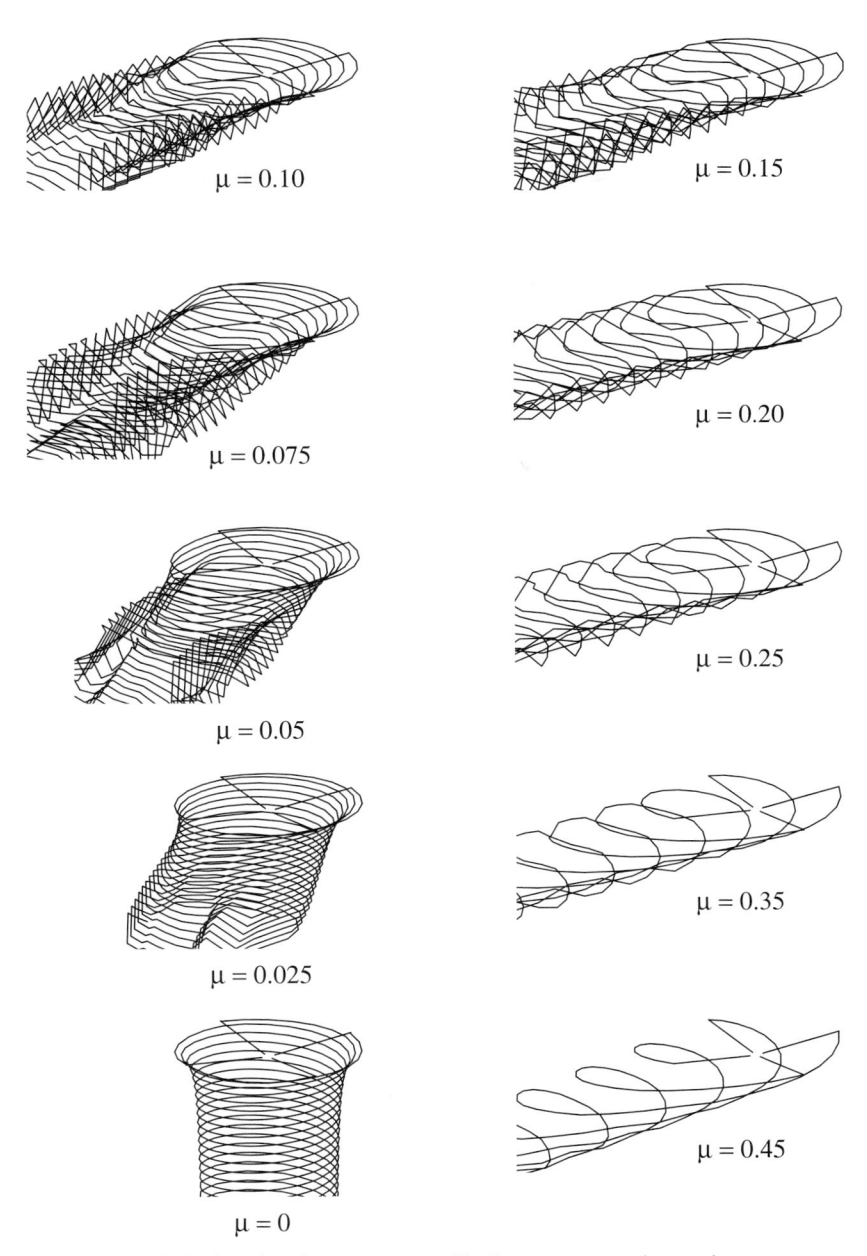

μ = 0.10

μ = 0.15

μ = 0.075

μ = 0.20

μ = 0.05

μ = 0.25

μ = 0.025

μ = 0.35

μ = 0

μ = 0.45

Figure 9.11. Calculated wake geometry of helicopter rotor tip vortices.

one behind each blade, skewed aft in forward flight. Figure 9.11 shows the calculated wake geometry of tip vortices for a helicopter rotor operating from hover to high-speed forward flight. The idealized wake geometry in hover is a contracted helix, as discussed in section 3.8.1. In forward flight, the distorted wake geometry exhibits an overall pattern in which the edges of the wake arising from the rotor disk roll up to form vortices, as behind a circular wing (Figure 9.11). These super-vortices consist of the helical tip vortices from individual blades. The consequence of this pattern is that near the rotor the tip vortices tend to move upward on the sides of the

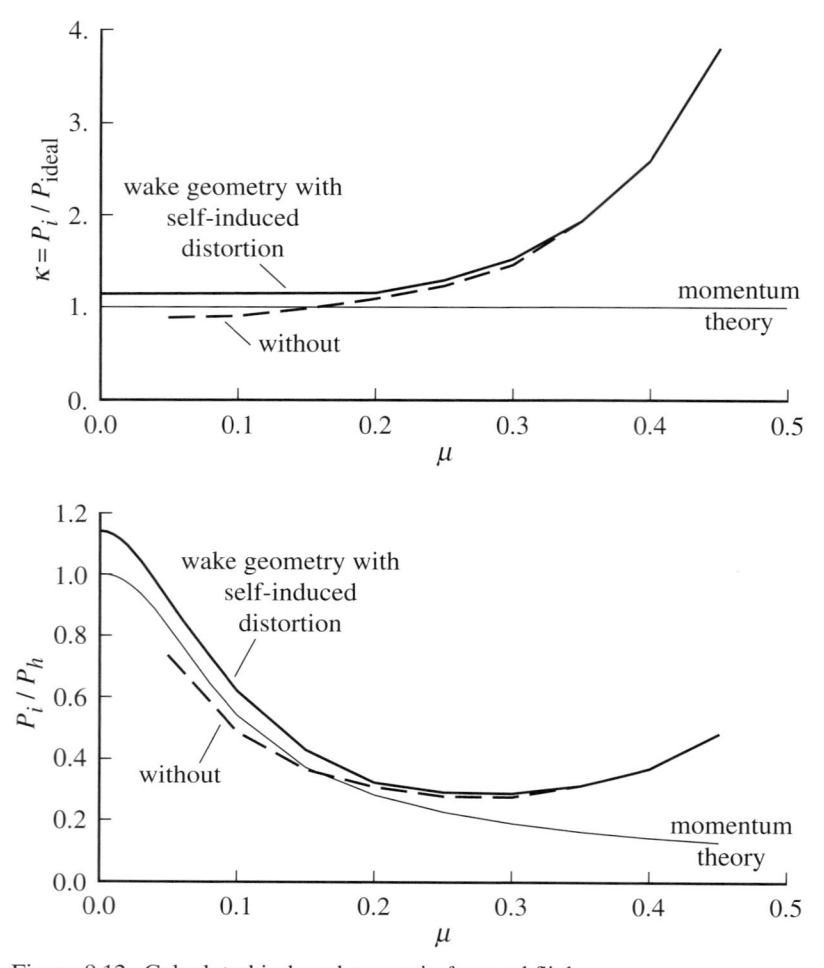

Figure 9.12. Calculated induced power in forward flight.

disk, and they tend to move downward in the middle of the disk. So the self-induced distortion moves the tip vortices closer to the blades on the advancing and retreating sides (compared to a rigid geometry), thereby increasing the blade-vortex interaction loads. The transition from hover-like to wing-like behavior occurs around $\mu \cong 0.05$ or $V/v_h \cong 1$. Simons, Pacifico, and Jones (1966) in a flow visualization experiment found that at low speed the trailing vortices from the leading edge of the rotor disk tend to pass upward through the disk first and then downward. At the rear of the disk the vortices tend to be convected downward at a rate higher than the mean inflow.

The wake geometry distortion has a large influence on the blade airloading in hover and low-speed flight, typically at advance ratios below about 0.20–0.25. At higher speeds, the helicopter rotor has a large tip-path-plane incidence relative to the free stream, typically several degrees forward, to provide the propulsive force. In such flight conditions the wake is convected rapidly aft as well as downward relative to the disk by the normal component of the free stream, and the distorted geometry is less important. Figure 9.12 shows the typical influence of the wake geometry on the

9.5 Wake Geometry

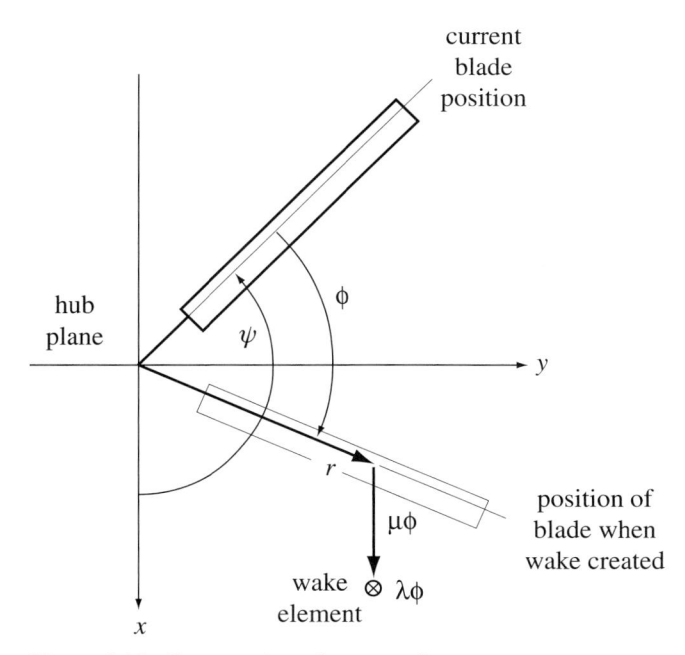

Figure 9.13. Construction of rotor wake geometry.

rotor induced power in forward flight. If the self-induced distortion is not included in the calculations, the induced power is underestimated at low speed.

The blade rotation together with the rotor flight speed creates the basic helical geometry of the wake. Figure 9.13 illustrates the construction of the wake geometry relative to the rotor hub plane. The figure shows a wake element created at radial station r on the blade, with current age ϕ. The blade is now at azimuth angle $\psi = \Omega t$ (dimensionless time). The wake element was created when the blade was at azimuth $(\psi - \phi)$, and the position of the blade at that time was $x = r\cos(\psi - \phi)$, $y = r\sin(\psi - \phi)$, and $z = r\beta(\psi - \phi)$ (including just the rigid flap motion of the blade). Since leaving the blade, the wake element has been convected with the local flow. Consider just the mean convection velocity, which has components λ and μ in the hub plane. The inflow ratio λ includes the mean induced velocity. Then the current position of the wake element is

$$x = r\cos(\psi - \phi) + \mu\phi \qquad (9.36)$$

$$y = r\sin(\psi - \phi) \qquad (9.37)$$

$$z = r\beta(\psi - \phi) - \lambda\phi \qquad (9.38)$$

which defines the basic skewed helix of the rotor wake in forward flight. Actually, there are N such helices, one behind each blade, which are obtained by substituting for ψ the azimuth angle of the m-th blade: $\psi_m = \psi + m\Delta\psi$ ($m = 1$ to N, $\Delta\psi = 2\pi/N$). As at the rotor disk, the induced velocity throughout the wake is highly nonuniform. When the nonuniform induced velocity is included in the convection of the wake element, the vertical position becomes

$$z = r\beta(\psi - \phi) - \int_0^\phi \lambda \, d\phi \qquad (9.39)$$

which defines the distorted wake geometry; the distortion in the x and y directions is required as well.

A number of wake geometry models are used in rotary-wing analyses. The rigid wake model is the undisturbed helical geometry, in which all the wake elements are convected with the same mean velocity. An elementary extension is the semi-rigid wake model, in which each element is convected downward with the induced velocity of the point on the rotor disk where it was created. The free wake or nonrigid wake model includes the distortion from the basic helix as each wake element is convected with the local flow, in particular the distortion caused by the wake self-induced velocities. The distorted geometry can be calculated or measured experimentally. When measured wake geometry information is used, it is often called a prescribed wake model.

The rigid wake model is the simplest geometry, requiring a negligible amount of computation. It is also the furthest from the true rotor wake geometry, which can involve significant distortion. If the flight condition is such that the wake is convected away from the disk (large tip-path-plane incidence at high speed, or high climb rates), and hence there is no significant vortex-blade interaction, then a rigid wake geometry can be a satisfactory model. The semi-rigid wake geometry does not require additional computation, using only the nonuniform inflow at the rotor disk. The assumption that the wake elements are convected with the velocity at the disk should be good for small age, but not very accurate by the time the vortex encounters the following blade. Thus the semi-rigid wake model generally offers little improvement over the rigid wake model. When the helicopter operating state is such that the wake remains close to the rotor disk, the distortion of the wake geometry has a large effect on the loading, and the free wake model is required. Calculating the distorted wake geometry requires evaluating the induced velocity throughout the wake rather than just at the rotor disk, and therefore it involves a very large amount of computation. The use of a prescribed wake model is limited by the necessity of performing measurements for the required rotor and flight condition. The choice of the wake geometry model is usually a balance between accuracy and economy.

The basic undistorted geometry is calculated from the position in the air at which the wake element was created, plus convection by the wind. Then the distortion produced by the self-induced velocities is added. Let ψ be the current time and ϕ the age of the element in the wake. Thus $\delta = \psi - \phi$ is the time when the vorticity was created and identifies a particular element in the wake. The position of a the wake element in space is

$$r_W(\psi, \phi) = r_Q(\psi - \phi) + W\phi + D(\psi, \phi) \tag{9.40}$$

or

$$r_W(\psi, \delta) = r_Q(\delta) + W(\psi - \delta) + D(\psi, \delta) \tag{9.41}$$

Here r_Q is the blade position, including rotor forward and axial speed, evaluated at past times. Convection by the wind or gusts gives the distortion $\int_{\psi-\phi}^{\psi} W \, d\psi = W\phi$ if the velocity W is constant. The wake geometry distortion $D(\psi, \phi)$ is the perturbation of the position from the undistorted geometry. By definition the wake geometry connects to the wing at $\phi = 0$, so $D(t, 0) = 0$. The undistorted part of the wake geometry can be evaluated at any time ψ and age ϕ as required. The geometry and distortion are functions of time ψ and of either wake age ϕ or the element creation

9.5 Wake Geometry

time δ. It is convenient to store the geometry information as a function of age ϕ, but following the motion of a wake element is done for fixed δ.

The rigid geometry is calculated by assuming that the wake elements are all convected by the average interference velocity λ_{mean}. This uniform convection gives $D(\psi, \phi) = \lambda_{\text{mean}} \phi$.

The prescribed distortion for a hovering rotor is described by two-stage vertical convection and exponential spanwise contraction; see section 3.8.1. The vertical and spanwise distortion (scaled with radius R) are

$$D_z = \begin{cases} K_1 \phi & \phi < \phi_1 \\ K_1 \phi_1 + K_2(\phi - \phi_1) & \phi > \phi_1 \end{cases} \quad (9.42)$$

$$D_r = \left(1 - e^{-K_3 \phi}\right)(1 - K_4) \quad (9.43)$$

where $\phi_1 = 2\pi/N$ is the age at the encounter with the following blade (N is the number of blades). Section 3.8.1 summarizes the prescribed wake geometry models of Landgrebe (1971, 1972) and Kocurek and Tangler (1977). This geometry is steady relative to the moving wing, so D_z and D_r are functions of only wake age ϕ, not time. The vertical convection is defined by K_1 and K_2, the rates before and after encountering the first following blade. The spanwise contraction is defined by K_3 and K_4, the rate of contraction and the maximum contraction ratio, respectively.

The major issue in a free wake analysis is developing an efficient yet accurate procedure for performing the calculations. Conceptually, the free wake calculation is simple. At each time step, the induced velocity is evaluated at every element in the wake by summing the contributions from all elements in the wake (as in a calculation of the nonuniform inflow at the rotor disk). Then the convection velocities are numerically integrated to obtain the positions of the wake elements at the next time step. Developing special procedures for economic calculation of the distorted wake is important. Among the techniques developed are the following:

a) When calculating the velocity at a collocation point, the wake is divided into "near wake" and "far wake" regions. The velocity contribution of the far wake is small, so when the velocity is required again in the algorithm, only the contribution of the near wake need be recalculated (Landgrebe (1969)).
b) During the basic time step, the existing wake is convected by the wind and the induced velocity with little relative distortion, while the blades move and generate new wake behind the trailing edges. Thus the minimal update of the induced velocities involves just adding the contribution of this new vorticity (Scully (1975)).
c) Integration of the blade-induced velocity in time is approximately equal to the average velocity from the bound vortex (a vortex line segment), which is equivalent to the velocity from a vortex sheet element (Scully (1975)).
d) For steady-state operating conditions, the distortion is constant or periodic (in an appropriate frame). Thus the convergence of the calculation can be improved by including an iteration between revolutions with distortion increment relaxation and propagation as well as an outer iteration with distortion relaxation (Scully (1975), Johnson (1995)).
e) The effect of the core vorticity distribution on the self-induced velocity of a vortex arc is obtained using an appropriate cutoff distance (Bliss, Teske, and Quackenbush (1987)).

f) Implicit time integration methods (with relaxed iteration) are used for stability. In the predictor-corrector form, these methods require only one evaluation of the velocity per time step (Bagai and Leishman (1995), Bhagwat and Leishman (2001)).

g) If a group of wake elements is a large distance from a group of collocation points where the induced velocity is required, a multipole representation of the wake elements can be developed and the velocity calculated at the collocation points using a Taylor series expansion (Wachspress, Quackenbush, and Boschitsch (2003)).

Efficiency of computation continues to be a priority, as advances in computational capability are used to enable the solution of larger and more complicated problems.

Following Helmholz, a point on a vortex wake is convected with the local fluid velocity. Thus the equation of motion of a wake node, identified by the time when the vorticity was created ($\delta = \psi - \phi$), is

$$\frac{dr_W(\psi, \delta)}{d\psi} = W + q(\psi, \delta) \tag{9.44}$$

with q the self-induced velocity and W the wind. Integrating over time from $\psi - \phi$ to ψ gives

$$r_W(\psi, \delta) = r_W(\psi - \phi, \delta) + W\phi + \int_{\psi - \phi}^{\psi} q(\psi = \sigma, \delta) \, d\sigma$$

$$= r_Q(\psi - \phi) + W\phi + D(\psi, \phi) \tag{9.45}$$

So the distortion is calculated by integrating just the self-induced velocity q over time:

$$D(\psi, \phi) = \int_{\psi - \phi}^{\psi} q(\psi = \sigma, \phi = \sigma - \delta) \, d\sigma \tag{9.46}$$

for fixed δ. The induced velocity at a given time is evaluated by integrating over all vorticity in the wake. For incompressible flow, the Biot-Savart law gives the velocity as an integral of the wake strength times an influence coefficient:

$$q(\psi) = \iint C(\psi, \phi; r) G(\psi, \phi; r) \, dr \, d\phi \tag{9.47}$$

where ϕ is the wake age and r the wing span variable, so the integral is over the wake surface. The influence coefficient C depends on the wake geometry. The wake strength G depends on the wing bound circulation at past times, $\Gamma(\psi - \phi; r)$. The wake age is discretized by using the geometry and strength only at a set of ages $\phi_k = k \Delta\phi$, $k = 0$ to K, for fixed wake age increment $\Delta\phi$. Time is discretized in the distortion calculation, with fixed increment $\Delta\psi$. Typically implementation of the integration algorithm requires that the time increment and the wake age increment be equal. The discretized integral is

$$D(\psi, \phi) = D(\psi - \Delta\psi, \phi - \Delta\psi) + \int_{\psi - \Delta\psi}^{\psi} q(\psi = \sigma, \phi = \sigma - \delta) \, d\sigma$$

$$= D(\psi - \Delta\psi, \phi - \Delta\psi) + \Delta D \tag{9.48}$$

where ΔD is the distortion increment, evaluated from the velocity at the current and past times, and constant δ. An integration algorithm that consists of a predictor followed by an iterated corrector with relaxation has the following form:

$$\text{predictor: } \Delta D_j = \Delta\psi\, P(q_{j-1}, \ldots)$$

$$\text{evaluate } q_j \text{ from } D_j$$

$$\text{relaxation: } q_j = \lambda_R q_j + (1 - \lambda_R) q_{j\text{old}}$$

$$\text{corrector: } \Delta D_j = \Delta\psi\, C(q_j, \ldots)$$

where subscript j indicates q or D at time $\psi = j\Delta\psi$. Several integration algorithms have been implemented. For all of the following, the predictor is $\Delta D_j = \Delta\psi\, q_{j-1}$.

a) Euler (first order): corrector $\Delta D_j = \Delta\psi\, q_j$. Using only the predictor gives explicit Euler integration (similar to Johnson (1995)).
b) Trapezoidal (second order): corrector $\Delta D_j = \frac{\Delta\psi}{2}(q_j + q_{j-1})$. A single pass (no iteration) gives the trapezoidal predictor-corrector algorithm. The result with one iteration and $\lambda_R = \frac{1}{2}$ is similar to the PIPC algorithm of Bagai and Leishman (1995).
c) Backward difference (second order): corrector $\Delta D_j = -\frac{2}{3}D_{j-1} + D_{j-2} - \frac{1}{3}D_{j-3} + \frac{2}{3}\Delta\psi(q_j + q_{j-1})$. The result with one iteration and $\lambda_R = \frac{1}{2}$ is similar to the PC2B algorithm of Bhagwat and Leishman (2001).

If the relative distortion of the wake from $\psi - \Delta\psi$ to ψ is ignored, then during this time the only change is the addition of the new wake of age $\phi = 0$ to $\Delta\psi$ directly behind the wings. Then $q(\psi, \phi) \cong q(\psi - \Delta\psi, \phi - \Delta\psi) + q_1(\psi)$, where q_1 is the velocity of the new wake generated behind the wing during $\psi - \Delta\psi$ to ψ. With this approximation, trapezoidal integration gives $\Delta D/\Delta\psi = q(\psi - \Delta\psi, \phi - \Delta\psi) + \frac{1}{2}q_1(\psi)$. Johnson (1995) used this integration method. Similarly, Scully (1975) used $q(\psi - \Delta\psi, \phi - \Delta\psi) \cong q(\psi, \phi) - q_1(\psi)$; hence $\Delta D/\Delta\psi = q(\psi, \phi) - \frac{1}{2}q_1(\psi)$.

Since the velocity and distortion are periodic in time, relaxation of the distortion increment can be introduced to improve convergence. In addition, propagation of the distortion information can be included in the trim solution. The distortion increment $\Delta D = q\,\Delta\psi$ is not the same as that calculated during the last revolution; the difference is $\delta D = \Delta D - \Delta D_{\text{old}}$. The difference δD affects all future values of D at this $\psi - \phi$. Since the trim distortion is periodic, δD also affects values of D at past times and larger age. Thus the propagation procedure adds δD to all values of $D(\psi, \phi)$ at future time ψ and fixed $\psi - \phi$.

The distortion can be required at an age ϕ beyond which it has been calculated. Just stopping the wake model at some age produces inaccurate results, characterized by spurious instabilities. At the last age where the distortion is required, the calculation of induced velocity must include several revolutions of wake beyond that point. Extrapolation of the geometry is accomplished by following the path of a specific wake element in time; hence by considering the geometry for a fixed value of the time that a wake element was created, $\delta = t - \phi$. Let ϕ_{last} be the maximum age of the available distortion. The distortion is extrapolated by assuming that the vortex element is convected for time $(\phi - \phi_{\text{last}})$ by a constant velocity:

$$D(\psi, \phi) = D\big(\psi - (\phi - \phi_{\text{last}}), \phi_{\text{last}}\big) + (\phi - \phi_{\text{last}})\lambda_{\text{conv}} \qquad (9.49)$$

The convection velocity λ_{conv} can be obtained from the average distortion increment at ϕ_{last}:

$$\lambda_{\text{conv}} = \frac{1}{2\pi} \sum_{\psi} \left(D(\psi, \phi_{\text{last}}) - D(\psi - \Delta\psi, \phi_{\text{last}} - \Delta\psi) \right) \frac{1}{\Delta\psi} \qquad (9.50)$$

where the average is over the last revolution of the free wake.

Generally the calculation of the wake-induced velocities on the nodes of the wake is similar to the calculation of nonuniform inflow at points on the rotor disk. However, special treatment is required for the induced velocity from vortex elements adjacent to a collocation point. For a collocation point on a rolled-up trailed vortex line, the induced velocity from the two adjacent line segments is calculated by replacing them by circular-arc vortex line segments. The effect of the core vorticity distribution on the self-induced velocity of a vortex arc is obtained using an appropriate cutoff distance (Bliss, Teske, and Quackenbush (1987); see section 9.7). The cutoff distance is proportional to the vortex core radius. If instead straight line segments are used adjacent to the collocation point, these segments do not contribute to the self-induced velocity, and the model is equivalent to a cutoff distance proportional to the wake age discretization.

Integration of the blade-induced velocities gives a distortion increment:

$$\Delta D = \int_{\psi - \Delta\psi}^{\psi} q_{\text{wing}}\, dt = q_B \Delta\psi \qquad (9.51)$$

so q_B is the average blade-induced velocity. For an instantaneous value of the blade-induced velocity, a good model of the flow about the wing would be required. For the average value, representing the blade by a bound vortex appears to be adequate. The average velocity from the bound vortex (a line vortex) as the blade moves from time $\psi - \Delta\psi$ to time ψ is equivalent to the velocity produced by a vortex sheet defined by the (convected) blade positions.

For a three-dimensional wing, the Kutta condition requires that the wake leave the trailing edge tangent to the wing surface (Hess (1974)). In the absence of a calculation of the detailed flow field near the wing, this requirement can be satisfied by using an initial convection velocity $q_K = \Gamma/\pi c$, where Γ and c are the section bound circulation and chord. With the bound vortex at the quarter chord, this is its induced velocity at the three-quarter chord. This result is obtained using the zero-lift chord line for the trailing-edge bisector, and a lift-curve slope of 2π. The velocity direction can be obtained from the geometry of the bound vortex and the collocation point. The initial velocity q_K is used at $\phi = 0$, and the wake-induced velocity q is used at ages greater than some ϕ_K, with ϕ_K selected based on correlation with measured wake geometry and performance.

In the undistorted wake geometry, the initial span station of the tip vortex can be obtained from Betz rollup, which is calculated assuming that the centroid of the rolled-up trailed vorticity is conserved (Betz (1932) and Rossow (1973)). Consider the bound vorticity from r_A to r_B, rolling up into a trailed line. The centroid of the trailed vorticity is at r_C:

$$r_C(\Gamma_A - \Gamma_B) = \int_{r_A}^{r_B} -\frac{\partial \Gamma}{\partial r} r\, dr = \Gamma_A r_A - \Gamma_B r_B + \int_{r_A}^{r_B} \Gamma\, dr$$
$$= \Gamma_A r_A - \Gamma_B r_B - (r_A - r_B)\Gamma_M \qquad (9.52)$$

in terms of the mean bound circulation Γ_M.

A factor in the wake geometry is the stability of the vortex elements. A vortex line is susceptible to short wavelength instabilities, and a helical vortex also exhibits long wavelength instabilities due to the interaction of successive turns of the helix. Such instabilities are generally of secondary importance for the loading, since they occur in the wake some distance from the rotor. Also, the concept of a unique geometry is an idealization, because in the real flow, turbulence and vortex instabilities can produce significant variations in the wake geometry with time, even for a nominally steady operating state.

A wake stability analysis begins with the equation for convection of vortex elements: $d\mathbf{r}/d\psi = \mathbf{V}(\mathbf{r})$, where $\mathbf{r}$ is a point on the vortex and $\mathbf{V}$ is the velocity induced there by all vorticity in the wake; see Bhagwat and Leishman (2000). The periodic solution of this equation for a given operating condition (convection velocity and wake strength) is $\mathbf{r}_0$. The perturbation of the wake geometry relative to this solution is governed by the equation

$$\frac{d}{d\psi}\delta\mathbf{r} = \mathbf{V}(\mathbf{r}_0 + \delta\mathbf{r}) - \mathbf{V}(\mathbf{r}_0) = \delta\mathbf{V}(\delta\mathbf{r}) \tag{9.53}$$

The model is discretized by considering a finite set of collocation points and expanding $\delta\mathbf{V}$ in terms of all the $\delta\mathbf{r}$ variables. The differential equations for the collection of $\delta\mathbf{r}$ generate an eigenvalue problem when it is assumed that the perturbation position is proportional to $e^{\alpha\psi + i\omega\phi}$, with ϕ the wake age. Here α is the growth rate and ω is the wave number (ω cycles per rotor revolution). The solution for the eigenvalues gives the maximum α as a function of ω. Typically α has peaks at half-integer values of ω/N ($\omega = (k + \frac{1}{2})N$, where N is the number of blades), for which the vortex filaments are out-of-phase.

Widnall (1972) examined the stability of a vortex filament of finite core size and infinite extent, considering small sinusoidal displacements of the vortex center-line. A helical vortex filament under its own influence rigidly translates and rotates. Three modes of instability were found. The first mode is a very short-wave instability, characteristic of curved line vortices. The principal length scale is the local radius of curvature. The second mode is a long-wave mode, typically appearing as a displacement of the helix boundary. The third mode is a mutual-inductance mode, which appears as the pitch of the helix decreases and the neighboring turns of the helix begin to interact strongly, analogous to the instability of a line of two-dimensional point vortices or a vortex sheet. Typically the maximum growth rate occurs when the perturbation is 180° out-of-phase on successive turns. Accounting for the vortex core is essential to the analysis of these instabilities. The wavelength of the short-wave mode decreases as core radius decreases. Decreasing the core radius increases the amplification rate of the long-wave mode, but decreases the amplification rate of the mutual-inductance mode. At large amplitudes, the mutual-induction mode results in adjacent turns of the helix wrapping around each other. In a vertical cross-section of the wake, vortex pairing occurs, with vortices rotating about each other until they change places, a phenomenon observed experimentally in hovering rotor tests by Landgrebe (1972), Tangler, Wohlfeld, and Miley (1973), Caradonna et al. (1999), and Martin, Bhagwat, and Leishman (1999).

9.6 Examples

As an example of the influence of the wake model on rotor aerodynamics, consider an articulated rotor with solidity $\sigma = 0.08$, Lock number $\gamma = 8$, and twist

$\theta_{tw} = -8°$. Section 6.13 presented results based on momentum theory inflow and ideal aerodynamics. The calculations here include nonuniform inflow with a free wake geometry. Also, airfoil tables for advanced helicopter airfoils give the effects of stall and compressibility (the airfoils used have a zero-lift angle-of-attack of about $-1.6°$). The section aerodynamic model includes attached flow, unsteady aerodynamic loads, but not dynamic stall. The rotor has four blades and a root cutout of $12\% R$. The only blade motion is rigid flap and pitch, with no hinge offsets from the center of rotation.

The influence of twist is examined by considering $\theta_{tw} = 0$ and $\theta_{tw} = -16°$. The baseline operating condition is thrust $C_T/\sigma = 0.08$ and propulsive force corresponding to fuselage drag $f/A = 0.008$. A thrust variation of $C_T/\sigma = 0.12$ is examined, as are propulsive force variations of $f/A = 0$ and $f/A = 0.016$. The calculations were performed by adjusting collective pitch and shaft incidence to match the required thrust and propulsive force, with cyclic pitch for zero flapping relative to the shaft.

Figure 9.14 shows the variation of the section angle-of-attack as the blade moves around the rotor disk, for the baseline condition ($C_T/\sigma = 0.08$, $f/A = 0.008$, $\theta_{tw} = -8°$) at advance ratios of $\mu = 0.15, 0.30$, and 0.45. Figure 6.18 shows the corresponding uniform inflow results. To compare the two figures, $1.6°$ must be added to the nonuniform inflow results, because of the positive lift at zero angle-of-attack. With nonuniform inflow, the angle-of-attack distribution exhibits blade-vortex interaction at low speed and somewhat more negative loading on the advancing tip at high speed. At high speed, uniform-inflow theory predicts that the maximum angle-of-attack occurs on the tip of the retreating blade ($r = 1$ and $\psi \cong 240°$), while nonuniform inflow shifts the stall inboard and into the third quadrant. Moreover, the vortex wake increases the rate of change of angle-of-attack, which influences the dynamic stall occurrence. Thus nonuniform inflow substantially alters the angle-of-attack distribution.

The blade angle-of-attack distribution depends on the operating condition (thrust and propulsive force) and twist, as shown in Figure 9.15. Figure 6.19 shows the corresponding uniform inflow results. With the effects of stall included, this rotor cannot achieve $C_T/\sigma = 0.16$ at high speed. With nonuniform inflow there is much less difference between the twisted and untwisted blades than with uniform inflow. With the vortex wake model, the angle-of-attack distribution for zero twist is very different from that obtained using uniform inflow, particularly on the retreating side of the disk. As with uniform inflow, large twist ($\theta_{tw} = -16°$) leads to negative angle-of-attack on the advancing tip even for $\mu = 0.30$.

Figure 9.16 shows the radial and azimuthal variation of the blade section lift at advance ratios of $\mu = 0.15, 0.30$, and 0.45, for the baseline condition ($C_T/\sigma = 0.08$, $f/A = 0.008$, $\theta_{tw} = -8°$). Figure 6.21 shows the corresponding uniform inflow results. Figure 9.17 shows the radial and azimuthal variation of the wake-induced downwash. Blade-vortex interaction on the advancing and retreating sides of the disk is evident in the loading and downwash at low speed. At high speed, the downwash is far from uniform over the rotor disk, with substantial influence on the loading (compare to Figure 6.21). For reference, momentum theory gives the inflow value $\lambda = 0.021, 0.011$, and 0.007 at $\mu = 0.15, 0.30$, and 0.45.

The section lift (Figures 9.16 and 6.21) is presented as $d(C_T/\sigma)/dr = F_z/(\rho(\Omega R)^2 c)$. In an aerodynamic investigation, commonly the section normal force is presented as $M^2 c_n = N/(\frac{1}{2}\rho c_s^2 c)$. Both forms are scaled using a constant velocity (ΩR or c_s). The lift coefficient $c_\ell = L/(\frac{1}{2}\rho U^2 c)$ would show the occurrence of stall,

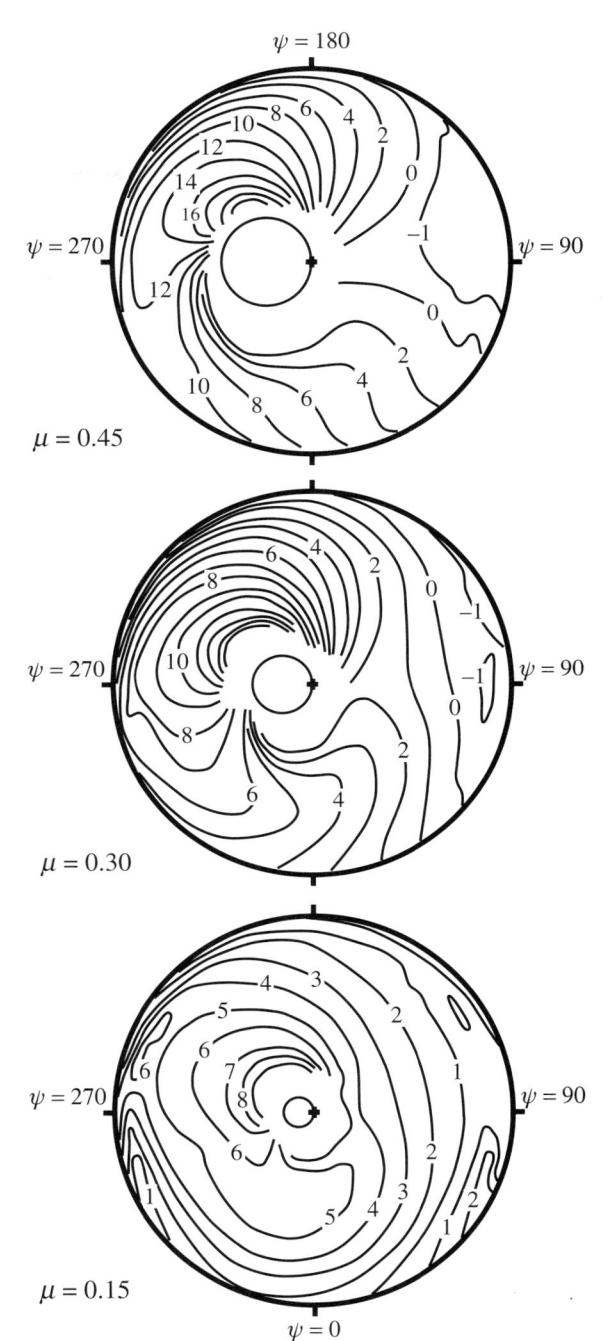

Figure 9.14. Blade angle-of-attack distribution (in degrees) for $C_T/\sigma = 0.08$, $f/A = 0.008$, and $\theta_{tw} = -8°$ (nonuniform inflow); $\mu = 0.15, 0.30, 0.45$.

but neither the lift L (local wind-axis force) nor the section resultant velocity U is available from measurements or high-fidelity analyses.

There is also a significant effect of the wake-induced velocities on the blade motion, even the mean and 1/rev flapping, and hence on the collective and cyclic pitch control angles required for a given flight state. A major influence of nonuniform inflow is to increase the lateral flapping and hence the lateral cyclic required to

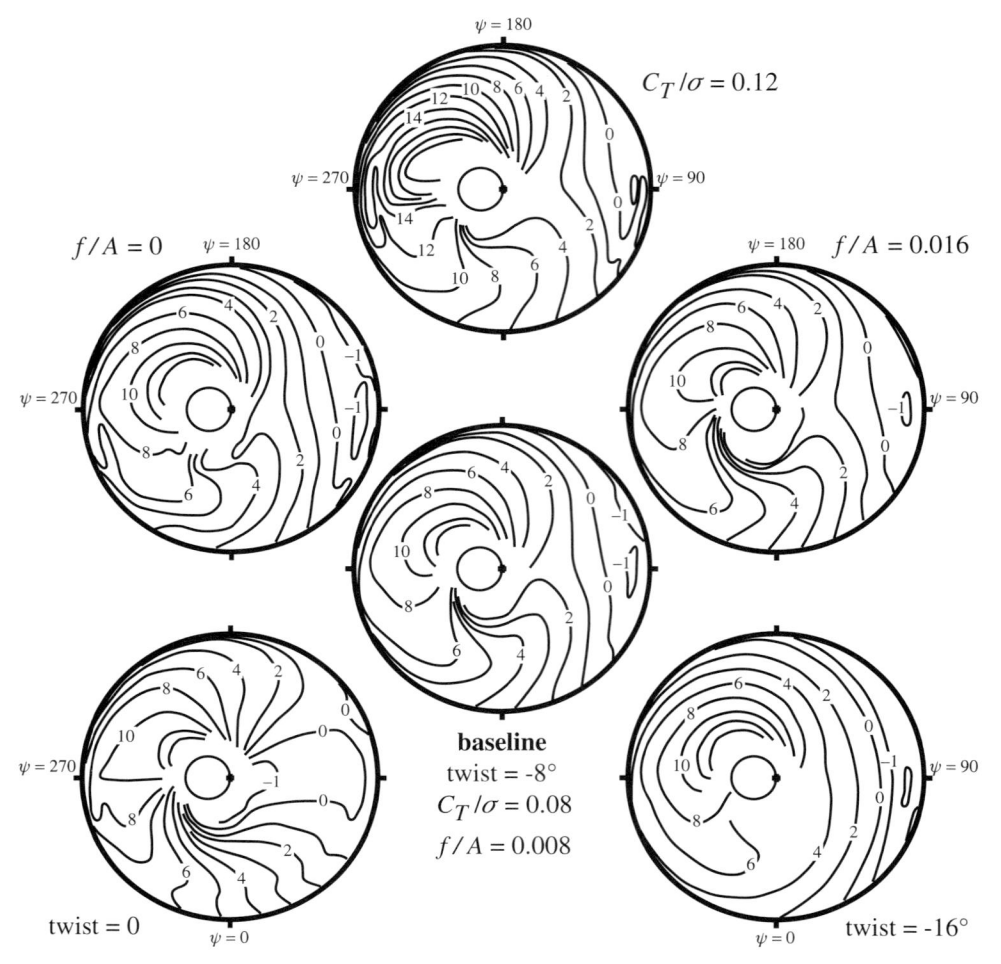

Figure 9.15. Variation of blade angle-of-attack distribution at $\mu = 0.30$ (in degrees) with thrust C_T/σ, propulsive force f/A, and twist θ_{tw} (nonuniform inflow).

trim the rotor. Figure 9.18 compares measured lateral flapping at low speed (from Harris (1972)) with calculations using free wake, rigid wake, and momentum theory inflow models. The rotor was articulated, with a small hinge offset (2.3%R) and solidity $\sigma = .0891$. The operating condition is $C_T/\sigma = 0.08$, fixed cyclic control, and shaft angle for $i_{\text{TPP}} = 1.0°$. Momentum theory (uniform inflow, without an empirical inflow gradient over the disk) predicts a small negative β_{1s} that increases steadily with μ. The experimental results show instead a large lateral flap magnitude at low speed, with a peak around $\mu = 0.08$ in Figure 9.18. A free wake geometry analysis is required to satisfactorily calculate the lateral flapping. In contrast, the longitudinal flapping angles are well predicted even with simple inflow models. Examination of the solution identifies the blade-vortex interaction on the advancing and retreating sides of the rotor disk as the source of a large longitudinal inflow variation (see Figure 9.17), the 1/rev component of which is responsible for a lateral flapping angle increment. Interaction with the rolled-up tip vortices produces large peak inflow values. The self-induced distortion of the wake moves the tip vortices close to the blades (much closer than suggested by rigid wake geometry), thereby increasing the strength of the blade-vortex interaction.

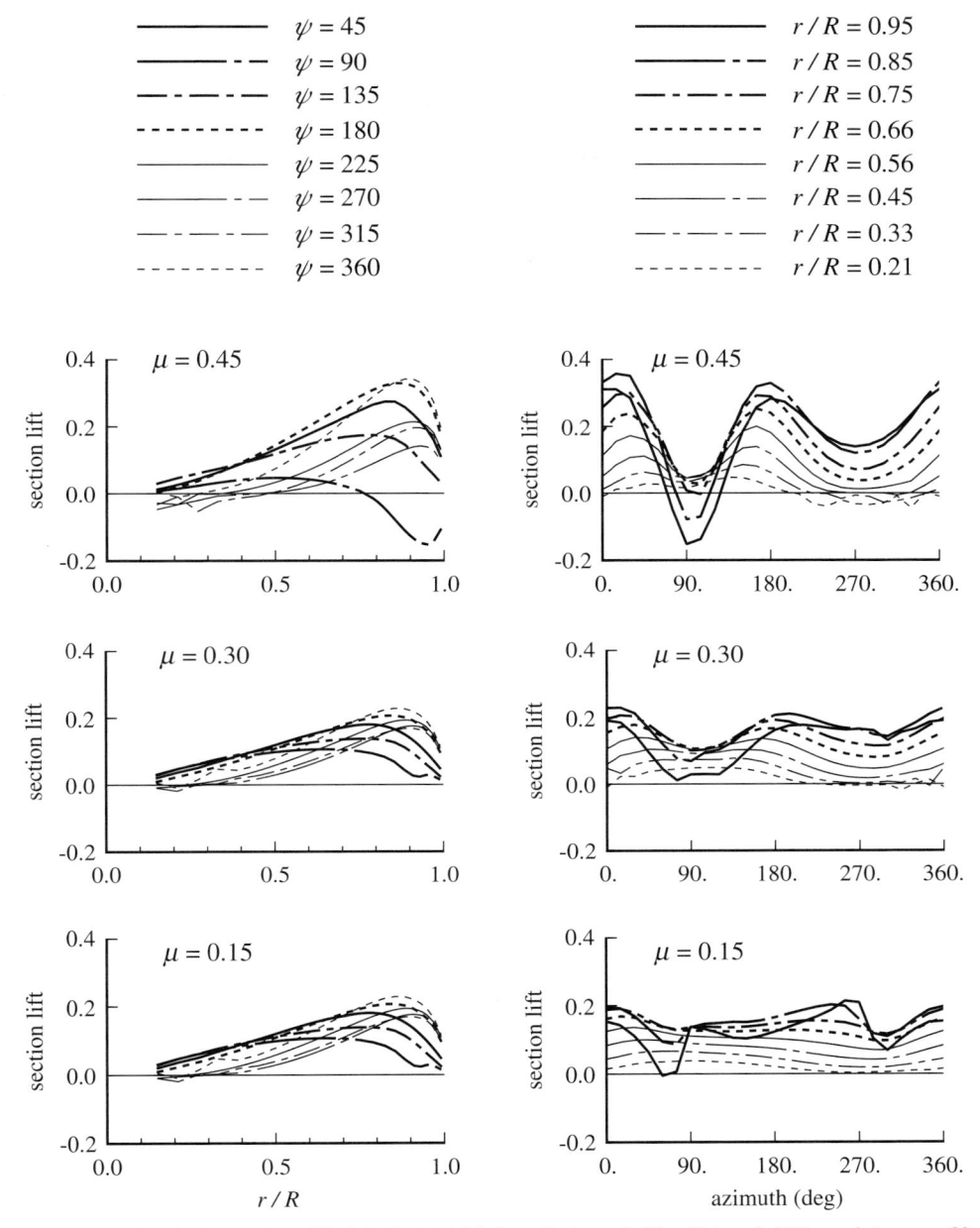

Figure 9.16. Blade section lift ($d(C_T/\sigma)/dr$) for $C_T/\sigma = 0.08$, $f/A = 0.008$, and $\theta_{tw} = -8°$ (nonuniform inflow); $\mu = 0.15, 0.30, 0.45$.

9.7 Vortex Core

A line vortex is an idealization in which a finite amount of vorticity is concentrated into a line of infinitesimal cross-section. There is a singularity at such a vortex line, with the induced velocity increasing as the inverse of the distance from the line. In the real fluid, viscosity eliminates this singularity by diffusing the vorticity over a small but finite region, called the vortex core. The maximum induced velocity occurs at some distance from the center of the line vortex, which is defined as the core radius. The vortex core is an important factor in the induced velocity character,

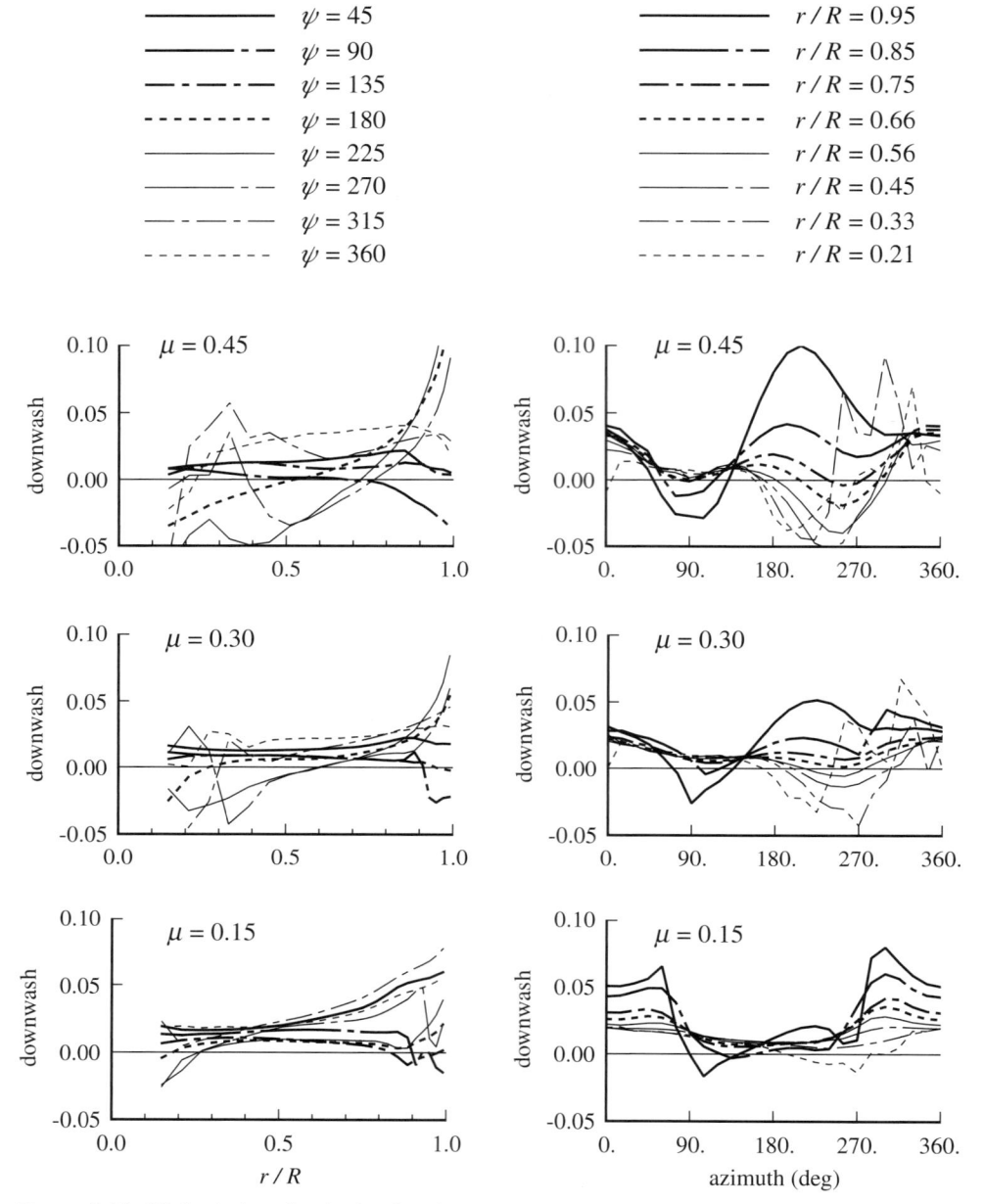

———	$\psi = 45$	———	$r/R = 0.95$
— - —	$\psi = 90$	— - —	$r/R = 0.85$
— - — - —	$\psi = 135$	— - — - —	$r/R = 0.75$
- - - - - -	$\psi = 180$	- - - - - -	$r/R = 0.66$
———	$\psi = 225$	———	$r/R = 0.56$
— - —	$\psi = 270$	— - —	$r/R = 0.45$
— - — - —	$\psi = 315$	— - — - —	$r/R = 0.33$
- - - - - -	$\psi = 360$	- - - - - -	$r/R = 0.21$

Figure 9.17. Wake-induced velocity for $C_T/\sigma = 0.08$, $f/A = 0.008$, and $\theta_{tw} = -8°$ (nonuniform inflow); $\mu = 0.15, 0.30, 0.45$.

determining the maximum velocities near the tip vortices. Because the rotor blade often passes very close to the tip vortices from preceding blades, the vortex core has a significant role in the wake-induced velocity of the rotor and must be included in the representation of the wake vorticity.

Consider the tangential or circumferential velocity v about a line vortex, at a distance r from the line. The vortex strength is given by the circulation Γ. For a potential line vortex (no core), $v = \Gamma/2\pi r$. For small r, viscosity reduces the magnitude of v by spreading the vorticity over a nonzero domain instead of a line. The

9.7 Vortex Core

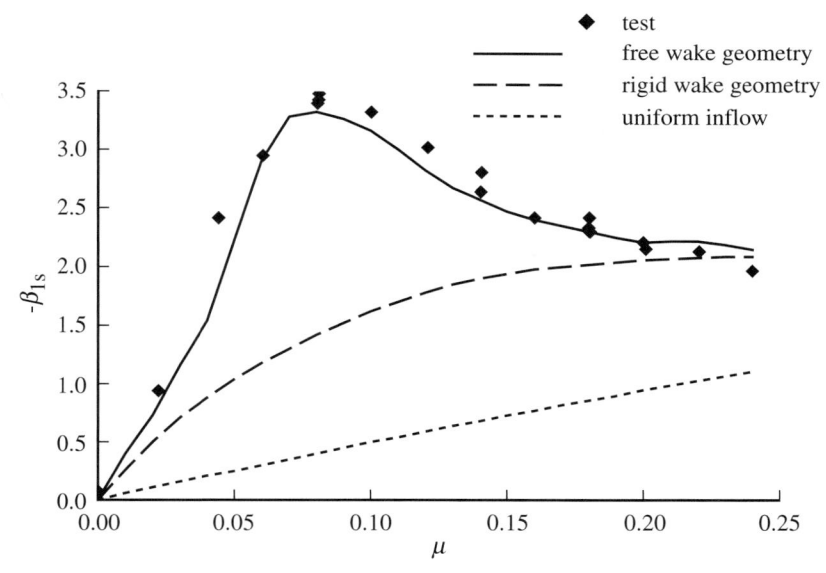

Figure 9.18. Influence of wake model on calculated lateral flapping; measured data from Harris (1972).

core radius is defined at the point of maximum tangential or circumferential velocity: the core radius r_c is the distance r at which the maximum value of v is encountered. Figure 9.19 shows the circumferential velocity as a function of r for the same vortex strength and several values of the core radius: 100%, 75%, 50%, and 33% of the nominal radius r_0. At large distances from the center, the velocity approaches that of a potential vortex, but the peak velocity is inversely proportional to r_c.

For fixed vortex strength Γ, the peak velocity depends on the distribution of vorticity in the core. From the swirl velocity v, the circulation is $\gamma = 2\pi r v$ and the vorticity is $\zeta = \frac{1}{r}\frac{d}{dr} r v$. For the potential vortex, $v = \Gamma/2\pi r$, so $\gamma = \Gamma$ and all the vorticity is concentrated at $r = 0$. With a finite core, let $\gamma = f(r)\Gamma$ be the circulation

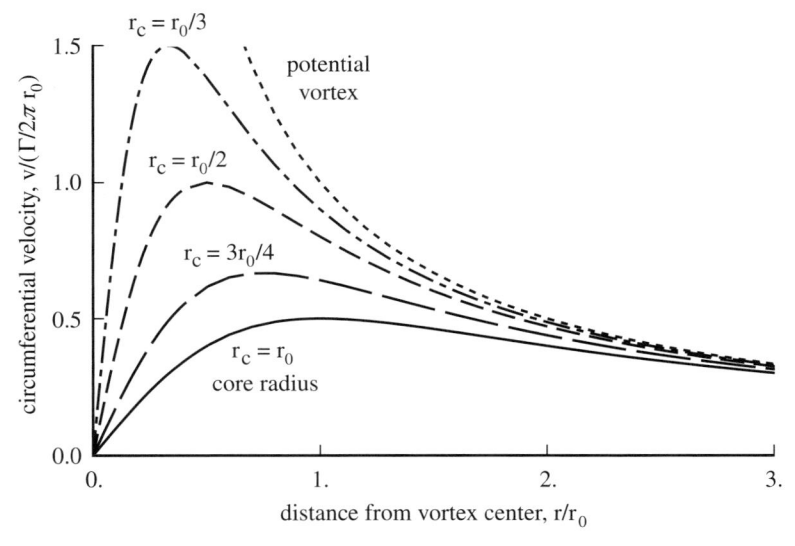

Figure 9.19. Tip vortex core radius and peak velocity.

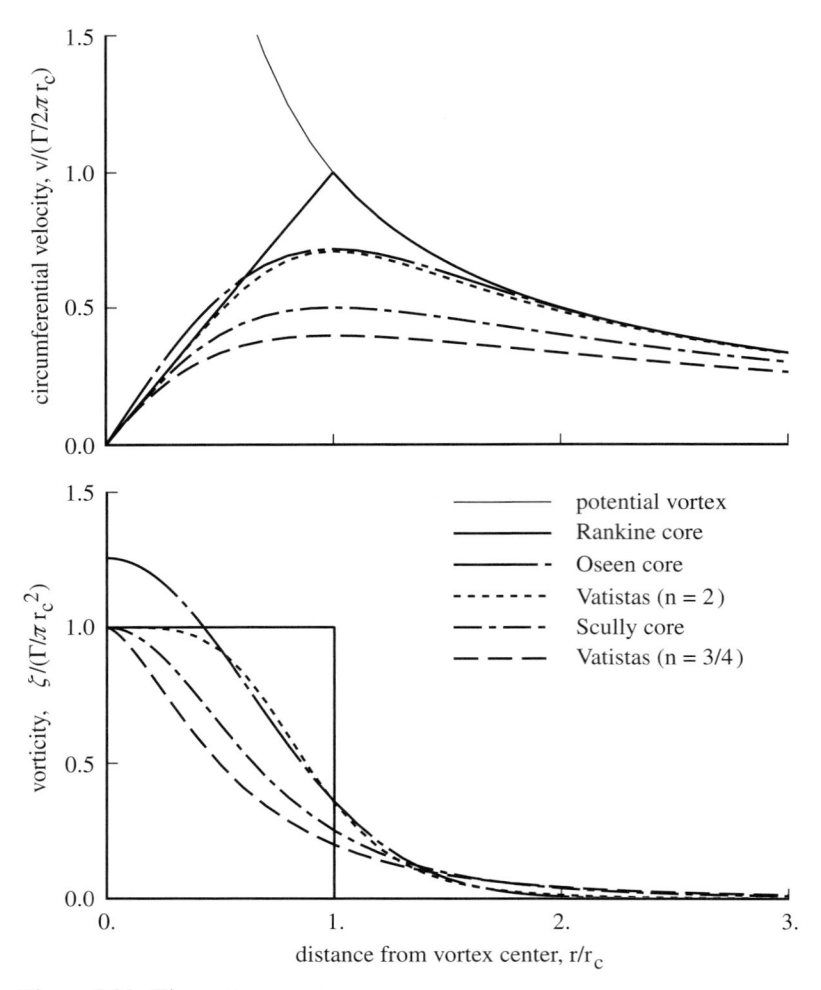

Figure 9.20. Tip vortex core types.

at $r(f(0) = 0$ and $f(\infty) = 1)$; so $v = f\Gamma/2\pi r$. Then for fixed strength Γ and core radius r_c, the maximum velocity about the vortex is determined by the value of f at r_c. The peak velocity magnitude is reduced as the vorticity outside of r_c is increased. Figure 9.20 compares the circumferential velocity for several distributions of vorticity in the core.

A Rankine vortex core has solid body rotation of the fluid inside r_c, produced by a uniform vorticity distribution concentrated entirely within the core radius. The velocity and circulation are

$$v = \frac{\Gamma}{2\pi r_c} \begin{cases} r/r_c & r < r_c \\ \dfrac{1}{r/r_c} & r > r_c \end{cases} \tag{9.54}$$

$$\gamma = \Gamma \begin{cases} (r/r_c)^2 & r < r_c \\ 1 & r > r_c \end{cases} \tag{9.55}$$

and the vorticity is $\Gamma/\pi r_c^2$ for $r < r_c$.

9.7 Vortex Core

Oseen (see Lamb (1932)) obtained the solution of the Navier-Stokes equations for decay of a laminar vortex:

$$v = \frac{\Gamma}{2\pi r}\left(1 - e^{-r^2/4\nu t}\right) = \frac{\Gamma}{2\pi r}\left(1 - e^{-ar^2/r_c^2}\right) \tag{9.56}$$

introducing the core radius $r_c^2 = a4\nu t$. Here $a \cong 1.2564$ is the solution of $e^a = 1 + 2a$, so $r_c = 2.2418\sqrt{\nu t}$ (ν is the kinematic viscosity). The vorticity is

$$\zeta = \frac{\Gamma}{\pi r_c^2}\left(ae^{-ar^2/r_c^2}\right) \tag{9.57}$$

At the core radius, $f = \gamma/\Gamma = 2a/(1+2a) \cong 0.7153$.

A simple model for a distributed vorticity core has the circulation proportional to $f = r^2/(r^2 + r_c^2)$, so half the vorticity is outside the core radius. Then

$$v = \frac{\Gamma}{2\pi r}\frac{r^2}{r^2 + r_c^2} \tag{9.58}$$

$$\zeta = \frac{\Gamma}{\pi r_c^2}\frac{r_c^4}{(r^2 + r_c^2)^2} \tag{9.59}$$

This is the Scully model, developed based on the form of the Biot-Savart solution for vortex-induced velocity. It was introduced for rotor wake analysis by Scully (1975), although there are earlier uses of the model in other fields (see Bhagwat and Leishman (2002)).

Vatistas, Kozel, and Mih (1991) defined a power-law core:

$$v = \frac{\Gamma}{2\pi r}\frac{r^2}{(r^{2n} + r_c^{2n})^{1/n}} \tag{9.60}$$

$$\zeta = \frac{\Gamma}{\pi r_c^2}\frac{r_c^{2n+2}}{(r^{2n} + r_c^{2n})^{(1+1/n)}} \tag{9.61}$$

The peak velocity $v_{max} = 2^{-1/n}(\Gamma/2\pi r_c)$ increases with n. The Scully model is obtained with $n = 1$, and the Rankine core with $n = \infty$. For $n = 2$ the Vatistas model is close to the Lamb-Oseen distribution (see Bagai and Leishman (1995)).

The peak circumferential velocity (at $r = r_c$) is proportional to Γ/r_c and a factor that depends on the vorticity distribution:

$$\text{Rankine} \quad v_{max} = \frac{\Gamma}{2\pi r_c} \tag{9.62}$$

$$\text{Oseen} \quad v_{max} = \frac{\Gamma}{2\pi r_c}\left(1 - e^{-a}\right) = \frac{\Gamma}{2\pi r_c}\frac{2a}{1+2a} = \frac{\Gamma}{2\pi r_c}0.7153 \tag{9.63}$$

$$\text{Scully} \quad v_{max} = \frac{\Gamma}{2\pi r_c}0.5 \tag{9.64}$$

$$\text{Vatistas} \quad v_{max} = \frac{\Gamma}{2\pi r_c}2^{-1/n} = \frac{\Gamma}{2\pi r_c}0.7071 \quad (n = 2) \tag{9.65}$$

The Rankine core has all the vorticity inside $r = r_c$, so has the largest possible peak velocity given Γ and r_c (equal to the potential vortex velocity). The Scully core has half the vorticity inside and half the vorticity outside of $r = r_c$. Vorticity outside the core reduces the peak velocity for fixed Γ and r_c. Measurements of the velocity distributions about tip vortices show that the maximum tangential velocity is much less than $\Gamma/2\pi r_c$, indicating that a substantial fraction of the vorticity is outside the core radius. Measurements of rotor tip vortices imply $n = 1$ to $n = 2$ in the Vatistas model.

The self-induced velocity at a collocation point on a curved vortex line is finite only with a nonzero core radius. Bliss, Teske, and Quackenbush (1987) showed that the correct induced velocity is obtained from a line of concentrated vorticity (zero core radius) if the integration over the line stops a cutoff distance d either side of the collocation point. Assuming that the ratio of the core radius to the vortex radius-of-curvature is small, the required cutoff distance is

$$d/r_c = \frac{1}{2} e^{\frac{1}{2}-A+C} \tag{9.66}$$

The quantities A and C follow from the kinetic energy in the viscous core:

$$A = \lim_{r \to \infty} \left[\int_0^r rv^2 dr - \ln r \right] \tag{9.67}$$

$$C = \int_0^\infty 2rw^2 dr \tag{9.68}$$

Here r is the distance from the center of the core (scaled with the core radius r_c), and v and w are the swirl and axial velocities (scaled with the circulation $\Gamma/2\pi r_c$). The integral A depends on the vortex core model:

Rankine $\quad A = \frac{1}{4} \quad$ from $\quad v = \min(r, 1/r)$ $\hfill(9.69)$

Oseen $\quad A = \frac{1}{2}\left(\gamma + \ln \frac{a}{2}\right) \cong 0.0562 \quad$ from $\quad v = \frac{1}{r}\left(1 - e^{-ar^2}\right)$ $\hfill(9.70)$

Scully $\quad A = -\frac{1}{2} \quad$ from $\quad v = r/(r^2+1)$ $\hfill(9.71)$

Vatistas $\quad A \cong \frac{1}{4}\frac{(n+3)(n-2)}{(n+1)n} \quad$ from $\quad v = r/(r^{2n}+1)^{1/n}$ $\hfill(9.72)$

Vatistas $\quad A = 0 \quad$ from $\quad v = r/\sqrt{r^4+1} \quad (n=2)$ $\hfill(9.73)$

(where $\gamma \cong 0.5772$ is Euler's constant). Hence the cutoff distance is

Rankine $\quad d/r_c = \frac{1}{2} e^{1/4} \cong 0.6420$ $\hfill(9.74)$

Oseen $\quad d/r_c = \frac{1}{\sqrt{2a}} e^{(1-\gamma)/2} \cong 0.7793$ $\hfill(9.75)$

Scully $\quad d/r_c = \frac{1}{2} e \cong 1.3591$ $\hfill(9.76)$

Vatistas $\quad d/r_c = \frac{1}{2} e^{1/2} \cong 0.8244 \quad (n=2)$ $\hfill(9.77)$

(ignoring the axial velocity, so $C = 0$). Use of this cutoff distance is equivalent to solving for the velocity field using the method of matched asymptotic expansions (Widnall, Bliss, and Zalay (1971)).

After the initial formation of the vortex at the trailing edge of the wing tip, the strength increases as trailed vorticity is entrained, and the core radius grows due to viscous diffusion. The Oseen solution of the Navier-Stokes equations for decay of a laminar vortex gives $r_c = \sqrt{a4\nu t} = 2.2418\sqrt{\nu t}$, where t is the time since

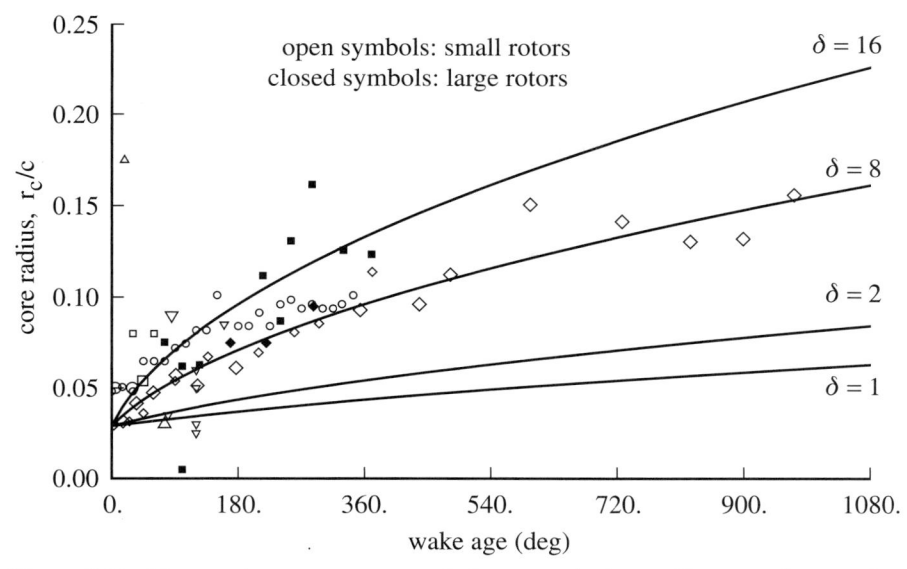

Figure 9.21. Measured vortex core growth for rotors in hover. Data are from various tests collected by Ramasamy and Leishman (2007) and McAlister, Schuler, Branum, and Wu (1995).

the vortex element was created (dimensional wake age). Squire (1965) introduced an eddy viscosity to account for turbulence, replacing ν with $\nu + \epsilon = \delta\nu$. Assuming the eddy viscosity is proportional to the vortex strength gives $\delta = 1 + k\Gamma/\nu$, where $\Gamma/2\pi\nu = (\Gamma/2\pi r_c)r_c/\nu$ is a Reynolds number based on the core radius and the peak swirl velocity. In terms of dimensionless wake age, $t = (\phi + \phi_0)/\Omega$, including effective origin since the core radius is finite at the trailing edge, so

$$r_c^2 = \left(\frac{4a\delta\nu}{\Omega}\right)(\phi + \phi_0) = r_{c0}^2 + \left(\frac{4a\delta\nu}{\Omega c^2}\right)c^2\phi = r_{c0}^2 + \left(\frac{4a\delta}{(c/R)Re}\right)c^2\phi \quad (9.78)$$

where $Re = (\Omega R)c/\nu$ is the blade Reynolds number. This result can be interpreted in terms of the wake age ϕ_1 required for the core to grow by 100% of the blade chord:

$$\phi_1 = \frac{(c/R)Re}{4a\delta} \quad (9.79)$$

Figure 9.21 shows the vortex core radius as a function of wake age, measured for a number of hovering rotors; the data are from Ramasamy and Leishman (2007) and McAlister, Schuler, Branum, and Wu (1995). Most of the rotors tested were small, with diameters up to about 6 feet. The Oseen and Squire results for the core size were calculated for a blade Reynolds number of $(c/R)Re = 30000$. For Figure 9.21, $\delta = 8$ to 16 matches the measurements, with $r_{c0} = 0.03c$. The core radius is about 5% of the chord at the blade trailing edge, 5–10% chord at the encounter with the first following blade, and about 10% chord after one revolution (for hover).

9.8 Blade-Vortex Interaction

The encounter of rotor blades with a tip vortex from a preceding blade is called blade-vortex interaction (BVI). The tip vortex induces a large aerodynamic loading

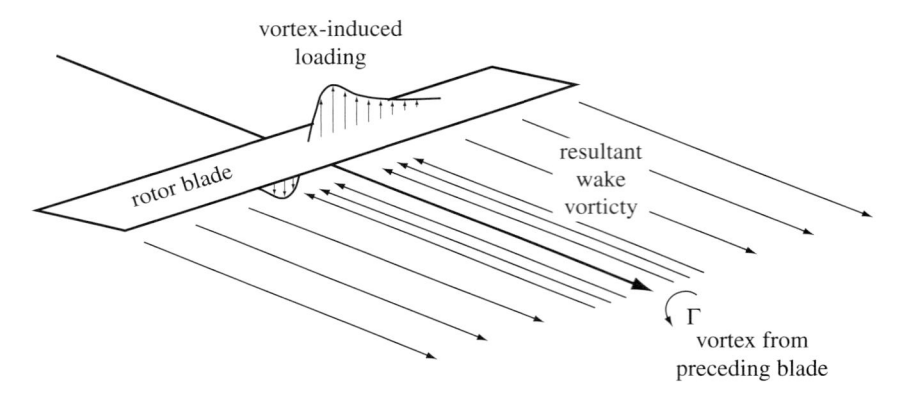

Figure 9.22. Vortex-induced blade loading.

for the small vortex-blade separations characteristic of the helicopter rotor in both hover and forward flight. A vortex that is a distance h below the blade induces a downwash velocity (the component normal to the blade surface) that is zero directly over the vortex and has positive and negative peaks a distance h on either side of the intersection. The induced bound circulation and loading have the same general form as the induced velocity distribution (Figure 9.22), although the peaks are somewhat further apart than $2h$ because of lifting-surface effects. The spanwise gradient of the bound circulation produces trailed vorticity in the wake behind the blade. Because this induced wake vorticity has a direction parallel to the free tip vortex, if the free vortex is not perpendicular to the blade span there is actually a radial component of the trailed vorticity (shed wake). If the vortex is not perpendicular to the span, the vortex-blade intersection sweeps radially along the blade as the vortex is convected with the free stream, and the aerodynamic-loading is unsteady.

A model problem for blade-vortex interaction is an infinite-aspect-ratio, nonrotating wing in a subsonic free stream, encountering a straight, infinite vortex at an angle Λ with the wing (Figure 9.23). The wing has chord c; the vortex lies in a plane parallel to the wing, a distance h below it. The vortex is convected past the wing by the free stream. For the linear aerodynamic solution the distortion of the vortex-line geometry by the interaction with the blade is neglected, and the blade and wake are represented by a planar distribution of vorticity. This model problem can be solved for the case of a sinusoidal induced velocity distribution with wave fronts parallel to the vortex line. The vortex-induced velocity distribution can be obtained by a suitable combination of sinusoidal waves of various wavelengths (a Fourier transform), so the same superposition applied to the loading solution gives the vortex-induced loading.

Figure 9.23 shows the limiting cases: the perpendicular encounter ($\Lambda = 90°$), which is a steady, three-dimensional aerodynamic problem, and the parallel encounter ($\Lambda = 0°$, also called airfoil-vortex interaction), which is an unsteady, two-dimensional problem. The perpendicular encounter is characteristic of the blade interaction with the tip vortex from the preceding blade, on the advancing and retreating sides of the disk in forward flight, and is a factor in vibration and structural loads. The parallel encounter is characteristic of the blade interaction with vortices one revolution or so old, especially in the first quadrant, and is a factor

9.8 Blade-Vortex Interaction

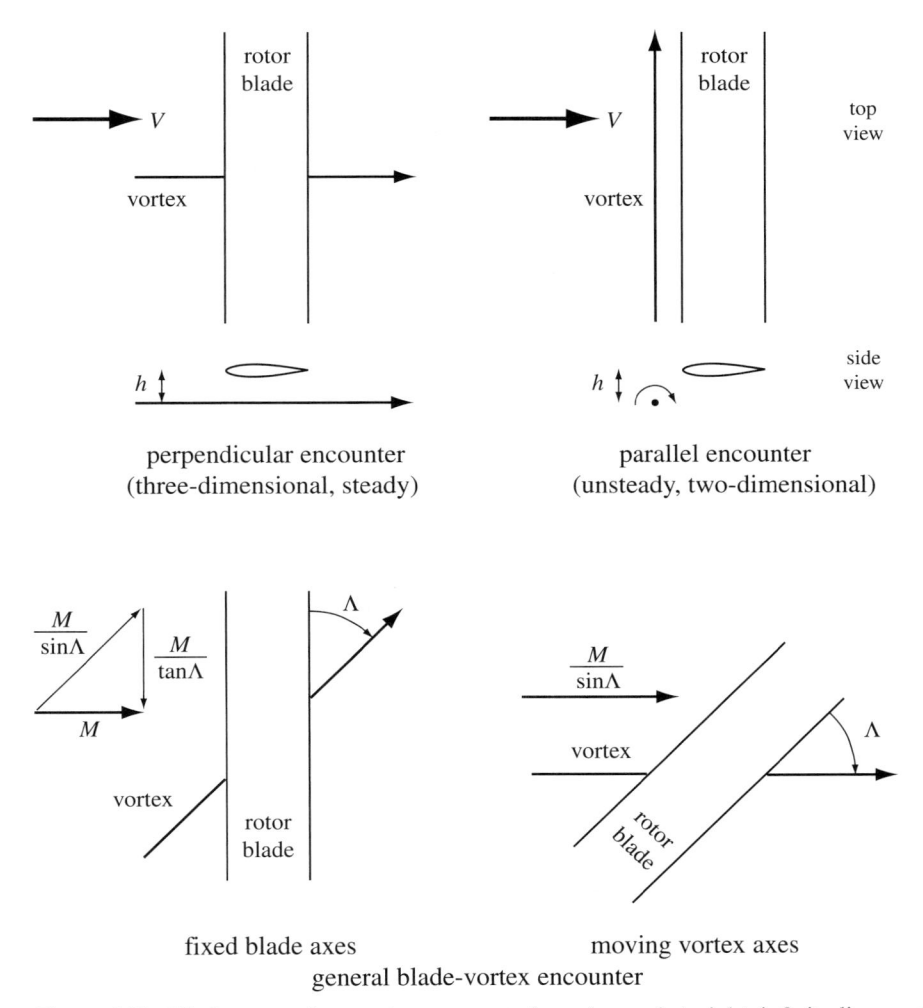

Figure 9.23. Blade-vortex interaction: non-rotating wing and straight, infinite line vortex.

in BVI noise generation. The general interaction has the vortex at angle Λ in a compressible free stream with Mach number M (in blade axes, Figure 9.23). As the vortex is convected, the intersection with the blade sweeps spanwise at speed $M/\tan\Lambda$. In axes moving with this intersection, the problem is stationary (reducing from four space-time dimensions to just three dimensions), with a speed parallel the vortex of $M/\sin\Lambda$. The aerodynamic equations in the moving axes are elliptical (more three-dimensional, including the perpendicular encounter) for $M < \sin\Lambda$. The equations are hyperbolic (more unsteady, including the parallel encounter) for $M > \sin\Lambda$.

Consider a straight, infinite line vortex of strength Γ, perpendicular to an infinite-aspect-ratio blade with spanwise coordinate y. The vortex lies in a plane parallel to the wing surface, a distance h below it. The normal w and spanwise u components of the vortex-induced velocity are required on the wing at y (at $y\sin\Lambda$ for the non-perpendicular case or ($y\sin\Lambda - Mt\cos\Lambda$) to include the unsteady, two-dimensional limit). For a potential vortex, the swirl velocity is $\Gamma/2\pi r$, where $r = \sqrt{y^2 + h^2}$ is the

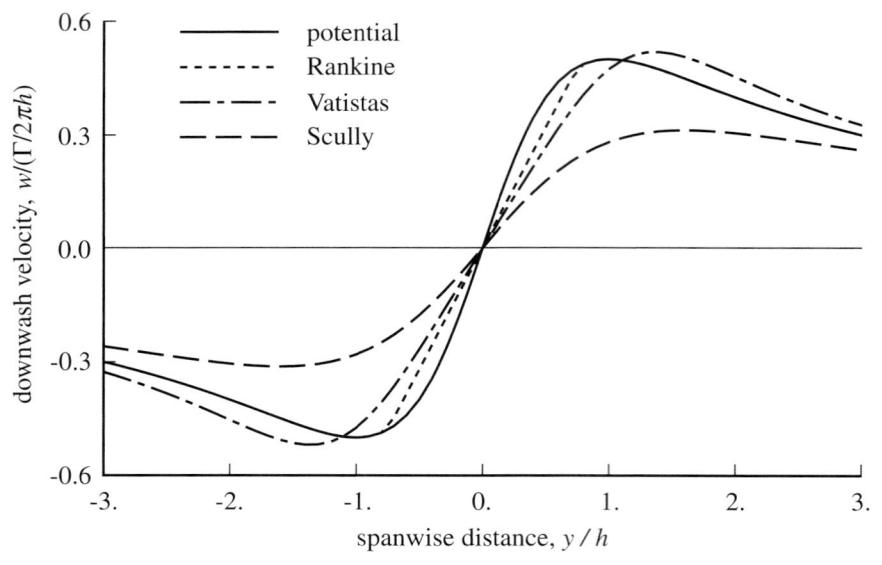

Figure 9.24. Normal velocity induced by a straight infinite line vortex.

distance from the vortex center to blade point. So

$$w = \frac{\Gamma}{2\pi r} \frac{y}{r} = \frac{\Gamma}{2\pi} \frac{y}{y^2 + h^2} \tag{9.80}$$

$$u = \frac{\Gamma}{2\pi r} \frac{h}{r} = \frac{\Gamma}{2\pi} \frac{h}{y^2 + h^2} \tag{9.81}$$

The downwash has peaks at $y = \pm h$, of magnitude $w_{\text{peak}} = \Gamma/4\pi h$ (Figure 9.24). This normal velocity induces a loading on the blade with a similar radial variation. The spanwise component of the induced velocity has a peak at $y = 0$ of $u_{\text{peak}} = \Gamma/2\pi h$.

For $h = 0$ the downwash is $w = \Gamma/2\pi y$, which is singular at the vortex line ($y = 0$). So for small h and y, a vortex core must be included in the wake model to obtain physically realistic calculations of the induced velocity. For the Rankine (equation 9.54), Scully (equation 9.58), and Vatistas (equation 9.60, with $n = 2$) core models, the downwash is

$$\text{Rankine} \quad w = \frac{\Gamma}{2\pi} \begin{cases} \dfrac{y}{r_c^2} & r < r_c \\ \dfrac{y}{r^2} & r > r_c \end{cases} \tag{9.82}$$

$$\text{Scully} \quad w = \frac{\Gamma}{2\pi} \frac{y}{r^2 + r_c^2} = \frac{\Gamma}{2\pi} \frac{y}{y^2 + h^2 + r_c^2} \tag{9.83}$$

$$\text{Vatistas} \quad w = \frac{\Gamma}{2\pi} \frac{y}{(r^4 + r_c^4)^{1/2}} \tag{9.84}$$

where $r = \sqrt{y^2 + h^2}$. With the Scully model, the effect of the core is equivalent to moving the vortex farther away from the blade, to $h_{\text{eq}} = (h^2 + r_c^2)^{1/2}$; such a simple

9.8 Blade-Vortex Interaction

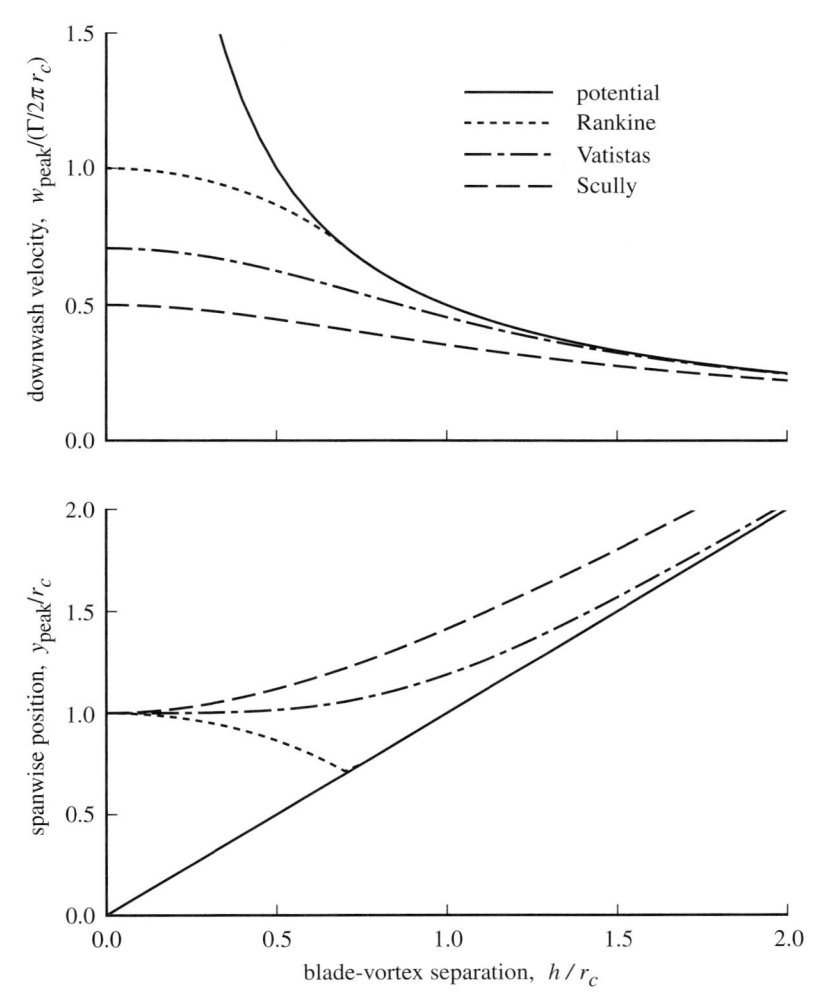

Figure 9.25. Magnitude and position of peak downwash velocity.

correction can be useful for analytical work. The downwash peaks are

$$\text{potential} \quad w_{\text{peak}} = \frac{\Gamma}{2\pi} \frac{1}{2h} \quad \text{at } y = \pm h \tag{9.85}$$

$$\text{Rankine} \quad w_{\text{peak}} = \frac{\Gamma}{2\pi} \frac{\sqrt{r_c^2 - h^2}}{r_c^2} \quad \text{at } y = \pm\sqrt{r_c^2 - h^2} \quad \left(h < \frac{r_c}{\sqrt{2}}\right) \tag{9.86}$$

$$\text{Scully} \quad w_{\text{peak}} = \frac{\Gamma}{2\pi} \frac{1}{2\sqrt{h^2 + r_c^2}} \quad \text{at } y = \pm\sqrt{h^2 + r_c^2} \tag{9.87}$$

$$\text{Vatistas} \quad w_{\text{peak}} = \frac{\Gamma}{2\pi} \frac{1}{\sqrt{2h^2 + 2\sqrt{h^4 + r_c^4}}} \quad \text{at } y = \pm\sqrt[4]{h^4 + r_c^4} \tag{9.88}$$

Figure 9.25 shows the spanwise location and magnitude of the downwash peaks as a function of separation h.

The airloads produced by blade-vortex interaction depend on numerous physical effects, including the extent of the tip vortex rollup; the tip vortex strength; the size of the viscous core; the distorted wake geometry; lifting-surface effects on the induced loading; and possibly even vortex bursting, vortex-induced stall on the blade, or blade-induced geometry changes. The peak-to-peak value of the vortex-induced loading dominates the measured and calculated airloads. Physical factors influence the peak-to-peak loading. The rollup process, at the generating blade and in the wake, produces the strength and core size of the tip vortex at the encounter with a following wing. The strength is less than or equal to the peak bound circulation, and the core size is typically 5–20% of the chord. If the strength is assumed to equal the peak bound circulation when in fact the strength is less, then the analysis over-predicts the loading. In close blade-vortex encounters, the induced loading varies rapidly along the span, so the interaction has a small effective aspect ratio. First-order lifting-line theory over-predicts such loading, especially if the radial and azimuthal resolution in the wake are not small enough. Second-order lifting-line theory or lifting-surface theory is needed for accurate prediction of BVI loads (as well as the airloading for swept tips, yawed flow, and low-aspect-ratio blades). Lifting-surface effects reduce the peak induced loading by 20-40% for a vortex-blade separation equal to 25% of the chord. This effect can be approximated by increasing the viscous core size by about 15% chord. There are compressibility and viscous effects involved in the interaction and possibly vortex bursting. The high radial pressure gradients on the blade might cause vortex-induced stall. The geometry of the vortex is distorted locally by the blade. The vortex interacts with the trailed wake it induces behind the blade, with the effect of diffusing and reducing the circulation in the vortex. Measured data exhibit unsteadiness and noise. If in nominally periodic loading the azimuth at the occurrence of the blade-vortex interaction changes from revolution to revolution, then an averaging process reduces the measured peak loads, and the analysis over-predicts the loading.

Computational factors also influence the loading. If the calculated blade-vortex separation is too large, as when a rigid wake geometry is used, then the analysis under-predicts the loading. Typically the radial and the azimuthal (age) resolution of the wake are too large for close blade-vortex interactions, so the analysis over-predicts the loading. If the radial and azimuthal (time) discretization of the calculated airloading are too large, the analysis under-predicts the loading.

The core size r_c is a convenient parameter with which to control the amplitude of the calculated blade-vortex interaction loads, since it determines the maximum tangential velocity about the vortex (inversely proportional to r_c). Moreover, the core size is often neither measured nor calculated, so is an input parameter of the analysis. The approach is to model all effects possible in the theory as accurately as possible and then to use the value of r_c to account for the actual viscous core radius, as well as for all phenomena of the interaction that are not otherwise modeled (or are inaccurately modeled). A goal for the development of better models is that the vortex core size represents the actual physical core (5–20% chord) and nothing else.

9.9 Vortex Elements

Calculations of the rotor nonuniform inflow and free wake geometry are based on the induced velocity produced by discrete elements of the vortex wake. The basic building blocks are straight line and quadrilateral sheet elements. A connected series

9.9 Vortex Elements

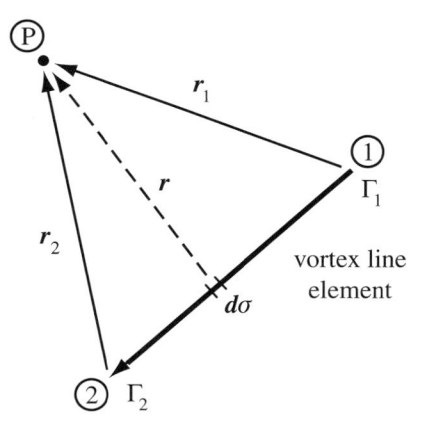

Figure 9.26. Vortex line element.

of finite-length vortex line segments can represent the tip vortex spirals, and a vortex lattice can represent the inboard shed and trailed vorticity.

9.9.1 Vortex Line Segment

Figure 9.26 shows the geometry for a finite-length, straight vortex line element. The line segment extends from point 1 to point 2 in space, and the velocity is required at point P. The vortex strength varies linearly from Γ_1 at point 1 to Γ_2 at point 2. The convention for positive circulation is about the vector from point 1 to point 2. The geometry is defined by the position vectors $\mathbf{r}_1$ and $\mathbf{r}_2$ from the ends of the line segment to P. The geometry is actually specified by vectors from a common origin to points 1, 2, and P, from which $\mathbf{r}_1$ and $\mathbf{r}_2$ can be calculated. The velocity is obtained in the same axes as used to describe the position vectors. The Biot-Savart law gives the induced velocity produced by this line segment:

$$\Delta \mathbf{v} = -\frac{1}{4\pi} \int \frac{\Gamma \mathbf{r} \times \mathbf{d\sigma}}{r^3} \tag{9.89}$$

where $\mathbf{r}$ is the vector from the element $\mathbf{d\sigma}$ on the segment to the point P, and $r = |\mathbf{r}|$. The coordinate σ is measured along the vortex segment, from s_1 to s_2:

$$s_1 = \frac{1}{s}(\mathbf{r}_1 \cdot \mathbf{r}_2 - r_1^2) = \frac{1}{s}\mathbf{r}_1 \cdot (\mathbf{r}_2 - \mathbf{r}_1) \tag{9.90}$$

$$s_2 = \frac{1}{s}(r_2^2 - \mathbf{r}_1 \cdot \mathbf{r}_2) = \frac{1}{s}\mathbf{r}_2 \cdot (\mathbf{r}_2 - \mathbf{r}_1) = s_1 + s \tag{9.91}$$

where s is the length of the segment:

$$s^2 = |\mathbf{r}_1 - \mathbf{r}_2|^2 = r_1^2 + r_2^2 - 2\mathbf{r}_1 \cdot \mathbf{r}_2 \tag{9.92}$$

Write $\mathbf{r} = \mathbf{r}_m - \sigma \hat{\epsilon}$, where $\mathbf{r}_m$ is the minimum distance from the vortex line (including its extension beyond the end points of the segment) to the point P, and $\hat{\epsilon}$ is the unit vector in the direction of the vortex:

$$\mathbf{r}_m = \frac{1}{s^2}\left(\mathbf{r}_1(r_2^2 - \mathbf{r}_1 \cdot \mathbf{r}_2) + \mathbf{r}_2(r_1^2 - \mathbf{r}_1 \cdot \mathbf{r}_2)\right) = \frac{1}{s}(\mathbf{r}_1 s_2 - \mathbf{r}_2 s_1) \tag{9.93}$$

$$\hat{\epsilon} = \frac{1}{s}(\mathbf{r}_1 - \mathbf{r}_2) \tag{9.94}$$

The vectors $\mathbf{r}_m$ and $\hat{\epsilon}$ are perpendicular, and
$$r_m^2 s^2 = |\mathbf{r}_1 s_2 - \mathbf{r}_2 s_1|^2 = r_1^2 r_2^2 - (\mathbf{r}_1 \cdot \mathbf{r}_2)^2 \tag{9.95}$$

The vortex strength varies linearly along the segment:
$$\begin{aligned}\Gamma &= \frac{1}{s}\Big(\Gamma_1(s_2 - \sigma) + \Gamma_2(\sigma - s_1)\Big) \\ &= \frac{1}{s}\Big((\Gamma_1 s_2 - \Gamma_2 s_1) + \sigma(\Gamma_2 - \Gamma_1)\Big) \\ &= \Gamma_m + \sigma \Gamma_s \end{aligned} \tag{9.96}$$

Hence
$$\begin{aligned}\Delta \mathbf{v} &= \frac{\mathbf{r}_1 \times \mathbf{r}_2}{4\pi s} \int_{s_1}^{s_2} \frac{\Gamma_m + \sigma \Gamma_s}{(r_m^2 + \sigma^2)^{3/2}} d\sigma \\ &= \frac{\mathbf{r}_1 \times \mathbf{r}_2}{4\pi s} \left[\frac{\Gamma_m \sigma / r_m^2 - \Gamma_s}{(r_m^2 + \sigma^2)^{1/2}} \right]_{\sigma=s_1}^{\sigma=s_2} \\ &= \frac{\mathbf{r}_1 \times \mathbf{r}_2}{4\pi s r_m^2} \left[\Gamma_m \left(\frac{s_2}{r_2} - \frac{s_1}{r_1} \right) - \Gamma_s r_m^2 \left(\frac{1}{r_2} - \frac{1}{r_1} \right) \right] \\ &= \frac{\mathbf{r}_1 \times \mathbf{r}_2}{4\pi s r_m^2} \left[\Gamma_1 \left\{ \frac{s_2}{s}\left(\frac{s_2}{r_2} - \frac{s_1}{r_1}\right) + \frac{r_m^2}{s}\left(\frac{1}{r_2} - \frac{1}{r_1}\right) \right\} \right. \\ &\quad \left. - \Gamma_2 \left\{ \frac{s_1}{s}\left(\frac{s_2}{r_2} - \frac{s_1}{r_1}\right) + \frac{r_m^2}{s}\left(\frac{1}{r_2} - \frac{1}{r_1}\right) \right\} \right] \\ &= \Gamma_1 \Delta \mathbf{v}_1 + \Gamma_2 \Delta \mathbf{v}_2 \end{aligned} \tag{9.97}$$

is the induced velocity of the vortex line segment with linearly varying circulation. The velocity increment from a constant-strength element is obtained using $\Gamma_m = \Gamma$ and $\Gamma_s = 0$. The induced velocity of a vortex line segment with stepped variation of circulation (Γ_1 from point 1 to the midpoint, Γ_2 from the midpoint to point 2) is obtained by applying the constant-strength result to the two pieces.

The influence of the vortex core is accounted for by multiplying the induced velocity of the line segment by the factor f:

$$\text{Rankine} \quad f = \min(r_m^2/r_c^2, 1) \tag{9.98}$$

$$\text{Scully} \quad f = r_m^2/(r_m^2 + r_c^2) \tag{9.99}$$

$$\text{Vatistas } (n=2) \quad f = r_m^2/\sqrt{r_m^4 + r_c^4} \tag{9.100}$$

$$\text{Vatistas} \quad f = r_m^2/(r_m^{2n} + r_c^{2n})^{1/n} \tag{9.101}$$

$$\text{Oseen} \quad f = \left(1 - e^{-a r_m^2 / r_c^2}\right) \tag{9.102}$$

The core radius r_c is the location of the maximum tangential velocity. Thus the vortex core is accounted for by using the factor f/r_m^2 instead of $1/r_m^2$ in the expression for $\Delta \mathbf{v}$. Using the Scully core is particularly simple: replace r_m^2 in the denominator with $r_m^2 + r_c^2$.

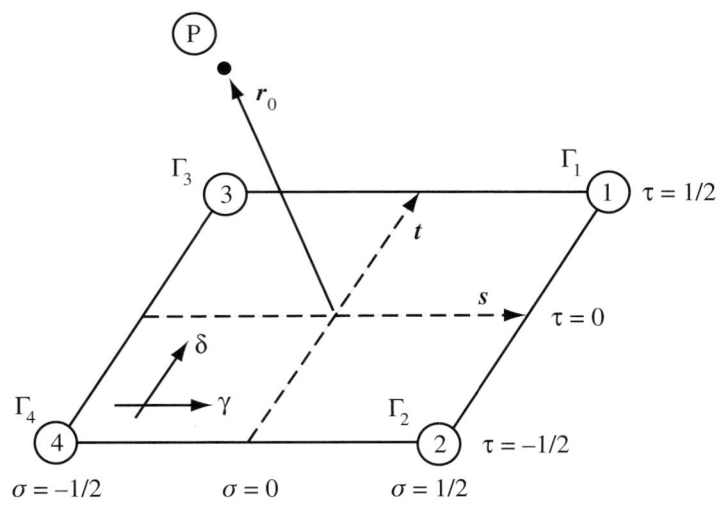

Figure 9.27. Vortex sheet element.

9.9.2 Vortex Sheet Element

A vortex sheet element can be used to model the inboard shed and trailed vorticity. Figure 9.27 shows the geometry for a nonplanar, quadrilateral vortex sheet element. The four corners of the sheet element are at points 1 to 4 in space, and the velocity is required at point P. The strength of the sheet vorticity is defined in terms of the wing bound circulation associated with the four corners (Γ_1, Γ_2, Γ_3, and Γ_4). The geometry is defined by the position vectors $\mathbf{r}_1$, $\mathbf{r}_2$, $\mathbf{r}_3$, and $\mathbf{r}_4$, from the corners to P. The geometry is actually specified by vectors from a common origin to points 1, 2, 3, 4, and P, from which the vectors from the corners to P can be calculated. The velocity is obtained in the same axes as used to describe the position vectors. The velocity induced by a vortex sheet element is required in the form

$$\Delta\mathbf{v} = \Gamma_1 \Delta\mathbf{v}_{t1} + \Gamma_2 \Delta\mathbf{v}_{t2} + \Gamma_3 \Delta\mathbf{v}_{t3} + \Gamma_4 \Delta\mathbf{v}_{t4}$$
$$= (\Gamma_1 - \Gamma_3)\Delta\mathbf{v}_{t1} + (\Gamma_2 - \Gamma_4)\Delta\mathbf{v}_{t2} \quad (9.103)$$

for the trailed vorticity, and

$$\Delta\mathbf{v} = \Gamma_1 \Delta\mathbf{v}_{s1} + \Gamma_2 \Delta\mathbf{v}_{s2} + \Gamma_3 \Delta\mathbf{v}_{s3} + \Gamma_4 \Delta\mathbf{v}_{s4}$$
$$= (\Gamma_1 - \Gamma_2)\Delta\mathbf{v}_{s1} + (\Gamma_3 - \Gamma_4)\Delta\mathbf{v}_{s3} \quad (9.104)$$

for the shed vorticity.

For the general case of a nonplanar-quadrilateral element, the induced velocity must be evaluated by numerical integration. For the case of a planar-rectangular element the integrations can be performed analytically. The planar-rectangular element can be used as an approximation for the general element. The edges of adjacent elements do not join with this approximation, but the amount of computation required is reduced. As a further approximation, the sheet can be replaced by trailed and shed line segments, with large core radii to avoid high induced velocities near the lines.

The sheet geometry is defined by the vectors **s** and **t** joining the midpoints of the sides (Figure 9.27). These vectors always intersect; the vector from the intersection to P is $\mathbf{r}_0$. The vector between the midpoints of the diagonals is **u**. The sheet surface is described by the coordinates σ and τ, each varying from $-1/2$ to $1/2$; the origin $\sigma = \tau = 0$ is at $\mathbf{r}_0$. Then the vector from (σ, τ) on the sheet to P is

$$\mathbf{r} = \mathbf{r}_0 - \sigma \mathbf{s} - \tau \mathbf{e}_t = \mathbf{r}_0 - \sigma \mathbf{e}_s - \tau \mathbf{t} = \mathbf{r}_0 - \sigma \mathbf{e}_s - \tau \mathbf{e}_t + 2\sigma\tau \mathbf{u} \quad (9.105)$$

where the vectors in the trailed and shed directions are $\mathbf{e}_t = \mathbf{t} + 2\sigma \mathbf{u}$ and $\mathbf{e}_s = \mathbf{s} + 2\tau \mathbf{u}$. The bound circulation corresponding to the four corners is known, but not the actual vorticity distribution over the sheet. A linear variation of the trailed vorticity $\delta(\tau)$ and shed vorticity $\gamma(\sigma)$ is assumed:

$$\delta = -\frac{1}{|\mathbf{e}_s|}(\Gamma_t + 2\tau \Gamma_u) \quad (9.106)$$

$$\gamma = -\frac{1}{|\mathbf{e}_t|}(\Gamma_s + 2\sigma \Gamma_u) \quad (9.107)$$

where $\Gamma_s = \frac{1}{2}(\Gamma_2 + \Gamma_4 - \Gamma_1 - \Gamma_3)$, $\Gamma_t = \frac{1}{2}(\Gamma_3 + \Gamma_4 - \Gamma_1 - \Gamma_2)$, and $\Gamma_u = \frac{1}{2}(\Gamma_2 + \Gamma_3 - \Gamma_1 - \Gamma_4)$. So the differential vorticity on the sheet is

$$\omega \, dA = \left[\gamma \frac{\mathbf{e}_s}{|\mathbf{e}_s|} + \delta \frac{\mathbf{e}_t}{|\mathbf{e}_t|} \right] |\mathbf{e}_s| \, d\sigma |\mathbf{e}_t| \, d\tau$$

$$= \left[(\Gamma_s + 2\sigma \Gamma_u) \mathbf{e}_s - (\Gamma_t + 2\tau \Gamma_u) \mathbf{e}_t \right] d\sigma \, d\tau \quad (9.108)$$

The Biot-Savart law gives the induced velocity produced by this sheet element:

$$\Delta \mathbf{v} = -\frac{1}{4\pi} \int \frac{\mathbf{r} \times \omega}{r^3} \, dA$$

$$= -\frac{1}{4\pi} \int_{-1/2}^{1/2} \int_{-1/2}^{1/2} [(\Gamma_s + 2\sigma \Gamma_u)(\mathbf{r} \times \mathbf{e}_s) - (\Gamma_t + 2\tau \Gamma_u)(\mathbf{r} \times \mathbf{e}_t)] \frac{d\sigma \, d\tau}{r^3}$$

$$(9.109)$$

and hence

$$\Delta \mathbf{v}_{t1} = \frac{1}{4\pi} \iint -\left(\frac{1}{2} + \tau\right) (\mathbf{r} \times \mathbf{e}_t) \frac{d\sigma \, d\tau}{r^3} \quad (9.110)$$

$$\Delta \mathbf{v}_{t2} = \frac{1}{4\pi} \iint -\left(\frac{1}{2} - \tau\right) (\mathbf{r} \times \mathbf{e}_t) \frac{d\sigma \, d\tau}{r^3} \quad (9.111)$$

$$\Delta \mathbf{v}_{s1} = \frac{1}{4\pi} \iint \left(\frac{1}{2} + \sigma\right) (\mathbf{r} \times \mathbf{e}_s) \frac{d\sigma \, d\tau}{r^3} \quad (9.112)$$

$$\Delta \mathbf{v}_{s3} = \frac{1}{4\pi} \iint \left(\frac{1}{2} - \sigma\right) (\mathbf{r} \times \mathbf{e}_s) \frac{d\sigma \, d\tau}{r^3} \quad (9.113)$$

The velocity from an infinitesimally thin vortex sheet has a logarithmic singularity near the edges. To avoid this singularity, r^3 in the integrand denominator is replaced by

$$(r^2 + d_{vs}^2)^{3/2} \quad (9.114)$$

The parameter d_{vs} can be considered a viscous core size (compare with the role of a core radius for a line segment). The introduction of d_{vs} also improves the convergence of the numerical integration when the collocation point is close to the sheet surface.

9.9 Vortex Elements

For a planar-rectangular element, the velocity integrals can be evaluated analytically. The planar assumption means $\mathbf{u} = 0$; the rectangular assumption means $\mathbf{s} \cdot \mathbf{t} = 0$. Hence define

$$\rho = \mathbf{r}_0 - \sigma \mathbf{s} - \tau \mathbf{t} = \mathbf{r}_m - (\sigma - \sigma_m)\mathbf{s} - (\tau - \tau_m)\mathbf{t} \quad (9.115)$$

where the minimum distance between P and the plane of the sheet gives $\sigma_m = \mathbf{r}_0 \cdot \mathbf{s}/|s|^2$, $\tau_m = \mathbf{r}_0 \cdot \mathbf{t}/|t|^2$, and $\mathbf{r}_m = \mathbf{r}_0 - \sigma_m \mathbf{s} - \tau_m \mathbf{t}$. Note that $\mathbf{r}_m \cdot \mathbf{s} = 0$ and $\mathbf{r}_m \cdot \mathbf{t} = 0$. So

$$\rho^2 = \rho_m^2 + (\sigma - \sigma_m)^2 s^2 + (\tau - \tau_m)^2 t^2 \quad (9.116)$$

$$\rho_m^2 = r_m^2 + d_{vs}^2 \quad (9.117)$$

$$\rho \times \mathbf{s} = \mathbf{r}_m \times \mathbf{s} - (\tau - \tau_m)\mathbf{t} \times \mathbf{s} \quad (9.118)$$

$$\rho \times \mathbf{t} = \mathbf{r}_m \times \mathbf{t} - (\sigma - \sigma_m)\mathbf{s} \times \mathbf{t} \quad (9.119)$$

Then the induced velocity can be written as the planar-rectangular result plus a correction term:

$$\Delta \mathbf{v}_{t1} = (\Delta \mathbf{v}_{t1})_{PR} + \frac{1}{4\pi} \iint -\left(\frac{1}{2} + \tau\right) \left(\frac{\mathbf{r} \times \mathbf{e}_t}{r^3} - \frac{\rho \times \mathbf{t}}{\rho^3}\right) d\sigma \, d\tau \quad (9.120)$$

and similarly for $\Delta \mathbf{v}_{t2}$, $\Delta \mathbf{v}_{s1}$, and $\Delta \mathbf{v}_{s3}$. If the element is actually a planar rectangle, then the correction term is zero ($\mathbf{r} \times \mathbf{e}_t/r^3 = \rho \times \mathbf{t}/\rho^3$). The planar-rectangular expressions can be integrated analytically:

$$(\Delta \mathbf{v}_{t1,2})_{PR} = \frac{1}{4\pi} \iint -\left(\frac{1}{2} \pm \tau\right) (\rho \times \mathbf{t}) \frac{d\sigma \, d\tau}{\rho^3}$$

$$= -\frac{\mathbf{r}_m \times \mathbf{t}}{4\pi} \left[\left(\frac{1}{2} \pm \tau_m\right) I_1 \pm I_2\right] + \frac{\mathbf{s} \times \mathbf{t}}{4\pi} \left[\left(\frac{1}{2} \pm \tau_m\right) I_3 \pm I_4\right] \quad (9.121)$$

$$(\Delta \mathbf{v}_{s1,3})_{PR} = \frac{1}{4\pi} \iint \left(\frac{1}{2} \pm \sigma\right) (\rho \times \mathbf{s}) \frac{d\sigma \, d\tau}{\rho^3}$$

$$= \frac{\mathbf{r}_m \times \mathbf{s}}{4\pi} \left[\left(\frac{1}{2} \pm \sigma_m\right) I_1 \pm I_3\right] - \frac{\mathbf{t} \times \mathbf{s}}{4\pi} \left[\left(\frac{1}{2} \pm \sigma_m\right) I_2 \pm I_4\right] \quad (9.122)$$

where

$$I_1 = \iint \frac{d\sigma \, d\tau}{\rho^3} = \frac{1}{\rho_m st} \tan^{-1} \frac{(\sigma - \sigma_m)(\tau - \tau_m)st}{\rho_m \rho} \quad (9.123)$$

$$I_2 = \iint \frac{(\tau - \tau_m) \, d\sigma \, d\tau}{\rho^3} = \frac{1}{st^2} \ln(\rho - s(\sigma - \sigma_m)) \quad (9.124)$$

$$I_3 = \iint \frac{(\sigma - \sigma_m) \, d\sigma \, d\tau}{\rho^3} = \frac{1}{s^2 t} \ln(\rho - t(\tau - \tau_m)) \quad (9.125)$$

$$I_4 = \iint \frac{(\sigma - \sigma_m)(\tau - \tau_m) \, d\sigma \, d\tau}{\rho^3} = -\frac{\rho}{s^2 t^2} \quad (9.126)$$

each evaluated for $\sigma = -1/2$ to $1/2$ and $\tau = -1/2$ to $1/2$. For the general case of a nonplanar quadrilateral, the correction terms are evaluated numerically.

The basic induced velocity of a vortex sheet is given by the arc-tangent term, which produces for the point P approach the surface

$$\Delta \mathbf{v} \rightarrow \pm\frac{1}{2}(\delta \hat{\epsilon}_s - \gamma \hat{\epsilon}_t) \qquad (9.127)$$

where $\hat{\epsilon}_s$ and $\hat{\epsilon}_t$ are unit vectors. The plus sign is for just above the sheet and the minus sign is for just below. So there is a jump across the sheet that is equal to the vorticity strength. For a point approaching an edge, there velocity is

$$\Delta \mathbf{v} \rightarrow \frac{1}{2\pi} \gamma_m \hat{\epsilon}_n \ln r_m \qquad (9.128)$$

(for a side edge), where ϵ_n is the unit normal. There is a logarithmic singularity at the edge of the vortex sheet. At the side edges of the rotor wake, the trailed vorticity produces a large velocity normal to the sheet, which is responsible for the rollup of the tip vortices. Elsewhere in the wake the logarithmic singularity is a result of the discreteness of the model, since replacing a curved sheet by a series of flat panels introduces infinite curvature where the edges join.

An economical approximation is to replace the vortex sheet by a line segments, with either a linear or a stepped circulation distribution, and a large core size to eliminate the high induced velocity near the lines. The strength and position of the line segments are determined from the circulation and position of the four corners of the sheet. The core radius can be specified arbitrarily, or $r_c = |\mathbf{s}|/2$ can be used for the trailed vorticity and $r_c = |\mathbf{t}|/2$ for the shed vorticity.

9.9.3 Circular-Arc Vortex Element

The entire wake can be modeled using curved vortex elements, as in Figure 9.10. Even when the wake is modeled using straight line elements, the curvature must be considered in evaluating the self-induced velocity for a free wake geometry calculation. There is no induced velocity contribution from elements adjacent to a collocation point on the vortex if these adjacent elements are straight line elements. This is equivalent to using a cutoff distance equal to the length of the vortex elements, instead of on the order of the vortex core radius. So a circular-arc element should be used for the adjacent elements. This section only considers the case of a collocation point at one end of the curved element.

Figure 9.28 shows the geometry for a circular-arc vortex line element. The line segment extends from point 1 to point 2 in space, and the velocity is required at point P. Here P is located at one end of the segment; hence the velocity is required at point 1 or at point 2. To define a curved line, a third point is required, either point 0 (before point 1) or point 3 (after point 2). A circular arc of radius R is defined by points (0,1,2) or (1,2,3), and the vortex line segment is that portion of the arc extending from point 1 to point 2. The vortex strength varies linearly from Γ_1 at point 1 to Γ_2 at point 2. The convention for positive circulation is about the vector from point 1 to point 2. The geometry is defined by the position vectors $\mathbf{r}_1$ and $\mathbf{r}_2$ from a common origin to the ends of the line segment, plus the vector $\mathbf{r}_0$ or $\mathbf{r}_3$ to a third point. The velocity is obtained in the same axes as used to describe the position vectors. The velocity induced by a vortex line segment is required in the form

$$\Delta v = \Gamma_1 \Delta \mathbf{v}_1 + \Gamma_2 \Delta \mathbf{v}_2 \qquad (9.129)$$

9.9 Vortex Elements

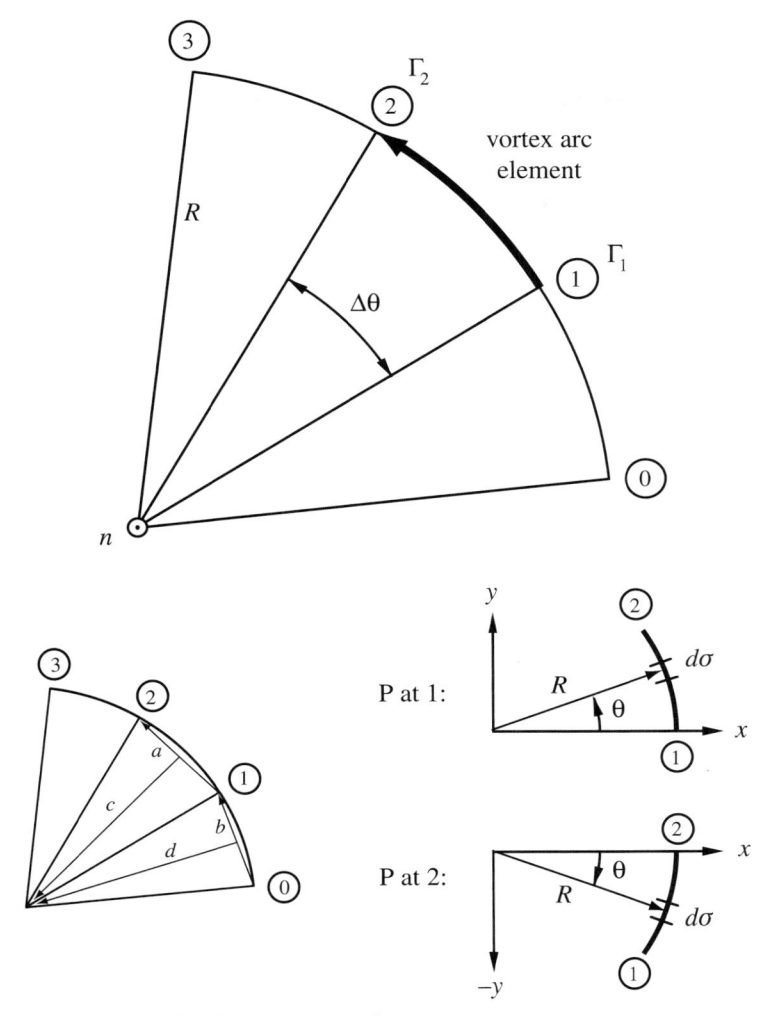

Figure 9.28. Circular-arc vortex element.

A non-zero core radius is needed to obtain a finite induced velocity on the line segment. The correct induced velocity is obtained from a line of concentrated vorticity (zero core radius) if the integration over the line stops a cutoff distance d before the collocation point; see section 9.7.

The Biot-Savart law gives the induced velocity produced by this line segment:

$$\Delta \mathbf{v} = -\frac{1}{4\pi} \int \frac{\Gamma \mathbf{r} \times d\boldsymbol{\sigma}}{r^3} \qquad (9.130)$$

where $\mathbf{r}$ is the vector from the element $d\boldsymbol{\sigma}$ on the segment to the point P, and $r = |\mathbf{r}|$. The coordinate σ is measured along the vortex segment. The circular arc has radius R and angular length $\Delta\theta$ (Figure 9.28). The vector $\mathbf{n}$ is normal the plane of the arc. Consider the case with the collocation point P at end point 1. Distance along the vortex segment is defined by the angle θ, measured from point 1. Then

$$\mathbf{r} = R \begin{pmatrix} 1 - \cos\theta \\ -\sin\theta \\ 0 \end{pmatrix} \qquad (9.131)$$

$$\mathbf{d\sigma} = R\, d\theta \begin{pmatrix} -\sin\theta \\ \cos\theta \\ 0 \end{pmatrix} \tag{9.132}$$

$$r = 2R \sin\frac{\theta}{2} \tag{9.133}$$

$$\mathbf{r} \times \mathbf{d\sigma} = -\mathbf{n} R^2 d\theta\, 2\sin^2\frac{\theta}{2} \tag{9.134}$$

The vortex strength along the segment is assumed to be

$$\Gamma = \Gamma_1 + (\Gamma_2 - \Gamma_1)\frac{\sin\theta/2}{\sin\Delta\theta/2} = \Gamma_m + \Gamma_\theta \sin\frac{\theta}{2} \tag{9.135}$$

which is a linear variation for small $\Delta\theta$ (for analytical integration, a factor of $\sin\theta/2$ is needed instead of $\theta/2$). Assuming small d/R,

$$\Delta\mathbf{v} \cong \frac{\mathbf{n}}{8\pi R}\int_{d/R}^{\Delta\theta}\left[\Gamma_m \frac{d\theta}{2\sin\theta/2} + \Gamma_\theta \frac{d\theta}{2}\right] = \frac{\mathbf{n}}{8\pi R}\left[\Gamma_m \ln\tan\frac{\theta}{4} + \Gamma_\theta \frac{\theta}{2}\right]\Big|_{\theta=d/R}^{\theta=\Delta\theta}$$

$$\cong \frac{\mathbf{n}}{8\pi R}\left[\Gamma_m \ln\left(\frac{4R}{d}\tan\frac{\Delta\theta}{4}\right) + \Gamma_\theta \frac{\Delta\theta}{2}\right]$$

$$= \frac{\mathbf{n}}{8\pi R}\left[\Gamma_1\left(\ln\left(\frac{4R}{d}\tan\frac{\Delta\theta}{4}\right) - \frac{\Delta\theta/2}{\sin\Delta\theta/2}\right) + \Gamma_2\left(\frac{\Delta\theta/2}{\sin\Delta\theta/2}\right)\right]$$

$$= \Gamma_1 \Delta\mathbf{v}_1 + \Gamma_2 \Delta\mathbf{v}_2 \tag{9.136}$$

is the induced velocity of the circular-arc vortex segment with linearly varying circulation (P at 1). The self-induced velocity of a vortex ring is given by twice the above result for $\Delta\theta = \pi$:

$$\Delta\mathbf{v} = \frac{\mathbf{n}\Gamma}{4\pi R}\ln\frac{4R}{d} \tag{9.137}$$

The result for the collocation point P at end point 2 is obtained in a similar manner.

To complete the evaluation of the induced velocity, the quantities $\mathbf{n}$, R, and $\Delta\theta$ must be obtained from the geometry of the circular arc. Consider the arc defined by points (0,1,2). The chord and normal vectors are $\mathbf{a} = \mathbf{r}_2 - \mathbf{r}_1$, $\mathbf{b} = \mathbf{r}_1 - \mathbf{r}_0$, $\mathbf{n} = \mathbf{b} \times \mathbf{a}/|\mathbf{b} \times \mathbf{a}|$ (Figure 9.28). The vectors $\mathbf{c}$ and $\mathbf{d}$ from the chord vectors to the center of the arc can be constructed from $\mathbf{b}$ and $\mathbf{n}$. Then

$$R = \frac{ab}{2}\frac{|\mathbf{a} + \mathbf{b}|}{|\mathbf{b} \times \mathbf{a}|} \tag{9.138}$$

$$\mathbf{n} = 2R \frac{\mathbf{b} \times \mathbf{a}}{ab|\mathbf{a} + \mathbf{b}|} \tag{9.139}$$

$$\sin\frac{\Delta\theta}{2} = \frac{a}{2R} = \frac{|\mathbf{b} \times \mathbf{a}|}{b|\mathbf{a} + \mathbf{b}|} \tag{9.140}$$

For the arc defined by points (1,2,3), the chord vectors are $\mathbf{a} = \mathbf{r}_3 - \mathbf{r}_2$ and $\mathbf{b} = \mathbf{r}_2 - \mathbf{r}_1$. Then the results for R and $\mathbf{n}$ are unchanged, and $\mathbf{a}$ and $\mathbf{b}$ are interchanged in $\sin\Delta\theta/2$.

9.10 History

Computation of harmonic airloading on a helicopter rotor blade in forward flight using realistic wake models began with Miller of the Massachusetts Institute of Technology (MIT) and Piziali and DuWaldt of the Cornell Aeronautical Laboratory (CAL). Miller (1962a, 1964b) described the problem:

> The determination of the air loads acting on rotor blades in forward flight presents an interesting and challenging problem in applied aerodynamics. Of particular importance for design purposes are the oscillatory components of this loading occurring at harmonics of the rotor speed. Unlike a wing, the trailing and shed vortex system of the blade generates a spiral wake that returns close to the blade. Because of its close proximity to the blade, the wake cannot be considered as rigid. Also, since the resulting loads are highly time-dependent, unsteady aerodynamic effects become important.... The oscillatory air loads occurring at harmonics of the rotor speed are the primary source of the blade stresses that establish the fatigue life of the structure and of the periodic hub loads that determine the fuselage vibration level.

With a uniform or linear inflow distribution, the predicted harmonic blade loading is of the order μ^n (where n is the harmonic number), in contrast to the large fifth or sixth harmonics that are measured in certain flight states such as transition or flare. This large harmonic loading is the source of the roughness and noise associated with such flight states and is due primarily to the wake-induced velocities. The work at MIT and CAL was the extension of non-rotating wing and two-dimensional airfoil unsteady aerodynamic theory to the complicated wake of the helicopter rotor in forward flight, which was made possible by the digital computer.

According to Miller (1963), experimental work at MIT in the 1950s made clear the importance of unsteady aerodynamics for rotor blades:

> The tests... clearly indicated the need for an analytical tool for computing blade downwash velocities which would take into account the individual blade wake geometry and also introduce the effects of unsteady aerodynamics. Attempts to obtain a closed form solution to this problem, or one based on tabulated integrals, were not successful and it was evident that extensive computer facilities would be required to explore this problem and, hopefully, to provide a basis for obtaining simplified solutions suitable for engineering applications. In 1960 the availability of an IBM 709 computer at the MIT Computation Center and funds from a Carnegie grant permitted initiation of such a program.

Miller's rotor model consisted of a sheet of distributed vorticity for the blade and a wake of shed and trailed vorticity. The Biot-Savart law gave the induced velocity increment caused by trailed vorticity (from radial change in bound circulation), simplified to the case of a lifting line in which the variation over the chord is neglected. Integrating from the blade to infinity down the spiral was necessary and was initially accomplished with numerical integration. From Miller (1964a),

> Computations of air loads is complicated by the existence of singularities in the solution. These occur as the shed wake approaches the trailing edge of the rotor and whenever the blade passes through a trailing vortex line generated by itself or another blade. The treatment of the singularities and of the non-uniform flow field presents no basic problem providing lifting-surface theory is used. However, this requires the numerical evaluation of the downwash at several chordwise as well as spanwise stations and hence, usually involves a prohibitive amount of machine computation time. Approximate methods

have therefore been used to evaluate the unsteady aerodynamic effects.... One of the most troublesome of the singularities is that associated with a shed vortex approaching the blade. In the simplest solution for the blade air loads it is convenient to replace the blade by a single vortex line and normally the high aspect ratio of conventional rotors would suggest that this is a reasonable approach.

The initial approach (Miller (1962a)) was to use a combined analytical and numeral procedure. Miller (1962b) developed a simpler method, in which the induced velocity was calculated at a single point on the airfoil chord, but only using the shed wake up to a small distance behind the collocation point. Thus in evaluating the induced velocity, the integral over wake age began at the bound vortex for the trailed wake, but a quarter chord behind the bound vortex for the shed wake. Miller (1964b) presented analytical work to support the assumptions and simplifications. Three-dimensional solutions for forward flight were obtained by numerical integration on a high-speed digital computer. Comparisons were made with airloads measured on rotor blades (Miller (1964b)):

> More recent experimental data, however, have supported the prediction of these abrupt changes in downwash near the 90° and 180° azimuth positions.... The computations of [Miller (1962a)] indicated that this abrupt load change is largely dominated by the vortex generated by the immediately preceding blade.

Scully (1965) dealt with the efficiency of the wake model, which led to replacing the numerical integration over the wake helices by a wake model consisting of finite straight-line vortex elements. Scully (1975) conducted an extensive investigation of the inboard trailed and shed wake representation and concluded that a model using a small number of line vortex elements, with a large core to better simulate the sheet vorticity, is the best in terms of both accuracy and economy.

The work by Piziali and DuWaldt was motivated in part by flight test experience with high oscillatory blade loads, attributed to nonuniform inflow velocities at blades. Piziali (1962, 1963) concentrated on the aerodynamic aspect of the problem:

> Any accurate method of computing the airloads must adequately predict the wake-induced velocities at the blades because the airloads are strongly influenced by these velocities. However, because the vortical wake of the rotating wing is extremely complex and difficult to adequately represent mathematically, the practical solution of the aeroelastic problem has been delayed. Early attempts to solve this problem analytically were based on relatively drastic simplifications of the wake of the rotor to make them computationally feasible. The modern high-speed computing machines of today have made it possible to account for much more of the detail of the wake than has been possible in the past and thus permit an adequate aerodynamic representation of the blades and wake to be formulated which will enable accurate computation of the rotor airload distributions.

Piziali and DuWaldt (1962) started with a vortex-lattice model of the wake. The Biot-Savart law gave the induced velocity at the collocation points (three-quarter chord) on the rotor disk. Simultaneous equations were formulated for the bound circulation at the collocation points and an iterative solution implemented. The early work did not consider the issue of the near shed wake, but Piziali (1966) dealt with the effects of discretization of the shed wake. The analysis permitted important observations regarding the aerodynamic environment and performance of rotors. "It is interesting to note that the computed induced velocity distributions are always such as to *oppose* the retreating blade tip stall which has always been predicted on the basis of an assumed uniform inflow distribution" (Piziali and DuWaldt (1962)). Piziali and DuWaldt (1963) also looked at the calculated induced drag of the blades: "Induced power calculated from the induced drag distributions was found to be more than three times that calculated on the assumption of a uniform induced velocity."

Crimi (1966) developed a method for calculating the induced velocity at any point in the flow field, including a calculation of the distorted wake geometry and the effects of a fuselage. The wake model consisted of just the tip vortices; the inboard shed and trailed wake were neglected entirely. The blade loading and circulation were assumed to be known, so that only the wake geometry had to be calculated.

Landgrebe (1969) developed a method for calculating the rotor distorted wake geometry. The wake model consisted of up to 10 trailed vortex lines; the shed wake vorticity was neglected. Only the geometry of the tip vortices was calculated. To reduce the computation required, Landgrebe divided the wake elements into far wake and near wake regions. The near wake elements were those that were found in the first iteration to contribute significantly to the induced velocity at a given point in the wake. For successive iterations, only the induced velocity contributions of the near wake elements were updated. The result is a reduction of about an order of magnitude in the computation required to obtain the free wake geometry.

Clark and Leiper (1970) developed a method for calculating the distorted wake geometry of a hovering rotor. Their wake model consisted of a number of constant-strength trailed vortex lines; in hover there is no shed vorticity in the wake. The far wake was approximated by segments of ring vortices. The distorted geometry of all the trailers was calculated. A substantial influence of the distorted geometry on the loading distribution was found, particularly near the tip, and hence on the hover performance of the rotor.

Landgrebe (1971, 1972) conducted an experimental investigation of the performance and wake geometry of a model hovering rotor; see section 3.8.1. The wake geometry was measured by flow visualization, and the data were used to develop expressions for the axial convection and radial contraction of the tip vortices and inboard vortex sheets. This generalized wake geometry information was used in calculations of the rotor performance and produced a significant improvement in correlation with measured performance, compared to the results based on an uncontracted, rigid wake model. Landgrebe also calculated the distorted tip vortex geometry for the hovering rotor, and concluded that, although the wake geometry calculation was qualitatively good, the important first blade-vortex interaction was not well predicted. Landgrebe found a reduction in the stability of the wake vortices with increasing distance from the rotor disk in both the measurements and the calculations. An instability in which successive coils of the helices rolled around one another was observed in many of the model rotor flow visualizations. Shortly beyond this instability, further observation of the tip vortices was difficult. In no case was a smoothly contracting tip vortex observed for a large enough distance below the rotor disk to definitely preclude the possibility of an instability. Usually three or four turns of the helices were clearly evident, and then nothing of the wake structure could be seen.

Scully (1967, 1975) developed a method for calculating the free wake geometry of a helicopter rotor. The emphasis in this work was on developing efficient yet accurate computation techniques for the wake geometry and inflow calculations. Since the case of steady-state flight was considered, the solution was periodic. Only the geometry of the tip vortices was obtained. Shed vorticity and inboard trailed vorticity were retained (circulation was conserved), with rigid geometry. An efficient calculation of the wake geometry requires many variations on the basic numerical integration procedure. Scully adopted Landgrebe's near wake and far wake scheme for reducing the computation. The other major consideration for minimizing the computation was the matter of updating the induced velocity calculation. As time increases, the entire wake is convected downstream and the rotor blades move forward, adding new trailed and shed vorticity to the beginning of the wake. If there were no distortion of the wake during this time increment, the induced velocity at a given wake element would not change except for the contributions from the newly created wake vorticity just behind the blade. Thus the induced velocity could be obtained by just adding at each step the contribution from the new wake just behind the blade. The wake does distort as it is convected and as the estimate of the distortion improves, so updating the calculation of the induced velocity in the wake was necessary. The distortion was extrapolated beyond the last calculated age as required for the accurate evaluation of the velocities. The induced velocity from tip vortex elements adjacent to a collocation point was calculated using the result for the self-induced velocity of a vortex ring, which gives the effect of the core radius. Scully found that the wake geometry has a significant influence on the predicted rotor aerodynamic loading, because the distorted wake tends to be much closer to the blades than the rigid wake model would indicate.

Bliss, Teske, and Quackenbush (1987) developed a free wake geometry calculation using curved vortex elements. Only the tip vortices were modeled. An approximate integration of the Biot-Savart law for the induced velocity of a parabolic arc was obtained. The parabolic arc is close to a circular arc passing through the same mid and end points, up to fairly large arc angles. Quackenbush, Bliss, and Wachspress (1989) developed a free wake geometry analysis of a hovering rotor, using curved elements. An influence coefficient method was used to find the wake geometry by solving directly for the positions (steady in the rotating frame) that satisfy the equilibrium conditions, without numerical integration. Quackenbush, Wachspress, and Boschitsch (1995) developed the constant vorticity contour (CVC) model for a rotor wake. The sheet of vorticity behind each blade was discretized by laying out vortex filaments along contours of constant sheet strength. These filaments were constructed of curved vortex elements that distort in response to the local velocity field. This approach captured the free geometry and nonuniform inflow of the full wake. A vortex lattice (lifting surface) model was used for the blade. To reduce computation time, Quackenbush, Boschitsch, and Wachspress (1996) introduced a new algorithm (generalized periodic relaxation) for finding the geometry that satisfies the equations for convection of the vortex nodes. They also used a fast hierarchical vortex method in which, if a group of wake elements is a large distance from a group of collocation points where the induced velocity is required, a multipole representation of the wake elements is developed and the velocity calculated at the collocation points using a Taylor series expansion.

Crouse and Leishman (1993) produced an improved method for calculating rotor free wake geometry, using a predictor-corrector algorithm for the numerical integration. The predictor was based on explicit integration and the corrector on implicit integration. Bagai and Leishman (1995) developed a high-order pseudo-implicit predictor-corrector (PIPC) relaxation method for solution of the equations of the wake self-induced motion. The equation of motion of a wake node (equation 9.44) is formulated in terms of ψ and ϕ instead of ψ and $\delta = \psi - \phi$:

$$\frac{dr_W(\psi,\phi)}{d\psi} + \frac{dr_W(\psi,\phi)}{d\phi} = W + q(\psi,\phi) \qquad (9.141)$$

so the integration algorithm involves discretization of the wake age ϕ as well as time ψ. A Vatistas core model was used, with $n = 2$. The actual growth of the viscous core is small, but a larger value of the Squire parameter δ can improve convergence of the solution. A value of $\delta = 10^4$ was used as a compromise between good convergence without suppressing the distortion. To improve computational efficiency, Bagai and Leishman (1998) introduced adaptive grid sequencing and velocity field interpolation, which reduced the number of induced velocity evaluations in the free wake solution. Bhagwat and Leishman (2001, 2003) developed a time-accurate free-vortex method for rotors during maneuvers, based on a predictor-corrector 2nd-backward (PC2B) integration algorithm.

9.11 REFERENCES

Ashley, H., and Landahl, M. *Aerodynamics of Wings and Bodies.* Reading, MA: Addison-Wesley Publishing Company, Inc., 1965.

Betz, A. "Behavior of Vortex Systems." Zeitschrift fuer Angewandte, Mathematik und Mechanik, *XII*:3 (1932); also NACA TM 713.

Bagai, A., and Leishman, J.G. "Rotor Free-Wake Modeling Using a Pseudo-Implicit Technique – Including Comparisons with Experimental Data." Journal of the American Helicopter Society, *40*:3 (July 1995).

Bagai, A., and Leishman, J.G. "Adaptive Grid Sequencing and Interpolation Schemes for Helicopter Rotor Wake Analyses." AIAA Journal, *36*:9 (September 1998).

Bhagwat, M.J., and Leishman, J.G. "Stability Analysis of Helicopter Rotor Wakes in Axial Flight." Journal of the American Helicopter Society, *45*:3 (July 2000).

Bhagwat, M.J., and Leishman, J.G. "Stability, Consistency and Convergence of Time-Marching Free-Vortex Rotor Wake Algorithms." Journal of the American Helicopter Society, *46*:1 (January 2001).

Bhagwat, M.J., and Leishman, J.G. "Generalized Viscous Vortex Model for Application to Free-Vortex Wake and Aeroacoustic Calculations." American Helicopter Society 58th Annual Forum, Montreal, Canada, June 2002.

Bhagwat, M.J., and Leishman, J.G. "Rotor Aerodynamics During Maneuvering Flight Using a Time-Accurate Free-Vortex Wake." Journal of the American Helicopter Society, *48*:3 (July 2003).

Bilanin, A.J., and Donaldson, C.DuP. "Estimation of Velocities and Roll-Up in Aircraft Vortex Wakes." Journal of Aircraft, *12*:7 (July 1975).

Bliss, D.B., Dadone, L., and Wachspress, D.A. "Rotor Wake Modeling for High Speed Applications." American Helicopter Society 43rd Annual Forum, St. Louis, MO, May 1987.

Bliss, D.B., Teske, M.E., and Quackenbush, T.R. "A New Methodology for Free Wake Analysis Using Curved Vortex Elements." NASA CR 3958, December 1987.

Brower, M. "Lifting Surface and Lifting Line Solutions for Rotor Blade Interaction with Curved and Straight Vortex Lines." Massachusetts Institute of Technology, ASRL TR 194-5, November 1981.

Caradonna, F., Hendley, E., Silva, M., Huang, S., Komerath, N., Reddy, U., Mahalingam, R., Funk, R., Wong, O., Ames, R., Darden, L., Villareal, L., and Gregory, J. "Performance Measurement and Wake Characteristics of a Model Rotor in Axial Flight." Journal of the American Helicopter Society, *44*:2 (April 1999).

Clark, D.R., and Leiper, A.C. "The Free Wake Analysis. A Method for the Prediction of Helicopter Rotor Hovering Performance." Journal of the American Helicopter Society, *15*:1 (January 1970).

Crimi, P. "Prediction of Rotor Wake Flows." CAL/AVLABS Symposium on Aerodynamic Problems Associated with V/STOL Aircraft, Buffalo, NY, June 1966.

Crouse, G.L., Jr., and Leishman, J.G. "A New Method for Improved Rotor Free-Wake Convergence." AIAA Paper No. 93-0872, January 1993.

DeYoung, J., and Harper, C.W. "Theoretical Symmetric Span Loading at Subsonic Speeds for Wings Having Arbitrary Plan Form." NACA Report 921, 1948.

Harris, F.D. "Articulated Rotor Blade Flapping Motion at Low Advance Ratio." Journal of the American Helicopter Society, *17*:1 (January 1972).

Hess, J.L. "The Problem of Three-Dimensional Lifting Potential Flow and Its Solution by Means of Surface Singularity Distribution." Computer Methods in Applied Mechanics and Engineering, *4*:3 (November 1974).

Hooper, W.E. "The Vibratory Airloading of Helicopter Rotors." Vertica, *8*:2 (1984).

Johnson, W. "Recent Developments in Rotary-Wing Aerodynamic Theory." AIAA Journal, *24*:8 (August 1986).

Johnson, W. "A General Free Wake Geometry Calculation for Wings and Rotors." American Helicopter Society 51st Annual Forum, Ft. Worth, TX, May 1995.

Kocurek, J.D., Berkowitz, L.F., and Harris, F.D. "Hover Performance Methodology at Bell Helicopter Textron." American Helicopter Society 36th Annual Forum, Washington, DC, May 1980.

Kocurek, J.D., and Tangler, J.L. "A Prescribed Wake Lifting Surface Hover Performance Analysis." Journal of the American Helicopter Society, *22*:1 (January 1977).

Lamb, H. Hydrodynamics. 6th Edition. Mineola, NY: Dover Publications, Inc., 1932 (section 334a).

Landgrebe, A.J. "An Analytical Method for Predicting Rotor Wake Geometry." Journal of the American Helicopter Society, *14*:4 (October 1969).

Landgrebe, A.J. "An Analytical and Experimental Investigation of Helicopter Rotor Hover Performance and Wake Geometry Characteristics." USAAMRDL TR 71-24, June 1971.

Landgrebe, A.J. "The Wake Geometry of a Hovering Helicopter Rotor and Its Influence on Rotor Performance." Journal of the American Helicopter Society, *17*:4 (October 1972).

Martin, P.B., Bhagwat, M.J., and Leishman, J.G. "Visualization of a Helicopter Rotor Wake in Hover." AIAA Paper No. 99-3225, June 1999.

McAlister, K.W., Schuler, C.A., Branum, L., and Wu, J.C. "3-D Wake Measurements Near a Hovering Rotor for Determining Profile and Induced Drag." NASA TP 3577, August 1995.

Miller, R.H. "Rotor Blade Harmonic Air Loading." IAS Paper No. 62-82, January 1962a.

Miller, R.H. "On the Computation of Airloads Acting on Rotor Blades in Forward Flight." Journal of the American Helicopter Society, *7*:2 (April 1962b).

Miller, R.H. "A Discussion of Rotor Blade Harmonic Airloading." CAL/TRECOM Symposium on Dynamic Load Problems Associated with Helicopters and V/STOL Aircraft, Buffalo, NY, June 1963.

Miller, R.H. "Unsteady Air Loads on Helicopter Rotor Blades." Journal of the Royal Aeronautical Society, *68*:640 (April 1964a).

Miller, R.H. "Rotor Blade Harmonic Air Loading." AIAA Journal, *2*:7 (July 1964b).

Miller, R.H. "Factors Influencing Rotor Aerodynamics in Hover and Forward Flight." Vertica, *9*:2 (1985).

Piziali, R.A. "Method for the Solution of the Aeroelastic Response Problem for Rotating Wings." Journal of Sound and Vibration, *4*:3 (1966).

Piziali, R.A., and DuWaldt, F.A. "A Method for Computing Rotary Wing Airload Distribution in Forward Flight." TCREC TR 62-44, November 1962.

Piziali, R., and DuWaldt, F. "Computed Induced Velocity, Induced Drag, and Angle of Attack Distribution for a Two-Bladed Rotor." American Helicopter Society 19th Annual National Forum, Washington, DC, May 1963.

Prandtl, L. "Applications of Modern Hydrodynamics to Aeronautics." NACA Report 116, 1921.

Quackenbush, T.R., Bliss, D.B., and Wachspress, D.A. "New Free-Wake Analysis of Rotorcraft Hover Performance Using Influence Coefficients." Journal of Aircraft, *26*:12 (December 1989).

Quackenbush, T.R., Boschitsch, A.H., and Wachspress, D.A. "Fast Analysis Methods for Surface-Bounded Flows with Applications to Rotor Wake Modeling." American Helicopter Society 52nd Annual Forum, Washington, DC, June 1996.

Quackenbush, T.R., Lam, C.-M.G., Wachspress, D.A., and Bliss, D.B. "Computational Analysis of High Resolution Unsteady Airloads for Rotor Aeroacoustics." NASA CR 194894, May 1994.

Quackenbush, T.R., Wachspress, D.A., and Boschitsch, A.H. "Rotor Aerodynamic Loads Computation Using a Constant Vorticity Contour Free Wake Model." Journal of Aircraft, *32*:5 (September–October 1995).

Ramasamy, M., and Leishman, J.G. "A Reynolds Number-Based Blade Tip Vortex Model." Journal of the American Helicopter Society, *52*:3 (July 2007).

Rossow, V.J. "On the Inviscid Rolled-Up Structure of Lift-Generated vortices." Journal of Aircraft, *10*:11 (November 1973).

Scully, M.P. "Approximate Solutions for Computing Helicopter Harmonic Airloads." MIT ASRL TR 123-2, December 1965.

Scully, M.P. "A Method of Computing Helicopter Vortex Wake Distortion." MIT ASRL TR 138-1, June 1967.

Scully, M.P. "Computation of Helicopter Rotor Wake Geometry and Its Influence on Rotor Harmonic Airloads." Massachusetts Institute of Technology, ASRL TR 178-1, March 1975.

Simons, I.A., Pacifico, R.E., and Jones, J.P. "The Movement, Structure and Breakdown of Trailing Vortices from a Rotor Blade." CAL/AVLABS Symposium on Aerodynamic Problems Associated with V/STOL Aircraft, Buffalo, NY, June 1966.

Squire, H.B. "The Growth of a Vortex in Turbulent Flow," Aeronautical Quarterly, *16*:3 (August 1965).

Tangler, J.L., Wohlfeld, R.M., and Miley, S.J. "An Experimental Investigation of Vortex Stability, Tip Shapes, Compressibility, and Noise for Hovering Model Rotors." NASA CR 2305, September 1973.

Van Dyke, M. "Lifting-Line Theory as a Singular-Perturbation Problem." Stanford University Report SUDAER 165, August 1963.

van Holten, T. "The Computation of Aerodynamic Loads on Helicopter Blades in Forward Flight, Using the Method of the Acceleration Potential." Technische Hogeschool Delft, Report VTH-189, March 1975.

van Holten, T. "On the Validity of Lifting Line Concepts in Rotor Analysis." Vertica, *1*:3 (1977).

Vatistas, G.H., Kozel, V., and Mih, W.C. "A Simpler Model for Concentrated Vortices." Experiments in Fluids, *11*:1 (April 1991).

Wachspress, D.A., Quackenbush, T.R., and Boschitsch, A.H. "Rotorcraft Interactional Aerodynamics with Fast Vortex/Fast Panel Methods." Journal of the American Helicopter Society, *48*:4 (October 2003).

Weissinger, J. "The Lift Distribution of Swept-Back Wings." NACA TM 1120, March 1947.

Widnall, S.E. "The Stability of a Helical Vortex Filament." Journal of Fluid Mechanics, *54*:4 (August 1972).

Widnall, S.E., Bliss, D.B., and Zalay, A. "Theoretical and Experimental Study of the Stability of a Vortex Pair." In *Aircraft Wake Turbulence and its Detection*, Olsen J., Goidberg, A. and Rogers, M. (Editors). New York: Plenum Press, 1971.

10 Unsteady Aerodynamics

10.1 Two-Dimensional Unsteady Airfoil Theory

Since the aerodynamic environment of the rotor blade in forward flight or during transient motion is unsteady, lifting-line theory requires an analysis of the unsteady aerodynamics of a two-dimensional airfoil. Consider the problem of a two-dimensional airfoil undergoing unsteady motion in a uniform free stream. Linear, incompressible aerodynamic theory represents the airfoil and its wake by thin surfaces of vorticity (two-dimensional vortex sheets) in a straight line parallel to the free stream velocity. For the linear problem the solution for the thickness and camber loads can be separated from the loads due to angle-of-attack and unsteady motion. In the development of unsteady thin-airfoil theory, the foundation is constructed for a number of extensions of the analysis for rotary wings, which are presented in later sections of this chapter.

The airfoil and shed wake in unsteady thin-airfoil theory are modeled by planar sheets of vorticity, as shown in Figure 10.1. An airfoil of chord $2b$ is in a uniform free stream with velocity U. Since the bound circulation of the section varies with time, there is shed vorticity in the wake downstream of the airfoil. The vorticity strength on the airfoil is γ_b, and in the wake γ_w. The blade motion (Figure 10.2) is described by a heaving motion h (positive downward) and a pitch angle α about an axis at $x = ab$ (positive for nose upward). The aerodynamic pitch moment is evaluated about the axis at $x = ab$. The airfoil motion produces an upwash velocity relative to the blade of

$$w_a = U\alpha + \dot{h} + (x - ab)\dot{\alpha} \tag{10.1}$$

In addition to the velocity w_a, at the blade section there is also a downwash velocity λ due to the shed wake, and w_b due to the vorticity representing the blade surface. From the strength of the vortex sheets representing the airfoil and shed wake, these induced velocities are

$$w_b(x) = \frac{1}{2\pi} \int_{-b}^{b} \frac{\gamma_b}{x - \xi}\, d\xi \tag{10.2}$$

$$\lambda(x) = \frac{1}{2\pi} \int_{b}^{\infty} \frac{\gamma_w}{x - \xi}\, d\xi \tag{10.3}$$

10.1 Two-Dimensional Unsteady Airfoil Theory

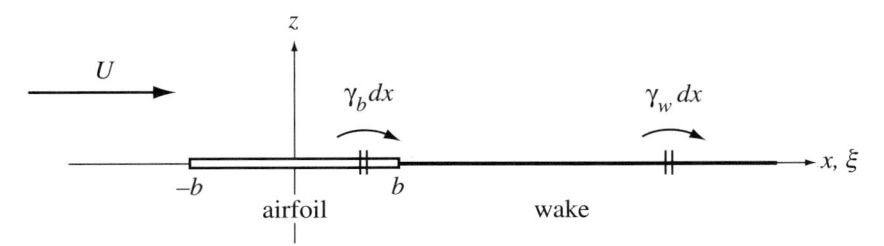

Figure 10.1. Unsteady thin airfoil theory model of the two-dimensional wing and wake.

The boundary condition of no flow through the wing surface, $w_b + \lambda - w_a = 0$, gives an integral equation for the bound vorticity γ_b:

$$\frac{1}{2\pi} \int_{-b}^{b} \frac{\gamma_b d\xi}{x - \xi} = w_a - \lambda \tag{10.4}$$

From the bound circulation γ_b the chordwise pressure loading can be found. The shed wake vorticity is given by the time rate of change in the total bound circulation $\Gamma = \int_{-b}^{b} \gamma_b dx$:

$$\gamma_w = -\frac{1}{U} \frac{d\Gamma}{dt} \tag{10.5}$$

evaluated at the time the element was shed, $t - (x - b)/U$. So the wake-induced velocity λ is also defined by the blade vorticity γ_b. The boundary condition of no pressure difference across the wake requires that the shed vorticity be convected with the free stream, so $\gamma_w = \gamma_w(x - Ut)$. Finally, the Kutta condition of finite velocity at the blade trailing edge requires $\gamma_b = 0$ at $x = b$.

With the Kutta condition, the integral equation inverts to

$$\gamma_b = -\frac{2}{\pi} \sqrt{\frac{b - x}{b + x}} \int_{-b}^{b} \sqrt{\frac{b + \xi}{b - \xi}} \frac{w_a - \lambda}{x - \xi} d\xi \tag{10.6}$$

Now write for the wake-induced velocity and the upwash due to the airfoil motion,

$$\lambda = \sum_{n=0}^{\infty} \lambda_n \cos n\theta \tag{10.7}$$

$$w_a = \sum_{n=0}^{\infty} w_n \cos n\theta \tag{10.8}$$

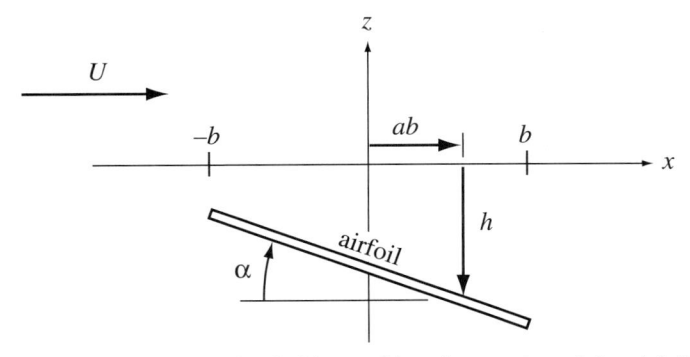

Figure 10.2. Unsteady pitching and heaving motion of the airfoil.

where $x = b\cos\theta$ ($\theta = 0$ at the trailing edge and $\theta = \pi$ at the leading edge). Then the solution for γ_b reduces to

$$\gamma_b = 2\sum_{n=0}^{\infty}(w_n - \lambda_n)f_n(\theta) \tag{10.9}$$

where f_n is the Glauert series:

$$f_n(\theta) = \begin{cases} \tan(\theta/2) & n = 0 \\ \sin n\theta & n \geq 1 \end{cases} \tag{10.10}$$

In terms of x rather than θ, the expansion of the normal velocity is

$$w_a = w_0 + w_1(x/b) + w_2(2x^2/b^2 - 1) + \ldots \tag{10.11}$$

For the blade motion considered, $w_0 = U\alpha + \dot{h} - ab\dot{\alpha}$ (w_a at the midchord), $w_1 = b\dot{\alpha}$, and $w_n = 0$ for $n \geq 2$. The first terms in the Glauert series are

$$f_0 = \sqrt{\frac{b-x}{b+x}} \tag{10.12}$$

$$f_1 = \sqrt{1 - (x/b)^2} \tag{10.13}$$

$$f_2 = 2(x/b)\sqrt{1 - (x/b)^2} \tag{10.14}$$

The coefficients w_n can be evaluated for a particular blade motion. To complete the solution, the wake-induced velocity λ is required.

On substituting for γ_b, the airfoil bound circulation becomes

$$\Gamma = \int_{-b}^{b}\gamma_b dx = 2\pi b\left[\left(w_0 + \frac{1}{2}w_1\right) - \left(\lambda_0 + \frac{1}{2}\lambda_1\right)\right] \tag{10.15}$$

Next divide γ_b into two parts: the circulatory vorticity γ_{b_C}, which gives Γ but corresponds to $w_b = 0$ and so has no effect on the boundary conditions; and the noncirculatory vorticity $\gamma_{b_{NC}}$, which satisfies the boundary conditions but gives $\Gamma = 0$. Hence $\gamma_b = \gamma_{b_C} + \gamma_{b_{NC}}$, and the expressions

$$\gamma_{b_C} = \frac{2}{\sin\theta}\left[\left(w_0 + \frac{1}{2}w_1\right) - \left(\lambda_0 + \frac{1}{2}\lambda_1\right)\right] \tag{10.16}$$

$$\gamma_{b_{NC}} = -\frac{2}{\sin\theta}\left[(w_0 - \lambda_0)\cos\theta + \frac{1}{2}(w_1 - \lambda_1)\cos 2\theta\right] + 2\sum_{n=2}^{\infty}(w_n - \lambda_n)f_n(\theta) \tag{10.17}$$

give

$$\int_{-b}^{b}\gamma_{b_C}dx = \Gamma \tag{10.18}$$

$$\int_{-b}^{b}\gamma_{b_{NC}}dx = 0 \tag{10.19}$$

10.1 Two-Dimensional Unsteady Airfoil Theory

$$\frac{1}{2\pi} \int_{-b}^{b} \frac{\gamma_{b_C}}{x-\xi} d\xi = 0 \tag{10.20}$$

$$\frac{1}{2\pi} \int_{-b}^{b} \frac{\gamma_{b_{NC}}}{x-\xi} d\xi = w_a - \lambda \tag{10.21}$$

as required. The relation

$$\frac{1}{\pi} \int_0^{\pi} \frac{\cos n\theta \, d\theta}{\cos\theta - \cos\phi} = \frac{\sin n\phi}{\sin\phi} \tag{10.22}$$

is used to establish the last two results.

The pressure is obtained by linearizing the unsteady Bernoulli equation:

$$p = -\rho \left(U \frac{\partial \phi}{\partial x} + \frac{\partial \phi}{\partial t} \right) \tag{10.23}$$

where ϕ is the velocity potential. The differential pressure on the airfoil surface is then

$$-\Delta p = \rho \left(U \frac{\partial \Delta\phi}{\partial x} + \frac{\partial \Delta\phi}{\partial t} \right) \tag{10.24}$$

where Δp is the upper surface pressure minus the lower surface pressure. The velocity parallel to the blade surface is $u = \partial\phi/\partial x$, and the blade vorticity strength is $\gamma_b = \Delta u$. Then

$$\frac{\partial \Delta\phi}{\partial x} = \Delta u = \gamma_b \tag{10.25}$$

$$\frac{\partial \Delta\phi}{\partial t} = \frac{\partial}{\partial t} \int_{-\infty}^{x} \Delta u \, dx = \frac{\partial}{\partial t} \int_{-b}^{x} \gamma_b dx \tag{10.26}$$

The differential pressure is thus

$$-\Delta p = \rho \left(U\gamma_b + \frac{\partial}{\partial t} \int_{-b}^{x} \gamma_{b_{NC}} dx \right) \tag{10.27}$$

Only the non-circulatory vorticity contributes pressure through the $\partial\phi/\partial t$ term. The unsteady circulatory vorticity produces pressure through the shed-wake-induced velocity λ. Substituting the expressions for γ_{b_C} and $\gamma_{b_{NC}}$ gives

$$-\Delta p = \rho U \gamma_b + \rho b \int_{\theta}^{\pi} \dot{\gamma}_{b_{NC}} \sin\theta \, d\theta$$

$$= 2\rho U \sum_{n=0} (w_n - \lambda_n) f_n + \rho b \left(2(\dot{w}_0 - \dot{\lambda}_0) \sin\theta + \frac{1}{2}(\dot{w}_1 - \dot{\lambda}_1) \sin 2\theta \right.$$

$$\left. - \sum_{n=1} (\dot{w}_{n+1} - \dot{\lambda}_{n+1}) \frac{\sin n\theta}{n} + \sum_{n=3} (\dot{w}_{n-1} - \dot{\lambda}_{n-1}) \frac{\sin n\theta}{n} \right)$$

$$= \sum_{n=0} p_n f_n(\theta) \tag{10.28}$$

where

$$p_0 = 2\rho U (w_0 - \lambda_0) \tag{10.29}$$

$$p_1 = 2\rho U (w_1 - \lambda_1) + \rho b \left(2(\dot{w}_0 - \dot{\lambda}_0) - (\dot{w}_2 - \dot{\lambda}_2) \right) \tag{10.30}$$

and

$$p_n = 2\rho U(w_n - \lambda_n) + \frac{\rho b}{n}\left((\dot{w}_{n-1} - \dot{\lambda}_{n-1}) - (\dot{w}_{n+1} - \dot{\lambda}_{n+1})\right) \qquad (10.31)$$

for $n \geq 2$.

The net aerodynamic forces on the airfoil are the lift L (positive upward) and moment M about the axis at $x = ab$ (positive nose upward):

$$L = \int_{-b}^{b} (-\Delta p)\,dx \qquad (10.32)$$

$$M = \int_{-b}^{b} (-\Delta p)(-x + ab)\,dx \qquad (10.33)$$

Substituting for Δp gives

$$L = \rho\left(U\Gamma - \frac{\partial}{\partial t}\Gamma_{NC}^{(1)}\right) \qquad (10.34)$$

$$M = -\rho\left(U\Gamma^{(1)} - \frac{1}{2}\frac{\partial}{\partial t}\Gamma_{NC}^{(2)}\right) \qquad (10.35)$$

where

$$\Gamma^{(n)} = \int_{-b}^{b} x^n \gamma_b\,dx \qquad (10.36)$$

$$\Gamma_{NC}^{(n)} = \int_{-b}^{b} x^n \gamma_{b_{NC}}\,dx \qquad (10.37)$$

The required circulations can be evaluated by substituting for γ_b:

$$\Gamma = 2\pi b\left[\left(w_0 + \frac{1}{2}w_1\right) - \left(\lambda_0 + \frac{1}{2}\lambda_1\right)\right] \qquad (10.38)$$

$$\Gamma^{(1)} = 2\pi b^2 \left[-\left(\frac{1}{2} + a\right)\left(\left(w_0 + \frac{1}{2}w_1\right) - \left(\lambda_0 + \frac{1}{2}\lambda_1\right)\right)\right.$$
$$\left. + \frac{1}{4}\left((w_1 + w_2) - (\lambda_1 + \lambda_2)\right)\right] \qquad (10.39)$$

$$\Gamma_{NC}^{(1)} = 2\pi b^2\left[-\frac{1}{2}\left(w_0 - \frac{1}{2}w_2\right) + \frac{1}{2}\left(\lambda_0 - \frac{1}{2}\lambda_2\right)\right] \qquad (10.40)$$

$$\Gamma_{NC}^{(2)} = 2\pi b^3\left[a\left(\left(w_0 - \frac{1}{2}w_2\right) - \left(\lambda_0 - \frac{1}{2}\lambda_2\right)\right) - \frac{1}{8}\left((w_1 - w_3) - (\lambda_1 - \lambda_3)\right)\right]$$
$$(10.41)$$

For the blade motion considered here,

$$w_0 + \frac{1}{2}w_1 = U\alpha + \dot{h} + \left(\frac{1}{2} - a\right)b\dot{\alpha} = w_{.75c} \qquad (10.42)$$

$$w_0 - \frac{1}{2}w_2 = U\alpha + \dot{h} - ab\dot{\alpha} = w_{.5c} \qquad (10.43)$$

$$w_1 + w_2 = b\dot{\alpha} \qquad (10.44)$$

$$w_1 - w_3 = b\dot{\alpha} \qquad (10.45)$$

where $w_{.75c}$ is the upwash at the three-quarter chord and $w_{.5c}$ is the upwash at the midchord. The coefficients λ_n in the expansion of the induced velocity over the chord can be written in terms of the wake vorticity as follows:

$$\begin{aligned}\lambda_n &= \frac{2}{\pi}\int_0^\pi \lambda \cos n\theta \, d\theta \\ &= \frac{2}{\pi}\int_0^\pi \left[\frac{1}{2\pi}\int_b^\infty \frac{\gamma_w d\xi}{x-\xi}\right]\cos n\theta \, d\theta \\ &= -\frac{1}{\pi}\int_b^\infty \gamma_w \left[\frac{1}{\pi}\int_0^\pi \frac{\cos n\theta}{\xi - b\cos\theta}d\theta\right]d\xi \\ &= -\frac{1}{\pi}\int_b^\infty \gamma_w \left[\frac{\left(\xi - \sqrt{\xi^2 - b^2}\right)^n}{b^n\sqrt{\xi^2-b^2}}\right]d\xi \end{aligned} \quad (10.46)$$

So

$$\lambda_0 + \frac{1}{2}\lambda_1 = -\frac{1}{2\pi b}\int_b^\infty \gamma_w \left(\sqrt{\frac{\xi+b}{\xi-b}} - 1\right)d\xi \quad (10.47)$$

$$\lambda_1 + \frac{1}{2}\lambda_2 = -\frac{1}{\pi b^2}\int_b^\infty \gamma_w \left(\xi - \sqrt{\xi^2 - b^2}\right)d\xi \quad (10.48)$$

$$\lambda_1 + \lambda_2 = -\frac{1}{\pi b^2}\int_b^\infty \gamma_w \left(\xi - \sqrt{\xi^2 - b^2}\right)\left(\sqrt{\frac{\xi+b}{\xi-b}} - 1\right)d\xi \quad (10.49)$$

$$\lambda_1 - \lambda_3 = -\frac{2}{\pi b^3}\int_b^\infty \gamma_w \left(\xi - \sqrt{\xi^2 - b^2}\right)^2 d\xi \quad (10.50)$$

The circulations required for the airfoil lift are then

$$\Gamma = 2\pi b\left(U\alpha + \dot{h} + \left(\frac{1}{2} - a\right)b\dot{\alpha}\right) + \int_b^\infty \left(\sqrt{\frac{\xi+b}{\xi-b}} - 1\right)\gamma_w d\xi \quad (10.51)$$

and

$$\begin{aligned}\frac{\partial}{\partial t}\Gamma_{NC}^{(1)} &= \frac{\partial}{\partial t}\left[-\pi b^2\left(U\alpha + \dot{h} - ab\dot{\alpha}\right) - \int_b^\infty \left(\xi - \sqrt{\xi^2 - b^2}\right)\gamma_w d\xi\right] \\ &= -\pi b^2\left(U\dot{\alpha} + \ddot{h} - ab\ddot{\alpha}\right) - U\int_b^\infty \frac{\partial}{\partial \xi}\left(\xi - \sqrt{\xi^2 - b^2}\right)\gamma_w d\xi \\ &= -\pi b^2\left(U\dot{\alpha} + \ddot{h} - ab\ddot{\alpha}\right) - U\int_b^\infty \left(1 - \frac{\xi}{\sqrt{\xi^2 - b^2}}\right)\gamma_w d\xi \end{aligned} \quad (10.52)$$

The airfoil lift now is

$$\begin{aligned}L &= 2\pi\rho Ub\left(U\alpha + \dot{h} + \left(\frac{1}{2} - a\right)b\dot{\alpha}\right) + \rho\pi b^2\left(U\dot{\alpha} + \ddot{h} - ab\ddot{\alpha}\right) \\ &\quad + \rho U\int_b^\infty \frac{b}{\sqrt{\xi^2 - b^2}}\gamma_w d\xi \\ &= L_Q + L_{NC} + L_W \end{aligned} \quad (10.53)$$

L_Q is the quasistatic lift, which is the only term present for the steady case ($L = 2\pi \rho U^2 b \alpha$); L_{NC} is the non-circulatory lift, which is due to $\partial \Gamma_{NC}^{(1)} / \partial t$; and L_W is the lift due to the shed-wake-induced velocity. For the unsteady case L_Q is due to the angle-of-attack at the three-quarter chord. From equations 10.34, 10.38, and 10.40, the terms in $L = L_Q + L_{NC} + L_W$ can be written

$$L_Q = 2\pi \rho U b \left(w_0 + \frac{1}{2} w_1 \right) \tag{10.54}$$

$$L_{NC} = \rho \pi b^2 \left(\dot{w}_0 - \frac{1}{2} \dot{w}_2 \right) \tag{10.55}$$

$$L_W = -2\pi \rho U b \left(\lambda_0 + \frac{1}{2} \lambda_1 \right) - \rho \pi b^2 \left(\dot{\lambda}_0 - \frac{1}{2} \dot{\lambda}_2 \right) \tag{10.56}$$

Now the bound circulation is

$$\Gamma = \frac{L_Q}{\rho U} + \int_b^\infty \left(\sqrt{\frac{\xi + b}{\xi - b}} - 1 \right) \gamma_w d\xi \tag{10.57}$$

and conservation of vorticity requires $\Gamma = -\int_b^\infty \gamma_w d\xi$; hence

$$L_Q = -\rho U \int_b^\infty \sqrt{\frac{\xi + b}{\xi - b}} \gamma_w d\xi \tag{10.58}$$

and

$$L_C = L_Q + L_W = -\rho U \int_b^\infty \frac{\xi}{\sqrt{\xi^2 - b^2}} \gamma_w d\xi \tag{10.59}$$

The lift can therefore be written as

$$L = \frac{\int_b^\infty \frac{\xi}{\sqrt{\xi^2 - b^2}} \gamma_w d\xi}{\int_b^\infty \sqrt{\frac{\xi + b}{\xi - b}} \gamma_w d\xi} L_Q + L_{NC} \tag{10.60}$$

The effect of the shed wake is to multiply the quasistatic lift L_Q by a factor that depends on γ_w, and hence on the airfoil motion. To evaluate this factor, a specific time history of motion must be considered. Assume that the airfoil has purely harmonic motion at frequency ω: $\alpha = \bar{\alpha} e^{i\omega t}$ and $h = \bar{h} e^{i\omega t}$. Then the wake vorticity γ_w must also be periodic in time and has the form $\gamma_w = \bar{\gamma}_w e^{i\omega(t - \xi/U)}$ when the requirement of convection with the free stream velocity is applied as well. Then $\gamma_w e^{i\omega t}$ factors out of the integrals over the wake, giving

$$L = C(k) L_Q + L_{NC}$$
$$= 2\pi \rho U b C(k) \left(U\alpha + \dot{h} + \left(\frac{1}{2} - a \right) b \dot{\alpha} \right) + \rho \pi b^2 \left(U\dot{\alpha} + \ddot{h} - ab\ddot{\alpha} \right) \tag{10.61}$$

where $C(k)$ is a function depending only on the dimensionless frequency $k = \omega b / U$. $C(k)$ is the Theodorsen lift deficiency function (Theodorsen (1935)). Since the magnitude of C varies from 1 at low frequency to 0.5 at high frequency, the effect of the shed wake is to reduce the circulatory lift below the quasistatic value.

10.1 Two-Dimensional Unsteady Airfoil Theory

The circulations required for the aerodynamic moment about the axis $x = ab$ are obtained by similar manipulations:

$$\Gamma^{(1)} = -b\left(\frac{1}{2}+a\right)\Gamma + \frac{1}{2}\pi b^3 \dot{\alpha} + \frac{1}{2}\int_b^\infty \left(\xi - \sqrt{\xi^2 - b^2}\right)\left(\sqrt{\frac{\xi+b}{\xi-b}} - 1\right)\gamma_w d\xi \qquad (10.62)$$

$$\frac{1}{2}\frac{\partial}{\partial t}\Gamma_{NC}^{(2)} = -b\left(\frac{1}{2}+a\right)\frac{\partial}{\partial t}\Gamma_{NC}^{(1)} - \frac{1}{2}\pi b^3 \left(U\dot{\alpha} + \ddot{h} + \left(\frac{1}{4} - a\right)b\ddot{\alpha}\right)$$

$$+ \frac{1}{2}U\int_b^\infty \left(\xi - \sqrt{\xi^2 - b^2}\right)\left(\sqrt{\frac{\xi+b}{\xi-b}} - 1\right)\gamma_w d\xi \qquad (10.63)$$

Thus the moment is

$$M = b\left(\frac{1}{2}+a\right)L + M_{QC}$$

$$= b\left(\frac{1}{2}+a\right)L - \frac{1}{2}\rho\pi b^3\left(2U\dot{\alpha} + \ddot{h} + \left(\frac{1}{4}-a\right)b\ddot{\alpha}\right)$$

$$= b\left(\frac{1}{2}+a\right)C(k)L_Q + \rho\pi b^3\left(a\ddot{h} - \left(\frac{1}{2}-a\right)U\dot{\alpha} - \left(\frac{1}{8}+a^2\right)b\ddot{\alpha}\right) \qquad (10.64)$$

M_{QC} is the moment about the quarter chord, which is the aerodynamic center predicted by thin-airfoil theory. With the pitch axis at the quarter chord ($a = -\frac{1}{2}$) there is no moment due to the lift. The virtual mass terms ($\ddot{h}$ and $\ddot{\alpha}$) arise both from M_{QC} and from the non-circulatory lift L_{NC}. The non-circulatory pitch damping moment is due to lift acting at the three-quarter chord; for $a = \frac{1}{2}$ this moment is zero.

Let us now examine the Theodorsen lift deficiency function $C(k)$, which defines the influence of the shed wake on the aerodynamic loads during unsteady motion. Recall that to evaluate the wake influence, harmonic motion at frequency ω was assumed, giving $\gamma_w = \overline{\gamma}_w e^{i\omega(t-\xi/U)}$. Hence

$$C(k) = \frac{\int_b^\infty \frac{\xi}{\sqrt{\xi^2-b^2}}\gamma_w d\xi}{\int_b^\infty \sqrt{\frac{\xi+b}{\xi-b}}\gamma_w d\xi} = \frac{\int_1^\infty \frac{\xi}{\sqrt{\xi^2-1}}e^{-ik\xi}d\xi}{\int_1^\infty \sqrt{\frac{\xi+1}{\xi-1}}e^{-ik\xi}d\xi} = \frac{H_1^{(2)}(k)}{H_1^{(2)}(k) + iH_0^{(2)}(k)} \qquad (10.65)$$

where $H_n^{(2)} = J_n - iY_n$ is the Hankel function, and the reduced frequency is $k = \omega b/U$. The real and imaginary parts are defined by $C = F + iG$. Figure 10.3 shows the magnitude and phase of the lift deficiency function for reduced frequencies up to $k = 1$. For $k = 0$, $C = 1$ as is required of the static limit; for large frequencies the magnitude approaches $|C| = 0.5$, so the shed wake reduces the circulatory lift to one-half the quasistatic value. There is a moderate phase shift that has a maximum just above 15° at about $k = 0.3$ and approaches zero again at high frequencies. For small frequencies, the lift deficiency function is approximately

$$C(k) \cong \left(1 - \frac{\pi}{2}k\right) + ik\left(\ln\frac{k}{2} + \gamma\right) \qquad (10.66)$$

where $\gamma = 0.5772156\ldots$ is Euler's constant. For a rotor, the frequency of the blade motion can be expressed in terms of the rotational speed Ω. Consider n/rev motion,

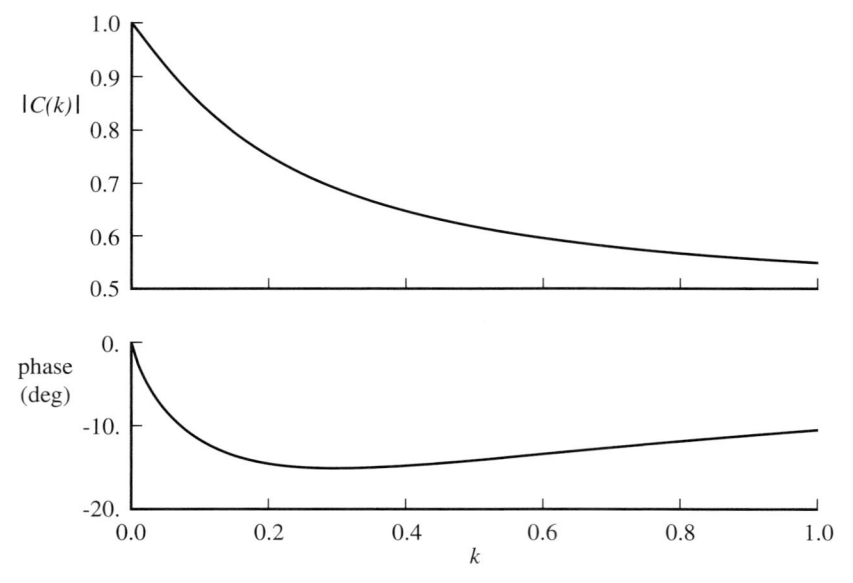

Figure 10.3. Theodorsen lift deficiency function.

where $\omega = n\Omega$. The free stream velocity in hover is Ωr, and the semi-chord is $c/2$, so the reduced frequency becomes $k = nc/2r$. For the high-aspect-ratio blades of rotors, typically $k \cong 0.05n$. For the lower harmonics, the reduced frequency is small, and the lift deficiency function is near unity. For 1/rev motion there is perhaps a 5% reduction in the lift due to the shed wake. Thus the neglect of the shed wake and other unsteady aerodynamic effects in the analysis of the rotor performance and flap motion of the earlier chapters is justified. For the higher harmonics, the reduced frequency is large enough that the shed wake effects must be accounted for to obtain an accurate estimate of the loads.

An alternative form of the unsteady thin-airfoil result is a Glauert series for the pressure, developed by Cicala (1951):

$$-\Delta p = \rho U^2 e^{i\omega t} \sum_{n=0}^{\infty} a_n f_n(\theta) \tag{10.67}$$

where $x = b\cos\theta$. Expand the upwash due to the blade motion as a cosine series:

$$w_a = Ue^{i\omega t}\left(A_0 + 2\sum_{n=1}^{\infty} A_n \cos n\theta\right) \tag{10.68}$$

Then the solution can be written as

$$a_0 = 2(A_0 + A_1)C(k) - 2A_1 \tag{10.69}$$

$$a_n = -\frac{2ik}{n}(A_{n+1} - A_{n-1}) + 4A_n \tag{10.70}$$

with the lift and moment given by

$$L = \rho U^2 b\pi \left(a_0 + \frac{1}{2}a_1\right)e^{i\omega t} \tag{10.71}$$

$$M_{QC} = -\rho U^2 b^2 \frac{\pi}{4}(a_1 + a_2)e^{i\omega t} \tag{10.72}$$

10.1 Two-Dimensional Unsteady Airfoil Theory

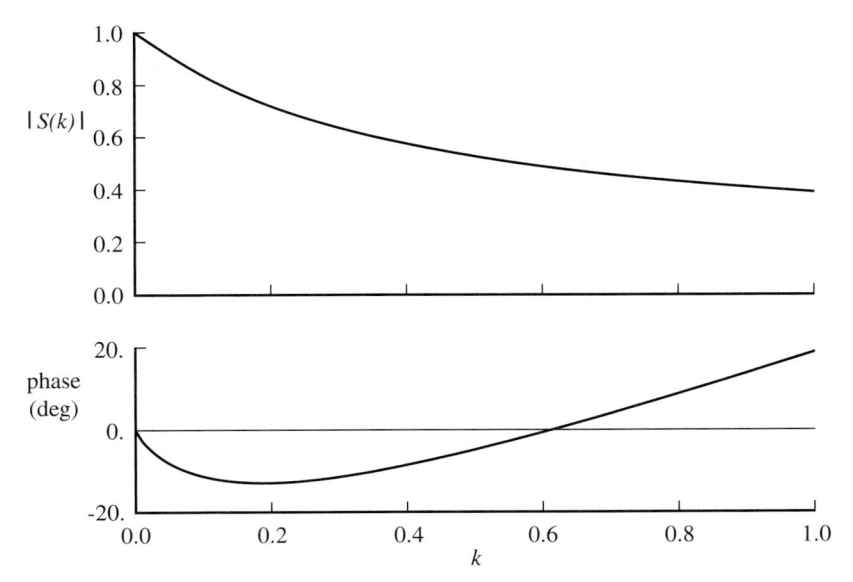

Figure 10.4. Sears function for sinusoidal gust loading.

For example, consider an encounter with a sinusoidal gust of wavelength $2\pi b/k$, so the airfoil sees the upwash velocity

$$w_a = w_0 e^{i\omega(t-x/U)} = w_0 e^{i\omega t} e^{-ikx/b} = w_0 e^{i\omega t} e^{-ik\cos\theta} \quad (10.73)$$

Expanding $e^{-ik\cos\theta}$ as a cosine series in θ gives

$$A_n = \frac{w_0}{U}(-1)^n J_n(k) \quad (10.74)$$

where J_n is the Bessel function. Thus

$$a_0 = \frac{w_0}{U} 2\left[(J_0(k) - iJ_1(k))C(k) + iJ_1(k)\right] \quad (10.75)$$

$$a_n = \frac{w_0}{U}\frac{2ik}{n}(-1)^{n-1}\left[J_{n+1} + J_{n-1} - \frac{2n}{k}J_n\right] = 0 \quad (10.76)$$

The pressure then has only the first term in the Glauert series:

$$-\frac{\Delta p}{\rho U^2} = e^{i\omega t}\frac{w_0}{U} 2S(k)\sqrt{\frac{b-x}{b+x}} \quad (10.77)$$

The lift is

$$\frac{L}{\rho U^2 b} = e^{i\omega t}\frac{w_0}{U} 2\pi S(k) \quad (10.78)$$

and $M_{QC} = 0$. Here $S(k)$ is the Sears function,

$$S(k) = (J_0(k) - iJ_1(k))C(k) + iJ_1(k) \quad (10.79)$$

which is shown in Figure 10.4; see Sears (1941). Since any gust can be Fourier analyzed, the resulting aerodynamic lift always acts at the quarter chord. At $k = 0$, the Sears function is $S = 1$. For large frequency, S is approximately

$$S(k) \sim \frac{1}{\sqrt{2\pi k}} e^{i(k-\pi/4)} \tag{10.80}$$

so the magnitude approaches zero (in contrast to the Theodorsen function), while the phase is linear with k.

10.2 Lifting-Line Theory and Near Shed Wake

Two-dimensional airfoil theory shows that the shed wake is an important factor in determining the unsteady aerodynamic loading at frequencies characteristic of rotor blade motion. Unlike the two-dimensional model, the rotary-wing shed wake is in a helical sheet behind the blade, but the major effects are produced by the shed wake nearest to the trailing edge. The near shed wake (extending 15° to 45° in wake age behind the blade) must be modeled appropriately in a calculation of induced velocity and airloading on the rotating wing. Helicopter airloads analyses generally use lifting-line theory to calculate the wake-induced velocity at the bound vortex. Although for the trailed vorticity the wing in lifting-line theory is collapsed to a bound vortex and the induced velocity evaluated at a single point on the chord, the near shed vorticity properly is part of the wing problem. The two-dimensional loads due to the shed wake are obtained from the distribution of induced velocity over the chord (in terms of the inflow coefficients λ_0, λ_1, and λ_2), the evaluation of which requires calculation of the inflow at many points along the chord. In a numerical implementation of lifting-line theory, treating both the shed and trailed vorticity in the wake problem is desirable, by evaluating the induced velocity from all wake elements at a single chordwise collocation point.

Miller (1964) considered a lifting-line theory approximation for the near shed wake. Since the lifting-line assumption of high aspect ratio also implies low reduced frequency, the result is expected to be equivalent to a low-frequency approximation. The approach is to determine what treatment of the shed wake in the lifting-line evaluation of the induced velocity correctly gives the unsteady loads on the two-dimensional airfoil, particularly the lift deficiency function. Evaluating the induced velocity at a single point on the airfoil, equation 10.56 becomes $L_W = -2\pi \rho U b \lambda$, and the circulatory lift is

$$L_C = L_Q - 2\pi \rho U b \lambda \tag{10.81}$$

where the induced velocity is obtained from the shed wake vorticity:

$$\lambda = \frac{1}{2\pi} \int_b^\infty \frac{\gamma_w}{x - \xi} d\xi \tag{10.82}$$

In the lifting-line approximation, the airfoil is collapsed to a bound vortex at the quarter chord ($x = -b/2$), and the wake vorticity is extended up to bound vortex. So

$$\lambda = -\frac{1}{2\pi} \int_{-\frac{b}{2}}^\infty \frac{\gamma_w}{\xi + \frac{b}{2}} d\xi \tag{10.83}$$

The wake vorticity is given by the time variation of the bound circulation:

$$\gamma_w = -\frac{1}{U} \frac{d\Gamma}{dt} \tag{10.84}$$

10.2 Lifting-Line Theory and Near Shed Wake

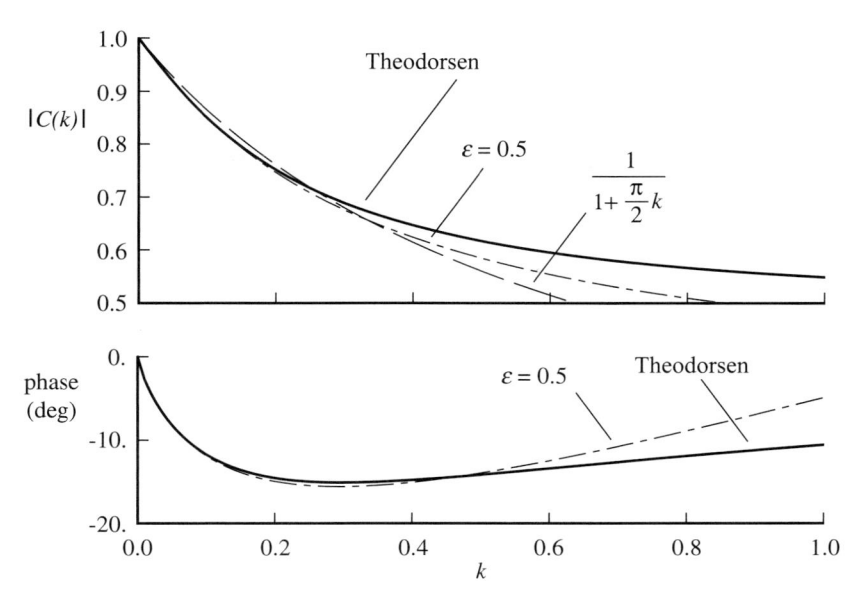

Figure 10.5. Lifting-line approximations for the Theodorsen lift deficiency function.

now at $t - (\xi + \frac{b}{2})/U$. Assuming harmonic motion so that $\Gamma = \overline{\Gamma} e^{i\omega t}$, γ_w is

$$\gamma_w = -\frac{i\omega}{U} \Gamma e^{-i\omega(\xi+\frac{b}{2})/U} \qquad (10.85)$$

and the induced velocity becomes

$$\lambda = \frac{i\omega}{U} \Gamma \frac{1}{2\pi} \int_{-\frac{b}{2}}^{\infty} \frac{e^{-i\omega(\xi+\frac{b}{2})/U}}{\xi + \frac{b}{2}} d\xi$$

$$= \Gamma \frac{ik}{2\pi b} \int_0^\infty \frac{e^{-ik\xi}}{\xi} d\xi$$

$$= \Gamma \frac{k}{2\pi b} \left(\int_0^\infty \frac{\sin k\xi}{\xi} d\xi + i \int_0^\infty \frac{\cos k\xi}{\xi} d\xi \right) \qquad (10.86)$$

The cosine integral is not finite, so is omitted for now. The remaining integral (the real part) is

$$\lambda = \Gamma \frac{k}{2\pi b} \int_0^\infty \frac{\sin k\xi}{\xi} d\xi = \Gamma \frac{k}{4b} = \frac{L_C}{2\pi \rho U b} \frac{\pi}{2} k \qquad (10.87)$$

Then the unsteady lift is $L_C = L_Q - 2\pi \rho U b \lambda = L_Q - L_C \frac{\pi}{2} k$, or

$$L_C = \frac{L_Q}{1 + \frac{\pi}{2}k} \qquad (10.88)$$

Thus an approximate lift deficiency function has been obtained:

$$C = \frac{1}{1 + \frac{\pi}{2}k} \qquad (10.89)$$

which is a correct approximation to order k for the Theodorsen lift deficiency function. Figure 10.5 compares this result with Theodorsen's function. The approximation

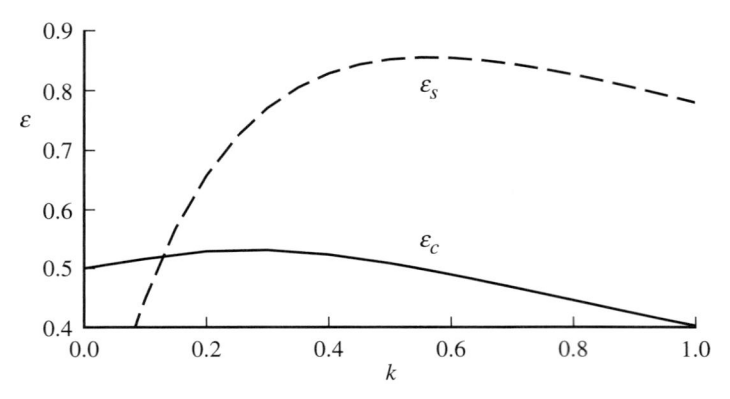

Figure 10.6. Integration limit for near shed wake model in lifting-line theory.

is good even for fairly large values of reduced frequency, but at high k the correct value for $C(k)$ is significantly underestimated.

The lifting-line approximation gives the proper results, except that actually the integral over the wake vorticity is divergent. The difficulty is due to the singularity in the induced velocity at the edge of the vortex sheet, which was extended up to the quarter chord. To correct the model, consider stopping the shed wake a distance $b\epsilon$ behind the quarter chord (where the induced velocity is evaluated). Hence λ is

$$\lambda = \frac{i\omega}{U}\Gamma\frac{1}{2\pi}\int_{-\frac{b}{2}+b\epsilon}^{\infty}\frac{e^{-i\omega(\xi+\frac{b}{2})/U}}{\xi+\frac{b}{2}}d\xi = \Gamma\frac{ik}{2\pi b}\int_{\epsilon}^{\infty}\frac{e^{-ik\xi}}{\xi}d\xi$$

$$= \Gamma\frac{k}{2\pi b}I = \frac{L_C}{2\pi\rho U b}kI \qquad (10.90)$$

where

$$I = \int_{\epsilon}^{\infty}\frac{\sin k\xi}{\xi}d\xi + i\int_{\epsilon}^{\infty}\frac{\cos k\xi}{\xi}d\xi \qquad (10.91)$$

The resulting lift deficiency function is

$$C = \frac{1}{1+kI} \qquad (10.92)$$

Requiring that this approximation give exactly the Theodorsen function determines the parameter ϵ; actually there are two values, ϵ_c and ϵ_s, for the cosine and sine integrals, respectively. The important parameter is ϵ_c, which prevents the divergence of the cosine integral. The limit for low frequency is $\epsilon_c = \frac{1}{2}$. Figure 10.6 shows the results for ϵ_c and ϵ_s over a range of frequencies. For $k = 0$ to 1, $\epsilon = \frac{1}{2}$ is a good approximation, particularly for the cosine integral.

Figure 10.5 shows the lift deficiency function obtained from equation 10.92 using $\epsilon = \frac{1}{2}$. To first order in k, equation 10.92 gives

$$C = \frac{1}{1+kI} \cong 1 - kI \cong \left(1 - \frac{\pi}{2}k\right) + ik(\ln k\epsilon + \gamma) \qquad (10.93)$$

which matches the expansion of Theodorsen's function (equation 10.66) if $\epsilon = \frac{1}{2}$. It is concluded that the near shed wake in the lifting-line model should be extended to a quarter chord ($b\epsilon = b/2 = c/4$) behind the point where the induced velocity is being calculated.

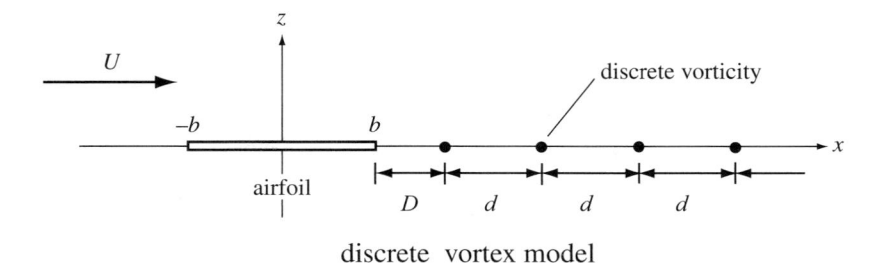

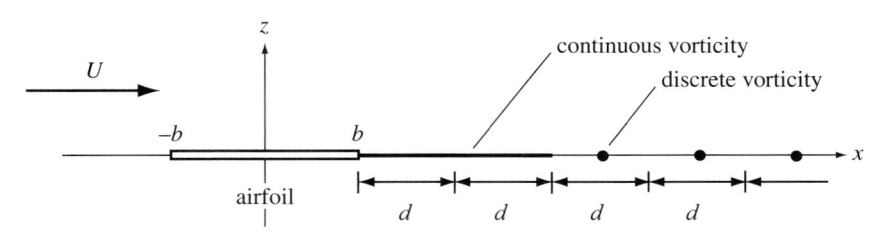

10.2 Lifting-Line Theory and Near Shed Wake

Figure 10.7. Models for shed wake of a two-dimensional airfoil.

For a sinusoidal gust, the lifting-line approximation uses the downwash at the three-quarter chord. Equation 10.73 becomes $w_a = w_0 e^{i\omega t} e^{-ik/2}$, and $A_0 = \frac{w_0}{U} e^{-ik/2}$ for the cosine series. Then

$$S = \frac{1}{2}\left(a_0 + \frac{1}{2}a_1\right)\frac{U}{w_0} = \frac{1}{2}(2A_0 C + ikA_0)\frac{U}{w_0} = A_0\left(C + \frac{ik}{2}\right)\frac{U}{w_0}$$

$$= e^{-ik/2}\left(C(k) + \frac{ik}{2}\right) \tag{10.94}$$

is the lifting-line approximation for the Sears function.

Piziali (1966) considered the effect of a discrete vortex representation of the rotor wake on the shed wake influence. The spirals of the rotary-wing wake are most easily represented by a lattice of finite-strength line vortex segments. In two-dimensional airfoil theory, the corresponding shed wake model is a series of point vortices (Figure 10.7). The distance between the vortices in the wake is $d = 2\pi U/N\omega$, for N vortices per cycle of oscillation. The induced velocity is calculated N times per cycle. The discrete shed vortices correspond to a step change in the airfoil bound circulation. The distance of the first vortex behind the trailing edge is D. Piziali calculated the ratio of the unsteady lift and moment to their quasisteady counterparts for this model and then compared this ratio with the Theodorsen function for pitch and heaving of the airfoil at various frequencies. It was found that $D = d$ does not give good results even with a large number of points per cycle. However, if the entire discrete wake model is advanced closer to the trailing edge so that $D = d/3$, reasonable results are obtained over the frequency range of interest. The conclusion was that with a vortex lattice wake model in a rotary-wing airloads analysis, the shed wake elements should be advanced by about 70% of the azimuthal spacing, so the first elements are closer to the blade trailing edge. A linear variation of bound circulation between the azimuthal collocation points generates vortex sheet elements in the wake. Collapsing each sheet element to a point vortex at the center of the element implies $D = d/2$. Daughaday and Piziali (1966) found that the calculation of the lift and moment at high frequency can be improved by representing the shed wake just behind the blade as a continuous distribution of vorticity, replacing the first few discrete elements in the vortex lattice (Figure 10.7).

Lifting-line theory calculates the blade loads from the velocity induced at the section by the shed and trailed vorticity in the wake. For the inflow calculation, the blade is modeled by the bound vortex at the quarter chord, and the trailed vorticity (due to the spanwise lift variation) is extended up to the bound vortex. The induced velocity is then evaluated along the bound vortex or at the three-quarter chord. The simplest and most economical representation of the complex wake structure is a lattice of finite-strength vortex-line elements. A line vortex is in fact a good representation of the rolled-up tip vortices of the blades. Based on the two-dimensional airfoil analyses discussed in this section, such a lifting-line model can be used for the shed wake as well, and the shed-wake-induced velocity can be calculated at a single chordwise point. To correct for the neglect of the chordwise variation of the induced velocity, the shed wake is not extended all the way to collocation point, but is stopped a quarter chord behind it.

10.3 Reverse Flow

Extending thin-airfoil theory to reverse flow ($U < 0$) introduces several sign changes in the results. The variable θ still runs from $\theta = 0$ at the trailing edge to $\theta = \pi$ at the leading edge, so $x = \pm b \cos\theta$. The convention is that in the double sign ($\pm$) the upper sign is for normal flow, and the lower sign is for reverse flow. The differential pressure $-\Delta p$ is still positive upward, but the vorticity (γ_b) changes direction in reverse flow. In reverse flow, equation 10.27 becomes

$$-\Delta p = \rho \left(-U\gamma_b + \frac{\partial}{\partial t} \int_x^b \gamma_{b_{NC}} dx \right)$$

$$= -\rho U \gamma_b + \rho b \int_\theta^\pi \dot{\gamma}_{b_{NC}} \sin\theta \, d\theta$$

$$= \sum_{n=0} p_n f_n(\theta) \qquad (10.95)$$

and the absolute value of U is used in equations 10.29, 10.30, and 10.31. The circulation and section loads are now

$$\Gamma = \pm \int_{-b}^{b} \gamma_b \, dx = \pm 2\pi b \left[\left(w_0 + \frac{1}{2} w_1 \right) - \left(\lambda_0 + \frac{1}{2} \lambda_1 \right) \right] \qquad (10.96)$$

$$L = \pm \int_{-b}^{b} (-\Delta p) dx = \pm \pi b \left(p_0 + \frac{1}{2} p_1 \right)$$

$$= \pm 2\pi \rho |U| b \left[\left(w_0 + \frac{1}{2} w_1 \right) - \left(\lambda_0 + \frac{1}{2} \lambda_1 \right) \right]$$

$$\pm \pi \rho b^2 \left[\left(\dot{w}_0 - \frac{1}{2} \dot{w}_2 \right) - \left(\dot{\lambda}_0 - \frac{1}{2} \dot{\lambda}_2 \right) \right] \qquad (10.97)$$

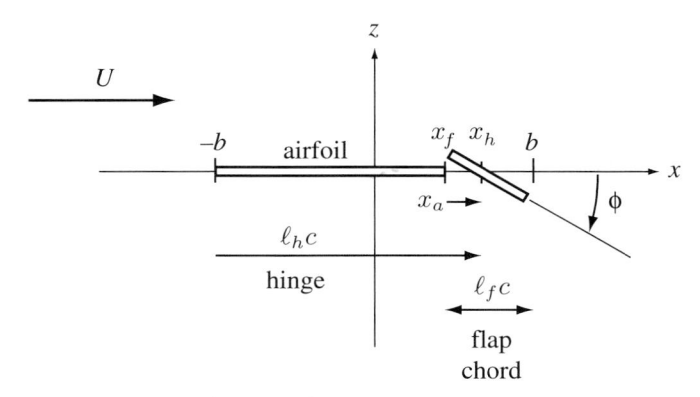

Figure 10.8. Trailing-edge flap geometry.

$$M_{QC} = \int_{-b}^{b} (-\Delta p)\left(-x - \frac{b}{2}\right) dx$$

$$= -\frac{\pi}{4} b^2 \left(p_0(2 \mp 2) + p_1 \pm p_2\right)$$

$$= -\frac{1}{2}\pi \rho |U| b^2 \left[(w_0(2 \mp 2) + w_1 \pm w_2) - (\lambda_0(2 \mp 2) + \lambda_1 \mp \lambda_2)\right]$$

$$-\frac{1}{2}\pi \rho b^3 \left[\left(\dot{w}_0 - \frac{1}{2}\dot{w}_2\right) - \left(\dot{\lambda}_0 - \frac{1}{2}\dot{\lambda}_2\right)\right]$$

$$\mp \frac{1}{8}\pi \rho b^3 \left[(\dot{w}_1 - \dot{w}_3) - (\dot{\lambda}_1 - \dot{\lambda}_3)\right] \tag{10.98}$$

These unsteady loads are used with static airfoil characteristics from a wind-tunnel test, in which the lift is positive upward as the angle-of-attack varies from $-180°$ to $180°$. Positive upwash w with reverse flow corresponds to a positive angle-of-attack near $180°$, for which the lift is downward; hence the $\pm$ signs on L. M_{QC} is the moment about the theoretical aerodynamic center (quarter chord), positive nose up.

10.4 Trailing-Edge Flap

For an airfoil with a trailing-edge flap, the unsteady loads are derived following Küssner and Schwarz (1941) and Theodorsen and Garrick (1942). Figure 10.8 shows the geometry. The flap chord is $\ell_f c$, with the hinge axis $\ell_h c$ aft of the leading edge. The flap angle is ϕ. The flap leading-edge coordinate is $x_f = b(1 - 2\ell_f)$, so $\cos\theta_f = (1 - 2\ell_f)$. The flap hinge coordinate is $x_h = b(-1 + 2\ell_h)$. The distance of the hinge aft of the leading edge is $x_a = x_h - x_f = 2b(\ell_f + \ell_h - 1)$. An aerodynamically balanced flap (flap leading edge forward of the hinge) has $x_a > 0$.

Thin airfoil theory is linear, so the effects of the trailing-edge flap can be examined separately from the other airfoil motion. The vertical deflection of the airfoil is $z = -(x - x_h)\phi = -(x - x_f)\phi + x_a\phi$ for $x > x_f$. So the upwash along the section is

$$w = -\left(U\frac{\partial z}{\partial x} + \frac{\partial z}{\partial t}\right) = U\phi + (x - x_h)\dot{\phi} - Ux_a\delta(x_f)\phi \tag{10.99}$$

The last term is present only with a sealed leading-edge gap. For an open gap, the step change in displacement at $x = x_f$ is assumed not to contribute to the upwash.

The coefficients in the cosine Fourier series for w are

$$w_0 = \frac{1}{\pi}\left[U\theta_f\phi + (bS_1 - x_h\theta_f)\dot{\phi} - U\frac{x_a/b}{S_1}\phi\right] \quad (10.100)$$

$$w_1 = \frac{2}{\pi}\left[US_1\phi + \left(\frac{b}{2}\left(\theta_f + \frac{S_2}{2}\right) - x_hS_1\right)\dot{\phi} - U\frac{(x_a/b)C_1}{S_1}\phi\right] \quad (10.101)$$

$$w_n = \frac{2}{\pi}\left[U\frac{S_n}{n}\phi + \left(\frac{b}{2}\left(\frac{S_{n-1}}{n-1} - \frac{S_{n+1}}{n+1}\right) - x_h\frac{S_n}{n}\right)\dot{\phi} - U\frac{(x_a/b)C_n}{S_1}\phi\right] \quad (10.102)$$

where $S_k = \sin k\theta_f$ and $C_k = \cos k\theta_f$. Reverse flow is not considered here. The total airfoil loads are obtained from equations 10.96 to 10.98. The flap lift and hinge moment are

$$L_f = \int_{x_f}^{b}(-\Delta p)dx$$

$$= b\left[p_0(\theta_f - S_1) + \frac{1}{2}p_1\left(\theta_f - \frac{S_2}{2}\right) + \sum_{n=2}\frac{1}{2}p_n\left(\frac{S_{n-1}}{n-1} + \frac{S_{n+1}}{n+1}\right)\right] \quad (10.103)$$

$$M_f = -\int_{x_f}^{b}(-\Delta p)(x - x_h)dx$$

$$= b^2\left[p_0\left(\left(\frac{1}{2} + x_h/b\right)\theta_f + S_1(1 - x_h/b) + \frac{1}{4}S_2\right)\right.$$
$$+ p_1\left(\frac{1}{4}\left(S_1 - \frac{S_3}{3}\right) + \frac{x_h/b}{2}\left(\theta_f - \frac{S_2}{2}\right)\right)$$
$$+ p_2\left(\frac{1}{4}\left(\theta_f - \frac{S_4}{4}\right) + \frac{x_h/b}{2}\left(S_1 - \frac{S_3}{3}\right)\right)$$
$$\left. + \sum_{n=3}p_n\left(\frac{1}{4}\left(\frac{S_{n-2}}{n-2} - \frac{S_{n+2}}{n+2}\right) + \frac{x_h/b}{2}\left(\frac{S_{n-1}}{n-1} - \frac{S_{n+1}}{n+1}\right)\right)\right] \quad (10.104)$$

with p_n given by equations 10.29 to 10.31. The series converge with about 10 pressure terms, except that the aerodynamic balance terms produced by static flap deflection ($x_a\phi$ terms) do not converge at all for flap lift and hinge moment. These static loads for a flap with sealed gap must be obtained from tests or a more appropriate theory.

10.5 Unsteady Airfoil Theory with a Time-Varying Free Stream

The rotating blade of a helicopter rotor in forward flight sees a periodically varying free stream velocity: $u_T = r + \mu\sin\psi = r(1 + (\mu/r)\sin\psi)$. For either high advance ratio or the inboard sections, the 1/rev variation of the velocity is a significant fraction of the mean. In such cases the time-varying free stream must be included in the unsteady airfoil theory, both for its direct effects and for its influence through the shed wake. Only the case $\mu/r < 1$ is considered. If $\mu/r > 1$, the blade section passes through the reverse flow region, and a simple wake model is not applicable.

Consider the two-dimensional airfoil and wake model described in section 10.1. Only a few modifications are required to account for a time-varying free stream

10.5 Unsteady Airfoil Theory with a Time-Varying Free Stream

velocity U. The time derivative acts on the velocity now, so equations 10.52 amd 10.63 become

$$\frac{\partial}{\partial t}\Gamma_{NC}^{(1)} = -\pi b^2 \left(\frac{d}{dt}(U\alpha + \dot{h}) - ab\ddot{\alpha} \right) - U \int_b^\infty \left(1 - \frac{\xi}{\sqrt{\xi^2 - b^2}} \right) \gamma_w d\xi \quad (10.105)$$

$$\frac{1}{2}\frac{\partial}{\partial t}\Gamma_{NC}^{(2)} = -b\left(\frac{1}{2}+a\right)\frac{\partial}{\partial t}\Gamma_{NC}^{(1)} - \frac{1}{2}\pi b^3 \left(\frac{d}{dt}(U\alpha + \dot{h}) + \left(\frac{1}{4}-a\right)b\ddot{\alpha}\right)$$

$$+ \frac{1}{2}U \int_b^\infty \left(\xi - \sqrt{\xi^2 - b^2} \right) \left(\sqrt{\frac{\xi+b}{\xi-b}} - 1 \right) \gamma_w d\xi \quad (10.106)$$

Then the lift and moment are

$$L = L_C + L_{NC}$$

$$= L_Q + \rho U \int_b^\infty \frac{b}{\sqrt{\xi^2 - b^2}} \gamma_w d\xi + \rho \pi b^2 \left(\frac{d}{dt}(U\alpha + \dot{h}) - ab\ddot{\alpha} \right) \quad (10.107)$$

$$M = b\left(\frac{1}{2}+a\right)L + M_{QC}$$

$$= b\left(\frac{1}{2}+a\right)L - \frac{1}{2}\rho\pi b^3 \left(\frac{d}{dt}(U\alpha + \dot{h}) + U\dot{\alpha} + \left(\frac{1}{4}-a\right)b\ddot{\alpha} \right) \quad (10.108)$$

with the quasistatic lift still

$$L_Q = 2\pi \rho U b \left(U\alpha + \dot{h} + \left(\frac{1}{2}-a\right)b\dot{\alpha} \right) \quad (10.109)$$

The only changes are $\dot{U}\alpha$ terms added to the non-circulatory lift and moment. The quasistatic lift and circulatory lift are still given in terms of the wake vorticity:

$$L_Q = -\rho U \int_b^\infty \sqrt{\frac{\xi+b}{\xi-b}} \gamma_w d\xi \quad (10.110)$$

$$L_C = -\rho U \int_b^\infty \frac{\xi}{\sqrt{\xi^2 - b^2}} \gamma_w d\xi \quad (10.111)$$

Relating L_Q and L_C in terms of a lift deficiency function requires a knowledge of the dependence of γ_w on ξ. The criterion that there be no pressure difference across the vortex sheet gives

$$-\Delta p = \rho \left(\frac{\partial}{\partial t} + U \frac{\partial}{\partial \xi} \right) \Delta \phi = 0 \quad (10.112)$$

which implies

$$\left(\frac{\partial}{\partial t} + U \frac{\partial}{\partial \xi} \right) \frac{\partial \Delta \phi}{\partial \xi} = \left(\frac{\partial}{\partial t} + U \frac{\partial}{\partial \xi} \right) \gamma_w = 0 \quad (10.113)$$

the solution of which has the form

$$\gamma_w = \gamma_w \left(\xi - \int^t U \, dt \right) \quad (10.114)$$

If the free stream velocity is constant, the wake vorticity is convected at a constant rate and γ_w is a function of $(\xi - Ut)$ as before. Considering the rotor blade in forward flight, the dimensionless free stream velocity is $U = r + \mu \sin \psi$, so

$$\gamma_w = \gamma_w(\xi - r\psi + \mu \cos \psi) \tag{10.115}$$

Here $\psi = \Omega t$ is the dimensionless time variable. Now assume periodic motion of the blade. For the flow field to be entirely periodic, the blade motion can consist only of harmonics of the fundamental frequency Ω of the free stream variation. The period of the flow is then $2\pi/\Omega$. The wake vorticity must be a periodic function of ξ, with a wavelength equal to the distance the wake is convected during the period: $\int_0^{2\pi} U \, d\psi = 2\pi r$. Next, write the periodic function γ_w as a Fourier series in ξ with period $2\pi r$:

$$\gamma_w = \sum_{m=-\infty}^{\infty} \gamma_{w_m}(\psi) e^{-im\xi/r} \tag{10.116}$$

Since γ_w must be a function of the quantity $(\xi - r\psi + \mu \cos \psi)$ alone,

$$\gamma_w = \sum_{m=-\infty}^{\infty} \overline{\gamma}_m e^{im(\psi - (\mu/r) \cos \psi) - im\xi/r} \tag{10.117}$$

where $\overline{\gamma}_m$ are constants. For $\mu = 0$ this reduces to $\gamma_w = \overline{\gamma}_w e^{i\omega(t - \xi/U)}$ as before.

With the structure of the wake vorticity established, the relation between the quasistatic and circulatory lift can be constructed. Substituting for γ_w gives

$$L_Q = -\rho U \sum_{m=-\infty}^{\infty} \overline{\gamma}_m e^{im(\psi - (\mu/r) \cos \psi)} \int_b^{\infty} \sqrt{\frac{\xi + b}{\xi - b}} e^{-im\xi/r} d\xi \tag{10.118}$$

$$L_C = -\rho U \sum_{m=-\infty}^{\infty} \overline{\gamma}_m e^{im(\psi - (\mu/r) \cos \psi)} \int_b^{\infty} \frac{\xi}{\sqrt{\xi^2 - b^2}} e^{-im\xi/r} d\xi \tag{10.119}$$

Noting that

$$\frac{1}{2\pi} \int_0^{2\pi} (1 + (\mu/r) \sin \psi) e^{in(\psi - (\mu/r) \cos \psi)} d\psi = 1 \tag{10.120}$$

if $n = 0$ and is zero otherwise, the harmonics γ_m can be evaluated:

$$\gamma_m = \frac{-\int_0^{2\pi} (1 + (\mu/r) \sin \psi) e^{-im(\psi - (\mu/r) \cos \psi)} L_Q d\psi}{2\pi \rho U \int_b^{\infty} \sqrt{\frac{\xi+b}{\xi-b}} e^{-im\xi/r} d\xi} \tag{10.121}$$

Thus the circulatory lift is

$$L_C = 2\pi \rho U b \sum_{m=-\infty}^{\infty} e^{im(\psi - (\mu/r) \cos \psi)} C(mb/r)$$

$$\frac{1}{2\pi} \int_0^{2\pi} (1 + (\mu/r) \sin \psi) e^{-im(\psi - (\mu/r) \cos \psi)} Q \, d\psi \tag{10.122}$$

where $C(mb/r)$ is the Theodorsen lift deficiency function at reduced frequency $k = mb/r$ ($\omega = m\Omega$ and an average velocity $\overline{U} = \Omega r$), and

$$Q = \frac{L_Q}{2\pi \rho U b} = U\alpha + \dot{h} + \left(\frac{1}{2} - a\right) b\dot{\alpha} \tag{10.123}$$

10.5 Unsteady Airfoil Theory with a Time-Varying Free Stream

If now the quasistatic circulation is written as a Fourier series, $Q = \sum_{n=-\infty}^{\infty} Q_n e^{in\psi}$, then the lift can be written in a form analogous to the constant velocity result:

$$L_C = 2\pi \rho U b \sum_{n=-\infty}^{\infty} Q_n e^{in\psi} C_\mu(n, \psi) \tag{10.124}$$

where $C_\mu(n, \psi)$ is the modified lift deficiency function for the n-th harmonic of the blade motion with free stream velocity $U = r + \mu \sin \psi$:

$$C_\mu(n, \psi) = \sum_{m=-\infty}^{\infty} e^{im(\psi - (\mu/r)\cos\psi) - in\psi} C(mb/r)$$

$$\frac{1}{2\pi} \int_0^{2\pi} (1 + (\mu/r)\sin\psi) e^{-im(\psi - (\mu/r)\cos\psi) + in\psi} d\psi \tag{10.125}$$

For a constant free stream velocity ($\mu = 0$), the integral is non-zero only for $m = n$, and hence $C_{\mu=0}(n, \psi) = C(nb/r)$ as required. An alternative form is

$$L_C = 2\pi \rho U b \sum_{n=-\infty}^{\infty} Q_n \left[\sum_{\ell=-\infty}^{\infty} C_{\ell n} e^{i\ell\psi} \right] \tag{10.126}$$

where

$$C_{\ell n} = \sum_{m=-\infty}^{\infty} \left[\frac{1}{2\pi} \int_0^{2\pi} e^{im(\psi - (\mu/r)\cos\psi) - i\ell\psi} d\psi \right] C(mb/r)$$

$$\left[\frac{1}{2\pi} \int_0^{2\pi} (1 + (\mu/r)\sin\psi) e^{-im(\psi - (\mu/r)\cos\psi) + in\psi} d\psi \right] \tag{10.127}$$

The coefficients $C_{\ell n}$ are the harmonics in a Fourier series expansion of $e^{in\psi} C_\mu(n, \psi)$. This form shows that the time-varying free stream couples the harmonics of the circulation and lift through the influence of the shed wake.

The integrals appearing in the lift deficiency function for a time-varying free stream can be evaluated in terms of Bessel functions:

$$I_{mn} = \frac{1}{2\pi} \int_0^{2\pi} (1 + (\mu/r)\sin\psi) e^{-im(\psi - (\mu/r)\cos\psi) + in\psi} d\psi$$

$$= \frac{n}{m} \frac{1}{\pi} \int_0^\pi e^{im(\mu/r)\cos\psi} \cos(n-m)\psi \, d\psi$$

$$= \begin{cases} \frac{n}{m} i^{|n-m|} J_{|n-m|}(m\mu/r) & m > 0 \\ \frac{n}{m} (-i)^{|n-m|} J_{|n-m|}(|m\mu/r|) & m < 0 \end{cases} \tag{10.128}$$

and

$$I_{mn} = \frac{1}{2\pi} \int_0^{2\pi} (1 + (\mu/r)\sin\psi) e^{in\psi} d\psi$$

$$= \begin{cases} 1 & n = 0 \\ \frac{i}{2}\mu/r & n = 1 \\ 0 & n \geq 2 \end{cases} \tag{10.129}$$

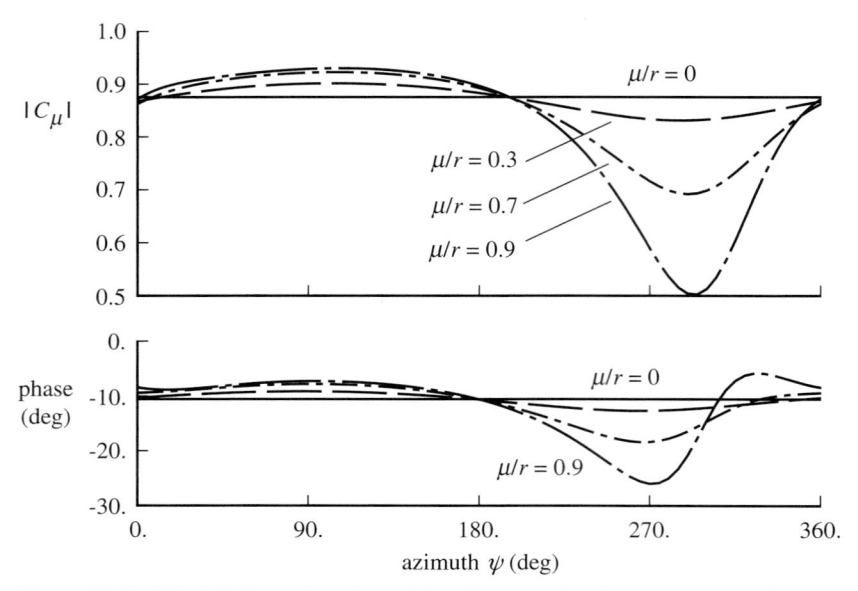

Figure 10.9. Lift deficiency function with a time-varying free stream, for the second harmonic ($n = 2$) and $b/r = 0.04$.

for $m = 0$. Figure 10.9 shows typical results for $C_\mu(n, \psi)$, with $n = 2$ and $b/r = 0.04$. The 1/rev variation in the free stream velocity produces a basic 1/rev variation of C with ψ. The largest influence occurs nearest the reverse flow boundary, at $\psi = 270°$. Most of the range of velocities and radial stations of interest are covered by $0 < \mu/r < 0.7$. The model breaks down for $\mu/r > 1$, when the section passes through the reverse flow region. For small μ/r, the lift deficiency function is approximately

$$C_\mu(n, \psi) \cong \sum_{m=-\infty}^{\infty} e^{i(m-n)\psi}(1 - im(\mu/r)\cos\psi)C(mb/r)$$

$$\frac{1}{2\pi}\int_0^{2\pi}(1 + (\mu/r)(\sin\psi + im\cos\psi))e^{i(n-m))\psi}\,d\psi$$

$$= C_n + (\mu/r)\frac{in}{2}\left[\cos\psi(C_{n+1} + C_{n-1} - 2C_n) + i\sin\psi(C_{n+1} - C_{n-1})\right]$$

(10.130)

where C_n means $C(nb/r)$. Assuming as well small values of b/r gives

$$C_\mu(n, \psi) \cong C(nb/r) - \left[(nb/r)(\mu/r)\sin\psi\right]C'(nb/r)$$

$$\cong C(nb/(r + \mu\sin\psi))$$

(10.131)

Thus for small variations in the free stream velocity (small $nb\mu/r^2$), the lift deficiency function is nearly the same as the Theodorsen function $C(k)$, with the reduced frequency based on the local velocity, $k = \omega b/U$. This approximation works well for moderate n. Figure 10.9 shows the basic dependence on the local reduced frequency. On the advancing side, the increased velocity lowers the reduced frequency, and hence the lift deficiency function is nearer unity. On the retreating side there is the greatest accumulation of shed vorticity in the wake near the trailing edge, and thus the greatest reduction in the lift.

In summary, a time-varying free stream has the following influence on the unsteady aerodynamics of a two-dimensional airfoil: there are additional non-circulatory lift and moment terms due to $d(U\alpha)/dt$; there is coupling by the wake of all the harmonics of the quasistatic and unsteady circulation; and there is a significant influence on the lift deficiency function due to stretching and compressing of the vorticity in the shed wake. For the free stream variation of the rotor blade in forward flight, all these effects basically produce 1/rev variations of the loads. The non-circulatory lift and moment terms are valid for a general time variation of U. The simple approximation $C_\mu(n, \psi) \cong C(k)$ using the local reduced frequency is good up to $\mu/r \cong 0.7$. For small enough μ/r the cruder approximation $C_\mu(n, \psi) \cong C(nb/r)$ using the mean reduced frequency can be chosen, neglecting entirely the influence of a time-varying free stream on the shed wake.

10.6 Unsteady Airfoil Theory for the Rotary Wing

Application of unsteady airfoil theory to rotary wings is often done in the context of lifting-line theory (see section 9.2). Thus the steady two-dimensional loads are obtained from measured airfoil data as a function of angle-of-attack and Mach number. The angle-of-attack is evaluated from the blade motion at the quarter chord. The influence of the near shed wake is in the induced velocity calculated from a vortex wake model, along with the influence of the tip vortices, trailed wake, and far shed wake. What is required from thin-airfoil theory are expressions for the unsteady loads in attached flow.

The airfoil upwash velocity for thin air-foil theory is written in terms of the upwash at the quarter chord, w_{QC}, and the gradient of the upwash along the chord, $w' = dw/dx$:

$$w_a = w_{QC} + \left(x + \frac{c}{4}\right) w' = w_0 + w_1 \cos\theta \tag{10.132}$$

with $x = \pm b \cos\theta$. Thus $w_0 = w_{QC} + \frac{c}{4} w'$ and $w_1 = \pm \frac{c}{2} w'$. Equations 10.97 and 10.98 for the lift and moment, which include reverse flow, become

$$L = \pm 2\pi\rho |U| b \left[w_{QC} - \lambda + \left(\frac{b}{2} \pm \frac{b}{2}\right) w' \right] \pm \pi\rho b^2 \left(\dot{w}_{QC} + \frac{b}{2} \dot{w}'\right) \tag{10.133}$$

$$M = b\left(\frac{1}{2} + a\right) L - \frac{1}{2}\pi\rho |U| b^2 \left[(w_{QC} - \lambda)(2 \mp 2) + \frac{b}{4} w' \right] - \frac{1}{2}\pi\rho b^3 \left(\dot{w}_{QC} + \frac{3b}{4} \dot{w}'\right) \tag{10.134}$$

where the upper sign in $\pm$ or $\mp$ is for normal flow, and the lower sign is for reverse flow. The lift term in brackets is the quasistatic lift L_Q, and the $\dot{w}$ terms are the non-circulatory lift L_{NC}. Multiplying by $a/2\pi$ introduces the real lift-curve slope a. The chord $c = 2b$ is used instead of the semi-chord, and the section velocity is written $U = u_T$. The distance of the aerodynamic center (quarter chord for thin-airfoil theory) aft of the pitch axis is $x_A = -b\left(\frac{1}{2} + a\right)$. The lift and moment about the pitch axis are then

$$L = \pm a \frac{1}{2}\rho |u_T| c \left(w_{QC} - \lambda\right) \pm a\rho \frac{c^2}{8} \left(\left(\frac{1}{2} \pm \frac{1}{2}\right) 2|u_T|w' + \dot{w}_{QC} + \frac{c}{4}\dot{w}'\right) \quad (10.135)$$

$$M = -x_A L - \frac{c}{2}\left(\frac{1}{2} \mp \frac{1}{2}\right)\left[\pm a\frac{1}{2}\rho |u_T| c \left(w_{QC} - \lambda\right)\right] - a\rho \frac{c^3}{32}\left(|u_T|w' + \dot{w}_{QC} + \frac{3c}{8}\dot{w}'\right) \quad (10.136)$$

$$= -\left(x_A + \frac{c}{2}\left(\frac{1}{2} \mp \frac{1}{2}\right)\right)\left[\pm a\frac{1}{2}\rho |u_T| c \left(w_{QC} - \lambda\right)\right]$$

$$\mp a\rho x_A \frac{c^2}{8}\left(\left(\frac{1}{2} \pm \frac{1}{2}\right) 2|u_T|w' + \dot{w}_{QC} + \frac{c}{4}\dot{w}'\right) - a\rho \frac{c^3}{32}\left(|u_T|w' + \dot{w}_{QC} + \frac{3c}{8}\dot{w}'\right) \quad (10.137)$$

The terms proportional to $(w_{QC} - \lambda)$ are the thin-airfoil theory results for the static lift and moment, the lift acting at the three-quarter chord in reverse flow. If these loads are accounted for in the measured static airfoil data, they are excluded from the unsteady loads. The lift and moment can also be expressed in terms of the upwash at the pitch axis:

$$L = \pm a \frac{1}{2}\rho |u_T| c \left(w_{PA} + x_A w' - \lambda\right)$$

$$\pm a\rho \frac{c^2}{8}\left(\left(\frac{1}{2} \pm \frac{1}{2}\right) 2|u_T|w' + \dot{w}_{PA} + x_A \dot{w}' + \frac{c}{4}\dot{w}'\right) \quad (10.138)$$

$$M = -x_A L - \frac{c}{2}\left(\frac{1}{2} \mp \frac{1}{2}\right)\left[\pm a\frac{1}{2}\rho |u_T| c \left(w_{PA} + x_A w' - \lambda\right)\right]$$

$$- a\rho \frac{c^3}{32}\left(|u_T|w' + \dot{w}_{PA} + x_A \dot{w}' + \frac{3c}{8}\dot{w}'\right) \quad (10.139)$$

$$= -\left(x_A + \frac{c}{2}\left(\frac{1}{2} \mp \frac{1}{2}\right)\right)\left[\pm a\frac{1}{2}\rho |u_T| c \left(w_{PA} + x_A w' - \lambda\right)\right]$$

$$\mp a\rho x_A \frac{c^2}{8}\left(\left(\frac{1}{2} \pm \frac{1}{2}\right) 2|u_T|w' + \dot{w}_{PA} + x_A \dot{w}' + \frac{c}{4}\dot{w}'\right)$$

$$- a\rho \frac{c^3}{32}\left(|u_T|w' + \dot{w}_{PA} + x_A \dot{w}' + \frac{3c}{8}\dot{w}'\right) \quad (10.140)$$

since $w_{QC} = w_{PA} + x_A w'$.

In thin-wing theory, the blade is defined by a surface a distance $z_b(r, x, t)$ above the disk plane, with r the radial coordinate and x the chordwise coordinate (from the pitch axis, positive toward the trailing edge). The blade has time-varying velocity relative to the air, with velocity perpendicular and radial components. The upwash is

$$w_a = -\frac{D}{Dt} z_b = -\left[\frac{\partial}{\partial t} + (\Omega r + \Omega R\mu \sin \psi)\frac{\partial}{\partial x} + (-\Omega x + \Omega R\mu \cos \psi)\frac{\partial}{\partial r}\right] z_b \quad (10.141)$$

10.6 Unsteady Airfoil Theory for the Rotary Wing

If the total derivative in equation 10.24 for the differential pressure on the blade surface includes a radial velocity term, the result is slender-body loads, which are best ignored in the context of lifting-line theory. Let the blade motion consist of vertical deflection of the elastic axis z_0 and nose-up pitch about the elastic axis by θ:

$$z_b = z_0(r, t) - x\theta(r, t) \tag{10.142}$$

Then the upwash is

$$\begin{aligned}w_a &= -\dot{z}_0 + x\dot{\theta} + (\Omega r + \Omega R\mu \sin\psi)\theta - (-\Omega x + \Omega R\mu \cos\psi)(z_0' - x\theta') \\ &= u_T\theta - (\dot{z}_0 + u_R z_0') + x(\dot{\theta} + \Omega z_0' + u_R\theta') - x^2\Omega\theta' \\ &= w_{PA} + xw' \end{aligned} \tag{10.143}$$

neglecting the x^2 term. For just rigid flap and rigid pitch motion, $z_b = r\beta - x\theta$, and the upwash is

$$\begin{aligned} w_a &= -r\dot{\beta} + x\dot{\theta} + (\Omega r + \Omega R\mu \sin\psi)\theta - (-\Omega x + \Omega R\mu \cos\psi)\beta \\ &= u_T\theta - (r\dot{\beta} + u_R\beta) + x(\dot{\theta} + \Omega\beta) \\ &= w_{PA} + xw' \end{aligned} \tag{10.144}$$

In terms of the classical description of the airfoil motion in terms of heaving h and pitch angle α (Figure 10.2), the rotor motion gives

$$U\alpha + \dot{h} = u_T\theta - (r\dot{\beta} + u_R\beta) = w_{PA} \tag{10.145}$$
$$\dot{\alpha} = \dot{\theta} + \Omega\beta = w' \tag{10.146}$$

For hover, w_a depends on the quantities $r(\Omega\theta - \dot{\beta})$ and $(\dot{\theta} + \Omega\beta)$, as required to maintain the equivalence of flapping and feathering. The term $r\Omega\theta$ is the component of upwash when the blade is pitched up relative to the free stream Ωr. The term $\Omega\beta$ is the component of the rotational speed Ω that is a pitch rate when the blade is flapped by β.

The distinction between angle-of-attack and pitch angle is important. In the context of flight dynamics, angle-of-attack is produced by a perturbation of the aircraft velocity, whereas pitch angle is produced by a perturbation of the aircraft orientation. For thin-airfoil theory, angle-of-attack $(\alpha + \dot{h}/U)$ gives a uniform distribution of upwash along the chord, whereas pitch rate $\dot{\alpha}$ gives the linear variation of upwash along the chord.

The moment due to pitch rate is particularly important. From the contributions $\Delta \dot{w}_{PA} = u_T\dot{\theta}$ and $\Delta w' = \dot{\theta}$,

$$\begin{aligned}\frac{\partial M}{\partial \dot{\theta}} &= -a\rho|u_T|\left[\pm\left(x_A + \frac{c}{2}\left(\frac{1}{2}\mp\frac{1}{2}\right)\right)\frac{c}{2}x_A \pm x_A\frac{c^2}{8}\left(\left(\frac{1}{2}\pm\frac{1}{2}\right)2\pm 1\right) + \frac{c^3}{32}(1\pm 1)\right] \\ &= \mp a\rho|u_T|\frac{c}{2}\left[\left(x_A + \frac{c}{4}\right)\left(x_A + \frac{c}{2}\left(\frac{1}{2}\pm\frac{1}{2}\right)\right)\right] \\ &= \begin{cases} -a\rho|u_T|\frac{c}{2}\left(x_A + \frac{c}{4}\right)\left(x_A + \frac{c}{2}\right) & \text{normal flow} \\ +a\rho|u_T|\frac{c}{2}\left(x_A + \frac{c}{4}\right)x_A & \text{reverse flow} \end{cases}\end{aligned} \tag{10.147}$$

or $\partial M/\partial\dot\theta = 2\pi\rho|U|b^3 a(\frac{1}{2}\mp a)$. This unsteady aerodynamic load is the source of damping for the blade pitch and torsion motion. For the pitch axis at the aerodynamic center ($x_A = 0$ or $a = -\frac{1}{2}$), the damping in normal flow is $\partial M/\partial\dot\theta = -a\rho u_T \frac{c^3}{16}$, from the non-circulatory moment (negative derivative is positive damping). For the pitch axis at the aerodynamic center, the damping in reverse flow is zero.

Incompressible thin-airfoil theory provides the principal unsteady load needed to model rotor blades. Closed-form solutions are not possible for the compressible case. A number of approximations have been developed for the thin-airfoil theory solutions in compressible flow.

The ONERA EDLIN (Equations Differentielles Lineaires) theory for unsteady loads was presented by Petot (1989). The extended model includes the effects of heave and pitch, as well as the time-varying free stream. In the absence of stall, thin-airfoil theory results compared well with measured behavior. To include the effects of compressibility, Küssner's coefficients are used, as tabulated by van der Vooren (1964) and curve-fit by Petot (1989). Omitting the static terms, the lift and quarter-chord moment are

$$L_{US} = \pm a\rho \frac{c^2}{8}\left(\left(\frac{1}{2}\pm\frac{1}{2}\right)2|u_T|w' + \dot w f_{L0} + \frac{c}{4}\dot w' f_{L1}\right) + a\frac{1}{2}\rho|u_T|L_1 \quad (10.148)$$

$$M_{US} = -a\rho\frac{c^3}{32}\left(|u_T|w' f_{M0} + \dot w f_{M0} + \frac{3c}{8}\dot w' f_{M1}\right) \quad (10.149)$$

$$\dot L_1 + \lambda L_1 = \mu\left[\pm\dot w + \dot w'\frac{c}{2}\left(\frac{1}{2}\pm\frac{1}{2}\right)\right] \quad (10.150)$$

where

$$f_{L0} = \beta\left[1 + 5(\beta^{0.57} - 1)\right] \quad (10.151)$$

$$f_{L1} = \beta\left[1 + 3.92(\beta - 1)\right] \quad (10.152)$$

$$f_{M0} = \beta\left[1 + 1.4M^2\right] \quad (10.153)$$

$$f_{M1} = \beta\left[-1.2625 + 1.5330\tan^{-1}(10.5 - 15M)\right] \quad (10.154)$$

$$\lambda = \frac{2|u_T|}{c}\lambda_0(1 - 0.76M) \quad (10.155)$$

$$\mu = -\frac{1}{4}(3 - \beta) \quad (10.156)$$

with $\beta = \sqrt{1 - M^2}$, and M is the section Mach number. These factors give a good representation of Küssner's coefficients, except that the moment produced by heave is always real, when it should exhibit a phase shift for nonzero Mach number. The L_1 term accounts for the airfoil shed wake effects (lift deficiency function). A constant was changed so the incompressible circulatory loads match the lift deficiency function value of $C = 0.5$ at high frequency. Petot (1989) used the Prandtl-Glauert correction for the lift-curve slope, $a = 2\pi/\beta$, and gave $\lambda_0 = 0.17$.

The Leishman-Beddoes theory for unsteady loads in attached flow was presented by Leishman (1988), Leishman and Beddoes (1989), Leishman and Nguyen (1990), and Hariharan and Leishman (1996). The theory is based on the indicial response of a thin airfoil in compressible flow to heave and pitch motions. The indicial response is a combination of impulsive (small time) and circulatory (long time) terms, each approximated by exponential functions of time. In this form, the equations for the loads can be transformed from indicial response to the Laplace domain, and thence to state equations (ordinary differential equations in time). The impulsive indicial response is derived using piston theory, which is valid for nonzero Mach number and small enough time. While giving nonsingular results at zero Mach number, this theory does not include the incompressible limit exactly. The amplitude of the circulatory indicial response is obtained from the quasistatic incompressible response, scaled with $\beta = \sqrt{1 - M^2}$. The resulting unsteady loads are obtained from first-order differential equations for both the impulsive and the circulatory terms. This theory includes the effects of the airfoil shed wake, but not entirely in the "circulatory" terms. Care must be taken with a vortex wake or dynamic inflow model to ensure that the shed wake effects are neither omitted nor duplicated. The equations for unsteady lift, moment, and drag were given in Johnson (1998).

10.7 Two-Dimensional Model for Hovering Rotor

The wake of a rotor in hover or vertical flight consists of helical vortex sheets below the disk, one from each blade. Even in hover, rotor stability and loads involve unsteady motion. Unsteady motion of the rotor blade produces shed vorticity in the wake spirals. With low disk loading the wake remains near the rotor disk and therefore passes close to the following blades. Thus the wake vorticity is not convected downstream of the airfoil as with fixed wings, and the shed vorticity sheets below the rotor disk must be accounted for to correctly estimate the unsteady loads. For high inflow or forward flight, the rotor wake is convected away from the blades, so the returning shed wake influence is primarily a concern of vertical flight. Assuming a high aspect ratio of the blade, lifting-line theory requires a knowledge of the loads on the blade section, and the returning shed wake of the rotor must be incorporated into the two-dimensional unsteady airfoil theory. The wake far from the blade section has little influence, so the emphasis is on modeling the wake near the blade, which for low inflow consists of vortex sheets that are nearly planar surfaces parallel to the disk plane. Based on these considerations, a two-dimensional model for the unsteady aerodynamics of the rotor can be constructed.

Loewy (1957) developed a two-dimensional model for the unsteady aerodynamics of the blade of a hovering rotor, including the effect of the returning shed wake; Figure 10.10 shows the model. Consider first a single-bladed rotor, so all the vorticity comes from the same blade. There is a two-dimensional thin airfoil, with a shed wake vortex sheet extending downstream from the trailing edge to infinity. The surfaces of the wake spiral below the blade are modeled as a series of planar, two-dimensional vortex sheets with vertical separation h, extending from infinity upstream to infinity downstream. All the wake vortex sheets are parallel to the free stream velocity. Except for the wake-induced velocity, the model and its analysis are the same as in section 10.1. The free stream velocity U is constant here.

Assuming harmonic blade motion at frequency ω, the strength of the vortex sheet directly behind the blade ($n = 0$; see Figure 10.10) is still $\gamma_w = \overline{\gamma}_w e^{i\omega(t-x/U)}$. Since the sheets below the blade represent the successive spirals of a single helix, the vorticity strength must be related to that of the first sheet. Consider a point x on a sheet. Moving one rotor revolution downstream must be equivalent to going

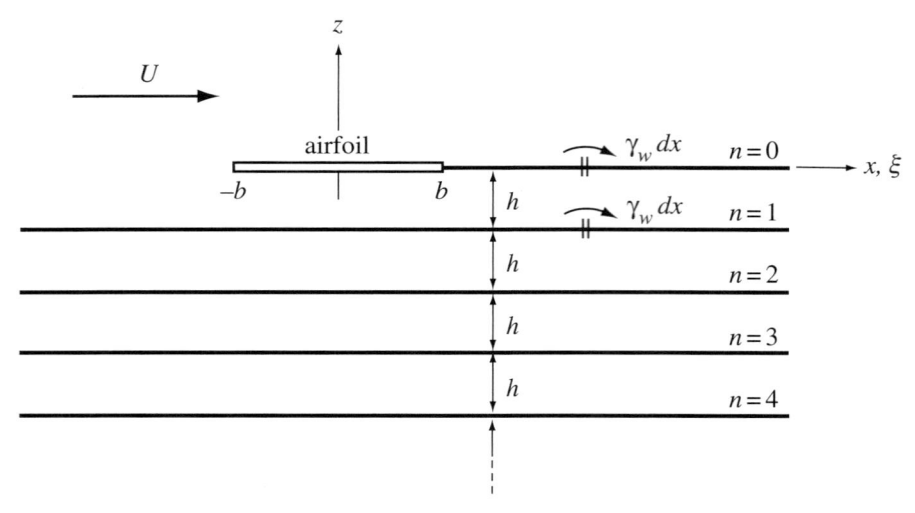

Figure 10.10. Two-dimensional model of hovering rotor (single-blade case).

directly downward to the next sheet at the same x. So, the vortex strength must be a function of the quantity $(x + n\Delta x)$, where n is the index of the wake sheets and Δx is the distance the wake is convected in a single revolution of the rotor: $\Delta x = 2\pi r = 2\pi(\Omega r)/\Omega = 2\pi U/\Omega$. Furthermore, since all the vorticity is convected downstream at the velocity U, the vortex strength in the sheets must also be a function of $(t - x/U)$. Assuming harmonic motion requires the time dependence of the wake strength to be $e^{i\omega t}$, and the wake structure is completely specified:

$$\gamma_{wn} = \overline{\gamma}_w e^{i\omega(t - x/U - n2\pi/\Omega)} = \overline{\gamma}_w e^{i\omega t} e^{-ikx/b} e^{-in2\pi(\omega/\Omega)} \qquad (10.157)$$

where γ_{wn} the strength of the n-th sheet. The reduced frequency is $k = \omega b/U$, and the wavelength is $2\pi b/k$. If ω/Ω is an integer, the loading is at a harmonic of the rotor speed. Since $e^{in2\pi(\omega/\Omega)} = 1$ in that case, the vorticity in all the sheets is exactly in phase.

The analysis of the unsteady aerodynamics proceeds as in section 10.1. With the present wake model, the wake-induced velocity is

$$\lambda(x) = \frac{1}{2\pi}\int_b^\infty \frac{\gamma_w}{x-\xi}d\xi + \sum_{n=1}^\infty \frac{1}{2\pi}\int_{-\infty}^\infty \frac{\gamma_{wn}(x-\xi)}{(x-\xi)^2 + h^2 n^2}d\xi \qquad (10.158)$$

The second term is the additional contribution of the returning shed wake. Substituting for γ_{wn} gives

$$\Delta\lambda = \sum_{n=1}^\infty \frac{1}{2\pi}\int_{-\infty}^\infty \frac{\gamma_{wn}(x-\xi)}{(x-\xi)^2 + h^2 n^2}d\xi$$

$$= \overline{\gamma}_w e^{i\omega t}\sum_{n=1}^\infty e^{-in2\pi(\omega/\Omega)}\frac{1}{2\pi}\int_{-\infty}^\infty \frac{e^{ik\xi/b}(x-\xi)}{(x-\xi)^2 + h^2 n^2}d\xi$$

$$= \overline{\gamma}_w e^{i\omega t}\sum_{n=1}^\infty \frac{i}{2}e^{-ikx/b}e^{-n((kh/b) + i2\pi(\omega/\Omega))}$$

$$= \frac{i}{2}\overline{\gamma}_w e^{i\omega t} e^{-ikx/b} W \qquad (10.159)$$

10.7 Two-Dimensional Model for Hovering Rotor

where

$$W = \sum_{n=1}^{\infty} e^{-n((kh/b)+i2\pi(\omega/\Omega))} = \frac{1}{e^{kh/b}e^{i2\pi(\omega/\Omega)} - 1} \quad (10.160)$$

To evaluate the unsteady loads on the airfoil, the wake-induced velocity is expanded as a series:

$$\lambda = \sum_{n=0}^{\infty} \lambda_n \cos n\theta \quad (10.161)$$

where $x = \cos\theta$. Using the following expression for the Bessel function,

$$J_n(k) = \frac{i^n}{\pi} \int_0^{\pi} e^{-ik\cos\theta} \cos n\theta \, d\theta \quad (10.162)$$

we obtain

$$\Delta\lambda_0 = \frac{1}{\pi}\int_0^{\pi} \Delta\lambda \, d\theta = \frac{i}{2}\overline{\gamma}_w e^{i\omega t} W J_0(k) \quad (10.163)$$

$$\Delta\lambda_1 = \frac{2}{\pi}\int_0^{\pi} \Delta\lambda \cos\theta \, d\theta = \overline{\gamma}_w e^{i\omega t} W J_1(k) \quad (10.164)$$

$$\Delta\lambda_2 = \frac{2}{\pi}\int_0^{\pi} \Delta\lambda \cos 2\theta \, d\theta = -i\overline{\gamma}_w e^{i\omega t} W J_2(k) \quad (10.165)$$

Then

$$\Delta\left(\lambda_0 + \frac{1}{2}\lambda_1\right) = \overline{\gamma}_w e^{i\omega t} W \frac{1}{2}(J_1(k) + iJ_0(k)) \quad (10.166)$$

$$\frac{b}{U}\frac{\partial}{\partial t}\Delta\left(\lambda_0 - \frac{1}{2}\lambda_2\right) = -\overline{\gamma}_w e^{i\omega t} W \frac{k}{2}(J_2(k) + J_0(k))$$

$$= -\overline{\gamma}_w e^{i\omega t} W J_1(k) \quad (10.167)$$

The changes in the bound circulation and lift are

$$\Delta\Gamma = -2\pi b\, \Delta\left(\lambda_0 + \frac{1}{2}\lambda_1\right)$$

$$= -2\pi b \overline{\gamma}_w e^{i\omega t} W \frac{1}{2}(J_1(k) + iJ_0(k)) \quad (10.168)$$

$$\Delta L = -2\pi\rho U b\left[\Delta\left(\lambda_0 + \frac{1}{2}\lambda_1\right) + \frac{1}{2}\frac{b}{U}\frac{\partial}{\partial t}\Delta\left(\lambda_0 - \frac{1}{2}\lambda_2\right)\right]$$

$$= -2\pi\rho U b \overline{\gamma}_w e^{i\omega t} W \frac{i}{2}J_0(k) \quad (10.169)$$

The total lift is now

$$L = L_Q + L_{NC} + \rho U \int_b^{\infty} \frac{b}{\sqrt{\xi^2 - b^2}}\gamma_w \xi - 2\pi\rho U b \overline{\gamma}_w e^{i\omega t} W \frac{i}{2}J_0 \quad (10.170)$$

with L_Q and L_{NC} as in section 10.1. The total circulation is

$$\Gamma = \frac{L_Q}{\rho U} + \int_b^{\infty}\left(\sqrt{\frac{\xi+b}{\xi-b}} - 1\right)\gamma_w \, d\xi - 2\pi b\overline{\gamma}_w e^{i\omega t} W \frac{1}{2}(J_1 + iJ_0) = -\int_b^{\infty}\gamma_w \, d\xi \quad (10.171)$$

which gives

$$L_Q = -\rho U \int_b^\infty \sqrt{\frac{\xi+b}{\xi-b}} \gamma_w d\xi + 2\pi \rho U b \overline{\gamma}_w e^{i\omega t} W \frac{1}{2}(J_1 + iJ_0) \tag{10.172}$$

On substituting $\gamma_w = \overline{\gamma}_w e^{i\omega(t-\xi/U)}$, the last expression gives $\overline{\gamma}_w$ in terms of L_Q, which can then be used to evaluate the circulatory lift in terms of L_Q. The result is

$$L = C'L_Q + L_{NC} \tag{10.173}$$

where

$$C'(k, \omega/\Omega, h) = 1 + \frac{-\int_1^\infty \frac{1}{\sqrt{\xi^2-1}} e^{-ik\xi} d\xi + \pi i J_0 W}{\int_1^\infty \sqrt{\frac{\xi+1}{\xi-1}} e^{-ik\xi} d\xi - \pi(J_1 + iJ_0)W} \tag{10.174}$$

$$= \frac{\int_1^\infty \frac{\xi}{\sqrt{\xi^2-1}} e^{-ik\xi} d\xi - \pi J_1 W}{\int_1^\infty \sqrt{\frac{\xi+1}{\xi-1}} e^{-ik\xi} d\xi - \pi(J_1 + iJ_0)W} \tag{10.175}$$

$$= \frac{H_1^{(2)}(k) + 2J_1(k)W}{H_1^{(2)}(k) + iH_0^{(2)}(k) + 2(J_1(k) + iJ_0(k))W} \tag{10.176}$$

is Loewy's lift deficiency function. The only influence of the returning shed wake on the two-dimensional unsteady loads of an airfoil is in the lift deficiency function, with Loewy's function replacing Theodorsen's. The modification of the lift deficiency function by the returning shed wake is determined by the quantity W. For a single blade

$$W(kh/b, \omega/\Omega) = \frac{1}{e^{kh/b}e^{i2\pi(\omega/\Omega)} - 1} \tag{10.177}$$

As h approaches infinity, W approaches zero, and hence Loewy's function C' reduces to the Theodorsen function $C(k)$. In addition to the reduced frequency k, the rotor model introduces the parameters h/b and ω/Ω. The wake spacing is given by h/b, and ω/Ω determines the relative phase of the vorticity in the successive wake sheets. When ω/Ω is equal to an integer, the strengths of all the sheets are exactly in phase. Only the fractional part of ω/Ω is important.

Now consider the case of an N-bladed rotor. Again the two-dimensional model of the rotor wake is a series of parallel vortex sheets with vertical spacing h arrayed below the blade. Here only every N-th sheet is due to a given blade, however. Let n be the index of the rotor revolutions as above, and let $m = 0, 1, 2, \ldots, N-1$ be the blade index (see Figure 10.11). When $n = 0$ and $m \geq 1$, all the wake sheets should extend upstream to a blade. Extending these sheets upstream to infinity is consistent with the two-dimensional model. The shed vorticity of sheets from a given blade (fixed m) must again be a function of $(x + n\Delta x) = (x + n\pi U/\Omega)$. To determine the relation between the vorticity strength in sheets from different blades, assume that all the blades have the same motion, but that the motion of each blade leads to the following one by the time $\Delta t = \Delta \psi \Omega$. Then moving directly down to the next sheet must be equivalent to moving downstream a distance $(\Delta x/N - U\Delta t)$, where $\Delta x/N$ is

10.7 Two-Dimensional Model for Hovering Rotor

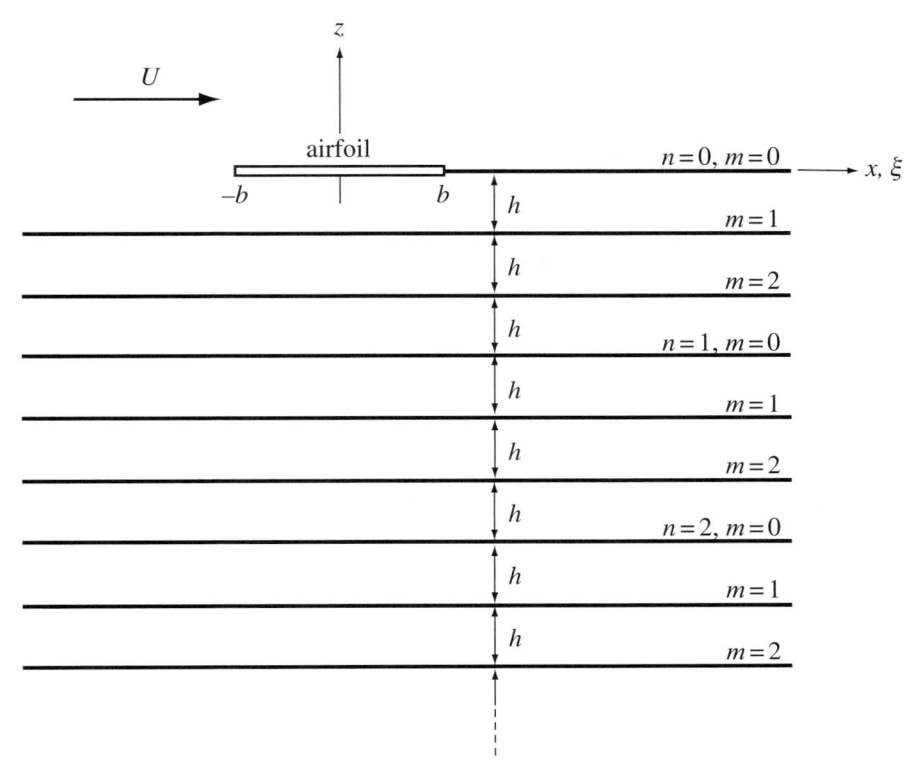

Figure 10.11. Two-dimensional wake model of an N-bladed rotor (three-bladed case shown).

the spacing between the blades. Therefore the shed vorticity must also be a function of $(x + m(\Delta x/N - U \Delta t))$. For harmonic blade motion at frequency ω

$$\gamma_{nm} = \overline{\gamma}_w e^{i\omega(t-x/U-n2\pi/\Omega+m\Delta t-2\pi m/N\Omega)}$$
$$= \overline{\gamma}_w e^{i\omega t} e^{-ikx/b} e^{-i2\pi(\omega/\Omega)(n+m/N)+im\omega\Delta t} \quad (10.178)$$

The wake-induced velocity for the N-bladed rotor is then

$$\lambda(x) = \frac{1}{2\pi}\int_b^\infty \frac{\gamma_w}{x-\xi}\,d\xi + \sum_{m=1}^{N-1}\frac{1}{2\pi}\int_{-\infty}^\infty \frac{\gamma_{0m}(x-\xi)}{(x-\xi)^2 + h^2 m^2}\,d\xi$$
$$+ \sum_{n=1}^\infty \sum_{m=0}^{N-1}\frac{1}{2\pi}\int_{-\infty}^\infty \frac{\gamma_{nm}(x-\xi)}{(x-\xi)^2 + h^2(Nn+m)^2}\,d\xi \quad (10.179)$$

Substituting for γ_{nm} gives

$$\Delta\lambda = \overline{\gamma}_w e^{i\omega t}\sum_{n=0}^\infty\sum_{m=0}^{N-1}\left\{e^{-i2\pi(\omega/\Omega)(n+m/N)+im\omega\Delta t}\frac{1}{2\pi}\int_{-\infty}^\infty \frac{e^{-ik\xi/b}(x-\xi)}{(x-\xi)^2 + h^2(Nn+m)^2}\,d\xi\right\}$$
$$- \overline{\gamma}_w e^{i\omega t}\frac{1}{2\pi}\int_b^\infty \frac{e^{-ik\xi/b}}{x-\xi}\,d\xi$$
$$= \frac{i}{2}\overline{\gamma}_w e^{i\omega t} e^{-ikx/b} W \quad (10.180)$$

where

$$W = -1 + \sum_{n=0}^{\infty}\sum_{m=0}^{N-1} e^{-i2\pi(\omega/\Omega)(n+m/N)+im\omega\Delta t} e^{-kh(Nn+m)/b}$$

$$= \frac{1 + \sum_{m=1}^{N-1}\left(e^{kNh/b}e^{i2\pi(\omega/\Omega)}\right)^{1-m/N} e^{im\omega\Delta t}}{e^{kNh/b}e^{i2\pi(\omega/\Omega)} - 1} \tag{10.181}$$

This is the same form for $\Delta\lambda$ as for the single blade, so the same result for the unsteady loads, and in particular for the lift deficiency function C', is obtained. The number of blades influences the solution only through the function W appearing in C'. For the special case $\omega\Delta t = (2\pi/N)\ell$, with ℓ an integer,

$$W = \frac{1}{\left(e^{kNh/b}e^{i2\pi(\omega/\Omega-\ell)}\right)^{1/N} - 1} = \frac{1}{e^{kh/b}e^{i2\pi(\omega/N\Omega - \ell/N)} - 1} \tag{10.182}$$

which is the single-blade result with the equivalent phase $(\omega/\Omega)_e = \omega/N\Omega - \ell/N$.

The multiblade coordinate transform (section 15.4) introduces degrees of freedom that describe the motion of the rotor as a whole. The rotating-frame degree of freedom of the m-th blade is obtained from the multiblade coordinates by

$$\beta^{(m)} = \beta_0 + \sum_n (\beta_{nc}\cos n\psi_m + \beta_{ns}\sin n\psi_m) + \beta_{N/2}(-1)^m \tag{10.183}$$

where $\psi_m = \Omega t + m2\pi/N$ is the azimuth angle of the m-th blade. Each of the non-rotating modes (collective β_0, cyclic β_{nc} and β_{ns}, and reactionless $\beta_{N/2}$) defines the relative motion of the N blades of the rotor, and hence the relationship between the shed vorticity in successive sheets of the wake. So for each non-rotating mode the function W and then the lift deficiency function C' can be evaluated.

In the collective mode the motions of all the blades are exactly in phase, and the only phase shift in the wake vorticity is due to the spacing between the blades. The time phase shift $\Delta t = 0$, for which

$$W = \frac{1}{e^{kh/b}e^{i2\pi(\omega/N\Omega)} - 1} \tag{10.184}$$

The collective mode thus is equivalent to a single-bladed rotor with wake spacing $h_e = h$ and $(\omega/\Omega)_e = \omega/N\Omega$. The relative phase between the vorticity in successive wake sheets is determined by the two parameters $\Delta\psi$ and ω/Ω for the N-bladed rotor, but only by ω/Ω with a single blade. All the wake sheets are in phase $((\omega/\Omega)_e$ is an integer) only if the collective mode oscillation occurs at a frequency that is a multiple of N/rev.

For the reactionless mode (the $\beta_{N/2}$ degree of freedom, which is present only with an even number of blades) successive blades have identical motion except for opposite signs. Reactionless motion $\beta_{N/2} = \overline{\beta}e^{i\omega t}$ gives

$$\beta^{(m)} = \overline{\beta}e^{i\omega t}(-1)^m = \overline{\beta}e^{i\omega t}e^{im\pi} \tag{10.185}$$

so $\omega\Delta t = \pi$, or $\ell = N/2$. The blades are 180° out-of-phase and

$$W = \frac{1}{e^{kh/b}e^{i2\pi(\omega/N\Omega - \frac{1}{2})} - 1} \tag{10.186}$$

The reactionless mode is also equivalent to a single-bladed rotor, with $h_e = h$ and $(\omega/\Omega)_e = \omega/N\Omega - \frac{1}{2}$.

10.7 Two-Dimensional Model for Hovering Rotor

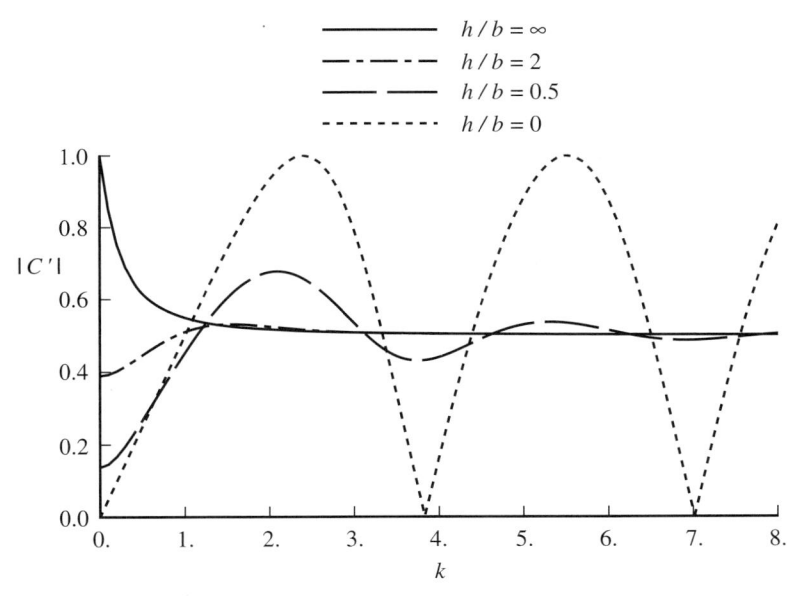

Figure 10.12. Magnitude of Loewy's lift deficiency function C' as a function of reduced frequency and wake spacing, for $\omega/\Omega =$ integer.

The cyclic degrees of freedom β_{nc} and β_{ns} are coupled. There are two modes, with eigenvectors $\beta_{ns} = \pm i\beta_{nc}$, the upper sign for the regressive mode and the lower sign for the progressive mode. The corresponding eigenvalues are $s = s_R \mp in$, where s_R is the rotating frame eigenvalue. Thus the cyclic motion $\beta_{nc} = \overline{\beta}e^{i\omega_{NR}t}$ gives

$$\beta^{(m)} = \overline{\beta}e^{i\omega_{NR}t}(\cos n\psi_m \pm i \sin n\psi_m) = \overline{\beta}e^{i\omega_{NR}t}e^{\pm in\psi_m}$$
$$= \overline{\beta}e^{i(\omega_{NR}\pm n\Omega)t}e^{\pm inm2\pi/N} = \overline{\beta}e^{i\omega t}e^{im\omega\Delta t} \quad (10.187)$$

So the rotating frequency gives $\omega = \omega_{NR} \pm n$, and the phase is $\omega\Delta t = \pm n2\pi/N$ or $\ell = \pm n$. The cyclic modes give W equivalent to a single-bladed rotor, with $h_e = h$ and $(\omega/\Omega)_e = (\omega/\Omega - n)/N$ for regressive modes or $(\omega/\Omega)_e = (\omega/\Omega + n)/N$ for progressive modes. For example, the first cyclic modes (β_{1c} and β_{1s}) of a four-bladed rotor have $(\omega/\Omega)_e = (\omega/\Omega \mp 1)/4$.

Next let us examine the behavior of Loewy's lift deficiency function,

$$C' = \frac{H_1^{(2)}(k) + 2J_1(k)W}{H_1^{(2)}(k) + iH_0^{(2)}(k) + 2(J_1(k) + iJ_0(k))W} \quad (10.188)$$

The N-bladed rotor case is equivalent to a single-bladed rotor, using the same wake spacing and a value of ω/Ω such that the successive wake sheets have the proper relative phase. It is sufficient therefore to consider the single-bladed rotor case, for which

$$W = \frac{1}{e^{kh/b}e^{i2\pi(\omega/\Omega)} - 1} \quad (10.189)$$

Figures 10.12 to 10.16 show the magnitude and phase of C' for the cases of $\omega/\Omega =$ integer, integer $+ \frac{1}{4}$, integer $+ \frac{1}{2}$, and integer $+ \frac{3}{4}$. Plots on the plane of real and imaginary C' are also interesting; see Anderson and Watts (1976). The limit as h approaches infinity is $W = 0$, and thus $C' = C(k)$, the Theodorsen function. The

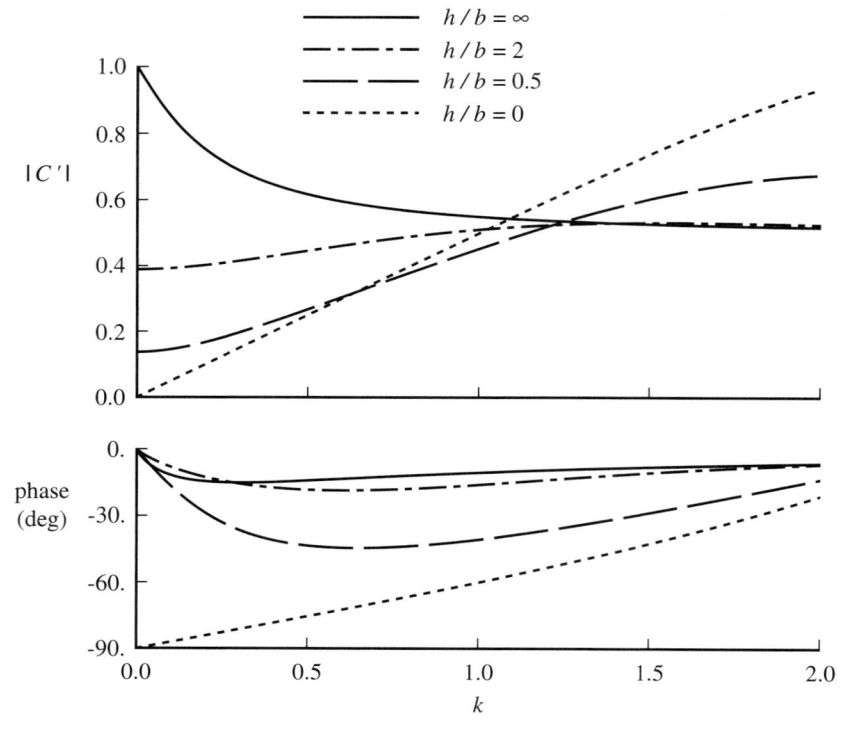

Figure 10.13. Loewy's lift deficiency function C' for $\omega/\Omega = $ integer.

case $h = 0$ is not a physically realistic limit, but does show the behavior of C' for small wake separation. With $h = 0$,

$$W = \frac{1}{e^{i2\pi(\omega/\Omega)} - 1} \tag{10.190}$$

Then

$$C' = \frac{J_1}{J_1 + iJ_0} \qquad \text{for } \omega/\Omega = \text{integer} \tag{10.191}$$

$$C' = \frac{Y_1}{Y_1 + iY_0} \qquad \text{for } \omega/\Omega = \text{integer} + \frac{1}{2} \tag{10.192}$$

These Bessel functions give C' an oscillatory behavior at large k (see Figure 10.12). The real part F' oscillates between 0 and 1, and the imaginary part between -0.5 and 0.5, with a period of π. Since $|C'| = \sqrt{F'}$ for these cases, the magnitude of C' goes to zero at certain frequencies. For large reduced frequency the lift deficiency function is

$$C' \sim \frac{1}{2}\left(1 + ie^{-kh/b}e^{-i2\pi(\omega/\Omega)}e^{i2k}\right) \tag{10.193}$$

Hence the oscillatory behavior observed in Figure 10.12 is a general result for large k. The period is π, and the amplitude of the oscillation diminishes with increasing wake spacing h.

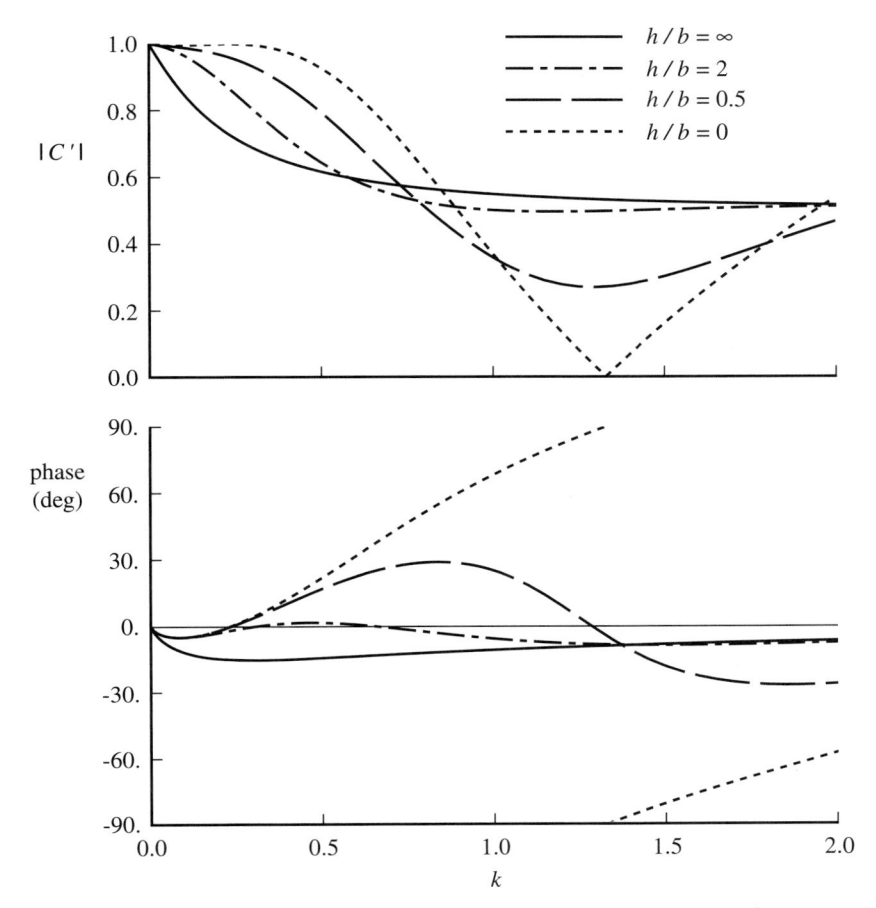

Figure 10.14. Loewy's lift deficiency function C' for $\omega/\Omega = $ integer $+ \frac{1}{4}$.

For small reduced frequency the lift deficiency function is approximately

$$C' \cong \frac{1 - i\frac{\pi}{2}k^2 W}{1 + \frac{\pi}{2}k - ik\left(\ln\frac{k}{2} + \gamma\right) + \left(1 - \frac{i}{2}k\right)\pi kW} \qquad (10.194)$$

where γ is Euler's constant. If ω/Ω is not equal to an integer, W is of order 1 or smaller for all h; then to order k,

$$C' \cong \frac{1}{1 + \frac{\pi}{2}k - ik\left(\ln\frac{k}{2} + \gamma\right) + \pi kW} \qquad (10.195)$$

This gives $C' = 1$ at $k = 0$, independent of the wake spacing h (see Figures 10.14 to 10.16). For $\omega/\Omega = $ integer, $W \cong b/kh$, so to order 1

$$C' \cong \frac{1}{1 + \pi kW} = \frac{1}{1 + \frac{\pi}{h/b}} \qquad (10.196)$$

These results for small reduced frequency are of most interest for helicopter rotors. If ω/Ω is not an integer, there is only an order k correction of the Theodorsen function due to the returning shed wake. For oscillations at harmonics of the rotor speed ($\omega/\Omega = $ integer), the shed wake reduces the lift deficiency function at low frequency to $C' = h/(h + \pi b)$. Figure 10.13 shows that this low-frequency result is

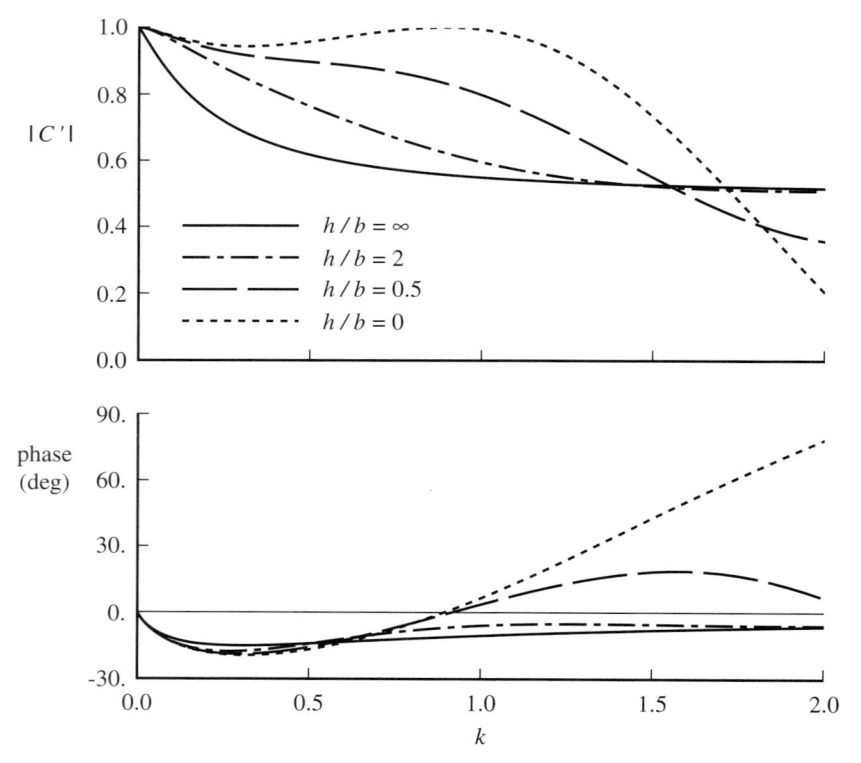

Figure 10.15. Loewy's lift deficiency function C' for $\omega/\Omega =$ integer $+ \frac{1}{2}$.

a good approximation even to $k = 0.2$ or so. $C'(0) \neq 1$ now; in fact, $C'(0) = 0$ when $h = 0$, because of the returning shed wake.

The wake spacing h/b is determined by the rate at which the helical vortex sheets of the rotor are convected downward. Using the mean induced velocity at the rotor disk for the convection velocity of the wake near the rotor, the distance the wake moves in a single rotor revolution is $Nh = v(2\pi/\Omega)$, or

$$\frac{h}{b} = \frac{v2\pi}{\Omega Nb} = \frac{4\lambda}{\sigma} \qquad (10.197)$$

where λ is the inflow ratio and σ is the rotor solidity. The lift deficiency function for low k and harmonic oscillations is then

$$C' \cong \frac{1}{1 + \frac{\pi\sigma}{4\lambda}} \qquad (10.198)$$

Typically $\lambda \cong 0.07$ for the hovering helicopter rotor, which gives $h/b \cong 3$ or 4, and so $C' \cong 0.5$.

Thus the reduction of the unsteady loads due to the returning shed wake can be large, with serious consequences for rotor control, loads, and stability in the critical circumstances of low inflow and harmonic oscillations. The vorticity in successive wake sheets below the blade is exactly in phase for critical values of ω/Ω, which depend on whether the motion occurs in collective, cyclic, or reactionless modes. Important examples of such motion in rotor dynamics are cyclic pitch control and flapping, which give 1/rev motion in the rotating frame, and flutter instabilities with a natural frequency at n/rev. The reduction of the circulatory lift at low frequency

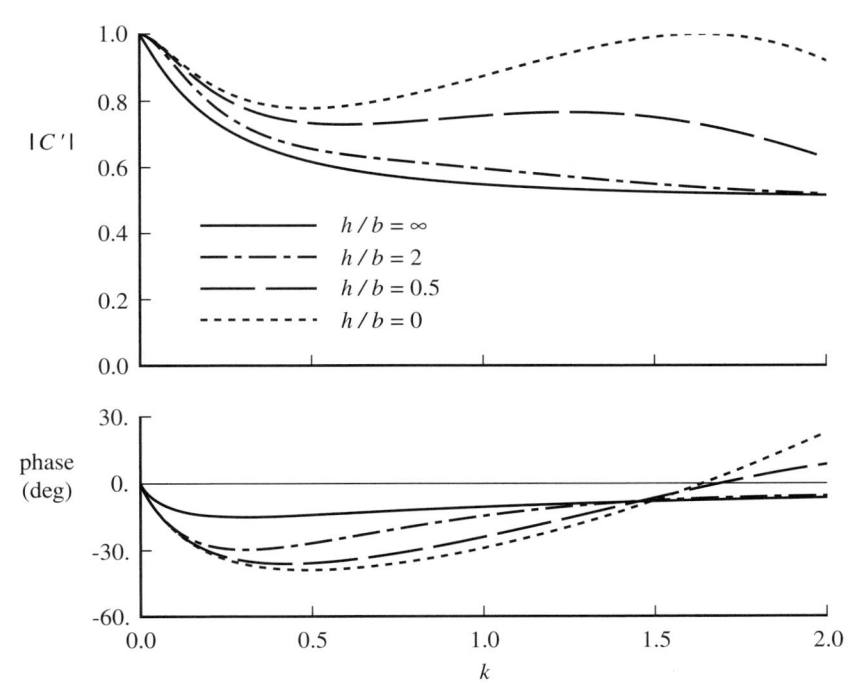

Figure 10.16. Loewy's lift deficiency function C' for $\omega/\Omega = \text{integer} + \frac{3}{4}$.

decreases the rotor response to collective and cyclic pitch control. At high reduced frequencies C' has an oscillatory behavior, with minima near zero for small wake spacing. The result is a decrease in the damping of the flap motion and the flapwise bending modes, thereby increasing the response of the blade vibration to harmonic airloads.

Wake-excited flutter has been observed on rotors operating in hover at low collective. The returning shed wake reduces the circulatory loads that are responsible for flap damping, which can lead to a pitch-flap flutter instability. The circulatory lift does not influence the pitch moment for oscillation about the quarter chord, but the blade pitch damping moments are also influenced by the returning shed wake if the pitch axis is off the aerodynamic center. For oscillation near certain harmonics of the rotor speed, about the leading edge or about the midchord, the damping in pitch can be negative. Thus a single-degree-of-freedom instability is possible.

Similar unsteady aerodynamic theories were developed by Timman and van de Vooren (1957) and by Jones (1958). Timman and van de Vooren (1957) considered the limit of no vertical convection, with all shed sheets in the plane of the airfoil. They performed two-degree-of-freedom flutter calculations, including comparisons with experiment. Jones (1958) considered the same two-dimensional aerodynamic model as Loewy, but did not obtain Loewy's result for the lift deficiency function.

Loewy's aerodynamic theory was verified experimentally. Daughaday, DuWaldt, and Gates (1957) conducted an experimental investigation of the bending load amplification on model helicopter rotor blades, with particular attention to the effect of the returning shed wake on the aerodynamic damping of the bending modes. They conducted a test of a one-bladed, teetering model rotor, varying the mass characteristics, pitch spring, and pitch-flap coupling. Using excitation through a moment at the flap hinge, forced response was measured and damping obtained from the free decay. For the damping of the first bending mode in hover, a reduction of damping ratio when the frequency was near 3/rev was observed experimentally and predicted using Loewy's theory. This wake effect was significant at zero collective, but reduced for 4° collective. The second bending mode damping was reduced near 7/rev and 8/rev. Pitch-flap flutter was investigated for a two-bladed rotor with large pitch-flap coupling, varying rotor speed, and blade chordwise center-of-gravity position. A stabilizing effect of the returning wake was observed for pitch mode frequency near 2/rev and predicted using Loewy's theory.

Ham, Moser, and Zvara (1958) measured blade flap response to collective pitch and to vertical hub motion on a two-bladed articulated rotor model with rectangular, untwisted blades. At low blade pitch, significant effects of the returning shed wake were observed for excitation frequencies near an integer. In the hover flap response to collective, for zero mean collective, the magnitude of the response was nearly quasistatic (good prediction using $C = 1$) except near 2/rev, where the reduced magnitude observed experimentally was predicted using Loewy's C'. The phase was nearly quasistatic at low frequency, with significant phase shift above 1.5/rev, predicted using Loewy's function. In the hover flap response to vertical hub motion, for zero collective, a large increase in the magnitude was observed experimentally at frequencies near a multiple of 2/rev and was predicted using Loewy's C'. At collectives of 5° and 10° in hover, the measured response was reduced. Forward flight ($\mu = 0.1$ and 0.2) produced little effect on the measured response to hub motion for zero collective.

Silveira and Brooks (1959) conducted an experimental study of the damping of the flapwise bending modes of a two-bladed hovering rotor at low pitch (and hence low inflow). They tested two-bladed teetering and articulated rotors in hover, for collective pitch angles of 0° and 3°. For the teetering rotor, the flap bending modes were excited by vertical motion of the hub and then the damping was obtained from free decay. The measured damping of the second and third elastic flap modes exhibited a reduction in damping at multiples of 2/rev, which was successfully predicted using the real part of Loewy's function, F'.

The lifting-line approximation was examined in section 10.2 for the near shed wake in unsteady two-dimensional airfoil theory. Following Miller (1964), this analysis is now extended to include the returning shed wake. The lifting-line approximation involves calculating the induced velocity at a single point on the chord, rather than using the distribution over the chord. Because the treatment required for the near shed wake in this approximation has already been derived, here the near shed wake is treated correctly and the lifting-line approximation is used only for the wake sheets below the airfoil. Equation 10.159 for the additional induced velocity due to the returning shed wake, evaluated at the quarter chord ($x = -b/2$), is

$$\Delta \lambda = \frac{i}{2} \overline{\gamma}_w e^{i\omega t} e^{-ikx/b} W = \frac{i}{2} \overline{\gamma}_w e^{i\omega t} e^{-ik/2} W \qquad (10.199)$$

Since the lifting-line approximation corresponds to low reduced frequency in the unsteady aerodynamics, the approximation $e^{-ik/2} \cong 1$ can also be used. The bound circulation and lift increments are then

$$\Delta \Gamma = -2\pi b \overline{\gamma}_w e^{i\omega t} W \frac{i}{2} \qquad (10.200)$$

$$\Delta L = -2\pi \rho U b \overline{\gamma}_w e^{i\omega t} W \frac{i}{2} \qquad (10.201)$$

from which the lift deficiency function is

$$C' = \frac{H_1^{(2)}(k)}{H_1^{(2)}(k) + iH_0^{(2)}(k) + 2iW} \qquad (10.202)$$

This result can be obtained from Loewy's function by using the order 1 approximations of the Bessel functions. It is a good approximation to Loewy's function at least up to $k = 0.5$ for the complete range of wake spacing. The greatest error is in the imaginary part of C' (hence the phase shift) at low h/b. Thus the lifting-line analysis is a satisfactory treatment for the returning shed wake as well as for the trailed wake. Only the near shed wake requires special treatment.

Miller (1964) also considered a further approximation to Loewy's analysis, modeling the wake sheets below the blade by a continuous vorticity distribution. This model is analogous to the actuator disk representation of the rotor and thus serves to illustrate the connection between the continuous vorticity analysis for the rotor and the discrete wake model of Loewy's two-dimensional theory. Only the case of harmonic blade motion is considered, $\omega/\Omega =$ integer. Since the loading on the blade is periodic, the vorticity strength in the wake is independent of the vertical position. Also assuming low reduced frequency, the variation of the induced velocity over the chord is neglected. The wake-induced velocity at $x = 0$ on the airfoil is then

$$\lambda = -\frac{1}{2\pi}\int_{-\infty}^{0}\int_{-\infty}^{\infty}\frac{\xi}{\xi^2 + z^2}\gamma\, d\xi\, dz \qquad (10.203)$$

where $\gamma(\xi, t)$ is the strength of the continuous vorticity distribution below the airfoil. Since each discrete wake sheet with strength γ_w has now been spread over the distance h, $\gamma = \gamma_w/h$. The shed wake strength is obtained from the time derivative of the bound circulation Γ. If sinusoidal motion at frequency ω is assumed,

$$\gamma = -\frac{1}{Uh}\frac{d}{dt}\Gamma(t - \xi/U) = -\frac{i\omega}{Uh}\overline{\Gamma}e^{i\omega(t-\xi/U)} = -\frac{ik\Gamma/b^2}{h/b}e^{-i\omega\xi/U} \qquad (10.204)$$

With the wake convected downward by the mean inflow ratio λ_0, the wake spacing is $h/b = 4\lambda_0/\sigma$ (equation 10.197). So

$$\gamma = -\frac{ik\Gamma/b^2}{4\lambda_0/\sigma}e^{-i\omega\xi/U} \qquad (10.205)$$

After substituting for γ, the velocity induced by the shed wake is

$$\lambda = \frac{ik\Gamma/b^2}{4\lambda_0/\sigma}\frac{1}{2\pi}\int_{-\infty}^{0}\int_{-\infty}^{\infty}\frac{\xi}{\xi^2+z^2}e^{-ik\xi/b}\,d\xi\,dz = \frac{k\Gamma/b^2}{4\lambda_0/\sigma}\frac{1}{2k/b} = \frac{\pi\sigma}{4\lambda_0}\frac{L}{2\pi\rho Ub} \qquad (10.206)$$

Then from $L = L_Q - 2\pi\rho Ub\lambda = L_Q - (\pi\sigma/4\lambda_0)L$, $L = C'L_Q$, with the lift deficiency function

$$C' = \frac{1}{1 + \frac{\pi\sigma}{4\lambda_0}} \qquad (10.207)$$

which is the low-frequency limit of Loewy's function for the case of harmonic oscillation (equation 10.198).

10.8 Blade-Vortex Interaction

The model problem for blade-vortex interaction is an infinite wing encountering a straight infinite vortex in a subsonic, compressible free stream (Figure 10.17). The

model geometry   coordinate systems

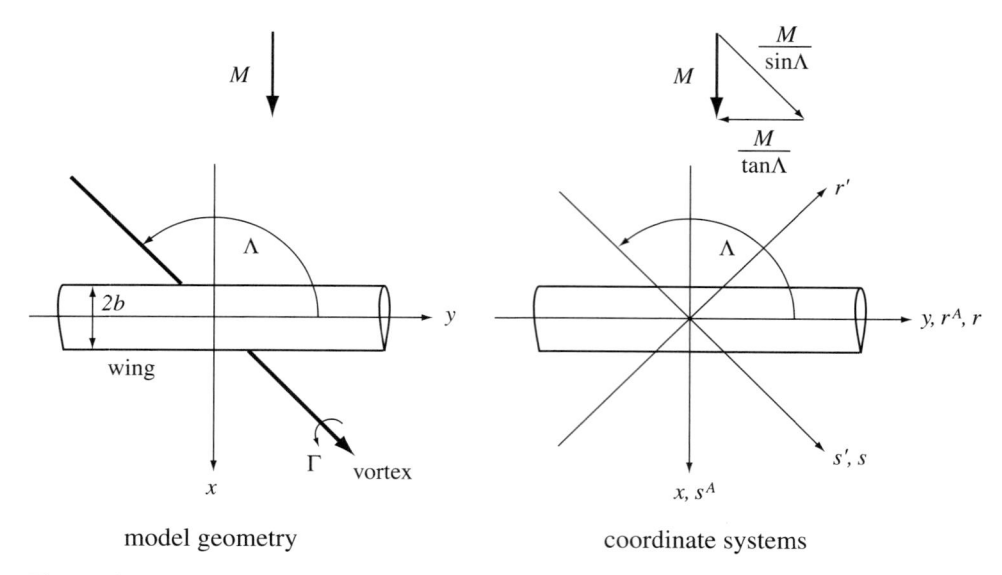

Figure 10.17. Blade-vortex interaction.

wing and vortex lie in parallel planes with separation h. The wing semi-chord is b. The angle between the wing axis and the vortex is Λ: $\Lambda = 90°$ for a perpendicular interaction, and $\Lambda = 180°$ for a parallel interaction. The wing spanwise variable r is measured from the vortex-midchord intersection. The vortex-induced downwash velocity at the midchord is

$$w = -\frac{\Gamma}{2\pi}\frac{r}{r^2 + h^2} \tag{10.208}$$

for the perpendicular interaction (equation 9.80). The effect of the viscous core can be included by using a larger effective separation, $h_{eq} = (h^2 + r_c^2)^{1/2}$, based on the distributed vorticity model of the Scully core (see section 9.8).

For linear aerodynamic analysis, applying the Fourier transform to the vortex-induced downwash and lift reduces the problem to finding the loading produced by a convected sinusoidal gust. For the perpendicular interaction, any downwash field can be represented by a superposition of sinusoidal gusts:

$$w(r) = \int_{-\infty}^{\infty} e^{i\nu r/b}\,\overline{w}(\nu)\,d\nu \tag{10.209}$$

Here ν is the dimensionless wave number, corresponding to wavelength $\lambda = 2\pi b/\nu$. For the vortex-induced downwash, the Fourier transform of equation 10.208 gives the spectrum

$$\frac{\overline{w}/U}{\Gamma/2\pi U b} = \frac{i}{2}\,\mathrm{sign}\nu\, e^{-|\nu|h/b} \tag{10.210}$$

The loading function $g_L(\nu)$ gives the section lift produced by a sinusoidal gust:

$$g_L(\nu) = -\frac{\overline{L}(\nu)/2\pi\rho U^2 b}{\overline{w}(\nu)/U} \tag{10.211}$$

The $\overline{w}$ spectrum is 10% of the long wavelength value at $vh/b \cong 2.3$, and for wing-vortex separation equal to the quarter chord ($h/b = 0.5$) the wave number is $v \cong 4.6$, with wavelength $\lambda \cong 1.4b$.

Lifting-line theory obtains the wing bound circulation Γ from the vortex-induced downwash w and the upwash from the trailed wake:

$$\Gamma = -2\pi b(w - w_i) = -2\pi bw - \frac{b}{2}\int_{-\infty}^{\infty}\frac{d\Gamma}{d\rho}\frac{d\rho}{r - \rho} \qquad (10.212)$$

For a sinusoidal gust, $\Gamma = \overline{\Gamma}e^{ivr/b}$ and $w = \overline{w}e^{ivr/b}$, so

$$\overline{\Gamma} = -2\pi b\overline{w} - \frac{iv}{2}\overline{\Gamma}\int_{-\infty}^{\infty}e^{iv(\rho-r)/b}\frac{d\rho}{r-\rho} = -2\pi b\overline{w} - \frac{\pi}{2}v\overline{\Gamma} \qquad (10.213)$$

and the loading function is

$$g_L = -\frac{\overline{\Gamma}}{2\pi b\overline{w}} = \frac{1}{1 + \frac{\pi}{2}v} \qquad (10.214)$$

with a continuous trailed wake sheet. Consider a discretized wake, with trailed lines uniformly spaced, a distance d apart. The bound circulation Γ_m is evaluated at the collocation points $r = md$. The trailers at $\rho = \left(n + \frac{1}{2}\right)d$ have strength $\delta = \Gamma_{n+1} - \Gamma_n$. Then

$$\Gamma_m = -2\pi bw_m - \frac{b}{2}\sum_{n=-\infty}^{\infty}\frac{\Gamma_{n+1} - \Gamma_n}{md - \left(n + \frac{1}{2}\right)d} \qquad (10.215)$$

Substituting $\Gamma = \overline{\Gamma}e^{ivr/b}$ and $w = \overline{w}e^{ivr/b}$ gives

$$\overline{\Gamma} = -2\pi b\overline{w} - \frac{1}{2}\overline{\Gamma}\sum_{n=-\infty}^{\infty}\frac{e^{iv(n+1-m)d/b} - e^{iv(n-m)d/b}}{\left(m - n - \frac{1}{2}\right)d/b}$$

$$= -2\pi b\overline{w} - \frac{\pi}{2}v\left|\frac{\sin vd/2b}{vd/2b}\right|\overline{\Gamma} \qquad (10.216)$$

and the loading function is

$$g_L = -\frac{\overline{\Gamma}}{2\pi b\overline{w}} = \frac{1}{1 + \frac{\pi}{2}v\left|\frac{\sin vd/2b}{vd/2b}\right|} \qquad (10.217)$$

This discretized wake solution is a good approximation for the continuous wake solution up to $vd \cong 2b$. So the trailer spacing should be $d \cong 2b/v = b\lambda/\pi$, or $d \cong 0.5b$ (quarter chord) for $v = 4$. For spacing equal to the wavelength ($d = \lambda$), $g_L = 1$. Figure 10.18 compares the lifting-line solutions for the loading function with the lifting-surface solution. For small separation between the wing and vortex, the lifting-line loading is significantly larger than the lifting-surface loading.

For the parallel interaction, the Fourier transform introduces the reduced frequency k instead of the wave number, and the incompressible loading function is the Sears function of two-dimensional, unsteady airfoil theory. The lifting-line approximation is

$$g_L = S = e^{-ik/2}\left(C(k) + \frac{ik}{2}\right) \qquad (10.218)$$

(equation 10.94). Figure 10.19 compares the lifting-line solution with the exact unsteady, two-dimensional solution. This is a low-frequency approximation, accurate

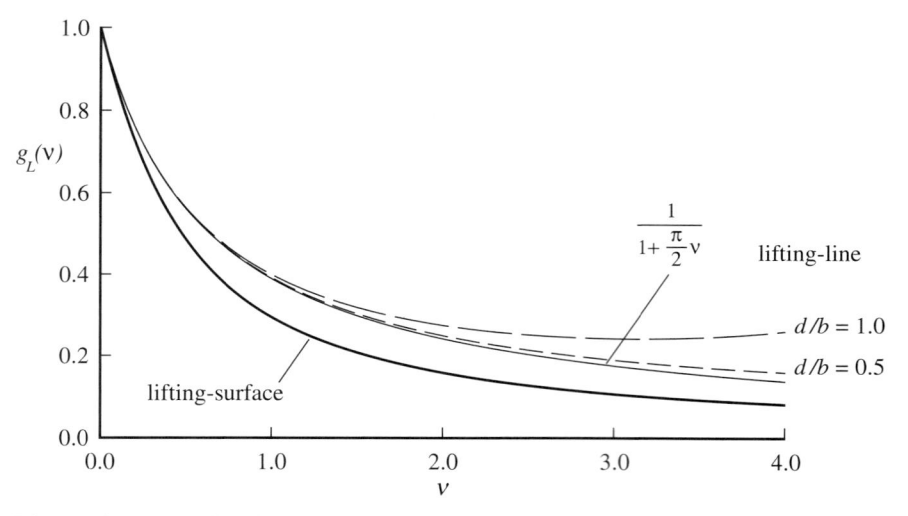

Figure 10.18. Loading function for perpendicular interaction.

to about $k \cong 0.5$. To order k,

$$S \cong C \cong \frac{1}{1 + \frac{\pi}{2}k} \qquad (10.219)$$

which approximates the magnitude well even for large k, but neglects the phase shift entirely. The phase shift of the Sears function S leads to significant asymmetry in the blade-vortex interaction loads for the parallel interaction, with the first peak (in time) larger than the second peak.

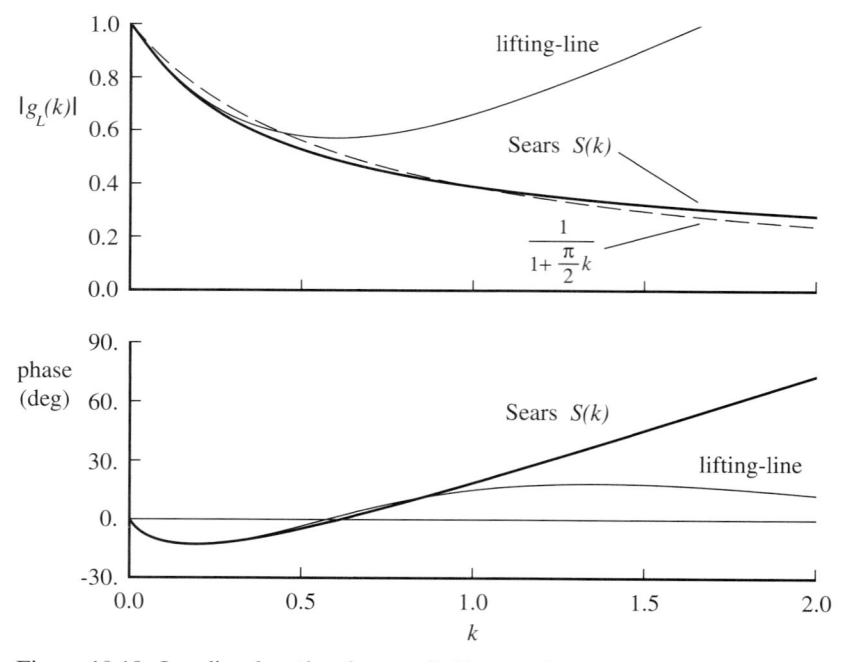

Figure 10.19. Loading function for parallel interaction.

10.8 Blade-Vortex Interaction

For the large variations of downwash along the wing that are produced by a close vortex, lifting-line theory does not produce accurate loading. Second-order lifting-line theory can give the section lift accurately for the vortex-blade separations typical of helicopter rotors, but in practical implementations the moment and pressure results are still essentially first order (section 9.2). Moreover, for the two-dimensional, parallel interaction, analytical solutions are only possible for incompressible flow. Thus compressible lifting-surface theory is next developed for the blade-vortex interaction problem of Figure 10.17. The solution is obtained in the form of an aerodynamic influence function for an infinite in an oblique, sinusoidal gust. The development follows Johnson (1970, 1971).

The vortex is convected past the blade by the free stream, so the vortex-induced downwash in the plane of the wing is a one-dimensional field, depending only on the perpendicular distance from the vortex. For $\Lambda = 90°$, the problem becomes the steady, three-dimensional flow due to a vortex perpendicular to the wing. For $\Lambda = 180°$, the problem is the unsteady, two-dimensional flow of a point vortex past an airfoil. Some singular behavior of the problem can be expected, because the first of these limits is an elliptic problem, whereas the second is hyperbolic (due to the time dependence). The transition between the elliptic and hyperbolic problems occurs at $M = \sin \Lambda$.

Figure 10.17 shows the coordinate systems used. The (x, y) system is fixed, and all the others are moving. Because of the infinite geometry of the model, the problem is steady in a coordinate system with its origin translating along the blade centerline, following the projection of the convected vortex. A natural coordinate system for such a translating frame of reference has one coordinate (s') aligned in the direction of the free vortex (Figure 10.17). In this frame the vortex is stationary, so the problem is steady and there is no shed vorticity (wake vorticity normal to the direction of the relative free stream, generated by time-varying blade circulation). Both the induced-wake vorticity behind the wing and the relative free stream velocity in this coordinate system must be in the direction of the vortex (the s' direction). The relative free stream Mach number in this translating coordinate system is $M/\sin \Lambda$, and the wing is swept at an angle Λ relative to this velocity component. The relative free stream attains a sonic value ($M/\sin \Lambda = 1$) at the transition between the elliptic and hyperbolic domains.

Linear lifting-surface theory is used to obtain the solution to the model problem in the form of an integral equation relating the pressure and downwash at the wing surface. This integral equation is a double integral over the wing surface; it is transformed into a single (chordwise) integral by the use of the Fourier transform with respect to the span variable. In what follows all quantities are dimensionless, based on the fluid density, the wing semi-chord, and the free stream velocity (ρ, b, and U). The several coordinate systems used are shown in Figure 10.17. The (s', r') system is the translating coordinate system, with s' aligned in the vortex direction. The problem is solved for the loads in the (s^A, r^A) system. One coordinate is along the span (so the Fourier transform can be used), and the other is in the chordwise direction (the integration direction to obtain the section airloading). The blade leading and trailing edges are given by $s^A = \pm 1$; this system is orthogonal. The problem is solved for the circulation in the (s, r) system. One coordinate is along the span (so the Fourier transform can be used), and the other is in the s' direction (the integration direction to obtain the circulation). The s metric has been stretched so that the blade leading and trailing edges are still given by $s = \pm 1$; this system is not orthogonal.

The solution is most conveniently formulated in terms of the acceleration potential ψ. The linearized equation of motion is

$$\left\{\nabla^2 - M^2 \left(\frac{\partial}{\partial x} + \frac{\partial}{\partial t}\right)^2\right\} \psi = 0 \tag{10.220}$$

where

$$\psi = \left(\frac{\partial}{\partial x} + \frac{\partial}{\partial t}\right)\phi = -p \tag{10.221}$$

Here ϕ is the velocity potential and p is the perturbation pressure. The boundary conditions are

$$\left.\frac{\partial \phi}{\partial z}\right|_{z=0} = w \quad \text{on the airfoil} \tag{10.222}$$

$$\Delta \psi = -\Delta p = 0 \quad \text{off the airfoil} \tag{10.223}$$

where Δ means the difference between the quantities at $z = 0^+$ and at $z = 0^-$. Here the downwash w is a function of r' only. In the (s', r') system the equation of motion becomes

$$\left\{[1 - (M/\sin\Lambda)^2]\frac{\partial^2}{\partial s'^2} + \frac{\partial^2}{\partial r'^2} + \frac{\partial^2}{\partial z^2}\right\} \psi = 0 \tag{10.224}$$

$$\psi = -p = \frac{1}{\sin\Lambda}\frac{\partial \phi}{\partial s'} \tag{10.225}$$

It is seen that in this system the problem is indeed steady and is elliptic or hyperbolic as $M/\sin\Lambda$ is less than or greater than one.

The elementary lifting solution for the acceleration potential is the dipole solution, denoted by ψ_d. Using superposition, the acceleration potential at an arbitrary point due to a lifting surface is written

$$\psi(s^A, r^A, z) = \int_{-\infty}^{\infty}\int_{-1}^{1} L^A(\sigma^A, \rho^A)\psi_d(s'_0, r'_0, z)\, d\sigma^A\, d\rho^A \tag{10.226}$$

where $s'_0 = s' - \sigma'$, $r'_0 = r' - \rho'$, and $L^A(\sigma^A, \rho^A) = -\Delta p = \Delta \psi$ is the differential pressure across the lifting surface. The advantage of the acceleration potential is that the boundary condition off the airfoil, $\Delta\psi = -\Delta p = 0$, is automatically satisfied by placing the elementary solutions only on the lifting surface. Integration of equation 10.225 and application of the boundary condition on the airfoil give

$$w(r') = \lim_{z \to 0} \frac{\partial}{\partial z}\int_0^{\infty} \psi|_{s_0 = s_0 - \lambda}\, d\lambda \tag{10.227}$$

where $s_0 = s - \sigma$; the notation in the integrand means that ψ is expressed as a function of s by means of the coordinate transformations between the (s^A, r^A), (s', r'), and (s, r) systems, and then the quantity s_0 is replaced by the quantity $(s_0 - \lambda)$. Substitution of ψ from equation 10.226 gives the integral equation:

$$w(r') = \int_{-\infty}^{\infty}\int_{-1}^{1} \frac{L^A(\sigma^A, \rho^A)}{2\pi/\alpha}\left\{\lim_{z \to 0}\frac{\partial}{\partial z}\int_0^{\infty}\psi_d|_{s_0=s_0-\lambda}\,d\lambda\frac{2\pi}{\alpha}\right\} d\sigma^A d\rho^A \tag{10.228}$$

where $\alpha = \sqrt{1-M^2}$ is the Prandtl-Glauert compressibility factor. Writing $L^A(s^A, r^A)$ as a Fourier integral

$$L^A(s^A, r^A) = \int_{-\infty}^{\infty} \overline{L}^A(s^A, \nu) e^{i\nu r^A} d\nu \tag{10.229}$$

and taking the Fourier transform of the integral equation give

$$\int_{-1}^{1} \overline{G}^A(\sigma^A, \hat{\nu}) K^A(\sigma_0^A, \hat{\nu}) d\sigma^A = -e^{i\hat{\nu} s^A \cos \Lambda} \tag{10.230}$$

where $s_0^A = s^A - \sigma^A$ and $\hat{\nu} = \nu/\sin \Lambda$. The downwash now is just an oblique, sinusoidal gust, $e^{i\hat{\nu} s^A \cos \Lambda}$. The kernel function is

$$K^A(\sigma_0^A, \hat{\nu}) = -\frac{2\pi}{\alpha} \int_{-\infty}^{\infty} e^{-i\nu r_0^A} \lim_{z \to 0} \frac{\partial}{\partial z} \int_0^{\infty} \psi_d \big|_{s_0 = s_0 - \lambda} d\lambda \, dr^A \tag{10.231}$$

and the aerodynamic influence function is

$$\overline{G}^A(\sigma^A, \hat{\nu}) = \frac{\overline{L}^A(\sigma^A, \hat{\nu})}{(2\pi/\alpha)\overline{w}(\hat{\nu})/\sin \Lambda} \tag{10.232}$$

Note that $2\pi/\alpha$ is the theoretical subsonic lift-curve slope. Here $\overline{w}(\hat{\nu})/\sin \Lambda$ is the Fourier transform of $w(r')$ with respect to r^A, at $s^A = 0$. The wave number has been written in the form $\hat{\nu} = \nu/\sin \Lambda$, which is the wave number of the sinusoidal gust. This quantity is the span wave number for $\Lambda = 90°$, and the reduced frequency for $\Lambda = 180°$. The vortex-induced downwash for general interaction angle is

$$w(r^A \sin \Lambda) = -\frac{\Gamma}{2\pi} \frac{r^A \sin \Lambda}{(r^A \sin \Lambda)^2 + h^2} \tag{10.233}$$

which has the Fourier transform

$$\overline{w}(\hat{\nu}) = \frac{\Gamma}{2\pi} \frac{i}{2} \operatorname{sign} \hat{\nu} \, e^{-|\hat{\nu}|h} \tag{10.234}$$

For the two-dimensional, unsteady limit $\Lambda = 180°$, the span variable $r^A \sin \Lambda$ becomes the time variable.

The kernels depend on the parameters M and Λ. In particular, they depend on the sign of the quantity

$$\beta^2 = -B^2 = 1 - (M/\sin \Lambda)^2 \tag{10.235}$$

hence whether the flow is in the elliptic or hyperbolic domain. The derivation of the kernel functions follows the standard techniques of aerodynamic theory, particularly Possio's method; see Bisplinghoff, Ashley, and Halfman (1955). The detailed derivation of the kernels has given by Johnson (1970).

The elliptic domain is defined by $M < \sin \Lambda$, or $\beta^2 = 1 - (M/\sin \Lambda)^2 > 0$. The elliptic kernel is

$$K_\beta^A(s_0^A, \hat{\nu}) = e^{i\hat{\nu} s_0^A \cos \Lambda} \left\{ e^{ia\mu s_0^A} \left[\pm a K_1(a|s_0^A|) + ia\mu K_0(a|s_0^A|) \right] \right.$$
$$\left. - \frac{i\hat{\nu}}{2\alpha} \left[i\pi + \ln \frac{\alpha - \cos \Lambda}{\alpha + \cos \Lambda} \right] \pm \frac{\hat{\nu}}{\beta \sin \Lambda} \int_0^{a|s_0^A|} e^{\pm i\mu \xi} K_0(\xi) d\xi \right\} \tag{10.236}$$

where

$$\pm = \operatorname{sign} s_0^A \qquad a = \hat{v}\frac{\beta \sin \Lambda}{\alpha^2} \qquad \mu = \frac{-\cos \Lambda}{\beta \sin \Lambda}$$

$$a\mu = \hat{v}\frac{-\cos \Lambda}{\alpha^2} \qquad \beta \sin \Lambda = \sqrt{\sin^2 \Lambda - M^2}$$

and K_0, K_1 are modified Bessel functions. For $\Lambda = 90°$, the kernel is

$$K_\beta^A(s_0^A, \hat{v}) = -\frac{1}{2\alpha}\int_{-\infty}^{\infty} e^{-i v r_0^A} \frac{1}{(r_0^A)^2}\left[1 + \frac{s_0^A}{\sqrt{(s_0^A)^2 + \alpha^2(r_0^A)^2}}\right] dr_0^A \qquad (10.237)$$

which is the Fourier transform of the steady, three-dimensional, lifting-surface kernel.

The hyperbolic domain is defined by $M > \sin \Lambda$, or $B^2 = (M/\sin \Lambda)^2 - 1 > 0$. The hyperbolic kernel is

$$K_B^A(s_0^A, \hat{v}) = e^{i\hat{v}s_0^A \cos \Lambda}\left\{\frac{i\pi}{2}e^{ia\mu s_0^A}\left[\mp a H_1^{(2)}(a|s_0^A|) - ia\mu H_0^{(2)}(a|s_0^A|)\right]\right.$$

$$\left. - \frac{i\hat{v}}{2\alpha}\left[\ln\frac{-\cos\Lambda + \alpha}{-\cos\Lambda - \alpha}\right] \mp \frac{i\pi\hat{v}}{2B\sin\Lambda}\int_0^{a|s_0^A|} e^{\pm i\mu\xi}H_0^{(2)}(\xi)d\xi\right\} \qquad (10.238)$$

where

$$\pm = \operatorname{sign} s_0^A \qquad a = \hat{v}\frac{B\sin\Lambda}{\alpha^2} \qquad \mu = \frac{-\cos\Lambda}{B\sin\Lambda}$$

$$a\mu = \hat{v}\frac{-\cos\Lambda}{\alpha^2} \qquad B\sin\Lambda = \sqrt{M^2 - \sin^2\Lambda}$$

and $H_0^{(2)}, H_1^{(2)}$ are Hankel functions. For $\Lambda = 180°$, and writing $\hat{v} = v/\sin\Lambda = k$ (the reduced frequency), the kernel function is Possio's form of the kernel for unsteady, compressible flow about a two-dimensional thin airfoil:

$$K_2^A(s_0^A, k) = e^{-iks_0^A}\left\{\frac{i\pi}{2}e^{iks_0^A/\alpha^2}\left[\mp a H_1^{(2)}(a|s_0^A|) - \frac{ik}{\alpha^2}H_0^{(2)}(a|s_0^A|)\right]\right.$$

$$\left. - \frac{ik}{2\alpha}\left[\ln\frac{1+\alpha}{1-\alpha}\right] \mp \frac{i\pi k}{2}\int_0^{k|s_0^A|/\alpha^2} e^{\pm i\xi}H_0^{(2)}(M\xi)d\xi\right\} \qquad (10.239)$$

where $a = kM/\alpha^2$; see Bisplinghoff, Ashley, and Halfman (1955). Also for this limit, the compressed span variable $r^A \sin\Lambda$ should be interpreted as the time variable.

The kernels in the two domains can be obtained from each other by noting that $\beta = iB$. For $\hat{v} = 0$ the kernel reduces to $K^A(s_0^A, 0) = 1/s_0^A$. For this kernel the integral equation inverts directly to give

$$\overline{G}^A(s^A, 0) = -\frac{1}{\pi}\sqrt{\frac{1-s^A}{1+s^A}} \qquad (10.240)$$

The limit $\hat{v} = 0$ corresponds to a wavelength of the sinusoidal gust that is very large compared with the wing chord. The problem reduces to a two-dimensional airfoil in a uniform downwash, and equation 10.240 is the standard result from thin-airfoil theory.

10.8 Blade-Vortex Interaction

The transitional case is $M = \sin \Lambda$, or $\beta = B = 0$. The equation of motion becomes Laplace's equation in two dimensions (z and r'). The limit of either the elliptic or the hyperbolic kernel at $M = \sin \Lambda$ gives the transitional kernel:

$$K_T^A(s_0^A, \hat{v}) = e^{-i\hat{v}s_0^A \alpha} \left\{ \frac{e^{i(\hat{v}/\alpha)s_0^A}}{s_0^A} - \frac{i\hat{v}}{\alpha}\left[\gamma + \frac{i\pi}{2} + \ln(\hat{v}/\alpha)|s_0^A|\right] - \frac{i\hat{v}}{\alpha} \int_0^{(\hat{v}/\alpha)|s_0^A|} \frac{e^{\pm it} - 1}{t} dt \right\}$$
(10.241)

This is the kernel for the integral equation that is the solution of the linear problem, but for most of the transitional domain the appropriate equation of motion is nonlinear. The equation for the first-order potential when $M \cong \sin \Lambda$ has the same form as the equation for three-dimensional, steady, transonic flow, which is exactly the problem in the $\Lambda = 90°$ case ($M \cong 1$). With $\Lambda > 90°$, the flow is not transonic for the transitional case $M = \sin \Lambda$, but there is a velocity that has become sonic: the vector sum of the Mach number and the sweep velocity of the vortex along the wing, $M/\sin \Lambda$ (Figure 10.17). When this combination of physical and geometric velocities is sonic, disturbances produced by the downwash follow it as both are convected along the blade. Thus the aerodynamic problem is nonlinear in an order M^2 region near $M = \sin \Lambda$. The two-dimensional, incompressible problem at $M = 0$ and $\Lambda = 180°$ is a special case for which the linear equations are appropriate. For this case alone the integral equation can be solved in closed form by classical techniques. The loading on a two-dimensional airfoil due to a sinusoidal gust in an incompressible flow is

$$\overline{G}^A(s^A, \hat{v}) = -\frac{1}{\pi}\sqrt{\frac{1 - s^A}{1 + s^A}} S(\hat{v}) \qquad (10.242)$$

where S is the Sears function.

To solve the integral equations, the influence function for the pressure can be written as a Glauert series:

$$\overline{G}^A(s^A, \hat{v}) = \sum_{n=0}^{\infty} \overline{g}_n^A(\hat{v}) f_n(\theta) \qquad (10.243)$$

where $s^A = \cos \theta$ and f_n is defined by equation 10.10. Substituting for this expansion of $\overline{G}^A$ and evaluating the integral equation at a set of collocation points along the chord, a set of algebraic equations is obtained. Given the wave number $\hat{v}$, these algebraic equations can be solved for $\overline{g}_n$. The section lift, moment, and circulation are then obtained by integration over the chord. The resulting loading influence functions are

$$\overline{g}_L(\hat{v}) = \frac{\overline{L}(\hat{v})}{(2\pi/\alpha)\overline{w}(\hat{v})/\sin \Lambda} = \pi\left[\overline{g}_0^A(\hat{v}) + \frac{1}{2}\overline{g}_1^A(\hat{v})\right] \qquad (10.244)$$

$$\overline{g}_M(\hat{v}) = \frac{\overline{M}_{QC}(\hat{v})}{(2\pi/\alpha)\overline{w}(\hat{v})/\sin \Lambda} = -\frac{\pi}{4}\left[\overline{g}_1^A(\hat{v}) + \overline{g}_2^A(\hat{v})\right] \qquad (10.245)$$

$$\overline{g}_C(\hat{v}) = \frac{\overline{\Gamma}(\hat{v})}{(2\pi/\alpha)\overline{w}(\hat{v})/\sin \Lambda} = \pi\left[\overline{g}_0(\hat{v}) + \frac{1}{2}\overline{g}_1(\hat{v})\right]e^{i\hat{v}\cos\Lambda} \qquad (10.246)$$

All these functions are Fourier transforms with respect to r^A, with $w(\hat{v})/\sin \Lambda$ the Fourier transform of the downwash $w(r')$ with respect to r^A, at $s^A = 0$.

An important character of the model problem is that it includes as limits in M and Λ several types of flow about a lifting wing. The elliptical domain includes the steady, three-dimensional loading. The hyperbolic domain includes the unsteady, two-dimensional loading. The transitional region, separating the elliptic and hyperbolic domains, joins transonic, three-dimensional (nonlinear) and incompressible, two-dimensional flows.

10.9 REFERENCES

Anderson, W.D., and Watts, G.A. "Rotor Blade Wake Flutter – A Comparison of Theory and Experiment." Journal of the American Helicopter Society, *21*:2 (April 1976).

Bisplinghoff, R.L., Ashley, H., and Halfman, R.L. *Aeroelasticity*. Reading, MA: Addison-Wesley Publishing Company, Inc., 1955.

Cicala, P. "Present State of Development in Nonsteady Motion of a Lifting Surface." NACA TM 1277, 1951.

Daughaday, H., DuWaldt, F., and Gates, C. "Investigation of Helicopter Blade Flutter and Load Amplification Problems." Journal of the American Helicopter Society, *2*:3 (July 1957).

Daughaday, H., and Piziali, R.A. "An Improved Computational Model for Predicting the Unsteady Aerodynamic Loads of Rotor Blades." Journal of the American Helicopter Society, *11*:4 (October 1966).

Ham, N.D., Moser, H.H., and Zvara, J. "Investigation of Rotor Response to Vibratory Aerodynamic Inputs. Part I. Experimental Results and Correlation with Theory." WADC TR 58-87, Part I., October 1958.

Hariharan, N., and Leishman, J.G. "Unsteady Aerodynamics of a Flapped Airfoil in Subsonic Flow by Indicial Concepts." Journal of Aircraft, *33*:5 (September-October 1996).

Johnson, W. "A Lifting Surface Solution for Vortex Induced Airloads and Its Application to Rotary Wing Airloads Calculations." NASA CR 192328, April 1970.

Johnson, W. "A Lifting-Surface Solution for Vortex-Induced Airloads." AIAA Journal, *9*:4 (April 1971).

Johnson, W. "Rotorcraft Aerodynamic Models for a Comprehensive Analysis." American Helicopter Society 54th Annual Forum, Washington, DC, May 1998.

Jones, J.P. "The Influence of the Wake on the Flutter and Vibration of Rotor Blades." The Aeronautical Quarterly, *9*:Part 3 (August 1958).

Küssner, H.G., and Schwarz, L. "The Oscillating Wing with Aerodynamically Balanced Elevator." NACA TM 991, October 1941.

Leishman, J.G. "Validation of Approximate Indicial Aerodynamic Functions for Two-Dimensional Subsonic Flow." Journal of Aircraft, *25*:10 (October 1988).

Leishman, J.G., and Beddoes, T.S. "A Semi-Empirical Model for Dynamic Stall." Journal of the American Helicopter Society, *34*:3 (July 1989).

Leishman, J.G., and Nguyen, K.Q. "State-Space Representation of Unsteady Airfoil Behavior." AIAA Journal, *28*:5 (May 1990).

Loewy, R.G. "A Two-Dimensional Approximation to the Unsteady Aerodynamics of Rotary Wings." Journal of the Aeronautical Sciences, *24*:2 (February 1957).

Miller, R.H. "Unsteady Air Loads on Helicopter Rotor Blades." Journal of the Royal Aeronautical Society, *68*:640 (April 1964).

Petot, D. "Differential Equation Modeling of Dynamic Stall." La Recherche Aerospatiale, Number 1989-5 (corrections dated October 1990).

Piziali, R.A. "Method for the Solution of the Aeroelastic Response Problem for Rotating Wings." Journal of Sound and Vibration, *4*:3 (1966).

Sears, W.R. "Some Aspects of Non-Stationary Airfoil Theory and Its Practical Application." Journal of the Aeronautical Sciences, *8*:3 (January 1941).

Silveira, M.A., and Brooks, G.W. "Dynamic-Model Investigation of the Damping of Flapwise Bending Modes of Two-Blade Rotors in Hovering and a Comparison with QuasiStatic and Unsteady Aerodynamic Theories." NASA TN D-175, December 1959.

Theodorsen, T. "General Theory of Aerodynamic Instability and the Mechanism of Flutter." NACA Report 496, 1935.

Theodorsen, T., and Garrick, I.E. "Nonstationary Flow About a Wing-Aileron-Tab Combination Including Aerodynamic Balance." NACA Report 736, 1942.

Timman, R., and van de Vooren, A.I. "Flutter of a Helicopter Rotor Rotating in Its Own Wake." Journal of the Aeronautical Sciences, 24:9 (September 1957).

van der Vooren, A.I. "The Theodorsen Circulation Function and Aerodynamic Coefficients." AGARD Manual on Aeroelasticity, Volume VI, January 1964.

11 Actuator Disk

The analysis of the wake is considerably simplified if the rotor is modeled as an actuator disk, which is a circular surface of zero thickness that can support a pressure difference and thus accelerate the air through the disk. The actuator disk neglects the discreteness in the rotor and wake associated with a finite number of blades, and it distributes the vorticity throughout the wake volume. The actuator disk model is the basis for momentum theory (sections 3.1.1 and 5.1.1). The simplest version of vortex theory uses an actuator disk model, which produces a tractable mathematical problem, at least for axial flight (section 3.7). In contrast to hover, the mathematical problem in forward flight is still not trivial, because of the skewed cylindrical geometry (section 5.2). Some results from actuator disk models were presented in section 5.2.1.

The focus of this chapter is the unsteady aerodynamics of the rotor associated with the three-dimensional wake. In particular, the dynamic inflow model is developed. This is a finite-state model, relating a set of inflow variables and loading variables by differential equations. Such a model is required for aeroelastic stability calculations and real time simulation. Vortex theory uses the Biot-Savart law for the velocity induced by the wake vorticity. Potential theory solves the fluid dynamic equations for the velocity potential or acceleration potential.

11.1 Vortex Theory

For the actuator disk in axial flow, the wake is a right circular cylinder (Figure 11.1). With uniform loading, the bound circulation is constant over the span, and the trailed vorticity is concentrated in root and tip vortices. The tip vortex spirals form a continuous distribution of ring vortices on the surface of this cylinder. Continuity of vorticity requires a root vortex on the axis of the wake and axial vorticity on the surface of the cylinder. Such axial vorticity does not contribute to the axial induced velocity at the rotor disk. Thus the wake consists of radial vorticity on the actuator disk (the bound circulation), a root vortex on the axis of symmetry, a surface distribution of axial vorticity on the cylinder, and a vortex tube of ring vortices. With nonuniform loading, there is trailed and shed vorticity within the wake cylinder due to radial and azimuthal variation of the bound circulation, respectively.

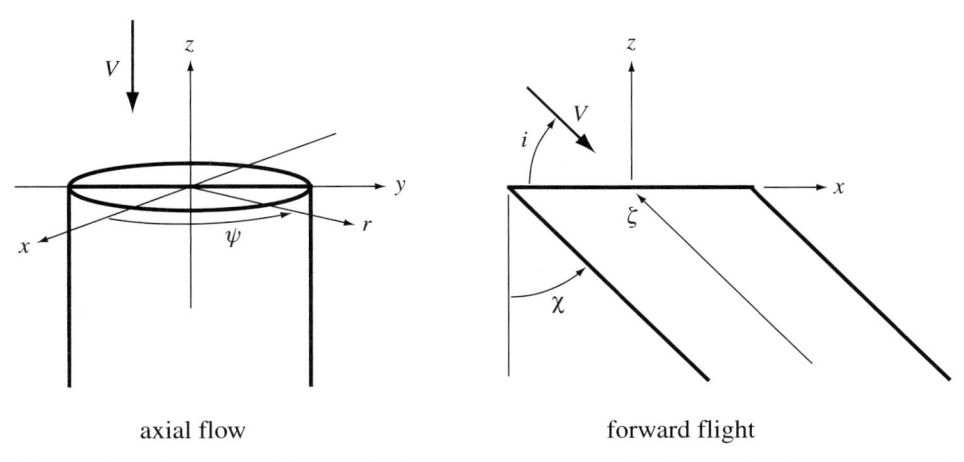

Figure 11.1. Actuator disk models for vortex theory. Axial flow wake is a right circular cylinder; forward flight wake is a skewed cylinder.

The velocity produced by a distribution of vorticity ω in incompressible, invisid flow is

$$\mathbf{u}(\mathbf{x}) = -\frac{1}{4\pi} \int \frac{\mathbf{s} \times \omega}{s^3} dV(\mathbf{x}') \tag{11.1}$$

where $\mathbf{s} = \mathbf{x} - \mathbf{x}'$. The velocity can also be obtained from the vector potential $\mathbf{B}$:

$$\mathbf{B}(\mathbf{x}) = \frac{1}{4\pi} \int \frac{\omega}{s} dV(\mathbf{x}') \tag{11.2}$$

such that $\nabla^2 \mathbf{B} = -\omega$, $\mathbf{u} = \nabla \times \mathbf{B}$. In two-dimensional or axisymmetric flow, the vector potential is the stream function. The velocity produced by a line vortex with circulation κ is

$$\mathbf{u}(\mathbf{x}) = -\frac{1}{4\pi} \int \frac{\kappa \, \mathbf{s} \times d\ell(\mathbf{x}')}{s^3} \tag{11.3}$$

$$\mathbf{B}(\mathbf{x}) = \frac{1}{4\pi} \int \frac{\kappa}{s} d\ell(\mathbf{x}') \tag{11.4}$$

substituting $\int \omega dV = \kappa d\ell$ ($d\ell$ is the tangent to the vortex at $\mathbf{x}'$). This result can also be written $\mathbf{u} = -(\kappa/4\pi)\nabla\Sigma$, where

$$\Sigma(\mathbf{x}) = \int \frac{\mathbf{s} \cdot \mathbf{n}}{s^3} dA(\mathbf{x}') \tag{11.5}$$

is the solid angle subtended by the line vortex at point $\mathbf{x}$ ($\mathbf{n}dA$ is the element of area on the surface bounded by the line vortex).

Consider a rotor with N blades, each with bound circulation Γ. The vertical convection of the wake is $(V + v)$, where V is the climb velocity of the rotor and v is the induced velocity at the disk. The wake is convected a distance $Nh = \frac{V+v}{\Omega/2\pi}$ in one revolution, where h is the distance between wake sheets. The wake pitch angle is $\phi = \tan^{-1}\frac{V+v}{\Omega r}$. For uniform loading (constant Γ), the strength of the circumferential or ring vorticity on the wake surface is $\gamma = \Gamma/h$, so the induced velocity at the disk is

$$v = \frac{1}{2}\gamma = \frac{1}{2}\frac{\Gamma}{h} = \frac{1}{2}\frac{N\Gamma}{2\pi/\Omega}\frac{1}{V+v} = \frac{T}{2\rho A(V+v)} \tag{11.6}$$

which is the momentum theory result; see section 3.7.2. This result is based on the assumptions of no contraction (a right circular cylinder for the wake); wake spacing from the velocity at the disk $(V + v)$ (which can be interpreted as the velocity at the edge of the far wake, $(V + \frac{w}{2})$); and vorticity strength γ independent of axial distance z. In fact, from mass conservation the wake free surface must contract, and the convection velocity and vorticity strength must depend on z. The cylindrical surface of wake vorticity in this model is not a streamline; rather there must be flow through the wake surface; see Castles (1957) for examples. For hover, half the flow in the far wake comes through the disk, and half through the wake cylinder.

For nonuniform loading, the vorticity within the wake cylinder is

$$\omega \, dV = \left(\gamma_s \mathbf{e}_r - \gamma_t (\mathbf{e}_\psi + \mathbf{e}_z \tan\phi)\right) r \, d\psi \, dr \, dz \tag{11.7}$$

where (r, ψ, z) are cylindrical coordinates (Figure 11.1), with unit vectors $\mathbf{e}_r, \mathbf{e}_\psi, \mathbf{e}_z$. The trailed, shed, and axial vorticity strengths are obtained by spreading the wake sheets over the vertical distance h:

$$\gamma_t = -\frac{1}{h} \frac{\partial \Gamma}{\partial r} \tag{11.8}$$

$$\gamma_s = -\frac{1}{hr} \frac{\partial \Gamma}{\partial \psi} \tag{11.9}$$

$$\gamma_z = \gamma_t \tan\phi = -\frac{\tan\phi}{h} \frac{\partial \Gamma}{\partial r} = -\frac{N}{2\pi r} \frac{\partial \Gamma}{\partial r} \tag{11.10}$$

For this general case,

$$\mathbf{s} = (r\cos\psi - r'\cos\psi')\mathbf{e}_x + (r\sin\psi - r'\sin\psi')\mathbf{e}_y + (z - z')\mathbf{e}_z \tag{11.11}$$

$$s^2 = r^2 + r'^2 - 2rr'C + (z-z')^2 = (r' - rC)^2 + (rS)^2 + (z-z')^2 \tag{11.12}$$

$$\omega = \gamma_s \mathbf{e}'_r - \gamma_t \mathbf{e}'_\psi - \gamma_z \mathbf{e}'_z$$
$$= (\gamma_t \sin\psi' + \gamma_s \cos\psi')\mathbf{e}_x + (-\gamma_t \cos\psi' + \gamma_s \sin\psi')\mathbf{e}_y - \gamma_z \mathbf{e}_z$$
$$= (-\gamma_t S + \gamma_s C)\mathbf{e}_r + (-\gamma_t C - \gamma_s S)\mathbf{e}_\psi - \gamma_z \mathbf{e}_z \tag{11.13}$$

where $C = \cos(\psi - \psi')$ and $S = \sin(\psi - \psi')$. The transformation between Cartesian and cylindrical coordinates gives $\mathbf{e}_x = \mathbf{e}_r \cos\psi - \mathbf{e}_\psi \sin\psi$ and $\mathbf{e}_y = \mathbf{e}_r \sin\psi + \mathbf{e}_\psi \cos\psi$. Then the velocity at an arbitrary point is

$$u_r = -\frac{1}{4\pi} \int_{-\infty}^{0} \int_{0}^{2\pi} \int_{0}^{R} \frac{1}{s^3}\left[(z-z')(\gamma_t C + \gamma_s S) - \gamma_z r'S\right] r' \, dr' \, d\psi' \, dz' \tag{11.14}$$

$$u_\psi = -\frac{1}{4\pi} \int_{-\infty}^{0} \int_{0}^{2\pi} \int_{0}^{R} \frac{1}{s^3}\left[(z-z')(-\gamma_t S + \gamma_s C) + \gamma_z(r - r'C)\right] r' \, dr' \, d\psi' \, dz' \tag{11.15}$$

$$u_z = -\frac{1}{4\pi} \int_{-\infty}^{0} \int_{0}^{2\pi} \int_{0}^{R} \frac{1}{s^3}\left[\gamma_t(r' - rC) - \gamma_s rS\right] r' \, dr' \, d\psi' \, dz' \tag{11.16}$$

$$\mathbf{B} = \frac{1}{4\pi} \int_{-\infty}^{0} \int_{0}^{2\pi} \int_{0}^{R} \frac{1}{s}\left[(-\gamma_t S + \gamma_s C)\mathbf{e}_r + (-\gamma_t C - \gamma_s S)\mathbf{e}_\psi - \gamma_z \mathbf{e}_z\right] r' \, dr' \, d\psi' \, dz' \tag{11.17}$$

11.1 Vortex Theory

The shed wake terms (γ_s) are zero if the bound circulation Γ is constant. The terms with the factor $S = \sin(\psi - \psi')$ are zero if γ does not depend on azimuth, since s is symmetric in $(\psi - \psi')$. Analytical results can be obtained for a number of special cases.

Consider the velocity produced by the circumferential or ring vorticity, for γ_t independent of azimuth. With the change of variables $\psi - \psi' = 2\theta - \pi$:

$$s^2 = (z-z')^2 + (r+r')^2 - 2rr'(1+\cos(\psi-\psi')) = A^2(1-k^2\sin^2\theta) \quad (11.18)$$

where $A^2 = (z-z')^2 + (r+r')^2$ and $k^2 = 4rr'/A^2$. Then the axial velocity, radial velocity, and stream function are

$$u_z = -\frac{1}{4\pi}\int_{-\infty}^{0}\int_{0}^{R}\left\{\int_{0}^{2\pi}\frac{r'-rC}{s^3}d\psi'\right\}\gamma_t r'\,dr'\,dz'$$

$$= -\frac{1}{2\pi}\int_{-\infty}^{0}\int_{0}^{R}\left\{\frac{2r'}{A^3}\int_{0}^{\pi/2}\frac{(r+r')-2r\sin^2\theta}{(1-k^2\sin^2\theta)^{3/2}}d\theta\right\}\gamma_t\,dr'\,dz'$$

$$= -\frac{1}{2\pi}\int_{-\infty}^{0}\int_{0}^{R}\frac{1}{Ar}\left\{Kr - E\frac{r(1-k^2/2)-r'k^2/2}{1-k^2}\right\}\gamma_t\,dr'\,dz' \quad (11.19)$$

$$u_r = -\frac{1}{4\pi}\int_{-\infty}^{0}\int_{0}^{R}\left\{\int_{0}^{2\pi}\frac{(z-z')C}{s^3}d\psi'\right\}\gamma_t r'\,dr'\,dz'$$

$$= -\frac{1}{2\pi}\int_{-\infty}^{0}\int_{0}^{R}\left\{\frac{2r'(z-z')}{A^3}\int_{0}^{\pi/2}\frac{2\sin^2\theta-1}{(1-k^2\sin^2\theta)^{3/2}}d\theta\right\}\gamma_t\,dr'\,dz'$$

$$= \frac{1}{2\pi}\int_{-\infty}^{0}\int_{0}^{R}\frac{z-z'}{Ar}\left\{K - E\frac{1-k^2/2}{1-k^2}\right\}\gamma_t\,dr'\,dz' \quad (11.20)$$

$$rB_\psi = -\frac{1}{4\pi}\int_{-\infty}^{0}\int_{0}^{R}\left\{\int_{0}^{2\pi}\frac{C}{s}d\psi'\right\}\gamma_t r'\,dr'\,dz'$$

$$= -\frac{1}{2\pi}\int_{-\infty}^{0}\int_{0}^{R}\left\{\frac{2rr'}{A}\int_{0}^{\pi/2}\frac{2\sin^2\theta-1}{(1-k^2\sin^2\theta)^{1/2}}d\theta\right\}\gamma_t\,dr'\,dz'$$

$$= -\frac{1}{2\pi}\int_{-\infty}^{0}\int_{0}^{R}A\left\{K(1-k^2/2)-E\right\}\gamma_t\,dr'\,dz' \quad (11.21)$$

where K and E are complete elliptic integrals:

$$K(k) = \int_{0}^{\pi/2}\frac{1}{\sqrt{1-k^2\sin^2\theta}}d\theta \quad (11.22)$$

$$E(k) = \int_{0}^{\pi/2}\sqrt{1-k^2\sin^2\theta}\,d\theta \quad (11.23)$$

It is also possible to transform the integration over azimuth into integrals of Bessel functions.

For the vorticity γ independent of axial distance, the integration over z' can be evaluated. Write $a^2 = r^2 + r'^2 - 2rr'C$, $s^2 = (z - z')^2 + a^2$. Then

$$u_r = -\frac{1}{4\pi} \int_0^{2\pi} \int_0^R \left\{ (\gamma_t C + \gamma_s S) \frac{1}{\sqrt{z^2 + a^2}} - \gamma_z r' S \frac{1}{a^2} \left[1 - \frac{z}{\sqrt{z^2 + a^2}} \right] \right\} r' \, dr' \, d\psi' \quad (11.24)$$

$$u_\psi = -\frac{1}{4\pi} \int_0^{2\pi} \int_0^R \left\{ (-\gamma_t S + \gamma_s C) \frac{1}{\sqrt{z^2 + a^2}} \right.$$
$$\left. + \gamma_z (r - r'C) \frac{1}{a^2} \left[1 - \frac{z}{\sqrt{z^2 + a^2}} \right] \right\} r' \, dr' \, d\psi' \quad (11.25)$$

$$u_z = -\frac{1}{4\pi} \int_0^{2\pi} \int_0^R \left\{ (\gamma_t (r' - rC) - \gamma_s rS) \frac{1}{a^2} \left[1 - \frac{z}{\sqrt{z^2 + a^2}} \right] \right\} r' \, dr' \, d\psi' \quad (11.26)$$

On the axis ($r = 0$, so $a = r'$), this becomes

$$u_r = -\frac{1}{4\pi} \int_0^{2\pi} \int_0^R \left\{ (\gamma_t C + \gamma_s S) \frac{r'}{\sqrt{z^2 + r'^2}} - \gamma_z S \left[1 - \frac{z}{\sqrt{z^2 + r'^2}} \right] \right\} dr' \, d\psi' \quad (11.27)$$

$$u_\psi = -\frac{1}{4\pi} \int_0^{2\pi} \int_0^R \left\{ (-\gamma_t S + \gamma_s C) \frac{r'}{\sqrt{z^2 + r'^2}} - \gamma_z C \left[1 - \frac{z}{\sqrt{z^2 + r'^2}} \right] \right\} dr' \, d\psi' \quad (11.28)$$

$$u_z = -\frac{1}{4\pi} \int_0^{2\pi} \int_0^R \left\{ \gamma_t \left[1 - \frac{z}{\sqrt{z^2 + r'^2}} \right] \right\} dr' \, d\psi'$$
$$= -\frac{1}{2} \int_0^R \overline{\gamma}_t \left[1 - \frac{z}{\sqrt{z^2 + r'^2}} \right] dr' \quad (11.29)$$

using $\overline{\gamma}_t = \frac{1}{2\pi} \int_0^{2\pi} \gamma_t d\psi$, since for the axial velocity the azimuth integration acts only on the vorticity strength. If the bound circulation is constant radially, $\gamma_t = -\frac{1}{h} \frac{\partial \Gamma}{\partial r} = \frac{\Gamma}{h} \delta(R)$ gives

$$u_z = -\frac{1}{2} \frac{\Gamma}{h} \left[1 - \frac{z}{\sqrt{z^2 + R^2}} \right] \quad (11.30)$$

See also section 3.7.2; here u_z and z are positive upward. At the disk ($z = 0$ and $r = 0$), the downwash is again $v = -u_z = \frac{1}{2} \frac{\Gamma}{h} = \frac{T}{2\rho A(V+v)}$. At the disk ($z = 0$) for arbitrary radial station,

$$u_r = -\frac{1}{4\pi} \int_0^{2\pi} \int_0^R \left\{ (\gamma_t C + \gamma_s S) \frac{1}{a} - \gamma_z \frac{r' S}{a^2} \right\} r' \, dr' \, d\psi' \quad (11.31)$$

$$u_\psi = -\frac{1}{4\pi} \int_0^{2\pi} \int_0^R \left\{ (-\gamma_t S + \gamma_s C) \frac{1}{a} + \gamma_z \frac{r - r'C}{a^2} \right\} r' \, dr' \, d\psi' \quad (11.32)$$

$$u_z = -\frac{1}{4\pi} \int_0^{2\pi} \int_0^R \left\{ (\gamma_t (r' - rC) - \gamma_s rS) \frac{1}{a^2} \right\} r' \, dr' \, d\psi' \quad (11.33)$$

11.1 Vortex Theory

If the bound circulation is constant radially (γ_t only at the tip) and azimuthally ($\gamma_s = 0$),

$$u_z = -\frac{1}{4\pi} \int_0^{2\pi} \int_0^R \gamma_t \frac{r' - rC}{a^2} r' \, dr' \, d\psi'$$

$$= -\frac{1}{2}\frac{\Gamma}{h} \frac{1}{2\pi} \int_0^{2\pi} \frac{R^2 - rRC}{r^2 + R^2 - 2rRC} \, d\psi'$$

$$= -\frac{1}{2}\frac{\Gamma}{h} \frac{1}{2\pi} \int_0^{2\pi} \sum_{m=0}^{\infty} \left(\frac{r}{R}\right)^m \cos m(\psi - \psi') \, d\psi'$$

$$= -\frac{1}{2}\frac{\Gamma}{h} \tag{11.34}$$

So for uniform loading (constant bound circulation), the downwash is uniform over the disk. Using the expansion

$$S = \frac{r' - rC + irS}{r^2 + r'^2 - 2rr'C} = \frac{r' - rC + irS}{(r' - rC)^2 + (rS)^2} = \frac{1}{r' - rC - irS} = \frac{1}{r' - re^{i(\psi - \psi')}}$$

$$= \frac{1}{r'}\frac{1}{1 - (r/r')e^{i(\psi - \psi')}} = \frac{1}{r'}\sum_{m=0}^{\infty} \left(\frac{r}{r'}\right)^m e^{im(\psi - \psi')} \tag{11.35}$$

for $r' > r$, and

$$S = -\frac{1}{re^{i(\psi - \psi')}} \frac{1}{1 - (r'/r)e^{-i(\psi - \psi')}} = -\frac{1}{re^{i(\psi - \psi')}} \sum_{m=0}^{\infty} \left(\frac{r'}{r}\right)^m e^{-im(\psi - \psi')}$$

$$= -\frac{1}{r'}\sum_{m=1}^{\infty} \left(\frac{r'}{r}\right)^m e^{-im(\psi - \psi')} \tag{11.36}$$

for $r' < r$, it follows that

$$\frac{1}{2\pi} \int_0^{2\pi} \frac{r'(r' - rC)}{r^2 + r'^2 - 2rr'C} \, d\psi'$$

$$= \begin{cases} \frac{1}{2\pi} \int_0^{2\pi} \sum_{m=0}^{\infty} \left(\frac{r}{r'}\right)^m \cos m(\psi - \psi') \, d\psi' = 1 & r' > r \\ \frac{1}{2\pi} \int_0^{2\pi} -\sum_{m=1}^{\infty} \left(\frac{r'}{r}\right)^m \cos m(\psi - \psi') \, d\psi' = 0 & r' < r \end{cases} \tag{11.37}$$

So with a radial variation of bound circulation,

$$u_z = -\frac{1}{2} \int_0^R \gamma_t \left\{ \frac{1}{2\pi} \int_0^{2\pi} \frac{r'(r' - rC)}{r^2 + r'^2 - 2rr'C} \, d\psi' \right\} dr' = -\frac{1}{2} \int_r^R \gamma_t \, dr'$$

$$= \frac{1}{2h} \int_r^R \frac{\partial \Gamma}{\partial r'} \, dr' = -\frac{1}{2}\frac{\Gamma(r)}{h} \tag{11.38}$$

Next consider harmonic variation of the bound circulation: $\Gamma_n = \Gamma_c \cos n\psi' + \Gamma_s \sin n\psi'$. Because the circulation is periodic, the sheets of wake vorticity are in phase, and γ can be taken as independent of z. With no radial variation, $\gamma_t = \frac{\Gamma_n}{h}\delta(R)$ and $\gamma_s = -\frac{1}{hr'}\frac{d\Gamma_n}{d\psi'}$. The axial velocity at the disk is

$$u_z = -\frac{1}{4\pi}\int_0^{2\pi}\left\{\int_0^R \gamma_t \frac{r'-rC}{r^2+r'^2-2rr'C}r'\,dr' - \int_0^R \gamma_s \frac{rS}{r^2+r'^2-2rr'C}r'\,dr'\right\}d\psi'$$

$$= -\frac{1}{4\pi}\int_0^{2\pi}\left\{\frac{\Gamma_n}{h}\frac{R^2-rRC}{r^2+R^2-2rRC} + \frac{1}{h}\frac{d\Gamma_n}{d\psi'}\int_0^R \frac{rS}{r^2+r'^2-2rr'C}dr'\right\}d\psi'$$

$$= -\frac{1}{4\pi}\int_0^{2\pi}\left\{\frac{\Gamma_n}{h}\sum_{m=0}^{\infty}\left(\frac{r}{R}\right)^m \cos m(\psi-\psi')\right.$$

$$\left. +\frac{1}{h}\frac{d\Gamma_n}{d\psi'}\int_0^r \frac{1}{r'}\sum_{m=1}^{\infty}\left(\frac{r'}{r}\right)^m \sin m(\psi-\psi')\,dr'\right.$$

$$\left. +\frac{1}{h}\frac{d\Gamma_n}{d\psi'}\int_r^R \frac{1}{r'}\sum_{m=0}^{\infty}\left(\frac{r}{r'}\right)^m \sin m(\psi-\psi')\,dr'\right\}d\psi'$$

$$= -\frac{1}{2}\left\{\frac{\Gamma_n}{h}\frac{1}{2}\left(\frac{r}{R}\right)^n + \frac{\Gamma_n}{h}\frac{n}{2}\left[\int_0^r \frac{1}{r'}\left(\frac{r'}{r}\right)^n dr' + \int_r^R \frac{1}{r'}\left(\frac{r}{r'}\right)^n dr'\right]\right\}$$

$$= -\frac{1}{2}\left\{\frac{\Gamma_n}{h}\frac{1}{2}\left(\frac{r}{R}\right)^n + \frac{\Gamma_n}{h}\frac{1}{2}\left[2-\left(\frac{r}{R}\right)^n\right]\right\}$$

$$= -\frac{1}{2}\frac{\Gamma_n(\psi)}{h} \tag{11.39}$$

using

$$\frac{1}{2\pi}\int_0^{2\pi}\Gamma_n(\psi')\cos m(\psi-\psi')\,d\psi' = \frac{1}{2}\Gamma_n(\psi) \tag{11.40}$$

$$\frac{1}{2\pi}\int_0^{2\pi}\frac{d}{d\psi'}\Gamma_n(\psi')\sin m(\psi-\psi')\,d\psi' = \frac{n}{2}\Gamma_n(\psi) \tag{11.41}$$

Finally, for both radial and harmonic variation of the bound circulation, $\gamma_t = -\frac{1}{h}\frac{\partial\Gamma}{\partial r'}$ and $\gamma_s = -\frac{1}{hr'}\frac{\partial\Gamma}{\partial\psi'}$, and

$$u_z = -\frac{1}{4\pi}\int_0^{2\pi}\left\{\int_0^R \gamma_t \frac{r'-rC}{r^2+r'^2-2rr'C}r'\,dr' - \int_0^R \gamma_s \frac{rS}{r^2+r'^2-2rr'C}r'\,dr'\right\}d\psi'$$

$$= -\frac{1}{4\pi}\int_0^{2\pi}\left\{-\int_0^R \frac{1}{h}\frac{\partial\Gamma}{\partial r'}\frac{r'-rC}{r^2+r'^2-2rr'C}r'\,dr'\right.$$

$$\left. +\int_0^R \frac{1}{h}\frac{\partial\Gamma}{\partial\psi'}\frac{rS}{r^2+r'^2-2rr'C}dr'\right\}d\psi'$$

$$= -\frac{1}{4\pi} \int_0^{2\pi} \left\{ \int_0^r \frac{1}{h} \frac{\partial \Gamma}{\partial r'} \sum_{m=1}^{\infty} \left(\frac{r'}{r}\right)^m \cos m(\psi - \psi') dr' \right.$$

$$- \int_r^R \frac{1}{h} \frac{\partial \Gamma}{\partial r'} \sum_{m=0}^{\infty} \left(\frac{r}{r'}\right)^m \cos m(\psi - \psi') dr'$$

$$+ \int_0^r \frac{1}{hr'} \frac{\partial \Gamma}{\partial \psi'} \sum_{m=1}^{\infty} \left(\frac{r'}{r}\right)^m \sin m(\psi - \psi') dr'$$

$$\left. + \int_r^R \frac{1}{hr'} \frac{\partial \Gamma}{\partial \psi'} \sum_{m=0}^{\infty} \left(\frac{r}{r'}\right)^m \sin m(\psi - \psi') dr' \right\} d\psi'$$

$$= -\frac{1}{2} \left\{ \int_0^r \frac{1}{h} \frac{\partial \Gamma}{\partial r'} \frac{1}{2} \left(\frac{r'}{r}\right)^n dr' - \int_r^R \frac{1}{h} \frac{\partial \Gamma}{\partial r'} \frac{1}{2} \left(\frac{r}{r'}\right)^n dr' \right.$$

$$\left. + \int_0^r \frac{\Gamma}{hr'} \frac{n}{2} \left(\frac{r'}{r}\right)^n dr' + \int_r^R \frac{\Gamma}{hr'} \frac{n}{2} \left(\frac{r}{r'}\right)^n dr' \right\}$$

$$= -\frac{1}{2} \left\{ \frac{\Gamma}{h} \frac{1}{2} \left(\frac{r'}{r}\right)^n \Big|_0^r - \frac{\Gamma}{h} \frac{1}{2} \left(\frac{r}{r'}\right)^n \Big|_r^R - \int_0^r \frac{\Gamma}{h} \frac{n}{2r'} \left(\frac{r'}{r}\right)^n dr' - \int_r^R \frac{\Gamma}{h} \frac{n}{2r'} \left(\frac{r}{r'}\right)^n dr' \right.$$

$$\left. + \int_0^r \frac{\Gamma}{hr'} \frac{n}{2} \left(\frac{r'}{r}\right)^n dr' + \int_r^R \frac{\Gamma}{hr'} \frac{n}{2} \left(\frac{r}{r'}\right)^n dr' \right\}$$

$$= -\frac{1}{2} \left\{ \frac{\Gamma}{h} \frac{1}{2} \left(\frac{r'}{r}\right)^n \Big|_0^r - \frac{\Gamma}{h} \frac{1}{2} \left(\frac{r}{r'}\right)^n \Big|_r^R \right\}$$

$$= -\frac{1}{2} \frac{\Gamma(r, \psi)}{h} \tag{11.42}$$

It is remarkable that $u_z = -\Gamma/2h$ is obtained for both uniform and nonuniform loading. These results were derived by Miller (1964).

In summary, for uniform loading, the vortex theory result for the induced velocity at the disk of a rotor in axial flight is uniform downwash:

$$v = -u_z = \frac{1}{2} \frac{\Gamma}{h} = \frac{1}{2} \frac{N\Gamma}{2\pi/\Omega} \frac{1}{V+v} = \frac{T}{2\rho A(V+v)} \tag{11.43}$$

using $(V + v)$ for the vertical convection h. For nonuniform loading, vortex theory gives

$$v_n(r, \psi) = -u_z = \frac{1}{2} \frac{\Gamma(r, \psi)}{h} = \frac{1}{2} \frac{N\Gamma_n}{2\pi/\Omega} \frac{1}{V+v} = \frac{dT_n(r, \psi)/dA}{2\rho(V+v)} \tag{11.44}$$

again using $h = \frac{V+v}{N\Omega/2\pi}$. This is the perturbation inflow v_n due to the perturbation loading dT_n/dA, using the mean axial velocity $(V + v)$ to define the vorticity distribution in the wake cylinder. Equation 11.44 is not the same as differential momentum theory, which obtains the total induced velocity $v(r, \psi)$ from the total local loading $dT(r, \psi)/dA$: $2\rho(V + v)v = dT/dA$. Using the local convection h for each wake cylinder at r leads to equation 3.157.

The induced velocity distribution can be evaluated now for a number of loading distributions. Consider steady loading with bound circulation distribution $\Gamma = \Gamma_0 g(r)$ (here the radial coordinate r is dimensionless). Using $T = \rho \Omega N R^2 \int \Gamma r \, dr$, the

induced velocity is related to the thrust:
$$v = \frac{1}{2}\frac{\Gamma}{h} = \frac{T}{2\rho A(V+v)}\frac{g}{2\int_0^1 gr\,dr} \tag{11.45}$$

Thus
$$v = \frac{T}{2\rho A(V+v)} \qquad \text{for constant } \Gamma, g = 1 \tag{11.46}$$

$$v = \frac{T}{2\rho A(V+v)}\frac{3}{2}\sqrt{1-r^2} \qquad \text{for } g = \sqrt{1-r^2} \tag{11.47}$$

$$v = \frac{T}{2\rho A(V+v)}\frac{15}{4}r^2\sqrt{1-r^2} \qquad \text{for } g = r^2\sqrt{1-r^2} \tag{11.48}$$

For 1/rev loading distribution, $\Gamma = (\Gamma_c \cos\psi + \Gamma_s \sin\psi)f(r)$, and the moment on the disk is $M_x = \rho\Omega N R^3 \frac{1}{2\pi}\int\int \Gamma(r\sin\psi)r\,dr\,d\psi$, giving
$$v = \frac{1}{2}\frac{\Gamma}{h} = \frac{-M_y\cos\psi + M_x\sin\psi}{\rho AR(V+v)}\frac{f}{2\int_0^1 fr^2\,dr} \tag{11.49}$$

Thus
$$v = \frac{-M_y\cos\psi + M_x\sin\psi}{\rho AR(V+v)}2r \qquad \text{for } f = r \tag{11.50}$$

$$v = \frac{-M_y\cos\psi + M_x\sin\psi}{\rho AR(V+v)}\frac{15}{4}r\sqrt{1-r^2} \qquad \text{for } f = r\sqrt{1-r^2} \tag{11.51}$$

These vortex theory results can be interpreted as a lift deficiency function. The n-th harmonic of the loading produces the n-th harmonic of the induced velocity, and bound circulation constant spanwise produces an induced velocity independent of r. The rotor thrust can be written $T_n = T_{nQ} - T_{nW}$, where T_{nQ} is the quasistatic thrust and T_{nW} is the thrust due to the harmonic inflow:
$$T_{nW} = -N\int_0^R \frac{1}{2}\rho(\Omega r)c\,2\pi v_n\,dr = -\frac{1}{2}N\rho A\Omega c v_n = -\frac{1}{4}N\rho A\Omega c\frac{\Gamma_n}{h} = -\frac{\pi}{h/b}T_n \tag{11.52}$$
where $b = c/2$. So $T_n = T_{nQ} - \frac{\pi}{h/b}T_n = C'T_{nQ}$, with the lift deficiency function
$$C' = \frac{1}{1 + \frac{\pi}{h/b}} = \frac{1}{1 + \frac{\pi\sigma}{4\lambda_0}} \tag{11.53}$$

for the rotary wing. Note that C' is independent of the harmonic number. The wake spacing is given by $h/b = \frac{V+v}{Nb(\Omega/2\pi)} = \frac{4\lambda_0}{\sigma}$, with λ_0 the mean flow through the disk. Remarkably, this lift deficiency function is identical to the function obtained using a two-dimensional, continuous wake model (equation 10.207) and is the low-frequency approximation to Loewy's function for harmonic loading (equations 10.196 and 10.198). For nonuniform loading (bound circulation varying radially as well as azimuthally), vortex theory gives nonuniform induced velocity that depends only on the local load. The total section lift is $L_n = L_{nQ} - L_{nW} = L_{nQ} - \frac{1}{2}\rho(\Omega r)c2\pi v_n = L_{nQ} - \frac{\pi\sigma}{4\lambda_0}L_n$, or $L_n = C'L_{nQ}$ with

$$C' = \frac{1}{1 + \frac{\pi\sigma}{4\lambda_0}} \tag{11.54}$$

11.1 Vortex Theory

again. Since the lift deficiency function is independent of r as well as of frequency, the loading can be integrated over the span to obtain $T_n = C' T_{nQ}$, as for the case of constant bound circulation.

The vortex theory model for the wake in forward flight is a skewed cylinder (figure 11.1); see Coleman, Feingold, and Stempin (1945) and Drees (1949). The wake skew angle χ is obtained from the velocity components at the disk:

$$\tan \chi = \frac{V \cos i}{V \sin i + v} = \frac{\mu}{\lambda} \tag{11.55}$$

So $\chi = 0$ in climb ($i = 90°$), and approaches $\chi = 90°$ in high-speed forward flight ($i \cong 0$). The convection along the wake axis gives $h = \frac{U}{N\Omega/2\pi}$, using the resultant velocity $U^2 = (V \cos i)^2 + (V \sin i + v)^2$.

For nonuniform loading, the vorticity within the wake cylinder is now

$$\omega \, dV = \left(\gamma_s \mathbf{e}_r - \gamma_t \mathbf{e}_\psi - \gamma_z (\mathbf{e}_z \cos \chi - \mathbf{e}_x \sin \chi) \right) r \, d\psi \, dr \, d\zeta \tag{11.56}$$

where $d\zeta = dz/\cos \chi$. For this general case,

$$\mathbf{s} = (r \cos \psi - r' \cos \psi' - (z - z') \tan \chi) \mathbf{e}_x + (r \sin \psi - r' \sin \psi') \mathbf{e}_y + (z - z') \mathbf{e}_z \tag{11.57}$$

$$s^2 = r^2 + r'^2 - 2rr'C - 2(r \cos \psi - r' \cos \psi')(z - z') \tan \chi + (z - z')^2 (1 + \tan^2 \chi) \tag{11.58}$$

$$\omega = (\gamma_t \sin \psi' + \gamma_s \cos \psi' + \gamma_z \sin \chi) \mathbf{e}_x + (-\gamma_t \cos \psi' + \gamma_s \sin \psi') \mathbf{e}_y - \gamma_z \cos \chi \mathbf{e}_z \tag{11.59}$$

and

$$(\mathbf{s} \times \omega)_z = \gamma_t (r' - rC + (z - z') \tan \chi \cos \psi') + \gamma_s (-rS - (z - z') \tan \chi \sin \psi')$$
$$+ \gamma_z (-r \sin \psi + r' \sin \psi') \sin \chi \tag{11.60}$$

The z-component of the induced velocity is

$$u_z = -\frac{1}{4\pi} \int_{-\infty}^{0} \int_{0}^{2\pi} \int_{0}^{R} \frac{1}{s^3 \cos \chi} \left[(\mathbf{s} \times \omega)_z \right] r' \, dr' \, d\psi' \, dz' \tag{11.61}$$

For the vorticity γ independent of axial distance, the integration over z' can be performed. The induced velocity at the disk, due to just γ_t, is

$$u_z = -\frac{1}{4\pi} \int_{0}^{2\pi} \int_{0}^{R} \gamma_t \left[1 - \frac{r^2 - rr' \cos(\psi - \psi') - rs \cos \psi \sin \chi}{s^2 - s(r \cos \psi - r' \cos \psi') \sin \chi} \right] dr' \, d\psi' \tag{11.62}$$

For uniform loading, $\gamma_t = \frac{\Gamma}{h} \delta(R)$:

$$u_z = -v_0 \frac{1}{2\pi} \int_{0}^{2\pi} \left[1 - \frac{r^2 - rR \cos(\psi - \psi') - rs \cos \psi \sin \chi}{s^2 - s(r \cos \psi - R \cos \psi') \sin \chi} \right] d\psi' \tag{11.63}$$

with $v_0 = \frac{1}{2} \frac{\Gamma}{h}$. On the longitudinal axis, $\psi = 0$:

$$u_z = -v_0 \frac{1}{2\pi} \int_{0}^{2\pi} \left[1 - \frac{r^2 - rR \cos \psi' - rs \sin \chi}{s^2 - s(r - R \cos \psi') \sin \chi} \right] d\psi' \tag{11.64}$$

where $s^2 = r^2 + R^2 - 2rR\cos\psi'$ now. So at the center of the disk ($r = 0$), $u_z = -v_0$. The gradient of the induced velocity along the longitudinal axis, at the center of the disk is

$$\frac{du_z}{dr} = -\frac{v_0}{R}\frac{1}{2\pi}\int_0^{2\pi}\frac{\cos\psi' + \sin\chi}{1 + \cos\psi'\sin\chi}d\psi' = -\frac{v_0}{R}\frac{1 - \cos\chi}{\sin\chi} = -\frac{v_0}{R}\tan\frac{\chi}{2} \quad (11.65)$$

See Coleman, Feingold, and Stempin (1945). Section 5.2.1 presented results from vortex theory in forward flight.

11.2 Potential Theory

Consider incompressible, potential flow about a rotor, which is modeled as an actuator disk that can support a pressure jump. An analytical solution can be obtained for a circular wing or disk by writing Laplace's equation in ellipsoidal coordinates and expanding the potential in associated Legendre functions (spherical harmonics). The derivation follows Kinner (1937), Mangler (1948a, 1948b), and Mangler and Squire (1950).

Let $\Phi = p/\rho$ be the perturbation pressure or acceleration potential and x_i the flow field coordinates. The acceleration potential satisfies Laplace's equation $\nabla^2\Phi = 0$, with boundary conditions that the perturbation pressure is zero at infinity and the pressure discontinuity on the rotor disk is the loading. Figure 11.2 shows the geometry. The rotor wake is a skewed cylinder in a free stream V. The wake skew angle is χ measured from the z-axis or i measured from the disk plane. All lengths are dimensionless, scaled with the rotor radius R. The Cartesian coordinates are (x, y, z). The origin is at the center of the rotor disk, and x–y is the disk plane. The axes are moving with the rotor disk, with the x-axis downstream and the z-axis in the positive thrust direction. The coordinate ξ is aligned with the velocity vector, positive upstream. The rotor wake is a skewed cylinder. Figure 11.2 also shows the ellipsoidal coordinates (ν, η, θ), defined as follows:

$$x = \sqrt{1 - \nu^2}\sqrt{1 + \eta^2}\cos\theta \quad (11.66)$$

$$y = \sqrt{1 - \nu^2}\sqrt{1 + \eta^2}\sin\theta \quad (11.67)$$

$$z = \nu\eta \quad (11.68)$$

$$\theta = \tan^{-1}(y/x) \quad (11.69)$$

$$\nu = \frac{\text{sign}z}{\sqrt{2}}\sqrt{1 - s + \sqrt{(1-s)^2 + 4z^2}} \quad (11.70)$$

$$s = x^2 + y^2 + z^2 = 1 - \nu^2 + \eta^2$$

$$\eta = z/\nu \quad (11.71)$$

The range of the ellipsoidal coordinates is $-1 \leq \nu \leq 1, 0 \leq \eta \leq \infty, 0 \leq \theta \leq 2\pi$, with sign$\nu = $ sign$z$. The variable θ is the disk azimuth angle, measured from the x-axis. On the rotor disk, $\eta = 0$ so $\nu = \pm\sqrt{1 - (x^2 + y^2)} = \pm\sqrt{1 - r^2}$. In the x–y plane outside the rotor disk, $\nu = 0$ so $\eta = \sqrt{r^2 - 1}$. On the z-axis, $\nu = \pm 1$ so $\eta = |z|$.

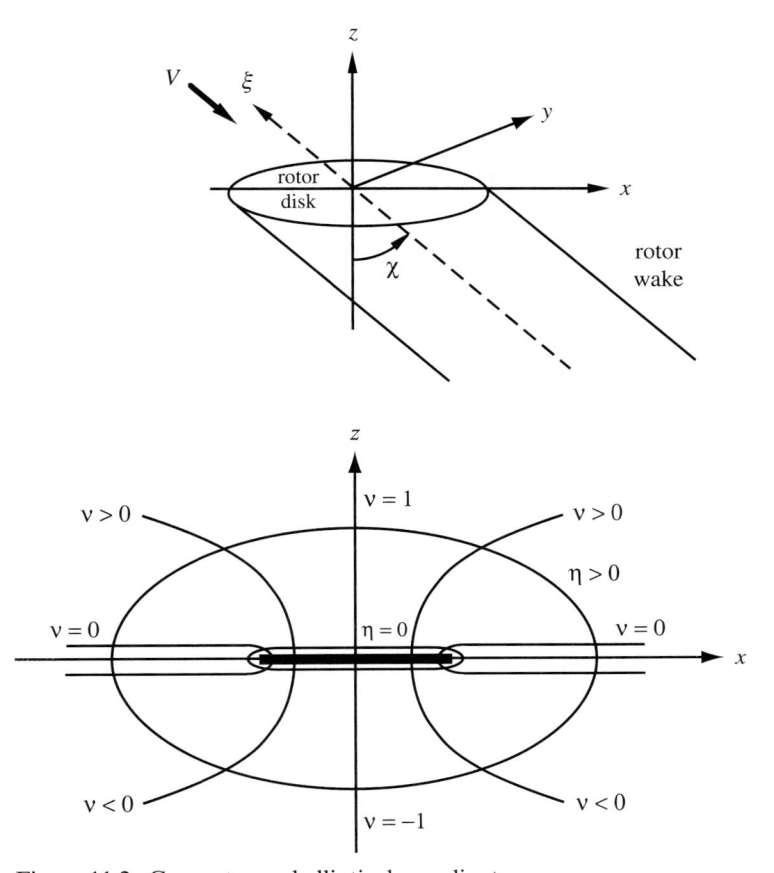

Figure 11.2. Geometry and elliptical coordinates.

In ellipsoidal coordinates (and dropping a factor of $(v^2 + \eta^2)^{-1}$), Laplace's equation $\nabla^2 \Phi = 0$ is

$$\frac{\partial}{\partial v}\left[(1-v^2)\frac{\partial \Phi}{\partial v}\right] + \frac{\partial}{\partial \eta}\left[(1+\eta^2)\frac{\partial \Phi}{\partial \eta}\right] + \frac{\partial}{\partial \theta}\left[\frac{v^2+\eta^2}{(1-v^2)(1+\eta^2)}\frac{\partial \Phi}{\partial \theta}\right] = 0 \quad (11.72)$$

Using the method of separation of variables, write $\Phi = \Phi_1(v)\Phi_2(\eta)\Phi_3(\theta)$. Then

$$\frac{\partial}{\partial v}\left[(1-v^2)\frac{\partial \Phi_1}{\partial v}\right] + \left[-\frac{m^2}{1-v^2} + n(n+1)\right]\Phi_1 = 0 \quad (11.73)$$

$$\frac{\partial}{\partial \eta}\left[(1+\eta^2)\frac{\partial \Phi_2}{\partial \eta}\right] + \left[\frac{m^2}{1+\eta^2} - n(n+1)\right]\Phi_2 = 0 \quad (11.74)$$

$$\frac{\partial^2 \Phi_3}{\partial \theta^2} + m^2 \Phi_3 = 0 \quad (11.75)$$

The solution for Φ_3 is periodic if m is an integer. The solution for Φ_1 is finite for $v = \pm 1$ (on the z-axis) if n is an integer ($n \geq m$). The solutions are $\Phi_1 = P_n^m(v)$ and $Q_n^m(v)$, the associated Legendre function of the first kind; $\Phi_2 = P_n^m(i\eta)$ and $Q_n^m(i\eta)$, the associated Legendre function of the second kind; and $\Phi_3 = \sin(m\theta)$ or $\cos(m\theta)$. This family of distributions does not encompass all solutions. In particular, a uniformly loaded disk (constant pressure) is not included.

The associated Legendre functions $P_n^m(v)$ ($n \geq m$) are related to Legendre polynomials $P_n(v)$:

$$P_n^m(v) = (-1)^m (1-v^2)^{m/2} \frac{d^m}{dv^m} P_n(v) \tag{11.76}$$

The following stable recurrence relation can be used to evaluate $P_n^m(v)$:

$$P_m^m = (-1)^m (2m-1)!! (1-v^2)^{m/2} \tag{11.77}$$

$$P_{m+1}^m = v(2m+1) P_m^m \tag{11.78}$$

$$(k-m) P_k^m = v(2k-1) P_{k-1}^m - (k+m-1) P_{k-2}^m \tag{11.79}$$

The double-factorial notation means that for n even/odd, $n!!$ is the product of all even/odd integers less than or equal to n:

$$n!! = \begin{cases} n(n-2)(n-4)\ldots 2 & n \text{ even} \\ n(n-2)(n-4)\ldots 1 & n \text{ odd} \end{cases} \tag{11.80}$$

with $0!! = 1$ and $1!! = 1$. The first few associated Legendre functions are

$$P_1^0(v) = v \tag{11.81}$$

$$P_3^0(v) = \frac{1}{2}(5v^3 - 3v) \tag{11.82}$$

$$P_2^1(v) = -3v\sqrt{1-v^2} \tag{11.83}$$

$$Q_1^0(i\eta) = \eta \left(\frac{\pi}{2} - \tan^{-1}\eta\right) - 1 \tag{11.84}$$

$$Q_3^0(i\eta) = -\frac{1}{2}\eta(5\eta^2 + 3)\left(\frac{\pi}{2} - \tan^{-1}\eta\right) + \frac{5}{2}\eta^2 + \frac{2}{3} \tag{11.85}$$

$$Q_2^1(i\eta) = -3\eta\sqrt{1+\eta^2}\left(\frac{\pi}{2} - \tan^{-1}\eta\right) + \frac{3\eta^2 + 2}{\sqrt{1+\eta^2}} \tag{11.86}$$

The functions are orthogonal:

$$\int_0^1 P_n^m(v) P_{n'}^m(v) dv = \begin{cases} 0 & n \neq n' \\ \frac{1}{2n+1} \frac{(n+m)!}{(n-m)!} & n = n' \end{cases} \tag{11.87}$$

The Legendre polynomial $P_n(v)$ contains even/odd powers of v for n even/odd. So $d^m P_n/dv^m$ has $(1,\ldots,v^{n-m})$ terms for n and m either both odd or both even and $(v,\ldots,v^{n-m})$ terms for only n or m odd ($n+m$ odd). The factor $(1-v^2)^{m/2} = r^m$ on the rotor disk, where $v = \pm\sqrt{1-r^2}$.

$Q_n^m(v)$ is infinite on $v = \pm 1$ (the z-axis), and $P_n^m(i\eta)$ is infinite for $\eta \to \infty$. So the solution is a series in the terms $P_n^m(v) Q_n^m(i\eta)$. On the rotor disk, $\eta = 0$ and $v = \pm\sqrt{1-r^2}$, with $v > 0$ above the disk and $v < 0$ below the disk. $Q_n^m(i\eta)$ is continuous at $\eta = 0$. So the pressure difference on the disk is

$$\frac{\Delta p}{\rho} = \Phi(v < 0, \eta = 0) - \Phi(v > 0, \eta = 0) \sim P_n^m(v < 0) - P_n^m(v > 0) \tag{11.88}$$

So $P_n^m(v)$ with $n+m$ even does not produce a pressure discontinuity on the disk, since it is an even function of v. Only terms for $n+m$ odd are retained, and $\Delta p/\rho = -2\Phi(v > 0, \eta = 0)$. Also, $P_n^m(v)$ is only defined for $n \geq m$.

11.2 Potential Theory

Thus the solution of Laplace's equation is a series for the acceleration potential:

$$\Phi = \frac{p}{\rho} = (\Omega R)^2 \sum_{m=0}^{\infty} \sum_{n=m+1,m+3,\ldots}^{\infty} P_n^m(\nu) Q_n^m(i\eta) \left(c_n^m \cos m\theta + d_n^m \sin m\theta \right) \quad (11.89)$$

and hence for the rotor loading:

$$\Delta p = -2\rho (\Omega R)^2 \sum_{m=0}^{\infty} \sum_{n=m+1,m+3,\ldots}^{\infty} P_n^m(\nu) Q_n^m(i0) \left(c_n^m \cos m\theta + d_n^m \sin m\theta \right) \quad (11.90)$$

The coefficients c_n^m and d_n^m are constant for steady loading and functions of time for unsteady loading. The coordinates are dimensionless, so c_n^m and d_n^m are scaled with the tip speed squared. The bound circulation is $N\Gamma = \frac{2\pi}{\rho\Omega} \frac{dT}{dA} = \frac{2\pi}{\Omega} \frac{-\Delta p}{\rho}$.

The thrust is obtained by integrating Δp over the rotor disk, for $m = 0$:

$$T = \int \Delta p \, dA = R^2 \int_0^{2\pi} \int_0^1 (p_{\text{lower}} - p_{\text{upper}}) r \, dr \, d\theta$$

$$= \rho(\Omega R)^2 2\pi R^2 \sum_{n=1,3,\ldots} Q_n^0(i0) c_n^0 \int_0^1 \left(P_n^0(\nu)_{\text{lower}} - P_n^0(\nu)_{\text{upper}} \right) r \, dr$$

$$= \rho(\Omega R)^2 2\pi R^2 \sum_{n=1,3,\ldots} Q_n^0(i0) c_n^0 \left(-\int_{-1}^0 P_n^0(\nu)_{\text{lower}} \nu \, d\nu + \int_1^0 P_n^0(\nu)_{\text{upper}} \nu \, d\nu \right)$$

$$= -\rho(\Omega R)^2 2\pi R^2 \sum_{n=1,3,\ldots} Q_n^0(i0) c_n^0 \int_{-1}^1 P_n^m(\nu) \nu \, d\nu$$

$$= -\rho(\Omega R)^2 2\pi R^2 Q_1^0(i0) c_1^0 \frac{2}{3} = \rho A (\Omega R)^2 \frac{4}{3} c_1^0 \quad (11.91)$$

or $c_1^0 = \frac{3}{4} C_T$. The thrust is produced solely by the $P_1^0(\nu) = \nu = \sqrt{1-r^2}$ loading term. It is conventional to consider a pressure distribution that is zero at the center, as well as at the tips, by including the $n = 3$ term. At $r = 0$ or $\nu = 1$,

$$P_1^0(1) Q_1^0(i0) c_1^0 + P_3^0(1) Q_3^0(i0) c_3^0 = 0 \quad (11.92)$$

So $c_3^0 = \frac{3}{2} c_1^0 = \frac{9}{8} C_T$, and

$$\frac{\Delta p}{\rho(\Omega R)^2} = 2 \left(\frac{3}{4} P_1^0(\nu) - P_3^0(\nu) \right) C_T = \frac{15}{4} \nu (1 - \nu^2) C_T = \frac{15}{4} r^2 \sqrt{1 - r^2} \, C_T \quad (11.93)$$

is the loading.

The pitch moment M_y (positive front edge up) and roll moment M_x (positive right edge up) are

$$-M_y = \int \Delta p \, x \, dA = R^3 \int_0^{2\pi} \int_0^1 (p_{\text{lower}} - p_{\text{upper}})(r \cos \theta) r \, dr \, d\theta$$

$$= \rho(\Omega R)^2 \pi R^3 \sum_{n=2,4,\ldots} Q_n^1(i0) c_n^1 \int_0^1 \left(P_n^1(\nu)_{\text{lower}} - P_n^1(\nu)_{\text{upper}} \right) r^2 \, dr$$

$$= -\rho(\Omega R)^2 \pi R^3 \sum_{n=2,4,\ldots} Q_n^1(i0) c_n^1 \int_{-1}^1 P_n^1(\nu) \sqrt{1-\nu^2} \, \nu \, d\nu$$

$$= \rho(\Omega R)^2 \pi R^3 Q_2^1(i0) c_2^1 \frac{4}{5} = \rho A R (\Omega R)^2 \frac{8}{5} c_2^1 \quad (11.94)$$

$$M_x = \int \Delta p\, y\, dA = R^3 \int_0^{2\pi} \int_0^1 (p_{\text{lower}} - p_{\text{upper}})(r \sin\theta) r\, dr\, d\theta$$

$$= \rho(\Omega R)^2 \pi R^3 \sum_{n=2,4,\ldots} Q_n^1(i0) d_n^1 \int_0^1 \left(P_n^1(\nu)_{\text{lower}} - P_n^1(\nu)_{\text{upper}}\right) r^2 dr$$

$$= -\rho(\Omega R)^2 \pi R^3 \sum_{n=2,4,\ldots} Q_n^1(i0) d_n^1 \int_{-1}^1 P_n^1(\nu)\sqrt{1-\nu^2}\,\nu\, d\nu$$

$$= \rho(\Omega R)^2 \pi R^3 Q_2^1(i0) d_2^1 \frac{4}{5} = \rho A R (\Omega R)^2 \frac{8}{5} d_2^1 \qquad (11.95)$$

or $c_2^1 = -\frac{5}{8} C_{My}$ and $d_2^1 = \frac{5}{8} C_{Mx}$. The moments are produced solely by the $P_2^1 = -3\nu\sqrt{1-\nu^2} = -3r\sqrt{1-r^2}$ loading term.

In summary, the solution that produces thrust is

$$\Phi = (\Omega R)^2 \left[P_1^0(\nu) Q_1^0(i\eta) + \frac{3}{2} P_3^0(\nu) Q_3^0(i\eta) \right] \frac{3}{4} C_T \qquad (11.96)$$

$$\Delta p = 2\rho(\Omega R)^2 \left[P_1^0(\nu) - P_3^0(\nu) \right] \frac{3}{4} C_T \qquad (11.97)$$

where $P_0^1 = \nu = \sqrt{1-r^2}$ and $P_1^0 - P_3^0 = \frac{5}{2}(\nu - \nu^3) = \frac{5}{2} r^2 \sqrt{1-r^2}$. The P_3^0 term is included so the loading is zero at the center of the disk. The solution that produces moments is

$$\Phi = (\Omega R)^2 \left[P_2^1(\nu) Q_2^1(i\eta) \right] \frac{5}{8} \left(-C_{My} \cos\theta + C_{Mx} \sin\theta \right) \qquad (11.98)$$

$$\Delta p = 2\rho(\Omega R)^2 \left[-P_2^1(\nu) \right] \frac{5}{4} \left(-C_{My} \cos\theta + C_{Mx} \sin\theta \right) \qquad (11.99)$$

where $-P_2^1 = 3\nu\sqrt{1-\nu^2} = 3r\sqrt{1-r^2}$.

The velocities are obtained from the momentum equation. For small disturbances relative to the velocity V (Figure 11.2), the linearized equation of momentum conservation is

$$\frac{\partial u_i}{\partial t} - V \frac{\partial u_i}{\partial \xi} = -\frac{\partial \Phi}{\partial x_i} \qquad (11.100)$$

where u_i are the components of the perturbation velocity; and x_i the flow field coordinates. The convection term has been written in terms of the coordinate ξ: $\zeta = x \cos\chi + z \sin\chi$ and $\xi = -x \sin\chi + z \cos\chi$, so

$$V_x \frac{\partial}{\partial x} - V_z \frac{\partial}{\partial z} = V\left(\sin\chi \frac{\partial}{\partial x} - \cos\chi \frac{\partial}{\partial z}\right) = -V \frac{\partial}{\partial \xi} \qquad (11.101)$$

For steady flow, the equation for the normal velocity is

$$V \frac{\partial u_z}{\partial \xi} = \frac{\partial \Phi}{\partial z} \qquad (11.102)$$

Integrating Φ along a streamline ξ from upstream to the disk gives the downwash $v = -u_z$:

$$v(x_0, y_0) = \frac{1}{V} \int_{-x_0 \sin\chi}^{\infty} \frac{\partial \Phi}{\partial z} d\xi \qquad (11.103)$$

11.2 Potential Theory

using coordinates (x_0, y_0) on the rotor disk: $x_0 = r\cos\psi$ and $y_0 = r\sin\psi$. The integration is over ξ for fixed $\zeta = x_0 \cos\chi$, so

$$x = x_0 \cos^2 \chi - \xi \sin \chi \tag{11.104}$$

$$y = y_0 \tag{11.105}$$

$$z = x_0 \cos\chi \sin\chi + \xi \cos\chi \tag{11.106}$$

from which elliptical coordinates (ν, η, θ) are determined, and then

$$R\frac{\partial \Phi}{\partial z} = \frac{1}{\nu^2 + \eta^2}\left[\eta(1-\nu^2)\frac{\partial}{\partial \nu} + \nu(1+\eta^2)\frac{\partial}{\partial \eta}\right]\Phi \tag{11.107}$$

The induced velocity $\lambda = v/\Omega R$ is represented by the truncated series:

$$v = v_0 + v_c r \cos\psi + v_s r \sin\psi \tag{11.108}$$

where

$$v_0 = \frac{1}{\pi}\int_0^{2\pi}\int_0^1 vr\,dr\,d\psi = \frac{1}{A}\int v\,dA \tag{11.109}$$

$$v_c = \frac{4}{\pi}\int_0^{2\pi}\int_0^1 vr^2\cos\psi\,dr\,d\psi = \frac{4}{A}\int vr\cos\psi\,dA \tag{11.110}$$

$$v_s = \frac{4}{\pi}\int_0^{2\pi}\int_0^1 vr^2\sin\psi\,dr\,d\psi = \frac{4}{A}\int vr\sin\psi\,dA \tag{11.111}$$

are the inflow variables.

For axial flow ($\chi = 0$), $\xi = z$ and at the disk $z = 0$ or $\eta = 0$. So

$$v = \frac{1}{V}\int_0^\infty \frac{\partial \Phi}{\partial z}dz = -\frac{1}{V}\Phi(\eta=0) = \frac{1}{2\rho V}\Delta p \tag{11.112}$$

This result does not depend on the solution for Φ, and hence is independent of the loading distribution. Compare with the vortex theory result

$$v(r, \psi) = \frac{1}{2}\frac{\Gamma(r,\psi)}{h} = \frac{dT(r,\psi)/dA}{2\rho(V+v)} = \frac{\Delta p}{2\rho(V+v)} \tag{11.113}$$

(equation 11.44). The inflow variables are thus

$$v_0 = \frac{1}{2\rho V}\frac{1}{A}\int \Delta p\,dA = \frac{T}{2\rho AV} \tag{11.114}$$

$$v_c = \frac{1}{2\rho V}\frac{4}{A}\int r\cos\psi\,\Delta p\,dA = -\frac{2M_y}{\rho ARV} \tag{11.115}$$

$$v_s = \frac{1}{2\rho V}\frac{4}{A}\int r\sin\psi\,\Delta p\,dA = \frac{2M_x}{\rho ARV} \tag{11.116}$$

or in coefficient form

$$\begin{pmatrix}\lambda_0 \\ \lambda_c \\ \lambda_s\end{pmatrix} = \frac{1}{V}\begin{bmatrix}\frac{1}{2} & 0 & 0 \\ 0 & 2 & 0 \\ 0 & 0 & 2\end{bmatrix}\begin{pmatrix}C_T \\ -C_{My} \\ C_{Mx}\end{pmatrix} \tag{11.117}$$

for axial flow ($V = \mu_z$). This result holds for the loading distributions from the separation of variables solution

$$\frac{\Delta p}{\rho(\Omega R)^2} = \frac{15}{4} r^2 \sqrt{1 - r^2} C_T + \frac{15}{2} r \sqrt{1 - r^2} \left(-C_{My} \cos\theta + C_{Mx} \sin\theta \right) \tag{11.118}$$

or

$$\frac{\Delta p}{\rho(\Omega R)^2} = \frac{3}{2} \sqrt{1 - r^2} C_T + \frac{15}{2} r \sqrt{1 - r^2} \left(-C_{My} \cos\theta + C_{Mx} \sin\theta \right) \tag{11.119}$$

and for the simpler loading distribution

$$\frac{\Delta p}{\rho(\Omega R)^2} = C_T + 4r \left(-C_{My} \cos\theta + C_{Mx} \sin\theta \right) \tag{11.120}$$

From both potential theory and vortex theory, the induced velocity depends only on the local pressure on the actuator disk in axial flight. So uniform plus linear loading distribution on the disk (equation 11.120) corresponds to uniform plus linear inflow distribution (equation 11.108). The potential theory result is a perturbation relative to the rotor velocity V. To cover low speed, including hover, vortex theory implies replacing V with $V + v$. Vortex theory gives the perturbation inflow due to the perturbation loading for fixed wake geometry, which is determined by the axial convection velocity $V + v$. That is an appropriate model for moment perturbations, but low-frequency thrust changes should approach the momentum theory result. Perturbing $(V + v)v = T/2\rho A$ gives

$$\delta v = \frac{\delta T}{2\rho A (V + 2v)} \tag{11.121}$$

Hence for the thrust loading, replacing V with $V + 2v$ extends the potential theory result to hover.

For edgewise flow ($\chi = 90°$), $\xi = -x$ and at the disk $z = 0$ or $\eta = 0$. The integration over ξ goes from $x = -\infty$ to $-\sqrt{1 - y_0^2}$ ahead of the rotor ($v = 0$) and from $x = -\sqrt{1 - y_0^2}$ to x_0 on the disk ($\eta = 0$). So

$$v = \frac{1}{V} \int_{-x_0}^{\infty} \frac{\partial \Phi}{\partial z} d\xi = \frac{1}{V} \int_{-\infty}^{x_0} \frac{\partial \Phi}{\partial z} dx$$

$$= \frac{1}{V} \left[\int_{-\infty}^{-\sqrt{1-y_0^2}} \frac{\partial \Phi}{\partial z} \bigg|_{v=0} dx + \int_{-\sqrt{1-y_0^2}}^{x_0} \frac{\partial \Phi}{\partial z} \bigg|_{\eta=0} dx \right] \tag{11.122}$$

Integrating over r and ψ, the inflow variables are

$$\begin{pmatrix} \lambda_0 \\ \lambda_c \\ \lambda_s \end{pmatrix} = \frac{1}{V} \begin{bmatrix} \frac{1}{2} & -\frac{15\pi}{64} & 0 \\ \frac{15\pi}{64} & 0 & 0 \\ 0 & 0 & 4 \end{bmatrix} \begin{pmatrix} C_T \\ -C_{My} \\ C_{Mx} \end{pmatrix} \tag{11.123}$$

using the loading from equation 11.118. If instead the loading from equation 11.119 is used, the only change is $\lambda_c = \frac{1}{V} \frac{3\pi}{8} C_T$.

For arbitrary flow angle, Mangler (1948b) found the Fourier series for the induced velocity produced by thrust:

$$v = \frac{1}{V} \left[\frac{3}{4} \sqrt{1 - r^2} + \frac{3\pi}{8} r \cos\psi \sqrt{\frac{1 - \cos\chi}{1 + \cos\chi}} \right] C_T \tag{11.124}$$

11.2 Potential Theory

for the loading of equation 11.119, or

$$v = \frac{1}{V}\left[\frac{15}{8}r^2\sqrt{1-r^2} + \frac{15\pi}{128}(9r^2-4)r\cos\psi\sqrt{\frac{1-\cos\chi}{1+\cos\chi}}\right]C_T \tag{11.125}$$

for the loading of equation 11.118. The longitudinal inflow variation, $\partial\lambda_c/\partial C_T$, is proportional to

$$\sqrt{\frac{1-\cos\chi}{1+\cos\chi}} = \tan\frac{\chi}{2} \tag{11.126}$$

as from vortex theory. Pitt (1980) found the inflow due to moments for arbitrary flow angle, completing the matrix:

$$\begin{pmatrix}\lambda_0 \\ \lambda_c \\ \lambda_s\end{pmatrix} = \frac{1}{V}\begin{bmatrix}\frac{1}{2} & -\frac{15\pi}{64}\sqrt{\frac{1-\cos\chi}{1+\cos\chi}} & 0 \\ \frac{15\pi}{64}\sqrt{\frac{1-\cos\chi}{1+\cos\chi}} & \frac{4\cos\chi}{1+\cos\chi} & 0 \\ 0 & 0 & \frac{4}{1+\cos\chi}\end{bmatrix}\begin{pmatrix}C_T \\ -C_{My} \\ C_{Mx}\end{pmatrix} \tag{11.127}$$

with $V = \sqrt{\mu^2 + \mu_z^2}$ and $\tan\chi = \mu/\mu_z$ (for large velocity). Note $\frac{2\cos\chi}{1+\cos\chi} = 1 - X^2$ and $\frac{2}{1+\cos\chi} = 1 + X^2$, where $X = \sqrt{\frac{1-\cos\chi}{1+\cos\chi}}$.

The potential theory result is a perturbation relative to the velocity $V = \sqrt{\mu^2 + \mu_z^2}$. To be useful for helicopter rotors, this result must be extended to the case of small rotor velocity, including hover, by using an effective velocity V_{eff} in place of V in the relation between inflow and loading variables. Momentum theory gives the mean induced velocity due to thrust:

$$\lambda_i = \frac{C_T}{2\sqrt{\mu^2 + \lambda^2}} \tag{11.128}$$

where $\lambda = \mu_z + \lambda_i$. Vortex theory interprets the denominator as the mean convection velocity or mean wake separation, so $\partial\lambda_i/\partial C_T = 1/(2\sqrt{\mu^2 + \lambda^2})$, which implies

$$V_{\text{eff}} = \sqrt{\mu^2 + \lambda^2} \tag{11.129}$$

using the mean induced velocity in $\lambda = \mu_z + \bar{\lambda}_i$. Then for hover $V_{\text{eff}} = |\bar{\lambda}_i|$ and for forward flight $V_{\text{eff}} = \mu$. Equation 11.129 is appropriate for harmonic variation of the loading, including hub moments. Differential momentum theory allows the convection velocity and wake separation to vary with thrust:

$$\frac{\partial\lambda_i}{\partial C_T} = \frac{1}{2\sqrt{\mu^2 + \lambda^2}} - \frac{C_T\lambda}{2(\mu^2 + \lambda^2)^{3/2}}\frac{\partial\lambda_i}{\partial C_T} = \frac{1}{2}\frac{\sqrt{\mu^2 + \lambda^2}}{\mu^2 + \lambda(\lambda + \lambda_i)} \tag{11.130}$$

which implies

$$V_{\text{eff}} = \frac{\mu^2 + \lambda(\lambda + \lambda_i)}{\sqrt{\mu^2 + \lambda^2}} \tag{11.131}$$

Then for hover $V_{\text{eff}} = 2|\bar{\lambda}_i|$, and for forward flight $V_{\text{eff}} = \mu$. Equation 11.131 is the correct limit for slow variation of the thrust. The wake skew angle is $\tan\chi = \mu/|\lambda|$.

Then $\cos \chi = |\lambda|/\sqrt{\mu^2 + \lambda^2}$, and

$$\tan \frac{\chi}{2} = \sqrt{\frac{1 - \cos \chi}{1 + \cos \chi}} = \frac{\mu}{\sqrt{\mu^2 + \lambda^2} + |\lambda|} \tag{11.132}$$

$$\frac{1}{1 + \cos \chi} = \frac{\sqrt{\mu^2 + \lambda^2}}{\sqrt{\mu^2 + \lambda^2} + |\lambda|} \tag{11.133}$$

$$\frac{\cos \chi}{1 + \cos \chi} = \frac{|\lambda|}{\sqrt{\mu^2 + \lambda^2} + |\lambda|} \tag{11.134}$$

Applying differential momentum theory $\delta\lambda = \frac{1}{2V_{\text{eff}}}\delta C_T$ to the loading of equation 11.120 gives

$$\begin{pmatrix} \lambda_0 \\ \lambda_c \\ \lambda_s \end{pmatrix} = \frac{1}{V_{\text{eff}}} \begin{bmatrix} \frac{1}{2} & 0 & 0 \\ 0 & 2 & 0 \\ 0 & 0 & 2 \end{bmatrix} \begin{pmatrix} C_T \\ -C_{My} \\ C_{Mx} \end{pmatrix} \tag{11.135}$$

It is still best to use $V_{\text{eff}} = \sqrt{\mu^2 + \lambda^2}$ for the moment terms. Differential momentum theory and potential theory give the same results for axial flow, but are different for forward flight.

For unsteady flow, the linearized momentum equation for $u_z = -v$ is

$$\frac{\partial v}{\partial t} - V\frac{\partial v}{\partial \xi} = \frac{\partial \Phi}{\partial z} \tag{11.136}$$

A result for the time delay of the inflow response to loading is obtained by writing the momentum equation for axial flow ($\chi = 0$) and hover ($V = 0$):

$$\dot{v} = \frac{\partial \Phi}{\partial z}\bigg|_{z=0} \tag{11.137}$$

on the rotor disk. Then the time derivatives of the inflow variables are

$$\dot{v}_0 = \frac{1}{\pi}\int_0^{2\pi}\int_0^1 \left(\frac{\partial \Phi}{\partial z}\bigg|_{z=0}\right) r\,dr\,d\psi = \frac{1}{M_0}C_T \tag{11.138}$$

$$\dot{v}_c = \frac{4}{\pi}\int_0^{2\pi}\int_0^1 \left(\frac{\partial \Phi}{\partial z}\bigg|_{z=0}\right) r^2 \cos\psi\,dr\,d\psi = -\frac{1}{M_c}C_{My} \tag{11.139}$$

$$\dot{v}_s = \frac{4}{\pi}\int_0^{2\pi}\int_0^1 \left(\frac{\partial \Phi}{\partial z}\bigg|_{z=0}\right) r^2 \sin\psi\,dr\,d\psi = \frac{1}{M_s}C_{Mx} \tag{11.140}$$

Although lacking rigor ($V = 0$ is not a meaningful case for this problem), the result gives the same values as the apparent mass of an impermeable disk: $M_0 = \frac{8}{3\pi}$ or $M_0 = \frac{128}{75\pi}$ (for the loadings of equations 11.119, and 11.118, respectively); and $M_c = M_s = \frac{64}{45\pi}$.

11.3 Dynamic Inflow

The wake-induced velocities at the rotor disk play an important role in rotor unsteady aerodynamics and therefore must be accounted for in determining both the periodic and transient blade loading. The best representation for the wake is a vortex model, but for stability analysis and real-time simulations a finite-state model of the wake is needed. A dynamic inflow model is a set of first-order differential equations relating

11.3 Dynamic Inflow

inflow variables and aerodynamic loading variables. The simplest model has three inflow states, consisting of uniform (λ_0) and linear (λ_c, λ_s) perturbations of the wake-induced downwash at the rotor disk:

$$\lambda = \lambda_0 + \lambda_c r \cos \psi + \lambda_s r \sin \psi \tag{11.141}$$

The loading variables are the integrated section lift of all blades: thrust C_T, pitch moment C_{My}, and roll moment C_{Mx} (aerodynamic contributions only). The dynamic inflow equations are

$$LM \begin{pmatrix} \dot{\lambda}_0 \\ \dot{\lambda}_c \\ \dot{\lambda}_s \end{pmatrix} + \begin{pmatrix} \lambda_0 \\ \lambda_c \\ \lambda_s \end{pmatrix} = L \begin{pmatrix} C_T \\ -C_{My} \\ C_{Mx} \end{pmatrix} \tag{11.142}$$

Dynamic inflow is a global, low frequency model for the wake-induced velocity. The model is generally sufficient to capture wake effects on the dynamic behavior of the lowest-frequency blade modes and the aircraft flight dynamics.

Pitt and Peters (1981) developed dynamic inflow based on the potential theory solution of Mangler and Squire (1950), giving the derivative matrix:

$$L = \frac{1}{V_{\text{eff}}} \begin{bmatrix} \frac{1}{2} & -\frac{15\pi}{64}\sqrt{\frac{1-\cos\chi}{1+\cos\chi}} & 0 \\ \frac{15\pi}{64}\sqrt{\frac{1-\cos\chi}{1+\cos\chi}} & \frac{4\cos\chi}{1+\cos\chi} & 0 \\ 0 & 0 & \frac{4}{1+\cos\chi} \end{bmatrix} \tag{11.143}$$

(see equation 11.127), with

$$V_{\text{eff}} = \frac{\mu^2 + \lambda(\lambda + \lambda_i)}{\sqrt{\mu^2 + \lambda^2}} \tag{11.144}$$

for the thrust terms and

$$V_{\text{eff}} = \sqrt{\mu^2 + \lambda^2} \tag{11.145}$$

for the moment terms; Peters (1974) and Pitt and Peters (1981) used equation 11.144 for all terms of L. The mass matrix was obtained from the virtual mass of an impermeable disk in translation or rotation:

$$M = \begin{bmatrix} \frac{128}{75\pi} & 0 & 0 \\ 0 & \frac{64}{45\pi} & 0 \\ 0 & 0 & \frac{64}{45\pi} \end{bmatrix} \tag{11.146}$$

The time lag is given by the matrix LM; the values are supported by experimental data. Omitting the time lag produces a quasistatic model, the effects of which are given by a constant lift deficiency function C'. From section 6.3, the aerodynamic thrust of the rotor is

$$\frac{C_T}{\sigma a} = \frac{1}{2\pi} \int_0^{2\pi} \int_0^1 \frac{F_z}{ac} \, dr \, d\psi \tag{11.147}$$

Substituting $\delta(F_z/ac) = -\frac{1}{2}u_T \delta\lambda \cong -\frac{1}{2}r\lambda_0$,

$$\frac{C_T}{\sigma a} = \left(\frac{C_T}{\sigma a}\right)_{QS} - \frac{1}{4}\lambda_0 = \left(\frac{C_T}{\sigma a}\right)_{QS} - \frac{1}{4}\frac{1}{2V_{\text{eff}}}C_T = C'\left(\frac{C_T}{\sigma a}\right)_{QS} \tag{11.148}$$

The subscript QS means the quasistatic loading, consisting of all loads except those due to the wake-induced velocities. The lift deficiency function is

$$C' = \frac{1}{1 + \frac{\sigma a}{8V_{\text{eff}}}} = \begin{cases} \frac{1}{1+\sigma a/16\lambda_i} & \text{hover} \\ \frac{1}{1+\sigma a/8\mu} & \text{forward flight} \end{cases} \quad (11.149)$$

The aerodynamic hub moments are

$$-\frac{2C_{My}}{\sigma a} = \frac{1}{\pi} \int_0^{2\pi} \int_0^1 \cos\psi \, \frac{F_z}{ac} r \, dr \, d\psi \quad (11.150)$$

$$\frac{2C_{Mx}}{\sigma a} = \frac{1}{\pi} \int_0^{2\pi} \int_0^1 \sin\psi \, \frac{F_z}{ac} r \, dr \, d\psi \quad (11.151)$$

Substituting $\delta(F_z/ac) = -\frac{1}{2} u_T \delta\lambda \cong -\frac{1}{2} r^2 (\lambda_x \cos\psi + \lambda_y \cos\psi)$,

$$\begin{pmatrix} -\frac{2C_{My}}{\sigma a} \\ \frac{2C_{Mx}}{\sigma a} \end{pmatrix} = \begin{pmatrix} -\frac{2C_{My}}{\sigma a} \\ \frac{2C_{Mx}}{\sigma a} \end{pmatrix}_{QS} - \frac{1}{8} \begin{pmatrix} \lambda_c \\ \lambda_s \end{pmatrix}$$

$$= \begin{pmatrix} -\frac{2C_{My}}{\sigma a} \\ \frac{2C_{Mx}}{\sigma a} \end{pmatrix}_{QS} - \frac{1}{8} \frac{1}{V_{\text{eff}}} \begin{pmatrix} -\frac{4\cos\chi}{1+\cos\chi} C_{My} \\ \frac{4}{1+\cos\chi} C_{Mx} \end{pmatrix}$$

$$= C' \begin{pmatrix} -\frac{2C_{My}}{\sigma a} \\ \frac{2C_{Mx}}{\sigma a} \end{pmatrix}_{QS} \quad (11.152)$$

The lift deficiency function for pitch moment is

$$C' = \frac{1}{1 + \frac{\sigma a}{8V_{\text{eff}}} \frac{2\cos\chi}{1+\cos\chi}} = \begin{cases} \frac{1}{1+\sigma a/8\lambda_i} & \text{hover} \\ 0 & \text{forward flight} \end{cases} \quad (11.153)$$

and for roll moment is

$$C' = \frac{1}{1 + \frac{\sigma a}{8V_{\text{eff}}} \frac{2}{1+\cos\chi}} = \begin{cases} \frac{1}{1+\sigma a/8\lambda_i} & \text{hover} \\ \frac{1}{1+\sigma a/4\mu} & \text{forward flight} \end{cases} \quad (11.154)$$

The aerodynamic thrust is reduced by the wake effects, but the influence of the wake in hover is greater for moments because of the shed vorticity. Typical values of these lift deficiency functions are $C' \cong 0.8$ for forward flight; $C' \cong .7$ for thrust changes in hover; and $C' \cong 0.5$ for moment changes in hover. If the dominant aerodynamic forces are lift perturbations caused by angle-of-attack changes, then the aerodynamic influence is described by the blade Lock number $\gamma = \rho a c R^4 / I_b$, which contains the lift-curve slope. Then the effects of the quasistatic dynamic inflow model are largely represented by an effective Lock number: $\gamma_e = \gamma C'$ (Curtiss and Shupe (1971)). However, often the time lag is needed to properly represent the effects of the unsteady aerodynamics.

The dynamic inflow model must also include the effects of the rotor motion:

$$LM \begin{pmatrix} \dot\lambda_0 \\ \dot\lambda_c \\ \dot\lambda_s \end{pmatrix} + \begin{pmatrix} \lambda_0 \\ \lambda_c \\ \lambda_s \end{pmatrix} = L \begin{pmatrix} C_T \\ -C_{My} \\ C_{Mx} \end{pmatrix} + U \begin{pmatrix} \delta\mu \\ \delta\mu_z \end{pmatrix} + A \begin{pmatrix} \dot\alpha_{y\text{TPP}} \\ -\dot\alpha_{x\text{TPP}} \end{pmatrix} \quad (11.155)$$

11.3 Dynamic Inflow

Perturbing the momentum theory result $\lambda_i = C_T/2\sqrt{\mu^2 + \lambda^2}$ gives

$$\delta\lambda = -\frac{\lambda_i(\mu\delta\mu + \lambda\delta\mu_z)}{\mu^2 + \lambda(\lambda + \lambda_i)} = \begin{cases} -\frac{1}{2}\delta\mu_z & \text{hover} \\ -\frac{\lambda_i}{\mu_z + 2\lambda_i}\delta\mu_z & \text{axial} \\ -\frac{\lambda_i}{\mu}\delta\mu - \frac{\lambda\lambda_i}{\mu^2}\delta\mu_z & \text{forward flight} \end{cases} \quad (11.156)$$

or

$$U = \begin{bmatrix} -\frac{\lambda_i\mu}{\mu^2+\lambda(\lambda+\lambda_i)} & -\frac{\lambda_i\lambda}{\mu^2+\lambda(\lambda+\lambda_i)} \\ 0 & 0 \\ 0 & 0 \end{bmatrix} \quad (11.157)$$

for the influence of in-plane and normal changes of the rotor velocity.

The last term in equation 11.155 is the effect of tip-path-plane pitch and roll rate: $\dot{\alpha}_{y\text{TPP}} = \dot{\alpha}_y - \dot{\beta}_{1c}$ and $\dot{\alpha}_{x\text{TPP}} = \dot{\alpha}_x + \dot{\beta}_{1s}$. Assuming axisymmetric response of the inflow to tip-path-plane angular motion gives

$$A = \begin{bmatrix} 0 & 0 \\ K_R & 0 \\ 0 & K_R \end{bmatrix} \quad (11.158)$$

These terms have been identified with the curvature of the wake geometry in hover and low-speed flight. $K_R > 1$ changes the sign of the low-frequency off-axis response of the flapping to shaft angular rate, a phenomenon identified in aircraft off-axis pitch and roll rate response to cyclic control in hover. In flight tests of helicopter low-frequency control, on-axis response (pitch rate due to longitudinal cyclic and roll rate due to lateral cyclic) has a phase of 0° whereas the off-axis response (pitch rate due to lateral cyclic and negative roll rate due to longitudinal cyclic) has a phase of 180°. Using momentum theory or dynamic inflow, the magnitude and phase of the on-axis response are predicted well, but the predicted off-axis phase is 0°. See Takahashi (1990), Harding and Bass (1990), Chaimovich, Rosen, Rand, Mansur, and Tischler (1992), and Arnold, Keller, Curtiss, and Reichert, (1998). Vortex theory provides a simple estimate of the effect. The induced velocity in hover is the ratio of the loading and the vertical convection of the wake: $v = \Gamma/2h = T/2\rho A v_{\text{conv}}$. From $v = \kappa\sqrt{T/2\rho A}$, with κ the factor increasing the induced velocity due to nonuniform loading, $v_{\text{conv}} = \frac{1}{\kappa}\sqrt{T/2\rho A}$. Here the loading is fixed and the vertical convection is perturbed by the tip-path-plane motion: $\delta v_{\text{conv}} = -\dot{\alpha}_y x + \dot{\alpha}_x y$. So

$$\lambda_c x + \lambda_s y = \delta v = -\frac{T}{2\rho A v_{\text{conv}}^2}\delta v_{\text{conv}} = -\kappa^2 \delta v_{\text{conv}} = -\kappa^2(-\dot{\alpha}_y x + \dot{\alpha}_x y) \quad (11.159)$$

This result, $K_R = \kappa^2$, gives the sign of the effect but underestimates the magnitude.

For hover, $K_R = 1.5$ to 2.2 has been found using vortex theory, free wake analysis, and parameter identification. Keller (1996) used vortex theory to calculate the induced velocity along the rotor center-line in hover and found an approximately linear variation (λ_c) due to pitch rate, with $K_R = 1.5$. Including this effect in the dynamic inflow model changed the sign of the off-axis response (pitch rate due to lateral cyclic and roll rate due to longitudinal cyclic), as found in flight test. Keller and Curtiss (1998) used the tip-path-plane rate instead of just the shaft rate. For forward flight, they found $K_R \cong 0$ by $\mu = 0.1$ to 0.15. Bagai, Leishman, and Park (1999) found $K_R = 1.75$ to 2.25 using a free wake analysis. Fletcher (1995) predicted the off-axis response by introducing a phase lag in the blade aerodynamic loads, equal to an azimuthal phase shift of ψ_a in the non-rotating frame. By identification from flight test data, $\psi_a = 39°$ in hover and $\psi_a = 15°$ at 80 knots. Mansur and Tischler (1998) identified $\psi_a = 39°$ in hover and $\psi_a = 16°$ at $\mu = 0.19$ for the UH-60A, and $\psi_a = 36°$ in hover and $\psi_a = 19°$ at $\mu = 0.14$ for the AH-64. The phase lag approaches the two-dimensional value ($\psi_a \cong 11°$) for $\mu = 0.25$. Schulein, Tischler, Mansur, and Rosen (2002) found by identification for the UH-60A either $\psi_a = 43$ to 45° or $K_R = 1.7$ to 2.0. Fixing $\psi_a = 11°$, they found $K_R = 1.4$. Arnold, Keller, Curtiss, and Reichert (1998) identified $\psi_a = 38°$ or $K_R = 2.2$ from UH-60A flight test data. Kramer, Gimonet, and von Grunhagen (2002) identified from Bo-105 flight test data $K_R = 2.5$ in hover and $K_R = 1.1$ at 80 knots for the roll response, and $K_R = 1.7$ in hover and $K_R = 1.6$ at 80 knots for the pitch response.

A higher-order dynamic wake model for rotor induced velocity was developed by Peters and He (1995). This is an unsteady wake theory for lifting rotors, based on the acceleration potential for an actuator disk. The induced inflow at the rotor disk is expressed in terms of a Fourier series azimuthally and polynomials radially, so the inflow velocity is defined by the degree of freedom vector α. The rotor loading is represented by the generalized force vector τ. The dynamic wake model is a system of first-order, ordinary differential equations for the inflow states α:

$$LM\dot{\alpha} + \alpha = L\tau \tag{11.160}$$

where the derivative matrix is $L = \partial \alpha / \partial \tau$ and the diagonal mass matrix M gives the time lag LM.

11.4 History

Carpenter and Fridovich (1953) examined rotor response to collective changes in hover, motivated by helicopter jump takeoff. They tested a three-bladed articulated rotor, measuring the transient thrust and flapping response. To predict the effect of a rapid blade-pitch increase, they extended momentum theory by adding a time lag: $T = m\dot{v} + 2\pi R^2 \rho v(v + V_v)$, where v is the instantaneous induced velocity. Regarding the mass m, they wrote,

> If the induced velocity is assumed to be uniform over the rotor disk, the initial flow field [for sudden increase in pitch angle] is exactly analogous to the flow field produced by an impervious disk which is moved normal to its plane. The "apparent additional mass" of fluid associated with an accelerating impervious disk is given in [Munk (1924)] as 63.7 percent of the mass of fluid in the circumscribed sphere.

or $\frac{2}{\pi}(\frac{4}{3}\pi a^3)\rho$; Munk cited Tuckerman (1925). Rebont, Valensi, and Soulez-Lariviere (1960a, 1960b, 1961) considered the response of hovering rotor thrust to collective pitch changes, making the distinction that their interest was descending flight (landing) and near autorotation, not takeoff as in Carpenter and Fridovich. They extended momentum theory "by adding an inertia term, $m\dot{\lambda}_i U$ [U is the tip speed], where m represents the virtual mass associated with the disk, which can be assumed equal to that associated with a solid disk in nonuniform translation perpendicular to its plane." The model was verified by experiments.

Curtiss and Shupe (1970) considered the influence of quasistatic inflow variation on hingeless rotor response, expressing the result as an effective Lock number. Essentially this was a derivation of the static lift deficiency function, appearing in the Lock number by way of the lift-curve slope.

11.4 History

Ormiston and Peters (1972) investigated the modeling requirements for accurate prediction of hingeless rotor response. "Simplified models of the nonuniform induced inflow were derived, using momentum and vortex theory, and found to be the most important factor in improving correlation with the data." The inflow variables were the mean and 1/rev terms in a Fourier series, uniform along the blade for a given azimuth (not uniform plus linear over the disk). The three inflow variables were expressed as a linear combination of the rotor aerodynamic thrust and hub moment perturbations. For high advance ratio it was necessary to identify the inflow derivatives with load from wind-tunnel test data of hingeless rotor response, consisting of steady thrust and hub moment response to collective and cyclic pitch control. These identified derivatives exhibited anomalous behavior around $\mu = 0.8$.

Hohenemser and Crews (1973) tested a two-blade hingeless rotor in hover and at $\mu = 0.2$, measuring the blade flap response to cyclic pitch (stick stir). Crews, Hohenemser, and Ormiston (1973) developed an analysis to predict this test data. Following Carpenter and Fridovich (1953), they introduced a time lag in the inflow model: $\tau \dot\lambda + \lambda = L(2\gamma/\sigma a)C$, for the cosine and sine 1/rev harmonics of the inflow (λ), in response to hub aerodynamic pitch and roll moments (C). For three-bladed rotors, the result is a lift deficiency function or effective Lock number that depends on excitation frequency. When appropriate values (dependent on collective) of τ and L were used, good prediction of the measured flap response was obtained. Banerjee, Crews, Hohenemser, and Yin (1977) identified values of τ and L. Although the tests were conducted by measuring rotor frequency response, the analysis was no longer in the frequency domain (as with Theodorsen and Loewy), but rather finite state models in the time domain.

Peters (1974) continued the investigation of modeling requirements for hingeless rotor response, using a dynamic rather than quasistatic inflow model, and comparing it with frequency response as well as static measurements. Peters characterized previous work in this way: "Unfortunately, while some success has been achieved using simple models of the rotor induced flow in hover, a completely satisfactory induced flow model for forward flight has not been found, not even for the condition of steady response. In addition, neither the physical values of the induced flow time constants nor the frequency range in which they are important is known." The three state dynamic inflow model was used, with inflow variables representing uniform and linear variation over the disk. Following Carpenter and Fridovich (1953), who obtained good agreement with transient thrust measurements, a time lag was included in the relation between inflow and loading variables. "An approximation to the apparent mass terms of a lifting rotor can be made in terms of the reaction forces (or moments) on an impermeable disk which is instantaneously accelerated (or rotated) in still air" (Peters (1974)). These apparent mass values were obtained from Tuckerman (1925). The resulting non-dimensional constants ($\tau = K_m/2v$ for uniform variable, $\tau = 2K_I/v$ for linear first-harmonic variables) were $K_m = \frac{8}{3\pi}$ and $K_I = \frac{16}{45\pi}$, the latter agreeing with the identified τ value of Crews, Hohenemser, and Ormiston (1973). Good agreement with hingeless rotor response measurements was obtained using inflow/loading derivatives from momentum theory for hover and from the empirical model of Ormiston and Peters (1972) for forward flight.

Kinner (1937) used the acceleration potential approach of Prandtl to solve for the flow around a circular wing. The paper begins, "This work was intended initially as a contribution to the autogiro theory. In order to limit the scope the cross-flow of the disk was assumed to be zero, thus the disk can be replaced as a fixed wing in stationary flow." A footnote to this paragraph stated that the idea originated from Dr. Küssner, AVA Göttingen. The work was Kinner's dissertation at the University of Göttingen under Prof. Ludwig Prandtl. Kinner developed a separation of variables solution for a circular wing.

Mangler (1948a, 1948b; Mangler and Squire 1950) used separation of variables in elliptical coordinates to solve Laplace's equation for the acceleration potential of a circular actuator disk. He had available an English translation of Kinner's paper. Mangler evaluated for hover the derivatives of uniform and linear inflow variables with thrust and hub moment loading variables. In an amazing analytical effort, he evaluated the integrals required to obtain the uniform and longitudinal inflow due to thrust for the actuator disk in forward flight (a skewed cylindrical wake).

The dynamic inflow model of Pitt and Peters (1981) was based on the actuator disk models of Kinner (1937) and Mangler and Squire (1950), as presented by Joglekar and Loewy (1970). Pitt and Peters described the objective of their work and its background:

> A linear, unsteady theory is developed that relates transient rotor loads (thrust, roll moment, and pitch moment) to the overall transient response of the rotor induced-flow field. The relationships are derived from an unsteady, actuator-disc theory; and some are obtained in closed form....
>
> It has been known for some thirty years that the induced-flow field associated with a lifting rotor responds in a dynamic fashion to changes in either blade pitch (i.e. pilot inputs) or rotor flapping angles (i.e. rotor or body dynamics) [Amer (1950), Sissingh (1951), Carpenter and Fridovich (1953)]. In recent years, it has been found that dynamic inflow for steady response in hover can be treated by an equivalent (i.e. reduced) Lock number [Shupe (1970)]. For more general conditions, such as transient motions or a rotor in forward flight, it has been determined that the induced flow can be treated by additional "degrees of freedom" of the system. Each degree of freedom represents a particular inflow distribution, and each has its own particular gain and time constant [Ormiston and Peters (1972), Peters (1974), Crews, Hohenemser, and Ormiston (1973)].
>
> Although the above results have provided some impressive correlation with experimental data, there is still no general theory to predict the gains and time-constants of dynamic inflow. Values from momentum theory give excellent results in hover, but are clearly inadequate in forward flight [Ormiston and Peters (1972), Peters (1974)]. A simple vortex model [Ormiston and Peters (1972)], gives some improvement in forward flight but is still not satisfactory. An empirical model based on the best fit of response data [Ormiston and Peters (1972), Peters (1974)] gives excellent results; but several peculiar singularities remain unexplained. Thus there is a need to determine the dynamic-flow behavior from fundamental, aerodynamic considerations.

Representing the inflow distribution over the rotor disk by uniform plus linear terms, $\lambda = \lambda_0 + \lambda_c r \cos\psi + \lambda_s r \sin\psi$, "the dynamic inflow models of Peters (1974) and Crews, Hohenemser, and Ormiston (1973) assume that the inflow is related to the aerodynamic loads in a linear, first-order fashion.... The purpose of this research is to find the elements of L and M from basic aerodynamic principles and to also investigate the validity of this linear, first-order form" (Pitt and Peters (1981)).

Pitt and Peters (1981) obtained a solution for the incompressible potential flow about the rotor, which was represented by an actuator disk. The coefficients of the loading distribution are related to the integrated forces on the disk, with thrust and pitch and roll moments obtained only from the lowest-order solutions. The induced velocity is represented by a series. Exact, analytical solutions are possible for axial and edgewise flow. Mangler (1948b) derived the Fourier series for the induced velocity (uniform and longitudinal gradient) from thrust, as analytical functions of the wake skew angle χ. Pitt numerically evaluated the roll and pitch moment response, matching the exact values for axial and edgewise flow, and, from these results, identified analytical functions for the moment terms in χ. In summary,

> An actuator-disc theory has been used to obtain gains and time constants (i.e. the L and M matrices) for both 3-degree-of-freedom and 5-degree-of-freedom dynamic-inflow models.... In axial flow (e.g. hover), the gains are identical to those obtained from simple momentum theory.... The apparent mass terms (the M matrix) for the simplest pressure distributions are identical to the apparent mass terms of an impermeable disc, but these values vary significantly with lift distribution. (Pitt and Peters (1981))

The result was the unsteady, actuator disk theory that is the basis of dynamic inflow models of the rotor wake:

Gaonkar and Peters (1986, 1988) summarized dynamic inflow research and the verification of the Pitt and Peters model by comparison of hingeless rotor response measurements (particularly the data from Kuczynski and Sissingh (1972)), both static and dynamic, in hover and in forward flight.

11.5 REFERENCES

Amer, K.B. "Theory of Helicopter Damping in Pitch or Roll and a Comparison with Flight Measurements." NACA TN 2136, October 1950.

Arnold, U.T.P., Keller, J.D., Curtiss, H.C., Jr., and Reichert, G. "The Effect of Inflow Models on the Predicted Response of Helicopters." Journal of the American Helicopter Society, *43*:1 (January 1998).

Bagai, A., Leishman, J.G., and Park, J. "Aerodynamic Analysis of a Helicopter in Steady Maneuvering Flight Using a Free-Vortex Rotor Wake Model." Journal of the American Helicopter Society, *44*:2 (April 1999).

Banerjee, D., Crews, S.T., Hohenemser, K.H., and Yin, S.K. "Identification of State Variables and Dynamic Inflow from Rotor Model Dynamic Tests." Journal of the American Helicopter Society, *22*:2 (April 1977).

Carpenter, P.J., and Fridovich, B. "Effect of a Rapid Blade-Pitch Increase on the Thrust and Induced-Velocity Response of a Full-Scale Helicopter Rotor." NACA TN 3044, November 1953.

Castles, W., Jr. "Approximate Solution for Streamlines About a Lifting Rotor Having Uniform Loading and Operating in Hovering or Low-Speed Vertical-Ascent Flight Conditions." NACA TN 3921, February 1957.

Chaimovich, M., Rosen, A., Rand, O., Mansur, M.H., and Tischler, M.B. "Investigation of the Flight Mechanics Simulation of a Hovering Helicopter." American Helicopter Society 48th Annual Forum, Washington, DC, June 1992.

Coleman, R.P., Feingold, A.M., and Stempin, C.W. "Evaluation of the Induced-Velocity Field of an Idealized Helicopter Rotor." NACA ARR L5E10, June 1945.

Crews, S.T., Hohenemser, K.H., and Ormiston, R.A. "An Unsteady Wake Model for a Hingeless Rotor." Journal of Aircraft, *10*:12 (December 1973).

Curtiss, H.C., Jr., and Shupe, N.K. "A Stability and Control Theory for Hingeless Rotors." American Helicopter Society 27th Annual National V/STOL Forum, Washington, DC, May 1971.

Drees, J.M. "A Theory of Airflow Through Rotors and Its Application to Some Helicopter Problems." Journal of the Helicopter Association of Great Britain, *3*:2 (July-August-September 1949).

Fletcher, J.W. "Identification of Linear Models of the UH-60 in Hover and Forward Flight." Twenty-First European Rotorcraft Forum, Saint Petersburg, Russia, September 1995.

Gaonkar, G.H., and Peters, D.A. "Effectiveness of Current Dynamic-Inflow Models in Hover and Forward Flight." Journal of the American Helicopter Society, *31*:2 (April 1986).

Gaonkar, G.H., and Peters, D.A. "Review of Dynamic Inflow Modeling for Rotorcraft Flight Dynamics." Vertica, *12*:3 (1988).

Harding, J.W., and Bass, S.M. "Validation of a Flight Simulation Model of the AH-64 Apache Attack Helicopter Against Flight Test Data." American Helicopter Society 46th Annual Forum, Washington, DC, May 1990.

Hohenemser, K.H., and Crews, S.T. "Model Tests on Unsteady Rotor Wake Tests." Journal of Aircraft, *10*:1 (January 1973).

Joglekar, M., and Loewy, R. "An Actuator-Disc Analysis of Helicopter Wake Geometry and the Corresponding Blade Response." USAAVLABS TR 69-66, December 1970.

Keller, J.D. "An Investigation of Helicopter Dynamic Coupling Using an Analytical Model." Journal of the American Helicopter Society, *41*:4 (October 1996).

Keller, J.D., and Curtiss, H.C., Jr. "A Critical Examination of the Methods to Improve the Off-Axis Response Prediction of Helicopters." American Helicopter Society 54th Annual Forum, Washington, DC, May 1998.

Kinner, W. "Die Kreisförmige Tragfläche auf Potentialtheoretischer Grundlage." Ingenieur-Archiv, *8*:1, 1937. ("Theory of the Circular Wing," R.T.P. Translation No. 2345, issued by the Ministry of Aircraft Production.)

Kramer, P., Gimonet, B., and von Grunhagen, W. "A Systematic Approach to Nonlinear Rotorcraft Model Identification." Aerospace Science and Technology, *6*:8 (December 2002).

Kuczynski, W.A., and Sissingh, G.J. "Characteristics of Hingeless Rotors with Hub Moment Feedback Controls Including Experimental Rotor Frequency Response." NASA CR 114427, January 1972.

Mangler, K.W. "Calculation of the Induced Velocity Field of a Rotor." RAE Report No. Aero 2247, February 1948a.

Mangler, K.W. "Fourier Coefficients for Downwash at a Helicopter Rotor." RAE Tech Note Aero. 1958, May 1948b.

Mangler, K.W., and Squire, H.B. "The Induced Velocity Field of a Rotor." ARC R&M 2642, May 1950.

Mansur, M.H., and Tischler, M.B. "An Empirical Correction for Improving Off-Axes Response in Flight Mechanics Helicopter Models." Journal of the American Helicopter Society, *43*:2 (April 1998).

Miller, R.H. "Unsteady Air Loads on Helicopter Rotor Blades." Journal of the Royal Aeronautical Society, *68*:640 (April 1964).

Munk, M.M. "Some Tables of the Factor of Apparent Additional Mass." NACA TN 197, July 1924.

Ormiston, R.A., and Peters, D.A. "Hingeless Helicopter rotor Response with Nonuniform Inflow and Elastic Blade Bending." Journal of Aircraft, *9*:10 (October 1972).

Peters, D.A. "Hingeless Rotor Frequency Response with Unsteady Inflow." American Helicopter Society Specialists' Meeting on Rotor Dynamics, Moffett Field, CA, February 1974. (NASA SP-352, February 1974).

Peters, D.A., and He, C.J. "Finite State Induced Flow Models – Part II: Three-Dimensional Rotor Disk." Journal of Aircraft, *32*:2 (March-April 1995).

Pitt, D.M. "Rotor Dynamic Inflow Derivatives and Time Constants from Various Inflow Models." USATSARCOM TR 81-2, December 1980.

Pitt, D.M., and Peters, D.A. "Theoretical Prediction of Dynamic-Inflow Derivatives." Vertica, *5*:1 (1981).

11.5 References

Rebont, J., Soulez-Lariviere, J., and Valensi, J. "Response of Rotor Lift to an Increase in Collective Pitch in the Case of Descending Flight, the Regime of the Rotor Being Near Autorotation." NASA TT F-18, April 1960a (original paper September 1958).

Rebont, J., Valensi, J., and Soulez-Lariviere, J. "Wind-Tunnel Study of the Response in Lift of a Rotor to an Increase in Collective Pitch in the Case of Vertical Flight Near the Autorotative Regime." NASA TT F-17, April 1960b (original paper September 1958).

Rebont, J., Valensi, J., and Soulez-Lariviere, J. "Response of a Helicopter Rotor to an Increase in Collective Pitch for the Case of Vertical Flight." NASA TT F-55, January 1961 (original paper May-June 1959).

Schulein, G.J., Tischler, M.B., Mansur, M.H., and Rosen, A. "Validation of Cross-Coupling Modeling Improvements for UH-60 Flight Mechanics Simulations." Journal of the American Helicopter Society, *47*:3 (July 2002).

Shupe, N.K. "A Study of the Dynamic Motions of Hingeless Rotor Helicopters." USAECOM TR ECOM-3323, August 1970.

Sissingh, G.J. "The Effect of Induced Velocity Variation on Helicopter Rotor Damping in Pitch or Roll." ARC CP 101, November 1951.

Takahashi, M.D. "A Flight-Dynamic Helicopter Mathematical Model with a Single Flap-Lag-Torsion Main Rotor." NASA TM 102267, February 1990.

Tuckerman, L.B. "Inertia Factors of Ellipsoids for Use in Airship Design." NACA Report 210, 1925.

12   Stall

To maintain low drag and high lift, the flow over an airfoil section must remain smooth and attached to the surface. This flow has a rapid acceleration around the nose of the airfoil to the point of maximum suction pressure, and then a slow deceleration along the remainder of the upper surface to the trailing edge. The deceleration must be gradual for the flow to remain attached to the surface. At a high enough angle-of-attack, stall occurs: the deceleration is too large for the boundary layer to support, and the flow separates from the airfoil surface. The maximum lift coefficient at stall is highly dependent on the Reynolds number, Mach number, and the airfoil shape. Figure 8.12 shows $c_{\ell\max}$ values from 1.0 to 1.6 for various airfoils, corresponding to stall angles-of-attack of 10° to 16°. The unstalled airfoil has a low drag and a lift coefficient linear with angle-of-attack. The airfoil in stall at high angles-of-attack has high drag, a loss of lift, and an increased nose-down pitch moment caused by a rearward shift of the center of pressure. The aerodynamic flow field of an airfoil or wing in stall is complex, and for the rotary wing there are important three-dimensional and unsteady phenomena as well.

Stall of a helicopter rotor blade is characteristically manifested as high control system loads and helicopter vibration, accompanied by an increase in profile power. The alternating control loads show a gradual increase with speed until stall occurs and then show a sharp and large rise in magnitude (Figure 12.1). The rapid growth of the blade torsion and control system loads associated with stall is a major constraint on the helicopter speed, lift, and maneuver capability. Because of the flow axisymmetry, stall of a hovering rotor occurs as a limit cycle torsional oscillation of the blade, called stall flutter. In the periodically varying aerodynamic environment of the rotor in forward flight, stall occurs on the retreating blade. This stall phenomenon of forward flight is often called stall flutter as well. Sometimes stall has a role in the dynamic stability of the rotor, but usually the problem is the extremely high loads on the blade and control system. Because of the high rate of increase of the stall loads with speed or load factor, increasing the structural flight envelope by strengthening the control system is not very effective. Stall also limits the aerodynamic performance of the helicopter. When the rotor blade stall in forward flight persists into the highly loaded fourth quadrant of the disk, there is an increase in power required, a loss of lift and propulsive force capability, and a loss of control power. Rotor stall produces a significant vibration of the helicopter, which serves as a signal to the pilot of the onset of stall. With a mechanical control

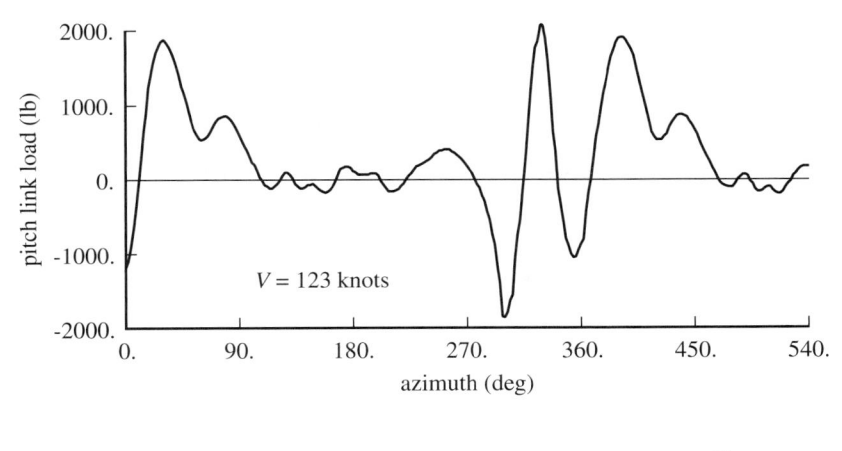

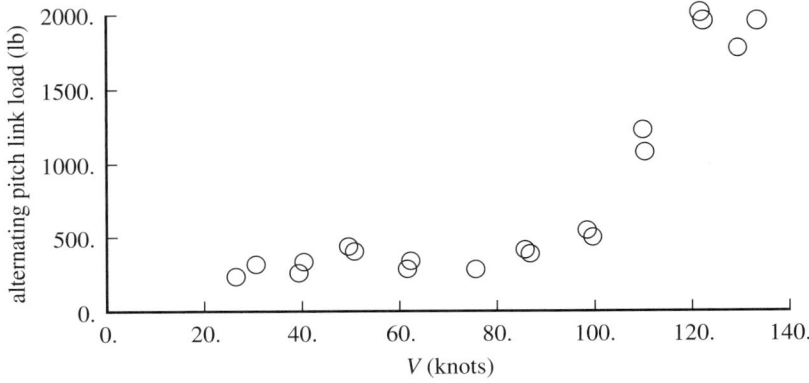

Figure 12.1. Influence of blade stall on rotor pitch link loads; from Tarzanin (1972).

system, the stall-induced torsional loads also produce a significant vibration in the cyclic and collective control sticks. Such vibration can be severe enough to be a limiting factor in itself. As a principal limitation of rotor performance and source of high loads and vibration in extreme operating conditions, stall is a major factor in the aerodynamic and structural design of the helicopter rotor and control system. Stall must be included in the analysis of the helicopter performance and aeroelastic behavior. It is difficult, however, to predict stall and all its effects, particularly in the complex three-dimensional and unsteady aerodynamic environment of the rotor blade.

12.1 Dynamic Stall

Dynamic stall is a flow phenomenon of rotor blades that involves large-scale, unsteady viscous effects. Although the yawed flow on a rotor blade does have an influence, the fundamental behavior is contained in the two-dimensional problem. Dynamic stall has been described by Carr, McAlister, and McCroskey (1977), McCroskey (1981), McCroskey and Pucci (1982), and Bousman (1998, 2001).

For an airfoil oscillating in pitch, a rapid increase in angle-of-attack can delay stall. When dynamic stall does occur, it is more severe and more persistent than static stall, with a large amount of hysteresis. The character of dynamic stall is determined primarily by the maximum angle-of-attack during the oscillation, with four regimes identifiable (McCroskey and Pucci (1982)). In the no-stall regime, the viscous-inviscid interaction is weak, and the vertical scale of the viscous zone is of the order of the boundary-layer thickness. For an NACA 0012 airfoil, this regime extends up to $\alpha_{\max} = 13°$ (for low Mach number). In this regime, there is little separation and the viscous effects are small. In the stall-onset regime (around $\alpha_{\max} = 14°$ for an NACA 0012 airfoil), there is some separation during part of the oscillation cycle. This regime gives the maximum lift possible with no significant penalty on section moment or drag. In the light-stall regime (around $\alpha_{\max} = 15°$ for an NACA 0012 airfoil), the viscous-inviscid interaction is strong, and the vertical scale of the viscous zone is of the order of the airfoil thickness. The airfoil loads show the usual static stall effects, plus phase lags and hysteresis. The pitch damping can be negative, a tendency strongest in this regime. The loads are most sensitive to the airfoil geometry, pitch rate, maximum angle-of-attack, and Mach number in the light-stall regime. In the deep-stall regime (around $\alpha_{\max} = 20°$ for an NACA 0012 airfoil), the flow is dominated by viscosity, with the vertical scale of the viscous zone on the order of the airfoil chord. The viscous zone exists over the upper surface during most of the cycle. The flow is characterized by shedding of a large vortex-like disturbance from the leading edge, which passes over the upper surface and produces section loads far in excess of the static values while the angle-of-attack is increasing; it is also characterized by a large amount of hysteresis.

Figure 12.2 illustrates the dynamic stall events on an oscillating airfoil (McCroskey (1981)). Point 1 is in the unseparated domain, with a thin, attached boundary layer. The section loads behave as predicted by linear, unsteady theory. At point 2, above the static stall angle, flow reversal within the boundary layer develops, but the lift continues to increase, extrapolated from the linear domain. For airfoils exhibiting trailing-edge stall (as in Figure 12.2), the reversed flow starts at the trailing edge and moves forward. For airfoils exhibiting leading-edge stall, the reversed flow develops quickly and very locally just downstream of the suction peak on the upper surface. For both types of airfoil, a vortex begins to evolve near the leading edge and spreads rearward over the upper surface (at less than half the free stream velocity), producing an area of high suction moving aft. At point 3, the pitching moment diverges (moment stall) and the drag begins to rise, but the lift is still increasing as a result of the vortex. At point 4, with the vortex near the trailing edge, the maximum lift (lift stall), drag, and moment occur (not simultaneously), followed by rapid drops. Secondary vortices can be shed, producing further load fluctuations (point 5). At point 6, the angle-of-attack is decreasing, with large hysteresis as the reattached flow develops from the leading edge.

The key characteristics of dynamic stall are the delay of the occurrence of stall and the vortex shed from the leading edge. Associated with the stall delay is significant hysteresis of the section lift and moment. The pressure disturbance produced by the leading-edge vortex as it passes over the airfoil upper surface first results in a large increase in lift and then in a large nose-down moment, followed rapidly by a large reduction of section loads and large-scale separation of the flow. The leading-edge vortex convection velocity is 25% to 40% of the free stream (Green, Galbraith, and Niven (1991)). The peak lift and moment depend primarily on the value of the angle-of-attack rate ($\dot{\alpha}c/V$) at the time stall occurs, as shown in Figure 12.3. At high pitch rates the peak lift coefficient can be as high as $c_{\ell\max} = 3.0$ and the peak nose-down moment coefficient as high as $c_{m\max} = -0.7$. Bousman (2001) showed that there is a general relation between the maximum lift coefficient,

12.1 Dynamic Stall

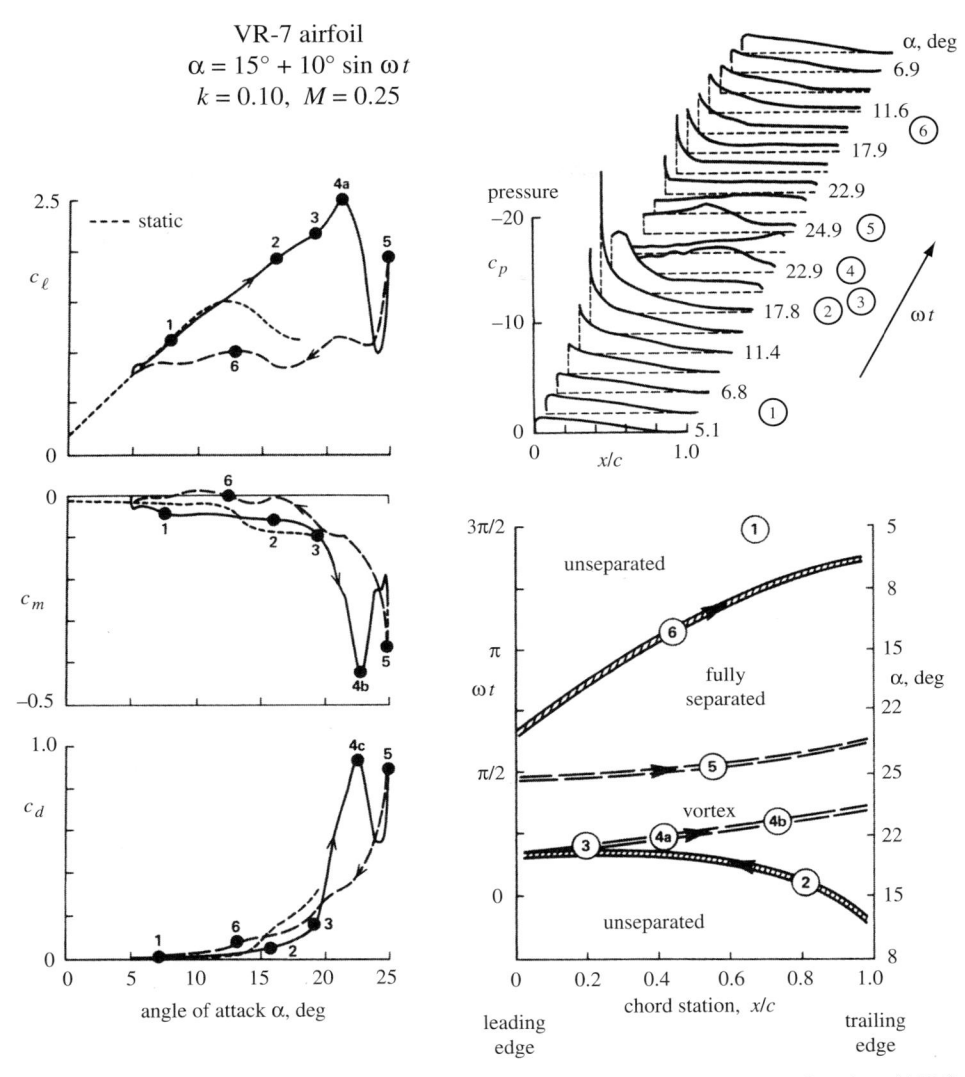

Figure 12.2. Dynamic stall events on an oscillating airfoil; adapted from McCroskey (1981).

minimum moment coefficient, and maximum drag coefficient measured during unsteady airfoil motion. The value of the unsteady $c_{\ell max}$ for small c_{mmin} is 0.05 to 0.12 higher than the static $c_{\ell max}$, reflecting the overshoot of lift in the light-stall regime. This is the lift increment that can be obtained during unsteady motion without a moment or drag penalty. The moment and additional lift produced by the leading-edge vortex are similar for all airfoils and exhibit only limited sensitivity to the mean pitch angle, alternating pitch angle, and reduced frequency. So increasing the airfoil static $c_{\ell max}$ is the most effective way to improve the airfoil (and hence rotor) dynamic lift capability. For $c_{mmin} < -0.1$, there is generally just one loading peak in lift and moment. For $c_{mmin} < -0.4$ or -0.5, two loading peaks are often observed, indicating secondary vortices have been shed.

Stall flutter can be examined in terms of the aerodynamic damping, or net aerodynamic work per cycle, of an airfoil oscillating in pitch. The data exhibit hysteresis

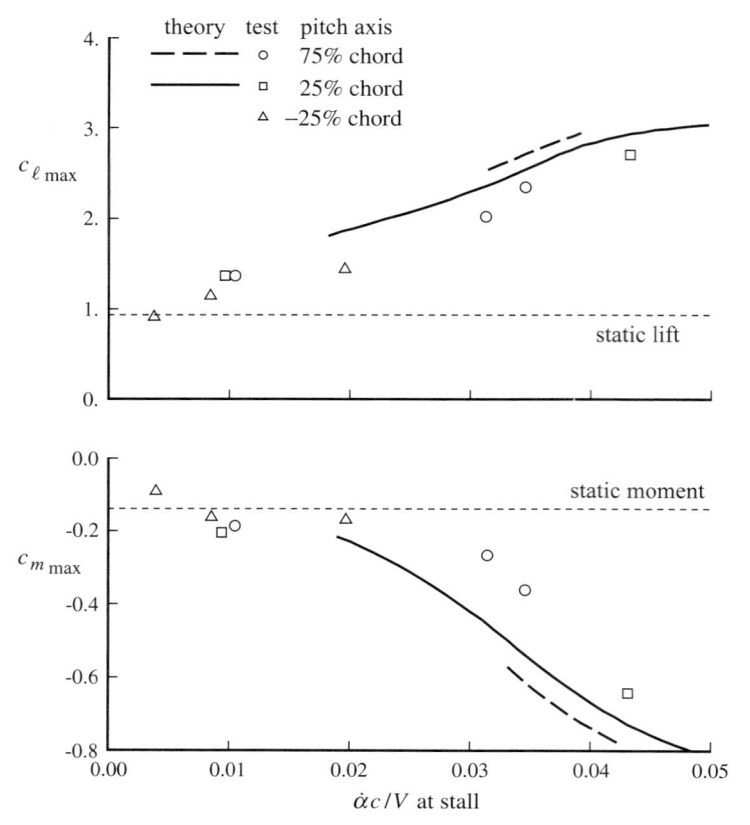

Figure 12.3. Peak lift and moment coefficients for a two-dimensional airfoil during a linear, transient increase in angle-of-attack through stall; from Ham and Garelick (1968).

loops; see Figure 12.4. The work done on the fluid by the airfoil is $W = -\oint c_m d\alpha$. For oscillations below or in stall, the damping is positive. For oscillations about a mean angle-of-attack near stall, the net pitch damping can be negative, and the airfoil extracts energy from the air. The largest negative damping typically occurs

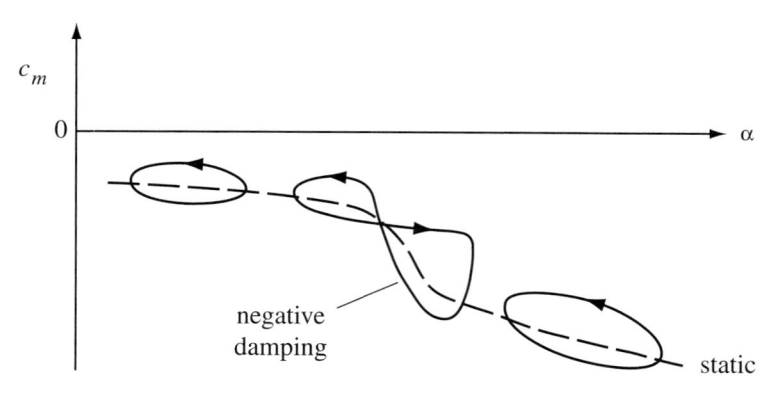

Figure 12.4. Typical unsteady moment data for an airfoil oscillating in pitch below, through, and in stall.

12.1 Dynamic Stall

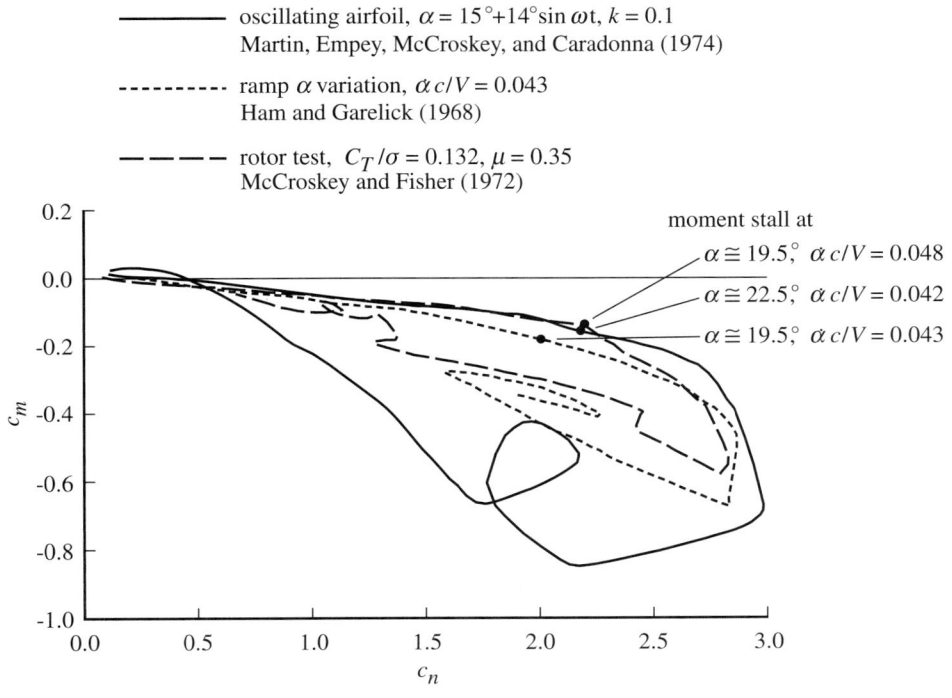

Figure 12.5. Dynamic stall loads measured in airfoil tests and on a rotor; from Martin, Empey, McCroskey, and Caradonna (1974).

at a reduced frequency $k = \omega c/2V \cong 0.3$ (Ham and Young (1966)). If the structural natural frequency is such that the airfoil oscillates at this frequency, a single-degree-of-freedom, limit cycle instability can occur. At a mean angle-of-attack near static stall, where the pitch damping is negative for small motion, the amplitude of the oscillatory motion increases until the net damping in pitch is zero.

Dynamic stall experiments have been conducted on two-dimensional airfoils, three-dimensional wings, and rotors. A convenient experimental arrangement consists of a two-dimensional airfoil oscillating about a pitch axis in a wind tunnel. The mean angle, oscillation amplitude, and oscillation frequency are typical of the aerodynamic environment of a rotary wing. The mean and oscillatory angles should be large and about equal, and the oscillation frequency should correspond to the rotor speed (for a 1/rev angle-of-attack variation). The pressure, section loads, and other quantities are measured during the oscillation cycle. Figure 12.2 is an example of such oscillatory airfoil data. There is actually a large scatter in the measured loads, particularly for decreasing angle-of-attack. The delay of stall due to the airfoil pitch rate is seen, as are the higher loads than in the static case. Such a presentation also shows the hysteresis of the unsteady loads: the lift and moment depend not just on the current angle-of-attack but also on the past history of the motions. The dynamic stall loads measured during oscillating airfoil tests and ramp angle-of-attack tests are similar, and both simulate well the loading measured on rotor blades in forward flight; see Figure 12.5. So the three-dimensional nature of the rotor flow does not fundamentally alter the section stall behavior.

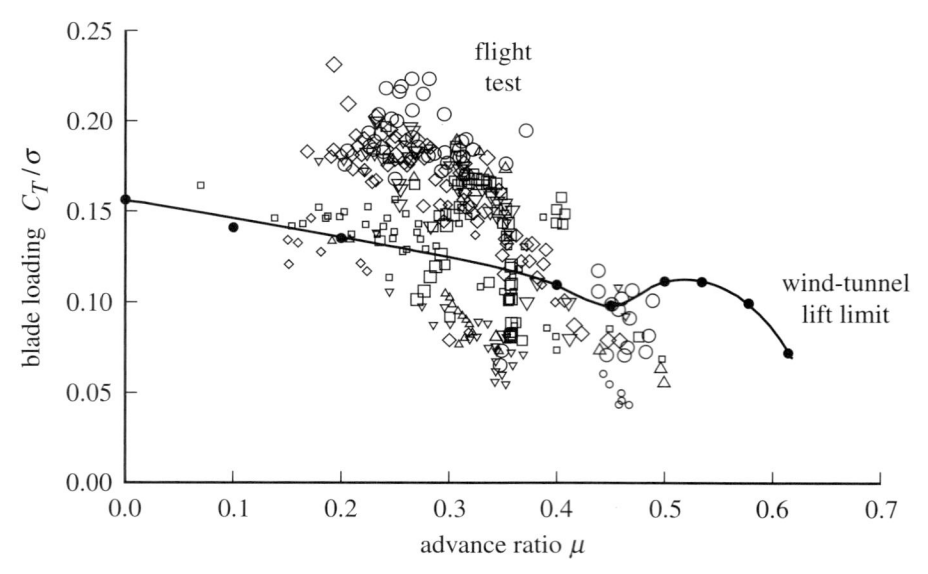

Figure 12.6. Flight test results (17 aircraft) for helicopter lift capability and steady wind-tunnel lift limit (McHugh (1978)).

12.2 Rotary-Wing Stall Characteristics

The effects of stall on the aerodynamics of the entire rotor follow from the effects on the blade section. The limit on the blade lift capability together with the drag increase results in a limit on the rotor thrust, a sharp increase in the rotor profile power at high loading, a reduced propulsive force capability, and high flapwise and chordwise bending stresses. The section center-of-pressure shift produces high blade torsional and control system loads. The helicopter vibration increase is due to the rapid change of the aerodynamic forces when the blade stalls. Stall over a limited portion of the disk can be tolerated, but as the stall area increases these effects eventually become objectionable.

Significant stall of the rotary wing is encountered when the helicopter is operating in a condition requiring the blades to work at high angles-of-attack, particularly high thrust and high speed. The rotor blade loading is a measure of the mean lift coefficient: $\bar{c}_\ell = 6C_T/\sigma$ for hover (section 3.6.5). The lateral asymmetry of the aerodynamic environment in forward flight increases with the advance ratio μ, so at a given thrust stall eventually occurs on the retreating side. Thus the limiting thrust of the rotor, C_T/σ, is a decreasing function of the advance ratio μ. Figure 12.6 illustrates the stall limit in terms of the demonstrated lift capability of various helicopters. The rotor can encounter stall during maneuvers of the helicopter or in aerodynamic turbulence. Turns, pull-ups, and similar maneuvers involve an increase in the rotor thrust and therefore are limited by stall. Both the maneuver and the gust encounter cases benefit from the transient nature of the angle-of-attack changes, which tend to delay the occurrence of stall. As a result, the transient maneuver capability of the rotor is greater than the static load capability. The highest C_T/σ points in Figure 12.6 were obtained in maneuvers. Nonuniform inflow has a major influence on the angle-of-attack distribution on rotor blades (section 9.4) and hence on the rotor stall behavior.

In hover the flow is axisymmetric, so stall is expected to occur in an annulus on the rotor disk (ignoring the effects of blade motion and unsteady aerodynamics). As the thrust is increased, the blade angle-of-attack change is greatest at the tip because the induced velocity increase with thrust produces the smallest inflow angle change there. The tip region stalls first on a hovering rotor, at least for low twist. Large twist shifts the peak angle-of-attack inboard (Figure 3.18). In autorotation, where the net inflow is upward through the disk, the angle-of-attack is largest at the root. For the autorotating rotor, the root sections are expected to stall first (Figure 4.10). Maximizing the hover C_T/σ capability requires large negative twist of the blade, so most sections reach the maximum lift coefficient at the same time. If a hovering rotor is operating at high lift, a gust or other disturbance can trigger dynamic stall of the blade. The resulting blade torsion motion can establish an oscillation in and out of stall. The energy to sustain the oscillation comes from the hysteresis of the moment coefficient as a function of angle-of-attack during dynamic stall (see Figure 12.4). The oscillation is a limit cycle in which the balance of the negative damping in stall and the positive damping below stall determine the oscillation amplitude. This single-degree-of-freedom limit cycle instability is called *stall flutter*.

In forward flight, the largest lift coefficients occur on the retreating blade and increase with speed and thrust. The stall angle-of-attack is reached in the third quadrant, inboard at moderate C_T/σ and μ, extending outward along the blade at high thrust and speed; see Figures 9.14 and 9.15. The maximum angle-of-attack increases with propulsive force and extends farther outward for small twist. With a uniform inflow model, the calculated peak angles-of-attack occur in the third quadrant and midspan, the stall region extending to the tip at high thrust and speed; see Figures 6.18 and 6.19. Uniform inflow calculations show the maximum angles-of-attack on the retreating tip for zero twist. Early work on rotor stall was based on such uniform inflow results. Generally, the induced velocity tends to be higher at the retreating tip than is indicated by the mean inflow value, which alleviates the high angles-of-attack at the tip. The stall region is thus moved inboard and into the third quadrant of the disk. Nonuniform inflow also tends to increase the maximum angle-of-attack on the disk and to increase the rate of change of a near stall.

The high blade loads, control loads, and vibration characteristic of helicopter rotor stall are manifestations of the blade aeroelastic response to the high aerodynamic loads in stall. The blade motion, in turn, influences the angle-of-attack, and hence the aerodynamic forces. In particular, the large nose-down pitch moments due to stall produce substantial torsional motion of the blade, which directly changes the angle-of-attack. A calculation of the effects of stall on the helicopter rotor cannot be concerned with the aerodynamic forces alone, but instead, a complete aeroelastic analysis that includes the dynamic response of the blades is required. The local angle-of-attack must be calculated from the actual wake-induced, nonuniform inflow and the elastic torsion motion of the blade.

The aerodynamic environment of the rotor blade is three-dimensional and unsteady, and for stall in particular, the effects of this complex environment must be accounted for. The radial velocity along the blade results in significant yaw angles. Yawed flow delays the occurrence of stall, and also influences the nature of the separated flow. The rotor blade section in forward flight has a large 1/rev variation in angle-of-attack, so the section loads involve dynamic stall. An unsteady increase in the angle-of-attack delays the occurrence of stall, so the section is capable of higher lift in unsteady conditions than can be sustained under static conditions. After dynamic stall does occur, the transient lift and nose-down moment are much greater than the static stall loads. From the delay of stall, in forward flight, the rotor is capable of a higher thrust with no more stall effects than is implied by the static airfoil characteristics. The large transient loads on the section when dynamic stall occurs are the source of the high vibration and loads associated with rotor stall, particularly the blade torsion and control system loads, in response to the pitch moment of dynamic stall. When stall occurs on the retreating blade in forward flight, the resulting large transient moment twists the blade nose down. If the blade is sufficiently flexible in pitch, this nose-down motion reduces the angle-of-attack enough for the flow to reattach. With the return of attached flow loads, the blade rebounds up in pitch, overshooting the static stall level. The overshoot in pitch increases the angle-of-attack so that the blade stalls again. A rotor operating at sufficiently high thrust or speed can exhibit two or three dynamic stall events.

Figure 12.7 illustrates dynamic stall on a helicopter rotor, showing the occurrence of lift and moment stall, and the regions of separated flow; the maps were developed by Bousman (1998). The top map is for the UH-60A in level flight at high blade loading. The bottom map is for the UH-60A executing the UTTAS maneuver, at 2.1 g load factor (revolution 14). Moment stall was identified from the drop in moment coefficient, associated with formation of the leading edge vortex. Lift stall was identified from the drop in normal force coefficient, produced when the vortex moves downstream of the airfoil. Boundary layer separation was identified from the upper surface pressure at 96.3% chord. These features were essentially the same as observed in two-dimensional wind-tunnel tests. Figure 12.8 shows the airloads and structural loads for the level flight condition. Lift stall lags moment stall by the time the vortex takes to pass over the airfoil. Large separated flows occur at the trailing edge when the vortex leaves the trailing edge. The pressure data also show shocks moving forward on the blade in the second quadrant, reaching the leading edge as the first dynamic stall vortex forms. The dynamic stall events are thus compressible flow phenomena, with $M = 0.3$–0.5 on the retreating side. Stall occurs largely outboard on the UH-60A blade, extending to the blade tip. In contrast, a three-dimensional oscillating wing does not show tip stall. The first stall event on the retreating blade is triggered by the basic 1/rev variation in angle-of-attack, occurring first inboard. The second and third stall events reflect the torsion motion of the blade, occurring over the outer part of the blade nearly simultaneously, at a frequency of 4–5/rev. In the high load-factor case, the blade pitch oscillation persists into the first quadrant, producing the third dynamic stall event. The resulting large blade torsion moments and control loads typically size the control system.

Rotor stall is a major consideration in the design of a helicopter. The limit on the thrust coefficient to solidity ratio C_T/σ is determined by the requirement for an adequate stall margin in forward flight. So for a given gross weight, the quantity $A_{\text{blade}}(\Omega R)^2$ is determined. The combination of an advancing-tip Mach number limit (due to compressibility effects on performance and noise) and an advance ratio limit (due to stall and other factors) constrains the tip speed ΩR (section 8.4). Then the minimum blade area that must be provided to meet the stall margin requirement is defined. The fact that the blade loading limit decreases with speed suggests using a fixed wing on the helicopter to reduce the rotor lift required in forward flight. Reducing the helicopter drag (hence the rotor propulsive requirement) is also effective in improving the rotor stall characteristics, raising the limits on C_T/σ and μ. Stall is also a major concern in selecting the blade airfoil section

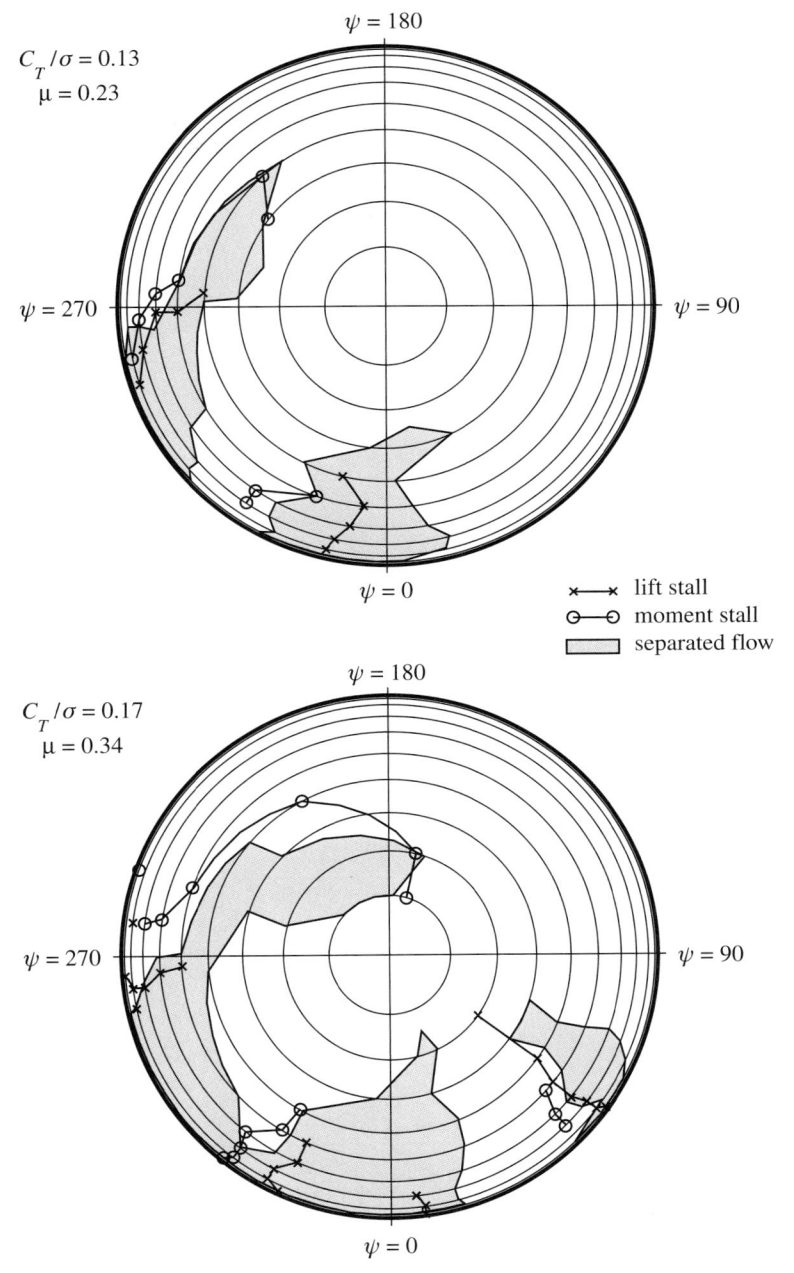

Figure 12.7. Dynamic stall rotor map, based on airloads measurements on UH-60A in flight; from Bousman (1998).

(section 8.7). An airfoil with high maximum lift coefficient at low to moderate Mach number is desired for the retreating blade stall environment.

12.3 Elementary Stall Criteria

The blade angle-of-attack or lift coefficient (the actual value or some representative value) is the primary criterion for stall of rotary wings. The rotor exhibits effects of

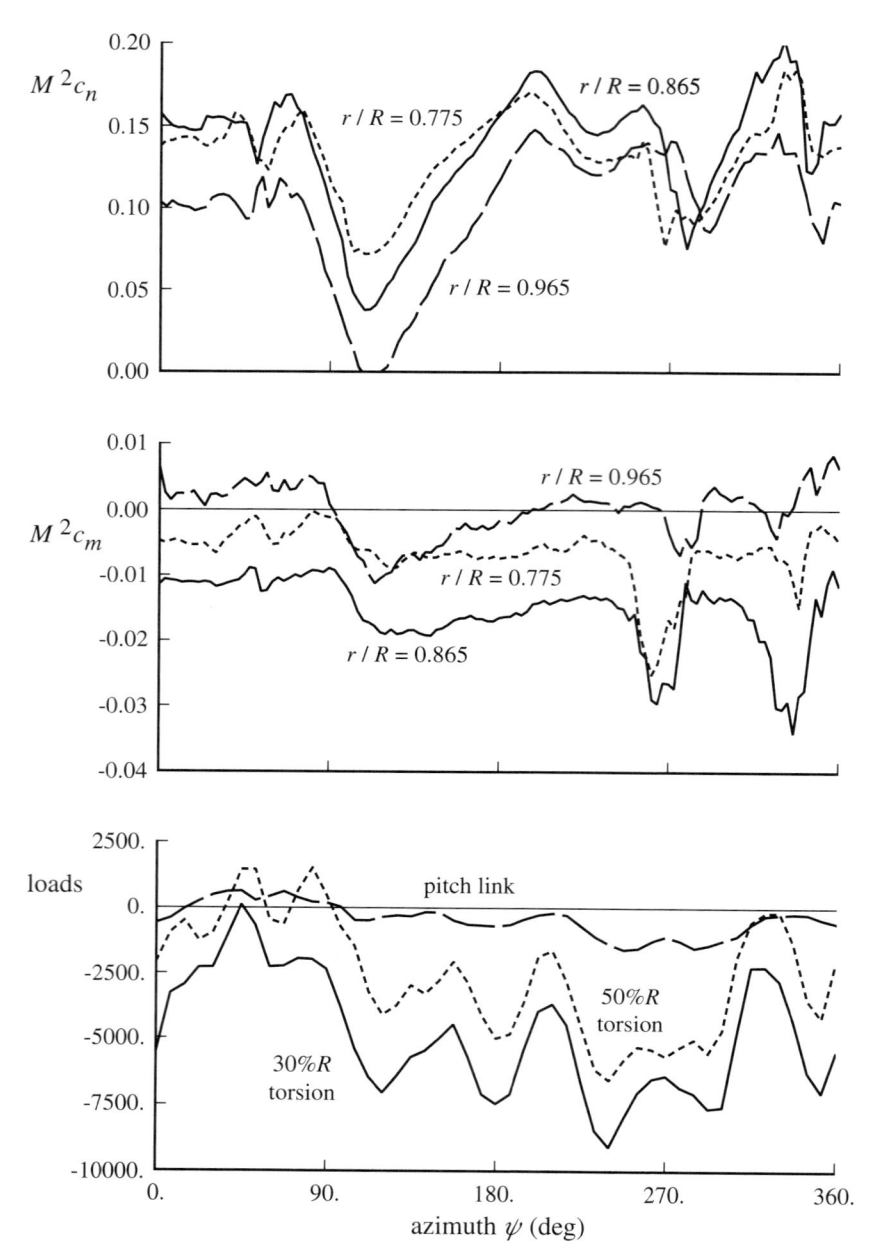

Figure 12.8. Influence of dynamic stall on airloads and structural loads of UH-60A at $C_T/\sigma = 0.13$ and $\mu = 0.23$; from Bousman (1998).

stall when the blade section angle-of-attack is above the stall angle over a significant portion of the rotor disk. Translation of a local aerodynamic criterion into rotor operating limits is complicated, and useful stall criteria have some empirical basis.

The NACA conducted a series of investigations directed at developing a means of predicting helicopter rotor stall; these investigations were summarized by Gessow and Myers (1952). The critical angle-of-attack on the rotor disk was identified by the following arguments. As the rotor thrust increases, the angle-of-attack change

12.3 Elementary Stall Criteria

due to the larger induced velocity is smallest at the tip. Consequently, the blade stalls first on the tip of the retreating blade, and the stall criterion is based on the angle-of-attack at $r/R = 1$ and $\psi = 270°$, denoted $\alpha_{1,270}$. For an unpowered rotor in forward flight (an autogyro), the rotor induced and profile losses increase as speed or thrust increases, requiring a larger rearward tilt of the tip-path plane to supply the rotor power. The resulting increased upward inflow velocity produces the greatest angle-of-attack increase on the inboard sections of the blade, and the autogyro rotor stalls first on the inboard sections of the retreating blade. The sections near the reverse flow region and at the root are at low dynamic pressure; the stall is of concern at radial stations where the normal velocity u_T is not too small. So for a rotor near autorotation, the stall criterion is based on the angle-of-attack at $u_T = 0.4$ and $\psi = 270°$, denoted $\alpha_{\mu+.4,270}$. The calculated maximum angle-of-attack is different when a nonuniform inflow model is used. The induced velocity tends to be higher at the retreating tip than is indicated by the mean inflow value, which alleviates the high angles-of-attack at the tip. The stall region is thus moved inboard and into the third quadrant of the disk. Nonuniform inflow also tends to increase the maximum angle-of-attack on the disk. Although nonuniform inflow is important for calculations of the actual rotor aerodynamics, a stall criterion based on uniform inflow analysis can still be useful, on the basis of correlation with measured stall effects.

Using an analysis such as Bailey (1941) to calculate the rotor performance and flapping in forward flight, the angle-of-attack on the retreating blade (the most critical of $\alpha_{1,270}$ and $\alpha_{\mu+.4,270}$) can be evaluated. This angle-of-attack depends primarily on C_T/σ and μ, with some dependence on total rotor power or tip-path-plane incidence. On the basis of rotor tuft behavior and the pilot's assessment of the vibration and control characteristics, Gustafson and Myers (1946) found that the angle-of-attack of the retreating blade tip correlated well with the helicopter stall behavior. This correlation with flight test results is the basis for using α calculated by a simple theory (uniform inflow, no stall in the airfoil lift and drag) as a criterion for the rotor stall. It was found that $\alpha_{1,270} = \alpha_{ss}$ corresponds to incipient stall and that $\alpha_{1,270} = \alpha_{ss} + 4°$ corresponds to excessive stall, where α_{ss} is the airfoil static stall angle-of-attack. For the NACA 0012 airfoil of the helicopter involved, $\alpha_{ss} = 12°$, and hence the boundaries of interest are $\alpha_{1,270} = 12°$ and $\alpha_{1,270} = 16°$. Below the boundary of incipient stall, there are no noticeable stall effects; above this boundary the vibration, loads, and power increase because of stall. At the boundary of excessive or objectionable stall, the helicopter is still controllable, but the loads and vibration have reached the limit for practical operation. Near autorotation these limits are applicable to $\alpha_{\mu+.4,270}$. Gustafson and Gessow (1947) found by flight tests that the increase in rotor profile power also correlated well with $\alpha_{1,270}$ (Figure 12.9). The ratio of the measured rotor profile power to the calculated value (obtained using a theory without stall effects) was around unity until $\alpha_{1,270}$ reached the static stall angle (12°). Above α_{ss} there was a sharp rise in the measured profile power, to about twice the predicted (no-stall) value at $\alpha_{1,270} = \alpha_{ss} + 4 = 16°$. At still higher angles-of-attack there are control difficulties as well. The observed effects of stall on the rotor were first a large increase in the rotor profile power and then vibration and loads large enough to limit the helicopter operation.

Fradenburgh (1960) introduced a criterion for rotor stall. Since stall effects can be identified by a rapid increase in the blade profile torque on the retreating side, the criterion is based on the maximum value of $NC_{Qo}/\sigma = \frac{1}{2} \int c_d u_T^2 r \, dr$ over $180° < \psi < 360°$; see Tanner (1964). A moderate degree of stall or the onset of

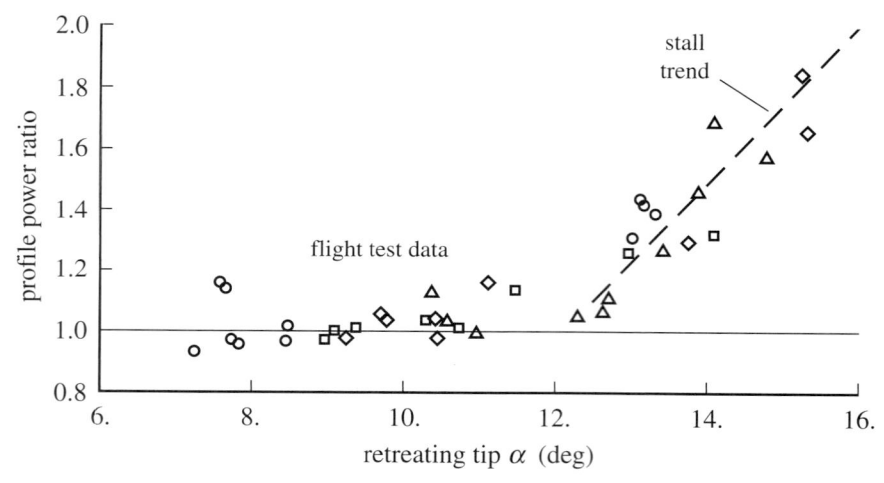

Figure 12.9. Variation of profile power ratio with calculated retreating tip angle-of-attack; from Gustafson and Gessow (1947).

significant stall phenomena occurs when $NC_{Q_o}/\sigma = 0.004$, corresponding to $c_d = 0.1$ and $u_T = 0.4$. The limit beyond which operation is undesirable is $NC_{Q_o}/\sigma = 0.008$ (Tanner (1964)).

McHugh (1978) conducted a wind-tunnel test to determine rotor aerodynamic lift limits in forward flight. Figure 12.10 shows the results for four values of propulsive force. The maximum rotor lift achievable for each speed and propulsive force was determined by blade stall, not power or structural load limitations. Wind-tunnel measurements provide the steady-state, sustained operation limit. A maneuvering helicopter can exceed this limit, as illustrated in Figure 12.6. Typical current helicopter drag levels correspond to $X/qD^2\sigma = C_D/(\frac{4}{\pi}\sigma) = 0.10$ (section 8.6). Reducing aircraft drag and hence rotor propulsive force requirement increases the lift capability. Yeo (2003) used comprehensive analysis to calculate the limits for this rotor. Good correlation with test data was obtained when the airfoil tables were corrected for the model-scale Reynolds number (lower calculation results in Figure 12.10). Using full-scale airfoil tables increased the lift limit by about $\Delta C_T/\sigma = 0.01$. Figure 12.10 also shows wind tunnel measurements for a model rotor (Floros, Gold, and Johnson (2004)), and the full-scale UH-60A rotor (Norman, Shinoda, Peterson, and Datta (2011)). McCloud and McCullough (1958) conducted wind-tunnel tests of the stall characteristics of a rotor in forward flight at $\mu = 0.3$ to 0.4. They investigated the performance gains possible when blade airfoil sections with increased maximum lift due to camber were used to delay retreating blade stall. The stall boundary was defined by a marked change in the rotor torque and in the characteristics of the blade torsional moment. These two criteria gave essentially the same boundaries of maximum C_T/σ as a function of μ. The improved stall characteristics of the cambered airfoil increased the rotor lift capability by about 15% (Figure 12.10). All the rotors involved in these tests had first-generation airfoils (see Figure 8.12).

Blade element theory for a hovering rotor provides estimates of the stall limit on rotor lift capability. The ideal rotor (section 3.6.7) has twist $\theta = \theta_t/r$ to obtain uniform inflow. The angle-of-attack is $\alpha = \alpha_t/r = (\theta_t - \lambda)/r$, so the ideal rotor always stalls at the blade root. From equation 3.132, $c_\ell = \frac{4}{B^2 r}(C_T/\sigma)$. As long as the stall region is confined to the root, where the dynamic pressure is low, stall effects are

12.3 Elementary Stall Criteria

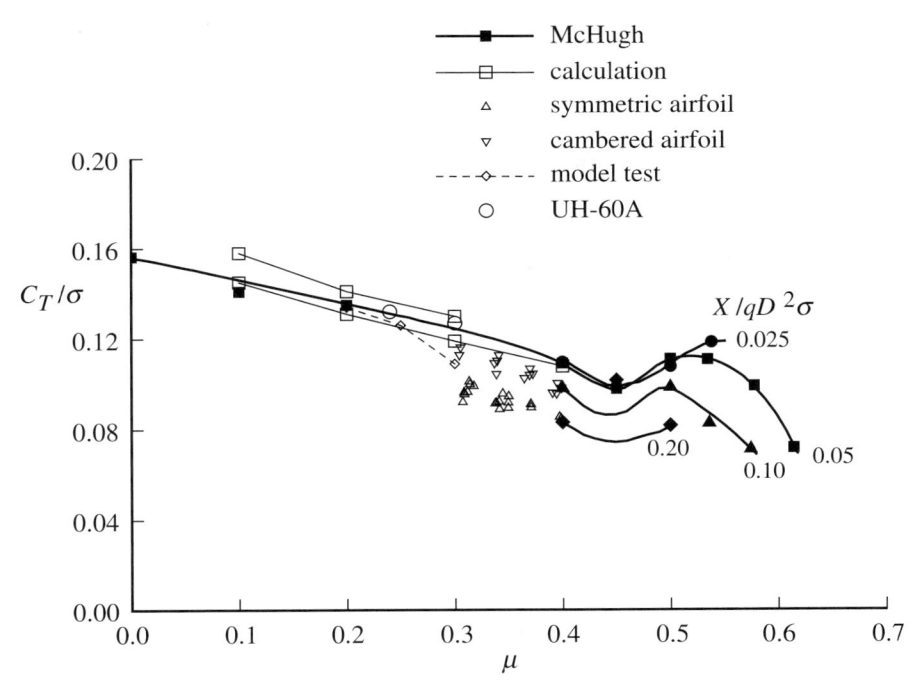

Figure 12.10. Measured rotor aerodynamic lift limits.

not significant. When stall occurs outboard of a critical radius r_crit, stall effects are objectionable. So the hover limit is

$$\left(\frac{C_T}{\sigma}\right)_\text{max hover} = \frac{B^2}{4} r_\text{crit}\, c_{\ell\text{max}} \qquad (12.1)$$

The observed limits in hover correspond to about $r_\text{crit} = 0.5$. The optimum rotor (section 3.6.8) has twist and taper for uniform inflow and constant angle-of-attack, so the entire blade stalls at once. From equation 3.137, $c_\ell = \frac{6}{B^2}(C_T/\sigma_e)$ (using equivalent solidity $\sigma_e = \frac{3}{2}\sigma_t$). Then the maximum hover lift capability is

$$\left(\frac{C_T}{\sigma_e}\right)_\text{max hover} = \frac{B^2}{6} c_{\ell\text{max}} \qquad (12.2)$$

This is the best possible performance, since at stall the entire blade is working at the maximum lift coefficient. For the optimum rotor both the twist and taper depend on the operating condition.

For a simple stall criterion in forward flight, the lift coefficient on the retreating blade is estimated in terms of the mean lift coefficient. The lateral asymmetry of the rotor angle-of-attack is due to the variation of dynamic pressure with azimuth, so the blade loading $u_T^2 c_\ell$ is equated to the mean loading $r^2 \bar{c}_\ell$, giving $c_\ell = \bar{c}_\ell (r/u_T)^2$. The rotor mean lift coefficient is proportional to the blade loading, $\bar{c}_\ell = 6(C_T/\sigma)K$, where from hover the empirical parameter K is expected to be approximately 1. The maximum lift coefficient occurs on the retreating blade at $\psi = 270°$. Thus the estimate of the rotor lift limit is

$$\frac{C_T}{\sigma} = \frac{c_{\ell\text{max}}}{6K}\left(1 - \frac{\mu}{r}\right)^2 \qquad (12.3)$$

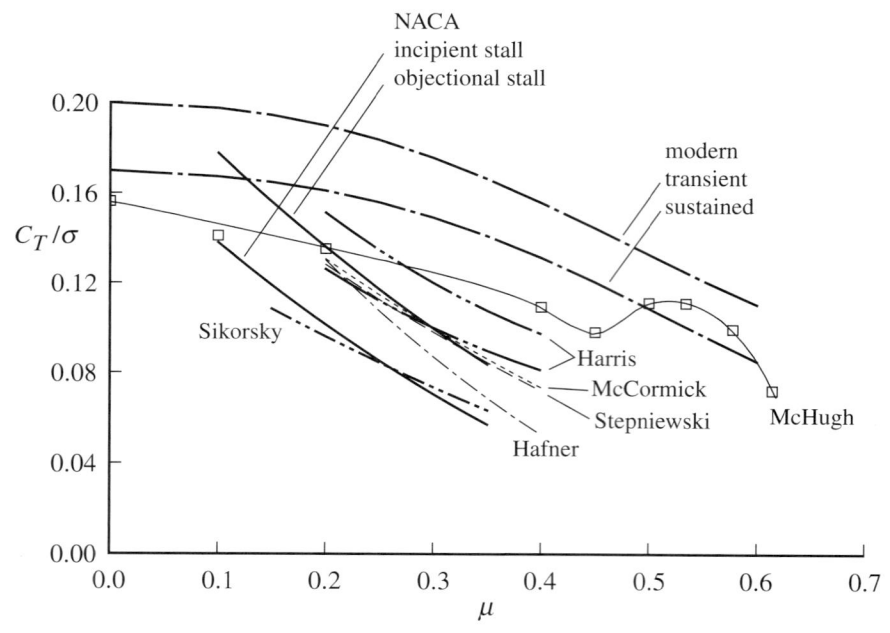

Figure 12.11. Elementary stall criteria.

The parameters in this criterion are chosen on the basis of experimental data to give the stall limit in forward flight at $\mu = 0.2$ to 0.4 typically. The resulting value at $\mu = 0$ is higher than the hover stall limit. A number of parameter values are found in the literature:

	r	$c_{\ell\max}/K$
Hafner (1949)	0.75	$c_{\ell\max}$
Stepniewski (1955)	1.0	1.2
NACA, incipient stall	0.87	1.08
NACA, objectionable stall	0.89	1.35
Sikorsky (1960), infinite blade life	1.0	0.9
McCormick (1967)	1.0	$c_{\ell\max}/(1.06 - 0.9\theta_{tw})$

Sikorsky's result is the limit of infinite blade life, based on helicopter flight tests for $\mu = 0.15$ to 0.35. Gessow and Myers (1952) presented curves for an untwisted rotor, based on the NACA stall research, which are approximated (for $D/L = 0.1$) using the parameters in the table. Figure 12.11 plots these criteria, using $c_{\ell\max} = 1.5$ for Hafner and $c_{\ell\max} = 1.45$ for McCormick. For reference, McHugh's wind-tunnel results are also shown. These criteria represent early helicopter technology. Figure 12.11 shows approximate modern technology lift limits for sustained and transient operation.

Harris (1987) developed a stall criterion for forward flight by considering a blade lift coefficient with a 1/rev variation: $c_\ell = c_{\ell 0} - c_{\ell s} \sin\psi - c_{\ell c}\cos\psi$. The thrust is $C_T = \frac{\sigma}{2}\int u_T^2 c_\ell dr$ (averaged over azimuth). Zero pitch moment requires $c_{\ell c} = 0$, and zero roll moment gives $c_{\ell s} = \frac{8}{3}\mu c_{\ell 0}/(1 + \frac{3}{2}\mu^2)$. Thus

$$\frac{C_T}{\sigma} = c_{\ell 0}\left(\frac{1}{6} + \frac{1}{4}\mu^2\right) - c_{\ell s}\frac{\mu}{4} = \frac{c_{\ell 0}}{6}\frac{1 - \mu^2 + \frac{9}{4}\mu^4}{1 + \frac{3}{2}\mu^2} \qquad (12.4)$$

The maximum lift coefficient is on the retreating blade at $\psi = 270°$: $c_{\ell\max} = c_{\ell 0} + c_{\ell s}$. Thus the rotor lift limit in forward flight is

$$\frac{C_T}{\sigma} = \frac{c_{\ell\max}}{6} \frac{1 - \mu^2 + \frac{9}{4}\mu^4}{1 + \frac{8}{3}\mu + \frac{3}{2}\mu^2} \qquad (12.5)$$

This limit is plotted in Figure 12.11 using $c_{\ell\max} = 1.25$ and $c_{\ell\max} = 1.5$, matching the symmetrical and cambered airfoil results of McCloud and McCullough (1958) (Figure 12.10). From $\int c_\ell u_T^2 dr = \bar{c}_\ell \int u_T^2 dr$, the mean lift coefficient in forward flight is

$$\bar{c}_\ell = 6 \frac{C_T}{\sigma} \frac{1}{1 + \frac{3}{2}\mu^2} \qquad (12.6)$$

So $\bar{c}_\ell$ decreases in forward flight because of the increased mean dynamic pressure.

12.4 Empirical Dynamic Stall Models

Calculation of the loads on an airfoil or wing or rotor blade undergoing dynamic stall is challenging. In vortex methods, discrete vortices are shed from the airfoil and tracked in the flow field, generally using the Biot-Savart equation. Vortex methods readily predict the effects of the leading-edge vortex, but the viscous aspects of the problem are prescribed, modeled, or approximated. Navier-Stokes calculations of dynamic stall are now often performed, but issues still remain in treating large separation regions, particularly concerning turbulence models.

Empirical models are used for efficient calculation of dynamic stall on rotors, particularly in rotorcraft comprehensive analyses. Typically these are finite-state differential-equation models for the section loads, based on measured dynamic and static data. Published empirical models always match the data (generally from oscillating airfoil tests) used in their development. The models are less accurate in the context of rotor loads calculations. Empirical models require many parameters, which must be identified for each airfoil from unsteady load measurements or calculations.

Dynamic stall is characterized by a delay in the occurrence of separated flow produced by the wing motion and by high transient loads induced by a vortex shed from the leading edge when stall does occur. Empirical models should still use airfoil tables for the steady characteristics, evaluated at an angle-of-attack that includes the dynamic stall delay, perhaps with a dynamic stall increment for the coefficients. By using the airfoil table data, the basic characteristics associated with airfoil shape are retained, and the dynamic effects are isolated from the static loads. Let α_d be the delayed angle-of-attack, calculated from the angle-of-attack α. Then the corrected coefficients are

$$c_\ell = \left(\frac{\alpha - \alpha_z}{\alpha_d - \alpha_z}\right) c_{\ell_{2D}}(\alpha_d) + \Delta c_{\ell_{DS}} \qquad (12.7)$$

$$c_d = \left(\frac{\alpha - \alpha_z}{\alpha_d - \alpha_z}\right)^2 \left(c_{d_{2D}}(\alpha_d) - c_{dz}\right) + c_{dz} + \Delta c_{d_{DS}} \qquad (12.8)$$

$$c_m = \left(\frac{\alpha - \alpha_z}{\alpha_d - \alpha_z}\right) \left(c_{m_{2D}}(\alpha_d) - c_{mz}\right) + c_{mz} + \Delta c_{m_{DS}} \qquad (12.9)$$

where α_z is the zero-lift angle-of-attack and c_{dz} and c_{mz} are the corresponding drag and moment. The form of the lift and moment corrections ensures that the coefficients below stall are unchanged. The Δc_{DS}'s are increments defined by the dynamic stall model, generally attributed to the leading-edge vortex. The angle-of-attack delay and coefficient increments are functions of the rate $\dot{\alpha}c/V$ or time delay $V\Delta t/c$.

Many empirical dynamic stall models have been developed. Three models are described here. The Boeing model is notable for its simplicity. The Leishman-Beddoes model requires only a small number of parameters, some obtained from the static airfoil data. The ONERA EDLIN model was from the beginning formulated as differential equations. Johnson (1998) gives more details of these models.

The Boeing dynamic stall model uses an angle-of-attack delay proportional to the square root of $\dot{\alpha}$, which produces the basic hysteresis effects. The coefficient increments produced by the leading-edge vortex are not used in this model. The Boeing model was developed by Harris, Tarzanin, and Fisher (1970), Tarzanin (1972), and Gormont (1973). The onset of dynamic stall is assumed to occur at $\alpha_{DS} = \alpha_{SS} + \Delta\alpha$, where α_{SS} is the static stall angle and $\Delta\alpha$ is proportional to $\sqrt{\dot{\alpha}}$. Lift and moment coefficients are obtained from static airfoil data using a delayed angle-of-attack:

$$\alpha_d = \alpha - \tau_d\sqrt{|\dot{\alpha}c/2U|}\,\text{sign}\,\dot{\alpha} \tag{12.10}$$

The time constant τ_d depends on airfoil geometry and Mach number and is different for lift and moment stall. Values of τ_d are obtained from oscillating airfoil tests. The model does not correct the section drag coefficient.

The Leishman-Beddoes dynamic stall model uses a delayed angle-of-attack, plus lift and moment increments from the leading-edge vortex. The model is documented in Leishman and Beddoes (1986, 1989) and Leishman and Crouse (1989). This model characterizes the airfoil static stall behavior by the trailing-edge separation point f (fraction of chord from leading edge) and a critical lift coefficient $c_{\ell_{CR}}$ at the separation onset boundary (leading-edge separation at low Mach number, shock reversal at high Mach number). The airfoil data for lift are used to identify constants s_1, s_2, and α_s that generate $f(\alpha)$ as follows:

$$f = \begin{cases} 1. - .3e^{(|\alpha-\alpha_z|-\alpha_s)/s_1} & |\alpha - \alpha_z| \leq \alpha_s \\ .04 + .66e^{(\alpha_s - |\alpha-\alpha_z|)/s_2} & |\alpha - \alpha_z| > \alpha_s \end{cases} \tag{12.11}$$

Then $|\alpha - \alpha_z| = \alpha_s$ or $f = .7$ is taken as the definition of stall. The parameters $c_{\ell_{CR}}$, s_1, s_2, and α_s for an airfoil depend on Mach number. The Leishman-Beddoes model for unsteady flow is based on $f_d = f(\alpha_d)$ at the delayed angle-of-attack.

Leishman and Beddoes (1989) write the static normal force, moment, and drag as functions of f:

$$c_{n_{2D}} = c_{n\alpha}(M)K_N(f)(\alpha - \alpha_z) \tag{12.12}$$

$$c_{m_{2D}} = c_{mz} + c_n K_M(f) \tag{12.13}$$

$$c_{d_{2D}} = c_{dz} + (1 - \eta K_D(f))(\alpha - \alpha_z)^2 \tag{12.14}$$

The Kirchhoff expression $K_N = \frac{1}{4}(1 + \sqrt{f})^2$ is used; several functional forms of K_M and K_D are found in the literature. In unsteady flow, a delayed separation point f_d is calculated from f, the loads are evaluated using f_d, and an increment for the leading-edge vortex is added. Since f_d corresponds to a delayed angle-of-attack, the analytical functions K can be replaced with the loads from airfoil tables:

$$\begin{aligned}
c_n &= c_{n\alpha}(M)K_N(f_d)(\alpha - \alpha_z) + \Delta c_{n_{DS}} \\
&= \left(\frac{\alpha - \alpha_z}{\alpha_d - \alpha_z}\right) c_{n\alpha}(M) K_N(f_d)(\alpha_d - \alpha_z) + \Delta c_{n_{DS}} \\
&= \left(\frac{\alpha - \alpha_z}{\alpha_d - \alpha_z}\right) c_{n_{2D}}(\alpha_d) + \Delta c_{n_{DS}}
\end{aligned} \quad (12.15)$$

The model can use the static loads directly from the airfoil tables, instead of fitting the static loads to analytical functions.

The delayed angle-of-attack α_d is calculated including static hysteresis around stall, a lag in the leading-edge pressure relative to c_ℓ, and an additional lag in the boundary layer response. There are separate α_d equations for lift and moment to allow different behavior during reattachment. Vortex lift accumulation begins at the onset of stall, driven by the difference between the linear and nonlinear lifts, $c_v = c_{\ell_L} - c_\ell$. The vortex loads Δc_{DS} are obtained from c_v with a time lag. Thus the model has first-order differential equations for the delayed angle-of-attack and the leading-edge vortex lift.

The ONERA EDLIN (Equations Differentielles Lineaires) dynamic stall model uses a stall delay plus lift, drag, and moment increments calculated from second-order differential equations. The model is documented in Petot (1989, 1997). The extended model of Petot includes the effects of heave and pitch, as well as time-varying free stream. Generalizations based on Petot's pitch model (Peters (1985)) lead to more complicated equations. The load is divided into two parts. The first part is the load in the absence of stall, which gives the unsteady load for attached flow. The second part of the load is driven by the difference between the linear load extrapolated to the unstalled domain and the real nonlinear static load. Tests show that dynamic stall occurs at a higher angle-of-attack than does static stall. The absence of stall is preserved in the model by forcing the difference between the linear and nonlinear loads to be zero for a time τ_d after exceeding the static stall angle.

The dynamic stall loads of the ONERA EDLIN model are calculated from second-order differential equations:

$$\ddot{L}_2 + a\dot{L}_2 + bL_2 = -bU\,\Delta c_\ell - eU\dot\alpha \quad (12.16)$$

$$\ddot{M}_2 + a\dot{M}_2 + bM_2 = -bU\,\Delta c_m - eU\dot\alpha \quad (12.17)$$

$$\ddot{D}_2 + a\dot{D}_2 + bD_2 = -bU\,\Delta c_d - eU\dot\alpha\,\text{sign}(\alpha - \alpha_z) \quad (12.18)$$

These equations are driven by the difference between the linear and non-linear loads: $\Delta c_\ell = c_{\ell_L} - c_\ell = c_{\ell\alpha}(\alpha - \alpha_z) - c_\ell$, $\Delta c_m = c_{m_L} - c_m = c_{m\alpha}(\alpha - \alpha_z) + c_{mz} - c_m$, $\Delta c_d = c_{d_L} - c_d = c_{dz} - c_d$; where c_ℓ, c_m, and c_d are the static coefficients, without the unsteady or leading-edge vortex terms. Then the load increments are

$$\Delta c_{\ell_{DS}} = \frac{1}{U}(L_2 + U\,\Delta c_\ell + dU\dot\alpha) \quad (12.19)$$

$$\Delta c_{m_{DS}} = \frac{1}{U}(M_2 + U\,\Delta c_m + dU\dot\alpha) + M_3 \quad (12.20)$$

$$\Delta c_{d_{DS}} = \frac{1}{U}(D_2 + U\,\Delta c_d + dU\dot\alpha\,\text{sign}(\alpha - \alpha_z)) \quad (12.21)$$

The stall delay is accounted for by setting the right-hand side of the differential equation to zero if $\tau_{SS} < \tau_d$, where $t_{SS} = \tau_{SS} c/2U$ is the time since the static stall angle was exceeded. Petot (1997) described a refined transition model, intended to accommodate airfoils that exhibit larger nose-down pitching moments at dynamic stall. The refined transition model assumes that the extra lift from dynamic stall is convected aft from the quarter chord after moment stall occurs, producing the extra moment term M_3. The coefficients in these equations (a, b, d, e) depend on the lift difference Δc_ℓ. Values of the parameters for various airfoils are given in Petot (1983, 1984).

12.5 REFERENCES

Bailey, F.J., Jr. "A Simplified Theoretical Method of Determining the Characteristics of a Lifting Rotor in Forward Flight." NACA Report 716, 1941.

Bousman, W.G. "A Qualitative Examination of Dynamic Stall from Flight Test Data." Journal of the American Helicopter Society, 43:4 (October 1998).

Bousman, W.G. "Evaluation of Airfoil Dynamic Stall Characteristics for Maneuverability." Journal of the American Helicopter Society, 46:4 (October 2001).

Carr, L.W., McAlister, K.W., and McCroskey, W.J. "Analysis of the Development of Dynamic Stall Based on Oscillating Airfoil Experiments." NASA TN D-8382, January 1977.

Floros, M.W., Gold, N.P., and Johnson, W. "An Exploratory Aerodynamic Limits Test with Analytical Correlation." American Helicopter Society 4th Decennial Specialist's Conference on Aeromechanics, San Francisco, CA, January 2004.

Fradenburgh, E.A. "Aerodynamic Efficiency Potential of Rotary Wing Aircraft." American Helicopter Society 16th Annual National Forum, Washington, DC, May 1960.

Gessow, A., and Myers, G.C., Jr. *Aerodynamics of the Helicopter*. New York: The Macmillan Company, 1952.

Gormont, R.E. "A Mathematical Model of Unsteady Aerodynamics and Radial Flow for Application to Helicopter Rotors." USAAVLABS TR 72-67, May 1973.

Green, R.B., Galbraith, R.A.M., and Niven, A.J. "Measurements of the Dynamic Stall Vortex Convection Speed." Seventeenth European Rotorcraft Forum, Berlin, Germany, September 1991.

Gustafson, F.B., and Gessow, A. "Effect of Blade Stalling on the Efficiency of a Helicopter Rotor as Measured in Flight." NACA TN 1250, April 1947.

Gustafson, F.B., and Myers, G.C., Jr. "Stalling of Helicopter Blades." NACA Report 840, 1946.

Hafner, R. "The Bristol 171 Helicopter." The Journal of the Royal Aeronautical Society, 53:460 (April 1949).

Ham, N.D., and Garelick, M.S. "Dynamic Stall Considerations in Helicopter Rotors." Journal of the American Helicopter Society, 13:2 (April 1968).

Ham, N.D., and Young, M.I. "Torsional Oscillation of Helicopter Blades due to Stall." Journal of Aircraft, 3:3 (May-June 1966).

Harris, F.D. "Rotary Wing Aerodynamics – Historical Perspective and Important Issues." American Helicopter Society National Specialists' Meeting on Aerodynamics and Aeroacoustics, Arlington, TX, February 1987.

Harris, F.D., Tarzanin, F.J., Jr., and Fisher, R.K., Jr. "Rotor High Speed Performance, Theory vs. Test." Journal of the American Helicopter Society, 15:3 (April 1970).

Johnson, W. "Rotorcraft Aerodynamic Models for a Comprehensive Analysis." American Helicopter Society 54th Annual Forum, Washington, DC, May 1998.

Leishman, J.G., and Beddoes, T.S. "A Generalised Model for Airfoil Unsteady Aerodynamic Behaviour and Dynamic Stall Using the Indicial Method." American Helicopter Society 42nd Annual Forum, Washington DC, June 1986.

Leishman, J.G., and Beddoes, T.S. "A Semi-Empirical Model for Dynamic Stall." Journal of the American Helicopter Society, 34:3 (July 1989).

Leishman, J.G., and Crouse, G.L., Jr. "State-Space Model for Unsteady Airfoil Behavior and Dynamic Stall." AIAA Paper No. 89-1319, April 1989.

McCloud, J.L., III, and McCullough, G.B. "Wind-Tunnel Tests of a Full-Scale Helicopter Rotor with Symmetrical and with Cambered Blade Sections at Advance Ratios from 0.3 to 0.4." NACA TN 4367, September 1958.

McCormick, B.W., Jr. *Aerodynamics of V/STOL Flight.* New York: Academic Press, Inc., 1967.

McCroskey, W.J. "The Phenomenon of Dynamic Stall." NASA TM 81264, March 1981.

McCroskey, W.J., and Fisher, R.K., Jr. "Detailed Aerodynamic Measurements on a Model Rotor in the Blade Stall Regime." Journal of the American Helicopter Society, *17*:1 (January 1972).

McCroskey, W.J., and Pucci, S.L. "Viscous-Inviscid Interaction on Oscillating Airfoils in Subsonic Flow." AIAA Journal, *20*:2 (February 1982).

McHugh, F.J. "What Are the Lift and Propulsive Force Limits of High Speed for the Conventional Rotor?" American Helicopter Society 34th Annual National Forum, Washington, DC, May 1978.

Martin, J.M., Empey, R.W., McCroskey, W.J., and Caradonna, F.X. "An Experimental Analysis of Dynamic Stall on an Oscillating Airfoil." Journal of the American Helicopter Society, *19*:1 (January 1974).

Norman, T.R., Shinoda P., Peterson, R.L., and Datta, A. "Full-Scale Wind Tunnel Test of the UH-60A Airloads Rotor." American Helicopter Society 67th Annual Forum, Virginia Beach, VA, May 2011.

Peters, D.A. "Toward a Unified Lift Model for Use in Rotor Blade Stability Analyses." Journal of the American Helicopter Society, *30*:3 (July 1985).

Petot, D. "Progress in the Semi-Empirical Prediction of the Aerodynamic Forces due to Large Amplitude Oscillations of an Airfoil in Attached or Separated Flow." Ninth European Rotorcraft Forum, Stresa, Italy, September 1983.

Petot, D. "Dynamic Stall Modeling of the NACA 0012 Profile." La Recherche Aerospatiale, *1984*:6 (1984).

Petot, D. "Differential Equation Modeling of Dynamic Stall." La Recherche Aerospatiale, *1989*:5 (1989), corrections dated October 1990.

Petot, D. "An Investigation of Stall on a 4.2m Diameter Experimental Rotor." Seventh International Workshop on Dynamics and Aeroelastic Modeling of Rotorcraft, St. Louis, MO, October 1997.

Sikorsky, I.A. "Aerodynamic Parameters Selection in Helicopter Design." Journal of the American Helicopter Society, *5*:1 (January 1960).

Stepniewski, W.Z. *Introduction to Helicopter Aerodynamics.* Morton, PA: Rotorcraft Publishing Committee, 1955 (first edition 1950).

Tanner, W.H. "Charts for Estimating Rotary Wing Performance in Hover and at High Forward Speeds." NASA CR 114, November 1964.

Tarzanin, F.J., Jr. "Prediction of Control Loads Due to Blade Stall." Journal of the American Helicopter Society, *17*:2 (April 1972).

Yeo, H. "Calculation of Rotor Performance and Loads Under Stalled Conditions." American Helicopter Society 59th Annual Forum, Phoenix, AZ, May 2003.

13 Computational Aerodynamics

Rotary-wing flow fields are as complex as any in aeronautics. The helicopter rotor in forward flight encounters three-dimensional, unsteady, transonic, viscous aerodynamic phenomena. Rotary-wing problems provide a stimulus for development and opportunities for application of the most advanced computational techniques.

Inviscid, potential aerodynamics is the starting point for many computational methods for rotors, allowing practical solutions of compressible and unsteady problems. Lifting-surface theory solves the linearized problem by using the result for a moving singularity, often of the acceleration potential. Panel methods use surface singularity distributions to solve problems with arbitrary geometry. Transonic rotor analyses use finite-difference techniques to solve the nonlinear flow equation.

The rotor wake is a factor in almost all helicopter problems. A major issue in advanced aerodynamic methods is how the wake can be included. Wake formation must at some level be considered a viscous phenomenon, and the helical geometry of the helicopter wake means that the detailed structure is important even at scales on the order of the rotor size. A useful rotor aerodynamic theory must account for the effects of viscosity, such as wake formation and blade stall, which are important for most operating conditions. Solution of Navier-Stokes equations for rotor flows is now common. Hybrid methods can be used for efficiency, typically using Navier-Stokes solutions near the blade and some vortex method for the rest of the flow field.

Sources for the derivations of the equations are Lamb (1932), Morse and Feshback (1953), Garrick (1957), Ashley and Landahl (1965), and Batchelor (1967).

13.1 Potential Theory

Potential theory requires the assumptions that the flow field is inviscid, irrotational, and isentropic. With concentrated vorticity in the wake and weak shocks on the rotor blades, these are generally good assumptions. In spite of the inviscid assumption, wake formation, stall, and drag must be addressed in some manner. Often the additional assumption of incompressibility is introduced for rotor analyses, particularly for unsteady problems.

13.1 Potential Theory

Consider a fixed reference frame (x, y, z, t), with the fluid at rest at infinity. The equations of mass and momentum conservation for inviscid flow without body forces are

$$\frac{D}{Dt}\rho + \rho \nabla \cdot \mathbf{v} = 0 \tag{13.1}$$

$$\frac{D}{Dt}\mathbf{v} = -\frac{1}{\rho}\nabla p \tag{13.2}$$

where $\mathbf{v}$ is the velocity, ρ the density, p the pressure, and $\frac{D}{Dt} = \frac{\partial}{\partial t} + \mathbf{v} \cdot \nabla$. An initially irrotational, uniform, inviscid fluid remains irrotational. Irrotationality ($\nabla \times \mathbf{v} = 0$) implies the existence of the velocity potential ϕ, such that velocity $\mathbf{v} = \nabla \phi$. A wing moving relative to a fluid at rest is considered because of the need to deal with both rotating and translating wings. Although $\mathbf{v}$ and ϕ can therefore be considered perturbations, they are not necessarily small. Introducing a coordinate system moving with the wing influences the equations and boundary conditions, but does not change the definition of ϕ as the potential relative to the fluid at rest.

For an irrotational flow, uniform and at rest at infinity, the inviscid equation of momentum conservation integrates to Kelvin's equation:

$$\phi_t + \frac{1}{2}v^2 + \int_{p_\infty}^{p} \frac{dp}{\rho} = 0 \tag{13.3}$$

since $\frac{1}{\rho}\nabla p = \nabla \int \frac{dp}{\rho}$ always, and $(\mathbf{v} \cdot \nabla)\mathbf{v} = \frac{1}{2}\nabla v^2$ for irrotational flow ($v = |\mathbf{v}|$). For isentropic flow, the local adiabatic speed of sound is $a^2 = \gamma RT$ (where γ is the ratio of specific heats), or $a^2 = dp/d\rho = \gamma p/\rho$. So $p/p_\infty = (\rho/\rho_\infty)^\gamma$, and the differential $da^2 = \gamma d(p/\rho) = (\gamma - 1)dp/\rho$ integrates to

$$\int_{p_\infty}^{p} \frac{dp}{\rho} = \frac{a_\infty^2}{\gamma - 1}\left(\frac{a^2}{a_\infty^2} - 1\right) \tag{13.4}$$

The pressure is

$$\frac{p}{p_\infty} = \left(\frac{\rho}{\rho_\infty}\right)^\gamma = \left(\frac{T}{T_\infty}\right)^{\gamma/(\gamma-1)} = \left(\frac{a^2}{a_\infty^2}\right)^{\gamma/(\gamma-1)} = \left[1 - \frac{\gamma - 1}{a_\infty^2}\left(\phi_t + \frac{1}{2}v^2\right)\right]^{\gamma/(\gamma-1)} \tag{13.5}$$

Then the equation for mass conservation gives

$$a^2 \nabla^2 \phi = -\frac{a^2}{\rho}\frac{D\rho}{Dt} = -\frac{1}{\gamma - 1}\frac{Da^2}{Dt} = \left(\frac{\partial}{\partial t} + \mathbf{v} \cdot \nabla\right)\left(\phi_t + \frac{1}{2}v^2\right) \tag{13.6}$$

with $\mathbf{v} = \nabla\phi$ and

$$a^2 = a_\infty^2 - (\gamma - 1)\left(\phi_t + \frac{1}{2}v^2\right) \tag{13.7}$$

So the equation for the velocity potential relative to still air, in tensor form, is

$$a^2 \nabla^2 \phi = \left(\frac{\partial}{\partial t} + \phi_{x_i}\frac{\partial}{\partial x_i}\right)\left(\phi_t + \frac{1}{2}\phi_{x_j}\phi_{x_j}\right) = \phi_{tt} + 2\phi_{x_i}\phi_{tx_i} + \phi_{x_i}\phi_{x_j}\phi_{x_i x_j} \tag{13.8}$$

The corresponding conservative form is $(\rho\phi_{x_i})_{x_i} = \rho_t$. The linearized equation of motion is the wave equation $a^2\nabla^2\phi = \phi_{tt}$, with $(p - p_\infty)/\rho_\infty = -\phi_t$. For incompressible flow, the potential equation reduces to the Laplace equation, $\nabla^2\phi = 0$, with $(p - p_\infty)/\rho_\infty = -\left(\phi_t + \frac{1}{2}v^2\right)$.

The acceleration potential ψ can be defined by considering the momentum equation, $D\mathbf{v}/Dt = -(\nabla p)/\rho = \nabla\psi$, so

$$\psi = -\int \frac{dp}{\rho} = \phi_t + \frac{1}{2}v^2 = \frac{D\phi}{Dt} \tag{13.9}$$

which is a nonlinear relation between the acceleration and velocity potentials. Small disturbances are assumed in order to relate ψ and ϕ in a practical manner. The linearized relation is

$$\psi \cong -\frac{p - p_\infty}{\rho} \cong \phi_t \tag{13.10}$$

and the acceleration potential is a solution of the wave equation, $a^2\nabla^2\psi = \psi_{tt}$.

The boundary condition at infinity is $\mathbf{v} = 0$ and $p = p_\infty$ (still air); hence $\nabla\phi = \phi_t = 0$, which implies $\phi = 0$. For an inviscid fluid, the boundary condition at a solid body defined by the surface $F(\mathbf{x}, t) = 0$ is $\frac{DF}{Dt} = F_t + \mathbf{v} \cdot F = 0$. The time derivative of $F = 0$ (on the body surface) yields $F_t = -\nabla F \cdot \mathbf{v}_b$, where $\mathbf{v}_b$ is the body surface velocity. So the boundary condition is that there be no flow normal to the surface: $\mathbf{n} \cdot \mathbf{v} = \frac{\partial \phi}{\partial n} = \mathbf{n} \cdot \mathbf{v}_b$, where the body normal is $\mathbf{n} = \nabla F/|\nabla F|$. For inviscid flow, there can be tangential velocity at the body surface. If the body is defined by $F = z - g = 0$ or $z = g(x, y, t)$, then the boundary condition is $v_z = g_t + v_x g_x + v_y g_y$ on $z = g$. The linearized boundary condition is $v_z \cong g_t$ on $z = 0$.

A wake is a vortex surface, which allows a tangential velocity jump but no normal velocity or pressure difference: $\Delta v_n = 0$ and $\Delta p = 0$ (where Δ means the upper surface value minus the lower surface value). Kelvin's equation gives

$$\Delta\phi_t + \frac{1}{2}\Delta(v^2) = \Delta\phi_t + \frac{1}{2}(\mathbf{v}_u + \mathbf{v}_l) \cdot \Delta\mathbf{v} = \Delta\phi_t + \mathbf{v}_w \cdot \nabla\Delta\phi = 0 \tag{13.11}$$

with $\mathbf{v}_w = \frac{1}{2}(\mathbf{v}_u + \mathbf{v}_l)$. The potential difference $\Delta\phi$ is constant for a point on the wake surface that is convected with the velocity $\mathbf{v}_w$.

13.2 Rotating Coordinate System

Consider the transformation from inertial axes $(t, \mathbf{r})$ relative to the still air, to moving axes $(t', \mathbf{r}')$ that contain the principal translation and rotation of the rotor (Figure 13.1). Typically the rotating y' coordinate is the rotor blade span axis, although in fact the rotor can be placed anywhere in the moving frame. The rotor is rotating at rate Ω, with the azimuth angle $\psi = \Omega t$ measured from downstream. The rotor has translation velocity V_∞ in the x–z plane, with the angle i relative to the x-axis (positive for climb). So the rotation and translation vectors are

$$\overline{\Omega} = (0 \quad 0 \quad \Omega)^T \tag{13.12}$$

$$\overline{\mu} = \mu\Omega R(-1 \quad 0 \quad \tan i)^T = (-\mu\Omega R \quad 0 \quad \lambda\Omega R)^T \tag{13.13}$$

where $\mu = V_\infty \cos i/\Omega R$ is the advance ratio and R the rotor radius. The inflow ratio $\lambda = \mu \tan i = \mu_z$, or $\lambda = \mu_z + \lambda_i$ if the induced velocity at the rotor disk is included. The coordinate transformation is $\mathbf{r} = \overline{\mu}t' + T\mathbf{r}'$ and $t = t'$, with the rotation matrix

$$T = Z_{90-\psi} = \begin{bmatrix} \sin\psi & \cos\psi & 0 \\ -\cos\psi & \sin\psi & 0 \\ 0 & 0 & 1 \end{bmatrix} \tag{13.14}$$

13.2 Rotating Coordinate System

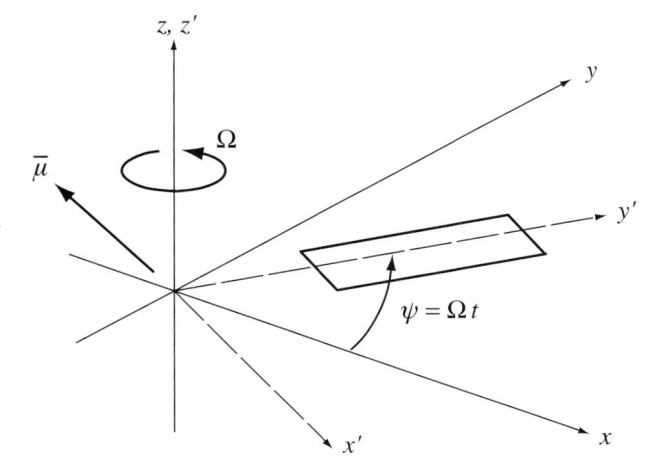

Figure 13.1. Rotating and translating axes.

The inverse transformation is $\mathbf{r}' = -\overline{\mu}'t + T^T\mathbf{r}$, with

$$\overline{\mu}' = T^T\overline{\mu} = \mu\Omega R(-\sin\psi \quad -\cos\psi \quad \tan i)^T \qquad (13.15)$$

Then the derivatives are $\nabla = \nabla'$ and

$$\frac{\partial}{\partial t} = \frac{\partial}{\partial t'} + \mathbf{V}\cdot\nabla' \qquad (13.16)$$

where $\mathbf{V}$ is the velocity of the air relative to the moving frame,

$$\mathbf{V} = \frac{\partial \mathbf{r}'}{\partial t} = -\overline{\Omega}'\times\mathbf{r}' - \overline{\mu}' = \begin{pmatrix} \Omega y' + \mu\Omega R\sin\psi \\ -\Omega x' + \mu\Omega R\cos\psi \\ -\mu\Omega R\tan i \end{pmatrix} \qquad (13.17)$$

$$\dot{\mathbf{V}} = \frac{d\mathbf{V}}{dt} = \left(\frac{\partial}{\partial t'} + \mathbf{V}\cdot\nabla'\right)\mathbf{V} = \begin{pmatrix} \Omega^2 x' + 2\Omega V_y \\ \Omega^2 y' - 2\Omega V_x \\ 0 \end{pmatrix} \qquad (13.18)$$

After the transformation to the rotating frame has been accomplished, the primes can be omitted for simplicity.

The potential equation is transformed by substituting for the derivatives. The scalar ϕ remains a perturbation potential, although not necessarily small, defined relative to the still air. Writing $\mathbf{U} = \mathbf{V} + \nabla\phi$, the potential equation in tensor form becomes

$$a^2\nabla^2\phi = \left(\frac{\partial}{\partial t} + U_i\frac{\partial}{\partial x_i}\right)\left(\phi_t + V_j\phi_{x_j} + \frac{1}{2}\phi_{x_j}\phi_{x_j}\right)$$
$$= \phi_{tt} + 2U_i\phi_{tx_i} + U_iU_j\phi_{x_ix_j} + \dot{V}_i\phi_{x_i} \qquad (13.19)$$

and

$$a^2 = a_\infty^2 - (\gamma - 1)\left(\phi_t + V_j\phi_{x_j} + \frac{1}{2}\phi_{x_j}\phi_{x_j}\right) \qquad (13.20)$$

As a result of rotation, the velocity $\mathbf{V}$ varies with time and space, and Coriolis and centrifugal terms are introduced in the form of $\dot{\mathbf{V}}$. The linearized potential equation is obtained by setting $\mathbf{U} \cong \mathbf{V}$. The incompressible equation is still $\nabla^2\phi = 0$.

The boundary condition on a solid body $F = 0$ becomes $F_t + \mathbf{U} \cdot \nabla F = 0$, or $\mathbf{n} \cdot \mathbf{v} = \frac{\partial \phi}{\partial n} = \mathbf{n} \cdot (\mathbf{v}_b - \mathbf{V})$, where now $\mathbf{v}_b$ is the body velocity relative to the moving frame. With $F = z - g$, the boundary condition is $U_z = g_t + U_x g_x + U_y g_y$ on $z = g$, which linearizes to $v_z = g_t + V_x g_x + V_y g_y - V_z$ on $z = 0$.

13.3 Lifting-Surface Theory

Lifting-surface theory solves the linearized equation of motion, generally using the acceleration potential. The development of the theory provides the result for a moving singularity, which is needed for acoustic theory as well.

The acceleration potential satisfies the wave equation. For a thin wing surface, $\Delta \psi = -\Delta p / \rho_\infty$, whereas on the wake surface $\Delta \psi = 0$, since the wake does not support a pressure jump. Hence the wing is represented by a distribution of singularities over the planform. The linearized boundary condition is $\partial \phi / \partial z = w$, with the velocity potential obtained from the integral of $\phi_t = \psi$. That the acceleration potential automatically accounts for the wake is an advantage, but precludes any simple incorporation of a rolled-up, distorted wake geometry. So acceleration-potential methods are not a basis for a general analysis, but can be a sound starting point for simplified models.

13.3.1 Moving Singularity

A solution of the wave equation is a stationary dipole at position $\mathbf{y}$

$$\psi_d = \frac{\partial}{\partial n_y} \frac{f(t - s/a)}{4\pi s} \tag{13.21}$$

where $\mathbf{s} = \mathbf{x} - \mathbf{y}$, $f(t)$ is an arbitrary function of time, and $\partial(\ldots)/\partial n_y = \mathbf{n} \cdot \nabla_y$ is the gradient in direction $\mathbf{n}$. Evaluating the acceleration potential near a surface distribution of dipoles ($\psi(\mathbf{x}) = \int \psi_d dA(\mathbf{y})$ in the limit $\mathbf{x} \to$ surface A) gives $\psi_u = f/2$ and $\psi_l = -f/2$, so $f = \Delta \psi = -\Delta p / \rho_\infty$, the force in the normal direction. Another solution is a stationary velocity potential source,

$$\phi_s = \frac{q(t - s/a)}{4\pi s} \tag{13.22}$$

A surface distribution of sources ($\partial \phi(\mathbf{x})/\partial n_x = \int (\partial \phi_s / \partial n_x) dA(\mathbf{y})$ in the limit $\mathbf{x} \to A$) gives $v_u = -q/2$ and $v_l = q/2$, so $q = -\Delta \partial \phi / \partial n = -\Delta v_n$ and hence can represent the wing thickness.

The derivation of a moving singularity follows Garrick (1957). For the potential at $\mathbf{x}$ produced by a source at $\mathbf{y}(\tau)$, replace $f(t)$ with $f(\tau) \delta(\tau - t)$ and integrate over τ:

$$\begin{aligned}\psi_s &= \int_{-\infty}^{t} \frac{f(\tau)\delta(\tau - t + s/a)}{4\pi s} d\tau = \int_{-\infty}^{s/a} \frac{f(\tau)\delta(\tau^*)}{4\pi s} \frac{d\tau}{d\tau^*} d\tau^* \\ &= \frac{f}{4\pi s} \frac{d\tau}{d\tau^*}\bigg|_{\tau^*=0} = \left[\frac{f}{4\pi s(1 - M_r)}\right]\end{aligned} \tag{13.23}$$

where the square brackets denote the retarded time $\tau^* = \tau + s/a - t = 0$ or the solution of $\tau = t - s/a$, and $aM_r = -ds/d\tau = \mathbf{s} \cdot (d\mathbf{y}/d\tau)/s = a(\mathbf{s} \cdot \mathbf{M})/s$. M_r is the

13.3 Lifting-Surface Theory

Mach number of the source in the observer direction. Then the moving dipole is

$$\psi_d = \left[\frac{\partial}{\partial n_y} \frac{f}{4\pi s(1-M_r)}\right] = \left[\frac{f\mathbf{n}\cdot\mathbf{s}}{4\pi s^3(1-M_r)} + \frac{\partial}{\partial t}\frac{f\mathbf{n}\cdot\mathbf{s}}{4\pi a s^2(1-M_r)}\right] \quad (13.24)$$

Including a potential source, the result is

$$\psi_d = \left[\frac{f\mathbf{n}\cdot\mathbf{s}}{4\pi s^3(1-M_r)} + \frac{\partial}{\partial t}\frac{f\mathbf{n}\cdot\mathbf{s}}{4\pi a s^2(1-M_r)} + \frac{\partial}{\partial t}\frac{q}{4\pi s(1-M_r)}\right] \quad (13.25)$$

which is the acoustic formulation used by Farassat (1975). The thickness term is not considered further here. For aerodynamic problems, the surface pressure is the unknown to be obtained from the boundary condition on the velocity potential. Integrating the acceleration potential gives the velocity potential for a moving dipole,

$$\phi_d(\mathbf{x}, t_0) = \int_{-\infty}^{t_0} \psi(\tau = t - s/a) dt = \int_{-\infty}^{\tau_0} \psi(\tau)(1 - M_r) d\tau$$

$$= \int_{-\infty}^{\tau_0} \frac{f\mathbf{n}\cdot\mathbf{s}}{4\pi s^3} d\tau + \left.\frac{f\mathbf{n}\cdot\mathbf{s}}{4\pi a s^2(1-M_r)}\right|_{\tau_0} \quad (13.26)$$

where τ_0 is the solution of $\tau_0 = t_0 - s(\tau_0)/a$. This is the aerodynamic formulation used by Dat (1973). Farassat (1982) noted the equivalence of the acoustic and aerodynamic formulations and discussed the aerodynamic applications.

Represent the wings by a surface of dipoles and apply the boundary condition to obtain a singular (Mangler-type) integral equation for the loading $f = -\Delta p/\rho_\infty$:

$$w(\mathbf{x}, t_0) = \lim_{x\to A}\frac{\partial \phi}{\partial n_x} = \lim_{x\to A}\frac{\partial}{\partial n_x}\int_{\text{wings}}\phi_d dA(\mathbf{y})$$

$$= \int_{\text{wings}}\lim_{x\to A}\frac{\partial}{\partial n_x}\left\{\int_{-\infty}^{\tau_0}\frac{f\mathbf{n}\cdot\mathbf{s}}{4\pi s^3}d\tau + \left.\frac{f\mathbf{n}\cdot\mathbf{s}}{4\pi a s^2(1-M_r)}\right|_{\tau_0}\right\} dA(\mathbf{y}) \quad (13.27)$$

where $\mathbf{s} = \mathbf{x}(t_0) - \mathbf{y}(\tau)$, $sM_r = \mathbf{s}\cdot\mathbf{M}_y$, and $\tau_0 = t_0 - s(\tau_0)/a$. The integral is over the surface of all wings. For incompressible flow, $\tau_0 = t_0$, the second term is zero, and the normal derivative can be evaluated; the integral equation becomes

$$w(\mathbf{x}, t_0) = \int_{\text{wings}}\lim_{x\to A}\frac{\partial}{\partial n_x}\left\{\int_{-\infty}^{\tau_0}\frac{f\mathbf{n}_y\cdot\mathbf{s}}{4\pi s^3}d\tau\right\} dA(\mathbf{y}) \quad (13.28)$$

$$= \int_{\text{wings}}\lim_{x\to A}\left\{\int_{-\infty}^{\tau_0}\frac{f}{4\pi}\left(\frac{\mathbf{n}_y\cdot\mathbf{n}_x}{s^3} - \frac{3\mathbf{n}_y\cdot\mathbf{s}\,\mathbf{n}_x\cdot\mathbf{s}}{s^5}\right)d\tau\right\} dA(\mathbf{y}) \quad (13.29)$$

For compressible flow, the derivative with respect to $\mathbf{n}_x$ can also be evaluated analytically, but the result is much more complicated. The required derivatives are of the form

$$\frac{\partial}{\partial n_x}G\mathbf{n}_y\cdot\mathbf{s} = G\mathbf{n}_y\cdot\mathbf{n}_x + \frac{\partial G}{\partial n_x}\mathbf{n}_y\cdot\mathbf{s} \quad (13.30)$$

for some quantity G. The vector $\mathbf{s}$ connects a point on the present wing surface ($\mathbf{x}$) with a point on the wing at a retarded time ($\mathbf{y}$). So $\mathbf{n}_y\cdot\mathbf{s} = 0$ and $\mathbf{n}_y\cdot\mathbf{n}_x = 1$ if the wing and wake are planar. For a fixed wing, neglecting any warp of the wing and wake surfaces is consistent with the linearization. For a rotary wing, neglecting the warp may also be possible. The important effect of rotation is that the denominator becomes periodically small as the wake passes under the blade. With a planar wing and wake, the integral equation is

$$w(\mathbf{x}, t_0) = \int_{\text{wings}} \left\{ \int_{-\infty}^{\tau_0} \frac{f}{4\pi s^3} d\tau + \frac{f}{4\pi a s^2 (1 - M_r)} \bigg|_{\tau_0} \right\} dA(\mathbf{y}) \quad (13.31)$$

For steady or harmonic loading, $f = e^{i\omega t} \ell / \rho_\infty$, and the magnitude of the loading factors from the integral over the wake. Then the integral equation can be written in terms of a kernel function:

$$w(\mathbf{x}, t_0) = e^{i\omega t_0} \int \frac{\ell}{\rho_\infty} K \, dA(\mathbf{y}) \quad (13.32)$$

where

$$K = \lim_{x \to A} \frac{\partial}{\partial n_x} \left\{ \int_{-\infty}^{\tau_0} \frac{e^{i\omega(\tau - t_0)} \mathbf{n} \cdot \mathbf{s}}{4\pi s^3} d\tau + \frac{e^{i\omega(\tau_0 - t_0)} \mathbf{n} \cdot \mathbf{s}}{4\pi a s^2 (1 - M_r)} \bigg|_{\tau_0} \right\} dA(\mathbf{y}) \quad (13.33)$$

In general, K is a function of t_0, so there is inter-harmonic coupling for the periodic loading of a rotor blade in forward flight (each loading harmonic gives many induced velocity harmonics). When $\mathbf{s}$ is a function of $(t_0 - \tau)$, as for fixed wings or the rotor in axial flight, $(\tau_0 - t_0)$ and the kernel K are independent of t_0, and there is no inter-harmonic coupling.

A lifting-line theory can be obtained by applying the chordwise integral to the wing loading only. An integral equation for the section lift is obtained, with the kernel function evaluated at one chordwise location only.

13.3.2 Fixed Wing

The fixed-wing problem is useful as a guide and a contrast to the rotor theory. Consider a thin wing on the $z = 0$ plane, moving in the negative x direction and undergoing harmonic loading. Following Garrick (1957), transform now to the moving frame: $\mathbf{x} = (x - Vt_0, y, z)^T$ and $\mathbf{y} = (\xi - V\tau, \eta, 0)^T$. The equation $a(t - \tau) = s$ for the retarded time is a quadratic, with solution

$$\tau = t + \frac{M\sigma - S}{\beta^2 a} \quad (13.34)$$

where $\sigma = x - \xi - V(t_0 - t)$, $S^2 = \sigma^2 + \beta^2(y - \eta)^2 + \beta^2 z^2$, and $\beta^2 = 1 - M^2$. The retarded distance is then $s(1 - M_r) = S$. Working directly from the acceleration-potential source ($\psi_s = f(\tau)/4\pi S$, $f = (l/\rho)e^{i\omega\tau}$) gives the integral equation

$$w = \lim_{z \to 0} \frac{\partial \phi}{\partial z} = \lim_{z \to 0} \frac{\partial}{\partial z} \int \int \phi_d d\xi \, d\eta = \lim_{z \to 0} \frac{\partial}{\partial z} \int \int \int_{-\infty}^{t_0} \frac{\partial \psi_s}{\partial z} dt \, d\xi \, d\eta$$

$$= \lim_{z \to 0} \frac{\partial}{\partial z} \int \int \frac{1}{V} \int_{-\infty}^{x-\xi} \frac{\partial \psi_s}{\partial z} d\sigma \, d\xi \, d\eta$$

$$= V e^{i\omega t_0} \int \int \frac{l}{\rho V^2} K \, d\xi \, d\eta \quad (13.35)$$

with the kernel

$$K = \lim_{z \to 0} \frac{\partial^2}{\partial z^2} e^{-i\omega(x-\xi)/V} \int_{-\infty}^{x-\xi} e^{i\omega(\sigma - MS)/V\beta^2} \frac{1}{4\pi S} d\sigma \quad (13.36)$$

which is the form given by Watkins, Runyan, and Woolston (1955).

13.3 Lifting-Surface Theory

Working from the moving dipole solution (equation 13.26), the retarded time is required only at t_0:

$$\tau_0 = t_0 + (M(x - \xi) - S)/a\beta^2 \quad (13.37)$$

So with $\sigma_0 = x - \xi - V(t_0 - \tau_0) = (x - \xi - MS)/\beta^2$, $s^2 = \sigma^2 + (y - \eta)^2$, and $S^2 = (x - \xi)^2 + \beta^2(y - \eta)^2 + \beta^2 z^2$, the kernel is

$$K = V \int_{-\infty}^{\tau_0} \frac{e^{i\omega(\tau - t_0)}}{4\pi s^3} d\tau + \frac{M e^{i\omega(\tau_0 - t_0)}}{4\pi s_0 S}$$

$$= e^{-i\omega(x-\xi)/V} \left\{ \int_{-\infty}^{\sigma_0} \frac{e^{i\omega\sigma/V}}{4\pi s^3} d\sigma + \frac{M e^{i\omega\sigma_0/V}}{4\pi s_0 S} \right\} \quad (13.38)$$

which is the form given by Landahl (1967). The integral in K can be reduced to a function of two parameters and is a good form for series evaluation. For the steady case

$$K = \int_{-\infty}^{\sigma_0} \frac{1}{4\pi s^3} d\sigma + \frac{M}{4\pi s_0 S} = \int_{-\infty}^{(x-\xi)/\beta} \frac{1}{4\pi s^3} d\sigma$$

$$= \frac{1}{4\pi} \frac{1}{(y - \eta)^2 + z^2} \left(1 + \frac{x - \xi}{S} \right) \quad (13.39)$$

which shows that the compressible kernel is obtained by scaling $(x - \xi)$ with β.

13.3.3 Rotary Wing

To complete the development of the theory for the rotary wing, the transformation to the rotating and translating coordinate system is introduced: $\mathbf{s} = \overline{\mu}(t - \tau) + T_t \mathbf{x}' - T_\tau \mathbf{y}'$ and $a\mathbf{M} = d\mathbf{y}/d\tau = \overline{\mu} + \overline{\Omega} \times T_\tau \mathbf{y}'$, where $\mathbf{x}' = (x, y, z)^T$ and $\mathbf{y}' = (\xi, \eta, 0)^T$. From

$$s^2 = |\overline{\mu}|^2(t - \tau)^2 + 2\overline{\mu} \cdot (T_t \mathbf{x}' - T_\tau \mathbf{y}')(t - \tau) + |T_t \mathbf{x}' - T_\tau \mathbf{y}'|^2 \quad (13.40)$$

the equation for the retarded time $(a(t_0 - \tau_0) = s$ at $z = 0)$, becomes

$$\left(a^2 - (\Omega R)^2(\mu^2 + \lambda^2)\right)(t_0 - \tau_0)^2$$
$$+ 2\mu\Omega R\left(x \sin \Omega t_0 + y \cos \Omega t_0 - \xi \sin \Omega \tau_0 - \eta \cos \Omega \tau_0\right)(t_0 - \tau_0)$$
$$- \left(x^2 + y^2 + \xi^2 + \eta^2 - 2(x\xi + y\eta) \cos \Omega(t_0 - \tau_0)\right.$$
$$\left. - 2(x\eta - y\xi) \sin \Omega(t_0 - \tau_0)\right) = 0 \quad (13.41)$$

and

$$aS = as(1 - M_r) = a^2(t_0 - \tau_0) - a\mathbf{s} \cdot \mathbf{M}$$
$$= \left(a^2 - (\Omega R)^2(\mu^2 + \lambda^2)\right)(t_0 - \tau_0) + \mu\Omega^2 R\left(\xi \cos \Omega \tau_0 - \eta \sin \Omega \tau_0\right)(t_0 - \tau_0)$$
$$+ \mu\Omega R\left(x \sin \Omega t_0 + y \cos \Omega t_0 - \xi \sin \Omega \tau_0 - \eta \cos \Omega \tau_0\right)$$
$$- \Omega\left((x\xi + y\eta) \sin \Omega(t_0 - \tau_0) - (x\eta - y\xi) \cos \Omega(t_0 - \tau_0)\right) \quad (13.42)$$

With these results the kernel can be evaluated, but because of the basic helical geometry, a transcendental equation must be solved for the retarded time, and even for steady or incompressible or axial flight, integrating analytically over the wake (τ) is not possible. The linearized wing is taken as the $z = 0$ plane, so $\mathbf{n} \cdot \mathbf{s} = \lambda \Omega R(t - \tau) + z \cong z$ for low inflow, and on the wing $\mathbf{n} \cdot \mathbf{s} \cong 0$ (planar wing and wake assumption). The rotor consists of N separate wing surfaces. In steady forward flight, the N blades have identical, periodic loading.

The mean wake-induced velocity λ_i should be included in the vertical convection, $\lambda = \mu \tan i + \lambda_i$, when evaluating s in the kernel (but not for the boundary condition). Integrating the acceleration potential along the undisturbed air path is equivalent to using the undistorted wake geometry defined by the translation and rotation of the coordinate system. Including λ_i in λ is a first approximation to the distorted wake geometry, needed to avoid unrealistically extreme interactions of the wing with the returning wake when the rotor incidence angle is small.

Then for harmonic loading of the rotary wing, the integral equation is

$$w = \Omega R e^{i\omega t_0} \int \int \frac{l}{\rho(\Omega R)^2} K \, d\xi \, d\eta \tag{13.43}$$

with the kernel function

$$K = \Omega R \int_{-\infty}^{\tau_0} \frac{e^{i\omega(\tau - t_0)}}{4\pi s^3} d\tau + \left. \frac{M_{\text{tip}} e^{i\omega(\tau_0 - t_0)}}{4\pi s^2 (1 - M_r)} \right|_{\tau_0} \tag{13.44}$$

The distance s is a function of t_0, τ, $\mathbf{x}'$, and $\mathbf{y}'$. Even in the steady case, there is no simple analytical evaluation of the integral over the helical wake. A transcendental equation must be solved for the retarded time τ_0. In forward flight the kernel is a function of t_0, but in axial flight the retarded time can be written $\tau_0 = t_0 - \Upsilon_0(\mathbf{x}', \mathbf{y}')$. In terms of polar coordinates ($x = r \sin \Theta$, $y = r \cos \Theta$, $\xi = \rho \sin \theta$, $\eta = \rho \cos \theta$),

$$\left(a^2 - (\Omega R)^2(\mu^2 + \lambda^2)\right)(t_0 - \tau_0)^2$$
$$+ 2\mu \Omega R \left(r \cos(\Theta - \Omega t_0) - \rho \cos(\theta - \Omega \tau_0)\right)(t_0 - \tau_0)$$
$$- \left(r^2 + \rho^2 - 2r\rho \cos(\Theta - \theta - \Omega(t_0 - \tau_0))\right) = 0 \tag{13.45}$$

$$aS = \left(a^2 - (\Omega R)^2(\mu^2 + \lambda^2)\right)(t_0 - \tau_0) + \mu \Omega^2 R\rho \sin(\theta - \Omega \tau_0)(t_0 - \tau_0)$$
$$+ \mu \Omega R \left(r \cos(\Theta - \Omega t_0) - \rho \cos(\theta - \Omega \tau_0)\right)$$
$$+ \Omega r\rho \sin(\Theta - \theta - \Omega(t_0 - \tau_0)) \tag{13.46}$$

Then the kernel for axial flow is

$$K = e^{-i\omega(\Theta - \theta)/\Omega} \left\{ \int_{-\infty}^{\sigma_0} \frac{e^{i\omega\sigma/\Omega}}{4\pi s^3} R \, d\sigma + \frac{M_{\text{tip}} e^{i\omega\sigma_0/\Omega}}{4\pi s_0 S} \right\} \tag{13.47}$$

where $\sigma_0 = \Theta - \theta - \Omega(t_0 - \tau_0)$ and

$$\beta_\lambda^2 R^2 (\sigma_0 - \Theta - \theta)^2 - M_{\text{tip}}^2 (r^2 + \rho^2 - 2r\rho \cos \sigma_0) = 0 \tag{13.48}$$

$$aS = -\beta_\lambda^2 (a^2/\Omega)(\sigma_0 - \Theta - \theta) + \Omega r\rho \sin \sigma_0 \tag{13.49}$$

$$s^2 = \lambda^2 R^2 (\sigma - \Theta - \theta)^2 + r^2 + \rho^2 - 2r\rho \cos \sigma \tag{13.50}$$

$$\beta_\lambda^2 = 1 - (\lambda M_{\text{tip}})^2 \tag{13.51}$$

This result is superficially similar to the fixed-wing kernel (equation 13.38), but the differences are significant. The use of rectangular coordinates is more appropriate for typical helicopter blade planforms. The integral in the kernel is a function of five parameters, not two, and a series evaluation does not work as well as for the fixed wing, because s is periodically small. Even in the steady, axial flow case, a transcendental equation must be solved for the retarded time. Even for incompressible flow, when the retarded time is not needed, the kernel involves an integral over the helical wake geometry. In forward flight, the kernel introduces inter-harmonic coupling. Although the lifting-surface problem as formulated is much simpler for axial flight, in hover the wake geometry is so important that the wake contraction, wake convection, and blade-vortex interaction have a first-order influence on the loading.

Under the assumption that the retarded time $(\tau_0 - t_0)$ is small, the equation is quadratic and an analytical solution can be obtained:

$$\tau_0 \cong t_0 + (\mathbf{V} \cdot (\mathbf{x}' - \mathbf{y}')/a - S)/a\beta^2 \quad (13.52)$$

where

$$S^2 = (\mathbf{V} \cdot (\mathbf{x}' - \mathbf{y}')/a)^2 + \beta^2 |\mathbf{x}' - \mathbf{y}'|^2 \quad (13.53)$$

$$\beta^2 = 1 - \left(|\mathbf{V}|^2 + \dot{\mathbf{V}} \cdot (\mathbf{x}' - \mathbf{y}')\right)/a^2 \quad (13.54)$$

and $s(1 - M_r) = S$. Here $\mathbf{V}$ and $\dot{\mathbf{V}}$ are the velocity and acceleration of the coordinate system at the observer point $\mathbf{x}$ (and are functions of t_0). Except for the acceleration term in β^2, this is the solution for a fixed wing at velocity $\mathbf{V}(\mathbf{x})$. Equation 13.52 illustrates the dependence of $(\tau_0 - t_0)$ on t_0 for forward flight, but not for axial flight. For axial flow and polar coordinates, the result is

$$\tau_0 = t_0 + \left((\Omega/a)r\rho \sin(\Theta - \theta) - S\right)/a\beta^2 \quad (13.55)$$

$$S^2 = \left((\Omega/a)r\rho \sin(\Theta - \theta)\right)^2 + \beta^2\left(r^2 + \rho^2 - 2r\rho \cos(\Theta - \theta)\right) \quad (13.56)$$

$$\beta^2 = 1 - (\Omega/a)^2\left(\lambda^2 R^2 + r\rho \cos(\Theta - \theta)\right) \quad (13.57)$$

The above approximations are based on $\Omega(t - \tau) \ll 1$, which with $a(t - \tau) = s$ implies $s\Omega/a \ll 1$, or $M_{\text{tip}}(s/R) \ll 1$. This requirement might be valid even at high tip Mach numbers, since the kernel is dominated by locations with small s/R.

Lifting-surface analyses of rotary wings have been developed based on the integral equation (Dat (1973), Runyan and Tail (1983, 1985)), and based on vortex-lattice models (Kocurek and Tangler (1973), Quackenbush, Bliss, and Wachspress (1989), and Wachspress, Quackenbush, and Boschitsch (2003a)).

13.4 Boundary Element Methods

Boundary element or panel methods use a surface singularity distribution to solve the linear potential equation for arbitrary geometry. The use of the velocity potential for inviscid, irrotational flow reduces the problem to the solution of an integral equation for a scalar function on a two-dimensional surface. The derivation is usually based on Green's theorem. Most of the work for helicopters has dealt with incompressible flow.

13.4.1 Surface Singularity Representations

The derivation of the incompressible integral equation follows Lamb (1932). The velocity potential satisfies Laplace's equation, $\nabla^2 \phi = 0$, in either the fixed or moving frame. The boundary condition is that $\partial \phi / \partial n = w$ on the body surface, and the potential jump in the wake is convected from the blade trailing edge. Thus for incompressible flow the partial differential equation is linear without further approximation, and unsteady effects enter only through the boundary conditions and the wake. Assume ϕ and ψ are single-valued functions satisfying Laplace's equation in a volume V bounded by a surface A. Using Green's theorem and substituting for Laplace's equation gives

$$\int \left(\phi \frac{\partial \psi}{\partial n} - \psi \frac{\partial \phi}{\partial n}\right) dA = -\int \left(\phi \nabla^2 \psi - \psi \nabla^2 \phi\right) dV = 0 \qquad (13.58)$$

For multiply connected regions (such as in two-dimensional problems or for an actuator disk), branch cuts and corresponding circulations can be introduced. Let ψ be a unit source at $\mathbf{x}$: $\psi = 1/(4\pi s)$, where $\mathbf{s} = \mathbf{x} - \mathbf{y}$. The volume V is the space external to the body and wake, excluding a small sphere around the singularity at $\mathbf{x}$. Evaluating the surface integral over this small sphere (radius $\epsilon \to 0$),

$$\frac{1}{4\pi} \int \left(-\phi \frac{\partial}{\partial n}\frac{1}{s} + \frac{\partial \phi}{\partial n}\frac{1}{s}\right) dA = \frac{1}{4\pi}\left(\phi \frac{1}{\epsilon^2} + \frac{\partial \phi}{\partial n}\frac{1}{\epsilon}\right) 4\pi \epsilon^2 = \phi(\mathbf{x}) \qquad (13.59)$$

gives

$$4\pi \phi(\mathbf{x}) = \int \left(\phi \frac{\partial}{\partial n}\frac{1}{s} - \frac{1}{s}\frac{\partial \phi}{\partial n}\right) dA(\mathbf{y}) \qquad (13.60)$$

which is a representation of ϕ as a surface distribution of sources and doublets. This is an integral equation, since specifying both ϕ and $\partial\phi/\partial n$ on the surface is not possible. If the point $\mathbf{x}$ is inside the body (outside V), the left-hand side of equation 13.60 is zero. The integral of the dipole distribution is singular as $\mathbf{x}$ approaches the surface. Accounting for the singular part (excluding a small circle about $\mathbf{x}$), the left-hand side becomes $2\pi \phi$.

To obtain a representation in terms of sources or doublets alone, consider the potential ϕ' of an arbitrary flow inside the body. For $\mathbf{x}$ outside the body,

$$0 = \int \left(\phi' \frac{\partial}{\partial n}\frac{1}{s} - \frac{1}{s}\frac{\partial \phi'}{\partial n}\right) dA(\mathbf{y}) \qquad (13.61)$$

so

$$4\pi \phi(\mathbf{x}) = \int \left[(\phi - \phi')\frac{\partial}{\partial n}\frac{1}{s} - \frac{1}{s}\left(\frac{\partial \phi}{\partial n} - \frac{\partial \phi'}{\partial n}\right)\right] dA(\mathbf{y}) \qquad (13.62)$$

For a source representation, let $\phi = \phi'$ on the surface:

$$4\pi \phi(\mathbf{x}) = -\int \frac{1}{s}\left(\frac{\partial \phi}{\partial n} - \frac{\partial \phi'}{\partial n}\right) dA(\mathbf{y}) = -\int \frac{\sigma}{s} dA(\mathbf{y}) \qquad (13.63)$$

so the tangential velocity is continuous and the normal velocity discontinuous at the surface. For a doublet representation, let $\partial \phi/\partial n = \partial \phi'/\partial n$ on the surface:

$$4\pi \phi(\mathbf{x}) = \int (\phi - \phi')\frac{\partial}{\partial n}\frac{1}{s} dA(\mathbf{y}) = \int \sigma \frac{\partial}{\partial n}\frac{1}{s} dA(\mathbf{y}) \qquad (13.64)$$

so the tangential velocity is discontinuous and the normal velocity continuous at the surface.

The surface must be collapsed into a thin layer in order to model a wake and for the thin-wing approximation. For a thin layer, the surface integral is taken over the upper surface only, and the integrand is the difference between the upper and lower surface singularity strengths. With a source distribution, $\Delta\sigma = \Delta\partial\phi/\partial n$ and $\Delta\phi = 0$. The tangential velocity is the same on both sides, so sources can represent the wing thickness, but cannot be used to model a wake. With a doublet distribution, $\Delta\sigma = \Delta\phi$ and $\Delta\partial\phi/\partial n = 0$. The normal velocity is the same on both sides, but the tangential velocity can jump. A doublet distribution is a vortex sheet, which can represent a wake or a lifting wing without thickness. For a thin layer, equation 13.60 involves only $\Delta\partial\phi/\partial n = \Delta w$, so the lifting boundary condition is lost. An integral equation based on equation 13.60 is not proper for a thin wing.

13.4.2 Integral Equation

Evaluating **x** in equation 13.60 on the surface produces an integral equation for ϕ on the surface:

$$2\pi\phi(\mathbf{x}_s) = \int_{body} \left(\phi(\mathbf{y}_s)\frac{\partial}{\partial n}\frac{1}{s} - \frac{1}{s}\frac{\partial\phi}{\partial n}\right) dA(\mathbf{y}) + \int_{wake} \Delta\phi(\mathbf{y}_s)\frac{\partial}{\partial n}\frac{1}{s} dA(\mathbf{y}) \quad (13.65)$$

The wing boundary condition gives $\partial\phi/\partial n$, and $\Delta\phi$ in the wake is convected from the trailing edge. Generally, both the integral equation and the dependent variable (potential rather than velocity) are well behaved, producing a well-conditioned numerical problem (Maskew (1982a)), even for cases such as a wake cutting another body (Clark and Maskew (1985)). Other integral equations (discussed later) involve higher-order singularities for lifting problems. Equation 13.65 is not applicable to thin wings, but practical thicknesses present no difficulties (Morino (1974)). The singular part of the doublet integral has already been included on the left-hand side. For a flat, constant-strength panel, there is no other contribution from the singular part produced by the local panel; hence, the singularity of the integral is of no concern.

A panel method is produced when the surface is approximated by a connected set of small elements. Typically, the panels are quadrilaterals, although sometimes they are approximated by flat quadrilaterals or triangles. Usually the singularity strength is constant over a panel. A constant-strength doublet panel representation is equivalent to a vortex lattice. Higher-order distributions are also used, but constant-strength panels are more efficient and are generally satisfactory for subsonic flow. The integral equation is evaluated at collocation points, normally the panel centers. Thus, a set of algebraic equations for the potential ϕ_k on the panels is obtained:

$$2\pi\phi_k = \sum_{body} A_{kj}\phi_j + \sum_{body} B_{kj}\frac{\partial\phi_j}{\partial n} + \sum_{wake} D_{kj}\Delta\phi_j \quad (13.66)$$

The influence coefficients depend only on the geometry (not the boundary conditions) and so are fixed for rigid-body motion and a prescribed wake geometry.

To evaluate the pressure, ϕ_t and the velocity $\nabla\phi$ are required (probably in the moving frame):

$$\frac{p - p_\infty}{\rho_\infty} = \begin{cases} -\left(\phi_t + \frac{1}{2}(\nabla\phi)^2\right) & \text{fixed frame} \\ -\left(\phi_t + \mathbf{V}\cdot\nabla\phi + \frac{1}{2}(\nabla\phi)^2\right) & \text{moving frame} \end{cases} \quad (13.67)$$

On the body, the boundary condition defines the normal velocity and the tangential velocity; ϕ_t can be obtained by numerical differentiation of the surface potential. A piecewise constant representation of ϕ must be fit to a polynomial distribution over several nearby panels before being differentiated.

In practice, the Kutta condition or Joukowski hypothesis is implemented by attaching the wake to the wing trailing edge (thereby specifying where the vorticity leaves the body), evaluating the wake strength from $\Delta\phi$ of the panels at the trailing edge, and convecting $\Delta\phi$ in the wake with the local velocity (for an undistorted wake geometry, approximated by the free stream). The theory behind the Kutta condition is more complex and subtle than practice would suggest. Mangler and Smith (1970) showed that at a finite angle trailing edge, the wake should be tangent to the upper or lower surface, depending on local flow conditions. The implications for a panel method have been discussed by Hess (1974), Summa (1975), and Morino, Kaprielian, and Sipcic (1985). The trailing-edge bisector can usually be used for the wake direction. The Kutta condition must be used in a form such that the results are not sensitive to the geometric details of its implementation.

Other integral equations can be obtained from equations 13.62 to 13.64 by evaluating the normal derivative of the potential, $\partial\phi/\partial n_x$, on the body surface:

$$4\pi \frac{\partial \phi}{\partial n_x} = -\int \sigma \frac{\partial}{\partial n_x} \frac{1}{s} dA(\mathbf{y}) = 2\pi\sigma(\mathbf{x}) - \int \sigma \frac{\partial}{\partial n_x} \frac{1}{s} dA(\mathbf{y}) \qquad (13.68)$$

$$4\pi \frac{\partial \phi}{\partial n_x} = \int \sigma \frac{\partial^2}{\partial n_x \partial n_y} \frac{1}{s} dA(\mathbf{y}) \qquad (13.69)$$

$$4\pi \frac{\partial \phi}{\partial n_x} = \int \left(\sigma_d \frac{\partial^2}{\partial n_x \partial n_y} \frac{1}{s} - \sigma_s \frac{\partial}{\partial n_x} \frac{1}{s} \right) dA(\mathbf{y}) \qquad (13.70)$$

In each case, $\partial\phi/\partial n_x$ is evaluated from the boundary condition, and the equation is solved for the singularity strength σ. A source distribution cannot represent a wake, so equation 13.68 is applicable only to non-lifting bodies. The second form of equation 13.68 shows the singular part of the integral. For a flat, constant-strength panel, there is no more contribution from the local panel, so the singularity is of no concern. The doublet distribution (equation 13.69) can represent a lifting wing, but not thickness if the wing surface is modeled as a thin layer. On a wake, the singularity difference is convected from the trailing edge. Equation 13.69 has a Mangler-type singularity, which is of a higher order than the singularities in equations 13.65 and 13.68 and requires more care to evaluate. However, a constant-strength doublet panel is equivalent to a vortex lattice, for which the singularity is of no concern. The combination of source and doublet singularities (equation 13.70) is not a unique representation. Prescribing the distribution of the doublet or source strength over the body or adding boundary conditions to the problem is necessary. Such flexibility can be used to improve the numerical conditioning of the integral equation. To obtain the velocity and pressure, even on the body surface, the singularity distribution must be integrated over the entire body (unless the thin-wing model is used).

13.4.3 Compressible Flow

Helicopter rotor blades normally operate at speeds involving compressible flow. For a helicopter fuselage, separation rather than compressibility effects is the primary concern. Consider subsonic potential flow with an arbitrary geometry. The small perturbation velocity potential (relative to still air) satisfies the wave equation: $a^2 \nabla^2 \phi = \phi_{tt}$. The derivation of the integral equation is guided by Morse and

Feshback (1953), where Green's function was used to solve the problem for a fixed body.

Using Green's theorem, substituting for the wave equation, and then bringing the time derivative outside the volume integral for a moving body, gives

$$\int \left(\phi \frac{\partial \psi}{\partial n} - \psi \frac{\partial \phi}{\partial n} \right) dA = -\int (\phi \nabla^2 \psi - \psi \nabla^2 \phi) dV$$

$$= -\frac{1}{a^2} \int (\phi \psi_{\tau\tau} - \psi \phi_{\tau\tau}) dV$$

$$= -\frac{1}{a^2} \int (\phi \psi_\tau - \psi \phi_\tau)_\tau dV$$

$$= -\frac{1}{a^2} \frac{\partial}{\partial \tau} \int (\phi \psi_\tau - \psi \phi_\tau) dV$$

$$+ \frac{1}{a^2} \int \frac{\partial F}{\partial \tau} \frac{1}{|\nabla F|} (\phi \psi_\tau - \psi \phi_\tau) dA \quad (13.71)$$

The body is defined by $F(\mathbf{y}, \tau) = 0$, so

$$\frac{\partial F}{\partial \tau} \frac{1}{|\nabla F|} = -\mathbf{n} \cdot \frac{\partial \mathbf{y}}{\partial \tau} = -a\mathbf{n} \cdot \mathbf{M} \quad (13.72)$$

where $\mathbf{n}$ is the body normal and $\mathbf{M}$ the surface Mach number. Let $\psi = \delta(\tau^*)/4\pi s$, where $\tau^* = \tau - t + s/a$ and $\mathbf{s} = \mathbf{x} - \mathbf{y}$, and $\delta(\tau^*)$ is the Dirac delta function. Then ψ is singular at $\mathbf{x}$, and the volume V must exclude a small sphere around $\mathbf{x}$. Evaluating the singular part of the left-hand side integral over the surface around $\mathbf{x}$ gives

$$\int \phi \frac{\partial}{\partial n} \frac{\delta(\tau^*)}{4\pi s} dA = \frac{\phi \delta(\tau^*)}{4\pi s}\bigg|_{s=0} \int \frac{\partial}{\partial n} \frac{1}{s} dA = \frac{\phi \delta(\tau - t)}{4\pi} (-4\pi) = -\phi(\mathbf{x}, \tau) \delta(\tau - t) \quad (13.73)$$

Next, the equation is integrated over time, from $\tau = -\infty$ to ∞. The remaining volume integral is zero, since $\phi = \phi_\tau = 0$ at $\tau = -\infty$ (fluid at rest), and $\psi = \psi_\tau = 0$ at $\tau = \infty$ (no effect if $t < \tau$). To proceed further, the order of the time and area integrations is interchanged. First the coordinates are transformed to a moving frame, $\mathbf{y}' = \mathbf{h}(\mathbf{y}, \tau)$ and $\tau' = \tau$, such that the body surface is at rest in the moving frame (F is not a function of τ'). The time derivative transforms as

$$\frac{\partial}{\partial \tau} = \frac{\partial}{\partial \tau'} + \mathbf{V} \cdot \nabla' \equiv \frac{d}{d\tau'} \quad (13.74)$$

where $\mathbf{V} = \partial \mathbf{h}/\partial \tau$ is the velocity of the air relative to the moving frame. For rigid-body motion, a rotating and translating (and in general accelerating) coordinate transformation is required, such as derived earlier for the rotor. For a flexible body the transformation is more complex. Hence

$$\phi(\mathbf{x}, t) = \int \int_{-\infty}^{\infty} \left[\left(\phi \frac{\partial \psi}{\partial n} - \psi \frac{\partial \phi}{\partial n} \right) - \frac{1}{a^2} \frac{\partial F}{\partial \tau'} \frac{1}{|\nabla F|} \left(\phi \frac{\partial \psi}{\partial \tau'} - \psi \frac{\partial \phi}{\partial \tau'} \right) \right] d\tau' \, dA(\mathbf{y}') \quad (13.75)$$

To transform ψ to the moving frame, note that $\tau^* = \tau' - t + s(\tau')/a$. So

$$\delta(\tau^*) = \delta(\tau' - \tau_r) \frac{\partial \tau'}{\partial \tau^*} = \delta(\tau' - \tau_r) \frac{1}{1 - M_r} \quad (13.76)$$

where the retarded time τ_r is the solution of $\tau^* = 0$, and $aM_r = -ds(\tau')/d\tau'$. Finally, $\psi = \delta(\tau' - \tau_r)/(4\pi s(1 - M_r))$.

Thus the integral equation for a moving body in compressible flow is

$$4\pi\phi(\mathbf{x},t) = \int\left[\left(\phi\frac{\partial}{\partial n}\frac{1}{S} - \frac{1}{1-M_r}\frac{\partial s}{\partial n}\frac{\partial}{\partial \tau'}\frac{\phi}{aS} - \frac{1}{S}\frac{\partial \phi}{\partial n}\right)\right.$$
$$-\mathbf{n}\cdot\mathbf{M}\left(\phi\mathbf{M}\cdot\nabla\frac{1}{S} + \frac{\partial}{\partial\tau'}\frac{\phi}{aS}\frac{M_r}{1-M_r}\right)$$
$$\left.-\frac{1}{S}\mathbf{M}\cdot\nabla\phi + \frac{2}{aS}\frac{\partial\phi}{\partial\tau'}\right)\Bigg]dA(\mathbf{y}') \qquad (13.77)$$

where the square brackets imply evaluation at the retarded time, and $S = s(1 - M_r)$. For $\mathbf{x}$ on the body surface, the left-hand side becomes $2\pi\phi$ when the singular part of the right-hand side is accounted for. The surface $A(\mathbf{y}')$ includes both wings and wakes. Collapsing the wake to a thin layer gives integrals of the potential jump $\Delta\phi$ over the wake surface. The first two terms on the right-hand side are recognized as the moving dipole, and the third term is a moving source. The operator ($\mathbf{n}\cdot\nabla - \mathbf{n}\cdot\mathbf{M}\mathbf{M}\cdot\nabla$) accounts for the normal on the moving body. For zero body slope in the flow direction, $\mathbf{n}\cdot\mathbf{M} = 0$, only the first three terms remain on the right-hand side, and the result is just an extension of the incompressible equation using moving singularities. For a thin wing, $\mathbf{n}\cdot\mathbf{M} \cong 0$ can be a good approximation. For a fixed body ($\mathbf{M} = 0$ and $M_r = 0$), the equation becomes

$$4\pi\phi(\mathbf{x},t) = \int\left[\phi\frac{\partial}{\partial n}\frac{1}{s} - \frac{\partial\phi}{\partial\tau}\frac{1}{as}\frac{\partial s}{\partial n} - \frac{1}{s}\frac{\partial\phi}{\partial n}\right]_{\tau=t-s/a}dA(\mathbf{y}') \qquad (13.78)$$

which is Kirchhoff's equation, a mathematical statement of Huygen's principle (Lamb (1932), Garrick (1957)).

For a fixed wing, $\mathbf{M} = (-M\ 0\ 0)^T$ (neglecting motion relative to the uniform flight speed), so $s(1 - M_r) = S$ and $\tau_r = t - \Upsilon$, with S and Υ constant:

$$S^2 = (x - \xi)^2 + \beta^2(y - \eta)^2 + \beta^2(z - \zeta)^2 \qquad (13.79)$$

$$\Upsilon = (S - M(x - \xi))/\beta^2 a \qquad (13.80)$$

The integral equation becomes

$$4\pi\phi(\mathbf{x},t) = \int\left[\left(\phi\frac{\partial}{\partial n}\frac{1}{S} - \frac{\dot\phi}{aS(1-M_r)}\frac{\partial s}{\partial n} - \frac{1}{S}\frac{\partial\phi}{\partial n}\right)\right.$$
$$\left.-\mathbf{n}\cdot i M\left(\phi M\frac{\partial}{\partial\xi}\frac{1}{S} - \frac{\dot\phi M_r}{aS(1-M_r)} - \frac{1}{S}M\frac{\partial\phi}{\partial\xi} - \frac{2\dot\phi}{aS}\right)\right]_{\tau=t-\Upsilon}dA(\mathbf{y}') \qquad (13.81)$$

For steady flow, the result becomes

$$4\pi\phi(\mathbf{x},t) = \int\left[\phi\left(\frac{\partial}{\partial n} - M^2\mathbf{i}\cdot\mathbf{n}\frac{\partial}{\partial\xi}\right)\frac{1}{S} - \frac{1}{S}\left(\frac{\partial\phi}{\partial n} - M^2\mathbf{i}\cdot\mathbf{n}\frac{\partial\phi}{\partial\xi}\right)\right]dA(\mathbf{y}') \qquad (13.82)$$

The substitution $x = \beta\hat{x}$, $\xi = \beta\hat{\xi}$ scales this to the solution of the incompressible problem $\hat\nabla^2\phi = 0$, with $S = \hat{s}/\beta$ and

$$\frac{\partial}{\partial n} - M^2\mathbf{i}\cdot\mathbf{n}\frac{\partial}{\partial\xi} = \frac{\partial}{\partial\hat{n}} \qquad (13.83)$$

identified as the body normal derivative in the scaled coordinate frame. Prandtl-Glauert scaling is obtained if the body slope is small in the flow direction, so $\mathbf{n} \cong \hat{\mathbf{n}}$.

The foundations of panel methods are in fixed-wing applications. Hess and Smith (1974) pioneered the use of equation 13.68 for steady, nonlifting bodies. Rubbert and Saaris (1969, 1972) and Hess (1974) developed equation 13.70 for steady lifting problems, using a surface distribution of sources and an internal or surface distribution of doublets. Djojodihardjo and Widnall (1969) used equation 13.69 for unsteady problems. Morino (1974) and Morino and Kuo (1974) developed a panel method for unsteady, compressible flow based on equation 13.81. Maskew (1982a, 1982b) developed a panel method based on equation 13.65.

Panel methods have been used to analyze helicopter fuselages, particularly to determine the body flow and the velocity induced at a rotor. Appropriately for helicopters, the methods used have been generally incompressible, but there have been many attempts to include a model of separated flow. Panel methods have been developed for rotating blades and the helicopter airframe; see Summa (1976), Summa and Maskew (1981), Summa (1985), Gennaretti and Morino (1992), Quackenbush, Lam, and Bliss (1994), Gennaretti, Luceri, and Morino (1997), Ahmed and Vidjaja (1998), Morino, Bernardini, and Gennaretti (2003), Wachspress, Quackenbush, and Boschitsch (2003b), Gennaretti and Bernardini (2007), and D'Andrea (2008). Issues for applications to rotary wings are incorporating drag and stall and implementation or integration of a free wake analysis.

13.5 Transonic Theory

Transonic rotor analyses use finite-difference techniques to solve the nonlinear equations for the flow about the blade. Transonic flow is normally encountered on rotors in high-speed forward flight, so the problem is unsteady. There is a once-per-revolution variation in the velocity seen by the blade (corresponding to a low reduced frequency), as well as higher harmonics from the blade surface boundary conditions. In spite of the fact that shocks are not isentropic, the potential equation is a good approximation up to a local normal-shock Mach number of about 1.3; when the error resulting from the isentropic assumption is large, the inviscid assumption is probably also inappropriate.

The equation for the velocity potential relative to the still air as derived for a rotating and translating coordinate system is

$$a^2 \nabla^2 \phi = \phi_{tt} + 2 U_i \phi_{tx_i} + U_i U_j \phi_{x_i x_j} + \dot{V}_i \phi_{x_i} \tag{13.84}$$

$$a^2 = a_\infty^2 - (\gamma - 1) \left(\phi_t + V_j \phi_{x_j} + \frac{1}{2} \phi_{x_j} \phi_{x_j} \right) \tag{13.85}$$

$$p/p_\infty = \left(a^2 / a_\infty^2 \right)^{\gamma/(\gamma-1)} \tag{13.86}$$

(equations 13.19 and 13.20), with the boundary condition $U_z = g_t + U_x g_x + U_y g_y$ on $z = g$. In full potential methods, the exact equation and tangent boundary condition are solved.

13.5.1 Small-Disturbance Potential

The equation for small disturbances is useful both to understand the essential character of the problem and to reduce the computational effort. The small-disturbance equation is simpler than the full potential equation, and the linearized boundary condition means that a simpler grid can be used. The small-disturbance approximation is generally valid for moderate airfoil thickness and angle-of-attack, but is always incorrect at the leading edge. The derivation of the small-disturbance equation is guided by Isom (1974).

The small parameter τ is a measure of the disturbance produced by the wing thickness and angle-of-attack. Hence, the wing surface is defined by $g \doteq c\tau$, where c is the blade chord (and "$\doteq$" indicates "order of"). Let the order of the velocity potential be defined by the small parameter δ, such that $\phi \doteq \Omega R c \delta$. The chord and tip speed are measures of the displacement and velocity in the chordwise direction: $x \doteq c$ and $V_x \doteq \Omega R$. Small disturbance implies a linearized boundary condition: $V_z + \phi_z = g_t + V_x g_x + V_y g_y$ on $z = 0$. Requiring $\phi_z \doteq V_x g_x$, it follows that $z \doteq c\delta/\tau$. For $V_z \doteq V_x g_x$, the tip-path-plane incidence $i \doteq \tau$ or less.

For subsonic flow, $\delta \doteq \tau$ is assumed, and the linearized potential equation is obtained:

$$\left(a_\infty^2 \delta_{ij} - V_i V_j\right) \phi_{x_i x_j} = \phi_{tt} + 2V_i \phi_{t x_i} + \dot{V}_i \phi_{x_i} \tag{13.87}$$

This equation can be simplified further for the rotor. For small tip-path-plane incidence, all the V_z terms are higher order and the Ωx term in V_y can be neglected. Relative to $a_\infty^2 \phi_{xx}$, the $\dot{V}_i \phi_{x_i}$ terms are order c/R small. Relative to $a_\infty^2 \phi_{xx}$, ϕ_{tt} is order $(\omega c/\Omega R)^2 \doteq k^2$ and $V_i \phi_{t x_i}$ is order k. So all of the unsteady terms must be retained if $k \doteq 1$. For a 1/rev variation, $k \doteq c/R$ is small and the equation is quasisteady. Hence, the equation for subsonic flow is

$$\left(a_\infty^2 - V_x^2\right) \phi_{xx} - 2V_x V_y \phi_{xy} + \left(a_\infty^2 - V_y^2\right) \phi_{yy} + a_\infty^2 \phi_{zz} = \phi_{tt} + 2V_x \phi_{tx} + 2V_y \phi_{ty} \tag{13.88}$$

where the right-hand side is neglected for frequencies $\omega \doteq \Omega$; $V_x = \Omega y + \mu \Omega R \sin \psi$ and $V_y = \mu \Omega R \cos \psi$.

With transonic flow at the advancing tip, $M_x \cong 1$ or $V_x \cong a_\infty$, and the subsonic ordering assumptions are contradicted. So for transonic flow, it is necessary to retain some of the nonlinear terms. For small disturbances, the $\phi_{x_j}^2$ term in a^2, the order ϕ^3 terms, and the nonlinear unsteady terms can always be neglected. Thus

$$\left[\left(a_\infty^2 - (\gamma-1)\left(\phi_t + V_k \phi_{x_k}\right)\right) \delta_{ij} - V_i V_j - V_j \phi_{x_i} - V_i \phi_{x_j}\right]\phi_{x_i x_j}$$
$$= \phi_{tt} + 2V_i \phi_{t x_i} + \dot{V}_i \phi_{x_i} \tag{13.89}$$

To derive the transonic ordering, assume that the advancing-tip Mach number is near 1, so $a_\infty^2 - V_x^2 \doteq (\Omega R)^2 \Delta$, for a small parameter Δ. Then the nonlinear terms are required only for the coefficient of ϕ_{xx}. Requiring that the largest terms in the coefficient of ϕ_{xx} be the same order ($a_\infty^2 - V_x^2 \doteq V_x \phi_x$) implies $\Delta \doteq \delta$. For a nontrivial problem, the ϕ_{zz} term must be the same order as the ϕ_{xx} term; hence $\delta \doteq \tau^{2/3}$ and $z \doteq c\tau^{-1/3}$. For the problem to be three-dimensional, the ϕ_{yy} term must be of the same order as the ϕ_{xx} term; hence $y \doteq c\tau^{-1/3}$. For yawed flow, the ϕ_{xy} term being the same order as the ϕ_{xx} term requires that the spanwise velocity be moderately small: $V_y \doteq \Omega R \tau^{1/3}$. For small tip-path-plane incidence, the remaining terms on the left-hand side are of a higher order.

For unsteady flow, requiring that the ϕ_{tx} term be the same order as the ϕ_{xx} term implies $k \doteq \tau^{2/3}$. For 1/rev time variation, $k \doteq c/R$; so the problem is unsteady if $c/R \doteq \tau^{2/3}$ and is marginally unsteady if $c/R \doteq \tau$. Alternatively, if $c/R \doteq \tau$ is assumed, the frequency must be $\omega/\Omega \doteq \tau^{-1/3}$ for an unsteady problem. If $k \doteq 1$, the problem is more unsteady than transonic, and the entire ordering scheme must change. The ϕ_t term (compared to $V_x\phi_x$) and the ϕ_{tt} term (compared to $V_x\phi_{tx}$) are order k small; nevertheless, they are sometimes retained to improve the numerical convergence of the solution. The $\dot{V}_i\phi_{x_i}$ terms (compared to the ϕ_{xx} term) and Ωx in V_y or V_x are of order $\tau^{-1/3}c/R$. For $c/R \doteq \tau$ these terms are of second order, but for $c/R \doteq \tau^{2/3}$ (required for the problem to be unsteady with 1/rev time variation), they are only moderately small.

In summary, the transonic scaling requires $\phi \doteq \Omega R c \tau^{2/3}$, $1 - M_x^2 \doteq \tau^{2/3}$, $x \doteq c, y$ and $z \doteq c\tau^{-1/3}$, $M_y \doteq \tau^{1/3}$, and $k \doteq \tau^{2/3}$; exactly as for fixed-wing problems.

In the boundary condition, the V_z term must be retained for $i \doteq \tau$, but $V_y g_y/V_x g_x \doteq \tau^{2/3}$ and $g_t/V_x g_x \doteq k$ are small. Hence $\phi_z = V_x g_x - V_z$ on $z = 0$. For small disturbances, the pressure is approximately

$$\frac{p - p_\infty}{\rho_\infty} \cong -\left(\phi_t + V_i\phi_{x_i} + \frac{1}{2}\phi_{x_i}\phi_{x_i}\right) \tag{13.90}$$

which with the transonic scaling (as for a^2) becomes $(p - p_\infty)/\rho_\infty \cong -V_x\phi_x$.

The transonic domain of the rotor is defined by $(1 - M_x^2) \doteq \tau^{2/3}$, where $V_x = \Omega y + \mu\Omega R \sin \psi$. The maximum Mach number $M_{at} = (1 + \mu)\Omega R/a$ is at the advancing tip ($y = R$ and $\psi = 90°$). Then for $M_{at} \cong 1$, the radial extent of transonic flow is $y/R \doteq \tau^{2/3}$. So $c/R \doteq \tau$ gives $y \doteq c\tau^{-1/3}$, which is consistent with the differential equation being three-dimensional; for $c/R \doteq \tau^{2/3}$ the transonic domain is somewhat larger, $y \doteq c$. The azimuthal extent of transonic flow is $\mu(\psi - \pi/2)^2 \doteq \tau^{2/3}$, or $k \doteq \mu^{1/2}\tau^{-1/3}c/R$ (using $\omega \doteq 1/(\psi - \pi/2)$). Then $\mu \doteq 1$ and $c/R \doteq \tau$, or $\mu \doteq \tau^{2/3}$ and $c/R \doteq \tau^{2/3}$, gives $k \doteq \tau^{2/3}$, which is consistent with the equation being unsteady. For the transonic domain to be consistent with the equation being quasisteady ($k \doteq \tau^{4/3}$) requires either a very small advance ratio or a very small c/R. Inside the transonic domain on the rotor tip, the radial velocity $V_y \cong \mu\Omega R \cos \psi \doteq \Omega R \mu^{1/2}\tau^{1/3}$, which is consistent with the requirement $V_y \doteq \Omega R\tau^{1/3}$.

The transonic small-disturbance equation for potential flow on a rotary wing is thus

$$\left(a_\infty^2 - V_x^2 - (\gamma + 1)V_x\phi_x - (\gamma - 1)\phi_t\right)\phi_{xx} - 2V_xV_y\phi_{xy} + a_\infty^2\phi_{yy} + a_\infty^2\phi_{zz}$$
$$= \phi_{tt} + 2V_x\phi_{tx} + \left(2V_y\Omega + \Omega^2 x\right)\phi_x + \left(-2V_x\Omega + \Omega^2 y\right)\phi_y \tag{13.91}$$

which is nonlinear, unsteady, and three-dimensional. The $\dot{V}_i\phi_{x_i}$ terms on the right-hand side are almost of a higher order. Without them, the equation is identical to that for a fixed wing, except that V_x and V_y vary over the rotor disk. So the transonic nature dominates the equation of motion, while the effects of rotation enter through the variation of the wing velocities. The small-disturbance equations for the boundary condition and pressure are quasistatic and not affected by radial flow. A number of versions of equation 13.91 have been used, such as the following:

a) Without the $\dot{V}_i\phi_{x_i}$ terms, and neglect Ωx in V_y for the ϕ_{xy} term
b) Without the ϕ_{tt} and $\phi_t\phi_{xx}$ terms, without the $\dot{V}_i\phi_{x_i}$ terms, and neglect Ωx in V_y for the ϕ_{xy} term
c) Hover ($V_x = \Omega y$ and $V_y = -\Omega x$) and steady
d) Hover, steady, and without the $\dot{V}_i\phi_{x_i}$ and ϕ_{xy} terms

Retaining all linear terms (the unsteady and radial flow terms) in the boundary condition and pressure poses no difficulties.

The nonlinear terms were retained in equation 13.91 based on the assumption that the flow is in the x direction in the rotating coordinates. In fact, a rotor blade has yawed flow, except exactly at $\psi = 90°$. Retaining the nonlinear terms for arbitrary yawed flow produces the following small-disturbance equation:

$$\left(a_\infty^2 - V_x^2 - (\gamma+1)V_x\phi_x - (\gamma-1)V_y\phi_y\right)\phi_{xx}$$
$$+ \left(a_\infty^2 - V_y^2 - (\gamma-1)V_x\phi_x - (\gamma+1)V_y\phi_y\right)\phi_{yy}$$
$$+ \left(a_\infty^2 - (\gamma-1)V_x\phi_x - (\gamma-1)V_y\phi_y\right)\phi_{zz} - 2\left(V_xV_y + V_x\phi_y + V_y\phi_x\right)\phi_{xy}$$
$$= 2V_x\phi_{tx} + 2V_y\phi_{ty} + \left(2V_y\Omega + \Omega^2 x\right)\phi_x + \left(-2V_x\Omega + \Omega^2 y\right)\phi_y \quad (13.92)$$

Type-dependent differencing is applied in the direction of the local flow, at yaw angle $\Lambda = \tan^{-1} V_y/V_x$. Ignoring the variation of V_x and V_y with y and x, the second derivative in the local flow direction is $\phi_{ss} = C^2\phi_{xx} + 2CS\phi_{xy} + S^2\phi_{yy}$, where $C = \cos\Lambda$ and $S = \sin\Lambda$. Rearranging the left-hand side of equation 13.92, and retaining the nonlinear terms only in the coefficient of ϕ_{ss}, gives

$$\left[a_\infty^2 - (V_x^2 + V_y^2) - (\gamma+1)(V_x\phi_x + V_y\phi_y)\right]\left(C^2\phi_{xx} + 2CS\phi_{xy} + S^2\phi_{yy}\right)$$
$$+ a_\infty^2\left(S^2\phi_{xx} - 2CS\phi_{xy} + C^2\phi_{yy} + \phi_{zz}\right)$$
$$= 2V_x\phi_{tx} + 2V_y\phi_{ty} + \left(2V_y\Omega + \Omega^2 x\right)\phi_x + \left(-2V_x\Omega + \Omega^2 y\right)\phi_y \quad (13.93)$$

When using equation 13.92 or equation 13.93, the radial flow terms are retained in the boundary condition and pressure as well.

The transonic equation for a rotor is solved by a finite-difference method adapted from fixed-wing research, using type-dependent differencing (central differencing in subsonic flow, backward differencing in supersonic flow) for the x derivative or in the local flow direction. The small-disturbance equation can be solved faster than the full potential or Euler equations, but the former is not valid for angles-of-attack above about 5° and never at the leading edge. To reduce the computation time, the quasistatic equations can be solved: all of the time derivatives in the potential equation are neglected, although the correct instantaneous velocities and boundary conditions are used. To define the loading on the advancing blade tip, the unsteady equation must be solved at many more azimuth locations than is necessary with the quasistatic equation. Based on comparison with measured pressures on a non-lifting rotor, quasistatic solutions are not accurate on the advancing side except near $\psi = 90°$, so solutions of the unsteady equations are required; see Chattot and Philippe (1980). Calculations and comparison with experiment have also demonstrated the fundamental three-dimensional nature of transonic flow on the advancing blade tip (Caradonna and Isom (1976)) and the importance of radial flow (Grant (1979)).

13.5.2 History

The first application of computational fluid dynamics to the rotary wing was by Caradonna and Isom (1972). They derived the rotating-frame equation for potential flow about a rotor blade and from it the equation for small-disturbance, transonic flow. Only hover was considered, so the equations were steady in the rotating frame. Solutions were obtained for a non-lifting rotor with rectangular blades and 6% thick biconvex airfoil sections. Isom (1974) extended the derivation of the small-disturbance potential equation to forward flight, using a rotating and translating coordinate system and transonic scaling as appropriate for rotor blades. The unsteady, three-dimensional small-disturbance equations were solved for a non-lifting, rectangular planform rotor at an advance ratio of $\mu = 0.4$ by Caradonna and Isom (1976). The differences between unsteady and quasisteady results were significant, particularly in the decelerating flow of the second quadrant of the rotor disk.

In 1975, ONERA conducted a test of an Alouette II tail rotor in the S2-Ch wind tunnel; see Caradonna and Philippe (1978). The rotor had two blades, no twist, and symmetrical NACA sections. A blade was instrumented with 30 upper-surface pressure transducers at three tip radial stations. Rectangular planform and swept tip blades were tested with transonic tip flow at high-speed forward flight ($\mu = 0.4$ to 0.55), non-lifting. The resulting data proved to be crucial to establishing the validity of the CFD analyses being developed. Chattot and Philippe (1980) showed correlation with-small disturbance calculations and compared results from a number of researchers with these test data. To obtain data on a lifting rotor, ONERA tested a three-bladed rotor in the S2-Ch wind tunnel; see Philippe and Chattot (1980). The blades were articulated, twisted, with straight or swept-parabolic tips. There were pressure transducers at three tip radial stations. The rectangular planform blade was tested in hover and forward flight, for advance ratios up to $\mu = 0.43$.

Chattot (1980) extended the method of Caradonna and Isom for solving the transonic small-disturbance equation, including application to a nearly arbitrary planform. Arieli and Tauber (1979) at NASA obtained quasistatic, full-potential solutions for rotors. Grant (1979) at RAE obtained quasistatic, small perturbation. Philippe and Chattot (1980) summarized the calculations of ONERA, U.S. Army, RAE, and NASA, compared with the non-lifting ONERA data.

Caradonna, Tung, and Desopper (1984) calculated the three-dimensional, unsteady, lifting flow on a rotor blade in forward flight, based on the transonic small-disturbance equations. The influence of the wake and the blade motion was accounted for by using an effective angle-of-attack in the blade boundary condition, calculated from a simple inflow model and the measured flapping and pitch control. The calculations were compared with the test data of Philippe and Chattot (1980). Desopper (1985) presented further comparisons of calculations with lifting rotor test data, for both rectangular and swept-tip blades. This was the start of work that included blade motion and the rotor wake effectively in the computational model and extended the physics modeled by the fluid equations.

13.6 Navier-Stokes Equations

Solutions of the compressible, turbulent Navier-Stokes equations are required for the three-dimensional, unsteady, transonic, viscous flows about helicopter rotors. Rotary-wing problems have always been challenging for computational methods. In particular, modeling the returning wake demands an accurate computational domain over a volume with dimensions on the order of the rotor diameter, while also modeling flow features on the scale of the boundary layer and vortex core. Assume about 10 grid points are required to resolve the fine-scale features, with the vortex core size about 10% chord. A typical aspect ratio of the blade is $R/c = 15$. The wake influence extends 4 diameters or so. Hence a uniform grid requires about 10^{12}

points. The complexity of a nonuniform grid is essential, permitting finer resolution near the bodies and capturing wake vorticity, with coarser resolution in the far field. Overset grids are needed to accommodate such geometries and to implement parallel solutions. Dynamic stall is important for rotor blades, and massive separation for helicopter airframes. Models of turbulence and transition must be appropriate for flows in the rotating frame. In addition to the basic rotation of the rotor, the blades have substantial rigid-body and elastic motion. The aerodynamic grids must follow the blade motion, and the airloads must be transferred to a structural model to calculate the motion.

By the mid-1980s, solutions for rotor blade flow were published using the full-potential equations: see Arieli and Tauber (1979), Chang and Tung (1985), Egolf and Sparks (1985). Sankar and Prichard (1985), Steinhoff and Ramachandran (1986), and Strawn and Tung (1986). By the late 1980s, the Euler equations were being solved for rotor blades: see Roberts and Murman (1985), Kroll (1986), Sankar and Tung (1986), Sankar, Wake, and Lekoudis (1986), Agarwal and Deese (1987), Chang (1987), Chen, McCroskey, and Ying (1988), and Kramer, Hertel, and Wagner (1988). The Euler equations retain the inviscid but not irrotational and isentropic assumptions. Solutions of the full Navier-Stokes equations appeared first around 1990, based on preliminary work in the 1980s: see Liu, Thomas, and Tung (1983), Agarwal and Deese (1988), Srinivasan and McCroskey (1988), Wake and Sankar (1989), Aoyama, Saito, and Kawachi (1990), Smith and Sankar (1991), Srinivasan and Baeder (1991), and Duque (1992).

Current solutions of Reynolds-averaged Navier-Stokes (RANS) equations offer significant improvements in the calculation of blade airloads, compared to methods based on lifting-line theory. Potsdam, Yeo, and Johnson (2006) presented results from RANS calculations for high-speed (transonic flow), low-speed (blade-vortex interaction), and high-thrust (dynamic stall) flight conditions, showing good correlation with flight test measurements of blade airloads. The computational fluid dynamics (CFD) code used high fidelity, overset grid methodology with first-principles-based wake capturing. A comprehensive analysis calculated blade motion and trim, coupled to the CFD code through the normal force, chord force, and pitching moment. Normal force and pitching moment magnitudes were accurately captured in the coupled solutions, and the shape of the airloads curves was well predicted. The phase of the airloads when compared with test data was very good for all flight conditions, resolving past problems of airloads phase prediction using comprehensive analysis. The blade-vortex interaction airloads were well predicted for the low-speed case, even though the far field grid resolution (10% or 5% chord) was not sufficient to resolve tip vortex cores. Dynamic stall events were evident in the calculations, although more improvement was needed.

13.6.1 Hover Boundary Conditions

Whether computational or experimental, a hovering rotor needs a very large operating domain to avoid significant effects of the walls, such as large-scale recirculation and major performance changes. With overset grids, solid walls (zero velocity boundary conditions) very far from the rotor can be used. An alternative is to prescribe velocity boundary conditions in the far field to keep the flow solution well behaved. See Srinivasan, Raghavan, Duque, and McCroskey (1993) and Strawn and Djomehri (2002).

13.6 Navier-Stokes Equations

From momentum theory, the outflow below the rotor is $w_{\text{out}} = \Omega R 2\sqrt{C_T/2}$. This outflow occurs on a circle where the contracted wake cuts the lower boundary, with radius $(1/\sqrt{2})R$ ideally, and about $0.78R$ from flow visualization. By mass conservation, there must be inflow velocities on the rest of the bounding surfaces. These inflow velocities far from the rotor can be estimated based on a sink at the rotor hub, with strength required to match the mass flux of the outflow:

$$w_{\text{in}} = \frac{\Omega R}{4}\sqrt{\frac{C_T}{2}}\left(\frac{R}{d}\right)^2 \tag{13.94}$$

where d is the distance from the hub center. The velocity direction is radially inward toward the hub, on the surfaces of the box bounding the computational domain. Alternatively, free stream pressure can be prescribed at the outflow area, with the mass outflow automatically adjusted to match the inflow. A sink at the rotor hub is not actually part of the computation, but is just the basis for estimating the velocity boundary conditions. With the far-field boundaries reasonably far from the rotor, the solution is not sensitive to the strength of the sink.

13.6.2 CFD/CSD Coupling

Computational fluid dynamics (CFD) offers advances in modeling the complex aerodynamics of rotors, but for forward flight the solution must include the structural dynamic motion of the blades and trim of the aircraft. Typically, the latter tasks are handled using a rotorcraft comprehensive analysis (designated CSD for symmetry). The information exchanged between the CFD and CSD codes consists of integrated section aerodynamic loads and the blade section motion or, more generally, blade pressures and deformations. To combine the CFD and CSD codes requires a fluid-structure interface definition and a coupling strategy. In rotorcraft terminology, the interface methods are classified as tight coupling or loose coupling. For tight coupling, information is exchanged at every time step, often with staggered integration of the aerodynamic and structural dynamic equations, with an outer loop to handle trim. Tight coupling is required for problems involving time history solutions, such as aircraft maneuvers or aeroelastic stability assessment. Use of a time history solution method for steady-state operating conditions poses problems, particularly finding the periodic motion and the trim controls in the presence of low-damped or unstable modes; these problems are exacerbated by the computationally intensive nature of CFD. In loose coupling, information is exchanged for the entire revolution of periodic loads or motion, with separate time integration in the CFD and CSD codes, and trim (adjusting controls to achieve target rotor state) is part of the comprehensive analysis (CSD) as usual. The loose coupling methodology was developed by Johnson for Tung, Caradonna, and Johnson (1986). The history and application of the loose and tight coupling methods have been summarized by Datta, Nixon, and Chopra (2007) and Tung and Ormiston (2009).

Loose coupling is based on an iteration between the CFD and CSD codes, each obtaining the solution over the entire revolution of the rotor. The procedure has the following steps:

a) The comprehensive analysis solves for the blade motion and rotorcraft trim, using its internal aerodynamic model. The blade motion is transmitted to the CFD analysis.

b) The CFD analysis solves for the flow field, using the periodic blade motion to move the aerodynamic surface and with it the computational grid.
c) The difference between the total loading from the CFD analysis and the total loading from the comprehensive analysis is a loading increment. The comprehensive analysis solves for the blade motion and rotorcraft trim, using its internal aerodynamic model, plus this increment as a prescribed loading.
d) Steps (b) and (c) are repeated until the solution converges, as measured by the change in blade loading, blade motion, and trim controls from one iteration to the next. At convergence, the total loading in the CSD analysis equals the total loading from the CFD analysis, yielding a completely consistent solution.

The blade motion is typically described by the linear and angular motion of points along the blade axis, hence assuming rigid chordwise motion, which is consistent with beam models of the comprehensive analysis and is easily accommodated in the CFD grid motion. Section loads are typically used in the comprehensive analysis, again for beam models, requiring integration of pressure loading from the CFD analysis. Consistent collocation points for the loading in the CFD and CSD analyses must be used, so the loads are conserved. Each iteration of the comprehensive analysis produces a converged periodic solution for the motion, but it is most efficient if each iteration of the computationally intensive CFD calculations is not carried to a periodic solution for the loading. In practice, a CFD iteration can solve for just $1/N$ revolution (where N is the number of blades), and full convergence of the coupled CFD/CSD solution is achieved with a total of two or three revolutions of the CFD calculation. Limiting the number of revolutions required in the CFD solution is a major advantage of loose coupling.

Simply setting the lift coefficient in the comprehensive analysis to the value calculated by the CFD analysis would not account for changes in the angle-of-attack as the wing motion and trim are updated. The key to the coupling is to keep the comprehensive analysis aerodynamics active, responding to changes in blade motion and trim. The comprehensive analysis aerodynamics function as an estimate of the CFD airloads due to these changes. As long as this estimate is a sufficiently accurate approximation of the CFD load in the next iteration, the process converges. This approach is also called the delta-coupling method, since the correction can be written

$$\begin{aligned} c_n(\alpha) &= c_{n_{\text{CFD}}}(\alpha_{\text{old}}) + c_{n_\alpha}(\alpha - \alpha_{\text{old}}) \\ &= c_{n_{\text{CFD}}}(\alpha_{\text{old}}) + c_{n_{\text{CA}}}(\alpha) - c_{n_{\text{CAold}}}(\alpha_{\text{old}}) \\ &= c_{n_{\text{CA}}}(\alpha) + \left(c_{n_{\text{CFD}}}(\alpha_{\text{old}}) - c_{n_{\text{CAold}}}(\alpha_{\text{old}})\right) \\ &= c_{n_{\text{CA}}} + \Delta c_n \end{aligned} \quad (13.95)$$

where $c_{n_{\text{CFD}}}$ is the normal force obtained from the CFD analysis, $c_{n_{\text{CA}}}$ is the normal force calculated by the comprehensive analysis, and $c_{n_{\text{CAold}}}$ is the normal force calculated by the comprehensive analysis in the previous cycle. The correction is formulated in a similar manner for chord force and pitching moment. Section axes (normal force and chord force) must be used since the CFD analysis (or blade pressure measurements) does not involve the section angle-of-attack. Loading coefficients are best based on the speed of sound ($M^2 c_n$ form) to avoid inconsistencies in the use of the section velocity. Better still is to use dimensional forces and moments (F) with components in a frame fixed to the rotating hub, thereby avoiding inconsistencies in the normalization and axes of the section coefficients. With a three-dimensional structural model, the delta-coupling method can use pressures and shear stresses on the surface, rather than section loading.

Equation 13.95 requires the loading that the comprehensive analysis calculates without the prescribed delta ($c_{n_{CA}}$). Using an updated delta formulation is more convenient:

$$(\Delta F)_{k+1} = (F_{CFD})_k - (F_{CAold})_k$$
$$= (F_{CFD})_k - (F_{CAtotal} - \Delta F)_k$$
$$= (\Delta F)_k + (F_{CFD} - F_{CAtotal})_k \quad (13.96)$$

since then the total comprehensive analysis load is required ($F_{CAtotal}$), not the load without the increment (F_{CAold}). The update of the loading increment is the current difference between the CFD loading (F_{CFD}) and the total comprehensive analysis loading ($F_{CAtotal}$, including the prescribed delta). Convergence is achieved when ($F_{CFD} - F_{CAtotal}$) approaches zero.

Early applications of loose coupling involved CFD methods with limited computational domains, which did not encompass the entire blade wake or even the entire blade surface. In such cases the boundary condition on the blade surface was implemented in terms of a partial angle-of-attack, accounting not only for the blade motion but also for the induced velocity from the wake outside the CFD domain. The accuracy and convergence of current implementations of loose coupling owe much to the use of CFD models that simulate all the blades and all the flow field. Even if the computationally intensive Navier-Stokes calculations are limited to the blade near field, with a hybrid or approximate model of the wake, the complete aerodynamic model belongs in the CFD part of the iteration, not the comprehensive analysis. Then the interface for the CFD boundary conditions consists of just the blade motion. A key to accurate calculation of airloading on rotor blades is moment coupling, since blade torsion motion has a direct influence on the loading. Improving the loading calculation (and avoiding convergence problems) depends on accurate pitch moment calculation by the CFD analysis. The loose coupling methodology is stable, convergent, and robust with full coupling of normal force, pitching moment, and chord force.

13.7 Boundary Layer Equations

The classical ordering assumptions applied to equations of viscous flow near a surface produce the boundary layer equations, with centrifugal and Coriolis terms due to the rotation of the wing. The Navier-Stokes equations for incompressible flow are

$$\nabla \cdot \mathbf{v} = 0 \quad (13.97)$$

$$\left(\frac{\partial}{\partial t} + \mathbf{v} \cdot \nabla\right)\mathbf{v} = -\frac{1}{\rho}\nabla p + \nu \nabla^2 \mathbf{v} \quad (13.98)$$

with $\nu = \mu/\rho$ the kinematic viscosity. Transforming to rotating and translating coordinates, with $\mathbf{v}$ now the velocity relative to the moving frame, gives

$$\nabla \cdot \mathbf{v} = 0 \quad (13.99)$$

$$\left(\frac{\partial}{\partial t} + \mathbf{v} \cdot \nabla\right)\mathbf{v} + \overline{\Omega} \times \overline{\Omega} \times \mathbf{r} + 2\overline{\Omega} \times \mathbf{v} = -\frac{1}{\rho}\nabla p + \nu \nabla^2 \mathbf{v} \quad (13.100)$$

where $\overline{\Omega} = (0 \ 0 \ \Omega)^T$, so

$$\overline{\Omega} \times \overline{\Omega} \times \mathbf{r} + 2\overline{\Omega} \times \mathbf{v} = \begin{pmatrix} -\Omega^2 x - 2\Omega v \\ -\Omega^2 y + 2\Omega u \\ 0 \end{pmatrix} \quad (13.101)$$

Following Fogarty (1951), we now introduce boundary layer coordinates: x the curvilinear coordinate parallel to the blade surface, and z normal to the blade surface. When the radius of curvature is large and the z-axis is nearly parallel to the axis of rotation, the surface curvature does not change the equations of motion. The assumption that the pressure gradient through the boundary layer is small holds if the boundary layer thickness is small relative to the curvature. The boundary layer assumptions imply that for large Reynolds number Re, the normal velocity w and normal coordinate z are order $Re^{1/2}$. Thus the equations for laminar boundary layers are

$$u_x + v_y + w_z = 0 \quad (13.102)$$

$$u_t + u u_x + v u_y + w u_z - \Omega^2 x - 2\Omega v = -\frac{1}{\rho} p_x + v u_{zz} \quad (13.103)$$

$$v_t + u v_x + v v_y + w v_z - \Omega^2 y + 2\Omega u = -\frac{1}{\rho} p_y + v v_{zz} \quad (13.104)$$

$$0 = -\frac{1}{\rho} p_z \quad (13.105)$$

As found by Prandtl, the pressure gradient through the boundary layer is negligible. The pressure and velocity boundary conditions are from the solution for the inviscid external or outer flow, and no-slip at the wall: $u = v = w = 0$ at $z = 0$, and u, v approach the outer solution as $z \to \infty$. In cylindrical coordinates ($x = r \sin \theta$ and $y = r \cos \theta$, with now u the tangential velocity and r the radial velocity),

$$\frac{1}{r} u_\theta + v_r + \frac{v}{r} + w_z = 0 \quad (13.106)$$

$$u_t + \frac{u}{r} u_\theta + v u_r + w u_z + \frac{uv}{r} - 2\Omega v = -\frac{1}{\rho r} p_\theta + v u_{zz} \quad (13.107)$$

$$v_t + \frac{u}{r} v_\theta + v v_r + w v_z - \frac{u^2}{r} - \Omega^2 r + 2\Omega u = -\frac{1}{\rho} p_r + v v_{zz} \quad (13.108)$$

$$0 = -\frac{1}{\rho} p_z \quad (13.109)$$

In hover the external flow gives $V_y/V_x = -x/y$, which is order c/y and hence is small except at the blade root. So in the boundary layer u/v is order c/y small, and x is order c. Assuming $c/y \ll 1$ and $u/v \doteq (c/y)$ (not near the reverse flow region or the blade root), the boundary layer equations become

$$u_x + w_z = 0 \quad (13.110)$$

$$u_t + u u_x + w u_z = -\frac{1}{\rho} p_x + v u_{zz} \quad (13.111)$$

$$v_t + u v_x + w v_z - \Omega^2 y + 2\Omega u = -\frac{1}{\rho} p_y + v v_{zz} \quad (13.112)$$

Now the mass conservation and u-momentum equations are the same as for two-dimensional flow, with the effects of rotation only entering through the outer flow and pressure; the cross-flow v-momentum equation is decoupled from the other equations. The unsteady terms are order $x/ut \doteq nc/y$ and hence are small for 1/rev variation ($n = 1$).

13.8 Static Stall Delay

Propeller tests of Himmelskamp (1945 tests cited in Schlichting (1979)) show very large measured values of the lift coefficient at the blade root, suggesting that rotation postpones stall. Stall at the propeller root is expected from the large pitch angles, but the three-dimensional flow results in pressure distributions much different from either attached or stalled two-dimensional flow; see Tung and Branum (1990). The effect depends on the local chord ratio c/r, which is a measure of the flow curvature, large at the blade root. The centrifugal forces on the air in the boundary layer can significantly affect the flow character, including separation.

Banks and Gadd (1963) used boundary layer calculations to examine the effect of rotation on separation. They assumed that the tangential velocity outside the boundary had the form $u_e = \Omega r(1 - K\theta)$ (in cylindrical coordinates), where K is a constant that gives a decelerating flow. Then $\frac{1}{\rho r} p_\theta = r\Omega^2 K(1 - K\theta)$ and $\frac{1}{\rho} p_r = r\Omega^2 K^2 \theta^2$. The onset of separation occurs when $\partial u/\partial z = 0$ at the surface. For K very large, the problem is two-dimensional flow with a linear adverse velocity gradient, which stalls at $K\theta = 0.136$. For $K \leq 0.55$, the separation condition is not reached, so the boundary layer is stabilized completely against separation.

Corrigan and Schillings (1994) solved the boundary layer equations with external velocity $u_e = U_\infty(1 - K\theta)$, to find the value of θ_{TE} at separation. Based on these results, they developed a correction as a stall delay: $c_\ell = c_{\ell_{2D}}(\alpha - \Delta\alpha) + c_{\ell\alpha}\Delta\alpha$, where $\Delta\alpha = f(c_{\ell\max}/c_{\ell\alpha})$. The factor is given as $f = (K\theta_{TE}/.136)^n - 1$, with $\theta_{TE} \cong c/r$ and $c/r = .1517K^{-1.084}$. This correction is equivalent to using $c_\ell = Kc_{\ell_{2D}}((\alpha - \alpha_z)/K + \alpha_z)$, with

$$K = f + 1 = \left(\frac{c/r}{.136}\left(\frac{.1517}{c/r}\right)^{1/1.084}\right)^n = \left(1.291(c/r)^{.0775}\right)^n \quad (13.113)$$

The exponent $n = 0.8$ to 1.8, with the larger values typically giving better correlation.

Du and Selig (1998) solved the boundary layer equations with external velocity $u_e = U_\infty(1 - Kr\theta)$. They found the laminar separation point and approximated its position as a function of c/r. The delay of separation was identified by comparing with two-dimensional, non-rotating solutions and was used to construct a stall washout correction:

$$c_\ell = c_{\ell_{2D}} + K_{sd}\left(c_{\ell\alpha}(\alpha - \alpha_z) - c_{\ell_{2D}}\right) \quad (13.114)$$

$$c_d = c_{d_{2D}} + K_{sd}\left(c_{dz} - c_{d_{2D}}\right) \quad (13.115)$$

For a rotor in axial flow, they give the following stall delay factor:

$$K_{sd} = \frac{1}{2\pi}\left[\frac{1.6 c/r}{0.1267}\frac{a - (c/r)^D}{b + (c/r)^D} - 1\right]$$

with $D = d/(\Lambda r/R)$ for lift and $D = d/(2\lambda r/R)$ for drag; $\Lambda = \Omega R/\sqrt{V^2 + (\Omega R)^2}$. The constants a, b, and d are approximately 1. Here c is the chord, R the blade radius, r the distance from the center of rotation, V the rotor axial velocity, and Ω the rotational speed. This expression is not valid for $c/r > 1$. Raj (2000) revised this model, particularly for an increase rather than a decrease in drag, using $D = (r/R)/(d\Lambda)$ and including an additional factor in K_{sd} of $(1 - r/R)$ for lift and $(2 - r/R)$ for drag.

Snel and van Holten (1994) proposed the stall washout model based on boundary layer analysis, with $K_{sd} = \tanh(3(c/r)^2)$.

The static stall delay phenomenon is important for tilt rotors in hover and for wind turbines. The phenomenon depends on the rotation of the blade, so can be active on all rotors in hover. A helicopter rotor has low twist, so has low angle-of-attack and low loading at the root and hence exhibits little effect of stall delay on performance. The large twist typical of tilt rotors means that in hover the angle-of-attack is high at the root. The loading is small inboard, but because of the static stall delay the blade can carry more loading at the root and hence less loading outboard, with a net reduction in power; see Corrigan and Schillings (1994). The static stall delay phenomenon must be accounted for in the analysis, either by solving the Navier-Stokes equations or by using one of the earlier empirical corrections, in order to accurately calculate hovering rotor performance. The stall delay shifts the drop in the hover figure of merit caused by stall to higher C_T/σ. The static stall delay also improves the performance of highly twisted tilt rotors in low-speed edgewise flight (Johnson (2000)).

13.9 REFERENCES

Agarwal, R.K., and Deese, J.E. "Euler Calculations for Flowfield of a Helicopter Rotor in Hover." Journal of Aircraft, 24:4 (April 1987).

Agarwal, R.K., and Deese, J.E. "Navier-Stokes Calculations of the Flowfield of a Helicopter Rotor in Hover." AIAA Paper No. 88-0106, January 1988.

Ahmed, S.R., and Vidjaja, V.T. "Unsteady Panel Method Calculation of Pressure Distribution on BO 105 Model Rotor Blades." Journal of the American Helicopter Society, 43:1 (January 1998).

Aoyama, T., Saito, S., and Kawachi, K. "Navier-Stokes Analysis of Blade Tip Shape in Hover." Sixteenth European Rotorcraft Forum, Glasgow, UK, September 1990.

Arieli, R., and Tauber, M.E. "Computation of Subsonic and Transonic Flow About Lifting Rotor Blades." AIAA Paper No. 79-1667, August 1979.

Ashley, H., and Landahl, M. Aerodynamics of Wings and Bodies. Reading, MA: Addison-Wesley Publishing Company, Inc., 1965.

Banks, W.H.H., and Gadd, G.E. "Delaying Effect of Rotation on Laminar Separation." AIAA Journal, 1:4 (April 1963).

Batchelor, G.K. An Introduction to Fluid Dynamics. London: Cambridge University Press, 1967.

Caradonna, F.X., and Isom, M.P. "Subsonic and Transonic Potential Flow over Helicopter Rotor Blades." AIAA Journal, 10:12 (December 1972).

Caradonna, F.X., and Isom, M.P. "Numerical Calculation of Unsteady Transonic Potential Flow over Helicopter Rotor Blades." AIAA Journal, 14:4 (April 1976).

Caradonna, F.X., and Philippe, J.J. "The Flow over a Helicopter Blade Tip in the Transonic Regime." Vertica, 2:1 (1978).

Caradonna, F.X., Tung, C., and Desopper, A. "Finite Difference Modeling of Rotor Flows Including Wake Effects." Journal of the American Helicopter Society, 29:2 (April 1984).

13.9 References

Chang, I.-C. "Unsteady Euler Solution of Transonic Helicopter Rotor Flow." American Helicopter Society National Specialists' Meeting on Aerodynamics and Aeroacoustics, Arlington, TX, February 1987.

Chang, I.-C., and Tung, C. "Numerical Solution of the Full-Potential Equation for Rotors and Oblique Wings Using a New Wake Model." AIAA Paper No. 85-0268, January 1985.

Chattot, J.J. "Calculation of Three-Dimensional Unsteady Transonic Flows Past Helicopter Blades." NASA TP 1721, October 1980.

Chattot, J.-J., and Philippe, J.-J. "Pressure Distribution Computation on a Non-Lifting Symmetrical Helicopter Blade in Forward Flight." La Recherche Aerospatiale, *1980*:5 (1980).

Chen, C.L., McCroskey, W.J., and Ying, S.X. "Euler Solution of Multiblade Rotor Flow." Vertica, *12*:3 (1988).

Clark, D.R., and Maskew, B. "Study for Prediction of Rotor/Wake/Fuselage Interference." NASA CR 177340, March 1985.

Corrigan, J.J., and Schillings, J.J. "Empirical Model for Stall Delay due to Rotation." American Helicopter Society 50th Annual Forum, Washington, DC, May 1994.

D'Andrea, A. "Development of a Multi-Processor Unstructured Panel Code Coupled with a CVC Free Wake Model for Advanced Analyses of Rotorcrafts and Tiltrotors." American Helicopter Society 64th Annual Forum, Montreal, Canada, April 2008.

Dat, R. "Lifting Surface Theory Applied to Fixed Wings and Propellers." La Recherche Aerospatiale, *1973*:4 (July-August 1973).

Datta, A., Nixon, M., and Chopra, I. "Review of Rotor Loads Prediction with the Emergence of Rotorcraft CFD." Journal of the American Helicopter Society, *52*:4 (October 2007).

Desopper, A. "Study of the Unsteady Transonic Flow on Rotor Blades with Different Tip Shapes." Vertica, *9*:3 (1985).

Djojodihardjo, R.H., and Widnall, S.E. "A Numerical Method for the Calculation of Nonlinear, Unsteady Lifting Potential Flow Problems." AIAA Journal, *7*:10 (October 1969).

Du, Z., and Selig, M.S. "A 3-D Stall-Delay Model for Horizontal Axis Wind Turbine Performance Prediction." AIAA Paper No. 98-0021, January 1998.

Duque, E.P.N. "A Numerical Analysis of the British Experimental Rotor Program Blade." Journal of the American Helicopter Society, *37*:1 (January 1992).

Egolf, T.A., and Sparks, S.P. "Hovering Rotor Airloads Prediction Using a Full Potential Flow Analysis with Realistic Wake Geometry." American Helicopter Society 41st Annual Forum, Ft. Worth, TX, May 1985.

Farassat, F. "Theory of Noise Generation from Moving Bodies with an Application to Helicopter Rotors." NASA TR R-451, December 1975.

Farassat, F. "Advanced Theoretical Treatment of Propeller Noise." von Kármán Institute Lecture Series 81-82/10, May 1982.

Fogarty, L.E. "The Laminar Boundary Layer on a Rotating Blade." Journal of the Aeronautical Sciences, *18*:4 (April 1951).

Garrick, I.E. "Nonsteady Wing Characteristics." In *Aerodynamic Components of Aircraft at High Speeds*, edited by A.E. Donovan and H.R. Lawrence. Princeton, NJ: Princeton University Press, 1957.

Gennaretti, M., and Bernardini, G. "Novel Boundary Integral Formulation for Blade-Vortex Interaction Aerodynamics of Helicopter Rotors." AIAA Journal, *45*:6 (June 2007).

Gennaretti, M., Luceri, L., and Morino, L. "A Unified Boundary Integral Methodology for Aerodynamics and Aeroacoustics of Rotors." Journal of Sound and Vibration, *200*:4 (1997).

Gennaretti, M., and Morino, L. "A Boundary Element Method for the Potential, Compressible Aerodynamics of Bodies in Arbitrary Motion." The Aeronautical Journal, *96*:951 (January 1992).

Grant, J. "The Prediction of Supercritical Pressure Distributions on Blade Tips of Arbitrary Shape over a Range of Advancing Blade Azumuth Angles." Vertica, *3*:3/4 (1979).

Hess, J.L. "The Problem of Three-Dimensional Lifting Potential Flow and Its Solution by Means of Surface Singularity Distribution." Computer Methods in Applied Mechanics and Engineering, *4*:3 (November 1974).

Hess, J.L., and Smith, A.M.O. "Calculation of Potential Flow About Arbitrary Bodies." In *Progess in Aeronautical Sciences*, Volume 8, edited by D. Kuchemann. Oxford, England: Pergamon Press, 1966.

Isom, M.P. "Unsteady Subsonic and Transonic Potential Flow over Helicopter Rotor Blades." NASA CR 2463, October 1974.

Johnson, W. "Calculation of Tilt Rotor Aeroacoustic Model (TRAM DNW) Performance, Airloads, and Structural Loads." American Helicopter Society Aeromechanics Specialists' Meeting, Atlanta, GA, November 2000.

Kocurek, J.D., and Tangler, J.L. "A Prescribed Wake Lifting Surface Hover Performance Analysis." Journal of the American Helicopter Society, 22:1 (January 1977).

Kramer, E., Hertel, J., and Wagner, S. "Computation of Subsonic and Transonic Helicopter Rotor Flow Using Euler Equations." Vertica, 12:3 (1988).

Kroll, N. "Computation of the Flow Fields of Propellers and Hovering Rotors Using Euler Equations." Twelfth European Rotorcraft Forum, Garmisch-Partenkirchen, Germany, September 1986.

Lamb, H. *Hydrodynamics*. New York: Dover Publications, 1932.

Landahl, M.T. "Kernel Function for Nonplanar Oscillating Surfaces in a Subsonic Flow." AIAA Journal, 5:5 (May 1967).

Liu, C.H., Thomas, J.L., and Tung, C. "Navier-Stokes Calculations for the Vortex Wake of a Rotor in Hover." AIAA Paper No. 83-1676, July 1983.

Mangler, K.W., and Smith, J.H.B. "Behavior of the Vortex Sheet at the Trailing Edge of a Lifting Wing." Aeronautical Journal, 74:719 (November 1970).

Maskew, B. "Prediction of Subsonic Aerodynamic Characteristics: A Case for Low-Order Panel Methods." Journal of Aircraft, 19:2 (February 1982a).

Maskew, B. "Program VSAERO – A Computer Program for Calculating the Nonlinear Aerodynamic Characteristics of Arbitrary Configurations." NASA CR 166476, December 1982b.

Morino, L. "A General Theory of Unsteady Compressible Potential Aerodynamics." NASA CR 2464, December 1974.

Morino, L., Bernardini, G., and Gennaretti, M. "A Boundary Element Method for Aerodynamics and Aeroacoustics of Bodies in Arbitrary Motions." International Journal of Aeroacoustics, 2:2 (April 2003).

Morino, L., Kaprielian, Z., Jr., and Sipcic, S.R. "Free Wake Analysis of Helicopter Rotors." Vertica, 9:2 (1985).

Morino, L., and Kuo, C.-C. "Subsonic Potential Aerodynamics for Complex Configurations: A General Theory." AIAA Journal 12:2 (February 1974).

Morse, P.M., and Feshback, H. *Methods of Theortical Physics*. New York: McGraw-Hill, Book Company, 1953.

Philippe, J.-J., and Chattot, J.-J. "Experimental and Theoretical Studies on Helicopter Blade Tips at ONERA." Sixth European Rotorcraft and Powered Lift Aircraft Forum, Bristol, United Kingdom, September 1980.

Potsdam, M., Yeo, H., and Johnson, W. "Rotor Airloads Prediction Using Loose Aerodynamic/Structural Coupling." Journal of Aircraft, 43:3 (May-June 2006).

Quackenbush, T.R., Bliss, D.B., and Wachspress, D.A. "New Free-Wake Analysis of Rotorcraft Hover Performance Using Influence Coefficients." Journal of Aircraft, 26:12 (December 1989).

Quackenbush, T.R., Lam, C.-M.G., and Bliss, D.B. "Vortex Methods for the Computational Analysis of Rotor/Body Interaction." Journal of the American Helicopter Society, 39:4 (October 1994).

Raj, N.V. "An Improved Semi-Empirical Model for 3-D Post-Stall Effects in Horizontal Axis Wind Turbines." Master of Science Thesis, University of Illinois, 2000.

Roberts, T.W., and Murman, E.M. "Solution Method for a Hovering Helicopter Rotor Using the Euler Equations." AIAA Paper No. 85-0436, January 1985.

Rubbert, P.E., and Saaris, G.R. "3-D Potential Flow Method Predicts V/STOL Aerodynamics." SAE Journal, 77:9 (September 1969).

Rubbert, P.E., and Saaris, G.R. "Review and Evaluation of a Three-Dimensional Lifting Potential Flow Analysis Method for Arbitrary Configurations." AIAA Paper No. 72-188, January 1972.

Runyan, H.L., and Tai, H. "Lifting Surface Theory for a Helicopter Rotor in Forward Flight." NASA CR 169997, 1983.

Runyan, H.L., and Tai, H. "Compressible, Unsteady Lifting-Surface Theory for a Helicopter Rotor in Forward Flight." NASA TP 2503, December 1985.

Sankar, L.N., and Prichard, D.S. "Solution of Transonic Flow Past Rotor Blades Using the Conservative Full Potential Equation." AIAA Paper No. 85-5012, October 1985.

Sankar, L.N., and Tung, C. "Euler Calculations for Rotor Configurations in Unsteady Forward Flight." American Helicopter Society 42nd Annual Forum, Washington DC, June 1986.

Sankar, N.L., Wake, B.E., and Lekoudis, S.G. "Solution of the Unsteady Euler Equations for Fixed and Rotor Wing Configurations." Journal of Aircraft, 23:4 (April 1986).

Schlichting, H. *Boundary-Layer Theory*. Seventh Edition. New York: McGraw-Hill Book Company, 1979.

Smith, M.J., and Sankar, L.N. "Evaluation of a Fourth-Order Compact Operator Scheme for Euler/Navier-Stokes Simulations of a Rotor in Hover." AIAA Paper No. 91-0766, January 1991.

Snel, H., and van Holten, T. "Review of Recent Aerodynamic Research on Wind Turbines with Relevance to Rotorcraft." Twentieth European Rotorcraft Forum, Amsterdam, The Netherlands, October 1994.

Srinivasan, G.R., and Baeder, J.D. "Recent Advances in Euler and Navier-Stokes Methods for Calculating Helicopter Rotor Aerodynamics and Acoustics." 4th International Symposium on Computational Fluid Dynamics, Davis, CA, September 1991.

Srinivasan, G.R., and McCroskey, W.J. "Navier-Stokes Calculations of Hovering Rotor Flowfields." Journal of Aircraft, 25:10 (October 1988).

Srinivasan, G.R., Raghavan, V., Duque, E.P.N., and McCroskey, W.J. "Flowfield Analysis of Modern Helicopter Rotors in Hover by Navier-Stokes Method." Journal of the American Helicopter Society, 38:3 (July 1993).

Steinhoff, J., and Ramachandran, K. "Free Wake Analysis of Compressible Rotor Flow Fields in Hover." Twelfth European Rotorcraft Forum, Garmisch-Partenkirchen, Germany, September 1986.

Strawn, R.C., and Djomehri, M.J. "Computational Modeling of Hovering Rotor and Wake Aerodynamics." Journal of Aircraft, 39:5 (September-October 2002).

Strawn, R.C., and Tung, C. "The Prediction of Transonic Loading on Advancing Helicopter Rotors." AGARD CP 412, April 1986.

Summa, J.M. "Potential Flow About Three-Dimensional Lifting Configurations with Application to Wings and Rotors." AIAA Paper No. 75-126, January 1975.

Summa, J.M. "Potential Flow About Impulsively Started Rotors." Journal of Aircraft, 13:4 (April 1976).

Summa, J.M. "Advanced Rotor Analysis Methods for the Aerodynamics of Vortex-Blade Interactions in Hover." Vertica, 9:4 (1985).

Summa, J.M., and Maskew, B. "A Surface Singularity Method for Rotors in Hover or Climb." USAAVRADCOM TR 81-D-23, December 1981.

Tung, C., and Branum, L. "Model Tilt-Rotor Hover Performance and Surface Pressure Measurement." American Helicopter Society 46th Annual Forum, Washington, DC, May 1990.

Tung, C., Caradonna, F.X., and Johnson, W. "The Prediction of Transonic Flows on an Advancing Rotor." Journal of the American Helicopter Society, 31:3 (July 1986).

Tung, C., and Ormiston, R.A. "History of Rotorcraft CFD/CSD Coupling." Rotor Korea 2009, 2nd International Forum on Rotorcraft Multidisciplinary Technology, Seoul, Korea, October 2009.

Wachspress, D.A., Quackenbush, T.R., and Boschitsch, A.H. "First-Principles Free-Vortex Wake Analysis for Helicopters and Tiltrotors." American Helicopter Society 59th Annual Forum, Phoenix, AZ, May 2003a.

Wachspress, D.A., Quackenbush, T.R., and Boschitsch, A.H. "Rotorcraft Interactional Aerodynamics with Fast Vortex/Fast Panel Methods." Journal of the American Helicopter Society, *48*:4 (October 2003b).

Wake, B.E., and Sankar, L.N. "Solutions of the Navier-Stokes Equations for the Flow About a Rotor Blade." Journal of the American Helicopter Society, *34*:2 (April 1989).

Watkins, C.E., Runyan, H.L., and Woolston, D.S. "On the Kernel Function of the Integral Equation Relating the Lift and Downwash Distribution of Oscillating Finite Wings in Subsonic Flow." NACA Report 1234, 1955.

14 Noise

14.1 Helicopter Rotor Noise

The helicopter is the quietest VTOL aircraft, but its noise level can still be high enough to compromise its utility unless specific attention is given to designing for low noise. As the restrictions on aircraft noise increase, the rotor noise becomes an increasingly important factor in helicopter design. The complex aerodynamics of rotors lead to a number of significant noise mechanisms. Helicopter rotor noise tends to be concentrated at harmonics of the blade passage frequency $N\Omega$, because of the periodic nature of the rotor as seen in the non-rotating frame. There is sound radiated because the mean thrust and drag forces rotate with the blades and because of the higher harmonic loading as well. The spectral lines are broadened at the higher harmonics because of the random character of the rotor flow, particularly variations in the wake-induced loads. The acoustic pressure signal is basically periodic in time (the period is $2\pi/N\Omega$), with sharp impulses due to localized aerodynamic phenomena such as compressibility effects and vortex-induced loads. Figure 14.1 illustrates the spectrum of rotor-generated sound. The contributions to helicopter rotor noise can be classified as vortex or broadband noise, rotational noise, and impulsive noise or blade slap. Although the distinction between these types of rotor noise is not as sharp as was once thought, the classification remains useful for purposes of exposition. Cox (1973), Burton, Schlinker, and Shenoy (1985), and Brentner and Farassat (1994, 2003) have presented summaries of helicopter rotor noise mechanisms and analysis.

The blade passage frequency $N\Omega$ is typically 10 to 20 Hz for a main rotor. $N\Omega$ is low for two-bladed rotors, as well as for large rotors unless the number of blades is large. It can be 30 Hz or higher for small helicopters or with a large number of blades. For tail rotors and propellers the blade passage frequency is higher, typically 50 to 100 Hz.

Broadband or vortex noise is a high-frequency swishing sound produced by the rotor and modulated in frequency and amplitude at the blade passage frequency. Broadband noise is random sound radiated as a result of random fluctuations of the forces on the blades, caused by random flow field fluctuations (turbulence). The sound energy is distributed over a substantial portion of the spectrum in the audible range, typically extending for main rotors from about 150 Hz to 1000 Hz, with a peak around 300 or 400 Hz. There is also a high-frequency range of main rotor broadband noise, occurring at around 2 to 5 kHz. Rotor broadband noise mechanisms are

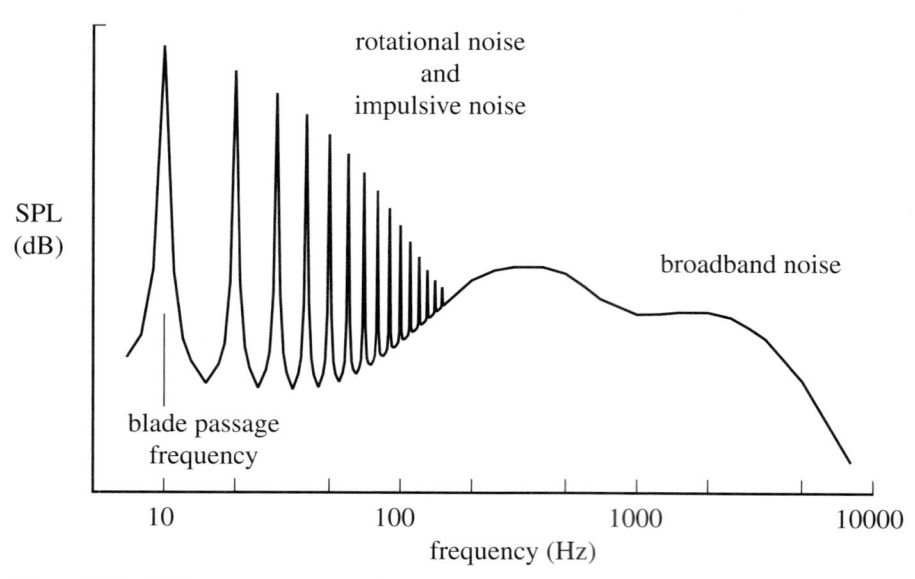

Figure 14.1. Helicopter rotor sound spectrum.

described by Burley and Brooks (2004). Dominant at frequencies up to 2 or 3 Hz is turbulence ingestion noise, from the blade encountering atmospheric and wake turbulence. Blade-wake interaction (BWI) noise is produced by the random lift fluctuations resulting from operation of the blade in the wake flow field, with random wake position and strength variations. Also contributing to BWI are recirculation when the rotor is near the ground, main rotor wake interaction with the tail rotor, and the interaction of twin main rotors. Blade self-noise, occurring at higher frequencies (3 to 6 kHz), is caused by the formation and shedding of self-induced boundary layer turbulence. Self-noise mechanisms include turbulent boundary layer, trailing-edge noise; boundary layer separation, especially large-scale separation in deep stall; laminar boundary layer, vortex shedding noise; blunt trailing-edge noise; and tip vortex formation and shedding noise. The random noise of rotating wings was originally associated with the shedding of vortices, as from a cylinder; hence the name "vortex noise."

Rotational noise is a thumping sound at the blade passage frequency, or at a multiple of $N\Omega$ if the fundamental is inaudible. As the higher harmonic content increases, the thumps sharpen into bangs and eventually into blade slap. The spectrum varies greatly with the rotor geometry and the operating condition. Rotational noise is a purely periodic sound pressure radiated as a result of the periodic forces exerted by the blade on the air. The spectrum of rotational noise thus consists of discrete lines at harmonics of the blade passage frequency $N\Omega$. Rotational noise dominates the low-frequency end of the spectrum, for the main rotor from below audible frequencies to about 150 Hz. It is found to extend even higher if a narrow band measurement is made, with a gradual progression to broadband noise. The fundamental frequency and perhaps also the first or second harmonic are below the threshold of hearing. The sound identified at the blade passage frequency is the higher harmonics and the vortex noise modulated at N/rev. Therefore the higher harmonics of main rotor rotational noise are important subjectively. For a tail rotor the fundamental frequency is higher, which so increases the subjective importance of the rotational noise that it becomes the dominant component. Rotational noise is produced by the force periodically exerted on the air at any fixed point on the rotor disk because of the rotation of the mean and unsteady loading with the blade. The higher harmonic blade loading is responsible for the large high-frequency content of helicopter rotor rotational noise. Rotational noise also includes thickness noise, from the periodic displacement of the fluid by the rotating blade.

Impulsive noise or blade slap is a sharp cracking, popping, or slapping sound occurring at the blade passage frequency and produced by the main rotor in certain flight conditions. Blade slap is a periodic, impulsive sound pressure disturbance, so is an extreme case of rotational noise. The impulsive character of blade slap results in a substantial increase of the sound level over the entire spectrum, covering a range of about 20 to 1000 Hz for a main rotor. When it occurs, blade slap is the dominant noise source, due to the high overall sound pressure level and objectionable impulsive character. Blade slap tends to occur most often in such maneuvers as flare to landing, shallow descents, and decelerating steep turns, as well as at high forward speeds. With some helicopters, blade slap occurs in level flight at moderate speeds. The principal sources of blade slap are blade-vortex interactions (BVI noise) and the effects of thickness and compressibility at high Mach number (high-speed impulsive, or HSI noise). Such aerodynamic phenomena produce a large, localized, transient force on the blade, which results in impulsive sound radiation. BVI noise is caused by lift on the blade and radiates forward and below the disk plane. BVI noise occurs in particular in descent, usually significant at the certification flight condition. A tandem helicopter has blade slap as a result of the interaction of the rear rotor blades with the tip vortices in the wake of the front rotor. HSI noise radiates noise in the plane of the rotor disk, due to transonic flow on the advancing blade. At high Mach numbers, thickness noise is an HSI noise source.

The sources of main rotor noise, in order of importance, are impulsive noise (when it occurs), broadband noise, and rotational noise. The rotational noise is most intense at very low frequency, but the first few harmonics are possibly below the threshold of hearing. Thus although rotational noise is the primary determinant of the overall sound pressure level, it is not the most important source in terms of subjective annoyance. When the sound level is corrected for frequency content based on human perception, the broadband noise often dominates. Rotational noise can be important when its level increases at high frequencies; hence in cases approaching blade slap. It can also produce acoustic fatigue and vibration of the helicopter structure. The low frequencies propagate best in the air, with the high frequencies being attenuated most with distance. Consequently, at very large distances from the helicopter the impulsive and rotational noise of the main rotor are most important. The acoustic detectability of the helicopter is often determined by the rotational noise.

On many helicopters the tail rotor is the source of the most noticeable and disturbing noise (next to blade slap), unless some effort is made to quiet it. The design limitations for a quiet tail rotor (mainly low tip speed) impose little aircraft performance degradation, because the tail rotor accounts for a small part of the total weight and power loss. The tail rotor noise mechanisms are the same as the main rotor mechanisms, but have a higher fundamental frequency. Hence tail rotor rotational noise is important subjectively. Interaction of the main rotor wake with the tail rotor produces noise at combinations of the main rotor and tail rotor blade passage frequencies. A fan-in-fin anti-torque system has a much higher blade passage frequency (300 to 800 Hz) and hence increased annoyance. The helicopter transmission and engine are sources of high-frequency sound, which is important mainly for the internal and near field noise. Since the high frequencies attenuate most with distance, the rotors are the primary source of far field noise.

The most important parameter influencing the noise level of a rotor is the blade tip speed. The vortex noise, rotational noise, and blade slap can all be significantly reduced by operating the rotor at a lower tip speed. At high Mach number, decreasing the tip speed is particularly effective in reducing the rotational noise and blade slap. The vortex noise generally decreases slowest, so at low tip speed becomes the dominant noise source. The vortex noise can also be reduced by lowering the rotor thrust or the blade loading T/A_b, and the rotational noise and blade slap can be reduced by lowering the thrust or the disk loading T/A. Decreasing the rotational speed of the rotor also lowers the frequency range of the rotor noise, which can be beneficial for large rotors. Increasing the number of blades tends to decrease the magnitude of the rotational noise harmonics, although the fundamental frequency also increases. Decreasing the higher harmonic content of the rotor airloads reduces the rotational noise; in particular, this requires decreasing the blade-vortex interaction loads, which also reduces blade slap.

The blade tip planform and section can influence the rotor noise by altering the tip aerodynamic loads and the structure of the trailed tip vortex. The airfoil section shape and thickness ratio at the tip should be chosen for good characteristics at high Mach number, because of the importance of compressibility effects in rotational noise and blade slap. An appropriate planform shape can minimize the tip vortex roll up and thus reduce the blade-vortex interaction loads.

The parameters influencing rotor noise, particularly tip speed, disk loading, and number of blades, are also major factors in determining rotor performance. So any improvement in the aerodynamic efficiency of the rotor can be used to design a quieter helicopter while maintaining the current level of performance and cost, rather than being used exclusively to improve the helicopter performance.

14.2 Rotor Sound Spectrum

Let us examine the statistics of the rotor sound field. Because sound consists of pressure waves radiated in a fluid, the noise analysis is concerned with the perturbation of the pressure from the atmospheric level. The rotor aerodynamic loading and hence the sound pressure are random signals with periodic (hence nonstationary) statistics. The fundamental frequency is the blade passage frequency, so the period is $2\pi/N\Omega$. The expected value of the sound pressure perturbation generated by the rotor is $\overline{p}(t) = E(p(t = \psi/\Omega))$. The expected value (designated by the operator $E(\ldots)$) is the sum or integration over all instances, weighted by the probability function. The Fourier transform of the auto-correlation R gives the spectrum S. For a stationary function $\tilde{p}$, the mean and auto-correlation do not depend on τ, so

$$R(\tau) = E\big(\tilde{p}(t)\tilde{p}(t+\tau)\big) = \int_{-\infty}^{\infty} e^{i\omega\tau} S(\omega) d\omega \tag{14.1}$$

$$S(\omega) = \frac{1}{2\pi} \int_{-\infty}^{\infty} e^{-i\omega\tau} R(\tau) d\tau \tag{14.2}$$

and the mean-squared value is $\sigma^2 = R(0) = \int_{-\infty}^{\infty} S(\omega) d\omega$.

14.2 Rotor Sound Spectrum

Since the pressure signal of a rotor is nonstationary, the expected value $\bar{p}$ is not constant. It is a periodic function of ψ, so can be written as a Fourier series:

$$\bar{p} = \sum_{m=-\infty}^{\infty} p_m e^{imN\Omega t} \tag{14.3}$$

where p_m are the harmonics of the discrete spectrum of $\bar{p}$. The auto-correlation of $\tilde{p} = p - \bar{p}$ is

$$R_p(t_1, t_2) = E(\tilde{p}(t_1)\tilde{p}(t_2)) \tag{14.4}$$

Since the rotor noise is nonstationary, R_p is not a function of $(t_1 - t_2)$ alone, but rather is a periodic function for constant $(t_1 - t_2)$. Define the variables $s_1 = \frac{1}{2}(t_1 + t_2)$ and $s_2 = t_2 - t_1$. The auto-correlation R_p is then periodic in s_1:

$$R_p(s_1, s_2) = E\bigl(\tilde{p}(t_1 = s_1 - s_2/2)\tilde{p}(t_2 = s_1 + s_2/2)\bigr)$$

$$= \sum_{m=-\infty}^{\infty} R_p^m(s_2) e^{imN\Omega s_1}$$

$$= \sum_{m=-\infty}^{\infty} e^{imN\Omega s_1} \int_{-\infty}^{\infty} e^{i\omega s_2} S_p^m(\omega) d\omega \tag{14.5}$$

The spectrum is obtained from the auto-correlation:

$$S_p^m(\omega) = \frac{N\Omega}{(2\pi)^2} \int_{-\infty}^{\infty} \int_{0}^{2\pi/N\Omega} e^{-imN\Omega s_1} e^{-i\omega s_2} R_p(s_1, s_2) ds_1 ds_2 \tag{14.6}$$

The standard deviation of the pressure at a particular instant in the period ($\psi = \Omega s_1$) can be obtained from the auto-correlation:

$$\sigma_p^2(\psi) = E\tilde{p}^2 = R_p(t_1 = t_2) = R_p(s_2 = 0) \tag{14.7}$$

Hence the spectral decomposition of σ_p^2 is

$$\sigma_p^2(\psi) = \sum_{m=-\infty}^{\infty} R_p^m(s_2 = 0) e^{imN\psi} = \sum_{m=-\infty}^{\infty} \left(\int_{-\infty}^{\infty} S_p^m(\omega) d\omega\right) e^{imN\psi} \tag{14.8}$$

So $Ep^2 = \bar{p}^2 + \sigma_p^2$, and the rms sound pressure is given by

$$(\text{rms } p)^2 = \lim_{T \to \infty} \frac{1}{T} \int_0^T p^2 dt = \lim_{K \to \infty} \frac{1}{K} \sum_{k=1}^{K} \frac{N\Omega}{2\pi} \int_{2\pi(k-1)/N\Omega}^{2\pi k/N\Omega} p^2 dt$$

$$= \frac{N}{2\pi} \int_0^{2\pi/N} Ep^2\, dt = \left(\bar{p}^2 + \sigma_p^2\right)_{\text{average over } \psi}$$

$$= \sum_{m=-\infty}^{\infty} |p_m|^2 + \int_{-\infty}^{\infty} S_p^0(\omega) d\omega \tag{14.9}$$

where

$$S_p^0(\omega) = \frac{1}{2\pi} \int_{-\infty}^{\infty} e^{-i\omega s_2} \left(\frac{N\Omega}{2\pi} \int_0^{2\pi/N\Omega} R_p(s_1, s_2) ds_1 \right) ds_2$$

$$= \frac{1}{2\pi} \int_{-\infty}^{\infty} e^{-i\omega s_2} \overline{R}_p(s_2) ds_2 \tag{14.10}$$

The expected value of the sound pressure $\overline{p}$ is the rotational noise (and blade slap if impulsive enough); the auto-correlation R_p is the broadband noise. Included in R_p is the standard deviation of the periodic pressure waveform, σ_p^2. The root-mean-square pressure, which is generally used for noise assessment, is then the average of $Ep^2 = \overline{p}^2 + \sigma_p^2$ over one period. The spectral decomposition of the rotational noise gives the discrete harmonics p_m. The spectral analysis of the rms pressure involves in addition the spectrum of the average of the auto-correlation, $S_p^0(\omega)$. Because the rotor noise is not stationary, there is more information about the broadband noise that is not considered by the rms pressure (such as the spectrum of σ_p^2). This additional information can be important for the subjective perception of rotor noise.

Sound is measured in units of decibels (dB), defined as

$$10 \log_{10} \frac{\text{sound power}}{\text{reference power}} \tag{14.11}$$

A logarithmic scale is used because of the need to handle orders-of-magnitude differences in the sound levels encountered and because the human ear has basically a logarithmic response to sound. The energy flux at a given point in the sound field is given by the acoustic intensity $I = E(pu)$. Here p is the perturbation pressure and u is the velocity due to the sound waves, so the instantaneous intensity pu is the power radiated per unit area. In the far field, the velocity and pressure disturbances of a sound wave are related by $u = p/\rho_0 c_s$, so the intensity is

$$I = E(pu) = \frac{E(p^2)}{\rho_0 c_s} = \frac{p_{\text{rms}}^2}{\rho_0 c_s} \tag{14.12}$$

where ρ_0 is the mean air density and c_s is the speed of sound. Thus the rms pressure is a measure of the acoustic intensity. Moreover, the human ear and the aircraft structure respond to the pressure deviations from the mean atmospheric value. Hence noise is measured in terms of the sound pressure level, defined as

$$\text{SPL} = 10 \log \frac{p_{\text{rms}}^2}{p_{\text{ref}}^2} = 20 \log \frac{p_{\text{rms}}}{p_{\text{ref}}} \tag{14.13}$$

in units of dB (re p_{ref}). For the reference pressure, $p_{\text{ref}} = 20\ \mu\text{Pa}$ is normally used. The spectrum of the rms pressure can then be regarded as the distribution of the sound energy over frequency.

The overall sound pressure level (OSPL) is the total rms pressure. Common practice is to measure and present the noise data in terms of its spectrum, either by octave band, third-octave band, or narrow band measurements. Since the subjective perception of sound depends on the frequency content, a number of frequency-weighted measures of the sound pressure level have been developed, notably the A-weighted sound level (dBA) and the perceived noise level. The A-weighted sound level is intended to account for the sensitivity of the human ear. The perceived noise level (PNL) is a metric developed for aircraft, using a weighting that depends on both the magnitude and the frequency of the spectrum, based on the sound annoyance level. Further corrections have been developed for aircraft noise. The effective perceived noise level (EPNL) accounts for the sound duration and the presence of discrete frequency tones. PNL and EPNL are defined in FAR Part 36 (U.S. Government (2012)).

The estimate of the subjective level of helicopter noise is difficult, however, because no standard has yet been formulated that is well suited to the noise characteristic of rotors: a random pressure disturbance with periodic statistics, composed of low-frequency rotational noise, high-frequency broadband noise, and often significant impulsive noise.

14.3 Broadband Noise

Rotor broadband or vortex noise is a high-frequency sound produced by random fluctuations of the forces on the blade. For an elementary analysis of vortex noise, consider a blade of length ℓ in a flow at speed V with a random lift force (per unit length) $F_z(t)$ induced by the turbulence and vorticity in the wake. Assuming impulsive loading over the chord, the section force is a vertical dipole, for which the sound pressure p is

$$p = -\frac{1}{4\pi}\frac{\partial}{\partial z}\frac{F_z(t - s/c_s)}{s} \tag{14.14}$$

where s is the distance from the dipole F_z to the observer and z is the vertical coordinate. From $\partial s/\partial z = z/s$, the sound pressure in the far field (s very large) is

$$p = -\frac{1}{4\pi c_s}\frac{z}{s^2}\left[\dot{F}_z(t - s/c_s) - \frac{c_s}{s}F_z(t - s/c_s)\right]$$

$$\cong -\frac{1}{4\pi c_s}\frac{z}{s_0^2}\dot{F}_z = -\frac{1}{4\pi c_s}\frac{\sin\theta_0}{s_0}\dot{F}_z \tag{14.15}$$

where s_0 is the distance from the blade to the observer and θ_0 is the angle of the observer from the disk plane ($\sin\theta_0 = z/s_0$). Integrating over the length of the rod gives the total sound pressure:

$$p = -\frac{1}{4\pi c_s}\frac{\sin\theta_0}{s_0}\int_0^\ell \dot{F}_z d\ell_1 \tag{14.16}$$

The rms pressure is then

$$p_{rms}^2 = E(p^2) = \left(\frac{\sin\theta_0}{4\pi c_s s_0}\right)^2 \int_0^\ell \int_0^\ell E\left(\dot{F}_z(\ell_1)\dot{F}_z(\ell_2)\right) d\ell_1 d\ell_2 \tag{14.17}$$

Now write $\dot{F}_z = \omega F_z$ for a force at frequency ω, and define

$$R_{ff} = E\left(F_z(\ell_1)F_z(\ell_2)\right) = E\left(F_z(\ell_2 + \ell_3)F_z(\ell_2)\right) \tag{14.18}$$

so

$$p_{rms}^2 = \left(\frac{\sin\theta_0}{4\pi c_s s_0}\right)^2 \omega^2 \int_0^\ell \int_0^\ell R_{ff} d\ell_1 d\ell_2 \tag{14.19}$$

Assuming that the force correlation is homogeneous over the length of the blade, it depends only on $\ell_3 = \ell_1 - \ell_2$: $R_{ff} = \overline{F}^2 R(\ell_3)$, where $\overline{F}$ is the rms level of the section lift (so $R(0) = 1$). Then

$$\int_0^\ell \int_0^\ell R_{ff} d\ell_1 d\ell_2 = \frac{1}{2}\overline{F}^2 \int_{-\ell}^\ell \int_{|\ell_3|}^{2\ell - |\ell_3|} R(\ell_3) d\ell_4 d\ell_3 = \overline{F}^2 \int_{-\ell}^\ell (\ell - |\ell_3|) R(\ell_3) d\ell_3 \quad (14.20)$$

where $\ell_4 = \ell_1 + \ell_2$. A perfect correlation of the force over the blade length would give $R(\ell_3) = 1$. A small correlation length over the blade is more appropriate, so assume $R(\ell_3)$ is small except for distances of order ℓ_c from the origin:

$$\int_0^\ell \int_0^\ell R_{ff} d\ell_1 d\ell_2 = \overline{F}^2 \ell \ell_c \quad (14.21)$$

where the correlation length $\ell_c = 2 \int_0^\ell (1 - \ell_3/\ell) R(\ell_3) d\ell_3 \cong 2 \int_0^\ell R(\ell_3) d\ell_3$ is here much less than ℓ. Hence

$$p_{\text{rms}}^2 = \left(\frac{\sin \theta_0}{4\pi c_s s_0}\right)^2 \omega^2 \overline{F}^2 \ell \ell_c \quad (14.22)$$

In terms of the rms lift coefficient $C_L = F/\frac{1}{2}\rho V^2 d$ and the Strouhal number $S = \omega d / 2\pi V$ (where d is a characteristic dimension of the section), the result is

$$p_{\text{rms}}^2 = \frac{\rho^2}{16 c_s^2} \left(\frac{\sin \theta_0}{s_0}\right)^2 C_L^2 S^2 V^6 \ell \ell_c \quad (14.23)$$

Yudin (1947) obtained essentially this result. He interpreted broadband noise on a rotating rod as being due to the oscillating forces induced on the section by the shed vortices, such as in a von Kármán vortex street. By considering the solution for a vortex street, he related the oscillatory lift force produced by the vortices to the steady drag force on the rod. Yudin quoted experimental data that indicate the sound is proportional to $(\sin \theta_0)^2 V^{5.5} \ell$, verifying the dipole character of the vortex noise and the assumption of small correlation length (a large correlation length would give $p^2 \sim \ell^2$). Yudin attributed the slightly smaller growth with speed to the dependence of the drag coefficient on V.

For the rotating blade, s_0 is the distance from the hub to the observer, at an angle θ_0 above the disk plane; the section lift coefficient is proportional to the blade loading C_T/σ; the tip speed ΩR is used for the speed V; and the Strouhal number is assumed constant (so the noise frequency scales with V/d). The blade radius R is used for the length ℓ, and the correlation length ℓ_c is assumed to be proportional to the blade chord c. The sound power must be multiplied by the number of blades N, so the sound is proportional to $N\ell\ell_c \sim NRc = A_b$, the total blade area. The rotor vortex noise is then

$$p_{\text{rms}}^2 = \text{constant} \times \frac{\rho^2}{c_s^2} \left(\frac{\sin \theta_0}{s_0}\right)^2 A_b (\Omega R)^6 (C_T/\sigma)^2 \quad (14.24)$$

A constant Strouhal number implies that the rotor vortex noise frequency scales with $\Omega R/c$, but since the blade velocity varies linearly along the span and varies in direction relative to the observer, the noise is spread over a considerable frequency range. The assumption that the sound is due to lift fluctuations implies the directivity of a vertical dipole: maximum on the rotor axis ($\theta_0 = 90°$) and zero in the rotor disk plane ($\theta_0 = 0°$). The far field sound varies with distance from the rotor as s_0^{-2}, which

14.3 Broadband Noise

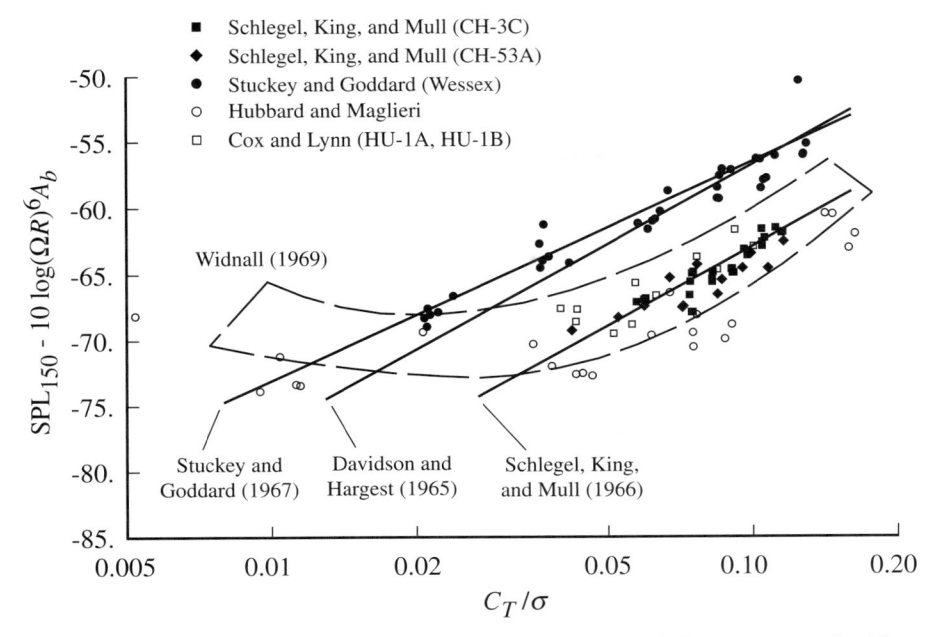

Figure 14.2. Measured hovering rotor noise as a function of C_T/σ compared with several empirical vortex noise expressions; adapted from Widnall (1969).

is required for constant total radiated power. For fixed C_T/σ and blade area, the vortex noise is proportional to the tip speed to the sixth power, because of the scaling of F_z with speed. Alternatively, in terms of the rotor thrust, $p^2 \sim T^2(\Omega R)^2/A_b$.

Equation 14.24 for the vortex noise can be generalized to

$$p_{\text{rms}}^2 = \text{constant} \times \frac{\rho^2}{c_s^2}\left(\frac{\sin\theta_0}{s_0}\right)^2 A_b (\Omega R)^6 f(C_T/\sigma) \tag{14.25}$$

Figure 14.2 shows measured hovering rotor noise in terms of

$$\text{SPL}_{150} - 10\log(\Omega R)^6 A_b \tag{14.26}$$

as a function of C_T/σ, where SPL_{150} is the sound pressure level on the rotor axis at 150 m from the hub (with ΩR in m/sec and A_b in m^2; subtract 41.3 dB for ft/sec and ft^2). The sound level at an arbitrary observer point is then

$$\text{SPL} = \text{SPL}_{150} + 20\log\left(\frac{\sin\theta_0}{s_0/150}\right) \tag{14.27}$$

At blade loadings typical of helicopter rotor operation, the relationship $p^2 \sim (C_T/\sigma)^2$ is reasonable. For low loading the noise is constant or even increases again, probably as a result of the proximity of the rotor wake.

A number of empirical expressions have been developed to predict vortex noise, based on correlation of measured rotor noise and on the functionality indicated by simple theories. Figure 14.2 shows the results of three such expressions, which predict the same overall sound pressure level to within a few decibels. Davidson and Hargest (1965) obtained from measurements of rotor noise in hover,

$$\text{SPL}_{150} = 10\log\left[(\Omega R)^6 A_b (C_T/\sigma)^2\right] - 36.7 \text{ dB} \tag{14.28}$$

They also suggested corrections for hovering in a wind, for forward flight, and for perceived noise level (PNdB). Schlegel, King, and Mull (1966) developed an empirical expression equivalent to

$$\text{SPL}_{150} = 10\log\left[(\Omega R)^6 A_b (C_T/\sigma)^2\right] - 42.9 \text{ dB} \tag{14.29}$$

for predicting vortex noise. The vortex noise spectrum extended from 150 to 9600 Hz, with the greatest intensity over 300 to 600 Hz. The peak frequency correlated as $f = 0.20\Omega R/h$, where h is the projected thickness of the blade. They gave generalized octave-band spectra for the vortex noise below and above stall conditions, scaled to this peak frequency. When the rotor stalled there was a sharp increase in the 1200 to 2400 Hz octave band, and the previous empirical expression under-predicted the vortex noise. Stuckey and Goddard (1967) found that the vortex noise measurements of a rotor on a whirl tower correlated best with the expression

$$\text{SPL}_{150} = 10\log\left[(\Omega R)^6 A_b (C_T/\sigma)^{1.66}\right] - 39.9 \text{ dB} \tag{14.30}$$

The broadband noise spectrum peak frequency was given by $f = 0.45\Omega R/c$ Hz (where c is the blade chord), and the spectrum fell off at about a constant rate of 7.5 dB per octave on either side of the peak. Widnall (1969) correlated existing data on main rotor vortex noise, identifying the trends shown in Figure 14.2. The measurements shown in Figure 14.2 are those used to develop these empirical expressions. Such simple methods are only accurate to about 5 dB at best.

Lowson and Ollerhead (1969) found that with very narrow band measurements the rotational noise harmonics are identifiable up to 400 Hz at least, in contrast with wide band measurements, the latter leading to the impression that random noise dominates the main rotor sound above about 150 Hz. Thus there is a gradual transition from harmonic (rotational) to broadband (vortex) noise as the frequency increases. They found that the vortex noise directivity is basically that of a dipole, but not exactly zero in the disk plane. They noted that there was increased noise at high collective because of stall and at low collective because of wake interference. Leverton and Pollard (1973) obtained noise measurements of a helicopter rotor on a whirl tower and found using a narrow band analysis that the spectrum in the frequency range generally associated with vortex noise was actually a combination of rotational and broadband components. They found that the broadband noise spectrum did not show a peak, but was flat out to a fundamental frequency of 325 to 450 Hz (depending on the thrust) and fell off at a roughly constant decibel-per-octave rate after that.

14.4 Rotational Noise

Rotor rotational noise is a periodic sound pressure disturbance. The rotational noise spectrum consists of discrete lines at the harmonics of the blade passage frequency $N\Omega$, which dominate the low-frequency portion of the rotor noise. With a narrow band analysis many harmonics of the rotational noise are identifiable. As the frequency increases, the random pressure fluctuations become more important until the discrete harmonics can no longer be discriminated from the broadband noise spectrum.

Rotational noise is produced by the periodic lift and drag forces acting on the blade. The blade exerts equal and opposite reaction forces on the air, and these rotate with the blade, so that at any fixed point in the rotor disk plane there is a periodic

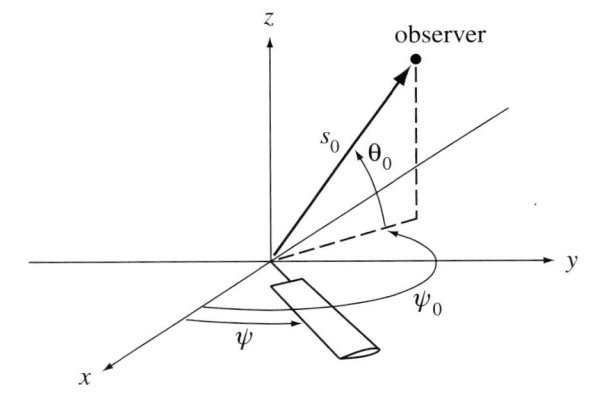

Figure 14.3. Rotor disk plane coordinate axes, and the observer position.

application of a force as the blades pass (at the fundamental frequency $N\Omega$). Such an unsteady force results in dipole radiation of a periodic pressure disturbance into the fluid, which is the rotational noise. The force acting on the air is unsteady because of the rotation of the blades and the periodic variation of the blade loading. For the rotational noise analysis, the section forces rotating with the blade (F_x, F_z, and F_r; respectively the chordwise, vertical, and radial forces) are replaced by an equivalent distribution of periodic forces on the surface of the rotor disk in the non-rotating frame (with components G_x, G_y, and G_z). Figure 14.3 shows the rotor coordinate axes used: the x-axis is aft, the y-axis is to the right, and the z-axis is upward. The rotor disk is in the x–y plane, with the blade at an azimuth angle $\psi = \Omega t$. The forces on the rotating blades can thus be represented by a distribution over the rotor disk of dipoles that do not rotate but have periodically varying strength. The rotor disk with the non-rotating dipole distribution then moves with the helicopter at a steady velocity in vertical or forward flight. Figure 14.3 also defines the position of an observer, where the rotor noise is detected, by the range s_0, azimuth angle ψ_0, and elevation θ_0.

14.4.1 Rotor Pressure Distribution

The differential pressure on the surface of the rotating blade, $\Delta p(r, x, \psi)$, is a periodic function of the blade azimuth angle ψ for steady-state operation of the rotor. Here x and r are rectangular coordinates rotating with the blade, respectively, chordwise (positive aft) and spanwise. Thus Δp is nonzero only within the area bounded by the blade leading and trailing edges ($x_{le} < x < x_{te}$) and by the root and tip ($r_R < r < R$). This pressure is written in terms of the section lift L and a chordwise loading function ℓ:

$$\Delta p(r, x, \psi) = L(r, \psi)\ell(x, \psi) \tag{14.31}$$

where by definition $\int_{x_{le}}^{x_{te}} \ell \, dx = 1$. The rotational noise analysis requires a knowledge of the pressure distribution over the rotor disk $\Delta p_1(r_1, \psi_1, t)$, where r_1 and ψ_1 are the polar coordinates of the point on the disk. With N blades, Δp_1 is periodic in time t with period $2\pi/N\Omega$. The pressure Δp_1, at a fixed point (r_1, ψ_1) on the rotor disk, as a function of time t, must be related to the pressure Δp at a point (x, r) on the rotating blade at azimuth angle ψ.

Each of the N blades produces an identical pressure pulse of duration $c/\Omega r$ as, one at a time, each sweeps past a given point on the rotor disk. So only the pressure produced by the reference blade when its azimuth angle ψ is in the vicinity of the disk point at ψ_1 must be considered. The pressure on the disk in rotating and non-rotating coordinates is $\Delta p_1(r_1, \psi_1, t) = \Delta p(r, x, \psi)$, and the blade coordinates and azimuth (r, x, ψ) must be determined for a given time and point on the rotor disk (r_1, ψ_1, t). The reference blade is at azimuth angle $\psi = \Omega(t - t_0)$, where t_0 is an initial time. Now given ψ and the disk point polar coordinates r_1 and ψ_1, the chordwise and radial coordinates of the corresponding point on the disk are

$$x = r_1 \sin(\psi - \psi_1) \qquad (14.32)$$

$$r = r_1 \cos(\psi - \psi_1) \qquad (14.33)$$

For a high-aspect-ratio blade x/r is small, so $x \cong r_1(\psi - \psi_1)$ and $r \cong r_1$. Then the distribution of pressure over the rotor disk is

$$\Delta p_1(r_1, \psi_1, t) = \Delta p\Big(r_1, x = r_1\big(\Omega(t - t_0) - \psi_1\big), \psi = \Omega(t - t_0)\Big) \qquad (14.34)$$

This determines Δp_1 for a time interval of length $T = 2\pi/N\Omega$ about $t = t_0 + \psi_1/\Omega$, and hence for all time since Δp_1 is periodic.

Now write the disk pressure distribution as a Fourier series in time:

$$\Delta p_1 = \sum_{m=-\infty}^{\infty} \Delta p_m e^{imN(\Omega t - \psi_1)} \qquad (14.35)$$

where the harmonics are obtained from

$$\Delta p_m = \frac{N\Omega}{2\pi} \int \Delta p\Big(r_1, x = r_1\big(\Omega(t - t_0) - \psi_1\big), \psi = \Omega(t - t_0)\Big) e^{-imN(\Omega t - \psi_1)} dt \qquad (14.36)$$

The integral is over the period $T = 2\pi/N\Omega$. Transforming the integration variable to $x = r_1(\Omega(t - t_0) - \psi_1)$ gives

$$\Delta p_m = \frac{N}{2\pi r_1} e^{-imN\Omega t_0} \int_{x_{le}}^{x_{te}} \Delta p\Big(r_1, x, \psi = \psi_1 + x/r_1\Big) e^{-imNx/r_1} dx \qquad (14.37)$$

Hence the harmonics of the pressure distribution on the rotor disk can be obtained from an integral of the blade pressure distribution over the chord. Separating the blade loading into the section lift and a chordwise distribution factor, write $\Delta p = L(r, \psi)\ell(x, \psi)$. For steady loading, L and ℓ are independent of ψ, so

$$\Delta p_m = \frac{N}{2\pi r_1} e^{-imN\Omega t_0} L(r_1) \ell_{mN} \qquad (14.38)$$

where

$$\ell_{mN} = \int_{x_{le}}^{x_{te}} \ell(x) e^{-imNx/r_1} dx \qquad (14.39)$$

Thus even a steady lift on the rotating blade produces an unsteady pressure on the air in the nonrotating frame. For the general case of periodic blade loading,

$$L = \sum_{n=-\infty}^{\infty} L_n(r) e^{in\psi} \qquad (14.40)$$

14.4 Rotational Noise

and the lift still factors out of the chordwise integration, giving

$$\Delta p_m = \frac{N}{2\pi r_1} e^{-imN\Omega t_0} \sum_{n=-\infty}^{\infty} L_n(r_1) \ell_{mN-n}(\psi_1) \qquad (14.41)$$

where now

$$\ell_{mN} = \int_{x_{le}}^{x_{te}} \ell(x, \psi = \psi_1 + x/r_1) e^{-imNx/r_1} dx \cong \int_{x_{le}}^{x_{te}} \ell(x, \psi_1) e^{-imNx/r_1} dx \qquad (14.42)$$

The factors ℓ_{mN} account for the influence of the blade chordwise loading distribution on the pressure at a disk point. If the chordwise loading factor $\ell(x, \psi)$ is independent of the blade azimuth position, then so is ℓ_{mN}; but ℓ_{mN} always varies with the radial station. Impulsive chordwise loading, $\Delta p = L(r, \psi)\delta(x)$, gives

$$\Delta p_m = \frac{N}{2\pi r_1} e^{-imN\Omega t_0} L(r_1, \psi_1) \qquad (14.43)$$

So the disk loading at the fixed point (r_1, ψ_1) is then simply an N/rev impulse of strength $NL/2\pi r_1$.

Consider the chordwise distribution factor ℓ_n for some specific cases. In general

$$\ell_0 = \int_{x_{le}}^{x_{te}} \ell(x) dx = 1 \qquad (14.44)$$

by the definition of $\ell(x)$, and for small nx/r, $\ell_n \cong 1 - inx_{CP}/r$, where $x_{CP} = \int \ell x \, dx$ is the location of the center of pressure. For impulsive loading or in the limit of small chord, $\ell(x) = \delta(x)$, so that $\ell_n = 1$ for all n. A rectangular distribution of pressure over the blade chord, $\ell(x) = 1/c$, gives

$$\ell_n = e^{-inx_m/r} \frac{\sin cn/2r}{cn/2r} \qquad (14.45)$$

where x_m is the coordinate of the blade midchord. From thin-airfoil theory, a loading distribution more appropriate for the lift on an airfoil is

$$\ell(x) = \frac{2}{\pi c} \sqrt{\frac{1-\xi}{1+\xi}} \qquad (14.46)$$

where $\xi = 2(x - x_m)/c$, which gives

$$\ell_n = e^{-inx_m/r} \left(J_0(nc/2r) + iJ_1(nc/2r) \right) \qquad (14.47)$$

with J_0 and J_1 Bessel functions. When the loading is distributed over a finite chord, the factors ℓ_n are reduced in magnitude relative to the impulsive loading result $\ell_n = 1$. The approximation of impulsive loading is simple and conservative, but for best accuracy the actual chordwise distribution must be accounted for.

14.4.2 Hovering Rotor with Steady Loading

Consider the rotational noise of a hovering rotor. The rotor disk is stationary in space, and with axisymmetric flow the loading on the blades is steady. The section aerodynamic forces are $F_z(r)$ normal to the disk plane and $F_x(r)$ in the disk plane.

For steady loading, write F_z and F_x in terms of the rotor thrust and torque:

$$F_z = \frac{1}{N}\frac{dT}{dr} \tag{14.48}$$

$$F_x = \frac{1}{Nr}\frac{dQ}{dr} \tag{14.49}$$

Since the sound is due to the pressure reaction on the fluid, the rotor profile torque should not be included in Q. Following section 14.4.1, the corresponding pressure distribution on the rotor disk in the nonrotating frame is

$$g_z = \sum_{m=-\infty}^{\infty} \frac{N}{2\pi r} F_z \ell_{mN} e^{imN(\Omega t - \psi)} \tag{14.50}$$

$$g_x = \sum_{m=-\infty}^{\infty} \frac{N}{2\pi r} F_x \ell_{mN} e^{imN(\Omega t - \psi)} \tag{14.51}$$

Here g_z and g_x are the normal and in-plane components of the pressure acting at the point (r, ψ) on the disk (g_x is still in the chordwise direction along the blade). An initial time $t_0 = 0$ has been used, so the reference blade azimuth is $\psi = 0$ at $t = 0$. The normal and in-plane pressures would have different chordwise distributions in general, but for now impulsive loading is assumed, which gives $\ell_{mN} = 1$. Considering just the m-th harmonic, the pressure on the rotor disk is thus

$$g_z = \frac{1}{2\pi r}\frac{dT}{dr} e^{imN(\Omega t - \psi)} \tag{14.52}$$

$$g_x = \frac{1}{2\pi r^2}\frac{dQ}{dr} e^{imN(\Omega t - \psi)} \tag{14.53}$$

The components of the pressure in the rectangular coordinate system (Figure 14.3) are then $G_x = -g_x \sin\psi$, $G_y = g_x \cos\psi$, and $G_z = g_z$. This pressure is equivalent to a distribution of dipoles over the surface of the rotor disk.

The dipole solution for the sound pressure p detected at an observer point (x, y, z) and due to a concentrated force with components G_x, G_y, and G_z at the point (x, y_1, z_1) is

$$p = -\frac{1}{4\pi}\left[\frac{\partial}{\partial x}\frac{G_x(t - s/c_s)}{s} + \frac{\partial}{\partial y}\frac{G_y(t - s/c_s)}{s} + \frac{\partial}{\partial z}\frac{G_z(t - s/c_s)}{s}\right] \tag{14.54}$$

where s is the distance from the dipole to the observer,

$$s^2 = (x - x_1)^2 + (y - y_1)^2 + (z - z_1)^2 \tag{14.55}$$

and $t - s/c_s$ is the retarded time, which accounts for the finite time s/c_s required for a sound wave emitted at the source to travel to the observer. In the present case, the forces acting on the rotor disk are periodic with fundamental frequency $N\Omega$. Thus the sound pressure is also periodic:

$$p = \sum_{m=-\infty}^{\infty} p_m e^{imN\Omega t} \tag{14.56}$$

14.4 Rotational Noise

Since G_x, G_y, and G_z have the time dependence $e^{imN\Omega t}$, the m-th harmonic of the sound pressure is

$$p_m = -\frac{1}{4\pi}\left[G_x\frac{\partial}{\partial x} + G_y\frac{\partial}{\partial y} + G_z\frac{\partial}{\partial z}\right]\frac{e^{-iks}}{s} \tag{14.57}$$

where $k = mN\Omega/c_s$, and the factor e^{-iks} is due to the retarded time. The force is at the disk point $x_1 = r\cos\psi$, $y_1 = r\sin\psi$, and $z_1 = 0$ so

$$-\left[G_x\frac{\partial}{\partial x} + G_y\frac{\partial}{\partial y}\right]\frac{e^{-iks}}{s} = g_x\left[\sin\psi\frac{\partial}{\partial x} - \cos\psi\frac{\partial}{\partial y}\right]\frac{e^{-iks}}{s}$$

$$= g_x\left[-\sin\psi\frac{\partial}{\partial x_1} + \cos\psi\frac{\partial}{\partial y_1}\right]\frac{e^{-iks}}{s}$$

$$= g_x\frac{1}{r}\frac{\partial}{\partial\psi}\frac{e^{-iks}}{s} \tag{14.58}$$

Then integrating over the rotor disk area $dS = r\,dr\,d\psi$ gives the total m-th harmonic of the rotational noise:

$$p_m = \frac{1}{8\pi^2}\int_0^{2\pi}\int_0^R \left[\frac{dT}{dr}\frac{\partial}{\partial z} + \frac{1}{r^2}\frac{dQ}{dr}\frac{\partial}{\partial\psi}\right]\frac{e^{-iks}}{s}e^{-imN\psi}\,dr\,d\psi \tag{14.59}$$

Now

$$\frac{\partial}{\partial z}\frac{e^{-iks}}{s} = \frac{e^{-iks}}{s}\left(-\frac{ikz}{s} - \frac{z}{s^2}\right) \tag{14.60}$$

and the torque term is integrated by parts with respect to ψ, giving

$$p_m = -\frac{imN\Omega}{8\pi^2 c_s}\int_0^{2\pi}\int_0^R \left[\frac{dT}{dr}\frac{z}{s}\left(1 - \frac{i}{ks}\right) - \frac{c_s}{\Omega r^2}\frac{dQ}{dr}\right]\frac{e^{-iks}}{s}e^{-imN\psi}\,dr\,d\psi \tag{14.61}$$

The rotational noise has a discrete spectrum, with harmonics at the frequencies $mN\Omega$. The sound pressure level is obtained by summing the contributions from all harmonics:

$$p_{rms}^2 = \frac{N\Omega}{2\pi}\int_0^{2\pi/N\Omega} p^2\,dt = \sum_{m=-\infty}^{\infty}|p_m|^2 \tag{14.62}$$

That discrete spectrum is two sided, extending from $m = -\infty$ to $m = \infty$. It is also common to work with a one-sided spectrum, defined for $m > 0$ only. Since p_m and p_{-m} are complex conjugates, the one-sided spectrum with the same p_{rms}^2 is obtained by multiplying the harmonics of the two-sided spectrum by $\sqrt{2}$:

$$|p_m| = \frac{mN\Omega}{4\sqrt{2}\pi^2 c_s}\left|\int_0^{2\pi}\int_0^R\left[\frac{dT}{dr}\frac{z}{s}\left(1-\frac{i}{ks}\right) - \frac{c_s}{\Omega r^2}\frac{dQ}{dr}\right]\frac{e^{-iks}}{s}e^{-imN\psi}\,dr\,d\psi\right| \tag{14.63}$$

This is the rotational noise spectrum of a hovering rotor with steady thrust and torque loading.

The far field approximation allows the integration over the rotor azimuth to be evaluated analytically. Assume that the observer is far from the rotor, so that $s \gg R$. Then to order R/s_0,

$$s = \sqrt{z^2 + (x - r\cos\psi)^2 + (y - r\sin\psi)^2} \cong s_0 - \frac{xr\cos\psi}{s_0} - \frac{yr\sin\psi}{s_0} \tag{14.64}$$

where $s_0^2 = z^2 + x^2 + y^2$ is the distance from the hub to the observer. For the rotor thrust term the following approximation is made:
$$\frac{z}{s}\left(1 - \frac{i}{ks}\right) \cong \frac{z}{s_0} \qquad (14.65)$$
which requires not just that $s \gg R$ but also that s be much greater than the wavelength of the sound, $ks \gg 1$. With $k = mN\Omega/c_s$, this criterion can be written as $s/R \gg 1/(mNM_{tip})$. Since M_{tip} is of order 1 for helicopter rotors, the criterion reduces to $s \gg R$ again. Furthermore, we make the approximation
$$\frac{1}{s}e^{-iks} \cong \frac{1}{s_0}e^{-iks_0}e^{ikr(x/s_0)\cos\psi} \qquad (14.66)$$

The last factor accounts for the difference in retarded time over the rotor disk and is the only influence of the order R/s_0 term in s to be retained. Since the hovering rotor is axisymmetric, the observer is assumed to be at $y = 0$, in the x–z plane. The sound pressure in the far field is thus

$$p_m = -\frac{imN\Omega e^{iks_0}}{8\pi^2 c_s s_0} \int_0^R \left[\frac{dT}{dr}\frac{z}{s_0} - \frac{c_s}{\Omega r^2}\frac{dQ}{dr}\right] \int_0^{2\pi} e^{ikr(x/s_0)\cos\psi - imN\psi} d\psi \, dr \qquad (14.67)$$

The integral over ψ can be evaluated in terms of Bessel functions using the relation
$$\int_0^{2\pi} e^{iz\cos\psi - in\psi} d\psi = 2\pi i^n J_n(z) \qquad (14.68)$$

Hence the far field sound pressure is
$$p_m = -\frac{imN\Omega i^{mN} e^{-imN\Omega s_0/c_s}}{4\pi c_s s_0} \int_0^R \left[\frac{dT}{dr}\frac{z}{s_0} - \frac{c_s}{\Omega r^2}\frac{dQ}{dr}\right] J_{mN}\left(\frac{mN\Omega r}{c_s}\frac{x}{s_0}\right) dr \qquad (14.69)$$

The far field approximation eliminates the integration over ψ and introduces the Bessel functions. In terms of the elevation angle θ_0 of the observer above the disk plane, $z = s_0 \sin\theta_0$ (see Figure 14.3), so

$$p_m = -\frac{imN\Omega i^{mN} e^{-imN\Omega s_0/c_s}}{4\pi c_s s_0} \int_0^R \left[\frac{dT}{dr}\sin\theta_0 - \frac{c_s}{\Omega r^2}\frac{dQ}{dr}\right] J_{mN}\left(\frac{mN\Omega r}{c_s}\cos\theta_0\right) dr$$
$$(14.70)$$

The far field approximation is usually valid beyond four or five rotor radii from the rotor hub.

As a further approximation, evaluate the integrand at an effective radius r_e, which is equivalent to assuming that the loading is all concentrated at r_e. Then the integration over the blade span can be eliminated, giving for the one-sided spectrum

$$|p_m| = \frac{mN\Omega}{2\sqrt{2\pi}c_s s_0}\left[T\sin\theta_0 - \frac{c_s}{\Omega r_e^2}Q\right] J_{mN}\left(\frac{mN\Omega r_e}{c_s}\cos\theta_0\right) \qquad (14.71)$$

or

$$|p_m| = \frac{mNM_{tip}}{2\sqrt{2\pi}Rs_0}\left[T\sin\theta_0 - \frac{R}{M_{tip}r_e^2}Q\right] J_{mN}\left(mNM_{tip}(r_e/R)\cos\theta_0\right) \qquad (14.72)$$

in terms of the tip Mach number $M_{tip} = \Omega R/c_s$.

The far field sound pressure level p_{rms}^2 is proportional to s_0^{-2}, as required for energy conservation. The rotational noise due to the thrust is zero on the rotor axis,

14.4 Rotational Noise

where $\cos\theta_0 = 0$ and therefore $J_{mN} = 0$, and in the disk plane, where $\sin\theta_0 = 0$. The sound thus has a broad maximum at an angle between the disk plane and rotor axis, typically around $\theta_0 = \pm 30°$. The noise due to the torque is zero on the rotor axis, with the same phase as the thrust noise below the disk ($\sin\theta_0 < 0$) and the opposite phase above the disk. The sound is thus greatest below the disk plane. For the helicopter rotor the effect of the torque term is small, $Qc_s/\Omega R^2 T \ll 1$, except near the disk plane where the thrust noise is small. The rotational noise of the hovering rotor has the functional form

$$|p_m|^2 = \frac{T^2/A}{s_0^2} f(mN, M_{\text{tip}}, \theta_0) \tag{14.73}$$

Thus the noise is proportional to the rotor thrust and disk loading. Because of the Bessel function behavior, the rotational noise harmonics fall off rapidly with harmonic number m (for steady loading of the blades). Increasing the number of blades N reduces the rotational noise harmonics, in addition to increasing the fundamental frequency $N\Omega$. At a constant thrust coefficient C_T, the sound pressure level is proportional to $(\Omega R)^6$ (neglecting the effect of the Bessel function). Hence the rotational noise increases with about the sixth power of the tip speed or tip Mach number.

To account for the actual chordwise distribution of the blade loading in the rotational noise expressions, the factor

$$\ell_{mN} = \int_{x_{le}}^{x_{te}} \ell(x) e^{-imNx/r} dx \tag{14.74}$$

is introduced inside the integral over the blade span (since ℓ_{mN} depends on r). Garrick and Watkins (1954) compared ℓ_{mN} for several elementary distributions. In section 14.4.1, expressions are given for rectangular and thin-airfoil chordwise pressure distributions. In general the factors ℓ_{mN} decrease in magnitude as $mNc/2r$ increases. Above about the 10-th harmonic the reduction in the rotational noise by the chordwise pressure distribution is significant. If a rectangular pressure distribution is assumed, the rotational noise is under-predicted, since the harmonics are reduced proportional to $(2r/cmN)$ at high harmonic number. With the more appropriate thin-airfoil theory pressure distribution, ℓ_{mN} decreases much more slowly, proportional to $(2r/cmN)^{1/2}$ at high harmonic number.

Gutin (1948) developed a theory for the rotational noise of propellers, obtaining the previous result for the far field noise of a stationary rotor with steady loading. Deming (1940) extended Gutin's result by considering the distribution of the loading over the blade span, instead of using an effective radius. Hubbard and Regier (1950) extended Gutin's analysis to include the near field noise calculation as well. They found that the far field result significantly underestimates the sound pressure in the near field, which extends to about $5R$ from the rotor hub.

The Gutin result is generally accurate for the rotational noise of static propellers. Reasonable agreement with the measured noise is found for the first few harmonics, and it is an adequate estimate of the overall sound pressure level. For the rotor of a hovering helicopter, however, the error is large. Stuckey and Goddard (1967) found that the Gutin result significantly underestimated all but the first harmonic of the rotational noise of a hovering rotor, although the trends with tip speed and thrust were given correctly. They found that eliminating the effective radius and far field approximations by integrating numerically over the rotor disk did not improve the predictions. The predicted rotational noise falls off rapidly with harmonic number, while the measured noise falls slowly or is even constant, apparently because of periodic blade loads that exist in even a nominally hovering condition. Schlegel, King, and Mull (1966) found that the rotational noise predicted by the Gutin theory was about 4 dB low for the fundamental harmonic of a helicopter rotor, and it deteriorated rapidly with harmonic number. Ollerhead and Lowson (1969) concluded that the Gutin theory significantly underestimates the rotational noise of a helicopter rotor because of the neglect of the higher harmonic airloads.

14.4.3 Vertical Flight and Steady Loading

Consider a helicopter rotor in vertical flight, still with steady blade loading. For a helicopter, the Mach number of the vertical velocity, $M = V_c/c_s = \lambda_c M_{\text{tip}}$ is small. This case also includes a propeller in forward flight, for which M is not small. The observer is assumed to be moving with the rotor. The rotor is represented by a distribution of forces over the disk, but now the solution for the sound generated by a dipole moving with uniform velocity in the z-direction must be used:

$$p = -\frac{1}{4\pi}\left[\frac{\partial}{\partial x}\frac{G_x(t-\sigma/c_s)}{S} + \frac{\partial}{\partial y}\frac{G_y(t-\sigma/c_s)}{S} + \frac{\partial}{\partial z}\frac{G_z(t-\sigma/c_s)}{S}\right] \quad (14.75)$$

where

$$S^2 = \beta^2(x-x_1)^2 + \beta^2(y-y_1)^2 + (z-z_1)^2 \quad (14.76)$$

$$\sigma = [S + M(z-z_1)]/\beta^2 \quad (14.77)$$

and $\beta^2 = 1 - M^2$. The m-th harmonic of the rotational noise is then

$$p_m = -\frac{1}{4\pi}\left[G_x\frac{\partial}{\partial x} + G_y\frac{\partial}{\partial y} + G_z\frac{\partial}{\partial z}\right]\frac{e^{-ik\sigma}}{S} = \frac{1}{4\pi}\left[g_z\frac{\partial}{\partial z} + g_x\frac{1}{r}\frac{\partial}{\partial \psi}\right]\frac{e^{-ik\sigma}}{S} \quad (14.78)$$

where $k = mN\Omega/c_s$. The total sound is obtained by integrating over the rotor disk area. Using

$$\frac{\partial}{\partial z}\frac{e^{-ik\sigma}}{S} = \frac{e^{-ik\sigma}}{S}\left(-\frac{ikM}{\beta^2} - \frac{ikz}{\beta^2 S} - \frac{z}{S^2}\right) \quad (14.79)$$

and integrating the torque term by parts with respect to ψ gives the sound pressure for the rotor with steady loading in axial flight:

$$p_m = -\frac{imN\Omega}{8\pi^2 c_s}\int_0^{2\pi}\int_0^R\left[\frac{dT}{dr}\left(\frac{M}{\beta^2} + \frac{z}{\beta^2 S} - \frac{iz}{kS^2}\right) - \frac{c_s}{\Omega r^2}\frac{dQ}{dr}\right]\frac{e^{-ik\sigma}}{S}e^{-imN\psi}\,dr\,d\psi \quad (14.80)$$

When $M = 0$, this reduces to the result of the last section.

The far field approximation now gives

$$S \cong S_0 - \frac{\beta^2 xr\cos\psi}{S_0} - \frac{\beta^2 yr\sin\psi}{S_0} \quad (14.81)$$

where $S_0^2 = z^2 + \beta^2 x^2 + \beta^2 y^2$. Assuming the observer is at $y = 0$, then

$$\frac{1}{S}e^{-ik\sigma} \cong \frac{1}{S_0}e^{-ik\sigma_0}e^{ikr(x/S_0)\cos\psi} \quad (14.82)$$

and the far field sound pressure is

$$p_m = -\frac{imN\Omega i^{mN} e^{-imN\Omega \sigma_0/c_s}}{4\pi c_s S_0} \int_0^R \left[\frac{dT}{dr}\left(M + \frac{z}{S_0}\right)\frac{1}{\beta^2} \right.$$
$$\left. - \frac{c_s}{\Omega r^2}\frac{dQ}{dr} \right] J_{mN}\left(\frac{mN\Omega r}{c_s}\frac{x}{S_0}\right) dr \quad (14.83)$$

Now in terms of the range and elevation angle of the observer (s_0 and θ_0 in Figure 14.3), $z = s_0 \sin\theta_0$ and $x = s_0 \cos\theta_0$. Hence $S_0^2 = s_0^2(1 - M^2 \cos^2\theta_0)$, and

$$p_m = -\frac{imN\Omega i^{mN} e^{-imN\Omega \sigma_0/c_s}}{4\pi c_s s_0 \sqrt{1 - M^2 \cos^2\theta_0}} \int_0^R \left[\frac{dT}{dr}\left(M + \frac{\sin\theta_0}{\sqrt{1 - M^2 \cos^2\theta_0}}\right)\frac{1}{\beta^2} \right.$$
$$\left. - \frac{c_s}{\Omega r^2}\frac{dQ}{dr} \right] J_{mN}\left(\frac{mN\Omega r}{c_s}\frac{\cos\theta_0}{\sqrt{1 - M^2 \cos^2\theta_0}}\right) dr \quad (14.84)$$

The principal influence of the rotor axial velocity is an order M increase in the thrust-induced noise above the disk when the helicopter is climbing. There is also a small (order M^2) increase in the magnitude of the noise since $S_0 < s_0$, and there is a shift in the directivity pattern.

Garrick and Watkins (1954) derived the rotational noise of a rotor in axial flight. They considered a propeller in forward flight, for which the axial Mach number can be large. Watkins and Durling (1956) extended the analysis to include more general chordwise loading distributions.

14.4.4 Stationary Rotor with Unsteady Loading

Consider next a stationary rotor with unsteady loading. The helicopter rotor in forward flight has periodic aerodynamic forces acting on the blades. If the effects of the helicopter translation on the sound radiation are neglected, the present model is obtained. In any case, it is useful to examine separately the influence of unsteady loads before the effects of the forward motion are included as well.

Assuming impulsive chordwise loading of the blades, section 14.4.1 gives the pressure distribution on the rotor disk as

$$g_z = \sum_{m=-\infty}^{\infty} \frac{N}{2\pi r} L(r, \psi) e^{imN(\Omega t - \psi)} \quad (14.85)$$

where $L(r, \psi)$ is the blade section lift, now depending on the blade azimuth as well as the radial station. The sound due to in-plane blade forces is no longer included in the analysis, since it is small compared to the sound due to the lift. Using the stationary dipole solution, the analysis proceeds as in section 14.4.2 to obtain the m-th harmonic of the sound pressure:

$$p_m = -\frac{imN^2\Omega}{8\pi^2 c_s} \int_0^{2\pi} \int_0^R L(r, \psi) \frac{z}{s}\left(1 - \frac{i}{ks}\right) \frac{e^{-iks}}{s} e^{-imN\psi} dr\, d\psi \quad (14.86)$$

The section lift is a periodic function of ψ, so

$$L = \sum_{n=-\infty}^{\infty} L_n(r) e^{in\psi} \quad (14.87)$$

For the far field approximation,

$$s \cong s_0 - \frac{r}{s_0}(x\cos\psi + y\sin\psi) = s_0 - r\cos\theta_0 \cos(\psi - \psi_0) \quad (14.88)$$

where ψ_0 is the azimuth angle of the observer (Figure 14.3). The loading is not axisymmetric now, so examining only observer points in the x–z plane is not sufficient. The rotational noise in the far field is then

$$p_m = -\frac{imN^2\Omega \sin\theta_0 e^{-imN\Omega s_0/c_s}}{8\pi^2 c_s s_0} \sum_{n=-\infty}^{\infty}$$

$$\int_0^R L_n \int_0^{2\pi} e^{ikr\cos\theta_0 \cos(\psi-\psi_0) - imN\psi + in\psi} d\psi \, dr$$

$$= -\frac{imN^2\Omega \sin\theta_0 e^{-imN\Omega s_0/c_s}}{4\pi c_s s_0} \sum_{n=-\infty}^{\infty} e^{i(n-mN)(\psi_0 - \pi/2)}$$

$$\int_0^R L_n J_{mN-n}\left(\frac{mN\Omega r}{c_s}\cos\theta_0\right) dr \quad (14.89)$$

So every harmonic of the blade loading contributes to the m-th harmonic of the sound pressure. In particular, the maximum sound produced by the loading harmonic L_n occurs at the sound harmonic $mN = n$. The higher harmonic loading contributes substantially more than the mean loading to the high-frequency rotational noise. To predict the rotational noise, which is significant up to $m = 20$ or 30, accurate data (measured or calculated) for the blade loads to very high harmonic number are required. At such high harmonics a deterministic loading does not really exist, however. A combined analysis of rotational and broadband noise is required to properly calculate the high frequency noise.

14.4.5 Forward Flight and Steady Loading

Consider next the helicopter rotor in steady forward flight at advance ratio μ. The higher harmonic loads on the blade are large and important to the rotational noise, but here only the mean loading is considered in order to examine the influence of forward motion of the rotor on the sound radiation. The rotor disk is represented by a distribution of vertical dipoles moving in the negative x-direction with Mach number $M = \mu M_{\text{tip}}$. The observer is moving with the rotor velocity also. The sound pressure due to a moving vertical dipole is

$$p = -\frac{1}{4\pi} \frac{\partial}{\partial z} \frac{G_z(t - \sigma/c_s)}{S} \quad (14.90)$$

where

$$S^2 = (x - x_1)^2 + \beta^2(y - y_1)^2 + \beta^2(z - z_1)^2 \quad (14.91)$$

$$\sigma = [S - M(x - x_1)]/\beta^2 \quad (14.92)$$

and $\beta^2 = 1 - M^2$. The m-th harmonic of the rotational noise is then

$$p_m = -\frac{1}{4\pi} G_z \frac{\partial}{\partial z} \frac{e^{-ik\sigma}}{S} = \frac{1}{4\pi} G_z \frac{e^{-ik\sigma}}{S}\left(\frac{ikz}{S} + \frac{\beta^2 z}{S^2}\right) \quad (14.93)$$

14.4 Rotational Noise

where

$$G_z = -\frac{1}{2\pi r}\frac{dT}{dr}e^{imN(\Omega t-\psi)} \qquad (14.94)$$

Integrated over the rotor disk, the sound pressure due to the rotor thrust in forward flight is

$$p_m = -\frac{imN\Omega}{8\pi^2 c_s}\int_0^{2\pi}\int_0^R \frac{dT}{dr}\frac{z}{S}\left(1-\frac{\beta^2 i}{kS}\right)\frac{e^{-ik\sigma}}{S}e^{-imN\psi}\,dr\,d\psi \qquad (14.95)$$

The far field approximation here gives

$$S \cong S_0 - \frac{xr\cos\psi}{S_0} - \frac{\beta^2 yr\sin\psi}{S_0} \qquad (14.96)$$

so in the retarded time

$$\sigma \cong \sigma_0 - \frac{1}{\beta^2}\left(\frac{x}{S_0}-M\right)r\cos\psi - \frac{y}{S_0}r\sin\psi \qquad (14.97)$$

where $\sigma_0 = (S_0 - Mx)/\beta^2$. Note that

$$\frac{1}{\beta^2}\left(\frac{x}{S_0}-M\right) = \frac{x-M\sigma_0}{S_0} \qquad (14.98)$$

$$(x-M\sigma_0)^2 + y^2 = \sigma_0^2 - z^2 \qquad (14.99)$$

Then

$$\sigma \cong \sigma_0 - \sqrt{\left(\frac{x-M\sigma_0}{S_0}\right)^2 + \left(\frac{y}{S_0}\right)^2}\,r\cos(\psi-\psi_r) = \sigma_0 - \frac{\sigma_0}{S_0}r\cos\theta_r\cos(\psi-\psi_r) \qquad (14.100)$$

where $\psi_r = \tan^{-1} y/(x-M\sigma_0)$ and $\theta_r = \sin^{-1} z/\sigma_0$. Here σ_0 is the distance from the rotor hub to the observer at the time the sound was emitted; that is, at the retarded time $t - \sigma_0/c_s$ (if the observer had been stationary while the rotor moved). Hence θ_r and ψ_r are the elevation and azimuth angle of the (fixed) observer at the retarded time. Finally, we can write

$$S_0 = \beta^2\sigma_0 + Mx = \sigma_0\left(1 - M\frac{-x+M\sigma_0}{\sigma_0}\right) = \sigma_0(1 - M\cos\delta_r) \qquad (14.101)$$

where δ_r is the angle between the observer and the forward velocity of the rotor at the retarded time, so that $M\cos\delta_r$ is the Mach number of the forward speed component in the direction of the observer. The sound pressure in the far field is thus

$$p_m = -\frac{imN\Omega}{4\pi c_s}\frac{\sin\theta_r}{\sigma_0(1-M\cos\delta_r)^2}e^{-imN\Omega\sigma_0/c_s - imN(\psi_r-\pi/2)}$$

$$\int_0^R \frac{dT}{dr}J_{mN}\left(\frac{mN\Omega r\cos\theta_r}{c_s(1-M\cos\delta_r)}\right)dr \qquad (14.102)$$

For the stationary rotor ($M = 0$), the elevation angle $\theta_r = \theta_0$, and the result of section 14.4.2 is recovered.

Since $\sin\theta_r/\sigma_0(1-M\cos\delta_r)^2 = z/S_0^2$ and $S_0 < s_0$, in forward flight there is an increase in the magnitude of the rotational noise harmonics. The effect of the $(1 - M\cos\delta_r)$ factor in the argument of the Bessel function is to increase the sound radiated forward of the rotor and to decrease the sound behind the rotor. Comparing the present result for the rotor in forward flight with the hovering rotor result in section 14.4.2, we observe that the forward flight noise is obtained if the hover expression is evaluated at the range $S_0 = \sigma_0(1 - M \cos \delta_r)$, the elevation θ_r, and an effective Mach number $M_{\text{eff}} = M_{\text{tip}}/(1 - M \cos \delta_r)$. Ahead of the rotor the effective tip Mach number is greater than M_{tip}, so the noise is increased, and behind the rotor $M_{\text{eff}} < M_{\text{tip}}$ and the noise is decreased.

14.4.6 Forward Flight and Unsteady Loading

Now let us consider the case of a helicopter rotor in forward flight, with periodic loading on the rotating blades. With impulsive chordwise loading, the normal pressure distribution on the rotor disk is the same as with steady loading:

$$G_z = -\frac{N}{2\pi r} L(r, \psi) e^{imN(\Omega t - \psi)} \tag{14.103}$$

except that now the section lift L varies with ψ. Hence the spectrum of the rotational noise in forward flight is

$$p_m = -\frac{imN^2\Omega}{8\pi^2 c_s} \int_0^{2\pi} \int_0^R L \frac{z}{S} \left(1 - \frac{\beta^2 i}{kS}\right) \frac{e^{-ik\sigma}}{S} e^{-imN\psi} \, dr \, d\psi \tag{14.104}$$

When the section lift is written as a Fourier series,

$$L = \sum_{n=-\infty}^{\infty} L_n(r) e^{in\psi} \tag{14.105}$$

the far field sound pressure is

$$p_m = -\frac{imN^2\Omega}{4\pi c_s} \frac{\sin\theta_r}{\sigma_0(1 - M\cos\delta_r)^2} e^{-imN\Omega\sigma_0/c_s} \sum_{n=-\infty}^{\infty} e^{-i(mN-n)(\psi_r - \pi/2)}$$

$$\int_0^R L_n J_{mN-n}\left(\frac{mN\Omega r \cos\theta_r}{c_s(1 - M\cos\delta_r)}\right) dr \tag{14.106}$$

Accounting for the actual chordwise pressure distribution gives instead

$$G_z = -\frac{N}{2\pi r} \sum_{n=-\infty}^{\infty} L_n \ell_{mN-n} e^{imN(\Omega t - \psi) + in\psi} \tag{14.107}$$

(see section 14.4.1), so the factor ℓ_{mN-n} must be included in the spanwise integration to evaluate the far field noise (assuming the chordwise distribution factor ℓ is independent of ψ). Alternatively, equation 14.37 for the m-th harmonic of the disk pressure can be used directly in the form

$$G_z = -\frac{N}{2\pi r} e^{imN(\Omega t - \psi)} \int_{x_{le}}^{x_{te}} \Delta p(r, x, \psi + x/r) e^{-imNx/r} dx$$

$$= -\frac{N}{2\pi r} e^{imN(\Omega t - \psi)} \sum_{n=-\infty}^{\infty} G_{zn} e^{in\psi} \tag{14.108}$$

where

$$G_{zn} = \frac{1}{2\pi} \int_0^{2\pi} e^{-in\psi} \int_{x_{le}}^{x_{te}} \Delta p(r, x, \psi + x/r) e^{-imNx/r} dx \, d\psi \tag{14.109}$$

which can be evaluated numerically, given the actual variation of the pressure over the surface of the blade and around the azimuth. With this form, L_n in the present result for impulsive loading is simply replaced by G_{zn}.

Loewy and Sutton (1966) developed a theory for helicopter rotor rotational noise in forward flight, including a treatment of unsteady airloads on the blades. They numerically integrated over the rotor disk to obtain the sound due to the dipole distribution at an arbitrary point in the near or far field. The blade flap motion and disk tilt were accounted for in determining the orientation of the dipoles. The rotor blade loading was assumed to be an input to the analysis. Simple chordwise distributions of the lift and drag, not impulsive loading, were used. An azimuth increment of 1° or less was required in the numerical integration, and the far field result significantly underestimated the noise in the near field. The principal effect of forward flight was to raise the level of the higher harmonics. The correlation with measured noise was good for the low harmonics, but the predicted noise harmonics (based on measured loadings) fell off rapidly with harmonic number while the measured values did not.

Schlegel, King, and Mull (1966) developed a theory for calculating helicopter rotor rotational noise in forward flight. They considered a stationary rotor (and hence used the stationary dipole solution), but included the unsteady airloads, as in section 14.4.4. The measured or calculated blade loading was assumed to be given, and a rectangular chordwise distribution of the lift was used. The sound pressure at an arbitrary field point was calculated by numerically integrating over the rotor disk. Comparison with flight test measurements of the rotational noise showed that the prediction of the first harmonic in forward flight had been improved (compared to predictions using the Gutin theory, which is accurate for the first harmonic in hover but underestimates the noise in forward flight). However, the prediction of the third, fourth, and higher harmonics was still poor. Schlegel and Bausch (1970) modified this analysis to use the actual chordwise loading distribution. Measured data for the pressure distribution over the rotating blade were converted to a distribution of pressure on the rotor disk, which was then harmonically analyzed. With this approach, good correlation with the measured noise was found up to at least the fourth harmonic, in both forward flight and hover. They presented examples of the theoretical influence that the higher harmonic airloads have on the noise and concluded that at least mN harmonics of the loads were required to obtain the m-th harmonic of the rotational noise.

Lowson and Ollerhead (1969) developed a theory for rotational noise of a rotor in forward flight, including the effects of unsteady airloads and the rotor motion. The derivation was based on the solution for the sound radiated by a rotating and translating dipole. The total rotational noise was calculated by representing the pressure distribution on the rotating blade as a distribution of such dipoles and then integrating over the surface of the blade. Assuming an impulsive chordwise loading reduces the calculation to an integral over the blade span of the section lift, drag, and radial forces. They examined the m-th harmonic of the sound pressure due to the n-th harmonic of the rotor loading and concluded that to calculate p_m the loading harmonics in the range

$$mN(1 - 0.8M_{\text{tip}}) < n < mN(1 + 0.8M_{\text{tip}}) \qquad (14.110)$$

are required. They concluded that the discrepancies in earlier analyses were due to the neglect of the very high harmonics of the loading, but that a practical calculation procedure should not rely on having such high-frequency aerodynamic data.

Lowson and Ollerhead also developed a simplified analysis that made the far field approximation and assumed a certain behavior of the higher harmonic airloads. An analytical evaluation of the integrals was then possible. In this simplified analysis, they considered impulsive chordwise loading (which is conservative); used an equivalent radius (the loads were concentrated at a single radial point as well); and from an examination of measured rotor loads assumed that the magnitude of the higher harmonics of the airloads varied with harmonic number n according to $F_n = F_0 n^{-k}$, where F_0 is the mean loading. The best correlation was obtained with $k = 2$ (for 10 harmonics of airloading, all outboard stations of the blade, and operating conditions from hover to $\mu = 0.2$). The spanwise variation of the loading was assumed to have a correlation length proportional to $1/n$, which had the approximate effect of increasing the exponent of the loading law to $k = 2.5$, giving $F_n = F_0 n^{-2.5}$. Since no trend was observed in the phase of the measured loading, the phase was assumed to be random; hence $|p_m|^2 = \sum_n |p_{mn}|^2$. They found that the calculated and measured rotational noise agreed fairly well for the first harmonic. In forward flight the predicted levels of the higher harmonics tended to be lower than measured, but the shape of the spectrum from about $m = 3$ to $m = 30$ was given well.

14.4.7 Doppler Shift

The rotational noise of the rotor in forward or vertical flight has been derived for an observer moving with the helicopter. For a fixed observer, these solutions can still be used to evaluate the time history of the sound pressure by using the instantaneous observer position relative to the rotor. The relative motion between the observer and the rotor produces a shift of the frequencies of the perceived sound, which is the Doppler effect. An acoustic source at frequency ω produces a sound pressure proportional to $e^{i\omega\tau}$, where $\tau = t - s(\tau)/c_s$ is the retarded time. The frequency of this sound at the observer then is

$$\omega_{\text{obs}} = \frac{\partial}{\partial t}\omega\tau = \omega\frac{\partial \tau}{\partial t} = \frac{\omega}{1 - M_r} = \frac{\omega}{1 - M\cos\delta_r} \quad (14.111)$$

since $dt/d\tau = d(\tau + s/c_s)/d\tau = 1 - M_r$, with $M_r = M\cos\delta_r$ the Mach number of the helicopter in the direction of the observer. A fixed observer hears a frequency increase as the helicopter approaches and a frequency decrease as the helicopter recedes.

14.4.8 Thickness Noise

The helicopter rotor produces both periodic (rotational) noise due to the thickness of the blades and noise due to the pressure forces. By periodically pushing the air aside, each blade produces a pressure disturbance. The sound pressure is linearly related to the blade lift and thickness, so the rotational noise due to the two sources can be evaluated separately and then added. Consider therefore a rotor blade with finite thickness but no lift. A symmetric section is assumed, so the pressure forces due to camber need not be considered. The thickness of the blade produces a velocity of the air normal to the section, first upward and then downward as the blade passes (considering only the air above the disk plane). This displacement velocity is determined by the boundary condition that the flow must be tangent to the airfoil surface. Let $v(r, x)$ be the velocity normal to the surface of the rotating blade due to

14.4 Rotational Noise

thickness. Such a velocity can be modeled by a distribution of sources over the blade, with strength proportional to v. For simplicity, v is assumed to be independent of ψ, although in forward flight v is periodic since it is proportional to the blade tangential velocity u_T. A source is a more effective radiator of sound than a dipole, but here the net source strength over the blade chord is zero (for a closed airfoil section), so the same order of noise as from a dipole is expected.

The sources on the rotating blade are now transformed to a distribution of fixed sources on the rotor disk, as required to give the periodic normal displacement velocity due to thickness. Following the approach of section 14.4.1, the distribution of normal velocity over the disk is

$$v_1(r,\psi,t) = \sum_{m=-\infty}^{\infty} \frac{N}{2\pi r} v_m(r) e^{imN(\Omega t - \psi)} \tag{14.112}$$

where

$$v_m = \int_{x_{le}}^{x_{te}} v(r,x) e^{-imNx/r} dx \tag{14.113}$$

For a hovering rotor, these sources on the rotor disk are stationary. The sound pressure due to a stationary but time-varying source of strength $2v_1\rho dA$ (considering both upper and lower surfaces of the rotor disk) is

$$p = \frac{\rho}{2\pi s} \frac{\partial}{\partial t} v_1(t - s/c_s) dA \tag{14.114}$$

On integrating over the rotor disk ($dA = r\,dr\,d\psi$), the m-th harmonic of the sound becomes

$$p_m = \rho \frac{imN^2\Omega}{4\pi^2} \int_0^{2\pi}\int_0^R v_m(r) \frac{e^{-iks}}{s} e^{-imN\psi} dr\,d\psi \tag{14.115}$$

In the far field, $s \cong s_0 - r\cos\theta_0 \cos(\psi - \psi_0)$ as usual, so

$$p_m = \rho \frac{imN^2\Omega}{4\pi^2 s_0} e^{-imN\Omega s_0/c_s} \int_0^R v_m \int_0^{2\pi} e^{-ikr\cos\theta_0 \cos(\psi-\psi_0) - imN\psi} d\psi\,dr$$

$$= \rho \frac{imN^2\Omega}{2\pi s_0} e^{-imN\Omega s_0/c_s} e^{-imN(\psi_0 - \pi/2)}$$

$$\int_0^R v_m J_{mN}\left(\frac{mN\Omega r}{c_s} \cos\theta_0\right) dr \tag{14.116}$$

is the spectrum of the rotational noise due to thickness. This noise is zero on the rotor axis, because of the $\cos\theta_0$ factor in the Bessel function.

To complete the analysis of the thickness noise, v_m must be evaluated. For thin sections, the velocity normal to the airfoil is given by the slope of the surface times the free stream velocity:

$$v(r,x) = \Omega r \frac{1}{2}\frac{dt}{dx} \tag{14.117}$$

where $t(x)$ is the thickness of the section. Then

$$v_m = \int_{x_{le}}^{x_{te}} \Omega r \frac{1}{2}\frac{dt}{dx} e^{-imNx/r} dx = imN\Omega \frac{1}{2} \int_{x_{le}}^{x_{te}} t e^{-imNx/r} dx = imN\Omega \frac{1}{2} A_{xs} a_{mN} \tag{14.118}$$

where

$$a_{mN} = \frac{1}{A_{xs}} \int_{x_{le}}^{x_{te}} t e^{-imNx/r} dx \qquad (14.119)$$

and A_{xs} is the area of the blade cross-section. Substituting for v_m then gives

$$p_m = -\frac{\rho(mN\Omega)^2 N}{4\pi s_0} e^{-imN\Omega s_0/c_s} e^{-imN(\psi_0 - \pi/2)}$$

$$\int_0^R A_{xs} a_{mN} J_{mN}\left(\frac{mN\Omega r}{c_s} \cos\theta_0\right) dr \qquad (14.120)$$

The coefficients a_{mN} can be evaluated for a given thickness distribution $t(x)$. For an impulsive thickness distribution (small chord), $t(x) = A_{xs}\delta(x)$, so $a_{mN} = 1$. Then only the section area is required. As an example, the area of the NACA 4-and 5-digit airfoils is $A_{xs} = 0.685\tau c^2$, where c is the chord and τ is the maximum thickness ratio (t_{max}/c). The thickness noise is usually small compared to the lift noise for the lower harmonics, but for high harmonic number and high tip Mach number the thickness noise can be large. The thickness noise consists of an impulsive type of pressure disturbance that is significant at high speeds.

Deming (1937, 1938) analyzed the rotational noise due to thickness, for a propeller at zero thrust. He found the theory predicted the propeller noise fairly well for the first five harmonics, and he found good agreement with the measured directivity.

14.5 Sound Generated Aerodynamically

The rotational noise analysis developed in section 14.4 follows classical linear acoustic analysis, using dipoles and sources that are stationary or moving with constant velocity. These singularities are distributed over the surface of the rotor disk, with time variation to model the periodic passage of the rotor blades. A more rigorous aeroacoustic theory is required to handle arbitrary motion of the rotor blades and the noise from nonlinear mechanisms.

14.5.1 Lighthill's Acoustic Analogy

Motivated by the problem of turbojet engine noise, Lighthill (1952) derived equations for analysis of sound generated aerodynamically. The Navier-Stokes equations for mass and momentum conservation are

$$\frac{\partial \rho}{\partial t} + \frac{\partial \rho u_j}{\partial x_j} = 0 \qquad (14.121)$$

$$\rho \frac{\partial u_i}{\partial t} + \rho u_j \frac{\partial u_i}{\partial x_j} = -\frac{\partial p}{\partial x_i} + \frac{\partial \tau_{ij}}{\partial x_j} \qquad (14.122)$$

where u_i is the velocity relative to still air, ρ is the density, and p is the pressure; see Batchelor (1967). The viscous stress tensor is

$$\tau_{ij} = \mu\left(\frac{\partial u_i}{\partial x_j} + \frac{\partial u_j}{\partial x_i} - \frac{2}{3}\delta_{ij}\frac{\partial u_k}{\partial x_k}\right) \qquad (14.123)$$

with μ the viscosity. The local adiabatic speed of sound is $c_s^2 = dp/d\rho$ (isentropic derivative). Here c_s is the still air value (also written $c_{s\infty}$, or a_∞, or a_0). For small (isentropic) disturbances the pressure and density perturbations are related by $\tilde{p} = c_s^2 \tilde{\rho}$.

Adding u_i times the mass conservation equation to the momentum conservation equation gives

$$\frac{\partial}{\partial t}\rho u_i = -\frac{\partial}{\partial x_j}\rho u_i u_j - \frac{\partial p}{\partial x_i} + \frac{\partial \tau_{ij}}{\partial x_j} = -\frac{\partial}{\partial x_j}\left(\rho u_i u_j + \delta_{ij}p - \tau_{ij}\right) \qquad (14.124)$$

or

$$\frac{\partial}{\partial t}\rho u_i + c_s^2 \frac{\partial \rho}{\partial x_i} = -\frac{\partial T_{ij}}{\partial x_j} \qquad (14.125)$$

where T_{ij} is Lighthill's turbulent stress tensor:

$$T_{ij} = \rho u_i u_j + \delta_{ij}(p - c_s^2 \rho) - \tau_{ij} \qquad (14.126)$$

Now combine the mass and momentum equations:

$$\frac{\partial^2 \rho}{\partial t^2} = -\frac{\partial}{\partial t}\frac{\partial \rho u_j}{\partial x_j} = -\frac{\partial}{\partial x_j}\frac{\partial \rho u_j}{\partial t} = \frac{\partial^2}{\partial x_i x_j}\left(\rho u_i u_j + \delta_{ij}p - \tau_{ij}\right) = c_s^2 \frac{\partial^2 \rho}{\partial x_j x_j} + \frac{\partial^2 T_{ij}}{\partial x_i x_j} \qquad (14.127)$$

which gives a wave equation for ρ, with a quadrupole source:

$$\frac{\partial^2 \rho}{\partial t^2} - c_s^2 \nabla^2 \rho = \frac{\partial^2 T_{ij}}{\partial x_i x_j} \qquad (14.128)$$

This is Lighthill's acoustic analogy. It is a rearrangement of the Navier-Stokes equations, derived without approximation. See also Curle (1955). In terms of perturbation quantities, $T_{ij} = \rho u_i u_j + \delta_{ij}(\tilde{p} - c_s^2 \tilde{\rho}) - \tau_{ij}$. For small disturbances (designated by the tilde),

$$\Box^2 \tilde{p} = \frac{1}{c_s^2}\frac{\partial^2 \tilde{p}}{\partial t^2} - \nabla^2 \tilde{p} = \frac{\partial^2 T_{ij}}{\partial x_i x_j} \qquad (14.129)$$

This is the wave equation for the acoustic (perturbation) pressure $\tilde{p}$. The operator is $\Box^2 = \frac{1}{c_s^2}\frac{\partial^2}{\partial t^2} - \nabla^2$. The approximation $\tilde{p} = c_s^2 \tilde{\rho}$ is only introduced on the left-hand side of the equation, requiring that the observer be in the linear region.

The tensor T_{ij} accounts for the generation of the sound and also for nonlinear propagation, reflection, and dissipation. The terms in T_{ij} are the convection of momentum, $\rho u_i u_j$; viscous stresses, τ_{ij}; and the effects of heat convection, $p - c_s^2 \rho = p_\infty - c_s^2 \rho_\infty + O(M^4)$. These quantities, hence T_{ij}, are known from the aerodynamic solution and are typically significant only within a limited volume of the fluid. For the jet noise problem, Lighthill used the approximation $T_{ij} \cong \rho_\infty u_i u_j$.

14.5.2 Ffowcs Williams-Hawkings Equation

Ffowcs Williams and Hawkings (1969) extended Lighthill's acoustic analogy to include moving surfaces. Consider a surface S defined by the function $f(x) = 0$. The normal to the surface is $n_i = \frac{1}{|\nabla f|}\partial f/\partial x_i$, and the normal velocity of the surface is $v_n = -\frac{1}{|\nabla f|}\partial f/\partial t$. The surface divides the volume into interior ($f < 0$) and exterior ($f > 0$) domains. The normal and normal velocity are outward, directed into the exterior domain. It is possible to define f such that $|\nabla f| = 1$ on the surface.

The Navier-Stokes equations are required in the volume V, bounded in the far field by the surface A and including the moving surface S. Mass conservation requires that the rate of change of the mass in V equal the mass flux through the surfaces A and S:

$$\frac{\partial}{\partial t}\int \rho\,dV = -\int \rho u_j n_j\,dA + \int \rho(u_j - v_j)n_j\,dS \tag{14.130}$$

in the exterior domain, and

$$\frac{\partial}{\partial t}\int \rho\,dV = -\int \rho u_j n_j\,dA - \int \rho(u_j - v_j)n_j\,dS \tag{14.131}$$

in the interior domain; $\mathbf{n}$ is the outward normal, and $\mathbf{v}$ is the surface velocity. Summing gives momentum conservation for the entire system:

$$\frac{\partial}{\partial t}\int \rho\,dV = -\int \rho u_j n_j\,dA + \int \Delta[\rho(u_j - v_j)]n_j\,dS$$

$$= -\int \frac{\partial \rho u_j}{\partial x_j}\,dV + \int \Delta[\rho(u_j - v_j)]n_j|\nabla f|\delta(f)\,dV \tag{14.132}$$

where $\Delta[\ldots]$ means the jump at the surface (exterior minus interior). Gauss's theorem has been used to transform the first term on the left-hand side to a volume integral, and the second term follows using:

$$\int Q\,dS = \int\left(\int Q\delta(f)\,df\right)dS = \int\left(\int Q|\nabla f|\delta(f)\,dn\right)dS = \int Q|\nabla f|\delta(f)\,dV \tag{14.133}$$

Here $\delta(f)$ is the Dirac delta function. The Heaviside function $H(f)$ is also used ($H = 1$ for $f > 0$, $H = 0$ for $f < 0$). The differential equation of mass conservation is then

$$\frac{\partial \rho}{\partial t} + \frac{\partial \rho u_j}{\partial x_j} = \Delta[\rho(u_j - v_j)]n_j|\nabla f|\delta(f) \tag{14.134}$$

Momentum conservation requires that the rate of change of the momentum in V equal the momentum flux through the surfaces A and S plus the normal forces acting on the surfaces:

$$\frac{\partial}{\partial t}\int \rho u_i\,dV = -\int (\rho u_i)u_j n_j\,dA + \int \sigma_{ij}n_j\,dA$$

$$+ \int \Delta[(\rho u_i)(u_j - v_j)]n_j\,dS - \int \Delta[\sigma_{ij}]n_j\,dS \tag{14.135}$$

where $\sigma_{ij} = -p\delta_{ij} + \tau_{ij}$. Then

$$\frac{\partial \rho u_i}{\partial t} + \frac{\partial}{\partial x_j}(\rho u_i u_j - \sigma_{ij}) = \Delta\bigl[-\sigma_{ij} + \rho u_i(u_j - v_j)\bigr]n_j|\nabla f|\delta(f) \tag{14.136}$$

is the differential equation of momentum conservation.

Now take the interior flow to be at rest: $u = 0$, $\rho = \rho_0$, $p = p_0$. Assume the surface S is impermeable, so the normal fluid velocity at the surface equals the surface velocity: $u_n = v_n$. Then the Navier-Stokes equations are

$$\frac{\partial \rho}{\partial t} + \frac{\partial \rho u_j}{\partial x_j} = \rho_0 v_n |\nabla f|\delta(f) \tag{14.137}$$

$$\frac{\partial \rho u_i}{\partial t} + \frac{\partial}{\partial x_j}(\rho u_i u_j - \sigma_{ij}) = (-\sigma_{ij} - p_0\delta_{ij})n_j|\nabla f|\delta(f) \tag{14.138}$$

14.5 Sound Generated Aerodynamically

Following Lighthill,

$$\frac{\partial}{\partial t}\rho u_i + c_s^2 \frac{\partial \rho}{\partial x_i} = -\frac{\partial T_{ij}}{\partial x_j} + \left((p - p_0)\delta_{ij} - \tau_{ij}\right) n_j |\nabla f| \delta(f) \qquad (14.139)$$

and the wave equation for ρ becomes

$$\frac{\partial^2 \rho}{\partial t^2} - c_s^2 \nabla^2 \rho = \frac{\partial^2}{\partial x_i \partial x_j}(T_{ij} H(f)) - \frac{\partial}{\partial x_i}\left((p - p_0)\delta_{ij} - \tau_{ij}\right) n_j |\nabla f| \delta(f)$$

$$+ \frac{\partial}{\partial t}\left(\rho_0 v_n |\nabla f| \delta(f)\right) \qquad (14.140)$$

This is the Ffowcs Williams-Hawkings (FW-H) equation, which is the foundation of current rotor noise calculations. It is a rearrangement of the Navier-Stokes equations (without approximation) to an inhomogeneous wave equation for the density, with two surface source terms and a volume source term. Equation 14.140 was derived by adding an artificial interior flow to the exterior flow, so the equations of motion are for the flow in unbounded space, with discontinuities at the moving surface. So the free-space Green's function can be used to solve the wave equation.

For small disturbances, the wave equation for the acoustic pressure $\tilde{p} = c_s^2 \tilde{\rho}$ is

$$\Box^2 \tilde{p} = \frac{\partial^2}{\partial x_i \partial x_j}(T_{ij} H(f)) - \frac{\partial}{\partial x_i}\left(\ell_i |\nabla f| \delta(f)\right) + \frac{\partial}{\partial t}\left(\rho_0 v_n |\nabla f| \delta(f)\right) \qquad (14.141)$$

where $\ell_i = P_{ij} n_j = \left((p - p_0)\delta_{ij} - \tau_{ij}\right) n_j$. The three terms on the right-hand side are quadrupole, dipole, and monopole sources of noise, respectively. The quadrupole term is Lighthill's turbulent stress tensor, in the volume exterior to the surface S. The dipole term is from the loading, the surface forces on S; the stress τ_{ij} is usually not a significant contribution to ℓ_i. The monopole term is from the thickness, the volume displacement of the surface.

Ffowcs Williams and Hawkings (1969) gave the mass and momentum equations (equations 14.134 and 14.136) for a general surface in the flow. From this result di Francescantonio (1997) developed the permeable surface FW-H equation. Let $f = 0$ now be a closed moving surface, permeable to the fluid and enclosing the solid surface that interacts with the fluid. Now the mass and momentum conservation equations are

$$\frac{\partial \rho}{\partial t} + \frac{\partial \rho u_j}{\partial x_j} = \left(\rho_0 v_n + \rho(u_n - v_n)\right) |\nabla f| \delta(f) \qquad (14.142)$$

$$\frac{\partial \rho u_i}{\partial t} + \frac{\partial}{\partial x_j}(\rho u_i u_j - \sigma_{ij}) = \left[\left((p - p_0)\delta_{ij} - \tau_{ij}\right) n_j\right.$$

$$\left. + \rho u_i (u_n - v_n)\right] |\nabla f| \delta(f) \qquad (14.143)$$

The wave equation for the acoustic pressure is

$$\Box^2 \tilde{p} = \frac{\partial^2}{\partial x_i \partial x_j}(T_{ij} H(f)) - \frac{\partial}{\partial x_i}\left(L_i |\nabla f| \delta(f)\right) + \frac{\partial}{\partial t}\left(\rho_0 V_n |\nabla f| \delta(f)\right) \qquad (14.144)$$

with

$$L_i = P_{ij} n_j + \rho u_i (u_n - v_n) = \ell_i + \rho u_i (u_n - v_n) \qquad (14.145)$$

$$\rho_0 V_n = \rho_0 v_n + \rho(u_n - v_n) \qquad (14.146)$$

Here **u** is the fluid velocity, and **v** is the surface velocity. The potential advantage of the permeable surface FW-H equation is when the surface encloses all quadrupole sources producing significant noise. Then the quadrupole term in the wave equation is not required in the volume outside the surface, just the dipole and monopole sources at the surface.

14.5.3 Kirchhoff Equation

The Kirchhoff method solves the wave equation for the propagation of the acoustic pressure $\tilde{p}$, from the pressure on a surface S that encloses all physical sources. The method is applicable to any phenomenon governed by the wave equation. For aerodynamically generated sound, the solution is an approximation appropriate in the linear domain. Given that $\tilde{p}$ satisfies the wave equation outside S, the surface is introduced by applying the wave equation to the generalized function $\tilde{p}H(f)$. The resulting Kirchhoff equation is

$$\Box^2 \tilde{p} = -\left(\frac{v_n}{c_s^2}\frac{\partial \tilde{p}}{\partial t} + \frac{\partial \tilde{p}}{\partial n}\right)\delta(f) - \frac{\partial}{\partial t}\left(\frac{v_n}{c_s^2}\tilde{p}\delta(f)\right) - \frac{\partial}{\partial x_i}(\tilde{p}n_i\delta(f)) \quad (14.147)$$

Brentner and Farassat (1998) showed that the permeable surface FW-H equation can be written as

$$\Box^2 \tilde{p} = -\left(\frac{v_n}{c_s^2}\frac{\partial \tilde{p}}{\partial t} + \frac{\partial \tilde{p}}{\partial n}\right)\delta(f) - \frac{\partial}{\partial t}\left(\frac{v_n}{c_s^2}\tilde{p}\delta(f)\right) - \frac{\partial}{\partial x_i}(\tilde{p}n_i\delta(f))$$

$$+ \frac{\partial}{\partial t}\left((\tilde{p} - c_s^2 \tilde{\rho})\frac{v_n}{c_s^2}\delta(f)\right) + \frac{\partial}{\partial t}(\tilde{p} - c_s^2 \tilde{\rho})\frac{v_n}{c_s^2}\delta(f)$$

$$- \frac{\partial}{\partial x_i}(\rho u_i u_n \delta(f)) - \frac{\partial}{\partial x_i}(\rho u_i u_j)n_i\delta(f) + \frac{\partial^2}{\partial x_i x_j}(T_{ij}H(f)) \quad (14.148)$$

This result is exact, so the last two lines show what is neglected by using the Kirchhoff equation and hence the linear solution for wave propagation. The additional terms are second order in perturbation quantities, so can reasonably be neglected in the linear domain ($\tilde{p} \cong c_s^2 \tilde{\rho}$ and small fluid velocities). The Kirchhoff equation is not adequate if the integration surface intersects shocks, vortices, inhomogeneous flow, or heat flux.

14.5.4 Integral Formulations

The free-space Green's function can be used to obtain the solution of the inhomogeneous wave equation as integrals of source terms over the surface and volume. For the acoustic analogy, the result is not an integral equation; rather the source flow field is considered to be an input to the acoustic solution. This approach allows separation of the aerodynamic problem (for the sources) from the acoustic problem (for the sound). Computational fluid dynamics calculations of the near field provide input needed for the acoustic integral formulations to calculate the radiated noise. The acoustic analogy is particularly effective if the volume sources can be ignored, requiring only integrals over surfaces. The alternative of solving the aeroacoustic problem out to the noise observation locations has numerical difficulties and requires significantly more computation. The integral formulation is not unique. The choice among

14.5 Sound Generated Aerodynamically

the numerous manipulations possible is based on implementation considerations, as well as facilitating exposition.

To illustrate the derivation of the integral forms, consider the inhomogeneous wave equation $\Box^2\phi = Q(\mathbf{x}, t)\delta(f)$, where Q is the source and $f(\mathbf{x}, t) = 0$ defines the source surface. The free-space Green's function for an observer at $(\mathbf{x}, t)$ and a source at $(\mathbf{y}, \tau)$ is

$$G = \begin{cases} 0 & \tau > t \\ \dfrac{\delta(g)}{4\pi r} & \tau \leq t \end{cases} \qquad (14.149)$$

where $r = |\mathbf{x} - \mathbf{y}|$ and $g = \tau - t + r/c_s$. The solution for a fixed observer is then

$$4\pi\phi(\mathbf{x}, t) = \int_{-\infty}^{t}\int Q(\mathbf{y}, \tau)\delta(f)\frac{\delta(g)}{r}d\mathbf{y}\,d\tau \qquad (14.150)$$

The integral is over all space and $\tau = -\infty$ to t. Now transform from $\mathbf{y}$ to moving coordinates η, such that $f = 0$ is independent of time. Then only $\delta(g)$ involves τ, not $\delta(f)$. A unit Jacobian $J = |\partial\eta/\partial\mathbf{y}| = 1$ is assumed, as for rigid-body motion. Usually the area change $(J \neq 1)$ of a more complicated transformation can be neglected. Next transform from τ to g. Then

$$4\pi\phi(\mathbf{x}, t) = \int\int Q(\mathbf{y}, \tau)\delta(f)\frac{\delta(g)}{r|\partial g/\partial\tau|}d\mathbf{y}\,dg$$

$$= \int\left[\frac{Q(\mathbf{y}, \tau)}{r|1 - M_r|}\delta(f)\right]_{g=0}d\mathbf{y}$$

$$= \int_{f=0}\left[\frac{Q(\mathbf{y}, \tau)}{r|1 - M_r|}\right]_{g=0}dS \qquad (14.151)$$

where S is the surface $f = 0$; $|\nabla f| = 1$ has been assumed. The position of the observer relative to the source is $\mathbf{r} = \mathbf{x} - \mathbf{y}$, and $r = |\mathbf{x} - \mathbf{y}|$, so $\hat{r} = \mathbf{r}/r$ is the unit vector in the radiation direction. The velocity of the source relative to the undisturbed fluid is $v_i = \partial y_i/\partial\tau$, and $M_r = \hat{r}_i v_i/c_s$ is the Mach number of the source (point $\mathbf{y}$) in the radiation direction. The derivatives of r are $\frac{\partial r}{\partial x_i} = \hat{r}_i$, $\frac{\partial r}{\partial y_i} = -\hat{r}_i$, and $\frac{\partial r}{\partial \tau} = \frac{\partial r}{\partial y_i}\frac{\partial y_i}{\partial \tau} = -\hat{r}_i v_i = -v_r$. The Doppler factor is

$$\frac{\partial g}{\partial \tau} = 1 + \frac{1}{c_s}\frac{\partial r}{\partial y_i}\frac{\partial y_i}{\partial \tau} = 1 - \frac{\hat{r}_i v_i}{c_s} = 1 - M_r \qquad (14.152)$$

The solution of $g = \tau - t + |\mathbf{x} - \mathbf{y}(\tau)|/c_s = 0$ is the retarded time, or emission time τ_e. At τ_e, the emission position is $\mathbf{y}_e$, and the emission distance is $r_e = |\mathbf{x} - \mathbf{y}_e| = c_s(t - \tau_e)$. For subsonic sources, $g = 0$ has only one solution, so there is only one emission point and time. Thus

$$4\pi\phi(\mathbf{x}, t) = \int_{f=0}\left[\frac{Q(\mathbf{y}, \tau)}{r(1 - M_r)}\right]_{\text{ret}}dS \qquad (14.153)$$

The subscript "ret" means the integrand is evaluated at the retarded time, $\tau = t - r/c_s$, the time when the sound was emitted. Often the retarded time is indicated simply by square brackets, omitting the subscript "ret." Solutions of the wave equation with time or space derivatives of the source follow directly, by taking the derivative of both the equation and the solution and redefining ϕ.

A number of formulations of the integration have been developed; see Farassat (1975) and Brentner (1997b). Equation 14.153 is the retarded time formulation. The numerical implementation of this solution is robust and efficient. One approach is to fix the observer time t and to solve for the retarded time τ as the integration over S is performed. An alternative is to fix the source time τ, find t to which the retarded surface element contributes (simple for a fixed observer), and then interpolate to get the required observer time. The Doppler singularity at $M_r = 1$ is integrable, but for supersonic source motion the retarded time equation can have multiple roots.

Writing $d\mathbf{y} = (c_s d\Gamma / \cos\theta) df\, dg$ gives the collapsing sphere formulation:

$$4\pi\phi(\mathbf{x},t) = \int_{-\infty}^{t} \int_{f=g=0} \frac{Q(\mathbf{y},\tau)}{r\sin\theta} c_s d\Gamma\, d\tau \tag{14.154}$$

where θ is the angle between the radiation direction $\hat{r}$ and the surface outward normal. The surface $g = \tau - t + r/c_s = 0$ is a collapsing sphere, and Γ is the intersection of the collapsing sphere and the source surface ($f = 0$). The collapsing sphere formulation does not exhibit the Doppler singularity, but is singular when the surface normal is parallel to the radiation vector ($\sin\theta = 0$).

Writing $d\mathbf{y}\, d\tau = (d\Sigma/\Lambda) dF\, dg$ gives the emission surface formulation:

$$4\pi\phi(\mathbf{x},t) = \int\int Q(\mathbf{y},\tau)\delta(F)\frac{\delta(g)}{r|\nabla F|} d\Sigma\, dF\, dg = \int_{F=0} \frac{1}{r}\left[\frac{Q(\mathbf{y},\tau)}{\Lambda}\right]_{\text{ret}} d\Sigma \tag{14.155}$$

Here $F = 0$ is the surface at the retarded time: $F = [f(\mathbf{y},\tau)]_{\text{ret}} = f(\mathbf{y}, t - r/c_s)$. From $\frac{\partial F}{\partial y_i} = \frac{\partial f}{\partial y_i} + \frac{1}{c_s}\frac{\partial f}{\partial t}\frac{\partial r}{\partial y_i} = n_i - M_n \hat{r}_i$, with M_n the surface normal Mach number, there follows $\Lambda = |\nabla F| = (1 + M_n^2 - 2M_n \cos\theta)^{1/2}$. The emission surface Σ is the set of points in space and time that emit signals that reach the observer at the same time. The emission surface formulation does not exhibit the Doppler singularity, but is singular when the surface normal is perpendicular to the radiation vector ($\cos\theta = 1$) and the surface normal velocity is sonic ($M_n = 1$).

Now let us apply these Green's function solutions to the Ffowcs Williams-Hawkings equation. Considering only the thickness and loading terms (surface sources), equation 14.141 is

$$\Box^2 \tilde{p} = \frac{\partial}{\partial t}(\rho_0 v_n \delta(f)) - \frac{\partial}{\partial x_i}(\ell_i \delta(f)) \tag{14.156}$$

(assuming $|\nabla f| = 1$). From equation 14.150, the solution is

$$4\pi \tilde{p}(\mathbf{x},t) = \frac{\partial}{\partial t}\int_{-\infty}^{t}\int \rho_0 v_n \delta(f)\frac{\delta(g)}{r} d\mathbf{y}\, d\tau - \frac{\partial}{\partial x_i}\int_{-\infty}^{t}\int \ell_i \delta(f)\frac{\delta(g)}{r} d\mathbf{y}\, d\tau \tag{14.157}$$

The spatial derivatives can be eliminated by using

$$\frac{\partial}{\partial t}\frac{\delta(g)}{r} = -\frac{\delta'(g)}{r} \tag{14.158}$$

in

$$\frac{\partial}{\partial x_i}\frac{\delta(g)}{r} = \left(-\frac{\delta(g)}{r^2} + \frac{\delta'(g)}{c_s r}\right)\frac{\partial r}{\partial x_i} = -\frac{\hat{r}_i \delta(g)}{r^2} - \frac{\partial}{\partial t}\frac{\hat{r}_i \delta(g)}{c_s r} \tag{14.159}$$

14.5 Sound Generated Aerodynamically

Then

$$4\pi \tilde{p}(\mathbf{x}, t) = \frac{\partial}{\partial t} \int_{-\infty}^{t} \int (\rho_0 v_n + \ell_i \hat{r}_i/c_s) \frac{\delta(f)\delta(g)}{r} d\mathbf{y}\, d\tau + \int_{-\infty}^{t} \int \ell_i \hat{r}_i \frac{\delta(f)\delta(g)}{r^2} d\mathbf{y}\, d\tau$$

$$= \frac{\partial}{\partial t} \int_{f=0} \left[\frac{\rho_0 v_n + \ell_r/c_s}{r(1-M_r)}\right]_{\text{ret}} dS + \int_{f=0} \left[\frac{\ell_r}{r^2(1-M_r)}\right]_{\text{ret}} dS \quad (14.160)$$

where $\ell_r = \ell_i \hat{r}_i$ is the loading in the radiation direction. This is Farassat's Formulation 1; see Farassat (1975, 1981). The loading is

$$\ell_i = P_{ij} n_j = \big((p-p_0)\delta_{ij} - \tau_{ij}\big) n_j \cong (p-p_0) n_i \quad (14.161)$$

so

$$\ell_r = P_{ij} n_j \hat{r}_i \cong (p-p_0) n_i \hat{r}_i = (p-p_0) \cos\theta \quad (14.162)$$

where θ is the angle between the surface normal and the radiation direction. Although the spatial derivatives have been eliminated, the time derivative remains, requiring numerical differential in implementations.

To move the time derivative inside the integral, note that

$$\frac{\partial}{\partial t}\Big[Q(\mathbf{y}, \tau)\Big]_{\text{ret}} = \left[\frac{\partial Q/\partial \tau}{|\partial t/\partial \tau|}\right]_{\text{ret}} = \left[\frac{1}{(1-M_r)} \frac{\partial Q}{\partial \tau}\right]_{\text{ret}} \quad (14.163)$$

Then

$$4\pi \tilde{p}(\mathbf{x}, t) = \int_{f=0} \left[\frac{1}{(1-M_r)} \frac{\partial}{\partial \tau} \frac{\rho_0 v_n + \ell_r/c_s}{r(1-M_r)}\right]_{\text{ret}} dS + \int_{f=0} \left[\frac{\ell_r}{r^2(1-M_r)}\right]_{\text{ret}} dS \quad (14.164)$$

Now define $\dot{v}_n = \partial v_n/\partial \tau = \partial(v_i n_i)/\partial \tau$, and $\dot{\ell}_r = \dot{\ell}_i \hat{r}_i = (\partial \ell_i/\partial \tau) \hat{r}_i$. From

$$\frac{\partial}{\partial \tau} \frac{1}{r(1-M_r)} = -\frac{1}{r^2(1-M_r)^2} \frac{\partial}{\partial \tau}(r - rM_r)$$

$$= -\frac{1}{r^2(1-M_r)^2} \frac{\partial}{\partial \tau}(r - r_i v_i/c_s)$$

$$= -\frac{1}{r^2(1-M_r)^2}(-\hat{r}_i v_i + v_i^2/c_s - r_i \dot{v}_i/c_s)$$

$$= \frac{1}{r^2(1-M_r)^2}(c_s M_r - c_s M^2 + r\dot{M}_r) \quad (14.165)$$

$$\frac{\partial \ell_r}{\partial \tau} = \frac{\partial(\ell_i \hat{r}_i)}{\partial \tau} = \frac{\partial \ell_i}{\partial \tau}\hat{r}_i + \ell_i \frac{\hat{r}_i v_r - v_i}{r}$$

$$= \dot{\ell}_i \hat{r}_i + \frac{c_s}{r}(\ell_r M_r - \ell_i M_i) \quad (14.166)$$

there follows

$$4\pi \tilde{p}(\mathbf{x}, t) = \int_{f=0} \left[\frac{\rho_0 \dot{v}_n + \dot{\ell}_r/c_s}{r(1-M_r)^2} + \frac{\ell_r - \ell_i M_i}{r^2(1-M_r)^2} \right.$$
$$\left. + \frac{\rho_0 v_n + \ell_r/c_s}{r^2(1-M_r)^3}(r\dot{M}_r + c_s M_r - c_s M^2)\right]_{\text{ret}} dS \quad (14.167)$$

This is Farassat's Formulation 1A, which is well suited for numerical evaluation of the noise from helicopter rotors with subsonic tip speeds; see Farassat and Succi

(1980) and Brentner (1986). The solution in the far field is

$$4\pi \tilde{p}(\mathbf{x}, t) \cong \int_{f=0} \left[\frac{1}{r(1 - M_r)} \frac{\partial}{\partial \tau} \frac{\rho_0 v_n + \ell_r/c_s}{1 - M_r} \right]_{\text{ret}} dS$$

$$\cong \int_{f=0} \left[\frac{\rho_0}{r(1 - M_r)^2} \left(\frac{\partial v_n}{\partial \tau} + \frac{v_n}{1 - M_r} \frac{\partial M_r}{\partial \tau} \right) \right]_{\text{ret}} dS$$

$$+ \int_{f=0} \left[\frac{\hat{r}_i}{c_s r(1 - M_r)^2} \left(\frac{\partial \ell_i}{\partial \tau} + \frac{\ell_i}{1 - M_r} \frac{\partial M_r}{\partial \tau} \right) \right]_{\text{ret}} dS$$

$$= \int_{f=0} \left[\frac{\rho_0}{r(1 - M_r)} \frac{\partial}{\partial \tau} \frac{v_n}{1 - M_r} \right]_{\text{ret}} dS$$

$$+ \int_{f=0} \left[\frac{\hat{r}_i}{c_s r(1 - M_r)} \frac{\partial}{\partial \tau} \frac{\ell_i}{1 - M_r} \right]_{\text{ret}} dS \qquad (14.168)$$

Lowson (1965) derived the far field loading result.

The thickness and loading terms of the Ffowcs Williams-Hawkings equations are responsible for most of the noise in many helicopter operating conditions, but when nonlinear flow phenomena are active, notably shocks that are responsible for high-speed impulsive noise, the quadupole terms become important. Considering the quadrupole term (volume sources), equation 14.141 is

$$\Box^2 \tilde{p} = \frac{\partial^2}{\partial x_i x_j} (T_{ij} H(f)) \qquad (14.169)$$

Farassat and Brentner (1988) wrote the solution as

$$4\pi \tilde{p}(\mathbf{x}, t) = \frac{\partial^2}{\partial x_i x_j} \int_{-\infty}^{t} \int T_{ij} H(f) \frac{\delta(g)}{r} d\mathbf{y} \, d\tau$$

$$= \frac{\partial^2}{\partial x_i x_j} \int_{F>0} \frac{T_{ij}}{r} c_s d\Omega \, d\tau$$

$$= \frac{1}{c_s} \frac{\partial^2}{\partial t^2} \int_{F>0} \frac{T_{rr}}{r} d\Omega \, d\tau + \frac{\partial}{\partial t} \int_{F>0} \frac{3 T_{rr} - T_{ii}}{r^2} d\Omega \, d\tau$$

$$+ c_s \int_{F>0} \frac{3 T_{rr} - T_{ii}}{r^3} d\Omega \, d\tau \qquad (14.170)$$

where $d\Omega$ is an element of the surface of the sphere $r = c_s(t - \tau)$, and $T_{rr} = T_{ij} \hat{r}_i \hat{r}_j$. The integration is over the volume external to the surface $F = [f(\mathbf{y}, \tau)]_{\text{ret}} = 0$. The time derivatives can be moved inside the integrals, if the $M_r = 1$ singularity is not an issue. See also Brentner (1997a). Some approximations are useful, particularly for in-plane observers or far field. In particular, if an approximate integration over the vertical coordinate is introduced, the result just requires a surface integral. If approximations are not appropriate, evaluating the quadrupole noise involves complicated volume integrals and requires the aerodynamic source solution throughout the volume.

Brentner, Burley, and Marcolini (1994) established the sensitivity of acoustic calculations to aspects of the numerical method. The blade motion is important, particularly flap and pitch motion. The compact chordwise loading approximation is reasonable, but the thickness noise requires integration over the chord in any case.

Blade-vortex interaction calculations require about 1024 points per revolution to accurately calculate the time derivative of the loading. In-plane noise calculation is very sensitive to the analysis parameters and requires the quadrupole sources at high speed.

With the motivation of avoiding the quadrupole integral, the noise can be obtained using a surface of acoustic sources off the body either through the Kirchhoff method or the permeable surface FW-H method. If significant quadrupole sources exist only in a finite (perhaps small) region around the body, the surface can be placed outside this region. Then quadrupole sources can be neglected, at the cost of evaluating the aerodynamic dipole and monopole sources out to this surface. For the Kirchhoff method, the solution of equation 14.147 is

$$4\pi \tilde{p}(\mathbf{x},t) = -\frac{1}{c_s}\frac{\partial}{\partial t}\int_{f=0}\left[\frac{\tilde{p}(M_n - \hat{r}_i n_i)}{r(1-M_r)}\right]_{\text{ret}} dS$$
$$-\int_{f=0}\left[\left(M_n\frac{1}{c_s}\frac{\partial \tilde{p}}{\partial \tau} + \frac{\partial \tilde{p}}{\partial n}\right)\frac{1}{r(1-M_r)}\right]_{\text{ret}} dS$$
$$+\int_{f=0}\left[\frac{\tilde{p}\hat{r}_i n_i}{r^2(1-M_r)}\right]_{\text{ret}} dS \quad (14.171)$$

For the permeable surface FW-H method, the solution of equation 14.144 follows by simply replacing ℓ_i and v_n in the integral solution of the Ffowcs Williams-Hawkings equation with L_i and V_n (equations 14.145 and 14.146).

The Kirchhoff method has limitations that present problems for rotor noise calculations. It is valid only in the linear propagation domain (inviscid, isentropic, small disturbances), the extent of which is not generally known. Consequently the solution is sensitive to placement of the surface. The Kirchhoff formulation depends only on the acoustic pressure, but space derivatives of the pressure are required and the pressure data must satisfy the wave equation.

The permeable surface Ffowcs Williams-Hawkings method is a unified theory for thickness, loading, and quadrupole noise, based on the conservation laws of the flow. The method requires the velocity, density, and pressure of the flow, all provided by computational fluid dynamics calculations of the near field. The solution should be insensitive to the placement of the surface, with the effect of quadrupole sources enclosed within the surface appearing in the surface sources. A non-rotating surface (enclosing the entire rotor) can be used, thereby avoiding the Doppler singularity. For both methods, moving the surface off the body decouples the acoustic calculations from the detailed geometry of the body. For the permeable surface FW-H method, the surface must still be placed outside significant quadrupole sources, while keeping the surface as close to the body as possible to minimize the domain required of the aerodynamic analysis. Quadrupole sources are not negligible if the surface is crossed by shocks or vorticity, which is difficult to avoid for flow fields producing impulsive noise.

14.5.5 Far Field Thickness and Loading Noise

The classical acoustic analyses of section 14.4 are based on acoustic sources that are stationary or moving with constant velocity, distributed over the rotor disk with time variation to simulate rotating blades. The Ffowcs Williams-Hawkings equation is the basis for analyses using rotating acoustic sources. The far field sound due to thickness and loading (equation 14.168) is

$$4\pi \tilde{p}(\mathbf{x}, t) = \int \left[\frac{\rho_0}{s(1 - M_r)} \frac{\partial}{\partial \tau} \frac{v_n}{1 - M_r} \right]_{\text{ret}} dS$$

$$+ \int \left[\frac{\hat{s}_i n_i}{c_s s(1 - M_r)} \frac{\partial}{\partial \tau} \frac{p - p_0}{1 - M_r} \right]_{\text{ret}} dS \quad (14.172)$$

writing now $\mathbf{s} = \mathbf{x} - \mathbf{y}$, $s = |\mathbf{s}|$ for consistency with section 14.4. The integrals are over the surfaces of all N blades, with the integrands evaluated at the retarded time $\tau = t - s(\tau)/c_s$. The first term is the thickness noise, produced by the velocity normal to the surface, v_n. The second term is the loading noise, produced by the surface normal force $\ell_i \cong (p - p_0)n_i$.

Consider a thin blade, so that the upper and lower surfaces coincide. Then only the difference in pressure and normal velocity between the upper and lower blade surfaces is required. The differential pressure on the blade surface, in rotating coordinates x and r, is written $\Delta p = L(r, \psi)\ell(x)$ (see section 14.4.1). The chordwise pressure distribution is assumed to not vary over the disk. The normal velocity due to the blade thickness (see section 14.4.8) has the same magnitude on the upper and lower surfaces. so

$$\Delta v_n = V_T A_{xs} \frac{da}{dx} \quad (14.173)$$

where $a(x) = t(x)/A_{xs}$ is the normalized thickness distribution, and $V_T = \Omega r + V_x \sin \psi$ is the free stream velocity seen by the blade. The velocity normal to the blade surface actually is

$$v_n = V_T \frac{t'/2}{\sqrt{1 + (t'/2)^2}} \quad (14.174)$$

at zero angle-of-attack. Since the arc length along the airfoil chord is $ds = \sqrt{1 + (t'/2)^2} dx$, the monopole strength is $\Delta v_n dS = V_T t' dx\, dr$, which gives $a(x) = t(x)/A_{xs}$ as earlier. The surface of the blade is defined by the root and tip and by the leading and trailing edges in the rotating coordinate system with axes r and x. The blade azimuth position is $\psi = \Omega \tau$. Instead of integrating over all the blades, it is equivalent to multiply the rotational noise due to one blade by N. The rotor is assumed to have a forward velocity V_x and a vertical velocity V_z. The position of a source on the blade surface is then

$$\mathbf{y}(\tau) = \begin{pmatrix} x_1 - V_x\tau \\ y_1 \\ z_1 - V_z\tau \end{pmatrix} = \begin{pmatrix} r\cos\Omega\tau + x\sin\Omega\tau - V_x\tau \\ r\sin\Omega\tau - x\cos\Omega\tau \\ V_z\tau \end{pmatrix}$$

$$= \begin{pmatrix} \sqrt{r^2 + x^2}\cos(\Omega\tau - \tan^{-1}(x/r)) - V_x\tau \\ \sqrt{r^2 + x^2}\sin(\Omega\tau - \tan^{-1}(x/r)) \\ V_z\tau \end{pmatrix} \quad (14.175)$$

where the components x_1, y_1, and z_1 represent the position relative to a tip-path-plane axis system moving with the rotor (as shown in Figure 14.3). The Mach number of a point on the blade is then

$$\mathbf{M} = \begin{pmatrix} -(\Omega/c_s)\sqrt{r^2 + x^2}\sin(\Omega\tau - \tan^{-1}(x/r)) - M_x \\ (\Omega/c_s)\sqrt{r^2 + x^2}\cos(\Omega\tau - \tan^{-1}(x/r)) \\ M_z \end{pmatrix} \quad (14.176)$$

14.5 Sound Generated Aerodynamically

The observer position is also defined relative to the axis system moving with the rotor hub:

$$\mathbf{x}(\tau) = \begin{pmatrix} x_0 - V_x t \\ y_0 \\ z_0 - V_z t \end{pmatrix}$$

The observer should actually be fixed in space, so that the time derivative does not operate on $\mathbf{x}$ when the sound pressure is evaluated, but for the present far field analysis a moving observer can be used with no difficulty. The observer position is evaluated at the present time t, not at the retarded time τ. The observer location relative to the rotor hub can be defined in terms of the range, elevation, and azimuth (see Figure 14.3); we have more need for these quantities at the retarded time.

The radial distance at the retarded time is $s = |\mathbf{x} - \mathbf{y}|$, or

$$s^2 = (x_0 - x_1 - V_x(t-\tau))^2 + (y_0 - y_1)^2 + (z_0 - z_1 - V_z(t-\tau))^2$$
$$= (x_0 - x_1 - M_x s)^2 + (y_0 - y_1)^2 + (z_0 - z_1 - M_z s)^2 \qquad (14.177)$$

since $s(\tau) = c_s(t - \tau)$. The solution of this quadratic equation for s is

$$s = \left(S - M_x(x_0 - x_1) + M_z(z_0 - z_1)\right)/\beta^2$$

where $\beta^2 = 1 - M_x^2 - M_z^2$ and

$$S^2 = \beta^2\left((x_0 - x_1)^2 + (y_0 - y_1)^2 + (z_0 - z_1)^2\right) + \left(M_x(x_0 - x_1) - M_z(z_0 - z_1)\right)^2 \qquad (14.178)$$

Define $\sigma_0 = s$ and $S_0 = S$ as the values from the center of the rotor ($x_1 = y_1 = z_1 = 0$), so

$$\sigma_0 = \sqrt{(x_0 - M_x\sigma_0)^2 + y_0^2 + (z_0 - M_z\sigma_0)^2}$$
$$= (S_0 - M_x x_0 + M_z z_0)/\beta^2 \qquad (14.179)$$

$$S_0^2 = \beta^2(x_0^2 + y_0^2 + z_0^2) + (M_x x_0 - M_z z_0)^2 \qquad (14.180)$$

Let $M \cos \delta_r$ be the value of $M_r = M_i \hat{s}_i$ at the center of the rotor:

$$M \cos \delta_r = M_x^2 + M_z^2 - \frac{M_x x_0 - M_z z_0}{\sigma_0} \qquad (14.181)$$

Note that $S_0 = \sigma_0(1 - M \cos \delta_r)$. Here σ_0 is the range of the observer from the rotor at the retarded time, and $M \cos \delta_r$ is the Mach number of the rotor toward the observer.

The acoustic far field is defined by the condition $S_0/R \gg 1$ ($\sigma_0 = S_0/(1 - M \cos \delta_r) > S_0$, so the condition $\sigma_0/R \gg 1$ is less critical). Expanding S and then s for large S_0/R gives

$$s \cong \sigma_0 - \frac{1}{S_0}\left((x_0 - M_x\sigma_0)x_1 + y_0 y_1 + (z_0 + M_z\sigma_0)z_1\right)$$

$$= \sigma_0 - \frac{1}{S_0}\sqrt{r^2 + x^2}\left((x_0 - M_x\sigma_0)\cos(\Omega r - \tan^{-1}(x/r))\right.$$
$$\left. + y_0 \sin(\Omega r - \tan^{-1}(x/r))\right)$$

$$= \sigma_0 - \frac{1}{S_0}\sqrt{r^2 + x^2}\sqrt{(x_0 - M_x\sigma_0)^2 + y_0^2}\cos(\Omega r - \tan^{-1}(x/r) - \psi_r)$$

$$= \sigma_0 - \frac{\cos\theta_r}{1 - M\cos\delta_r}\sqrt{r^2 + x^2}\cos(\Omega r - \tan^{-1}(x/r) - \psi_r) \qquad (14.182)$$

where $\psi_r = \tan^{-1} y_0/(x_0 - M_x\sigma_0)$ and $\theta_r = \sin^{-1}(z_0 + M_z\sigma_0)/\sigma_0$ are the azimuth and elevation of the observer at the retarded time. The far field approximation also gives

$$n_i \hat{s}_i \cong \frac{z_0 + V_z(t-\tau)}{\sigma_0} = \frac{z_0 + M_z\sigma_0}{\sigma_0} = \sin\theta_r \qquad (14.183)$$

The relative Mach number $M_r = M_i \hat{s}_i$ is

$$M_r = \frac{\Omega}{c_s}\sqrt{r^2+x^2}\left(-\frac{x_0 - M_x s}{s}\sin(\Omega\tau - \tan^{-1}(x/r))\right.$$
$$\left. + \frac{y_0}{s}\cos(\Omega\tau - \tan^{-1}(x/r))\right)$$
$$- \frac{M_x}{s}\sqrt{r^2+x^2}\cos(\Omega\tau - \tan^{-1}(x/r))$$
$$- M_z \frac{x_0 - M_x s}{s} + M_z \frac{z_0 + M_z s}{s} \qquad (14.184)$$

$$\cong M\cos\delta_r - \frac{\Omega}{c_s}\sqrt{r^2+x^2}\cos\theta_r \sin(\Omega\tau - \tan^{-1}(x/r) - \psi_r) \qquad (14.185)$$

in the far field. In summary, the far field approximation gives

$$s \cong \sigma_0 \qquad (14.186)$$
$$1 - M_r \cong (1 - M\cos\delta_r)(1 + \alpha \sin\psi) \qquad (14.187)$$
$$n_i \hat{s}_i \cong \sin\theta_r \qquad (14.188)$$
$$\tau = t - s/c_s \cong t - \sigma_0/c_s + (\alpha/\Omega)\cos\psi \qquad (14.189)$$

where

$$\psi = \Omega\tau - \tan^{-1}(x/r) - \psi_r \qquad (14.190)$$
$$\alpha = \frac{\Omega\sqrt{r^2+x^2}}{c_s}\frac{\cos\theta_r}{1 - M\cos\delta_r} \qquad (14.191)$$

and so

$$\Omega t = \psi - \alpha\cos\psi + \frac{\Omega\sigma_0}{c_s} + \tan^{-1}(x/r) + \psi_r \qquad (14.192)$$

is the dimensionless time.

The far field rotational noise due to the thickness and lift of the rotating blade is thus

$$\tilde{p}(\mathbf{x},t) = \frac{N\Omega\rho_0}{4\pi\sigma_0(1 - M\cos\delta_r)^2}\int_0^R \int_{x_{le}}^{x_{te}} \frac{A_{xs}a'}{1 + \alpha\sin\psi}\frac{\partial}{\partial\psi}\frac{V_T}{1 + \alpha\sin\psi} dx\, dr$$
$$- \frac{N\Omega\sin\theta_r}{4\pi c_s\sigma_0(1 - M\cos\delta_r)^2}\int_0^R \int_{x_{le}}^{x_{te}} \frac{\ell}{1 + \alpha\sin\psi}\frac{\partial}{\partial\psi}\frac{L}{1 + \alpha\sin\psi} dx\, dr \qquad (14.193)$$

14.5 Sound Generated Aerodynamically

The sound pressure is periodic with fundamental frequency $N\Omega$, so

$$\tilde{p} = \sum_{m=-\infty}^{\infty} p_m e^{imN\Omega t} \tag{14.194}$$

where the m-th harmonic is

$$p_m = \frac{\Omega}{2\pi} \int_0^{2\pi/\Omega} e^{-imN\Omega t} \tilde{p} \, dt = \frac{1}{2\pi} \int_0^{2\pi} e^{-imN\Omega t} \tilde{p}(1+\alpha \sin\psi) d\psi \tag{14.195}$$

since $\Omega dt = (1+\alpha \sin\psi)d\psi$. The blade section lift L and velocity V_T can be expanded as a Fourier series:

$$L = \sum_{n=-\infty}^{\infty} L_n(r) e^{in\Omega \tau} \tag{14.196}$$

$$V_T = \Omega r + V_x \sin \Omega \tau = \Omega r \sum_{n=-\infty}^{\infty} V_n(r) e^{in\Omega \tau} \tag{14.197}$$

(where $V_0 = 1$ and $V_{\pm 1} = \pm V_x/(2i\Omega r)$). The chordwise loading distributions ℓ and a are assumed to be independent of time. The m-th harmonic of the rotational noise is thus

$$p_m = \frac{N\Omega}{4\pi\sigma_0(1-M\cos\delta_r)^2} \int_0^R \int_{x_{le}}^{x_{te}} e^{-imN(\Omega\sigma_0/c_s + \psi_r + \tan^{-1}(x/r))}$$

$$\sum_{n=-\infty}^{\infty} e^{in(\psi_r + \tan^{-1}(x/r))} \left(\rho_0 A_{xs} a' \Omega r V_n - \frac{\sin\theta_r}{c_s} \ell L_n \right)$$

$$\frac{1}{2\pi} \int_0^{2\pi} e^{-imN(\psi - \alpha \cos\psi)} \frac{\partial}{\partial \psi} \frac{e^{in\psi}}{1+\alpha \sin\psi} d\psi \, dx \, dr \tag{14.198}$$

Integrating by parts gives

$$\frac{1}{2\pi} \int_0^{2\pi} e^{-imN(\psi - \alpha \cos\psi)} \frac{\partial}{\partial \psi} \frac{e^{in\psi}}{1+\alpha \sin\psi} d\psi$$

$$= imN \frac{1}{2\pi} \int_0^{2\pi} e^{-imN(\psi - \alpha \cos\psi) + in\psi} d\psi$$

$$= imN i^{mN-n} J_{mN-n}(mN\alpha) \tag{14.199}$$

using the Bessel function relation

$$\frac{1}{2\pi} \int_0^{2\pi} e^{iz \cos\psi - in\psi} d\psi = i^n J_n(z) \tag{14.200}$$

Next assume $x/r \ll 1$ so that the chordwise and spanwise integrals separate. Since $\tan^{-1} x/r \cong x/r$ and $\sqrt{r^2 + x^2} \cong r$, the Bessel functions do not depend on x. In addition, the thickness term is integrated by parts with respect to x. Define the chordwise loading factors ℓ_n and a_n as before:

$$\ell_n = \int_{x_{le}}^{x_{te}} \ell(x) e^{-inx/r} dx \tag{14.201}$$

$$a_n = \int_{x_{le}}^{x_{te}} a(x) e^{-inx/r} dx \tag{14.202}$$

Impulsive loading gives $\ell_n = a_n = 1$ for all n. Then the far field rotational noise due to the rotor blade thickness and lift is

$$p_m = -\frac{(mN\Omega)^2 N\rho_0}{4\pi\sigma_0(1 - M\cos\delta_r)^2} e^{-imN\Omega\sigma_0/c_s} \sum_{n=-\infty}^{\infty} e^{-i(mN-n)(\psi_r - \pi/2)} \left(1 - \frac{n}{mN}\right)$$

$$\int_0^R A_{xs} a_{mN-n} V_n J_{mN-n} dr$$

$$-\frac{imN^2\Omega \sin\theta_r}{4\pi c_s \sigma_0 (1 - M\cos\delta_r)^2} e^{-imN\Omega\sigma_0/c_s} \sum_{n=-\infty}^{\infty} e^{-i(mN-n)(\psi_r - \pi/2)}$$

$$\int_0^R \ell_{mN-n} L_n J_{mN-n} dr \qquad (14.203)$$

where the Bessel function has the argument

$$\frac{mN\Omega r}{c_s} \frac{\cos\theta_r}{1 - M\cos\delta_r} \qquad (14.204)$$

This result agrees with the solutions for the lift and thickness rotational noise that were derived in section 14.4. Working directly with the loading and motion of the rotating blade as here is more convenient for developing advanced analyses of the rotor noise.

14.5.6 Broadband Noise

Analysis of broadband noise requires a statistical approach, as in section 14.2. The far field loading noise (equation 14.168) is

$$4\pi \tilde{p}(\mathbf{x}, t) = \int \left[\frac{\hat{r}_i}{c_s r(1 - M_r)} \frac{\partial}{\partial \tau} \frac{\ell_i}{1 - M_r}\right]_{\tau = t - r/c_s} dS \qquad (14.205)$$

Considering the broadband noise produced by random loading with deterministic motion, in this section $\tilde{p}$ and ℓ_i are random perturbations relative to the periodic expected value: $\tilde{p} - E(\tilde{p})$ and $\ell_i - E(\ell_i)$. The relation between the auto-correlation of the acoustic pressure, $R_p(t_1, t_2) = E(\tilde{p}(t_1)\tilde{p}(t_2))$, and the auto-correlation of the loading, $R_{\ell ij}(\tau_1, \tau_2) = E(\ell_i(\tau_1)\ell_j(\tau_2))$, is

$$R_p(t_1, t_2) = \iint C_{ij}(\tau_1, \tau_2) R_{\ell ij}(\tau_1, \tau_2) dS_1 dS_2 \qquad (14.206)$$

since the expected value operator only acts on the random loading. The influence function is

$$C_{ij}(\tau_1, \tau_2) = \left(\frac{1}{4\pi c_s}\right)^2 \left[\frac{\hat{r}_i}{r(1 - M_r)} \frac{\partial}{\partial \tau} \frac{1}{1 - M_r}\right]_{\tau_1} \left[\frac{\hat{r}_i}{r(1 - M_r)} \frac{\partial}{\partial \tau} \frac{1}{1 - M_r}\right]_{\tau_2} \qquad (14.207)$$

From equation 14.6, the sound spectrum is

$$S_p^m(\omega) = \frac{N\Omega}{(2\pi)^2} \int_{-\infty}^{\infty} \int_0^{2\pi/N\Omega} e^{-imN\Omega s_1} e^{-i\omega s_2}$$

$$\left(\iint C_{ij}(\tau_1, \tau_2) R_{\ell ij}(\tau_1, \tau_2) dS_1 dS_2\right) ds_1 ds_2 \qquad (14.208)$$

14.5 Sound Generated Aerodynamically

The integration can be changed from (s_1, s_2) to (τ_1, τ_2), with the Jacobian $\frac{\partial(s_1,s_2)}{\partial(\tau_1,\tau_2)} = \frac{\partial t_1}{\partial \tau_1}\frac{\partial t_2}{\partial \tau_2} = (1 - M_r)_1(1 - M_r)_2$. From equation 14.5, the loading spectrum gives

$$R_{\ell i j}(\tau_1, \tau_2) = \sum_{n=-\infty}^{\infty} e^{in\Omega\sigma_1} \int_{-\infty}^{\infty} e^{iv\sigma_2} S_{\ell i j}^n(v) dv \tag{14.209}$$

where $\sigma_1 = \frac{1}{2}(\tau_2 + \tau_1)$ and $\sigma_2 = \tau_2 - \tau_1$. Substituting,

$$S_p^m(\omega) = \iint \sum_{n=-\infty}^{\infty} \int_{-\infty}^{\infty} S_{\ell i j}^n(v) \left(\frac{N\Omega}{(2\pi)^2} \int_{-\infty}^{\infty} \int_{0}^{2\pi/N\Omega} e^{-imN\Omega s_1} e^{-i\omega s_2} \right.$$
$$\left. C_{ij}(\tau_1, \tau_2) e^{in\Omega\sigma_1} e^{iv\sigma_2} ds_1 ds_2 \right) dv dS_1 dS_2 \tag{14.210}$$

or

$$S_p^0(\omega) = \iint \int_{-\infty}^{\infty} S_{\ell i j}^0(v) \left(\frac{1}{2\pi} \int_{-\infty}^{\infty} e^{-i\omega s_2} C_{ij}(\sigma_2) e^{iv\sigma_2} ds_2 \right) dv dS_1 dS_2 \tag{14.211}$$

for rms pressure ($m = 0$) due to average loading ($n = 0$). Spectral decomposition of $S_{\ell i j}$ in space can be introduced as well, perhaps to model a convected gust or blade-vortex interaction.

To proceed further, consider hover and one blade (no significant blade-to-blade correlation of the random loading) and compact chordwise loading so $\Delta p = L(r, t)\delta(x)$. Following section 14.5.5, the random acoustic pressure due to random loading is

$$\tilde{p}(\mathbf{x}, t) = -\frac{\Omega \sin \theta_r}{4\pi c_s s_0} \int_0^R \left[\frac{1}{1 + \alpha \sin \psi} \frac{\partial}{\partial \psi} \frac{L}{1 + \alpha \sin \psi} \right]_\tau dr \tag{14.212}$$

and for the far field geometry

$$s_0 = \sqrt{x_0^2 + y_0^2 + z_0^2} \tag{14.213}$$

$$\psi = \Omega\tau - \psi_r \tag{14.214}$$

$$\Omega t = \Omega\tau + \frac{\Omega s_0}{c_s} - \alpha \cos \psi = \psi - \alpha \cos \psi + \frac{\Omega s_0}{c_s} - \psi_r \tag{14.215}$$

$$\alpha = \frac{\Omega r}{c_s} \cos \theta_r \tag{14.216}$$

with $\theta_r = \sin^{-1} z_0/s_0$. For hover, $\psi_r = \tan^{-1} y_0/x_0$ can be ignored. The sound spectrum ($m = 0$ for rms noise) is

$$S_p^0(\omega) = \left(\frac{\Omega \sin \theta_r}{4\pi c_s s_0} \right)^2 \frac{\Omega}{(2\pi)^2} \int_{-\infty}^{\infty} \int_0^{2\pi/\Omega} e^{-i\omega s_2}$$
$$\left(\iint \left[\frac{1}{1 + \alpha \sin \psi} \frac{\partial}{\partial \psi} \frac{L}{1 + \alpha \sin \psi} \right]_{\tau_1} \left[\frac{1}{1 + \alpha \sin \psi} \frac{\partial}{\partial \psi} \frac{L}{1 + \alpha \sin \psi} \right]_{\tau_2} \right.$$
$$\left. R_L(\tau_1, \tau_2) dr_1 dr_2 \right) ds_1 ds_2 \tag{14.217}$$

Changing the integration from (s_1, s_2) to (ψ_1, ψ_2), and integrating by parts with respect to ψ, gives

$$S_p^0(\omega) = \left(\frac{(\omega/\Omega)\sin\theta_r}{4\pi c_s s_0}\right)^2 \frac{\Omega}{(2\pi)^2} \iint \left(\iint e^{-i\omega s_2} R_L(\tau_1, \tau_2) d\psi_2 d\psi_1\right) dr_1 dr_2 \tag{14.218}$$

Next introduce the loading spectrum:

$$S_p^0(\omega) = \left(\frac{(\omega/\Omega)\sin\theta_r}{4\pi c_s s_0}\right)^2 \frac{\Omega}{(2\pi)^2} \iint \sum_{n=-\infty}^{\infty}$$
$$\int_{-\infty}^{\infty} S_L^n(\nu) \left(\iint e^{-i\omega s_2 + i\nu\sigma_2 + in\Omega\sigma_1} d\psi_2 d\psi_1\right) d\nu dr_1 dr_2 \tag{14.219}$$

Using the expansion $e^{-iz\cos\beta} = \sum_k (-i)^k J_k(z) e^{-ik\beta}$, the integrals over ψ can be evaluated:

$$\int_{-\infty}^{\infty} \int_{-\psi_1}^{-\psi_1+4\pi} e^{-i\omega s_2 + i\nu\sigma_2 + in\Omega\sigma_1} d\psi_2 d\psi_1$$

$$= \int_{-\infty}^{\infty} e^{-i(\omega/\Omega)\alpha\cos\psi_1 + i(\omega/\Omega - \nu/\Omega + n/2)\psi_1}$$

$$\left(\int_{-\psi_1}^{-\psi_1+4\pi} e^{i(\omega/\Omega)\alpha\cos\psi_2 - i(\omega/\Omega - \nu/\Omega - n/2)\psi_2} d\psi_2\right) d\psi_1$$

$$= \int_{-\infty}^{\infty} \sum_k (-i)^k J_k(\alpha\omega/\Omega) e^{i(\omega/\Omega - \nu/\Omega + n/2 - k)\psi_1}$$

$$\left(\int_{-\psi_1}^{-\psi_1+4\pi} \sum_j i^j J_j(\alpha\omega/\Omega) e^{-i(\omega/\Omega - \nu/\Omega - n/2 - j)\psi_2} d\psi_2\right) d\psi_1$$

$$= 2\pi \sum_k (-i)^k J_k(\alpha\omega/\Omega) \delta(\omega/\Omega - \nu/\Omega + n/2 - k)$$

$$4\pi i^j J_j(\alpha\omega/\Omega)\big|_{j=\omega/\Omega - \nu/\Omega - n/2}$$

$$= 8\pi^2 \sum_k (-i)^n J_k(\alpha\omega/\Omega) J_{k-n}(\alpha\omega/\Omega) \delta(\omega/\Omega - \nu/\Omega + n/2 - k) \tag{14.220}$$

So the broadband noise spectrum is

$$S_p^0(\omega) = \left(\frac{\omega\sin\theta_r}{4\pi c_s s_0}\right)^2 \sum_{n=-\infty}^{\infty} \sum_{k=-\infty}^{\infty}$$
$$\iint (-i)^n J_k\left(\frac{\alpha\omega}{\Omega}\right) J_{k-n}\left(\frac{\alpha\omega}{\Omega}\right) 2S_L^n(\omega - \Omega(k - n/2)) dr_1 dr_2 \tag{14.221}$$

or

$$S_p^0(\omega) = \left(\frac{\omega\sin\theta_r}{4\pi c_s s_0}\right)^2 \sum_{k=-\infty}^{\infty} \iint \left(J_k\left(\frac{\omega r}{c_s}\cos\theta_r\right)\right)^2 2S_L^0(\omega - k\Omega) dr_1 dr_2 \tag{14.222}$$

for rms pressure ($m = 0$) due to average loading ($n = 0$). For small loading correlation length, this spectrum has peaks at low harmonics of the blade passage frequency, transitioning to a smooth spectrum at high frequency.

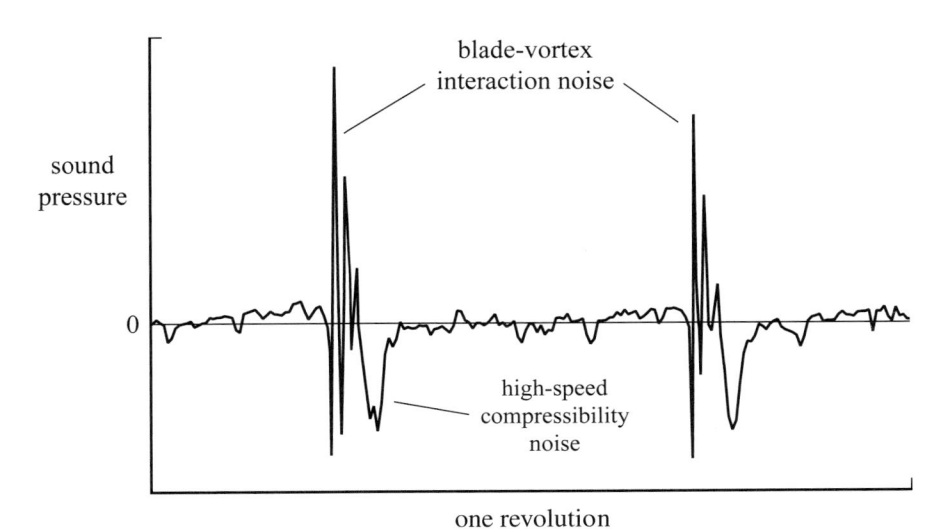

Figure 14.4. Flight test measurement of blade-vortex interaction noise; from Boxwell and Schmitz (1982).

14.6 Impulsive Noise

Blade-vortex interaction (BVI) noise is characterized by a set of sound pressure pulses occurring at the blade passage frequency, as illustrated in Figure 14.4. BVI noise is caused by the vortex-induced loading pulses on the rotor blade. The impulsive loading and resulting impulsive noise are strong at low speed and descent, as the noise is radiated forward and below the disk plane.

Let us examine the locus of blade-vortex interaction on the rotor disk. Consider the undistorted tip vortices of an N-bladed rotor. The blade is at radial station r and azimuth ψ; hence at disk plane coordinates

$$x = r\cos\psi \tag{14.223}$$

$$y = r\sin\psi \tag{14.224}$$

The vortex from the n-th preceding blade, with age ϕ, is at

$$x = \cos(\psi + n\Delta\psi - \phi) + \mu\phi \tag{14.225}$$

$$y = \sin(\psi + n\Delta\psi - \phi) \tag{14.226}$$

where $\Delta\psi = 2\pi/N$ is the inter-blade spacing. At the intersection of the blade and vortex these coordinates are equal. Eliminating r gives

$$\mu\phi\sin\psi + \sin(\phi - n\Delta\psi) = 0 \tag{14.227}$$

Taking wake age ϕ as the parameter of the intersection, solve for ψ and then r:

$$\sin\psi = \frac{\sin(n\Delta\psi - \phi)}{\mu\phi} \tag{14.228}$$

$$r = \frac{\sin(\psi + n\Delta\psi - \phi)}{\sin\psi} = \frac{\cos(\psi + n\Delta\psi - \phi) + \mu\phi}{\cos\psi} \tag{14.229}$$

Given $\sin\psi$, there are two values for the azimuth in the range $\psi = 0$ to 2π. The solution represents an intersection if $0 \leq r \leq 1$ and $|\sin\psi| \leq 1$. The direction **v** of the vortex and the direction **b** of the blade are

$$\mathbf{v} = \begin{pmatrix} x'/\sqrt{x'^2 + y'^2} \\ y'/\sqrt{x'^2 + y'^2} \end{pmatrix} \quad (14.230)$$

$$x' = \frac{\partial x}{\partial \phi} = \sin(\psi + n\Delta\psi - \phi) + \mu \quad (14.231)$$

$$y' = \frac{\partial y}{\partial \phi} = -\cos(\psi + n\Delta\psi - \phi) \quad (14.232)$$

$$\mathbf{b} = \begin{pmatrix} \cos\psi \\ \sin\psi \end{pmatrix} \quad (14.233)$$

Then the angle between the blade and the vortex at the intersection is $\gamma = \cos^{-1}(\mathbf{v} \cdot \mathbf{b})$. The trace speed or phase speed of the intersection is obtained using

$$\frac{dx}{d\psi} = -\sin(\psi + n\Delta\psi - \phi)\left(1 - \frac{\partial\phi}{\partial\psi}\right) + \mu\frac{\partial\phi}{\partial\psi} \quad (14.234)$$

$$\frac{dy}{d\psi} = \cos(\psi + n\Delta\psi - \phi)\left(1 - \frac{\partial\phi}{\partial\psi}\right) \quad (14.235)$$

The derivative of the equation 14.227 gives $\mu\phi\cos\psi + \mu\phi'\sin\psi + \cos(\phi - n\Delta\psi)\phi' = 0$, so

$$\phi' = \frac{\partial\phi}{\partial\psi} = -\frac{\mu\phi\cos\psi}{\mu\sin\psi + \cos(\phi - n\Delta\psi)} \quad (14.236)$$

and the trace Mach number is $M_{\text{trace}} = M_{\text{tip}}\sqrt{(dx/d\psi)^2 + (dy/d\psi)^2}$. Noise radiation is strong for supersonic trace speed, with directivity $\gamma \pm \cos^{-1}(1/M_{\text{trace}})$.

Figure 14.5 shows the loci of blade-vortex intersection on the rotor disk for a four-bladed rotor at advance ratios of $\mu = 0.15$ and $\mu = 0.3$. The heavy lines indicate supersonic trace speed (for $M_{\text{tip}} = 0.65$). The vortex direction tends to be aligned with the intersection locus. On the front of the disk and on the advancing ($\psi = 90°$) and retreating ($\psi = 270°$) tips, the interaction with the blade is nearly perpendicular. On the back of the disk the interaction is nearly parallel. These parallel interactions lead to supersonic trace speeds. Thus a four-bladed rotor at $\mu = 0.15$ is expected to produce BVI noise due to interaction of the blade with the vortices from the 5th, 6th, and 7th preceding blades. A pattern similar to Figure 14.5 is observed in blade pressure and airloads measurements, although the BVI loading should be zero directly above the intersection, with peaks occurring a distance h (the blade-vortex separation) either side of the intersection. Figure 14.6 illustrates the blade-vortex interaction airloads on the advancing and retreating sides of the rotor that are responsible for strong BVI noise.

High-speed impulsive (HSI) noise occurs at high advancing-tip Mach number, with directivity primarily in the disk plane. In the rotating frame, the sonic cylinder is where the rotational velocity equals the undisturbed sound speed: at $r/R = 1/M_{\text{tip}}$ for hover. At low tip speed the flow is subsonic near the blade, and the sonic cylinder is far from the disk. As the tip Mach number increases, the effects of compressibility and nonlinear aerodynamics increase, and the sonic cylinder comes closer to the blade tip. Pockets of supersonic flow form on the blade, and eventually the shocks extend beyond the tip. At a large enough tip Mach number, the supersonic flow reaches the sonic cylinder, an event called "delocalization"; see Yu, Caradonna,

14.6 Impulsive Noise

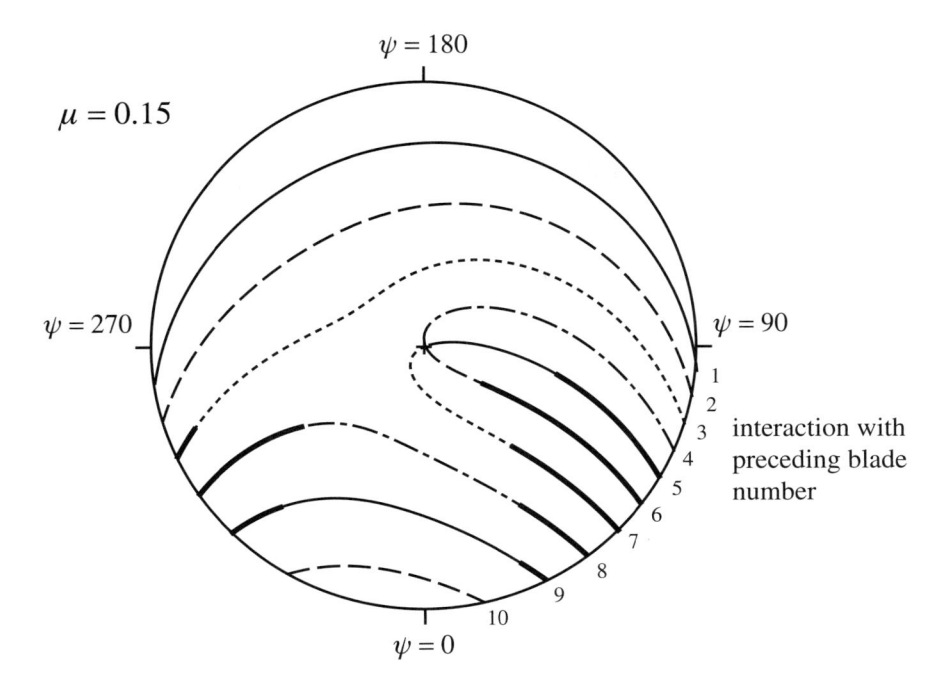

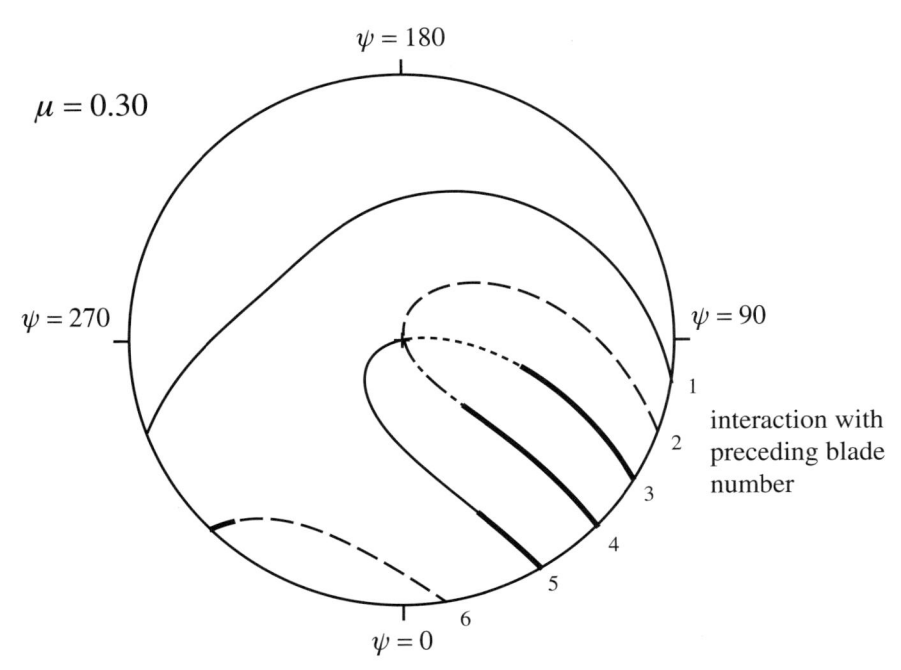

Figure 14.5. Trace of blade-vortex interaction on the rotor disk plane; heavy lines indicate supersonic trace speed.

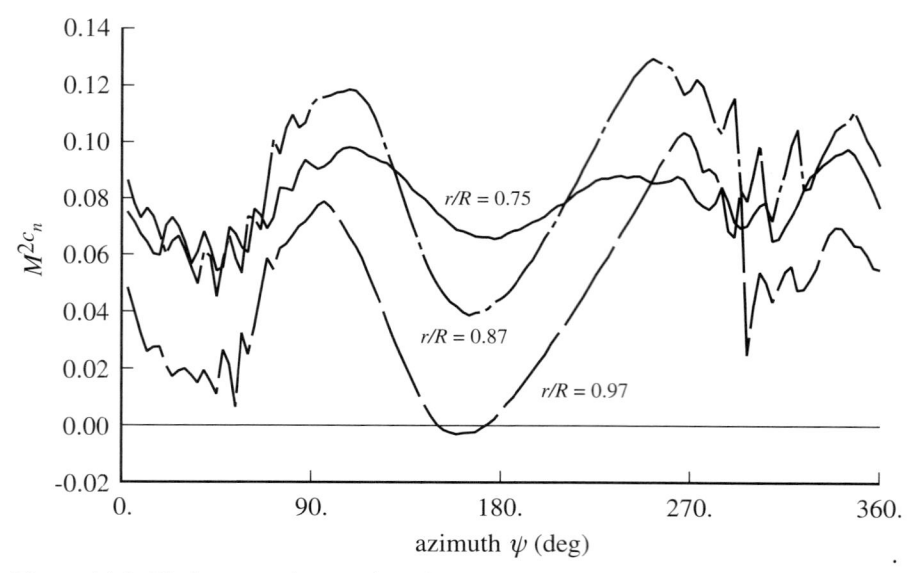

Figure 14.6. Blade-vortex interaction airloads, measured on a four-bladed model rotor at $\mu = 0.15$, $C_T/\sigma = 0.057$, and $i_{TPP} = -4.2°$; from Gmelin, Heller, Mercker, Philippe, Preisser, and Yu (1995).

and Schmitz (1978). Disturbances at the sonic cylinder propagate strongly, so at delocalization the amplitude of the noise increases and becomes more impulsive. Before delocalization, the thickness and loading noise sources account for most of the rotor noise. After delocalization, the quadrupoles are a major contributor to HSI noise.

Boxwell, Yu, and Schmitz (1979) conducted a hover test of a 1/7-scale UH-1H two-bladed rotor, measuring noise for high tip Mach numbers. The sound pressure was measured in the tip-path plane, 1.5 diameters from the hub. Figure 14.7 shows the measured high-speed impulsive noise for $M_{tip} = 0.8$ to $M_{tip} = 0.962$, presented for a common time increment that corresponds to an azimuth increment of 30° to 36°. Note the order-of-magnitude change in the pressure scale. At low M_{tip} the sound pressure pulse is small and symmetric. The waveform changes from symmetric to saw-tooth over a small tip speed range around $M_{tip} = 0.89$; this is delocalization. The large, sharp rise in pressure is produced by a radiated shock wave. Beyond $M_{tip} = 0.9$, the rise of the peak pressure is slower.

Figure 14.8 shows the measured peak negative sound pressure level, as well as calculations using several models. Linear thickness noise theory (monopole) under-predicts the peak, and the theoretical pressure pulse is symmetric even at $M_{tip} = 0.9$. Nonlinear models, sufficient to capture the noise associated with shocks, are required for HSI noise calculation. Schmitz and Yu (1981) obtained good correlation up to $M_{tip} = 0.9$ by using an approximate integration of the quadrupole term in the Ffowcs Williams-Hawkings equation. At higher tip Mach numbers the quadrupole calculations over-predicted the peak. Baeder (1991) obtained good correlation up to $M_{tip} = 1.0$ by solving the Euler equations out to the noise measurement location. The Euler equations can model shocks correctly (since the isentropic assumption of full potential methods is eliminated) and can model nonlinear propagation of

14.6 Impulsive Noise

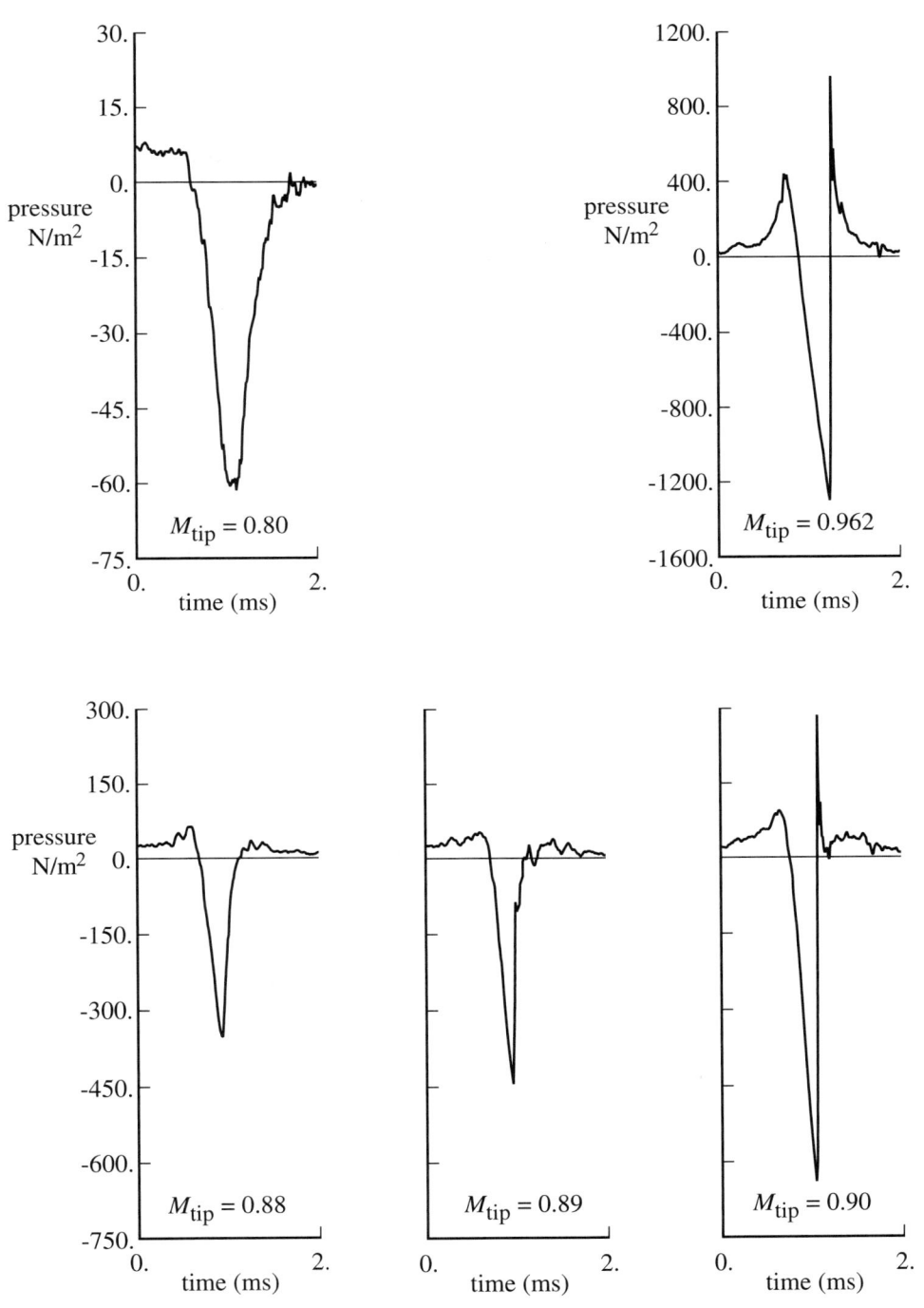

Figure 14.7. Measured high-speed impulsive sound pressure of a two-bladed rotor in hover; from Boxwell, Yu, and Schmitz (1979).

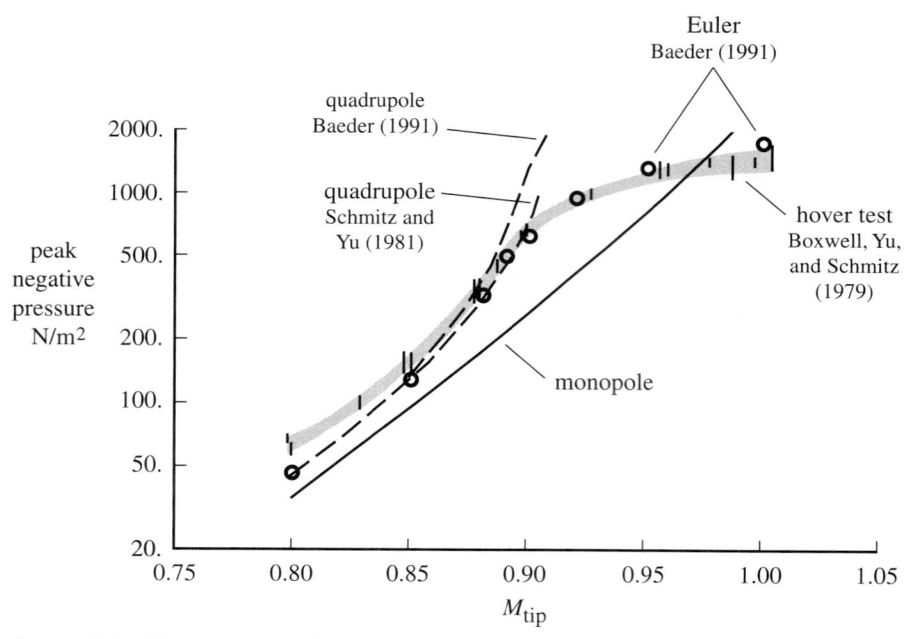

Figure 14.8. Measured and calculated peak negative sound pressure for high-speed impulsive noise.

acoustic waves as well as convention of entropy and vorticity. The FW-H equation is equivalent to the Navier-Stokes equations, so the difference between the quadrupole and Euler calculations reflects the approximate representation used for the quadrupole strength. Using computational fluid dynamic solutions to the acoustic far field is not efficient and currently not practical. The challenge is to accurately calculate the flow variables at a surface, as well as the quadrupole sources external to the surface, accurately enough to use the FW-H equation for the sound pressure calculation.

14.7 Noise Certification

Noise limits for certification of civil helicopters are established by the ICAO (International Civil Aviation Organization) and national regulations (such as FAR Part 36 (2012)). Noise limits have been established for three reference flight profiles: approach, takeoff, and flyover. Figure 14.9 shows the stage 2 limits (the U.S. requirements as of 2012) and the results of certification flight tests as summarized by Marze (1982) and Cox (1993).

The flight tests are conducted at maximum certified weight. Noise in EPNdB (effective perceived noise level) is measured at three microphone locations on the ground, on the centerline of the flight path, and ±150 m to each side. The arithmetic mean of the three microphone measurements is used. At least six measurements are made, such that the mean has less than 1.5 EPNdB error with 90% confidence level. The reference condition is sea level pressure, 25° C temperature, 70% humidity, and zero wind; corrections are permitted for actual atmospheric conditions. The flight path must be flown to specified limits in speed, altitude, and lateral position.

14.7 Noise Certification

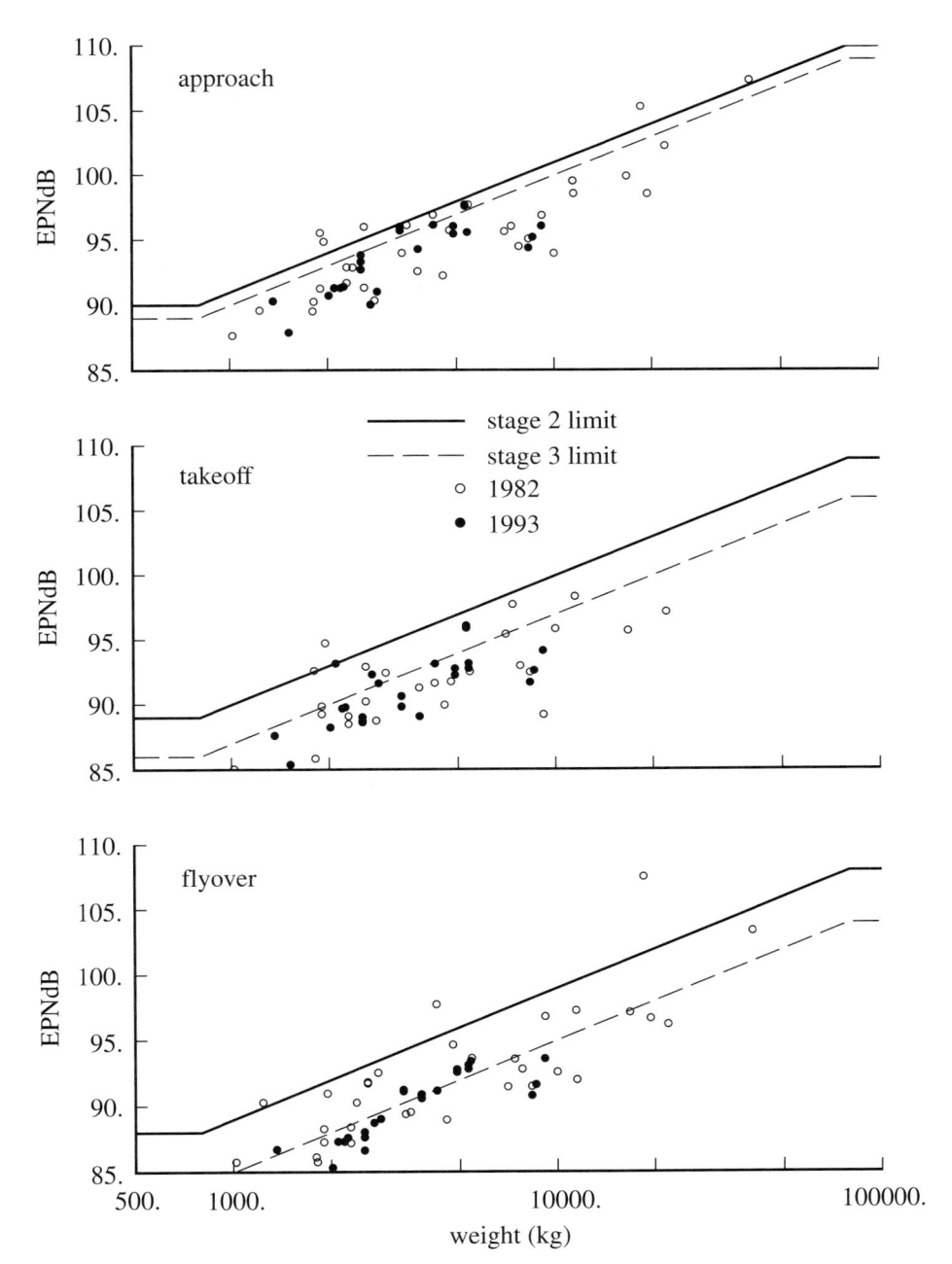

Figure 14.9. Helicopter noise certification limits and measurements.

The approach flight condition is a 6° flight path angle to 120-m altitude at the microphones. The speed is the greater of the best climb speed or the lowest approved approach speed. The principal concern is blade-vortex interaction noise. The takeoff flight condition starts at an altitude of 20 m, 500 m from the microphones. The test is conducted using maximum takeoff power, at speed for the best rate of climb. Tail rotor and engine noise are typically significant. The flyover flight condition is at an altitude of 150 m and at a specified speed based on level flight speed at maximum continuous power and the never-exceed speed. The noise source of concern is high-speed impulsive noise.

As shown in Figure 14.9, the stage 2 noise limits for each profile are a function of aircraft weight, in EPNdB:

weight:	≤ 800 kg	800 to 80000 kg	≥ 80000 kg
approach	90		110
takeoff	89	3.01 EPNdB per double	109
flyover	89		108

Figure 14.9 also shows the stage 3 limits. Negative and positive variations can be traded, subject to no exceedance greater than 3 EPNdB and the sum of the exceedances not more than 4 EPNdB. There is an alternate certification procedure for helicopters with a maximum certified takeoff weight of less than 7000 lb.

14.8 REFERENCES

Baeder, J.D. "Euler Solutions to Nonlinear Acoustics of Non-Lifting Rotor Blades." American Helicopter Society and Royal Aeronautical Society International Technical Specialists Meeting: Rotorcraft Acoustics and Rotor Fluid Dynamics, Valley Forge, PA, October 1991.

Batchelor, G.K. *An Introduction to Fluid Dynamics.* London: Cambridge University Press, 1967.

Boxwell, D.A., and Schmitz, F.H. "Full-Scale Measurements of Blade-Vortex Interaction Noise." Journal of the American Helicopter Society, *27*:4 (October 1982).

Boxwell, D.A., Yu, Y.H., and Schmitz, F.H. "Hovering Impulsive Noise – Some Measured and Calculated Results." Vertica, *3*:1 (1979).

Brentner, K.S. "Prediction of Helicopter Rotor Discrete Frequency Noise." NASA TM 87721, October 1986.

Brentner, K.S. "An Efficient and Robust Method for Predicting Helicopter Rotor High-Speed Impulsive Noise" Journal of Sound and Vibration, *203*:1 (1997a).

Brentner, K.S. "Numerical Algorithms for Acoustic Integrals with Examples for Rotor Noise Prediction." AIAA Journal, *35*:4 (April 1997b).

Brentner, K.S., Burley, C.L., and Marcolini, M.A. "Sensitivity of Acoustic Predictions to Variation of Input Parameters." Journal of the American Helicopter Society, *39*:3 (July 1994).

Brentner, K.S., and Farassat, F. "Helicopter Noise Prediction: The Current Status and Future Direction." Journal of Sound and Vibration, *170*:1 (1994).

Brentner, K.S., and Farassat, F. "Analytical Comparison of the Acoustic Analogy and Kirchhoff Formulation for Moving Surfaces." AIAA Journal, *36*:8 (August 1998).

Brentner, K.S., and Farassat, F. "Modeling Aerodynamically Generated Sound of Helicopter Rotors." Progress in Aerospace Sciences, *39*:2-3 (February 2003).

Burley, C.L., and Brooks, T.F. "Rotor Broadband Noise Prediction with Comparison to Model Data." Journal of the American Helicopter Society, *49*:1 (January 2004).

Burton, D.E., Schlinker, R.H., and Shenoy, R. "The Status of Analytical Helicopter Noise Prediction Methods." NASA CR 172606, August 1985.

Cox, C.R. "Aerodynamic Sources of Rotor Noise." Journal of the American Helicopter Society, *18*:1 (January 1973).

Cox, C.R. "Helicopter Noise Certification Experience and Compliance Cost Reductions." Nineteenth European Rotorcraft Forum, Cernobbio, Italy, September 1993.

Cox, C.R., and Lynn, R.R. "A Study of the Origin and Means of Reducing Helicopter Noise." TCREC TR 62-73, November 1962.

Curle, N. "The Influence of Solid Boundaries upon Aerodynamic Sound." Proceedings of the Royal Society of London, Series A, *231*:1187 (September 1955).

14.8 References

Davidson, I.M., and Hargest, T.J. "Helicopter Noise." The Journal of the Royal Aeronautical Society, *69*:653 (May 1965).

Deming, A.F. "Noise from Propellers with Symmetrical Sections at Zero Blade Angle." NACA TN 605, July 1937.

Deming, A.F. "Noise from Propellers with Symmetrical Sections at Zero Blade Angle, II." NACA TN 679, December 1938.

Deming, A.F. "Propeller Rotation Noise due to Torque and Thrust." NACA TN 747, January 1940.

di Francescantonio, P. "A New Boundary Integral Formulation for the Prediction of Sound Radiation." Journal of Sound and Vibration, *202*:4 (1997).

Farassat, F. "Theory of Noise Generation from Moving Bodies with an Application to Helicopter Rotors." NASA TR R-451, December 1975.

Farassat, F. "Linear Acoustic Formulas for Calculation of Rotating Blade Noise." AIAA Journal, *19*:9 (September 1981).

Farassat, F., and Brentner, K.S. "The Use and Abuses of the Acoustic Analogy in Helicopter Noise Prediction." Journal of the American Helicopter Society, *33*:1 (January 1988).

Farassat, F., and Succi, G.P. "A Review of Propeller Discrete Frequency Noise Prediction Technology with Emphasis on Two Current Methods for Time Domain Calculations." Journal of Sound and Vibration, *71*:3 (1980).

Ffowcs Williams, J.E., and Hawkings, D.L. "Sound Generation by Turbulence and Surfaces in Arbitrary Motion." Philosophical Transactions of the Royal Society of London, *264A*:1151 (May 1969).

Garrick, I.E., and Watkins, C.E. "A Theoretical Study of the Effect of Forward Speed on the Free-Space Sound-Pressure Field Around Propellers." NACA Report 1198, 1954.

Gmelin, B.L., Heller, H., Mercker, E., Philippe, J.-J., Preisser, J.S., and Yu, Y.H. "The HART Programme: A Quadrilateral Cooperative Research Effort." American Helicopter Society 51st Annual Forum, Ft. Worth, TX, May 1995.

Gutin, L. "On the Sound Field of a Rotating Propeller." NACA TM 1195, October 1948.

Hubbard, H.H., and Maglieri, D.J. "Noise Characteristics of Helicopter Rotors at Tip Speeds up to 900 Feet Per Second." Journal of the Acoustical Society of America, *32*:9 (September 1960).

Hubbard, H.H., and Regier, A.A. "Free-Space Oscillating Pressures Near the Tips of Rotating Propellers." NACA Report 996, 1950.

Leverton, J.W., and Pollard, J.S. "A Comparison of the Overall and Broadband Noise Characteristics of Full-Scale and Model Helicopter Rotors." Journal of Sound and Vibration, *30*:2 (1973).

Lighthill, M.J. "On Sound Generated Aerodynamically. I. General Theory." Proceedings of the Royal Society of London, Series A, *211*:1107 (March 1952).

Loewy, R.G., and Sutton, L.R. "A Theory for Predicting the Rotational Noise of Lifting Rotors in Forward Flight, Including a Comparison with Experiment." Journal of Sound and Vibration, *4*:3 (1966).

Lowson, M.V. "The Sound Field for Singularities in Motion." Proceedings of the Royal Society of London, Series A, *286*:1407 (August 1965).

Lowson, M.V., and Ollerhead, J.B. "A Theoretical Study of Helicopter Rotor Noise." Journal of Sound and Vibration, *9*:2 (1969).

Marze, H.J. "Helicopter External Noise: ICAO Standards and Operational Regulations." Eighth European Rotorcraft Forum, Aix-en-Provence, France, September 1982.

Ollerhead, J.B., and Lowson, M.V. "Problems of Helicopter Noise Estimation and Reduction." AIAA Paper No. 69-195, February 1969.

Schlegel, R.G., and Bausch, W.E. "Helicopter Rotor Rotational Noise Prediction and Correlation." USAAVLABS TR 70-1, November 1970.

Schlegel, R., King, R., and Mull, H. "Helicopter Rotor Noise Generation and Propagation." USAAVLABS TR 66-4, October 1966.

Schmitz, F.H., and Yu, Y.H. "Transonic Rotor Noise – Theoretical and Experimental Comparisons." Vertica, *5*:1 (1981).

Stuckey, T.J., and Goddard, J.O. "Investigation and Prediction of Helicopter Rotor Noise. Part I. Wessex Whirl Tower Results." Journal of Sound and Vibration, 5:1 (1967).

United States Government, "Part 36 – Noise Standards, Aircraft Type and Airworthiness Certification." Code of Federal Regulations, Title 14, Aeronautics and Space, 2012.

Watkins, C.E., and Durling, B.J. "A Method for Calculation of Free-Space Sound Pressures Near a Propeller in Flight Including Consideration of the Chordwise Blade Loading." NACA TN 3809, November 1956.

Widnall, S.E. "A Correlation of Vortex Noise Data from Helicopter Main Rotors." Journal of Aircraft, 6:3 (May-June 1969).

Yu, Y.H., Caradonna, F.X., and Schmitz, F.H. "The Influence of the Transonic Flow Field on High-Speed Helicopter Impulsive Noise." Fourth European Rotorcraft and Powered Lift Aircraft Forum, Stresa, Italy, September 1978.

Yudin, E.Y. "On the Vortex Sound from Rotating Rods." NACA TM 1136, March 1947.

15 Mathematics of Rotating Systems

This chapter presents some mathematics that are useful in the analysis of periodic dynamic systems, specifically an N-bladed helicopter rotor rotating at speed Ω. The period for a single blade is $T = 2\pi/\Omega$. In terms of the dimensionless time, measured by the azimuth angle ψ, the period is 2π. For the entire rotor, viewed in the non-rotating frame, the period is $T = 2\pi/N\Omega$. We are interested in the steady-state behavior of a rotating system, which in the rotating frame must be periodic, and thus a Fourier series analysis is appropriate. We are also concerned with the transient behavior of a rotating system, particularly the dynamic stability.

15.1 Fourier Series

A Fourier series is a representation of a periodic function $\beta(\psi)$ as a linear combination of sine and cosine functions with fundamental period 2π:

$$\beta(\psi) = \beta_0 + \beta_{1c}\cos\psi + \beta_{1s}\sin\psi + \beta_{2c}\cos 2\psi + \beta_{2s}\sin 2\psi + \ldots$$

$$= \beta_0 + \sum_{n=1}^{\infty}(\beta_{2n}\cos n\psi + \beta_{ns}\sin n\psi) \tag{15.1}$$

It is assumed that the time scale has been normalized so that the dimensionless period is 2π. The Fourier coefficients or harmonics are constants, which can be evaluated from integrals of $\beta(\psi)$ as follows:

$$\beta_0 = \frac{1}{2\pi}\int_0^{2\pi}\beta\,d\psi \tag{15.2}$$

$$\beta_{nc} = \frac{1}{\pi}\int_0^{2\pi}\beta\cos n\psi\,d\psi \tag{15.3}$$

$$\beta_{ns} = \frac{1}{\pi}\int_0^{2\pi}\beta\sin n\psi\,d\psi \tag{15.4}$$

A more concise representation is given by the complex form of the Fourier series:

$$\beta(\psi) = \sum_{n=-\infty}^{\infty}\beta_n e^{in\psi} \tag{15.5}$$

where

$$\beta_n = \frac{1}{2\pi} \int_0^{2\pi} \beta e^{-in\psi} d\psi \qquad (15.6)$$

Since β is real, β_n and β_{-n} are complex conjugates. The real and complex harmonics are related by

$$\beta_n = \frac{1}{2}(\beta_{nc} - i\beta_{ns}) \qquad (15.7)$$

for $n \geq 1$; β_0 has the same definition in both forms. The complex form can be useful in manipulating the equations of a periodic system, since a single expression defines all the harmonics. To interpret the results, the real form must still be considered.

The Fourier series is a linear transformation between a representation of a periodic motion by a continuous function $\beta(\psi)$ over one period, and a representation by an infinite set of constants $(\beta_0, \beta_{1c}, \beta_{1s}, \ldots)$. The Fourier coefficients represent the motion in the non-rotating frame, as for the flap and lag motion discussed in section 6.1.2. The usefulness of the Fourier series description of steady-state rotor motion is based on the fact that only the lowest few harmonics have significant magnitude, so that the complete periodic motion is described by a small set of numbers.

The Fourier coefficients describing the blade motion are the steady-state solution of the linear differential equation for the motion. An example is equation 6.130 for the flap motion:

$$\ddot{\beta} + \beta = \gamma \left(M_\theta \theta + M_\lambda \lambda + M_{\dot{\beta}} \dot{\beta} + M_\beta \beta \right) \qquad (15.8)$$

In general, the coefficients of the equations of motion (in this case the aerodynamic flap moments M_θ, M_λ, $M_{\dot{\beta}}$, and M_β) are periodic functions of ψ. There are two approaches to solving the equations of motion for the Fourier coefficients: the substitutional method and the operational method. In the substitutional method, the degrees of freedom (and their time derivatives) as well as the equation coefficients are written as Fourier series. Then products of sines and cosines are reduced to sums of sines and cosines by using trigonometric relations. Next, all the coefficients of like harmonics in the equation (that is, the coefficients of 1, $\cos \psi$, $\sin \psi$, $\cos 2\psi$, $\sin 2\psi, \ldots$) are collected. Finally, the collected coefficients of $\sin n\psi$ and $\cos n\psi$ are individually set to zero. The result is an infinite set of linear algebraic equations for the harmonics $(\beta_0, \beta_{1c}, \beta_{1s}, \ldots)$. The Fourier series representation must be truncated to obtain a finite set of algebraic equations, which can then be solved for the required harmonics.

In the operational method, the following operators are applied to the differential equation of motion:

$$\frac{1}{2\pi} \int_0^{2\pi} (\ldots) d\psi \qquad \frac{1}{\pi} \int_0^{2\pi} (\ldots) \cos n\psi \, d\psi \qquad \frac{1}{\pi} \int_0^{2\pi} (\ldots) \sin n\psi \, d\psi \qquad (15.9)$$

The periodic coefficients are again written as Fourier series, and products of harmonics are reduced to sums of harmonics. The operation here is simpler than in the substitutional method since Fourier series have not been introduced for the degrees of freedom. The integral operators only act on the product of the degrees of freedom and a cosine or sine harmonic; hence on terms of the form $\beta \cos k\psi$ or $\beta \sin k\psi$. The definitions of the Fourier coefficients are then used to replace these integrals by the appropriate harmonics of the blade motion. The result is the set of linear algebraic equations, which can be solved for the required harmonics. The substitutional and operational methods produce the same algebraic equations. The operational method has the advantage of obtaining the equations one at a time. The method can be interpreted as resolving into the non-rotating frame the moment equilibrium that produced the equation of motion.

15.2 Sum of Harmonics

To determine the total influence of a rotor with N blades undergoing identical periodic motion, sums of harmonics of the form $\sum_{m=1}^{N} \cos n\psi_m$ or $\sum_{m=1}^{N} \sin n\psi_m$ must be evaluated. Here the azimuth angle of each blade is $\psi_m = \psi + m\Delta\psi$, with ψ the dimensionless time variable (and the azimuth angle of the reference blade), and $\Delta\psi = 2\pi/N$ is the azimuthal spacing between the blades. The summation is over all the blades: $m = 1$ to N. The result for the sum of these harmonics is

$$\frac{1}{N} \sum_{m=1}^{N} \cos n\psi_m = f_n \cos n\psi \qquad (15.10)$$

$$\frac{1}{N} \sum_{m=1}^{N} \sin n\psi_m = f_n \sin n\psi \qquad (15.11)$$

$$\frac{1}{N} \sum_{m=1}^{N} e^{in\psi_m} = f_n e^{in\psi} \qquad (15.12)$$

where $f_n = 1$ only if $n = pN$ for p some integer, and $f_n = 0$ otherwise. Hence the sum is zero unless the harmonic number is a multiple of the number of blades.

To prove this result, consider the sum

$$S = \sum_{m=1}^{N} e^{inm\Delta\psi} = \sum_{m=1}^{N} e^{2\pi i m(n/N)} \qquad (15.13)$$

After factoring $e^{in\psi}$ from $\sum_{m=1}^{N} e^{in\psi_m}$, what must be proved is that $S = Nf_n$. If n/N is an integer, then

$$\left(e^{2\pi i}\right)^{(n/N)m} = (1)^{(n/N)m} = 1 \qquad (15.14)$$

for all m, and so $S = \sum_{m=1}^{N} 1 = N$. For the case of n/N not an integer, multiplying the series S by $e^{2\pi i n/N}$ is equivalent to subtracting the first ($m = 1$) term and adding an $m = N + 1$ term:

$$Se^{2\pi i n/N} = S + e^{2\pi i(n/N)(N+1)} - e^{2\pi i(n/N)} = S + e^{2\pi i n} e^{2\pi i(n/N)} - e^{2\pi i(n/N)} = S \quad (15.15)$$

since $e^{2\pi i n} = 1$. But $e^{2\pi i(n/N)} \neq 1$ if n/N is not an integer, so necessarily $S = 0$. Hence $S = Nf_n$ as required.

In rotor dynamics, sums of the following form are also encountered:

$$\frac{1}{N} \sum_{m=1}^{N} (-1)^m \cos n\psi_m = g_n \cos n\psi \qquad (15.16)$$

$$\frac{1}{N} \sum_{m=1}^{N} (-1)^m \sin n\psi_m = g_n \sin n\psi \qquad (15.17)$$

$$\frac{1}{N}\sum_{m=1}^{N}(-1)^m e^{in\psi_m} = g_n e^{in\psi} \tag{15.18}$$

where $g_n = 1$ only if $n = N/2 + pN$ for p some integer, and $g_n = 0$ otherwise. Thus the sums are zero unless the harmonic number equals an odd multiple of $N/2$, which also requires that the rotor have an even number of blades. The proof of this result is similar to that given earlier; note that $(-1)^m = e^{i(N/2)m\Delta\psi}$.

15.3 Harmonic Analysis

In numerical work, a periodic function $f(\psi)$ is typically evaluated at j equally spaced points around the azimuth: $f_j = f(\psi_j)$, where $\psi_j = j2\pi/J$ for $j = 1$ to J. The function f can be estimated at points between the known values using the Fourier interpolation formula:

$$\hat{f}(\psi) = \sum_{k=-K}^{K} F_k e^{ik\psi} \tag{15.19}$$

for $K \leq (J-1)/2$, where

$$F_k = \frac{1}{J}\sum_{j=1}^{J} f_j e^{-ik\psi_j} \tag{15.20}$$

is a numerical evaluation of the harmonics of a Fourier series representation of $f(\psi)$. If $L < (J-1)/2$, this expression is a least-squared-error representation of f. If $L = (J-1)/2$, it gives $\hat{f}(\psi_j) = f_j$ exactly.

Although if matches the periodic function exactly at the known values, the Fourier interpolation formula can be a poor representation elsewhere. Fourier interpolation can give large excursions because of the higher harmonics and does not estimate derivatives of the function well. For numerical harmonic analysis, using linear interpolation is better:

$$\hat{f}(\psi) = f(\psi_j) + \frac{\psi - \psi_j}{\psi_{j+1} - \psi_j}\left(f(\psi_{j+1}) - f(\psi_j)\right) \tag{15.21}$$

for $\psi_j \leq \psi \leq \psi_{j+1}$. This interpolation is equivalent to

$$\hat{f}(\psi) = \sum_{k=-\infty}^{\infty} F_k e^{ik\psi} \tag{15.22}$$

with the harmonics

$$F_k = \left(\frac{\sin \pi k/J}{\pi k/J}\right)^2 \frac{1}{J}\sum_{j=1}^{J} f_j e^{-ik\psi_j} \tag{15.23}$$

The factor $(\sin(\pi k/J)/(\pi k/J))^2$ reduces the magnitude of the higher harmonics, but now an infinite number of harmonics are required. By truncating the Fourier series ($k = -K$ to K) the corners of the linear interpolation are rounded off. Usually $K \cong J/3$ is satisfactory.

15.4 Multiblade Coordinates

Typically the rotor equations of motion are derived in the rotating frame, with degrees of freedom describing the motion of each blade separately. An example is the flapping equation as derived in section 6.5. In fact, the rotor responds as a whole to excitation from the non-rotating frame, such as aerodynamic gusts, control inputs, or shaft motion. Working with degrees of freedom that reflect this behavior is desirable. Such a representation of the rotor motion simplifies the analysis and facilitates understanding its behavior. For the steady-state solution, the appropriate representation of the blade motion is a Fourier series, the harmonics of which describe the motion of the rotor as a whole; see section 6.1.2. The equations of motion in the non-rotating frame are simply algebraic equations for the harmonics. Now, instead of the steady-state solution we are concerned with the general dynamic behavior, including the transient response of the rotor. The degrees of freedom that describe the motion of the rotor in the non-rotating frame are called multiblade coordinates (MBC).

The appropriate transformation of the degrees of freedom and the equations of motion to the non-rotating frame is of the Fourier type. There are many similarities between this coordinate change and Fourier series, Fourier interpolation, and discrete Fourier transforms. The common factor is the periodic nature of the system. Multiblade coordinates were introduced to rotor dynamics analysis by Coleman (1943), who used the transformation for the blade lag motion in ground resonance analysis, and they were fully developed by Hohenemser and Yin (1972).

The use of multiblade coordinates is crucial for problems involving the rotor motion coupled with the fixed frame, such as hub motion, swashplate control, or dynamic inflow. MBC are a physically relevant, non-rotating frame representation of the rotor motion; for example, coning and tip-path-plane tilt for blade flapping. Consequently, introduction of MBC separates the coupling of the rotor and fixed frame into subsets and, most importantly eliminates periodic coefficients (except for two-bladed rotors). MBC also reduce the periodicity of the equations resulting from edgewise flight aerodynamics.

15.4.1 Transformation of the Degrees of Freedom

Consider a rotor with N blades equally spaced around the azimuth at $\psi_m = \psi + m\Delta\psi$, where ψ is the dimensionless time variable ($\psi = \Omega t$ for constant rotational speed) and $\Delta\psi = 2\pi/N$ is the azimuthal spacing between blades. The blade index m ranges from 1 to N. Let $\beta^{(m)}$ be the degree of freedom in the rotating frame for the m-th blade. The multiblade coordinate transformation is a linear transformation of the degrees of freedom from the rotating to the non-rotating frame. The following new degrees of freedom are introduced:

$$\beta_0 = \frac{1}{N} \sum_{m=1}^{N} \beta^{(m)} \qquad (15.24)$$

$$\beta_{nc} = \frac{2}{N} \sum_{m=1}^{N} \beta^{(m)} \cos n\psi_m \qquad (15.25)$$

$$\beta_{ns} = \frac{2}{N} \sum_{m=1}^{N} \beta^{(m)} \sin n\psi_m \qquad (15.26)$$

$$\beta_{N/2} = \frac{1}{N} \sum_{m=1}^{N} \beta^{(m)} (-1)^m \tag{15.27}$$

These degrees of freedom describe the motion of the rotor in the non-rotating frame. As an example, for the rotor flap motion, β_0 is the coning degree of freedom, whereas β_{1c} and β_{1s} are the tip-path-plane tilt degrees of freedom. The remaining degrees of freedom are called reactionless modes, since they involve no net force or moment on the rotor hub. The corresponding inverse transformation is

$$\beta^{(m)} = \beta_0 + \sum_n (\beta_{nc} \cos n\psi_m + \beta_{ns} \sin n\psi_m) + \beta_{N/2} (-1)^m \tag{15.28}$$

which gives the motion of the individual blades again. The summation over the harmonic index n goes from $n = 1$ to $(N-1)/2$ for N odd, and from $n = 1$ to $(N-2)/2$ for N even. The degree of freedom $\beta_{N/2}$ appears in the transformation only if N is even.

The variables β_0, β_{nc}, β_{ns}, and $\beta_{N/2}$ are degrees of freedom and hence are functions of time, just as the variables $\beta^{(m)}$ are. These degrees of freedom describe the motion of the rotor as a whole in the non-rotating frame, whereas $\beta^{(m)}$ describes the motion of an individual blade in the rotating frame. Thus we have a linear, reversible transformation between the N degrees of freedom $\beta^{(m)}$ in the rotating frame ($m = 1 \ldots N$) and the N degrees of freedom $\beta_0, \beta_{nc}, \beta_{ns}, \beta_{N/2}$ in the non-rotating frame. Compare this coordinate transformation with a Fourier series representation of the steady-state solution. In the latter case, where $\beta^{(m)}$ is a periodic function of ψ_m, the motions of all the blades are identical. So the motion in the rotating frame can be represented by a Fourier series, the coefficients of which are constant in time but infinite in number. Thus there are similarities between multiblade coordinates and the Fourier series, but they are by no means identical.

The collective and cyclic modes (β_0, β_{1c}, and β_{1s}, where in general β can be any degree of freedom of the blade) are of particular importance because of their fundamental role in the coupled motion of the rotor and the non-rotating system. For axial flow only, the collective and cyclic modes of the rotor degrees of freedom couple with the fixed system, whereas the reactionless modes ($\beta_{2c}, \beta_{2s}, \ldots, \beta_{nc}, \beta_{ns}, \beta_{N/2}$) correspond to purely internal rotor motion. Nonaxial flow to some extent couples all the rotor degrees of freedom and the fixed system variables, but still the collective and cyclic motions dominate the rotor dynamic behavior.

The multiblade coordinate transform can be applied to a configuration with blades that are not equally spaced or not identical. If the blade-to-blade differences are small, the benefits of the transform can still be expected.

Now consider the transformation of time derivatives of the motion. The derivatives of equation 15.28 are

$$\dot{\beta}^{(m)} = \dot{\beta}_0 + \sum_n \left[(\dot{\beta}_{nc} + n\Omega \beta_{ns}) \cos n\psi_m + (\dot{\beta}_{ns} - n\Omega \beta_{nc}) \sin n\psi_m \right] + \dot{\beta}_{N/2} (-1)^m$$

$$\tag{15.29}$$

$$\ddot{\beta}^{(m)} = \ddot{\beta}_0 + \sum_n \left[(\ddot{\beta}_{nc} + 2n\Omega \dot{\beta}_{ns} + n\dot{\Omega} \beta_{ns} - n^2 \Omega^2 \beta_{nc}) \cos n\psi_m \right.$$

$$\left. + (\ddot{\beta}_{ns} - 2n\Omega \dot{\beta}_{nc} - n\dot{\Omega} \beta_{nc} - n^2 \Omega^2 \beta_{ns}) \sin n\psi_m \right] + \ddot{\beta}_{N/2} (-1)^m \tag{15.30}$$

15.4 Multiblade Coordinates

where $\Omega = \dot{\psi}$. The Ω's are omitted for dimensionless equations, and usually the trim rotor speed is constant (or its perturbations are represented by a separate degree of freedom), so $\dot{\Omega} = 0$. Then the harmonics of the time derivatives are

$$\frac{1}{N} \sum_{m=1}^{N} \dot{\beta}^{(m)} = \dot{\beta}_0 \tag{15.31}$$

$$\frac{2}{N} \sum_{m=1}^{N} \dot{\beta}^{(m)} \cos n\psi_m = \dot{\beta}_{nc} + n\beta_{ns} \tag{15.32}$$

$$\frac{2}{N} \sum_{m=1}^{N} \dot{\beta}^{(m)} \sin n\psi_m = \dot{\beta}_{ns} - n\beta_{nc} \tag{15.33}$$

$$\frac{1}{N} \sum_{m=1}^{N} \dot{\beta}^{(m)} (-1)^m = \dot{\beta}_{N/2} \tag{15.34}$$

and

$$\frac{1}{N} \sum_{m=1}^{N} \ddot{\beta}^{(m)} = \ddot{\beta}_0 \tag{15.35}$$

$$\frac{2}{N} \sum_{m=1}^{N} \ddot{\beta}^{(m)} \cos n\psi_m = \ddot{\beta}_{nc} + 2n\dot{\beta}_{ns} - n^2\beta_{nc} \tag{15.36}$$

$$\frac{2}{N} \sum_{m=1}^{N} \ddot{\beta}^{(m)} \sin n\psi_m = \ddot{\beta}_{ns} - 2n\dot{\beta}_{nc} - n^2\beta_{ns} \tag{15.37}$$

$$\frac{1}{N} \sum_{m=1}^{N} \ddot{\beta}^{(m)} (-1)^m = \ddot{\beta}_{N/2} \tag{15.38}$$

The transformation of the velocity and acceleration from the rotating frame introduces Coriolis and centrifugal terms in the non-rotating frame.

The steady-state blade motion is periodic and identical for all blades; hence it can be represented by a Fourier series. Using a complex representation of the harmonics, $\beta^{(m)} = \sum_n \beta_n e^{in\psi_m}$, gives

$$\beta_0 = \frac{1}{N} \sum_m \sum_n \beta_n e^{in\psi_m} = \sum_n \beta_n \frac{1}{N} \sum_m e^{in\psi_m} = \sum_n \beta_n e^{in\psi} \quad \text{for } n = pN \tag{15.39}$$

$$\beta_{kc} = \frac{1}{N} \sum_m \sum_n \beta_n 2 e^{in\psi_m} \cos k\psi_m = \sum_n \beta_n \frac{1}{N} \sum_m \left(e^{i(n+k)\psi_m} + e^{i(n-k)\psi_m} \right)$$
$$= \sum_n \beta_n \left(e^{i(n+k)\psi} + e^{i(n-k)\psi} \right) \quad \text{for } n \pm k = pN \tag{15.40}$$

$$\beta_{ks} = \frac{1}{N} \sum_m \sum_n \beta_n 2 e^{in\psi_m} \sin k\psi_m = \sum_n \beta_n \frac{1}{N} \sum_m (-i) \left(e^{i(n+k)\psi_m} - e^{i(n-k)\psi_m} \right)$$
$$= \sum_n \beta_n (-i) \left(e^{i(n+k)\psi} - e^{i(n-k)\psi} \right) \quad \text{for } n \pm k = pN \tag{15.41}$$

$$\beta_{N/2} = \frac{1}{N} \sum_m \sum_n \beta_n e^{in\psi_m}(-1)^m = \sum_n \beta_n \frac{1}{N} \sum_m e^{in\psi_m} e^{i(N/2)m\Delta\psi}$$
$$= \sum_n \beta_n e^{in\psi} \qquad \text{for } n = N/2 + pN \tag{15.42}$$

So the Fourier series for the multiblade coordinates have only harmonics at multiples of the number of blades (except for $\beta_{N/2}$).

15.4.2 Matrix Form

Let $\beta_{\text{rot}} = \left(\beta^{(m)}\right)$ be the vector of rotating frame variables (length N), and $\beta_{\text{non}} = \left(\beta_0 \ \beta_{nc} \ \beta_{ns} \ \beta_{N/2}\right)^T$ be the vector of multiblade coordinates (also length N). The multiblade coordinate transformation is

$$\beta_{\text{rot}} = T\beta_{\text{non}} \tag{15.43}$$
$$\dot{\beta}_{\text{rot}} = T\dot{\beta}_{\text{non}} + \dot{T}\beta_{\text{non}} \tag{15.44}$$
$$\ddot{\beta}_{\text{rot}} = T\ddot{\beta}_{\text{non}} + 2\dot{T}\dot{\beta}_{\text{non}} + \ddot{T}\beta_{\text{non}} \tag{15.45}$$

where the m-th row of T is

$$t_m = \begin{pmatrix} 1 & \cos k\psi_m & \sin k\psi_m & (-1)^m \end{pmatrix} \tag{15.46}$$

This is a linear, time-varying transformation of the variables. Note the following properties of the transformation:

$$DT^T T = I \tag{15.47}$$
$$DT^T \dot{T} = E_1 \tag{15.48}$$
$$DT^T \ddot{T} = E_2 = E_1^2 \tag{15.49}$$

or

$$T^{-1} = DT^T = S^T \tag{15.50}$$
$$\dot{T} = TE_1 \tag{15.51}$$
$$\ddot{T} = TE_2 \tag{15.52}$$

where

$$D = \begin{bmatrix} \frac{1}{N} & & & \\ & \frac{2}{N} & & \\ & & \frac{2}{N} & \\ & & & \frac{1}{N} \end{bmatrix} \tag{15.53}$$

$$E_1 = \begin{bmatrix} 0 & & & \\ & 0 & k\Omega & \\ & -k\Omega & 0 & \\ & & & 0 \end{bmatrix} \tag{15.54}$$

15.4 Multiblade Coordinates

$$E_2 = \begin{bmatrix} 0 & & \\ & -k^2\Omega^2 & \\ & & -k^2\Omega^2 \\ & & & 0 \end{bmatrix} \tag{15.55}$$

Thus the transformation from rotating to non-rotating variables is

$$\beta_{\text{non}} = T^{-1}\beta_{\text{rot}} = DT^T\beta_{\text{rot}} \tag{15.56}$$

$$\dot{\beta}_{\text{non}} = DT^T\dot{\beta}_{\text{rot}} - E_1 DT^T \beta_{\text{rot}} \tag{15.57}$$

$$\ddot{\beta}_{\text{non}} = DT^T\ddot{\beta}_{\text{rot}} - 2E_1 DT^T\dot{\beta}_{\text{rot}} + E_2 DT^T \beta_{\text{rot}} \tag{15.58}$$

It is also necessary to transform the motion or output equations, which is conventionally accomplished by multiplying the equations by S^T.

The MBC transformation matrix T can be factored into a multiblade part T_0 and a rotating to non-rotating frame part R:

$$T = T_0 R = \begin{bmatrix} & \vdots & & \\ 1 & \cos km\Delta\psi & \sin k\Delta\psi & (-1)^m \\ & \vdots & & \end{bmatrix} \begin{bmatrix} 1 & & & \\ & \cos k\psi & \sin k\psi & \\ & -\sin k\psi & \cos k\psi & \\ & & & 1 \end{bmatrix} \tag{15.59}$$

So $\beta_{\text{rot}} = T_0 R \beta_{\text{non}}$ and $\beta_{\text{non}} = R^T D T_0^T \beta_{\text{rot}}$.

That the transformation is reversible, hence that the rotating and non-rotating degrees of freedom describe the same motion, follows from $T^{-1} = DT^T$, or $T^T T = D^{-1}$. Using a complex representation of the multiblade coordinates, $\beta_n = \frac{1}{N}\sum_m \beta^{(m)} e^{-in\psi_m}$, gives

$$T^T T = \begin{bmatrix} & 1 & & \\ \cdots & e^{-i\ell\psi_m} & \cdots \\ & (-1)^m & & \end{bmatrix} \begin{bmatrix} & \vdots & & \\ 1 & e^{ik\psi_m} & (-1)^m \\ & \vdots & & \end{bmatrix}$$

$$= \sum_m \begin{bmatrix} 1 & e^{ik\psi_m} & (-1)^m \\ e^{-i\ell\psi_m} & e^{i(k-\ell)\psi_m} & (-1)^m e^{-i\ell\psi_m} \\ (-1)^m & (-1)^m e^{ik\psi_m} & 1 \end{bmatrix}$$

$$= \begin{bmatrix} N & 0 & 0 \\ 0 & N/2 & 0 \\ 0 & 0 & N \end{bmatrix} \tag{15.60}$$

using the results of section 15.2 to sum the harmonics.

15.4.3 Conversion of the Equations of Motion

The multiblade coordinate transformation must be accompanied by a conversion of the differential equations of motion from the rotating to the non-rotating frame. This conversion is accomplished by operating on the rotating-frame equation of motion with the following summation operators:

$$\frac{1}{N}\sum_{m=1}^{N}(\ldots) \quad \frac{2}{N}\sum_{m=1}^{N}(\ldots)\cos n\psi_m \quad \frac{2}{N}\sum_{m=1}^{N}(\ldots)\sin n\psi_m \quad \frac{1}{N}\sum_{m=1}^{N}(\ldots)(-1)^m \tag{15.61}$$

The result is N differential equations in the non-rotating frame, obtained by summing the rotating equation over all N blades. These same operators are involved in transforming the degrees of freedom. The conversion of the equations is not complete, however, until the summation operator is eliminated by using it to transform the degrees of freedom to multiblade coordinates.

A procedure analogous to the substitutional method for Fourier series consists of the following steps. The periodic coefficients in the rotating-frame equation of motion are written as Fourier series, and the multiblade coordinates are introduced for the degrees of freedom and their time derivatives. Then products of harmonics are written as sums of harmonics using trigonometric relations. Next, all coefficients of 1, $\cos\psi_m$, $\sin\psi_m$, ..., $\cos n\psi_m$, $\sin n\psi_m$, $(-1)^m$ are collected and individually set to zero, producing the required differential equations. A difficulty with this approach arises because, unlike the Fourier series case, only N equations are to be obtained. Thus any harmonics $\cos k\psi_m$ and $\sin k\psi_m$ with $k > N/2$ must be rewritten as products of harmonics in the proper range ($k < N/2$) and harmonics of N/rev. For example, consider a second harmonic appearing in the equations for a three-bladed rotor. By writing

$$\cos 2\psi_m = \cos 3\psi_m \cos \psi_m + \sin 3\psi_m \sin \psi_m \qquad (15.62)$$

$$\sin 2\psi_m = \sin 3\psi_m \cos \psi_m - \cos 3\psi_m \sin \psi_m \qquad (15.63)$$

the second harmonics contribute 3/rev terms to the $\cos \psi_m$ and $\sin \psi_m$ equations. A better approach is to apply the summation operators given earlier instead of trying to collect coefficients of like harmonics. Then, since the summation over all blades acts only on the harmonics, only terms of the form

$$\sum_{m=1}^{N} \cos k\psi_m \quad \sum_{m=1}^{N} \sin k\psi_m \quad \sum_{m=1}^{N} (-1)^m \cos k\psi_m \quad \sum_{m=1}^{N} (-1)^m \sin k\psi_m \qquad (15.64)$$

must be evaluated to complete the equations. These sums can be evaluated using the results of section 15.2. Recall that the first two sums give harmonics of N/rev if k is a multiple of N, and the sums involving $(-1)^m$ give harmonics of $(N/2)$/rev if k is an odd multiple of $N/2$.

An operational method, which requires less manipulation of the harmonics, proceeds as follows. Again the periodic coefficients of the rotating equations are written as Fourier series, and the summation operators are applied to the equations. Products of harmonics are reduced to sums of harmonics. Since the rotating degrees of freedom are still present, terms of the form

$$\frac{2}{N} \sum_{m=1}^{N} \beta^{(m)} \cos k\psi_m \qquad \frac{2}{N} \sum_{m=1}^{N} \beta^{(m)} \sin k\psi_m \qquad (15.65)$$

$$\frac{2}{N} \sum_{m=1}^{N} \beta^{(m)} (-1)^m \cos k\psi_m \qquad \frac{2}{N} \sum_{m=1}^{N} \beta^{(m)} (-1)^m \sin k\psi_m \qquad (15.66)$$

must be evaluated. If $k < N/2$, the first two sums are simply the definitions of the non-rotating degrees of freedom β_{kc} and β_{ks}. For the general case, write $k = n + pN$, where p is an integer and n is the principal value of the harmonic, such that $n < N/2$. Then if the complex form is used and the definition of the non-rotating degrees of

15.4 Multiblade Coordinates

freedom (for n, which now has the proper range) is applied,

$$\frac{1}{N}\sum_{m=1}^{N}\beta^{(m)}e^{-ik\psi_m} = \frac{1}{N}\sum_{m=1}^{N}\beta^{(m)}e^{-in\psi_m}e^{-ipN\psi_m}$$

$$= e^{-ipN\psi}\frac{1}{N}\sum_{m=1}^{N}\beta^{(m)}e^{-in\psi_m}$$

$$= e^{-ipN\psi}\beta_n \tag{15.67}$$

since $e^{-ipNm\Delta\psi} = e^{2\pi ipm} = 1$. If N is even, the case $k = n + pN$ with $n = N/2$ must also be considered, for which

$$\frac{1}{N}\sum_{m=1}^{N}\beta^{(m)}e^{-ik\psi_m} = e^{-i(p+1/2)N\psi}\beta_{N/2} \tag{15.68}$$

The real form is as follows. Writing $k = n + pN$ where $n < N/2$, then

$$\frac{2}{N}\sum_{m=1}^{N}\beta^{(m)}\cos k\psi_m = \beta_{nc}\cos pN\psi - \beta_{ns}\sin pN\psi \tag{15.69}$$

$$\frac{2}{N}\sum_{m=1}^{N}\beta^{(m)}\sin k\psi_m = \beta_{nc}\sin pN\psi + \beta_{ns}\cos pN\psi \tag{15.70}$$

or if $n = N/2$

$$\frac{1}{N}\sum_{m=1}^{N}\beta^{(m)}\cos k\psi_m = \beta_{N/2}\cos(p+1/2)N\psi \tag{15.71}$$

$$\frac{1}{N}\sum_{m=1}^{N}\beta^{(m)}\sin k\psi_m = \beta_{N/2}\sin(p+1/2)N\psi \tag{15.72}$$

Similarly, for the summations involving $(-1)^m$, write $k = n + (p - 1/2)N$, where $n < N/2$. Then

$$\frac{2}{N}\sum_{m=1}^{N}\beta^{(m)}(-1)^m\cos k\psi_m = \beta_{nc}\cos(p-1/2)N\psi - \beta_{ns}\sin(p-1/2)N\psi \tag{15.73}$$

$$\frac{2}{N}\sum_{m=1}^{N}\beta^{(m)}(-1)^m\sin k\psi_m = \beta_{nc}\sin(p-1/2)N\psi + \beta_{ns}\cos(p-1/2)N\psi \tag{15.74}$$

or if $n = N/2$

$$\frac{1}{N}\sum_{m=1}^{N}\beta^{(m)}(-1)^m\cos k\psi_m = \beta_{N/2}\cos pN\psi \tag{15.75}$$

$$\frac{1}{N}\sum_{m=1}^{N}\beta^{(m)}(-1)^m\sin k\psi_m = \beta_{N/2}\sin pN\psi \tag{15.76}$$

With these results the construction of the differential equations in the nonrotating frame is straightforward.

Two assumptions have been made in outlining these procedures for the conversion of the equations to the non-rotating frame: first, that the number of degrees of freedom involved is small enough for an analytical construction to be practical; and second, that analytical expressions are available for the periodic coefficients as Fourier series. For comprehensive dynamics analyses neither assumption is valid, and a procedure better suited to numerical work is required. The rotating degrees of freedom are written in terms of the multiblade coordinates:

$$\beta^{(m)} = \beta_0 + \sum_n (\beta_{nc} \cos n\psi_m + \beta_{ns} \sin n\psi_m) + \beta_{N/2}(-1)^m \tag{15.77}$$

with similar expressions for the time derivatives, and the summation operators are applied to the rotating equation. Then it is necessary to evaluate summations over all N blades that involve the periodic coefficient multiplied by two factors: a factor of 1, $\cos n\psi_m$, $\sin n\psi_m$, or $(-1)^m$ from the multiblade coordinates, and a factor of 1, $\cos k\psi_m$, $\sin k\psi_m$, or $(-1)^m$ from the summation operators. The construction of the non-rotating equations in this manner is simple, and the evaluation is easily performed numerically. The value of an analytical approach is that with simple periodic coefficients many of these summations are exactly zero, which simplifies the non-rotating equations of motion.

With a constant coefficient differential equation, the conversion to the non-rotating frame is elementary. The summation operators then act only on the degrees of freedom, not on the equation coefficients, and the definitions of the non-rotating degrees of freedom (and their derivatives) can be applied directly. Consider for example a mass-spring-damper system of the form

$$\ddot{\beta}^{(m)} + \frac{\gamma}{8}\dot{\beta}^{(m)} + \nu^2 \beta^{(m)} = \frac{\gamma}{8}\theta^{(m)} \tag{15.78}$$

(the flapping equation in hover). The resulting non-rotating equations are

$$\ddot{\beta}_0 + \frac{\gamma}{8}\dot{\beta}_0 + \nu^2 \beta_0 = \frac{\gamma}{8}\theta_0 \tag{15.79}$$

for β_0,

$$\begin{pmatrix} \ddot{\beta}_{nc} \\ \ddot{\beta}_{ns} \end{pmatrix} + \begin{bmatrix} \frac{\gamma}{8} & 2n \\ -2n & \frac{\gamma}{8} \end{bmatrix} \begin{pmatrix} \dot{\beta}_{nc} \\ \dot{\beta}_{ns} \end{pmatrix} + \begin{bmatrix} \nu^2 - n^2 & n\frac{\gamma}{8} \\ -n\frac{\gamma}{8} & \nu^2 - n^2 \end{bmatrix} \begin{pmatrix} \beta_{nc} \\ \beta_{ns} \end{pmatrix} = \frac{\gamma}{8} \begin{pmatrix} \theta_{nc} \\ \theta_{ns} \end{pmatrix} \tag{15.80}$$

for β_{nc} and β_{ns}, and

$$\ddot{\beta}_{N/2} + \frac{\gamma}{8}\dot{\beta}_{N/2} + \nu^2 \beta_{N/2} = \frac{\gamma}{8}\theta_{N/2} \tag{15.81}$$

for $\beta_{N/2}$. These equations show the basic manner in which inertia, damping, and spring terms transform to the non-rotating frame. The conversion introduces Coriolis and centrifugal terms to the β_{nc} and β_{ns} equations. The only coupling of the non-rotating degrees of freedom occurs in the β_{nc} and β_{ns} equations. The number of blades influences only the number of degrees of freedom and equations that must be analyzed.

The equations and the procedure by which they are obtained are much more complicated with periodic coefficients. Consider the differential equation for the flapping motion in forward flight:

$$\ddot{\beta}^{(m)} + \left(\frac{\gamma}{8} + \frac{\gamma}{6}\mu \sin \psi_m\right) \dot{\beta}^{(m)} + \left(v^2 + \frac{\gamma}{6}\mu \cos \psi_m + \frac{\gamma}{8}\mu^2 \sin 2\psi_m\right) \beta^{(m)}$$
$$= \left(\frac{\gamma}{8}(1+\mu^2) + \frac{\gamma}{3}\mu \sin \psi_m - \frac{\gamma}{8}\mu^2 \cos 2\psi_m\right) \theta^{(m)} - \left(\frac{\gamma}{6} + \frac{\gamma}{4}\mu \sin \psi_m\right) \lambda \quad (15.82)$$

(see section 6.5). The inertia and centrifugal-structural spring terms $\left(\ddot{\beta}^{(m)} + v^2\beta^{(m)}\right)$ transform as hover. The transformation of the aerodynamic terms to the non-rotating frame is given in section 16.8.4 for the cases of two, three, and four blades. As the number of blades increases, the periodic coefficients are cleared from the lower degrees of freedom and equations. There are always periodic coefficients in the complete set of equations, though, regardless of the number of blades. The higher harmonics of the coefficients in the rotating frame contribute to the mean values of the coefficients in the non-rotating frame (see section 16.8.4). Only multiples of N/rev harmonics appear in the equations for the N-bladed rotor. This follows from the results for $\sum_m \beta^{(m)} \cos k\psi_m$ and $\sum_m \beta^{(m)} \sin k\psi_m$ given earlier. Hence, although the rotating equations have period 2π, the equations in the non-rotating frame have a period $T = 2\pi/N$, as expected with identical blades. The exception is that $(N/2)$/rev harmonics (in general, odd multiples of $(N/2)$/rev) appear in the matrix elements coupling the $\beta_{N/2}$ mode with the other degrees of freedom. So for a rotor with an even number of blades, when the $\beta_{N/2}$ degree of freedom is included in the analysis, the period is $T = 4\pi/N$. The period is twice the expected result because the blades are no longer identical: the $\beta_{N/2}$ mode identifies alternate blades with a plus or minus amplitude by the $(-1)^m$ factor. Thus the period of $4\pi/N$ follows from the mathematical description of the rotor motion; the solution must still correspond to a physical system with period $2\pi/N$.

The equations of motion for the rotor in edgewise forward flight must always have periodic coefficients, whether they are written in the rotating or non-rotating frame. The solutions of such equations have distinctive behavior and are also more difficult to obtain than the solutions of constant coefficient equations (see section 15.6). When the equations are only weakly periodic, there might be some constant coefficient system that closely represents the behavior of the true system. An example is the periodic coefficients arising from the aerodynamics of forward flight, which have higher harmonics only of order μ and smaller. It is necessary to establish the best means to construct such a constant coefficient approximation and to determine its range of validity. The constant coefficient system can be constructed by retaining only the mean values of the original periodic coefficients. Clearly a better approximation results if the coefficients are averaged in the non-rotating frame, because the higher harmonics in the rotating equation contribute to the mean values of the coefficients in the non-rotating equations. By working in the non-rotating frame, more equations must be solved. The constant coefficient approximation is an important tool in the analysis of rotor dynamics.

15.4.4 Reactionless Mode and Two-Bladed Rotors

The reactionless mode $\beta_{N/2} = \frac{1}{N}\sum_m \beta^{(m)}(-1)^m$, present for rotors with an even number of blades, introduces special behavior and some difficulties. In this mode all the blades have identical motion, but the displacement alternates the sign from one blade to the next around the azimuth. Identification of the blades by ± 1 is artificial and leads to $\frac{N}{2}\Omega$ periodic terms in the non-rotating frame equations, when only $N\Omega$ periodicity is expected for identical blades. For a two-bladed rotor the non-rotating degrees of freedom are the coning and teetering modes:

$$\beta_0 = \frac{1}{2}\left(\beta^{(2)} + \beta^{(1)}\right) \tag{15.83}$$

$$\beta_1 = \frac{1}{2}\left(\beta^{(2)} - \beta^{(1)}\right) \tag{15.84}$$

In this case β_1 replaces the cyclic modes β_{1c}, and β_{1s} and therefore couples with the fixed system. Because of the absence of the cyclic modes, two-bladed rotor dynamics are fundamentally different from the dynamics of rotors with three or more blades. Specifically, there is no tip-path-plane representation of the flap motion, and coupling with the fixed frame exhibits 1/rev periodicity.

Defining degrees of freedom $\beta_{(N/2)c}$ and $\beta_{(N/2)s}$ as for β_{nc} and β_{ns}, would give

$$\beta_{(N/2)c} = \frac{2}{N}\sum_{m=1}^{N} \beta^{(m)} \cos\frac{N}{2}\psi_m = \frac{2}{N}\sum_{m=1}^{N} \beta^{(m)}(-1)^m \cos\frac{N}{2}\psi = 2\beta_{N/2}\cos\frac{N}{2}\psi \tag{15.85}$$

$$\beta_{(N/2)s} = \frac{2}{N}\sum_{m=1}^{N} \beta^{(m)} \sin\frac{N}{2}\psi_m = \frac{2}{N}\sum_{m=1}^{N} \beta^{(m)}(-1)^m \sin\frac{N}{2}\psi = 2\beta_{N/2}\sin\frac{N}{2}\psi \tag{15.86}$$

So $\beta_{(N/2)c}$ and $\beta_{(N/2)s}$ defined in this manner are not independent variables and are not useful. Consider a more general definition:

$$\beta_{(N/2)c} = \beta_{N/2}C - \alpha_{N/2}S \tag{15.87}$$

$$\beta_{(N/2)s} = \beta_{N/2}S + \alpha_{N/2}C \tag{15.88}$$

where $C = \cos\frac{N}{2}\psi$ and $S = \sin\frac{N}{2}\psi$, which inverts to

$$\beta_{N/2} = \beta_{(N/2)c}C + \beta_{(N/2)s}S \tag{15.89}$$

$$\alpha_{N/2} = -\beta_{(N/2)c}S + \beta_{(N/2)s}C \tag{15.90}$$

and

$$\dot{\beta}_{N/2} = \dot{\beta}_{(N/2)c}C + \dot{\beta}_{(N/2)s}S + \frac{N}{2}\Omega\alpha_{N/2} \tag{15.91}$$

$$\dot{\alpha}_{N/2} = -\dot{\beta}_{(N/2)c}S + \dot{\beta}_{(N/2)s}C - \frac{N}{2}\Omega\beta_{N/2} \tag{15.92}$$

The new degree of freedom $\alpha_{N/2}$ is being added to the system (total $N+1$ degrees of freedom now), which requires a new equation as well. Following Hoffman (1976), the variable $\frac{N}{2}\Omega\alpha_{N/2} = \dot{\beta}_{N/2}$ is used, so the states $\beta_{N/2}$ and $\dot{\beta}_{N/2}$ are replaced by $\beta_{(N/2)c}$ and $\beta_{(N/2)s}$. Equation 15.91 gives the constraint equation

$$\dot{\beta}_{(N/2)c}C + \dot{\beta}_{(N/2)s}S = 0 \tag{15.93}$$

There follows

$$\dot{\beta}_{N/2} = \frac{N}{2}\Omega(-\beta_{(N/2)c}S + \beta_{(N/2)s}C) \tag{15.94}$$

$$\ddot{\beta}_{N/2} = \frac{N}{2}\Omega(-\dot{\beta}_{(N/2)c}S + \dot{\beta}_{(N/2)s}C) - \left(\frac{N}{2}\Omega\right)^2 \beta_{N/2} \quad (15.95)$$

Peters (1994) described other possible definitions of $\alpha_{N/2}$.

The equation of motion for $\beta_{N/2}$ is obtained from $E_{N/2} = \frac{1}{N}\sum_m E^{(m)}(-1)^m$. Together with a constraint equation E_C, two equations of motion for $\beta_{(N/2)c}$ and $\beta_{(N/2)s}$ are obtained from

$$\begin{pmatrix} E_{(N/2)c} \\ E_{(N/2)s} \end{pmatrix} = \begin{bmatrix} 2C & 2S \\ 2S & -2C \end{bmatrix} \begin{pmatrix} E_{N/2} \\ E_C \end{pmatrix} \quad (15.96)$$

For a second-order equation of motion ($E_{N/2} = M\ddot{\beta}_{N/2} + \ldots$), using equation 15.95 for the acceleration and equation 15.93 for the constraint, this becomes

$$\begin{pmatrix} E_{(N/2)c} \\ E_{(N/2)s} \end{pmatrix} = \begin{bmatrix} 2C & 2S \\ 2S & -2C \end{bmatrix} \left\{ M\frac{N}{2}\Omega \begin{bmatrix} -S & C \\ C & S \end{bmatrix} \begin{pmatrix} \dot{\beta}_{(N/2)c} \\ \dot{\beta}_{(N/2)s} \end{pmatrix} \right.$$

$$\left. -M\left(\frac{N}{2}\Omega\right)^2 \begin{bmatrix} C & S \\ 0 & 0 \end{bmatrix} \begin{pmatrix} \beta_{(N/2)c} \\ \beta_{(N/2)s} \end{pmatrix} \right\} + \ldots$$

$$= M\frac{N}{2}\Omega \begin{bmatrix} 0 & 2 \\ -2 & 0 \end{bmatrix} \begin{pmatrix} \dot{\beta}_{(N/2)c} \\ \dot{\beta}_{(N/2)s} \end{pmatrix}$$

$$- M\left(\frac{N}{2}\Omega\right)^2 \begin{bmatrix} 2C^2 & 2CS \\ 2CS & 2S^2 \end{bmatrix} \begin{pmatrix} \beta_{(N/2)c} \\ \beta_{(N/2)s} \end{pmatrix} + \ldots \quad (15.97)$$

So the second-order equation for $\beta_{N/2}$ plus the constraint equation becomes two first-order equations for $\beta_{(N/2)c}$ and $\beta_{(N/2)s}$.

This approach gives a representation of the tip-path-plane motion of a two-bladed rotor:

$$\beta_1 = \beta_{1c}C + \beta_{1s}S \quad (15.98)$$

$$\dot{\beta}_1 = -\beta_{1c}\Omega S + \beta_{1s}\Omega C \quad (15.99)$$

where $C = \cos\psi$ and $S = \sin\psi$; with the constraint equation

$$\dot{\beta}_{1c}C + \dot{\beta}_{1s}S = 0 \quad (15.100)$$

These equations do not imply any approximation. They provide an adequate representation of the teeter motion β_1 for slow variation of β_{1c} and β_{1s}. The inverse is

$$\beta_{1c} = \beta_1 C - \frac{1}{\Omega}\dot{\beta}_1 S \quad (15.101)$$

$$\beta_{1s} = \beta_1 S + \frac{1}{\Omega}\dot{\beta}_1 C \quad (15.102)$$

The states β_1 and $\dot{\beta}_1$ are replaced by β_{1c} and β_{1s}. The acceleration is

$$\ddot{\beta}_1 = -\dot{\beta}_{1c}\Omega S + \dot{\beta}_{1s}\Omega C - \Omega^2 \beta_1 \quad (15.103)$$

The equation of motion for the teeter flap degree of freedom in hover,

$$\ddot{\beta}_1 + \frac{\gamma}{8}\dot{\beta}_1 + \nu^2 \beta_1 = \frac{\gamma}{8}(\theta_{1c}\cos\psi + \theta_{1s}\sin\psi) \quad (15.104)$$

becomes

$$\begin{bmatrix} 0 & 2 \\ -2 & 0 \end{bmatrix} \begin{pmatrix} \dot{\beta}_{1c} \\ \dot{\beta}_{1s} \end{pmatrix} + \begin{bmatrix} -\frac{\gamma}{8}2CS + (\nu^2 - 1)2C^2 & \frac{\gamma}{8}2C^2 + (\nu^2 - 1)2CS \\ -\frac{\gamma}{8}2S^2 + (\nu^2 - 1)2CS & \frac{\gamma}{8}2CS + (\nu^2 - 1)2S^2 \end{bmatrix} \begin{pmatrix} \beta_{1c} \\ \beta_{1s} \end{pmatrix}$$

$$= \frac{\gamma}{8} \begin{bmatrix} 2C^2 & 2CS \\ 2CS & 2S^2 \end{bmatrix} \begin{pmatrix} \theta_{1c} \\ \theta_{1s} \end{pmatrix} \quad (15.105)$$

which are first-order equations with periodic coefficients. Averaging the coefficients gives

$$\begin{bmatrix} 0 & 2 \\ -2 & 0 \end{bmatrix} \begin{pmatrix} \dot{\beta}_{1c} \\ \dot{\beta}_{1s} \end{pmatrix} + \begin{bmatrix} \nu^2 - 1 & \frac{\gamma}{8} \\ -\frac{\gamma}{8} & \nu^2 - 1 \end{bmatrix} \begin{pmatrix} \beta_{1c} \\ \beta_{1s} \end{pmatrix} = \frac{\gamma}{8} \begin{pmatrix} \theta_{1c} \\ \theta_{1s} \end{pmatrix} \quad (15.106)$$

which does produce the correct static solution for the tip-path-plane tilt.

15.4.5 History

In developing the equations of motion for ground resonance of a rotor with three or more blades, Coleman (1943) noted "a proper choice of coordinates leads to equations with constant coefficients." For an N-bladed rotor, each blade with lag degree of freedom $\beta^{(m)}$, he introduced "special linear combinations of the hinge deflections":

$$\theta_n = \frac{bi}{N} \sum_{m=1}^{N} \beta^{(m)} e^{inm\Delta\psi} \quad (15.107)$$

$$\zeta_n = \theta_n e^{in\Omega t} \quad (15.108)$$

where b is the distance from the lag hinge to the blade center of mass. The θ_n variables represent the hinge motion in the rotating coordinate system, whereas the ζ_n variables are in the fixed system. "Geometrically, θ_1 or ζ_1 is the complex vector representing the displacement due to hinge deflection of the center of mass of all the blades." Only ζ_1 couples with the in-plane shaft motion. "The physical meaning of this partial separation of variables is that a blade motion represented by ζ_1 involves a motion of the common center of mass of the blades and, thus, a coupling effect with the pylon. Blade motions in which the common center of mass does not move are represented by $\zeta_2, \ldots, \zeta_n$." The variables ζ_0 and ζ_1 are recognized as the mean and first-harmonic multiblade coordinates:

$$\beta_0 = \frac{1}{N} \sum_{m=1}^{N} \beta^{(m)} = \frac{1}{bi} \zeta_0 \quad (15.109)$$

$$\frac{\beta_{1c} + i\beta_{1s}}{2} = \frac{1}{N} \sum_{m=1}^{N} \beta^{(m)} e^{in\psi_m} = e^{in\psi} \frac{1}{N} \sum_{m=1}^{N} \beta^{(m)} e^{inm\Delta\psi} = \frac{1}{bi} \zeta_1 \quad (15.110)$$

Coleman used complex combinations of the lag degrees of freedom and the shaft motion degrees of freedom to facilitate derivation and solution of the ground resonance equations, for an axisymmetric system working with two complex equations instead of four real equations. Coleman and Feingold (1947) analyzed ground

15.4 Multiblade Coordinates

resonance of two-bladed rotors, using the collective and reactionless multiblade coordinates: $\theta_0 = \frac{b}{2}\left(\beta^{(2)} + \beta^{(1)}\right)$ and $\theta_1 = \frac{b}{2}\left(\beta^{(2)} - \beta^{(1)}\right)$.

Coleman recognized the physical relevance of MBC, the separation of the coupling of rotor and shaft motion, and the fact that MBC lead to equations with constant coefficients. Unfortunately, the character of the multiblade coordinates was obscured by the use of complex coordinates. Appropriate notation proved elusive in the development of multiblade coordinates.

Miller (1948) conducted an evaluation of the stability and control characteristics of several different types of helicopters, in which multiblade coordinates were used for the blade flap motion:

> The equations of motion are derived by considering the displacement of the helicopter and its blades relative to a system of axes fixed in space.... x is the displacement at any time t of the rotor hub in the X direction and α_1, β_1, the corresponding angular displacements of the helicopter and tip path plane.... The blade flapping can be expressed as $\beta_\psi = \beta_0 + \beta_1 \cos \Omega t + \beta_2 \sin \Omega t$ higher harmonics of flapping having no effect on the stability of the helicopter as a whole. β_1 and β_2 are functions of time. β_0 is constant since the thrust is constant.

Thus β_1 and β_2 were the longitudinal and lateral tip-path-plane tilt relative to space, whereas $\beta_1 - \alpha_1$ and $\beta_2 - \alpha_2$ were the tip-path-plane tilt relative to the hub. The rotating frame flap equation of motion was derived from the equilibrium of inertial, aerodynamic, and hinge spring moments. Then the fixed frame flap equations were obtained by setting to zero separately the coefficients of 1, $\cos \Omega t$, and $\sin \Omega t$. Citing Coleman, Miller also used complex combinations of variables for the airframe in-plane and angular motion and the rotor flap motion, to reduce the number of equations from six to three.

Grodko (1968) in the ground resonance chapter of Mil's book cited Coleman, but used multiblade coordinates instead of Coleman's variables:

> Investigations conducted by Coleman and B.Ya. Zherebtsov showed that, for a rotor with a number of blades $n \geq 3$, this system of equations can be reduced to a system of linear equations with constant coefficients, if we replace [the lag variables] $\xi_k(t)$ by new variables $x_c(t)$ and $z_c(t)$ representing the coordinates of the center of gravity of the blade system.

The work presented the corresponding transformation for the time derivatives of $\xi_k(t)$ and the approach for transforming the equations of motion.

Young and Lytwyn (1967) used multiblade coordinates in an examination of the dynamic stability of low disk loading propeller-rotors. The motion analyzed consisted of nacelle pitch and yaw degrees of freedom, plus flapping freedoms for an N-bladed rotor ($N \geq 3$). They noted that "the N blade freedoms are reduced to two quasi coordinates by observing that of all possible patterns of blade motion, only the two representing longitudinal and lateral tilting of the tip path plane provide a possibility of an unstable coupling with the nacelle freedoms." The notation β_c and β_s was used for the quasi-coordinates.

Johnson and Hohenemser (1970) investigated tilting moment feedback control on a hingeless rotor. They defined cyclic pitch as $\theta_{\text{cyc}} = -\theta_\text{I} \sin \psi_k + \theta_\text{II} \cos \psi_k$, and the feedback control as

$$\dot{\theta}_\mathrm{I} = -k_{11}\sum_{k=1}^{n}\beta_k\cos\psi_k - k_{12}\sum_{k=1}^{n}\beta_k\sin\psi_k$$

$$\dot{\theta}_\mathrm{II} = -k_{21}\sum_{k=1}^{n}\beta_k\cos\psi_k - k_{22}\sum_{k=1}^{n}\beta_k\sin\psi_k$$

where β_k is the flap degree of freedom of the k-th blade. Then they observed,

> Because of the inclusion of azimuth position, a two-axis resolution of the equations of motion is necessary. For this purpose the following definitions are made.
>
> $$\beta_\mathrm{I} = \sum_{k=1}^{n}\beta_k\cos\psi_k$$
>
> $$\beta_\mathrm{II} = \sum_{k=1}^{n}\beta_k\sin\psi_k$$
>
> ... By use of the preceding expressions, the equations of motion in terms of the redefined variables are obtained by first multiplying the blade equations of motion by $\cos\psi_k$, summing from $k=1$ to n and then making the appropriate substitutions. The process is then repeated for $\sin\psi_k$.

The use of Roman numeral subscripts "I" and "II" for the first-harmonic multiblade coordinates followed from the nomenclature for cyclic control.

Hohenemser and Yin (1972) introduced the terminology "multiblade coordinates." They cited Coleman (1943) and Young and Lytwyn (1967), and then generalized MBC for an N-bladed rotor:

> The multiblade coordinates represent collective flapping or coning, differential collective flapping (only for even bladed rotors), and cyclic flapping of various orders, defining tilting or warping of the rotor plane.
>
> The flapping angle β_k of the k-th blade in terms of multiblade flapping coordinates is
>
> $$\beta_k = \beta_0 + \beta_d(-1)^k + \beta_\mathrm{I}\cos\psi_k + \beta_\mathrm{II}\sin\psi_k$$
> $$+ \beta_\mathrm{III}\cos 2\psi_k + \beta_\mathrm{IV}\sin 2\psi_k$$
> $$+ \beta_\mathrm{V}\cos 3\psi_k + \beta_\mathrm{VI}\sin 3\psi_k + \ldots$$
>
> where $\psi_k = t + (2\pi/N)(k-1)$ is the azimuth angle of the k-th blade. For an N-bladed rotor only the first N terms are retained, whereby β_d occurs only in even-bladed rotors.

Hohenemser and Yin gave the inverse transformation and described the method for transforming the rotating frame equations of motion to the non-rotating frame. They also analyzed the periodic differential equations of flapping motion in edgewise forward flight using Floquet theory and demonstrated the utility of the constant coefficient equations obtained by dropping periodic terms after applying the MBC transformation.

15.5 Eigenvalues and Eigenvectors of the Rotor Motion

We now examine the characteristics of the rotor motion, in particular the eigenvalues and eigenvectors of the system described by the non-rotating frame degrees of

15.5 Eigenvalues and Eigenvectors of the Rotor Motion

freedom and equations. Consider a constant-coefficient, mass-spring-damper system in the rotating frame; for example, the rotor flapping equation in hover:

$$\ddot{\beta}^{(m)} + \frac{\gamma}{8}\dot{\beta}^{(m)} + v^2\beta^{(m)} = 0 \tag{15.111}$$

The homogeneous equation is sufficient, since only the roots and mode shapes are required here. The uncoupled motion of all the rotor degrees of freedom (lag, pitch, elastic bending, and so on) is described by similar equations. To be general, an arbitrary level of damping ($\gamma/8$) is allowed, and a natural frequency (v) is considered that unlike the flap motion is not necessarily near 1/rev. The eigenvalues of the rotating equation are the solution of the quadratic equation $s_R^2 + (\gamma/8)s_R + v^2 = 0$, or

$$s_R = -\frac{\gamma}{16} + i\sqrt{v^2 - \left(\frac{\gamma}{16}\right)^2} \tag{15.112}$$

and its conjugate.

In the non-rotating frame, the equations for β_0 and $\beta_{N/2}$ are identical to the rotating equation:

$$\ddot{\beta}_0 + \frac{\gamma}{8}\dot{\beta}_0 + v^2\beta_0 = 0 \tag{15.113}$$

$$\ddot{\beta}_{N/2} + \frac{\gamma}{8}\dot{\beta}_{N/2} + v^2\beta_{N/2} = 0 \tag{15.114}$$

The roots of both equations are then the same as the rotating roots: $s = s_R$ and its conjugate. The differential equations for β_{nc} and β_{ns} are

$$\begin{pmatrix}\ddot{\beta}_{nc}\\\ddot{\beta}_{ns}\end{pmatrix} + \begin{bmatrix}\frac{\gamma}{8} & 2n \\ -2n & \frac{\gamma}{8}\end{bmatrix}\begin{pmatrix}\dot{\beta}_{nc}\\\dot{\beta}_{ns}\end{pmatrix} + \begin{bmatrix}v^2 - n^2 & n\frac{\gamma}{8} \\ -n\frac{\gamma}{8} & v^2 - n^2\end{bmatrix}\begin{pmatrix}\beta_{nc}\\\beta_{ns}\end{pmatrix} = 0 \tag{15.115}$$

(equation 15.80), or

$$\begin{bmatrix}s^2 + \frac{\gamma}{8}s + v^2 - n^2 & 2ns + n\frac{\gamma}{8} \\ -\left(2ns + n\frac{\gamma}{8}\right) & s^2 + \frac{\gamma}{8}s + v^2 - n^2\end{bmatrix}\begin{pmatrix}\beta_{nc}\\\beta_{ns}\end{pmatrix} = 0 \tag{15.116}$$

The transformation to the non-rotating frame introduces centrifugal and Coriolis terms that couple the β_{nc} and β_{ns} equations. The roots are the solution of the characteristic equation:

$$\left(s^2 + \frac{\gamma}{8}s + v^2 - n^2\right)^2 + \left(2ns + n\frac{\gamma}{8}\right)^2 = 0 \tag{15.117}$$

or

$$s = -\frac{\gamma}{16} \pm in + i\sqrt{v^2 - \left(\frac{\gamma}{16}\right)^2} = s_R \pm in \tag{15.118}$$

and their conjugates. Hence the non-rotating eigenvalues for the β_{nc} and β_{ns} degrees of freedom are simply the rotating roots shifted in frequency by n/rev: $s = s_R \pm in$. The corresponding eigenvectors are $\beta_{nc}/\beta_{ns} = i$ for $s = s_R + in$, and $\beta_{nc}/\beta_{ns} = -i$ for $s = s_R - in$.

The eigenvalues $s = s_R \pm in$ correspond to a coupled motion of β_{nc} and β_{ns}, which is a damped oscillation at frequency $\text{Im} s = \text{Im} s_R \pm n$/rev or $\omega = \omega_R \pm n$/rev. The exponential decay rate, $\text{Re} s = \text{Re} s_R = -\gamma/16$, is the same as for the rotating roots. The $s = s_R + in$ root has frequency $\omega = \omega_R + n$/rev, and $\beta_{nc} = i\beta_{ns}$ implies that the motion of β_{nc} leads that of β_{ns} by a phase of 90°, meaning by one-quarter of the oscillation period $2\pi/(\omega_R + n)$. Thus $s = s_R + in$ is a high-frequency, progressive mode. The frequency $\omega_R + n$/rev is always greater than the rotor speed. The $s = s_R - in$ root has frequency $|\omega_R - n|$. If $\omega_R > n$/rev, then $\beta_{nc} = -i\beta_{ns}$ implies that the motion of β_{nc} lags β_{ns} by 90°. If, however, $\omega_R < n$/rev, so that the frequency $\omega_R - n$/rev is negative, then $\beta_{nc} = -i\beta_{ns}$ implies that β_{nc} leads β_{ns} by 90°. Thus $s = s_R - in$ is a low-frequency mode (the frequency can be below 1/rev if the rotating frequency is near n/rev), regressive if $\omega_R > n$/rev and progressive if $\omega_R < n$/rev.

Consider the important case of the cyclic modes ($n = 1$) for the flap and lag motion of the rotor. For the flap motion, the rotating natural frequency ω_R is usually slightly below 1/rev for articulated rotors and perhaps slightly above 1/rev for hingeless rotors. Then for the high-frequency mode $s = s_R + i$, β_{1c} leading β_{1s} means that the tip-path plane is wobbling in the same direction as the rotor rotation, at a speed around 2/rev. For the low-frequency mode $s = s_R - i$, the tip-path plane wobbles at a low rate, again in the same direction as the rotor rotation if the rotating frequency is below 1/rev, but in the opposite direction if ω_R is above 1/rev.

For the lag motion, articulated and soft in-plane hingeless rotors have a rotating frequency below 1/rev. The high-frequency lag mode is a progressive mode in which the rotor center-of-gravity whirls in the same direction as the rotor rotation at a speed above 1/rev. The low-frequency lag mode is also a progressive whirling, but with a low frequency in the non-rotating frame. For stiff in-plane rotors the rotating lag frequency is above 1/rev, and the low frequency lag mode is a regressive mode in which the rotor center-of-gravity whirls in the opposite direction to the rotor rotation.

Figure 15.1 summarizes the transformation of the eigenvalues describing the rotor dynamics from the rotating to the non-rotating frame. The case of a three-bladed rotor with a rotating frequency just below 1/rev is shown. In the rotating frame there are triple roots at s_R and its conjugate corresponding to the three independent blades, or in general there are N pairs of roots for an N-bladed rotor. In the non-rotating frame there are still N pairs of roots at s_R and its conjugate again for the β_0 and $\beta_{N/2}$ modes and at $s_R \pm in$ and their conjugates for the coupled β_{nc} and β_{ns} motion. Thus the transformation leaves the real part of the roots unchanged and shifts the frequency by $\pm n$/rev. Figure 15.1 shows the collective, high-frequency, and low-frequency modes for the three-bladed rotor case. When the individual blades of the rotor are not independent, but rather are coupled through the fixed system (such as by the control system or shaft motion), the non-rotating modes are not all influenced in the same manner, and the real parts of the roots are not necessarily identical nor are the frequencies separated by exactly n/rev. The basic character illustrated by Figure 15.1 still dominates the roots in the non-rotating frame.

15.6 Analysis of Linear, Periodic Systems

The aeroelastic behavior of the rotor or helicopter is described in many cases by linear differential equations with periodic coefficients. The periodic coefficients arise as a result of the aerodynamic forces in forward flight or a basic asymmetry in the rotor system (such as with a two-bladed rotor). Helicopter analysis therefore requires a means for obtaining the dynamic behavior of periodic systems, in particular the eigenvalues describing the stability.

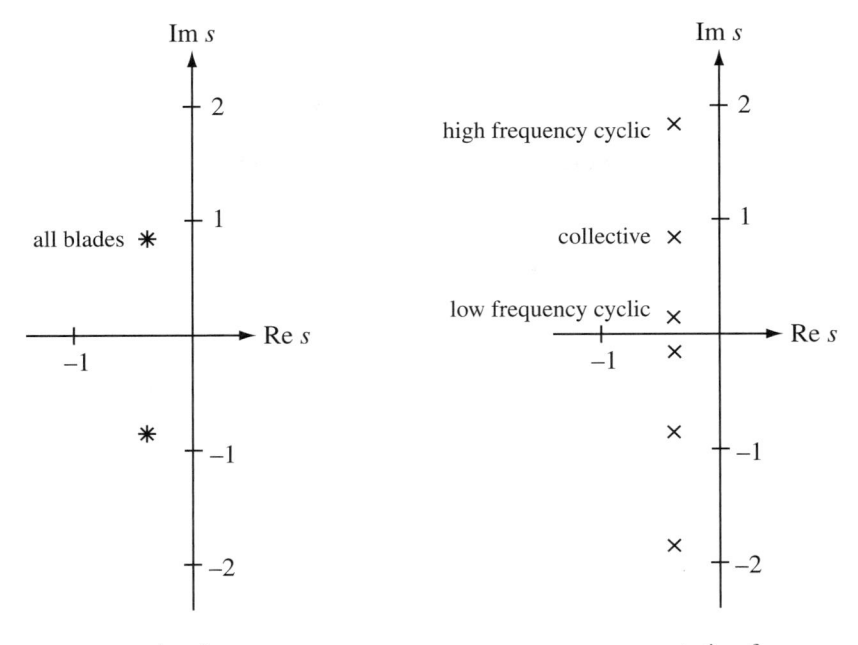

15.6 Analysis of Linear, Periodic Systems

Figure 15.1. Transformation of eigenvalues from the rotating to the non-rotating frame (shown for $N = 3$).

Consider a physical system described by linear ordinary differential equations of second order:

$$A_2 \ddot{x}_1 + A_1 \dot{x}_1 + A_0 x_1 = B_0 v \qquad (15.119)$$

Here x_1 is the vector of degrees of freedom; v is the vector of input variables; and A_2, A_1, A_0, and B_0 are matrices of the coefficients of the equations of motion. For a time-invariant system, the coefficient matrices are constant. Of interest for rotorcraft problems is the more general case of time-varying coefficients, especially periodic coefficients. Dealing with these equations in a standard first-order form is convenient, both in the mathematical development of the theory and the actual computation of the dynamic response. Thus we define $x_2 = \dot{x}_1$, so

$$\dot{x}_2 = \ddot{x}_1 = -A_2^{-1}(A_1 \dot{x}_1 + A_0 x_1 - B_0 v) \qquad (15.120)$$

Then the equations of motion become

$$\begin{pmatrix} \dot{x}_2 \\ \dot{x}_1 \end{pmatrix} = \begin{bmatrix} -A_2^{-1}A_1 & -A_2^{-1}A_0 \\ I & 0 \end{bmatrix} \begin{pmatrix} x_2 \\ x_1 \end{pmatrix} + \begin{bmatrix} A_2^{-1}B_0 \\ 0 \end{bmatrix} v \qquad (15.121)$$

or

$$\dot{x} = Ax + Bv \qquad (15.122)$$

where

$$x = \begin{pmatrix} x_2 \\ x_1 \end{pmatrix} = \begin{pmatrix} \dot{x}_1 \\ x_1 \end{pmatrix} \qquad (15.123)$$

is the state variable vector, consisting of the displacement and velocity of all second-order degrees of freedom. By transforming from second order to first order, the degree of the system (the dimension of x) has been doubled. The spring terms can be absent for a degree of freedom (zero column in A_0), so the degree of freedom is actually first order. Such lower order variables should be represented as a single state to avoid spurious zero eigenvalues. The degrees of freedom are reordered so that the second-order variables x_1 appear first and the first-order variables x_0 appear last in the vector. Then the last columns of A_0 corresponding to x_0 are zero: $A_0 = [\widetilde{A}_0 \; 0]$. The differential equations are

$$\begin{pmatrix} \ddot{x}_1 \\ \dot{x}_0 \\ \dot{x}_1 \end{pmatrix} = \begin{bmatrix} -A_2^{-1}A_1 & -A_2^{-1}\widetilde{A}_0 \\ I & 0 & 0 \end{bmatrix} \begin{pmatrix} \dot{x}_1 \\ x_0 \\ x_1 \end{pmatrix} + \begin{bmatrix} A_2^{-1}B_0 \\ 0 \end{bmatrix} v \qquad (15.124)$$

or $\dot{x} = Ax + Bv$, which is again the standard first-order form.

15.6.1 Linear, Constant Coefficient Equations

The analysis of a linear, time-invariant system is developed first, as a background for the periodic system analysis. Moreover, the time-invariant case is more practical to solve and hence more widely used. Consider the system described by ordinary differential equations of the form $\dot{x} = Ax + Bv$, where A and B are constant matrices. The state vector x has dimension n. The dynamic behavior of this system is determined by the eigenvalues and eigenvectors of the matrix A. For a system of order n, there are n eigenvalues λ_i ($i = 1, \ldots, n$) and corresponding eigenvectors u_i, which are the solution of the algebraic equations $(A - \lambda_i I)u_i = 0$. These homogeneous equations have a non-zero solution for u only if the determinant of the coefficients is zero: $\det(A - \lambda I) = 0$. This determinant defines a polynomial of order n in λ, called the characteristic equation. Define Λ as the diagonal eigenvalue matrix, and define the modal matrix M as the matrix with the eigenvectors as columns (ordered to correspond to the eigenvalues in Λ). Then the eigenvalue equation becomes $AM - M\Lambda = 0$, or $A = M\Lambda M^{-1}$.

To show the relation between the eigenvalues and the linear differential equation, consider the homogeneous equation $\dot{x} = Ax$. Expand the state vector x in terms of the eigenvectors of A:

$$x(t) = \sum_{i=1}^{n} \alpha_i(t) u_i \qquad (15.125)$$

where the α_i are scalar functions of time. This expansion is possible because the eigenvectors u_i form a complete linearly independent set, so the constants α_i can be found for any x. Substituting for x in $\dot{x} = Ax$, and using $Au_i = u_i \lambda_i$, gives $\dot{\alpha}_i = \lambda_i \alpha_i$. The solution is $\alpha_i = c_i e^{\lambda_i t}$, where the c_i are scalar constants. The solution of the differential equation has now been obtained in terms of the eigenvalues and eigenvectors:

$$x(t) = \sum_{i=1}^{n} c_i e^{\lambda_i t} u_i \qquad (15.126)$$

The constants c_i are obtained from the initial conditions $x(0) = \sum_{i=1}^{n} c_i u_i$. This expression is called the normal mode expansion of the response.

In matrix form, the expansion of the state vector in terms of the eigenvectors is accomplished by the linear transformation $x = Mq$, where M is the modal matrix and q is the vector of the normal coordinates (equivalent to $\alpha_i(t)$). Then using $AM = M\Lambda$, the differential equation $\dot{x} = Ax$ becomes $\dot{q} = \Lambda q$. Since the eigenvalue matrix Λ is diagonal, the differential equations for q_i are decoupled and easily integrated, giving

$$q = e^{\Lambda t} q(0) \tag{15.127}$$

or

$$x = M e^{\Lambda t} q(0) \tag{15.128}$$

Solving the initial conditions $x(0) = Mq(0)$ for $q(0)$ and substituting gives

$$x = M e^{\Lambda t} M^{-1} x(0) = e^{At} x(0) \tag{15.129}$$

which is the solution of the homogeneous differential equation.

Analysis of a linear time-invariant system thus requires an evaluation of the eigenvalues and eigenvectors of A. The normal mode expansion shows that the solution is unstable if $\text{Re}(\lambda_i) > 0$ for any mode, since in that case $e^{\lambda_i t}$ increases without bound as time increases. The eigenvalues determine the stability of the system, a fact that is often shown graphically as the variation of the roots with some parameter in the plane of Imλ vs. Reλ. The system is stable if all the roots are in the left half-plane of the root locus diagram. The eigenvectors u_i describe the mode shape of the state variable x corresponding to each eigenvalue. The eigenvalues of the real matrix A must be either real or occur in complex conjugate pairs. Complex roots are described in terms of the frequency $\omega = \text{Im}\lambda$, the natural frequency $\omega_n = |\lambda|$, and the damping ratio $\zeta = \text{Re}\lambda/|\lambda|$. The motion is a decaying oscillation at frequency ω. The roots are on the imaginary axis (neutrally damped) when $\zeta = 0$ and on the real axis when $\zeta = 1$, so ζ is thus the fraction of critical damping. For an unstable oscillation, $\zeta < 0$. Real roots are described by the time constant $\tau = -1/\lambda$ or by the time to decay to one-half amplitude $\tau_{1/2} = 0.693\tau$. The eigenvectors corresponding to complex eigenvalues must be complex conjugates also, and the corresponding initial values of the normal coordinates ($q(0) = M^{-1}x(0)$) are conjugates. Hence the total contribution of the pair of complex roots to the state vector,

$$\Delta x = u_1 e^{\lambda_1 t} q_1(0) + u_2 e^{\lambda_2 t} q_2(0) = 2\text{Re}\left(u_1 e^{\lambda_1 t} q_1(0)\right) \tag{15.130}$$

is real, as required of a physical system.

Now consider the response to input v. Using $x = Mq$, the normal form of the differential equation $\dot{x} = Ax + Bv$ becomes $\dot{q} = \Lambda q + M^{-1} Bv$. Since Λ is diagonal, these equations are readily integrated to obtain

$$q(t) = e^{\Lambda(t-t_0)} q(t_0) + \int_{t_0}^{t} e^{\Lambda(t-\tau)} M^{-1} Bv \, d\tau \tag{15.131}$$

The first term is the transient response and depends on the initial conditions, whereas the second term is the forced response to the input v occurring after t_0. In a stable system, the transient dies out as t increases. In terms of the state vector, the solution is

$$x(t) = e^{A(t-t_0)} x(t_0) + \int_{t_0}^{t} e^{A(t-\tau)} Bv \, d\tau \tag{15.132}$$

The matrix $\Phi(t, t_0) = e^{A(t-t_0)}$ is called the state transition matrix; it relates the state at t to the state at t_0. As an example of forced response of the system, consider a sinusoidal input $v = \hat{v} e^{i\omega t}$. A property of a linear, time-invariant system is that the forced response must also be a sinusoid at frequency ω: $x = \hat{x} e^{i\omega t}$. Integrating the expression for $x(t)$ (with $t_0 = -\infty$), or substituting into the differential equation and solving directly, gives

$$\hat{x} = -(A - i\omega I)^{-1} B\hat{v} = -(A + i\omega I)\left(A^2 + \omega^2 I\right)^{-1} B\hat{v} \tag{15.133}$$

This is in the form $\hat{x} = H\hat{v}$, where $H(\omega)$ is the matrix of the transfer functions of the system response. Also useful is the step response, obtained by integrating $v = 0$ for $t < 0$ and $v = \hat{v}$ for $t > 0$:

$$x = A^{-1}\left(e^{At} - I\right) B\hat{v} \tag{15.134}$$

The limit as time approaches infinity is the steady-state response, $x = -A^{-1}B\hat{v}$.

15.6.2 Linear, Periodic Coefficient Equations

Now let us consider a linear, time-varying dynamic system described by ordinary differential equations of the form $\dot{x} = A(t)x + B(t)v$. The coefficient matrices A and B are functions of time. We are particularly interested in periodic systems, for which $A(t + T) = A(t)$, where T is the period. The analysis of periodic coefficient equations is called Floquet-Lyapunov theory. Floquet theory is a numerical method for extracting the stability of a linear, periodic dynamic system in terms of eigenvalues and periodic eigenvectors, and it was described as a numerical recipe in texts on linear systems in the 1960s (such as DeRusso, Roy, and Close (1965)). As a numerical solution method, Floquet theory was made practical by the digital computer, which made possible investigations of the unique dynamic behavior of periodic systems.

The solution of $\dot{x} = A(t)x$ must be of the form $x(t) = \Phi(t, t_0)x(t_0)$, since for a linear system the degrees of freedom at t must always be a linear combination of the degrees of freedom at t_0. The matrix $\Phi(t, t_0)$ is called the state transition matrix. By definition, $\Phi(t_0, t_0) = I$ and $\Phi(t_2, t_0) = \Phi(t_2, t_1)\Phi(t_1, t_0)$, and letting $t_2 = t_0$ gives $\Phi(t_1, t_0) = \Phi^{-1}(t_0, t_1)$. By substituting $x(t) = \Phi x(t_0)$ into $\dot{x} = Ax$, the differential equation for Φ is obtained: $\dot{\Phi} = A\Phi$, with initial conditions $\Phi(t_0, t_0) = I$. When the response to the input v is included, the state transition matrix gives the complete solution:

$$x(t) = \Phi(t, t_0)x(t_0) + \int_{t_0}^{t} \Phi(t, \tau)B(\tau)v(\tau)\, d\tau \tag{15.135}$$

Thus the analysis of a linear system involves finding the state transition matrix. For a time-invariant system, Φ must have the further property of depending only on the difference $t - t_0$. The result for a constant coefficient equation is $\Phi = e^{A(t-t_0)}$.

Now restrict the system to the periodic coefficient case, $A(t + T) = A(t)$. The differential equation for Φ becomes

$$\frac{d}{dt}\Phi(t, t_0) = A(t)\Phi(t, t_0) \tag{15.136}$$

and

$$\frac{d}{dt}\Phi(t + T, t_0) = A(t + T)\Phi(t + T, t_0) = A(t)\Phi(t + T, t_0) \tag{15.137}$$

So $\Phi(t + T, t_0)$ must be a linear combination of $\Phi(t, t_0)$, since both are solutions of the same equation:

$$\Phi(t + T, t_0) = \Phi(t, t_0)\alpha \tag{15.138}$$

15.6 Analysis of Linear, Periodic Systems

where α is a constant matrix, depending on the system. Write the state transition matrix as $\Phi(t, 0) = P(t)e^{\beta t}$ or, more generally,

$$\Phi(t, t_0) = P(t)e^{\beta(t-t_0)}P^{-1}(t_0) \tag{15.139}$$

where β is a constant matrix defined by $\alpha = e^{\beta T}$. Now

$$P(t+T) = \Phi(t+T, 0)e^{-\beta(t+T)} = \Phi(t, 0)\alpha e^{-\beta T}e^{-\beta t} = \Phi(t, 0)e^{-\beta t} = P(t) \tag{15.140}$$

Hence the matrix P is periodic, with initial conditions $P(0) = I$. Thus the solution of a periodic system must take the form of an exponential factor with decay or growth determined by the constant matrix β, multiplied by a purely periodic factor P. This is the principal result of Floquet-Lyapunov theory.

From $\Phi(t+T, t_0) = \Phi(t, t_0)\alpha$, it follows that $\Phi(t+NT, t_0) = \Phi(t, t_0)\alpha^N$. Consequently, all the information about the solution is contained in the state transition matrix for a single period. Since by definition $\alpha = \Phi(t_0+T, t_0)$, the solution for all other times can be constructed from that data. Let Θ be the eigenvalue matrix of α, and S the corresponding modal matrix, so that $\alpha = S\Theta S^{-1}$. Then $\alpha^N = S\Theta^N S^{-1}$; hence the system is unstable, with the state transition matrix increasing without bound as time increases if $|\theta_i| > 1$ for any eigenvalue of α. The more conventional roots of the system are the eigenvalues of β. Let Λ be the eigenvalue matrix of β, and S the modal matrix, so that $\beta = S\Lambda S^{-1}$ (α and β have the same eigenvectors). From the definition $\alpha = e^{\beta T}$, the eigenvalues are related by $\Theta = e^{\Lambda T}$ or

$$\Lambda = \frac{1}{T}\ln\Theta \tag{15.141}$$

The solution is thus unstable if $\text{Re}\,\lambda_i > 0$ for any eigenvalue. The logarithm of a complex function has many branches, giving values for λ_i that differ in frequency by multiples of $2\pi/T$. The principal value of λ_i can be used, or we can use the value with the frequency expected from physical considerations.

The state transition matrix for a periodic system can be written in a normal form analogous to that of a time-invariant system. Using the eigenvalues of β gives

$$\Phi(t, t_0) = P(t)e^{\beta(t-t_0)}P^{-1}(t_0) = [P(t)S]\,e^{\Lambda(t-t_0)}\,[P(t_0)S]^{-1} \tag{15.142}$$

which can be compared with the result for a time-invariant system,

$$\Phi(t, t_0) = e^{A(t-t_0)} = Me^{\Lambda(t-t_0)}M^{-1} \tag{15.143}$$

The periodic matrix PS can therefore be considered the modal matrix (the eigenvectors) of the periodic system, with the eigenvalues Λ determining the principal frequency and damping of the modes. Thus the expansion in normal coordinates q is defined as $x = PSq$. The transient solution $x(t) = \Phi(t, t_0)x(t_0)$ gives the solution for the normal coordinates, simply $q(t) = e^{\Lambda(t-t_0)}q(t_0)$ as for the time-invariant case. When u is written for the columns of PS, the normal form of the solution is

$$x(t) = P(t)Se^{\Lambda t}q(0) = \sum_i u_i(t)e^{\lambda_i t}q_i(0) \tag{15.144}$$

with the initial conditions being obtained from $q(0) = S^{-1}x(0)$. Compared to the time-invariant system, the periodic system is also described by normal modes u_i and roots λ_i, but now the eigenvectors are periodic functions rather than constants: $u_i(t+T) = u_i(t)$ follows from the periodicity of P. If the substitution $\Phi = Pe^{\beta t}$ is made, the differential equation for Φ gives

$$\dot{P} = AP - P\beta \tag{15.145}$$

From this there follows a differential equation for the eigenvectors u_i, which are the columns of PS:

$$\dot{u}_i = (A - \lambda_i I)u_i \tag{15.146}$$

The requirement that u_i be periodic is then sufficient to determine the eigenvalues λ_i. For the time-invariant case (A constant) the only "periodic" solution is $u_i =$ constant, and the problem reduces to $(A - \lambda_i I)u_i = 0$ as earlier.

The analysis of the dynamic behavior of a system of linear, periodic coefficient equations therefore requires that the state transition matrix Φ be obtained over one period, $t = 0$ to T, by integrating $\dot{\Phi} = A\Phi$ with $\Phi(0) = I$. The eigenvalues and eigenvectors of the matrix $\alpha = \Phi(T)$ are then obtained, and the roots of the system are $\Lambda = \frac{1}{T}\ln\Theta$. The mode shapes are given by $PS = \Phi S e^{-\Lambda t}$, or $u_i = e^{\lambda_i t}\Phi v_i$ (where v_i are the eigenvectors of α). The system is unstable if $|\theta_i| > 1$ or $\mathrm{Re}\lambda_i > 0$ for any mode. Because the time-varying eigenvectors of a periodic system involve a great deal of information, often the analysis is only concerned with the eigenvalues.

Including the forced response to the input v, the solution for x can be obtained from the state transition matrix. Alternatively, by using $x = PSq$, the normal equations can be integrated to obtain

$$q(t) = e^{\Lambda(t-t_0)}q(t_0) + \int_{t_0}^{t} e^{\Lambda(t-\tau)}(PS)^{-1}B(\tau)v(\tau)\,d\tau \tag{15.147}$$

Although this is formally similar to the solution for the time-invariant case, here PS and B are periodic matrices. In addition to making it difficult to evaluate the response, this periodicity has a fundamental influence on its character. For example, the response to sinusoidal excitation at frequency ω is not at that same frequency alone, but rather is composed of harmonics at frequencies $\omega \pm n2\pi/T$ for all integers n, where $2\pi/T$ is the fundamental frequency of the system. Thus the frequency response of a periodic system is not described by a single transfer-function matrix, but rather by a transfer function $H_n(\omega)$ for each of the harmonics $\omega + n2\pi/T$.

Finally, let us examine in more detail the behavior of the eigenvalues of a periodic system. The eigenvalues θ_i of the matrix $\alpha = \Phi(T)$ are either real or occur in complex conjugate pairs. The roots λ_i are obtained from $\lambda = \frac{1}{T}\ln\theta$, or

$$\lambda = \frac{1}{T}(\ln|\theta| + i\angle\theta) + in\frac{2\pi}{T} \tag{15.148}$$

where $\angle\theta$ is the argument or phase angle of θ. The principal part of the eigenvalue is

$$\lambda_P = \frac{1}{T}(\ln|\theta| + i\angle\theta) \tag{15.149}$$

and a multiple of the fundamental frequency $2\pi/T$ can be added, depending on the branch of the logarithm that the root is on. A complex conjugate pair for θ gives a conjugate pair for the roots λ_P also. A real, positive θ gives a principal root λ_P with zero imaginary part, so that the frequency of λ is a multiple of the fundamental frequency of the system (i.e., n/rev). For a real and negative θ, the frequency of the principal root λ_P is π/T, one-half the fundamental frequency; the frequency of

15.6 Analysis of Linear, Periodic Systems

λ is $(n + 1/2)$/rev. So when θ is real, the λ roots are complex, but do not have a corresponding conjugate. To interpret these roots, two questions must be answered: how is the branch of the logarithm selected, that is, what multiple of the fundamental frequency is added to the frequency of λ_P; and what is the meaning of the λ roots associated with real θ? As for the interpretation of complex roots of a time-invariant system, these concerns are resolved by considering the actual physical response $x(t)$ rather than the eigenvalues and eigenvectors separately. The principal value λ_P is uniquely determined from θ, and there is a corresponding principal value of the mode shape u. The physical response of the system depends on the product $ue^{\lambda t}$. Hence adding a multiple of the fundamental, $n2\pi/T$, to the frequency of the root corresponds to multiplying the mode by the periodic function $e^{-in(2\pi/T)t}$. Since the theory only requires that the mode shape $u(t)$ be periodic, no guidance is offered on apportioning this periodicity between the eigenvalue and eigenvector. If the system being analyzed is time invariant for some limit, then the frequencies of the roots are determined by the requirement that the roots be continuous as the periodicity is introduced. For example, the periodic coefficients due to the rotor aerodynamic forces in forward flight drop out in the hover limit, $\mu = 0$. One way to mechanize this choice of frequencies is to require that the mean value of the eigenvector have the largest magnitude; then the harmonic of largest magnitude in the eigenvector corresponding to the principal value of the eigenvalue gives the frequency $n2\pi/T$. This criterion gives the correct results for the time-invariant case. The frequencies of the roots can also be established by using a knowledge of the uncoupled natural frequencies of the system or of other considerations of the physical characteristics of the response.

For a real and positive θ root, there is a single complex λ root with a frequency equal to a multiple of the system fundamental frequency. The principal value λ_P is on the real axis, however, so requiring that the contribution to $x(t)$ be real means that the corresponding principal value of the eigenvector is also real. Giving λ a frequency $n2\pi/T$ then corresponds to multiplying the mode shape by $e^{-in(2\pi/T)t}$ without changing the product $ue^{\lambda t}$. For a real and negative θ root, the principal value λ_P has a frequency of one-half the system fundamental, $\lambda_P = \frac{1}{T}(\ln|\theta| + i\pi)$. Requiring that $ue^{\lambda t}$ be real implies that the function $w(t) = u(t)e^{i(\pi/T)t}$ is real, and since u is periodic it follows that w is anti-periodic: $w(t + T) = -w(t)$. Thus the implication of the $(1/2)$/rev frequency of λ is that the contribution to the response is of the form $\Delta x = c_i w(t)e^{(t/T)\ln|\theta|}$, where $w(t)$ is a real, anti-periodic function. Therefore, while as eigenvalues of the real matrix α the roots θ must appear as real numbers or complex conjugate pairs, the λ roots are under no such restriction. A real θ gives a single λ root with a frequency equal to a multiple of one-half the fundamental frequency of the system. The property of the solution that allows such behavior is the corresponding periodicity of the eigenvectors.

Figure 15.2 sketches a root locus that might be encountered with a periodic system. The behavior illustrated is typical of systems with strongly periodic coefficients. If the parameter being varied, for example the advance ratio μ, is such that at $\mu = 0$ the system is time invariant, the roots start out as complex conjugates on both the θ and λ planes (point A). As μ increases, the system periodicity increases, and the roots change. The λ roots remain complex conjugates, though, as long as the θ roots are complex. If the θ roots reach the real axis (point B), one increases along the real axis while the other decreases. On the λ plane the roots have reached an n/rev frequency for some critical μ (or $(n + 1/2)$/rev for negative, real θ), and as μ increases

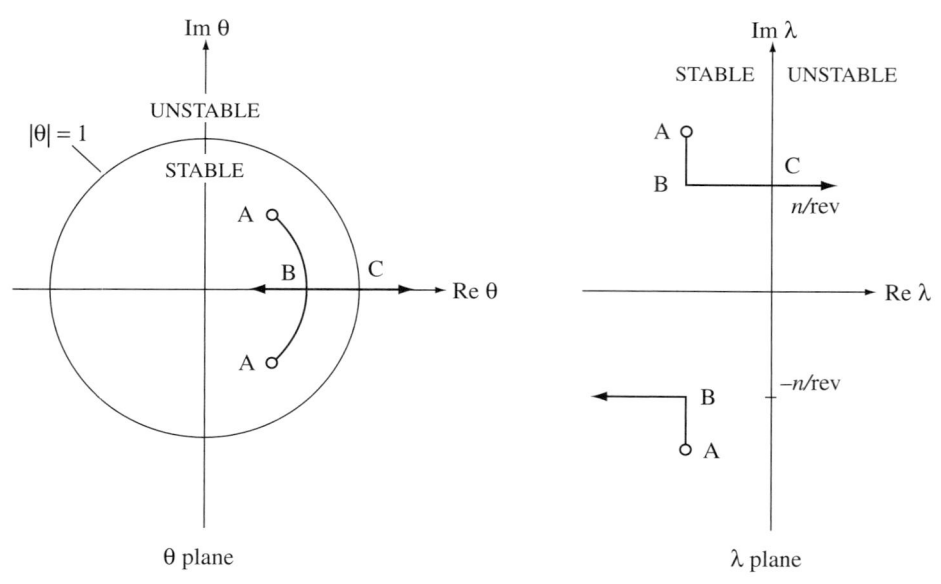

Figure 15.2. Sketch of a possible root locus for a periodic system.

further, the real part of one root increases and that of the other decreases while the frequency remains fixed at n/rev. The criterion for instability is $|\theta| > 1$ or $\text{Re}\lambda > 0$, so a stability boundary is crossed when the locus moves outside the $|\theta| = 1$ circle on the θ plane or into the right half-plane on the λ plane. With a time-invariant system, two types of instabilities are possible: a complex conjugate pair of roots can cross the Imλ axis at a non-zero frequency, or a single root on the real axis can go through the origin into the right half-plane. With periodic systems a third type of instability is introduced and, in fact, dominates the behavior for strong periodicity. Figure 15.2 illustrates this instability of periodic systems. After the θ roots reach the real axis, one becomes less stable and the other more stable. Often the root being destabilized eventually crosses over the stability boundary. For a time-invariant system, such a splitting of the branches of the root loci on the λ plane can only occur at the real axis. With periodic systems this behavior is generalized so that it can occur at any frequency that is a multiple of one-half the fundamental frequency of the system. The interpretation of this behavior is that the instability occurs with the oscillatory motion locked to the frequency of the system.

Floquet theory is a subject that was often entwined with multiblade coordinates (section 15.4) in early work, the connection being the periodic coefficients of rotor equations of motion, particularly with edgewise aerodynamics or two-bladed rotors. There are problems in rotor analysis that do not require both Floquet theory and MBC. For example, multiblade coordinates are needed to represent the blade motion of a rotor in axial flow when coupling with the non-rotating system is involved (such as shaft motion or control inputs), but the rotor is then a constant coefficient system. Alternatively, for the shaft-fixed dynamics of a rotor in forward flight, a single blade representation in the rotating frame can be used, but there are periodic coefficients due to the forward flight aerodynamics, and Floquet analysis is needed to determine the system stability.

15.7 Solution of the Equations of Motion

Solving the rotor equations of motion for any but the simplest of problems requires a numerical integration technique. The steady-state (periodic) response and the maneuver response must be obtained. Integration in time is accomplished using the standard methods of numerical analysis (Runge-Kutta), structural dynamics (Newmark, Hilber-Hughes-Taylor), or multibody dynamics. Special techniques have been developed to solve rotor equations of motion in steady-state flight, taking advantage of the periodicity of the equations and the solution. Simply integrating the equations until the solution converges to periodic motion presents convergence and accuracy issues in general and cannot be expected to work for low-damped or unstable systems. Solution methods based on the assumption of periodicity are best for the problem of the rotor in steady-state flight.

15.7.1 Early Methods

Gessow (1956) developed a harmonic analysis method for integrating the differential equation for the blade flap motion. The equation of motion for flapping in the rotating frame is

$$\ddot{\beta} + \nu^2 \beta = \gamma M_F \tag{15.150}$$

where M_F is the aerodynamic flap moment and the flap natural frequency ν is near 1/rev. Gessow's procedure is to calculate M_F at a finite number of points around the azimuth from the current estimate of the blade motion. Then the harmonics of a Fourier expansion of M_F can be evaluated:

$$M_F = \sum_{n=0}^{\infty} \left(M_{F_{nc}} \cos n\psi + M_{F_{ns}} \sin n\psi \right) \tag{15.151}$$

Assuming periodic motion, the solution of the flap equation is then

$$\beta_{nc} = \frac{\gamma M_{F_{nc}}}{\nu^2 - n^2} \tag{15.152}$$

$$\beta_{ns} = \frac{\gamma M_{F_{ns}}}{\nu^2 - n^2} \tag{15.153}$$

where β_{nc} and β_{ns} are the harmonics of the flap motion. With this new estimate of the blade motion, the flap moments can be recalculated. The successive calculations of the flap moments and blade motion are repeated until the solution converges, which is indicated when the change in blade motion from one iteration to the next falls below a specified tolerance level. With the converged solution for the blade motion, the rotor forces and performance can then be calculated. The only difficulty lies with the first harmonics of the flap motion, β_{1c} and β_{1s}. For $n = 1$ the flap equation gives

$$(\nu^2 - 1)\beta_{1c} = \gamma M_{F_{1c}} \tag{15.154}$$

$$(\nu^2 - 1)\beta_{1s} = \gamma M_{F_{1s}} \tag{15.155}$$

For an articulated rotor ($\nu = 1$) the left-hand side vanishes, and in general a different approach is required because the tip-path-plane tilt is primarily determined by the balance of aerodynamic moments on the blade. Expand the lateral flap moment about the current (k-th) iteration:

$$\frac{\nu^2-1}{\gamma}\beta_{1s} = M_{F_{1s}} \cong (M_{F_{1s}})_k + \frac{\partial M_{F_{1s}}}{\partial \beta_{1c}}(\beta_{1c}-\beta_{1c_k}) \tag{15.156}$$

Recall that the balance of lateral moments on the disk determines the longitudinal tip-path-plane tilt β_{1c}. So

$$(\beta_{1c})_{k+1} = \left[\beta_{1c} - \frac{M_{F_{1s}}-\beta_{1s}(\nu^2-1)/\gamma}{\partial M_{F_{1s}}/\partial \beta_{1c}}\right]_k \tag{15.157}$$

$$(\beta_{1s})_{k+1} = \left[\beta_{1s} - \frac{M_{F_{1c}}-\beta_{1c}(\nu^2-1)/\gamma}{\partial M_{F_{1c}}/\partial \beta_{1s}}\right]_k \tag{15.158}$$

The derivatives of the flap moment can be estimated from a simple analysis, since they do not affect the final solution, but only the convergence to it. Gessow gives

$$\frac{\partial M_{F_{1s}}}{\partial \beta_{1c}} = \frac{1}{8}\left(1-\frac{1}{2}\mu^2\right) \tag{15.159}$$

$$\frac{\partial M_{F_{1c}}}{\partial \beta_{1s}} = -\frac{1}{8}\left(1+\frac{1}{2}\mu^2\right) \tag{15.160}$$

With these expressions the 1/rev flap motion can be updated from the current calculation of the flap moments.

Tanner (1964), Berman (1965), and others have used a numerical integration method designed specifically for the periodic, steady-state case. If the 2/rev and higher harmonics are neglected, so that $\beta = \beta_0 + \beta_{1c}\cos\psi + \beta_{1s}\sin\psi$, then the motion at time-step $\psi_{k+1} = \psi_k + \Delta\psi$ is obtained exactly from the motion at ψ_k:

$$\beta_{k+1} = \beta_k + \dot{\beta}_k \sin\Delta\psi + \ddot{\beta}_k(1-\cos\Delta\psi) \tag{15.161}$$

$$\dot{\beta}_{k+1} = \dot{\beta}_k \cos\Delta\psi + \ddot{\beta}_k \sin\Delta\psi \tag{15.162}$$

This technique can be extended to the case of transient motion of an arbitrary degree of freedom, as follows. Consider the equation of motion

$$\ddot{\beta} + \nu^2\beta = g(\beta,\dot{\beta},\psi) \tag{15.163}$$

It is assumed that the forcing function is constant over the interval from ψ_k to ψ_{k+1}: $g \cong g_k$. Then the linear differential equation $\ddot{\beta}+\nu^2\beta = g_k$ can be integrated analytically, with initial conditions β_k and $\dot{\beta}_k$ at ψ_k. Evaluating the solution at $\psi_{k+1}=\psi_k+\Delta\psi$ gives

$$\beta_{k+1} = \beta_k \cos\nu\Delta\psi + \dot{\beta}_k\frac{\sin\nu\Delta\psi}{\nu} + g_k\frac{1-\cos\nu\Delta\psi}{\nu^2}$$

$$= \beta_k + \dot{\beta}_k\frac{\sin\nu\Delta\psi}{\nu} + \ddot{\beta}_k\frac{1-\cos\nu\Delta\psi}{\nu^2} \tag{15.164}$$

$$\dot{\beta}_{k+1} = \dot{\beta}_k\cos\Delta\nu\psi - \beta_k\nu\sin\nu\Delta\psi + g_k\frac{\sin\Delta\psi}{\nu}$$

$$= \dot{\beta}_k\cos\Delta\nu\psi + \ddot{\beta}_k\frac{\sin\Delta\psi}{\nu} \tag{15.165}$$

For small ($\nu\Delta\psi$) these equations reduce to the Taylor series expansion. A damping estimate $C\dot{\beta}$ can be added to both sides of the equation to improve the solution behavior.

15.7.2 Harmonic Analysis

Harmonic analysis or harmonic balance methods are based on representing the motion solution as Fourier series. Application of the inverse Fourier transform then leads to algebraic equations for the harmonics. The algebraic equations can be solved by a Newton-Raphson method. A frequency domain approach not only enforces periodicity of the solution but also allows the use of a large time-step, if the solution is adequately described by a small number of harmonics.

Johnson (1981) developed a harmonic solution method. The equations to be solved take the form $Hu = R$, where $H = M(d^2/dt^2) + C(d/dt) + K$ is a differential operator (mass, damper, and spring). In general R depends on time and on u and its derivatives, and the problem is nonlinear. The solution is assumed to be periodic, so the motion u can be represented by a Fourier series:

$$u = u_0 + \sum_{n=1}^{N} u_{nc} \cos n\psi + u_{ns} \sin n\psi \tag{15.166}$$

In the derivation that follows, a complex Fourier series representation is used for simplicity. The force $R_j = R(\psi_j)$ is evaluated at azimuth steps $\psi_j = j\Delta\psi$, $\Delta\psi = 2\pi/J$, for $j = 1$ to J. The harmonics R_n are obtained from R_j over one period by

$$R_n = \frac{K_n}{J} \sum_{j=1}^{J} e^{-in\psi_j} R_j \tag{15.167}$$

The factor K_n determines the interpolation option: $K_n = 1$ for Fourier interpolation, and

$$K_n = \left(\frac{\sin \pi n/J}{\pi n/J} \right)^2 \tag{15.168}$$

for linear interpolation (harmonics of a linear interpolation between the data at R_j; see section 15.3). The harmonics obtained by Fourier interpolation represent the force as a function of time that matches the discrete points R_j at ψ_j exactly, but the behavior is uncontrolled in between. The harmonics obtained by linear interpolation represent a smoothed curve, close to the discrete points. Linear interpolation of R is usually required for convergence. The harmonics of the left-hand side of the equation Hu are obtained by Fourier analysis. To improve convergence of the method, an estimate of the damping can be added to both sides of the equation of motion:

$$H_n u_n = \left[H(in) + in K_n D \right] u_n = (R + D\dot{u})_n = R_n \tag{15.169}$$

The damping term is added to R before the harmonic analysis is performed. The factor of K_n on the left-hand side is included so that this added damping is treated exactly the same on both sides of the equation. Consequently the value of D affects convergence but not the final solution for u, its primary function being to prevent problems near resonances.

By assuming a Fourier series representation of the solution, the differential equations are converted to algebraic equations, to be solved for the harmonics. An iterative solution is required. Each iteration finds the solution over a period and hence has a time loop. At each time step over the period, the equation force R_j is evaluated. To ensure convergence of nonlinear problems, this force is relaxed: $R_j = \lambda R_j + (1-\lambda) R_{j\text{old}}$, where the old force is from the previous iteration. Then the damping term is added to R_j and the harmonics R_n evaluated. The solution for the degrees of freedom is $u_n = H_n^{-1} R_n$. At the end of each iteration (the end of each period), convergence is tested. Since the correct solution for u is not known, convergence must be tested by comparing the values of two successive iterations: error = $\|u - u_{\text{old}}\| \leq$ tolerance, where the error is some norm of the difference between iterations, typically the root-mean-squared value of the response.

15.7.3 Time Finite Element

A time finite element solution method has been widely applied to solving rotor equations, notably by Chopra and Sivaneri (1982) and Bir, Chopra, and Nguyen (1990). The equations to be solved take the form $Hu = R$, where $H = M(d^2/dt^2) + C(d/dt) + K$ is a differential operator (mass, damper, and spring). In general R depends on time and on u and its derivatives, and the problem is nonlinear. For a periodic structural dynamic system (period T), Hamilton's principle in weak form gives

$$0 = \delta \int_0^T L\, dt = \int_0^T \left[\delta \dot{u}^T M \dot{u} + \delta u^T (R - C\dot{u} - Ku) \right] dt \quad (15.170)$$

Integrating the kinetic energy term by parts, and using the assumption of periodicity, gives

$$0 = \int_0^T \delta u^T [R - (M\ddot{u} + C\dot{u} + Ku)]\, dt = \int_0^T \delta u^T [R - Hu]\, dt = \int_0^T \delta u^T \widehat{R}\, dt \quad (15.171)$$

The finite element method expands the response as $u(t) = h(t)^T q$, where h are the shape functions in time and q are the finite element variables. Then the equations

$$0 = \int_0^T h^T \widehat{R}\, dt \quad (15.172)$$

are to be solved for q, typically by a Newton-Raphson algorithm. As with spatial finite elements, the period can be divided into N_t elements, with polynomial shape functions over each element:

$$u_i(t) = h(s) q_i \quad (15.173)$$

with $s = (t - t_{i-1})/(t_i - t_{i-1})$, for $i = 1$ to N_t. Typically h are Lagrange polynomials, and Gaussian integration is used over each element. The variables q_i are identified so u is continuous at the ends of each element and from the end to the beginning of the period.

It is also possible to use harmonics for the shape functions, with the advantage that only a few variables q are required to represent low-frequency motion. Let the variables q be the coefficients in a Fourier series representation of u:

$$u = h^T q = u_0 + \sum_{n=1}^{N} u_{nc} \cos n\psi + u_{ns} \sin n\psi \quad (15.174)$$

and then

$$0 = \int_0^T h^T \widehat{R}\, dt = \int_0^T h^T [R - Hu]\, dt \quad (15.175)$$

15.7 Solution of the Equations of Motion

are the harmonics of the equations of motion. The equations to be solved take the form $\widehat{R} = R - Hu = 0$, where H is a time-invariant operator. The force $R_j = R(\psi_j)$ is evaluated at azimuth steps $\psi_j = j \Delta\psi$, $\Delta\psi = 2\pi/J$, for $j = 1$ to J. The harmonics R_n are obtained from R_j over one period. Then the equation residuals are evaluated: $\widehat{R}_n = R_n - H_n u_n$. By introducing time finite elements, the differential equations are converted to algebraic equations, $\widehat{R}_n(u_n) = 0$. In this form a Newton-Raphson method is applicable. Including a relaxation factor λ and writing D for the derivative matrix, the iteration is

$$u_{n_{k+1}} = u_{n_k} - \lambda D^{-1} \widehat{R}(u_{n_k}) \tag{15.176}$$

The derivative matrix is

$$D = \frac{\partial \widehat{R}_n}{\partial u_n} = \frac{\partial R_n}{\partial u_n} - H_n \tag{15.177}$$

The derivative matrix is usually not available analytically, so must be obtained by an identification process. A simple estimate follows the assumption that most of the dependence on the motion is in the operator H, so $D \cong -H_n = -(H(in) + D_e in)$, including the damping estimate D_e to avoid singularities at resonant harmonics. With this result for D, the iteration is

$$u_{n_{m+1}} = u_{n_m} + \lambda H_n^{-1}(R_n - H_n u_{n_m}) = \lambda H_n^{-1} R_n + (1 - \lambda) u_{n_m} \tag{15.178}$$

At the end of each iteration (the end of each period), convergence is tested. Since the correct solution for u is not known, convergence must be tested by comparing the values of two successive iterations: error $= \|u - u_{\text{old}}\| \leq$ tolerance, where the error is some norm of the difference between iterations, typically the root-mean-squared value of the response.

15.7.4 Periodic Shooting

Periodic shooting is a solution procedure with a foundation in Floquet theory; see Peters and Izadpanah (1981). Consider the equation of motion in state variable form:

$$\dot{x} = Ax + b \tag{15.179}$$

with A and b periodic. Following section 15.6, the solution of the homogeneous equation ($b = 0$) is described by the state transition matrix Φ: $x(t) = \Phi(t)x(0)$, for $0 \leq t \leq T$. The state transition matrix is obtained by integrating $\dot{\Phi} = A\Phi$. What is required is the periodic solution of the inhomogeneous equation. Equation 15.179 can be integrated with zero initial conditions to obtain the motion x_F. Then from linearity, $x(t) = x_F(t) + \Phi(t)x(0)$ is also a solution of equation 15.179, and requiring $x(T) = x(0)$ gives the initial conditions $x(0) = (I - \Phi(T))^{-1}x_F(T)$, so

$$x(t) = x_F(t) + \Phi(t)(I - \Phi(T))^{-1}x_F(T) \tag{15.180}$$

is the periodic solution.

For nonlinear equations, $\dot{x} = f(x)$, an iterative solution is necessary. With the k-th estimate of the initial conditions, $x_k(0)$, the equations are integrated to obtain $x_k(t)$. Then a perturbation δx is needed, such that $x_k + \delta x$ is periodic. The perturbation satisfies the linearized equation

$$\delta \dot{x} = \frac{\partial f}{\partial x} \delta x = A(t)x \tag{15.181}$$

which has the state transition matrix Φ. Rather than linearizing f and integrating $\dot{\Phi} = A\Phi$, the initial conditions $x_k(0)$ can be perturbed and $\dot{x} = f(x)$ integrated (the "shooting" step) to generate Φ. Then $\delta x(0)$ is obtained from $x(t) = x_k(t) + \Phi(t)\delta x(0)$ and the periodicity requirement $x(T) = x(0)$, and

$$x_{k+1}(0) = x_k(0) + (I - \Phi(T))^{-1}(x_k(T) - x_k(0)) \tag{15.182}$$

is the new estimate of the initial conditions.

Friedmann (1983, 1986) developed a solution method more directly related to Floquet theory. From equation 15.135), the general solution of the linear problem $\dot{x} = Ax + b$ is

$$x(t) = \Phi(t)x(0) + \Phi(t)\int_0^t [\Phi(\tau)]^{-1}b(\tau)d\tau \tag{15.183}$$

The state transition matrix is the solution of $\dot{\Phi} = A\Phi$ with initial conditions $\Phi(0) = I$. Then $x(T) = x(0)$ gives $x(0) = (I - \Phi(T))^{-1}\Phi(T)\int_0^T \Phi^{-1}b\,d\tau$, so

$$x(t) = \Phi(t)\left[(I - \Phi(T))^{-1}\Phi(T)\int_0^T \Phi^{-1}b\,d\tau + \int_0^t \Phi^{-1}b\,d\tau\right] \tag{15.184}$$

is the periodic solution. The equations for the rotor are usually nonlinear: $M\ddot{y} + C\dot{y} + Ky = F(y, \dot{y}, \ddot{y}, t)$, or in state variable form $\dot{x} = f(x, \dot{x}, t)$. The equation of motion is linearized about the current solution estimate $x_k(t)$:

$$\dot{x}_{k+1} \cong f_k + \frac{\partial f}{\partial x}(x_{k+1} - x_k) + \frac{\partial f}{\partial \dot{x}}(\dot{x}_{k+1} - \dot{x}_k) = A_k x_{k+1} + b_k \tag{15.185}$$

which is solved for x_{k+1}. The iterative method thus solves a sequence of linearized equations. When the iteration has converged, the stability of the system is obtained from the eigenvalues of the last state transition matrix Φ.

15.7.5 Algebraic Equations

The analysis of rotors leads to nonlinear algebraic equations, used either directly such as to find the trim controls needed to achieve a steady-state flight condition or to balance the induced velocity and blade loading, or indirectly such as to find the periodic solution of differential equations. Nonlinear algebraic equations can be written in two forms: fixed point, $x = G(x)$, and zero point, $f(x) = 0$. Here x, G, and f are vectors. Efficient and convergent methods are required to find the solution $x = \alpha$ of these equations. Note that $f'(\alpha) = 0$ or $G'(\alpha) = 1$ means that α is a higher-order root. For nonlinear problems, the method is iterative: $x_{n+1} = F(x_n)$. The operation F depends on the solution method. The solution error is

$$\epsilon_{n+1} = \alpha - x_{n+1} = F(\alpha) - F(x_n) = (\alpha - x_n)F'(\xi_n) \cong \epsilon_n F'(\alpha) \tag{15.186}$$

Thus the iteration converges if F is not too sensitive to errors in x: $|F'(\alpha)| < 1$ for scalar x. When x is a vector, the criterion is that all the eigenvalues of the derivative matrix $\partial F/\partial x$ have magnitude less than one. The equations in this section are generally written for scalar x; the extension to vector x is straightforward. Convergence is linear for F' nonzero and is quadratic for $F' = 0$.

15.7.6 Successive Substitution

The successive substitution method (with relaxation) is an example of a fixed point solution. A direct iteration is simply $x_{n+1} = G(x_n)$, but $|G'| > 1$ for many practical problems. A relaxed iteration uses $F = (1 - \lambda)x + \lambda G$:

$$x_{n+1} = (1 - \lambda)x_n + \lambda G(x_n) \tag{15.187}$$

with relaxation factor λ. The convergence criterion is then

$$|F'(\alpha)| = |1 - \lambda + \lambda G'| < 1 \tag{15.188}$$

so a value of λ can be found to ensure convergence for any finite G'. Specifically, the iteration converges if the magnitude of λ is less than the magnitude of $2/(1 - G')$ (and λ has the same sign as $(1 - G')$). Quadratic convergence ($F' = 0$) is obtained with $\lambda = 1/(1 - G')$. Over-relaxation ($\lambda > 1$) can be used if $|G'| < 1$.

Since the correct solution $x = \alpha$ is not known, convergence must be tested by comparing the values of two successive iterations: error $= \|x_{n+1} - x_n\| \leq$ tolerance, where the error is some norm of the difference between iterations (typically absolute value for scalar x). The effect of the relaxation factor is to reduce the difference between iterations: $x_{n+1} - x_n = \lambda(G(x_n) - x_n)$. Hence a reduction of λ must be accompanied by a reduction in the tolerance to maintain the same convergence accuracy.

The successive substitution method fails if $G'(\alpha) = \infty$. In such a case, the iteration typically oscillates about the correct solution, the magnitude of the oscillation decreasing as λ approaches zero. But at $\lambda = 0$ the iteration is turned off, so the correct solution can never be found.

15.7.7 Newton-Raphson

The Newton-Raphson method (with relaxation and identification) is an example of a zero point solution. A problem specified as $h(x) = h_{\text{target}}$ becomes a zero point problem with $f = h - h_{\text{target}}$. The Taylor series expansion of $f(x) = 0$ leads to the iteration operator $F = x - f/f'$:

$$x_{n+1} = x_n - [f'(x_n)]^{-1} f(x_n) \tag{15.189}$$

which gives quadratic convergence. The behavior of this iteration depends on the accuracy of the derivative f'. Typically the analysis can evaluate directly f, but not f'. So f' must be evaluated by numerical perturbation of f, and for efficiency the derivatives should not be evaluated for each x_n. These approximations compromise the convergence of the method, so a relaxation factor λ is introduced to compensate. Hence a modified Newton-Raphson iteration is used, $F = x - Cf$:

$$x_{n+1} = x_n - Cf(x_n) = x_n - \lambda D^{-1} f(x_n) \tag{15.190}$$

where the derivative matrix D is an estimate of f'. The convergence criterion is then

$$|F'(\alpha)| = |1 - Cf'| = |1 - \lambda D^{-1} f'| < 1 \tag{15.191}$$

since $f(\alpha) = 0$. The iteration converges if the magnitude of λ is less than the magnitude of $2D/f'$ (and λ has the same sign as D/f'). Quadratic convergence is obtained with $\lambda = D/f'$. The Newton-Raphson method ideally uses the local derivative in the gain factor, $C = 1/f'$, so has quadratic convergence:

$$F'(\alpha) = \frac{ff''}{f'^2} = 0 \tag{15.192}$$

since $f(\alpha) = 0$ (if $f' \neq 0$ and f'' is finite; if $f' = 0$ then there is a multiple root, $F' = 1/2$, and the convergence is only linear). A relaxation factor is still useful, since the convergence is quadratic only sufficiently close to the solution.

The derivative matrix D is obtained by an identification process, either perturbation or recursive. In the secant method, the derivative of f is evaluated numerically at each step:

$$f'(x_n) \cong \frac{f(x_n) - f(x_{n-1})}{x_n - x_{n-1}} \tag{15.193}$$

It can be shown then that the error reduces during the iteration according to

$$|\epsilon_{n+1}| \cong |f''/2f'| |\epsilon_n| |\epsilon_{n-1}| \cong |f''/2f'|^{.62} |\epsilon_n|^{1.62} \tag{15.194}$$

which is slower than the quadratic convergence of the Newton-Raphson method (ϵ_n^2), but still better than linear convergence. In practical problems, whether the iteration converges at all is often more important than the rate of convergence. A Newton-Raphson method has good convergence when x is sufficiently close to the solution, but frequently has difficulty converging elsewhere. Hence the initial estimate x_0 that starts the iteration is a important parameter affecting convergence. Convergence of the solution for x can be tested in terms of the required value (zero) for f: error $= \|f\| \leq$ tolerance, where the error is some norm of f (typically an absolute value for scalar f).

15.8 REFERENCES

Berman, A. "A New Approach to Rotor Blade Dynamic Analysis." Journal of the American Helicopter Society, *10*:3 (July 1965).

Bir, G., Chopra, I., and Nguyen, K. "Development of UMARC (University of Maryland Advanced Rotorcraft Code)." American Helicopter Society 46th Annual Forum, Washington, DC, May 1990.

Chopra, I., and Sivaneri, N.T. "Aeroelastic Stability of Rotor Blades Using Finite Element Analysis." NASA CR 166389, August 1982.

Coleman, R.P. "Theory of Self-Excited Mechanical Oscillations of Hinged Rotor Blades." NACA ARR 3G29, July 1943.

Coleman, R.P., and Feingold, A.M. "Theory of Ground Vibrations of a Two-Blade Helicopter Rotor on Anisotropic Flexible Supports." NACA TN 1184, January 1947.

DeRusso, P.M., Roy, R.J., and Close, C.M. *State Variables for Engineers*. New York: John Wiley and Sons, Inc., 1965.

Friedmann, P.P. "Formulation and Solution of Rotary-Wing Aeroelastic Stability and Response Problems." Vertica, 7:2 (1983).

Friedmann, P.P. "Numerical Methods for Determining the Stability and Response of Periodic Systems with Application to Helicopter Rotor Dynamics and Aeroelasticity." Computers and Mathematics with Applications, *12A*:1 (1986).

Gessow, A. "Equations and Procedures for Numerically Calculating the Aerodynamic Characteristics of Lifting Rotors." NACA TN 3747, October 1956.

Grodko, L.N. "Ground Resonance." In *Helicopter, Calculation and Design*, Mil, M.L., et al. Moscow: Izdatel'stvo Mashinostroyeniye, 1966. Translation "Vibrations and Dynamic Stability", NASA TT F-519, May 1968.

15.8 References

Hoffman, J.A. "A Multi-Blade Coordinate Transformation Procedure for Rotors with Two Blades." Paragon Pacific Inc., Report PPI-1014-5, September 1976.

Hohenemser, K.H., and Yin, S.-K. "Some Applications of the Method of Multiblade Coordinates." Journal of the American Helicopter Society, *17*:3 (July 1972).

Johnson, R.L., and Hohenemser, K.H. "On the Dynamics of Lifting Rotors with Thrust or Tilting Moment Feedback Controls." Journal of the American Helicopter Society, *15*:1 (January 1970).

Johnson, W. "Development of a Comprehensive Analysis for Rotorcraft. Part II – Aircraft Model, Solution Procedure, and Applications." Vertica, *5*:3 (1981).

Miller, R.H. "Helicopter Control and Stability in Hovering Flight." Journal of the Aeronautical Sciences, *15*:8 (August 1948).

Peters, D.A. "Fast Floquet Theory and Trim for Multi-Bladed Rotorcraft." Journal of the American Helicopter Society, *39*:4 (October 1994).

Peters, D.A., and Izadpanah, A.P. "Helicopter Trim by Periodic Shooting with Newton-Raphson Iteration." American Helicopter Society 37th Annual Forum, New Orleans, LA, May 1981.

Tanner, W.H. "Charts for Estimating Rotary Wing Performance in Hover and at High Forward Speeds." NASA CR 114, November 1964.

Young, M.I., and Lytwyn, R.T. "The Influence of Blade Flapping Restraint on the Dynamic Stability of Low Disk Loading Propeller-Rotors." Journal of the American Helicopter Society, *12*:4 (October 1967).

16 Blade Motion

The differential equations of motion for the rotor blade are derived in this chapter. First the focus is on the inertial and structural forces on the blade, with the aerodynamics represented by the net forces and moments on the blade section. Then the aerodynamic loads are analyzed in more detail to complete the equations. In subsequent chapters the equations are solved for a number of fundamental rotor problems, including flap response, aeroelastic stability, and aircraft flight dynamics. In Chapter 6 the flap and lag dynamics of an articulated rotor were analyzed for only the rigid motion of the blade, including hinge spring or offset. The present chapter extends the derivation of the equations of motion to include a hingeless rotor, higher blade bending modes, blade torsion, and pitch motion. The corresponding hub reactions and blade loads are derived, and the rotor shaft motion is included in the analysis.

The rotor blade equations of motion are derived using the Newtonian approach, with a normal mode representation of the blade motion. The chapter begins with a discussion of the other approaches by which the dynamics can be analyzed. Engineering beam theory is commonly used in helicopter blade analyses. The blade section is assumed to be rigid, so its motion is represented by the bending and rotation of a slender beam. This is normally a good model for the rotor blade, although a more detailed structural analysis is required to obtain the effective beam parameters for some portions of the blade, such as flexbeams and at the root.

16.1 Sturm-Liouville Theory

The results of Sturm-Liouville theory are required to deal with the normal modes of the blade bending and torsion motion. Consider an ordinary differential equation of the form $\mathcal{L}y + \lambda Ry = 0$, where $\mathcal{L}$ is a linear differential operator of the form

$$\mathcal{L} = \frac{d^2}{dx^2} S \frac{d^2}{dx^2} + \frac{d}{dx} P \frac{d}{dx} + Q \tag{16.1}$$

Here S, P, Q, and R are symmetric operators. (An operator S is symmetric if $\phi_1 S \phi_2 = \phi_2 S \phi_1$ for all functions ϕ_1 and ϕ_2.) With the appropriate boundary conditions at the end points $x = a$ and $x = b$, this is an eigenvalue problem for λ.

Consider any two distinct eigenvalues λ_1 and λ_2 and their corresponding eigenfunctions ϕ_1 and ϕ_2. Using the differential equations satisfied by these functions, and integrating twice by parts, we obtain

$$(\lambda_2 - \lambda_1) \int_a^b \phi_1 R \phi_2 \, dx = \int_a^b (\phi_2 \mathcal{L} \phi_1 - \phi_1 \mathcal{L} \phi_2) dx$$

$$= \left[\phi_2 \left(\frac{d}{dx} S \frac{d^2 \phi_1}{dx^2} + P \frac{d\phi_1}{dx} \right) - \phi_1 \left(\frac{d}{dx} S \frac{d^2 \phi_2}{dx^2} + P \frac{d\phi_2}{dx} \right) \right] \Big|_a^b$$

$$- \left[\frac{d\phi_2}{dx} S \frac{d^2 \phi_1}{dx^2} - \frac{d\phi_1}{dx} S \frac{d^2 \phi_2}{dx^2} \right] \Big|_a^b \quad (16.2)$$

The right-hand side is zero for boundary conditions of the following form:

$$\frac{d}{dx} S \frac{d^2 y}{dx^2} = K_1 y \quad \text{and} \quad S \frac{d^2 y}{dx^2} = K_2 \frac{dy}{dx} \quad (16.3)$$

or $S = 0$, and

$$P \frac{dy}{dx} = K_3 y \quad (16.4)$$

or $P = 0$, where K_1, K_2, and K_3 are constants. With such boundary conditions

$$\int_a^b \phi_1 R \phi_2 \, dx = 0 \quad (16.5)$$

so the eigen-solutions are orthogonal over the interval from a to b, with weighting function R. For a beam in bending, the following end restraints satisfy these boundary conditions:

a) a free end, for which $d^2 y/dx^2 = d^3 y/dx^3 = 0$ and $P = 0$
b) a hinged end, for which $y = 0$ and $S d^2 y/dx^2 = K dy/dx$, where K is the hinge spring constant ($d^2 y/dx^2 = 0$ with no spring)
c) a cantilever end, for which $y = 0$ and $dy/dx = 0$ (which is also the limit of $K = \infty$ with a spring)

For a rod in torsion (so $S = 0$), the following end restraints satisfy these boundary conditions:

a) a free end with $dy/dx = 0$
b) a fixed end with $y = 0$
c) a restrained end with $P dy/dx = Ky$, where K is the spring constant

A proper Sturm-Liouville problem has boundary conditions of the form given earlier, and R and P of opposite sign to S and Q. These conditions are satisfied for the blade bending and torsion problems encountered in this chapter. Then the eigen-solutions are orthogonal, the eigenvalues λ are real and positive, and an expansion of an arbitrary function over the interval $x = a$ to $x = b$ as a series in the eigen-solutions converges.

The eigenvalue λ can be obtained from the eigen-solutions as follows:

$$-\lambda \int_a^b \phi R\phi \, dx = \int_a^b \phi \mathcal{L}\phi \, dx$$

$$= \left[\phi \frac{d}{dx} S \frac{d^2\phi}{dx^2} - \frac{d\phi}{dx} S \frac{d^2\phi}{dx^2} + \phi P \frac{d\phi}{dx}\right]\bigg|_a^b$$

$$+ \int_a^b \left[\frac{d^2\phi}{dx^2} S \frac{d^2\phi}{dx^2} - \frac{d\phi}{dx} P \frac{d\phi}{dx} + \phi Q\phi\right] dx \quad (16.6)$$

For example, for a beam with a free end at $x = b$ and a general restrained end at $x = a$,

$$-\lambda \int_a^b \phi R\phi \, dx = \frac{d\phi}{dx} K \frac{d\phi}{dx}\bigg|_{x=a} + \int_a^b \left[\frac{d^2\phi}{dx^2} S \frac{d^2\phi}{dx^2} - \frac{d\phi}{dx} P \frac{d\phi}{dx} + \phi Q\phi\right] dx \quad (16.7)$$

and for a rod with a free end at $x = b$ and a restrained end at $x = a$,

$$\lambda \int_a^b \phi R\phi \, dx = \phi K\phi\big|_{x=a} + \int_a^b \left[\frac{d\phi}{dx} P \frac{d\phi}{dx} - \phi Q\phi\right] dx \quad (16.8)$$

The exact value for λ is obtained if the exact eigen-function is used. These expressions are also useful to estimate λ by using approximate mode shapes.

16.2 Derivation of Equations of Motion

The equations of motion for the rotor blade are derived in this chapter using an integral Newtonian method to obtain the partial differential equations for bending or torsion and using a normal mode expansion to obtain ordinary differential equations for the normal coordinates. The choice of this approach was based on the physical insight that is gained by working directly with the forces and accelerations of the blade. Other methods are also used to derive the equations of motion for analyses of rotor dynamics. As a guide to what may be encountered in the literature, this section outlines several of these alternative approaches.

Rotor dynamics analyses are frequently based on Lagrange's equations,

$$\frac{d}{dt}\frac{\partial T}{\partial \dot{q}_i} - \frac{\partial T}{\partial q_i} + \frac{\partial U}{\partial q_i} = Q_i \quad (16.9)$$

where T and U are the kinetic and potential energies of the entire system, q_i are the generalized coordinates (degrees of freedom), and Q_i are generalized forces. Usually T gives the inertial terms, U the structural terms, and Q_i the aerodynamic terms of the equations of motion. The derivation of the equations of motion by Lagrange's equations is simply formulated and routinely (if somewhat laboriously) executed. Consequently, Lagrangian methods are often used in the development of the most comprehensive models found in the literature.

To illustrate the various derivation methods, consider the bending of a cantilever beam (Figure 16.1). A distributed loading $p(r)$, tip force F_T, and tip moment M_T are included. The spanwise variable is r and the bending deflection is z. The beam is not rotating, since the present purpose is to examine the methods of analysis rather than the rotor blade behavior. The objective is to obtain a set of ordinary differential equations (in time) describing the bending motion $z(r, t)$ of the cantilever beam.

16.2 Derivation of Equations of Motion

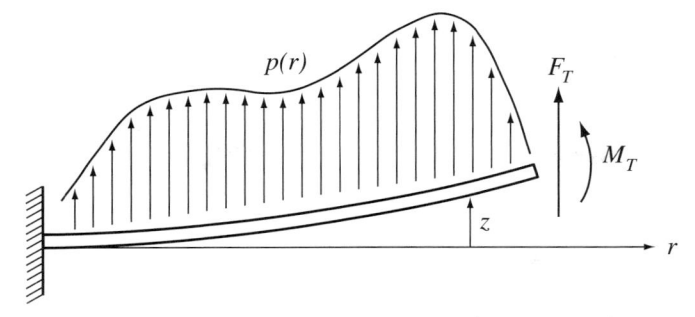

Figure 16.1. Bending of a cantilever beam (non-rotating).

16.2.1 Integral Newtonian Method

Newtonian methods derive the equations of motion from the equilibrium of forces on the body. In the case of the cantilever beam, the moment on the section at r consists of the tip moment M_T, the tip force F_T with moment arm $(R - r)$, and the section loading $(p - m\ddot{z})$ integrated over the portion outboard of r with moment arm $(\rho - r)$:

$$M(r) = \int_r^R (p - m\ddot{z})(\rho - r)d\rho + M_T + (R - r)F_T \tag{16.10}$$

Equating this to $M(r) = EIz''$ from engineering beam theory, and taking the second derivative, gives the required partial differential equation for bending:

$$\left(EIz''\right)'' + m\ddot{z} = p \tag{16.11}$$

Boundary conditions are needed to complete the problem formulation. Evaluating $M(r)$ and $M'(r)$ at the tip gives $EIz'' = M_T$ and $(EIz'')' = -F_T$ at $r = R$. The boundary conditions for the cantilever root are just $z = z' = 0$ at $r = 0$.

16.2.2 Differential Newtonian Method

The equation of motion can also be derived from the equilibrium of forces and moments on the differential beam element extending from r to $r + dr$. Let S and M be the shear and moment on the section at r, and $S + dS = S + S'dr$ and $M + dM = M + M'dr$ the reactions on the section at $r + dr$. Force balance on the differential element gives

$$p\,dr + (S + S'dr) - S = m\ddot{z}\,dr \tag{16.12}$$

or

$$p + S' = m\ddot{z} \tag{16.13}$$

Moment balance gives

$$(M + M'dr) - M + S\,dr = 0 \tag{16.14}$$

or

$$M' + S = 0 \tag{16.15}$$

Eliminating S then gives the partial differential equation

$$M'' + m\ddot{z} = p \tag{16.16}$$

which with $M = EIz''$ becomes

$$(EIz'')'' + m\ddot{z} = p \tag{16.17}$$

16.2.3 Lagrangian Method

Lagrangian methods derive the equation of motion from energy considerations instead of from the equilibrium of forces. Hamilton's principle states that the motion of a dynamic system is determined by the condition

$$\int_{t_1}^{t_2} (\delta T - \delta U + \delta W) dt = 0 \tag{16.18}$$

where T is the system kinetic energy, U the potential energy, and δW the virtual work of non-conservative forces. Since for a conservative system the criterion is that $\int_{t_1}^{t_2}(T - U)dt$ has a minimum value, this principle is also called the principle of least action.

For the cantilever beam, the kinetic and potential energy are

$$T = \int_0^R \frac{1}{2} m \dot{z}^2 dr \tag{16.19}$$

$$U = \int_0^R \frac{1}{2} EI z''^2 dr \tag{16.20}$$

and the virtual work by the distributed force and tip loads is

$$\delta W = M_T \delta z'(R) + F_T \delta z(R) + \int_0^R p \delta z \, dr \tag{16.21}$$

Hamilton's principle thus requires

$$\int_{t_1}^{t_2} \left[\int_0^R \left(p - m\ddot{z} - (EIz'')'' \right) \delta z \, dr + \left(-EIz'' + M_T \right) \delta z'(R) \right.$$
$$\left. + \left((EIz'')' + F_T \right) \delta z(R) \right] dt = 0 \tag{16.22}$$

It is necessary to integrate the kinetic energy term by parts with respect to t and the potential energy term twice by parts with respect to r. Since the variational δz is arbitrary,

$$(EIz'')'' + m\ddot{z} = p \tag{16.23}$$

$$(EIz'')|_{r=R} = M_T \tag{16.24}$$

$$(EIz'')'|_{r=R} = -F_T \tag{16.25}$$

These expressions are the partial differential equation for bending of the beam, and the boundary conditions at the tip.

16.2.4 Normal Mode Method

This chapter uses an expansion of the bending deflection as a series in the normal modes of free vibration to obtain the ordinary differential equations of motion. Consider the cantilever beam, but with no tip force or moment. The differential equation for free vibration is

$$(EIz'')'' + m\ddot{z} = 0 \tag{16.26}$$

Assuming $z = \eta(r)e^{i\nu t}$, the modal equation

$$(EI\eta'')'' - \nu^2 m\eta = 0 \tag{16.27}$$

with boundary conditions $\eta = \eta' = 0$ at $r = 0$, and $\eta'' = \eta''' = 0$ at $r = R$, is obtained. This is a proper Sturm-Liouville eigenvalue problem, with a series of eigen-solutions η_k and eigenvalues ν_k^2. The mode shapes are orthogonal with weighting function m, so if $i \neq k$,

$$\int_0^R \eta_i \eta_k m \, dr = 0 \tag{16.28}$$

Sturm-Liouville theory also gives for the eigenvalues

$$\nu^2 = \frac{\int_0^R EI\eta''^2 dr}{\int_0^R \eta^2 m \, dr} \tag{16.29}$$

Finally, Sturm-Liouville theory shows that an expansion of a function as a series in q_k converges.

The bending deflection is expanded as a series in the free vibration modes,

$$z(r,t) = \sum_{k=1}^{\infty} \eta_k(r) q_k(t) \tag{16.30}$$

where q_k is the degree of freedom for the k-th mode. Substituting this expansion in the partial differential equation gives

$$\sum_{k=1}^{\infty} \left[(EI\eta_k'')'' q_k + m\eta_k \ddot{q}_k \right] = p \tag{16.31}$$

Using the equation satisfied by q_k, the structural term is replaced by the natural frequency:

$$\sum_{k=1}^{\infty} \left[m\eta_k \left(\ddot{q}_k + \nu_k^2 q_k \right) \right] = p \tag{16.32}$$

Operating with $\int_0^R (\ldots) \eta_k dr$, and using the orthogonality of the modes, gives

$$\left(\int_0^R \eta_k^2 m \, dr \right) (\ddot{q}_k + \nu_k^2 q_k) = \int_0^R \eta_k p \, dr \tag{16.33}$$

which is the ordinary differential equation of motion for the k-th bending mode. Using the energy expression for ν_k^2, this equation can be written as

$$\left(\int_0^R \eta_k^2 m \, dr \right) \ddot{q}_k + \left(\int_0^R EI\eta_k''^2 \, dr \right) q_k = \int_0^R \eta_k p \, dr \tag{16.34}$$

By using the orthogonal modes of free vibration, the structural and inertial terms of the equations of motion are uncoupled.

The limitation of using normal modes is that since each of the modes satisfies a homogeneous boundary condition, the solution z must do the same. Thus directly including the tip force and moment in this approach is not possible.

16.2.5 Galerkin Method

The Galerkin method also uses a modal expansion of the bending deflection to obtain the ordinary differential equations of motion, but not necessarily the normal modes of free vibration. Let $z = \sum_k \eta_k(r) q_k(t)$, where q_k are the generalized coordinates describing the motion and η_k are a series of modes. Each of the modes η_k must satisfy the boundary conditions at the root, and the total deflection z must satisfy the boundary conditions at the tip. The true solution satisfies the differential equation

$$(EIz'')'' + m\ddot{z} = p \tag{16.35}$$

The Galerkin solution does not in general satisfy this equation exactly, since a finite number of modes is used. Therefore, define an error function

$$\epsilon = p - (EIz'')'' - m\ddot{z} \tag{16.36}$$

The equations of motion are obtained from the requirement that the equation error ϵ be small and specifically that

$$\int_0^R \eta_i \epsilon \, dr = 0 \tag{16.37}$$

Substituting for ϵ and employing the modal expansion for z gives

$$\sum_k \int_0^R m \eta_i \eta_k dr \, \ddot{q}_k + \sum_k \int_0^R \eta_i \left(EI\eta_k''\right)'' dr \, q_k = \int_0^R \eta_i p \, dr \tag{16.38}$$

Now integrating twice by parts and using the boundary conditions, the structural term can be written:

$$\sum_k \int_0^R \eta_i \left(EI\eta_k''\right)'' dr \, q_k$$

$$= \sum_k \left[\eta_i \left(EI\eta_k''\right)' q_k - \eta_i' \left(EI\eta_k''\right) q_k\right]_0^R + \sum_k \int_0^R EI\eta_i'' \eta_k'' dr \, q_k$$

$$= \left[\eta_i \left(EIz''\right)' - \eta_i' \left(EIz''\right)\right]_0^R + \sum_k \int_0^R EI\eta_i'' \eta_k'' dr \, q_k$$

$$= -\eta_i(R) F_T - \eta_i'(R) M_T + \sum_k \int_0^R EI\eta_i'' \eta_k'' dr \, q_k \tag{16.39}$$

Thus the equation of motion for the i-th mode is

$$\sum_k M_{ik} \ddot{q}_k + \sum_k K_{ik} q_k = \int_0^R \eta_i p \, dr + \eta_i(R) F_T + \eta_i'(R) M_T \tag{16.40}$$

where $M_{ik} = \int_0^R m \eta_i \eta_k dr$ and $K_{ik} = \int_0^R EI \eta_i'' \eta_k'' dr$. The normal mode expansion gave a similar result. With the Galerkin method the mass and spring matrices are not diagonal because the free vibration modes are not necessarily used, but the excitation by the tip force and moment are now included. The Galerkin method is equivalent to the Rayleigh-Ritz method (discussed later) when the proper weighting function is used for the equation error integral (η_i in this case). The Rayleigh-Ritz procedure has a stronger physical and mathematical basis, but the Galerkin procedure allows the use of a Newtonian approach to derive the equation of motion.

16.2 Derivation of Equations of Motion

If the free vibration modes are used in the Galerkin method, the mass and spring matrices are diagonal and the equations of motion become

$$\left(\int_0^R \eta_k^2 m\, dr\right) (\ddot{q}_k + v_k^2 q_k) = \int_0^R \eta_k p\, dr + \eta_k(R) F_T + \eta_k'(R) M_T \quad (16.41)$$

This is just the normal mode result, but now with the excitation due to the tip loading included. This suggests that the equations of motion for the rotor can be derived using the normal mode procedure, and then the influence of point loads on the blade can be accounted for by adding terms according to the Galerkin procedure. Such an approach is useful, for example, in adding a lag damper to the normal mode analysis or for the control system force and moment at the pitch bearing.

16.2.6 Rayleigh-Ritz Method

If the energy and virtual work are expressed in terms of the generalized coordinates q_i

$$T = T(q_i, \dot{q}_i) \quad (16.42)$$

$$U = U(q_i) \quad (16.43)$$

$$\delta W = \sum_i Q_i \delta q_i \quad (16.44)$$

then application of Hamilton's principle leads to Lagrange's equations,

$$\frac{d}{dt}\frac{\partial T}{\partial \dot{q}_i} - \frac{\partial T}{\partial q_i} + \frac{\partial U}{\partial q_i} = Q_i \quad (16.45)$$

By means of Lagrange's equations the ordinary differential equations of motion for the generalized coordinates describing the motion are obtained directly from the expressions for the system energy, without going through the partial differential equation.

Consider again a modal expansion for the bending deflection, as in the Galerkin method: $z = \sum_k \eta_k q_k$. Substituting for z gives the energy and virtual work in terms of the generalized coordinates:

$$T = \sum_i \sum_k \frac{1}{2}\int_0^R \eta_i \eta_k m\, dr\, \dot{q}_i \dot{q}_k = \sum_i \sum_k \frac{1}{2} M_{ik} \dot{q}_i \dot{q}_k \quad (16.46)$$

$$U = \sum_i \sum_k \frac{1}{2}\int_0^R EI \eta_i'' \eta_k''\, dr\, q_i q_k = \sum_i \sum_k \frac{1}{2} K_{ik} q_i q_k \quad (16.47)$$

$$\delta W = \sum_i Q_i \delta q_i = \left[\int_0^R \eta_i p\, dr + \eta_i(R) F_T + \eta_i'(R) M_T\right] \delta q_i \quad (16.48)$$

Application of Lagrange's equations then gives directly

$$\sum_k M_{ik} \ddot{q}_k + \sum_k K_{ik} q_k = \int_0^R \eta_i p\, dr + \eta_i(R) F_T + \eta_i'(R) M_T \quad (16.49)$$

which is identical to the result obtained by the Galerkin method.

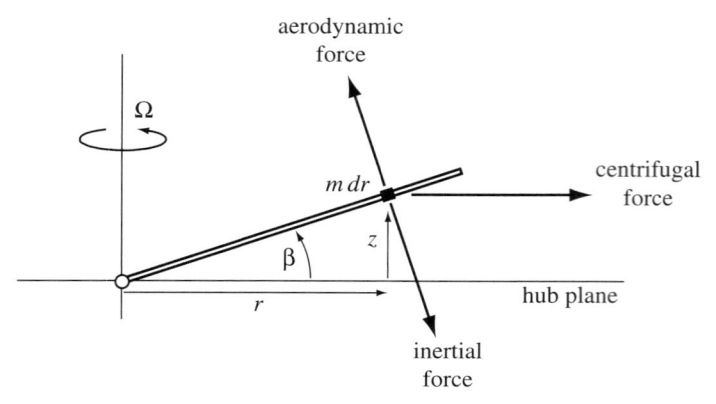

Figure 16.2. Rotor blade flapping moments.

16.2.7 Lumped Parameter and Finite Element Methods

In lumped parameter or finite element methods the continuous physical system is modeled by a series of discrete elements. For example, the cantilever beam considered in Figure 16.1 might be represented by finite masses located at a series of points and be connected by massless elastic elements with uniform properties (lumped mass model) or represented by finite-length beam segments (finite element model). The equations of motion are usually derived by Lagrangian methods. The greatest advantage of finite element methods is that they generally have the flexibility to treat complex configurations. The problem for a new system is to specify its geometry and properties in the manner required by the method used, rather than developing an entirely new analysis.

16.3 Out-of-Plane Motion

16.3.1 Rigid Flapping

We begin the development of the equations of motion for a rotor blade by considering the rigid flap motion of an articulated rotor, a derivation presented in detail in Chapter 6. The degree of freedom β is the angle of rotation about the flap hinge (Figure 16.2), so the out-of-plane deflection is $z = \beta r$. There is no flap-hinge offset or spring restraint. The equation of motion is obtained from the equilibrium of moments about the flap hinge. Based on the results of section 6.9, the gravitational moments are neglected in this chapter. The section forces and their moment arms about the flap hinge are as follows:

i) an inertial force $m\ddot{z} = mr\ddot{\beta}$ opposing the flap motion, with moment arm r about the flap hinge
ii) a centrifugal force $m\Omega^2 r$ directed radially outward, with moment arm $z = r\beta$
iii) an aerodynamic force F_z normal to the blade, with moment arm r

Here m is the blade mass per unit length at radial station r. The equilibrium of moments about the flap hinge gives

$$\left(\int_0^R r^2 m\, dr\right)(\ddot{\beta} + \Omega^2 \beta) = \int_0^R r F_z\, dr \qquad (16.50)$$

16.3 Out-of-Plane Motion

On dividing by the flap moment of inertia $I_b = \int_0^R r^2 m\, dr$ and using dimensionless quantities,

$$\ddot{\beta} + \beta = \frac{1}{I_b}\int_0^1 rF_z\, dr = \gamma \int_0^1 r\frac{F_z}{ac}\, dr \qquad (16.51)$$

where $\gamma = \rho acR^4/I_b$ is the blade Lock number. This is the equation of motion for the rigid flapping of an articulated rotor blade. The centrifugal spring gives a natural frequency of $\nu = 1$/rev in the rotating frame.

With an offset flap hinge, the blade out-of-plane deflection due to rigid rotation about the hinge becomes $z = \beta\eta$, where β is the degree of freedom and the mode shape is $\eta = (r - e)/(1 - e)$ (e is the flap-hinge offset). Since the mode shape has been normalized to $\eta = 1$ at the tip, β is the angle that a line from the center of rotation to the blade tip makes with the hub plane. The section forces are now

i) an inertial force $m\ddot{z} = m\eta\ddot{\beta}$, with moment arm $(r - e)$
ii) a centrifugal force $m\Omega^2 r$, with moment arm $z = \eta\beta$
iii) an aerodynamic force F_z, with moment arm $(r - e)$

Including a hinge spring (with precone angle β_p), the moment equilibrium about the flap hinge becomes

$$\int_e^R \eta(r - e)m\, dr\, \ddot{\beta} + \int_e^R \eta rm\, dr\, \Omega^2\beta + K_\beta(\beta - \beta_p) = \int_e^R (r - e)F_z\, dr \qquad (16.52)$$

Divide by $(1 - e)$ and write $I_\beta = \int_e^R \eta^2 m\, dr$ for the generalized mass of the flap mode. Then

$$I_\beta(\ddot{\beta} + \nu^2\beta) = \frac{K_\beta}{\Omega^2(1 - e)}\beta_p + \int_e^1 \eta F_z\, dr \qquad (16.53)$$

where the natural frequency of the flap motion is

$$\nu^2 = 1 + \frac{e}{1 - e}\frac{\int_e^1 \eta m\, dr}{\int_e^1 \eta^2 m\, dr} + \frac{K_\beta}{I_\beta\Omega^2(1 - e)} \qquad (16.54)$$

Finally, divide by the characteristic inertia I_b to obtain

$$\widehat{I}_\beta(\ddot{\beta} + \nu^2\beta) = \frac{K_\beta}{I_b\Omega^2(1 - e)}\beta_p + \gamma\int_e^1 \eta\frac{F_z}{ac}\, dr \qquad (16.55)$$

where $\widehat{I}_\beta = I_\beta/I_b$ and the Lock number is again $\gamma = \rho acR^4/I_b$. The flap frequency $\nu = 1$/rev if there is no hinge offset or spring. For a uniform mass distribution

$$\nu^2 = 1 + \frac{3}{2}\frac{e}{1 - e} + \frac{K_\beta}{I_\beta\Omega^2(1 - e)} \qquad (16.56)$$

so in general $\nu > 1$/rev.

The practice in this chapter is to use I_b for the characteristic inertia of the rotor blade. This parameter normalizes the generalized masses of the blade motion (such as $\widehat{I}_\beta = I_\beta/I_b$) and represents the blade inertial forces in the Lock number, $\gamma = \rho acR^4/I_b$. The actual value of I_b has no influence on the numerical solution, since the entire equation of motion is divided by I_b. A good choice for the blade inertia is $I_b = \int_0^R r^2 m\, dr$, even for blades with hinge offset or no flap hinge at all. This is a well-defined parameter of the blade that can be obtained from the rotary inertia about the shaft and avoids any dependence on the flap mode shape.

16.3.2 Out-of-Plane Bending

Consider now the out-of-plane bending of a rotor blade with arbitrary root constraint. This model includes the higher bending modes of articulated blades and covers the case of a hingeless rotor blade with cantilever root restraint. In Chapter 6, hingeless rotor dynamics were discussed in terms of the fundamental flapping mode. The present analysis adds the equation by which the frequency and mode shape can be calculated, as well as a derivation of the differential equation of motion. The equation of motion is obtained by considering the equilibrium of aerodynamic, inertial, and structural bending moments on the blade portion outboard of radial station r. Let $z(r)$ be the out-of-plane deflection of the blade. The forces acting on the blade section at radial station ρ, with their moment arm about the radial station at r, are as follows:

i) an inertial force $m\ddot{z}(\rho)$, with moment arm $(\rho - r)$
ii) a centrifugal force $m\Omega^2\rho$, with moment arm $(z(\rho) - z(r))$
iii) an aerodynamic force F_z, with moment arm $(\rho - r)$

The moment on the blade section at r due to the forces acting on the blade outboard of r is then

$$M(r) = \int_r^R \left[(F_z - m\ddot{z})(\rho - r) - m\Omega^2\rho(z(\rho) - z(r))\right] d\rho \qquad (16.57)$$

Now engineering beam theory relates the structural moment to the bending curvature of the blade:

$$M(r) = EI \frac{d^2z}{dr^2} \qquad (16.58)$$

where E is the modulus of elasticity of the blade section and I is the modulus-weighted area moment about the chordwise principal axis. The equilibrium distribution of the structural, inertial, and aerodynamic moments on the section gives

$$EI \frac{d^2z}{dr^2} + \int_r^R m\Omega^2\rho(z(\rho) - z(r))d\rho + \int_r^R m\ddot{z}(\rho - r)d\rho = \int_r^R F_z(\rho - r)d\rho \qquad (16.59)$$

The second derivative then gives the partial differential equation for the out-of-plane bending of a rotor blade:

$$\frac{d^2}{dr^2} EI \frac{d^2z}{dr^2} - \frac{d}{dr}\left[\int_r^R m\Omega^2\rho\, d\rho \frac{dz}{dr}\right] + m\ddot{z} = F_z \qquad (16.60)$$

The boundary conditions are as follows. The blade tip is a free end, with zero moment and shear force, so $d^2z/dr^2 = d^3z/dr^3 = 0$ at $r = R$. The root of an articulated blade has a hinge, with zero displacement and moment, so $z = d^2z/dr^2 = 0$ at $r = e$ (allowing for a hinge offset). The root of a hingeless rotor has cantilever restraint, with zero displacement and slope, so $z = dz/dr = 0$ at $r = e$ (allowing for a very stiff hub). The root restraint can be generalized by considering a hinge with spring constant K_β, so that $EI(d^2z/dr^2) = K_\beta(dz/dr)$ at $r = e$. For $K_\beta = 0$ this reduces to the articulated rotor case, and for $K_\beta = \infty$ it reduces to the hingeless rotor case.

The partial differential equation for the blade bending is solved by the method of separation of variables, which leads to ordinary differential equations (in time) for the degrees of freedom, as for rigid flapping. Thus the out-of-plane deflection $z(r, t)$ is expanded as a series in mode shapes describing the spanwise deformation.

16.3 Out-of-Plane Motion

A single equation of motion is obtained for the degree of freedom corresponding to each mode. First, an appropriate series of mode shapes for the rotating blades must be obtained. When the mode shapes are chosen such that the forced response of the blade is well described by the first few modes, the rotor dynamics problems can be solved by considering the smallest number of degrees of freedom. Consider the free vibration of the rotating blade at frequency v. In the homogeneous partial differential equation for the bending (without the aerodynamic force F_z in this case), write $z = \eta(r)e^{ivt}$, where η is the spanwise mode shape. The result is

$$\frac{d^2}{dr^2}EI\frac{d^2\eta}{dr^2} - \frac{d}{dr}\left[\int_r^R m\Omega^2 \rho \, d\rho \, \frac{d\eta}{dr}\right] - v^2 m\eta = 0 \qquad (16.61)$$

with the same boundary conditions on η as given earlier for z, for a hinged or cantilevered blade as appropriate. Since the aerodynamic force has been dropped, this result can be viewed as the equation for vibration in a vacuum, involving the equilibrium of structural, centrifugal, and inertial moments alone. This modal equation and its boundary conditions constitute an eigenvalue problem for the natural frequencies v and mode shapes $\eta(r)$. According to section 16.1 it is a proper Sturm-Liouville problem, so there exists a series of eigen-solutions $\eta_k(r)$ and corresponding eigenvalues v_k^2. The mode shapes are orthogonal with weighting function m:

$$\int_0^R \eta_k \eta_i m \, dr = 0 \qquad (16.62)$$

if $i \neq k$. Moreover, an expansion of an arbitrary function of r (such as the actual blade bending deflection) as a series in these modes converges. The modal equation is linear, so the solutions are only defined to within a multiplicative factor. The mode shapes are normalized to unit deflection at the blade tip: $\eta(1) = 1$ (or $\eta(R) = R$ with dimensional quantities). The series of natural frequencies v_1, v_2, v_3, and so on, are ordered by magnitude, such that the fundamental mode v_1, has the lowest frequency. When the modes are ordered in this fashion, the k-th mode shape has $k - 1$ nodes where $\eta(r) = 0$ (not counting the root, where $\eta = 0$ always).

Now the bending deflection z is expanded as a series in the rotating natural mode shapes:

$$z(r, t) = \sum_{k=1}^{\infty} \eta_k(r) q_k(t) \qquad (16.63)$$

The degrees of freedom of the bending motion are $q_k(t)$. With the modes normalized to unit deflection at the tip, q_k represents the angle from the hub plane made by a line from the center of rotation to the tip for the k-th mode. Since orthogonal modes are being used, the equations for q_k are simple. Substitute this expansion for z into the partial differential equation for bending:

$$\sum_k \left((EI\eta_k'')'' - \left[\int_r^R m\Omega^2 \rho \, d\rho \, \eta_k'\right]'\right) q_k + \sum_k m\eta_k \ddot{q}_k = F_z \qquad (16.64)$$

The differential equation satisfied by the mode shape q_k states that the terms in brackets equal $(v_k^2 m \eta_k)$, giving

$$\sum_{k=1}^{\infty} m\eta_k \left(\ddot{q}_k + v_k^2 q_k\right) = F_z \qquad (16.65)$$

Next, operate on this equation with $\int_0^R (\ldots) \eta_k dr$. Define the generalized mass of the k-th bending mode

$$I_{qk} = \int_0^R \eta_k^2 m\, dr \tag{16.66}$$

and recall that $\int_0^R \eta_k \eta_i m\, dr = 0$ if $i \neq k$. Then the bending equation becomes

$$I_{qk}\left(\ddot{q}_k + v_k^2 q_k\right) = \int_0^R \eta_k F_z dr \tag{16.67}$$

Using the free vibration modes of the rotating blade has allowed the structural and centrifugal terms to be replaced by the natural frequencies v_k, and because these modes are orthogonal the differential equation for the k-th mode is not coupled with other bending modes (except through the aerodynamic force). Dividing by I_b, and using dimensionless quantities, gives

$$\widehat{I}_{qk}\left(\ddot{q}_k + v_k^2 q_k\right) = \gamma \int_0^1 \eta_k \frac{F_z}{ac} dr \tag{16.68}$$

where $\widehat{I}_{qk} = I_{qk}/I_b$. This is the differential equation of motion for the k-th out-of-plane bending mode of the elastic rotor blade. A further result of Sturm-Liouville theory (section 16.1) is that the natural frequencies can be obtained from the mode shapes by

$$v^2 = \frac{K_\beta [\eta'(e)]^2 + \int_0^R \left[EI\eta''^2 + \int_r^R m\Omega^2 \rho\, d\rho\, \eta'^2\right] dr}{\int_0^R \eta^2 m\, dr} \tag{16.69}$$

The dimensionless frequency is obtained by dividing by Ω^2. This relation can be interpreted as an energy balance: $v^2 \int \eta^2 m\, dr$ is the maximum kinetic energy of the vibrating blade, $\int EI\eta''^2 dr$ is the maximum potential energy of bending, $K_\beta [\eta'(e)]^2$ is the potential energy in the hinge spring, and $\iint m\Omega^2 \rho\, d\rho\, \eta'^2 dr$ is the potential energy in the centrifugal spring. This relation can be written as $v^2 = K_1 + K_2 \Omega^2$, which is the Southwell form. The Southwell coefficients K_1 and K_2 (representing the structural and centrifugal stiffening, respectively) are constants involving integrals of the blade mode shape, which in fact are also somewhat sensitive to the rotor speed Ω. However, the Southwell form gives the basic dependence of the blade bending frequencies on the rotor speed (for a further discussion see section 18.1). This energy relation gives the exact frequency when the correct mode shape (which must be obtained by solving the modal equation) is used. Equation 16.69 is also the basis for obtaining estimates of the natural frequencies by using approximate mode shapes. Since the modes are integrated, the accuracy of the frequency estimate is good as long as the modes are fairly close to the correct shape.

The fundamental flapping mode is the lowest frequency solution of the modal equation. For an articulated rotor with no hinge offset or spring, $\eta = r$ satisfies the differential equation with dimensionless natural frequency $v = 1$/rev, which is also obtained from

$$v^2 = \frac{\int_0^1 \int_r^1 m\rho\, d\rho\, dr}{\int_0^1 r^2 m\, dr} = \frac{\int_0^1 \rho m \int_0^\rho dr\, d\rho}{\int_0^1 r^2 m\, dr} = 1 \tag{16.70}$$

The equation thus reduces to that for rigid flapping. With a hinge offset and spring, the mode shape $\eta = (r-e)/(1-e)$ gives the equation of motion and the natural frequency as in section 16.3.1:

$$\begin{aligned}
v^2 &= \frac{\int_e^1 \int_r^1 m\rho \, d\rho \, dr}{(1-e)^2 \int_e^1 \eta^2 m \, dr} + \frac{K_\beta}{I_\beta \Omega^2 (1-e)^2} \\
&= \frac{\int_e^1 m\rho(\rho-e) d\rho}{(1-e)^2 \int_e^1 \eta^2 m \, dr} + \frac{K_\beta}{I_\beta \Omega^2 (1-e)^2} \\
&= 1 + \frac{e}{1-e} \frac{\int_e^1 \eta m \, dr}{\int_e^1 \eta^2 m \, dr} + \frac{K_\beta}{I_\beta \Omega^2 (1-e)^2}
\end{aligned} \qquad (16.71)$$

except for an additional factor of $(1-e)$ in the spring term, due to a different definition of the spring constant K_β. The modal equation is actually not quite satisfied if $\eta = (r-e)/(1-e)$, but the bending involved in the fundamental mode of an articulated blade is small. For a hingeless rotor, there must be bending at the blade root, where the cantilever restraint requires zero slope. However, the centrifugal stiffening dominates the fundamental mode of even the hingeless blade, as indicated by the fact that the natural frequency is only slightly above 1/rev (typically $v = 1.10$ to 1.15). Except in the root region, therefore, the mode shape of the hingeless blade does not differ substantially from that of the articulated rotor. The natural frequency is the dominant parameter of the blade bending mode, not the mode shape. Even the small increase of the fundamental frequency above 1/rev for the hingeless blade has a major impact on the root loads of the blade and on the behavior of the rotor in general.

The second flapwise bending mode has a rotating natural frequency typically around 2.6 to 2.8/rev. As the modal number increases, so do the number of nodes and the curvature of the mode shape. The higher modes thus play an important role in the blade bending loads and their calculation. For an articulated blade, the second out-of-plane mode is often called the first bending mode, since the fundamental flap mode does not involve elastic motion of the blade. If no better estimate is available, $\eta = 4r^2 - 3r$ can be used as an approximation for the second out-of-plane mode shape of an articulated rotor blade. This expression is orthogonal to the first mode, $\eta = r$; however, the boundary conditions of zero moment at the root and tip are not satisfied. The expression $\eta = r - (\pi/3)\sin \pi r$ satisfies all the conditions except for zero shear at the tip. Such approximate mode shapes are useful for evaluating the inertial and aerodynamic coefficients in a dynamics analysis, particularly for estimating the natural frequency of the second mode from the energy relation.

The utility of the normal mode representation of the blade motion depends on being able to use only a small number of modes to solve most rotor problems. The frequency content of the forces exciting the blade provides a good guide to the number of modes that must be included. In many cases, the fundamental flap mode is a sufficient representation of the blade for both articulated and hingeless rotors. For problems such as the calculation of oscillatory rotor loads or helicopter vibration, up to three to five out-of-plane modes can be required.

16.3.3 Non-Rotating Frame

The degrees of freedom and equations of motion in the nonrotating frame are obtained using multiblade coordinates (section 15.4). The equations derived for the out-of-plane bending are for each blade of an N-bladed rotor in the rotating frame. The multiblade coordinate transformation introduces N degrees of freedom (β_0, β_{1c}, β_{1s}, ... β_{nc}, β_{ns}, $\beta_{N/2}$) to describe the rotor motion in the non-rotating frame. The corresponding N equations of motion are obtained by operating on the rotating equation with

$$\frac{1}{N}\sum_{m=1}^{N}(\ldots) \quad \frac{2}{N}\sum_{m=1}^{N}(\ldots)\cos n\psi_m \quad \frac{2}{N}\sum_{m=1}^{N}(\ldots)\sin n\psi_m \quad \frac{1}{N}\sum_{m=1}^{N}(\ldots)(-1)^m \quad (16.72)$$

as appropriate. The inertial and structural terms of the equation that are encountered in this chapter have constant coefficients. These summation operators therefore act only on the rotating degrees of freedom and their time derivatives. By using the definitions of the non-rotating degrees of freedom (and the corresponding transformations of the time derivatives, given in section 15.4.1), the conversion of the equations of motion to the non-rotating frame is straightforward.

With independent blades the equations of motion in the rotating frame can be used directly. Unless there is some coupling of the blades through the fixed frame, there is no reason to use multiblade coordinates, except that the constant coefficient approximation for the aerodynamics in forward flight is better made in the non-rotating frame. The usefulness of multiblade coordinates is more apparent later in this chapter, when rotor shaft motion is involved.

Consider the fundamental flap mode of an articulated or hingeless rotor blade. The equation of motion for the m-th blade ($m = 1$ to N) in the rotating frame is

$$\widehat{I}_\beta \left(\ddot{\beta}^{(m)} + \nu^2 \beta^{(m)} \right) = \gamma \int_0^1 \eta \frac{F_z}{ac} dr = \gamma M_F^{(m)} \tag{16.73}$$

Applying the summation operators, which act only on $\ddot{\beta}^{(m)}$ and $\beta^{(m)}$, gives

$$\widehat{I}_\beta \left(\ddot{\beta}_0 + \nu^2 \beta_0 \right) = \frac{1}{N}\sum_{m=1}^{N} \gamma M_F^{(m)} = \gamma M_{F_0} \tag{16.74}$$

$$\widehat{I}_\beta \left(\ddot{\beta}_{nc} + 2n\dot{\beta}_{ns} + (\nu^2 - n^2) \beta_{nc} \right) = \frac{2}{N}\sum_{m=1}^{N} \gamma M_F^{(m)} \cos n\psi_m = \gamma M_{F_{nc}} \tag{16.75}$$

$$\widehat{I}_\beta \left(\ddot{\beta}_{ns} - 2n\dot{\beta}_{nc} + (\nu^2 - n^2) \beta_{ns} \right) = \frac{2}{N}\sum_{m=1}^{N} \gamma M_F^{(m)} \sin n\psi_m = \gamma M_{F_{ns}} \tag{16.76}$$

$$\widehat{I}_\beta \left(\ddot{\beta}_{N/2} + \nu^2 \beta_{N/2} \right) = \frac{1}{N}\sum_{m=1}^{N} \gamma M_F^{(m)}(-1)^m = \gamma M_{F_{N/2}} \tag{16.77}$$

The influence of this transformation on the eigenvalues and eigenvectors of the rotor dynamics is discussed in section 15.5.

16.3.4 Bending Moments

The flapwise bending moment on the blade was obtained in section 16.3.2 as

$$M(r) = \int_r^R \left[(F_z - m\ddot{z})(\rho - r) - m\Omega^2 \rho (z(\rho) - z(r)) \right] d\rho \qquad (16.78)$$

Substituting for the modal expansion of z and using dimensionless quantities gives

$$M(r) = \int_r^1 F_z(\rho - r) d\rho - \sum_k \left[\ddot{q}_k \int_r^1 m\eta_k(\rho - r) d\rho + q_k \int_r^1 m\rho \big(\eta_k(\rho) - \eta_k(r)\big) d\rho \right] \qquad (16.79)$$

This is a force-balance formula for the bending moment, integrating the applied (aerodynamic) and body (inertial and centrifugal) forces acting on the blade sections. Now expand the aerodynamic loading as a series in the bending mode shapes: $F_z = \sum_k F_{zk} m \eta_k(r)$. The constants are $F_{zk} = \int_0^1 \eta_k F_z dr / I_{qk}$. On substituting for the expansion of F_z into the bending moment, and noting that the equation of motion for the k-th bending mode gives $F_{zk} = \ddot{q}_k + v_k^2 q_k$, the bending moment becomes

$$M(r) = \sum_k q_k \left[v_k^2 \int_r^1 m\eta_k(\rho - r) d\rho - \int_r^1 m\rho \big(\eta_k(\rho) - \eta_k(r)\big) d\rho \right] \qquad (16.80)$$

Thus the bending moment can be evaluated from the response of the blade modes and from the corresponding mode shapes and frequencies. The bending moment can also be obtained from the blade curvature:

$$M(r) = EI \frac{d^2 z}{dr^2} = \sum_k q_k (EI \eta_k'') \qquad (16.81)$$

which is equivalent to equation 16.80, as can be shown by integrating the differential equation for q_k twice.

Generally the force-balance approach (equation 16.79) gives the best results. The structural dynamics of the blade filter the blade response, so a small number of modes is usually adequate to describe the inertial and centrifugal loading, but representing the aerodynamic loading by a modal expansion usually requires a large number of modes. Truncation of the series representation of the aerodynamic loading leads to significant differences between bending moments calculated using equations 16.79 and 16.80. Calculating the bending moment from the curvature (equation 16.81) often is not acceptable because the beam representation of rotor blades usually involves step changes in the stiffness EI. The bending moment must be a continuous function of r (except where there are discrete loads acting on the section), so a step change in stiffness must be accompanied by a step in curvature. A step in curvature cannot be modeled well with a small number of modes. A finite element model has a similar problem within the element, since a small number of polynomial shape functions cannot model well a step in curvature, but at least nodes can be put at step changes in properties.

The blade loads need not be expressed in terms of the modal response. The mode shapes and frequencies may not be available, and better accuracy can be possible by calculating the bending moments directly from the aerodynamic loading. The partial differential equation for out-of-plane deflection (equation 16.60) can be integrated along the span for a given loading F_z. Integrating a fourth-order equation and then differentiating the deflection twice to obtain the moment is not a good approach for numerical work. Working directly in terms of the bending moment is preferable. The equilibrium of forces outboard of r gives equation 16.57, which can be written

$$M(r) = \int_r^1 F_z(\rho - r)d\rho - \int_r^1 \left[\ddot{z}(\rho - r) + \rho(z(\rho) - z(r))\right]m\,d\rho \quad (16.82)$$

where z is obtained from the integral of $M = EIz''$. If periodic loading is considered, the exciting force F_z and the blade response are expanded as Fourier series: $F_z = \sum F_n e^{in\psi}$, $M = \sum M_n e^{in\psi}$, and $z = \sum z_n e^{in\psi}$. Then the equations for the n-th harmonic of the bending moment are

$$M_n = \int_r^1 F_n(\rho - r)d\rho + \int_r^1 \left[n^2 z_n(\rho - r) - \rho(z_n(\rho) - z_n(r))\right]m\,d\rho \quad (16.83)$$

$$z_n = z_n(1) - (1 - r)z_n'(1) + \int_r^1 \frac{M_n}{EI}(\rho - r)d\rho \quad (16.84)$$

These equations are numerically integrated, starting at the tip where the boundary conditions are $M_n(1) = M_n'(1) = 0$ are automatically satisfied by the equation for M_n. The values of $z_n(1)$ and $z_n'(1)$ must be chosen to satisfy the two boundary conditions at the root. These equations can be linear or nonlinear, depending on the aerodynamic model for F_z. The linear problem can be solved by superposition, whereas the nonlinear problem can be solved using some search algorithm. This bending moment equation is the same as equation 16.79, but here depends directly on the deflection z rather than on the modal response, so no modal truncation is involved. The aerodynamic force might be calculated using a simpler model (such as just rigid flapping motion), but the aerodynamic damping is important for the high-frequency response, so the lift due to $\dot{z}$ must be included in F_z.

Cierva (1926) gave an approximate method for calculating blade bending moments, using the airloading and motion obtained when rigid flapping alone is considered. Elastic bending of the blade significantly reduces the loads and so must be accounted for. Consider the limit of a rigid blade. For an articulated rotor the blade motion then is just rigid flapping, $z = \beta r$, and the bending moment on the rigid blade is

$$M_R = \int_r^1 F_z(\rho - r)d\rho - (\ddot{\beta} + \beta)\int_r^1 \rho(\rho - r)m\,d\rho \quad (16.85)$$

which implies a radius of curvature

$$r_R = \frac{1}{d^2z/dr^2} = \frac{EI}{M_R} \quad (16.86)$$

In the limit of zero structural stiffness ($EI = 0$), the blade has only centrifugal stiffness, and the partial differential equation of bending reduces to

$$-\frac{d}{dr}\left[\int_r^1 m\rho\,d\rho\,\frac{dz}{dr}\right] = F_z - m\ddot{z} \quad (16.87)$$

or

$$-\frac{d}{dr}\left[\int_r^1 m\rho\,d\rho\,\frac{dz_e}{dr}\right] = F_z - mr(\ddot{\beta} + \beta) \quad (16.88)$$

where z_e is the blade elastic deflection ($z = \beta r + z_e$) and the inertial force $m\ddot{z}_e$ has been neglected. This equation integrates to

$$\int_r^1 m\rho \, d\rho \frac{dz_e}{dr} = \int_r^1 F_z d\rho - (\ddot{\beta} + \beta) \int_r^1 m\rho \, d\rho \qquad (16.89)$$

or $T(dz_e/dr) = S_R$, where S_R is the vertical shear force calculated for the rigid blade and T is the centrifugal tension force. The radius of curvature is then

$$r_F = \frac{1}{d^2 z/dr^2} = \frac{1}{d^2 z_e/dr^2} = \frac{1}{d(S_R/T)/dr} \qquad (16.90)$$

Define M_F as the moment on the blade with stiffness EI, but with the curvature of the zero stiffness solutions: $M_F = EI/r_F = EI d(S_R/T)/dr$. Now construct a composite solution for the radius of curvature of the blade with stiffness EI, valid for both the limits $EI = \infty$ and $EI = 0$:

$$r_C = r_R + r_F = \frac{EI}{M_R} + \frac{1}{d(S_R/T)/dr} \qquad (16.91)$$

Then the bending moment on the actual blade is

$$M = \frac{EI}{r_C} = \frac{EI}{\frac{EI}{M_R} + \frac{1}{d(S_R/T)/dr}} = \frac{M_R M_F}{M_R + M_F} = M_R \frac{1}{1 + \frac{M_R}{EI \, d(S_R/T)/dr}} \qquad (16.92)$$

The last form is a correction of the rigid blade moment for the effects of bending. Thus the bending moment on the blade can be obtained from the moment M_R and shear force S_R calculated considering rigid flap motion alone. Flax (1947) discussed the analytical basis of this method.

16.4 In-Plane Motion

16.4.1 Rigid Flap and Lag

Chapter 6 introduced the lag dynamics of an articulated rotor. Here the coupled equations for rigid flap and lag motion are derived in more detail. Consider an articulated rotor with both flap and lag hinges. Hinge offsets and springs are included, and the flap and lag offsets are not necessarily equal. The degree of freedom for rigid rotation about the flap hinge is again β, with mode shape $\eta_\beta = (r-e)/(1-e)$. The in-plane motion consists of rigid rotation about the lag hinge, generating an in-plane displacement $x = \zeta \eta_\zeta$, where ζ is the lag degree of freedom with mode shape $\eta_\zeta = (r-e)/(1-e)$. The flap motion is positive upward, and the lag motion is positive in the direction opposite the rotor rotation. The equations of motion are obtained from the equilibrium of moments about the hinges.

The section forces producing flap moments remain the same as in section 16.3.1, with the addition of a Coriolis force due to the lag motion. The Coriolis acceleration is twice the cross-product of the angular velocity vector and the velocity vector relative to the rotating frame. The inertial force in the d'Alembert sense is then in the opposite direction. The product of the rotor rotational velocity Ω and the in-plane velocity of the section $\dot{x}$ gives a Coriolis force $2\Omega \dot{x} m = 2\Omega \dot{\zeta} \eta_\zeta m$, directed radially inward. This force has a moment arm $z = \eta_\beta \beta$ about the flap hinge, producing a total moment of

$$-\int_e^R (2\Omega \dot{\zeta} \eta_\zeta m)(\eta_\beta \beta) \, dr \qquad (16.93)$$

600   Blade Motion

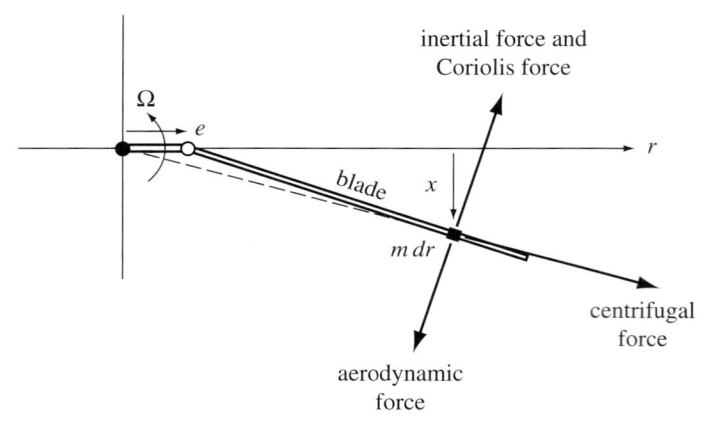

Figure 16.3. Rotor blade lagging moments.

Including this term in the moment equilibrium gives for the flap equation of motion

$$\widehat{I}_\beta(\ddot\beta + \nu_\beta^2 \beta) - \widehat{I}_{\beta\zeta} 2\beta\dot\zeta = \frac{K_\beta}{I_\beta \Omega^2 (1-e)} \beta_p + \gamma \int_e^1 \eta_\beta \frac{F_z}{ac} dr \qquad (16.94)$$

where $\widehat{I}_{\beta\zeta} = \int_e^1 \eta_\beta \eta_\zeta m\, dr / ((1-e) I_b)$.

The in-plane forces acting on the blade section (Figure 16.3) and their moment arms about the offset lag hinge are as follows:

i) an inertial force $m\ddot x = m\eta_\zeta \ddot\zeta$ opposing the lag motion, with moment arm $(r-e)$ about the lag hinge
ii) a centrifugal force $m\Omega^2 r$ directed radially outward from the center of rotation, with moment arm $x(e/r) = \eta_\zeta \zeta (e/r)$ about the lag hinge
iii) an aerodynamic force F_x in the drag direction, with moment arm $(r-e)$
ii) a Coriolis force $2\Omega \dot z z' m = 2\Omega \dot\beta \beta \eta_\beta \eta'_\beta m$ in the same direction as the inertial force, with moment arm $(r-e)$

Since the out-of-plane velocity $\dot z$ has a radially inward component $\dot z (dz/dr)$ when the blade is flapped up, the Coriolis force arises from the product of the rotational speed of the rotor and this radial velocity of the blade. The equilibrium of moments about the lag hinge gives

$$\left(\int_e^R (r-e) \eta_\zeta m\, dr \right) \ddot\zeta + \left(\int_e^R e \eta_\zeta m\, dr \right) \Omega^2 \zeta$$
$$+ \left(\int_e^R \eta_\beta \eta_\zeta m\, dr \right) 2\Omega\beta\dot\beta + K_\zeta \zeta = \int_e^R (r-e) F_x dr \qquad (16.95)$$

Dividing by $(1-e)$ and using dimensionless quantities gives

$$\left(\int_e^1 \eta_\zeta^2 m\, dr \right) \ddot\zeta + \left(\frac{e}{1-e} \int_e^1 \eta_\zeta m\, dr \right) \zeta + \frac{K_\zeta}{\Omega^2 (1-e)} \zeta$$
$$+ \left(\frac{1}{1-e} \int_e^1 \eta_\beta \eta_\zeta m\, dr \right) 2\beta\dot\beta = \int_e^1 \eta_\zeta F_x dr \qquad (16.96)$$

16.4 In-Plane Motion

Next, define the lag inertia as $I_\zeta = \int_e^R \eta_\zeta^2 m \, dr$ and divide by I_b, to obtain

$$\widehat{I}_\zeta \left(\ddot{\zeta} + v_\zeta^2 \zeta \right) + \widehat{I}_{\beta\zeta} 2\beta\dot{\beta} = \gamma \int_e^1 \eta_\zeta \frac{F_x}{ac} dr \qquad (16.97)$$

which is the equation of motion for the rigid lag of an articulated blade.

The rotating natural frequency of the lag motion is

$$v_\zeta^2 = \frac{e}{1-e} \frac{\int_e^1 \eta_\zeta m \, dr}{\int_e^1 \eta_\zeta^2 m \, dr} + \frac{K_\zeta}{I_\zeta \Omega^2 (1-e)} \qquad (16.98)$$

The lag hinge must have an offset or spring to obtain a nonzero lag frequency. For a uniform mass distribution and no spring,

$$v_\zeta^2 = \frac{3}{2} \frac{e}{1-e} \qquad (16.99)$$

More generally, the lag frequency is given by $v_\zeta^2 = eS_\zeta/I_\zeta$, where I_ζ is the second moment of inertia about the lag hinge and S_ζ is the first moment. S_ζ equals the product of the blade mass and the radial distance of the center of gravity from the lag hinge. Assuming the same mode shapes and spring constants for the flap and lag motion, the expressions for the natural frequencies here and in section 16.3.1 give

$$v_\beta^2 = 1 + v_\zeta^2 \qquad (16.100)$$

For an articulated blade with coincident flap and lag hinges, the mode shapes are in fact identical and this result is correct. In general, this relation is an expression of the fundamentally different role of centrifugal forces in flap and lag dynamics. The centrifugal force always acts as a spring on the flap motion to produce a natural frequency of at least 1/rev. However, since the centrifugal force goes through the center of rotation, the lag motion must depend on the hinge offset to obtain a centrifugal spring.

The flap and lag equations of motion are coupled by nonlinear terms due to the blade Coriolis forces: $-\widehat{I}_{\beta\zeta} 2\beta\dot{\zeta}$ in the flap equation and $\widehat{I}_{\beta\zeta} 2\beta\dot{\beta}$ in the lag equation. For a linear stability analysis, these terms are linearized about the trim motion:

$$\Delta(\beta\dot{\zeta}) = \beta_{\text{trim}} \Delta\dot{\zeta} + \dot{\zeta}_{\text{trim}} \Delta\beta \cong \beta_0 \Delta\dot{\zeta} \qquad (16.101)$$

$$\Delta(\beta\dot{\beta}) = \beta_{\text{trim}} \Delta\dot{\beta} + \dot{\beta}_{\text{trim}} \Delta\beta \cong \beta_0 \Delta\dot{\beta} \qquad (16.102)$$

where β_0 is the trim coning angle. The last approximation is based on using the mean values of the periodic trim lag and flap motion; the result is exact for hover. Thus a blade with a finite coning angle has a flap moment due to lag velocity and a lag moment due to flap velocity. Since the Coriolis terms are nonlinear, these coupling moments are small. However, all the lag moments are small compared to the flap moments, so the Coriolis force due to flapping velocity is an important factor in the lag dynamics.

Since an articulated rotor has a mechanical lag damper, the term $\widehat{C}_\zeta \dot{\zeta}$ should be added to the lag equation of motion. Here $\widehat{C}_\zeta = C_\zeta/I_b\Omega$, and C_ζ is the lag moment due to angular velocity about the lag hinge. For hingeless rotors the structural damping of the blade should be included by adding the term $\widehat{I}_\zeta g_s v_\zeta \dot{\zeta}$ to the lag equation, where g_s is the structural damping coefficient; typically $g_s = 0.01$ to 0.03 (0.5 to 1.5% critical damping). The structural damping is small, but can be important to the lag dynamics because the in-plane forces are all small.

16.4.2 Structural Coupling

Structural coupling between the flap and lag motion can have an important effect, because a small amount of out-of-plane motion can significantly increase the aerodynamic damping of the lag mode. Ormiston and Hodges (1972) introduced a simple model for structural coupling of rigid flap and rigid lag motion. The degrees of freedom β and ζ still represent purely out-of-plane and purely in-plane motion, respectively, so the inertial and aerodynamic loads are unchanged. The structural restraint at the flap and lag hinges consists now of flexibility both inboard and outboard of the pitch bearing (all at a single point). The inboard springs $K_{\beta h}$ and $K_{\zeta h}$ (subscript "h" for hub) are parallel and normal to the hub plane. The outboard springs $K_{\beta b}$ and $K_{\zeta b}$ (subscript "b" for blade) are in axes rotated by the pitch angle θ_s. The total rotation of the blade is the sum of the rotations in the hub and blade springs:

$$\begin{pmatrix} \beta \\ \zeta \end{pmatrix} = \begin{pmatrix} \beta_h \\ \zeta_h \end{pmatrix} + \begin{bmatrix} C & -S \\ S & C \end{bmatrix} \begin{pmatrix} \beta_b \\ \zeta_b \end{pmatrix} \qquad (16.103)$$

or $x = x_h + T x_b$, where $C = \cos\theta_s$ and $S = \sin\theta_s$. The moment transmitted through the springs must be equal:

$$\begin{pmatrix} M_\beta \\ M_\zeta \end{pmatrix} = \begin{bmatrix} K_{\beta h} & 0 \\ 0 & K_{\zeta h} \end{bmatrix} \begin{pmatrix} \beta_h \\ \zeta_h \end{pmatrix} = \begin{bmatrix} C & -S \\ S & C \end{bmatrix} \begin{bmatrix} K_{\beta b} & 0 \\ 0 & K_{\zeta b} \end{bmatrix} \begin{pmatrix} \beta_b \\ \zeta_b \end{pmatrix} \qquad (16.104)$$

or $M = K_h x_h = T K_b x_b$. Hence $x_b = K_b^{-1} T^T K_h x_h$, the total rotation is $x = (I + K_b^{-1} T^T K_h) x_h$, and $M = (K_h^{-1} + T K_b^{-1} T^T)^{-1} x$. For matched stiffness (equal flap and lag stiffnesses in each spring set), this reduces to $M = (K_h + K_b)x$; there is no influence of θ_s and no structural coupling. For the uncoupled case ($\theta_s = 0$), the stiffness is $K = (K_h^{-1} + K_b^{-1})^{-1}$. Now define

$$\frac{1}{K_\beta} = \frac{1}{K_{\beta h}} + \frac{1}{K_{\beta b}} \qquad \frac{1}{K_{\beta b}} = \frac{\chi_\beta}{K_\beta} \qquad \frac{1}{K_{\beta h}} = \frac{1 - \chi_\beta}{K_\beta} \qquad (16.105)$$

$$\frac{1}{K_\zeta} = \frac{1}{K_{\zeta h}} + \frac{1}{K_{\zeta b}} \qquad \frac{1}{K_{\zeta b}} = \frac{\chi_\zeta}{K_\zeta} \qquad \frac{1}{K_{\zeta h}} = \frac{1 - \chi_\zeta}{K_\zeta} \qquad (16.106)$$

where χ_β and χ_ζ are measures of the distribution of stiffness. For $\chi = 0$, all the flexibility is inboard (uncoupled); for $\chi = 1$ all the flexibility is outboard of the pitch bearing (fully coupled). A total coupling parameter is $\chi(K_\zeta - K_\beta) = \chi_\beta K_\zeta - \chi_\zeta K_\beta$. Then the stiffness matrix can be written:

$$K = (K_h^{-1} + T K_b^{-1} T^T)^{-1} = \begin{bmatrix} \dfrac{1}{K_{\beta h}} + \dfrac{C^2}{K_{\beta b}} + \dfrac{S^2}{K_{\zeta b}} & SC\left(\dfrac{1}{K_{\beta b}} - \dfrac{1}{K_{\zeta b}}\right) \\ SC\left(\dfrac{1}{K_{\beta b}} - \dfrac{1}{K_{\zeta b}}\right) & \dfrac{1}{K_{\zeta h}} + \dfrac{S^2}{K_{\beta b}} + \dfrac{C^2}{K_{\zeta b}} \end{bmatrix}$$

$$= \frac{1}{\Delta} \begin{bmatrix} K_\beta + S^2 \chi(K_\zeta - K_\beta) & -SC\chi(K_\zeta - K_\beta) \\ -SC\chi(K_\zeta - K_\beta) & K_\zeta - S^2 \chi(K_\zeta - K_\beta) \end{bmatrix} \qquad (16.107)$$

where $\Delta = 1 + S^2 \chi(1-\chi)(K_\zeta - K_\beta)^2 /(K_\zeta K_\beta)$. For small θ_s this is

$$K \cong \begin{bmatrix} K_\beta & -\chi(K_\zeta - K_\beta)\theta_s \\ -\chi(K_\zeta - K_\beta)\theta_s & K_\zeta \end{bmatrix} \qquad (16.108)$$

The off-diagonal terms can be included in equations 16.94 and 16.97 to account for structural coupling.

16.4.3 In-Plane Bending

Consider the pure in-plane motion of a rotating blade, including now the blade bending and an arbitrary root restraint. The in-plane forces due to out-of-plane motion are important, but are neglected for now to concentrate on the in-plane natural frequencies and mode shapes. The section forces and their moment arms about the blade section at r are as follows:

i) an inertial force $m\ddot{x}(\rho)$, with moment arm $(\rho - r)$
ii) a centrifugal force $m\Omega^2\rho$, with moment arm $(r/\rho)x(\rho) - x(r)$
iii) an aerodynamic force F_x, with moment arm $(\rho - r)$

The lag moment at r, due to the inertial and aerodynamic forces on the section outboard of r, is thus

$$M(r) = \int_r^R \left[(F_x - m\ddot{x})(\rho - r) - m\Omega^2\rho \left(x(\rho)\frac{r}{\rho} - x(r) \right) \right] d\rho \qquad (16.109)$$

Engineering beam theory gives the structural bending moment as $M_x(r) = EI_{xx}d^2x/dr^2$, where E is the modulus of elasticity and I_{xx} the modulus-weighted area moment about the vertical principal axis of the section. Equating the structural moment to the inertial and aerodynamic moments, and taking the second derivative, gives the partial differential equation for the in-plane bending motion of the rotating blade:

$$\frac{d^2}{dr^2} EI_{xx} \frac{d^2x}{dr^2} - \frac{d}{dr}\left[\int_r^R m\Omega^2\rho\, d\rho \frac{dx}{dr} \right] - \Omega^2 mx + m\ddot{x} = F_x \qquad (16.110)$$

The boundary conditions for articulated and hingeless blades are as discussed for out-of-plane bending in section 16.3.2. The modal equation is obtained by assuming free vibration of the rotating blade. Substituting $x = \eta(r)e^{i\nu t}$ in the homogeneous equation then gives

$$\frac{d^2}{dr^2} EI_{xx} \frac{d^2\eta}{dr^2} - \frac{d}{dr}\left[\int_r^R m\Omega^2\rho\, d\rho \frac{d\eta}{dr} \right] - \Omega^2 m\eta - \nu^2 m\eta = 0 \qquad (16.111)$$

This is again a proper Sturm-Liouville eigenvalue problem, for which there exists a series of orthogonal eigen-solutions η_{xk} and corresponding eigenvalues ν_{xk}^2.

The in-plane displacement can now be expanded as a series in the normal modes:

$$x(r, t) = \sum_{k=1}^{\infty} \eta_{xk}(r) q_{xk}(t) \qquad (16.112)$$

where q_{xk} are the in-plane bending degrees of freedom. This modal expansion is substituted into the partial differential equation, and the modal equation is used to replace the structural and centrifugal spring terms by the natural frequency ν_{xk}. Operating with $\int_0^R (\ldots)\eta_{xk} dr$ and using the orthogonality of the modes gives

$$\widehat{I}_{\zeta k}(\ddot{q}_{xk} + \nu_{xk}^2 q_{xk}) = \gamma \int_0^1 \eta_{xk}\frac{F_x}{ac}dr \qquad (16.113)$$

where $\widehat{I}_{\zeta k} = \int_0^R \eta_{xk}^2 dr/I_b$. This is the equation of motion for pure in-plane bending of the blade.

The natural frequency can be obtained from the mode shape by using the energy relation from Sturm-Liouville theory:

$$\nu^2 = \frac{K_\zeta \left[\eta'(e)\right]^2 + \int_0^R \left[EI\eta''^2 + \int_r^R m\Omega^2 \rho \, d\rho \, \eta'^2 - \Omega^2 m\eta^2\right] dr}{\int_0^R \eta^2 m \, dr} \quad (16.114)$$

Assuming the same mass and stiffness distributions, this is formally equivalent to $\nu_{\text{flap}}^2 = 1 + \nu_{\text{lag}}^2$, which can also be deduced by comparing the modal equations for in-plane and out-of-plane bending (see section 16.3.2). However, the chordwise bending stiffness (EI_{xx}) is much greater than the flapwise bending stiffness (EI_{zz}), typically by a factor of 20 to 40. Moreover, the in-plane and out-of-plane mode shapes are not the same. Thus the relation $\nu_{\text{flap}}^2 = 1 + \nu_{\text{lag}}^2$ really is only applicable to the fundamental modes of an articulated blade with coincident hinges. The similarity between the out-of-plane and in-plane modal problems can be used to advantage in numerical solutions for the modes.

16.4.4 In-Plane and Out-of-Plane Bending

Now the equations of motion for in-plane and out-of-plane bending are derived. This derivation is a generalization of the rigid flap and lag results. Assuming that there is no structural coupling of the bending motion, the displacement z is still purely out of plane and the displacement x is purely in plane. With this assumption, the only coupling of the equations of motion is due to the Coriolis forces, and it is only necessary to add the Coriolis terms to the results of sections 16.3.2 and 16.4.3.

For out-of-plane bending, there is a Coriolis force $2\Omega \dot{x} m$ directed radially inward, with moment arm $(z(\rho) - z(r))$ about the blade station at r. The flapwise bending moment at r then becomes

$$M(r) = \int_r^R \left[(F_z - m\ddot{z})(\rho - r) - (m\Omega^2 \rho - 2\Omega \dot{x} m)(z(\rho) - z(r))\right] d\rho \quad (16.115)$$

and the partial differential equation for out-of-plane bending is

$$(EI_{zz} z'')'' - \left[\int_r^R m\Omega^2 \rho \, d\rho \, z'\right]' + m\ddot{z} + \left[z' \int_r^R 2\Omega \dot{x} m \, d\rho\right]' = F_z \quad (16.116)$$

When the aerodynamic force and the Coriolis term are dropped, the same modal equation as in section 16.3.2 is obtained. The out-of-plane deflection is now expanded as a series in the modes η_{zk}:

$$z = \sum_z \eta_{zk}(r) q_{zk}(t) \quad (16.117)$$

where q_{zk} are the degrees of freedom. This expansion is substituted into the partial differential equation, and the modal equation is used to replace the structural and centrifugal spring terms with the natural frequency ν_{zk}. Then the operation $\int_0^R (\ldots) \eta_{zk} dr$ produces the ordinary differential equation for the k-th out-of-plane bending mode of the rotating blade:

$$I_{\beta k} \left(\ddot{q}_{zk} + \nu_{zk}^2 q_{zk}\right) + \int_0^1 \eta_{zk} \left[z' \int_r^1 2\dot{x} m \, d\rho\right]' dr = \int_0^1 \eta_{zk} F_z dr \quad (16.118)$$

16.4 In-Plane Motion

Integrating by parts and changing the order of integration converts the Coriolis term to

$$\int_0^1 \eta_{zk}\left[z'\int_r^1 2\dot{x}m\,d\rho\right]' dr = -2\int_0^1 \dot{x}m\int_0^r \eta'_{zk}z'\,d\rho\,dr \cong -2\beta_0\int_0^1 \dot{x}\eta_{zk}m\,dr \quad (16.119)$$

The last approximation follows from linearizing $\dot{x}z'$ about the trim condition, using the mean trim values of $\dot{x}$ and z' and assuming that the trim blade slope is principally due to the coning angle β_0. For rigid flap and lag this Coriolis force reduces to the previous result, $-2I_{\beta\zeta}\beta_0\dot{\zeta}$ (the extra factor of $(1-e)$ was lost when we assumed $z' \cong \beta_0$ instead of $z' = \eta'\beta = \beta/(1-e)$).

There are two Coriolis forces that must be considered for in-plane bending. First, the lag velocity $\dot{x}$ and rotor speed Ω give a radially inward Coriolis force $2\Omega\dot{x}m$. This is the same force that produces a flapwise bending moment. It also produces a chordwise moment, with moment arm $(x(\rho) - x(r))$ about the blade station at r. Second, the in-plane and out-of-plane deflection produces a nonlinear radial shortening of the blade by

$$-\frac{1}{2}\int_0^\rho \left(x'^2 + z'^2\right) d\rho^* \quad (16.120)$$

and thus there is a radially inward velocity of the blade section equal to

$$-\int_0^\rho \left(x'\dot{x}' + z'\dot{z}'\right) d\rho^* \quad (16.121)$$

The cross-product of this velocity and the rotor rotational speed gives an in-plane Coriolis force with moment arm $(\rho - r)$ about the blade station at r (see Figure 16.3). The total lag bending moment is thus

$$M(r) = \int_r^R \Bigg[(F_x - m\ddot{x})(\rho - r) - m\Omega^2\rho\left(x(\rho)\frac{r}{\rho} - x(r)\right)$$

$$+ 2\Omega\dot{x}m(x(\rho) - x(r)) - 2\Omega m\int_0^\rho (x'\dot{x}' + z'\dot{z}')\,d\rho^*(\rho - r)\Bigg]d\rho \quad (16.122)$$

and the partial differential equation for in-plane bending becomes

$$(EI_{xx}x'')'' - \left[\int_r^R m\Omega^2\rho\,d\rho\,x'\right]' - \Omega^2 mx + m\ddot{x}$$

$$+ \left[2\Omega x'\int_r^R \dot{x}m\,d\rho\right]' + 2\Omega m\int_0^r (x'\dot{x}' + z'\dot{z}')\,d\rho = F_x \quad (16.123)$$

After the in-plane deflection is expressed in terms of the normal modes, $x = \sum_k \eta_{xk}q_{xk}$, the ordinary differential equation for the k-th in-plane bending mode can be obtained by the usual steps:

$$I_{\zeta k}\left(\ddot{q}_{xk} + v_{xk}^2 q_{xk}\right) + \int_0^1 \eta_{xk}2m\int_0^r (x'\dot{x}' + z'\dot{z}')\,d\rho\,dr$$

$$+ \int_0^1 \eta_{xk}\left[x'\int_r^1 2\dot{x}m\,d\rho\right]' dr = \int_0^1 \eta_{xk}F_x\,dr \quad (16.124)$$

The two Coriolis terms can then be written:

$$2\int_0^1 \eta_{xk} m \int_0^r (x'\dot{x}' + z'\dot{z}')\,d\rho\,dr - 2\int_0^1 \dot{x}m \int_0^r \eta'_{xk}x'\,d\rho\,dr$$

$$\cong 2\int_0^1 \eta_{xk} m \int_0^r z'\dot{z}'\,d\rho\,dr$$

$$\cong 2\beta_0 \int_0^1 \dot{z}\eta_{xk}m\,dr \qquad (16.125)$$

For rigid lag motion, where x' is independent of r, the two in-plane velocity terms cancel exactly. Similarly, they cancel if the trim lag displacement is primarily due to the rigid mode. Therefore these two terms have been neglected for the general case.

Substituting the modal expansions for $\dot{x}$ in the flap equation and for $\dot{z}$ in the lag equation then completes the coupled equations of motion for out-of-plane and in-plane bending:

$$\widehat{I}_{\beta k}(\ddot{q}_{zk} + \nu_{zk}^2 q_{zk}) - \sum_{i=1}^{\infty} \widehat{I}_{\beta_k \zeta_i} 2\beta_0 \dot{q}_{xi} = \gamma \int_0^1 \eta_{zk}\frac{F_z}{ac}dr \qquad (16.126)$$

$$\widehat{I}_{\zeta k}(\ddot{q}_{xk} + \nu_{xk}^2 q_{xk}) + \sum_{i=1}^{\infty} \widehat{I}_{\beta_i \zeta_k} 2\beta_0 \dot{q}_{zi} = \gamma \int_0^1 \eta_{xk}\frac{F_x}{ac}dr \qquad (16.127)$$

where $\widehat{I}_{\beta k} = \int_0^R \eta_{zk}^2 dr/I_b$, $\widehat{I}_{\zeta k} = \int_0^R \eta_{xk}^2 dr/I_b$, and $\beta_0 \widehat{I}_{\beta_k \zeta_i} = z'_{\text{trim}} \int_0^R \eta_{zk}\eta_{xi}dr/I_b$.

However, this set of equations is not a sufficient model for the out-of-plane and in-plane bending of a rotor blade in most cases. Unless the blade is untwisted and operating at zero pitch, there is considerable structural coupling of the in-plane and out-of-plane deflections. The structural principal axes are rotated by the blade pitch angle, whereas the centrifugal forces always act relative to the shaft axes. Therefore, when the blade pitch is non-zero the axes of structural and centrifugal stiffening do not coincide, and the free vibration modes of the blade are not purely out-of-plane or purely in-plane as was assumed here. Section 16.4.2 described a simple model for flap-lag coupling of the fundamental modes. A better analysis must use a single series of coupled flap-lag bending modes to represent the blade deflection. The blade torsional motion must be included in such an analysis, since the coupling between bending and pitch can have a major influence on the dynamics.

16.5 Torsional Motion

16.5.1 Rigid Pitch and Flap

The rotor dynamics analysis is now extended to include the blade pitch degree of freedom. Consider an articulated rotor, with no flap-hinge offset (Figure 16.4). A general flap frequency can be obtained by using a hinge spring. In addition, we now consider the blade pitch motion, consisting of rigid rotation about the feathering axis, restrained by the rotor control system. If there is flexibility in the control system, the blade rigid pitch motion is a degree of freedom, not just a control input (as in Chapter 6). The pitch bearing is assumed to be outboard of the flap hinge, and there is no pitch-flap coupling (δ_3). The chordwise position of the blade section of gravity is a distance x_I behind the feathering axis (Figure 16.4).

16.5 Torsional Motion

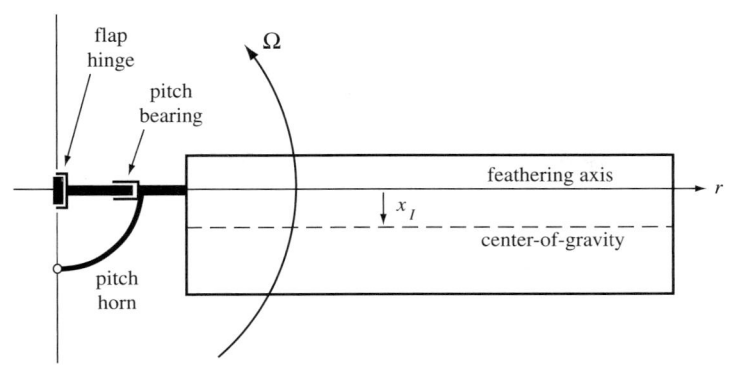

Figure 16.4. Articulated rotor blade with flapping and feathering motion.

The flap degree of freedom β is the angle of rigid rotation about the flap hinge. The out-of-plane deflection of the blade is thus $z = r\beta$. Let θ be the degree of freedom for the pitch motion, defined as the nose-up angle of rigid rotation about the feathering axis. The built-in twist of the blade is not considered here, since it is only involved in the trim forces on the blade. The rotor control system commands a pitch angle θ_{con}, while the actual blade pitch angle is θ. The difference $(\theta - \theta_{con})$ is due to control system flexibility and produces a restoring moment about the feathering axis equal to $K_\theta(\theta - \theta_{con})$, where K_θ is the control system spring constant.

The flapping equation of motion is obtained as usual from the equilibrium of moments about the flap hinge. The forces acting on the blade section center-of-gravity are now

i) an inertial force $m(\ddot{z} - x_I\ddot{\theta}) = m(r\ddot{\beta} - x_I\ddot{\theta})$, with moment arm r
ii) a centrifugal force $m\Omega^2 r$, with moment arm $z - x_I\theta = r\beta - x_I\theta$
iii) an aerodynamic force F_z, with moment arm r

Including the hinge spring moment, the equation of motion becomes

$$\int_0^R m(r\ddot{\beta} - x_I\ddot{\theta})r\,dr + \int_0^R m\Omega^2 r(r\beta - x_I\theta)\,dr + K_\beta\beta = \int_0^R rF_z\,dr \qquad (16.128)$$

or

$$\left(\int_0^R r^2 m\,dr\right)(\ddot{\beta} + v^2\beta) - \left(\int_0^R x_I rm\,dr\right)(\ddot{\theta} + \Omega^2\theta) = \int_0^R rF_z\,dr \qquad (16.129)$$

where v is the rotating natural frequency of the flap motion. Define $I_b = \int_0^R r^2 m\,dr$ and $I_x = \int_0^R x_I rm\,dr$. Dividing by I_b and using dimensionless quantities then gives

$$\ddot{\beta} + v^2\beta - \widehat{I_x}(\ddot{\theta} + \theta) = \gamma \int_0^1 r\frac{F_z}{ac}dr \qquad (16.130)$$

where $\widehat{I_x} = I_x/I_b$. Thus the pitch motion introduces inertial and centrifugal flap moments when the center-of-gravity is offset from the feathering axis.

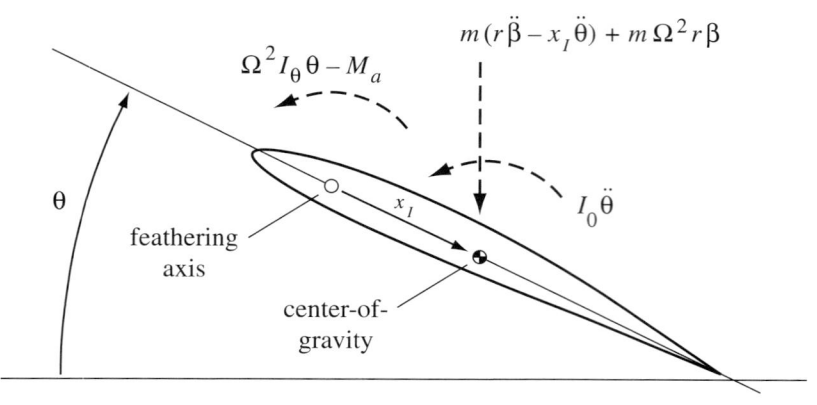

Figure 16.5. Blade section pitch moments.

The pitch equation of motion is obtained from the equilibrium of moments about the feathering axis (Figure 16.5). The forces acting on the blade section and their moment arms about the feathering axis are as follows:

i) an inertial moment $I_0\ddot{\theta}$ about the section center-of-gravity
ii) an inertial force $m(r\ddot{\beta} - x_I\ddot{\theta})$ acting on the center-of-gravity, with moment arm x_I about the feathering axis
iii) a propeller moment $I_\theta \Omega^2 \theta$ about the feathering axis, acting to oppose the pitch motion
iv) a flapping centrifugal spring force $m\Omega^2 r\beta$ acting at the center-of-gravity, with moment arm x_I about the feathering axis
v) a nose-up aerodynamic moment M_a about the feathering axis

Here I_0 is the pitch moment of inertia of the section about the center-of-gravity, and $I_\theta = I_0 + x_I^2 m$ is the section moment of inertia about the feathering axis. When the blade flaps up, the centrifugal force has a component $m\Omega^2 r\beta$ normal to the blade. This force is responsible for the centrifugal flap moment, and when the center-of-gravity is offset from the feathering axis it also produces a pitch moment. The propeller moment is also due to centrifugal forces. The centrifugal force on a blade mass element dm acts on a line through the center of rotation (Figure 16.6). For an element a distance x behind the feathering axis, there is then a chordwise component of this centrifugal force equal to

$$\left(\sqrt{r^2 + x^2}\, \Omega^2 dm\right) \frac{x}{\sqrt{r^2 + x^2}} = x\Omega^2 dm \qquad (16.131)$$

When the blade is pitched up by the angle θ, this chordwise force acts on a line a distance $x\theta$ below the feathering axis (see Figure 16.6). For mass elements forward of the feathering axis, the centrifugal force component is directed forward and acts on a line above the feathering axis. Thus there is a centrifugal feathering moment opposing the pitch motion. The propeller moment is obtained by integrating over the blade section:

$$\int_{\text{section}} (x\theta)(x\Omega^2 dm) = \theta\Omega^2 \int_{\text{section}} x^2 dm = \theta\Omega^2 I_\theta \qquad (16.132)$$

where I_θ is the section moment of inertia about the feathering axis.

16.5 Torsional Motion

Figure 16.6. Origin of the propeller moment.

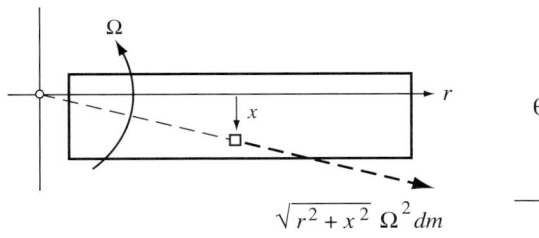

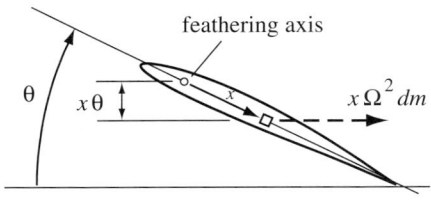

(left) centrifugal force on mass element dm
(right) resulting moment about feathering axis

The equilibrium of moments about the feathering axis gives

$$\int_0^R \left[I_\theta \ddot\theta - (r\ddot\beta - x_I\ddot\theta)x_I m + I_\theta \Omega^2 \theta - m\Omega^2 r\beta x_I\right] dr + K_\theta(\theta - \theta_{\text{con}}) = \int_0^R M_a \, dr \quad (16.133)$$

or

$$\left(\int_0^R I_\theta \, dr\right)(\ddot\theta + \Omega^2 \theta) - \left(\int_0^R x_I rm \, dr\right)(\ddot\beta + \Omega^2 \beta) + K_\theta(\theta - \theta_{\text{con}}) = \int_0^R M_a \, dr \quad (16.134)$$

The restoring moment from the control system, $K_\theta(\theta - \theta_{\text{con}})$, has been included; θ_{con} is the pitch angle commanded by the control system, and K_θ is the effective spring constant of the flexible control system. Now define the total moment of inertia about the feathering axis as $I_f = \int_0^R I_\theta \, dr$, and write

$$\omega^2 = \frac{K_\theta}{I_f \Omega^2} \quad (16.135)$$

where ω is the dimensionless natural frequency of the blade pitch motion. Dividing by I_b and using dimensionless quantities gives the pitch equation of motion,

$$\widehat{I}_f(\ddot\theta + (\omega^2 + 1)\theta) - \widehat{I}_x(\ddot\beta + \beta) = \gamma \int_0^R \frac{M_a}{ac} dr + \widehat{I}_f \omega^2 \theta_{\text{con}} \quad (16.136)$$

where $\widehat{I}_f = I_f/I_b$. Here ω is the non-rotating natural frequency of the pitch motion, and the propeller moment gives a spring equivalent to a 1/rev natural frequency. The rotating pitch natural frequency is therefore $(\omega^2 + 1)^{1/2}$. Typically the control system stiffness gives $\omega = 3$ to 5/rev, so the propeller moment is small compared to the structural spring.

To summarize, the equations of motion for rigid flapping and rigid pitch about the feathering axis are

$$\ddot\beta + \nu^2 \beta - \widehat{I}_x(\ddot\theta + \theta) = \gamma \int_0^1 r\frac{F_z}{ac} dr \quad (16.137)$$

$$\widehat{I}_f(\ddot\theta + (\omega^2 + 1)\theta) - \widehat{I}_x(\ddot\beta + \beta) = \gamma \int_0^1 \frac{M_a}{ac} dr + \widehat{I}_f \omega^2 \theta_{\text{con}} \quad (16.138)$$

where $\widehat{I}_x = \int_0^R x_I rm \, dr/I_b$ and $I_f = \int_0^R I_\theta \, dr/I_b$. The flap and pitch motions are coupled by inertial and centrifugal forces when the blade center-of-gravity is offset from the feathering axis. Here ν is the rotating natural frequency of the flap motion, and

ω is the non-rotating pitch natural frequency due to the control system stiffness. I_θ is the pitch moment of inertia of the section about the feathering axis, and x_I is the offset of the section center-of-gravity behind the feathering axis. For constant x_I we have $\widehat{I}_x = x_I \int_0^R rm\,dr/I_b = x_I \widehat{S}_b \cong \frac{3}{2} x_I$. Since the center-of-gravity offset is usually a small fraction of the chord, x_I (which is normalized using the rotor radius) is a second-order-small quantity. The ratio of the pitch inertia to the flap inertia is roughly $\widehat{I}_f \cong 0.1(c/R)^2$. In general, all the pitch moments are two orders smaller than the flap moments.

In the limit of a very stiff control system, the restoring moment $K_\theta(\theta - \theta_{\text{con}})$ must remain finite while $K_\theta \to \infty$, since the restoring moment is equal to the sum of the inertial and aerodynamic pitch moments on the blade. Then $\theta \to \theta_{\text{con}}$ in this limit, and the pitch motion is just the input commanded by the control system. Alternatively, in the limit of $\omega \to \infty$ the equation for the pitch motion reduces to $\widehat{I}_f \omega^2 \theta = \widehat{I}_f \omega^2 \theta_{\text{con}}$, or $\theta = \theta_{\text{con}}$ again. Kinematic pitch-flap coupling due to the control system geometry is a feedback of the flap angle to the commanded pitch of the form $\Delta \theta_{\text{con}} = -K_P \beta$. The equation of motion becomes

$$\widehat{I}_f(\ddot{\theta} + (\omega^2 + 1)\theta) - \widehat{I}_x(\ddot{\beta} + \beta) + K_P \widehat{I}_f \omega^2 \beta = \gamma \int_0^1 \frac{M_a}{ac} dr + \widehat{I}_f \omega^2 \theta_{\text{con}} \quad (16.139)$$

and in the limit of infinite control, system stiffness reduces to $\theta = \theta_{\text{con}} - K_P \beta$ as required.

16.5.2 Structural Pitch-Flap and Pitch-Lag Coupling

The order of the flap and pitch hinges or, for a hingeless rotor, the distribution of bending inboard and outboard of the pitch bearing has an important influence on the blade dynamics. The preceding analysis assumed that the pitch bearing was outboard of the flap hinge, so that flap motion tilts the feathering axis along with the blade. If the pitch bearing is inboard of the flap hinge, the feathering axis remains in the hub plane when the blade flaps, resulting in different moment arms of the section forces about the feathering axis.

Consider the rigid flap and rigid pitch of an articulated rotor blade, now with the pitch bearing inboard of the flap and lag hinges. The flapping equation is not changed, at least for small angles of flap and pitch. However, there is a change in the manner in which the centrifugal forces produce pitch moments. The centrifugal force $m\Omega^2 r$ does not now have a component about the feathering axis when the blade flaps, because the centrifugal force and feathering axis are both parallel to the hub plane. However, the chordwise component of the centrifugal force has a moment arm of $(x\theta - r\beta)$ about the feathering axis, so the propeller moment becomes

$$\int_{\text{section}} (x\theta - r\beta)(x\Omega^2 dm) = \theta\Omega^2 I_\theta - (m\Omega^2 r\beta)x_I \quad (16.140)$$

Thus there is no net change in the pitch moment due to the centrifugal forces, but there are a number of nonlinear effects of the flap and lag motion that must be considered when the pitch bearing is inboard. The trim flap and lag motion displaces the blade section from the feathering axis, so that all in-plane and out-of-plane forces have a moment arm to produce pitch moments. In particular, the pitching motion produces an in-plane acceleration when the blade is flapped up and an out-of-plane acceleration when the blade is lagged back. Hence the effective pitch inertia with

16.5 Torsional Motion

the flap and lag hinges outboard is increased to

$$I = \int_0^R \left(I_\theta + z_{\text{trim}}^2 m + x_{\text{trim}}^2 m\right) dr$$

$$= \int_0^R I_\theta \, dr + \beta_{\text{trim}}^2 \int_0^R \eta_\beta^2 m \, dr + \zeta_{\text{trim}}^2 \int_0^R \eta_\zeta^2 m \, dr$$

$$\cong I_f + (\beta_0^2 + \zeta_0^2) I_b \qquad (16.141)$$

The resulting increase in pitch inertia and decrease in the effective pitch natural frequency can be substantial.

If the flap and lag motion occur outboard of the pitch bearing, there is a coupling of the pitch moment with the flap and lag moments that is particularly important for hingeless rotors. Consider the pitch moment resulting from the flap and lag motion of a rigid blade, with hinge springs to obtain general frequencies. The hinge spring moments are zero with the blade at the precone angle β_p and the sweep angle ζ_s. The pitch axis is coned at $\beta_p + \beta_d$; the droop β_d and sweep ζ_s occur outboard of the pitch bearing. The forces on the deflected blade section and their moment arms about the feathering axis are as follows:

i) the normal force $F_z - m r \ddot{\beta} - m r \Omega^2 \beta$, with moment arm $r\zeta$ due to the lag motion
ii) the in-plane force $F_x - m r \ddot{\zeta} - 2\Omega m r \beta \dot{\beta}$, with moment arm $r(\beta - \beta_p - \beta_d)$ due to the flap motion

Then using the flap and lag equations of motion from sections 16.3.1 and 16.4.1, the nose-down moment about the feathering axis is

$$\Delta M_\theta = \zeta \left[\int_0^1 r F_z dr - (\ddot{\beta} + \beta) \int_0^1 r^2 m \, dr\right]$$

$$- (\beta - \beta_p - \beta_d) \left[\int_0^1 r F_x dr - (\ddot{\zeta} + 2\beta\dot{\beta}) \int_0^1 r^2 m \, dr\right]$$

$$= \zeta K_\beta (\beta - \beta_p) - (\beta - \beta_p - \beta_d) K_\zeta (\zeta - \zeta_s)$$

$$= \zeta M_\beta - (\beta - \beta_p - \beta_d) M_\zeta \qquad (16.142)$$

where $\widehat{K}_\beta = K_\beta/I_b\Omega^2 = \nu_\beta^2 - 1$ and $\widehat{K}_\zeta = K_\zeta/I_b\Omega^2 = \nu_\zeta^2$ are the hinge spring constants. This result can be interpreted as follows. The net flap moment at the root $M_\beta = K_\beta(\beta - \beta_p)$ has a nose-down component about the feathering axis when the blade is lagged by ζ. Similarly, the lag moment $M_\zeta = K_\zeta(\zeta - \zeta_s)$ has a nose-up pitch component when the blade is flapped by $(\beta - \beta_p - \beta_d)$ relative to the pitch bearing. Then the total pitch moment is $\Delta M_\theta = \zeta M_\beta - (\beta - \beta_p - \beta_d) M_\zeta$.

Here β and ζ are the total flap and lag angles. The pitch moment in terms of the elastic flap deflection $\beta_e = \beta - \beta_p$ and the elastic lag deflection $\zeta_e = \zeta - \zeta_s$ is

$$\Delta M_\theta = \zeta K_\beta (\beta - \beta_p) - (\beta - \beta_p - \beta_d) K_\zeta (\zeta - \zeta_s)$$

$$= (K_\beta - K_\zeta)(\beta - \beta_p)(\zeta - \zeta_s) + K_\beta(\beta - \beta_p)\zeta_s + K_\zeta \beta_d (\zeta - \zeta_s)$$

$$= (K_\beta - K_\zeta)\beta_e \zeta_e + K_\beta \beta_e \zeta_s + K_\zeta \beta_d \zeta_e \qquad (16.143)$$

Although nonlinear in the flap and lag motion, this pitch moment can be significant. The principal effect of this moment is to produce a static pitch deflection due to the control system flexibility, $\Delta\theta = -\Delta M_\theta/K_\theta$. Thus the effect on the linearized dynamics is to introduce an effective pitch-flap and pitch-lag coupling. The pitch-flap coupling is

$$K_{P_\beta} = -\frac{\partial\theta}{\partial\beta} = \frac{1}{K_\theta}\left[(K_\beta - K_\zeta)\zeta_e + K_\beta\zeta_s\right] \quad (16.144)$$

for a given trim lag deflection, and the pitch-lag coupling is

$$K_{P_\zeta} = -\frac{\partial\theta}{\partial\zeta} = \frac{1}{K_\theta}\left[(K_\beta - K_\zeta)\beta_e + K_\zeta\beta_d\right] \quad (16.145)$$

for a given trim flap deflection. These couplings depend on the trim elastic coning and lag angles, which depend on the rotor thrust and torque and also on the precone (see Chapter 6). The couplings are proportional to the difference in flap and lag stiffnesses, $\widehat{K}_\beta - \widehat{K}_\zeta = v_\beta^2 - 1 - v_\zeta^2$. For an articulated rotor with no springs but with coincident flap and lag hinges, $v_\beta^2 = 1 + v_\zeta^2$, the pitch moment is zero and this coupling disappears.

A similar result can be derived for the torsional moment at an arbitrary blade section. Consider bending of a blade with out-of-plane deflection $z(r)$ and in-plane deflection $x(r)$. The forces acting on the blade outboard of r produce a torsional moment on the section at r:

$$\Delta M_r = \int_r^R \Big([z(\rho) - z(r) - (\rho - r)z'(r)]G_x - [x(\rho) - x(r) - (\rho - r)x'(r)]G_z\Big)d\rho \quad (16.146)$$

where G_x is the total section in-plane force, including both inertial and aerodynamic contributions, and G_z is the total section out-of-plane force. Then the nose-down torsional loading on the section is

$$\Delta T = \frac{\partial M_r}{\partial r} = x''\int_r^R (\rho - r)G_z d\rho - z''\int_r^R (\rho - r)G_x d\rho \quad (16.147)$$

Now $M_x = \int_r^R (\rho - r)G_z d\rho$ and $M_z = \int_r^R (\rho - r)G_x d\rho$ are, respectively, the flapwise and chordwise bending moments on the section at r, so

$$\Delta T = M_x x'' - M_z z'' \quad (16.148)$$

In terms of the flapwise and chordwise bending stiffnesses, the torsional loading is thus

$$\Delta T = M_x M_z \left(\frac{1}{EI_{xx}} - \frac{1}{EI_{zz}}\right) = x''z''(EI_{zz} - EI_{xx}) \quad (16.149)$$

The coupling is proportional to the product of the in-plane and out-of-plane deflections and the difference between the flapwise and chordwise bending stiffnesses. For a blade with $EI_{zz} = EI_{xx}$, the torsion-bending coupling disappears. This is called the matched-stiffness case and corresponds to the condition $v_\beta^2 = 1 + v_\zeta^2$ for a rigid blade. For a hingeless rotor with $v_\beta = 1.10$ to 1.15/rev, matched stiffness implies $v_\zeta = 0.46$ to 0.57/rev. The matched-stiffness blade has equal non-rotating flap and lag frequencies. Usually the chordwise stiffness of a rotor blade is much greater than the flapwise stiffness. However, the matched-stiffness condition can be achieved with a soft in-plane hingeless rotor, at least at the root, where it is most important for the fundamental modes.

The effects that have been discussed in this section are primarily important for a hingeless rotor, which requires a more complete model of the bending and torsion

16.5 Torsional Motion

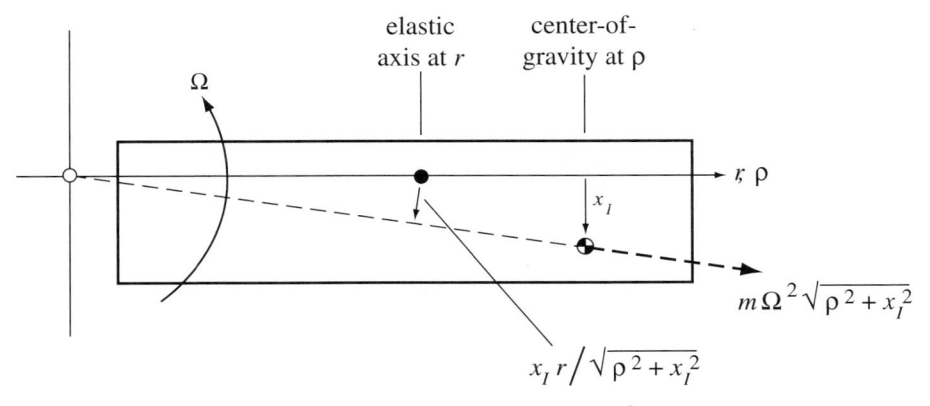

Figure 16.7. Origin of the centrifugal bending moment $m\Omega^2 r x_I \theta$.

dynamics for an accurate analysis. Flap or lag bending outboard of the feathering bearing produces substantial torsional moments. The resulting effective pitch-lag and pitch-flap couplings are important factors in hingeless rotor dynamics.

16.5.3 Torsion and Out-of-Plane Bending

Consider now the torsion and out-of-plane bending motion of an elastic blade. Excluding the in-plane motion from such an analysis is not entirely consistent. For example, the in-plane forces on the blade produce torsional moments when there is out-of-plane bending, as was seen from the preceding section. These forces are relieved by the blade lag motion, however, so they should not be considered unless the model includes the in-plane motion as well. For hingeless rotors in particular, a fully coupled flap-lag-torsion analysis is required to adequately represent the dynamics. Thus we are primarily concerned here with extending the rigid flap and rigid pitch analysis of section 16.5.1 to include the higher bending modes and elastic torsion and with laying the foundation for the development of more complete models.

The blade is assumed to have a straight elastic axis coincident with the feathering axis. The blade pitch now consists of the rigid pitch angle p_0 due to control system flexibility and a deflection θ_e due to elastic torsion of the blade: $\theta = p_0 + \theta_e$. The built-in twist only influences the trim forces and so can be ignored. The notation for the rigid pitch angle is chosen to be consistent with the modal expansion that is introduced for the elastic torsion θ_e.

The equation of motion for bending is obtained from the equilibrium of moments on the blade outboard of r. The section forces at radial station ρ and their moment arms about the elastic axis at r are as follows:

i) an inertial force $m(\ddot{z} - x_I \ddot{\theta})$, with moment arm $(\rho - r)$
ii) a centrifugal force $m\Omega^2 \rho$ acting on the center of gravity, with moment arm $(z - x_I \theta - z(r))$ about the elastic axis at r
iii) a centrifugal force $(m\Omega^2 r x_I)\theta(r)$
iv) an aerodynamic force F_z, with moment arm $(\rho - r)$

The centrifugal moment $m\Omega^2 r x_I \theta(r)$ at r due to the forces at ρ arises as follows. Figure 16.7 shows that the centrifugal force $m\Omega^2(\rho^2 + x_I^2)^{1/2}$ acting on the section center-of-gravity has a moment arm $x_I r/(\rho^2 + x_I^2)^{1/2}$ about the elastic axis at r, producing an in-plane bending moment $m\Omega^2 r x_I$. The section at r has a pitch angle $\theta(r)$, so the flapwise component of this bending moment on the blade is $m\Omega^2 r x_I \theta(r)$.

Thus the total moment on the blade section at r is

$$M(r) = \int_r^R \left[(F_z - m(\ddot{z} - x_I\ddot{\theta}))(\rho - r) \right.$$
$$\left. - m\Omega^2 \rho (z(\rho) - x_I\theta - z(r)) - \theta(r)m\Omega^2 r x_I \right] d\rho \quad (16.150)$$

By equating this expression to $M(r) = EI d^2z/dr^2$ and taking the second derivative, the partial differential equation for bending of the elastic axis is obtained:

$$\frac{d^2}{dr^2} EI \frac{d^2z}{dr^2} - \frac{d}{dr}\left[\int_r^R m\Omega^2 \rho\, d\rho \frac{dz}{dr} \right] + m\ddot{z} - mx_I\ddot{\theta} + \frac{d}{dr}\left[\frac{d(r\theta)}{dr} \int_r^R \Omega^2 x_I m\, d\rho \right] = F_z \quad (16.151)$$

The out-of-plane deflection can be expanded as a series in the normal modes, $z(r,t) = \sum_k \eta_k(r) q_k(t)$, where q_k are the bending degrees of freedom. The modal equation is the same as in section 16.3.2. By substituting for z and operating with $\int_0^R (\ldots) \eta_k dr$, the equation of motion for the k-th bending mode is obtained:

$$I_{qk}(\ddot{q}_k + v_k^2 q_k) - \int_0^R \eta_k x_I \ddot{\theta} m\, dr + \int_0^R \eta_k \frac{d}{dr}\left[\frac{d(r\theta)}{dr} \int_r^R \Omega^2 x_I m\, d\rho \right] dr = \int_0^R \eta_k F_z dr \quad (16.152)$$

The θ term can be written as

$$\int_0^R \eta_k \frac{d}{dr}\left[\frac{d(r\theta)}{dr} \int_r^R \Omega^2 x_I m\, d\rho \right] dr = -\int_0^R x_I \Omega^2 m \int_0^r \eta_k'(\rho\theta)' d\rho\, dr \quad (16.153)$$

For rigid pitch and flap ($\eta = r$ and θ independent of r) this equation of motion reduces to equation 16.129.

The equations of motion for rigid pitch and elastic torsion are obtained from equilibrium of torsion moments about the elastic axis. The forces acting on the blade section at ρ and their moment arms about the elastic axis at r are as follows:

i) an inertial moment $I_0 \ddot{\theta}$ about the section center of gravity;
ii) an inertial force $m(r\ddot{z} - x_I\ddot{\theta})$ acting on the center-of-gravity, with moment arm x_I about the elastic axis
iii) a nose-down propeller moment $I_\theta \Omega^2 \theta - x_I \Omega^2 m(z - z(r))$ about the elastic axis
iv) a nose-up centrifugal moment $(m\Omega^2 x_I r) z'(r)$
v) a nose-up aerodynamic moment M_a about the elastic axis

The propeller moment is due to the in-plane centrifugal force component $x\Omega^2 dm$ (Figure 16.6) acting with moment arm $(z(r) - (z - x\theta))$ about the elastic axis at r, so

$$\int_{\text{section}} (z(r) - (z - x\theta)) x \Omega^2 dm = I_\theta \Omega^2 \theta - x_I \Omega^2 m(z - z(r)) \quad (16.154)$$

The centrifugal moment $(m\Omega^2 x_I r) z'(r)$ is due to the in-plane bending moment $m\Omega^2 x_I r$ discussed for the flapping equation (see Figure 16.7). When the blade is flapped up by the angle $z'(r)$, this moment has a torsional component $(m\Omega^2 x_I r) z'(r)$ about the elastic axis at r. The total nose-up torsional moment on the blade section at r is thus

$$M_r = \int_r^R \left[M_a - I_\theta \ddot{\theta} - I_\theta \Omega^2 \theta + mx_I \ddot{z} + x_I \Omega^2 m(z - z(r) + r z'(r)) \right] d\rho \quad (16.155)$$

16.5 Torsional Motion

The equation of motion for rigid pitch is obtained from the equilibrium of moments about the pitch bearing at $r = 0$. The inertial and aerodynamic pitch moments of the blade are reacted by a moment from the control system:

$$M_r(0) = K_\theta \left(\theta(0) - \theta_{\text{con}}\right) \tag{16.156}$$

where K_θ is the control system stiffness, θ_{con} is the root pitch angle commanded by the control system, and $\theta(0)$ is the actual root pitch. We define the elastic torsion of the blade to be zero at the pitch bearing, $\theta_e(0) = 0$, so that the root pitch equals the rigid pitch degree of freedom; that is, $\theta(0) = p_0$. The differential equation of motion for rigid pitch is thus

$$\int_0^R \left[I_\theta \ddot{\theta} + I_\theta \Omega^2 \theta - m x_I \ddot{z} - m x_I z \Omega^2\right] dr + K_\theta (p_0 - \theta_{\text{con}}) = \int_0^R M_a \, dr \tag{16.157}$$

For rigid flap and pitch this reduces to the result of section 16.5.1.

Engineering beam theory relates the torsional moment to the elastic torsion deflection by

$$M_r = GJ \frac{d\theta_e}{dr} \tag{16.158}$$

where GJ is the torsional rigidity of the blade section. If the expressions for the structural moment on the section and the total inertial and aerodynamic moment are equated and the derivative taken with respect to r,

$$-\frac{d}{dr} GJ \frac{d\theta_e}{dr} + I_\theta \ddot{\theta} + I_\theta \Omega^2 \theta - m x_I \ddot{z} + r \frac{d}{dr} \left[\frac{dz}{dr} \int_r^R m \Omega^2 x_I \, d\rho\right] = M_a \tag{16.159}$$

This is the partial differential equation for elastic torsion of the rotating blade. The boundary conditions are $d\theta_e/dr = 0$ at $r = R$ (a free end at the tip) and $\theta_e = 0$ at $r = 0$ (a fixed end at the root). Consider free torsional vibration of the non-rotating blade:

$$-\frac{d}{dr} GJ \frac{d\theta_e}{dr} + I_\theta \ddot{\theta}_e = 0 \tag{16.160}$$

Solving this equation by separation of variables, write $\theta_e = \xi(r) e^{i\omega t}$, which gives

$$\frac{d}{dr} GJ \frac{d\xi}{dr} + \omega^2 I_\theta \xi = 0 \tag{16.161}$$

with the boundary conditions $\xi(0) = 0$ and $\xi'(R) = 0$. This is a proper Sturm-Liouville eigenvalue problem, for which there exists a series of eigen-solutions $\xi_k(r)$ and corresponding eigenvalues ω_k^2. Since the mode shapes are orthogonal with weighting function I_θ:

$$\int_0^R \xi_k \xi_i I_\theta \, dr = 0 \tag{16.162}$$

if $k \neq i$. The eigenvalues are ordered according to size (ω_1 is the smallest torsion frequency) and the mode shapes are normalized to unit deflection at the tip, $\xi(R) = 1$. According to Sturm-Liouville theory, the natural frequencies can be obtained from the mode shapes by

$$\omega^2 = \frac{\int_0^R GJ \xi'^2 \, dr}{\int_0^R I_\theta \xi^2 \, dr} \tag{16.163}$$

The free vibration of a non-rotating blade with uniform GJ and I_θ distributions has the exact solution,

$$\xi_k = \sin\left[\left(k - \frac{1}{2}\right)\pi \frac{r}{R}\right] \tag{16.164}$$

with the corresponding natural frequencies

$$\omega_k = \left(k - \frac{1}{2}\right)\pi \sqrt{\frac{GJ}{I_\theta R^2}} \tag{16.165}$$

for $k = 1$ to ∞. These functions are useful in solving the modal equation for the true mode shapes, such as by the Galerkin method, and serve as approximate mode shapes when better estimates are not available. The simple function $\xi_1 = r/R$ can also be used as an approximation to the first mode shape. For torsion the non-rotating free vibration modes are used. Rotating modes could be used instead by retaining the centrifugal spring term (the propeller moment) in the modal equation. For the torsional stiffness typical of rotor blades, the rotation has little effect on the free vibration frequencies and mode shapes. Thus it is reasonable to use the non-rotating modes, which are generally simpler to calculate.

Now the torsional deflection is expanded as a series in the normal modes:

$$\theta_e(r, t) = \sum_{k=1}^{\infty} \xi_k(r) p_k(t) \tag{16.166}$$

where p_k are the degrees of freedom of elastic torsion. With the mode shapes normalized to $\xi_k = 1$ at the tip, p_k is the pitch angle at the tip for the k-th mode. Using a mode shape $p_0 = 1$ for rigid pitch, the total blade pitch can be written as

$$\theta = p_0 + \theta_e = \sum_{k=0}^{\infty} \xi_k(r) p_k(t) \tag{16.167}$$

Next, the expansion for θ_e is substituted into the partial differential equation for torsion, the modal equation satisfied by ξ_k is used to replace the torsional stiffness term by the natural frequency ω_k^2, and the equation is operated on with $\int_0^R (\ldots)\xi_k dr$. Using the orthogonality of the elastic torsion modes (the rigid pitch and elastic torsion modes are not orthogonal), the following differential equation is obtained for the k-th mode:

$$I_{pk}(\ddot{p}_k + (\omega_k^2 + \Omega^2) p_k) + \left(\int_0^R I_\theta \xi_k dr\right)(\ddot{p}_0 + \Omega^2 p_0) - \int_0^R \xi_k m x_I \ddot{z}\, dr$$

$$+ \int_0^R \xi_k r \frac{d}{dr}\left[\frac{dz}{dr}\int_r^R \Omega^2 m x_I d\rho\right] dr = \int_0^R \xi_k M_a dr \tag{16.168}$$

where $I_{pk} = \int_0^R \xi_k^2 I_\theta dr$ is the generalized mass of the mode. The bending term can be written as

$$\int_0^R \xi_k r \frac{d}{dr}\left[\frac{dz}{dr}\int_r^R \Omega^2 m x_I d\rho\right] dr = -\int_0^R x_I \Omega^2 m \int_0^r z'(\rho \xi_k)' d\rho\, dr \tag{16.169}$$

16.5 Torsional Motion

Now substituting the expansion $\theta = \sum_{k=0}^{\infty} \xi_k p_k$ into the equation of motion for rigid pitch gives

$$I_{p0}(\ddot{p}_0 + (\omega_0^2 + 1)\Omega^2 p_0) + \sum_{j=1}^{\infty} \left(\int_0^R I_\theta \xi_j dr \right) (\ddot{p}_j + \Omega^2 p_j) - \int_0^R m x_I \ddot{z} \, dr$$

$$- \int_0^R m x_I z \Omega^2 dr = \int_0^R M_a dr + I_{p0}\omega_0^2 \Omega^2 \theta_{\text{con}} \qquad (16.170)$$

where $I_{p0} = \int_0^R I_\theta dr$ is the pitch moment of inertia of the blade and ω_0 is the natural frequency of the rigid pitch motion due to control system flexibility: $\omega_0^2 = K_\theta/(I_{p0}\Omega^2)$.

Finally, substitute the modal expansion for z into the torsion and pitch equations, and the expansion for θ into the bending equation; divide by I_b, and use dimensionless quantities. The equations of motion for out-of-plane bending, rigid pitch, and elastic torsion of the rotating blade result:

$$\widehat{I}_{qk}(\ddot{q}_k + \nu_k^2 q_k) - \sum_{j=0}^{\infty} \left(\widehat{I}_{q_k\ddot{p}_j} \ddot{p}_j + \widehat{I}_{q_k p_j} p_j \right)$$

$$= \gamma \int_0^1 \eta_k \frac{F_z}{ac} dr \qquad (16.171)$$

$$\widehat{I}_{p0}(\ddot{p}_0 + (\omega_0^2 + 1)p_0) + \sum_{j=1}^{\infty} \widehat{I}_{p_0 p_j}(\ddot{p}_j + p_j) - \sum_{j=1}^{\infty} \left(\widehat{I}_{q_j\ddot{p}_0} \ddot{q}_j + \widehat{I}_{q_j p_0} q_j \right)$$

$$= \gamma \int_0^1 \frac{M_a}{ac} dr + \widehat{I}_{p0}\omega_0^2 \theta_{\text{con}} - \sum_{j=1}^{\infty} \widehat{I}_{p0}\omega_0^2 K_{Pj} q_j \qquad (16.172)$$

$$\widehat{I}_{pk}(\ddot{p}_k + (\omega_k^2 + 1)p_k) + \widehat{I}_{p_0 p_k}(\ddot{p}_0 + p_0) - \sum_{j=1}^{\infty} \left(\widehat{I}_{q_j\ddot{p}_k} \ddot{q}_j + \widehat{I}_{q_j p_k} q_j \right)$$

$$= \gamma \int_0^1 \xi_k \frac{M_a}{ac} dr \qquad (16.173)$$

where the inertial coefficients are

$$\widehat{I}_{qk} = \frac{1}{I_b} \int_0^1 \eta_k^2 m \, dr \qquad (16.174)$$

$$\widehat{I}_{q_k\ddot{p}_j} = \frac{1}{I_b} \int_0^1 \eta_k \xi_j x_I m \, dr \qquad (16.175)$$

$$\widehat{I}_{q_k p_j} = \frac{1}{I_b} \int_0^1 x_I m \int_0^r \eta_k'(\rho \xi_j)' d\rho \, dr \qquad (16.176)$$

$$\widehat{I}_{pk} = \frac{1}{I_b} \int_0^1 \xi_k^2 I_\theta \, dr \qquad (16.177)$$

$$\widehat{I}_{p_0 p_k} = \frac{1}{I_b} \int_0^1 \xi_k I_\theta \, dr \qquad (16.178)$$

The bending and torsion equations are coupled by inertial and centrifugal forces if the section center-of-gravity is offset from the elastic axis. Kinematic pitch-bending coupling of the form $\Delta\theta_{\text{con}} = -\sum_j K_{Pj}q_j$ has been included. For rigid flap and pitch, these equations reduce to those obtained in section 16.5.1.

Because the rigid pitch mode is not orthogonal to the elastic bending modes, the equations for p_0 and p_k ($k \geq 1$) are coupled by inertial and centrifugal forces. The problem can also be formulated without the separate rigid pitch degree of freedom. Then the p_0 degree of freedom and equation of motion are dropped, and θ_e represents the complete pitch motion, including that due to control system flexibility. The boundary condition for the torsion equation becomes

$$GJ\frac{d\theta_e}{dr} = K_\theta \left(\theta_e - \theta_{\text{con}} + \sum_k K_{Pk}q_k\right) \tag{16.179}$$

at $r = 0$. The modal equation for free vibration can be solved with the boundary condition

$$GJ\frac{d\xi}{dr} = K_\theta \xi \tag{16.180}$$

for a general restrained end. The solution is a single series of orthogonal modes including both control system flexibility and blade torsional flexibility. However, this series of modes always gives $GJ\theta'_e = K_\theta \theta_e$ at the pitch bearing, which implies that the commanded pitch control and the pitch-bending feedback are zero. This is a typical result for normal modes, implying that discrete forces and moments applied at the end points cannot be handled. The problem also arises in treating the lag damper of an articulated blade, where the normal modes imply that the moment at the hinge is always zero. For this reason the rigid pitch and elastic torsion motion were separated in the present normal modes analysis. This is a rigorous approach and is easily implemented in a numerical solution. Moreover, rigid pitch alone is a sufficient model of the blade pitch motion for many rotors. The coupled rigid pitch/elastic torsion modes can be used in the rotor analysis, including a proper representation of the end conditions, with the Rayleigh-Ritz or Galerkin methods (see section 16.2).

16.5.4 Non-Rotating Frame

The rotor control system couples the pitch motion of the individual blades. Each non-rotating mode of pitch motion has a different load path in the fixed control system, and hence a different effective stiffness. This coupling can be accounted for by using a separate natural frequency for each non-rotating degree of freedom. Consider the pitch equation of motion for the m-th blade in the rotating frame:

$$\widehat{I}_f\left(\ddot{\theta}^{(m)} + (\omega^2 + 1)\theta^{(m)}\right) = \gamma \int_0^1 \frac{M_a}{ac}dr = \gamma M_f^{(m)} \tag{16.181}$$

The corresponding equations of motion in the non-rotating frame are

$$\widehat{I}_f\left(\ddot{\theta}_0 + (\omega_0^2 + 1)\theta_0\right) = \gamma M_{f_0} \tag{16.182}$$

$$\widehat{I}_f\left(\ddot{\theta}_{nc} + 2n\dot{\theta}_{ns} + (\omega_{nc}^2 + 1 - n^2)\theta_{nc}\right) = \gamma M_{f_{nc}} \tag{16.183}$$

$$\widehat{I}_f\left(\ddot{\theta}_{ns} - 2n\dot{\theta}_{nc} + (\omega_{ns}^2 + 1 - n^2)\theta_{ns}\right) = \gamma M_{f_{ns}} \tag{16.184}$$

$$\widehat{I}_f\left(\ddot{\theta}_{N/2} + (\omega_{N/2}^2 + 1)\theta_{N/2}\right) = \gamma M_{f_{N/2}} \tag{16.185}$$

where a separate natural frequency has been introduced for each equation. This is equivalent to assuming that the restoring moment provided by the control system responds to the nonrotating modes in such a way that

$$M_\theta^{(m)} = K_0\left(\theta_0 - \theta_0^{\text{con}}\right) + \sum_n K_{nc}\left(\theta_{nc} - \theta_{nc}^{\text{con}}\right)\cos n\psi_m$$
$$+ \sum_n K_{ns}\left(\theta_{ns} - \theta_{ns}^{\text{con}}\right)\sin n\psi_m + K_{N/2}\left(\theta_{N/2} - \theta_{N/2}^{\text{con}}\right)(-1)^m \quad (16.186)$$

instead of $M_\theta = K_\theta(\theta - \theta_{\text{con}})$ as in section 16.5.1. Thus ω_0 is the stiffness of the collective control system, whereas ω_{1c} and ω_{1s} are the stiffnesses of the cyclic control system. The higher multiblade coordinates produce no net force in the non-rotating control system and so are only due to flexibility in the pitch horn and pitch link and to swashplate bending. Thus for the reactionless modes ($\omega_{2c}, \omega_{2s}, \ldots, \omega_{N/2}$) the frequencies are usually much higher than for the collective and cyclic modes.

This technique of using different natural frequencies in the non-rotating frame is also useful for the flap and lag motion. A gimballed rotor can be modeled by using $\nu = 1$ for the β_{1c} and β_{1s} rigid flap degrees of freedom and by using the appropriate cantilever frequency and mode shape for the coning and other degrees of freedom.

16.6 Hub Reactions

The net forces and moments at the root of the rotating blade are transmitted to the helicopter airframe. The steady and low-frequency components of these hub reactions in the non-rotating frame are the forces and moments required to trim and control the aircraft. The higher frequency components are responsible for helicopter vibration. When the shaft motion is included in the model, these rotor forces and moments determine the helicopter flight dynamics. Figure 16.8 shows the root shears and moments on the rotating blade, as well as the rotor forces and moments acting on the hub in the non-rotating frame. The vertical shear force S_z produces the rotor thrust, and the in-plane shear forces S_x and S_r produce the rotor side and drag forces. The flapwise root moment N_F produces the rotor pitch and roll moments, and the lagwise moment N_L produces the rotor shaft torque. Positive rotor hub reactions are acting on the helicopter, with the exception of the rotor torque Q, which is defined as the moment on the rotor (the torque reaction of the rotor on the hub is positive in the direction opposing the rotor rotation). Figure 16.8 indicates the positive directions of the rotor thrust T, drag force H, side force Y, pitch moment M_y, and roll moment M_x.

16.6.1 Rotating Loads

The net root forces and moments on the rotating blade are obtained by integrating the section inertial and aerodynamic forces, as in the derivation of the blade equations of motion. Consider an articulated rotor with no hinge offset, as in section 16.3.1. The vertical forces acting on the blade section are the inertial force $m\ddot{z} = mr\ddot{\beta}$ and the aerodynamic force F_z. The centrifugal force is always parallel to the hub plane; see Figure 16.9. The vertical shear force at the blade root is therefore

$$S_z = \int_0^R F_z\,dr - \ddot{\beta}\int_0^R rm\,dr \quad (16.187)$$

Blade Motion

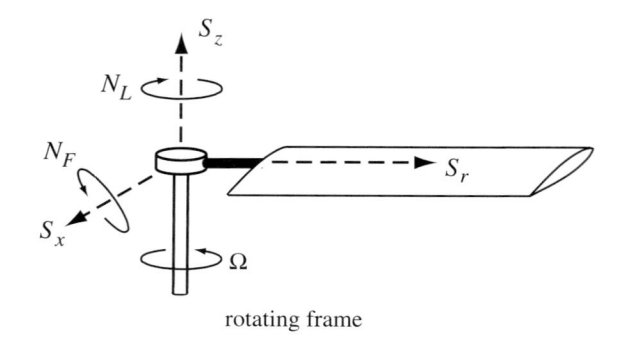

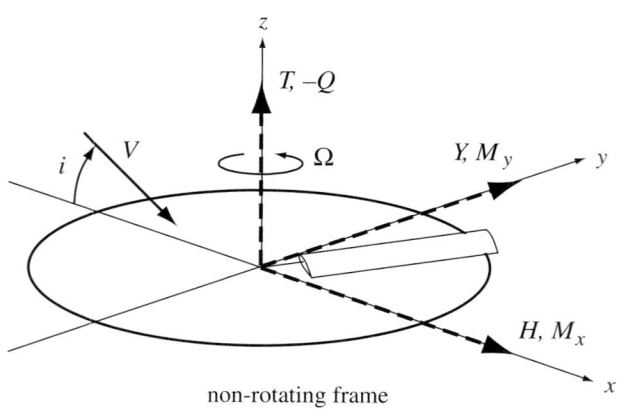

Figure 16.8. Rotor forces and moments acting on the hub.

In section 16.3.1 for the equation of flapping motion, the root flap moment was obtained from integration of the inertial, centrifugal, and aerodynamic forces on the section:

$$N_F = \int_0^R rF_z dr - (\ddot{\beta} + \Omega^2\beta) \int_0^R r^2 m\, dr \qquad (16.188)$$

The root moment is simply the hinge moment in this case, since there is no offset of the flap hinge; the moment can be nonzero only with a hinge spring. The moment

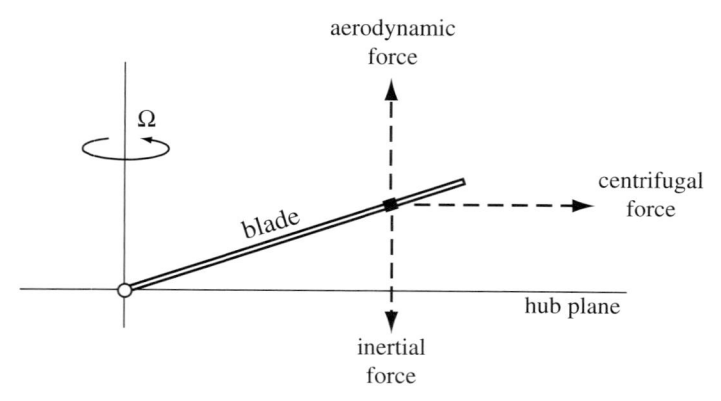

Figure 16.9. Blade section forces producing the vertical shear and flapwise moment at the root.

transmitted to the hub through the hinge spring is $N_F = K_\beta(\beta - \beta_p)$, or since $v_\beta^2 = 1 + K_\beta/(I_b\Omega^2)$,

$$N_F = I_b\Omega^2(v_\beta^2 - 1)(\beta - \beta_p) \qquad (16.189)$$

From section 6.15, this relation applies when there is a hinge offset as well.

Next consider the case of general out-of-plane bending motion, including both articulated and hingeless rotors. The forces acting on the blade section are described in section 16.3.2, and the equation of motion for the normal bending modes is derived. The vertical shear force at the root is obtained by integrating the aerodynamic and inertial forces on the blade:

$$S_z = \int_0^R (F_z - m\ddot{z}) dr \qquad (16.190)$$

Substituting for the modal expansion $z = \sum_k \eta_k q_k$ gives

$$S_z = \int_0^R F_z dr - \sum_k \ddot{q}_k \int_0^R \eta_k m \, dr \qquad (16.191)$$

The root moment is obtained from the flap moments due to the aerodynamic, inertial, and centrifugal forces on the blade section (Figure 16.9) or simply by evaluating the flapwise bending moment expression given in section 16.3.2 at the root:

$$N_F = \int_0^R [(F_z - m\ddot{z})r - m\Omega^2 rz] dr$$

$$= \int_0^R rF_z dr - \sum_k (\ddot{q}_k + \Omega^2 q_k) \int_0^R r\eta_k m \, dr \qquad (16.192)$$

Recall that the differential equation of motion for q_k is

$$I_{qk}(\ddot{q}_k + v_k^2 q_k) = \int_0^R \eta_k F_z dr \qquad (16.193)$$

The aerodynamic loading F_z therefore contributes directly to the root shear and moment, but it also excites the blade bending motion, which then cancels part of the hub reaction. Indeed, the flap hinge was introduced so that the blade motion rather than the structure would absorb the root moments. Since the mode shapes q_k form a complete series, the aerodynamic loading can be expanded as $F_z = \sum_k F_{zk} \eta_k m$, where the coefficients are $F_{zk} = \int_0^R \eta_k F_z dr / \int_0^R \eta_k^2 m \, dr$. When the expansion for F_z is substituted, the root moment becomes

$$N_F = \sum_k (F_{zk} - \ddot{q}_k - \Omega^2 q_k) \int_0^R r\eta_k m \, dr \qquad (16.194)$$

The equation of motion for q_k gives $F_{zk} = \ddot{q}_k + v_k^2 q_k$, so

$$N_F = \sum_k q_k \Omega^2 (v_k^2 - 1) \int_0^R r\eta_k m \, dr \qquad (16.195)$$

For an articulated rotor with no hinge offset, $v_1 = 1$ and $\eta_1 = r$ for the first mode, and all the higher mode shapes are orthogonal to $\eta_1 = r$; hence $N_F = 0$, as required. If only a single flap mode is used and the mode shape is approximated by $\eta \cong r$,

the expression reduces to
$$N_F = I_b\Omega^2(\nu_\beta^2 - 1)\beta \qquad (16.196)$$
as earlier. Thus the hub moment can be obtained from the flap deflection and the natural frequency of the fundamental flap mode. The simplicity of this result makes it very useful. In a similar fashion the vertical shear force of the root can be expressed as
$$S_z = \sum_k (F_{zk} - \ddot{q}_k) \int_0^R \eta_k m\, dr = \sum_k q_k \Omega^2 \nu_k^2 \int_0^R \eta_k m\, dr \qquad (16.197)$$

It is more convenient, though, to relate the vertical shear, and hence the rotor thrust, directly to the aerodynamic force.

If the number of modes is large, the same result should be obtained for the hub moment regardless of whether the forces are integrated along the blade or equation 16.195 is used. With the latter approach, using a finite number of modes is equivalent to truncating the expansion $F_z = \sum_k F_{zk} \eta_k m$, which may not be an adequate representation of the loading if only a small number of modes are used. Thus better results are generally to be expected from using the integrals of the blade section forces to obtain the hub reactions, although in some cases the improved accuracy may not be as valuable as a simple equation.

Next let us examine the in-plane shear forces and torque moment at the blade root, including the in-plane blade motion. Consider an articulated blade with an in-plane displacement given by $x = \eta_\zeta \zeta$. There are three forces acting in the radial direction:

i) a centrifugal force $m\Omega^2 r$
ii) a radially inward Coriolis force $2\Omega \dot{x} m = 2\Omega \eta_\zeta \dot{\zeta} m$
iii) a radial aerodynamic force F_r, due to the radial drag and an in-plane component of the lift when the blade flaps

The Coriolis force is due to the product of the rotor rotational speed Ω and the in-plane velocity $\dot{x}$; this is the force responsible for the $\beta\dot{\zeta}$ flap moment (see section 16.4.1). Thus the radial shear force at the root is
$$S_r = \int_0^R (F_r + m\Omega^2 r - 2\Omega \eta_\zeta \dot{\zeta} m)\, dr$$
$$= \int_0^R F_r\, dr + \Omega^2 \int_0^R rm\, dr - 2\Omega\dot{\zeta} \int_0^R \eta_\zeta m\, dr \qquad (16.198)$$

The centrifugal force is constant and is reacted by the identical centrifugal forces on the other blades. Hence only the aerodynamic and Coriolis forces contribute to the hub reactions in the non-rotating frame.

Figure 16.10 shows the in-plane forces normal to the r axis that act on the blade section:

i) an inertial force $m\ddot{x} = m\eta_\zeta \ddot{\zeta}$
ii) a centrifugal force $m\Omega^2 x = m\Omega^2 \eta_\zeta \zeta$
iii) an aerodynamic force F_x, consisting of profile and induced drag terms

The second force arises because the centrifugal force $m\Omega^2 r$ has a component $(m\Omega^2 r)(x/r)$ normal to the r axis that acts in the same direction as the lag motion

16.6 Hub Reactions

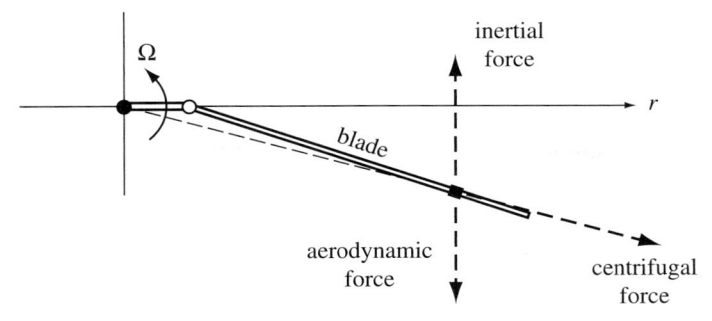

Figure 16.10. Blade section forces producing the in-plane shear at the root.

(Figure 16.10). The in-plane Coriolis force due to the flapping motion is small compared to the centrifugal force and has been neglected. Thus the total in-plane shear force at the blade root is

$$S_x = \int_0^R \left(F_x - m\eta_\zeta \ddot{\zeta} + m\Omega^2 \eta_\zeta \zeta \right) dr$$
$$= \int_0^R F_x dr - \left(\ddot{\zeta} - \Omega^2 \zeta \right) \int_0^R \eta_\zeta m \, dr \qquad (16.199)$$

The torque moment acting on the rotor hub is due to the section in-plane forces as discussed in deriving the lag equation of motion:

i) an inertial force $m\ddot{x}$
ii) a Coriolis force $2\Omega \dot{z} z' m$
iii) an aerodynamic force F_x

These forces have moment arm r about the center of rotation. The centrifugal force always passes through the shaft axis and so does not contribute to the torque. The moment at the root is thus

$$N_L = \int_0^R r \left(F_x - m\eta_\zeta \ddot{\zeta} - 2\Omega m \beta \dot{\beta} \eta_\beta \eta'_\beta \right) dr$$
$$= \int_0^R r F_x dr - \ddot{\zeta} \int_0^R r \eta_\zeta m \, dr - 2\Omega \beta \dot{\beta} \int_0^R \eta_\beta \eta'_\beta r m \, dr \qquad (16.200)$$

These results are readily extended to the case of general in-plane bending. As in section 16.4.3, expand the in-plane deflection as a series in the normal modes: $x = \sum_k \eta_{xk} q_{xk}$. Then the radial and in-plane shear forces are

$$S_r = \int_0^R \left(F_r + m\Omega^2 r - 2\Omega \dot{x} m \right) dr$$
$$= \int_0^R F_r dr + \Omega^2 \int_0^R r m \, dr - 2\Omega \sum_k \dot{q}_{xk} \int_0^R \eta_{xk} m \, dr \qquad (16.201)$$

$$S_x = \int_0^R \left(F_x - m\ddot{x} + m\Omega^2 x \right) dr$$
$$= \int_0^R F_x dr - \sum_k (\ddot{q}_{xk} - \Omega^2 q_{xk}) \int_0^R \eta_{xk} m \, dr \qquad (16.202)$$

Neglecting the Coriolis terms, the torque moment is

$$N_L = \int_0^R r(F_x - m\ddot{x})\,dr$$
$$= \int_0^R rF_x\,dr - \sum_k \ddot{q}_{xk} \int_0^R r\eta_{xk} m\,dr \qquad (16.203)$$

By expanding the aerodynamic loading F_x as a series in the bending mode shapes as for the hub moment, the in-plane shear force and the torque moment can be written as

$$S_x = \sum_k q_{xk}\Omega^2(v_{xk}^2 + 1) \int_0^R \eta_{xk} m\,dr \qquad (16.204)$$

$$N_L = \sum_k q_{xk}\Omega^2 v_{xk}^2 \int_0^R r\eta_{xk} m\,dr \qquad (16.205)$$

These results are not as useful as the corresponding expression for the flap moment, however, since the blade lag motion q_{xk} must be found in order to evaluate S_x and N_L in this manner. If the in-plane shear and torque are left in terms of the integrated aerodynamic forces, they can be evaluated even if the analysis neglects the lag motion.

16.6.2 Non-Rotating Loads

The total forces and moments acting on the rotor hub are obtained by resolving the rotating forces in the non-rotating frame and summing over all N blades:

$$T = \sum_{m=1}^N S_z \qquad (16.206)$$

$$H = \sum_{m=1}^N (S_r \cos\psi_m + S_x \sin\psi_m) \qquad (16.207)$$

$$Y = \sum_{m=1}^N (S_r \sin\psi_m - S_x \cos\psi_m) \qquad (16.208)$$

$$M_x = \sum_{m=1}^N N_F \sin\psi_m \qquad (16.209)$$

$$M_y = -\sum_{m=1}^N N_F \cos\psi_m \qquad (16.210)$$

$$Q = \sum_{m=1}^N N_L \qquad (16.211)$$

16.6 Hub Reactions

where ψ_m is the azimuth angle of the m-th blade. Working with the hub reactions in rotor coefficient form is convenient. In terms of dimensionless quantities,

$$\frac{T}{NI_b} = \frac{T/\rho R^4 \Omega^2}{NI_b/\rho R^5} = \frac{\rho a c R^4}{I_b} \frac{T/\rho A(\Omega R)^2}{(Nc/\pi R)a} = \gamma \frac{C_T}{\sigma a} \quad (16.212)$$

Thus the equations for the hub reactions are divided by NI_b, giving

$$\gamma \frac{C_T}{\sigma a} = \frac{1}{N} \sum_{m=1}^{N} \frac{S_z}{I_b} = \frac{\gamma}{N} \sum_{m=1}^{N} \frac{S_z}{ac} \quad (16.213)$$

and similar results for the other forces and moments.

Consider first the rotor thrust. Equation 16.197 for the vertical shear force of the m-th blade in terms of dimensionless quantities is

$$\frac{S_z}{I_b} = \gamma \int_0^1 \frac{F_z}{ac} dr - \sum_k \widehat{S}_{qk} \ddot{q}_k^{(m)} \quad (16.214)$$

where $\widehat{S}_{qk} = \int_0^1 \eta_k m \, dr/I_b$. So

$$\gamma \frac{C_T}{\sigma a} = \frac{\gamma}{N} \sum_{m=1}^{N} \int_0^1 \frac{F_z}{ac} dr - \sum_k \widehat{S}_{qk} \left(\frac{1}{N} \sum_{m=1}^{N} \ddot{q}_k^{(m)} \right) \quad (16.215)$$

Now from the definition of multiblade coordinates (section 15.4.1), the acceleration of the coning degree of freedom of the k-th bending mode is

$$\ddot{\beta}_0^{(k)} = \frac{1}{N} \sum_{m=1}^{N} \ddot{q}_k^{(m)} \quad (16.216)$$

Hence the rotor thrust becomes

$$\gamma \frac{C_T}{\sigma a} = \frac{\gamma}{N} \sum_{m=1}^{N} \int_0^1 \frac{F_z}{ac} dr - \sum_k \widehat{S}_{qk} \ddot{\beta}_0^{(k)} \quad (16.217)$$

The first term is the net aerodynamic lift on the rotor, and the second term is the vertical acceleration due to the coning motion of the blades.

Equation 16.195 for the root flapwise moment on the rotating blade is

$$\frac{N_F}{I_b} = \sum_k \widehat{I}_{qk\alpha} q_k^{(m)} (v_k^2 - 1) \quad (16.218)$$

where $\widehat{I}_{qk\alpha} = \int_0^1 r \eta_k m \, dr/I_b$. The pitch and roll moments on the rotor hub are then

$$-\gamma \frac{2C_{My}}{\sigma a} = \sum_k \widehat{I}_{qk\alpha}(v_k^2 - 1) \frac{2}{N} \sum_{m=1}^{N} q_k^{(m)} \cos \psi_m \quad (16.219)$$

$$\gamma \frac{2C_{Mx}}{\sigma a} = \sum_k \widehat{I}_{qk\alpha}(v_k^2 - 1) \frac{2}{N} \sum_{m=1}^{N} q_k^{(m)} \sin \psi_m \quad (16.220)$$

For a rotor with three or more blades, the definitions of the cyclic degrees of freedom $\beta_{1c}^{(k)}$ and $\beta_{1s}^{(k)}$ then give

$$\begin{pmatrix} -\gamma \dfrac{2C_{My}}{\sigma a} \\ \gamma \dfrac{2C_{Mx}}{\sigma a} \end{pmatrix} = \sum_k \widehat{I}_{q_k\alpha}(\nu_k^2 - 1) \begin{pmatrix} \beta_{1c}^{(k)} \\ \beta_{1s}^{(k)} \end{pmatrix} \qquad (16.221)$$

Using just a single flap mode, the hub moments are then simply proportional to the rotor tip-path-plane tilt:

$$\begin{pmatrix} -\gamma \dfrac{2C_{My}}{\sigma a} \\ \gamma \dfrac{2C_{Mx}}{\sigma a} \end{pmatrix} = \widehat{I}_\beta(\nu_\beta^2 - 1) \begin{pmatrix} \beta_{1c} \\ \beta_{1s} \end{pmatrix} \qquad (16.222)$$

This result was obtained in sections 6.14 and 6.15 for more limited models of the blade motion. If instead

$$\frac{N_F}{I_b} = \gamma \int_0^1 r \frac{F_z}{ac} dr - \sum_k \widehat{I}_{q_k\alpha}(\ddot{q}_k^{(m)} + q_k^{(m)}) \qquad (16.223)$$

(equation 16.192) is used as the expression for the moment, then

$$-\gamma \frac{2C_{My}}{\sigma a} = \frac{2\gamma}{N}\sum_{k=1}^N \cos\psi_m \int_0^1 r\frac{F_z}{ac}dr - \sum_k \widehat{I}_{q_k\alpha}\left(\ddot{\beta}_{1c}^{(k)} + 2\dot{\beta}_{1s}^{(k)}\right) \qquad (16.224)$$

$$\gamma \frac{2C_{Mx}}{\sigma a} = \frac{2\gamma}{N}\sum_{k=1}^N \sin\psi_m \int_0^1 r\frac{F_z}{ac}dr - \sum_k \widehat{I}_{q_k\alpha}\left(\ddot{\beta}_{1s}^{(k)} - 2\dot{\beta}_{1c}^{(k)}\right) \qquad (16.225)$$

In the steady-state case, the tip-path-plane tilt is constant. Then only the aerodynamic forces contribute to the hub pitch and roll moment, which is the result derived in section 6.3.

The rotor drag and side forces are obtained by resolving into the non-rotating frame the in-plane and radial shear forces on the root of the rotating blade:

$$\frac{S_x}{I_b} = \gamma \int_0^1 \frac{F_x}{ac}dr - \widehat{S}_\zeta(\ddot{\zeta}^{(m)} - \zeta^{(m)}) \qquad (16.226)$$

$$\frac{S_r}{I_b} = \gamma \int_0^1 \frac{F_r}{ac}dr - 2\widehat{S}_\zeta\dot{\zeta}^{(m)} \qquad (16.227)$$

where $\widehat{S}_\zeta = \int_0^1 \eta_\zeta m\, dr/I_b$ (equations 16.199 and 16.198). Only the fundamental lag mode has been considered for simplicity, and the centrifugal force has been dropped from S_r since it does not contribute to the total hub forces. Now the definitions of the cyclic lag degrees of freedom in multiblade coordinates give

$$\frac{2}{N}\sum_{m=1}^N [(\ddot{\zeta}^{(m)} - \zeta^{(m)})\sin\psi_m + 2\dot{\zeta}^{(m)}\cos\psi_m]$$

$$= (\ddot{\zeta}_{1s} - 2\dot{\zeta}_{1c} - \zeta_{1s} - \zeta_{1s}) + 2(\dot{\zeta}_{1c} + \zeta_{1s}) = \ddot{\zeta}_{1s} \qquad (16.228)$$

$$\frac{2}{N}\sum_{m=1}^N [(\ddot{\zeta}^{(m)} - \zeta^{(m)})\cos\psi_m - 2\dot{\zeta}^{(m)}\sin\psi_m]$$

$$= (\ddot{\zeta}_{1c} + 2\dot{\zeta}_{1s} - \zeta_{1c} - \zeta_{1c}) - 2(\dot{\zeta}_{1s} - \zeta_{1c}) = \ddot{\zeta}_{1c} \qquad (16.229)$$

(again assuming the rotor has three or more blades). Hence the rotor drag and side forces are

$$\gamma \frac{2C_H}{\sigma a} = \frac{2\gamma}{N} \sum_{m=1}^{N} \left[\cos\psi_m \int_0^1 \frac{F_r}{ac} dr + \sin\psi_m \int_0^1 \frac{F_x}{ac} dr \right] - \widehat{S}_\zeta \ddot{\zeta}_{1s} \quad (16.230)$$

$$\gamma \frac{2C_Y}{\sigma a} = \frac{2\gamma}{N} \sum_{m=1}^{N} \left[\sin\psi_m \int_0^1 \frac{F_r}{ac} dr - \cos\psi_m \int_0^1 \frac{F_x}{ac} dr \right] + \widehat{S}_\zeta \ddot{\zeta}_{1c} \quad (16.231)$$

The in-plane forces on the rotor have inertial reactions due to the longitudinal and lateral shifts of the rotor center-of-gravity that are associated with the cyclic lag degrees of freedom. Recall that in Chapter 6 the steady-state rotor forces were expressed as $H = \beta_{1c}T + H_{\text{TPP}}$ and $Y = -\beta_{1s}T + Y_{\text{TPP}}$. To write the present results in terms of the tilt of the thrust vector with the tip-path plane requires a detailed consideration of the aerodynamic forces F_x and F_r, which is given in section 16.8.

Finally, equation 16.200 for the torque on a single blade (neglecting the Coriolis term), is

$$\frac{N_L}{I_b} = \gamma \int_0^1 r \frac{F_x}{ac} dr - \widehat{I}_{\zeta\alpha} \ddot{\zeta}^{(m)} \quad (16.232)$$

where $\widehat{I}_{\zeta\alpha} = \int_0^1 r\eta_\zeta m\, dr/I_b$. Then the total rotor torque is

$$\gamma \frac{C_Q}{\sigma a} = \frac{\gamma}{N} \sum_{m=1}^{N} \int_0^1 r \frac{F_x}{ac} dr - \widehat{I}_{\zeta\alpha} \ddot{\zeta}_0 \quad (16.233)$$

since the collective lag degree of freedom is $\frac{1}{N}\sum_m \ddot{\zeta}^{(m)} = \ddot{\zeta}_0$.

For a two-bladed rotor the cyclic flap and lag degrees of freedom in the non-rotating frame do not exist, so different results are obtained. Instead of the cyclic degrees of freedom such as β_{1c} and β_{1s}, the two-bladed rotor has a single teetering degree of freedom, β_1. Determining the hub moment requires an evaluation of the sums:

$$\frac{2}{N}\sum_{m=1}^{N} \beta^{(m)} \sin\psi_m = 2\sin\psi \frac{1}{N}\sum_{m=1}^{N} \beta^{(m)} (-1)^m = 2\beta_1 \sin\psi \quad (16.234)$$

$$\frac{2}{N}\sum_{m=1}^{N} \beta^{(m)} \cos\psi_m = 2\cos\psi \frac{1}{N}\sum_{m=1}^{N} \beta^{(m)} (-1)^m = 2\beta_1 \cos\psi \quad (16.235)$$

where $\beta_1 = \frac{1}{2}(\beta^{(2)} - \beta^{(1)})$ is the teetering degree of freedom. Then the hub moments are

$$-\gamma \frac{2C_{My}}{\sigma a} = \widehat{I}_\beta (\nu_\beta^2 - 1) 2\beta_1 \cos\psi \quad (16.236)$$

$$\gamma \frac{2C_{Mx}}{\sigma a} = \widehat{I}_\beta (\nu_\beta^2 - 1) 2\beta_1 \sin\psi \quad (16.237)$$

628   **Blade Motion**

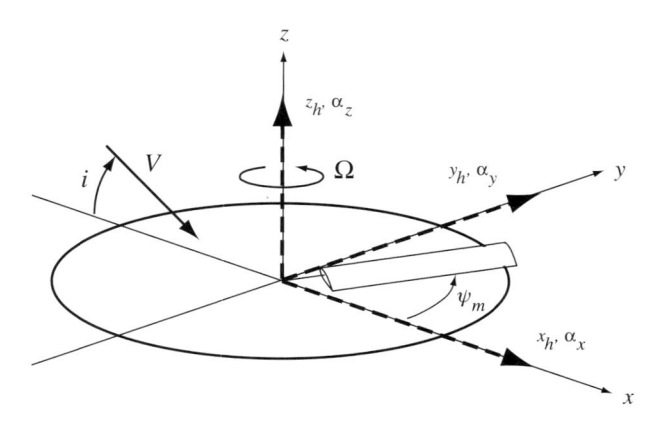

Figure 16.11. Definition of the linear and angular shaft motion.

Similarly, the rotor drag and side force depend on the differential lag degree of freedom $\zeta_1 = \frac{1}{2}(\zeta^{(2)} - \zeta^{(1)})$:

$$\gamma \frac{2C_H}{\sigma a} = \frac{2\gamma}{N} \sum_{m=1}^{N} \left[\cos\psi_m \int_0^1 \frac{F_r}{ac} dr + \sin\psi_m \int_0^1 \frac{F_x}{ac} dr \right]$$
$$- 2\widehat{S}_\zeta \left((\ddot{\zeta}_1 - \zeta_1)\sin\psi + 2\dot{\zeta}_1 \cos\psi \right) \tag{16.238}$$

$$\gamma \frac{2C_Y}{\sigma a} = \frac{2\gamma}{N} \sum_{m=1}^{N} \left[\sin\psi_m \int_0^1 \frac{F_r}{ac} dr - \cos\psi_m \int_0^1 \frac{F_x}{ac} dr \right]$$
$$+ 2\widehat{S}_\zeta \left((\ddot{\zeta}_1 - \zeta_1)\cos\psi - 2\dot{\zeta}_1 \sin\psi \right) \tag{16.239}$$

Although the steady-state, periodic motion of the two-bladed rotor is the same as for rotors with three or more blades, the transient dynamics are fundamentally different because of the absence of the cyclic degrees of freedom.

16.7 Shaft Motion

So far only the motion of the rotor has been considered. The shaft motion is also an important factor, both for helicopter flight dynamics involving the rigid-body degrees of freedom and for aeroelastic problems involving the coupled motion of the airframe and rotor. Figure 16.11 defines the linear and angular hub motion. The perturbation of the hub position from the equilibrium flight path is given by the displacements x_h, y_h, and z_h. The perturbed orientation is given by the hub rotations α_x, α_y, and α_z. For now an inertial reference frame is used, so that the coordinate frame remains fixed in space during the perturbed motion of the hub.

The shaft motion introduces additional out-of-plane and in-plane acceleration terms that must be included in the bending equations of motion. Consider the rigid flap and lag model developed in sections 16.3.1 and 16.4.1. The additional section accelerations producing flap moments are as follows:

   i) an angular acceleration $r(\ddot\alpha_x \sin\psi_m - \ddot\alpha_y \cos\psi_m)$
  ii) a Coriolis acceleration $2\Omega r(\dot\alpha_x \cos\psi_m + \dot\alpha_y \sin\psi_m)$
 iii) a vertical acceleration $\ddot{z}_h$

Each of these terms gives a downward inertial force on the section, with moment arm $(r - e)$ about the offset flap hinge. The angular acceleration $(\ddot{\alpha}_x \sin \psi_m - \ddot{\alpha}_y \cos \psi_m)$ is the flapwise component of the pitch and roll acceleration of the hub. The Coriolis acceleration arises from the cross-product of the angular velocity $(\dot{\alpha}_x \cos \psi_m + \dot{\alpha}_y \sin \psi_m)$ of the rotor disk about the blade radial axis and the rotational velocity of the section Ωr. Integrating these forces over the blade span gives the additional flap moment

$$\left(\int_0^R \eta_\beta r m \, dr \right) \left((\ddot{\alpha}_x + 2\Omega \dot{\alpha}_y) \sin \psi_m - (\ddot{\alpha}_y - 2\Omega \dot{\alpha}_x) \cos \psi_m \right) + \left(\int_0^R \eta_\beta m \, dr \right) \ddot{z}_h \tag{16.240}$$

The flap equation of motion becomes

$$\widehat{I}_\beta (\ddot{\beta} + v_\beta^2 \beta) - \widehat{I}_{\beta\zeta} 2\beta \dot{\zeta} + \widehat{I}_{\beta\alpha} \left((\ddot{\alpha}_x + 2\dot{\alpha}_y) \sin \psi_m - (\ddot{\alpha}_y - 2\dot{\alpha}_x) \cos \psi_m \right)$$

$$+ \widehat{S}_\beta \ddot{z}_h = \gamma \int_e^1 \eta_\beta \frac{F_z}{ac} dr = \gamma M_F \tag{16.241}$$

where $\widehat{I}_{\beta\alpha} = \int_0^1 r \eta_\beta m \, dr / I_b$ and $\widehat{S}_\beta = \int_0^1 \eta_\beta m \, dr / I_b$. The shaft motion appears in the blade equation of motion with periodic coefficients, because it is defined in the non-rotating frame.

The additional in-plane accelerations producing lag moments on the blade are

i) a hub angular acceleration $r \ddot{\alpha}_z$
ii) a hub in-plane linear acceleration $(\ddot{x}_h \sin \psi_m - \ddot{y}_h \cos \psi_m)$

The angular acceleration term gives an inertial force in the lag direction, and the linear acceleration gives a force opposing the lag motion; both have moment arms $(r - e)$ about the lag hinge. Integrating over the span gives the lag moments

$$-\left(\int_0^R r \eta_\zeta m \, dr \right) \ddot{\alpha}_z + \left(\int_0^R \eta_\zeta m \, dr \right) (\ddot{x}_h \sin \psi_m - \ddot{y}_h \cos \psi_m) \tag{16.242}$$

so that the lag equation of motion becomes

$$\widehat{I}_\zeta (\ddot{\zeta} + v_\zeta^2 \zeta) + \widehat{I}_{\beta\zeta} 2\beta \dot{\beta} - \widehat{I}_{\zeta\alpha} \ddot{\alpha}_z + \widehat{S}_\zeta (\ddot{x}_h \sin \psi_m - \ddot{y}_h \cos \psi_m) = \gamma \int_0^1 \eta_\zeta \frac{F_x}{ac} dr = \gamma M_L \tag{16.243}$$

where $\widehat{I}_{\zeta\alpha} = \int_0^1 r \eta_\zeta m \, dr / I_b$ and $\widehat{S}_\zeta = \int_0^1 \eta_\zeta m \, dr / I_b$.

Next, transform the flap and lag equations of motion to the non-rotating frame. The hub acceleration and velocity are independent of the blade index, so the summation operators act only on the $\sin \psi_m$ and $\cos \psi_m$ factors. The hub motion contributes only to the collective and cyclic equations in the non-rotating frame (at least for the inertial forces). The result for the flap motion is

$$\widehat{I}_\beta (\ddot{\beta}_0 + v_\beta^2 \beta_0) - \widehat{I}_{\beta\zeta} 2\beta_{\text{trim}} \dot{\zeta}_0 + \widehat{S}_\beta \ddot{z}_h = \gamma M_{F_0} \tag{16.244}$$

$$\widehat{I}_\beta (\ddot{\beta}_{1c} + 2\dot{\beta}_{1s} + (v_\beta^2 - 1) \beta_{1c}) - \widehat{I}_{\beta\zeta} 2\beta_{\text{trim}} (\dot{\zeta}_{1c} + \zeta_{1s}) - \widehat{I}_{\beta\alpha} (\ddot{\alpha}_y - 2\dot{\alpha}_x) = \gamma M_{F_{1c}} \tag{16.245}$$

$$\widehat{I}_\beta (\ddot{\beta}_{1s} - 2\dot{\beta}_{1c} + (v_\beta^2 - 1) \beta_{1s}) - \widehat{I}_{\beta\zeta} 2\beta_{\text{trim}} (\dot{\zeta}_{1s} - \zeta_{1c}) + \widehat{I}_{\beta\alpha} (\ddot{\alpha}_x + 2\dot{\alpha}_y) = \gamma M_{F_{1s}} \tag{16.246}$$

and for the lag motion:

$$\widehat{I}_\zeta \left(\ddot{\zeta}_0 + v_\zeta^2 \zeta_0\right) + \widehat{I}_{\beta\zeta} 2\beta_{\text{trim}} \dot{\beta}_0 - \widehat{I}_{\zeta\alpha} \ddot{\alpha}_z = \gamma M_{L_0} \qquad (16.247)$$

$$\widehat{I}_\zeta \left(\ddot{\zeta}_{1c} + 2\dot{\zeta}_{1s} + \left(v_\zeta^2 - 1\right)\zeta_{1c}\right) + \widehat{I}_{\beta\zeta} 2\beta_{\text{trim}} \left(\dot{\beta}_{1c} + \beta_{1s}\right) - \widehat{S}_\zeta \ddot{y}_h = \gamma M_{L_{1c}} \qquad (16.248)$$

$$\widehat{I}_\zeta \left(\ddot{\zeta}_{1s} - 2\dot{\zeta}_{1c} + \left(v_\zeta^2 - 1\right)\zeta_{1s}\right) + \widehat{I}_{\beta\zeta} 2\beta_{\text{trim}} \left(\dot{\beta}_{1s} - \beta_{1c}\right) + \widehat{S}_\zeta \ddot{x}_h = \gamma M_{L_{1s}} \qquad (16.249)$$

In the non-rotating frame the inertial coupling between the rotor and shaft motion is thus quite limited. The coning mode responds to vertical acceleration, the cyclic flap modes respond to the pitch and roll motion, the collective lag responds to yaw acceleration of the shaft, and the cyclic lag modes respond to the longitudinal and lateral hub acceleration. There is no coupling at all of the shaft motion with the equations of motion for the reactionless degrees of freedom ($2c, 2s, \ldots, nc, ns, N/2$).

All three of the vertical inertial forces due to the shaft motion that produce flap moments must also be included in the root shear force, which becomes

$$S_z = \int_0^R F_z \, dr - \ddot{\beta} \int_0^R \eta_\beta m \, dr - \ddot{z}_h \int_0^R m \, dr$$

$$- \left(\left(\ddot{\alpha}_x + 2\dot{\alpha}_y\right) \sin \psi_m - \left(\ddot{\alpha}_y - 2\dot{\alpha}_x\right) \cos \psi_m\right) \int_0^R rm \, dr \qquad (16.250)$$

Similarly, the root hub moment is

$$N_F = \int_0^R r F_z \, dr - \left(\ddot{\beta} + \Omega^2 \beta\right) \int_0^R r \eta_\beta m \, dr - \ddot{z}_h \int_0^R rm \, dr$$

$$- \left(\left(\ddot{\alpha}_x + 2\dot{\alpha}_y\right) \sin \psi_m - \left(\ddot{\alpha}_y - 2\dot{\alpha}_x\right) \cos \psi_m\right) \int_0^R r^2 m \, dr \qquad (16.251)$$

Alternatively, the expression $N_F = I_b \Omega^2 (\eta_\beta^2 - 1)\beta$ can be used here.

The radial acceleration ($\ddot{x}_h \cos \psi_m + \ddot{y}_h \sin \psi_m$) due to the in-plane hub motion must be added to the radial shear force:

$$S_r = \int_0^R F_r \, dr - 2\Omega \dot{\zeta} \int_0^R \eta_\zeta m \, dr - \left(\ddot{x}_h \cos \psi_m + \ddot{y}_h \sin \psi_m\right) \int_0^R m \, dr \qquad (16.252)$$

The in-plane inertial forces due to the shaft motion that produce lag moments must be included in the in-plane root shear force and the blade torque:

$$S_x = \int_0^R F_x \, dr - \left(\ddot{\zeta} - \Omega^2 \zeta\right) \int_0^R \eta_\zeta m \, dr + \ddot{\alpha}_x \int_0^R rm \, dr$$

$$- \left(\ddot{x}_h \sin \psi_m - \ddot{y}_h \cos \psi_m\right) \int_0^R m \, dr \qquad (16.253)$$

$$N_L = \int_0^R r F_x \, dr - \ddot{\zeta} \int_0^R r \eta_\zeta m \, dr - 2\Omega \beta \dot{\beta} I_{\beta\zeta} + \ddot{\alpha}_x \int_0^R r^2 m \, dr$$

$$- \left(\ddot{x}_h \sin \psi_m - \ddot{y}_h \cos \psi_m\right) \int_0^R rm \, dr \qquad (16.254)$$

The results for the rotor forces and moments, obtained by summing the root reaction over all N blades, are simpler because many of the new terms cancel.

16.7 Shaft Motion

The rotor thrust (equation 16.217) is

$$\gamma \frac{C_T}{\sigma a} = \frac{1}{N} \sum_{m=1}^{N} \frac{S_z}{I_b} = \left(\gamma \frac{C_T}{\sigma a}\right)_{\text{aero}} - \widehat{S}_\beta \ddot{\beta}_0 - \widehat{M}_b \ddot{z}_h \quad (16.255)$$

where $\widehat{S}_\beta = \int_0^R \eta_\beta m\, dr/I_b$ and $\widehat{M}_b = \int_0^R m\, dr/I_b$. $\widehat{M}_b$ is the normalized mass of a single blade. For the pitch and roll moments, we can still use equation 16.222, or equations 16.224 and 16.225 become

$$-\gamma \frac{2C_{My}}{\sigma a} = -\left(\gamma \frac{2C_{My}}{\sigma a}\right)_{\text{aero}} - \widehat{I}_{\beta\alpha}\left(\ddot{\beta}_{1c} + 2\dot{\beta}_{1s}\right) + \widehat{I}_0\left(\ddot{\alpha}_y - 2\dot{\alpha}_x\right) \quad (16.256)$$

$$\gamma \frac{2C_{Mx}}{\sigma a} = \left(\gamma \frac{2C_{Mx}}{\sigma a}\right)_{\text{aero}} - \widehat{I}_{\beta\alpha}\left(\ddot{\beta}_{1s} - 2\dot{\beta}_{1c}\right) - \widehat{I}_0\left(\ddot{\alpha}_x + 2\dot{\alpha}_y\right) \quad (16.257)$$

where $\widehat{I}_{\beta\alpha} = \int_0^R r\eta_\beta m\, dr/I_b$ and $\widehat{I}_0 = \int_0^R r^2 m\, dr/I_b$. The rotor drag and side forces (equations 16.230 and 16.231) become

$$\gamma \frac{2C_H}{\sigma a} = \frac{2}{N} \sum_{m=1}^{N} \left(\frac{S_r}{I_b}\cos\psi_m + \frac{S_x}{I_b}\sin\psi_m\right) = \left(\gamma \frac{2C_H}{\sigma a}\right)_{\text{aero}} - \widehat{S}_\zeta \ddot{\zeta}_{1s} - 2\widehat{M}_b \ddot{x}_h$$

$$(16.258)$$

$$\gamma \frac{2C_Y}{\sigma a} = \frac{2}{N} \sum_{m=1}^{N} \left(\frac{S_r}{I_b}\sin\psi_m - \frac{S_x}{I_b}\cos\psi_m\right) = \left(\gamma \frac{2C_Y}{\sigma a}\right)_{\text{aero}} + \widehat{S}_\zeta \ddot{\zeta}_{1c} - 2\widehat{M}_b \ddot{y}_h$$

$$(16.259)$$

Finally, the rotor torque (equation 16.233) becomes

$$\gamma \frac{C_Q}{\sigma a} = \frac{1}{N} \sum_{m=1}^{N} \frac{N_L}{I_b} = \left(\gamma \frac{C_Q}{\sigma a}\right)_{\text{aero}} - \widehat{I}_{\zeta\alpha}\ddot{\zeta}_0 + \widehat{I}_0 \ddot{\alpha}_z \quad (16.260)$$

The only inertial contributions to the thrust, drag, and side forces of the rotor are the reactions to the linear acceleration of the total rotor mass. The angular acceleration reactions of the entire rotor produce moments on the hub.

The utility of multiblade coordinates is apparent in the interaction between the rotor and the non-rotating system. The shaft motion appears in the rotating equations of motion with periodic coefficients, which are eliminated by conversion to the non-rotating frame. Summing the blade root forces to obtain the total rotor hub reactions naturally leads to the non-rotating degrees of freedom for the rotor motion. Moreover, the coupling between the rotor and the fixed system is limited, because the non-rotating degrees of freedom define the motion of the rotor as a whole in certain patterns that naturally lead to an association with only certain components of the shaft motion and hub forces. Specifically, the rotor tip-path-plane tilt appears with the shaft pitch and roll motion and also with the hub pitch and roll moments. The cyclic lag degrees of freedom, which produce an in-plane shift of the rotor center-of-gravity, are associated with the hub in-plane displacements and forces. The rotor coning motion appears with the vertical shaft displacement and the rotor thrust, whereas the collective lag motion appears with the shaft yaw and rotor torque. Finally, the reactionless rotor modes do not couple with the shaft motion and hub forces at all. In axial flow, there is some additional coupling of the dynamics by the aerodynamic forces, but the motion still separates into a vertical system (z_h and α_z), a lateral-longitudinal system (x_h, y_h, α_x, and α_y), and the reactionless modes. In forward flight the aerodynamic forces couple all the degrees of freedom of the helicopter, but this basic decomposition remains a dominant characteristic of the behavior.

The influence of the shaft motion is different for the case of a two-bladed rotor because of the absence of the cyclic modes. The equations for β_{1c} and β_{1s} are replaced by the teetering equation, which including the shaft motion is

$$\widehat{I}_\beta(\ddot{\beta}_1 + v_\beta^2 \beta_1) - \widehat{I}_{\beta\zeta} 2\beta_{\text{trim}}\dot{\zeta}_1 + \widehat{I}_{\beta\alpha}\left((\ddot{\alpha}_x + 2\dot{\alpha}_y)\sin\psi - (\ddot{\alpha}_y - 2\dot{\alpha}_x)\cos\psi\right) = \gamma M_{F_1} \tag{16.261}$$

Similarly, an equation for the differential lag motion ζ_1 replaces the ζ_{1c} and ζ_{1s} equations:

$$\widehat{I}_\zeta(\ddot{\zeta}_1 + v_\zeta^2 \zeta_1) + \widehat{I}_{\beta\zeta} 2\beta_{\text{trim}}\dot{\beta}_1 + \widehat{S}_\zeta(\ddot{x}_h \sin\psi - \ddot{y}_h \cos\psi) = \gamma M_{L_1} \tag{16.262}$$

The equations of motion for β_0 and ζ_0 given earlier are valid for both $N \geq 3$ and the two-bladed rotor case, as are the results for the rotor thrust and torque. For the in-plane hub forces, the inertial reaction to the rotor acceleration must be added to equations 16.238 and 16.239:

$$\Delta\left(\gamma\,\frac{2C_H}{\sigma a}\right) = -2\widehat{M}_b \ddot{x}_h \tag{16.263}$$

$$\Delta\left(\gamma\,\frac{2C_Y}{\sigma a}\right) = -2\widehat{M}_b \ddot{y}_h \tag{16.264}$$

The pitch and roll moments of the two-bladed rotor were obtained in terms of the tip-path-plane motion and so are unchanged; the shaft motion influences the hub moments through the solution for β_1. The most notable feature of the coupling between the fixed frame and the two-bladed rotor in both the hub forces and the shaft motion is the periodic coefficients due to the lack of axisymmetry of the rotor. As a result, the analysis of the dynamics of a two-bladed rotor is distinctly different from that for rotors with three or more blades.

In flight dynamics analyses, for helicopters as well as for airplanes, a body axis system is most frequently used. With a body axis system the coordinate axes remain fixed in the body during its perturbed motion, whereas an inertial axis system remains fixed relative to space. Since the trim velocity of the aircraft is defined relative to the reference axes, the angular velocity of the body axes must rotate the velocity vector as well, which implies a centrifugal acceleration relative to inertial space:

$$\left(\frac{d\mathbf{v}}{dt}\right)_{\text{inertial axes}} = \left(\frac{d\mathbf{v}}{dt}\right)_{\text{body axes}} + \omega \times \mathbf{v} \tag{16.265}$$

The results that have been derived in this section for the shaft motion require knowing the hub acceleration in inertial space. The rotor in steady flight has trim velocity components μ in the disk plane and ($\mu \tan i$) normal to the disk plane. Then the inertial accelerations in terms of the body axis motion are

$$(\ddot{x}_h)_{\text{inertial}} = (\ddot{x}_h + \dot{\alpha}_y \mu \tan i)_{\text{body axes}} \tag{16.266}$$

$$(\ddot{y}_h)_{\text{inertial}} = (\ddot{y}_h - \dot{\alpha}_z \mu - \dot{\alpha}_x \mu \tan i)_{\text{body axes}} \tag{16.267}$$

$$(\ddot{z}_h)_{\text{inertial}} = (\ddot{z}_h + \dot{\alpha}_y \mu)_{\text{body axes}} \tag{16.268}$$

In hover, where the trim velocity of the helicopter is zero, there is no difference between the inertial axes and body axes for a linear analysis of the inertial forces. The use of body axes affects as well the formulation of the rotor aerodynamic forces due to shaft motion.

16.8 Aerodynamic Loads

This section derives the aerodynamic forces required to complete the differential equations describing the helicopter rotor blade motion. So far in this chapter, the inertial and structural terms of the equations of motion have been derived, and the net aerodynamic forces required have been defined in terms of integrals of the section forces and pitch moment over the blade span. Now these aerodynamic forcing terms are expressed as functions of the blade and shaft motion and the rotor loading. Blade element theory is the basis of the rotor aerodynamic model, so the section loads depend on the aerodynamic environment at that radial station alone, with the wake influence given by the induced velocity. For the blade torsion dynamics in particular, unsteady aerodynamic effects must be considered. The forces on the blade section are derived for the general case of large pitch and inflow, but small angles must be assumed in order to obtain analytical expressions for the aerodynamic coefficients. For the same reason the effects of stall, compressibility, and reverse flow are neglected. The small angle assumption is usually very good for the low disk loading helicopter rotor, and compressibility can be accounted for in a rough fashion by using the lift-curve slope at the Mach number of a representative blade radius. The neglect of reverse flow restricts the model to advance ratios below about $\mu = 0.5$, which covers the speed range of most current helicopters. When the aerodynamic coefficients are to be evaluated numerically, a more complex aerodynamic model is easily implemented. The linear differential equations describing the rotor dynamics are completed by expressing the aerodynamic forces in terms of the perturbation blade motion. The resulting equations are solved in subsequent chapters. The analysis begins with a derivation of the forces acting on the blade section.

16.8.1 Section Aerodynamics

Consider the air velocity and aerodynamic forces at the rotor blade section, as shown in Figure 16.12. A hub plane reference axis system is used for the aerodynamic analysis. The hub plane is fixed relative to the shaft and thus is tilted and displaced by the shaft motion. The pitch angle θ is measured from the reference plane. The velocities u_T, u_P, and u_R are the components of the air velocity seen by the blade, resolved in the hub plane axis system. The tangential velocity u_T is in the hub plane, positive in the blade drag direction; the radial velocity u_R is positive when directed radially outward; and the perpendicular velocity u_P is normal to the reference plane, positive when directed down through the disk. The resultant velocity in the section is $U = (u_T^2 + u_P^2)^{1/2}$ and the inflow angle is $\phi = \tan^{-1} u_P/u_T$. Then the blade section angle-of-attack is $\alpha = \theta - \phi$. The aerodynamic lift and drag forces (L and D) are, respectively, normal to and parallel to the resultant velocity U. F_x and F_z are the components of the section lift and drag resolved into the hub plane axes. The radial force F_r is positive outward (in the same direction as u_R); F_r consists of a radial drag force and an in-plane component of the blade lift due to flapwise bending of the blade. The section aerodynamic moment at the elastic axis is M_a, positive in

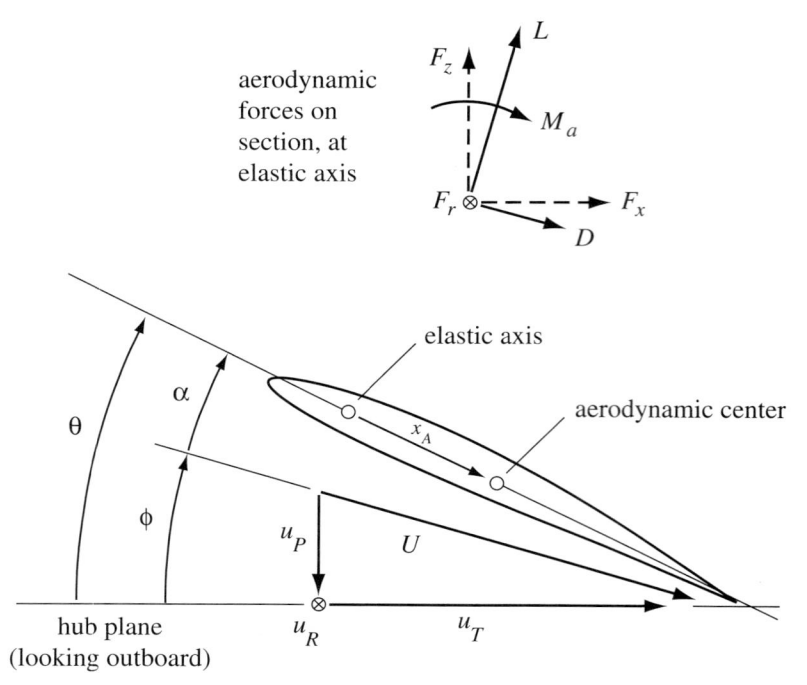

Figure 16.12. Rotor blade section aerodynamics.

the nose-upward direction. The aerodynamic center of the section is a distance x_A behind the elastic axis.

The blade lift and drag forces can be written in terms of the section coefficients:

$$L = \frac{1}{2}\rho U^2 c c_\ell \qquad (16.269)$$

$$D = \frac{1}{2}\rho U^2 c c_d \qquad (16.270)$$

where ρ is the air density and c is the rotor chord. Dimensionless quantities are used from this point on in the analysis, so the air density ρ is omitted. The section forces resolved relative to the hub plane axes are

$$F_z = L\cos\phi - D\sin\phi = (Lu_T - Du_P)/U \qquad (16.271)$$

$$F_x = L\sin\phi + D\cos\phi = (Lu_P + Du_T)/U \qquad (16.272)$$

The section lift and drag coefficients, $c_\ell = c_\ell(\alpha, M)$ and $c_d = c_d(\alpha, M)$, are functions of the angle-of-attack and Mach number:

$$\alpha = \theta - \phi \qquad (16.273)$$

$$M = M_{\text{tip}} U \qquad (16.274)$$

where M_{tip} is the tip Mach number (the tip speed ΩR divided by the speed of sound) in hover. In fact, the lift and drag of the rotor blade depend on other parameters as well, such as the local yaw angle of the flow and unsteady angle-of-attack changes. Such effects can be included in a numerical analysis, but are neglected here. The nose-up moment about the elastic axis is

$$M_a = -x_A L + M_{AC} + M_{US} = -x_A \frac{1}{2} U^2 c c_\ell + \frac{1}{2} U^2 c^2 c_{m_{ac}} + M_{US} \quad (16.275)$$

where x_A is the distance that the aerodynamic center is behind the elastic axis, M_{AC} is the moment about the aerodynamic center, and M_{US} is the unsteady aerodynamic moment.

The radial force on the section is

$$F_r = \frac{u_R}{U} D - z' F_z = \frac{1}{2} U u_R c c_d - z' F_z \quad (16.276)$$

The first term is the radial drag force, obtained by assuming that the viscous drag force on the section has the same yaw angle as the local velocity (see section 6.22). The second term in F_r is the radial component of the normal force F_z due to the local flapwise bending slope z'. An exact analysis would transform the aerodynamics forces (L, D, and radial drag) on the section tilted by flap and lag bending (z' and x') to forces relative to the hub plane axes (F_z, F_x, F_r). The bending slopes are assumed to be small, so F_z and F_x are obtained by just rotating the lift and drag by the inflow angle ϕ, neglecting the section tilt. The radial component of the section lift and drag is then $-z'(L \cos \phi - D \sin \phi) = -z' F_z$. The corresponding term due to lag bending ($x' F_x$) is neglected.

Next, substitute for L and D in terms of the section coefficients, and divide by the two-dimensional lift curve slope a and the section chord c. The result is

$$\frac{F_z}{ac} = U \left(u_T \frac{c_\ell}{2a} - u_P \frac{c_d}{2a} \right) \quad (16.277)$$

$$\frac{F_x}{ac} = U \left(u_P \frac{c_\ell}{2a} + u_T \frac{c_d}{2a} \right) \quad (16.278)$$

$$\frac{F_r}{ac} = U u_R \frac{c_d}{2a} - z' \frac{F_z}{ac} \quad (16.279)$$

$$\frac{M_a}{ac} = -x_A U^2 \frac{c_\ell}{2a} + U^2 c \frac{c_m}{ac} + \frac{M_{US}}{2a} \quad (16.280)$$

The integrals of these forces over the blade span are required in the rotor equations of motion. The section pitch moment M_a is not considered further until section 16.8.10.

The objective is to obtain the aerodynamic forces in the rotor blade equations in terms of the perturbed motion of the blade. Thus the perturbation section forces must be expressed in terms of the perturbations of the velocities and pitch angle. Each component of the velocity seen by the blade has a trim term, due to operation of the rotor in its equilibrium state, and a perturbation due to the perturbed motion of the system. The latter term is due to the system degrees of freedom and is assumed to be small in deriving linear differential equations describing the rotor dynamics. Thus the blade pitch and section velocities are written as trim plus perturbation terms:

$$\theta = (\theta)_{\text{trim}} + \delta\theta \quad (16.281)$$

$$u_T = (u_T)_{\text{trim}} + \delta u_T \quad (16.282)$$

$$u_P = (u_P)_{\text{trim}} + \delta u_P \tag{16.283}$$
$$u_R = (u_R)_{\text{trim}} + \delta u_R \tag{16.284}$$

After this substitution has been made, the subscript "trim" can be dropped. The perturbation of the angle-of-attack, resultant velocity, and Mach number are

$$\delta\alpha = \delta\theta - (u_T \delta u_P - u_P \delta u_T)/U^2 \tag{16.285}$$
$$\delta U = (u_T \delta u_T + u_P \delta u_P)/U \tag{16.286}$$
$$\delta M = M_{\text{tip}} \delta U \tag{16.287}$$

and the perturbations of the section coefficients are

$$\delta c_\ell = \frac{\partial c_\ell}{\partial \alpha}\delta\alpha + \frac{\partial c_\ell}{\partial M}\delta M = c_{\ell\alpha}\delta\alpha + c_{\ell M}\delta M \tag{16.288}$$
$$\delta c_d = \frac{\partial c_d}{\partial \alpha}\delta\alpha + \frac{\partial c_d}{\partial M}\delta M = c_{d\alpha}\delta\alpha + c_{dM}\delta M \tag{16.289}$$

The perturbations of the section aerodynamic forces are obtained by carrying out the differential operation on the relations for F_z, F_x, and F_r, using the above results to express the perturbation forces in terms of $\delta\theta$, δu_T, δu_P, and δu_R. The coefficients of the perturbation quantities are evaluated at the trim state. The result for the section forces is as follows:

$$\delta\frac{F_z}{ac} = U\left(u_T \frac{c_{\ell\alpha}}{2a} - u_P \frac{c_{d\alpha}}{2a}\right)\delta\theta$$
$$+ \left[-\frac{u_T}{U}\left(u_T \frac{c_{\ell\alpha}}{2a} - u_P \frac{c_{d\alpha}}{2a}\right) + \frac{u_T u_P}{U}\left(\frac{c_\ell}{2a} + M\frac{c_{\ell M}}{2a}\right)\right.$$
$$\left. - \frac{u_P^2}{U}\left(\frac{c_d}{2a} + M\frac{c_{dM}}{2a}\right) - U\frac{c_d}{2a}\right]\delta u_P$$
$$+ \left[\frac{u_P}{U}\left(u_T \frac{c_{\ell\alpha}}{2a} - u_P \frac{c_{d\alpha}}{2a}\right) + \frac{u_T^2}{U}\left(\frac{c_\ell}{2a} + M\frac{c_{\ell M}}{2a}\right)\right.$$
$$\left. - \frac{u_T u_P}{U}\left(\frac{c_d}{2a} + M\frac{c_{dM}}{2a}\right) + U\frac{c_\ell}{2a}\right]\delta u_T$$
$$= F_{z\theta}\delta\theta + F_{zP}\delta u_P + F_{zT}\delta u_T \tag{16.290}$$

$$\delta\frac{F_x}{ac} = U\left(u_P \frac{c_{\ell\alpha}}{2a} + u_T \frac{c_{d\alpha}}{2a}\right)\delta\theta$$
$$+ \left[-\frac{u_T}{U}\left(u_P \frac{c_{\ell\alpha}}{2a} + u_T \frac{c_{d\alpha}}{2a}\right) + \frac{u_P^2}{U}\left(\frac{c_\ell}{2a} + M\frac{c_{\ell M}}{2a}\right)\right.$$
$$\left. + \frac{u_T u_P}{U}\left(\frac{c_d}{2a} + M\frac{c_{dM}}{2a}\right) + U\frac{c_\ell}{2a}\right]\delta u_P$$
$$+ \left[\frac{u_P}{U}\left(u_P \frac{c_{\ell\alpha}}{2a} + u_T \frac{c_{d\alpha}}{2a}\right) + \frac{u_T u_P}{U}\left(\frac{c_\ell}{2a} + M\frac{c_{\ell M}}{2a}\right)\right.$$
$$\left. + \frac{u_T^2}{U}\left(\frac{c_d}{2a} + M\frac{c_{dM}}{2a}\right) + U\frac{c_d}{2a}\right]\delta u_T$$
$$= F_{x\theta}\delta\theta + F_{xP}\delta u_P + F_{xT}\delta u_T \tag{16.291}$$

16.8 Aerodynamic Loads

$$\delta \frac{F_r}{ac} = U\left(u_R \frac{c_{d\alpha}}{2a}\right)\delta\theta$$
$$+ \left[-\frac{u_T u_R}{U}\frac{c_{d\alpha}}{2a} + \frac{u_P u_R}{U}\left(\frac{c_d}{2a} + M\frac{c_{dM}}{2a}\right)\right]\delta u_P$$
$$+ \left[\frac{u_P u_R}{U}\frac{c_{d\alpha}}{2a} + \frac{u_T u_R}{U}\left(\frac{c_d}{2a} + M\frac{c_{dM}}{2a}\right)\right]\delta u_T$$
$$+ \left[U\frac{c_d}{2a}\right]\delta u_R - \left(\frac{F_z}{ac}\right)\delta z' - (z')\delta\frac{F_z}{ac}$$
$$= F_{r\theta}\delta\theta + F_{rP}\delta u_P + F_{rT}\delta u_T + F_{rR}\delta u_R \qquad (16.292)$$

These equations are applicable to low or high inflow and to axial or edgewise flight.

Since the low disk loading helicopter rotor in edgewise flight is characterized by low inflow, small angles can be assumed in the aerodynamic analysis. Specifically, θ, ϕ, and c_d/c_ℓ are all assumed to be small compared to unity. So α and u_P/u_T are also small, $U \cong u_T$, and $\phi \cong u_P/u_T$. The section forces are then

$$\frac{F_z}{ac} \cong \frac{L}{ac} \cong u_T^2 \frac{c_\ell}{2a} \qquad (16.293)$$

$$\frac{F_x}{ac} \cong u_T u_P \frac{c_\ell}{2a} + u_T^2 \frac{c_d}{2a} \qquad (16.294)$$

$$\frac{F_r}{ac} \cong u_T u_R \frac{c_d}{2a} - z' \frac{F_z}{ac} \qquad (16.295)$$

It is usually consistent with the small angle approximation to also assume a constant lift-curve slope and neglect stall and the zero-lift angle. Then the lift coefficient is simply $c_\ell = a\alpha = a(\theta - u_P/u_T)$, and

$$\frac{F_z}{ac} = \frac{1}{2}u_T^2\alpha = \frac{1}{2}(u_T^2\theta - u_T u_P) \qquad (16.296)$$

$$\frac{F_x}{ac} = \frac{1}{2}u_T u_P\alpha + u_T^2\frac{c_d}{2a} = \frac{1}{2}(u_T u_P\theta - u_P^2) + u_T^2\frac{c_d}{2a} \qquad (16.297)$$

With the small angle assumption, the perturbation forces become

$$\delta\frac{F_z}{ac} = \frac{1}{2}u_T^2\delta\theta - \frac{1}{2}u_T\delta u_P + \frac{1}{2}(u_P + 2u_T\alpha)\delta u_T \qquad (16.298)$$

$$\delta\frac{F_x}{ac} = \frac{1}{2}u_T u_P\delta\theta + \frac{1}{2}(u_T\alpha - u_P)\delta u_P + \left(\frac{1}{2}u_P\theta + 2u_T\frac{c_d}{2a}\right)\delta u_T \qquad (16.299)$$

$$\delta\frac{F_r}{ac} = u_R\frac{c_d}{2a}\delta u_T + u_T\frac{c_d}{2a}\delta u_R - \left(\frac{F_z}{ac}\right)\delta z' - (z')\delta\frac{F_z}{ac} \qquad (16.300)$$

In the process of making the small angle approximation we have also neglected reverse flow, and the terms due to $c_{d\alpha}$, $c_{\ell M}$, and c_{dM} have been dropped. In this form, the section forces can be integrated analytically, while retaining the basic characteristics of the rotor aerodynamics.

To complete the specification of the perturbation forces, the trim and perturbation velocities are required. The trim terms are as follows. The components of the

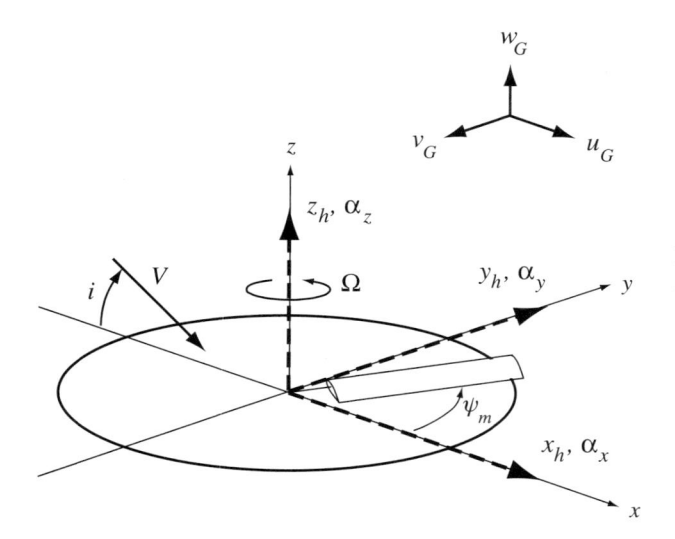

Figure 16.13. Rotor hub motion and aerodynamic gust components.

helicopter trim forward speed in the hub plane axes are the advance ratio and inflow ratio:

$$\mu = \frac{V \cos i_{\text{HP}}}{\Omega R} \tag{16.301}$$

$$\lambda_{\text{HP}} = \lambda_i + \mu \tan i_{\text{HP}} \tag{16.302}$$

The in-plane trim velocity seen by the rotor blade is due to the rotor rotation and the advance ratio, giving $u_T = r + \mu \sin \psi$ and $u_R = \mu \cos \psi$ (section 6.1.1). Any aerodynamic forces involving u_T and u_R alone depend only on the advance ratio μ. The blade pitch θ and normal velocity u_P depend on the operating condition of the rotor, in particular on the thrust coefficient, as well as on the advance ratio. So the aerodynamic forces involving the trim values of θ or u_P require the solution for the blade angle-of-attack and loading at the given operating state. The velocity and loading of the rotor blade in edgewise forward flight are periodic because of the rotation of the blade relative to the forward speed of the helicopter. Hence the aerodynamics of forward flight introduce periodic coefficients in the equations of motion describing the dynamics. For hover, or vertical or axial flight, the aerodynamic environment is axisymmetric, so the differential equations have constant coefficients.

The perturbation velocities depend on the degrees of freedom considered in the model of the blade motion. The following sections of this chapter are concerned with deriving the perturbation velocities corresponding to the various models of the rotor motion and with integrating the resulting perturbation forces to obtain the required aerodynamic terms in the equations of motion. The rotor shaft motion is also included in the blade velocity perturbations. The definitions of the displacements and rotations of the rotor hub are shown in Figure 16.13. The dimensionless shaft motion variables are assumed to be small. In addition, the excitation of the rotor by aerodynamic turbulence is considered. The gust velocity components defined relative to the non-rotating hub plane axes are u_G, v_G, and w_G (longitudinal, lateral, and vertical components, respectively, as shown in Figure 16.13). The gust velocity components are assumed to be both uniform throughout space and small (in terms of the dimensionless velocities, based on the rotor tip speed ΩR). The aerodynamic forces due to a uniform perturbation of the wake-induced inflow velocity are also found. Such an inflow perturbation can be used to model the unsteady aerodynamics of the rotor.

16.8.2 Flap Motion

Consider the rigid flap motion of a blade with no hinge offset (section 16.3.1). The aerodynamic flap moment is

$$M_F = \int_0^1 r \frac{F_z}{ac} dr \qquad (16.303)$$

For small angles $F_z/ac \cong L/ac \cong \frac{1}{2}u_T^2 \alpha$, so the perturbation force is

$$\delta \frac{F_z}{ac} = \left(\frac{1}{2}u_T^2\right)\delta\theta - \left(\frac{1}{2}u_T\right)\delta u_P \qquad (16.304)$$

The only degree of freedom for the blade is the rigid flap motion β. The blade pitch control is included, as is a uniform perturbation of the inflow velocity. Hence $\delta\theta$ and δu_P are here

$$\delta\theta = \theta - K_P \beta \qquad (16.305)$$

$$\delta u_P = \lambda + r\dot\beta + \beta u_R \qquad (16.306)$$

The perturbation pitch angle consists of the control input and the kinematic pitch-flap coupling. The perturbation normal velocity consists of the inflow perturbation, flapping velocity, and the normal component of the radial velocity u_R when the blade is flapped up. Here θ, λ, and β are small perturbations of the trim quantities. Recall from Chapter 6 that the pitch angle and normal velocity for the steady-state flapping solution are

$$\theta = \theta_{\text{con}} + \theta_{tw} r - K_P \beta \qquad (16.307)$$

$$u_P = \lambda_{\text{HP}} + r\dot\beta + \beta\mu\cos\psi \qquad (16.308)$$

Since θ and u_P are linear functions, the perturbation quantities have the same form. Substituting for δF_z in the flap moment gives

$$M_F = M_\theta(\theta - K_P \beta) + M_\lambda \lambda + M_{\dot\beta} \dot\beta + M_\beta \beta \qquad (16.309)$$

where the aerodynamic coefficients are

$$M_\theta = \int_0^1 \frac{1}{2} r u_T^2 dr = \frac{1}{8} + \frac{\mu}{3}\sin\psi + \frac{\mu^2}{4}\sin^2\psi \qquad (16.310)$$

$$M_\lambda = -\int_0^1 \frac{1}{2} r u_T dr = -\left(\frac{1}{6} + \frac{\mu}{4}\sin\psi\right) \qquad (16.311)$$

$$M_{\dot\beta} = -\int_0^1 \frac{1}{2} r^2 u_T dr = -\left(\frac{1}{8} + \frac{\mu}{6}\sin\psi\right) \qquad (16.312)$$

$$M_\beta = -\int_0^1 \frac{1}{2} r u_T u_R dr = u_R M_\lambda \qquad (16.313)$$

Including the tip loss factor, the upper limit of integration should be $r = B$ rather than $r = 1$. All of these flap moments are due to the lift increment produced by the blade angle-of-attack change. The identical coefficients were derived in section 6.5, where the steady-state solution for the flap motion was obtained by assuming a Fourier series for the flap response. Here we have a linear differential equation for the perturbed flap motion. In hover ($\mu = 0$) this differential equation has constant coefficients. In forward flight the aerodynamic coefficients in the equation of motion are periodic functions of the azimuth angle ψ.

For a fundamental flapping mode that is applicable generally to offset-hinge articulated rotors and hingeless rotors, the required flap moment is

$$M_F = \int_0^1 \eta \frac{F_z}{ac} dr \tag{16.314}$$

where $\eta(r)$ is the out-of-plane mode shape. With the deflection $z = \eta\beta$ now, the normal velocity is

$$\delta u_P = \lambda + \dot{z} + z'u_R = \lambda + \eta\dot{\beta} + \eta'\beta u_R \tag{16.315}$$

The flap moment is again written as

$$M_F = M_\theta(\theta - K_P\beta) + M_\lambda \lambda + M_{\dot{\beta}}\dot{\beta} + M_\beta \beta \tag{16.316}$$

where the aerodynamic coefficients are

$$M_\theta = \int_0^1 \frac{1}{2}\eta u_T^2 dr = \frac{1}{8}c_2 + \frac{\mu}{3}c_1 \sin\psi + \frac{\mu^2}{4}c_0 \sin^2\psi \tag{16.317}$$

$$M_\lambda = -\int_0^1 \frac{1}{2}\eta u_T dr = -\left(\frac{1}{6}c_1 + \frac{\mu}{4}c_0 \sin\psi\right) \tag{16.318}$$

$$M_{\dot{\beta}} = -\int_0^1 \frac{1}{2}\eta^2 u_T dr = -\left(\frac{1}{8}d_1 + \frac{\mu}{6}d_0 \sin\psi\right) \tag{16.319}$$

$$M_\beta = -\int_0^1 \frac{1}{2}\eta\eta' u_T u_R dr = -\mu \cos\psi \left(\frac{1}{6}f_1 + \frac{\mu}{4}f_0 \sin\psi\right) \tag{16.320}$$

with $c_n = (n+2)\int_0^1 \eta r^n dr$, $d_n = (n+3)\int_0^1 \eta^2 r^n dr$, and $f_n = (n+2)\int_0^1 \eta\eta' r^n dr$ (see also section 6.15).

For the case of out-of-plane bending of the blade (section 16.3.2) the aerodynamic force is

$$M_{Fk} = \int_0^1 \eta_k \frac{F_z}{ac} dr \tag{16.321}$$

and $z = \sum_i \eta_i q_i$ gives

$$\delta\theta = \theta - \sum_i K_{Pi} q_i \tag{16.322}$$

$$\delta u_P = \lambda + \sum_i \eta_i \dot{q}_i + \sum_i \eta'_i q_i u_R \tag{16.323}$$

Then

$$M_{Fk} = M_{q_k\theta}\left(\theta - \sum_i K_{Pi} q_i\right) + M_{q_k\lambda}\lambda + \sum_i \left(M_{q_k\dot{q}_i}\dot{q}_i + M_{q_k q_i}q_i\right) \tag{16.324}$$

where the aerodynamic coefficients are

$$M_{q_k \theta} = \int_0^1 \frac{1}{2} \eta_k u_T^2 \, dr \tag{16.325}$$

$$M_{q_k \lambda} = -\int_0^1 \frac{1}{2} \eta_k u_T \, dr \tag{16.326}$$

$$M_{q_k \dot{q}_i} = -\int_0^1 \frac{1}{2} \eta_k \eta_i u_T \, dr \tag{16.327}$$

$$M_{q_k q_i} = -\int_0^1 \frac{1}{2} \eta_k \eta_i' u_T u_R \, dr \tag{16.328}$$

The influence of the rotor wake on the unsteady aerodynamic forces of the blade can be accounted for by using an appropriate model for the inflow perturbation λ. Alternatively, the quasistatic lift can be multiplied by a lift deficiency function $C'(k)$. Then the factor C' must be included in the integrands of the aerodynamic coefficients; for example

$$M_\theta = \int_0^1 C'(k) \frac{1}{2} r u_T^2 \, dr \tag{16.329}$$

The reduced frequency and therefore C' vary along the blade span, but often the lift deficiency function can be evaluated just at an effective radius (typically $0.75R$), in which case

$$M_\theta \cong C'(k_e) \int_0^1 \frac{1}{2} r u_T^2 \, dr \tag{16.330}$$

The lift deficiency function is based on harmonic motion and hence is applicable to the frequency response or flutter boundary calculation. In forward flight, Theodorsen's function should be used for $C'(k)$. If the lift deficiency function is to be integrated numerically, the reduced frequency should be based on the local free stream velocity: $k = \omega b / u_T$. For the low harmonics of flap motion the reduced frequency is small, and the near shed wake effects are small (Theodorsen's function $C \cong 1$). In hover at low thrust, the returning shed wake effects can be significant, and Loewy's lift deficiency function should be used for C' (see section 10.7). When the spacing between the wake spirals is small and oscillation occurs at a harmonic of the rotor speed so that the layers of wake vorticity are in phase, Loewy's function substantially reduces the blade lift.

16.8.3 Flap and Lag Motion

The aerodynamic forces required for the fundamental flap and lag motions of a rotor blade are

$$M_F = \int_0^1 \eta_\beta \frac{F_z}{ac} \, dr \tag{16.331}$$

$$M_L = \int_0^1 \eta_\zeta \frac{F_x}{ac} \, dr \tag{16.332}$$

(see section 16.4.1). Here η_β and η_ζ are the mode shapes for the flap (purely out-of-plane) and lag (purely in-plane) motion, respectively. The perturbations of the section forces are

$$\delta \frac{F_z}{ac} = \frac{1}{2}u_T^2 \delta\theta - \frac{1}{2}u_T \delta u_P + \frac{1}{2}(u_P + 2u_T\alpha)\delta u_T \tag{16.333}$$

$$\delta \frac{F_x}{ac} = \frac{1}{2}u_T u_P \delta\theta + \frac{1}{2}(u_T\alpha - u_P)\delta u_P + \left(\frac{1}{2}u_P\theta + 2u_T\frac{c_d}{2a}\right)\delta u_T \tag{16.334}$$

In the case of the lag degree of freedom, the in-plane force and velocity must be considered. The in-plane velocity perturbation δu_T produces dynamic pressure and small angle-of-attack changes. The resulting lift perturbation is much smaller than the lift due to $\delta\theta$ and δu_P, which directly produce angle-of-attack changes. The in-plane forces are due to the induced drag perturbations and hence are much smaller than the out-of-plane forces. The δu_T term of δF_x also has a profile drag change due to the dynamic pressure perturbation. The blade pitch and normal velocity perturbations are

$$\delta\theta = \theta - K_P\beta \tag{16.335}$$

$$\delta u_P = \lambda + \eta_\beta \dot\beta + \eta_\beta' \beta u_R \tag{16.336}$$

as earlier. The lag motion produces a perturbation of the in-plane blade velocity:

$$\delta u_T = -\eta_\zeta \dot\zeta - \eta_\zeta' \zeta u_R \tag{16.337}$$

Substituting for the force and velocity perturbations then gives the flap and lag moments:

$$M_F = M_\theta(\theta - K_P\beta) + M_\lambda\lambda + M_{\dot\beta}\dot\beta + M_\beta\beta + M_{\dot\zeta}\dot\zeta + M_\zeta\zeta \tag{16.338}$$

$$M_L = Q_\theta(\theta - K_P\beta) + Q_\lambda\lambda + Q_{\dot\beta}\dot\beta + Q_\beta\beta + Q_{\dot\zeta}\dot\zeta + Q_\zeta\zeta \tag{16.339}$$

where the aerodynamic coefficients are

$$M_\theta = \int_0^1 \frac{1}{2}\eta_\beta u_T^2 dr \tag{16.340}$$

$$M_\lambda = -\int_0^1 \frac{1}{2}\eta_\beta u_T dr \tag{16.341}$$

$$M_{\dot\beta} = -\int_0^1 \frac{1}{2}\eta_\beta^2 u_T dr \tag{16.342}$$

$$M_\beta = -u_R \int_0^1 \frac{1}{2}\eta_\beta \eta_\beta' u_T dr \tag{16.343}$$

$$M_{\dot\zeta} = -\int_0^1 \frac{1}{2}\eta_\beta \eta_\zeta (u_P + 2u_T\alpha) dr \tag{16.344}$$

$$M_\zeta = -u_R \int_0^1 \frac{1}{2}\eta_\beta \eta_\zeta' (u_P + 2u_T\alpha) dr \tag{16.345}$$

$$Q_\theta = \int_0^1 \frac{1}{2}\eta_\zeta u_T u_P dr \tag{16.346}$$

$$Q_\lambda = \int_0^1 \frac{1}{2}\eta_\zeta (u_T\alpha - u_P) dr \tag{16.347}$$

16.8 Aerodynamic Loads

$$Q_{\dot{\beta}} = \int_0^1 \frac{1}{2}\eta_\zeta \eta_\beta (u_T\alpha - u_P) dr \tag{16.348}$$

$$Q_\beta = u_R \int_0^1 \frac{1}{2}\eta_\zeta \eta'_\beta (u_T\alpha - u_P) dr \tag{16.349}$$

$$Q_{\dot{\zeta}} = -\int_0^1 \eta_\zeta^2 \left(\frac{1}{2}u_P\theta + 2u_T\frac{c_d}{2a}\right) dr \tag{16.350}$$

$$Q_\zeta = -u_R \int_0^1 \eta_\zeta \eta'_\zeta \left(\frac{1}{2}u_P\theta + 2u_T\frac{c_d}{2a}\right) dr \tag{16.351}$$

The coefficients M_θ, M_λ, $M_{\dot{\beta}}$, and M_β are the flap moments produced by the lift changes due to angle-of-attack perturbations. These coefficients are therefore defined entirely by the rotor advance ratio and the flapping mode shape of the blade. The remaining coefficients involve either in-plane velocity or in-plane force or both, and therefore they require the solution for the trim blade motion and loading (that is, θ, u_P, and α as well as u_T and u_P) to be evaluated. These coefficients then depend on the rotor operating state, particularly the thrust coefficient.

In hover or vertical flight, where the trim solution for the blade loading is axisymmetric, the aerodynamic coefficients are more easily evaluated than in forward flight. The trim velocities in vertical flight are $u_T = r$, $u_P = \lambda_{HP}$, and $u_R = 0$; the blade angle-of-attack becomes $\alpha = \theta - \lambda_{HP}/r$. To evaluate the integrals analytically, assume that the induced velocity λ_{HP} is uniform over the rotor disk and that the blade mode shapes are $\eta_\beta = \eta_\zeta = r$. If the coefficients are evaluated numerically, the actual mode shapes can be used, and perhaps also the combined blade element and momentum theory result for nonuniform inflow. For hover or vertical flight,

$$M_\theta = \int_0^1 \frac{1}{2} r^3 dr = \frac{1}{8} \tag{16.352}$$

$$M_\lambda = -\int_0^1 \frac{1}{2} r^2 dr = -\frac{1}{6} \tag{16.353}$$

$$M_{\dot{\beta}} = -\int_0^1 \frac{1}{2} r^3 dr = -\frac{1}{8} \tag{16.354}$$

$$M_{\dot{\zeta}} = -\int_0^1 \frac{1}{2} r^2 (\lambda_{HP} + 2r\alpha) dr = -\left(\frac{\lambda_{HP}}{6} + 2\int_0^1 \frac{1}{2} r^3 \alpha \, dr\right) \tag{16.355}$$

$$Q_\theta = \int_0^1 \frac{1}{2} r^2 \lambda_{HP} dr = \frac{\lambda_{HP}}{6} \tag{16.356}$$

$$Q_\lambda = \int_0^1 \frac{1}{2} r (r\alpha - \lambda_{HP}) dr = \int_0^1 \frac{1}{2} r^2 \alpha \, dr - \frac{\lambda_{HP}}{4} \tag{16.357}$$

$$Q_{\dot{\beta}} = \int_0^1 \frac{1}{2} r^2 (r\alpha - \lambda_{HP}) dr = \int_0^1 \frac{1}{2} r^3 \alpha \, dr - \frac{\lambda_{HP}}{6} \tag{16.358}$$

$$Q_{\dot{\zeta}} = -\int_0^1 r^2 \left(\frac{1}{2}\lambda_{HP}\theta + 2r\frac{c_d}{2a}\right) dr = -\left(\lambda_{HP}\int_0^1 \frac{1}{2} r^2 \theta \, dr + \frac{c_d}{4a}\right) \tag{16.359}$$

and $M_\beta = M_\zeta = Q_\beta = Q_\zeta = 0$. These coefficients can be expressed directly in terms of the rotor thrust. Momentum theory relates the induced velocity and thrust: for example, in hover $\lambda_{HP} = \kappa\sqrt{C_T/2}$. Recall that the definition of the thrust coefficient is

$$\frac{C_T}{\sigma a} = \int_0^1 \frac{F_z}{ac} dr = \int_0^1 \frac{1}{2} r^2 \alpha \, dr \tag{16.360}$$

So integrals of the blade angle-of-attack distribution can be related to the thrust. With $\alpha = \theta - \lambda_{HP}/r = \theta_0 + \theta_{tw} r - \lambda_{HP}/r$,

$$\frac{C_T}{\sigma a} = \frac{\theta_{.75}}{6} - \frac{\lambda_{HP}}{4} \tag{16.361}$$

So

$$\int_0^1 \frac{1}{2} r^3 \alpha \, dr = \frac{\theta_{.8}}{8} - \frac{\lambda_{HP}}{6} = \frac{3}{4} \frac{C_T}{\sigma a} + \frac{\lambda_{HP}}{48} + \frac{\theta_{tw}}{160} \tag{16.362}$$

$$\int_0^1 \frac{1}{2} r^2 \theta \, dr = \frac{C_T}{\sigma a} + \frac{\lambda_{HP}}{4} \tag{16.363}$$

The results for the aerodynamic coefficients are

$$M_{\dot{\zeta}} = -\left(\frac{3}{2}\frac{C_T}{\sigma a} + \frac{5}{24}\lambda_{HP} + \frac{1}{80}\theta_{tw}\right) \tag{16.364}$$

$$Q_\lambda = \frac{C_T}{\sigma a} - \frac{\lambda_{HP}}{4} \tag{16.365}$$

$$Q_{\dot{\beta}} = \frac{3}{4}\frac{C_T}{\sigma a} - \frac{7}{48}\lambda_{HP} + \frac{1}{160}\theta_{tw} \tag{16.366}$$

$$Q_{\dot{\zeta}} = -\left[\lambda_{HP}\left(\frac{C_T}{\sigma a} + \frac{\lambda_{HP}}{4}\right) + \frac{c_d}{4a}\right] \tag{16.367}$$

The rotor thrust coefficient thus defines these aerodynamic coefficients.

16.8.4 Non-Rotating Frame

The aerodynamic flap moment on the m-th blade of an N-bladed rotor has been obtained in the following form:

$$M_F^{(m)} = M_\theta \left(\theta^{(m)} - K_P \beta^{(m)}\right) + M_\lambda \lambda + M_{\dot{\beta}} \dot{\beta}^{(m)} + M_\beta \beta^{(m)} \tag{16.368}$$

The aerodynamic forces in the non-rotating frame are obtained by introducing multi-blade coordinates and evaluating the summations,

$$M_{F_0} = \frac{1}{N} \sum_{m=1}^{N} M_F^{(m)} \tag{16.369}$$

$$M_{F_{nc}} = \frac{2}{N} \sum_{m=1}^{N} M_F^{(m)} \cos n\psi_m \tag{16.370}$$

$$M_{F_{ns}} = \frac{2}{N} \sum_{m=1}^{N} M_F^{(m)} \sin n\psi_m \tag{16.371}$$

$$M_{F_{N/2}} = \frac{1}{N} \sum_{m=1}^{N} M_F^{(m)} (-1)^m \tag{16.372}$$

16.8 Aerodynamic Loads

where $\psi_m = \psi + m\Delta\psi$ is the azimuth angle of the m-th blade ($\Delta\psi = 2\pi/N$). When the aerodynamic coefficients are independent of ψ_m, as for the hovering rotor, the summation operators act only on the degrees of freedom, not on the aerodynamic coefficients. The summations are easily evaluated using the definitions of the non-rotating degrees of freedom and the corresponding results for their time derivatives (see section 15.4.1). The result for constant coefficients is

$$M_{F_0} = M_\theta \left(\theta_0 - K_P\beta_0\right) + M_\lambda\lambda + M_{\dot\beta}\dot\beta_0 \tag{16.373}$$

$$M_{F_{nc}} = M_\theta \left(\theta_{nc} - K_P\beta_{nc}\right) + M_{\dot\beta}\left(\dot\beta_{nc} + n\beta_{ns}\right) \tag{16.374}$$

$$M_{F_{ns}} = M_\theta \left(\theta_{ns} - K_P\beta_{ns}\right) + M_{\dot\beta}\left(\dot\beta_{ns} - n\beta_{nc}\right) \tag{16.375}$$

$$M_{F_{N/2}} = M_\theta \left(\theta_{N/2} - K_P\beta_{N/2}\right) + M_{\dot\beta}\dot\beta_{N/2} \tag{16.376}$$

(recall that $M_\beta = 0$ in hover). In forward flight the aerodynamic coefficients are periodic functions of ψ_m, and the evaluation of the summations is more complicated. In section 15.4.3 the techniques for converting the equations of motion to the non-rotating frame were discussed. The solutions of differential equations with periodic coefficients have a number of unique characteristics. A transformation of the degrees of freedom cannot change the physical behavior of a system, although it can make the analysis easier. Therefore, periodic coefficients must appear in the rotor equations in the non-rotating frame if they appear in the rotating frame. Moreover, with periodic coefficients the differential equations in the non-rotating frame depend on the number of blades. Using the forward flight aerodynamic coefficients given in section 16.8.2, the flap moments in the non-rotating frame are as follows. In these matrices, $C_n = \cos n\psi$ and $S_n = \sin n\psi$. For a two-bladed rotor ($N = 2$),

$$\begin{pmatrix} M_{F_0} \\ M_{F_1} \end{pmatrix} = \begin{bmatrix} \frac{1}{8}(1+\mu^2) - \frac{\mu^2}{8}C_2 & \frac{\mu}{3}S_1 \\ \frac{\mu}{3}S_1 & \frac{1}{8}(1+\mu^2) - \frac{\mu^2}{8}C_2 \end{bmatrix} \begin{pmatrix} \theta_0 - K_P\beta_0 \\ \theta_1 - K_P\beta_1 \end{pmatrix}$$

$$- \begin{bmatrix} \frac{1}{8} & \frac{\mu}{6}S_1 \\ \frac{\mu}{6}S_1 & \frac{1}{8} \end{bmatrix} \begin{pmatrix} \dot\beta_0 \\ \dot\beta_1 \end{pmatrix}$$

$$- \begin{bmatrix} \frac{\mu^2}{8}S_2 & \frac{\mu}{6}C_1 \\ \frac{\mu}{6}C_1 & \frac{\mu^2}{8}S_2 \end{bmatrix} \begin{pmatrix} \beta_0 \\ \beta_1 \end{pmatrix} + \begin{bmatrix} -\frac{1}{6} \\ -\frac{\mu}{4}S_1 \end{bmatrix} \lambda \tag{16.377}$$

For a three-bladed rotor ($N = 3$),

$$\begin{pmatrix} M_{F_0} \\ M_{F_{1c}} \\ M_{F_{1s}} \end{pmatrix}$$

$$= \begin{bmatrix} \frac{1}{8}(1+\mu^2) & -\frac{\mu^2}{16}C_3 & \frac{\mu}{6} - \frac{\mu^2}{16}S_3 \\ -\frac{\mu^2}{8}C_3 & \frac{1}{8}(1+\frac{1}{2}\mu^2) + \frac{\mu}{6}S_3 & -\frac{\mu}{6}C_3 \\ \frac{\mu}{3} - \frac{\mu^2}{8}S_3 & -\frac{\mu}{6}C_3 & \frac{1}{8}(1+\frac{3}{2}\mu^2) - \frac{\mu}{6}S_3 \end{bmatrix} \begin{pmatrix} \theta_0 - K_P\beta_0 \\ \theta_{1c} - K_P\beta_{1c} \\ \theta_{1s} - K_P\beta_{1s} \end{pmatrix}$$

$$- \begin{bmatrix} \frac{1}{8} & 0 & \frac{\mu}{12} \\ 0 & \frac{1}{8} + \frac{\mu}{12}S_3 & -\frac{\mu}{12}C_3 \\ \frac{\mu}{6} & -\frac{\mu}{12}C_3 & \frac{1}{8} - \frac{\mu}{12}C_3 \end{bmatrix} \begin{pmatrix} \dot\beta_0 \\ \dot\beta_{1c} \\ \dot\beta_{1s} \end{pmatrix}$$

$$-\begin{bmatrix} 0 & \frac{\mu^2}{16}S_3 & -\frac{\mu^2}{16}C_3 \\ \frac{\mu}{6}+\frac{\mu^2}{8}S_3 & \frac{\mu}{6}C_3 & \frac{1}{8}(1+\frac{1}{2}\mu^2)+\frac{\mu}{6}S_3 \\ -\frac{\mu^2}{8}C_3 & -\frac{1}{8}(1-\frac{1}{2}\mu^2)+\frac{\mu}{6}S_3 & -\frac{\mu}{6}C_3 \end{bmatrix}\begin{pmatrix}\beta_0\\ \beta_{1c}\\ \beta_{1s}\end{pmatrix}$$

$$+\begin{bmatrix}-\frac{1}{6}\\ 0\\ -\frac{\mu}{4}\end{bmatrix}\lambda \tag{16.378}$$

For a four-bladed rotor ($N = 4$),

$$\begin{pmatrix}M_{F_0}\\ M_{F_{1c}}\\ M_{F_{1s}}\\ M_{F_2}\end{pmatrix}$$

$$=\begin{bmatrix}\frac{1}{8}(1+\mu^2) & 0 & \frac{\mu}{6} & -\frac{\mu^2}{8}C_2 \\ 0 & \frac{1}{8}(1+\frac{1}{2}\mu^2)-\frac{\mu^2}{16}C_4 & -\frac{\mu^2}{16}S_4 & \frac{\mu}{3}S_2 \\ \frac{\mu}{3} & -\frac{\mu^2}{16}S_4 & \frac{1}{8}(1+\frac{3}{2}\mu^2)+\frac{\mu^2}{16}C_4 & -\frac{\mu}{3}C_2 \\ -\frac{\mu^2}{8}C_2 & \frac{\mu}{6}S_2 & -\frac{\mu}{6}C_2 & \frac{1}{8}(1+\mu^2)\end{bmatrix}$$

$$\times\begin{pmatrix}\theta_0-K_P\beta_0\\ \theta_{1c}-K_P\beta_{1c}\\ \theta_{1s}-K_P\beta_{1s}\\ \theta_2-K_P\beta_2\end{pmatrix}$$

$$-\begin{bmatrix}\frac{1}{8} & 0 & \frac{\mu}{12} & 0 \\ 0 & \frac{1}{8} & 0 & \frac{\mu}{6}S_2 \\ \frac{\mu}{6} & 0 & \frac{1}{8} & -\frac{\mu}{6}C_2 \\ 0 & \frac{\mu}{12}S_2 & -\frac{\mu}{12}C_2 & \frac{1}{8}\end{bmatrix}\begin{pmatrix}\dot{\beta}_0\\ \dot{\beta}_{1c}\\ \dot{\beta}_{1s}\\ \dot{\beta}_2\end{pmatrix}$$

$$-\begin{bmatrix}0 & 0 & 0 & \frac{\mu^2}{8}S_2 \\ \frac{\mu}{6} & \frac{\mu^2}{16}S_4 & \frac{1}{8}(1+\frac{1}{2}\mu^2)-\frac{\mu^2}{16}C_4 & \frac{\mu}{6}C_2 \\ 0 & -\frac{1}{8}(1-\frac{1}{2}\mu^2)-\frac{\mu^2}{16}C_4 & -\frac{\mu^2}{16}S_4 & \frac{\mu}{6}S_2 \\ \frac{\mu^2}{8}S_2 & \frac{\mu}{6}C_2 & \frac{\mu}{6}S_2 & 0\end{bmatrix}\begin{pmatrix}\beta_0\\ \beta_{1c}\\ \beta_{1s}\\ \beta_2\end{pmatrix}$$

$$+\begin{bmatrix}-\frac{1}{6}\\ 0\\ -\frac{\mu}{4}\\ 0\end{bmatrix}\lambda \tag{16.379}$$

For the two-bladed rotor the pitch control variables are collective pitch $\theta_0 = \frac{1}{2}(\theta^{(2)} + \theta^{(1)})$ and differential pitch $\theta_1 = \frac{1}{2}(\theta^{(2)} - \theta^{(1)})$. The usual swashplate control gives $\theta_1 = \theta_{1c}\cos\psi + \theta_{1s}\sin\psi$. Also observe that increasing the number of blades has the effect of clearing the periodic coefficients from the lower degrees of freedom and equations, although the periodicity always remains in the higher elements of the matrices.

The analysis of the time-invariant system response is much simpler than the analysis of the periodic system response, and more powerful tools are available. Thus we are interested in the possibility of an accurate constant coefficient representation of the rotor dynamics. Such a representation can only be approximate, since it can never correctly model all the behavior of a periodic system. From these equations for the flap moments, it is clear that the constant coefficient approximation should be introduced in the non-rotating frame, after the introduction of multiblade coordinates. If the mean values of the aerodynamic coefficients in the rotating frame are used, all the influence of forward flight is lost except for an increase of order μ^2 in M_θ. The mean values of the coefficients in the non-rotating frame include some of the higher harmonics of the coefficients in the rotating frame. From the earlier results for three-bladed and four-bladed rotors, the constant coefficient approximation for the flap moments is

$$\begin{pmatrix} M_{F_0} \\ M_{F_{1c}} \\ M_{F_{1s}} \end{pmatrix} = \begin{bmatrix} \frac{1}{8}(1+\mu^2) & 0 & \frac{\mu}{6} \\ 0 & \frac{1}{8}\left(1+\frac{1}{2}\mu^2\right) & 0 \\ \frac{\mu}{3} & 0 & \frac{1}{8}\left(1+\frac{3}{2}\mu^2\right) \end{bmatrix} \begin{pmatrix} \theta_0 - K_P\beta_0 \\ \theta_{1c} - K_P\beta_{1c} \\ \theta_{1s} - K_P\beta_{1s} \end{pmatrix}$$

$$- \begin{bmatrix} \frac{1}{8} & 0 & \frac{\mu}{12} \\ 0 & \frac{1}{8} & 0 \\ \frac{\mu}{6} & 0 & \frac{1}{8} \end{bmatrix} \begin{pmatrix} \dot\beta_0 \\ \dot\beta_{1c} \\ \dot\beta_{1s} \end{pmatrix}$$

$$- \begin{bmatrix} 0 & 0 & 0 \\ \frac{\mu}{6} & 0 & \frac{1}{8}\left(1+\frac{1}{2}\mu^2\right) \\ 0 & -\frac{1}{8}\left(1-\frac{1}{2}\mu^2\right) & 0 \end{bmatrix} \begin{pmatrix} \beta_0 \\ \beta_{1c} \\ \beta_{1s} \end{pmatrix}$$

$$+ \begin{bmatrix} -\frac{1}{6} \\ 0 \\ -\frac{\mu}{4} \end{bmatrix} \lambda \qquad (16.380)$$

There are additional degrees of freedom and equations for $N \geq 4$. Since increasing the number of blades tends to sweep the periodic coefficients to the higher degrees of freedom, it is expected that for a rotor with a large number of blades the constant coefficient approximation is a good representation of the dynamics involving primarily the collective and cyclic degrees of freedom (β_0, β_{1c}, and β_{1s} here).

Alternatively, the constant coefficient approximation is readily obtained directly from the rotating equation. Consider a typical term in the non-rotating equation, of the form

$$\frac{2}{N} \sum_{m=1}^{N} \left(M_\theta^{(m)} \theta^{(m)}\right) \cos\psi_m \qquad (16.381)$$

Substituting the multiblade coordinates for $\theta^{(m)}$ gives

$$\frac{2}{N} \sum_{m=1}^{N} M_\theta^{(m)} \left[\theta_0 \cos\psi_m + \theta_{1c}\frac{1}{2}(1+\cos 2\psi_m) + \theta_{1s}\frac{1}{2}\sin 2\psi_m\right] \qquad (16.382)$$

If the complete (periodic) coefficient is required, the evaluation of the summation over $M_\theta^{(m)}$ is rather complicated and also depends on N. If only the mean value of the coefficient is required, the summation operator simply picks out the corresponding harmonic in the Fourier series expansion of $M_\theta^{(m)}$. Hence in the present example we obtain

$$\theta_0 M_\theta^{1c} + \theta_{1c}\left(M_\theta^0 + \frac{1}{2}M_\theta^{2c}\right) + \theta_{1s}\frac{1}{2}M_\theta^{2s} \qquad (16.383)$$

where M_θ^{nc} and M_θ^{ns} are the Fourier series harmonics of M_θ. The complete result for the β_0, β_{1c}, and β_{1s} flap moments is

$$\begin{pmatrix} M_{F_0} \\ M_{F_{1c}} \\ M_{F_{1s}} \end{pmatrix} = \begin{bmatrix} M_\theta^0 & 0 & \frac{1}{2}M_\theta^{1s} \\ 0 & M_\theta^0 + \frac{1}{2}M_\theta^{2c} & 0 \\ M_\theta^{1s} & 0 & M_\theta^0 - \frac{1}{2}M_\theta^{2c} \end{bmatrix} \begin{pmatrix} \theta_0 - K_P\beta_0 \\ \theta_{1c} - K_P\beta_{1c} \\ \theta_{1s} - K_P\beta_{1s} \end{pmatrix}$$

$$+ \begin{bmatrix} M_{\dot\beta}^0 & 0 & \frac{1}{2}M_{\dot\beta}^{1s} \\ 0 & M_{\dot\beta}^0 + \frac{1}{2}M_{\dot\beta}^{2c} & 0 \\ M_{\dot\beta}^{1s} & 0 & M_{\dot\beta}^0 - \frac{1}{2}M_{\dot\beta}^{2c} \end{bmatrix} \begin{pmatrix} \dot\beta_0 \\ \dot\beta_{1c} \\ \dot\beta_{1s} \end{pmatrix}$$

$$+ \begin{bmatrix} 0 & \frac{1}{2}\left(M_\beta^{1c} - M_{\dot\beta}^{1s}\right) & 0 \\ M_\beta^{1c} & 0 & M_{\dot\beta}^0 + \frac{1}{2}\left(M_\beta^{2c} + M_{\dot\beta}^{2s}\right) \\ 0 & -M_{\dot\beta}^0 + \frac{1}{2}\left(M_\beta^{2c} + M_{\dot\beta}^{2s}\right) & 0 \end{bmatrix} \begin{pmatrix} \beta_0 \\ \beta_{1c} \\ \beta_{1s} \end{pmatrix}$$

$$+ \begin{bmatrix} M_\lambda^0 \\ 0 \\ M_\lambda^{1s} \end{bmatrix} \lambda \qquad (16.384)$$

Use has been made of the fact that all the odd-cosine and even-sine harmonics of M_θ, $M_{\dot\beta}$, and M_λ are zero, as are the even-cosine and odd-sine harmonics of M_β. Expressions are available for the harmonics of the coefficients M_θ, $M_{\dot\beta}$, M_β, and M_λ in forward flight. Substituting the results for the harmonics calculated with $\eta_\beta = r$ gives the previous result for the constant coefficient approximation.

Often it is neither necessary nor possible to obtain explicit expressions for the periodic coefficients in the non-rotating frame, as was done here for the flap moments. Because the conversion of the equations to the non-rotating frame is rather tedious and must be repeated for every value of N, such an approach is justified only for analytical investigations with a small number of degrees of freedom. Moreover, with any but the simplest models the harmonics of the coefficients in the rotating frame must themselves be evaluated numerically. Thus a general procedure is desired for converting the rotor blade equations of motion to the non-rotating frame, one that can easily be implemented in numerical investigations. Consider again the flap moment (equation 16.368), where the aerodynamic coefficients are periodic functions of ψ_m. Substituting multiblade coordinates for $\theta^{(m)}$ and $\beta^{(m)}$, and applying the summation operators to convert the equations to the non-rotating frame, the equations are

$$\begin{pmatrix} M_{F_0} \\ M_{F_{1c}} \\ M_{F_{1s}} \end{pmatrix} = \frac{1}{N}\sum_{m=1}^{N} \begin{bmatrix} M_\theta & M_\theta C & M_\theta S \\ M_\theta 2C & M_\theta 2C^2 & M_\theta 2CS \\ M_\theta 2S & M_\theta 2CS & M_\theta 2S^2 \end{bmatrix} \begin{pmatrix} \theta_0 - K_P\beta_0 \\ \theta_{1c} - K_P\beta_{1c} \\ \theta_{1s} - K_P\beta_{1s} \end{pmatrix}$$

$$+ \frac{1}{N}\sum_{m=1}^{N} \begin{bmatrix} M_{\dot\beta} & M_{\dot\beta} C & M_{\dot\beta} S \\ M_{\dot\beta} 2C & M_{\dot\beta} 2C^2 & M_{\dot\beta} 2CS \\ M_{\dot\beta} 2S & M_{\dot\beta} 2CS & M_{\dot\beta} 2S^2 \end{bmatrix} \begin{pmatrix} \dot\beta_0 \\ \dot\beta_{1c} \\ \dot\beta_{1s} \end{pmatrix}$$

$$+ \frac{1}{N}\sum_{m=1}^{N} \begin{bmatrix} M_\beta & M_\beta C - M_{\dot\beta} S & M_\beta S + M_{\dot\beta} C \\ M_\beta 2C & M_\beta 2C^2 - M_{\dot\beta} 2CS & M_\beta 2CS + M_{\dot\beta} 2C^2 \\ M_\beta 2S & M_\beta 2CS - M_{\dot\beta} 2S^2 & M_\beta 2S^2 + M_{\dot\beta} 2CS \end{bmatrix} \begin{pmatrix} \beta_0 \\ \beta_{1c} \\ \beta_{1s} \end{pmatrix}$$

$$+ \frac{1}{N}\sum_{m=1}^{N} \begin{bmatrix} M_\lambda \\ M_\lambda 2C \\ M_\lambda 2S \end{bmatrix} \lambda \qquad (16.385)$$

where $C = \cos\psi_m$ and $S = \sin\psi_m$. The summation of the coefficients over all the blades ($m = 1$ to N) is to be performed numerically now. The corresponding rows and columns of these matrices are easily obtained for the β_{nc}, β_{ns}, and $\beta_{N/2}$ degrees of freedom as required, depending on the number of blades. Each row of a matrix has one of the factors 1, $2\cos k\psi_m$, $2\sin k\psi_m$, or $(-1)^m$ from the summation operator. Each column has one of the factors 1, $\cos n\psi_m$, $\sin n\psi_m$, or $(-1)^m$ from the multiblade coordinates (or 0, $-n\sin n\psi_m$, $n\cos n\psi_m$, 0 for the Coriolis terms resulting from the transformation of time derivatives). The form of this result does not depend on the number of blades, except for the size of the matrices. It is the result of the summations over all the blades that depends on N. See also the discussion in section 15.4.3.

The constant coefficient approximation requires the mean values of the equation coefficients in the non-rotating frame, which are obtained by applying the operator

$$\frac{1}{2\pi}\int_0^{2\pi}(\ldots)d\psi \tag{16.386}$$

to the matrices above. This operator produces harmonics of the coefficients, which can be evaluated using numerical integration. The result is the replacement of the summation over the number of blades by a summation over the rotor azimuth:

$$\frac{1}{N}\sum_{m=1}^N\{\ldots\}M(\psi_m) \to \frac{1}{J}\sum_{j=1}^J\{\ldots\}M(\psi_j) \tag{16.387}$$

with the coefficient evaluated at J points equally spaced around the azimuth, so $\psi_j = j\Delta\psi$ and $\Delta\psi = 2\pi/J$.

16.8.5 Hub Reactions in Rotating Frame

To evaluate the net rotor forces acting on the hub, first the shears and moments at the root of an individual blade must be found. The blade motion considered is the fundamental flap and lag modes. The perturbation forces and velocities for this case were given in section 16.8.3. The vertical shear forces at the root are

$$\int_0^1 \frac{F_z}{ac}dr = T_\theta(\theta - K_P\beta) + T_\lambda\lambda + T_{\dot\beta}\dot\beta + T_\beta\beta + T_{\dot\zeta}\dot\zeta + T_\zeta\zeta \tag{16.388}$$

where the aerodynamic coefficients are the same as for the flap moments, but without the factor η_β in the integrands:

$$T_\theta = \int_0^1 \frac{1}{2}u_T^2 dr = \frac{1}{6} + \frac{\mu}{2}\sin\psi + \frac{\mu^2}{2}\sin^2\psi \tag{16.389}$$

$$T_\lambda = -\int_0^1 \frac{1}{2}u_T dr = -\left(\frac{1}{4} + \frac{\mu}{2}\sin\psi\right) \tag{16.390}$$

$$T_{\dot\beta} = -\int_0^1 \frac{1}{2}\eta_\beta u_T dr = -\left(\frac{1}{6} + \frac{\mu}{4}\sin\psi\right) \tag{16.391}$$

$$T_\beta = -u_R\int_0^1 \frac{1}{2}\eta'_\beta u_T dr = -u_R\left(\frac{1}{4} + \frac{\mu}{2}\sin\psi\right) \tag{16.392}$$

$$T_{\dot\zeta} = -\int_0^1 \frac{1}{2}\eta_\zeta(u_P + 2u_T\alpha)\,dr \tag{16.393}$$

$$T_\zeta = -u_R\int_0^1 \frac{1}{2}\eta'_\zeta(u_P + 2u_T\alpha)\,dr \tag{16.394}$$

Here $T_{\dot\beta}$ and T_β have been evaluated using $\eta_\beta = r$. For hover, $T_{\dot\zeta} = 0$ and

$$T_\zeta = -\left(\frac{\lambda_{HP}}{4} + 2\int_0^1 \frac{1}{2}r^2\alpha\,dr\right) = -\left(\frac{\lambda_{HP}}{4} + \frac{2C_T}{\sigma a}\right) \tag{16.395}$$

assuming that the induced velocity is uniform and $\eta_\zeta = r$. The behavior of these aerodynamic coefficients for thrust is similar to the behavior of the flap moments, with just a change in the numerical constants because the factor η_β has been removed from the integrands.

The flapwise aerodynamic moment at the rotor hub is $\int_0^1 r(F_z/ac)\,dr$, which can be evaluated from the flap moment M_F by simply replacing the mode shape η_β by r. Alternatively, the hub moment can be obtained directly from the flap response using $N_F = I_b\Omega^2(\nu_\beta^2 - 1)\beta$.

The perturbation of the section radial force is

$$\delta\frac{F_r}{ac} = u_R\frac{c_d}{2a}\delta u_T + u_T\frac{c_d}{2a}\delta u_R - \left(\frac{F_z}{ac}\right)\delta z' - (z')\delta\frac{F_z}{ac} \tag{16.396}$$

$$= \left(-z'\frac{1}{2}u_T^2\right)\delta\theta + \left(z'\frac{1}{2}u_T\right)\delta u_P + \left(u_R\frac{c_d}{2a} - z'\frac{1}{2}(u_P + 2u_T\alpha)\right)\delta u_T$$

$$+ \left(u_T\frac{c_d}{2a}\right)\delta u_R - \left(\frac{F_z}{ac}\right)\delta z' \tag{16.397}$$

The perturbation velocities are given in section 16.8.3. Here we also require $\delta u_R = \eta'_\zeta\zeta\mu\sin\psi$ and $\delta z' = \eta'_\beta\beta$. Then the radial aerodynamic shear force at the blade root is

$$\int_0^1 \frac{F_r}{ac}\,dr = R_\theta(\theta - K_P\beta) + R_\lambda\lambda + R_{\dot\beta}\dot\beta + R_\beta\beta + R_{\dot\zeta}\dot\zeta + R_\zeta\zeta \tag{16.398}$$

where

$$R_\theta = -\int_0^1 z'\frac{1}{2}u_T^2\,dr \tag{16.399}$$

$$R_\lambda = \int_0^1 z'\frac{1}{2}u_T\,dr \tag{16.400}$$

$$R_{\dot\beta} = \int_0^1 z'\frac{1}{2}\eta_\beta u_T\,dr \tag{16.401}$$

$$R_\beta = -\int_0^1 \frac{F_z}{ac}\eta'_\beta\,dr + u_R\int_0^1 z'\frac{1}{2}\eta'_\beta u_T\,dr \tag{16.402}$$

$$R_{\dot\zeta} = -\int_0^1 \eta_\zeta\left(u_R\frac{c_d}{2a} - z'\frac{1}{2}(u_P + 2u_T\alpha)\right)dr \tag{16.403}$$

$$R_\zeta = \int_0^1 \eta'_\zeta\left[-u_R\left(u_R\frac{c_d}{2a} - z'\frac{1}{2}(u_P + 2u_T\alpha)\right) + u_T\frac{c_d}{2a}\mu\sin\psi\right]dr \tag{16.404}$$

16.8 Aerodynamic Loads

Most of the terms in these aerodynamic coefficients are due to the radial tilt of the thrust vector. The coefficient R_β, which gives the radial force due to flap displacement, is particularly important. Assuming $\eta_\beta = r$, the first term in R_β is

$$-\int_0^1 \frac{F_z}{ac} dr = -\frac{S_z}{ac} = -\frac{C_T}{\sigma a} \tag{16.405}$$

where $S_z = T/N$ is the thrust of a single blade. So this term gives a radial force $\Delta R = -(T/N)\beta$, which is just the in-plane component of the blade thrust when it is tilted by flap deflection of the blade. Because of the importance of this term, we write

$$R_\beta = -\frac{C_T}{\sigma a} + \widehat{R}_\beta \tag{16.406}$$

where

$$\widehat{R}_\beta = \int_0^1 \left(\frac{C_T}{\sigma a} - \frac{1}{2}u_T^2 \alpha \eta'_\beta\right) dr + u_R \int_0^1 z' \frac{1}{2}\eta'_\beta u_T dr \tag{16.407}$$

$\widehat{R}_\beta$ is non-zero if $\eta'_\beta \neq 1$ or in forward flight where the radial velocity u_R is nonzero and the blade lift varies around the azimuth.

Assuming that the trim slope of the blade z' is independent of r (hence written $z' = \eta'_\beta \beta = \beta_{\text{trim}}$), the radial force aerodynamic coefficients can be related to the corresponding vertical force coefficients. For example,

$$R_\theta \cong -z' \int_0^1 \frac{1}{2}u_T^2 dr = -z' T_\theta \tag{16.408}$$

The coefficients can be readily evaluated for the hover case. Assuming uniform inflow, $\eta_\beta = \eta_\zeta = r$, and constant z', we obtain

$$R_\theta = -\frac{z'}{6} \tag{16.409}$$

$$R_\lambda = \frac{z'}{4} \tag{16.410}$$

$$R_{\dot\beta} = \frac{z'}{6} \tag{16.411}$$

$$R_\zeta = z'\left(\frac{\lambda_{\text{HP}}}{4} + \frac{2C_T}{\sigma a}\right) \tag{16.412}$$

and $\widehat{R}_\beta = R_{\dot\zeta} = 0$.

The in-plane aerodynamic shear force at the blade root is obtained from the section drag force F_x as was the lag moment, giving

$$\int_0^1 \frac{F_x}{ac} dr = H_\theta(\theta - K_P \beta) + H_\lambda \lambda + H_{\dot\beta}\dot\beta + H_\beta \beta + H_{\dot\zeta}\dot\zeta + H_\zeta \zeta \tag{16.413}$$

with the aerodynamic coefficients as follows:

$$H_\theta = \int_0^1 \frac{1}{2} u_T u_P dr \tag{16.414}$$

$$H_\lambda = \int_0^1 \frac{1}{2}(u_T \alpha - u_P) dr \tag{16.415}$$

$$H_{\dot\beta} = \int_0^1 \frac{1}{2}\eta_\beta(u_T\alpha - u_P)dr \tag{16.416}$$

$$H_\beta = u_R \int_0^1 \frac{1}{2}\eta'_\beta(u_T\alpha - u_P)dr \tag{16.417}$$

$$H_{\dot\zeta} = -\int_0^1 \eta_\zeta\left(\frac{1}{2}u_P\theta + 2u_T\frac{c_d}{2a}\right)dr \tag{16.418}$$

$$H_\zeta = -u_R \int_0^1 \eta'_\zeta\left(\frac{1}{2}u_P\theta + 2u_T\frac{c_d}{2a}\right)dr \tag{16.419}$$

The coefficient $H_{\dot\beta}$, which gives the in-plane force due to flapping velocity, is of particular importance. For hover,

$$H_{\dot\beta} = \int_0^1 \frac{1}{2}(r^2\alpha - r\lambda_{HP})dr = \frac{C_T}{\sigma a} - \frac{\lambda_{HP}}{4} \tag{16.420}$$

assuming $\eta_\beta = r$. The first term is an in-plane force $\Delta H = (T/N)\dot\beta$. The flapping velocity gives an angle-of-attack change $\delta\alpha = -\eta_\beta\dot\beta/r = -\dot\beta$ that tilts the thrust vector backward and thus produces an in-plane shear force on the blade. This coefficient is therefore written as

$$H_{\dot\beta} = \frac{C_T}{\sigma a} + \widehat{H}_{\dot\beta} \tag{16.421}$$

where

$$\widehat{H}_{\dot\beta} = \int_0^1 \left(\frac{1}{2}\eta_\beta u_T\alpha - \frac{C_T}{\sigma a} - \frac{1}{2}\eta_\beta u_P\right)dr \tag{16.422}$$

Even for hover there is a non-zero inflow contribution, $\widehat{H}_{\dot\beta} = -\lambda_{HP}/4$, which can be a significant fraction of the thrust term in $H_{\dot\beta}$. The aerodynamic coefficients for hover are

$$H_\theta = \frac{\lambda_{HP}}{4} \tag{16.423}$$

$$H_\lambda = \frac{3}{2}\frac{C_T}{\sigma a} - \frac{5}{8}\lambda_{HP} - \frac{\theta_{tw}}{48} \tag{16.424}$$

$$H_{\dot\beta} = -\frac{\lambda_{HP}}{4} \tag{16.425}$$

$$H_\zeta = -\left[\lambda_{HP}\left(\frac{3}{2}\frac{C_T}{\sigma a} + \frac{3}{8}\lambda_{HP} - \frac{\theta_{tw}}{48}\right) + \frac{c_d}{3a}\right] \tag{16.426}$$

and $H_\beta = H_{\dot\zeta} = 0$.

The aerodynamic torque moment acting on the rotor hub at the center of rotation can be obtained from the lag moment M_L (derived in section 16.8.3) by replacing the mode shape η_ζ by r:

$$\int_0^1 r\frac{F_x}{ac}dr = Q_\theta(\theta - K_P\beta) + Q_\lambda\lambda + Q_{\dot\beta}\dot\beta + Q_\beta\beta + Q_{\dot\zeta}\dot\zeta + Q_\zeta\zeta \tag{16.427}$$

Since we have been using the approximation $\eta_\zeta \cong r$ to evaluate the aerodynamic coefficients, there is no need to distinguish in the notation between the lag and

16.8.6 Hub Reactions in Non-Rotating Frame

The total aerodynamic forces and moments acting on the rotor hub were derived in section 16.6.2. The aerodynamic thrust, torque, drag force, and side force of the rotor are obtained by summing the root reactions of all N blades:

$$\left(\frac{C_T}{\sigma a}\right)_{\text{aero}} = \frac{1}{N} \sum_{m=1}^{N} \int_0^1 \frac{F_z}{ac} dr \tag{16.428}$$

$$\left(\frac{C_Q}{\sigma a}\right)_{\text{aero}} = \frac{1}{N} \sum_{m=1}^{N} \int_0^1 r\frac{F_x}{ac} dr \tag{16.429}$$

$$\left(\frac{2C_H}{\sigma a}\right)_{\text{aero}} = \frac{2}{N} \sum_{m=1}^{N} \left[\cos\psi_m \int_0^1 \frac{F_r}{ac} dr + \sin\psi_m \int_0^1 \frac{F_x}{ac} dr\right] \tag{16.430}$$

$$\left(\frac{2C_Y}{\sigma a}\right)_{\text{aero}} = \frac{2}{N} \sum_{m=1}^{N} \left[\sin\psi_m \int_0^1 \frac{F_r}{ac} dr - \cos\psi_m \int_0^1 \frac{F_x}{ac} dr\right] \tag{16.431}$$

(equations 16.217, 16.233, 16.230, and 16.231). Expressions for the root forces and moments were derived in the preceding paragraphs as linear functions of the rotating degrees of freedom of the blade. In hover, since the aerodynamic coefficients in these expressions are constants, the summation operators in the total hub reactions act only on the blade degrees of freedom. Hence for this constant coefficient case the summations are easily evaluated using the definitions of the degrees of freedom in the non-rotating frame (multiblade coordinates). In hover the thrust and torque only involve the rotor collective degrees of freedom (here the coning and collective lag modes). The result is

$$\frac{C_T}{\sigma a} = T_\theta(\theta_0 - K_P\beta_0) + T_\lambda \lambda + T_{\dot\beta}\dot\beta_0 + T_{\dot\zeta}\dot\zeta_0 \tag{16.432}$$

$$\frac{C_Q}{\sigma a} = Q_\theta(\theta_0 - K_P\beta_0) + Q_\lambda \lambda + Q_{\dot\beta}\dot\beta_0 + Q_{\dot\zeta}\dot\zeta_0 \tag{16.433}$$

For hover, the rotor in-plane hub forces involve only the cyclic degrees of freedom in the non-rotating frame. Neglecting the forces due to the blade lag motion, which are much smaller than those due to the flap motion, the rotor drag and side forces in hover are

$$\begin{pmatrix} \frac{2C_H}{\sigma a} \\ \frac{2C_Y}{\sigma a} \end{pmatrix} = -\frac{2C_T}{\sigma a}\begin{pmatrix} \beta_{1c} \\ \beta_{1s} \end{pmatrix} + \begin{bmatrix} R_{\dot\beta} & H_{\dot\beta} \\ -H_{\dot\beta} & R_{\dot\beta} \end{bmatrix}\begin{pmatrix} \dot\beta_{1c} \\ \dot\beta_{1s} \end{pmatrix} + \begin{bmatrix} -\widehat{H}_{\dot\beta} & R_{\dot\beta} \\ -R_{\dot\beta} & -\widehat{H}_{\dot\beta} \end{bmatrix}\begin{pmatrix} \beta_{1c} \\ \beta_{1s} \end{pmatrix}$$

$$+ \begin{bmatrix} R_\theta & H_\theta \\ -H_\theta & R_\theta \end{bmatrix}\begin{pmatrix} \theta_{1c} - K_P\beta_{1c} \\ \theta_{1s} - K_P\beta_{1s} \end{pmatrix} \tag{16.434}$$

using $R_\beta = -C_T/\sigma a$ and $H_{\dot\beta} = C_T/\sigma a + \widehat{H}_{\dot\beta}$. Substituting for the aerodynamic coefficients gives

$$\begin{pmatrix} \dfrac{2C_H}{\sigma a} \\ \dfrac{2C_Y}{\sigma a} \end{pmatrix} = -\dfrac{2C_T}{\sigma a}\begin{pmatrix} \beta_{1c} \\ \beta_{1s} \end{pmatrix} + \begin{bmatrix} 0 & \dfrac{C_T}{\sigma a} \\ -\dfrac{C_T}{\sigma a} & 0 \end{bmatrix}\begin{pmatrix} \dot\beta_{1c} \\ \dot\beta_{1s} \end{pmatrix}$$
$$+ \begin{bmatrix} -\dfrac{z'}{6} & \dfrac{\lambda_{HP}}{4} \\ -\dfrac{\lambda_{HP}}{4} & -\dfrac{z'}{6} \end{bmatrix}\begin{pmatrix} -\dot\beta_{1c} - \beta_{1s} + \theta_{1c} - K_P\beta_{1c} \\ -\dot\beta_{1s} + \beta_{1c} + \theta_{1s} - K_P\beta_{1s} \end{pmatrix} \quad (16.435)$$

This result is of the form

$$\begin{pmatrix} H \\ Y \end{pmatrix} = -T\begin{pmatrix} \beta_{1c} \\ \beta_{1s} \end{pmatrix} + \begin{pmatrix} H_{TPP} \\ Y_{TPP} \end{pmatrix} \quad (16.436)$$

Thus the rotor drag and side hub forces consist of the in-plane component of the thrust vector tilted with the tip-path plane, plus the in-plane forces relative to the tip-path plane. Consequently, the flapping response is a principal factor in the rotor hub reactions. Recall that the rotor hub moments can also be related to the tip-path-plane tilt. The total moment about the helicopter center-of-gravity a distance h below the hub is then

$$\begin{pmatrix} -\dfrac{2C_{My}}{\sigma a} - h\dfrac{2C_H}{\sigma a} \\ \dfrac{2C_{Mx}}{\sigma a} - h\dfrac{2C_Y}{\sigma a} \end{pmatrix} = \left[\dfrac{\widehat{I}_\beta(\nu_\beta^2 - 1)}{\gamma} + h\dfrac{2C_T}{\sigma a}\right]\begin{pmatrix} \beta_{1c} \\ \beta_{1s} \end{pmatrix} \quad (16.437)$$

plus the moments due to the in-plane forces relative to the tip-path plane.

The in-plane force due to the tilt of the rotor thrust vector with the tip-path plane arises from two sources, one-half from R_β and one-half from $H_{\dot\beta}$. The slope of the blade due to flap deflection tilts the blade lift radially, producing an in-plane component of the thrust (R_β). The rotating-frame flap velocity due to tip-path-plane tilt changes the blade angle-of-attack, which tilts the blade lift chordwise and thereby produces an in-plane component of the thrust ($H_{\dot\beta}$). Although R_β acts only on the flap displacement, the $H_{\dot\beta}$ coefficient produces forces due to tip-path-plane tilt rate ($\dot\beta_{1c}$ and $\dot\beta_{1s}$) as well. Also, any blade pitch change, flap displacement, or flap velocity changes the blade lift magnitude. Since the lift has an in-plane component due to the trim induced velocity, in-plane hub forces are produced by these lift magnitude changes (through $-\widehat{H}_{\dot\beta} = H_\theta = \lambda_{HP}/4$).

For a two-bladed rotor the cyclic flap degrees of freedom β_{1c} and β_{1s} are replaced by the teetering mode β_1, so the above results are applicable only when $N > 3$. When $N = 2$ the in-plane hub forces become

$$\begin{pmatrix} \dfrac{2C_H}{\sigma a} \\ \dfrac{2C_Y}{\sigma a} \end{pmatrix} = \begin{pmatrix} R_\beta 2C \\ R_\beta 2S \end{pmatrix}\beta_1 + \begin{bmatrix} R_{\dot\beta}2C + H_{\dot\beta}2S \\ R_{\dot\beta}2S - H_{\dot\beta}2C \end{bmatrix}\dot\beta_1$$
$$+ \begin{bmatrix} R_\theta 2C + H_\theta 2S \\ R_\theta 2S - H_\theta 2C \end{bmatrix}(\theta_1 - K_P\beta_1) \quad (16.438)$$

where $C = \cos\psi$ and $S = \sin\psi$. Thus even in hover the two-bladed rotor dynamics are described by periodic coefficient differential equations.

The derivation of the hub reactions in forward flight follows the derivation for the flap moment in the non-rotating frame (section 16.8.4). For the constant coefficient approximation we obtain

$$\begin{pmatrix} \dfrac{C_T}{\sigma a} \\ \dfrac{2C_H}{\sigma a} \\ \dfrac{2C_Y}{\sigma a} \end{pmatrix}$$

$$= \begin{bmatrix} T_\beta^0 & \frac{1}{2}T_\beta^{1c} & \frac{1}{2}T_\beta^{1s} \\ R_\beta^{1c}+H_\beta^{1s} & R_\beta^0+\frac{1}{2}\left(R_\beta^{2c}+H_\beta^{2s}\right) & H_\beta^0+\frac{1}{2}\left(R_\beta^{2s}-H_\beta^{2c}\right) \\ R_\beta^{1s}-H_\beta^{1c} & -H_\beta^0+\frac{1}{2}\left(R_\beta^{2s}-H_\beta^{2c}\right) & R_\beta^0-\frac{1}{2}\left(R_\beta^{2c}+H_\beta^{2s}\right) \end{bmatrix} \begin{pmatrix} \dot{\beta}_0 \\ \dot{\beta}_{1c}+\beta_{1s} \\ \dot{\beta}_{1s}-\beta_{1c} \end{pmatrix}$$

$$+ \begin{bmatrix} T_\beta^0 & \frac{1}{2}T_\beta^{1c} & \frac{1}{2}T_\beta^{1s} \\ R_\beta^{1c}+H_\beta^{1s} & R_\beta^0+\frac{1}{2}\left(R_\beta^{2c}+H_\beta^{2s}\right) & H_\beta^0+\frac{1}{2}\left(R_\beta^{2s}-H_\beta^{2c}\right) \\ R_\beta^{1s}-H_\beta^{1c} & -H_\beta^0+\frac{1}{2}\left(R_\beta^{2s}-H_\beta^{2c}\right) & R_\beta^0-\frac{1}{2}\left(R_\beta^{2c}+H_\beta^{2s}\right) \end{bmatrix} \begin{pmatrix} \beta_0 \\ \beta_{1c} \\ \beta_{1s} \end{pmatrix}$$

$$+ \begin{bmatrix} T_\theta^0 & \frac{1}{2}T_\theta^{1c} & \frac{1}{2}T_\theta^{1s} \\ R_\theta^{1c}+H_\theta^{1s} & R_\theta^0+\frac{1}{2}\left(R_\theta^{2c}+H_\theta^{2s}\right) & H_\theta^0+\frac{1}{2}\left(R_\theta^{2s}-H_\theta^{2c}\right) \\ R_\theta^{1s}-H_\theta^{1c} & -H_\theta^0+\frac{1}{2}\left(R_\theta^{2s}-H_\theta^{2c}\right) & R_\theta^0-\frac{1}{2}\left(R_\theta^{2c}+H_\theta^{2s}\right) \end{bmatrix} \begin{pmatrix} \theta_0-K_P\beta_0 \\ \theta_{1c}-K_P\beta_{1c} \\ \theta_{1s}-K_P\beta_{1s} \end{pmatrix}$$

$$+ \begin{bmatrix} T_\lambda^0 \\ R_\lambda^{1c}+H_\lambda^{1s} \\ R_\lambda^{1s}-H_\lambda^{1c} \end{bmatrix} \lambda \tag{16.439}$$

The superscripts denote the harmonics of the Fourier series expansions of the aerodynamic coefficients for forward flight. With more than three blades there are additional degrees of freedom and equations, but the coupled dynamics of the fixed frame and rotor are dominated by these collective and cyclic modes. The forces due to the lag motion have also been neglected. For hover only the mean terms remain in the matrices, and these equations reduce to the previous results. Perhaps the most important effect of forward flight is the coupling of the vertical and the lateral-longitudinal dynamics.

16.8.7 Shaft Motion

The linear and angular shaft motions were defined in Figure 16.13. The perturbation linear velocity of the hub has components $\dot{x}_h$, $\dot{y}_h$, and $\dot{z}_h$, whereas the orientation of the shaft relative to the inertial reference frame is given by the perturbation angles α_x, α_y, and α_z. We also consider aerodynamic turbulence with velocity components u_G, v_G, and w_G (normalized by the rotor tip speed ΩR). Including the shaft motion and gust, the perturbation velocities of the rotor blade section become

$$\delta u_P = (\lambda + \dot{z}_h - w_G - \mu\alpha_y) \\ + r(\dot{\beta} + \dot{\alpha}_x \sin\psi - \dot{\alpha}_y \cos\psi) + \mu\cos\psi\,\beta \tag{16.440}$$

$$\delta u_T = -r(\dot\zeta - \dot\alpha_z) - \mu \cos\psi(\zeta - \alpha_z)$$
$$- (\dot x_h - u_G - \lambda_{HP}\alpha_y)\sin\psi + (\dot y_h + v_G + \lambda_{HP}\alpha_x)\cos\psi \quad (16.441)$$
$$\delta u_R = \mu \sin\psi(\zeta - \alpha_z)$$
$$- (\dot x_h - u_G - \lambda_{HP}\alpha_y)\cos\psi - (\dot y_h + v_G + \lambda_{HP}\alpha_x)\sin\psi \quad (16.442)$$

The vertical velocity of the hub contributes to δu_P, and the in-plane velocity resolved in the rotating frame contributes to δu_T and δu_R. The influence of the aerodynamic gust components is analogous to the hub velocities. The angular rates of pitch and roll of the rotor disk give a normal velocity of the blade section (δu_P), whereas the yaw motion of the hub produces velocity perturbations in a manner similar to the blade lag motion. Finally, the trim velocity of the rotor (with components μ and λ_{HP}) is defined relative to the unperturbed inertial reference frame. Pitch and roll rotations of the shaft (α_y and α_x) therefore produce perturbation components of these velocities relative to the hub plane. Since the resulting $\lambda_{HP}\alpha_x$ and $\lambda_{HP}\alpha_y$ terms in the velocity perturbations are an order smaller than the other terms, they can usually be neglected for low inflow helicopter rotors. The blade pitch is measured from the hub plane, so $\delta\theta = \theta - K_P\beta$ still. Only first mode flap and lag motion has been considered for the rotor blade. Since the equivalent mode shape for the angular motion of the hub is exactly $\eta = r$, the blade mode shapes have been approximated by $\eta_\beta = \eta_\zeta = r$. Then the same aerodynamic coefficients can be used in many cases for both the blade and shaft motion, simplifying the analysis. For numerical work the actual blade mode shapes can be used, which modifies the aerodynamic coefficients of the rotor degrees of freedom slightly, but does not greatly influence the basic behavior of the rotor.

With these velocity perturbations, the aerodynamic flap and lag moments now become

$$M_F = M_\theta(\theta - K_P\beta) + M_\lambda(\lambda + \dot z_h - w_G - \mu\alpha_y)$$
$$+ M_{\dot\beta}(\dot\beta + \dot\alpha_x\sin\psi - \dot\alpha_y\cos\psi) + M_\beta\beta + M_{\dot\zeta}(\dot\zeta - \dot\alpha_z) + M_\zeta(\zeta - \alpha_z)$$
$$+ M_\mu((-\dot x_h + u_G)\sin\psi + (\dot y_h + v_G)\cos\psi) \quad (16.443)$$
$$M_L = Q_\theta(\theta - K_P\beta) + Q_\lambda(\lambda + \dot z_h - w_G - \mu\alpha_y)$$
$$+ Q_{\dot\beta}(\dot\beta + \dot\alpha_x\sin\psi - \dot\alpha_y\cos\psi) + Q_\beta\beta + Q_{\dot\zeta}(\dot\zeta - \dot\alpha_z) + Q_\zeta(\zeta - \alpha_z)$$
$$+ Q_\mu((-\dot x_h + u_G)\sin\psi + (\dot y_h + v_G)\cos\psi) \quad (16.444)$$

Since the velocity produced by the shaft motion is similar to that produced by the blade motion already considered, only two new aerodynamic coefficients appear:

$$M_\mu = \int_0^1 r\frac{1}{2}(u_P + 2u_T\alpha)dr \quad (16.445)$$

$$Q_\mu = \int_0^1 r\left(\frac{1}{2}u_P\theta + 2u_T\frac{c_d}{2a}\right)dr \quad (16.446)$$

16.8 Aerodynamic Loads

which are the flap and lag moments due to in-plane velocity of the blade. For hover these coefficients are

$$M_\mu = \frac{2C_T}{\sigma a} + \frac{\lambda_{HP}}{4} \tag{16.447}$$

$$Q_\mu = \lambda_{HP}\left(\frac{3}{2}\frac{C_T}{\sigma a} + \frac{3}{8}\lambda_{HP} - \frac{\theta_{tw}}{48}\right) + \frac{c_d}{3a} \tag{16.448}$$

On transforming to the non-rotating frame, the flap moments in hover become

$$M_{F_0} = M_\theta(\theta_0 - K_P\beta_0) + M_\lambda(\lambda + \dot{z}_h - w_G) + M_{\dot{\beta}}\dot{\beta}_0 - M_{\dot{\zeta}}\dot{\alpha}_z \tag{16.449}$$

$$M_{F_{1c}} = M_\theta(\theta_{1c} - K_P\beta_{1c}) + M_{\dot{\beta}}(\dot{\beta}_{1c} + \beta_{1s} - \dot{\alpha}_y) + M_\mu(\dot{y}_h + v_G) \tag{16.450}$$

$$M_{F_{1s}} = M_\theta(\theta_{1s} - K_P\beta_{1s}) + M_{\dot{\beta}}(\dot{\beta}_{1s} - \beta_{1c} + \dot{\alpha}_x) + M_\mu(-\dot{x}_h + u_G) \tag{16.451}$$

for a rotor with three or more blades. If $N > 4$ there are additional degrees of freedom and equations, but in hover they are not influenced by the shaft motion. The characteristic pattern of limited interaction between the shaft motion and non-rotating degrees of freedom, already found in the inertial terms, is also observed in the hover aerodynamics. There are coning moments due to the vertical velocity and yaw rate of the hub and due to the vertical gusts. There is a longitudinal flap moment due to the lateral in-plane velocity and pitch rate of the hub and due to the lateral gusts. Finally there are lateral flap moments due to the longitudinal in-plane velocity and roll rate of the hub and due to longitudinal gusts. For a two-bladed rotor the aerodynamic flap moment for the teetering mode is instead

$$\begin{aligned} M_{F1} &= M_\theta(\theta_1 - K_P\beta_1) + M_{\dot{\beta}}(\dot{\beta}_1 + \dot{\alpha}_x \sin\psi - \dot{\alpha}_y \cos\psi) \\ &\quad + M_\mu\big((-\dot{x}_h + u_G)\sin\psi + (\dot{y}_h + v_G)\cos\psi\big) \end{aligned} \tag{16.452}$$

Thus there are periodic coefficients coupling the rotor and shaft motion, even in hover.

The vertical and in-plane aerodynamic shear forces at the blade root due to the shaft motion are

$$\begin{aligned} \Delta\int_0^1 \frac{F_z}{ac}dr &= T_\lambda(\dot{z}_h - w_G - \mu\alpha_y) + T_{\dot{\beta}}(\dot{\alpha}_x \sin\psi - \dot{\alpha}_y \cos\psi) - T_{\dot{\zeta}}\dot{\alpha}_z - T_\zeta\alpha_z \\ &\quad + T_\mu\big((-\dot{x}_h + u_G)\sin\psi + (\dot{y}_h + v_G)\cos\psi\big) \end{aligned} \tag{16.453}$$

$$\begin{aligned} \Delta\int_0^1 \frac{F_x}{ac}dr &= H_\lambda(\dot{z}_h - w_G - \mu\alpha_y) + H_{\dot{\beta}}(\dot{\alpha}_x \sin\psi - \dot{\alpha}_y \cos\psi) - H_{\dot{\zeta}}\dot{\alpha}_z - H_\zeta\alpha_z \\ &\quad + H_\mu\big((-\dot{x}_h + u_G)\sin\psi + (\dot{y}_h + v_G)\cos\psi\big) \end{aligned} \tag{16.454}$$

As for the flap and lag moments, there are only two new aerodynamic coefficients, which are due to the in-plane velocity perturbations:

$$T_\mu = \int_0^1 \frac{1}{2}(u_P + 2u_T\alpha)dr \tag{16.455}$$

$$H_\mu = \int_0^1 \left(\frac{1}{2}u_P\theta + 2u_T\frac{c_d}{2a}\right)dr \tag{16.456}$$

and which for hover are

$$T_\mu = \frac{3C_T}{\sigma a} + \frac{\lambda_{HP}}{4} - \frac{\theta_{tw}}{24} \qquad (16.457)$$

$$H_\mu = \lambda_{HP}\left(\frac{3C_T}{\sigma a} + \frac{3}{4}\lambda_{HP} - \frac{\theta_{tw}}{8}\right) + \frac{c_d}{2a} \qquad (16.458)$$

The radial aerodynamic force due to the shaft motion is

$$\Delta \int_0^1 \frac{F_r}{ac} dr = R_\lambda(\dot{z}_h - w_G - \mu\alpha_y) + R_{\dot{\beta}}(\dot{\alpha}_x \sin\psi - \dot{\alpha}_y \cos\psi) - R_{\dot{\zeta}}\dot{\alpha}_z - R_\zeta \alpha_z$$
$$+ R_\mu\big((-\dot{x}_h + u_G)\cos\psi - (\dot{y}_h + v_G)\sin\psi\big)$$
$$+ R_r\big((-\dot{x}_h + u_G)\sin\psi + (\dot{y}_h + v_G)\cos\psi\big) \qquad (16.459)$$

The two new aerodynamic coefficients in the radial force are due to the in-plane velocity perturbations resolved in the radial direction (R_μ) and in the chordwise direction (R_r):

$$R_\mu = \int_0^1 \frac{c_d}{2a} u_T dr = \frac{c_d}{2a}\left(\frac{1}{2} + \mu\sin\psi\right) \qquad (16.460)$$

$$R_r = \int_0^1 \left[\frac{c_d}{2a} u_R - z'\frac{1}{2}(u_P + 2u_T\alpha)\right] dr = \frac{c_d}{2a}\mu\cos\psi - z'T_\mu \qquad (16.461)$$

In hover R_μ is the single contribution of the radial drag force to the rotor dynamics. All the other radial forces are due to the tilt of the thrust vector by the blade flap deflection. The torque moment at the hub center of rotation is here identical to the lag moment (since $\eta_\zeta = r$ has been assumed), giving

$$\Delta \int_0^1 r\frac{F_x}{ac} dr = Q_\lambda(\dot{z}_h - w_G - \mu\alpha_y) + Q_{\dot{\beta}}(\dot{\alpha}_x\sin\psi - \dot{\alpha}_y\cos\psi) - Q_{\dot{\zeta}}\dot{\alpha}_z - Q_\zeta\alpha_z$$
$$+ Q_\mu\big((-\dot{x}_h + u_G)\sin\psi + (\dot{y}_h + v_G)\cos\psi\big) \qquad (16.462)$$

Summing the root forces for all N blades gives the total rotor hub reactions in the non-rotating frame. For hover, the thrust and torque perturbations including the shaft motion are

$$\frac{C_T}{\sigma a} = T_\theta(\theta_0 - K_P\beta_0) + T_\lambda(\lambda + \dot{z}_h - w_G) + T_{\dot{\beta}}\dot{\beta}_0 + T_{\dot{\zeta}}(\dot{\zeta}_0 - \dot{\alpha}_z) \qquad (16.463)$$

$$\frac{C_Q}{\sigma a} = Q_\theta(\theta_0 - K_P\beta_0) + Q_\lambda(\lambda + \dot{z}_h - w_G) + Q_{\dot{\beta}}\dot{\beta}_0 + Q_{\dot{\zeta}}(\dot{\zeta}_0 - \dot{\alpha}_z) \qquad (16.464)$$

The rotor drag and side forces due to the shaft motion are

$$\Delta\begin{pmatrix}\dfrac{2C_H}{\sigma a} \\ \dfrac{2C_Y}{\sigma a}\end{pmatrix} = \begin{bmatrix} -R_{\dot\beta} & H_{\dot\beta} \\ H_{\dot\beta} & R_{\dot\beta} \end{bmatrix}\begin{pmatrix}\dot{\alpha}_y \\ \dot{\alpha}_x\end{pmatrix}$$
$$+ \begin{bmatrix} -(H_\mu + R_\mu) & R_r \\ -R_r & -(H_\mu + R_\mu) \end{bmatrix}\begin{pmatrix}\dot{x}_h - u_G \\ \dot{y}_h + v_G\end{pmatrix} \qquad (16.465)$$

16.8 Aerodynamic Loads

(for three or more blades). The flap response to the rotor shaft motion tilts the rotor thrust vector and by this means also contributes to the hub in-plane forces.

With a two-bladed rotor, the summation of the root shears over both blades to obtain the hub in-plane forces does not eliminate the sinusoidal variation of the coefficients. The contributions of the shaft motion to the hub forces are for this case

$$\Delta \begin{pmatrix} \frac{2C_H}{\sigma a} \\ \frac{2C_Y}{\sigma a} \end{pmatrix} = \begin{bmatrix} -R_{\dot\beta}2C^2 - H_{\dot\beta}2CS & R_{\dot\beta}2CS + H_{\dot\beta}2S^2 \\ -R_{\dot\beta}2CS + H_{\dot\beta}2C^2 & R_{\dot\beta}2S^2 - H_{\dot\beta}2CS \end{bmatrix} \begin{pmatrix} \dot\alpha_y \\ \dot\alpha_x \end{pmatrix}$$
$$+ \begin{bmatrix} -H_\mu 2S^2 - R_\mu 2C^2 - R_r 2CS & H_\mu 2CS - R_\mu 2CS + R_r 2C^2 \\ -H_\mu 2CS - R_\mu 2CS - R_r 2S^2 & -H_\mu 2C^2 - R_\mu 2S^2 + R_r 2CS \end{bmatrix} \begin{pmatrix} \dot x_h - u_G \\ \dot y_h + v_G \end{pmatrix}$$
(16.466)

where $C = \cos\psi$ and $S = \sin\psi$.

For aircraft flight dynamics analyses, a body-axis reference frame is most frequently used. With the inertial axis system considered so far, angular motion of the shaft tilts the axes relative to the trim velocity components μ and λ_{HP}, which are fixed in space, producing perturbations of the air velocity as seen in the reference frame. With body axes, however, the helicopter trim velocity vector remains fixed relative to the reference axes when the shaft is tilted. Thus for body axes the velocity perturbations are

$$\delta u_P = (\lambda + \dot z_h - w_G) + r(\dot\beta + \dot\alpha_x \sin\psi - \dot\alpha_y \cos\psi) + \mu \cos\psi \beta \quad (16.467)$$
$$\delta u_T = -r(\dot\zeta - \dot\alpha_z) - \mu\cos\psi\zeta - (\dot x_h - u_G)\sin\psi + (\dot y_h + v_G)\cos\psi \quad (16.468)$$
$$\delta u_R = \mu\sin\psi\zeta - (\dot x_h - u_G)\cos\psi - (\dot y_h + v_G)\sin\psi \quad (16.469)$$

So the $\mu\alpha_y$, $\mu\alpha_z$, $\lambda_{HP}\alpha_y$, and $\lambda_{HP}\alpha_x$ terms are dropped from the rotor equations of motion and the hub reactions. The use of body axes adds corresponding terms to the inertia forces.

The shaft motion contributions to the non-rotating equations of motion in forward flight can be derived following section 16.8.4. The constant coefficient approximation for the aerodynamic flap moments in forward flight is

$$\Delta \begin{pmatrix} M_{F_0} \\ M_{F_{1c}} \\ M_{F_{1s}} \end{pmatrix} = \begin{bmatrix} 0 & \frac{1}{2}M_{\dot\beta}^{1s} & M_\zeta^0 \\ M_{\dot\beta}^0 + \frac{1}{2}M_{\dot\beta}^{2c} & 0 & M_\zeta^{1c} \\ 0 & M_{\dot\beta}^0 - \frac{1}{2}M_{\dot\beta}^{2c} & M_\zeta^{1s} \end{bmatrix} \begin{pmatrix} -\dot\alpha_y \\ \dot\alpha_x \\ -\dot\alpha_z \end{pmatrix}$$
$$+ \begin{bmatrix} M_\lambda^0 & \frac{1}{2}M_\mu^{1s} & \frac{1}{2}M_\mu^{1c} \\ 0 & \frac{1}{2}M_\mu^{2s} & M_\mu^0 + \frac{1}{2}M_\mu^{2c} \\ M_\lambda^{1s} & M_\mu^0 - \frac{1}{2}M_\mu^{2c} & \frac{1}{2}M_\mu^{2s} \end{bmatrix} \begin{pmatrix} \dot z_h - w_G \\ -\dot x_h + u_G \\ \dot y_h + v_G \end{pmatrix} \quad (16.470)$$

This result is for body axes, since the problem considered in this text involving the shaft motion is that of the helicopter flight dynamics. Similarly, the constant coefficient approximation for the hub forces is

$$\Delta \begin{pmatrix} \dfrac{C_T}{\sigma a} \\ \dfrac{2C_H}{\sigma a} \\ \dfrac{2C_Y}{\sigma a} \end{pmatrix}$$

$$= \begin{bmatrix} \tfrac{1}{2}T^{1c}_{\dot\beta} & \tfrac{1}{2}T^{1s}_{\dot\beta} & T^0_{\dot\zeta} \\ R^0_{\dot\beta} + \tfrac{1}{2}\left(R^{2c}_{\dot\beta} + H^{2s}_{\dot\beta}\right) & H^0_{\dot\beta} + \tfrac{1}{2}\left(R^{2s}_{\dot\beta} - H^{2c}_{\dot\beta}\right) & R^{1c}_{\dot\zeta} + H^{1s}_{\dot\zeta} \\ -H^0_{\dot\beta} + \tfrac{1}{2}\left(R^{2s}_{\dot\beta} - H^{2c}_{\dot\beta}\right) & R^0_{\dot\beta} - \tfrac{1}{2}\left(R^{2c}_{\dot\beta} + H^{2s}_{\dot\beta}\right) & R^{1s}_{\dot\zeta} - H^{1c}_{\dot\zeta} \end{bmatrix} \begin{pmatrix} -\dot\alpha_y \\ \dot\alpha_x \\ -\dot\alpha_z \end{pmatrix}$$

$$+ \begin{bmatrix} T^0_\lambda & \tfrac{1}{2}T^{1s}_\mu & \tfrac{1}{2}T^{1c}_\mu \\ R^{1c}_\lambda + H^{1s}_\lambda & R^0_\mu + H^0_\mu + \tfrac{1}{2}\left(R^{2c}_\mu - H^{2c}_\mu + R^{2s}_r\right) & R^0_r - \tfrac{1}{2}\left(R^{2s}_\mu - H^{2s}_\mu - R^{2c}_r\right) \\ R^{1s}_\lambda - H^{1c}_\lambda & R^0_r + \tfrac{1}{2}\left(R^{2s}_\mu - H^{2s}_\mu - R^{2c}_r\right) & -(R^0_\mu + H^0_\mu) + \tfrac{1}{2}\left(R^{2c}_\mu - H^{2c}_\mu + R^{2s}_r\right) \end{bmatrix}$$

$$\times \begin{pmatrix} \dot z_h - w_G \\ -\dot x_h + u_G \\ \dot y_h + v_G \end{pmatrix} \qquad (16.471)$$

Again observe that forward flight fully couples the dynamics of the rotor and shaft motion.

16.8.8 Summary

Let us summarize the results derived for the hover aerodynamics, including the hub reactions and shaft motion. For simplicity, the cyclic lag degrees of freedom are omitted, and the special case of a two-bladed rotor is not considered. The axisymmetry of the aerodynamics in vertical flight separates the dynamics into a vertical group, consisting of the coning moment and the rotor thrust and torque:

$$M_{F_0} = M_\theta(\theta_0 - K_P\beta_0) + M_\lambda(\lambda + \dot z_h - w_G) + M_{\dot\beta}\dot\beta_0 + M_{\dot\zeta}(\dot\zeta_0 - \dot\alpha_z) \qquad (16.472)$$

$$\left(\frac{C_T}{\sigma a}\right)_{aero} = T_\theta(\theta_0 - K_P\beta_0) + T_\lambda(\lambda + \dot z_h - w_G) + T_{\dot\beta}\dot\beta_0 + T_{\dot\zeta}(\dot\zeta_0 - \dot\alpha_z) \qquad (16.473)$$

$$\left(\frac{C_Q}{\sigma a}\right)_{aero} = Q_\theta(\theta_0 - K_P\beta_0) + Q_\lambda(\lambda + \dot z_h - w_G) + Q_{\dot\beta}\dot\beta_0 + Q_{\dot\zeta}(\dot\zeta_0 - \dot\alpha_z) \qquad (16.474)$$

and a lateral-longitudinal group, consisting of the pitch and roll flap moments and the rotor in-plane hub forces:

$$\begin{pmatrix} M_{F_{1c}} \\ M_{F_{1s}} \end{pmatrix} = M_\theta \begin{pmatrix} \theta_{1c} - K_P\beta_{1c} \\ \theta_{1s} - K_P\beta_{1s} \end{pmatrix} + M_{\dot\beta}\begin{pmatrix} \dot\beta_{1c} + \beta_{1s} - \dot\alpha_y \\ \dot\beta_{1s} - \beta_{1c} + \dot\alpha_x \end{pmatrix}$$

$$+ M_\mu \begin{pmatrix} \dot y_h + v_G \\ -\dot x_h + u_G \end{pmatrix} \qquad (16.475)$$

$$\begin{pmatrix} \dfrac{2C_H}{\sigma a} \\ \dfrac{2C_Y}{\sigma a} \end{pmatrix}_{aero} = R_\beta \begin{pmatrix} \beta_{1c} \\ \beta_{1s} \end{pmatrix} + \begin{bmatrix} R_{\dot\beta} & H_{\dot\beta} \\ -H_{\dot\beta} & R_{\dot\beta} \end{bmatrix} \begin{pmatrix} \dot\beta_{1c} + \beta_{1s} - \dot\alpha_y \\ \dot\beta_{1s} - \beta_{1c} + \dot\alpha_x \end{pmatrix}$$

16.8 Aerodynamic Loads

$$+ \begin{bmatrix} R_\theta & H_\theta \\ -H_\theta & R_\theta \end{bmatrix} \begin{pmatrix} \theta_{1c} - K_P\beta_{1c} \\ \theta_{1s} - K_P\beta_{1s} \end{pmatrix}$$

$$+ \begin{bmatrix} -(H_\mu + R_\mu) & R_r \\ -R_r & -(H_\mu + R_\mu) \end{bmatrix} \begin{pmatrix} \dot{x}_h - u_G \\ \dot{y}_h + v_G \end{pmatrix} \quad (16.476)$$

The aerodynamic coefficients for hover can be evaluated analytically assuming uniform induced velocity, $\eta_\beta = \eta_\zeta = r$ and neglecting the tip losses:

$$M_\theta = -M_{\dot\beta} = \frac{1}{8} \quad (16.477)$$

$$M_\lambda = -\frac{1}{6} \quad (16.478)$$

$$M_\mu = \frac{2C_T}{\sigma a} + \frac{\lambda_{HP}}{4} \quad (16.479)$$

$$M_{\dot\zeta} = -\left(\frac{3}{2}\frac{C_T}{\sigma a} + \frac{5}{24}\lambda_{HP} + \frac{1}{80}\theta_{tw}\right) \quad (16.480)$$

$$T_\theta = -T_{\dot\beta} = \frac{1}{6} \quad (16.481)$$

$$T_\lambda = -\frac{1}{4} \quad (16.482)$$

$$T_{\dot\zeta} = -\left(\frac{2C_T}{\sigma a} + \frac{\lambda_{HP}}{4}\right) \quad (16.483)$$

$$Q_\theta = \frac{\lambda_{HP}}{6} \quad (16.484)$$

$$Q_\lambda = \frac{C_T}{\sigma a} - \frac{\lambda_{HP}}{4} \quad (16.485)$$

$$Q_{\dot\beta} = \frac{3}{4}\frac{C_T}{\sigma a} - \frac{7}{48}\lambda_{HP} + \frac{1}{160}\theta_{tw} \quad (16.486)$$

$$Q_{\dot\zeta} = -\left(\lambda_{HP}\left(\frac{C_T}{\sigma a} + \frac{\lambda_{HP}}{4}\right) + \frac{c_d}{4a}\right) \quad (16.487)$$

$$H_\theta = \frac{\lambda_{HP}}{4} \quad (16.488)$$

$$H_{\dot\beta} = \frac{C_T}{\sigma a} - \frac{\lambda_{HP}}{4} \quad (16.489)$$

$$H_\mu + R_\mu = \lambda_{HP}\left(\frac{3C_T}{\sigma a} + \frac{3}{4}\lambda_{HP} - \frac{\theta_{tw}}{8}\right) + \frac{3c_d}{4a} \quad (16.490)$$

$$R_\theta = -R_{\dot\beta} = -\frac{\beta_{trim}}{6} \quad (16.491)$$

$$R_r = -\beta_{trim}\left(\frac{3C_T}{\sigma a} + \frac{\lambda_{HP}}{4} - \frac{\theta_{tw}}{24}\right) \quad (16.492)$$

$$R_\beta = -\frac{C_T}{\sigma a} \quad (16.493)$$

The behavior of a particular aerodynamic coefficient depends primarily on whether it is an out-of-plane or in-plane force and whether it is due to the blade pitch or an out-of-plane or in-plane velocity. Hence a set of six aerodynamic coefficients is sufficient to establish the basic behavior of the forces; for example,

$$M_\theta = \int_0^1 r F_{z\theta} dr = \int_0^1 \frac{1}{2} r u_T^2 dr \tag{16.494}$$

$$M_{\dot\beta} = \int_0^1 r^2 F_{zP} dr = -\int_0^1 \frac{1}{2} r^2 u_T dr \tag{16.495}$$

$$M_\mu = \int_0^1 r F_{zT} dr = \int_0^1 r \frac{1}{2}(u_P + 2u_T\alpha) dr \tag{16.496}$$

$$H_\theta = \int_0^1 F_{x\theta} dr = \int_0^1 \frac{1}{2} u_T u_P dr \tag{16.497}$$

$$H_{\dot\beta} = \int_0^1 r F_{xP} dr = \int_0^1 \frac{1}{2} r (u_T\alpha - u_P) dr \tag{16.498}$$

$$H_\mu = \int_0^1 F_{xT} dr = \int_0^1 \left(\frac{1}{2} u_P\theta + 2u_T \frac{c_d}{2a}\right) dr \tag{16.499}$$

The remaining coefficients are all similar to ones from this set, as can be seen in the expressions for hover given earlier.

Evaluating the aerodynamic coefficients in forward flight is a more involved task than evaluating them in hover. The trim pitch and velocities are then periodic functions of the rotor azimuth:

$$u_T = r + \mu \sin\psi \tag{16.500}$$

$$u_R = \mu \cos\psi \tag{16.501}$$

$$\begin{aligned} u_P &= \lambda_{\text{HP}} + r\dot\beta + \beta\mu\cos\psi \\ &= \lambda_{\text{HP}} + r(\beta_{1s}\cos\psi - \beta_{1c}\sin\psi) \\ &\quad + (\beta_0 + \beta_{1c}\cos\psi + \beta_{1s}\sin\psi)\mu\cos\psi \end{aligned} \tag{16.502}$$

$$\theta = \theta_0 + r\theta_{tw} + \theta_{1c}\cos\psi + \theta_{1s}\sin\psi \tag{16.503}$$

(see Chapter 6). A complete trim solution is thus required, not just a specification of the rotor thrust coefficient. Helicopter force and moment equilibrium gives the tip-path-plane tilt, and the hub-plane inflow ratio $\lambda_{\text{HP}} = \lambda_{\text{TPP}} - \mu\beta_{1c}$. The thrust coefficient and flapping equations can then be solved for the collective and cyclic pitch control and the coning angle. As for hover, the coefficients can be integrated analytically over the span assuming uniform inflow and $\eta_\beta = r$. It is simplest to leave the expressions in terms of the harmonics of the trim pitch and flap motion, rather than trying to obtain the explicit dependence on the parameters of the operating condition (such as thrust coefficient) as for hover. The results for the basic set of six aerodynamic coefficients are as follows:

$$M_\theta = \frac{1}{8} + \frac{\mu}{3}\sin\psi + \frac{\mu^2}{4}\sin^2\psi \tag{16.504}$$

$$M_{\dot\beta} = -\left(\frac{1}{8} + \frac{\mu}{6}\sin\psi\right) \tag{16.505}$$

16.8 Aerodynamic Loads

$$M_\mu = \theta \left(\frac{1}{3} + \frac{\mu}{2}\sin\psi\right) + \theta_{tw}\left(\frac{1}{4} + \frac{\mu}{3}\sin\psi\right)$$
$$- \left(\frac{1}{4}\lambda_{HP} + \frac{1}{6}\dot{\beta} + \frac{1}{4}\beta u_R\right) \quad (16.506)$$

$$H_\theta = \lambda_{HP}\left(\frac{1}{4} + \frac{\mu}{2}\sin\psi\right) + \dot{\beta}\left(\frac{1}{6} + \frac{\mu}{4}\sin\psi\right)$$
$$+ \beta u_R\left(\frac{1}{4} + \frac{\mu}{2}\sin\psi\right) \quad (16.507)$$

$$H_{\dot\beta} = \theta\left(\frac{1}{6} + \frac{\mu}{4}\sin\psi\right) + \theta_{tw}\left(\frac{1}{8} + \frac{\mu}{6}\sin\psi\right)$$
$$- \left(\frac{1}{2}\lambda_{HP} + \frac{1}{3}\dot{\beta} + \frac{1}{2}\beta u_R\right) \quad (16.508)$$

$$H_\mu = \theta\left(\frac{1}{2}\lambda_{HP} + \frac{1}{4}\dot{\beta} + \frac{1}{2}\beta u_R\right) + \theta_{tw}\left(\frac{1}{4}\lambda_{HP} + \frac{1}{6}\dot{\beta} + \frac{1}{4}\beta u_R\right)$$
$$+ \frac{c_d}{2a}\left(1 + 2\mu\sin\psi\right) \quad (16.509)$$

In forward flight all the aerodynamic coefficients are periodic functions of the rotor azimuth.

A uniform perturbation of the wake-induced velocity has been included in the aerodynamic analysis for use in a model of the rotor unsteady aerodynamics. Dynamic inflow (see section 11.3) relates the uniform and linear inflow perturbations to the transient changes in the aerodynamic thrust and hub moments on the rotor:

$$LM\begin{pmatrix}\dot\lambda_0\\\dot\lambda_c\\\dot\lambda_s\end{pmatrix} + \begin{pmatrix}\lambda_0\\\lambda_c\\\lambda_s\end{pmatrix} = L\begin{pmatrix}C_T\\-C_{My}\\C_{Mx}\end{pmatrix}_{aero} \quad (16.510)$$

Rotor velocity changes also produce a perturbation of the inflow:

$$\delta\lambda = -\frac{\lambda_i(\mu\delta\mu + \lambda\delta\mu_z)}{\mu^2 + \lambda(\lambda + \lambda_i)} \quad (16.511)$$

(equation 11.156), where here $\delta\mu = -\dot{x}_h + u_G$ and $\delta\mu_z = \dot{z}_h - w_G$. For hover this relation reduces to $\delta\lambda = -\frac{1}{2}\delta\mu_z = -\frac{1}{2}(\dot{z}_h - w_G)$. Hence the coning moment, thrust, and torque due to the rotor vertical velocity perturbations ($\dot{z}_h - w_G$) in hover are reduced by a factor of one-half by the effect of this inflow perturbation: $\lambda + \dot{z}_h - w_G = \frac{1}{2}(\dot{z}_h - w_G)$. Without the time lag, these inflow equations reduce to linear algebraic equations for the induced velocity perturbations in terms of the system degrees of freedom. Eliminating $\lambda_0, \lambda_c, \lambda_s$ from the model leads to a lift deficiency function representation of the wake effects, as shown in section 11.3. With large order systems it is more practical to accomplish this substitution numerically, and if the time lag is included the inflow perturbations are degrees of freedom. The effects of the wake, as represented by the dynamic inflow model, are often important to the dynamic behavior of the rotor.

16.8.9 Large Angles and High Inflow

General results for the aerodynamic loads are obtained using the exact expressions for the force perturbations (equations 16.290, 16.291, and 16.292), which are valid for large angles. Still assuming that the blade flap motion is pure out-of-plane and the blade lag motion is pure in-plane, and approximating the mode shapes by $\eta_\beta = \eta_\zeta = r$, the perturbation velocities of the blade section are given by equations 16.440, 16.441, and 16.442. The aerodynamic flap and lag moments are

$$M_F = M_\theta(\theta - K_P\beta) + M_\lambda(\lambda + \dot{z}_h - w_G - \mu\alpha_y)$$
$$+ M_{\dot\beta}(\dot\beta + \dot\alpha_x \sin\psi - \dot\alpha_y \cos\psi) + M_\beta\beta + M_{\dot\zeta}(\dot\zeta - \dot\alpha_z) + M_\zeta(\zeta - \alpha_z)$$
$$+ M_\mu\big((-\dot{x}_h + u_G + \lambda_{HP}\alpha_y)\sin\psi + (\dot{y}_h + v_G + \lambda_{HP}\alpha_x)\cos\psi\big) \quad (16.512)$$

$$M_L = Q_\theta(\theta - K_P\beta) + Q_\lambda(\lambda + \dot{z}_h - w_G - \mu\alpha_y)$$
$$+ Q_{\dot\beta}(\dot\beta + \dot\alpha_x \sin\psi - \dot\alpha_y \cos\psi) + Q_\beta\beta + Q_{\dot\zeta}(\dot\zeta - \dot\alpha_z) + Q_\zeta(\zeta - \alpha_z)$$
$$+ Q_\mu\big((-\dot{x}_h + u_G + \lambda_{HP}\alpha_y)\sin\psi + (\dot{y}_h + v_G + \lambda_{HP}\alpha_x)\cos\psi\big) \quad (16.513)$$

and the hub forces are

$$\int_0^1 \frac{F_z}{ac}dr = T_\theta(\theta - K_P\beta) + T_\lambda(\lambda + \dot{z}_h - w_G - \mu\alpha_y)$$
$$+ T_{\dot\beta}(\dot\beta + \dot\alpha_x \sin\psi - \dot\alpha_y \cos\psi) + T_\beta\beta + T_{\dot\zeta}(\dot\zeta - \dot\alpha_z) + T_\zeta(\zeta - \alpha_z)$$
$$+ T_\mu\big((-\dot{x}_h + u_G + \lambda_{HP}\alpha_y)\sin\psi + (\dot{y}_h + v_G + \lambda_{HP}\alpha_x)\cos\psi\big) \quad (16.514)$$

$$\int_0^1 \frac{F_x}{ac}dr = H_\theta(\theta - K_P\beta) + H_\lambda(\lambda + \dot{z}_h - w_G - \mu\alpha_y)$$
$$+ H_{\dot\beta}(\dot\beta + \dot\alpha_x \sin\psi - \dot\alpha_y \cos\psi) + H_\beta\beta + H_{\dot\zeta}(\dot\zeta - \dot\alpha_z) + H_\zeta(\zeta - \alpha_z)$$
$$+ H_\mu\big((-\dot{x}_h + u_G + \lambda_{HP}\alpha_y)\sin\psi + (\dot{y}_h + v_G + \lambda_{HP}\alpha_x)\cos\psi\big) \quad (16.515)$$

$$\int_0^1 \frac{F_r}{ac}dr = R_\theta(\theta - K_P\beta) + R_\lambda(\lambda + \dot{z}_h - w_G - \mu\alpha_y)$$
$$+ R_{\dot\beta}(\dot\beta + \dot\alpha_x \sin\psi - \dot\alpha_y \cos\psi) + R_\beta\beta + R_{\dot\zeta}(\dot\zeta - \dot\alpha_z) + R_\zeta(\zeta - \alpha_z)$$
$$+ R_\mu\big((-\dot{x}_h + u_G + \lambda_{HP}\alpha_y)\cos\psi - (\dot{y}_h + v_G + \lambda_{HP}\alpha_x)\sin\psi\big)$$
$$+ R_r\big((-\dot{x}_h + u_G + \lambda_{HP}\alpha_y)\sin\psi + (\dot{y}_h + v_G + \lambda_{HP}\alpha_x)\cos\psi\big) \quad (16.516)$$

Without the small angle assumption, the coefficients are

$$M_\theta = \int_0^1 rF_{z\theta}\,dr \quad (16.517)$$

$$M_\lambda = \int_0^1 rF_{zP}\,dr \qquad M_{\dot\beta} = \int_0^1 r^2 F_{zP}\,dr \qquad M_\beta = \int_0^1 ru_R F_{zP}\,dr \quad (16.518)$$

$$M_\mu = \int_0^1 rF_{zT}\,dr \qquad M_{\dot\zeta} = -\int_0^1 r^2 F_{zT}\,dr \qquad M_\zeta = -\int_0^1 ru_R F_{zT}\,dr \quad (16.519)$$

$$Q_\theta = \int_0^1 rF_{xP}dr \tag{16.520}$$

$$Q_\lambda = \int_0^1 rF_{xP}dr \qquad Q_{\dot\beta} = \int_0^1 r^2 F_{xP}dr \qquad Q_\beta = \int_0^1 ru_R F_{xP}dr \tag{16.521}$$

$$Q_\mu = \int_0^1 rF_{xT}dr \qquad Q_{\dot\zeta} = -\int_0^1 r^2 F_{xT}dr \qquad Q_\zeta = -\int_0^1 ru_R F_{xT}dr \tag{16.522}$$

$$R_\theta = \int_0^1 F_{r\theta}dr \tag{16.523}$$

$$R_\lambda = \int_0^1 F_{rP}dr \qquad R_{\dot\beta} = \int_0^1 rF_{rP}dr \qquad R_\beta = \int_0^1 u_R F_{rP}dr \tag{16.524}$$

$$R_r = \int_0^1 F_{rT}dr \qquad R_{\dot\zeta} = -\int_0^1 rF_{rT}dr \tag{16.525}$$

$$R_\zeta = -\int_0^1 u_R F_{rT}dr + \int_0^1 (\mu \sin\psi)F_{rR}dr \qquad R_\mu = \int_0^1 F_{rR}dr \tag{16.526}$$

The T and H coefficients follow the pattern of the M and Q coefficients, but with one less factor of r in the integrand. The coefficients are evaluated numerically. Sometimes evaluating the integrands at an effective radius r_e or effective inflow angle $\phi_e = \tan^{-1} u_P/u_T$ is sufficient.

Analytical integration is possible for axial flow at large inflow: $u_T = r$, $u_R = 0$, and $u_P = \lambda \cong \mu_z$. These conditions are appropriate for whirl flutter of propellers or tiltrotors, for which the lift-curve slope terms ($c_{\ell\alpha} = a$) dominate. Then with $U = \sqrt{r^2 + \lambda^2}$ here,

$$M_\theta = \frac{1}{2}\int_0^1 Ur^2 dr = g_2 \tag{16.527}$$

$$M_\lambda = -\frac{1}{2}\int_0^1 \frac{r^3}{U}dr = -f_3 \qquad M_{\dot\beta} = -\frac{1}{2}\int_0^1 \frac{r^4}{U}dr = -f_4 \tag{16.528}$$

$$M_\mu = \frac{1}{2}\int_0^1 \frac{\lambda r^2}{U}dr = \lambda f_2 \qquad M_{\dot\zeta} = -\frac{1}{2}\int_0^1 \frac{\lambda r^3}{U}dr = -\lambda f_3 \tag{16.529}$$

$$T_\theta = \frac{1}{2}\int_0^1 Ur\,dr = g_1 \tag{16.530}$$

$$T_\lambda = -\frac{1}{2}\int_0^1 \frac{r^2}{U}dr = -f_2 \qquad T_{\dot\beta} = -\frac{1}{2}\int_0^1 \frac{r^3}{U}dr = -f_3 \tag{16.531}$$

$$T_\mu = \frac{1}{2}\int_0^1 \frac{\lambda r}{U}dr = \lambda f_1 \qquad T_{\dot\zeta} = -\frac{1}{2}\int_0^1 \frac{\lambda r^2}{U}dr = -\lambda f_2 \tag{16.532}$$

$$Q_\theta = \frac{1}{2}\int_0^1 U\lambda r\,dr = \lambda g_1 \tag{16.533}$$

$$Q_\lambda = -\frac{1}{2}\int_0^1 \frac{\lambda r^2}{U}dr = -\lambda f_2 \qquad Q_{\dot\beta} = -\frac{1}{2}\int_0^1 \frac{\lambda r^3}{U}dr = -\lambda f_3 \tag{16.534}$$

$$Q_\mu = \frac{1}{2}\int_0^1 \frac{\lambda^2 r}{U}dr = \lambda^2 f_1 \qquad Q_\zeta = -\frac{1}{2}\int_0^1 \frac{\lambda^2 r^2}{U}dr = -\lambda^2 f_2 \qquad (16.535)$$

$$H_\theta = \frac{1}{2}\int_0^1 U\lambda dr = \lambda g_0 \qquad (16.536)$$

$$H_\lambda = -\frac{1}{2}\int_0^1 \frac{\lambda r}{U}dr = -\lambda f_1 \qquad H_{\dot\beta} = -\frac{1}{2}\int_0^1 \frac{\lambda r^2}{U}dr = -\lambda f_2 \qquad (16.537)$$

$$H_\mu = \frac{1}{2}\int_0^1 \frac{\lambda^2}{U}dr = \lambda^2 f_0 \qquad H_{\dot\zeta} = -\frac{1}{2}\int_0^1 \frac{\lambda^2 r}{U}dr = -\lambda^2 f_1 \qquad (16.538)$$

where $g_n = \int_0^1 (Ur^n/2)dr$ and $f_n = \int_0^1 (r^n/2U)dr$,

$$f_0 = \frac{1}{2}\ln\frac{1+\sqrt{1+\lambda^2}}{\lambda} \qquad (16.539)$$

$$f_1 = \frac{1}{2}\left(\sqrt{1+\lambda^2}-V\right) \qquad (16.540)$$

$$f_2 = \frac{1}{4}\sqrt{1+\lambda^2}-\frac{1}{2}\lambda^2 f_0 \qquad (16.541)$$

$$f_3 = \frac{1}{6}\left(\sqrt{1+\lambda^2}(1-2\lambda^2)+2\lambda^3\right) \qquad (16.542)$$

$$f_4 = \frac{1}{16}\sqrt{1+\lambda^2}(2-3\lambda^2)+\frac{3}{8}\lambda^4 f_0 \qquad (16.543)$$

$$g_0 = \frac{1}{4}\sqrt{1+\lambda^2}+\frac{1}{2}\lambda^2 f_0 \qquad (16.544)$$

$$g_1 = \frac{1}{6}\left((\sqrt{1+\lambda^2})^3-\lambda^3\right) \qquad (16.545)$$

$$g_2 = \frac{1}{16}\sqrt{1+\lambda^2}(2+\lambda^2)-\frac{1}{8}\lambda^4 f_0 \qquad (16.546)$$

The R coefficients and the β and ζ derivatives are zero. With high inflow ($\lambda = \mu_z$ order 1), the μ and $\dot\zeta$ derivatives are not small; rather they are the same order as the λ and $\dot\beta$ derivatives, because now in-plane velocity perturbations directly change angle-of-attack and lift. The in-plane rate derivative $H_{\dot\beta} = C_T/\sigma a + \widehat{H}_{\dot\beta} = C_T/\sigma a - \lambda_{HP}/4$ for hover and approaches $H_{\dot\beta} = -\frac{1}{6}$ for large λ.

16.8.10 Pitch and Flap Motion

Consider the rigid flap and rigid pitch motion of a rotor blade, as in section 16.5.1. For the feathering moments on the blade, the effects of unsteady aerodynamics must be included. The aerodynamic forces required for the equations of motion are the flap moment M_F and the pitch moment about the feathering axis M_f:

$$M_F = \int_0^1 r\frac{F_z}{ac}dr \cong \int_0^1 r\frac{L}{ac}dr \qquad (16.547)$$

16.8 Aerodynamic Loads

$$M_f = \int_0^1 \frac{M_a}{ac} dr \tag{16.548}$$

The section lift and pitch moment are obtained from the unsteady aerodynamic theory developed for rotary wings in section 10.6:

$$\frac{L}{ac} = \frac{1}{2} u_T \left(w_{PA} + x_A w' - \lambda \right) C'(k)$$
$$+ \frac{c}{8} \left(2u_T w' C'(k) + \dot{w}_{PA} + x_A \dot{w}' + \frac{c}{4} \ddot{w}' \right) \tag{16.549}$$

$$\frac{M}{ac} = -x_A \frac{L}{ac} - \frac{c^2}{32} \left(u_T w' + \dot{w}_{PA} + x_A \dot{w}' + \frac{3c}{8} \ddot{w}' \right) \tag{16.550}$$

The lift deficiency function $C'(k)$ has been included to account for shed wake effects. The inflow perturbation λ is an alternative to the lift deficiency function. The effects of reverse flow have been neglected. Here c is the blade chord and x_A is the distance the aerodynamic center is behind the elastic axis. The upwash velocity at the pitch axis and the upwash gradient are

$$w_{PA} = u_T \theta - (r\dot{\beta} + u_R \beta) \tag{16.551}$$

$$w' = \theta + \beta \tag{16.552}$$

for rigid flap and rigid pitch degrees of freedom. The degree of freedom θ is the actual blade pitch angle, whereas in the preceding sections of this chapter θ has been the pitch control variable. Here the commanded pitch and kinematic pitch-flap coupling enter the solution through the pitch equation of motion (see section 16.5.1). The derivative required for the non-circulatory terms is

$$\dot{w}_{PA} = u_T \dot{\theta} - (r\ddot{\beta} + u_R \dot{\beta}) + u_R \theta + (\mu \sin \psi) \beta \tag{16.553}$$

which in hover is just $r(\dot{\theta} - \ddot{\beta})$ and hence a source of pitch damping. The virtual mass terms ($\ddot{\theta}$ and $\ddot{\beta}$) are neglected now. Substituting for the perturbation forces and velocities gives

$$M_F = M_\theta \theta + M_{\dot{\theta}} \dot{\theta} + M_\lambda \lambda + M_{\dot{\beta}} \dot{\beta} + M_\beta \beta \tag{16.554}$$

$$M_f = m_\theta \theta + m_{\dot{\theta}} \dot{\theta} + m_\lambda \lambda + m_{\dot{\beta}} \dot{\beta} + m_\beta \beta \tag{16.555}$$

where the aerodynamic coefficients are

$$M_\theta = \int_0^1 r \left[C'(k) \frac{1}{2} u_T^2 + \frac{c}{8} u_R \right] dr \tag{16.556}$$

$$M_{\dot{\theta}} = \int_0^1 r \left[u_T c \frac{1}{8} \left(1 + 2C'(k)(1 + 2\xi_A) \right) \right] dr \tag{16.557}$$

$$M_\lambda = -\int_0^1 r \left[C'(k) \frac{1}{2} u_T \right] dr \tag{16.558}$$

$$M_{\dot{\beta}} = -\int_0^1 r \left[C'(k) \frac{1}{2} r u_T + \frac{c}{8} u_R - \frac{c^2}{32} (1 + 4\xi_A) \right] dr \tag{16.559}$$

$$M_\beta = -\int_0^1 r \left[C'(k) \frac{1}{2} u_R u_T - \frac{c}{4} C'(k) u_T - \frac{c}{8} \mu \sin \psi \right] dr \tag{16.560}$$

$$m_\theta = -\int_0^1 \left[x_A C'(k) \frac{1}{2} u_T^2 + \frac{c^2}{32} u_R (1 + 4\xi_A) \right] dr \tag{16.561}$$

$$m_{\dot\theta} = -\int_0^1 \left[\frac{c^2}{16} u_T \left(1 + 4\xi_A \left(1 + 2C'(k)(1 + 2\xi_A)\right)\right) \right] dr \tag{16.562}$$

$$m_\lambda = \int_0^1 \left[x_A C'(k) \frac{1}{2} u_T \right] dr \tag{16.563}$$

$$m_{\dot\beta} = \int_0^1 \left[x_A C'(k) \frac{1}{2} r u_T + \frac{c^2}{32} u_R (1 + 4\xi_A) \right.$$
$$\left. - \frac{c^3}{32} \left(\frac{3}{8} + 2\xi_A (1 + 2\xi_A)\right) \right] dr \tag{16.564}$$

$$m_\beta = \int_0^1 \left[x_A C'(k) \frac{1}{2} u_R u_T - \frac{c^2}{32} u_T (1 + 8\xi_A C'(k)) \right.$$
$$\left. - \frac{c^2}{32} \mu \sin\psi (1 + 4\xi_A) \right] dr \tag{16.565}$$

with $\xi_A = x_A/c$. The aerodynamic coefficients can be integrated analytically assuming constant chord and aerodynamic center offset and by evaluating the lift deficiency function at an effective radius (such as $r_e = 0.75$), so that the reduced frequency is $k_e = \omega b/(r_e + \mu \sin\psi)$. The results are

$$M_\theta = C'(k_e) \left(\frac{1}{8} + \frac{\mu}{3}\sin\psi + \frac{\mu^2}{4}\sin^2\psi\right) + \frac{c}{16}\mu\cos\psi \tag{16.566}$$

$$M_{\dot\theta} = \frac{c}{4}(1 + 2C'(k_e)(1 + 2\xi_A))\left(\frac{1}{6} + \frac{\mu}{4}\sin\psi\right) \tag{16.567}$$

$$M_\lambda = -C'(k_e)\left(\frac{1}{6} + \frac{\mu}{4}\sin\psi\right) \tag{16.568}$$

$$M_{\dot\beta} = -C'(k_e)\left(\frac{1}{8} + \frac{\mu}{6}\sin\psi\right) - \frac{c}{8}\mu\cos\psi + \frac{c^2}{64}(1 + 4\xi_A) \tag{16.569}$$

$$M_\beta = -\mu\cos\psi\, C'(k_e)\left(\frac{1}{6} + \frac{\mu}{4}\sin\psi\right)$$
$$+ \frac{c}{2}C'(k_e)\left(\frac{1}{6} + \frac{\mu}{4}\sin\psi\right) + \frac{c}{16}\mu\sin\psi \tag{16.570}$$

$$m_\theta = -x_A C'(k_e)\left(\frac{1}{6} + \frac{\mu}{2}\sin\psi + \frac{\mu^2}{2}\sin^2\psi\right)$$
$$- \frac{c^2}{32}\mu\cos\psi(1 + 4\xi_A) \tag{16.571}$$

$$m_{\dot\theta} = -\frac{c^2}{16}(1 + 4\xi_A(1 + 2C'(k_e)(1 + 2\xi_A)))\left(\frac{1}{2} + \mu\sin\psi\right) \tag{16.572}$$

$$m_\lambda = x_A C'(k_e)\left(\frac{1}{4} + \frac{\mu}{2}\sin\psi\right) \tag{16.573}$$

16.8 Aerodynamic Loads

$$m_{\dot{\beta}} = x_A C'(k_e) \left(\frac{1}{6} + \frac{\mu}{4} \sin \psi \right) + \frac{c^2}{32} \mu \cos \psi \, (1 + 4\xi_A)$$
$$- \frac{c^3}{32} \left(\frac{3}{8} + 2\xi_A \, (1 + 2\xi_A) \right) \tag{16.574}$$

$$m_{\beta} = x_A C'(k_e) \mu \cos \psi \left(\frac{1}{4} + \frac{\mu}{2} \sin \psi \right)$$
$$- \frac{c^2}{32} (1 + 8\xi_A C'(k_e)) \left(\frac{1}{2} + \mu \sin \psi \right)$$
$$- \frac{c^2}{32} \mu \sin \psi \, (1 + 4\xi_A) \tag{16.575}$$

The non-circulatory lift terms are an order c/R smaller than the flap moments due to the circulatory lift. The rotor wake can significantly reduce the circulatory lift forces through the lift deficiency function, however. The circulatory lift also produces feathering moments through the pitch axis-aerodynamic center offset x_A. The non-circulatory forces are the source of the aerodynamic pitch damping moment of the blade ($m_{\dot{\theta}}$). In hover, the aerodynamic coefficients reduce to

$$M_{\theta} = \frac{1}{8} C'(k_e) \tag{16.576}$$

$$M_{\dot{\theta}} = \frac{c}{24} \left(1 + 2C'(k_e) \, (1 + 2\xi_A) \right) \tag{16.577}$$

$$M_{\lambda} = -\frac{1}{6} C'(k_e) \tag{16.578}$$

$$M_{\dot{\beta}} = -\frac{1}{8} C'(k_e) + \frac{c^2}{64} (1 + 4\xi_A) \tag{16.579}$$

$$M_{\beta} = \frac{c}{12} C'(k_e) \tag{16.580}$$

$$m_{\theta} = -\frac{x_A}{6} C'(k_e) \tag{16.581}$$

$$m_{\dot{\theta}} = -\frac{c^2}{32} \left(1 + 4\xi_A \, (1 + 2C'(k_e) \, (1 + 2\xi_A)) \right) \tag{16.582}$$

$$m_{\lambda} = \frac{x_A}{4} C'(k_e) \tag{16.583}$$

$$m_{\dot{\beta}} = \frac{x_A}{6} C'(k_e) - \frac{c^3}{32} \left(\frac{3}{8} + 2\xi_A \, (1 + 2\xi_A) \right) \tag{16.584}$$

$$m_{\beta} = -\frac{c^2}{64} (1 + 8\xi_A C'(k_e)) \tag{16.585}$$

The circulatory lift produces flap moments due to θ, $\dot{\beta}$, and λ and the corresponding pitch moments through x_A. The non-circulatory forces produce flap and pitch moments due to $\dot{\theta}$ and β.

16.9 REFERENCES

de la Cierva, J. "The Development of the Autogyro." The Journal of the Royal Aeronautical Society, *30*:181 (January 1926).

Flax, A.H. "The Bending of Rotor Blades." Journal of the Aeronautical Sciences, *14*:1 (January 1947).

Ormiston, R.A., and Hodges, D.H. "Linear Flap-Lag Dynamics of Hingeless Helicopter Rotor Blades in Hover." Journal of the American Helicopter Society, *17*:2 (April 1972).

17 Beam Theory

17.1 Beams and Rotor Blades

An adequate blade structural model is essential for the prediction of rotor loads and stability. Rotor blades almost universally have a high structural fineness ratio and thus are well idealized as beams. The complexities of rotation, and now multiple load paths and composite construction, have required extensive and continuing efforts to develop appropriate beam models for the solution of rotor problems. For exposition of beam theory, particularly relevant to rotor blade analyses, see Hodges (2006) and Bauchau (1985).

A beam is a structure that has small cross-section dimensions relative to an axial line. Based on the slender geometry, beam theory develops a one-dimensional model of the three-dimensional structure. The deflection of the structure is described as functions of the axial coordinate, obtained from ordinary differential equations (in the axial coordinate). The equations depend on cross-section properties, including two-dimensional elastic stiffnesses. The three-dimensional stress field is determined from the deflection variables. Beam theory combines kinematic equations relating strain measures to deflection variables, constitutive equations relating stress resultants to strain measures, and equilibrium equations relating stress resultants to applied loads. When inertial loads are included, the motion is described by partial differential equations, in time and the axial coordinate.

Although a beam model offers major simplifications of the structural analysis, the design and operation of rotor blades introduce complications. For a rotor blade, the undeformed state is a twisted line, perhaps curved or kinked, and rotation generates large axial forces. Current blades have built-up construction using composite materials. Consequently, significant structural coupling of bending, torsion, and extension is possible. The deflections are not necessarily small, so a geometrically nonlinear description is required. Most beam models for rotor blades do assume small strain and linearly elastic materials.

Fundamental to applicability of beam models is the Saint-Venant Principle: distributions of stress on the cross-section with the same resultant loads exhibit only local differences, producing the same effect far enough away. The difference between the exact stress and an engineering approximation decays exponentially over a short distance along the beam. So the effects of loading can be determined without details of the stress distribution, and that end effects are only important near the beam end.

With the Saint-Venant Principle, elementary solutions for extension, bending, and torsion of beams (notably deformation of homogeneous, isotropic, prismatic beams) provide the basis for engineering beam theory.

Engineering beam theory has its origins in the work of Euler and Bernoulli. For bending and extension deformations, it is assumed that plane cross-sections remain plane and normal to the neutral axis, and axial stress dominates the section loading. Euler-Bernoulli beam theory is applicable for large wavelength deflection of slender beams (low-frequency modes), but not thin-walled open sections. There are four strain measures: two bending curvatures, torsion, and extension strain. Timoshenko beam theory incorporates shear deformation and rotational inertia, relaxing the assumption that plane sections remain normal to the deformed beam axis. There are six strain measures, adding the two shear strains. The added mechanisms of deformation lower the global stiffness of the beam, resulting in a larger deflection under a static load and lower predicted eigen-frequencies for a given set of boundary conditions. Vlasov theory is applicable to thin-walled beams with open cross-sections, adding strain measures to account for the constraint of warping at the end of the beam, which produces a significant increase in local torsion stiffness.

For a beam constructed of isotropic materials, there are six fundamental stiffnesses: extension, two bending, torsion, and two shear. The extension and bending stiffnesses are obtained from simple integrals over the cross-section, whereas the torsion and shear stiffnesses depend on warping of the section. For anisotropic materials of composite beams, there can be elastic coupling of the global deformations, implying 21 stiffnesses (symmetric 6×6 matrix), and simple integrals are not sufficient. Linear, two-dimensional, finite-element analyses for the cross-section stiffnesses have been developed; see Giavotto, Borri, Mantegazza, Ghiringhelli, Carmaschi, Maffioli, and Mussi (1983) and Cesnik and Hodges (1997).

17.2 Engineering Beam Theory for a Twisted Rotor Blade

Engineering beam theory was used in Chapter 16 to derive the equations of motion for an elastic rotor blade, with the simplifying assumption of no structural coupling of bending and torsion and purely in-plane and out-of-plane bending motion. Blade pitch and twist introduce structural coupling of the out-of-plane and in-plane bending deflection. The free vibration modes in the centrifugal force field of the rotating blade hence involve coupled flap and lag motion, which has a major impact on the rotor dynamics. Following Houbolt and Brooks (1958), this section derives engineering beam theory for bending and torsion of a rotating blade. The focus is still on the linear terms.

The undeformed elastic axis is assumed to be a straight line, and the blade is assumed to have a high structural aspect ratio so that engineering beam theory is applicable. Figure 17.1 shows the geometry considered. The spanwise variable r is measured from the center of rotation along the elastic axis. The coordinates x and z are the principal axes of the section, with origin at the elastic axis. Then by definition $\int xzE\,dA = 0$. This integral is weighted with the modulus of elasticity. The modulus-weighted centroid, or tension center, is assumed to be located on the x-axis at a distance x_C aft of the elastic axis; hence $\int zE\,dA = 0$ and $\int xE\,dA = x_C$. The angle of the major principal axis (the x-axis) with respect to the hub plane is the blade pitch θ. The existence of the elastic axis means that torsion about the elastic axis occurs without bending of the blade. Thus the pitch angle consists of the root pitch

17.2 Engineering Beam Theory for a Twisted Rotor Blade

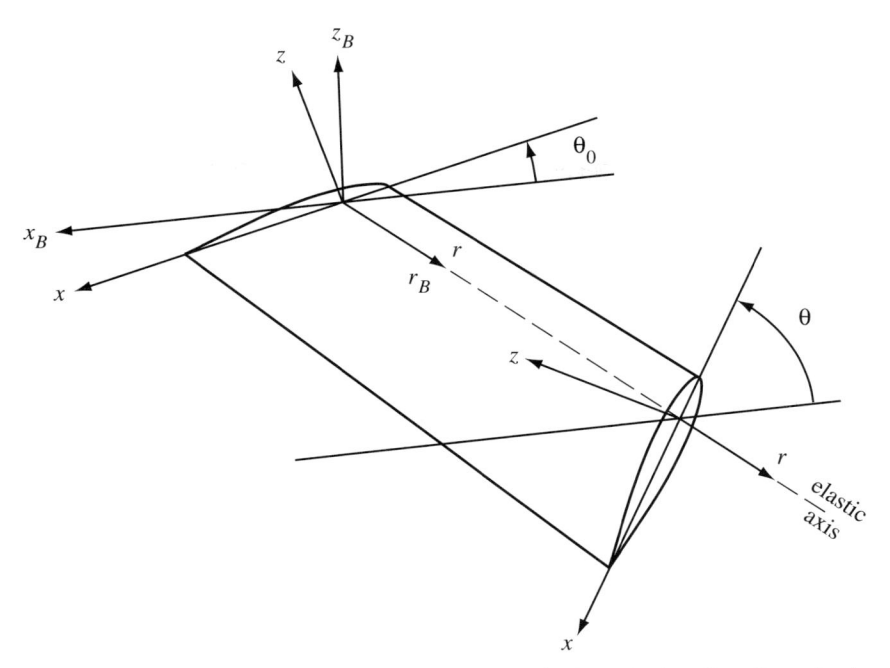

Figure 17.1. Geometry of the twisted rotor blade (before the bending distortion).

θ_0, the built-in twist θ_{tw}, and an elastic torsion deflection θ_e: $\theta = \theta_0 + \theta_{tw} + \theta_e$. The twist θ_{tw} is a function of r and is defined to be zero at the root. Shear stress in the blade is due to θ_e only. The elastic twist θ_e is assumed to be small, but the trim pitch angles θ_0 and θ_{tw} can be large.

The unit vectors of the rotating hub plane axis system are $\mathbf{i}_B, \mathbf{j}_B$, and $\mathbf{k}_B$ (respectively, the x_B, r_B, and z_B axes; see Figure 17.1). The unit vectors of the principal axes of the section are $\mathbf{i}, \mathbf{j}$, and $\mathbf{k}$ (respectively, the x, r, and z axes), which are rotated by the angle θ from the hub plane:

$$\begin{pmatrix} \mathbf{i} \\ \mathbf{j} \\ \mathbf{k} \end{pmatrix} = \begin{bmatrix} \cos\theta & 0 & -\sin\theta \\ 0 & 1 & 0 \\ \sin\theta & 0 & \cos\theta \end{bmatrix} \begin{pmatrix} \mathbf{i}_B \\ \mathbf{j}_B \\ \mathbf{k}_B \end{pmatrix} \qquad (17.1)$$

Elastic torsion is included in the definition of $\mathbf{i}$ and $\mathbf{k}$, but not the blade bending. So $\partial \mathbf{i}/\partial r = -\theta' \mathbf{k}$ and $\partial \mathbf{k}/\partial r = \theta' \mathbf{i}$.

The distortion of the blade is described by a deflection of the elastic axis with components x_0, r_0, and z_0 (Figure 17.2). The bending of the elastic axis produces a rotation of the section by the angles ϕ_x and ϕ_z. The elastic torsion θ_e has already been included in θ. Engineering beam theory assumes that plane sections perpendicular to the elastic axis remain so after bending. This description of the blade motion is sufficient to define the distortion of all elements of the section. The quantities x_0, r_0, z_0, and θ_e are assumed to be small. The unit vectors of the deformed cross section $\mathbf{i}_S$, $\mathbf{j}_S$, and $\mathbf{k}_S$ are rotated by ϕ_x and ϕ_z from the undeformed cross-section:

$$\begin{pmatrix} \mathbf{i}_S \\ \mathbf{j}_S \\ \mathbf{k}_S \end{pmatrix} = \begin{bmatrix} 1 & \phi_z & 0 \\ -\phi_z & 1 & \phi_x \\ 0 & -\phi_x & 1 \end{bmatrix} \begin{pmatrix} \mathbf{i} \\ \mathbf{j} \\ \mathbf{k} \end{pmatrix} \qquad (17.2)$$

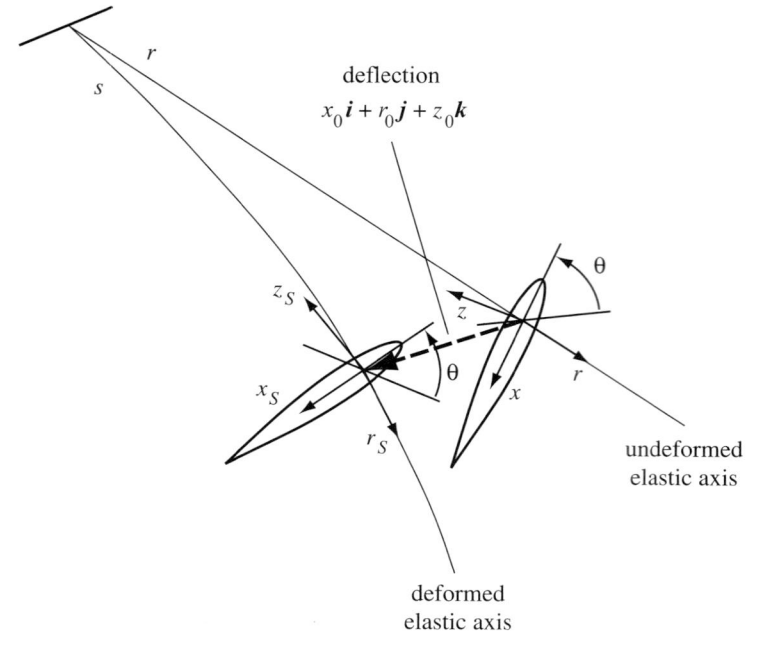

Figure 17.2. Definition of the blade deformation.

The vector $\mathbf{j}_S$ is the tangent to the deformed elastic axis. Hence by definition $\mathbf{j}_S = d\mathbf{r}/ds$, where $\mathbf{r} = x_0\mathbf{i} + (r + r_0)\mathbf{j} + z_0\mathbf{k}$ is the deflected position and s is the arc length along the deformed elastic axis. Then to first order,

$$\mathbf{j}_S = \mathbf{j} + (x_0\mathbf{i} + z_0\mathbf{k})' = \mathbf{j} + (x_0' + z_0\theta')\mathbf{i} + (z_0' - x_0\theta')\mathbf{k} \qquad (17.3)$$

Comparing the two definitions of $\mathbf{j}_S$ shows that the rotation of the section is $-\phi_z = x_0' + z_0\theta'$ and $\phi_x = z_0' - x_0\theta'$. The undeflected position of a blade element is $\mathbf{r} = x\mathbf{i} + r\mathbf{j} + z\mathbf{k}$, and the deflected position is

$$\begin{aligned}\mathbf{R} &= (r + r_0)\mathbf{j} + x_0\mathbf{i} + z_0\mathbf{k} + x\mathbf{i}_S + z\mathbf{k}_S \\ &= (x + x_0)\mathbf{i} + (r + r_0 + x\phi_z - z\phi_x)\mathbf{j} + (z + z_0)\mathbf{k}\end{aligned} \qquad (17.4)$$

For now the elastic extension r_0 is neglected, which simplifies the strain analysis since then to first order $s = r$. The extension r_0 produces a uniform strain over the section, which can be introduced later.

The metric tensor g_{mn} of the undistorted blade is defined by

$$(ds)^2 = d\mathbf{r} \cdot d\mathbf{r} = \left(\frac{\partial \mathbf{r}}{\partial x_m} dx_m\right) \cdot \left(\frac{\partial \mathbf{r}}{\partial x_n} dx_n\right) = g_{mn} dx_m dx_n \qquad (17.5)$$

where ds is the differential length in the material and x_m are general curvilinear coordinates. Similarly, the metric tensor G_{mn} of the deformed blade is

$$(dS)^2 = d\mathbf{R} \cdot d\mathbf{R} = \left(\frac{\partial \mathbf{R}}{\partial x_m} dx_m\right) \cdot \left(\frac{\partial \mathbf{R}}{\partial x_n} dx_n\right) = G_{mn} dx_m dx_n \qquad (17.6)$$

Then the strain tensor γ_{mn} is defined by the differential length increment, $2\gamma_{mn} dx_m dx_n = (dS)^2 - (ds)^2$, or $\gamma_{mn} = \frac{1}{2}(G_{mn} - g_{mn})$. For engineering beam theory, only the axial component of the strain is required. For the specific case of the

17.2 Engineering Beam Theory for a Twisted Rotor Blade

twisted rotor blade, the metric g_{mn} is obtained from the undistorted position vector, $\mathbf{r} = x\mathbf{i} + r\mathbf{j} + z\mathbf{k}$. The axial component is

$$g_{rr} = \frac{\partial \mathbf{r}}{\partial r} \cdot \frac{\partial \mathbf{r}}{\partial r} = 1 + \theta_{tw}'^2(x^2 + z^2) \tag{17.7}$$

The metric G_{mn} is obtained from equation 17.4 for the position vector of the deformed blade:

$$G_{rr} = \frac{\partial \mathbf{R}}{\partial r} \cdot \frac{\partial \mathbf{R}}{\partial r} = \left(1 + x\phi_z' - z\phi_x'\right)^2 + \left(x_0' + \theta'(z + z_0)\right)^2 + \left(z_0' - \theta'(x + x_0)\right)^2 \tag{17.8}$$

Hence the axial component of the strain tensor is

$$\gamma_{rr} = \frac{1}{2}(G_{rr} - g_{rr}) = \frac{1}{2}\Big[\left(1 + x\phi_z' - z\phi_x'\right)^2 - 1 + \left(x_0' + \theta'(z + z_0)\right)^2 - \theta_{tw}'^2 z^2 \\ + \left(z_0' - \theta'(x + x_0)\right)^2 - \theta_{tw}'^2 x^2\Big] \tag{17.9}$$

The linear strain is then

$$\gamma_{rr} \cong \epsilon_{rr} = x\phi_z' - z\phi_x' + \theta_{tw}'^2(xx_0 + zz_0) + \theta_{tw}'\left(zx_0' - xz_0' + \theta_e'(x^2 + z^2)\right) \tag{17.10}$$

since x_0, z_0, ϕ_x, ϕ_z, and θ_e are small.

The strain due to the blade tension, e_T, is a constant given by $T = \int E\epsilon_{rr}dA = \epsilon_T \int E\,dA$, where T is the tension force on the section. Substituting for ϵ_{rr} and including the strain due to the blade extension r_0 again, we obtain

$$\epsilon_T = \phi_z' x_C + \theta_{tw}'^2 x_0 x_C - \theta_{tw}' z_0' x_C + \theta_{tw}' \theta_e' k_P^2 + r_0' \tag{17.11}$$

where $\int zE\,dA = 0$, $\int xE\,dA = x_C$, and $\int(x^2 + z^2)E\,dA = k_P^2 \int E\,dA$; k_P is the modulus-weighted radius of gyration about the elastic axis. So the strain can be written as

$$\epsilon_{rr} = \epsilon_T + (x - x_C)\left(\phi_z' - \theta_{tw}'\phi_x\right) - z\left(\phi_x' + \theta_{tw}'\phi_z\right) + \theta_{tw}'\theta_e'\left(x^2 + z^2 - k_P^2\right) \tag{17.12}$$

where e_T is obtained from the tension force.

Engineering beam theory assumes that all the stresses due to bending are negligible except for the axial component, $\sigma_{rr} = E\epsilon_{rr}$. Assuming that the lines of stress follow the twisted filaments of the beam, the direction of σ_{rr} is given by the unit vector

$$\mathbf{e} = \frac{\partial \mathbf{R}/\partial r}{|\partial \mathbf{R}/\partial r|} \tag{17.13}$$

where $\mathbf{R}$ is the deformed position of the section; see Figure 17.3. The moment about the elastic axis due to the elemental force $\sigma_{rr}dA$ on the cross-section is

$$d\mathbf{M} = (x\mathbf{i}_S + z\mathbf{k}_S) \times (\sigma_{rr}\mathbf{e}\,dA) = \left(-z\mathbf{i}_S + x\mathbf{k}_S + \theta_{tw}'(x^2 + z^2)\mathbf{j}_S\right)\sigma_{rr}dA \tag{17.14}$$

Then the total moment, with components M_x, M_r, and M_z (Figure 17.3), is obtained by integrating over the section:

$$\int_{\text{section}} d\mathbf{M} = M_x\mathbf{i}_S + M_r\mathbf{j}_S + M_z\mathbf{k}_S \tag{17.15}$$

hence

$$M_{x_{EA}} = -\int z\sigma_{rr}dA \tag{17.16}$$

Beam Theory

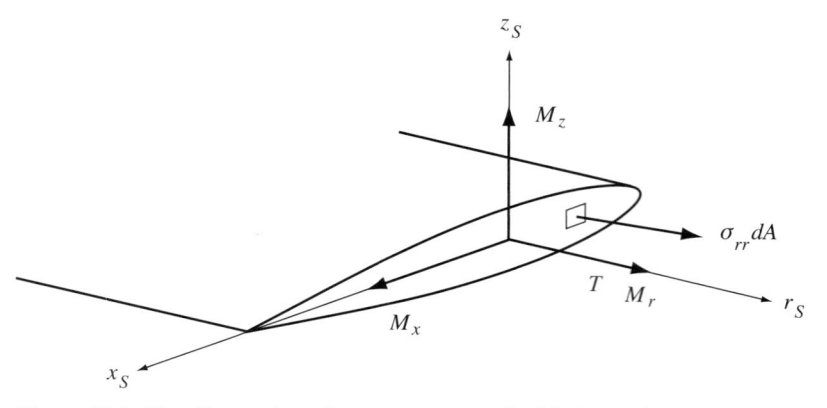

Figure 17.3. Bending and torsion moments on the blade section.

$$M_{z_{EA}} = \int x\sigma_{rr}dA \tag{17.17}$$

$$M_r = GJ\theta'_e + \int (x^2 + z^2)\theta'\sigma_{rr}dA \tag{17.18}$$

The torsion moment $GJ\theta'_e$ due to the shear forces of elastic torsion has been added to M_r. For bending, working with moments about the tension center at x_C is more appropriate:

$$M_x = -\int z\sigma_{rr}dA \tag{17.19}$$

$$M_z = \int (x - x_C)\sigma_{rr}dA \tag{17.20}$$

After substituting for σ_{rr} and integrating, the bending and torsion moments are as follows:

$$M_x = EI_{zz}\left(\phi'_x + \theta'\phi_z\right) - \theta'_{tw}EI_{zp}\theta'_e \tag{17.21}$$

$$M_z = EI_{xx}\left(\phi'_z - \theta'\phi_x\right) + \theta'_{tw}EI_{xp}\theta'_e \tag{17.22}$$

$$M_r = \left(GJ + k_P^2 T + \theta'^2_{tw}EI_{pp}\right)\theta'_e + \theta'_{tw}k_P^2 T$$
$$+ \theta'_{tw}\left(EI_{xp}\left(\phi'_z - \theta'\phi_x\right) - EI_{zp}\left(\phi'_x + \theta'\phi_z\right)\right) \tag{17.23}$$

where

$$EI_{zz} = \int z^2 E\, dA \tag{17.24}$$

$$EI_{xx} = \int (x - x_C)^2 E\, dA \tag{17.25}$$

$$k_P^2 \int E\, dA = \int (x^2 + z^2) E\, dA \tag{17.26}$$

$$EI_{xp} = \int (x - x_C)(x^2 + z^2) E\, dA \tag{17.27}$$

$$EI_{zp} = \int z(x^2 + z^2) E\, dA \tag{17.28}$$

17.2 Engineering Beam Theory for a Twisted Rotor Blade

$$EI_{pp} = \int (x^2 + z^2 - k_P^2)^2 E\, dA \qquad (17.29)$$

Since the tension T acts at the tension center x_C, the bending moments about the elastic axis can be obtained from those about the tension center by $M_{z_{EA}} = M_z + x_C T$ and $M_{x_{EA}} = M_x$.

The structural moments (equations 17.21 and 17.22) couple the flap and lag bending motion. Higher-order stiffnesses (EI_{xp} and EI_{zp}) couple bending and torsion moments. Equation 17.23 has a torsion moment due to the tension T acting on a twisted blade: $\Delta M_r = k_P^2 T (\theta'_{tw} + \theta'_e)$. Called the trapeze effect, the $k_P^2 T \theta'_{tw}$ term acts to untwist the blade, whereas $k_P^2 T \theta'_e$ increases the effective torsion stiffness. These torsion moments arise because of the assumption that the axial stress follows the twisted filaments of the beam, equation 17.13. This assumption is reasonable for a rotor blade, which has a chordwise dimension of the structure much larger than the normal dimension. The assumption is not correct for an axisymmetric beam, where with isotropic materials twist has no meaning. A more careful derivation replaces the coefficient k_P^2 with k_T^2, defined so $k_T = k_P$ for a flat beam and $k_T = 0$ for an axisymmetric beam.

This result can be written using a vector representation of the blade bending. Define the two-dimensional section bending moment vector $M = (M_x\ M_z)^T$ and the flap-lag deflection vector $u = (z_0\ -x_0)^T$ in principal axes. A superscript B is used for quantities relative to the hub axes. The derivatives of u are

$$u' = \begin{pmatrix} z'_0 - x_0 \theta' \\ -x'_0 - z_0 \theta' \end{pmatrix} = \begin{pmatrix} \phi_x \\ \phi_z \end{pmatrix} \qquad (17.30)$$

$$u'' = \begin{pmatrix} \phi'_x + \theta' \phi_z \\ \phi'_z - \theta' \phi_z \end{pmatrix} \qquad (17.31)$$

Then the bending and torsion moments can be written as

$$M = EI u'' + \theta'_{tw} EI_P \theta_e \qquad (17.32)$$

$$M_r = \left(GJ + k_P^2 T + \theta'^2_{tw} EI_{pp}\right) \theta'_e + \theta'_{tw} k_P^2 T + \theta'_{tw} (EI_P)^T u'' \qquad (17.33)$$

with the stiffness matrices

$$EI = \begin{bmatrix} EI_{zz} & 0 \\ 0 & E_{xx} \end{bmatrix} \qquad (17.34)$$

$$EI_P = \begin{bmatrix} -EI_{zp} \\ EI_{xp} \end{bmatrix} \qquad (17.35)$$

in principal axes. This is the engineering beam theory result relating the structural moments to the deflections of the rotor blade section. For a blade with zero pitch, it reduces to the usual results for uncoupled in-plane and out-of-plane bending of a beam:

$$M_x = EI_{zz} z''_0 \qquad (17.36)$$

$$M_z = EI_{xx} x''_0 \qquad (17.37)$$

$$M_r = \left(GJ + k_P^2 T\right) \theta'_e \qquad (17.38)$$

The vector form allows a simultaneous treatment of the coupled in-plane and out-of-plane bending of the blade.

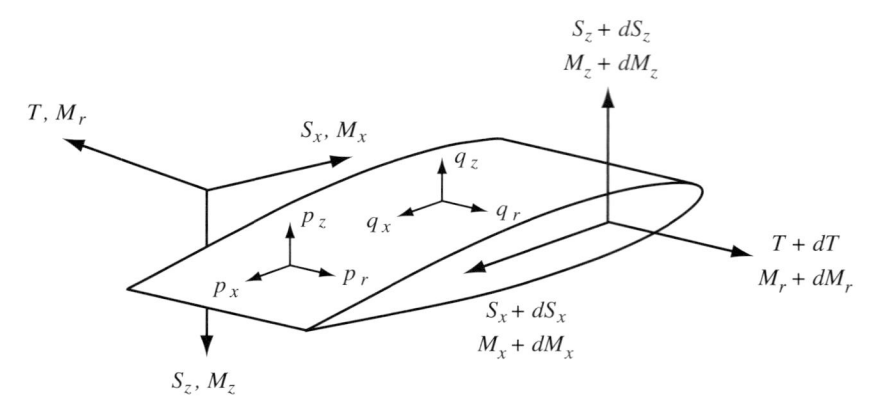

Figure 17.4. Forces and moments acting on the blade element from r to $r + dr$.

Houbolt and Brooks (1958) considered the bending displacement defined in terms of the hub plane axes. Their result can be obtained from the present relations by writing $u^B = (w \quad -v)^T$. Then the curvature in hub plane axes and principal axes is

$$(u^B)'' = \begin{pmatrix} w'' \\ -v'' \end{pmatrix} \tag{17.39}$$

$$u'' = \begin{pmatrix} w'' \cos\theta + v'' \sin\theta \\ w'' \sin\theta - v'' \cos\theta \end{pmatrix} \tag{17.40}$$

and the bending stiffness matrix is

$$EI^B = \begin{bmatrix} EI_{zz}\cos^2\theta + EI_{xx}\sin^2\theta & (EI_{xx} - EI_{zz})\cos\theta\sin\theta \\ (EI_{xx} - EI_{zz})\cos\theta\sin\theta & EI_{zz}\sin^2\theta + EI_{xx}\cos^2\theta \end{bmatrix} \tag{17.41}$$

in hub plane axes.

The differential equations for blade bending and torsion can be derived from the equilibrium of forces and moments on a differential element of the beam extending from r to $r + dr$. Consider the shear forces, bending moments, tension, and torsion moment on the blade section (as shown in Figure 17.4), defined relative to the hub plane axes so the axes are the same at r and at $r + dr$. The blade also has distributed forces (components p_x, p_r, and p_z, force/length) and moments (components q_x, q_r, and q_z, moment/length) acting on the section element. The equilibrium of forces and moments on this blade element give

$$S'_x + p_x = 0 \tag{17.42}$$

$$S'_z + p_z = 0 \tag{17.43}$$

$$T' + p_r = 0 \tag{17.44}$$

$$M'_x - Tw' + S_z + q_x = 0 \tag{17.45}$$

$$M'_z + Tv' - S_x + q_a = 0 \tag{17.46}$$

$$M'_r + S_x w' - S_z v' + q_r = 0 \tag{17.47}$$

17.2 Engineering Beam Theory for a Twisted Rotor Blade

where w and v are the bending deflections in hub plane axes. Eliminating the shears, these equations become

$$M''_x - (Tw')' + q'_x - p_z = 0 \qquad (17.48)$$

$$M''_z + (Tv')' + q'_z + p_x = 0 \qquad (17.49)$$

$$M'_r - w' \int^r p_x dr + v' \int^r p_z dr + q_r = 0 \qquad (17.50)$$

where the tension is $T = -\int^r p_r dr$. The structural analysis gives the section moments M_x, M_z, and M_r. The distributed forces and moments are due to the inertial and aerodynamic forces acting on the blade. To combine the in-plane and out-of-plane equations again, define the following two-dimensional vectors: $M = (M_x \ M_z)^T$, $u = (w \ -v)^T$, $q = (q_x \ q_z)^T$, and $p = (p_z \ -p_x)^T$ (in hub plane axes). Then the equilibrium equation for bending is

$$M'' - (Tu')' + q' - p = 0 \qquad (17.51)$$

With no torsion, engineering beam theory gives $M = EIu''$ (equation 17.32), so

$$(EIu'')'' - (Tu')' + q' - p = 0 \qquad (17.52)$$

is the differential equation for the blade bending.

To complete the equations of motion, the distributed loadings must be derived. For the bending modal equations, only the centrifugal and inertial loadings due to bending deflection are required. The tension is due to the centrifugal force: $T = \Omega^2 \int_r^R \rho m \, d\rho$. The out-of-plane and in-plane accelerations give an inertial force on the blade section: $p = -m\ddot{u}$. The section centrifugal force $m\Omega^2 \rho$ has an in-plane component $(m\Omega^2 \rho)(v/\rho)$ in the lag direction due to the in-plane deflection v, giving a lag moment on the section at r of $q_z = -\int_r^R m\Omega^2 v \, d\rho$. Hence

$$q' = -m \begin{bmatrix} 0 & 0 \\ 0 & \Omega^2 \end{bmatrix} u = -m\overline{\Omega}\,\overline{\Omega}^T u \qquad (17.53)$$

where $\overline{\Omega} = (0 \ \Omega)^T$ is the rotational speed vector in hub plane axes.

The partial differential equation of the rotating blade in a vacuum is then

$$(EIu'')'' - \Omega^2 \frac{d}{dr}\left(\int_r^R \rho m \, d\rho \, u'\right) - m\overline{\Omega}\,\overline{\Omega}^T u + m\ddot{u} = 0 \qquad (17.54)$$

Using the method of separation of variables, write the bending deflection as $u = \eta(r)e^{i\nu t}$. Then the equation for the vector mode shape η is

$$(EI\eta'')'' - \Omega^2 \frac{d}{dr}\left(\int_r^R \rho m \, d\rho \, \eta'\right) - m\overline{\Omega}\,\overline{\Omega}^T \eta - m\nu^2 \eta = 0 \qquad (17.55)$$

This is the modal equation for coupled flap-lag bending of the rotating blade. For a blade with no pitch or twist, this vector equation separates into two equations for out-of-plane deflection and in-plane deflections: equations 16.61 and 16.111. The boundary conditions are

i) The blade tip is a free end, with zero moment and shear force, so $EI\eta'' = (EI\eta'')' = 0$ at $r = R$.

ii) The root of an articulated blade has a hinge, with zero displacement $\eta = 0$ and spring moment $EI\eta'' = K_s\eta'$ at $r = e$ (allowing for a hinge offset), where K_s is the hinge spring matrix.

iii) The root of a hingeless rotor has cantilever restraint, with zero displacement and slope, so $\eta = \eta' = 0$ at $r = e$ (allowing for a very stiff hub).

The modal equation and its boundary conditions form a proper Sturm-Liouville eigenvalue problem (see section 16.1) for which there is a series of eigensolutions $\eta_k(r)$ and corresponding eigenvalues ν_k. The modes are orthogonal with weighting function m, so

$$\int_0^R \eta_i^T \eta_k m \, dr = 0 \tag{17.56}$$

if $i \neq k$. The natural frequencies ν_k can be obtained from the mode shapes using the energy relation

$$\nu^2 = \frac{\eta'(e)^T K_s \eta'(e) + \int_0^R \left[\eta''^T EI\eta'' + \Omega^2 \int_r^R \rho m \, d\rho \, \eta'^2 - m\left(\overline{\Omega}^T \eta\right)^2\right] dr}{\int_0^R \eta^2 m \, dr} \tag{17.57}$$

which can also be used to estimate the frequencies from approximate mode shapes.

17.3 Nonlinear Beam Theory

Nonlinear beam theory for rotor blades is developed in this section. The theory is derived first for exact kinematics of the elastic motion for small strain and linear elastic materials but without any limitation on the magnitude of the deformation. This development draws on the work of Hodges (1990), Bauchau and Hong (1988), Smith and Chopra (1993), and Yuan, Friedmann, and Venkatesan (1992). With the assumption of small strain, beam theory produces a linear relation between the section structural loads and the strain measures (such as curvature). Some interesting (if not important) nonlinear terms are retained as well, coupling extension and torsion. Following Hodges and Dowell (1974), the equations can retain only second-order effects of elastic motion in the strain energy and kinetic energy, restricting the elastic motion to moderate deflection.

Two structural models are considered. The first model is the Euler-Bernoulli beam theory for isotropic materials with a straight elastic axis. The second model is beam theory for anisotropic or composite materials, including transverse shear deformation, with a straight beam axis. Beam theories have been developed for a curved axis, but here the undistorted axis is assumed to be straight. When modeled by finite elements, the blade can be reasonably represented by a series of small elements with straight axes, with jumps and kinks of the axis occurring at nodes.

Classical beam theory works best if the reference axis is the elastic axis. The elastic axis is defined as a line along which transverse loads produce bending only, with no torsion at any section. Such a line often is not independent of the loading distribution and may not exist at all. So often the elastic axis is taken as the locus of shear centers. The shear center or flexural center is a point on a section where a tranverse load must be applied to produce bending only, with no twist of the section; it is a property of the section. The torsional center or center of twist is the point about which the section rotates under torsion load. The neutral axis is the point on the section with zero stress produced by bending. In addition to treating anisotropic or composite materials, the present general model eliminates the assumption that an elastic axis exists. Identification of the beam axis is not entirely arbitrary, since the structure must be slender relative to the beam axis, allowing application of the beam theory assumptions; structural and inertial properties are defined in planes perpendicular to the beam axis, and the elastic motion is described by extension, bending, and torsion of the beam axis.

The effects of cross-section warping and transverse shear are included in the section structural properties. Their effects on the inertial forces and surface geometry are neglected. Any variables describing the warp amplitude are eliminated by expressing them in terms of the strain measures. This treatment of warp is not sufficient for open sections or for restrained warping at end conditions. Transverse shear is introduced by variables that describe the rotation of the cross-section relative to the plane perpendicular to the bent beam axis. Good results have been obtained using reduced section properties, so a quasistatic reduction is used to eliminate the transverse shear variables from the section structural relations; see Hodges, Atilgan, Cesnik, and Fulton (1992), Cesnik and Hodges (1997), and Smith and Chopra (1991). Alternate approaches are a global (rather than section) quasistatic elimination of the transverse shear and warping variables or use of an integration or solution method that suppresses the high-frequency, short wavelength modes associated with such variables.

The description of the beam motion includes large rigid-body motion, which for the helicopter encompasses rotation of the rotor as well as rigid and elastic motion of the airframe. Depending on the shape functions for the elastic deflection, this rigid-body motion can be either frame motion for the beam or the motion of one end of the beam. In the context of multibody dynamics, joints can be included at any point on the beam.

17.3.1 Beam Cross-Section Motion

The elastic motion is represented by the deflection, extension, and torsion of the beam axis. The rigid motion describes the position and orientation of a frame at one end of the beam. The elastic motion is measured relative to the rigid motion. The undeflected structure has a straight beam axis of length ℓ. The beam axis is the positive x-axis, with the beam extending from $x = 0$ to $x = \ell$. Pitch angles are measured from the x–y plane, positive for rotation about the x-axis. Figure 17.5 illustrates the geometry. The structural and inertial properties of the undeflected beam are defined as follows.

a) θ_C: pitch of the structural principal axes
b) y_C and z_C: offset of the tension center (modulus-weighted centroid) from the beam axis, relative to the principal axes (at θ_C)
c) k_P: modulus-weighted radius of gyration about the beam axis
d) θ_I: pitch of the inertial principal axes
e) y_I and z_I: offset of the center of gravity (mass-weighted centroid) from the beam axis relative to the principal axes

The pitch angles can be large. In the following, the notation $C_\beta = \cos\beta$, $S_\beta = \sin\beta$ is used. For a rotation matrix, the notation $C = X_\alpha Y_\beta Z_\gamma$ means a rotation by the angles γ, then β, and then α about the z, y, and x-axes, respectively (see Chapter 2).

Beam Theory

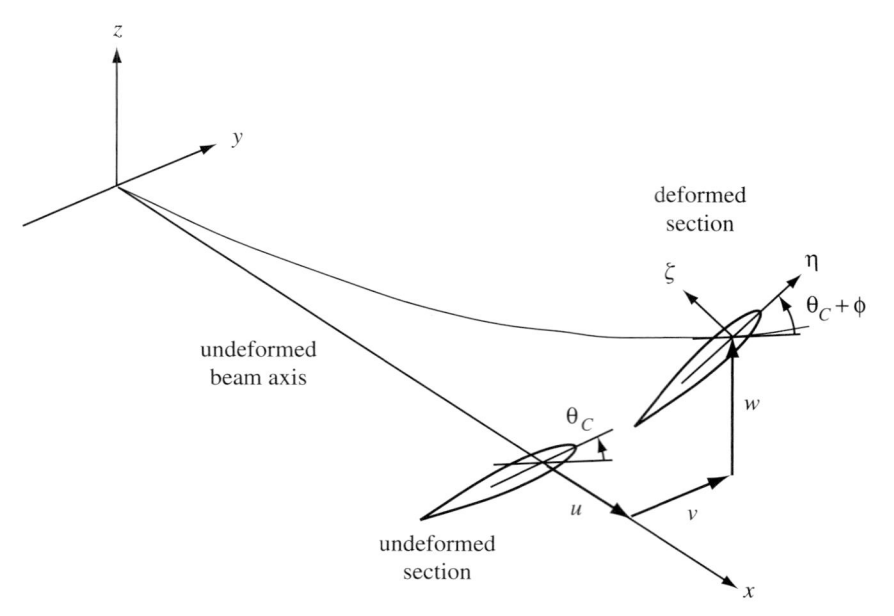

Figure 17.5. Definition of the blade deformation for the nonlinear beam model.

The theory requires the motion of a point on the beam cross-section. For the structural contributions to the equations of motion, the effects of warp and transverse shear are included. The position of a cross-section point relative to the body axes is constructed as follows:

a) Constant axial position x
b) Then elastic axial deflection u along the x-axis
c) Then rotation of the cross-section by v then ω, produced by transverse shear deformation
d) Then elastic bending deflections v then w, along the y and z-axes, respectively, which also produce rotation of the cross-section axes by the angles ζ and β
e) Then elastic torsion and a constant rotation θ_X about the x-axis
f) Then the position relative to the bent and rotated cross-section axes, including warp W of the cross-section

The pitch angle $\theta_X = \theta_C$ for the structural analysis, and $\theta_X = \theta_I$ for the inertial analysis (section principal axes). The warp displacement W can have three components. Thus the position on the cross-section is

$$r^B = \begin{pmatrix} x+u \\ v \\ w \end{pmatrix} + C \begin{pmatrix} 0 \\ \eta \\ \zeta \end{pmatrix} + C \begin{pmatrix} W_1 \\ W_2 \\ W_3 \end{pmatrix} \quad (17.58)$$

where the variables η and ζ identify the cross-section point, relative to the section principal axes at θ_X (Figure 17.5). The variables x, η, and ζ are curvilinear coordinates of the beam. The section is rotated by the matrix C: $C = (Z_{-v}Y_{\omega})C^{BE} = (Z_{-v}Y_{\omega})(Z_{-\zeta}Y_{\beta}X_{-\theta})$. The position and orientation of the deformed beam axis (E) relative to the undeformed beam frame (B) are given by $x^{EB/B} = (x+u \ v \ w)^T$ and $C^{BE} = Z_{-\zeta}Y_{\beta}X_{-\theta}$.

17.3 Nonlinear Beam Theory

The section warping displacement can in general be described by a set of warping functions S_i and scalar amplitudes A_i: $W = (W_1, W_2, W_3)^T = \sum_i S_i(\eta, \zeta) A_i(x)$ (for example, Hodges, Atilgan, Cesnik, and Fulton (1992)). From the virtual displacement δW, differential equations (in x) are obtained for the amplitudes A_i (static equations if the inertial effects of warping are neglected). Here it is assumed that these equations are solved to eliminate the warping variables, so the effects of warp are accounted for in the section elastic constants. For an isotropic beam with an elastic axis, Saint Venant's torsional warping function can be used: $W_1 = \lambda \phi'$ (where ϕ is the elastic torsion). For simplicity, this expression is also used in the equations presented here for an anisotropic beam, although the analysis that obtains the section elastic constants must fully consider the effects of warp.

The order of the bending (v then w) follows from the order of the Euler angles to describe the rotation of the section, so the order is not unique. If Rodrigues parameters were used instead, the bending deflections would be treated identically. Here the cross-section is still perpendicular to the bent beam axis, and β and ζ are the rotations of the cross-section produced by bending deflection. The rotation angles are obtained from the kinematics of the elastic deflection. Let r be the arc length along the deflected beam axis. The notation $(\ldots)^+$ is used for the derivative with respect to r, whereas $(\ldots)'$ is the derivative with respect to x. The tangent to the beam axis is

$$t^B = \frac{dx^{EB/B}}{dr} = \frac{1}{r'}\frac{dx^{EB/B}}{dx} = \begin{pmatrix} x^+ + u^+ \\ v^+ \\ w^+ \end{pmatrix} = \frac{1}{r'}\begin{pmatrix} 1 + u' \\ v' \\ w' \end{pmatrix} \quad (17.59)$$

where $r' = dr/dx = |dx^{EB/B}/dx| = \sqrt{(1+u')^2 + v'^2 + w'^2}$. Since the first row of $Y_{-\beta} Z_\zeta$ equals t^B:

$$t^B = \begin{pmatrix} \cos\beta \cos\zeta \\ \cos\beta \sin\zeta \\ \sin\beta \end{pmatrix} = \begin{pmatrix} x^+ + u^+ \\ v^+ \\ w^+ \end{pmatrix} \quad (17.60)$$

the rotation angles can be obtained from the bending:

$$\sin\beta = w^+ \quad (17.61)$$

$$\sin\zeta = v^+/\sqrt{1 - w^{+2}} \quad (17.62)$$

$$\cos\beta = \sqrt{1 - \sin^2\beta} \quad (17.63)$$

$$\cos\zeta = \sqrt{1 - \sin^2\zeta} \quad (17.64)$$

So β is a rotation about the negative y-axis, produced by bending w^+, and ζ is a rotation about the z-axis, produced by bending v^+. For moderate deflections, the relations can be simplified, consistent with second-order accuracy of the equations of motion. The second-order approximation for the geometry uses the following expressions: $S_\beta = w'$, $S_\zeta = v'$, $C_\beta = \sqrt{1 - S_\beta^2}$, and $C_\zeta = \sqrt{1 - S_\zeta^2}$. With such approximations, the rotation matrix is still proper.

The magnitudes of $\sin\beta$ and $\sin\zeta$ are less than one for values of u, v, and w describing a realizable deflection of the beam. The expressions used for $\cos\beta$ and $\cos\zeta$ assume that the magnitudes of β and ζ are less than 90°. Hence $|w'| < |r'|$ and $w'^2 + v'^2 < r'^2$ are required. The elastic extension is small for realistic motion, so the requirement is $|w'| < 1$ and $w'^2 + v'^2 < 1$. The shape functions that describe the bending deflection can violate these requirements, giving an inconsistent geometric model. So very large elastic motion of a beam must be modeled using several elements. A simple, extreme test case is a uniform beam bent in a circle. With exact kinematics for both rigid-body and elastic motion, accurate representation of the circle is obtained with as few as six elements. With a second-order model for the elastic motion and relative rigid motion with exact kinematics, 24 or more elements are needed. For practical rotor blade problems, the beam curvature is much smaller than the beam length, and second-order elastic motion gives accurate results.

The angular motion of the beam axis, in terms of the bending and pitch variables, is thus

$$C^{EB} = X_\theta Y_{-\beta} Z_\zeta = X_\theta \begin{bmatrix} C_\beta C_\zeta & C_\beta S_\zeta & S_\beta \\ -S_\zeta & C_\zeta & 0 \\ -S_\beta C_\zeta & -S_\beta S_\zeta & C_\beta \end{bmatrix}$$

$$= X_\theta \begin{bmatrix} b & v^+ & w^+ \\ -v^+/a & b/a & 0 \\ -w^+ b/a & -w^+ v^+/a & a \end{bmatrix} \quad (17.65)$$

$$\omega^{EB/E} = R \begin{pmatrix} \dot\theta \\ \dot\beta \\ \dot\zeta \end{pmatrix} \quad (17.66)$$

with

$$X_\theta = \begin{bmatrix} 1 & 0 & 0 \\ 0 & C_\theta & S_\theta \\ 0 & -S_\theta & C_\theta \end{bmatrix} \quad (17.67)$$

$$R = X_\theta \begin{bmatrix} 1 & 0 & S_\beta \\ 0 & -1 & 0 \\ 0 & 0 & C_\beta \end{bmatrix} = \begin{bmatrix} 1 & 0 & S_\beta \\ 0 & -C_\theta & S_\theta C_\beta \\ 0 & S_\theta & C_\theta C_\beta \end{bmatrix} \quad (17.68)$$

and $a = \sqrt{1 - w^{+2}}$, $b = \sqrt{1 - v^{+2} - w^{+2}}$. Similarly, the virtual rotation and the torsion/curvature are

$$\delta\psi = R \begin{pmatrix} \delta\theta \\ \delta\beta \\ \delta\zeta \end{pmatrix} \quad (17.69)$$

$$\kappa = R \begin{pmatrix} \theta^+ \\ \beta^+ \\ \zeta^+ \end{pmatrix} \quad (17.70)$$

Note that $\widetilde{\omega}^{EB/E} = C^{EB} \dot{C}^{BE}$, and $\widetilde{\kappa} = C^{EB}(C^{BE})^+$. For moderate deflections, these relations can be simplified, consistent with second-order accuracy of the equations of motion:

$$C^{EB} \cong X_\theta \begin{bmatrix} 1 - \tfrac{1}{2}(v'^2 + w'^2) & v' & w' \\ -v' & 1 - \tfrac{1}{2}v'^2 & 0 \\ -w' & -w'v' & 1 - \tfrac{1}{2}w'^2 \end{bmatrix} \quad (17.71)$$

$$\omega^{EB/E} \cong X_\theta \begin{bmatrix} 1 & 0 & w' \\ 0 & -1 & 0 \\ 0 & 0 & 1 \end{bmatrix} \begin{pmatrix} \dot\theta \\ \dot w' \\ \dot v' \end{pmatrix} \quad (17.72)$$

and $\dot\theta \cong \dot\phi - \int_0^x (\dot w' v'' + w' \dot v'') dx$.

17.3.2 Extension and Torsion Produced by Bending

Bending of the beam (v and w deflection) produces axial and torsional displacements. The extension u and pitch angle θ of a bent beam are nonzero even for large axial and torsional stiffnesses. These variables are therefore defined as the sum of elastic motion and motion produced by bending: $u = u_e + U$ and $\theta = \theta_C + \phi + \Theta$. Here u_e and ϕ are quasi-coordinates for the elastic extension motion and elastic torsion motion, respectively. For large axial and torsional stiffnesses, u_e and ϕ approach zero. Bending deflection produces the extension U and rotation Θ. The first term in θ is the pretwist of the structural principal axes (which can be replaced by θ_I or zero, depending on the geometry required). The elastic torsion ϕ is defined considering the curvature of the beam about the x-axis: $\kappa_x = \theta^+ + S_\beta \zeta^+ = (\theta_C + \phi)^+$. Hence the torsional displacement produced by bending is

$$\Theta = -\int_0^r S_\beta \zeta^+ \, dr = -\int_0^x S_\beta \zeta^+ r' \, dx = -\int_0^x S_\beta \zeta' \, dx = -\int_0^x w' \zeta^+ \, dx \quad (17.73)$$

If there is no elastic extension of the beam, then $r' = dr/dx = 1$, which gives the axial displacement produced by bending: $u'_{\text{bend}} = \sqrt{1 - (v'^2 + w'^2)} - 1$. Typically therefore the total axial displacement is written as

$$u = u_e + \int_0^x \left[\sqrt{1 - (v'^2 + w'^2)} - 1 \right] dx = u_e + U \quad (17.74)$$

It is simpler (and equivalent to second order) to instead define the elastic extension as $r' = 1 + u'_e$, so $u' = \sqrt{(1 + u'_e)^2 - (v'^2 + w'^2)} - 1$ and

$$u = u_e + \int_0^x \left[\sqrt{(1 + u'_e)^2 - (v'^2 + w'^2)} - (1 + u'_e) \right] dx = u_e + U \quad (17.75)$$

To second order in the displacement (or third order if $u_e = 0$), the extension and torsion produced by bending are

$$U_2 = -\frac{1}{2} \int_0^x (v'^2 + w'^2) \, dx \quad (17.76)$$

$$\Theta_2 = -\int_0^x w' v'' \, dx \quad (17.77)$$

These approximations are accurate for moderate deflection, specifically as long as v'^2, w'^2, and u'_e are small compared to 1. For the exact geometric model, the extension and torsion produced by bending are written $U = U_2 + \Delta U$ and $\Theta = \Theta_2 + \Delta \Theta$, and the increments ΔU and $\Delta \Theta$ are evaluated by numerical integration.

17.3.3 Elastic Variables and Shape Functions

The elastic motion of the beam is described by the variables u_e, v, w, and ϕ, as a function of beam axial station x. This motion is discretized using generalized coordinates $q(t)$ and shape functions $h(x)$:

$$u_e = h_u^T q_u \quad (17.78)$$

$$v = h_v^T q_v \quad (17.79)$$

$$w = h_w^T q_w \quad (17.80)$$

$$\phi = h_\phi^T q_\phi \quad (17.81)$$

The rigid motion of the entire beam is contained in the motion of the body axes or frame. The coordinates q represent motion measured relative to that frame motion. The axial and torsion variables exclude the motion produced by bending kinematics, so suppressing u_e or ϕ is equivalent to the limit of infinite axial or torsional stiffness. With the second-order approximation, U and Θ can be expressed as quadratic functions of the bending degrees of freedom q_v and q_w: $U_2 = q_v^T H_{vv} q_v + q_w^T H_{ww} q_w$ and $\Theta_2 = q_w^T H_{wv} q_v$.

A finite element analysis typically uses Hermite polynomials for the bending shape functions, so the degrees of freedom are displacement and rotation at the nodes. Then the boundary condition of the continuity of bending displacement and slope, extension, and torsion between elements requires simply equating nodal coordinates. The generalized coordinates q represent both rigid and elastic motion of the structure. The frame motion is either prescribed, or constraint equations are introduced to tie the frame to the structure. The frame motion must capture the large part of the rigid motion of the beam. Chopra and Sivaneri (1982) developed a 15 degree-of-freedom beam discretization for a finite element analysis of rotor blades: four coordinates each for u, v, and w, and three coordinates for ϕ. The degrees of freedom correspond to bending displacement and slope, extension, and torsion at each end node, plus three interior nodes: two for extension and one for torsion. The vector of generalized coordinates is

$$q^T = (u_1 \quad u_2 \quad u_3 \quad u_4 \quad v_1 \quad v_1' \quad v_2 \quad v_2' \quad w_1 \quad w_1' \quad w_2 \quad w_2' \quad \phi_1 \quad \phi_2 \quad \phi_3) \tag{17.82}$$

where subscripts 1 and 2 refer to the end nodes. The shape functions are Hermite polynomials for bending and Lagrange polynomials for extension and torsion:

$$H_v^T = H_w^T = \begin{pmatrix} 2\xi^3 - 3\xi^2 + 1 \\ \ell(\xi^3 - 2\xi^2 + \xi) \\ -2\xi^3 + 3\xi^2 \\ \ell(\xi^3 - \xi^2) \end{pmatrix} \tag{17.83}$$

$$H_u^T = \frac{1}{2} \begin{pmatrix} -9\xi^3 + 18\xi^2 - 11\xi + 1 \\ 27\xi^3 - 45\xi^2 + 18\xi \\ -27\xi^3 + 36\xi^2 - 9\xi \\ 9\xi^3 - 9\xi^2 + 2\xi \end{pmatrix} \tag{17.84}$$

$$H_\phi^T = \begin{pmatrix} 2\xi^2 - 3\xi + 1 \\ -4\xi^2 + 4\xi \\ 2\xi^2 - \xi \end{pmatrix} \tag{17.85}$$

with $\xi = x/\ell$; ℓ is the length of the beam element. This representation of the motion corresponds to linear variation of the bending moment and torsion moment and to quadratic variation of axial force.

An alternative is to have the frame attached to the end of the beam ($x = 0$). The shape functions can be orthogonal polynomials:

$$H_v^T = H_w^T = \begin{pmatrix} \xi^2 \\ 6\xi^3 - 5\xi^2 \\ 28\xi^4 - 42\xi^3 + 15\xi^2 \end{pmatrix} \tag{17.86}$$

for bending, $h(0) = h'(0) = 0$ to exclude the rigid motion; and

$$H_u^T = H_\phi^T = \begin{pmatrix} \xi \\ 4\xi^2 - 3\xi \\ 15\xi^3 - 20\xi^2 + 6\xi \\ 56\xi^4 - 105\xi^3 + 60\xi^2 - 10\xi \end{pmatrix} \qquad (17.87)$$

for extension and torsion, $h(0) = 0$ to exclude the rigid motion. The elastic motion is not orthogonal to the rigid motion, and the elastic degrees of freedom are not the total motion (nodal coordinates) for the other end of the beam. A 15 degree-of-freedom model uses three shape functions for extension and two shape functions each for bending and torsion, plus the six rigid-body degrees of freedom. This representation gives cubic displacements and quadratic rotation; hence, quadratic tension and linear moments along the beam.

The effects of transverse shear are introduced by variables ω and ν that rotate the cross-section (in addition to the rotation produced by bending). These variables can be nonzero at both ends. By a static reduction, the structural analysis accounts for the transverse shear effects in the section elastic constants, so ω and ν do not remain as degrees of freedom for the element.

17.3.4 Hamilton's Principle

The equations of motion are obtained using Hamilton's principle:

$$\delta \int_{t_1}^{t_2} L \, dt = \delta \int_{t_1}^{t_2} (T - U + W) \, dt = 0 \qquad (17.88)$$

where the terms in the Lagrangian L are the kinetic energy, the strain energy, and the work of external loads. In a finite element analysis, the Lagrangian is discretized. Let u be the displacement, represented by generalized coordinates a using $u = Na$ (N are the shape functions). The strain energy can be written as an integral over the structure volume: $\delta U = \int \delta \epsilon^T \sigma \, d\Omega$. The material properties define the stress σ as a function of the strain ϵ; and the geometry gives σ from u: $\epsilon = Lu = LNa = Ba$, $\sigma = D\epsilon + \sigma_0$. Hence the strain energy becomes

$$\delta U = \delta a^T \left[\int B^T D B \, d\Omega \, a + \int B^T \sigma_0 \, d\Omega \right] \qquad (17.89)$$

The generalization to nonlinear material properties and geometry is straightforward ($\sigma(\epsilon)$ and $\epsilon(u)$). The work of the external loads can be written as integrals of the body forces b, surface forces t, and discrete force F:

$$\delta W = \int \delta u^T b \, d\Omega + \int \delta u^T t \, d\Gamma + \delta u^T F = \delta a^T \left[\int N^T b \, d\Omega + \int N^T t \, d\Gamma + N^T F \right] \qquad (17.90)$$

The surface forces are assumed to be discretized, and the only body force is gravity, which is handled with the inertial loads. So the only work terms needed are those from the discrete forces acting on the structure (and similar terms for the discrete moments).

The equations of motion can be obtained from Hamilton's principle as the coefficient of δa^T inside an integral over time. The kinetic energy can be written as an integral over the structure volume:

$$\delta T = \delta \int \frac{1}{2} \rho v^2 \, d\Omega \qquad (17.91)$$

where ρ and v are the material density and velocity. Hence $\delta\dot{a}$ terms arise from the kinetic energy. These can be eliminated by integrating by parts (in time), with the assumption $\delta a = 0$ at t_1 and t_2. The integration and δ operation are complicated for the nonlinear and time-varying case. The appearance of $\delta\dot{a}$ can be avoided, and so the integration over time ignored, by using the d'Alembert approach: inertial acceleration is treated as a body force. So the kinetic energy can be written as

$$\delta T = \int \delta u^T (-\ddot{r} + g) \rho \, d\Omega = \delta a^T \left[\int N^T (-N\ddot{a} + g) \rho \, d\Omega \right] \tag{17.92}$$

where for the linear case the acceleration relative to inertial space is $\ddot{r} = \ddot{u} = N\ddot{a}$. The gravity force acting on the body has been included. Thus

$$\left(\int N^T N \rho \, d\Omega \right) \ddot{a} + \left(\int B^T D B \, d\Omega \right) a = \left(\int N^T \rho \, d\Omega \right) g + \int N^T t \, d\Gamma$$

$$+ N^T F - \int B^T \sigma_0 \, d\Omega \tag{17.93}$$

are the equations of motion for the linear problem. The derivation of the equations of motion follows this general approach.

17.3.5 Strain Energy

Evaluation of the strain energy begins with the analysis of strain; see Bisplinghoff, Mar, and Pian (1965) and Washizu (1964, 1975). The Green-Lagrange strain tensor is obtained from the metric tensors of the undistorted and distorted beams (g_{mn} and G_{mn}). In terms of curvilinear coordinates $y_m = (x, \eta, \zeta)$, the undistorted and distorted position vectors are

$$r = \begin{pmatrix} x \\ 0 \\ 0 \end{pmatrix} + X_{-\theta_C} \begin{pmatrix} 0 \\ \eta \\ \zeta \end{pmatrix} \tag{17.94}$$

$$R = \begin{pmatrix} x+u \\ v \\ w \end{pmatrix} + C \begin{pmatrix} 0 \\ \eta \\ \zeta \end{pmatrix} + C \begin{pmatrix} W_1 \\ W_2 \\ W_3 \end{pmatrix} \tag{17.95}$$

Here x is the distance along the straight beam axis, whereas η and ζ specify a position on the cross-section plane (parallel to the structural principal axes, but the origin is not necessarily at the tension center). Assuming small strain, the section loads can be expressed as linear combinations of the moment strain measure κ and force strain measure γ:

$$\kappa = K - k \tag{17.96}$$

$$\gamma = C^T \begin{pmatrix} 1+u' \\ v' \\ w' \end{pmatrix} - \begin{pmatrix} 1 \\ 0 \\ 0 \end{pmatrix} = \begin{pmatrix} \bar{\epsilon}_{11} \\ 2\bar{\epsilon}_{12} \\ 2\bar{\epsilon}_{13} \end{pmatrix} \tag{17.97}$$

where $\widetilde{K} = C^T C'$, $\widetilde{k} = X_{\theta_C} X'_{-\theta_C}$, and $k = (\theta'_C \ 0 \ 0)^T$ (Hodges (1985)). Now $K_x = \theta'_C + \phi'$, so $\kappa_x = \phi'$; $\gamma_x = \bar{\epsilon}_{11} = u'_e$. Writing $C = C_s C^{BE}$, the curvature K is the sum of shear and bending/torsion terms:

$$\widetilde{K} = C^T C' = C^{EB} (C_s^T C_s') C^{BE} + C^{EB} (C^{BE})' = C^{EB} \widetilde{K}_s C^{BE} + \widetilde{K}^{EB/E} \tag{17.98}$$

17.3 Nonlinear Beam Theory

Hence to second order,

$$\kappa = X_\theta \begin{pmatrix} 0 \\ -\omega' \\ v' \end{pmatrix} + X_\theta \begin{pmatrix} \phi' \\ -\beta' \\ C_\beta \zeta' \end{pmatrix} = X_\theta \begin{pmatrix} \phi' \\ -w'' - \omega' \\ v'' + v' \end{pmatrix} \quad (17.99)$$

$$\gamma = X_\theta \begin{bmatrix} 1 & v' + v & w' + \omega \\ -(v' + v) & 1 & 0 \\ -(w' + \omega) & 0 & 1 \end{bmatrix} \begin{pmatrix} 1 + u' \\ v' \\ w' \end{pmatrix} - \begin{pmatrix} 1 \\ 0 \\ 0 \end{pmatrix} = X_\theta \begin{pmatrix} u'_e \\ -v \\ -\omega \end{pmatrix} \quad (17.100)$$

with $\theta \cong \theta_C + \phi$ here. Without the shear deformation, the second-order moment strain measure is

$$\kappa = \begin{pmatrix} \phi' \\ -C_\theta w'' + S_\theta v'' \\ S_\theta w'' + C_\theta v'' \end{pmatrix} \quad (17.101)$$

For the geometrically exact model, rotation of the deformed section gives

$$K = R \begin{pmatrix} \theta' \\ \beta' \\ \zeta' \end{pmatrix} = X_\theta \begin{pmatrix} \theta' + S_\beta \zeta' \\ -\beta' \\ C_\beta \zeta' \end{pmatrix} = X_\theta \begin{pmatrix} \theta'_C + \phi' \\ -\beta' \\ C_\beta \zeta' \end{pmatrix} \quad (17.102)$$

From $\Theta' = -S_\beta \zeta'$, there follows $K_x = \theta' + S_\beta \zeta' = \theta'_C + \phi'$, and so $\kappa_x = \phi'$ exactly. Then the moment strain measure is

$$\kappa = X_\theta \begin{pmatrix} \phi' \\ -\beta' \\ C_\beta \zeta' \end{pmatrix} \quad (17.103)$$

to second order.

The basis vectors of the undistorted and distorted beam are $\mathbf{g}_m = \partial r / \partial y_m$ and $\mathbf{G}_m = \partial R / \partial y_m$, respectively. Then the metric tensors are $g_{mn} = \mathbf{g}_m \cdot \mathbf{g}_n$ and $G_{mn} = \mathbf{G}_m \cdot \mathbf{G}_n$, and the Green-Lagrange strain tensor is obtained from $f_{mn} = \frac{1}{2}(G_{mn} - g_{mn})$. The basis vector $\mathbf{g}_1$ is tangent to the line described by constant η and ζ, which is a helix for a beam with pretwist ($\theta'_C \neq 0$). So $\mathbf{g}_1$ is not perpendicular to $\mathbf{g}_2$ and $\mathbf{g}_3$. Using the strain $\gamma_{mn} = f_{mn}$ is equivalent to assuming that the axial stress follows the basis vectors in the twisted beam. Assuming that the constitutive relation is defined in local rectangular Cartesian coordinates z_k is generally more appropriate. The unit vectors of z_k are $\mathbf{e}_k = (\mathbf{i}, \mathbf{g}_2, \mathbf{g}_3)$. Thus the strain tensor γ_{mn} is related to f_{mn} by $f_{mn} = (\partial z_k / \partial y_m)(\partial z_l / \partial y_n) \gamma_{kl}$, where $\partial z_k / \partial y_m = \mathbf{e}_k \cdot \mathbf{g}_m$. The transformation

$$\left[\frac{\partial z_k}{\partial y_m} \right] = \begin{bmatrix} 1 & 0 & 0 \\ -\theta'_C \zeta & 1 & 0 \\ \theta'_C \eta & 0 & 1 \end{bmatrix} \quad (17.104)$$

gives $\gamma_{11} = f_{11} + 2\theta'_C(\zeta f_{12} - \eta f_{13})$. With the assumption of small strain, $\gamma_{mn} \cong \epsilon_{mn}$, where ϵ_{mn} is linear in the strain measures. Thus

$$\epsilon_{11} = \frac{1}{2}(G_{11} - g_{11}) + 2\theta'_C(\zeta \epsilon_{12} - \eta \epsilon_{13})$$

$$\cong u'_e - \kappa_z \eta + \kappa_y \zeta + (\theta'_C \phi' + \frac{1}{2}\phi'^2)(\eta^2 + \zeta^2) + W'_1$$

$$+ 2\theta'_C(\zeta \bar{\epsilon}_{12} - \eta \bar{\epsilon}_{13}) - \theta'_C \phi'(\zeta^2 + \eta^2)$$

$$+ \theta'_C(\zeta(W_{1\eta} + W'_2) - \eta(W_{1\zeta} + W'_3))$$

$$\cong u'_e - \kappa_z \eta + \kappa_y \zeta + \frac{1}{2}\phi'^2(\eta^2 + \zeta^2)$$
$$+ 2\theta'_C(\zeta \bar{\epsilon}_{12} - \eta \bar{\epsilon}_{13}) + \theta'_C \phi'(\zeta \lambda_\eta - \eta \lambda_\zeta) \quad (17.105)$$

$$2\epsilon_{12} = G_{12} - g_{12} \cong 2\bar{\epsilon}_{12} - (K_x - \theta'_C)\zeta + W_{1\eta} + W'_2$$
$$\cong 2\bar{\epsilon}_{12} + (\lambda_\eta - \zeta)\phi' \quad (17.106)$$

$$2\epsilon_{13} = G_{13} - g_{13} \cong 2\bar{\epsilon}_{13} + (K_x - \theta'_C)\eta + W_{1\zeta} + W'_3$$
$$\cong 2\bar{\epsilon}_{13} + (\lambda_\zeta + \eta)\phi' \quad (17.107)$$

are the required strains. The nonlinear term producing coupling between extension and torsion is conventionally retained in ϵ_{11}. In the final expression for each strain, the representative warping function $W_1 = \lambda \phi'$ has been used. The complete effects of warp must be considered when the section elastic constants are evaluated. Thus

$$\begin{pmatrix} \epsilon_{11} \\ 2\epsilon_{12} \\ 2\epsilon_{13} \end{pmatrix} = \begin{bmatrix} 1 & \theta'_C \zeta & -\theta'_C \eta & \theta'_C(\zeta\lambda_\eta - \eta\lambda_\zeta) + \frac{1}{2}\phi'(\eta^2 + \zeta^2) & \zeta & -\eta \\ 0 & 1 & 0 & \lambda_\eta - \zeta & 0 & 0 \\ 0 & 0 & 1 & \lambda_\zeta + \eta & 0 & 0 \end{bmatrix} \begin{pmatrix} u'_e \\ 2\bar{\epsilon}_{12} \\ 2\bar{\epsilon}_{13} \\ \phi' \\ \kappa_y \\ \kappa_z \end{pmatrix}$$
(17.108)

are the relations between the strain and the strain measures of the beam.

From Hamilton's principle, the strain energy is the integral over the structure of the product of the stress and strain:

$$\delta U = \int \delta \epsilon^T \sigma \, d\Omega = \int (\sigma_{11}\delta\epsilon_{11} + \sigma_{22}\delta\epsilon_{22} + \sigma_{33}\delta\epsilon_{33}$$
$$+ 2\sigma_{12}\delta\epsilon_{12} + 2\sigma_{13}\delta\epsilon_{13} + 2\sigma_{23}\delta\epsilon_{23}) \, d\Omega \quad (17.109)$$

The stress is obtained from the strain by the constitutive law $\sigma_{ij} = E_{ijkl}\epsilon_{kl}$. Beam theory assumes that only the stresses acting on the plane perpendicular to the beam axis are important. So σ_{22}, σ_{33}, and σ_{23} are neglected, and the constitutive law reduces to

$$\begin{pmatrix} \sigma_{11} \\ \sigma_{12} \\ \sigma_{13} \end{pmatrix} = \begin{bmatrix} Q_{11} & Q_{15} & Q_{16} \\ Q_{51} & Q_{55} & Q_{56} \\ Q_{65} & Q_{65} & Q_{66} \end{bmatrix} \begin{pmatrix} \epsilon_{11} \\ 2\epsilon_{12} \\ 2\epsilon_{13} \end{pmatrix} \quad (17.110)$$

The strain energy can now be written in terms of the section loads:

$$\delta U = \iint \delta \epsilon^T \sigma \, dA \, dx$$
$$= \int_0^\ell \left[F_x \delta u'_e + F_y 2\delta\bar{\epsilon}_{12} + F_z 2\delta\bar{\epsilon}_{13} + M_x \delta\phi' + M_y \delta\kappa_y + M_z \delta\kappa_z \right] dx \quad (17.111)$$

17.3 Nonlinear Beam Theory

The section loads are obtained from the stress, and hence from the section strain measures:

$$\begin{pmatrix} F_x \\ F_y \\ F_z \\ M_x \\ M_y \\ M_z \end{pmatrix} = \int \begin{bmatrix} 1 & 0 & 0 \\ \theta'_C \zeta & 1 & 0 \\ -\theta'_C \eta & 0 & 1 \\ \theta'_C(\zeta\lambda_\eta - \eta\lambda_\zeta) + \phi'(\eta^2 + \zeta^2) & \lambda_\eta - \zeta & \lambda_\zeta + \eta \\ \zeta & 0 & 0 \\ -\eta & 0 & 0 \end{bmatrix} \begin{pmatrix} \sigma_{11} \\ \sigma_{12} \\ \sigma_{13} \end{pmatrix} dA$$

$$= T \begin{pmatrix} u'_e \\ 2\overline{\epsilon}_{12} \\ 2\overline{\epsilon}_{13} \\ \phi' \\ \kappa_y \\ \kappa_z \end{pmatrix} \qquad (17.112)$$

At this point the effects of transverse shear are statically eliminated, reducing the 6×6 matrix T to the 4×4 matrix S. Neglecting the shear force (not the shear strain) is appropriate, so T is inverted, the F_y and F_z rows and columns of T^{-1} are eliminated to produce S^{-1}, and the inverse of the resulting matrix gives S. At a constrained end, assuming zero shear strain might be more appropriate; then S is simply obtained by eliminating the F_y and F_z rows and columns of T.

Including the nonlinear terms that couple extension and torsion, the section loads are

$$\begin{pmatrix} F_x \\ M_x \\ M_y \\ M_z \end{pmatrix} = \begin{bmatrix} S_{uu} & S_{u\phi} + \frac{1}{2}\phi' S_{uu}k_P^2 & S_{uw} & S_{uv} \\ S_{\phi u} + \phi' S_{uu}k_P^2 & S_{\phi\phi} & S_{\phi w} & S_{\phi v} \\ S_{wu} & S_{w\phi} & S_{ww} & S_{wv} \\ S_{vu} & S_{v\phi} & S_{vw} & S_{vv} \end{bmatrix} \begin{pmatrix} u'_e \\ \phi' \\ \kappa_y \\ \kappa_z \end{pmatrix} = S \begin{pmatrix} u'_e \\ \phi' \\ \kappa_y \\ \kappa_z \end{pmatrix} \qquad (17.113)$$

Using the beam theory for anisotropic or composite materials including transverse shear deformation, the section elastic constants S are the stiffnesses required. With

$$\begin{pmatrix} \delta u'_e \\ \delta \phi' \\ \delta \kappa_y \\ \delta \kappa_z \end{pmatrix} = C \begin{pmatrix} \delta q_u \\ \delta q_v \\ \delta q_w \\ \delta q_\phi \end{pmatrix} = \begin{bmatrix} C_f^T \\ C_x^T \\ C_y^T \\ C_z^T \end{bmatrix} \begin{pmatrix} \delta q_u \\ \delta q_v \\ \delta q_w \\ \delta q_\phi \end{pmatrix} \qquad (17.114)$$

the strain energy is

$$\delta U = \int_0^\ell \left[F_x \delta u'_e + M_x \delta\phi' + M_y \delta\kappa_y + M_z \delta\kappa_z \right] dx$$

$$= \delta q_u^T K_u + \delta q_v^T K_v + \delta q_w^T K_w + \delta q_\phi^T K_\phi = \delta q^T K \qquad (17.115)$$

Thus the structural terms for the equations of motion of the elastic degrees of freedom are:

$$K = \int_0^\ell \left[C_f F_x + C_x M_x + C_y M_y + C_z M_z \right] dx = \begin{bmatrix} K_u \\ K_v \\ K_w \\ K_\phi \end{bmatrix} \qquad (17.116)$$

The result is

$$K_u = \int_0^\ell h'_u F_x \, dx \tag{17.117}$$

$$K_v = \int_0^\ell h''_v (S_\theta M_y + C_\theta M_z) \, dx \tag{17.118}$$

$$K_w = \int_0^\ell h''_w (-C_\theta M_y + S_\theta M_z) \, dx \tag{17.119}$$

$$K_\phi = \int_0^\ell \left(h'_\phi M_x + h_\phi (\kappa_z M_y - \kappa_y M_z) \right) dx \tag{17.120}$$

for the second-order model.

Finally, the strain energy is obtained using Euler-Bernoulli theory for a beam of isotropic materials with an elastic axis. Transverse shear effects are neglected, and the elastic axis is the beam axis. The constitutive law is now

$$\begin{pmatrix} \sigma_{11} \\ \sigma_{12} \\ \sigma_{13} \end{pmatrix} = \begin{bmatrix} E & 0 & 0 \\ 0 & G & 0 \\ 0 & 0 & G \end{bmatrix} \begin{pmatrix} \epsilon_{11} \\ 2\epsilon_{12} \\ 2\epsilon_{13} \end{pmatrix} \tag{17.121}$$

Then the section loads are obtained from the stress, and hence from the section strain measures:

$$\begin{pmatrix} F_x \\ M_x \\ M_y \\ M_z \end{pmatrix} = \int dA \begin{bmatrix} 1 & 0 & 0 & 0 \\ \theta'_C(\zeta\lambda_\eta - \eta\lambda_\zeta) + \phi'(\eta^2 + \zeta^2) & \lambda_\eta - \zeta & \lambda_\zeta + \eta \\ \zeta & 0 & 0 \\ -\eta & 0 & 0 \end{bmatrix}$$

$$\begin{bmatrix} E & E\theta'_C(\zeta\lambda_\eta - \eta\lambda_\zeta) + \tfrac{1}{2}E\phi'(\eta^2 + \zeta^2) & E\zeta & -E\eta \\ 0 & G(\lambda_\eta - \zeta) & 0 & 0 \\ 0 & G(\lambda_\zeta + \eta) & 0 & 0 \end{bmatrix} \begin{pmatrix} u'_e \\ \phi' \\ \kappa_y \\ \kappa_z \end{pmatrix}$$

$$= S \begin{pmatrix} u'_e \\ \phi' \\ \kappa_y \\ \kappa_z \end{pmatrix} \tag{17.122}$$

with

$$S = \begin{bmatrix} EA & \theta'_C EA\, k_T^2 + \tfrac{1}{2}\phi' EA\, k_P^2 & EAz_C & -EAy_C \\ \theta'_C EA\, k_T^2 + \phi' EA\, k_P^2 & GJ + \theta'^2_C EI_{pp} & \theta'_C EI_{zp} & -\theta'_C EI_{yp} \\ EAz_C & \theta'_C EI_{zp} & \widehat{EI}_{zz} & -\widehat{EI}_{yz} \\ -EAy_C & -\theta'_C EI_{yp} & -\widehat{EI}_{yz} & \widehat{EI}_{yy} \end{bmatrix} \tag{17.123}$$

The section integrals are here evaluated relative to the elastic axis:

$$EI_{yp} = \int Ey(z\lambda_y - y\lambda_z) \, dA \tag{17.124}$$

$$EI_{zp} = \int Ez(z\lambda_y - y\lambda_z) \, dA \tag{17.125}$$

$$EI_{pp} = \int E(z\lambda_y - y\lambda_z)^2 \, dA \tag{17.126}$$

17.3 Nonlinear Beam Theory

$$\widehat{EI}_{yy} = \int Ey^2\, dA = \int E(y - y_C)^2\, dA + EAy_C^2 = EI_{yy} + EAy_C^2 \tag{17.127}$$

$$\widehat{EI}_{zz} = \int Ez^2\, dA = \int E(z - z_C)^2\, dA + EAz_C^2 = EI_{zz} + EAz_C^2 \tag{17.128}$$

$$\widehat{EI}_{yz} = \int Eyz\, dA = \int E(y - y_C)(z - z_C)\, dA + EAy_Cz_C = EAy_Cz_C \tag{17.129}$$

but are conventionally defined relative to the tension center instead. The higher-order section integrals EI_{xp}, EI_{zp}, and EI_{pp} are seldom available. Thus the stiffness matrix S is

$$\begin{bmatrix} EA & \theta_C' EA\, k_T^2 + \tfrac{1}{2}\phi' EA\, k_P^2 & EAz_C & -EAy_C \\ \theta_C' EA\, k_T^2 + \phi' EA\, k_P^2 & GJ & 0 & 0 \\ EAz_C & 0 & EI_{zz} + EAz_C^2 & -EAy_Cz_C \\ -EAy_C & 0 & -EAy_Cz_C & EI_{yy} + EAy_C^2 \end{bmatrix} \tag{17.130}$$

Beam theory for isotropic materials with an elastic axis requires the following section structural properties (modulus weighted):

$$EAy_C = \int Ey\, dA \tag{17.131}$$

$$EAz_C = \int Ez\, dA \tag{17.132}$$

$$EAk_P^2 = \int E(y^2 + z^2)\, dA \tag{17.133}$$

$$EA = \int E\, dA \tag{17.134}$$

$$EAk_T^2 = \int E(z\lambda_y - y\lambda_z)\, dA \tag{17.135}$$

$$GJ = \int G\big((\lambda_y - z)^2 + (\lambda_z + y)^2\big)\, dA \tag{17.136}$$

$$EI_{yy} = \int E(y - y_C)^2\, dA \tag{17.137}$$

$$EI_{zz} = \int E(z - z_C)^2\, dA \tag{17.138}$$

The integral is over the beam cross-section in structural principal axes with origin at the beam axis. Note that the section loads are approximately

$$F_x \cong EA(u_e' + \theta_C' \phi' k_T^2 - y_C \kappa_z + z_C \kappa_y)$$

$$M_x = (GJ + F_x k_P^2)\phi' + F_x k_T^2 \theta_C'$$

$$M_y \cong EI_{zz}\kappa_y + F_x z_C$$

$$M_z \cong EI_{yy}\kappa_z - F_x y_C$$

which illustrates the value of using the tension center for the beam axis, since then the bending moments M_y and M_z do not depend on the tension force F_x.

17.3.6 Extension-Torsion Coupling

In the previous structural analysis, the constitutive relation was assumed to be defined in local rectangular Cartesian coordinates. The consequence of this assumption is a distinction between the torsion moments produced by elastic torsion ϕ' and those by pretwist θ'_C, in the presence of a tension force:

$$M_x = (GJ + F_x k_P^2)\phi' + F_x k_T^2 \theta'_C \tag{17.139}$$

See Hodges (1980). For a circular cross-section, k_T must be zero, since pretwist is not then meaningful with isotropic materials.

If instead the constitutive law is applied in the curvilinear coordinates, the axial strain is

$$\epsilon_{11} = \frac{1}{2}(G_{11} - g_{11}) \cong u'_e - \kappa_z \eta + \kappa_y \zeta + (\theta'_C \phi' + \frac{1}{2}\phi'^2)(\eta^2 + \zeta^2) + W'_1 \tag{17.140}$$

so the relations between the strain and the strain measures of the beam become

$$\begin{pmatrix} \epsilon_{11} \\ 2\epsilon_{12} \\ 2\epsilon_{13} \end{pmatrix} = \begin{bmatrix} 1 & 0 & 0 & (\theta'_C + \tfrac{1}{2}\phi')(\eta^2 + \zeta^2) & \zeta & -\eta \\ 0 & 1 & 0 & \lambda_\eta - \zeta & 0 & 0 \\ 0 & 0 & 1 & \lambda_\zeta + \eta & 0 & 0 \end{bmatrix} \begin{pmatrix} u'_e \\ 2\overline{\epsilon}_{12} \\ 2\overline{\epsilon}_{13} \\ \phi' \\ \kappa_y \\ \kappa_z \end{pmatrix} \tag{17.141}$$

The structural properties for an isotropic beam, including the extension-torsion coupling term, are then

$$EI_{yp} = \int Ey(y^2 + z^2)\, dA \tag{17.142}$$

$$EI_{zp} = \int Ez(y^2 + z^2)\, dA \tag{17.143}$$

$$EI_{pp} = \int E(y^2 + z^2)^2\, dA \tag{17.144}$$

$$EAk_T^2 = \int E(y^2 + z^2)\, dA = EAk_P^2 \tag{17.145}$$

So $k_T = k_P$, and the torsion moment is

$$M_x = (GJ + F_x k_P^2)\phi' + F_x k_P^2 \theta'_C \tag{17.146}$$

This result often has been obtained (as in Houbolt and Brooks (1958)) by explicitly making the assumption that the axial stress is tangent to the line described by constant η and ζ, which is a helix for a pretwisted beam. Such an assumption might be appropriate for an anisotropic beam. In general, the extension-torsion coupling is defined by the section constant k_T for an isotropic beam or $S_{u\phi}$ for an anisotropic beam.

17.3.7 Kinetic Energy

Using the d'Alembert approach, in which inertial acceleration is treated as a body force, the virtual work of the inertial and gravitational forces is

$$\delta T = \int\!\!\int (\delta r^I)^T (\ddot{r}^I - g^I)\, dm\, dx \tag{17.147}$$

17.3 Nonlinear Beam Theory

The integration is over the section mass and then the beam length. The position relative to the inertial frame is the sum of the beam frame rigid motion relative to the inertial frame ($x^{BI/I}$ and C^{BI}) and the position of a point on the beam section relative to the beam frame (r^B): $r^I = x^{BI/I} + C^{IB} r^B$. So the inertial acceleration is

$$\ddot{r}^I = C^{IB}\left(\dot{v}^{BI/B} + \tilde{\omega}^{BI/B} r^B + 2\tilde{\omega}^{BI/B}\dot{r}^B + \ddot{r}^B + \tilde{\omega}^{BI/B} v^{BI/B} + \tilde{\omega}^{BI/B}\tilde{\omega}^{BI/B} r^B\right) \quad (17.148)$$

The virtual displacement can be expressed in terms of the generalized coordinates:

$$\delta r^I = \delta(x^{BI/I} + C^{IB} r^B) = \delta x^{BI/I} + \delta C^{IB} r^B + C^{IB} \delta r^B$$
$$= \delta x^{BI/I} - C^{IB}\tilde{r}^B \delta \psi^{BI/B} + C^{IB} \delta r^B$$
$$= C^{IB}\begin{bmatrix} I & -\tilde{r}^B & R_u^T & R_v^T & R_w^T & R_\phi^T \end{bmatrix} \delta q = C^{IB} R^T \delta q \quad (17.149)$$

where $\delta x^{BI/I}$ and $\delta \psi^{BI/B}$ are the virtual coordinates of the rigid-body motion of the beam frame. Then the virtual work is

$$\delta T = \delta q^T \iint R C^{BI}(\ddot{r}^I - g^I) \, dm \, dx$$
$$= (\delta x^{BI/B})^T M_x + (\delta \psi^{BI/B})^T M_\psi + \delta q_u^T M_u + \delta q_v^T M_v + \delta q_w^T M_w + \delta q_\phi^T M_\phi$$
$$= \delta q^T M \quad (17.150)$$

with M the inertial terms in the beam equations of motion.

The displacement of a point on the beam, relative to the beam frame, is $r^B = r_0 + r_y \eta_b + r_z \zeta_b$. The position on the beam axis is r_0. The variables η and ζ identify the cross-section point, relative to the section principal axes at θ_I. The variables η_b and ζ_b are relative to section axes that are bent but not twisted:

$$\begin{pmatrix} \eta_b \\ \zeta_b \end{pmatrix} = \begin{bmatrix} C_\theta & -S_\theta \\ S_\theta & C_\theta \end{bmatrix} \begin{pmatrix} \eta \\ \zeta \end{pmatrix} \quad (17.151)$$

with $\theta = \theta_I + \phi + \Theta$. The position at the center of gravity is $r_I = r_0 + r_y \eta_{Ib} + r_z \zeta_{Ib} = r_0 + \Delta r_I$. The matrix giving the virtual displacements can be written $R = R_0 + R_y \eta_b + R_z \zeta_b$. Thus the inertial terms in the beam equations of motion are evaluated from

$$M = \iint R C^{BI}(\ddot{r}^I - g^I) \, dm \, dx$$
$$= \iint R \left(\dot{v}^{BI/B} + \tilde{\omega}^{BI/B} v^{BI/B} - C^{BI} g^I \right.$$
$$\left. + \ddot{r}^B + \dot{\tilde{\omega}}^{BI/B} r^B + 2\tilde{\omega}^{BI/B}\dot{r}^B + \tilde{\omega}^{BI/B}\tilde{\omega}^{BI/B} r^B\right) dm \, dx$$
$$= \int_0^\ell \Big[R_I(\dot{v} + \tilde{\omega} v - Cg) m$$
$$+ R_I(\ddot{r}_0 + \dot{\tilde{\omega}} r_0 + 2\tilde{\omega}\dot{r}_0 + \tilde{\omega}\tilde{\omega} r_0) m$$
$$+ R_0(\Delta \ddot{r}_I + \dot{\tilde{\omega}} \Delta r_I + 2\tilde{\omega}\Delta \dot{r}_I + \tilde{\omega}\tilde{\omega}\Delta r_I) m$$
$$+ R_y(\ddot{r}_y + \dot{\tilde{\omega}} r_y + 2\tilde{\omega}\dot{r}_y + \tilde{\omega}\tilde{\omega} r_y) I_{\eta\eta}$$
$$+ R_z(\ddot{r}_z + \dot{\tilde{\omega}} r_z + 2\tilde{\omega}\dot{r}_z + \tilde{\omega}\tilde{\omega} r_z) I_{\zeta\zeta}$$
$$+ R_y(\ddot{r}_z + \dot{\tilde{\omega}} r_z + 2\tilde{\omega}\dot{r}_z + \tilde{\omega}\tilde{\omega} r_z) I_{\eta\zeta}$$
$$+ R_z(\ddot{r}_y + \dot{\tilde{\omega}} r_y + 2\tilde{\omega}\dot{r}_y + \tilde{\omega}\tilde{\omega} r_y) I_{\eta\zeta} \Big] dx \quad (17.152)$$

where

$$I_{\eta\eta} = \int \eta_b^2 \, dm = \frac{1}{2}(I_\theta + I_P C_{2\theta}) - y_I z_I m S_{2\theta} \tag{17.153}$$

$$I_{\zeta\zeta} = \int \zeta_b^2 \, dm = \frac{1}{2}(I_\theta - I_P C_{2\theta}) + y_I z_I m S_{2\theta} \tag{17.154}$$

$$I_{\eta\zeta} = \int \eta_b \zeta_b \, dm = \frac{1}{2} I_P S_{2\theta} + y_I z_I m C_{2\theta} \tag{17.155}$$

with $C_{2\theta} = \cos 2\theta = C_\theta^2 - S_\theta^2$ and $S_{2\theta} = \sin 2\theta = 2 S_\theta C_\theta$. The required section inertial properties (mass weighted) are

$$m = \int dm \tag{17.156}$$

$$m y_I = \int y \, dm \tag{17.157}$$

$$m z_I = \int z \, dm \tag{17.158}$$

$$I_\theta = \int (y^2 + z^2) \, dm \tag{17.159}$$

$$I_P = \int (y^2 - z^2) \, dm \tag{17.160}$$

The integral is over the beam cross-section, in inertial principal axes with origin at the beam axis. For a circular cross-section, $I_P = 0$; for a cross-section with small z dimension, $I_P \cong I_\theta$.

The kinematics of the rigid-body motion are exact. For the geometrically exact model, the motion of a point on the beam cross-section, relative to the beam frame, is

$$r^B = x^{EB/B} + C^{BE} \begin{pmatrix} 0 \\ \eta \\ \zeta \end{pmatrix} = x^{EB/B} + (C^{BE} X_\theta) \begin{pmatrix} 0 \\ \eta_b \\ \zeta_b \end{pmatrix}$$

$$= \begin{pmatrix} x+u \\ v \\ w \end{pmatrix} + \begin{pmatrix} -S_\zeta \\ C_\zeta \\ 0 \end{pmatrix} \eta_b + \begin{pmatrix} -S_\beta C_\zeta \\ -S_\beta S_\zeta \\ C_\beta \end{pmatrix} \zeta_b \tag{17.161}$$

The variables η_b and ζ_b identify the cross-section point, relative to section axes that are bent but not twisted. With the second-order model,

$$r^B = \begin{pmatrix} x+u \\ v \\ w \end{pmatrix} + \begin{pmatrix} -v' \\ 1 \\ 0 \end{pmatrix} \eta_b + \begin{pmatrix} -w' \\ 0 \\ 1 \end{pmatrix} \zeta_b \tag{17.162}$$

is the position on the cross-section. Since η_b and ζ_b depend on the pitch angle $\theta = \theta_I + \phi + \Theta$, they contribute to time derivatives ($\dot\eta_b = -\zeta_b \dot\theta$ and $\dot\zeta_b = \eta_b \dot\theta$), virtual displacements, and linearization relative to θ_I.

17.3.8 Equations of Motion

The equations for the rigid-body motion are obtained from the virtual work of the inertial and applied forces. The linear and angular equations are $M_x = F^B$ and

$M_\psi = M^B$, where F^B and M^B are the total force and moment acting on the beam, in the beam frame axes.

The equations for the elastic motion are obtained from the virtual work of the structural, inertial, and applied forces. The equations for the elastic generalized coordinates are

$$\begin{pmatrix} M_u + K_u \\ M_v + K_v \\ M_w + K_w \\ M_\phi + K_\phi \end{pmatrix} = \sum \left(v_q^T F^E + \omega_q^T M^E \right) \tag{17.163}$$

where F^E and M^E are the sum of all loads acting on the beam and the partial velocities v_q and ω_q are obtained from the expressions for the linear and angular velocities in terms of the generalized coordinates.

The equations of motion require integration of the beam properties (such as mass and stiffness) along its length. This integration is performed numerically, typically using Gaussian quadrature for a finite element analysis. Thus

$$\int_0^\ell f(x)\, dx = \frac{\ell}{2} \sum_{i=1}^N w_i f(x_i) + \mathrm{O}(f^{(2N)}) \tag{17.164}$$

where $x_i = \frac{\ell}{2}(\xi_i + 1)$, for Gaussian points ξ_i and weights w_i. Gaussian integration implies a polynomial approximation to the variation of the properties, which is accurate only if the variation is sufficiently smooth. Also, the shape functions for the elastic motion are continuous and cannot accurately represent large changes in the curvature or slope. Thus a beam having properties that vary rapidly along its length must be modeled by defining nodes to break the beam into segments, with the major jumps in properties at the nodes. If very short beam segments are required to accommodate the properties, then beam theory is probably not applicable.

17.3.9 Structural Loads

The section load at axial station x_L consists of the torsion and bending moments, the axial tension, and the section shear forces. The load acts on the beam segment extending inboard of x_L, at the tension center, in structural principal axes. The section load can be calculated from the deflection or by force balance.

The deflection method obtains the section load from the elastic motion and structural coefficients. Essentially the load is evaluated from the stiffness and elastic displacement at x_L: for uncoupled bending, moment = $EI \times$ curvature. The accuracy of this calculation depends on the accuracy of the representation of the curvature or slope (the product of the degrees of freedom and shape functions). At a step in stiffness there should be a corresponding step in curvature or slope, such that the load remains continuous. With a small number of shape functions, such a step cannot be simulated well, so the results for the reaction are not accurate near a step in stiffness. Also, the theory does not imply continuity of curvature on the two sides of a node.

The structural analysis provides expressions for the reactions at the beam axis, from which bending moments at the tension center can be obtained. Thus the section load from the deflection method is

$$\begin{pmatrix} F_x \\ M_x \\ M_{yTC} \\ M_{zTC} \end{pmatrix} = S \begin{pmatrix} u'_e \\ \phi' \\ \kappa_y \\ \kappa_z \end{pmatrix} \quad (17.165)$$

with

$$S = \begin{bmatrix} S_{uu} & S_{u\phi} + \tfrac{1}{2}\phi' S_{uu} k_P^2 & S_{uw} & S_{uv} \\ S_{\phi u} + \phi' S_{uu} k_P^2 & S_{\phi\phi} & S_{\phi w} & S_{\phi v} \\ S_{wu} - S_{uu} z_C & S_{w\phi} - (S_{u\phi} + \tfrac{1}{2}\phi' S_{uu} k_P^2) z_C & S_{ww} - S_{uw} z_C & S_{wv} - S_{uv} z_C \\ S_{vu} + S_{uu} y_C & S_{v\phi} + (S_{u\phi} + \tfrac{1}{2}\phi' S_{uu} k_P^2) y_C & S_{vw} + S_{uw} y_C & S_{vv} + S_{uv} y_C \end{bmatrix} \quad (17.166)$$

for the anisotropic structure, or

$$S = \begin{bmatrix} EA & \theta'_C EA\, k_T^2 + \tfrac{1}{2}\phi' EA\, k_P^2 & EAz_C & -EAy_C \\ \theta'_C EA\, k_T^2 + \phi' EA\, k_P^2 & GJ & 0 & 0 \\ 0 & -(\theta'_C EA\, k_T^2 + \tfrac{1}{2}\phi' EA\, k_P^2) z_C & EI_{zz} & 0 \\ 0 & (\theta'_C EA\, k_T^2 + \tfrac{1}{2}\phi' EA\, k_P^2) y_C & 0 & EI_{yy} \end{bmatrix} \quad (17.167)$$

for the isotropic structure. The torsion moment is M_x, the bending moments are M_y and M_z, and the axial tension force is F_x; the shear forces are not available with the deflection method.

The force-balance method obtains the section load from the difference between the applied forces and inertial forces acting on the beam segment. For symmetry, the section loads calculated using the forces on either side of x_L are combined. The force-balance method can capture the steps in the section load produced by discrete loads on the beam. The position of the tension center at span station x_L, from the origin of the beam frame, is r_L. The difference between the applied forces and the inertial forces, acting on the segment of beam outboard of x_L, is

$$F_{L+} = \left[\sum_{x > x_L} F - \int_{x_L}^{\ell} \int (a - g)\, dm\, dx \right] \quad (17.168)$$

$$M_{L+} = \left[\sum_{x > x_L} (M + (\tilde{x} - \tilde{r}_L) F) - \int_{x_L}^{\ell} \int (\tilde{r} - \tilde{r}_L)(a - g)\, dm\, dx \right] \quad (17.169)$$

The difference between the applied forces and the inertial forces, acting on the segment of beam inboard of x_L, is

$$F_{L-} = -\left[\sum_{x < x_L} F - \int_0^{x_L} \int (a - g)\, dm\, dx \right] \quad (17.170)$$

$$M_{L-} = -\left[\sum_{x < x_L} (M + (\tilde{x} - \tilde{r}_L) F) - \int_0^{x_L} \int (\tilde{r} - \tilde{r}_L)(a - g)\, dm\, dx \right] \quad (17.171)$$

All terms are transformed to the bent cross-section axes. The first term is the summation of all applied loads (forces F and moments M, acting at position x relative to the origin of the beam frame) outboard or inboard of x_L; the summation becomes an integral for distributed loads. The second term is the integral of the inertial acceleration a and gravitational acceleration g acting on the element of mass ($dm\,dx$) at position r (relative to the origin of the beam frame). The section loads calculated using the forces on either side of x_L are combined, such that the loads at the beam ends are the same as a nodal reaction. Thus

$$F_L = (x_L/\ell)F_{L+} + (1 - x_L/\ell)F_{L-} \tag{17.172}$$

$$M_L = (x_L/\ell)M_{L+} + (1 - x_L/\ell)F_{M-} \tag{17.173}$$

or

$$F_L = \left[\sum WF - \iint W(a-g)\,dm\,dx\right] \tag{17.174}$$

$$M_L = \left[\sum W(M + \tilde{x}F) - \iint W\tilde{r}(a-g)\,dm\,dx\right] - \tilde{r}_L F_L \tag{17.175}$$

with the weighting function

$$W = \begin{cases} x_L/\ell & x > x_L \\ x_L/\ell - 1 & x \le x_L \end{cases} \tag{17.176}$$

The required sums and integrals can be evaluated as for the rigid-body equations of motion, with the addition of the weighting function W. However, Gaussian integration of the inertial forces does not treat the step in W accurately. The integrated inertial load is continuous with x_L if the integrand at x_L is handled analytically. So the quadrature becomes

$$F = \int_0^\ell W f(x)\,dx = \int_0^\ell W\bigl(f(x) - f(x_L)\bigr)\,dx + \int_0^\ell W f(x_L)\,dx$$

$$= \frac{\ell}{2}\sum_{i=1}^N w_i W(x_i)\bigl(f(x_i) - f(x_L)\bigr) \tag{17.177}$$

where $x_i = (\xi_i + 1)(\ell/2)$ for Gaussian points ξ_i and weights w_i. The force-balance method can capture the steps in the section load produced by discrete loads on the beam. Distributed loads (as from aerodynamics) must be treated as such for good results.

17.4 Equations of Motion for Elastic Rotor Blade

The equations of motion in implicit form with finite element discretization, as developed in the preceding section, are best for computational applications. Differential equations of motion are also useful, particularly for an analysis of the entire blade. Consider bending and torsion of an elastic blade, without shear. Equation 17.111 for the strain energy becomes

$$\delta U = \iint \delta\epsilon^T \sigma\,dA\,dx = \int_0^\ell \left[F_x \delta u_e' + M_x \delta\phi' + M_y \delta\kappa_y + M_z \delta\kappa_z\right]dx \tag{17.178}$$

The virtual extension is $\delta u'_e = \delta u' + v'\delta v' + w'\delta w'$ (equation 17.77), and the curvature is

$$\begin{pmatrix} \delta\kappa_y \\ \delta\kappa_x \end{pmatrix} = \begin{bmatrix} C_\theta & S_\theta \\ -S_\theta & C_\theta \end{bmatrix} \begin{pmatrix} -\delta w'' + v''\delta\phi \\ \delta v'' + w''\delta\phi \end{pmatrix} \quad (17.179)$$

(equation 17.101). So

$$\delta U = \int_0^\ell \Big[F_x(\delta u' + v'\delta v' + w'\delta w') + M_x\delta\phi'$$
$$+ (M_z C_\theta + M_y S_\theta)(\delta v'' + w''\delta\phi)$$
$$+ (M_z S_\theta - M_y C_\theta)(\delta w'' - v''\delta\phi) \Big] dx \quad (17.180)$$

Integrating by parts gives

$$\delta U = \int_0^\ell \Big[-F'_x \delta u - \big(M'_x - (M_z C_\theta + M_y S_\theta) w'' + (M_z S_\theta - M_y C_\theta) v''\big)\delta\phi$$
$$+ \big((M_z C_\theta + M_y S_\theta)'' - (F_x v')'\big)\delta v$$
$$+ \big((M_z S_\theta - M_y C_\theta)'' - (F_x w')'\big)\delta w \Big] dx \quad (17.181)$$

See Hodges and Dowell (1974). Using the curvature in pitched axes,

$$\begin{pmatrix} \kappa_y \\ \kappa_x \end{pmatrix} = \begin{bmatrix} C_\theta & S_\theta \\ -S_\theta & C_\theta \end{bmatrix} \begin{pmatrix} -w'' \\ v'' \end{pmatrix} \quad (17.182)$$

equation 17.130 gives the section resultant loads:

$$F_x = EA\Big[u'_e + \theta'_C k_T^2 \phi' - y_C(v''C_\theta + w''S_\theta) + z_C(v''S_\theta - w''C_\theta)\Big] \quad (17.183)$$

$$M_x = GJ\phi' + EAu'_e(k_T^2 \theta'_C + k_P^2 \phi')$$
$$\cong GJ\phi' + F_x(k_T^2 \theta'_C + k_P^2 \phi') \quad (17.184)$$

$$M_y = EI_{zz}(v''S_\theta - w''C_\theta) + z_C EA\Big[u'_e - y_C(v''C_\theta + w''S_\theta) + z_C(v''S_\theta - w''C_\theta)\Big]$$
$$\cong EI_{zz}(v''S_\theta - w''C_\theta) + z_C F_x \quad (17.185)$$

$$M_z = EI_{yy}(v''C_\theta + w''S_\theta) - y_C EA\Big[u'_e - y_C(v''C_\theta + w''S_\theta) + z_C(v''S_\theta - w''C_\theta)\Big]$$
$$\cong EI_{yy}(v''C_\theta + w''S_\theta) - y_C F_x \quad (17.186)$$

Equation 17.181 has an elastic torsion moment due to bending:

$$\Delta T = -M_y(v''C_\theta + w''S_\theta) + M_z(v''S_\theta - w''C_\theta)$$
$$= (EI_{yy} - EI_{zz})(v''C_\theta + w''S_\theta)(v''S_\theta - w''C_\theta) \quad (17.187)$$

17.4 Equations of Motion for Elastic Rotor Blade

which is a generalization of equation 16.149. Torsion is complicated, even at second order, and often higher-order terms must be included. Equation 17.181 also shows the coupling of bending by pitch angle:

$$\begin{pmatrix} M_z C_\theta + M_y S_\theta \\ M_z S_\theta - M_y C_\theta \end{pmatrix} = \begin{bmatrix} C_\theta & -S_\theta \\ S_\theta & C_\theta \end{bmatrix} \begin{pmatrix} M_z \\ -M_y \end{pmatrix}$$

$$= \begin{bmatrix} C_\theta & -S_\theta \\ S_\theta & C_\theta \end{bmatrix} \begin{bmatrix} EI_{yy} & 0 \\ 0 & EI_{zz} \end{bmatrix} \begin{bmatrix} C_\theta & S_\theta \\ -S_\theta & C_\theta \end{bmatrix} \begin{pmatrix} v'' \\ w'' \end{pmatrix}$$

$$= X_\theta^T (EI) X_\theta \begin{pmatrix} v'' \\ w'' \end{pmatrix} \qquad (17.188)$$

The position of a point on the blade section is

$$r^B = \begin{pmatrix} x + u - v'\eta_b - w'\zeta_b \\ v + \eta_b \\ w + \zeta_b \end{pmatrix} \qquad (17.189)$$

(equation 17.162). The kinetic energy $\delta T = \iint (\delta r^I)^T \ddot{r}^I \, dm \, dx$ requires the virtual displacement (equation 17.149),

$$\delta r^I = C^{IB} \delta r^B = \begin{pmatrix} \delta u \\ \delta v \\ \delta w \end{pmatrix} + \eta_b \begin{pmatrix} -\delta v' - w'\delta\phi \\ 0 \\ \delta\phi \end{pmatrix} + \zeta_b \begin{pmatrix} -\delta w' + v'\delta\phi \\ -\delta\phi \\ 0 \end{pmatrix} \qquad (17.190)$$

(using $\delta\eta_b = -\zeta_b\delta\phi$ and $\delta\zeta_b = \eta_b\delta\phi$), as well as and the inertial acceleration (equation 17.148):

$$a = \ddot{r}^I = C^{IB}\left(\ddot{r}^B + 2\widetilde{\omega}\dot{r}^B + \widetilde{\omega}\widetilde{\omega}r^B\right)$$

$$= \begin{pmatrix} \ddot{u} - 2\Omega\dot{v} - \Omega^2(x+u) \\ \ddot{v} + 2\Omega\dot{u} - \Omega^2 v \\ \ddot{w} \end{pmatrix}$$

$$+ \eta_b \begin{pmatrix} -\ddot{v}' - 2\dot{w}'\dot{\phi} - w'\ddot{\phi} + \Omega^2 v' \\ -2\Omega(\dot{v}' + w'\dot{\phi}) - \Omega^2 \\ \ddot{\phi} \end{pmatrix}$$

$$+ \zeta_b \begin{pmatrix} -\ddot{w}' + 2\dot{v}'\dot{\phi} + v'\ddot{\phi} + 2\Omega\dot{\phi} + \Omega^2 w' \\ -\ddot{\phi} - 2\Omega(\dot{w}' - v'\dot{\phi}) \\ 0 \end{pmatrix} = \begin{pmatrix} A_x \\ A_y \\ A_z \end{pmatrix} \qquad (17.191)$$

with rotation at constant rate about the z-axis, so $\omega = (0 \ 0 \ \Omega)^T$. Then to lowest order, the kinetic energy is

$$\delta T = \iint \Big[A_z \delta u + A_y \delta v - A_x \eta_b \delta v' + A_z \delta w - A_x \zeta_b \delta w' $$
$$+ \left(\eta_b A_z - \zeta_b A_y - \Omega^2 x(-\eta_b w' + \zeta_b v')\right) \delta\phi \Big] dm \, dx \qquad (17.192)$$

where

$$A_x = -\Omega^2 x - 2\Omega\dot{v} \qquad (17.193)$$

$$A_y = \ddot{v} + 2\Omega\dot{u} - \Omega^2 v + \eta_b(-2\Omega\dot{v}' - \Omega^2) + \zeta_b(-\ddot{\phi} - 2\Omega\dot{w}') \qquad (17.194)$$

$$A_z = \ddot{w} + \eta_b\ddot{\phi} \qquad (17.195)$$

See Hodges and Dowell (1974). Integrating by parts, and introducing the section integrals of the mass, gives

$$\delta T = \int_0^\ell \Big[A_x m\, \delta u + \big(A_{yI} - \Omega^2 \zeta_{bI}\phi - (A_x \eta_{bI} + \Omega^2 x\zeta_{bI}\phi)'\big) m\, \delta v$$
$$+ \big(A_{zI} - (A_x \zeta_{bI} - \Omega^2 x\eta_{bI}\phi)'\big) m\, \delta w$$
$$+ \big(\eta_{bI} m(\ddot{w} + \Omega^2 xw') - \zeta_{bI} m(\ddot{v} + 2\Omega\dot{u} - \Omega^2 v + \Omega^2 xv')\big)$$
$$+ I_{\eta\eta}\ddot{\phi} + I_{\zeta\zeta}(\ddot{\phi} + 2\Omega\dot{w}') + I_{\zeta\eta}(2\Omega\dot{v}' + \Omega^2)$$
$$+ \Omega^2 (I_P C_{2\theta} + 2y_I z_I m S_{2\theta})\phi \Big) \delta\phi \Big] dx \qquad (17.196)$$

The centrifugal terms have been linearized about θ_I, using $\Delta\eta_b = -\zeta_b\phi$, $\Delta\zeta_b = \eta_b\phi$, and $\Delta I_{\zeta\eta} = (I_P C_{2\theta} + 2y_I z_I m S_{2\theta})\phi$ (see equation 17.155). The polar inertia is $I_P = I_{\eta\eta} + I_{\zeta\zeta}$ (equations 17.153 and 17.154), and the center-of-gravity position is $\eta_{bI} = y_I C_\theta - z_I S_\theta$ and $\zeta_{bI} = y_I S_\theta + z_I C_\theta$.

The equations of motion of the elastic rotor blade are obtained from $\delta U + \delta T = \delta W$. Hodges and Dowell (1974) also give a Newtonian derivation. To lowest order, the quasistatic axial force equation is

$$F_x' = mA_x = -m(\Omega^2 x + 2\Omega\dot{v}) \qquad (17.197)$$

For a single load path, this equation can be integrated from the tip to obtain the tension force at a section, including the centrifugal force T:

$$F_x = \int_x^R (\Omega^2 x + 2\Omega\dot{v}) m\, dx = T + \int_x^R 2\Omega\dot{v} m\, dx \qquad (17.198)$$

Equation 17.77 gives the axial velocity: $\dot{u} = -\int_0^x (v'\dot{v}' + w'\dot{w}') dx$ (neglecting $\dot{u}_e$). The torsion equation is

$$-\Big((GJ + F_x k_P^2)\phi' + F_x k_T^2 \theta_C'\Big)' - (EI_{yy} - EI_{zz})(v''C_\theta + w''S_\theta)(v''S_\theta - w''C_\theta)$$
$$+ I_\theta \ddot{\phi} + \Omega^2(I_P C_{2\theta} + 2y_I z_I m S_{2\theta})\phi + I_{\zeta\zeta} 2\Omega\dot{w}' + I_{\zeta\eta}(2\Omega\dot{v}' + \Omega^2)$$
$$+ \eta_{bI} m(\ddot{w} + \Omega^2 xw') - \zeta_{bI} m(\ddot{v} + 2\Omega\dot{u} - \Omega^2 v + \Omega^2 xv') = M_\phi \qquad (17.199)$$

where M_ϕ is the section applied moment. Notable terms are the effective torsion stiffness $(GJ + F_x k_P^2)$, trapeze effect $(F_x k_T^2 \theta_C')$, pitch inertia $(I_\theta \ddot{\phi})$, propeller moment $(I_P \Omega^2 \phi)$, and coupling with bending $(y_I m(\ddot{w} + \Omega^2 xw'))$; see section 16.5.3. The bending equations are

$$\Big(EI_{yy}(v''C_\theta + w''S_\theta)C_\theta + EI_{zz}(v''S_\theta - w''C_\theta)S_\theta\Big)'' - (F_x v')'$$
$$+ m(\ddot{v} + 2\Omega\dot{u} - \Omega^2 v) + m\eta_{bI}(-2\Omega\dot{v}' - \Omega^2) - m\zeta_{bI}\Omega^2\phi + m\zeta_{bI}(-\ddot{\phi} - 2\Omega\dot{w}')$$
$$+ m\Big(\eta_{bI}(-\Omega^2 x - 2\Omega\dot{v}) + \zeta_{bI}\Omega^2 x\phi\Big)' = F_v \qquad (17.200)$$

$$\Big(EI_{yy}(v''C_\theta + w''S_\theta)S_\theta - EI_{zz}(v''S_\theta - w''C_\theta)C_\theta\Big)'' - (F_x w')'$$
$$+ m\ddot{w} + m\eta_{bI}\ddot{\phi}$$
$$+ m\Big(\zeta_{bI}(-\Omega^2 x - 2\Omega\dot{v}) - \eta_{bI}\Omega^2 x\phi\Big)' = F_w \qquad (17.201)$$

With no center-of-gravity offsets and no torsion, the bending equations simplify to

$$\left(EI_{yy}(v''C_\theta + w''S_\theta)C_\theta + EI_{zz}(v''S_\theta - w''C_\theta)S_\theta\right)'' - (Tv')'$$
$$+ m(\ddot{v} - \Omega^2 v)$$
$$- m2\Omega \int_o^x (v'\dot{v}' + w'\dot{w}')dx - \left(v' \int_x^R 2\Omega \dot{v} m\, dx\right)' = F_v \qquad (17.202)$$

$$\left(EI_{yy}(v''C_\theta + w''S_\theta)S_\theta - EI_{zz}(v''S_\theta - w''C_\theta)C_\theta\right)'' - (Tw')'$$
$$+ m\ddot{w} - \left(w' \int_x^R 2\Omega \dot{v} m\, dx\right)' = F_w \qquad (17.203)$$

which is an extension of equations 16.116 and 16.123 to include structural coupling due to the blade pitch. The equations for free vibration in bending omit the Coriolis terms:

$$\left(\begin{bmatrix} C_\theta & S_\theta \\ -S_\theta & C_\theta \end{bmatrix}\begin{bmatrix} EI_{zz} & 0 \\ 0 & EI_{yy} \end{bmatrix}\begin{bmatrix} C_\theta & -S_\theta \\ S_\theta & C_\theta \end{bmatrix}\begin{pmatrix} w'' \\ v'' \end{pmatrix}\right)''$$
$$- \left(T\begin{pmatrix} w' \\ v' \end{pmatrix}\right)' + m\begin{pmatrix} \ddot{w} \\ \ddot{v} - \Omega^2 v \end{pmatrix} = 0 \qquad (17.204)$$

Writing the bending as a two-dimension vector $z = (w\ v)^T$ (in hub plane axes), the equation is

$$\left(X_\theta (EI) X_\theta^T z''\right)'' - (Tz')' - m\overline{\Omega}\,\overline{\Omega}^T z + m\ddot{z} = 0 \qquad (17.205)$$

with $\overline{\Omega} = (0\ \Omega)^T$; see equation 17.54. The centrifugal force is in the hub plane axes and the stiffness is in the section axes, so pitch couples in-plane and out-of-plane bending.

17.5 History

There was notable early work on beam theory for rotor blades by Morduchow (1950), Leone (1954, 1957), Yntema (1954), Targoff (1955), Miller and Ellis (1956), Daughaday, DuWaldt, and Gates (1957), and Brooks (1958).

Houbolt and Brooks (1958) brought together beam theory for bending and torsion deflection of rotor blades. The principal assumptions were a single loadpath, permitting determination of the tension directly from the centrifugal force; isotropic material; and structural and inertial terms retained only to first order in bending and torsion deflection. The differential equations were developed using both Newtonian and Lagrangian methods.

By the end of the 1960s, the hingeless rotor was maturing rapidly, notably the Lockheed XH-51A, the AH-56A Cheyenne compound helicopter, the MBB Bö-105, and the Westland Lynx. The Cheyenne, Bö-105, and Lynx were developed using relatively simple rotor dynamics models (Johnston and Cook (1971), Huber (1973), Hansford and Simons (1973)).

There was notable work on beam theory by Mil' et al. (1966), Chang (1967), Arcidiacono (1969), Piziali (1970), and Hansford and Simons (1973). The development of the Bö-105 and the Lynx was accompanied by an understanding of the torsion-bending coupling arising from nonlinear structural dynamics of hingeless rotors (Huber (1973), Hansford and Simons (1973)). Bending deflections can result in a torsion moment component of trim bending moments, leading to effective pitch-lag and pitch-flap couplings that significantly influence blade stability and aircraft flight dynamics; see section 16.5.2.

Developing even a theory of the stability of the flap and lag motion of a hingeless rotor proved difficult. Hohenemser and Heaton (1967) presented a careful derivation and an analysis of the linearized, second-order equations of flap-lag motion. However, this work had an error in the treatment of the effect of lag motion on the in-plane velocity for the aerodynamic model. Ormiston and Hodges (1972) presented a very careful derivation of the flap-lag equations of motion, as well as a thorough exploration of the influence on blade stability of parameters such as thrust, flap and lag frequency, pitch-lag coupling, and structural flap-lag coupling. Ormiston developed a useful simulation of variable elastic coupling by introducing springs both inboard and outboard of the pitch rotation. Regarding the original issue of hingeless rotor stability, the importance of nonlinear inertial, structural, and aerodynamic terms was clear. The conclusion was that instabilities were possible, largely due to pitch-lag or pitch-flap coupling, and that flap-lag structural coupling was potentially stabilizing.

Motivated by the government and industry activities in hingeless rotor development, particularly the AH-56A Cheyenne development, the U.S. Army laboratory at Ames Research Center initiated research in rotor dynamics, including systematic development of a theoretical basis for hingeless rotor blade stability and dynamics and a series of careful experiments to provide data to guide and substantiate the analysis. Hodges and Ormiston (1973a) worked on elastic flap-lag-torsion equations of a rotor blade, including effects of nonlinearity. Friedmann and Tong (1972) and Friedman (1973a, 1973b) developed a model for bending-torsion motion of elastic blades.

Hodges and Dowell (1974) developed a rigorous derivation of the nonlinear bending-torsion equations of motion for a rotor blade. The theory was intended for application to long, straight, slender, homogeneous, isotropic beams with moderate displacements. The theory was accurate to second order, based on the restriction that squares of bending slopes, twist, thickness-to-radius, and chord-to-radius are small. The equations of motion were derived by the variational method (based on Hamilton's principle) and the Newtonian method (based on the summation of forces and moments acting on a differential blade element). They summarized the evidence regarding the importance of nonlinearities in rotor blade dynamics: inertial nonlinearities in flap-lag dynamics, based on rigid blade models (Ormiston and Hodges (1972)) or elastic models (Hodges and Ormiston (1973b)); structural bending-torsion nonlinearities (Huber (1973)); and elastic bending-torsion coupling in the torsion equation (Mil' et al. (1966)). A single load-path was assumed, permitting elimination of the axial extension variable. A systematic, self-consistent approach was developed for determining which terms in the equations to retain and which to ignore. In particular, it was assumed that with bending and torsion deflections and cross-section dimensions all of order ϵ, the extension and section warping were order ϵ^2. This approach and the assumptions made regarding order of various quantities formed the basis for future extensions as well. Hodges and Dowell concluded,

> In the resulting system of equations, several important nonlinear terms are identified. First, the centrifugal term proportional to lead-lag velocity in the tension equation

combines with the centrifugal coupling terms in the bending equations to produce nonlinear flap-lag inertial terms. The longitudinal velocity in the lead-lag equation, a Coriolis terms, is expressed in terms of bending quantities as another nonlinear flap-lag inertial term. These terms, when linearized with respect to equilibrium, are antisymmetic gyroscopic terms and significantly influence hingeless rotor stability and forced response phenomena. Second, the nonlinear bending-torsion coupling term in the torsion equation is written in a form similar to the one identified by Mil'. The twisting moment arises from bending in two directions and is proportional to the difference in bending stiffness and the product of curvatures. The counterpart nonlinear bending-torsion coupling terms in the bending equations appear in the form of a change in elastic coupling due to elastic twist. These bending-torsion coupling terms are also important in determining the aeroelastic stability of hingeless rotors.

There followed work by Hodges (1976) and Hodges and Ormiston (1976, 1977) exploring the dynamic stability and response of elastic rotor blades.

Subsequent work included numerous investigations devoted to extending the equations of motion, either to higher order or with different ordering assumptions. In the 1980s, beam models were developed using exact kinematics and introducing implicit formulations, notably by Hodges, Ormiston, and Peters (1980) and Hodges (1985). A key step was the development of beam models for anisotropic or composite materials by Bauchau and Hong (1988), Hodges (1990), Yuan, Friedmann, and Venkatesan (1992), and Smith and Chopra (1993).

Finite elements are needed to model the complexity of rotor structures. Finite element models were developed for rotor blade analysis in the early 1980s. Multibody dynamics technology is needed to model the mechanisms found in rotors. Finite element and multi-body dynamics modeling capability, including input-driven definition of the geometry, was fully integrated into comprehensive analysis with the introduction of CAMRAD II (Johnson (1994)) and DYMORE (Bauchau, Bottasso, and Nikishkov (2001)). With the development of rotor models combining finite elements and multi-body dynamics, large rigid-body motion of small individual elements can be handled with exact kinematics. Then for most problems of rotor dynamics, the second-order model of Hodges and Dowell is satisfactory for the motion within the element.

The use of finite elements in rotor blade analysis is crucial to modeling the true complexity of the mechanical and structural systems being presented by designers. The use of finite element techniques to obtain free vibration modes of rotor blades is common (such as by Bratanow and Ecer (1974), and Yasue (1977)). Friedmann and Straub (1980) and Straub and Friedman (1981, 1982) discretized the partial differential equations of a rotor blade using a local Galerkin method of weight residuals. Although recognizing that "the finite element method is ideally suited for modeling the complicated and redundant structural system encountered in bearingless rotors," (Friedmann and Straub (1980)) they considered the equations of a single load-path rotating beam.

Borri, Lanz, and Mantegazza (1981) and Borri, Lanz, Mantegazza, Orlandi, and Russo (1982) developed an analysis with the "blade motion represented by finite elements in space-time domain," based on Hamilton's variational principle. These papers specifically focused on the time-finite-element development and observed the advantage of leaving the integration problems and tedious algebraic manipulations to the computer. Hodges and Rutkowski (1981) applied variable order finite elements to blade analysis. Sivaneri and Chopra (1982) analyzed the bending-torsion stability of a rotor blade using "a finite element formulation based on the principle of virtual work." The subject was still a single load-path configuration, with the extension displacement eliminated by substitution using the centrifugal force.

Application of the finite element method to a bearingless rotor was finally made by Sivaneri and Chopra (1984). "The finite element formulation allows the multi-beams of the flexure to be considered individually. The multibeams of the flexure and the single beam of the outboard are discretized into beam elements, each with fifteen nodal degrees of freedom." With multiple load-paths, "the distribution of the centrifugal force in the multibeams of the inboard blade is not known a priori and hence the axial deflection can not be eliminated."

Flexible and accurate modeling of the mechanisms that comprise the hub, blade root, and control system of rotors requires the technology of multibody dynamics. Hodges, Hopkins, Kunz, and Hinnant (1987) brought to rotorcraft the combination of finite elements and multibody dynamics in a project during the early 1980s to develop the GRASP software. The code "was developed to perform aeroelastic stability analysis of rotorcraft in steady, axial flight and ground contact conditions" and thus was implemented with a limited aerodynamic model applicable only to hover and axial flow. Elliott and McConville (1989) applied the general purpose multibody system analysis tool ADAMS to rotary-wing aeroelastic problems. Bauchau and Kang (1993) implemented and validated "a multibody formulation for helicopter nonlinear structural dynamic analysis," focusing on the appropriate coordinates to represent the element motion and development of the corresponding constraint equations. Although no aerodynamic model was included, the method was applied to the problem of ground resonance.

17.6 REFERENCES

Arcidiacono, P.J. "Prediction of Rotor Instability at High Forward Speeds. Volume I. Steady Flight Differential Equations of Motion for a Flexible Helicopter Blade with Chordwise Mass Unbalance." USAAVLABS TR 68-18A, February 1969.

Bauchau, O.A. "A Beam Theory for Anisotropic Materials." Journal of Applied Mechanics, 52:2 (June 1985).

Bauchau, O.A., Bottasso, C.L., and Nikishkov, Y.G. "Modeling Rotorcraft Dynamics with Finite Element Multibody Procedures." Mathematical and Computer Modeling, 33:10–11 (May-June 2001).

Bauchau, O.A., and Hong, C.H. "Nonlinear Composite Beam Theory." Journal of Applied Mechanics, 55:1 (March 1988).

Bauchau, O.A., and Kang, N.K. "A Multibody Formulation for Helicopter Structural Dynamic Analysis." Journal of the American Helicopter Society, 38:2 (April 1993).

Bisplinghoff, R.L., Mar, J.W., and Pian, T.H.H. Statics of Deformable Solids. Reading, MA: Addison-Wesley Publishing Company, Inc., 1965.

Borri, M., Lanz, M., and Mantegazza, P. "A General Purpose Program for Rotor Blade Dynamics." Seventh European Rotorcraft and Powered Lift Aircraft Forum, Garmisch-Partenkirchen, Germany, 1981.

Borri, M., Lanz, M., Mantegazza, P., Orlandi, D., and Russo, A. "STAHR: A Program for Stability and Trim Analysis of Helicopter Rotors." Eighth European Rotorcraft Forum, Aix-en-Provence, France, September 1982.

Bratanow, T., and Ecer, A. "Sensitivity of Rotor Blade Vibration Characteristics to Torsional Oscillations." Journal of Aircraft, 11:7 (July 1974).

Brooks, G.W. "On the Determination of the Chordwise Bending Frequencies of Rotor Blades." Journal of the American Helicopter Society, 3:3 (July 1958).

Cesnik, C.E.S., and Hodges, D.H. "VABS: A New Concept for Composite Rotor Blade Cross-Sectional Modeling." Journal of the American Helicopter Society, 42:1 (January 1997).

Chang, T.T. "A Method for Predicting the Trim Constants and the Rotor-Blade Loadings and Responses of a Single-Rotor Helicopter." USAAVLABS TR 67-71, November 1967.

Chopra, I., and Sivaneri, N.T. "Aeroelastic Stability of Rotor Blades Using Finite Element Analysis." NASA CR 166389, August 1982.

Daughaday, H., DuWaldt, F., and Gates, C. "Investigation of Helicopter Blade Flutter and Load Amplification Problems." Journal of the American Helicopter Society, 2:3 (July 1957).

Elliott, A.S., and McConville, J.B. "Application of a General-Purpose Mechanical Systems Analysis Code to Rotorcraft Dynamics Problems." American Helicopter Society National Specialists' Meeting on Rotorcraft Dynamics, Arlington, TX, November 1989.

Friedmann, P. "Aeroelastic Instabilities of Hingeless Helicopter Blades." Journal of Aircraft, 10:10 (October 1973b).

Friedmann, P. "Some Conclusions Regarding the Aeroelastic Stability of Hingeless Helicopter Blades in Hover and in Forward Flight." Journal of the American Helicopter Society, 18:4 (October 1973a).

Friedmann, P.P., and Straub, F. "Application of the Finite Element Method to Rotary-Wing Aeroelasticity." Journal of the American Helicopter Society, 25:1 (January 1980).

Friedmann, P., and Tong, P. "Dynamic Nonlinear Elastic Stability of Helicopter Rotor Blades in Hover and in Forward Flight." NASA CR 114485, May 1972.

Giavotto, V., Borri, M., Mantegazza, P., Ghiringhelli, G., Carmaschi, V., Maffioli, G.C., and Mussi, F. "Anisotropic Beam Theory and Applications." Computers and Structures, 16:1–4 (1983).

Hansford, R.E., and Simons, I.A. "Torsion-Flap-Lag Coupling on Helicopter Rotor Blades." Journal of the American Helicopter Society, 18:4 (October 1973).

Hodges, D.H. "Nonlinear Equations of Motion for Cantilever Rotor Blades in Hover with Pitch Link Flexibility, Twist, Precone, Droop, Sweep, Torque Offset, and Blade Root Offset." NASA TM X-73112, May 1976.

Hodges, D.H. "Torsion of Pretwisted Beams due to Axial Loading." Journal of Applied Mechanics, 47:2 (June 1980).

Hodges, D.H. "Nonlinear Equations for Dynamics of Pretwisted Beams Undergoing Small Strains and Large Rotations." NASA TP 2470, May 1985.

Hodges, D.H. "A Mixed Variational Formulation Based on Exact Intrinsic Equations for Dynamics of Moving Beams." International Journal of Solids and Structures, 26:11 (1990).

Hodges, D.H. Nonlinear Composite Beam Theory. Reston, VA: American Institute of Aeronautics and Astronautics, 2006.

Hodges, D.H., Atilgan, A.R., Cesnik, C.E.S., and Fulton, M.V. "On a Simplified Strain Energy Function for Geometrically Nonlinear Behavior of Anisotropic Beams." Composites Engineering, 2:5–7 (1992).

Hodges, D.H., and Dowell, E.H. "Nonlinear Equations of Motion for the Elastic Bending and Torsion of Twisted Nonuniform Rotor Blades." NASA TN D-7818, December 1974.

Hodges, D.H., Hopkins, A.S., Kunz, D.L., and Hinnant, H.E. "Introduction to GRASP – General Rotorcraft Aeromechanical Stability Program – A Modern Approach to Rotorcraft Modeling." Journal of the American Helicopter Society, 32:2 (April 1987).

Hodges, D.H., and Ormiston, R.A. "Stability of Elastic Bending and Torsion of Uniform Cantilevered Rotor Blades in Hover." AIAA Paper No. 73-405, March 1973a.

Hodges, D.H., and Ormiston, R.A. "Nonlinear Equations for Bending of Rotating Beams with Application to Linear Flap-Lag Stability of Hingeless Rotors." NASA TM X-2770, May 1973b.

Hodges, D.H., and Ormiston, R.A. "Stability of Elastic Bending and Torsion of Uniform Cantilever Rotor Blades in Hover with Variable Structural Coupling." NASA TN D-8192, April 1976.

Hodges, D.H., and Ormiston, R.A. "Stability of Hingeless Rotor Blades in Hover with Pitch-Link Flexibility." AIAA Journal, 15:4 (April 1977).

Hodges, D.H., Ormiston, R.A., and Peters, D.A. "On the Nonlinear Deformation Geometry of Euler-Bernoulli Beams." NASA TP 1566, April 1980.

Hodges, D.H., and Rutkowski, M.J. "Free-Vibration Analysis of Rotating Beams by a Variable-Order Finite-Element Method." AIAA Journal, 19:11 (November 1981).

Hohenemser, K.H., and Heaton, P.W., Jr. "Aeroelastic Instability of Torsionally Rigid Helicopter Blades." Journal of the American Helicopter Society, 12:2 (April 1967).

Houbolt, J.C., and Brooks, G.W. "Differential Equations of Motion for Combined Flapwise Bending, Chordwise Bending, and Torsion of Twisted Nonuniform Rotor Blades." NACA Report 1346, 1958.

Huber, H.B. "Effect of Torsion-Flap-Lag Coupling on Hingeless Rotor Stability." American Helicopter Society 29th Annual National Forum, Washington, DC, May 1973.

Johnson, W. "Technology Drivers in the Development of CAMRAD II." American Helicopter Society Aeromechanics Specialists Conference, San Francisco, CA, January 1994.

Johnston, J.F., and Cook, J.R. "AH-56A Vehicle Development." American Helicopter Society 27th Annual National V/STOL Forum, Washington, DC, May 1971.

Leone, P.F. "Theory of Rotor Blade Uncoupled Flap Bending Aero-Elastic Vibrations." American Helicopter Society 10th Annual Forum, Washington, DC, 1954.

Leone, P.F. "Theoretical and Experimental Study of the Coupled Flap Bending and Torsion Aero-Elastic Vibrations of a Helicopter Rotor Blade." American Helicopter Society 13th Annual National Forum, Washington, DC, May 1957.

Mil', M.L., Nekrasov, A.V., Braverman, A.S., Grodko, L.N., and Leykand, M.A. *Helicopter, Calculation and Design.* Moscow: Izdatel'stvo Mashinostroyeniye, 1966 (Volume I: Aerodynamics, NASA TT F-494, September 1967; Volume II: Vibrations and Dynamic Stability, TT F-519, May 1968).

Miller, R.H., and Ellis, C.W. "Helicopter Blade Vibration and Flutter." Journal of the American Helicopter Society, 1:3 (July 1956).

Morduchow, M. "A Theoretical Analysis of Elastic Vibrations of Fixed-End and Hinged Helicopter Blades in Hovering and Vertical Flight." NACA TN 1999, January 1950.

Ormiston, R.A., and Hodges, D.H. "Linear Flap-Lag Dynamics of Hingeless Helicopter Rotor Blades in Hover." Journal of the American Helicopter Society, 17:2 (April 1972).

Piziali, R.A. "An Investigation of the Structural Dynamics of Helicopter Rotors." USAAVLABS TR 70-24, 1970.

Sivaneri, N.T., and Chopra, I. "Dynamic Stability of a Rotor Blade Using Finite Element Analysis." AIAA Journal, 20:5 (May 1982).

Sivaneri, N.T., and Chopra, I. "Finite Element Analysis for Bearingless Rotor Blade Aeroelasticity." Journal of the American Helicopter Society, 29:2 (April 1984).

Smith, E.C., and Chopra, I. "Formulation and Evaluation of an Analytical Model for Composite Box-Beams." Journal of the American Helicopter Society, 36:3 (July 1991).

Smith, E.C., and Chopra, I. "Aeroelastic Response, Loads, and Stability of a Composite Rotor in Forward Flight." AIAA Journal, 31:7 (July 1993).

Straub, F.K., and Friedmann, P.P. "A Galerkin Type Finite Element Method for Rotary-Wing Aeroelasticity in Hover and Forward Flight." Vertica, 5:1 (1981).

Straub, F.K., and Friedmann, P.P. "Application of the Finite Element Method to Rotary Wing Aeroelasticity." NASA CR 165854, February 1982.

Targoff, W.P. "The Bending Vibrations of a Twisted Rotating Beam." WADC TR 56-27, December 1955.

Washizu, K. "Some Considerations on a Naturally Curved and Twisted Slender Beam." Journal of Mathematics and Physics, 43:2 (June 1964).

Washizu, K. *Variational Methods in Elasticity and Plasticity.* Second Edition, Pergamon Press, New York, 1975.

Yasue, M. "Gust Response and Its Alleviation for a Hingeless Helicopter Rotor in Cruising Flight." MIT ASRL TR 189-1, 1977.

Yntema, R.T. "Rapid Estimation of Bending Frequencies of Rotating Beams." NACA Conference on Helicopters, Langley Field, VA, May 1954.

Yuan, K.-A., Friedmann, P.P., and Venkatesan, C. "Aeroelastic Behavior of Composite Rotor Blades with Swept Tips." American Helicopter Society 48th Annual Forum, Washington, DC, June 1992.

18 Dynamics

18.1 Blade Modal Frequencies

The modal equations for blade bending and torsion are derived in sections 16.3.2, 16.4.3, 16.5.3, and 17.2. The results for the blade natural frequencies can be written in terms of the rotor rotational speed Ω:

$$\nu^2 = K_1 + K_2 \Omega^2 \tag{18.1}$$

where the coefficients K_1 and K_2 are the structural and centrifugal stiffening, respectively. For example, equation 17.57 for flap and lag bending gives

$$K_1 = \frac{\eta'(e)^T K_s \eta'(e) + \int_0^R \eta''^T E I \eta'' dr}{\int_0^R \eta^2 m\, dr} \tag{18.2}$$

$$K_2 = \frac{\int_0^R \left(\eta'^2 \int_r^R \rho m\, d\rho - m(k^T \eta)^2 \right) dr}{\int_0^R \eta^2 m\, dr} \tag{18.3}$$

Southwell and Gough (1921) obtained this expression from Rayleigh energy considerations. It applies to fully coupled flap bending, lag bending, and torsion modes of a rotating beam. The coefficients K_1 and K_2 are constants involving the integrals of the blade mode shape, which do depend on Ω. However, the Southwell form gives the basic dependence of the blade frequencies on the rotor speed and emphasizes the relative strengths of structural and centrifugal stiffening.

Figure 18.1 shows the variation of the bending frequency with rotor speed that the Southwell form implies. The natural frequencies of the blade must be kept away from resonance with harmonics of the rotor speed, so the n/rev lines are also usually shown on the frequency diagram. In the limit $\Omega = 0$, the blade frequency is $\nu^2 = K_1$. So $\sqrt{K_1}$ is the non-rotating natural frequency, due to the structural stiffness. In the limit of large Ω, the blade frequency approaches $(\nu/\Omega)^2 = K_2$. So $\sqrt{K_2}$ is the dimensionless (per-revolution) natural frequency at high speed, due to the centrifugal forces.

Figure 18.2 shows a typical plot of blade frequencies as a function of rotor speed, for articulated and soft in-plane hingeless rotors. This presentation is called a Southwell plot, or a fan-plot, from the constant per-revolution lines. For specific rotors, the fan-plot is commonly dimensional, showing blade frequencies in Hz as

18.1 Blade Modal Frequencies

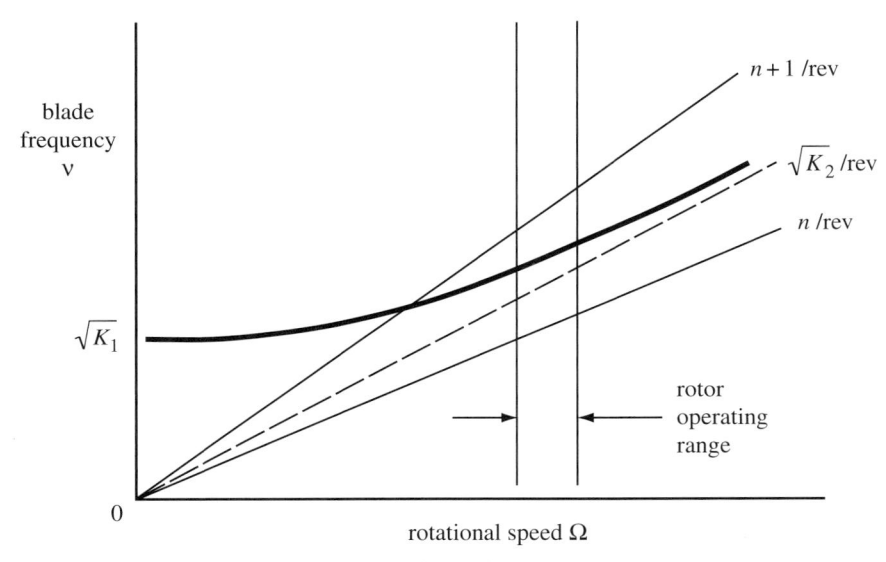

Figure 18.1. Rotating blade frequency behavior.

a function of rotor speed in rpm. The centrifugal stiffness dominates the first flap modes, so the frequency approaches a constant per-rev line. For an articulated rotor, with or without hinge offset but no hinge spring, the first flap mode frequency is always a constant per-rev value. The centrifugal stiffness also dominates the first lag modes, but the lag bending stiffness is much larger than the flap bending stiffness, so the frequency curves for lag modes are flatter with rotor speed and a larger rotor speed is required to approach a constant per-rev line. For a torsion mode, the uncoupled frequency is $\omega^2 = \omega_{NR}^2 + \Omega^2$, where ω_{NR} is the non-rotating natural frequency; see sections 16.5.1 and 16.5.3. Usually the structural stiffness is an order of magnitude greater than the centrifugal stiffness, so the frequency curve is very flat with rpm. Such different behavior with rotor speed can be used to identify the primary motion of a mode. As a consequence of the different variation with rotor speed, uncoupled frequencies cross. The frequencies of the coupled modes do not cross at such resonances, indicating strong coupling of the motions and a switch of the modal identities. Except for the first flap and lag of an articulated rotor (rigid motion, with just centrifugal stiffening), all modes have flap, lag, and torsion deflection. Flap and lag bending are coupled by blade pitch and twist, torsion and bending are coupled by chordwise center-of-gravity offset, and there can be structural coupling of all motions.

The frequencies of the fundamental flap and lag modes depend on the rotor hub configuration; see section 8.1. Teetering rotors have a flap frequency of 1/rev, whereas articulated rotors have hinge offsets of 3% to 6%; hence flap frequencies of 1.02 to 1.04/rev typically. For hingeless rotors, the structural stiffness is still small compared to the centrifugal stiffening of the blade, so the flap frequency is not far above 1/rev, typically 1.10 to 1.15/rev. Modern bearingless rotors have flap frequencies between 1.05 and 1.07/rev. The coning frequency of teetering and gimballed rotors is above 1/rev. A lift-offset rotor requires a stiff blade to carry the roll moment and hence has a flap frequency above 1.4/rev. With a hinge offset of 3% to 6%, the lag frequency of an articulated rotor is around 0.20 to 0.30/rev. Depending on the structural design of the root, hingeless and bearingless blades can be either soft in-plane (lag frequency

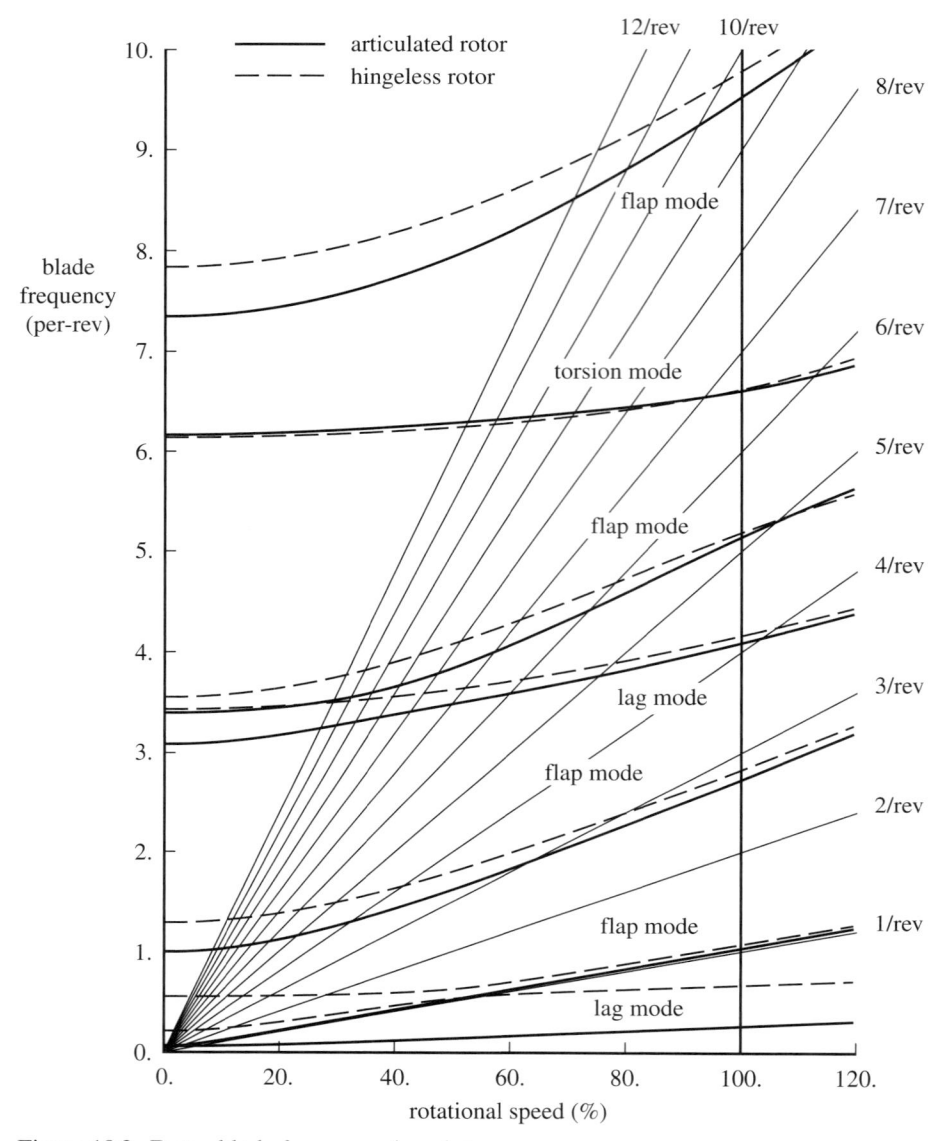

Figure 18.2. Rotor blade frequency fan plot.

below 1/rev, typically 0.65 to 0.75/rev) or stiff in-plane (lag frequency above 1/rev). The second flapwise bending mode has a rotating natural frequency typically around 2.6 to 2.8/rev. The frequencies of the higher blade modes depend substantially on the lag, flap, torsion, and pitch link stiffnesses.

For a hingeless rotor blade, bending at the root is important even for the fundamental modes. The principal axes for the centrifugal stiffening are always the hub plane axes, whereas the principal axes for the structural stiffening are determined by the blade pitch. Only if these axes coincide are the free vibration modes of the blade purely in-plane and purely out-of-plane. Pitch of the blade, particularly at the root, introduces significant coupling of the flap and lag motion in the fundamental modes. For the fundamental bending modes of stiff in-plane hingeless rotors, the centrifugal stiffening dominates the out-of-plane motion, whereas structural stiffening dominates the in-plane motion. Even a small root pitch then greatly influences the mode shapes and frequencies. Soft in-plane blades tend to have matched stiffnesses at the root, which reduces the coupling caused by collective pitch. For the higher bending modes, the structural stiffening becomes increasingly important, and consequently twist influences the mode shape.

A basic requirement for minimum vibration and loads is that the natural frequencies of the blade modes avoid resonances with harmonics of the rotor speed. The design criterion is that any intersections of the blade frequencies and the n/rev lines should not occur within the normal rotor speed operating range. Typically resonances at least up to 5 or 6/rev must be considered. Except for the rigid modes of an articulated rotor, the structural stiffness dominates at low rotor speed, and therefore the natural frequencies cross the n/rev lines. Such resonances can produce load amplification during the rotor run-up, but are not a major problem as long as they occur at low rotor speed.

For articulated rotors, the fundamental flap and lag modes are almost entirely rigid rotation about the hinges, and the mode shape is nearly the rigid blade solution, $\eta = (r-e)/(1-e)$. If no better estimate is available, $\eta = 4r^2 - 3r$ can be used as an approximation to the second out-of-plane mode shape of an articulated rotor blade. This expression is orthogonal to the first mode, $\eta = r$, but the boundary conditions of zero moment at the root and tip are not satisfied.

The zero-stiffness limit gives the minimum possible bending frequency due to centrifugal stiffening alone. When $EI = 0$, equation 17.55 for the bending modes separates into purely out-of-plane and purely in-plane equations:

$$\left(\int_r^1 m\rho\, d\rho\, \eta_z' \right)' + m v_z^2 \eta_z = 0 \tag{18.4}$$

$$\left(\int_r^1 m\rho\, d\rho\, \eta_x' \right)' + m(v_x^2 + 1)\eta_x = 0 \tag{18.5}$$

See equations 16.61 and 16.111. Without the structural terms, there is no coupling of the flap and lag modes. Moreover, the in-plane and out-of-plane motion have identical mode shapes, with corresponding frequencies given by $v_{\text{flap}}^2 = 1 + v_{\text{lag}}^2$. This is a singular limit, since dropping the structural terms reduces the order of the equations. For small EI, the boundary conditions are satisfied in small regions near the ends of the blade. For $EI = 0$, the two boundary conditions at the tip must be dropped. For a uniform mass distribution, the equation for η reduces to Legendre's equation,

$$((1-r^2)\eta')' + 2v^2 \eta = (1-r^2)\eta'' - 2r\eta' + 2v^2 \eta = 0 \tag{18.6}$$

The solutions that satisfy the boundary conditions and are finite at the tip are the odd Legendre polynomials, $\eta_k(r) = P_{2k-1}(r)$, with eigenvalues $v_k^2 = k(2k-1)$. These polynomials can be obtained from

$$P_n = \frac{1}{2^n n!} \frac{d^n}{dr^n} (r^2 - 1)^n \tag{18.7}$$

which gives $\eta_1 = P_1 = r$, $\eta_2 = P_3 = \frac{1}{2}(5r^2 - 3r)$, and so on. The corresponding frequencies are $v_1 = 1$, $v_2 = \sqrt{6} \cong 2.45$, $v_3 = \sqrt{15} \cong 3.87$ for flap, and $v_1 = 0$,

$\nu_2 = \sqrt{5} \cong 2.24$, $\nu_3 = \sqrt{14} \cong 3.74$ for lag. For the third modes and above, the curvature is large enough that the structural stiffening begins to dominate the solution, so beginning with ν_3 (or even ν_2 for lag), this lower bound on the frequency is very conservative.

With uniform mass and stiffness, the modal equation for flap bending (equation 16.61) becomes

$$EI\eta'''' - \frac{m\Omega^2}{2}\left((R^2 - r^2)\eta'\right)' - \nu^2 m\eta = 0 \tag{18.8}$$

Solutions in terms of elementary functions are found only for the non-rotating case, $EI\eta'''' - \nu^2 m\eta = 0$. For an articulated blade the boundary conditions are $\eta(0) = \eta''(0) = \eta''(R) = \eta'''(R) = 0$, and the modes are

$$\eta = \frac{\sin ar}{2\sin a} + \frac{\sinh ar}{2\sinh a} \tag{18.9}$$

where a is a solution of $\tan a = \tanh a$: $a = 0, 1.2499\pi, \ldots, (3n + \frac{5}{4})\pi$. The corresponding frequencies are $\nu = a^2\sqrt{EI/mR^4}$. For a hingeless rotor, the boundary conditions $\eta(0) = \eta'(0) = \eta''(R) = \eta'''(R) = 0$ give

$$\eta = \frac{(\sin a + \sinh a)(\cosh ar - \cos ar) - (\cos a + \cosh a)(\sinh ar - \sin ar)}{2(\sin a \cosh a - \cos a \sinh a)} \tag{18.10}$$

where a is a solution of $\cos a \cosh a = -1$: $a = 0.5969\pi, 1.4942\pi, 2.5002\pi, \ldots, (n + \frac{1}{2})\pi$. These solutions give the non-rotating frequencies, and the mode shapes are useful as a series in the solution for the modes of a rotating blade, since they are orthogonal and satisfy the boundary conditions.

With uniform inertia and stiffness, the modal equation for torsion (section 16.5.3) becomes

$$GJ\xi'' + (\omega^2 - \Omega^2)I_\theta \xi = 0 \tag{18.11}$$

with boundary conditions $\xi(0) = \xi'(R) = 0$. The modes are

$$\xi = \frac{\sin ar}{\sin a} = \sin \pi \left(n - \frac{1}{2}\right)r \tag{18.12}$$

where a is the solution of $\cos a = 0$: $a = (n - \frac{1}{2})\pi$. The corresponding frequencies are $\omega^2 = \Omega^2 + a^2(GJ/I_\theta R^2)$.

Duncan polynomials (Duncan (1937)) are useful for trial functions in the modal solution. For bending,

$$y_n = \frac{1}{6}(n+2)(n+3)r^{n+1} - \frac{1}{3}n(n+3)r^{n+2} + \frac{1}{6}n(n+1)r^{n+3} \tag{18.13}$$

which gives $y_0 = r$, $y_1 = 2r^2 - \frac{4}{3}r^3 + \frac{1}{3}r^4$. For $n = 1, 2, 3, \ldots$ these polynomials satisfy the fixed-free boundary conditions of a hingeless rotor: $y(0) = y'(0) = y''(1) = y'''(1) = 0$. For $n = 0, 2, 3, \ldots$ they satisfy the pinned-free boundary conditions of an articulated rotor: $y(0) = y''(0) = y''(1) = y'''(1) = 0$. They are not orthogonal, but the set of polynomials is complete, and they can be orthogonalized by the Gram-Schmidt process. Duncan polynomials are useful as approximations for the first bending modes.

The Duncan polynomials for torsion are $y_n = (n+1)r^n - nr^{n+1}$. The function $y_0 = 1$ satisfies pinned-free boundary conditions. The function $y_1 = 2r - r^2$ satisfies fixed-free boundary conditions.

18.2 Rotor Structural Loads

The blade, hub, and control loads produced by the aerodynamic and inertial forces acting on the rotor are needed to design the helicopter structural components to the specified strength and fatigue criteria. Designing the structure requires the stresses in the blade, which in the context of beam theory are obtained from the bending and torsional moments acting on the blade section. With articulated blades the critical bending moment is usually the oscillatory load somewhere around the blade midspan. For hingeless rotors the highest bending moments are at the blade root. The net reactions at the blade root are needed to determine the loads in the rotor hub. The feathering moments on the blades lead to loads in the rotor control system, which are often a limiting factor in extreme operating conditions. The designer is usually concerned with the periodic or nearly periodic loads occurring in steady-state or maneuvering flight. Since the periodic aerodynamic environment of the helicopter rotor produces high oscillatory loads in the blades, hub, and control system, the fatigue analysis is a major part of rotor structural design. Because it depends critically on the details of the stress distribution, the fatigue life must normally be verified by tests. This is particularly true for helicopter rotors since many components are designed for finite fatigue life because of the high load levels.

Figure 18.3 illustrates the rotor loads measured in flight and in a wind tunnel on full-scale articulated rotors. It shows the flap bending moment at midspan, as the oscillatory $10^5 C_M/\sigma = 10^5 M/(\rho N c \Omega^2 R^4)$ (mean removed), for low-speed and high-speed conditions. Figure 9.0 showed the corresponding airloads. These flap bending loads are all dominated by the 3/rev component and hence exhibit a similar behavior. The flap bending motion consists primarily of the response of the first elastic flap bending mode, which for most rotor blades has a frequency just below 3/rev. The first elastic lag bending mode and the torsion mode have frequencies that vary greatly between blades, so the lag and torsion structural loads do not exhibit universal behavior.

Bousman (1990) compared structural blade loads measured on seven full-scale articulated rotors in high-speed flight conditions ($\mu = 0.37$ to 0.39) without effects of stall or maneuvers. The oscillatory flap bending moment and pitch link force increased by a factor of two or three beginning at $\mu = 0.25$–0.30, due to the asymmetric aerodynamic environment and compressibility effects (the thrust C_T/σ was well below the stall boundary). For all the rotors, the dominant flap bending load component was 3/rev. The flap oscillatory load magnitude showed an initial peak in the transition regime, then decreased, and then increased with speed. The blade torsion moments and pitch link loads showed a large positive moment in the first quadrant and then a negative moment in the second quadrant, followed in some cases by oscillation at the first torsion frequency. For several rotors, the blade root oscillatory chord bending moment exhibited a basically 1/rev variation, apparently associated with the lag damper. Otherwise, the vibratory chord bending moment behavior of the seven rotors differed significantly. Torsion and pitch link load behavior of a teetering rotor were different from that of the articulated rotors.

Equations for the structural loads acting on the rotor blade cross-section were derived in sections 16.3.4 and 17.3.9 for modal and finite element analyses, respectively. The force-balance method obtains the section load from the difference between the applied forces and inertial forces acting on the blade. The deflection method obtains the section load from the elastic displacement and structural

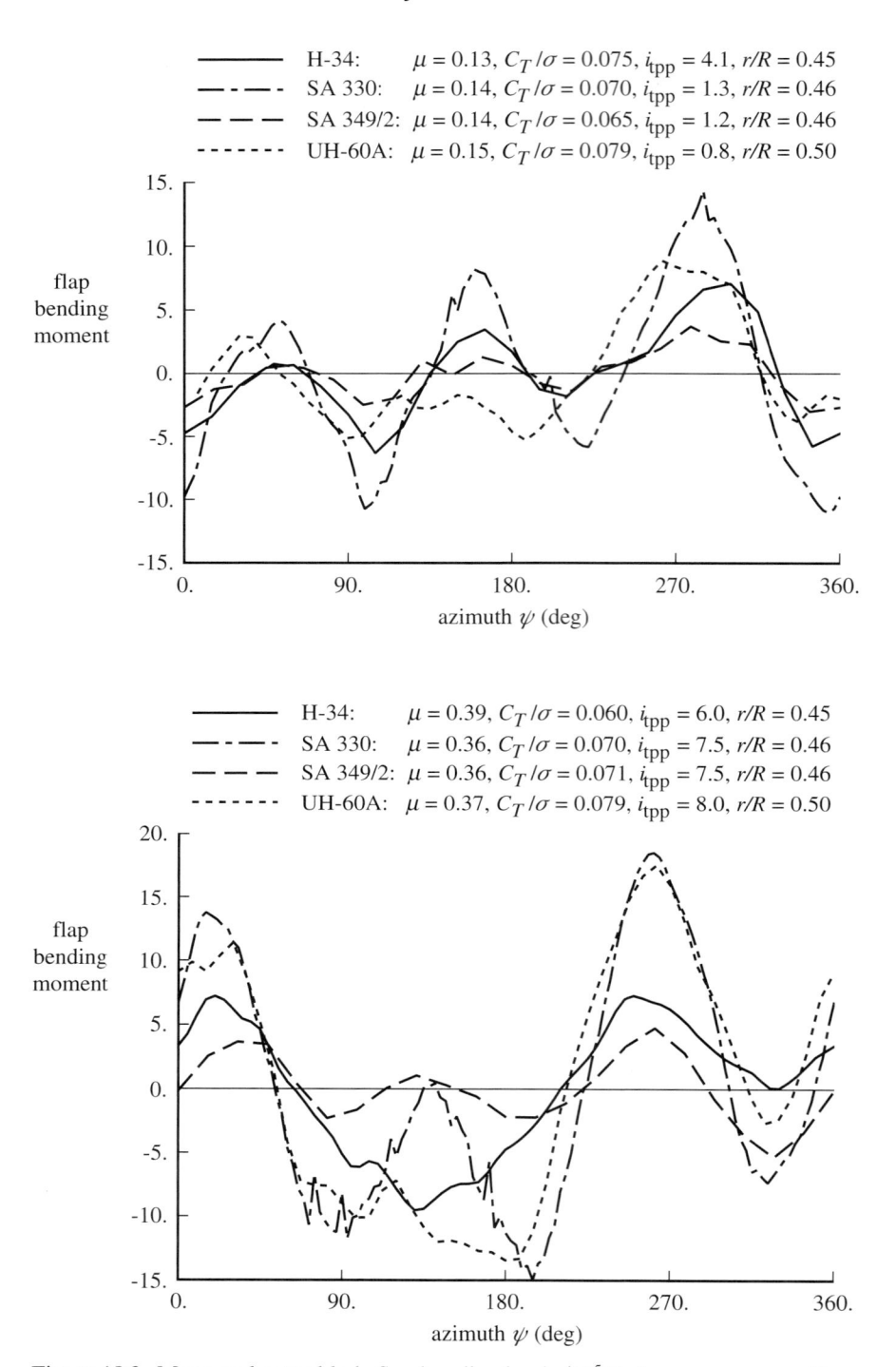

Figure 18.3. Measured rotor blade flap bending loads ($10^5 C_M/\sigma$, mean removed).

stiffness. Generally a force-balance approach (equations 16.79 or 17.174–17.175) gives the best results. The structural dynamics of the blade filter the blade response, so a small number of modes or shape functions are usually adequate to describe the inertial and centrifugal loading.

Calculating the loads by a deflection method (equations 16.81 or 17.165) depends on an accurate representation of the curvature or slope, derived from the product of the degrees of freedom and the modes or shape functions. However, a deflection method is often not acceptable because the beam representation of rotor blades usually involves step changes in the stiffness. The loads must be a continuous function of radial station (except where there are discrete shear forces acting on the section), so at a step in stiffness there should be a corresponding step in curvature or slope such that the load remains continuous. A step in curvature cannot be well modeled with a small number of modes. Moreover, if the equations of motion are obtained by the Galerkin or Rayleigh-Ritz approach, the deflection method may not even be applicable, since the boundary conditions of the modes need not be consistent with the loads applied at the blade root (such as those due to a lag damper or control system inputs). A finite element model has a similar problem within the element, since a small number of polynomial shape functions cannot model well a step in curvature. In the finite element analysis, nodes can be put at step changes in properties, but the theory does not imply continuity of curvature on the two sides of a node.

Accurate calculation of rotor blade structural loads requires good models for the aerodynamic forces and moments acting on the blade surface and for the resulting inertial loads and structural response. Calculating accurately the mean and oscillatory (one-half peak-to-peak) loads is often sufficient for fatigue and strength determination. The critical operating conditions include maneuvers, as well as level flight at high thrust or high speed. The design loads for the rotor blades and hub are typically encountered in pull-out, push-over, and roll-reversal maneuvers at moderate flight speeds. Inaccurate calculation of the time history of the loads implies limitations in the fidelity of the models, decreasing confidence in prediction of loads beyond the correlation base.

Calculation of flap bending moments is most accurate and is generally good for time history as well as mean and oscillatory load components. The flap loads depend on the blade lift forces and flap bending deflection. Calculating lag bending moments is more difficult, since the in-plane loads are much smaller than the out-of-plane loads on the blade. The in-plane aerodynamic loads are the induced drag, depending on the rotor wake, and the viscous drag, sensitive to stall and compressibility effects. Because the in-plane loads are small, inertial and structural nonlinear terms are important. Lag damper loads are also reflected in lag bending moments. Blade torsion loads and hence the control system loads are difficult to calculate, particularly since they become complicated in high-thrust and high-speed flight conditions (see Chapter 12).

18.3 Vibration

Vibration is the oscillatory response of the helicopter airframe (and other components in the non-rotating frame) to the rotor hub forces and moments. The aerodynamic loads on the fuselage and tail produced by the rotor wake can also contribute to airframe vibration. There are other important sources of helicopter vibration,

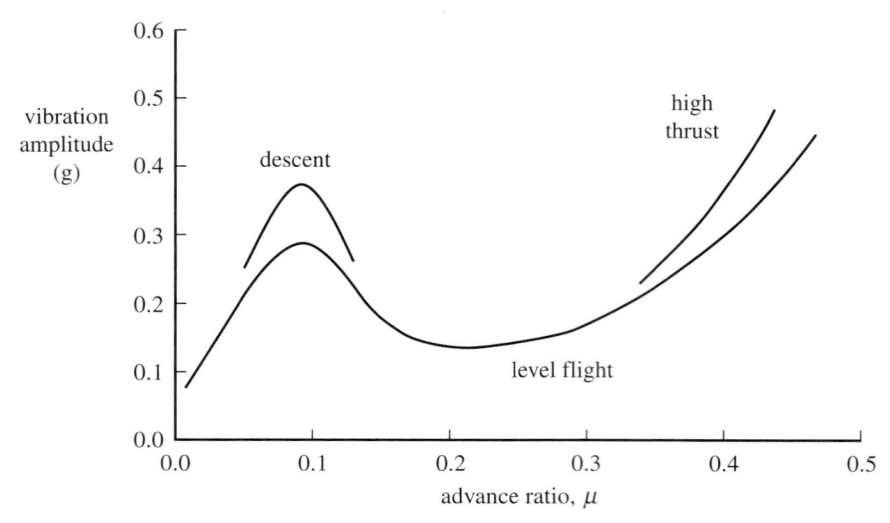

Figure 18.4. Typical helicopter vibration variation with speed, without an absorber or control to reduce vibration.

notably the engine and transmission, but the rotor influence is usually most significant. In steady-state forward flight, the periodic forces at the root of the blade are transmitted to the helicopter, producing a periodic vibratory response. Thus helicopter vibration is characterized by harmonic excitation in the non-rotating frame, primarily at 1/rev and N/rev (where N is the number of blades).

As illustrated in Figure 18.4, the vibration is generally low (but not zero) in hover and increases with forward flight to high levels at the maximum speed of the aircraft. There is also a high level of vibration in transition ($\mu \cong 0.1$) because of the rotor wake influence on the blade airloads. The vibration increases with descent at low speed and increases with thrust at high speed. With the appropriate scale, Figure 18.4 could be as well a plot of blade oscillatory structural loads or the rotor noise, since the same aerodynamic sources are involved: blade-vortex interaction in the transition regime, compressibility at high speed, and stall at high thrust. The levels of vibration shown in Figure 18.4 are characteristic of a helicopter without absorbers or control to reduce vibration.

Let us examine how the periodic rotor forces are transmitted through the hub to the aircraft. The root reaction of the m-th blade ($m = 1$ to N) is assumed to be a periodic function of $\psi_m = \psi + m\Delta\psi$ ($\Delta\psi = 2\pi/N$). Therefore, all the blades have identical loading and motion. Section 16.6 and Figure 16.8 define the rotating frame blade root loads and the non-rotating frame hub loads. Consider first the vertical shear force $S_z^{(m)}$ at the root of the m-th blade, written as a complex Fourier series in ψ_m:

$$S_z^{(m)} = \sum_{n=-\infty}^{\infty} S_{z_n} e^{in\psi_m} \qquad (18.14)$$

The total thrust force of the rotor is obtained by summing the root vertical shears over all N blades:

$$T = \sum_{m=1}^{N} S_z^{(m)} \qquad (18.15)$$

18.3 Vibration

Using the results of section 15.2 for the summation of harmonics,

$$T = \sum_{m=1}^{N} \sum_{n=-\infty}^{\infty} S_{z_n} e^{in\psi_m} = \sum_{n=-\infty}^{\infty} S_{z_n} \sum_{m=1}^{N} e^{in\psi_m} = \sum_{p=-\infty}^{\infty} NS_{z_{pN}} e^{ipN\psi} \quad (18.16)$$

The forces from all the blades exactly cancel at the hub, except for those harmonics at multiples of N/rev, which are transmitted to the aircraft. The in-plane shear forces on the rotating blade are $S_x^{(m)}$ in the blade drag direction and $S_r^{(m)}$ radially. The in-plane hub forces in the non-rotating frame, the rotor drag force H and side force Y, are given by

$$H = \sum_{m=1}^{N} \left(S_r^{(m)} \cos \psi_m + S_x^{(m)} \sin \psi_m \right) \quad (18.17)$$

$$Y = \sum_{m=1}^{N} \left(S_r^{(m)} \sin \psi_m - S_x^{(m)} \cos \psi_m \right) \quad (18.18)$$

Writing the rotating shear forces as Fourier series in ψ_m, we obtain

$$H = \sum_{n=-\infty}^{\infty} \left[S_{r_n} \sum_{m=1}^{N} e^{in\psi_m} \cos \psi_m + S_{x_n} \sum_{m=1}^{N} e^{in\psi_m} \sin \psi_m \right]$$

$$= \sum_{n=-\infty}^{\infty} \left[S_{r_n} \frac{1}{2} \sum_{m=1}^{N} \left(e^{i(n+1)\psi_m} + e^{i(n-1)\psi_m} \right) \right.$$

$$\left. + S_{x_n} \frac{1}{2i} \sum_{m=1}^{N} \left(e^{i(n+1)\psi_m} - e^{i(n-1)\psi_m} \right) \right]$$

$$= \sum_{p=-\infty}^{\infty} \frac{N}{2} \left(S_{r_{pN-1}} + S_{r_{pN+1}} - iS_{x_{pN-1}} + iS_{x_{pN+1}} \right) e^{ipN\psi} \quad (18.19)$$

and similarly

$$Y = \sum_{p=-\infty}^{\infty} \frac{N}{2} \left(-iS_{r_{pN-1}} + iS_{r_{pN+1}} - S_{x_{pN-1}} - S_{x_{pN+1}} \right) e^{ipN\psi} \quad (18.20)$$

Thus for the in-plane hub forces as well, only the harmonics of N/rev appear in the non-rotating frame, produced by the $(pN \pm 1)$/rev harmonics of the rotating shear forces. The rotor torque transmitted through the hub is obtained from the root lagwise moment $N_L^{(m)}$ in a fashion similar to the rotor thrust, giving

$$Q = \sum_{m=1}^{N} N_L^{(m)} = \sum_{p=-\infty}^{\infty} N N_{L_{pN}} e^{ipN\psi} \quad (18.21)$$

Finally, the hub pitch and roll moments are obtained from the flapwise moment $N_F^{(m)}$ at the root of the rotating blade:

$$M_y = -\sum_{m=1}^{N} N_F^{(m)} \cos \psi_m = \sum_{p=-\infty}^{\infty} \frac{N}{2} \left(-N_{F_{pN-1}} - N_{F_{pN+1}} \right) e^{ipN\psi} \quad (18.22)$$

$$M_x = \sum_{m=1}^{N} N_F^{(m)} \sin \psi_m = \sum_{p=-\infty}^{\infty} \frac{N}{2} \left(-iN_{F_{pN-1}} + iN_{F_{pN+1}} \right) e^{ipN\psi} \quad (18.23)$$

Table 18.1. *Transmission of rotor hub forces and moments*

Non-rotating frame, at pN/rev		Rotating frame
thrust	from	vertical shear at pN/rev
torque	from	lagwise moment at pN/rev
rotor drag and side forces	from	in-plane shears at $(pN \pm 1)$/rev
pitch and roll moments	from	flapwise moment at $(pN \pm 1)$/rev
collective control system loads	from	feathering moments at pN/rev
cyclic control system loads	from	feathering moments at $(pN \pm 1)$/rev

So the rotor transmits forces and moments to the non-rotating frame only at harmonics of N/rev, as summarized in Table 18.1. The transmission of the blade feathering moments to the collective and cyclic control systems has also been included in Table 18.1. If the control system is entirely mechanical, these control loads produce vibration in the pilot's collective and cyclic sticks.

Thus the rotor hub acts as a filter, transmitting to the helicopter only harmonics of the rotor forces at multiples of N/rev. This result is based on the assumption that all the blades are identical and have the same periodic motion. Although this is not perfectly true, still the N/rev harmonics dominate the vibration produced by real rotors. Figure 18.5 shows the 1/rev and N/rev frequencies of existing helicopter rotors.

The helicopter N/rev vibration is due to the higher harmonic loading of the rotor. The sources of this loading are the rotor wake and the effects of stall and compressibility at high speed. The helicopter vibration is low in hover where the aerodynamic environment is nearly axisymmetric. The only sources of higher harmonic loading are the small asymmetries such as those due to aerodynamic interference with the fuselage and other rotors. In transition, at advance ratios around $\mu = 0.1$, there is a peak in the vibration level due to the wake-induced loads on the rotor. Since the helicopter drag is small at low speeds, the tip-path-plane incidence remains small,

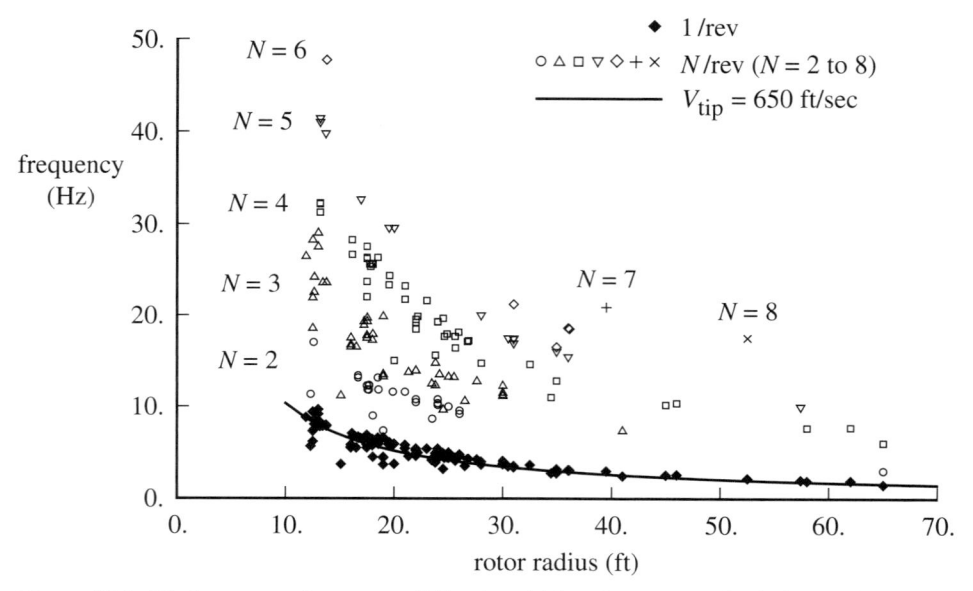

Figure 18.5. Blade passage frequency (N/rev) and 1/rev frequency of existing helicopters.

and the tip vortices in the wake remain close to the disk plane. The advance ratio is high enough so that the blades sweep past the tip vortices from preceding blades. Such close blade-vortex encounters produce significant higher harmonic airloading at the harmonics transmitted through the hub as vibration. This vibration is increased by operations that keep the wake near the plane of the disk, such as decelerating or descending flight. As the speed increases, the helicopter rotor tip-path plane tilts forward to provide the propulsive force, which means that the wake is convected away from the disk plane and the wake-induced vibration decreases. At still higher speeds the vibration increases again, primarily as a result of the higher harmonic loading produced by stall and compressibility effects. Such vibration can limit the maximum speed of the aircraft.

Tiltrotors have inherently low vibration in cruise, with the rotors nominally operating in axial flow. However, the aircraft usually trims with the rotor shaft not exactly aligned with the free stream velocity; hence with some asymmetry in the rotor aerodynamic environment. The rotor blades encounter a flow disturbance as they pass in front of the wing and near the fuselage, which can produce significant vibratory hub loads.

Helicopter rotors can produce a significant 1/rev vibration as well, because of the large 1/rev variation of the loading in forward flight and the fact that any aerodynamic or inertial dissimilarity between the blades primarily generates 1/rev vibration. A major effort is made with every rotor to eliminate the differences between the blades in the tracking and balancing operations. The inertial properties of the blades can be adjusted using balance weights, particularly at the tips; and the aerodynamic properties can be matched using aerodynamic trim tabs and by adjusting the pitch links. However, enough 1/rev vibration often remains that it must be considered in the helicopter design.

Track and balance are accomplished through a whirl test to compare each blade with a reference blade set and through field adjustments in service. Balance refers to the inertial similarity of the blades. Tip weights can be added to match the moment of inertia, hence root centrifugal force. Weights can be added at the root or on the hub to cancel residual 1/rev inertial loads. Track refers to the height at the blade tip. Pitch link length can be changed to lower or raise the blade tip for all collectives. Control loads are adjusted by changing the aerodynamic pitch moment. A trailing-edge tab can be deflected to lower or raise the moment for all collectives. Moving a balance weight off the pitch axis increases the propeller moment (from centrifugal forces) at high collective, affecting both track and control load.

Calculating rotorcraft vibration is difficult. Added to the modeling requirements for rotor performance and blade structural load calculation is the need for an accurate model of the airframe structure. Moreover, the higher harmonics of the blade root loads ($N-1$, N, and $N+1$/rev), not just mean and peak-to-peak loads, must be accurately calculated, and the high frequency airframe modes (near N/rev) are important.

Airframe structural dynamic analysis requires a large-order finite element model. Rotorcraft airframes are difficult to model because of the presence of large concentrated masses (transmission and engine), large openings for doors, and complex load-paths from the rotor to the primary structure. The airframe frequencies are required to design the structure to avoid resonances. Typically initial models of the helicopter airframe produce frequency calculations accurate to about $\pm 20\%$ (for the modes at N/rev and above, ± 5–10% for fundamental modes), which can be reduced to $\pm 5\%$ through refinement of the model; see Dompka (1990) and Cronkhite

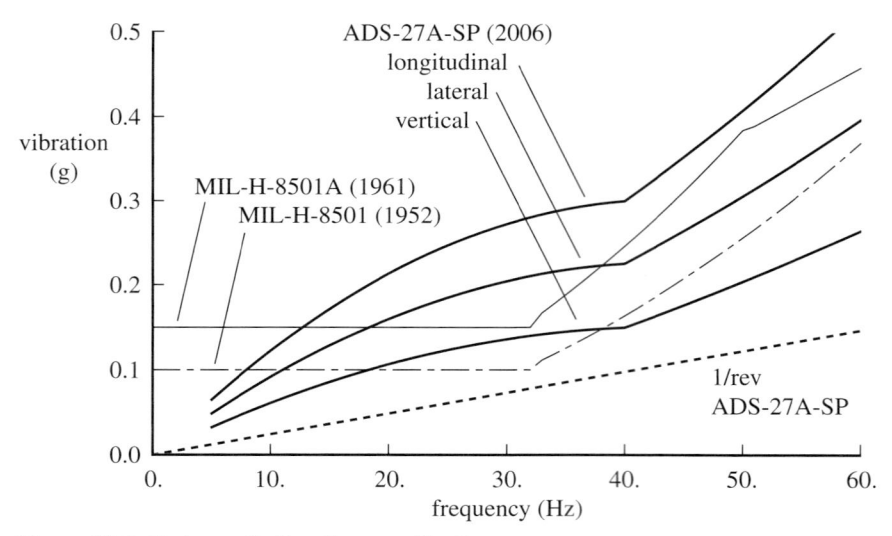

Figure 18.6. Rotorcraft vibration specifications.

(1992). Calculating the transmissibility between the rotor hub and the cockpit and cabin requires as well the modal mass and modal damping of the airframe. There are no good methods to predict modal damping. Thus accurate airframe frequencies and modal response information must be obtained by a shake test. From shake test measurements, the modal damping of metal helicopter airframes is 1–3% critical (Dompka (1990), Cronkhite (1992)); for composite airframes it is 0.5–5% critical damping (Dompka and Calapodas (1991)) or 1.5–4% critical damping (Cronkhite (1992)). So typical airframe modal damping is 2% critical, but the actual value of damping (and hence the modal response at resonance) can differ by a factor of 2 or more.

18.4 Vibration Requirements and Vibration Reduction

Without vibration treatment, helicopter rotor-induced vibration can be 0.5g or more at N/rev, as illustrated in Figure 18.4. Reducing the vibration below 0.1g or so is essential. The long-term goal is rotorcraft with vibration levels below 0.02g.

Figure 18.6 shows early and recent specifications for rotorcraft vibration. The specification is in terms of acceleration, velocity, or displacement, which can be converted to acceleration in g's as a function of frequency, as shown here. The MIL-H-8501 and MIL-H-8501A limits shown are for all locations (pilot, crew, passengers, and litters) and for steady flight up to cruise speed. The MIL-H-8501A specification (1961) refined the limit at high frequency, but also increased the limit by 50% relative to MIL-H-8501 (1952), reflecting an assessment of what the technology could provide. The design standard ADS-27A-SP (2006) defines an intrusion index, based on the normalized vibration spectra for the longitudinal, lateral, and vertical directions at a location. The intrusion index is the square root of the sum of the squares of the four largest peaks (excluding 1/rev), for each of the three normalized spectra:

$$I = \sqrt{\sum (g/g_{\text{norm}})^2} \qquad (18.24)$$

18.4 Vibration Requirements and Vibration Reduction

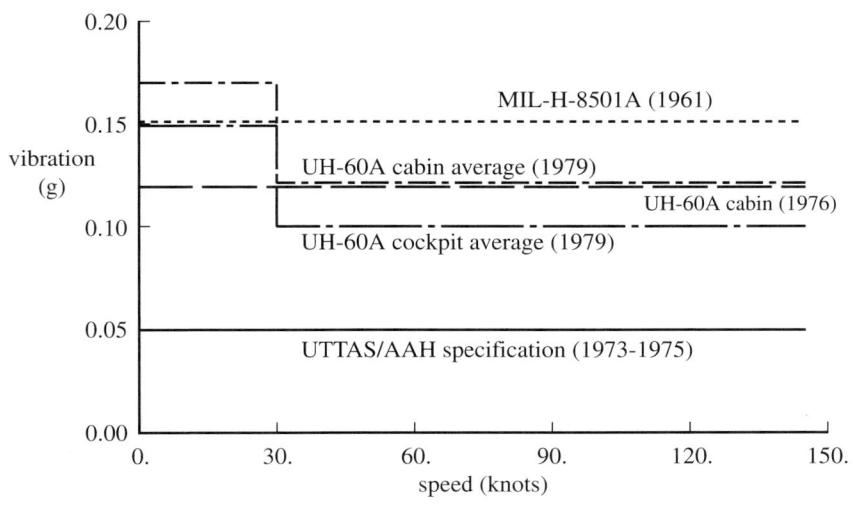

Figure 18.7. UTTAS specification for N/rev vibration in steady flight.

The normalization is defined in terms of velocity (appropriate for defining the effects of vibration on humans) as a linear function of frequency, shown in Figure 18.6 in terms of g's (g_{norm}). The intrusion index limit I_{limit} is defined for various operating conditions and positions in the aircraft. Thus the vibration requirement is

$$\sum \left(\frac{g}{g_{\text{norm}} I_{\text{limit}}}\right)^2 < 1 \tag{18.25}$$

If $I_{\text{limit}} = 1$ and there is only one component of vibration, then g_{norm} is the requirement. Considering just N/rev for longitudinal, lateral, and vertical components, the limit would be $g_{\text{norm}}/\sqrt{3}$. For steady flight conditions up to cruise speed, the limit in ADS-27A-SP is $I = 1.2$ at the pilot and $I = 2.0$ in the cabin. Specifications for particular aircraft have used the intrusion index, but with increased limits. ADS-27A-SP also has a specification for 1/rev vibration: 0.15 inch/sec for steady flight conditions up to cruise speed (Figure 18.6), which is 0.010–0.015g at 4–6 Hz. The development of ADS-27 was discussed by Crews (1987).

Specifications must reflect what is possible with the technology available. Figure 18.7 shows the vibration specification for the UTTAS procurement, which led to the UH-60A helicopter development. The UTTAS/AAH specification (1973–1975) was for less than 0.05g up to N/rev frequencies, at all stations (pilot, copilot, passenger, and litter), for steady flight up to cruise speed. In 1976 this limit was increased to 0.12g for N/rev vibration in the cabin. By 1979 the specification had become more elaborate, increasing the vibration level permitted, as well as distinguishing between cockpit and cabin and increasing the vibration limit at low speed.

Vibration reduction can be accomplished by minimizing the source from the rotor, particularly the oscillatory hub forces and moments but also including the aerodynamic loads acting on the fuselage and tail, and by reducing the response of the structure. Vibration reduction methods can be passive or active. Passive methods use pendulum or bifilar absorbers on the rotor or structural mode tuning, isolators, or absorbers on the airframe. Active methods include open and closed loop, fixed and adaptive techniques. Active control of the rotor reduces vibration by reducing the oscillatory hub forces and moments. Rotor active control methods can be implemented with higher harmonic control, individual blade control, trailing-edge flaps, and other approaches. Active control of the airframe reduces vibration by controlling the airframe structural response, using actuators on the fuselage or tuned absorbers.

The first principle in designing a rotorcraft to minimize vibration is to avoid structural resonances with the frequencies of the exciting forces. The helicopter airframe must be designed to avoid resonances with the harmonics of the rotor speed, particularly near 1/rev and N/rev. Resonances must be avoided as well with the speeds of other rotating components, including the engine, transmission, and tail rotor. The analysis of the vibration modes of a helicopter is a difficult task because of the complexity of the structure, but reasonable accuracy is possible with sufficiently complex finite element models. A shake test of the actual structure is necessary to determine the true natural frequencies. Adjusting the airframe frequencies to avoid resonances is also complicated by the large number of exciting frequencies that must be considered. Resonances in the rotor amplify the root loads, and hence the transmitted vibration. Therefore, the blades must also be designed to avoid resonances with N/rev and $(N \pm 1)$/rev. If the distinction is relevant, namely for teetering and gimballed rotors, the collective modes of the rotor should avoid N/rev resonances and the cyclic modes should avoid $(N \pm 1)$/rev. Considering the blade loads and the fact that the rotor hub is not a perfect filter of the root forces, it is generally necessary to avoid resonances of the rotating natural frequencies of the blade with all harmonics of the rotor speed. As more is learned about helicopter dynamic response and how to predict it, designing for low vibration with more sophisticated approaches than frequency placement will become possible.

The filtering of the blade vibratory forces by the rotor hub helps the task of vibration reduction or avoidance, because only a few frequencies need be considered, and because the low harmonics with the largest magnitude are not transmitted to the helicopter. The exception is the case of a two-bladed rotor, for which all harmonics in the rotating frame are transmitted to the airframe, including the highest loads at 1/rev and 2/rev. Two-bladed rotors are usually stiff in-plane as well and hence have larger blade root loads. The rotor pylon flexibility is a significant factor in frequency placement and vibration reduction on a two-bladed helicopter.

Design of the airframe and rotor can be tailored for vibration minimization. A good practice is to take advantage of nodes (points of zero motion) in the structural vibration modes of the helicopter airframe to minimize the vibration at critical points. Soft mounting of the rotor and transmission to the airframe is common for two-bladed rotors. For articulated and soft in-plane hingeless rotors, ground resonance considerations can, however, require a stiff mounting. Blade structural and inertial coupling of bending and torsion modes can reduce the root vibratory loads. For example, with a torsionally soft blade, the torsion motion can be coupled with the first flapwise bending mode to reduce the vibratory loads at the root. Often it is simpler to design an entirely separate device to act as a dynamic vibration isolation system.

Passive or active vibration reduction techniques are needed to achieve a low level of helicopter vibration. The design approach for vibration alleviation must be insensitive to a lack of knowledge about the vibration source and strength. That the vibration occurs at known, discrete frequencies helps. Passive methods rely on dynamic isolation tuned so a particular frequency is highly attenuated. Passive methods are responsible for achieving reasonable levels of vibration in current designs,

18.4 Vibration Requirements and Vibration Reduction

Bell 412

Mi-17

BK-117

S-92

AS 332

UH-60A

pendulum

bifilar

Figure 18.8. Vibration absorbers on rotor blades and hubs; photos courtesy Burkhard Domke.

but have the disadvantages of significant weight penalty and lack of flexibility. Welsh (2004) summarized the state of vibration reduction: vibration levels of 0.10g to 0.15g were achieved, in 1970 using absorbers and bifilars for about 1.5% of the gross weight and in 2000 using bifilars and active devices on the airframe for about 1.2% of the gross weight. The goal of active methods is to achieve levels of 0.05g for the same weights as passive methods.

A dynamic vibration isolation system, consisting of a mass and spring system attached between the rotor blades and the airframe, can be used in either the rotating or non-rotating frame. Such an isolator is tuned so that a particular frequency of vibration, usually N/rev, is highly attenuated. Then energy of the blade root loads at this frequency goes into the isolator rather than into airframe motion. Figure 18.8 shows some of the vibration absorbers found on rotor blades and hubs.

Consider a system with mass M_R acted on by an exciting force F, and an absorber mass M_A attached to the system with spring K_A. Let z_R be the position of M_R relative to inertial space and z_A the position of M_A relative to M_R. The equation of motion

in-plane
pendulum

vertical
pendulum

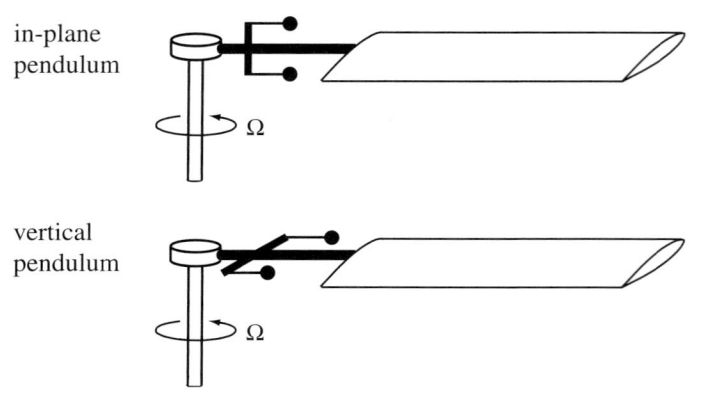

Figure 18.9. Blade pendulum absorbers.

for the absorber mass is $M_A(\ddot{z}_R + \ddot{z}_A) = -K_A z_A$, or $(M_A s^2 + K_A)z_A = -M_A s^2 z_R$ in Laplace form. The equation of motion for the system mass, $M_R \ddot{z}_R = F + K_A z_A$, then gives for the acceleration due to F:

$$M_R a_R = F - \frac{K_A M_A}{M_A s^2 + K_A} a_R = \frac{(M_A s^2 + K_A) M_R}{(M_A s^2 + K_A) M_R + K_A M_A} F$$

$$= \frac{M_A s^2 + K_A}{M_A s^2 + K_A(1 + M_A/M_R)} F \qquad (18.26)$$

So there is a zero in the response at frequency $\omega = \sqrt{K_A/M_A}$. The resonance is at a slightly larger frequency; hence not at the frequency of the exciting force. Absorbers mounted on the blade can use the centrifugal force for the spring, which has the advantage that the tuning is maintained as rotor speed Ω varies.

Figure 18.9 illustrates blade pendulum absorbers. An in-plane pendulum rotates about a vertical axis, tuned to a frequency $\omega = (N \pm 1)\Omega$, usually $(N-1)$/rev. The pendulum has a mass M, and the pivot is a distance L from the center of rotation. The center-of-gravity of the pendulum is a distance ℓ from the pivot, and $I = k^2 M$ is the moment of inertia about the center-of-gravity. Let X and Y be the position relative to the center of rotation, x and y the position of a pendulum mass element dm in pendulum axes, and ϕ the rotation of the pendulum from the spanwise axis. The centrifugal forces $dF_x = \Omega^2 X \, dm$ and $dF_y = \Omega^2 Y \, dm$ act on the mass dm, located at $X = x\cos\phi + y\sin\phi$ and $Y = L - x\sin\phi + y\cos\phi$. The centrifugal restoring moment is $dM_z = X \, dF_y - (Y - L)dF_x = \Omega^2 LX \, dm$, which integrates to

$$M_z = \int dM_z = \Omega^2 L \int (x\cos\phi + y\sin\psi)dm = \Omega^2 M L \ell \sin\phi \qquad (18.27)$$

So the in-plane pendulum natural frequency is

$$\omega^2 = \Omega^2 \frac{ML\ell}{M\ell^2 + I} = \Omega^2 \frac{L/\ell}{1 + (k/\ell)^2} \qquad (18.28)$$

For a given size and shape of the absorber mass (fixed radius of gyration k), a maximum frequency of $\omega^2 = \Omega^2(L/2k)$ is obtained, at $\ell = k$.

A vertical pendulum rotates about a horizontal axis, tuned to a frequency $\omega = N\Omega$ to attenuate vertical shear force or to $\omega = (N \pm 1)\Omega$ to attenuate hub moments. The centrifugal force $dF_y = \Omega^2 Y \, dm$ acts on the mass dm, located at

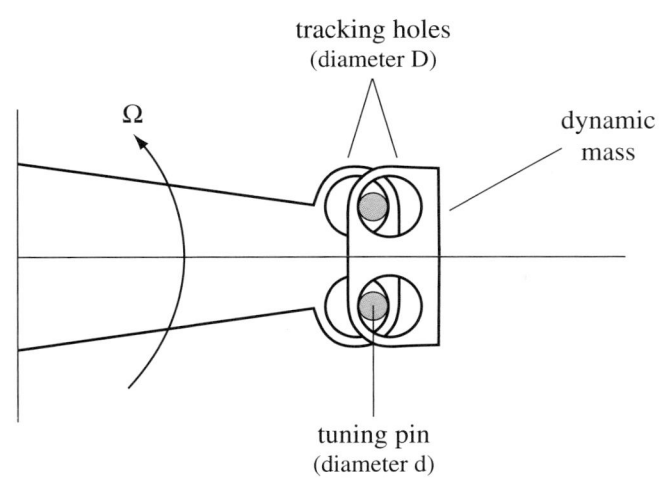

18.4 Vibration Requirements and Vibration Reduction

Figure 18.10. Bifilar vibration absorber for rotor hub.

$Y = L + y\cos\phi - z\sin\phi$ and $Z = y\sin\phi + z\cos\phi$. The centrifugal restoring moment is $dM_x = Z\,dF_y = \Omega^2 ZY\,dm$, which integrates to

$$M_x = \int dM_x = \Omega^2 \int (y\sin\phi + z\cos\phi)(L + y\cos\phi - z\sin\phi)dm$$

$$= \Omega^2 ML\ell\sin\phi + \Omega^2 \sin\phi\cos\phi \int (\ell^2 + ((y-\ell)^2 - z^2))dm$$

$$= \Omega^2 \left(ML\ell + \cos\phi(M\ell^2 + I_y - I_z)\right)\sin\phi \qquad (18.29)$$

where $I_y = Mk_y^2 = \int y^2 dm$ and $I_z = Mk_z^2 = \int z^2 dm$. The moment of inertia about the pivot is $(M\ell^2 + I_y + I_z)$. So the vertical pendulum natural frequency is

$$\omega^2 = \Omega^2 \frac{M\ell(L+\ell) + I_y - I_z}{M\ell^2 + I_y + I_z} = \Omega^2 \frac{\ell(L+\ell) + k_y^2 - k_z^2}{\ell^2 + k_y^2 + k_z^2} \cong \Omega^2 \left(1 + \frac{L/\ell}{1 + (k/\ell)^2}\right) \qquad (18.30)$$

The pendulum must be designed to null the oscillatory force at the hub node, not at the absorber. A pendulum can significantly reduce vertical hub shears (see Gabel and Reichert (1975)), by controlling the first and second elastic flap modes. These modes typically have frequencies near 3/rev and 5/rev and thus dominate 4/rev hub forces.

For high frequencies, as with a large number of blades, the required pendulum length ℓ can be impractically small. Paul (1969) described a bifilar absorber for rotor hubs to reduce in-plane vibration. Figure 18.10 shows the bifilar geometry. Typically there is an arm and dynamic mass for each blade, mounted on the hub; see Figure 18.8. The arm and dynamic mass have tracking holes of diameter D. The motion of the mass is guided by two pins, with diameter d. The dynamic mass can move, without rotation, in a circle of radius $\ell = D - d$ (the distance between the centers of the two tracking holes). So ℓ can be made small by increasing d. The center of that circle is the effective pivot point of the center-of-gravity, at a distance L from the center of rotation. With no motion, the center-of-gravity is a distance $L + \ell$ from the center of rotation. The centrifugal forces $dF_x = \Omega^2 X\,dm$ and $dF_y = \Omega^2 Y\,dm$ act on the mass dm, located at $X = x + \ell\sin\phi$ and $Y = L + y - \ell(1 - \cos\phi)$. The centrifugal

Dynamics

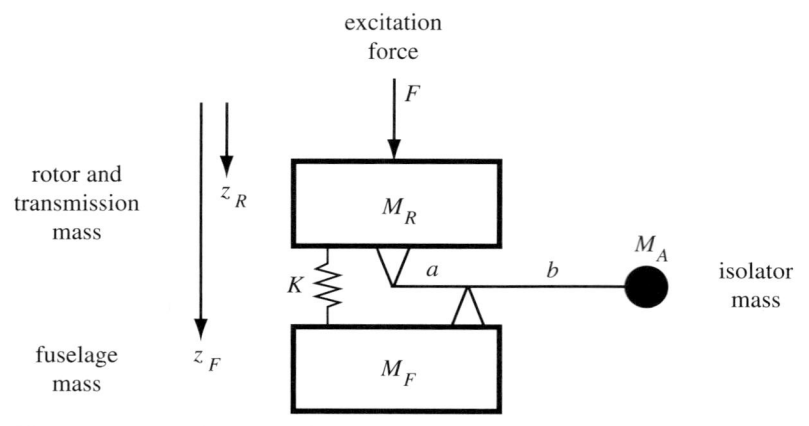

Figure 18.11. Anti-resonant isolator concept.

restoring moment is $dM_z = X\,dF_y - (Y - L)dF_x = \Omega^2 LX\,dm$, which integrates to

$$M_z = \int dM_z = \Omega^2 L \int (x + \ell \sin\phi)dm = \Omega^2 ML\ell \sin\phi \qquad (18.31)$$

as for the in-plane pendulum. Since there is no rotation of the dynamic mass as it moves, the inertial moment is $M\ell^2\ddot{\phi}$. So the in-plane pendulum natural frequency is

$$\omega^2 = \Omega^2(L/\ell) \qquad (18.32)$$

The moment of inertia of the dynamic mass does not affect the frequency of the bifilar. A nonlinear analysis of the effect of the bifilar is required for motion beyond $\phi = 20$ to 30 degrees.

An anti-resonant isolator is effective in the non-rotating frame, typically located between the transmission and airframe. Although there are many practical implementations, the concept of the anti-resonant isolator is shown in Figure 18.11. The rotor and transmission mass M_R are connected to the fuselage mass M_F through a spring K and damper C. Motion of the fuselage relative to the rotor ($z_F - z_R$) drives vertical motion of the isolator mass M_A, through a level with mechanical advantage b/a. The position of the isolator mass is $z_A = z_F + \frac{b}{a}(z_F - z_R)$. The force F acts on the rotor. The equations of motion are

$$M_R \ddot{z}_R = F + K(z_F - z_R) + C(\dot{z}_F - \dot{z}_R) - f_a \qquad (18.33)$$

$$M_F \ddot{z}_F = -K(z_F - z_R) - C(\dot{z}_F - \dot{z}_R) + f_b \qquad (18.34)$$

$$M_A \ddot{z}_A = f_a - f_b \qquad (18.35)$$

and $(a + b)f_a = bf_b$ from moment equilibrium at the isolator mass. Eliminating the pivot forces f_a and f_b gives the isolator transmissibility z_F/z_R, and the isolator mass response z_A/z_R:

$$\frac{z_F}{z_R} = \frac{(1 + b/a)(b/a)M_A s^2 + Cs + K}{((1 + b/a)^2 M_A + M_F)s^2 + Cs + K} \qquad (18.36)$$

$$\frac{z_A}{z_R} = \frac{-(b/a)M_F s^2 + Cs + K}{((1 + b/a)^2 M_A + M_F)s^2 + Cs + K} \qquad (18.37)$$

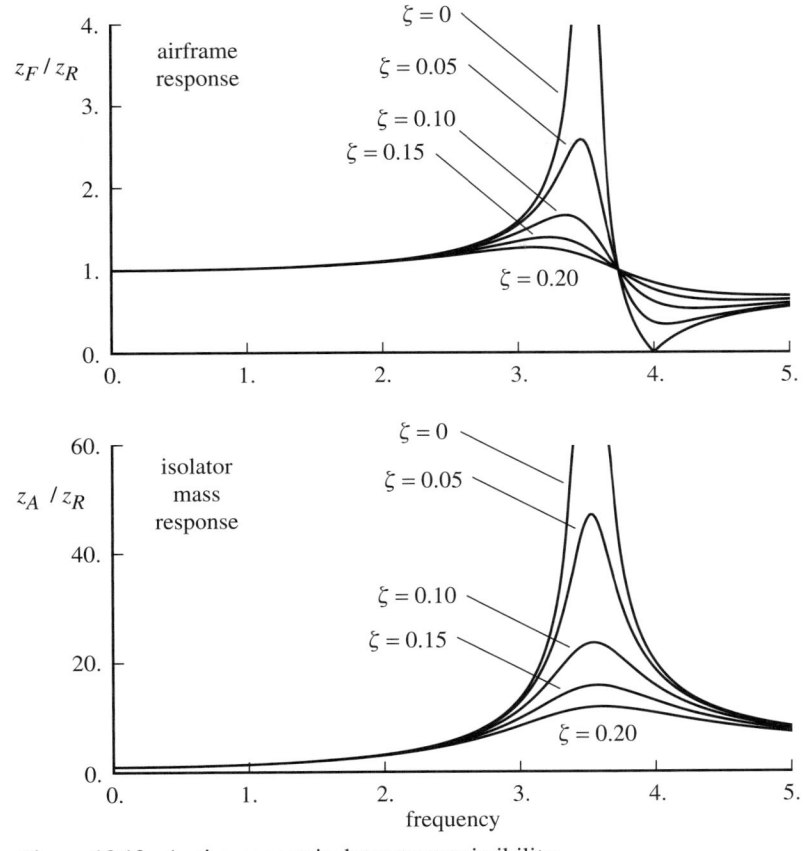

Figure 18.12. Anti-resonant isolator transmissibility.

The isolator results in fuselage response z_F/z_R with a zero and a pole at frequencies

$$\omega_z^2 = \frac{K}{(1+b/a)(b/a)M_A} \tag{18.38}$$

$$\omega_p^2 = \frac{K}{(1+b/a)^2 M_A + M_F} \tag{18.39}$$

for zero damping; the resonant frequency ω_p is always below the attenuation frequency ω_z. The isolator can be tuned to N/rev (for a specific rotor speed Ω), using a large value of b/a to keep the isolator mass M_A small. Figure 18.12 shows the fuselage and isolator response for the parameters $b/a = 20$, $M_A/M_F = 0.01$, and $\omega_z = 4$. The resulting resonant frequency is $\omega_p = 3.52$. Damping (ζ is the critical damping ratio) reduces the effectiveness of the isolation at $\omega = 4$, but is needed to control the transmissibility near the resonance.

18.5 Higher Harmonic Control

Primary control of the rotor is conventionally accomplished using mean and 1/rev (collective and cyclic) blade pitch motion. Thus higher harmonics of the pitch motion (2/rev and above) are potentially available to control the rotor dynamic behavior.

Active control of the rotor or aircraft has the objective of reducing vibration, rotor or airframe structural loads, and noise or even enhancing the rotor performance or flight envelope. Such control can be open loop or closed loop. Feedback control can also be used to improve blade stability or augment aeromechanical stability, as well as for rotor track and balance. There are many possible controls for these purposes, including swashplate motion, blade root twist, blade active twist change, blade servoflaps or trailing-edge flaps, and forces applied directly to the transmission or airframe. Quantities to be controlled can include blade loads, airframe vibration, noise (blade-vortex interaction or high-speed impulsive), stall flutter, maximum blade loading, and lift and drag distribution over the rotor disk. With a closed-loop system, there must be measurements that provide an estimate of the quantity to be controlled, such as blade pressures for noise and stall, microphones for noise, accelerometers for vibration, pitch link loads for stall flutter and blade loading, blade strain and acceleration for root loads, and blade strain for stability.

The term "higher harmonic control" (HHC) generally refers to oscillatory non-rotating-frame control through the swashplate. Individual blade control (IBC) generally refers to rotating-frame control, often blade root pitch control through a pitch link actuator. Non-rotating-frame control is adequate for vibration reduction. Oscillating the swashplate collective and cyclic at N/rev (or a multiple of N/rev, where N is the number of blades) produces N/rev and $N \pm 1$/rev blade pitch motion, which can reduce N/rev hub forces and moments. With only two or three blades, the swashplate mechanism provides complete control of the pitch motion of all the blades. With four or more blades, swashplate motion cannot control all modes and all harmonics of blade pitch. In particular, 2/rev is not accessible then, which limits the capability to control noise and enhance performance.

There are many variations in the literature on the meanings of higher harmonic control and individual blade control. Some early HHC investigations used rotating-frame control. The term "multicyclic control" was an alternative to HHC.

The original use of the term "individual blade control" (Ham (1980)) had a broader meaning of control, encompassing sensors, feedback, and actuation of the blade in the rotating frame. Ham's concept of IBC was based on using accelerometers on the blade to sense the appropriate modal response, and using time-domain feedback to control the blade. Control was implemented using blade root pitch or a trailing-edge flap, or even the swashplate. The approach was to first control the blade modes and then command modal motion to achieve the desired behavior of the rotor and aircraft.

18.5.1 Control Algorithm

The typical HHC control problem is to determine the harmonics of an input control to minimize the harmonics of the output quantity. Particular importance is attached to control in equilibrium flight, when the input and output are periodic. Dealing with periodic signals does not require a harmonic representation, rather just that the time history be parameterized. The controller must also be able to accommodate changes in the flight condition, both slow changes and maneuvers.

A class of algorithms for the multicyclic control of helicopter dynamic behavior is characterized by a linear, quasistatic, frequency-domain model of the helicopter response to control; identification of the helicopter model by least-squared error or Kalman-filter methods; and a minimum variance or quadratic performance function

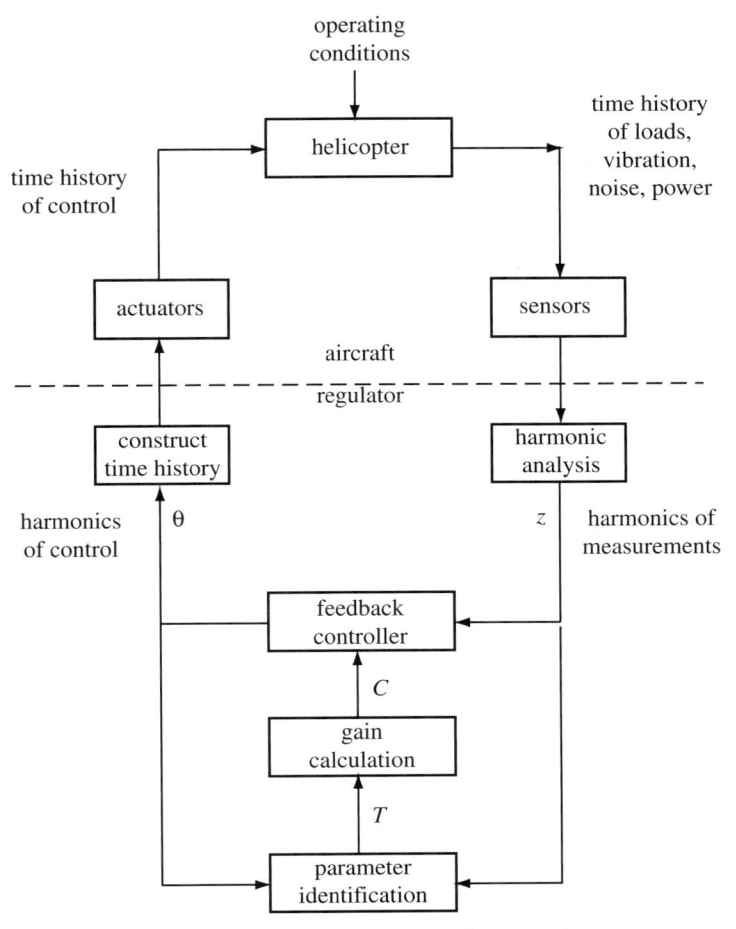

Figure 18.13. Schematic of frequency-domain control system.

controller. Such a control system combining recursive parameter estimation with linear feedback is called a self-tuning regulator.

Figure 18.13 outlines the control task. The regulator algorithm consists of parameter estimation, gain calculation, and the control feedback. Some of these steps can be performed off-line. A digital control system operating on the harmonics of the input and output is considered here. Hence, the regulator also includes transformations between the time and frequency domains and between analog and digital representations of the signals.

18.5.2 Helicopter Model

It is assumed that the helicopter can be represented by a linear, quasistatic frequency-domain model relating the output z to the input θ (see Figure 18.13) at time $t_n = n\,\Delta t$. Here z is a vector of the harmonics (both sine and cosine components) of the quantities to be controlled, such as loads and vibration, in either the rotating or the non-rotating frame. The input θ is a vector of the harmonics of the higher harmonic control, in either the rotating or non-rotating frame. The helicopter model depends on the operating condition, which is defined by the rotor lift, propulsive force, and forward speed (at least).

A local model of the helicopter response is a linearization about the current control value:

$$z_n = z_{n-1} + T(\theta_n - \theta_{n-1}) \tag{18.40}$$

or $\Delta z_n = T \Delta \theta_n$. The matrix T is the transfer function, relating changes of harmonics of output z to changes of harmonics of input θ. Control algorithms have also been developed for a global model, which is linear over the entire range of control:

$$z_n = z_0 + T\theta_n \tag{18.41}$$

with z_0 the uncontrolled vibration level. The local model leads to feedback of the current measurement of the output, z_n. The global model requires identification or measurement of z_0, perhaps from the previous value of z through $z_0 = z_{n-1} - T\theta_{n-1}$. For further information on the use of the global model in developing HHC algorithms, see Johnson (1982). The quasistatic assumption requires that the sampling or update time-step Δt be long enough for transients produced by control changes to die out and for the harmonics to be measured. Typically, this requires an interval of at least one rotor revolution.

The assumption of linear response to control is expected to be reasonable, since experimental data imply that only a small multicyclic control amplitude (of the order of 0.5° to 1.5°) is required for vibration alleviation and noise reduction. The uncontrolled vibration level (z_0) is a highly nonlinear function of the helicopter operating condition and involves nonlinear aerodynamic and dynamic phenomena. With the local model only, the response to changes in control input is linearized. The T-matrix is expected to vary with operating condition, especially the aircraft speed.

McCloud and Kretz (1974) and Kretz, Aubrun, and Larche (1973) tested multicyclic control on a full-scale jet-flap rotor in a wing tunnel. They examined the response of the blade loads and vibration to control in the rotating frame. They introduced the concept of a linear, quasistatic representation of the rotor response, including the notation "T" for the transfer function. This transfer function representation was attributed to Aubrun (McCloud and Kretz (1974)). The T-matrix was calculated from the wind-tunnel data by the least-squares method. Then the open-loop control required to minimize a quadratic performance function was calculated.

18.5.3 Identification

Consider identification of the T-matrix. Including measurement noise v_n, equation 18.40 becomes $\Delta z_n = T \Delta \theta_n + v_n$. The identification algorithms are derived for the j-th measurement,

$$\Delta z_{jn} = \Delta \theta_n^T t_{jn} + v_{jn} \tag{18.42}$$

where t_j^T is the j-th row of T. Here z_j and v_j are scalars. The task is to identify t from the measurements of z. The measurement noise is assumed to have zero mean and variance $E(v_n v_m) = r_n \delta_{nm}$ for the j-th measurement (where $E(\ldots)$ means the expected value).

18.5.3.1 Least-Squares Method

Off-line or on-line identification can be done by the method of least squares. Off-line identification implies constant parameters. A set of N measurements Δz is made, using a prescribed schedule of independent control perturbations $\Delta\theta$. The number of measurements N (the dimension of z_j below) must be greater than the number of parameters to be identified L (the dimension of t_j). Consider the sum of the squares of errors:

$$S = \sum_{n=1}^{N} \left(\Delta z_{jn} - \Delta\theta_n^T t_j\right)^2 = (z_j - \Theta t_j)^T (z_j - \Theta t_j) \tag{18.43}$$

where the vector z_j and matrix Θ are assembled from the measurement and control perturbations:

$$z_j = \begin{pmatrix} \vdots \\ \Delta z_{jn} \\ \vdots \end{pmatrix} \qquad \Theta = \begin{pmatrix} \vdots \\ \Delta\theta_n^T \\ \vdots \end{pmatrix} \tag{18.44}$$

The solution that minimizes S is the least-squares estimate: $\hat{t}_j = \left(\Theta^T \Theta\right)^{-1} \Theta^T z_j$ or $\hat{t}_j^T = z_j^T \Theta \left(\Theta^T \Theta\right)^{-1}$. Putting the rows together again gives

$$\widehat{T} = Z\Theta \left(\Theta^T \Theta\right)^{-1} \tag{18.45}$$

where

$$Z = \begin{bmatrix} \vdots \\ z_j^T \\ \vdots \end{bmatrix} = \begin{bmatrix} \cdots & \Delta z_n & \cdots \end{bmatrix} \tag{18.46}$$

$\widehat{T}$ is a linear estimate; that is, a linear function of the data Z. The measurement noise v_n is assumed to be stationary, with zero mean, and uncorrelated at different times ($E(v_n v_m) = r_n \delta_{nm}$). There is no noise in the measurement of θ. Then the least-squares estimate is unbiased, expected value $E(\hat{t}_j) = t_j$, and the error variance is

$$P = E\left((\hat{t}_j - t_j)(\hat{t}_j - t_j)^T\right) = r\left(\Theta^T \Theta\right)^{-1} \tag{18.47}$$

An unbiased estimate of r is

$$\hat{r} = \frac{1}{N-L} (z_j - \Theta\hat{t}_j)^T (z_j - \Theta\hat{t}_j) \tag{18.48}$$

With this type of measurement noise, the least-squares estimate is equivalent to the unbiased minimum error variance estimate, so it has the minimum error variance of all linear, unbiased estimators.

18.5.3.2 Generalized Least-Squares Method

The generalized least-squares estimate is obtained by minimizing the weighted sum of squares:

$$S_w = (z_j - \Theta t_j)^T W (z_j - \Theta t_j) \tag{18.49}$$

The solution is

$$\hat{t}_j = \left(\Theta^T W \Theta\right)^{-1} \Theta^T W z_j \tag{18.50}$$

The matrix W can be used to introduce weights based on the level of $\Delta\theta$ or Δz; for example, to emphasize the measurements at low vibration levels in order to improve the identification in the vicinity of the optimum response.

If the measurement noise has zero mean and variance $E(v_n v_m) = R$, then the generalized least-squares estimate with $W = R^{-1}$ is equivalent to the unbiased minimum variance estimate. Hence, when the noise is not stationary or is correlated, the weighting matrix is chosen to emphasize the more precise data. Using $W = R^{-1}$, the error variance is $P = (\Theta^T R^{-1} \Theta)^{-1}$. If, in addition, the noise has a normal probability distribution, then the minimum variance estimate is equivalent to the maximum-likelihood estimate.

18.5.3.3 Recursive Parameter Identification

Recursive algorithms can be used for on-line identification from a sequence of measurements of the response to control. These algorithms can be used when the parameters are constant or when they vary with time. The algorithms are still derived for the j-th measurements, but the subscripts j are omitted to simplify the notation. Since the parameters can be time varying now, the equation of the helicopter model is $\Delta z_n = \Delta \theta_n^T t_n + v_n$.

18.5.3.4 Recursive Generalized Least Squares

A recursive form of the generalized least-squares estimate can be used for on-line parameter identification. The weighted sum of squares

$$S_w = \sum_{n=1}^{N} \left(\Delta z_n - \Delta \theta_n^T t_n \right)^2 w_n \tag{18.51}$$

is to be minimized. The solution is equation 18.50. Here the weighting matrix is diagonal, and the notation $w_n = 1/r_n$ is used (where r_n can be interpreted as the noise variance). The error matrix is defined as

$$P_N = \left(\Theta_N^T W_N \Theta_N \right)^{-1} \tag{18.52}$$

The effect of adding one more measurement, Δz_{N+1}, is obtained by applying the matrix inversion formula to

$$P_{N+1} = \left(P_N^{-1} + \Delta \theta_{N+1} w_{N+1} \Delta \theta_{N+1}^T \right)^{-1} \tag{18.53}$$

The result is the recursive algorithm

$$\hat{t}_{n+1} = \hat{t}_n + k_{n+1} \left(\Delta z_{n+1} - \Delta \theta_{n+1}^T \hat{t}_n \right) \tag{18.54}$$

where the gain vector k_{n+1} is obtained from

$$P_{n+1} = P_n - P_n \Delta \theta_{n+1} \Delta \theta_{n+1}^T P_n / \left(r_{n+1} + \Delta \theta_{n+1}^T P_n \Delta \theta_{n+1} \right) \tag{18.55}$$

$$k_{n+1} = P_{n+1} \Delta \theta_{n+1} / r_{n+1} \tag{18.56}$$

This is the estimate for the j-th measurement. In general there is a different weight r for each measurement and hence a different solution for P and k. If the time behavior of r_n is the same for all measurements, the solution for k_{n+1} is the same, and the rows can be combined to form

$$\widehat{T}_{n+1} = \widehat{T}_n + \left(\Delta z_{n+1} - \widehat{T}_n \Delta \theta_{n+1} \right) k_{n+1}^T \tag{18.57}$$

(here Δz is the vector of all measurements). So the entire matrix can be identified in a single step, with k_{n+1} calculated only once.

If $r_n = 1$ (or any other constant), the recursive least-squares algorithm is obtained. The solution is the same as that from the batch least-squares algorithm. The recursive implementation might be useful to track the estimates and error as the data are acquired. Eventually the old data dominate (k approaches zero), so the recursive least-squares algorithm is not appropriate with time-varying parameters.

18.5.3.5 Exponential Window

A recursive estimate applicable to the case of time-varying parameters can be obtained using an exponential window for the weighting function. By setting $r_n = \alpha^n$, where $0 < \alpha < 1$, the current data are emphasized. Since r_n is continuously decreasing, solving for the gain k_{n+1} in terms of $P_n^* = P_n/\alpha^n$ is best:

$$P_{n+1}^* = \alpha^{-1}\left[P_n^* - P_n^*\Delta\theta_{n+1}\Delta\theta_{n+1}^T P_n^*/\left(\alpha + \Delta\theta_{n+1}^T P_n^*\Delta\theta_{n+1}\right)\right] \quad (18.58)$$

$$k_{k+1} = P_{n+1}^*\Delta\theta_{n+1} \quad (18.59)$$

This algorithm can be obtained directly by minimizing the sum

$$S_{N+1}^* = \alpha S_N^* + \left(\Delta z_{N+1} - \Delta\theta_{N+1}^T t\right)^2 \quad (18.60)$$

18.5.3.6 Kalman-Filter Identification

A Kalman filter can be used for on-line identification of time-varying parameters. The equation for the j-th measurement is again $\Delta z_n = \Delta\theta_n^T t_n + v_n$. The measurement noise has zero mean, variance $E(v_n v_m) = r_n \delta_{nm}$, and Gaussian probability distribution. The variation of the parameters is modeled as a random process:

$$t_{n+1} = t_n + u_n \quad (18.61)$$

where u_n is a random variable with zero mean, variance $E(u_n u_m) = Q_n \delta_{nm}$, and Gaussian probability distribution. This equation implies that t varies and that the order of the change in one time-step can be estimated, but no information is available about the specific dynamics governing the variation of t. The minimum error variance estimate of t_n is then obtained from a Kalman filter:

$$\hat{t}_n = \hat{t}_{n-1} + k_n\left(\Delta z_n - \Delta\theta_n^T \hat{t}_{n-1}\right) \quad (18.62)$$

where

$$M_n = P_{n-1} + Q_{n-1} \quad (18.63)$$

$$P_n = \left(M_n^{-1} + \Delta\theta_n\Delta\theta_n^T/r_n\right)^{-1}$$
$$= M_n - M_n\Delta\theta_n\Delta\theta_n^T M_n/\left(r_n + \Delta\theta_n^T M_n\Delta\theta_n\right)$$
$$= \left(I - k_n\Delta\theta^T\right) M_n \left(I - k_n\Delta\theta^T\right)^T + k_n k_n^T r_n \quad (18.64)$$

$$k_n = P_n\Delta\theta_n/r_n$$
$$= M_n\Delta\theta_n/\left(r_n + \Delta\theta_n^T M_n\Delta\theta_n\right) \quad (18.65)$$

Here M_n is the variance of the error in the estimate of t_n before the measurement, and P_n is the variance after the measurement. P depends on the control input θ, but not on the measurement z; no matrix inversion is required, because t_n is related to only one measured variable. The Kalman filter can be considered a time-variant dynamic system with state $\hat{t}$:

$$\hat{t}_n = \left(I - k_n \Delta \theta_n^T\right) \hat{t}_{n-1} + k_n \Delta z_n = \left(I - k_n \Delta \theta_n^T\right) \hat{t}_{n-1} + k_n \Delta \theta_n^T t_n + k_n v_n \quad (18.66)$$

where

$$I - k_n \Delta \theta_n^T = I - P_n \Delta \theta_n \Delta \theta_n^T / r_n = P_n M_n^{-1} \quad (18.67)$$

If there are no process dynamics ($Q_n = 0$), the Kalman filter is equivalent to the generalized least-squares algorithm with $w_n = 1/r_n$ (for minimum error variance). The variances Q_n, r_n, and P_0 are different for the various measurements. If Q and r have the same time variation for all measurements, and Q, r, and P_0 are proportional to the same function f_j, then the solution for k_{n+1} is the same, and the rows can be combined to form

$$\widehat{T}_n = \widehat{T}_{n-1} + \left(\Delta z_n - \widehat{T}_{n-1} \Delta \theta_n\right) k_n^T \quad (18.68)$$

(here Δz is the vector of all measurements). So the entire matrix $\widehat{T}$ can be identified in a single step, with P_n and k_n calculated only once. The basis of this result is that the ratio of the parameter and measurement noise variances, Q_{jn}/r_{jn}, is the same for every measurement.

18.5.4 Control

The control algorithm is based on the minimization of a performance index J that is a quadratic function of the input and output variables. This function depends on the input and output at the n-th time-step and perhaps at past times, but not on the future values. If all the parameters in the model are known, a deterministic controller is obtained. With unknown, estimated parameters, the certainty-equivalence principle is applied: the deterministic control solution is used with the estimated parameter values. Alternatively, a cautious controller can be obtained by minimizing the expected value of the performance function. Such control systems are called passive-adaptive or non-dual controllers. The performance index does not consider that future measurements are made, so ignores the possibility of learning from the measurements. A dual or active-adaptive controller actively probes the system to reduce the parameter errors. The control is used for learning, to improve the parameter estimates, but the improvement is achieved at the expense of short-term deterioration of the closed-loop performance.

The quadratic performance function used is

$$J = z_n^T W_z z_n + \theta_n^T W_\theta \theta_n + \Delta \theta_n^T W_{\Delta\theta} \Delta \theta_n \quad (18.69)$$

where $\Delta \theta_n = \theta_n - \theta_{n-1}$. The vectors θ and z contain the harmonics of the input and output. Typically the weighting matrices are diagonal and have the same value for all harmonics of a particular quantity. Then J is a weighted sum of the mean-squares of the vibration, loads, noise (etc.) and control. The matrix W_θ constrains the amplitude of the control, whereas $W_{\Delta\theta}$ constrains the rate of change of the control.

The control required is found by substituting for z in the performance function, using the helicopter model and then solving for θ_n that minimizes J. Setting

$\partial J/\partial \theta_{jn} = 0$ (for each component in the vector θ_n), gives a set of equations that can be solved for θ_n. The result is

$$\theta_n = -Cz_{n-1} + (C_{\Delta\theta} + CT)\theta_{n-1} \qquad (18.70)$$

or

$$\Delta\theta_n = -Cz_{n-1} - C_\theta \theta_{n-1} \qquad (18.71)$$

where

$$C = DT^T W_z \qquad (18.72)$$

$$C_\theta = DW_\theta \qquad (18.73)$$

$$C_{\Delta\theta} = DW_{\Delta\theta} \qquad (18.74)$$

$$D = \left(T^T W_z T + W_\theta + W_{\Delta\theta}\right)^{-1} \qquad (18.75)$$

Equation 18.71 defines closed-loop control obtained by feedback of the measured response z_{n-1}. The control weight W_θ constrains the amplitude of the control relative to the best value; hence it also constrains the reduction of z. The control increment weight $W_{\Delta\theta}$ reduces the control change in each step, but does not change the final value of z. When $W_\theta = 0$, the solution reduces to $\Delta\theta_n = -Cz_{n-1}$. If both W_θ and $W_{\Delta\theta}$ are zero, then $CT = I$. If $W_\theta = W_{\Delta\theta} = 0$, and the number of controls equals the number of measurements (so T is a square matrix), then $C = T^{-1}$ (regardless of W_z).

Using the largest value of $W_{\Delta\theta}$ that does not make the response too sluggish is appropriate to improve the transient response, the sensitivity to measurement noise, and the sensitivity of the stability to parameter errors. The rate limit $W_{\Delta\theta}$ should always be used during the start of the recursive identification, and a small value of $W_{\Delta\theta}$ can be used to avoid the possibility of control divergence, should the estimated T-matrix be too small. The control magnitude constraint W_θ is not very useful, since it limits the control relative to the ideal value rather than relative to an absolute value. Absolute limits on the control magnitude are probably better applied by uniformly reducing the elements of θ before the command signals are sent to the actuators.

18.5.5 Time-Domain Controllers

The T-matrix algorithm is based on the assumption of a steady or quasistatic operating condition; hence periodic input and response. The aircraft model relates sampled harmonics of response to harmonics of input:

$$z_{n+1} = z_n + T(\theta_{n+1} - \theta_n) \qquad (18.76)$$

The discrete-time controller is

$$\Delta\theta_{n+1} = -Cz_n \qquad (18.77)$$

(see Figure 18.14). For an equal number of controls and response quantities, and no control weights, $C = T^{-1}$. From this frequency-domain, discrete controller a time-domain, continuous compensator can be derived.

The approach follows Hall and Wereley (1989, 1993). The variables θ and z consist of the harmonics of the periodic input and response. For scalar y and u,

$$z = \begin{pmatrix} y_c \\ y_s \end{pmatrix} \qquad \theta = \begin{pmatrix} u_c \\ u_s \end{pmatrix} \qquad (18.78)$$

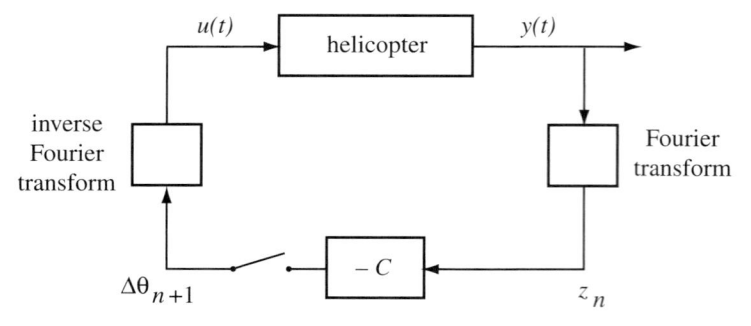

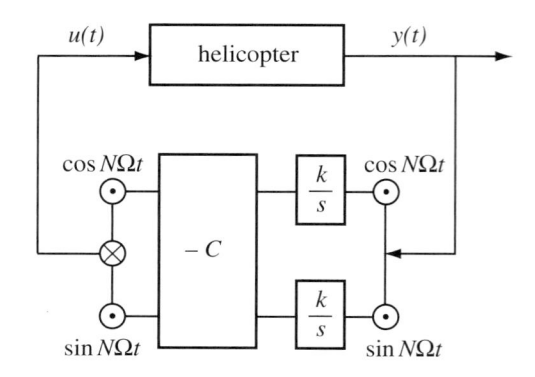

Dynamics

Figure 18.14. Time-domain higher harmonic control system.

For a time-invariant system, the input and response are related by a transfer function: $y = Gu$. Considering vibration at frequency N/rev, the transfer function is evaluated at $s = iN\Omega$: $(y_c - iy_s) = G_{N\Omega}(u_c - iu_s)$. Hence

$$\begin{pmatrix} y_c \\ y_s \end{pmatrix} = \begin{bmatrix} \text{Re}G_{N\Omega} & \text{Im}G_{N\Omega} \\ -\text{Im}G_{N\Omega} & \text{Re}G_{N\Omega} \end{bmatrix} \begin{pmatrix} u_c \\ u_s \end{pmatrix} = T \begin{pmatrix} u_c \\ u_s \end{pmatrix} \qquad (18.79)$$

so

$$C = \frac{1}{|G_{N\Omega}|^2} \begin{bmatrix} \text{Re}G_{N\Omega} & -\text{Im}G_{N\Omega} \\ \text{Im}G_{N\Omega} & \text{Re}G_{N\Omega} \end{bmatrix} = \begin{bmatrix} \alpha & -\beta \\ \beta & \alpha \end{bmatrix} \qquad (18.80)$$

Harmonic analysis gives the N/rev response by integrating over the period $T = 2\pi/N\Omega$, and the control is obtained from the Fourier series:

$$z_n = \frac{2}{T} \int_{(n-1)T}^{nT} \begin{pmatrix} \cos N\Omega\tau \\ \sin N\Omega\tau \end{pmatrix} y(\tau) d\tau \qquad (18.81)$$

$$\theta_{n+1} = \theta_n - Cz_n \qquad (18.82)$$

$$u(t) = \begin{bmatrix} \cos N\Omega t & \sin N\Omega t \end{bmatrix} \theta_{n+1} \qquad (18.83)$$

18.5 Higher Harmonic Control

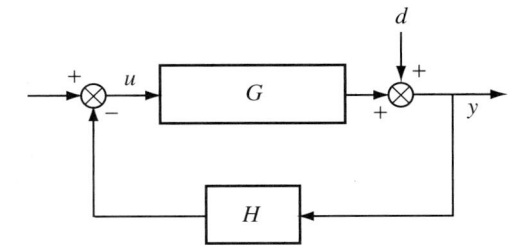

Figure 18.15. Disturbance rejection controller.

Substituting and extending the integration back in time,

$$\theta_{n+1} = \theta_n - C\frac{2}{T}\int_{(n-1)T}^{nT}\begin{pmatrix}\cos N\Omega\tau\\ \sin N\Omega\tau\end{pmatrix} y(\tau)d\tau$$

$$= \theta_{n-j} - C\frac{2}{T}\int_{(n-1-j)T}^{nT}\begin{pmatrix}\cos N\Omega\tau\\ \sin N\Omega\tau\end{pmatrix} y(\tau)d\tau$$

$$= -C\frac{2}{T}\int_{-\infty}^{nT}\begin{pmatrix}\cos N\Omega\tau\\ \sin N\Omega\tau\end{pmatrix} y(\tau)d\tau \tag{18.84}$$

Then eliminating the sample and hold,

$$u = -\begin{bmatrix}\cos N\Omega t & \sin N\Omega t\end{bmatrix} Ck \int_{-\infty}^{t}\begin{pmatrix}\cos N\Omega\tau\\ \sin N\Omega\tau\end{pmatrix} y(\tau)d\tau \tag{18.85}$$

with $k = 2/T$. Equation 18.85 is a time-domain, continuous controller (Figure 18.14) consisting of modulation, integration, and demodulation; see Hall and Wereley (1989, 1993). A transfer function form is obtained by writing equation 18.85 as

$$z_c(s) = \frac{k}{2s}(y(s - iN\Omega) + y(s + iN\Omega)) \tag{18.86}$$

$$z_s(s) = \frac{k}{2is}(y(s - iN\Omega) - y(s + iN\Omega)) \tag{18.87}$$

$$\begin{pmatrix}\theta_c(s)\\ \theta_s(s)\end{pmatrix} = -\begin{bmatrix}\alpha & -\beta\\ \beta & \alpha\end{bmatrix}\begin{pmatrix}z_s(s)\\ z_c(s)\end{pmatrix} \tag{18.88}$$

$$u(s) = \frac{1}{2}(\theta_c(s - iN\Omega) + \theta_c(s + iN\Omega))$$

$$+ \frac{1}{2i}(\theta_s(s - iN\Omega) - \theta_s(s + iN\Omega)) \tag{18.89}$$

Substituting gives $u(s) = -H(s)y(s)$, where

$$H = \frac{k(\alpha s + \beta N\Omega)}{s^2 + (N\Omega)^2} \tag{18.90}$$

This is a disturbance rejection compensator, illustrated in Figure 18.15, with d the uncontrolled response level. This system provides asymptotic rejection of the $N\Omega$ harmonics of y because of the poles at $\pm iN\Omega$ in the regulator, which give zeros at $\pm iN\Omega$ in the closed-loop transfer function: $y = d/(1 + GH)$. The gain k determines the stability and response to transients. A more complicated regulator can add poles and zeros to improve the stability, transient response, and robustness.

The compensator gives $y/d \to 0$ from $H \to \infty$ at the frequency $N\Omega$ of the disturbance. The poles at $\pm iN\Omega$ constitute a notch filter, or an oscillator tuned to the disturbance frequency. If the helicopter response to control can be treated as time invariant, then classical control design techniques can be applied after a shaping filter has been added to the model. A frequency-shaped cost function can be used, or the measurement can be shaped by a resonant filter:

$$z = \frac{(N\Omega)^2 s}{s^2 + (N\Omega)^2} y \qquad (18.91)$$

(see Du Val, Gregory, and Gupta (1984)). Equivalently the helicopter model can be augmented with

$$\ddot{z} + (N\Omega)^2 z = (N\Omega)^2 \dot{y} \qquad (18.92)$$

and with a classical compensator design for good behavior.

18.5.6 Effectiveness of HHC and IBC

There have been numerous tests and demonstrations of HHC and IBC: wind tunnel and flight, open and closed loop, 2/rev to 5/rev. Typically 1–2° of blade pitch amplitude can influence the rotor behavior significantly, including a factor of 10 reduction of vibration at high speed (from 0.3–0.6 g down to less than 0.05 g), up to 8–10 dB reduction of noise, or 5–7% reduction of power at high speed. Simultaneous reduction of noise and vibration has been achieved, using multiple harmonics of control. Rotor power often increases when the objective is just vibration or noise reduction. The pitch link loads, swashplate control loads, and blade torsion loads are increased for most cases. An HHC or IBC system adds weight to the aircraft and consumes power when active, but the goal is to have weight and power requirements that are less than for passive control (by vibration absorbers or blade design), with increased flexibility and effectiveness.

Tables 18.2 to 18.4 summarize the test experience with higher harmonic control and individual blade control of the rotor. The HHC algorithm has also been used in active vibration control, based on actuators that apply forces to the airframe. Flight tests have been reported for the UH-60L (Millott and Welsh (1999)), the S-92 (Goodman and Millott (2000)), and the Bell 430 (Heverly, Singh, and Pappas (2009)).

18.6 Lag Damper

The lag damper can be an important factor in rotor blade dynamic behavior. The aerodynamic forces due to lag motion are small for low-inflow helicopter rotor blades. So unlike the flap and pitch motion, the aerodynamic contribution to damping is small for lag motion. Soft in-plane rotors (lag frequency below 1/rev) are susceptible to ground resonance and require mechanical damping of the lag motion for stability. Current rotor designs often incorporate elastomeric bearings, dampers, or snubbers.

Analyses of rotor dynamics, particularly for stability, typically assume that springs and dampers are linear. Most dampers and snubbers are in fact nonlinear mechanisms. Physics-based models of damper components have been developed, some including separate degrees of freedom or states. A simpler approach is to use a parametric description, in which the force generated by the device is a function

Table 18.2a. *Vibration reduction and blade loads*

Test	Aircraft	Control		Improvement	Reference
WT	model	HHC	CL	up to 90% hub force; increased pitch link loads, especially at low speeds	Shaw and Albion (1981)
WT	model	HHC	CL	70–90% hub vertical force; increased edgewise bending, torsion moment, pitch link loads	Hammond (1983)
WT	model	HHC	CL	80–90% hub loads	Molusis, Hammond, and Cline (1983)
FLT	OH-6A	HHC	CL	30–90% pilot vibration; but longitudinal vibration increase at high speed; 90% in second test, below 0.05g except longitudinal; increased pitch link loads	Wood, Powers, Cline, and Hammond (1985)
FLT	S-76	HHC		40–80% cockpit vibration; with increased pushrod vibratory load	Miao, Kottapalli, and Frye (1986)
FLT	SA 349	HHC		90% cabin, 70-90% cockpit vibration, below 0.1g; increased flap bending and pitch rod load	Polychroniadis and Achache (1986)
WT	XV-15	HHC	CL	50% hub loads; increased control loads	Nguyen, Betzina, and Kitaplioglu (2001)
WT	Bo-105	IBC		85% hub vibration; simultaneous noise and vibration with multiple harmonics	Jacklin, Blaas, Teves, and Kube (1995)
FLT	Bo-105	IBC		2/rev 50%, simultaneous with noise reduction; 50% hub vertical force, 90% hub moment; cabin vibration to about 0.05g	Kube, van der Wall, Schultz, and Splettstoesser (1999)
WT	UH-60A	IBC		70% vibration	Jacklin, Haber, de Simone, Norman, Kitaplioglu, and Shinoda (2002)

Notes: WT = wind tunnel; FLT = flight; HHC = higher harmonic control (swashplate) IBC = individual blade control (pitch link); CL = closed loop.

of the displacement and velocity: $F(x, \dot{x})$. The parameters in such a model must be identified from test data. Because of the nonlinearity, the test conditions must be representative of the actual operation. Typically that means a dual-frequency test, combining motions at 1/rev and at the lag mode frequency.

Table 18.2b. *Vibration reduction and blade loads*

Test	Aircraft	Control		Improvement	Reference
FLT	CH-53G	IBC		90% one axis, 60% all directions in certain cases	Arnold (2003)
FLT	CH-53G	IBC	CL	up to 90% one component, 60% 3 components; with 2 harm (5/rev and 6/rev) up to 85% for 3 components	Arnold and Furst (2005)
WT	UH-60A	IBC	CL	eliminate single-component hub load	Norman, Theodore, Shinoda, Fuerst, Arnold, Makinen, Lorber, and O'Neill (2009)
WT	model	ATW	CL	40–60% vibratory hub forces	Bernhard and Wong (2005)
WT	model	TEF	CL	70–90% vertical force and hub moment	Koratkar and Chopra (2002)
FLT	BK 117	TEF	CL	50–90% vertical force and hub moment; cabin vibration 0.05g	Roth, Enenkl, and Dieterich (2006)
WT	MD 900	TEF	CL	95% vertical force, 70% hub moment	Straub, Anand, Birchette, and Lau (2009)

Notes: WT = wind tunnel; FLT = flight; IBC = individual blade control (pitch link); TEF = trailing-edge flap; ATW = active twist; CL = closed loop.

The spring/damper can be characterized by the in-phase and out-of-phase response to sinusoidal motion $x = \delta \cos \omega t$:

$$F = K'\delta \cos \omega t - K''\delta \sin \omega t \qquad (18.93)$$

K' is the spring term and K'' is the damping term. The loss tangent is defined as $L = K''/K' = \tan \Delta\phi$, such that $F = \overline{F} \cos(\omega t + \Delta\phi)$. A linear spring damper is described by $F = Kx + C\dot{x}$, so $K = K'$ and $C = K''/\omega = LK'/\omega$. The parameters are obtained by harmonic analysis of the force measured or calculated for motion $x = \delta \cos \omega t$:

$$K' = \frac{1}{\pi\delta} \int_0^{2\pi} F \cos \omega t \, d\omega t \qquad (18.94)$$

$$K'' = -\frac{1}{\pi\delta} \int_0^{2\pi} F \sin \omega t \, d\omega t \qquad (18.95)$$

which can be written in terms of energy:

$$K' = \frac{\omega}{\delta^2\pi} \int_0^{2\pi/\omega} Fx \, dt = \frac{4\omega}{\delta^2} \text{ (average potential energy)} \qquad (18.96)$$

$$K'' = \frac{1}{\delta^2\pi} \int_0^{2\pi/\omega} F\dot{x} \, dt = \frac{2}{\delta^2} \text{ (average energy loss)} \qquad (18.97)$$

Table 18.3. *Noise reduction*

Test	Aircraft	Control		Improvement	Reference
WT	model	HHC		5–6 dB mid-frequency (BVI) noise in low speed descent; low-frequency noise and vibration increased	Brooks, Booth, Jolly, Yeager, and Wilbur (1990)
WT	Bo-105 model	HHC		up to 6 dB BVI noise	Brooks, Booth, Boyd, Splettstoesser, Schultz, Kube, Niesl, and Streby (1994)
WT	Bo-105 model	HHC		6 dB advancing side, 2 dB retreating side BVI noise; minimum vibration control, increased BVI noise 1–2 dB	Splettstoesser, Kube, Wagner, Seelhorst, Boutier, Micheli, Mercker, and Pengel (1997)
WT	XV-15	HHC	CL	12 dB BVI noise; increase or decrease in vibratory hub loads	Nguyen, Betzina, and Kitaplioglu (2001)
WT	Bo-105	IBC		7 dB BVI noise	Niesl, Swanson, Jacklin, Blaas, and Kube (1994)
WT	Bo-105	IBC		85% BVI noise; simultaneous noise and vibration with multiple harmonics	Jacklin, Blaas, Teves, and Kube (1995)
FLT	Bo-105	IBC		2/rev BVI noise 4dB; 6 dB with increased vibration	Kube, van der Wall, Schultz, and Splettstoesser (1999)
WT	UH-60A	IBC		BVI noise up to 12 dB with 2/rev	Jacklin, Haber, de Simone, Norman, Kitaplioglu, and and Shinoda (2002)
FLT	CH-53G	IBC		BVI noise up to 3 dB with 2/rev	Arnold (2003)
WT	MD 900	TEF	CL	up to 6 dB BVI and in-plane noise	Straub, Anand, Birchette, and Lau (2009)

Notes: WT = wind tunnel; FLT = flight; HHC = higher harmonic control (swashplate)
IBC = individual blade control (pitch link); TEF = trailing-edge flap
CL = closed loop; BVI = blade-vortex interaction.

An equivalent linear spring/damper can be defined on the basis of equal work during a cycle of motion. For a linear damper, $F = C\dot{x}$, the work is $W = \oint F\dot{x}\,dt = C\pi\omega\delta^2$. Thus

$$C_{\text{equiv}} = \frac{\oint F\dot{x}\,dt}{\pi\omega\delta^2} \qquad (18.98)$$

Table 18.4. *Performance improvement utilizing 2/rev control*

Test	Aircraft	Control		Improvement	Reference
WT	model	HHC		20–40% L/D_e at high speed	McHugh and Shaw (1976)
WT	model	HHC	CL	6% power at 135 knots, 4% at 160 knots	Shaw and Albion (1981)
WT	Bo-105	IBC		7% power at high-speed condition	Jacklin, Blaas, Teves, and Kube (1995)
FLT	CH-53G	IBC		6% power at 130 knots; decreased pitch link loads	Arnold (2003)
WT	UH-60A	IBC	CL	5% power, 9% L/D_e at $\mu = 0.4$	Norman, Theodore, Shinoda, Fuerst, Arnold, Makinen, Lorber, and O'Neill (2009)
WT	model	ATW	CL	1–2% power	Bernhard and Wong (2005)

Notes: WT = wind tunnel; FLT = flight; HHC = higher harmonic control (swashplate) IBC = individual blade control (pitch link); ATW = active twist; CL = closed loop.

where here F is the force produced by motion at amplitude δ and frequency ω. For a dual-frequency test, the equivalent damping acting on the lag mode is obtained by using $\dot{x}$ and $\omega \delta^2$ of just that mode, while F is the total force. Similarly

$$K_{\text{equiv}} = \frac{\oint Fx\,dt}{\pi \delta^2/\omega} \qquad (18.99)$$

is the equivalent linear spring constant. With a nonlinear spring damper, the equivalent spring and damping constants depend on the amplitude and frequency of the motion.

Consider a nonlinear spring of the form

$$F = k_0 \text{sign} x + x\left(k_1 + k_2|x| + k_3 x^2 + \ldots\right) \qquad (18.100)$$

For $x = \delta \cos \omega t = \delta \cos \theta = \delta C$, the in-phase constant is

$$\begin{aligned}
K' &= \frac{1}{\pi \delta} \int_0^{2\pi} \left(k_0 \text{sign} C + k_1 \delta C + k_2 \delta^2 C|C| + k_3 \delta^3 C^3 + \ldots\right) C\,d\theta \\
&= \frac{4}{\pi \delta} \int_0^{\pi/2} \left(k_0 + k_1 \delta C + k_2 \delta^2 C^2 + k_3 \delta^3 C^3 + \ldots\right) C\,d\theta \\
&= \frac{4}{\pi \delta} \sum_{n=0} k_n \delta^n \int_0^{\pi/2} C^{n+1}\,d\theta \\
&= k_0 \frac{4}{\pi \delta} + k_1 + k_2 \delta \frac{8}{3\pi} + k_3 \delta^2 \frac{3}{4} + \ldots
\end{aligned} \qquad (18.101)$$

Each term also gives the equivalent linear spring constant. Consider a nonlinear damper of the form

$$F = c_0 \text{sign}\dot{x} + \dot{x}\left(c_1 + c_2|\dot{x}| + c_3\dot{x}^2 + \ldots\right) \quad (18.102)$$

For $\dot{x} = -\omega\delta \sin \omega t = -\omega\delta \sin \theta = -\delta S$, the out-of-phase constant is

$$K'' = \frac{1}{\pi\delta} \int_0^{2\pi} \left(c_0\text{sign}S + c_1\delta\omega S + c_2\delta^2\omega^2 S|S| + c_3\delta^3\omega^3 S^3 + \ldots\right) S\, d\theta$$

$$= \frac{4}{\pi\delta} \int_0^{\pi/2} \left(c_0 + c_1\delta\omega S + c_2\delta^2\omega^2 S^2 + c_3\delta^3\omega^3 S^3 + \ldots\right) S\, d\theta$$

$$= \frac{4}{\pi\delta} \sum_{n=0} c_n \delta^n \omega^n \int_0^{\pi/2} S^{n+1} d\theta$$

$$= c_0 \frac{4}{\pi\delta} + c_1\omega + c_2\delta\omega^2 \frac{8}{3\pi} + c_3\delta^2\omega^3 \frac{3}{4} + \ldots \quad (18.103)$$

Each term also gives the equivalent linear damping constant, $C_{\text{equiv}} = K''/\omega$. The coefficient c_1 gives linear, viscous damping, while c_0 is friction damping, and c_2 is hydraulic damping. Functional forms that combine x and $\dot{x}$ dependence are also useful, such as

$$F = \text{sign}x \left(k_0 + k_1|x| + k_2|x|^2 + \ldots + \ell_1|\dot{x}| + \ell_2|\dot{x}|^2 + \ldots\right)$$
$$+ \text{sign}\dot{x} \left(c_0 + c_1|\dot{x}| + c_2|\dot{x}|^2 + \ldots + d_1|x| + d_2|x|^2 + \ldots\right) \quad (18.104)$$

The d_n constants give nonlinear friction damping; d_1 can be considered structural damping.

An elastomeric damper can typically be modeled as a combination of nonlinear friction damping and nonlinear spring. For a hydraulic damper, the force depends only on the rate of motion, with quadratic variation for small amplitude. A relief valve is required to limit the damper loads during large 1/rev motion in forward flight. Thus the simplest model for the damper force is

$$F = \min((\dot{x}/\dot{x}_{\text{limit}})^2, 1) F_{\text{limit}} \text{sign}\dot{x} \quad (18.105)$$

Considering single-frequency motion, the equivalent damping increases with amplitude for small amplitude: $C_{\text{equiv}} \sim \delta\omega$. For large amplitude, the force limit produces friction damping behavior, and the damping decreases with amplitude: $C_{\text{equiv}} \sim 1/(\delta\omega)$. So for hover a hydraulic damper is very effective in stabilizing ground resonance. In forward flight, the steady-state 1/rev lag motion is enough to open the relief valve, so the damper does not produce excessive blade loads when the aircraft is off the ground. For tests of an articulated rotor in a wind tunnel, the lag damper must stabilize ground resonance in forward flight as well, so the damping must be active for perturbation motion at the lag mode frequency, while undergoing finite-amplitude motion at 1/rev. For such dual-frequency operation of a hydraulic damper, the friction damping due to the force limit dominates the behavior. Typically then the equivalent linear damping of the lag mode is roughly independent of the lag mode amplitude, decreasing as the 1/rev amplitude increases.

18.7 REFERENCES

Arnold, U.T.P. "Recent IBC Flight Test Results from the CH-53G Helicopter." Twenty-Ninth European Rotorcraft Forum, Friedrichshafen, Germany, September 2003.

Arnold, U.T.P., and Furst, D. "Closed Loop IBC Results from CH-53G Flight Tests." Aerospace Science and Technology, 9:5 (July 2005).

Bernhard, A.P.F., and Wong, J. "Wind-Tunnel Evaluation of a Sikorsky Active Rotor Controller Implemented on the NASA/Army/MIT Active Twist Rotor." Journal of the American Helicopter Society, 50:1 (January 2005).

Bousman, W.G. "The Response of Helicopter Rotors to Vibratory Airloads." Journal of the American Helicopter Society, 35:4 (October 1990).

Brooks, T.F., Booth, E.R., Jr., Boyd, D.D., Jr., Splettstoesser, W.R., Schultz, K.-J., Kube, R., Niesl, G., and Streby, O. "Analysis of a Higher Harmonic Control Test to Reduce Blade Vortex Interaction Noise." Journal of Aircraft, 31:6 (November-December 1994).

Brooks, T.F., Booth, E.R., Jr., Jolly, J.R., Jr., Yeager, W.T., Jr., and Wilbur, M.L. "Reduction of Blade-Vortex Interaction Noise Through Higher Harmonic Pitch Control." Journal of the American Helicopter Society, 35:1 (January 1990).

Crews, S.T. "Rotorcraft Vibration Criteria, A New Perspective." American Helicopter Society 43rd Annual Forum, St. Louis, MO, May 1987.

Cronkhite, J.D. "The NASA/Industry Design Analysis Methods for Vibration (DAMVIBS) Program – Bell Helicopter Textron Accomplishments." AIAA Paper No. 92-2201, April 1992.

Dompka, R.V. "Investigation of Difficult-Component Effects on Finite-Element-Model Vibration Prediction for the AH-1G Helicopter." Journal of the American Helicopter Society, 35:1 (January 1990).

Dompka, R.V., and Calapodas, N.J. "Finite Element Correlation of the U.S. Army/BHTI ACAP Composite Airframe Helicopter." American Helicopter Society 47th Annual Forum, Phoenix, AZ, May 1991.

Duncan, W.J. "Galerkin's Method in Mechanics and Differential Equations." ARC R&M 1798, August 1937.

Du Val, R.W., Gregory, C.Z., Jr., and Gupta, N.K. "Design and Evaluation of a State-Feedback Vibration Controller." Journal of the American Helicopter Society, 29:3 (July 1984).

Gabel, R., and Reichert, G. "Pendulum Absorbers Reduce Transition Vibration." American Helicopter Society 31st Annual National Forum, Washington, DC, May 1975.

Goodman, R.K., and Millott, T.A. "Design, Development, and Flight Testing of the Active Vibration Control System for the Sikorsky S-92." American Helicopter Society 56th Annual Forum, Virginia Beach, VA, May 2000.

Hall, S.R., and Wereley, N.M. "Linear Control Issues in the Higher Harmonic Control of Helicopter Vibrations." American Helicopter Society 45th Annual Forum, Boston, MA, May 1989.

Hall, S.R., and Wereley, N.M. "Performance of Higher Harmonic Control Algorithms for Helicopter Vibration Reduction." Journal of Guidance, Control, and Dynamics, 16:4 (July-August 1993).

Ham, N.D. "A Simple System for Helicopter Individual-Blade-Control Using Modal Decomposition." Vertica, 4:1 (1980).

Hammond, C.E. "Wind Tunnel Results Showing Rotor Vibratory Loads Reduction Using Higher Harmonic Blade Pitch." Journal of the American Helicopter Society, 28:1 (January 1983).

Heverly, D.E., II, Singh, R., and Pappas, J. "Adaptive Algorithms for Rotorcraft Active Vibration Control." American Helicopter Society 65th Annual Forum, Grapevine, TX, May 2009.

Jacklin, S.A., Blaas, A., Teves, D., and Kube, R. "Reduction of Helicopter BVI Noise, Vibration, and Power Consumption Through Individual Blade Control." American Helicopter Society 51st Annual Forum, Ft. Worth, TX, May 1995.

Jacklin, S.A., Haber, A., de Simone, G., Norman, T.R., Kitaplioglu, C., and Shinoda, P. "Full-Scale Wind Tunnel Test of an Individual Blade Control System for a UH-60 Helicopter." American Helicopter Society 58th Annual Forum, Montreal, Canada, June 2002.

Johnson, W. "Self-Tuning Regulators for Multicyclic Control of Helicopter Vibration." NASA TP 1996, March 1982.

Koratkar, N.A., and Chopra, I. "Wind Tunnel Testing of a Smart Rotor Model with Trailing-Edge Flaps." Journal of the American Helicopter Society, 47:4 (October 2002).

Kretz, M., Aubrun, J.-N., and Larche, M. "Wind Tunnel Tests of the Dorand DH 2011 Jet Flap Rotor." NASA CR-114693 and CR-114694, June 1973.

Kube, R., van der Wall, B.G., Schultz, K.J., and Splettstoesser, W.R. "IBC Effects on BVI Noise and Vibrations â: A Combined Numerical and Experimental Investigation." American Helicopter Society 55th Annual Forum, Montreal, Canada, May 1999.

McCloud, J. L., III, and Kretz, M. "Multicyclic Jet-Flap Control for Alleviation of Helicopter Blade Stresses and Fuselage Vibration." NASA SP-352, February 1974.

McHugh, F.J., and Shaw, J., Jr. "Benefits of Higher-Harmonic Blade Pitch: Vibration Reduction, Blade-Load Reduction, and Performance Improvement." American Helicopter Society Mideast Region Symposium on Rotor Technology, Essington, PA, August 1976.

Miao, W., Kottapalli, S.B.R., and Frye, H.M. "Flight Demonstration of Higher Harmonic Control (HHC) on S-76." American Helicopter Society 42nd Annual Forum, Washington DC, June 1986.

Millott, T.A., and Welsh, W.A. "Helicopter Active Noise and Vibration Reduction." Twenty-Fifth European Rotorcraft Forum, Rome, Italy, September 1999.

Molusis, J.A., Hammond, C.E., and Cline, J.H. "A Unified Approach to the Optimal Design of Adaptive and Gain Scheduled Controllers to Achieve Minimum Helicopter Rotor Vibration." Journal of the American Helicopter Society, 28:2 (April 1983).

Nguyen, K., Betzina, M., and Kitaplioglu, C. "Full-Scale Demonstration of Higher Harmonic Control for Noise and Vibration Reduction on the XV-15 Rotor." Journal of the American Helicopter Society, 46:3 (July 2001).

Niesl, G., Swanson, S.M., Jacklin, S.A., Blaas, A., and Kube, R. "Effect of Individual Blade Control on Noise Radiation." AGARD CP 552, October 1994.

Norman, T.R., Theodore, C., Shinoda, P., Fuerst, D., Arnold, U.T.P., Makinen, S., Lorber, P., and O'Neill, J. "Full-Scale Wind Tunnel Test of a UH-60 Individual Blade Control System for Performance Improvement and Vibration, Loads, and Noise Control." American Helicopter Society 65th Annual Forum, Grapevine, TX, May 2009.

Pául, W.F. "Development and Evaluation of the Main Rotor Bifilar Absorber." American Helicopter Society 25th Annual National Forum, Washington, DC, May 1969.

Polychroniadis, M., and Achache, M. "Higher Harmonic Control: Flight Tests of an Experimental System on SA 349 Research Gazelle." American Helicopter Society 42nd Annual Forum, Washington DC, June 1986.

Roth, D., Enenkl, B., and Dieterich, O. "Active Rotor Control by Flaps for Vibration Reduction – Full Scale Demonstrator and First Flight Test Results." Thirty-Second European Rotorcraft Forum, Maastricht, The Netherlands, September 2006.

Shaw, J., and Albion, N. "Active Control of the Helicopter Rotor for Vibration Reduction." Journal of the American Helicopter Society, 26:3 (July 1981).

Southwell, R.V., and Gough, B.S. "On the Free Transverse Vibrations of Airscrew Blades." ARC R&M 766, October 1921.

Splettstoesser, W.R., Kube, R., Wagner, W., Seelhorst, U., Boutier, A., Micheli, F., Mercker, E., and Pengel, K. "Key Results from a Higher Harmonic Control Aeroacoustic Rotor Test (HART)." Journal of the American Helicopter Society, 42:1 (January 1997).

Straub, F.K., Anand, V.R., Birchette, T.S., and Lau, B.H. "Wind Tunnel Test of the SMART Active Flap Rotor." American Helicopter Society 65th Annual Forum, Grapevine, TX, May 2009.

United States Government. "Requirements for Helicopter Flying Qualities." Military Specification MIL-H-8501, November 1952.

United States Government. "General Requirements for Helicopter Flying and Ground Handling Qualities." Military Specification MIL-H-8501A, September 1961.

U.S. Army Aviation and Missile Command, Aviation Engineering Directorate. "Requirements for Rotorcraft Vibration Specifications, Modeling and Testing." Aeronautical Design Standard, Standard Practice, ADS-27A-SP, May 2006.

Welsh, W.A. "Evolution of Active Vibration Control Technology." American Helicopter Society 4th Decennial Specialist's Conference on Aeromechanics, San Francisco, CA, January 2004.

Wood, E.R., Powers, R.W., Cline, J.H., and Hammond, C.E. "On Developing and Flight Testing a Higher Harmonic Control System." Journal of the American Helicopter Society, *30*:1 (January 1985).

19 Flap Motion

The blade flap hinge or, equivalently, flapping flexibility for hingeless and bearingless rotors is a crucial aspect of practical rotor design, allowing inertial forces due to flapping to counter applied airloads, thereby reducing structural loads. Hence flap motion of the blades plays a key role in the aeroelastic behavior of the rotor. Chapter 6 dealt with the steady-state solution for the flap response in forward flight. This chapter is concerned with the dynamic behavior of the flap motion. We consider the eigenvalues in the rotating and non-rotating frame and the flap response to control, gust, and shaft motion inputs. The hub reactions in response to shaft motion, including the effects of the flapping dynamics, are also examined. The equations derived here are used in Chapter 21 in the investigation of helicopter flight dynamics. For the shaft-fixed problems, a single independent blade in the rotating frame, which is a single-degree-of-freedom system, can be considered. With coupling of the blades, through either shaft motion or the rotor wake, the motion of the entire rotor must be considered; hence N degrees of freedom, one for each blade.

19.1 Rotating Frame

The equation of motion for the fundamental flapping mode of a rotor blade in the rotating frame is

$$\widehat{I}_\beta(\ddot{\beta} + v_\beta^2 \beta) = \gamma \int_0^1 \eta_\beta \frac{F_z}{ac} dr = \gamma M_F \tag{19.1}$$

from equation 16.55, without the precone term. The flapping degree of freedom is β. An arbitrary rotor blade is considered, described by the rotating natural frequency of the flap motion v_β and the out-of-plane mode shape $\eta_\beta(r)$. The normalized flap inertia $\widehat{I}_\beta = I_\beta/I_b$ has a value of approximately 1. The Lock number $\gamma = \rho a c R^4/I_b$ characterizes the relative magnitudes of the aerodynamic and inertial forces acting on the blade. Equation 16.309 gives the aerodynamic flap moment:

$$M_F = M_\theta(\theta - K_P \beta) + M_{\dot{\beta}}\dot{\beta} + M_\beta \beta + M_\lambda(\lambda - w_G) \tag{19.2}$$

In addition to the aerodynamic forces caused by the flap motion, those due to the blade pitch control and a vertical gust velocity are included. Kinematic pitch-flap coupling is introduced in terms of $K_P = \tan \delta_3$; see section 6.18. Assuming $\eta_\beta = r$

and neglecting reverse flow and tip losses, the aerodynamic coefficients are

$$M_\theta = \frac{1}{8} + \frac{\mu}{3}\sin\psi + \frac{\mu^2}{4}\sin^2\psi \tag{19.3}$$

$$M_\lambda = -\left(\frac{1}{6} + \frac{\mu}{4}\sin\psi\right) \tag{19.4}$$

$$M_{\dot\beta} = -\left(\frac{1}{8} + \frac{\mu}{6}\sin\psi\right) \tag{19.5}$$

$$M_\beta = -\mu\cos\psi\left(\frac{1}{6} + \frac{\mu}{4}\sin\psi\right) \tag{19.6}$$

Thus the flap equation of motion is

$$\widehat{I}_\beta\ddot\beta - \gamma M_{\dot\beta}\dot\beta + \left(\widehat{I}_\beta v_\beta^2 + K_P\gamma M_\theta - \gamma M_\beta\right)\beta = \gamma M_\theta\theta + \gamma M_\lambda(\lambda - w_G) \tag{19.7}$$

which is a linear ordinary differential equation that has periodic coefficients in forward flight. The aerodynamic forces provide the flap damping, flap springs in forward flight (M_β) and through the pitch-flap coupling, and the moments due to the control and gust inputs.

19.1.1 Hover Roots

In hover ($\mu = 0$) the aerodynamic environment is axisymmetric, and the aerodynamic coefficients are constants. In addition, $M_\beta = 0$ in hover. Thus the homogeneous equation is

$$\widehat{I}_\beta\ddot\beta - \gamma M_{\dot\beta}\dot\beta + \left(\widehat{I}_\beta v_\beta^2 + K_P\gamma M_\theta\right)\beta = 0 \tag{19.8}$$

The characteristic equation is

$$\widehat{I}_\beta s^2 - \gamma M_{\dot\beta} s + \left(\widehat{I}_\beta v_\beta^2 + K_P\gamma M_\theta\right) = 0 \tag{19.9}$$

which can be solved for the eigenvalues or roots of the flap dynamics in hover:

$$s = \frac{\gamma M_{\dot\beta}}{2\widehat{I}_\beta} \pm i\sqrt{v_\beta^2 + K_P\frac{\gamma M_\theta}{\widehat{I}_\beta} - \left(\frac{\gamma M_{\dot\beta}}{2\widehat{I}_\beta}\right)^2} \tag{19.10}$$

Substituting $\widehat{I}_\beta = 1$ and $-M_{\dot\beta} = M_\theta = \frac{1}{8}$ gives

$$s = \frac{\gamma}{16} \pm i\sqrt{v_\beta^2 + K_P\frac{\gamma}{8} - \left(\frac{\gamma}{16}\right)^2} \tag{19.11}$$

Unless the Lock number γ is very large, the transient flap motion in the rotating frame is a damped oscillation, with frequency, natural frequency, and damping ratio as follows:

$$\omega = \mathrm{Im}\,s = \sqrt{v_\beta^2 + K_P\frac{\gamma}{8} - \left(\frac{\gamma}{16}\right)^2} \tag{19.12}$$

$$\omega_n = |s| = \sqrt{v_\beta^2 + K_P\frac{\gamma}{8}} = v_{\beta e} \tag{19.13}$$

$$\zeta = -\frac{\mathrm{Re}\,s}{|s|} = \frac{\gamma}{16 v_{\beta e}} \tag{19.14}$$

19.1 Rotating Frame

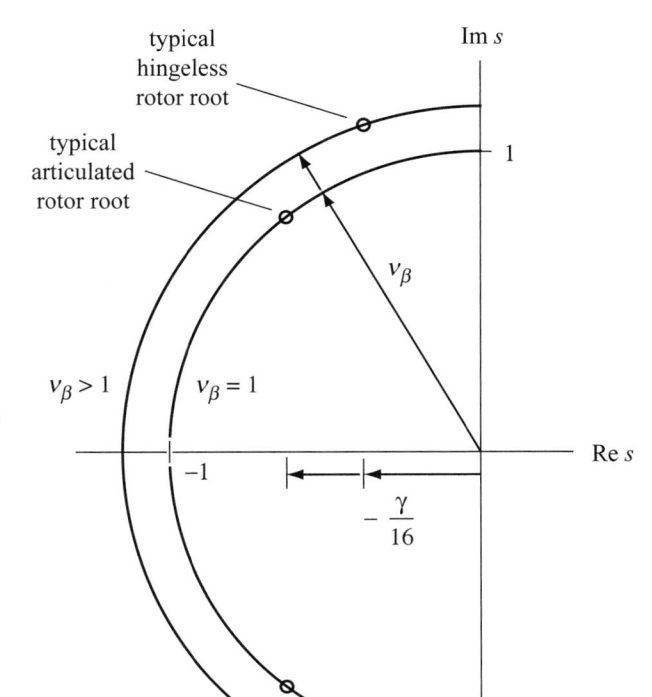

Figure 19.1. Flapping roots in hover (no pitch-flap coupling).

The pitch-flap coupling K_P introduces an aerodynamic spring on the flap motion through M_θ, giving the effective flapping natural frequency $\nu_{\beta e}$. The damping ratio is typically around 50% critical damping, so the flap motion is highly damped. The source of this damping is the aerodynamic lift forces on the blade due to the angle-of-attack change produced by a flapping velocity. Figure 19.1 shows the hover eigenvalues, with typical roots for articulated and hingeless rotors. For the articulated rotor the frequency is below 1/rev, whereas for the hingeless rotor ($\nu_\beta > 1$ and small γ) the frequency is likely to be above 1/rev. The location of the roots is determined by the natural frequency ν_β (which gives the distance from the origin) and by the damping $\text{Re}\, s = -\gamma/16$ (which gives the distance from the imaginary axis).

For $\gamma = 0$, the roots are $s = \pm i\nu_\beta$. Since there are no aerodynamic forces in this case, the motion is an undamped oscillation at the frequency ν_β determined by the centrifugal and structural stiffness of the blade. For $\gamma > 0$, the locus of roots describes a circle with radius $\sqrt{\nu_\beta^2 + K_P^2}$ and center at $-K_P$ on the real axis. The location of the two complex conjugate roots on this circle can be determined from the real part $\text{Re}\, s = -\gamma/16$, which depends on the Lock number alone. For large enough γ, specifically $\gamma/16 = K_P + \sqrt{\nu_\beta^2 + K_P^2}$, the loci intercept the real axis at $s = -\gamma/16$. In the absence of pitch-flap coupling the intercept occurs at $\gamma = 16\nu_\beta$, which is rather large. Unless the pitch-flap coupling ($K_P = \tan \delta_3$) is significantly negative, the flap roots are a complex conjugate pair, implying oscillatory transient motion. As the Lock number becomes still larger, one of the roots approaches $s = -\infty$ on the real axis, and the other goes to $s = -K_P$. If $K_P < 0$ (negative δ_3, flap-up/pitch-up), this latter root on the real axis goes through the origin into the right half-plane as γ increases. The criterion for stable motion (Re $s < 0$) is then

$$\frac{\gamma}{16} < \frac{v_\beta^2}{2|K_P|} \tag{19.15}$$

for $K_P < 0$. The main rotors of the helicopter are generally well away from this boundary, but the boundary can be a concern for rotors with large negative pitch-flap coupling and small Lock number. Since this instability is a static divergence, the boundary is simply determined by the spring terms in the flap equation. For stable motion a net positive flap spring is required, or $v_{\beta e}^2 > 0$, which gives the above boundary. This flap divergence is primarily a limit on the allowable negative pitch-flap coupling, so the stability requirement can be written instead as $K_P > -8v_\beta^2/\gamma$.

19.1.2 Forward Flight Roots

The homogeneous equation of the flap motion in forward flight is

$$\widehat{I}_\beta \ddot{\beta} - \gamma M_{\dot\beta}\dot{\beta} + (\widehat{I}_\beta v_\beta^2 + K_P\gamma M_\theta - \gamma M_\beta)\beta = 0 \tag{19.16}$$

The flap inertia I_β can be used in the Lock number here, so with no loss of generality $\widehat{I}_\beta = 1$ can be assumed. The aerodynamic coefficients are given in Section 19.1 for $\eta_\beta = r$ and neglecting reverse flow and tip losses. Above about $\mu = 0.5$, reverse flow must be included in the aerodynamic coefficients, as in section 6.8. The hover solution was found in the last section:

$$s = \frac{\gamma}{16} \pm i\sqrt{v_\beta^2 + K_P\frac{\gamma}{8} - \left(\frac{\gamma}{16}\right)^2} \tag{19.17}$$

Forward flight ($\mu > 0$) introduces periodic coefficients due to the rotation of the blade relative to the helicopter forward velocity. These coefficients radically influence the behavior of the root loci and also the analysis techniques required. The root loci of a time-invariant system can exhibit behavior in which two roots start as complex conjugates, meet at the real axis, and then proceed in opposite directions on the real axis. With periodic coefficients this behavior is generalized so that it can occur at any frequency that is a multiple of $\frac{1}{2}$/rev, not just on the real axis. The property of the solution that allows this behavior is the fact that the eigenvectors are themselves periodic, instead of constant as for a time-invariant system. In section 15.6 the behavior of roots of periodic-coefficient differential equations was discussed, and procedures for calculating the roots were developed.

Consider the stability of the rotor flapping motion in forward flight (see Johnson (1973, 1974)). For small μ, analytical solutions for the roots can be obtained, but numerical methods are required at moderate and high advance ratios. At $\mu = 0$ the roots are complex conjugate pairs (or perhaps two real roots) determined by v_β, γ, and K_P. The frequency of the hover roots has an important influence on the behavior at low μ. For values of v_β and γ such that the hover root frequency is not too close to a multiple of $\frac{1}{2}$/rev, the roots for low advance ratio only exhibit an order μ^2 change in frequency, which is quite small even at $\mu = 0.5$; there is no change in the real part of the root. When the hover frequency is near a multiple of one-half the fundamental frequency of the system, there can occur a degradation of the stability, perhaps even an instability, which is characteristic of periodic coefficient equations. If the hover root frequency is near n/rev, then as μ increases, the roots approach n/rev while remaining a complex conjugate pair. The roots reach $\text{Im}\, s = n$ for some critical μ, and then for still larger μ the frequency remains fixed while the real part of one root is increased and that of the other root decreased. The root being destabilized can cross into the right half-plane for large enough μ, indicating that the system has become unstable because of the periodic coefficients. Similar behavior can occur if the hover root frequency is near $(n + \frac{1}{2})$/rev. For the hover root frequency near $\frac{1}{2}$/rev there are order μ influences of the periodic coefficients. Initially there is an order μ change in the frequency, with the real part of the root remaining near the hover value $\text{Re}\, s = -\gamma/16$. The roots reach $\text{Im}\, s = \frac{1}{2}$ for a value of the advance ratio that decreases as the hover root moves closer to $\frac{1}{2}$/rev. For larger μ there are order μ changes in the real part of the roots while the frequency is fixed at $\frac{1}{2}$/rev. The order μ reduction in damping is small compared to the large aerodynamic damping in hover, so the flapping stability remains high for small advance ratio, even with the influence of the periodic coefficients. The roots exhibit a similar behavior when the hover frequency is near 1/rev, except that all the changes are of order μ^2, and hence are much smaller than those near $\frac{1}{2}$/rev. At $\mu = 2.25$ or so (there is some dependence on ν_β, γ, and K_P) a flapping instability is encountered because of the periodic forces on the rotor blade in forward flight. This instability usually occurs in a region where the frequency is fixed at 1/rev and the real part of one root has been decreased enough for the root to go into the right half-plane. For advance ratios high enough to encounter this instability, reverse flow effects must be included in the aerodynamic coefficients. Other degrees of freedom (such as elastic bending, lag motion, and torsion motion) significantly reduce the advance ratio at the stability boundary. A representation of the rotor blade motion by just the fundamental flap mode is not adequate at very high advance ratio.

Figure 19.2 shows typical root loci of the rotor blade flap motion from hover ($\mu = 0$) up to $\mu = 0.6$. Three cases are shown: (a) a typical articulated rotor with $\gamma = 12$ and $\nu_\beta = 1.0$, for which the hover frequency is near $\frac{1}{2}$/rev; (b) a typical hingeless rotor with $\gamma = 6$ and $\nu_\beta = 1.15$, for which the hover frequency is near 1/rev; and (c) an intermediate case with $\gamma = 6$ and $\nu_\beta = 1.0$, for which the hover frequency is not near a multiple of $\frac{1}{2}$/rev. There is a pair of roots for each case. The articulated rotor (case a) illustrates the order μ behavior near $\frac{1}{2}$/rev, and the hingeless rotor (case b) illustrates the order μ^2 behavior near 1/rev. Case (c) shows just the small frequency change of order μ^2 for roots away from a multiple of $\frac{1}{2}$/rev. The eigenvalues depend primarily on the Lock number and advance ratio, so the results can be presented as contours of constant real and imaginary parts of the roots on the γ–μ plane. Figure 19.3 is such a plot for the case $\nu_\beta = 1$ and $K_P = 0$. The regions in which the frequency is fixed at $\frac{1}{2}$/rev or 1/rev are due to the periodic coefficients. Since a horizontal line in Figure 19.3 corresponds to constant γ, the variation of $\text{Im}\, s$ and $\text{Re}\, s$ as such a line is traversed gives the root locus for varying μ.

For example, consider the articulated rotor with $\gamma = 12$ (case a in Figure 19.2). As μ increases the $\frac{1}{2}$/rev region comes closer, indicating that the frequency is approaching $\frac{1}{2}$/rev. When the $\gamma = 12$ line goes into the $\frac{1}{2}$/rev region, the frequency of the root remains fixed while for each point in the region there are two values of the real part, one more stable and one less stable than the hover root. When μ is of order 1 the real parts of the roots approach each other again, and at about $\mu = 1.7$ the roots move quickly from $\frac{1}{2}$/rev to 1/rev as complex conjugates, hence with the

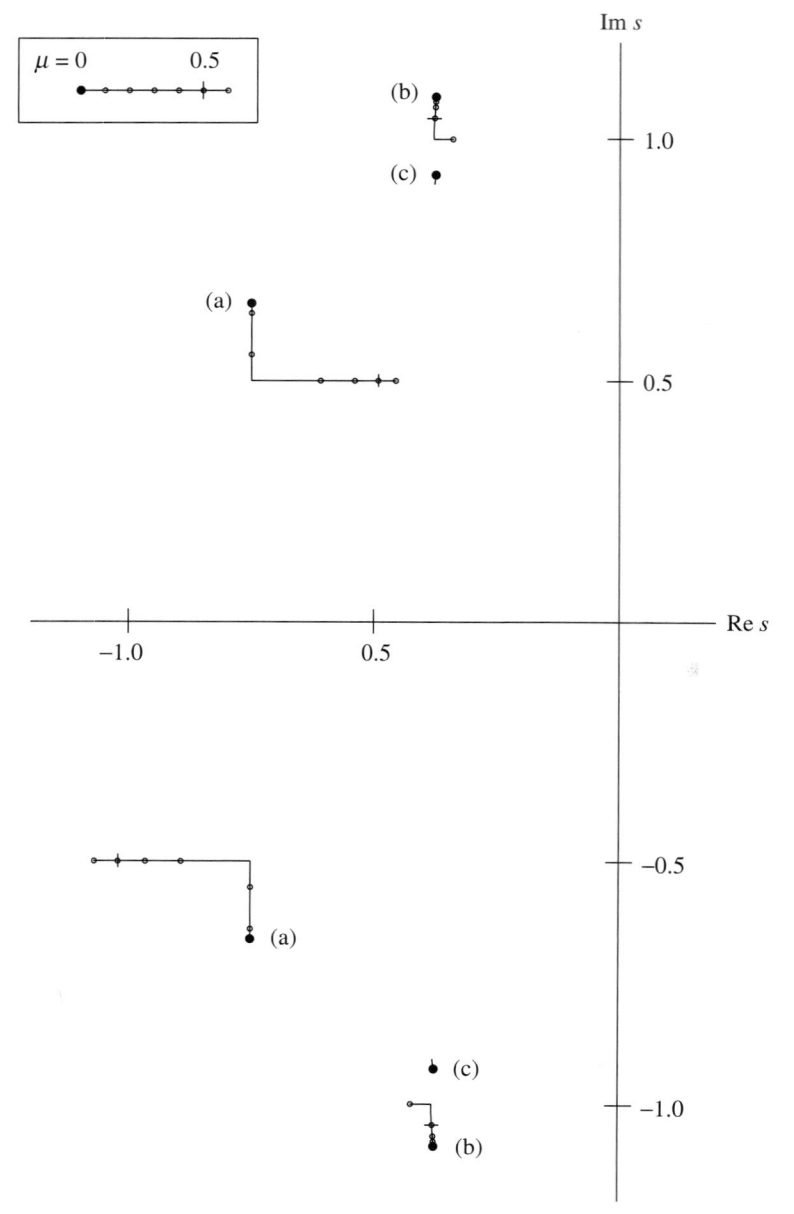

Figure 19.2. Influence of forward flight on flapping roots, for (a) $v_\beta = 1$ and $\gamma = 12$, (b) $v_\beta = 1.15$ and $\gamma = 6$, and (c) $v_\beta = 1$ and $\gamma = 6$.

same real parts. After the 1/rev region is entered, the damping of one root decreases again while that of the other increases. The branch being destabilized finally crosses into the right half-plane at about $\mu = 2.3$. Figure 19.3 shows that the critical regions where the frequency is fixed at a multiple of $\frac{1}{2}$/rev increase in importance as the periodic forces increase with μ, eventually dominating the behavior of the root loci at high advance ratio.

Because of the greater ease and scope of the analysis of constant-coefficient differential equations, it is useful to have a time-invariant model of the rotor dynamics

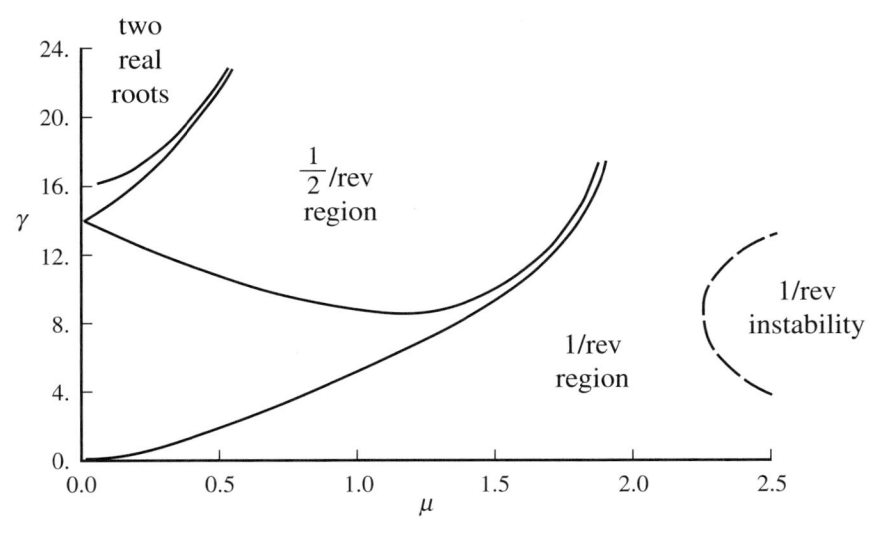

Figure 19.3. Flap roots in forward flight ($v_\beta = 1$ and $K_P = 0$).

in forward flight. Such a model must be approximate, since periodic systems have unique behavior, but the approximation is satisfactory for many conditions. If the mean values of the coefficients in the rotating frame are used, the only influence of forward flight on the flap moments that remains is an order μ^2 change in M_θ. If there is no pitch-flap coupling, forward flight has no influence at all on the eigenvalues. Unless μ is very small, this is not a satisfactory approximation. However, by using the mean values of the coefficients in the non-rotating frame (after introducing multiblade coordinates), much more of the influence of forward flight is retained; see section 16.8.4.

Consider, for example, the case of the articulated rotor with $\gamma = 12$ and $v_\beta = 1.0$ (case a of Figure 19.2). In the rotating frame the roots encounter the $\frac{1}{2}$/rev critical region as μ increases. Recall from section 15.5 that in the transformation to the non-rotating frame, the coning roots are unchanged while the low-frequency and high-frequency flap mode roots are shifted in frequency by 1/rev from the rotating roots, as shown in Figure 19.4. There are additional roots for rotors with more than three blades. Figure 19.4 also shows the results of the constant coefficient approximation in the non-rotating frame, which gives the influence of forward flight on the flap roots remarkably well. The approximation does not work in the rotating frame because without periodic coefficients the two flap roots must always remain complex conjugates. The transformation to the non-rotating frame places four roots near $\frac{1}{2}$/rev, two from the coning mode and two from the low-frequency flap mode. These four roots can behave in a fashion similar to the roots of a periodic system in a critical region: the frequency is fixed at a multiple of $\frac{1}{2}$/rev while the real parts decrease for one root and increase for the other, yet remain complex conjugates as required of a constant coefficient system. For the hingeless rotor example ($\gamma = 6$ and $v_\beta = 1.15$, which places the rotating frequency near 1/rev as shown in Figure 19.2), the transformation to the non-rotating frame shifts the roots of the low-frequency flap mode to the real axis, where the constant coefficient approximation can model the correct behavior of the roots.

Flap Motion

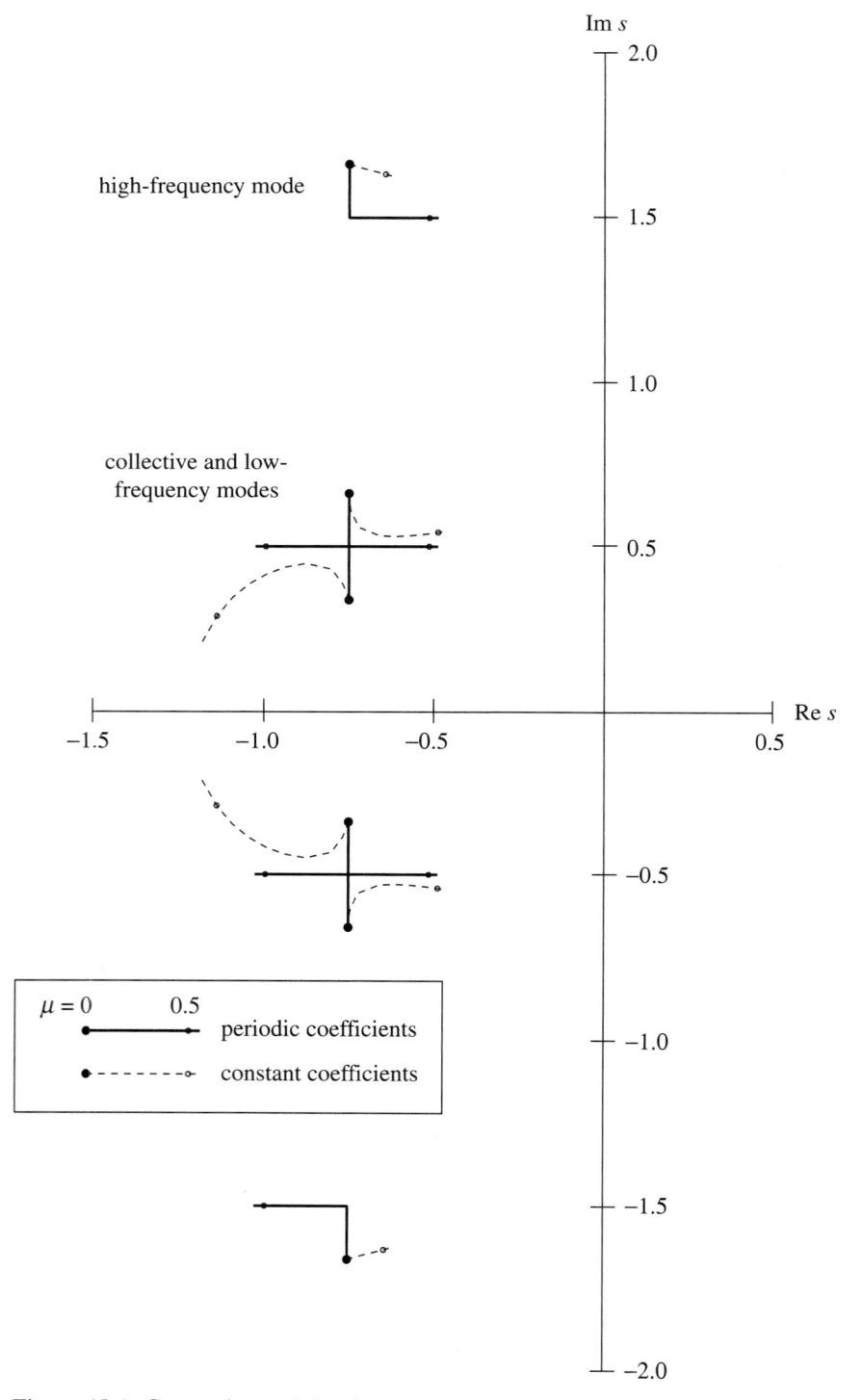

Figure 19.4. Comparison of the flapping roots of a three-bladed rotor from the periodic coefficient solution and the constant coefficient approximation.

In general, the characteristic behavior of the roots of a periodic system can be exhibited by the roots of a time-invariant system only when there are two roots on the real axis or four complex roots (two at positive frequency and two at negative frequency). The transformation to the non-rotating frame produces such loci because it shifts the frequency of the rotating roots by $\pm n/\text{rev}$ for the β_{nc} and β_{ns} mode). The high-frequency modes in the non-rotating frame are always isolated pairs (one at positive frequency and one at negative frequency), which must remain complex conjugates in the constant coefficient approximation. The constant coefficient approximation is expected therefore to be least satisfactory when the rotor high-frequency dynamics must be modeled. Increasing the number of blades improves the approximation by increasing the number of coupled degrees of freedom of the model in the non-rotating frame. In summary, the constant coefficient approximation to the rotor dynamics in forward flight produces differential equations that can be more easily and more thoroughly analyzed, but no longer completely describe the rotor behavior. The results from the time-invariant model must always be approximate. Generally the constant coefficient approximation in the non-rotating frame gives results that are close to the correct solution, particularly for the behavior involving the lower frequency modes, as long as the advance ratio is not too large. However, the validity of the constant coefficient approximation should always be checked by comparing the exact and approximate solutions for the particular problem being considered.

The flap equation in forward flight can be transformed to Mathieu's equation. Without pitch-flap coupling, the homogeneous equation of motion is

$$\ddot{\beta} + \left(\frac{\gamma}{8} + \frac{\gamma}{6}\mu \sin\psi\right)\dot{\beta} + \left(v_\beta^2 + \frac{\gamma}{6}\mu \cos\psi + \frac{\gamma}{8}\mu^2 \sin 2\psi\right)\beta = 0 \qquad (19.18)$$

for low μ (no reverse flow). The transformation $\beta = xe^{(-\frac{\gamma}{16}\psi + \frac{\gamma}{12}\mu \cos\psi)}$ eliminates the velocity term:

$$\ddot{x} + \left(v_\beta^2 + \frac{\gamma}{6}\mu \cos\psi + \frac{\gamma}{8}\mu^2 \sin 2\psi - \left(\frac{\gamma}{16} + \frac{\gamma}{12}\mu \sin\psi\right)^2 - \frac{\gamma}{12}\mu \cos\psi\right)x$$

$$= \ddot{x} + \left(v_\beta^2 - \frac{1}{4}\left(\frac{\gamma}{8}\right)^2 - \frac{2}{9}\left(\frac{\gamma}{8}\right)^2 \mu^2 + \frac{2}{3}\frac{\gamma}{8}\mu \cos\psi - \frac{2}{3}\left(\frac{\gamma}{8}\right)^2 \mu \sin\psi \right.$$

$$\left. + \frac{\gamma}{8}\mu^2 \sin 2\psi + \frac{2}{9}\left(\frac{\gamma}{8}\right)^2 \mu^2 \cos 2\psi\right)x = 0 \qquad (19.19)$$

which to order μ is

$$\ddot{x} + \left(v_\beta^2 - \frac{1}{4}\left(\frac{\gamma}{8}\right)^2 + \frac{2}{3}\frac{\gamma}{8}\mu\left(\cos\psi - \frac{\gamma}{8}\sin\psi\right)\right)x = 0 \qquad (19.20)$$

Substituting $\psi + \delta = 2z$ with $\tan\delta = \frac{\gamma}{8}$ gives

$$\frac{d^2x}{dz^2} + \left(4v_\beta^2 - \left(\frac{\gamma}{8}\right)^2 + \frac{8}{3}\mu\frac{\gamma}{8}\sqrt{1 + \left(\frac{\gamma}{8}\right)^2} \cos 2z\right)x = \frac{d^2x}{dz^2} + (a + 2q\cos 2z)x = 0 \qquad (19.21)$$

which is a form of Mathieu's equation. Here $a = 4\omega_h^2$, where ω_h is the frequency of the hover root, and q is proportional to the advance ratio μ. This equation has critical regions with boundaries $a(q)$. At $q = 0$ (hover), the critical values are $a = n^2$ for positive integers n; hence the frequency is at $\frac{n}{2}$/rev. To lowest order in q, the critical regions are described by $a = n^2 + bq^2 \pm cq^n$.

The prototypical problem of a periodic system in rotor dynamics is the solution of the equation of blade flap motion in edgewise flight. The stability of blade flap motion has been examined by numerous investigators, beginning with Glauert and Shone (1933). The methods used included infinite determinant, perturbation expansions and successive approximations, numerical integration, and analog computation. Horvay (1947) was the first to plot the frequency and stability regions on the parameter plane of Lock number (as $n = \gamma/8$) and advance ratio. Peters and Hohenemser (1971) brought the digital computer to the task of implementing Floquet theory as an analysis tool. The technique proved to be fast, although this investigation did not capture the order μ^2 behavior of the 1/rev region boundary at small advance ratio.

19.1.3 Hover Transfer Function

The response of the blade flapping motion to pitch and gust inputs can be defined by a transfer function. Only the hovering case is considered, since the periodic coefficients in forward flight introduce inter-harmonic coupling. In forward flight a sinusoidal input at a single frequency ω does not produce an output at that frequency alone, but rather at all frequencies $\omega \pm n$/rev. For a small advance ratio at least, the dominant response is still at the input frequency ω.

For hover, the flapping equation of motion in the rotating frame is

$$\widehat{I}_\beta \ddot{\beta} - \gamma M_{\dot{\beta}} \dot{\beta} + (\widehat{I}_\beta v_\beta^2 + K_P \gamma M_\theta)\beta = \gamma M_\theta \theta - \frac{1}{2}\gamma M_\lambda w_G \tag{19.22}$$

Recall from section 16.8.8 that in hover the inflow perturbation reduces the effect of the vertical velocity by a factor of one-half, $\lambda - w_G = -\frac{1}{2}w_G$, which has been used in equation 19.22. The transfer function of the flap motion in hover is then

$$\beta = \frac{\gamma M_\theta \theta - \frac{1}{2}\gamma M_\lambda w_G}{\widehat{I}_\beta \left(s^2 - (\gamma M_{\dot{\beta}}/\widehat{I}_\beta)s + v_{\beta e}^2\right)} \tag{19.23}$$

Here s is the Laplace variable, and $v_{\beta e}^2 = v_\beta^2 + K_P \gamma M_\theta/\widehat{I}_\beta$. The poles (the roots of the denominator polynomial) are the hover eigenvalues, and there are no zeros. Substituting for the coefficients gives

$$\beta = \frac{\frac{\gamma}{8}\theta - \frac{\gamma}{12}w_G}{s^2 - \frac{\gamma}{8}s + v_{\beta e}^2} \tag{19.24}$$

The frequency response is obtained from $s = i\omega$. Figure 19.5 shows the magnitude and phase of the frequency response of the blade flap motion to pitch control inputs for an articulated rotor, with $v_\beta = 1$, $K_P = 0$, and $\gamma = 8$. The response is that of a highly damped second-order system with natural frequency $v_{\beta e}$. The static response ($\omega = 0$) is $\beta/\theta = \gamma/8v_{\beta e}^2$, and at high frequency the magnitude decreases and the phase shifts to $-180°$ as the inertia dominates the system.

19.2 Non-Rotating Frame

The equations for the flap motion in the non-rotating frame are obtained by using multiblade coordinates to transform the rotating equation. Here only the case of hover and three or more blades is considered, so the equations have constant

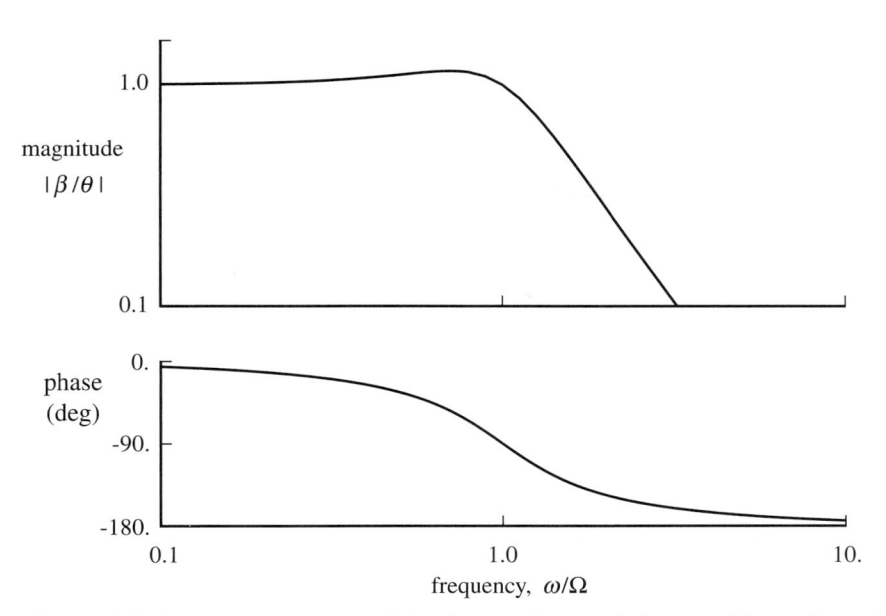

Figure 19.5. Frequency response of the flap motion to pitch control input ($v_\beta = 1$, $K_P = 0$, $\gamma = 8$).

coefficients. Excitation by blade pitch control, shaft motion, and aerodynamic gusts is included. The equations of motion are

$$\widehat{I}_\beta \ddot{\beta}_0 - \gamma M_{\dot{\beta}} \dot{\beta}_0 + \left(\widehat{I}_\beta v_\beta^2 + K_P \gamma M_\theta\right) \beta_0$$
$$= \gamma M_\theta \theta_0 + \gamma M_\lambda (\lambda + \dot{z}_h - w_G) - \widehat{S}_\beta \ddot{z}_h \quad (19.25)$$

$$\widehat{I}_\beta \begin{pmatrix} \ddot{\beta}_{1c} \\ \ddot{\beta}_{1s} \end{pmatrix} + \begin{bmatrix} -\gamma M_{\dot{\beta}} & 2\widehat{I}_\beta \\ -2\widehat{I}_\beta & -\gamma M_{\dot{\beta}} \end{bmatrix} \begin{pmatrix} \dot{\beta}_{1c} \\ \dot{\beta}_{1s} \end{pmatrix}$$
$$+ \begin{bmatrix} \widehat{I}_\beta(v_\beta^2 - 1) + K_P \gamma M_\theta & -\gamma M_{\dot{\beta}} \\ \gamma M_{\dot{\beta}} & \widehat{I}_\beta(v_\beta^2 - 1) + K_P \gamma M_\theta \end{bmatrix} \begin{pmatrix} \beta_{1c} \\ \beta_{1s} \end{pmatrix}$$
$$= \gamma M_\theta \begin{pmatrix} \theta_{1c} \\ \theta_{1s} \end{pmatrix} + \gamma M_\mu \begin{pmatrix} \dot{y}_h + v_G \\ -\dot{x}_h + u_G \end{pmatrix}$$
$$+ \widehat{I}_{\beta\alpha} \begin{pmatrix} \ddot{\alpha}_y \\ -\ddot{\alpha}_x \end{pmatrix} + \begin{bmatrix} -\gamma M_{\dot{\beta}} & 2\widehat{I}_{\beta\alpha} \\ -2\widehat{I}_{\beta\alpha} & -\gamma M_{\dot{\beta}} \end{bmatrix} \begin{pmatrix} \dot{\alpha}_y \\ -\dot{\alpha}_x \end{pmatrix} \quad (19.26)$$

$$\widehat{I}_\beta \begin{pmatrix} \ddot{\beta}_{nc} \\ \ddot{\beta}_{ns} \end{pmatrix} + \begin{bmatrix} -\gamma M_{\dot{\beta}} & 2n\widehat{I}_\beta \\ -2n\widehat{I}_\beta & -\gamma M_{\dot{\beta}} \end{bmatrix} \begin{pmatrix} \dot{\beta}_{nc} \\ \dot{\beta}_{ns} \end{pmatrix}$$
$$+ \begin{bmatrix} \widehat{I}_\beta(v_\beta^2 - n^2) + K_P \gamma M_\theta & -n\gamma M_{\dot{\beta}} \\ n\gamma M_{\dot{\beta}} & \widehat{I}_\beta(v_\beta^2 - n^2) + K_P \gamma M_\theta \end{bmatrix} \begin{pmatrix} \beta_{nc} \\ \beta_{ns} \end{pmatrix}$$
$$= \gamma M_\theta \begin{pmatrix} \theta_{nc} \\ \theta_{ns} \end{pmatrix} \quad (19.27)$$

$$\widehat{I}_\beta \ddot{\beta}_{N/2} - \gamma M_{\dot{\beta}} \dot{\beta}_{N/2} + \left(\widehat{I}_\beta v_\beta^2 + K_P \gamma M_\theta\right) \beta_{N/2}$$
$$= \gamma M_\theta \theta_{N/2} \quad (19.28)$$

as derived in section 16.7 (equations 16.244, 16.245, and 16.246) and section 16.8.7 (equations 16.449, 16.450, and 16.451). The flapping degrees of freedom in the non-rotating frame are the coning mode β_0, the tip-path-plane tilt modes β_{1c} and β_{1s}, and the reactionless modes (β_{nc}, β_{ns}, and $\beta_{N/2}$) as required to give a total of N degrees of freedom for an N-bladed rotor. The only coupling of the flap degrees of freedom is between β_{1c} and β_{1s} and between β_{nc} and β_{ns}. Moreover, only the coning and tip-path-plane tilt degrees of freedom respond to the shaft motion and gusts, so these three degrees of freedom are of the most interest. For the shaft-fixed case, these equations also apply to the two-bladed rotor in hover, where the degrees of freedom are the coning mode β_0 and the teetering mode β_1.

19.2.1 Hover Roots and Modes

The homogeneous equations give the following characteristic equations for the flap motion in the non-rotating frame:

$$\left(\widehat{I}_\beta s^2 - \gamma M_{\dot\beta} s + \widehat{I}_\beta v_\beta^2 + K_P \gamma M_\theta\right) \beta_0 = 0 \tag{19.29}$$

$$\begin{bmatrix} \widehat{I}_\beta s^2 - \gamma M_{\dot\beta} s + \widehat{I}_\beta (v_\beta^2 - n^2) + K_P \gamma M_\theta & n\left(2\widehat{I}_\beta s - \gamma M_{\dot\beta}\right) \\ -n\left(2\widehat{I}_\beta s - \gamma M_{\dot\beta}\right) & \widehat{I}_\beta s^2 - \gamma M_{\dot\beta} s + \widehat{I}_\beta (v_\beta^2 - n^2) + K_P \gamma M_\theta \end{bmatrix} \begin{pmatrix} \beta_{nc} \\ \beta_{ns} \end{pmatrix} = 0 \tag{19.30}$$

$$\left(\widehat{I}_\beta s^2 - \gamma M_{\dot\beta} s + \widehat{I}_\beta v_\beta^2 + K_P \gamma M_\theta\right) \beta_{N/2} = 0 \tag{19.31}$$

Since the characteristic equations for β_0 and $\beta_{N/2}$ are the same as for the single blade in the rotating frame, the non-rotating eigenvalues s_{NR} for the coning and reactionless modes are equal to the rotating eigenvalues:

$$s_{NR} = s_R = \frac{\gamma M_{\dot\beta}}{2\widehat{I}_\beta} \pm i \sqrt{v_\beta^2 + K_P \frac{\gamma M_\theta}{\widehat{I}_\beta} - \left(\frac{\gamma M_{\dot\beta}}{2\widehat{I}_\beta}\right)^2} \tag{19.32}$$

For the β_{nc} and β_{ns} degrees of freedom, the determinant of the matrix gives

$$\left(\widehat{I}_\beta s^2 - \gamma M_{\dot\beta} s + \widehat{I}_\beta (v_\beta^2 - n^2) + K_P \gamma M_\theta\right)^2 + n^2 \left(2\widehat{I}_\beta s - \gamma M_{\dot\beta}\right)^2 = 0 \tag{19.33}$$

which has the solution $s_{NR} = s_R \pm in$. Thus the transformation to the non-rotating frame simply shifts the frequency of the β_{nc} and β_{ns} roots by $\pm n$/rev, while the real part remains unchanged. The corresponding eigenvectors (using the characteristic equation to replace the numerator) are

$$\frac{\beta_{nc}}{\beta_{ns}} = \frac{\widehat{I}_\beta s^2 - \gamma M_{\dot\beta} s + \widehat{I}_\beta (v_\beta^2 - n^2) + K_P \gamma M_\theta}{n\left(2\widehat{I}_\beta s - \gamma M_{\dot\beta}\right)} = \pm i \tag{19.34}$$

The high-frequency mode $s_{NR} = s_R + in$ has the eigenvector $\beta_{nc}/\beta_{ns} = i$, and the low-frequency mode $s_{NR} = s_R - in$ has the eigenvector $\beta_{nc}/\beta_{ns} = -i$. So for both modes, β_{nc} and β_{ns} have equal magnitude, but are 90° apart in phase. For the high-frequency mode β_{nc} leads β_{ns} by one-quarter of an oscillation cycle, whereas for the low-frequency mode β_{ns} leads β_{nc}. See section 15.5 for a further discussion of the eigenvalues and eigenvectors in the non-rotating frame.

For the normal modes of the flap motion in hover, the only coupling is between the β_{nc} and β_{ns} degrees of freedom. The coning β_0 and reactionless $\beta_{N/2}$ modes are highly damped oscillations, with the same eigenvalues as the rotating flap response. The β_{nc} and β_{ns} degrees of freedom have two modes, each a damped oscillation with a frequency equal to the rotating flap frequency plus or minus n/rev. The high-frequency mode with eigenvalue $s_{NR} = s_R + in$ (assuming that the frequency of s_R is positive) is a whirling or wobbling motion of β_{nc} and β_{ns} at a frequency $\text{Im}\, s_R + n$ in the same direction as the rotor rotation. The low-frequency mode with eigenvalue $s_{NR} = s_R - in$ is a motion of β_{nc} and β_{ns} at a frequency $|\text{Im}\, s_R - n|$, in the same direction as the rotor rotation if the rotating frequency is below n/rev, and in the opposite direction if $\text{Im}\, s_R > n$. The high-frequency flap motion can thus be called a progressive mode, and the low frequency motion a regressive mode.

Since the β_{1c} and β_{1s} degrees of freedom represent tilt of the tip-path plane, their coupled motion is a wobble of the tip-path plane. The high-frequency flap mode is a wobble in the same direction as the rotor rotation, but at frequency $\text{Im}\, s_R + 1$, approximately 2/rev. The low-frequency mode is a wobble at the small frequency $|\text{Im}\, s_R - 1|$, in a direction depending on whether the rotating flap frequency is above or below 1/rev. For an articulated rotor the rotating frequency is below 1/rev, so the low-frequency mode is a wobble in the same direction as the rotor. For a hingeless rotor $\text{Im}\, s_R$ is likely above 1/rev, in which case the low-frequency motion is truly a regressive mode, the tip-path plane wobbling opposite the direction of rotor rotation.

19.2.2 Hover Transfer Functions

The equation of motion for the coning mode β_0 of a hovering rotor gives the following transfer function:

$$\beta_0 = \frac{\gamma M_\theta \theta_0 - \tfrac{1}{2}\gamma M_\lambda w_G + (-\widehat{S}_\beta s + \tfrac{1}{2}\gamma M_\lambda)\dot{z}_h}{\widehat{I}_\beta \left(s^2 - (\gamma M_{\dot\beta}/\widehat{I}_\beta)s + v_{\beta e}^2\right)} \tag{19.35}$$

using for the hover inflow perturbation $\lambda + \dot{z}_h - w_G = \tfrac{1}{2}(\dot{z}_h - w_G)$ (section 16.8.8). The response to collective pitch θ_0 and to vertical gusts is the same as the flap response of the blade in the rotating frame. The coning response to vertical shaft motion involves the rotor inertia as well as aerodynamic forces and introduces a zero at $s = \tfrac{1}{2}\gamma M_\lambda/\widehat{S}_\beta$ on the negative real axis. For low-frequency shaft motion, the aerodynamic forces dominate, and the response is like that to the vertical gusts. For high-frequency shaft motion the inertia dominates, so the coning response approaches $\beta_0/z_h = -\widehat{S}_\beta/\widehat{I}_\beta$.

The response of the tip-path-plane tilt to cyclic pitch and shaft motion is defined by the transfer function

$$\begin{bmatrix} \widehat{I}_\beta s^2 - \gamma M_{\dot\beta} s + \widehat{I}_\beta(v_\beta^2 - 1) + K_P \gamma M_\theta & 2\widehat{I}_\beta s - \gamma M_{\dot\beta} \\ -(2\widehat{I}_\beta s - \gamma M_{\dot\beta}) & \widehat{I}_\beta s^2 - \gamma M_{\dot\beta} s + \widehat{I}_\beta(v_\beta^2 - 1) + K_P \gamma M_\theta \end{bmatrix} \begin{pmatrix} \beta_{1c} \\ \beta_{1s} \end{pmatrix}$$

$$= \gamma M_\theta \begin{pmatrix} \theta_{1c} \\ \theta_{1s} \end{pmatrix} + \gamma M_\mu \begin{pmatrix} \dot{y}_h + v_G \\ -\dot{x}_h + u_G \end{pmatrix} + \begin{bmatrix} \widehat{I}_{\beta\alpha} s - \gamma M_{\dot\beta} & 2\widehat{I}_{\beta\alpha} \\ -2\widehat{I}_{\beta\alpha} & \widehat{I}_{\beta\alpha} s - \gamma M_{\dot\beta} \end{bmatrix} \begin{pmatrix} \dot\alpha_y \\ -\dot\alpha_x \end{pmatrix}$$

$$\tag{19.36}$$

Inverting the matrix gives for the response to cyclic

$$\begin{pmatrix} \beta_{1c} \\ \beta_{1s} \end{pmatrix} = \frac{1}{\Delta} \begin{bmatrix} \widehat{I}_\beta \left(s^2 - \dfrac{\gamma M_{\dot\beta}}{\widehat{I}_\beta} s + v_{\beta e}^2 - 1\right) & -(2\widehat{I}_\beta s - \gamma M_{\dot\beta}) \\ 2\widehat{I}_\beta s - \gamma M_{\dot\beta} & \widehat{I}_\beta \left(s^2 - \dfrac{\gamma M_{\dot\beta}}{\widehat{I}_\beta} s + v_{\beta e}^2 - 1\right) \end{bmatrix} \gamma M_\theta \begin{pmatrix} \theta_{1c} \\ \theta_{1s} \end{pmatrix}$$

$$\tag{19.37}$$

where

$$\Delta = \widehat{I}_\beta^2 \left(s^2 - \frac{\gamma M_{\dot\beta}}{\widehat{I}_\beta} s + v_{\beta e}^2 - 1 \right)^2 + \widehat{I}_\beta^2 \left(2s - \frac{\gamma M_{\dot\beta}}{\widehat{I}_\beta} \right)^2$$

is the characteristic equation. Now we introduce the parameter

$$N_\star = \frac{\widehat{I}_\beta(v_{\beta e}^2 - 1)}{-\gamma M_{\dot\beta}} = \frac{\widehat{I}_\beta(v_\beta^2 - 1)}{-\gamma M_{\dot\beta}} + K_P \frac{M_\theta}{M_{\dot\beta}} \cong \frac{v_\beta^2 - 1}{\gamma/8} + K_P \tag{19.38}$$

which defines the phase shift of the flap response due to the structural stiffening of the blade ($v_\beta > 1$) and the pitch-flap coupling; see sections 6.14, 6.15, and 6.18. Then the transfer function can be written:

$$\begin{pmatrix} \beta_{1c} \\ \beta_{1s} \end{pmatrix} = \frac{1}{\Delta} \begin{bmatrix} \frac{2\widehat{I}_\beta}{-\gamma M_{\dot\beta}} s + 1 & \frac{\widehat{I}_\beta}{-\gamma M_{\dot\beta}} s^2 + s + N_\star \\ -\left(\frac{\widehat{I}_\beta}{-\gamma M_{\dot\beta}} s^2 + s + N_\star\right) & \frac{2\widehat{I}_\beta}{-\gamma M_{\dot\beta}} s + 1 \end{bmatrix} \frac{M_\theta}{-M_{\dot\beta}} \begin{pmatrix} -\theta_{1s} \\ \theta_{1c} \end{pmatrix} \tag{19.39}$$

with

$$\Delta = \left(\frac{2\widehat{I}_\beta}{-\gamma M_{\dot\beta}} s + 1 \right)^2 + \left(\frac{\widehat{I}_\beta}{-\gamma M_{\dot\beta}} s^2 + s + N_\star \right)^2 \tag{19.40}$$

The static response ($s = 0$) is

$$\begin{pmatrix} \beta_{1c} \\ \beta_{1s} \end{pmatrix} = \frac{1}{1 + N_\star^2} \begin{bmatrix} 1 & N_\star \\ -N_\star & 1 \end{bmatrix} \frac{M_\theta}{-M_{\dot\beta}} \begin{pmatrix} -\theta_{1s} \\ \theta_{1c} \end{pmatrix} \tag{19.41}$$

For an articulated rotor ($v_\beta = 1$, $K_P = 0$, and $-M_\theta/M_{\dot\beta} = 1$) this reduces to

$$\begin{pmatrix} \beta_{1c} \\ \beta_{1s} \end{pmatrix} = \begin{pmatrix} -\theta_{1s} \\ \theta_{1c} \end{pmatrix} \tag{19.42}$$

which states that the tip-path plane remains exactly parallel to the control plane. In general, $-M_\theta/M_{\dot\beta}$ is the static gain of the flap response to cyclic, and $N_\star$ defines the phase shift between the tip-path-plane and control-plane tilt (see also sections 6.14 and 6.18).

After substituting for the coefficients, the transfer functions of the direct and cross response of the tip-path-plane tilt to cyclic pitch are

$$\frac{\beta_{1c}}{-\theta_{1s}} = \frac{\beta_{1s}}{\theta_{1c}} = \frac{\frac{16}{\gamma} s + 1}{\left(\frac{16}{\gamma} s + 1\right)^2 + \left(\frac{8}{\gamma} s^2 + s + N_\star\right)^2} \tag{19.43}$$

$$\frac{\beta_{1c}}{\theta_{1c}} = \frac{\beta_{1s}}{\theta_{1s}} = \frac{\frac{8}{\gamma} s^2 + s + N_\star}{\left(\frac{16}{\gamma} s + 1\right)^2 + \left(\frac{8}{\gamma} s^2 + s + N_\star\right)^2} \tag{19.44}$$

The poles (roots of the denominator polynomial) are the eigenvalues of the β_{1c} and β_{1s} motion in the non-rotating frame: $s_{NR} = s_R \pm i$. The direct transfer function has a single zero at

$$s = \frac{\gamma M_{\dot\beta}}{2\widehat{I}_\beta} \cong -\frac{\gamma}{16} \tag{19.45}$$

This zero is on the negative real axis, with the same real part as the poles. The cross transfer function has two zeros, which are the solution of the quadratic

$$\frac{\widehat{I}_\beta}{-\gamma M_{\dot\beta}}s^2 + s + N_\star = 0 \tag{19.46}$$

namely

$$s = \frac{\gamma M_{\dot\beta}}{2\widehat{I}_\beta} \pm \sqrt{1 - v_{\beta e}^2 + \left(\frac{\gamma M_{\dot\beta}}{2\widehat{I}_\beta}\right)^2} = \operatorname{Re} s_R \pm \sqrt{1 - (\operatorname{Im} s_R)^2} \tag{19.47}$$

If the rotating flap root frequency is below 1/rev, there are two real zeros, at equal distances on either side of the real part of the poles. If $\operatorname{Im} s_R > 1$, there are two complex conjugate zeros, with the same real part as the poles. For an articulated rotor ($v_\beta = 1$, $K_P = 0$) the two real zeros are $s = 0$ and $s = \gamma M_{\dot\beta}/\widehat{I}_\beta$, which are at the origin and twice the real part of the poles. The zero at the origin is responsible for the static response of β_{1c}/θ_{1c} and of β_{1s}/θ_{1s} being zero in this case. With negative pitch-flap coupling such that $v_{\beta e} < 1$, the zeros are shifted farther away from the real part of the poles, the zero at the origin moving into the right half-plane. For $v_{\beta e} > 1$, the two zeros move instead toward $\operatorname{Re} s_R$. At

$$v_{\beta e}^2 = 1 + \left(\frac{\gamma M_{\dot\beta}}{2\widehat{I}_\beta}\right)^2 \tag{19.48}$$

where the rotating frequency $\operatorname{Im} s_R = 1/\text{rev}$, the two zeros coincide at $s = \operatorname{Re} s_R$. Moreover, the two poles of the low-frequency flap mode are also at $s = \operatorname{Re} s_R$ on the real axis for this case. For still larger $v_{\beta e}$, such that $\operatorname{Im} s_R > 1$ (as is likely with a hingeless rotor), there are two complex conjugate zeros. The zeros have the same real part as the poles and a larger frequency than the low-frequency flap mode poles.

Substituting $s = i\omega$ gives the frequency response. Figure 19.6 presents the direct ($-\beta_{1c}/\theta_{1s} = \beta_{1s}/\theta_{1c}$) and cross ($\beta_{1c}/\theta_{1c} = \beta_{1s}/\theta_{1s}$) response of the tip-path plane to cyclic control inputs for typical articulated and hingeless rotors. For the hingeless rotor, the swashplate phase $\tan^{-1} N_\star$ has been included. The frequency response shows the resonance with the high-frequency flap mode around 2/rev. The low-frequency mode has a very large damping ratio, so is more evident in the phase than in the magnitude of the response.

The response of the tip-path-plane tilt to the in-plane shaft motion and gusts is similar to the response to cyclic pitch, but with a static gain of $-M_\mu/M_{\dot\beta} \cong 8(2C_T/\sigma a + \lambda_{\text{HP}}/4)$ instead of $-M_\theta/M_{\dot\beta} \cong 1$. The response to the shaft pitch and roll rate is more complicated, involving inertial and Coriolis forces as well as the aerodynamic forces. After substituting for the coefficients, the transfer functions of the direct and cross response of the tip-path-plane tilt to shaft pitch and roll rate are

$$\frac{\beta_{1c}}{\dot\alpha_y} = \frac{\beta_{1s}}{-\dot\alpha_x} = \frac{\frac{16}{\gamma}\left(\frac{16}{\gamma}s+1\right) + \left(\frac{8}{\gamma}s+1\right)\left(\frac{8}{\gamma}s^2+s+N_\star\right)}{\left(\frac{16}{\gamma}s+1\right)^2 + \left(\frac{8}{\gamma}s^2+s+N_\star\right)^2} \tag{19.49}$$

$$\frac{\beta_{1c}}{-\dot\alpha_x} = \frac{\beta_{1s}}{-\dot\alpha_y} = \frac{-\frac{8}{\gamma}s + \frac{16}{\gamma}N_\star - 1}{\left(\frac{16}{\gamma}s+1\right)^2 + \left(\frac{8}{\gamma}s^2+s+N_\star\right)^2} \tag{19.50}$$

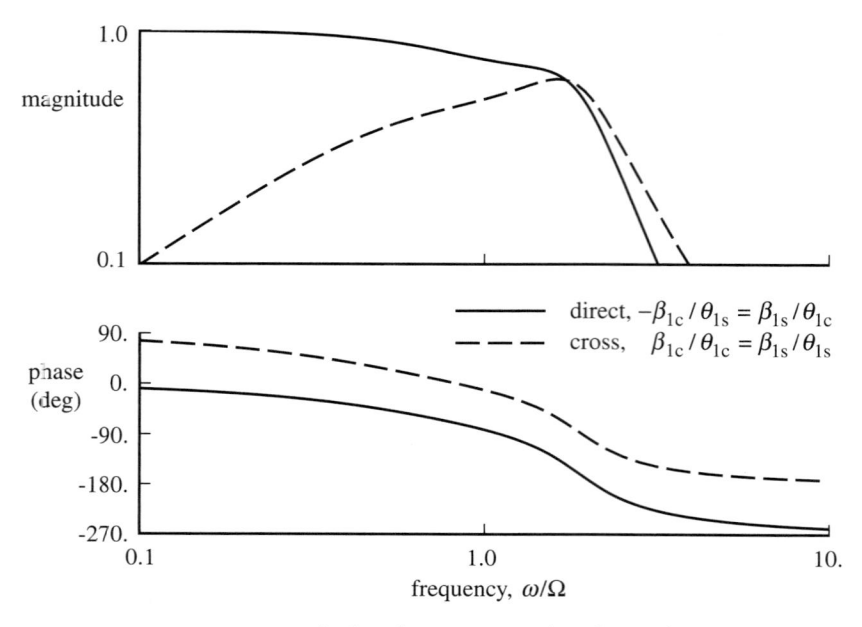

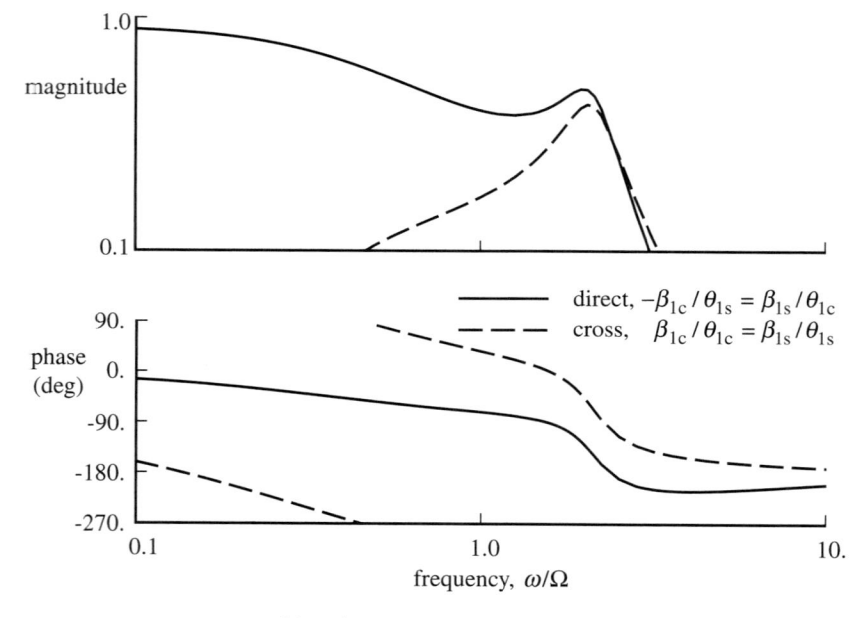

Figure 19.6. Frequency response of the tip-path-plane tilt to cyclic pitch in hover.

Figure 19.7 presents the direct and cross response to shaft motion for typical articulated and hingeless rotors. The cross response has a zero at $s = 2N_\star - \frac{\gamma}{8}$.

The reactionless modes β_{nc} and β_{ns} respond only to the pitch control inputs θ_{nc} and θ_{ns}, which are not present for conventional control systems. In hover the reactionless rotor modes are therefore not coupled with the non-rotating system,

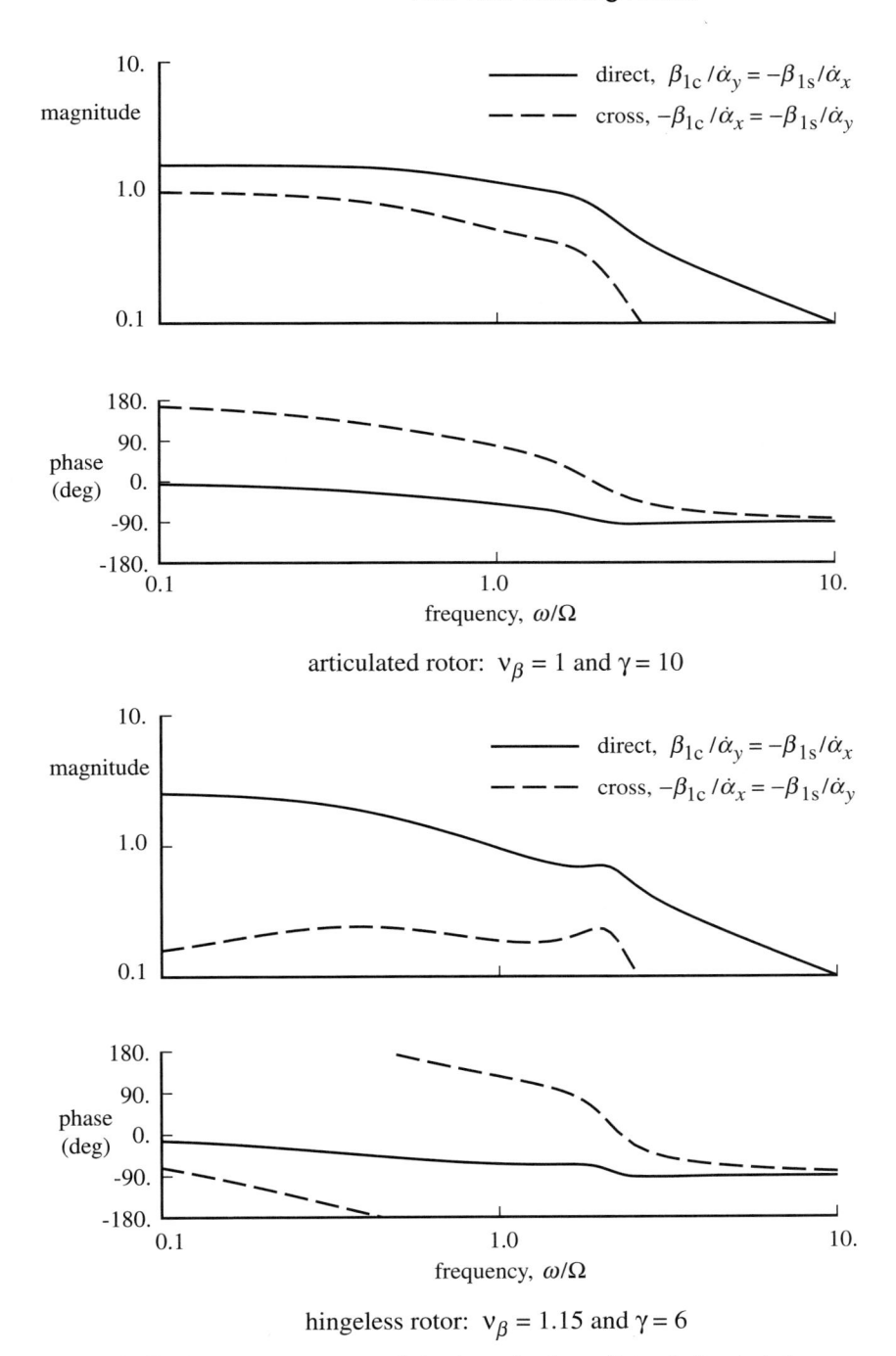

Figure 19.7. Frequency response of the tip-path-plane tilt to shaft rate in hover.

(upper plots) articulated rotor: $\nu_\beta = 1$ and $\gamma = 10$

(lower plots) hingeless rotor: $\nu_\beta = 1.15$ and $\gamma = 6$

Legend: direct, $\beta_{1c}/\dot\alpha_y = -\beta_{1s}/\dot\alpha_x$; cross, $-\beta_{1c}/\dot\alpha_x = -\beta_{1s}/\dot\alpha_y$

either through shaft motion, gusts, or control inputs. These degrees of freedom (β_{nc}, β_{ns}, and $\beta_{N/2}$) then represent purely internal rotor motion. In forward flight all the rotor degrees of freedom are coupled, but the coning and tip-path-plane tilt response still dominate the flap dynamics.

19.3 Low-Frequency Response

The transient flap motion is a damped oscillation that decays proportional to $e^{\text{Re}\, s_R \psi}$, where

$$\text{Re}\, s_R = \frac{\gamma M_{\dot{\beta}}}{2\widehat{I}_\beta} \cong -\frac{\gamma}{16} \qquad (19.51)$$

(dimensionless, based on the rotor rotational speed Ω). The time constant of the response is

$$\tau = \frac{1}{-\text{Re}\, s_R} = \frac{2\widehat{I}_\beta}{-\gamma M_{\dot{\beta}}} \cong \frac{16}{\gamma} \qquad (19.52)$$

radians. The time to half amplitude, $\tau_{1/2} = 0.693\tau$, is then typically 90° of the rotor azimuth. Because of the high flap damping, the flapping transients die out in less than one revolution of the rotor. In dimensional terms, the time to half amplitude is of the order of 0.05 sec. Hence the rotor flap motion responds on a much shorter time scale than the inputs from the pilot, from shaft motion due to the helicopter rigid-body degrees of freedom, or from aerodynamic gusts. For problems such as helicopter flight dynamics, it is sufficient therefore to consider only the low-frequency or even the steady-state response of the rotor, neglecting the transient flapping dynamics.

Consider the response of the tip-path-plane tilt for a hovering rotor with three or more blades. To lowest order in the Laplace variable s, equation 19.26 for the response of β_{1c} and β_{1s} is

$$\begin{bmatrix} \widehat{I}_\beta(\nu_\beta^2 - 1) + K_P\gamma M_\theta & -\gamma M_{\dot{\beta}} \\ \gamma M_{\dot{\beta}} & \widehat{I}_\beta(\nu_\beta^2 - 1) + K_P\gamma M_\theta \end{bmatrix} \begin{pmatrix} \beta_{1c} \\ \beta_{1s} \end{pmatrix}$$

$$= \gamma M_\theta \begin{pmatrix} \theta_{1c} \\ \theta_{1s} \end{pmatrix} + \gamma M_\mu \begin{pmatrix} \dot{y}_h + v_G \\ -\dot{x}_h + u_G \end{pmatrix} + \begin{bmatrix} -\gamma M_{\dot{\beta}} & 2\widehat{I}_{\beta\alpha} \\ -2\widehat{I}_{\beta\alpha} & -\gamma M_{\dot{\beta}} \end{bmatrix} \begin{pmatrix} \dot{\alpha}_y \\ -\dot{\alpha}_x \end{pmatrix} \qquad (19.53)$$

Assuming uniform trim induced velocity, $\eta_\beta = r$, and neglecting tip losses, the coefficients are

$$M_\theta = -M_{\dot{\beta}} = \frac{1}{8} \qquad (19.54)$$

$$M_\mu = \frac{2C_T}{\sigma a} + \frac{\lambda_{\text{HP}}}{4} \qquad (19.55)$$

$$\widehat{I}_\beta = \widehat{I}_{\beta\alpha} = 1$$

(section 16.8.8). In terms of $N_\star$ (equation 19.38), the low-frequency tip-path-plane tilt response is

$$\begin{pmatrix} \beta_{1c} \\ \beta_{1s} \end{pmatrix} = \frac{1}{1 + N_\star^2} \begin{bmatrix} 1 & N_\star \\ -N_\star & 1 \end{bmatrix} \left\{ \frac{M_\theta}{-M_{\dot{\beta}}} \begin{pmatrix} -\theta_{1s} \\ \theta_{1c} \end{pmatrix} + \frac{M_\mu}{-M_{\dot{\beta}}} \begin{pmatrix} \dot{x}_h - u_G \\ \dot{y}_h + v_G \end{pmatrix} \right.$$

$$\left. + \frac{2\widehat{I}_{\beta\alpha}}{-\gamma M_{\dot{\beta}}} \begin{pmatrix} \dot{\alpha}_y \\ -\dot{\alpha}_x \end{pmatrix} + \begin{pmatrix} \dot{\alpha}_x \\ \dot{\alpha}_y \end{pmatrix} \right\} \qquad (19.56)$$

19.3 Low-Frequency Response

The parameter $N_\star$ determines the lateral-longitudinal coupling of the rotor flap response. In this result most of the Lock number factors have cancelled, indicating that the flap response is primarily a balance of aerodynamic forces. The exception is the third term, which is a balance of Coriolis inertial forces due to the shaft angular velocity and the aerodynamic forces due to the flap motion. Blade pitch control produces an aerodynamic flap moment through M_θ. A 1/rev pitch input from longitudinal cyclic θ_{1s} produces a lateral aerodynamic moment on the disk. The rotor responds with a 90° lag (less if $N_\star > 0$); hence with longitudinal tilt of the tip-path plane. The flapping velocity in the rotating frame due to longitudinal tip-path-plane tilt β_{1c} produces a lateral aerodynamic moment on the disk through $M_{\dot\beta}$, which opposes the moment due to the cyclic pitch input. The tip-path plane tilts until equilibrium of moments is achieved, which gives the steady-state response. The effectiveness of cyclic pitch in producing flapping is thus governed by $-M_\theta/M_{\dot\beta}$; assuming small angle and $\eta_\beta = r$, this quantity has the value 1, implying that the tip-path plane remains parallel to the control plane. Longitudinal shaft velocity $\dot x_h$ and gust velocity u_G produce a lateral moment on the rotor disk through M_μ. This moment is due to the lateral asymmetry in the air velocity seen by the blades, which is similar to the effect of advance ratio in forward flight. Thus the steady-state response to longitudinal velocity is also longitudinal tip-path-plane tilt, but with effectiveness

$$\frac{M_\mu}{-M_{\dot\beta}} \cong 8\left(\frac{2C_T}{\sigma a} + \frac{\lambda_{\text{HP}}}{4}\right) \tag{19.57}$$

which typically has a value 0.35 in hover. A shaft roll rate $\dot\alpha_x$ also produces a lateral aerodynamic moment on the disk, through $M_{\dot\beta}$. The rotor responds with longitudinal tip-path-plane tilt until the flapping velocity in the rotating frame just cancels the velocity due to shaft roll. In this case the effectiveness is given by $M_{\dot\beta}/M_{\dot\beta} = 1$, again assuming $\eta_\beta = r$, so that the mode shapes of flapping and shaft tilt are identical. Similarly, lateral cyclic pitch θ_{1c}, lateral shaft velocity $\dot y_h$ and gust velocity v_G, and shaft pitch rate $\dot\alpha_y$ produce longitudinal aerodynamic moments on the rotor disk, and hence lateral tip-path-plane tilt.

Angular velocity of the shaft also produces tip-path-plane tilt proportional to

$$\frac{2\widehat{I}_{\beta\alpha}}{-\gamma M_{\dot\beta}} \cong \frac{16}{\gamma} \tag{19.58}$$

This is the lag of the tip-path plane required to precess the rotor to follow the shaft motion. For the rotor to follow the shaft with a pitch rate $\dot\alpha_y$, there must be an angular roll acceleration $2\Omega\widehat{I}_{\beta\alpha}\dot\alpha_y$ on the disk due to the Coriolis forces on the rotating blades. There must then be a lateral moment on the disk to provide this acceleration to precess the rotor. This moment is supplied by the aerodynamic forces on the blade. The rotor tip-path plane tilts back, lagging the shaft tilt, until the lateral moment due to the flap velocity in the rotating frame ($\gamma M_{\dot\beta}\beta_{1c}$) is sufficient to provide the required moment. Similarly, the rotor disk follows shaft roll angular velocity $\dot\alpha_x$, but with a lateral tip-path-plane tilt lag to provide the aerodynamic moment to precess the disk.

The parameter $N_\star$ is a measure of the structural and aerodynamic springs on the flap motion. When $N_\star = 0$, the rotor disk responds to applied moments with exactly a 90° phase shift (which is 90° in azimuth because the flap natural frequency is at 1/rev in this case). When $N_\star > 0$ because $\nu_\beta > 1$ or because of pitch-flap coupling, the flap response is quickened, and hence the lag in the response is less than 90°. Consider the tip-path-plane tilt due to cyclic pitch alone,

$$\begin{pmatrix} \beta_{1c} \\ \beta_{1s} \end{pmatrix} = \frac{1}{1+N_\star^2} \begin{bmatrix} 1 & N_\star \\ -N_\star & 1 \end{bmatrix} \frac{M_\theta}{-M_{\dot\beta}} \begin{pmatrix} -\theta_{1s} \\ \theta_{1c} \end{pmatrix} \qquad (19.59)$$

The off-diagonal terms represent the lateral-longitudinal coupling, which is zero only if $N_\star = 0$. The magnitude of this response is

$$\frac{|\beta|}{|\theta|} = \frac{M_\theta}{-M_{\dot\beta}} \frac{1}{\sqrt{1+N_\star^2}} \qquad (19.60)$$

and the azimuthal phase shift is $\Delta\psi = \tan^{-1} N_\star$. When $N_\star > 0$, the magnitude of the flap response is reduced slightly, and the phase lag has been reduced from 90° to $90° - \Delta\psi$. When $N_\star < 0$ (because of negative pitch-flap coupling), the magnitude of the flap response is again reduced, but the phase lag is increased. This behavior can also be viewed as due to removing the natural frequency of the flap motion from resonance with the 1/rev exciting forces, as discussed in section 6.14.

There is a similar phase shift and magnitude reduction in the flap response to shaft motion and gusts. When $N_\star = 0$, the response to lateral or longitudinal shaft motion is purely lateral or purely longitudinal tip-path-plane tilt, respectively. The phase shift when $N_\star \neq 0$ couples the lateral and longitudinal motions of the helicopter. The coupling of the response to cyclic control can be cancelled by a corresponding phase shift in the swashplate rigging, but the lateral-longitudinal coupling of the rotor response to shaft motion remains and can be troublesome if large. As an example, consider an articulated main rotor with offset hinges and no pitch-flap coupling. Using $\nu_\beta = 1.03$, $\gamma = 10$, and $K_P = 0$ gives $N_\star = 0.05$ and thus a negligible change in response magnitude and phase shift of only $\Delta\psi = \tan^{-1} N_\star = 3°$. For a hingeless rotor with $\nu_\beta = 1.15$, $\gamma = 6$, and $K_P = 0$ we obtain $N_\star = 0.43$. Then the response magnitude is reduced by 8%, and the phase shift is an appreciable $\Delta\psi = \tan^{-1} N_\star = 23°$. For a teetering tail rotor with $\nu_\beta = 1$ and $K_P = 1$, the phase shift of $\Delta\psi = 45°$ due to $N_\star = 1$ is large but not important, and the pitch-flap coupling reduces the flap response magnitude by 29%.

The rotor dynamics in forward flight are described by periodic-coefficient differential equations, but the constant coefficient approximation in the non-rotating frame does provide a good representation of the flap dynamics as long as the advance ratio is not too high. The constant coefficient approximation is particularly good for the low-frequency modes of the rotor. Consider a rotor with three or more blades in forward flight, with the flap motion described by the coning and tip-path-plane tilt modes. The inertial terms in the equations of motion are the same as in hover, and equations 16.384 and 16.470 give the constant coefficient approximation for the aerodynamic forces. Body axes are used for the aerodynamics here, since this result is intended for the helicopter flight dynamics. When only the lowest order terms in the Laplace variable s are retained, the low-frequency response of the rotor flap motion in forward flight is

$$\begin{bmatrix} \hat{I}_\beta v_\beta^2 + K_P \gamma M_\theta^0 & -\tfrac{1}{2}\gamma\left(M_\beta^{1c} - M_{\dot\beta}^{1s}\right) & K_P \gamma \tfrac{1}{2} M_\theta^{1s} \\ -K_P \gamma M_\theta^{1s} & -\gamma\left(M_{\dot\beta}^0 - \tfrac{1}{2}\left(M_\beta^{2c}+M_\beta^{2s}\right)\right) & -\hat{I}_\beta(v_\beta^2-1) - K_P\gamma\left(M_\theta^0 - \tfrac{1}{2} M_\theta^{2c}\right) \\ -\gamma M_\beta^{1c} & \hat{I}_\beta(v_\beta^2-1) + K_P\gamma\left(M_\theta^0 + \tfrac{1}{2} M_\theta^{2c}\right) & -\gamma\left(M_{\dot\beta}^0 + \tfrac{1}{2}\left(M_\beta^{2c}+M_\beta^{2s}\right)\right) \end{bmatrix} \begin{pmatrix} \beta_0 \\ \beta_{1c} \\ \beta_{1s} \end{pmatrix}$$

$$= \begin{bmatrix} \gamma M_\theta^0 & -\gamma \tfrac{1}{2} M_\theta^{1s} & 0 \\ -\gamma M_\theta^{1s} & \gamma\left(M_\theta^0 - \tfrac{1}{2} M_\theta^{2c}\right) & 0 \\ 0 & 0 & \gamma\left(M_\theta^0 + \tfrac{1}{2} M_\theta^{2c}\right) \end{bmatrix} \begin{pmatrix} \theta_0 \\ -\theta_{1s} \\ \theta_{1c} \end{pmatrix}$$

$$+ \begin{bmatrix} \gamma M_\lambda^0 & -\gamma \tfrac{1}{2} M_\mu^{1s} & \gamma \tfrac{1}{2} M_\mu^{1c} \\ -\gamma M_\lambda^{1s} & \gamma\left(M_\mu^0 - \tfrac{1}{2} M_\mu^{2c}\right) & -\gamma \tfrac{1}{2} M_\mu^{2s} \\ 0 & -\gamma \tfrac{1}{2} M_\mu^{2s} & \gamma\left(M_\mu^0 + \tfrac{1}{2} M_\mu^{2c}\right) \end{bmatrix} \begin{pmatrix} \dot z_h - w_G \\ \dot x_h - u_G \\ \dot y_h + v_G \end{pmatrix}$$

$$+ \begin{bmatrix} -\gamma M_\zeta^0 & 0 & -\gamma \tfrac{1}{2} M_\beta^{1s} \\ \gamma M_\zeta^{1s} & 2\hat{I}_{\beta\alpha} & \gamma\left(M_\beta^0 - \tfrac{1}{2} M_\beta^{2c}\right) \\ -\gamma M_\zeta^{1c} & -\gamma\left(M_\beta^0 + \tfrac{1}{2} M_\beta^{2c}\right) & 2\hat{I}_{\beta\alpha} \end{bmatrix} \begin{pmatrix} \dot\alpha_z \\ \dot\alpha_y \\ -\dot\alpha_x \end{pmatrix} \quad (19.61)$$

Forward speed of the helicopter has the effect of coupling the vertical and the lateral-longitudinal motions by aerodynamic forces of order μ, resulting in more complex behavior than for hover. Moreover, the task of calculating the dynamic behavior is more difficult because of the higher order of the system that must be considered and the additional aerodynamic coefficients that must be obtained. Of particular significance is the lateral aerodynamic moment due to vertical velocity of the helicopter:

$$-\gamma M_\lambda^{1s}(\dot z_h - w_G) = \gamma \frac{\mu}{4}(\dot z_h - w_G) \quad (19.62)$$

from $M_\lambda = -(\tfrac{1}{6} + \tfrac{\mu}{4}\sin\psi)$. This lateral moment on the disk primarily produces a longitudinal tip-path-plane tilt of

$$\Delta\beta_{1c} = \frac{M_\lambda^{1s}}{M_{\dot\beta}}(\dot z_h - w_G) = 2\mu(\dot z_h - w_G) \quad (19.63)$$

The downward velocity through the disk of $(\dot z_h - w_G)$ decreases the angle-of-attack of the blades. In forward flight the resulting lift decrease is largest on the advancing side and smallest on the retreating side, so there is a lateral moment of the rotor disk, in response to which the rotor flaps forward.

19.4 Hub Reactions

Next let us examine the forces and moments acting on the rotor hub, including the effects of the flap response. For use in the flight dynamics analysis (Chapter 21), the low-frequency response is of particular interest. Consider first the hovering rotor, for which the analysis is simplified not only by the constant coefficients but also by a complete decoupling of the vertical and longitudinal-lateral dynamics because of the axisymmetry.

From equations 16.255 and 16.463, the perturbation thrust due to the rotor is

$$\gamma \frac{C_T}{\sigma a} = \gamma T_\theta \theta_0 + \gamma T_\lambda (\lambda + \dot{z}_h - w_G) - \widehat{M}_b \ddot{z}_h - \widehat{S}_\beta \ddot{\beta}_0 + \gamma T_{\dot{\beta}} \dot{\beta}_0 - K_P \gamma T_\theta \beta_0 \quad (19.64)$$

and the coning equation is

$$\widehat{I}_\beta \ddot{\beta}_0 - \gamma M_{\dot{\beta}} \dot{\beta}_0 + \left(\widehat{I}_\beta v_\beta^2 + K_P \gamma M_\theta \right) \beta_0 = \gamma M_\theta \theta_0 + \gamma M_\lambda (\lambda + \dot{z}_h - w_G) - \widehat{S}_\beta \ddot{z}_h \quad (19.65)$$

Assuming $\eta_\beta = r$ and neglecting tip losses, the hover aerodynamic coefficients are $M_\theta = -M_{\dot{\beta}} = \frac{1}{8}, -M_\lambda = T_\theta = -T_{\dot{\beta}} = \frac{1}{6}$, and $-T_\lambda = \frac{1}{4}$. The effect of the axial velocity on the inflow gives again $\lambda + \dot{z}_h - w_G = \frac{1}{2}(\dot{z}_h - w_G)$. To the lowest order in s, the low-frequency thrust and coning equations are

$$\frac{C_T}{\sigma a} = T_\theta \theta_0 + \frac{1}{2} T_\lambda (\dot{z}_h - w_G) - K_P T_\theta \beta_0 \quad (19.66)$$

$$\beta_0 = \frac{1}{\widehat{I}_\beta v_{\beta e}^2} \left[\gamma M_\theta \theta_0 + \frac{1}{2} \gamma M_\lambda (\dot{z}_h - w_G) \right] \quad (19.67)$$

It is assumed that the rotor mass has been included in the helicopter gross weight, so the $\widehat{M}_b \ddot{z}_h$ term in C_T can be dropped. The coning motion influences the rotor low-frequency thrust only through the change in collective with pitch-cone coupling. Eliminating β_0 gives

$$\frac{C_T}{\sigma a} = \left(T_\theta - K_P T_\theta \frac{\gamma M_\theta}{\widehat{I}_\beta v_{\beta e}^2} \right) \theta_0 + \frac{1}{2} \left(T_\lambda - K_P T_\theta \frac{\gamma M_\lambda}{\widehat{I}_\beta v_{\beta e}^2} \right) (\dot{z}_h - w_G)$$

$$\cong \left[T_\theta \theta_0 + \frac{1}{2} T_\lambda (\dot{z}_h - w_G) \right] \frac{v_\beta^2}{v_{\beta e}^2} \quad (19.68)$$

where

$$\frac{v_\beta^2}{v_{\beta e}^2} = \frac{v_\beta^2}{v_\beta^2 + K_P \gamma M_\theta / \widehat{I}_\beta} \quad (19.69)$$

The thrust is given principally by the direct response to collective and vertical velocity perturbations. Pitch-flap coupling reduces the thrust response by the factor $v_\beta^2 / v_{\beta e}^2$ due to the collective pitch reduction with coning. The inflow also depends on the aerodynamic thrust. From section 11.3, the dynamic inflow model for the uniform induced velocity perturbation in hover is

$$\tau \dot{\lambda}_0 + \lambda_0 = \frac{1}{4\lambda_i} \delta C_T - \frac{1}{2} \delta \mu_z = \frac{1}{4\lambda_i} (C_T)_{\text{aero}} - \frac{1}{2} (\dot{z}_h - w_G) \quad (19.70)$$

Without pitch-cone coupling ($K_P = 0$), the low-frequency thrust response is then

$$\frac{C_T}{\sigma a} = T_\theta \theta_0 + T_\lambda (\dot{z}_h - w_G) + T_\lambda \left(\frac{1}{4\lambda_i} C_T - \frac{1}{2} (\dot{z}_h - w_G) \right)$$

$$= C' T_\theta \theta_0 + \frac{1}{2} C' T_\lambda (\dot{z}_h - w_G) \quad (19.71)$$

with the hover lift deficiency function

$$C' = \frac{1}{1 - \sigma a T_\lambda / 4\lambda_i} = \frac{1}{1 + \sigma a / 16 \lambda_i} \quad (19.72)$$

accounting for the effects of the rotor wake in the unsteady aerodynamics.

The in-plane forces on the hub, C_H and C_Y, and the hub moments C_{Mx} and C_{My} are closely related to the rotor tip-path-plane tilt. From equations 16.258, 16.259, 16.434, and 16.465, the in-plane hub forces of the hovering rotor are

$$\begin{pmatrix} \dfrac{2C_H}{\sigma a} \\ \dfrac{2C_Y}{\sigma a} \end{pmatrix} = -\frac{2\widehat{M}_b}{\gamma}\begin{pmatrix} \ddot{x}_h \\ \ddot{y}_h \end{pmatrix} - \frac{C_T}{\sigma a}\begin{pmatrix} \beta_{1c} \\ \beta_{1s} \end{pmatrix} + \begin{bmatrix} R_{\dot{\beta}} & H_{\dot{\beta}} \\ -H_{\dot{\beta}} & R_{\dot{\beta}} \end{bmatrix}\begin{pmatrix} \dot{\beta}_{1c} + \beta_{1s} - \dot{\alpha}_y \\ \dot{\beta}_{1s} - \beta_{1c} + \dot{\alpha}_x \end{pmatrix}$$

$$+ \begin{bmatrix} R_{\theta} & H_{\theta} \\ -H_{\theta} & R_{\theta} \end{bmatrix}\begin{pmatrix} \theta_{1c} - K_P\beta_{1c} \\ \theta_{1s} - K_P\beta_{1s} \end{pmatrix}$$

$$+ \begin{bmatrix} -(H_{\mu} + R_{\mu}) & R_r \\ -R_r & -(H_{\mu} + R_{\mu}) \end{bmatrix}\begin{pmatrix} \dot{x}_h - u_G \\ \dot{y}_h + v_G \end{pmatrix} \quad (19.73)$$

Assuming that the rotor mass $\widehat{M}_b$ is included in the helicopter gross weight, and writing $H_{\dot{\beta}} = \widehat{H}_{\dot{\beta}} + C_T/\sigma a$, the low-frequency response of the hub forces is

$$\begin{pmatrix} \dfrac{2C_H}{\sigma a} \\ \dfrac{2C_Y}{\sigma a} \end{pmatrix} = -\frac{2C_T}{\sigma a}\begin{pmatrix} \beta_{1c} \\ \beta_{1s} \end{pmatrix} + \begin{bmatrix} -\widehat{H}_{\dot{\beta}} - K_P R_{\theta} & R_{\dot{\beta}} - K_P H_{\theta} \\ -R_{\dot{\beta}} + K_P H_{\theta} & -\widehat{H}_{\dot{\beta}} - K_P R_{\theta} \end{bmatrix}\begin{pmatrix} \beta_{1c} \\ \beta_{1s} \end{pmatrix}$$

$$+ \begin{bmatrix} -H_{\theta} & R_{\theta} \\ -R_{\theta} & -H_{\theta} \end{bmatrix}\begin{pmatrix} -\theta_{1s} \\ \theta_{1c} \end{pmatrix} + \begin{bmatrix} -R_{\dot{\beta}} & -H_{\dot{\beta}} \\ H_{\dot{\beta}} & -R_{\dot{\beta}} \end{bmatrix}\begin{pmatrix} \dot{\alpha}_y \\ -\dot{\alpha}_x \end{pmatrix}$$

$$+ \begin{bmatrix} -(H_{\mu} + R_{\mu}) & R_r \\ -R_r & -(H_{\mu} + R_{\mu}) \end{bmatrix}\begin{pmatrix} \dot{x}_h - u_G \\ \dot{y}_h + v_G \end{pmatrix} \quad (19.74)$$

The first term is the in-plane force due to tilt of the thrust vector with the tip-path plane. The rotor hub moments can be obtained directly from the tip-path-plane tilt:

$$\begin{pmatrix} -\dfrac{2C_{My}}{\sigma a} \\ \dfrac{2C_{Mx}}{\sigma a} \end{pmatrix} = \frac{\widehat{I}_{\beta}(v_{\beta}^2 - 1)}{\gamma}\begin{pmatrix} \beta_{1c} \\ \beta_{1s} \end{pmatrix} \quad (19.75)$$

The total moment about the helicopter center-of-gravity a distance h below the rotor hub is then

$$\begin{pmatrix} -\dfrac{2C_{My}}{\sigma a} \\ \dfrac{2C_{Mx}}{\sigma a} \end{pmatrix}_{CG} = \left[\frac{\widehat{I}_{\beta}(v_{\beta}^2 - 1)}{\gamma} + h\frac{2C_T}{\sigma a}\right]\begin{pmatrix} \beta_{1c} \\ \beta_{1s} \end{pmatrix} \quad (19.76)$$

(equation 16.437), plus some in-plane forces due to tilt of the thrust vector relative to the tip-path plane. For an articulated rotor without hinge offset there are no hub moments, so all moments about the center-of-gravity come from the thrust vector tilt. With no hub moment capability, a helicopter must avoid flight at low load levels, where the control and damping from the rotor are lost because they are proportional to the rotor thrust. The moment-producing capability of an articulated rotor can be roughly doubled by using flap hinge offset, and the hub moment term does not depend on the thrust magnitude. With a hingeless rotor the hub moment term dominates, being typically three or four times the thrust tilt term. Thus a hingeless rotor has much greater control power and damping than the articulated rotor, but also higher gust response. See also the discussion in sections 6.14 and 6.15.

The rotor forces and moments acting on the helicopter are thus basically proportional to the tip-path-plane tilt. Longitudinal flapping produces a drag force C_H and pitch moment C_{My} on the helicopter; lateral flapping produces a side force C_Y and roll moment C_{Mx}. Cyclic pitch tilts the tip-path plane and consequently produces pitch and roll moments about the helicopter center-of-gravity. Thus the pilot can use cyclic pitch inputs to control the helicopter. Hub in-plane velocity ($\dot{x}_h$ or $\dot{y}_h$) produces flapping and thus an in-plane component of the thrust vector that opposes the motion. Hence there is a damping force on the helicopter speed perturbations. The corresponding moments due to $\dot{x}_h$ and $\dot{y}_h$ couple the linear and angular motion of the helicopter and are a major factor in the flight dynamics. Longitudinal and lateral aerodynamic gusts produce forces and moments on the helicopter by the same means. The rotor responds to shaft angular velocity with a tip-path-plane lag in order to precess the rotor. The result of this flapping response is a moment opposing the shaft motion; hence the rotor provides angular damping of the helicopter pitch and roll motion. Substituting for the low-frequency flap response, the hub forces relative to the tip-path plane are

$$\begin{pmatrix} \dfrac{2C_H}{\sigma a} \\ \dfrac{2C_Y}{\sigma a} \end{pmatrix}_{\text{TPP}} = \begin{bmatrix} -\widehat{H}_{\dot{\beta}} - H_\theta & R_{\dot{\beta}} + R_\theta \\ -R_{\dot{\beta}} - R_\theta & -\widehat{H}_{\dot{\beta}} - H_\theta \end{bmatrix} \begin{pmatrix} -\theta_{1s} \\ \theta_{1c} \end{pmatrix}$$

$$+ \begin{bmatrix} -\widehat{H}_{\dot{\beta}} 8M_\mu - (H_\mu + R_\mu) & R_{\dot{\beta}} 8M_\mu + R_r \\ -R_{\dot{\beta}} 8M_\mu - R_r & -\widehat{H}_{\dot{\beta}} 8M_\mu - (H_\mu + R_\mu) \end{bmatrix} \begin{pmatrix} \dot{x}_h - u_G \\ \dot{y}_h + v_G \end{pmatrix}$$

$$+ \begin{bmatrix} -\widehat{H}_{\dot{\beta}} \dfrac{16}{\gamma} & -\dfrac{C_T}{\sigma a} + R_{\dot{\beta}} \dfrac{16}{\gamma} \\ \dfrac{C_T}{\sigma a} - R_{\dot{\beta}} \dfrac{16}{\gamma} & -\widehat{H}_{\dot{\beta}} \dfrac{16}{\gamma} \end{bmatrix} \begin{pmatrix} \dot{\alpha}_y \\ -\dot{\alpha}_x \end{pmatrix} \qquad (19.77)$$

for an articulated rotor ($\nu_\beta = 1$ and $K_P = 0$). From the aerodynamic coefficients given in section 16.8.8, $-\widehat{H}_{\dot{\beta}} - H_\theta = 0$ and $R_{\dot{\beta}} + R_\theta = 0$, and

$$-\widehat{H}_{\dot{\beta}} 8M_\mu - (H_\mu + R_\mu) = \lambda_{\text{HP}} \left(\frac{C_T}{\sigma a} - \frac{\lambda_{\text{HP}}}{4} \right) - \frac{3c_d}{4a} \qquad (19.78)$$

$$R_{\dot{\beta}} 8M_\mu + R_r = \beta_{\text{trim}} \left(-\frac{1}{9} \frac{C_T}{\sigma a} + \frac{\lambda_{\text{HP}}}{12} \right) \qquad (19.79)$$

are small. Only for the hub forces due to shaft angular velocity is there an important contribution relative to the tip-path plane. For an articulated rotor, the direct response is

$$\Delta \begin{pmatrix} \dfrac{2C_H}{\sigma a} \\ \dfrac{2C_Y}{\sigma a} \end{pmatrix} = - \left(\frac{2C_T}{\sigma a} + \widehat{H}_{\dot{\beta}} \right) \frac{16}{\gamma} \begin{pmatrix} \dot{\alpha}_y \\ -\dot{\alpha}_x \end{pmatrix} \qquad (19.80)$$

Thus the effectiveness of the shaft angular velocity is not given by the thrust vector tilt alone, but instead by

$$H_{\dot{\beta}} - R_\beta = \frac{2C_T}{\sigma a} + \widehat{H}_{\dot{\beta}} = \frac{2C_T}{\sigma a} - \frac{\lambda_{\text{HP}}}{4} \qquad (19.81)$$

The $\widehat{H}_{\dot{\beta}}$ contribution reduces the pitch and roll damping from the rotor.

The constant coefficient approximation for the hub forces in forward flight is given in section 16.8, equations 16.439 and 16.471. As in hover, the low-frequency response requires only the aerodynamic terms. The hub in-plane forces and moments are still dominated by the flap response. Forward flight introduces a longitudinal tip-path-plane tilt due to vertical velocity of the helicopter:

$$\Delta \beta_{1c} \cong 2\mu(\dot{z}_h - w_G) \tag{19.82}$$

(equation 19.63). A downward velocity perturbation ($\dot{z}_h < 0$) is an increase in the aircraft angle-of-attack in forward flight. The rotor flaps back in response, thereby producing a pitch-up moment on the helicopter, which tends to increase the angle-of-attack still further. Thus in forward flight the rotor is a source of an angle-of-attack instability that is important for the helicopter flight dynamics.

19.5 Wake Influence

The rotor wake can have a strong influence on the blade flap response. Dynamic inflow (section 11.3) provides a global, low-frequency model for the wake-induced velocity. The simplest model has three inflow states, consisting of uniform (λ_0) and linear (λ_c, λ_s) perturbations of the wake-induced downwash at the rotor disk:

$$\lambda = \lambda_0 + \lambda_c r \cos\psi + \lambda_s r \sin\psi \tag{19.83}$$

First-order differential equations relate the inflow variables to the loading variables (thrust C_T, pitch moment C_{My}, and roll moment C_{Mx}, aerodynamic contributions only). The influence of the wake on the flap dynamics (equations 16.245 and 16.246) can be expressed in terms of a lift deficiency function C':

$$\widehat{I}_\beta \begin{pmatrix} \ddot{\beta}_{1c} + 2\dot{\beta}_{1s} + (v_\beta^2 - 1)\beta_{1c} \\ \ddot{\beta}_{1s} - 2\dot{\beta}_{1c} + (v_\beta^2 - 1)\beta_{1s} \end{pmatrix} = \gamma \begin{pmatrix} M_{F_{1c}} \\ M_{F_{1s}} \end{pmatrix}$$

$$= \gamma \begin{pmatrix} M_{F_{1c}} \\ M_{F_{1s}} \end{pmatrix}_{QS} - \frac{1}{8}\begin{pmatrix} \lambda_c \\ \lambda_s \end{pmatrix} = C'\gamma \begin{pmatrix} M_{F_{1c}} \\ M_{F_{1s}} \end{pmatrix} \tag{19.84}$$

as for equation 11.152. Combining the lift deficiency function and the Lock number gives an effective Lock number, $\gamma_e = \gamma C'$, which captures the major effects of the wake (Curtiss and Shupe (1971)).

The equations for the tip-path-plane tilt response to cyclic pitch, shaft tilt, and inflow in hover are

$$\begin{bmatrix} \widehat{I}_\beta s^2 - \gamma M_{\dot\beta} s + \widehat{I}_\beta(v_\beta^2-1) + K_P\gamma M_\theta & 2\widehat{I}_\beta s - \gamma M_{\dot\beta} \\ -(2\widehat{I}_\beta s - \gamma M_{\dot\beta}) & \widehat{I}_\beta s^2 - \gamma M_{\dot\beta} s + \widehat{I}_\beta(v_\beta^2-1) + K_P\gamma M_\theta \end{bmatrix}\begin{pmatrix} \beta_{1c} \\ \beta_{1s} \end{pmatrix}$$

$$= \gamma M_\theta \begin{pmatrix} \theta_{1c} \\ \theta_{1s} \end{pmatrix} + \gamma M_{\dot\beta}\begin{pmatrix} \lambda_c \\ \lambda_s \end{pmatrix} + \begin{bmatrix} \widehat{I}_{\beta\alpha}s - \gamma M_{\dot\beta} & 2\widehat{I}_{\beta\alpha} \\ -2\widehat{I}_{\beta\alpha} & \widehat{I}_{\beta\alpha}s - \gamma M_{\dot\beta} \end{bmatrix}\begin{pmatrix} \dot\alpha_y \\ -\dot\alpha_x \end{pmatrix} \tag{19.85}$$

(see equation 19.36). Substituting for the aerodynamic coefficients and introducing $N_\star$ (equation 19.38) gives

$$\begin{bmatrix} \frac{16}{\gamma}s + 1 & -(\frac{8}{\gamma}s^2 + s + N_\star) \\ \frac{8}{\gamma}s^2 + s + N_\star & \frac{16}{\gamma}s + 1 \end{bmatrix}\begin{pmatrix} \beta_{1c} \\ \beta_{1s} \end{pmatrix}$$

$$= \begin{pmatrix} -\theta_{1s} \\ \theta_{1c} \end{pmatrix} + \begin{pmatrix} \lambda_s \\ -\lambda_c \end{pmatrix} + \begin{bmatrix} \frac{16}{\gamma} & -(\frac{8}{\gamma}s+1) \\ \frac{8}{\gamma}s+1 & \frac{16}{\gamma} \end{bmatrix}\begin{pmatrix} \dot\alpha_y \\ -\dot\alpha_x \end{pmatrix} \tag{19.86}$$

Recognizing that the aerodynamic terms in these flap equations are the aerodynamic hub pitch and roll moments, the dynamic inflow equations for the inflow gradients (equation 11.155) in hover are

$$\left(\frac{2}{V_{\text{eff}}}ms+1\right)\begin{pmatrix}\lambda_c\\\lambda_s\end{pmatrix}=\frac{2}{V_{\text{eff}}}\begin{pmatrix}-C_{My}\\C_{Mx}\end{pmatrix}_{\text{aero}}+K_R\begin{pmatrix}\dot{\alpha}_y-s\beta_{1c}\\-\dot{\alpha}_x-s\beta_{1s}\end{pmatrix}$$

$$=\frac{\sigma a}{8V_{\text{eff}}}\left\{-\begin{bmatrix}s+K_P&1\\-1&s+K_P\end{bmatrix}\begin{pmatrix}\beta_{1c}\\\beta_{1s}\end{pmatrix}+\begin{pmatrix}\theta_{1c}\\\theta_{1s}\end{pmatrix}-\begin{pmatrix}\lambda_c\\\lambda_s\end{pmatrix}+\begin{pmatrix}\dot{\alpha}_y\\-\dot{\alpha}_x\end{pmatrix}\right\}$$

$$+K_R\begin{pmatrix}\dot{\alpha}_y-s\beta_{1c}\\-\dot{\alpha}_x-s\beta_{1s}\end{pmatrix}\qquad(19.87)$$

where $V_{\text{eff}}=\lambda_i$ and $m=\frac{64}{45\pi}$. Tip-path-plane pitch and roll rate produce an inflow gradient through the wake curvature change, with $K_R=1.5$ to 2.2 for hover. Thus

$$\begin{pmatrix}\lambda_s\\-\lambda_c\end{pmatrix}=(1-C)\left\{\begin{bmatrix}1&-(s+K_P)\\s+K_P&1\end{bmatrix}\begin{pmatrix}\beta_{1c}\\\beta_{1s}\end{pmatrix}-\begin{pmatrix}-\theta_{1s}\\\theta_{1c}\end{pmatrix}-\begin{bmatrix}0&-1\\1&0\end{bmatrix}\begin{pmatrix}\dot{\alpha}_y\\-\dot{\alpha}_x\end{pmatrix}\right\}$$

$$+\begin{bmatrix}0&CK_RD\\-CK_RD&0\end{bmatrix}\begin{pmatrix}\dot{\alpha}_y-s\beta_{1c}\\-\dot{\alpha}_x-s\beta_{1s}\end{pmatrix}\qquad(19.88)$$

with the lift deficiency function

$$C=1-\frac{\sigma a/8\lambda_i}{1+\sigma a/8\lambda_i+(2m/\lambda_i)s}=\frac{1}{1+(\sigma a/8\lambda_i)/(1+(2m/\lambda_i)s)}\qquad(19.89)$$

and $D=1/(1+(2m/\lambda_i)s)$. Substituting for the inflow, the equation for the flap response becomes

$$\begin{bmatrix}\frac{16}{\gamma}s+C&-(\frac{8}{\gamma}s^2+C(1-K_RD)s+N_\star-(1-C)K_P)\\\frac{8}{\gamma}s^2+C(1-K_RD)s+N_\star-(1-C)K_P&\frac{16}{\gamma}s+C\end{bmatrix}\begin{pmatrix}\beta_{1c}\\\beta_{1s}\end{pmatrix}$$

$$=C\begin{pmatrix}-\theta_{1s}\\\theta_{1c}\end{pmatrix}+\begin{bmatrix}\frac{16}{\gamma}&-(\frac{8}{\gamma}s+C(1-K_RD))\\\frac{8}{\gamma}s+C(1-K_RD)&\frac{16}{\gamma}\end{bmatrix}\begin{pmatrix}\dot{\alpha}_y\\-\dot{\alpha}_x\end{pmatrix}\qquad(19.90)$$

or

$$\begin{bmatrix}\frac{16}{\gamma_e}s+1&-(\frac{8}{\gamma_e}s^2+(1-K_e)s+N_e)\\\frac{8}{\gamma_e}s^2+(1-K_e)s+N_e&\frac{16}{\gamma_e}s+1\end{bmatrix}\begin{pmatrix}\beta_{1c}\\\beta_{1s}\end{pmatrix}$$

$$=\begin{pmatrix}-\theta_{1s}\\\theta_{1c}\end{pmatrix}+\begin{bmatrix}\frac{16}{\gamma_e}&-(\frac{8}{\gamma_e}s+1-K_e)\\\frac{8}{\gamma_e}s+1-K_e&\frac{16}{\gamma_e}\end{bmatrix}\begin{pmatrix}\dot{\alpha}_y\\-\dot{\alpha}_x\end{pmatrix}\qquad(19.91)$$

in terms of the effective Lock number $\gamma_e=C\gamma$, stiffness $N_e=(\nu_\beta^2-1)/\frac{\gamma_e}{8}+K_P$, and coupling $K_e=K_RD$. The static response is

$$\begin{bmatrix}1&-N_e\\N_e&1\end{bmatrix}\begin{pmatrix}\beta_{1c}\\\beta_{1s}\end{pmatrix}=\begin{pmatrix}-\theta_{1s}\\\theta_{1c}\end{pmatrix}+\begin{bmatrix}\frac{16}{\gamma_e}&-(1-K_R)\\1-K_R&\frac{16}{\gamma_e}\end{bmatrix}\begin{pmatrix}\dot{\alpha}_y\\-\dot{\alpha}_x\end{pmatrix}\qquad(19.92)$$

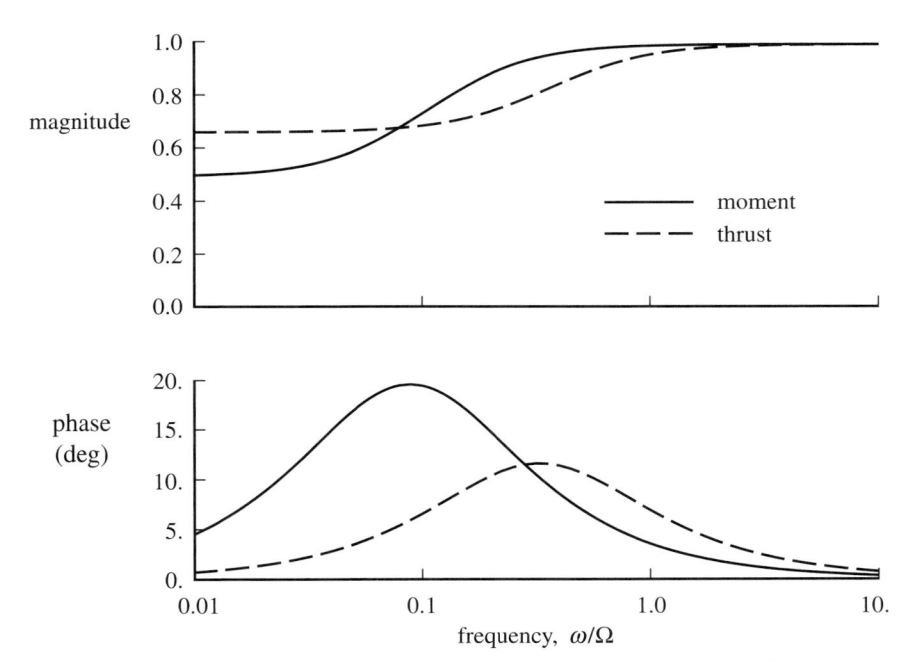

Figure 19.8. Lift deficiency functions introduced by dynamic inflow, for rotor response in hover; $C_T/\sigma = 0.08$.

The response to cyclic is determined by the balance of aerodynamic forces, so is not changed by the wake except for the increase in phase shift through N_e. For angular motion, the effect of the wake is to increase the damping $(16/\gamma_e)$ and to change the sign of the off-axis response (since $K_R > 1$). Figure 19.8 shows the lift deficiency function C as a function of frequency.

Figure 19.9 presents the direct and cross response of the tip-path plane to cyclic control inputs, including the effects of the wake. For the hingeless rotor, the swashplate phase $\tan^{-1} N_e$ has been included. Figure 19.10 presents the direct and cross response of the tip-path plane to shaft angular motion. Figures 19.6 and 19.7 show the corresponding results without the wake influence. The inflow gradient from wake curvature $(K_R = 1.5)$ primarily affects the sign of the static off-axis response for rotors with small hinge offset. For this hingeless rotor example, K_R has little effect on the response compared to the large coupling produced by $\nu_\beta = 1.15$. The time lags of the dynamic inflow model wash out the wake effects at high frequency. A quasistatic dynamic inflow model affects the response at all frequencies, as simply an effective (smaller) Lock number $\gamma_e = C\gamma$.

The equations for thrust, coning, and uniform inflow perturbation (without pitch-cone coupling) are for hover,

$$\frac{C_T}{\sigma a} = \frac{1}{6}\theta_0 - \frac{1}{4}\lambda_0 - \frac{1}{6}s\beta_0 \tag{19.93}$$

$$\left(s^2 + \frac{\gamma}{8}s + \nu_\beta^2\right)\beta_0 = \frac{\gamma}{8}\theta_0 - \frac{\gamma}{6}\lambda_0 \tag{19.94}$$

$$\left(\frac{1}{2V_{\text{eff}}}ms + 1\right)\lambda_0 = \frac{1}{2V_{\text{eff}}}(C_T)_{\text{aero}} \tag{19.95}$$

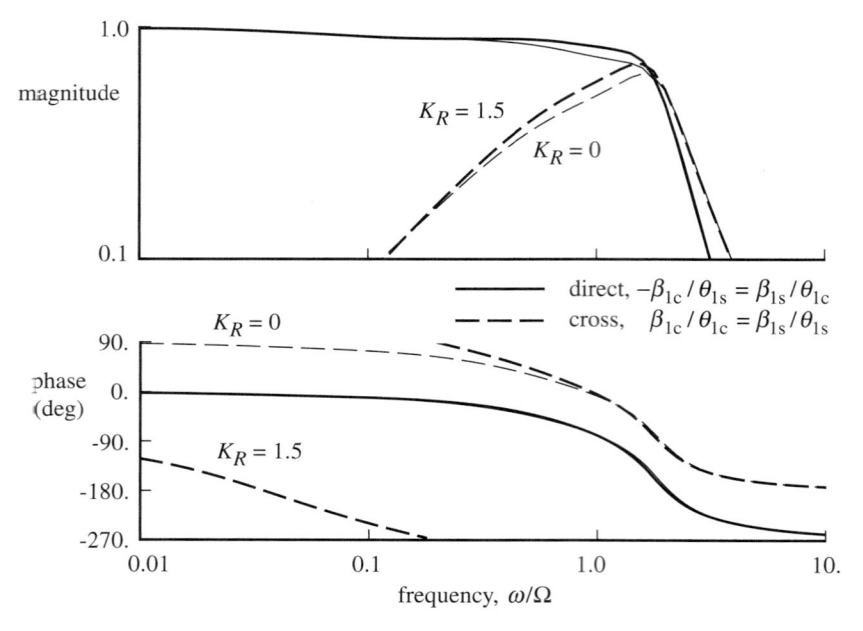

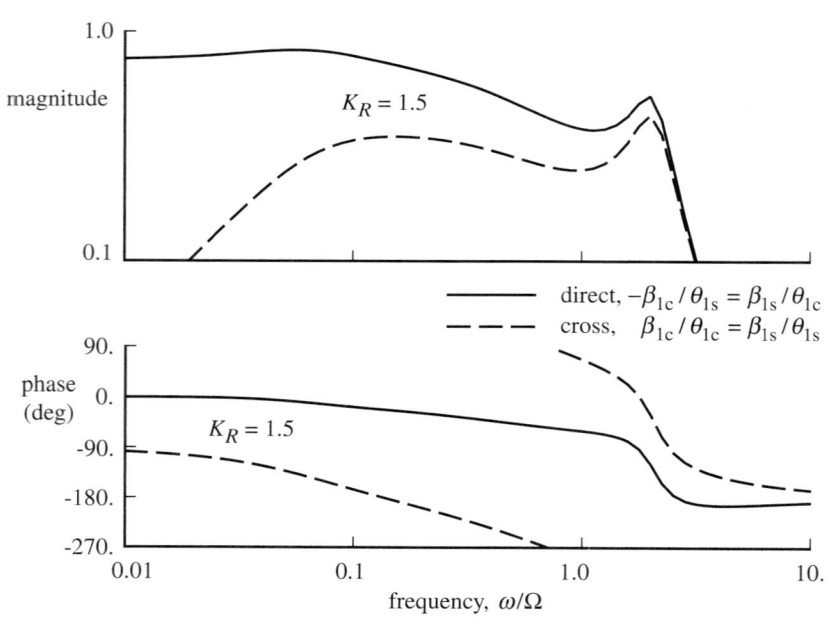

Figure 19.9. Frequency response of the tip-path-plane tilt to cyclic pitch in hover, including wake influence; $C_T/\sigma = 0.08$.

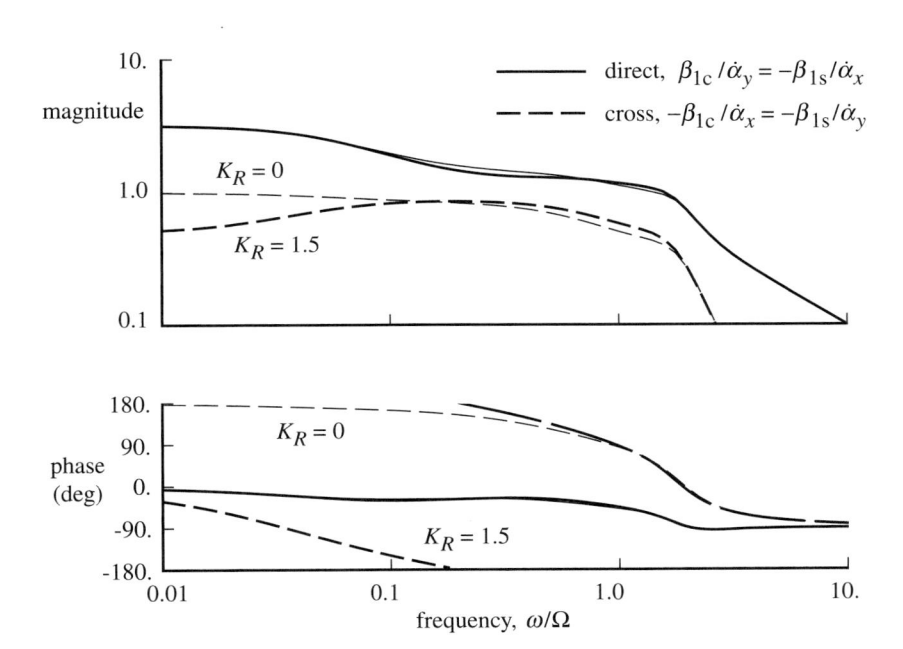

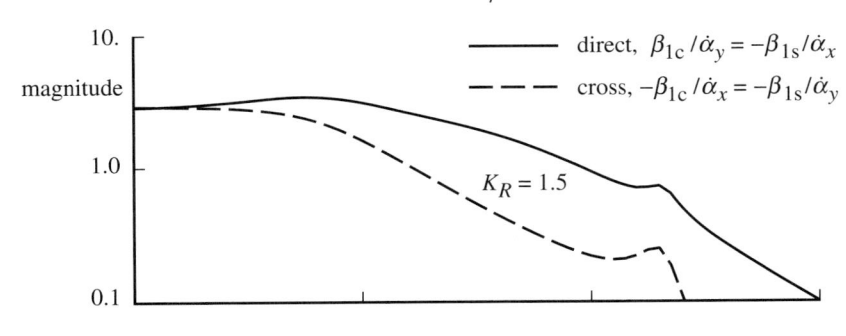

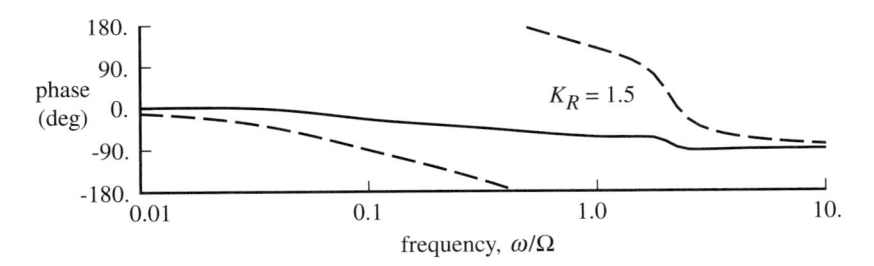

19.5 Wake Influence

Figure 19.10. Frequency response of the tip-path-plane tilt to shaft rate in hover, including wake influence; $C_T/\sigma = 0.08$.

where here $V_{\text{eff}} = 2\lambda_i$ and $m = \frac{8}{3\pi}$. Substituting for λ_0 gives

$$\frac{C_T}{\sigma a} = C\left(\frac{1}{6}\theta_0 - \frac{1}{6}s\beta_0\right) \tag{19.96}$$

$$\left(s^2 + \frac{\gamma}{8}C_0 s + \nu_\beta^2\right)\beta_0 = \frac{\gamma}{8}C_0\theta_0 \tag{19.97}$$

with the lift deficiency function

$$C = \frac{1}{1 + (\sigma a/16\lambda_i)/(1 + (m/4\lambda_i)s)} \tag{19.98}$$

and $C_0 = (8C + 1)/9$. The effective Lock number for the coning response is $\gamma_e = C_0\gamma$. Figure 19.8 shows the lift deficiency function for thrust as a function of frequency.

The rotor hub moments are obtained from the tip-path-plane tilt:

$$\begin{pmatrix} -\dfrac{2C_{My}}{\sigma a} \\ \dfrac{2C_{Mx}}{\sigma a} \end{pmatrix} = \frac{\nu_\beta^2 - 1}{\gamma}\begin{pmatrix} \beta_{1c} \\ \beta_{1s} \end{pmatrix} \tag{19.99}$$

Using $H_\theta = -\widehat{H}_{\dot\beta}$ and $R_{\dot\beta} = -R_\theta$ from section 16.8.8, the low-frequency hub forces due to cyclic pitch and shaft motion in hover (equation 19.74) are

$$\begin{aligned}
\begin{pmatrix} \dfrac{2C_H}{\sigma a} \\ \dfrac{2C_Y}{\sigma a} \end{pmatrix} &= -\frac{2C_T}{\sigma a}\begin{pmatrix} \beta_{1c} \\ \beta_{1s} \end{pmatrix} + \begin{bmatrix} -\widehat{H}_{\dot\beta} & R_{\dot\beta} \\ -R_{\dot\beta} & -\widehat{H}_{\dot\beta} \end{bmatrix}\begin{pmatrix} \beta_{1c} \\ \beta_{1s} \end{pmatrix} \\
&\quad + \begin{bmatrix} -H_\theta & R_\theta \\ -R_\theta & -H_\theta \end{bmatrix}\begin{pmatrix} -\theta_{1s} + K_P\beta_{1s} \\ \theta_{1c} - K_P\beta_{1c} \end{pmatrix} \\
&\quad + \begin{bmatrix} H_{\dot\beta} & -R_{\dot\beta} \\ R_{\dot\beta} & H_{\dot\beta} \end{bmatrix}\left\{\begin{pmatrix} \dot\alpha_x \\ \dot\alpha_y \end{pmatrix} + \begin{pmatrix} \lambda_s \\ -\lambda_c \end{pmatrix}\right\} \\
&= -\frac{C_T}{\sigma a}\begin{pmatrix} \beta_{1c} \\ \beta_{1s} \end{pmatrix} + \begin{bmatrix} -H_{\dot\beta} & R_{\dot\beta} \\ -R_{\dot\beta} & -H_{\dot\beta} \end{bmatrix}\left\{\begin{pmatrix} \beta_{1c} \\ \beta_{1s} \end{pmatrix} - \begin{pmatrix} \dot\alpha_x \\ \dot\alpha_y \end{pmatrix} - \begin{pmatrix} \lambda_s \\ -\lambda_c \end{pmatrix}\right\} \\
&\quad + \begin{bmatrix} -H_\theta & R_\theta \\ -R_\theta & -H_\theta \end{bmatrix}\begin{pmatrix} -\theta_{1s} + K_P\beta_{1s} \\ \theta_{1c} - K_P\beta_{1c} \end{pmatrix} \\
&= -\frac{C_T}{\sigma a}\begin{pmatrix} \beta_{1c} \\ \beta_{1s} \end{pmatrix} - \frac{C_T}{\sigma a}\begin{pmatrix} -\theta_{1s} + K_P\beta_{1s} \\ \theta_{1c} - K_P\beta_{1c} \end{pmatrix} \\
&\quad + \begin{bmatrix} -H_{\dot\beta} & R_{\dot\beta} \\ -R_{\dot\beta} & -H_{\dot\beta} \end{bmatrix}\left\{\begin{pmatrix} \beta_{1c} \\ \beta_{1s} \end{pmatrix} - \begin{pmatrix} -\theta_{1s} + K_P\beta_{1s} \\ \theta_{1c} - K_P\beta_{1c} \end{pmatrix} - \begin{pmatrix} \dot\alpha_x \\ \dot\alpha_y \end{pmatrix} - \begin{pmatrix} \lambda_s \\ -\lambda_c \end{pmatrix}\right\}
\end{aligned} \tag{19.100}$$

For low frequency the dynamic inflow equations are

$$\begin{pmatrix} \lambda_s \\ -\lambda_c \end{pmatrix} = (1 - C)\left\{\begin{pmatrix} \beta_{1c} \\ \beta_{1s} \end{pmatrix} - \begin{pmatrix} -\theta_{1s} + K_P\beta_{1s} \\ \theta_{1c} - K_P\beta_{1c} \end{pmatrix}\right\} - (1 - C + CK_R))\begin{pmatrix} \dot\alpha_x \\ \dot\alpha_y \end{pmatrix} \tag{19.101}$$

19.5 Wake Influence

(equation 19.88). Substituting for the inflow, and using $H_{\dot\beta} = \frac{C_T}{\sigma a} + \widehat{H}_{\dot\beta}$, gives

$$\begin{pmatrix} \frac{2C_H}{\sigma a} \\ \frac{2C_Y}{\sigma a} \end{pmatrix} = -\frac{C_T}{\sigma a}\begin{pmatrix} \beta_{1c} \\ \beta_{1s} \end{pmatrix} - \frac{C_T}{\sigma a}\begin{pmatrix} -\theta_{1s} + K_P\beta_{1s} \\ \theta_{1c} - K_P\beta_{1c} \end{pmatrix}$$

$$+ C\begin{bmatrix} -H_{\dot\beta} & R_{\dot\beta} \\ -R_{\dot\beta} & -H_{\dot\beta} \end{bmatrix}\left\{\begin{pmatrix} \beta_{1c} \\ \beta_{1s} \end{pmatrix} - \begin{pmatrix} -\theta_{1s} + K_P\beta_{1s} \\ \theta_{1c} - K_P\beta_{1c} \end{pmatrix} - (1 - K_R)\begin{pmatrix} \dot\alpha_x \\ \dot\alpha_y \end{pmatrix}\right\}$$

$$= \left\{-\frac{2C_T}{\sigma a} + \begin{bmatrix} (1-C)\frac{C_T}{\sigma a} - C\widehat{H}_{\dot\beta} & CR_{\dot\beta} \\ -CR_{\dot\beta} & (1-C)\frac{C_T}{\sigma a} - C\widehat{H}_{\dot\beta} \end{bmatrix}\begin{bmatrix} 1 & -K_P \\ K_P & 1 \end{bmatrix}\right\}\begin{pmatrix} \beta_{1c} \\ \beta_{1s} \end{pmatrix}$$

$$- \begin{bmatrix} (1-C)\frac{C_T}{\sigma a} - C\widehat{H}_{\dot\beta} & CR_{\dot\beta} \\ -CR_{\dot\beta} & (1-C)\frac{C_T}{\sigma a} - C\widehat{H}_{\dot\beta} \end{bmatrix}\begin{pmatrix} -\theta_{1s} \\ \theta_{1c} \end{pmatrix}$$

$$- C\begin{bmatrix} -H_{\dot\beta} & R_{\dot\beta} \\ -R_{\dot\beta} & -H_{\dot\beta} \end{bmatrix}(1 - K_R)\begin{pmatrix} \dot\alpha_x \\ \dot\alpha_y \end{pmatrix} \qquad (19.102)$$

Finally, substituting for the low-frequency flap response,

$$\begin{pmatrix} \beta_{1c} \\ \beta_{1s} \end{pmatrix} = \frac{1}{1 + N_e^2}\begin{bmatrix} 1 & N_e \\ -N_e & 1 \end{bmatrix}\left\{\begin{pmatrix} -\theta_{1s} \\ \theta_{1c} \end{pmatrix} + \begin{bmatrix} \frac{16}{\gamma_e} & -(1 - K_R) \\ 1 - K_R & \frac{16}{\gamma_e} \end{bmatrix}\begin{pmatrix} \dot\alpha_y \\ -\dot\alpha_x \end{pmatrix}\right\} \qquad (19.103)$$

(equation 19.92), the hub forces are

$$\begin{pmatrix} \frac{2C_H}{\sigma a} \\ \frac{2C_Y}{\sigma a} \end{pmatrix} = \left\{-\frac{2C_T}{\sigma a} + \begin{bmatrix} -CR_{\dot\beta} & (1-C)\frac{C_T}{\sigma a} - C\widehat{H}_{\dot\beta} \\ -((1-C)\frac{C_T}{\sigma a} - C\widehat{H}_{\dot\beta}) & -CR_{\dot\beta} \end{bmatrix}(N_e - K_P)\right\}$$

$$\frac{1}{1 + N_e^2}\begin{bmatrix} 1 & N_e \\ -N_e & 1 \end{bmatrix}\begin{pmatrix} -\theta_{1s} \\ \theta_{1c} \end{pmatrix}$$

$$+ \left\{-\frac{2C_T}{\sigma a} + \begin{bmatrix} (1-C)\frac{C_T}{\sigma a} - C\widehat{H}_{\dot\beta} & CR_{\dot\beta} \\ -CR_{\dot\beta} & (1-C)\frac{C_T}{\sigma a} - C\widehat{H}_{\dot\beta} \end{bmatrix}\begin{bmatrix} 1 & -K_P \\ K_P & 1 \end{bmatrix}\right\}$$

$$\frac{1}{1 + N_e^2}\begin{bmatrix} 1 & N_e \\ -N_e & 1 \end{bmatrix}\frac{16}{\gamma_e}\begin{pmatrix} \dot\alpha_y \\ -\dot\alpha_x \end{pmatrix}$$

$$+ \left\{-\frac{2C_T}{\sigma a} + \frac{C_T}{\sigma a}\begin{bmatrix} 1 & -K_P \\ K_P & 1 \end{bmatrix} + C\begin{bmatrix} -R_{\dot\beta} & -H_{\dot\beta} \\ H_{\dot\beta} & -R_{\dot\beta} \end{bmatrix}(N_e - K_P)\right\}$$

$$\frac{1}{1 + N_e^2}\begin{bmatrix} 1 & N_e \\ -N_e & 1 \end{bmatrix}(1 - K_R)\begin{pmatrix} \dot\alpha_x \\ \dot\alpha_y \end{pmatrix} \qquad (19.104)$$

where $N_e - K_P = \frac{\nu_\beta^2 - 1}{\gamma_e/8}$.

For $\nu_\beta = 1$, cyclic control produces in-plane hub forces simply by tilting the thrust vector with the tip-path plane. The flapping and hub moment amplitude produced by cyclic pitch are

$$|\beta| = \frac{1}{\sqrt{1 + N_e^2}} |\theta| \qquad (19.105)$$

$$8\frac{2|C_M|}{\sigma a} = C \frac{N_e}{\sqrt{1 + N_e^2}} |\theta| \qquad (19.106)$$

with $N_e = \frac{\nu_\beta^2 - 1}{C_\gamma/8} + K_P$. Figure 19.11 shows the rotor response to cyclic pitch in hover for a range of flap frequencies for $K_P = 0$ and neglecting the $R_{\dot\beta}$ terms. Flap frequency $\nu_\beta > 1$ reduces the flap amplitude and increases the hub moment. The lift deficiency function $C = 1/(1 + \sigma a/8\lambda_i)$ introduces a dependency on the rotor thrust. For $\nu_\beta \to \infty$, the tip-path-plane tilt is zero, the hub moment per unit cyclic approaches a constant value, and the hub force is

$$\begin{pmatrix} \frac{2C_H}{\sigma a} \\ \frac{2C_Y}{\sigma a} \end{pmatrix} \to -\left((1-C)\frac{C_T}{\sigma a} + C\frac{\lambda_{HP}}{4} \right) \begin{pmatrix} -\theta_{1s} \\ \theta_{1c} \end{pmatrix} \qquad (19.107)$$

The difference between tip-path-plane tilt and thrust vector tilt (Figure 19.11) is caused by tilt of the thrust vector relative to the tip-path plane. So even for very large flap frequency, there is a finite in-plane hub force due to cyclic pitch, equivalent to tilt of the thrust vector on the order of half the articulated rotor value. Thus the tandem and side-by-side helicopter configurations can use differential cyclic pitch for yaw control, regardless of the rotor flapping stiffness.

The hub in-plane forces produced by shaft pitch rate and roll rate, for $K_P = 0$ and neglecting the $R_{\dot\beta}$ terms, are

$$\begin{pmatrix} \frac{2C_H}{\sigma a} \\ \frac{2C_Y}{\sigma a} \end{pmatrix} = -\left((1+C)\frac{C_T}{\sigma a} + C\widehat{H}_{\dot\beta} \right)\left(\frac{16}{\gamma_e} + N_e(1-K_R) \right) \frac{1}{1+N_e^2} \begin{pmatrix} \dot\alpha_y \\ -\dot\alpha_x \end{pmatrix}$$

$$+ \left\{ \left((1+C)\frac{C_T}{\sigma a} + C\widehat{H}_{\dot\beta} \right)\left(N_e \frac{16}{\gamma_e} - (1-K_R) \right) \frac{1}{1+N_e^2} \right.$$

$$\left. + C\widehat{H}_{\dot\beta}(1-K_R) \right\} \begin{pmatrix} \dot\alpha_x \\ \dot\alpha_y \end{pmatrix} \qquad (19.108)$$

Equation 19.80 gives this result for an articulated rotor ($\nu_\beta = 1$) without the wake influence ($C = 1$). If the thrust vector tilted with the tip-path plane, the in-plane forces would be

$$\begin{pmatrix} \frac{2C_H}{\sigma a} \\ \frac{2C_Y}{\sigma a} \end{pmatrix} = -\frac{2C_T}{\sigma a}\begin{pmatrix} \beta_{1c} \\ \beta_{1s} \end{pmatrix} = -\frac{2C_T}{\sigma a}\left(\frac{16}{\gamma_e} + N_e(1-K_R) \right) \frac{1}{1+N_e^2} \begin{pmatrix} \dot\alpha_y \\ -\dot\alpha_x \end{pmatrix}$$

$$+ \frac{2C_T}{\sigma a}\left(N_e\frac{16}{\gamma_e} - (1-K_R) \right) \frac{1}{1+N_e^2} \begin{pmatrix} \dot\alpha_x \\ \dot\alpha_y \end{pmatrix} \qquad (19.109)$$

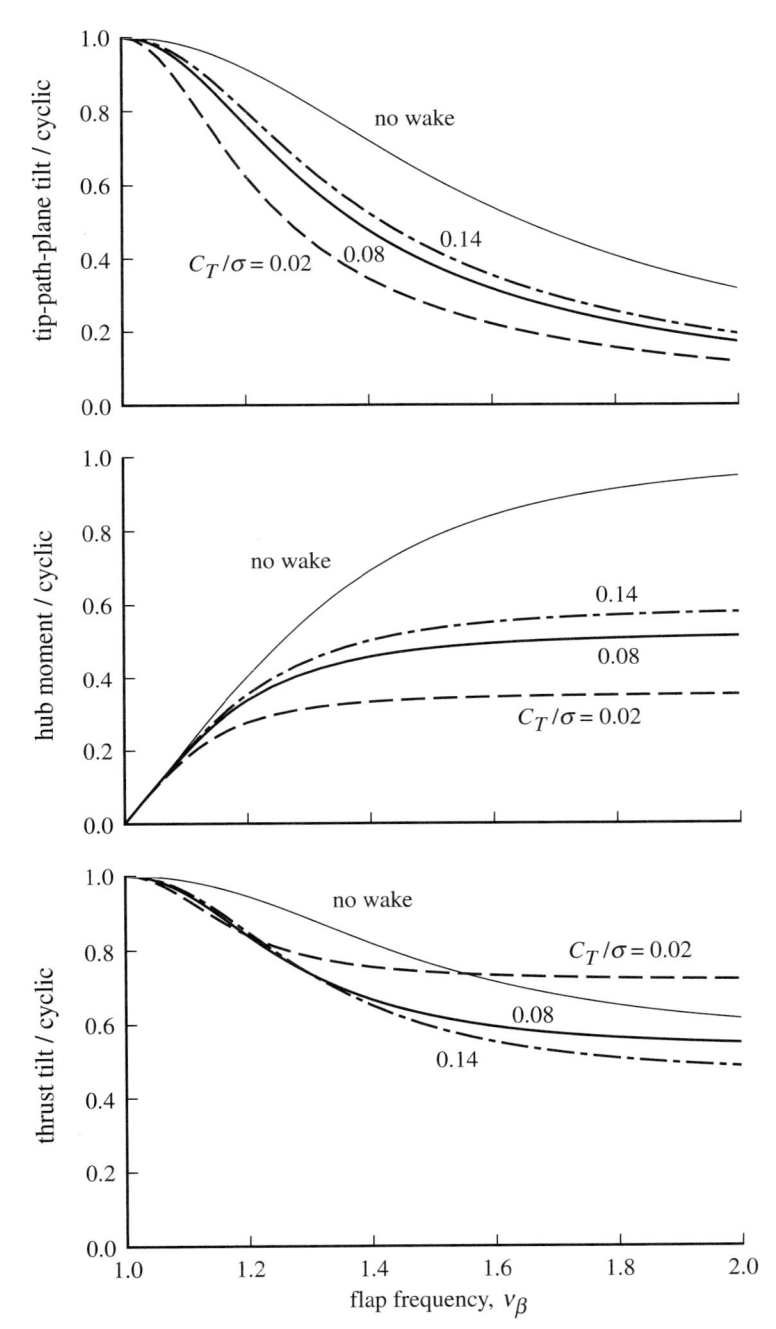

Figure 19.11. In-plane hub force due to cyclic pitch as a fraction of thrust tilt; $\gamma = 8$ and $\sigma = 0.08$.

So the tilt of the thrust vector due to shaft rate, relative to the tip-path-plane tilt, is reduced by the factor

$$f = \frac{(1+C)C_T/\sigma a + C\widehat{H}_{\dot{\beta}}}{2C_T/\sigma a} = \frac{1+C}{2} - C\frac{\lambda_{\text{HP}}}{8C_T/\sigma a} \qquad (19.110)$$

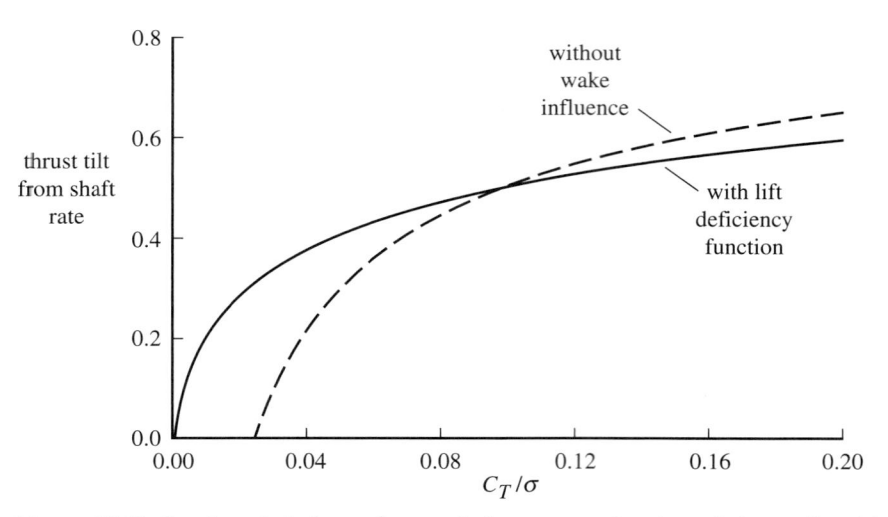

Figure 19.12. In-plane hub force due to shaft rate as a fraction of thrust tilt with tip-path plane; $\gamma = 8$ and $\sigma = 0.08$.

which is shown in Figure 19.12. The $\widehat{H}_{\dot{\beta}}$ contribution reduces the pitch and roll damping from the rotor, particularly at low thrust.

Miller (1948) identified this damping reduction and included it in a stability and control analysis. Amer (1950) also derived the damping reduction. He substituted for the inflow in terms of collective and thrust coefficient to obtain $f = \frac{3}{2}(1 - \frac{B^3 a}{18} \frac{\theta}{C_T/\sigma})$, although collective and thrust are not independent variables for hover. Amer and Gustafson (1951) showed that assuming the resultant rotor force is perpendicular to the tip-path plane leads to very incorrect stability derivatives. Sissingh (1951) added the wake influence, in terms of an inflow gradient.

19.6 Pitch-Flap Coupling and Feedback

Pitch-flap coupling ($K_P = \tan \delta_3$) was introduced in section 6.18, and its effects have been derived in this chapter. A more general representation of pitch-flap feedback includes rate feedback (K_D) and Oehmichen coupling (K_O):

$$\Delta \theta^{(m)} = -K_P \beta^{(m)} - K_D \dot{\beta}^{(m)} \mp K_O \beta^{(m \pm 1)} \qquad (19.111)$$

where m is the blade number. Oehmichen patented in 1929 a mechanism to change the blade pitch proportional to the flap motion of the preceding blade; see Leishman (2006). For 1/rev motion of a four-bladed rotor, $\beta^{(m-1)} \cong \dot{\beta}^{(m)}$, so Oehmichen coupling corresponds to rate feedback. The multiblade coordinate transformation of the feedback is

$$\Delta \theta_0 = -K_P \beta_0 - K_D \dot{\beta}_0 \mp K_O \beta_0 \qquad (19.112)$$

$$\Delta \theta_{nc} = -K_P \beta_{nc} - K_D (\dot{\beta}_{nc} + n\beta_{ns}) - K_O (\sin n\Delta\psi \, \beta_{ns} \pm \cos n\Delta\psi \, \beta_{nc}) \qquad (19.113)$$

$$\Delta \theta_{ns} = -K_P \beta_{ns} - K_D (\dot{\beta}_{ns} - n\beta_{nc}) - K_O (-\sin n\Delta\psi \, \beta_{nc} \pm \cos n\Delta\psi \, \beta_{ns}) \qquad (19.114)$$

where $\Delta\psi = 2\pi/N$ is the spacing between blades. Including now rate feedback and Oehmichen coupling, equations 19.25 and 19.26 become

$$\ddot{\beta}_0 + \frac{\gamma}{8}(1+K_D)\dot{\beta}_0 + \left(v_\beta^2 + \frac{\gamma}{8}(K_P \pm K_O)\right)\beta_0 = \frac{\gamma}{8}\theta_0 \qquad (19.115)$$

$$\begin{pmatrix}\ddot{\beta}_{1c}\\ \ddot{\beta}_{1s}\end{pmatrix} + \begin{bmatrix}\frac{\gamma}{8}(1+K_D) & 2\\ -2 & \frac{\gamma}{8}(1+K_D)\end{bmatrix}\begin{pmatrix}\dot{\beta}_{1c}\\ \dot{\beta}_{1s}\end{pmatrix}$$
$$+ \begin{bmatrix}v_\beta^2 - 1 + \frac{\gamma}{8}K_P & \frac{\gamma}{8}(1+K_D+K_O)\\ -\frac{\gamma}{8}(1+K_D+K_O) & v_\beta^2 - 1 + \frac{\gamma}{8}K_P\end{bmatrix}\begin{pmatrix}\beta_{1c}\\ \beta_{1s}\end{pmatrix} = \frac{\gamma}{8}\begin{pmatrix}\theta_{1c}\\ \theta_{1s}\end{pmatrix} \qquad (19.116)$$

for a four-bladed rotor ($\Delta\psi = 90°$). Pitch-flap coupling K_P introduces an aerodynamic spring, and rate feedback K_D adds to the aerodynamic flap damping. Oehmichen coupling adds or subtracts from pitch-cone coupling. For the low-frequency response, K_O acts the same as K_D, so Oehmichen coupling corresponds to rate feedback. Whereas Oehmichen coupling increases the stability of the low-frequency flap modes, it decreases the stability of the high-frequency modes.

19.7 Complex Variable Representation of Motion

Because of the symmetry of the rotor equations and motion, complex variables can be useful in the analysis of rotor dynamics. Coleman (1943) used complex combinations of the lag degrees of freedom and the shaft motion degrees of freedom to facilitate derivation and solution of the ground resonance equations, for an axisymmetric system working with two complex equations instead of four real equations. Miller (1948) conducted an evaluation of the stability and control characteristics of several different types of helicopters. Citing Coleman, Miller also used complex combinations of variables for the airframe in-plane and angular motion and the rotor flap motion, thereby reducing the number of equations from six to three. Curtiss (1973) provided an exposition on the use of complex coordinates for hovering rotor dynamics.

For the blade flap dynamics, define complex variables for the tip-path-plane tilt, cyclic pitch control, shaft angular velocity, and inflow gradient: $b = \beta_{1c} + i\beta_{1s}$, $t = \theta_{1c} + i\theta_{1s}$, $a = \dot{\alpha}_y - i\dot{\alpha}_x$, $l = \lambda_c + i\lambda_s$. Combining the β_{1c} equation and i times the β_{1s} equation gives

$$\widehat{I}_\beta \ddot{b} - (\gamma M_{\dot\beta} + i2\widehat{I}_\beta)\dot{b} + (\widehat{I}_\beta(v_\beta^2 - 1) + K_P\gamma M_\theta + i\gamma M_{\dot\beta})b$$
$$= \gamma M_\theta t + \gamma M_{\dot\beta} l + \widehat{I}_{\beta\alpha}\dot{a} - (\gamma M_{\dot\beta} + i2\widehat{I}_\beta)a \qquad (19.117)$$

(from equations 19.26 and 19.85), or

$$\ddot{b} + \left(\frac{\gamma}{8} - 2i\right)\dot{b} + \frac{\gamma}{8}(N_\star - i)b = \frac{\gamma}{8}\left(t - l + \frac{8}{\gamma}\dot{a} + \left(1 - \frac{16}{\gamma}i\right)a\right) \qquad (19.118)$$

The low-frequency response is

$$b = \frac{t - l + \left(1 - \frac{16}{\gamma}i\right)a}{N_\star - i} \qquad (19.119)$$

The $-i$ term in the denominator is the 90° phase shift of the articulated rotor. The solution of the characteristic equation,

$$s^2 + \left(\frac{\gamma}{8} - 2i\right)s + \frac{\gamma}{8}(N_\star - i) = 0 \tag{19.120}$$

is provided by the eigenvalues

$$s = -\frac{\gamma}{16} - i \pm i\sqrt{v^2 + \frac{\gamma}{8}K_P - \left(\frac{\gamma}{16}\right)^2} = s_R - i \tag{19.121}$$

and its conjugate. The eigen-vector is an oscillation of the complex variable $b = \beta_{1c} + i\beta_{1s}$, with i implying a 90° phase difference between β_{1c} and β_{1s}. The oscillation is at frequency $\omega_{NR} = \pm\omega_R - 1$, which is a progressive wobble (same direction as the rotor) for positive ω_{NR} and a regressive wobble (opposite to the rotor direction) for negative ω_{NR}.

The equation for the flap response to cyclic and shaft motion in hover is

$$\left(\frac{8}{\gamma}s^2 + \left(1 - \frac{16}{\gamma}i\right)s + N_\star - i\right)b = t + \left(\frac{8}{\gamma}s + 1 - \frac{16}{\gamma}i\right)a - l \tag{19.122}$$

Equation 19.88 for the dynamic inflow model becomes

$$l = (1-C)\bigl((-s - K_P + i)b + t + a\bigr) + CK_R D(a - sb) \tag{19.123}$$

Substituting for the inflow gives

$$\left(\frac{8}{\gamma}s^2 + \left(C(1-K_R)D - \frac{16}{\gamma}i\right)s + N_\star - (1-C)K_P - Ci\right)b$$
$$= Ct + \left(\frac{8}{\gamma}s + C(1-K_R)D - \frac{16}{\gamma}i\right)a \tag{19.124}$$

or

$$\left(\frac{8}{\gamma_e}s^2 + \left(1 - K_e - \frac{16}{\gamma_e}i\right)s + N_e - i\right)b = t + \left(\frac{8}{\gamma_e}s + 1 - K_e - \frac{16}{\gamma_e}i\right)a \tag{19.125}$$

The static equation is just $(N_e - i)b = t + (1 - K_e - \frac{16}{\gamma_e}i)a$

So this is a compact form for the expression and solution of the rotor dynamics, but the form introduces the issue of interpreting the motion and control implied by the complex variables.

19.8 Two-Bladed Rotor

The case of a two-bladed rotor requires special consideration because, unlike rotors with three or more blades, the rotor motion cannot be represented by tip-path-plane tilt degrees of freedom β_{1c} and β_{1s}. Instead, the dynamic behavior of the two-bladed rotor is described by the motion of the teetering degree of freedom β_1, which leads to periodic coefficient equations coupling the rotor and the non-rotating frame. For helicopter flight dynamics, the primary concern is with the low-frequency response of the flap motion and hub reactions to control inputs, shaft motion, and gusts. This section shows that the low-frequency response of the two-bladed rotor is nearly identical to that for the $N \geq 3$ case.

19.8 Two-Bladed Rotor

The frequency response of a linear, time-invariant dynamic system is described by a transfer function $H(\omega)$ relating the magnitude and phase of the input and output at frequency ω: $\overline{F} = H(\omega)\overline{\alpha}$. The implication of the periodic coefficients in the equations of motion for the two-bladed rotor is that such a transfer does not exist, since an input at frequency ω produces in general a response at all frequencies $\omega \pm n\Omega$ (n an integer). Then the input-output relation for sinusoidal excitation takes the form

$$F = \left(\sum_{n=-\infty}^{\infty} H_n(\omega) e^{in\Omega t}\right) \overline{\alpha} e^{i\omega t} \qquad (19.126)$$

Specifically, consider the flapping equation of motion for the two-bladed rotor:

$$\widehat{I}_\beta \ddot{\beta}_1 - \gamma M_{\dot\beta} \dot{\beta}_1 + \left(\widehat{I}_\beta v_\beta^2 + K_P \gamma M_\theta\right) \beta_1$$
$$= \left[\gamma M_\theta \theta_{1c} + \widehat{I}_{\beta\alpha}(\ddot{\alpha}_y - 2\dot{\alpha}_x) - \gamma M_{\dot\beta}\dot{\alpha}_y + \gamma M_\mu(\dot{y}_h + v_G)\right]\cos\psi$$
$$+ \left[\gamma M_\theta \theta_{1s} - \widehat{I}_{\beta\alpha}(\ddot{\alpha}_x + 2\dot{\alpha}_y) + \gamma M_{\dot\beta}\dot{\alpha}_x + \gamma M_\mu(-\dot{x}_h + u_G)\right]\sin\psi \qquad (19.127)$$

(equations 16.261 and 16.452). Assuming that the pitch control input is from the swashplate, we have written $\theta_1 = \theta_{1c}\cos\psi + \theta_{1s}\sin\psi$. For this equation, only the transfer functions $H_1(\omega)$ and $H_{-1}(\omega)$ are nonzero, meaning that the teetering response to sinusoidal inputs occurs only at the frequencies $\omega \pm \Omega$. Since the response to low-frequency inputs occurs at the frequencies $\pm\Omega$, the low-frequency flap motion can be written $\beta_1 = \beta_{1c}\cos\psi + \beta_{1s}\sin\psi$. The low-frequency response of the two-bladed rotor can thus be described by the steady-state motion of the tip-path plane. On substituting for β_1, the solution for β_{1c} and β_{1s} is identical to the low-frequency flap response obtained for rotors with three or more blades.

From equations 16.238, 16.239, 16.263, 16.264, 16.438, and 16.466, the in-plane hub forces for the rotor with two blades are

$$\begin{pmatrix} \dfrac{2C_H}{\sigma a} \\ \dfrac{2C_Y}{\sigma a} \end{pmatrix} = -\dfrac{2\widehat{M}_b}{\gamma}\begin{pmatrix} \ddot{x}_h \\ \ddot{y}_h \end{pmatrix} + R_\beta \begin{pmatrix} 2C \\ 2S \end{pmatrix}\beta_1 + \begin{bmatrix} R_{\dot\beta}2C + H_{\dot\beta}2S \\ R_{\dot\beta}2S - H_{\dot\beta}2C \end{bmatrix}\dot{\beta}_1$$

$$+ \begin{bmatrix} R_\theta 2C + H_\theta 2S \\ R_\theta 2S - H_\theta 2C \end{bmatrix}(\theta_1 - K_P\beta_1)$$

$$+ \begin{bmatrix} -H_\mu 2S^2 - R_\mu 2C^2 - R_r 2CS & H_\mu 2CS - R_\mu 2CS + R_r 2C^2 \\ -H_\mu 2CS - R_\mu 2CS - R_r 2S^2 & -H_\mu 2C^2 - R_\mu 2S^2 + R_r 2CS \end{bmatrix}\begin{pmatrix} \dot{x}_h - u_G \\ \dot{y}_h + v_G \end{pmatrix}$$

$$+ \begin{bmatrix} -R_{\dot\beta}2C^2 - H_{\dot\beta}2CS & R_{\dot\beta}2CS + H_{\dot\beta}2S^2 \\ -R_{\dot\beta}2CS + H_{\dot\beta}2C^2 & R_{\dot\beta}2S^2 - H_{\dot\beta}2CS \end{bmatrix}\begin{pmatrix} \dot{\alpha}_y \\ \dot{\alpha}_x \end{pmatrix} \qquad (19.128)$$

where $C = \cos\psi$ and $S = \sin\psi$. From equations 16.236 and 16.237, the hub moment is

$$\begin{pmatrix} -\dfrac{2C_{My}}{\sigma a} \\ \dfrac{2C_{Mx}}{\sigma a} \end{pmatrix} = \dfrac{\widehat{I}_\beta(v_\beta^2 - 1)}{\gamma}\begin{pmatrix} 2C \\ 2S \end{pmatrix}\beta_1 \qquad (19.129)$$

Substituting $\beta_1 = \beta_{1c} \cos\psi + \beta_{1s} \sin\psi$, the low-frequency response of the hub reactions is

$$\begin{pmatrix} \dfrac{2C_H}{\sigma a} \\ \dfrac{2C_Y}{\sigma a} \end{pmatrix} = -\dfrac{2C_T}{\sigma a}\begin{pmatrix} \beta_{1c} \\ \beta_{1s} \end{pmatrix}$$

$$+ \begin{bmatrix} -\widehat{H}_{\dot\beta}2S^2 - R_{\dot\beta}2CS & \widehat{H}_{\dot\beta}2CS + R_{\dot\beta}2C^2 \\ -K_P H_\theta 2CS - K_P R_\theta 2C^2 & -K_P H_\theta 2S^2 - K_P R_\theta 2CS \\ \widehat{H}_{\dot\beta}2CS - R_{\dot\beta}2S^2 & -\widehat{H}_{\dot\beta}2C^2 + R_{\dot\beta}2CS \\ +K_P H_\theta 2C^2 - K_P R_\theta 2CS & +K_P H_\theta 2CS - K_P R_\theta 2S^2 \end{bmatrix}\begin{pmatrix} \beta_{1c} \\ \beta_{1s} \end{pmatrix}$$

$$+ \begin{bmatrix} -H_\theta 2S^2 - R_\theta 2CS & H_\theta 2CS + R_\theta 2C^2 \\ H_\theta 2CS - R_\theta 2S^2 & -H_\theta 2C^2 + R_\theta 2CS \end{bmatrix}\begin{pmatrix} -\theta_{1s} \\ \theta_{1c} \end{pmatrix}$$

$$+ \begin{bmatrix} -H_\mu 2S^2 - R_\mu 2C^2 - R_r 2CS & H_\mu 2CS - R_\mu 2CS + R_r 2C^2 \\ -H_\mu 2CS - R_\mu 2CS - R_r 2S^2 & -H_\mu 2C^2 - R_\mu 2S^2 + R_r 2CS \end{bmatrix}\begin{pmatrix} \dot{x}_h - u_G \\ \dot{y}_h + v_G \end{pmatrix}$$

$$+ \begin{bmatrix} -H_{\dot\beta}2CS - R_{\dot\beta}2C^2 & -H_{\dot\beta}2S^2 - R_{\dot\beta}2CS \\ H_{\dot\beta}2C^2 - R_{\dot\beta}2CS & H_{\dot\beta}2CS - R_{\dot\beta}2S^2 \end{bmatrix}\begin{pmatrix} \dot\alpha_y \\ -\dot\alpha_x \end{pmatrix} \qquad (19.130)$$

and

$$\begin{pmatrix} -\dfrac{2C_{My}}{\sigma a} \\ \dfrac{2C_{Mx}}{\sigma a} \end{pmatrix} = \dfrac{\widehat{I}_\beta(\nu_\beta^2 - 1)}{\gamma}\begin{bmatrix} 2C^2 & 2CS \\ 2CS & 2S^2 \end{bmatrix}\begin{pmatrix} \beta_{1c} \\ \beta_{1s} \end{pmatrix} \qquad (19.131)$$

The average of these coefficients gives exactly the same expressions for the hub reactions as those obtained in section 19.4 for the low-frequency response of rotors with three or more blades. However, although this constant coefficient result is exact for the rotor with three or more blades in hover, because of the inertial and aerodynamic axisymmetry of the rotor, for the two-bladed rotor there are periodic variations of the coefficients in the hub reactions. The asymmetry of the rotor with two blades leads to large 2/rev variations of the coefficients even in hover. The thrust tilt term is obtained without periodic coefficients even with $N = 2$. Recall that the thrust dominates the hub in-plane forces except for the response to shaft angular velocity, where the $\widehat{H}_{\dot\beta}$ term due to tip-path-plane tilt is also important. Thus the primary influence of the periodic coefficients on the in-plane hub forces is found in the rotor pitch and roll damping. There is also a large 2/rev variation in the hub moment if $\nu_\beta > 1$, which is why two-bladed rotors are not often designed with a hub spring.

In summary, the two-bladed rotor is indeed a special case. The description of the flap dynamics is unique, involving the teetering degree of freedom β_1, which is fundamentally in the rotating frame, rather than the tip-path-plane tilt degrees of freedom. The frequency response of the two-bladed rotor motion is not given by the usual transfer function relation because the system is not time invariant. The low-frequency flap response does reduce to a tip-path-plane representation identical to the result for $N \geq 3$, but only for the steady-state limit ($\omega = 0$), which allows writing $\beta_1 = \beta_{1c} \cos\psi + \beta_{1s} \sin\psi$. The equations for the rotor low-frequency response that are used in the analysis of the helicopter flight dynamics are the same as for $N \geq 3$

if the averaged coefficients are used, but in fact the hub reactions involve large-amplitude periodic coefficients even in hover. The 2/rev variation of the coefficients due to the lack of axisymmetry with a two-bladed rotor is expected to influence primarily the rotor pitch and roll damping, and the hub moments if $v_\beta > 1$.

The special characteristics of the two-bladed rotor influence several aspects of the analysis of the aeroelastic behavior. In general periodic coefficient equations must be analyzed more often than for a rotor with three or more blades. A special procedure is also required to derive the rotor low-frequency response and implement the quasistatic approximation. For a rotor with three or more blades the low-frequency response can be obtained by dropping the flapping acceleration and velocity terms from the equations in the non-rotating frame. Such a procedure does not work with a two-bladed rotor because the equation of motion for β_1 is still in the rotating frame, so the β_1 response to low-frequency inputs from the non-rotating frame is not at low frequency also, but rather at 1/rev. Furthermore, a constant coefficient approximation cannot be used directly for the helicopter flight dynamics, since averaging the periodic coefficients of the two-bladed rotor equations of motion eliminates the coupling between the rotor and the shaft motion.

19.9 REFERENCES

Amer, K.B. "Theory of Helicopter Damping in Pitch or Roll and a Comparison with Flight Measurements." NACA TN 2136, October 1950.

Amer, K.B., and Gustafson, F.B. "Charts for Estimation of Longitudinal-Stability Derivatives for a Helicopter Rotor in Forward Flight." NACA TN 2309, March 1951.

Coleman, R.P. "Theory of Self-Excited Mechanical Oscillations of Hinged Rotor Blades." NACA ARR 3G29, July 1943.

Curtiss, H.C., Jr. "Complex Coordinates in Near Hovering Rotor Dynamics." Journal of Aircraft, *10*:5 (May 1973).

Curtiss, H.C., Jr., and Shupe, N.K. "A Stability and Control Theory for Hingeless Rotors." American Helicopter Society 27th Annual National V/STOL Forum, Washington, DC, May 1971.

Glauert, H., and Shone, G. "The Disturbed Motion of the Blades of a Gyroplane." A.R.C. 995, 1933.

Horvay, G. "Rotor Blade Flapping Motion." Quarterly of Applied Mathematics, *5*:2 (July 1947).

Johnson, W. "A Perturbation Solution of Helicopter Rotor Flapping Stability." Journal of Aircraft, *10*:5 (May 1973).

Johnson, W. "Perturbation Solutions for the Influence of Forward Flight on Helicopter Rotor Flapping Stability." NASA TM 62361, August 1974.

Leishman, J.G. "Etienne Oehmichen: Scientist, Engineer and Helicopter Pioneer." American Helicopter Society 62th Annual Forum, Phoenix, AZ, May 2006.

Miller, R.H. "Helicopter Control and Stability in Hovering Flight." Journal of the Aeronautical Sciences, *15*:8 (August 1948).

Peters, D.A., and Hohenemser, K.H. "Application of the Floquet Transition Matrix to Problems of Lifting Rotor Stability." Journal of the American Helicopter Society, *16*:2 (April 1971).

Sissingh, G.J. "The Effect of Induced Velocity Variation on Helicopter Rotor Damping in Pitch or Roll." ARC CP 101, November 1951.

20 Stability

The aeroelastic equations of motion for the rotor were derived in Chapter 16. The present chapter examines the solutions of these equations for a number of fundamental stability problems in rotor dynamics. To obtain analytical solutions, each problem must be restricted to a small number of degrees of freedom and to only the fundamental blade motion. Rotorcraft engineering currently has the capability to routinely calculate the dynamic behavior for much more detailed and complex models of the rotor and airframe. Thus elementary analyses are less necessary for actual numerical solutions, but are even more important as the basis for understanding the rotor dynamics.

20.1 Pitch-Flap Flutter

Traditionally, the term "flutter" refers to an aeroelastic instability involving the coupled bending and torsion motion of a wing. For the rotary wing, flutter refers to the pitch-flap motion of the blade. Often the term is generalized to include any aeroelastic instability of the rotor or aircraft, but the subject of this section is the blade pitch-flap stability. The classical problem considers two degrees of freedom: the rigid flap and rigid pitch motion of an articulated rotor blade. Since the control system is usually the softest element in the torsion motion, the rigid pitch degree of freedom is a good representation of the blade dynamics. A general fundamental flap mode with natural frequency v_β is considered. A thorough analysis of the flutter of a hingeless rotor blade usually requires that the in-plane motion be modeled as well. The rotation of the wing introduces a number of effects that make rotor blade flutter much different from the fixed-wing phenomenon. The centrifugal forces couple the flap and pitch motion if the center-of-gravity is offset from the feathering axis. Moreover, the returning shed wake has an important influence on the blade aerodynamic forces, as does the periodic aerodynamic environment of the blade in forward flight.

20.1.1 Pitch-Flap Equations

The differential equations for the rigid flap and rigid pitch motion of a rotor blade were derived in section 16.5.1:

$$\widehat{I}_\beta(\ddot{\beta} + v_\beta^2 \beta) - \widehat{I}_x(\ddot{\theta} + \theta) = \gamma \int_0^1 r \frac{F_z}{ac} dr = \gamma M_F \qquad (20.1)$$

$$\widehat{I}_f\big(\ddot{\theta} + (\omega_\theta^2 + 1)\theta\big) - \widehat{I}_x(\ddot{\beta} + \beta) + K_P\widehat{I}_f\omega_\theta^2\beta = \gamma \int_0^1 \frac{M_a}{ac} dr = \gamma M_f \qquad (20.2)$$

(equations 16.137 and 16.138). Here β is the degree of freedom of the perturbation flap motion, with rotating natural frequency ν_β, and θ is the pitch degree of freedom, with non-rotating natural frequency ω_θ. The inertial coefficients are $\widehat{I}_\beta = \int_0^1 r^2 m\,dr/I_b$, $I_f = \int_0^1 I_\theta dr/I_b$, and $\widehat{I}_x = \int_0^1 x_I rm\,dr/I_b$, with x_I the distance that the blade center-of-gravity is behind the feathering axis. The aerodynamic flap and pitch moments,

$$M_F = M_\theta \theta + M_{\dot\theta} \dot\theta + M_{\dot\beta} \dot\beta + M_\beta \beta \qquad (20.3)$$

$$M_f = m_\theta \theta + m_{\dot\theta} \dot\theta + m_{\dot\beta} \dot\beta + m_\beta \beta \qquad (20.4)$$

are derived in section 16.8.10. For hover, the aerodynamic coefficients are

$$M_\theta = \frac{1}{8} C'(k_e) \qquad (20.5)$$

$$M_{\dot\theta} = \frac{c}{24}\big(1 + 2C'(k_e)(1 + 2\xi_A)\big) \qquad (20.6)$$

$$M_{\dot\beta} = -\frac{1}{8} C'(k_e) + \frac{c^2}{64}(1 + 4\xi_A) \qquad (20.7)$$

$$M_\beta = \frac{c}{12} C'(k_e) \qquad (20.8)$$

$$m_\theta = -\frac{x_A}{6} C'(k_e) \qquad (20.9)$$

$$m_{\dot\theta} = -\frac{c^2}{32}\big(1 + 4\xi_A(1 + 2C'(k_e)(1 + 2\xi_A))\big) \qquad (20.10)$$

$$m_{\dot\beta} = \frac{x_A}{6} C'(k_e) - \frac{c^3}{32}\left(\frac{3}{8} + 2\xi_A(1 + 2\xi_A)\right) \qquad (20.11)$$

$$m_\beta = -\frac{c^2}{64}\big(1 + 8\xi_A C'(k_e)\big) \qquad (20.12)$$

where c is the chord and x_A is the distance the aerodynamic center is behind the feathering axis ($\xi_A = x_A/c$). The lift deficiency function C' is evaluated at an effective radius (typically 75% R), so that $k_e = \omega c/2r_e$. Section 16.8.10 gives the aerodynamic coefficients in forward flight as well.

The coupled differential equations for the flap and pitch motion are thus

$$\begin{bmatrix} \widehat{I}_\beta & -\widehat{I}_x \\ -\widehat{I}_x & \widehat{I}_f \end{bmatrix} \begin{pmatrix} \ddot\beta \\ \ddot\theta \end{pmatrix} + \begin{bmatrix} -\gamma M_{\dot\beta} & -\gamma M_{\dot\theta} \\ -\gamma m_{\dot\beta} & -\gamma m_{\dot\theta} \end{bmatrix} \begin{pmatrix} \dot\beta \\ \dot\theta \end{pmatrix}$$
$$+ \begin{bmatrix} \widehat{I}_\beta \nu_\beta^2 - \gamma M_\beta & -\widehat{I}_x - \gamma M_\theta \\ -\widehat{I}_x + K_P \widehat{I}_f \omega_\theta^2 - \gamma m_\beta & \widehat{I}_f(\omega_\theta^2 + 1) - \gamma m_\theta \end{bmatrix} \begin{pmatrix} \beta \\ \theta \end{pmatrix} = 0 \qquad (20.13)$$

No forcing terms have been included since only the stability of the motion is of interest here. For hover, for which aerodynamic coefficients are constant, and in terms of the Laplace variable s, the equations of motion are

$$\begin{bmatrix} \widehat{I}_\beta s^2 - \gamma M_{\dot\beta} s + \widehat{I}_\beta \nu_\beta^2 - \gamma M_\beta & -\widehat{I}_x s^2 - \gamma M_{\dot\theta} s - \widehat{I}_x - \gamma M_\theta \\ -\widehat{I}_x s^2 - \gamma m_{\dot\beta} s - \widehat{I}_x + K_P \widehat{I}_f \omega_\theta^2 - \gamma m_\beta & \widehat{I}_f s^2 - \gamma m_{\dot\theta} s + \widehat{I}_f(\omega_\theta^2 + 1) - \gamma m_\theta \end{bmatrix} \begin{pmatrix} \beta \\ \theta \end{pmatrix}$$
$$= 0 \qquad (20.14)$$

The eigenvalues are the roots of the characteristic equation:

$$\left(\widehat{I}_\beta s^2 - \gamma M_{\dot\beta} s + \widehat{I}_\beta v_\beta^2 - \gamma M_\beta\right)\left(\widehat{I}_f s^2 - \gamma m_{\dot\theta} s + \widehat{I}_f(\omega_\theta^2 + 1) - \gamma m_\theta\right)$$
$$- \left(\widehat{I}_x s^2 + \gamma M_{\dot\theta} s + \widehat{I}_x + \gamma M_\theta\right)\left(\widehat{I}_x s^2 + \gamma m_{\dot\beta} s + \widehat{I}_x - K_P \widehat{I}_f \omega_\theta^2 + \gamma m_\beta\right) = 0 \quad (20.15)$$

Although the four roots of this equation must in general be found numerically, the stability boundary can be determined analytically. A plane of the system parameters has regions in which all the roots have negative real parts, so that the motion is stable, and regions in which one or more roots have positive real parts, so that the motion is unstable. On the stability boundary one root must be on the imaginary axis of the s-plane, crossing from the left half-plane into the right half-plane. There are two ways a root can cross the imaginary axis into the right half-plane, producing an unstable system: as a real root along the real axis and as a complex conjugate pair at finite frequency. The instability associated with a real root going through the origin into the right half-plane is called divergence. It is a static instability, since with zero frequency no velocity or acceleration forces are involved. The instability associated with a complex conjugate pair of roots crossing the imaginary axis is called flutter. This instability involves an oscillatory motion of the system.

The most significant parameters for the rotor blade flutter stability are the pitch natural frequency ω_θ, determined by the control system stiffness, and the offsets of the center-of-gravity and aerodynamic center from the feathering axis. The separation of the center-of-gravity and aerodynamic center $(x_I - x_A)$ is more important than their distance from the feathering axis, but x_A must usually be kept small to avoid large oscillatory control loads in forward flight. Thus the principal parameters controlling the blade flutter stability are the control stiffness (ω_θ) and the chordwise mass balance (x_I).

20.1.2 Divergence Instability

A divergence instability occurs when a real root goes through the origin of the s-plane into the right half-plane. The divergence stability boundary is defined by the requirement that one root be $s = 0$, for which the characteristic equation becomes

$$\left(\widehat{I}_\beta v_\beta^2 - \gamma M_\beta\right)\left(\widehat{I}_f(\omega_\theta^2 + 1) - \gamma m_\theta\right) - \left(\widehat{I}_x + \gamma M_\theta\right)\left(\widehat{I}_x - K_P \widehat{I}_f \omega_\theta^2 + \gamma m_\beta\right) = 0 \quad (20.16)$$

which is a balance of the spring terms alone. Since increasing the flap or pitch springs should produce static stability, the criterion for a stable system is that this quantity be positive. Neglecting $\widehat{I}_x$ relative to γM_θ, and γM_β relative to $I_\beta v_\beta^2$ in the flap equation, the stability criterion can be written:

$$\left(\widehat{I}_\beta v_\beta^2 + K_P \gamma M_\theta\right)\left(\widehat{I}_f(\omega_\theta^2 + 1) - \gamma m_\theta\right) > \gamma M_\theta\left(\widehat{I}_x + \gamma m_\beta + K_P\left(\widehat{I}_f - \gamma m_\theta\right)\right) \quad (20.17)$$

The left-hand side is the product of the net flap and pitch springs, whereas the right-hand side is the product of the moments coupling the flap and pitch motion, primarily the aerodynamic flap moment due to pitch (M_θ), and the centrifugal pitch moment due to flapping (I_x). The flap and pitch springs are certainly positive. Negative pitch-flap coupling ($K_P < 0$) or a forward aerodynamic center ($x_A < 0$) contributes a negative spring, but these terms are unlikely to be larger than even the centrifugal springs alone. Divergence stability thus requires that the quantity $\left(\widehat{I}_x + \gamma m_\beta + K_P\left(\widehat{I}_f - \gamma m_\theta\right)\right)$ be small or negative; hence a forward center-of-gravity position is desired. Divergence stability can also be ensured by a large enough pitch spring (ω_θ).

Emphasizing the stability as a function of the parameters ω_θ (control system stiffness) and I_x (chordwise mass balance), the divergence stability criterion is

$$\omega_\theta^2 > \frac{1}{\widehat{I}_\beta v_{\beta e}^2}\left[\gamma M_\theta\left(\frac{\widehat{I}_x}{\widehat{I}_f}+\frac{\gamma m_\beta}{\widehat{I}_f}\right)-\widehat{I}_\beta v_\beta^2\left(1-\frac{\gamma m_\theta}{\widehat{I}_f}\right)\right] \quad (20.18)$$

which is a straight line on the plane of ω_θ^2 vs. I_x/I_f. In terms of the mass balance required, the criterion is

$$\frac{I_x}{I_f} < \frac{\widehat{I}_\beta v_\beta^2(\omega_\theta^2+1)}{\gamma M_\theta}-\frac{\gamma m_\beta}{\widehat{I}_f}+K_P\omega_\theta^2-\widehat{I}_\beta v_\beta^2\frac{m_\theta}{M_\theta \widehat{I}_f} \quad (20.19)$$

Using $\widehat{I}_x \cong \frac{3}{2}x_I$ and substituting for the aerodynamic coefficients (section 20.1.1), this becomes

$$x_I-\frac{8v_\beta^2}{9}x_A < \frac{16}{3\gamma}v_\beta^2(\omega_\theta^2+1)\widehat{I}_f+\frac{\gamma c^2}{96}(1+8\xi_A C')+\frac{2}{3}K_P\omega_\theta^2\widehat{I}_f \quad (20.20)$$

which shows that divergence depends on the distance the center-of-gravity is aft of the aerodynamic center $(x_I - x_A)$, since $8v_\beta^2/9 \cong 1$. The boundary is relatively insensitive to the pitch axis location for a fixed $x_I - x_A$. Since the right-hand side of this criterion is almost always positive, divergence stability is assured regardless of the pitch spring if the blade is mass balanced in such a way that the center-of-gravity is ahead of the aerodynamic center $(x_I - x_A < 0)$.

20.1.3 Flutter Instability

A flutter instability occurs when a pair of complex conjugate roots crosses the imaginary axis into the right half-plane. The flutter stability boundary is thus defined by the requirement that one root be on the imaginary axis, $s = i\omega$, where ω is a real and positive frequency. On substituting $s = i\omega$, the real and imaginary parts of the characteristic equation are

$$\begin{aligned}&\left(-\widehat{I}_\beta\omega^2+\widehat{I}_\beta v_\beta^2-\gamma M_\beta\right)\left(-\widehat{I}_f\omega^2+\widehat{I}_f(\omega_\theta^2+1)-\gamma m_\theta\right)-\omega^2\gamma M_{\dot\beta}\gamma m_{\dot\theta}\\ &\quad-\left(-\widehat{I}_x\omega^2+\widehat{I}_x+\gamma M_\theta\right)\left(-\widehat{I}_x\omega^2+\widehat{I}_x-K_P\widehat{I}_f\omega_\theta^2+\gamma m_\beta\right)+\omega^2\gamma M_{\dot\theta}\gamma m_{\dot\beta}\\ &=(\omega^2-1)^2\left(\widehat{I}_\beta\widehat{I}_f-\widehat{I}_x^2\right)\\ &\quad+(\omega^2-1)\left[-\left(\widehat{I}_\beta(v_\beta^2-1)-\gamma M_\beta\right)\widehat{I}_f-\widehat{I}_\beta\left(\widehat{I}_f\omega_\theta^2-\gamma m_\theta\right)\right.\\ &\quad\left.+\widehat{I}_x\left(-K_P\widehat{I}_f\omega_\theta^2+\gamma m_\beta\right)+\gamma M_\theta\widehat{I}_x-\gamma M_{\dot\beta}\gamma m_{\dot\theta}+\gamma M_{\dot\theta}\gamma m_{\dot\beta}\right]\\ &\quad+\left(\widehat{I}_\beta(v_\beta^2-1)-\gamma M_\beta\right)\left(\widehat{I}_f\omega_\theta^2-\gamma m_\theta\right)-\gamma M_\theta\left(-K_P\widehat{I}_f\omega_\theta^2+\gamma m_\beta\right)\\ &\quad-\gamma M_{\dot\beta}\gamma m_{\dot\theta}+\gamma M_{\dot\theta}\gamma m_{\dot\beta}=0 \quad (20.21)\end{aligned}$$

and

$$\begin{aligned}&M_{\dot\beta}\left(-\widehat{I}_f\omega^2+\widehat{I}_f(\omega_\theta^2+1)-\gamma m_\theta\right)+m_{\dot\theta}\left(-\widehat{I}_\beta\omega^2+\widehat{I}_\beta v_\beta^2-\gamma M_\beta\right)\\ &\quad+M_{\dot\theta}\left(-\widehat{I}_x\omega^2+\widehat{I}_x-K_P\widehat{I}_f\omega_\theta^2+\gamma m_\beta\right)+m_{\dot\beta}\left(-\widehat{I}_x\omega^2+\widehat{I}_x+\gamma M_\theta\right)\\ &=-(\omega^2-1)\left(M_{\dot\beta}\widehat{I}_f+m_{\dot\theta}\widehat{I}_\beta+M_{\dot\theta}\widehat{I}_x+m_{\dot\beta}\widehat{I}_x\right)+M_{\dot\beta}\left(\widehat{I}_f\omega_\theta^2-\gamma m_\theta\right)\\ &\quad+m_{\dot\theta}\left(\widehat{I}_\beta(v_\beta^2-1)-\gamma M_\beta\right)+M_{\dot\theta}\left(-K_P\widehat{I}_f\omega_\theta^2+\gamma m_\beta\right)+m_{\dot\beta}\gamma M_\theta=0 \quad (20.22)\end{aligned}$$

Eliminating $(\omega^2 - 1)$ from these two equations gives a single relation defining the flutter boundary in terms of the blade parameters. In a numerical solution, for a given value of ω_θ these two equations can be solved for ω and $\widehat{I}_x$.

To proceed analytically, the order of magnitude of the terms must be considered: c/R, x_A/c, and x_I/c are small, and $\widehat{I}_f$ is order $(c/R)^2$. Then to lowest order in c/R, the characteristic equations are

$$(\omega^2 - 1)^2 \widehat{I}_\beta \widehat{I}_f + (\omega^2 - 1)\left(-\widehat{I}_\beta(v_\beta^2 - 1)\widehat{I}_f - \widehat{I}_\beta \left(\widehat{I}_f \omega_\theta^2 - \gamma m_\theta\right) + \gamma M_\theta \widehat{I}_x - \gamma M_{\dot\beta}\gamma m_{\dot\theta}\right)$$
$$+ \widehat{I}_\beta(v_\beta^2 - 1)\left(\widehat{I}_f \omega_\theta^2 - \gamma m_\theta\right) - \gamma M_\theta \left(-K_P \widehat{I}_f \omega_\theta^2 + \gamma m_\beta\right) - \gamma M_{\dot\beta}\gamma m_{\dot\theta} = 0 \quad (20.23)$$

and

$$-(\omega^2 - 1)\left(M_{\dot\beta}\widehat{I}_f + m_{\dot\theta}\widehat{I}_\beta\right) + M_{\dot\beta}\widehat{I}_f \omega_\theta^2 + m_{\dot\theta}\widehat{I}_\beta(v_\beta^2 - 1) + \gamma \left(M_\theta m_{\dot\beta} - M_{\dot\beta}m_\theta\right) = 0 \quad (20.24)$$

The imaginary part is solved for the flutter frequency:

$$\omega^2 - 1 = \frac{M_{\dot\beta}\widehat{I}_f \omega_\theta^2 + m_{\dot\theta}\widehat{I}_\beta(v_\beta^2 - 1) + \gamma(M_\theta m_{\dot\beta} - M_{\dot\beta}m_\theta)}{M_{\dot\beta}\widehat{I}_f + m_{\dot\theta}\widehat{I}_\beta}$$

$$= \frac{1}{a}\left((a - 1)\omega_\theta^2 + (v_\beta^2 - 1)\right) + b \quad (20.25)$$

or

$$\omega^2 = \left(1 - \frac{1}{a}\right)(\omega_\theta^2 + 1) + \frac{1}{a}v_\beta^2 + b \quad (20.26)$$

where $a = 1 + (M_{\dot\beta}\widehat{I}_f)/(m_{\dot\theta}\widehat{I}_\beta)$. Since $a \cong 1.4$, the flutter frequency ω is usually significantly lower than the pitch natural frequency ω_θ. The term

$$b = \frac{\gamma(M_\theta m_{\dot\beta} - M_{\dot\beta}m_\theta)}{M_{\dot\beta}\widehat{I}_f + m_{\dot\theta}\widehat{I}_\beta} \quad (20.27)$$

is order c/R small, but is needed for small ω_θ. Substituting for the flutter frequency in the real part gives the criterion for flutter stability:

$$\omega_\theta^4 + A\omega_\theta^2 \frac{I_x}{I_f} + B\omega_\theta^2 + C\frac{I_x}{I_f} + D > 0 \quad (20.28)$$

where the coefficients are

$$A = -\frac{\gamma M_\theta}{\widehat{I}_\beta} a \quad (20.29)$$

$$B = \left(-\frac{\gamma m_\theta}{\widehat{I}_f} + \frac{\gamma M_{\dot\beta}\gamma m_{\dot\theta}}{\widehat{I}_\beta \widehat{I}_f}\right)a - 2(v_\beta^2 - 1)^2 - K_P \frac{M_\theta \gamma m_{\dot\theta}}{M_{\dot\beta}\widehat{I}_f}a^2 - ba\frac{a-2}{a-1} \quad (20.30)$$

$$C = -\frac{M_\theta \gamma m_{\dot\theta}}{M_{\dot\beta}\widehat{I}_f}\left((v_\beta^2 - 1)a + ba^2\right) \quad (20.31)$$

$$D = \frac{\gamma m_{\dot\theta}}{\widehat{I}_f}\left(\frac{\gamma m_{\dot\theta}}{\widehat{I}_f} + \frac{M_\theta \gamma m_\beta}{M_{\dot\beta}\widehat{I}_f}\right)a^2 + (v_\beta^2 - 1)^2 + \left(\frac{\gamma m_\theta}{\widehat{I}_f} + \left(\frac{\gamma m_{\dot\theta}}{\widehat{I}_f}\right)^2\right)(v_\beta^2 - 1)a$$

$$+ \frac{a^2}{a-1}b\left(b - \frac{\gamma m_\theta}{\widehat{I}_f} + \frac{\gamma M_{\dot\beta}\gamma m_{\dot\theta}}{\widehat{I}_\beta \widehat{I}_f}\right) + (v_\beta^2 - 1)b\frac{a-2}{a-1} \quad (20.32)$$

20.1 Pitch-Flap Flutter

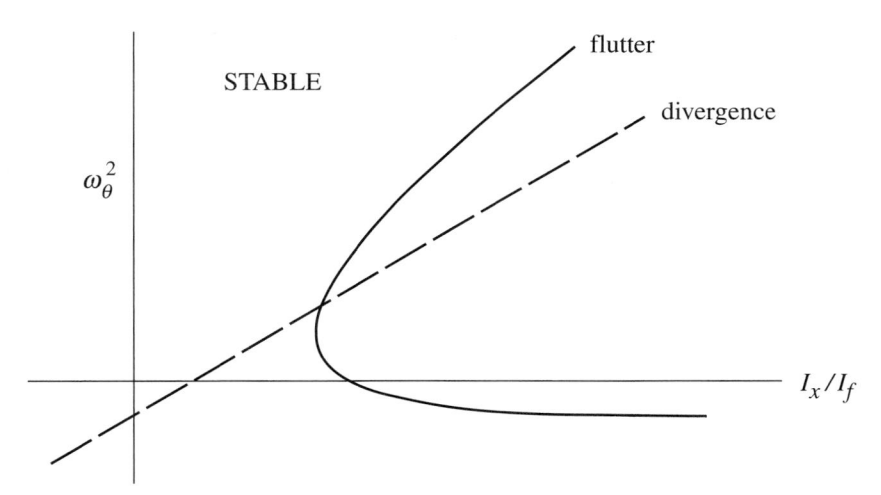

Figure 20.1. Sketch of flutter and divergence stability boundaries.

On the plane of ω_θ^2 vs. I_x/I_f, the flutter stability boundary is a quadratic function, as sketched in Figure 20.1. In terms of the required chordwise mass balance, the criterion for flutter stability is

$$\frac{I_x}{I_f} < -\frac{\omega_\theta^4 + B\omega_\theta^2 + D}{A\omega_\theta^2 + C} \qquad (20.33)$$

The asymptotes of the flutter boundary have slopes of zero and

$$\frac{\partial \omega_\theta^2}{\partial I_x/I_f} = -A \qquad (20.34)$$

The ratio of the slope of the upper asymptote of the flutter boundary to the slope of the divergence boundary (equation 20.19) is $v_{\beta e}^2 a$, which is always greater than 1. Hence for large enough I_x/I_f, flutter is always the critical instability. The lower asymptote of the flutter boundary is a line of zero slope at $\omega_\theta = -C/A$, which is usually negative. The solution at $\omega_\theta = 0$ is $I_x/I_f = -D/C$. The minimum I_x of the flutter boundary occurs at

$$\frac{I_x}{I_f} = -\frac{1}{A}\left(B - 2C/A + 2\sqrt{D + (C/A)^2 - BC/A}\right) \qquad (20.35)$$

which is approximately

$$\widehat{I}_x + \frac{\widehat{I}_\beta m_\theta}{M_\theta} = \frac{2\widehat{I}_\beta}{M_\theta}\sqrt{m_\theta^2 + \frac{M_\theta}{M_{\dot\beta}}m_{\dot\theta}m_\beta + \frac{\gamma M_{\dot\beta}m_{\dot\theta}}{M_\theta}} - K_P \frac{\widehat{I}_\beta m_{\dot\theta}}{M_{\dot\beta}} a \qquad (20.36)$$

Using $\widehat{I}_x \cong \frac{3}{2}x_I$ and substituting for the aerodynamic coefficients (section 20.1.1), this becomes

$$x_I - \frac{8}{9}x_A = c^2\left(\frac{1}{3\sqrt{2}} + \frac{\gamma}{48} - \frac{1}{4}K_P\right) \qquad (20.37)$$

Consequently, if the blade is mass balanced in such a way that the center-of-gravity is no farther aft than this distance, flutter stability is assured regardless of the pitch stiffness ω_θ. As for divergence, the distance between the center-of-gravity and aerodynamic center is the primary parameter in the pitch-flap dynamics.

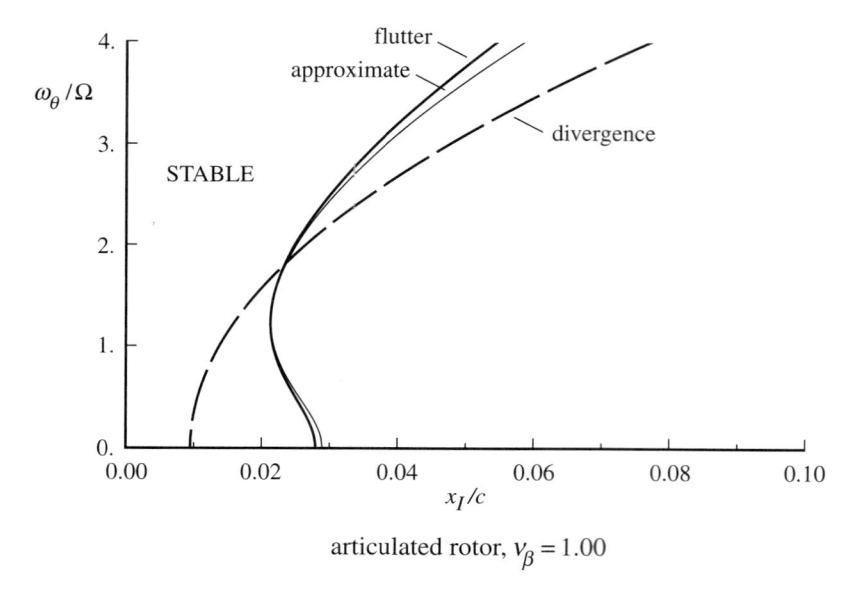

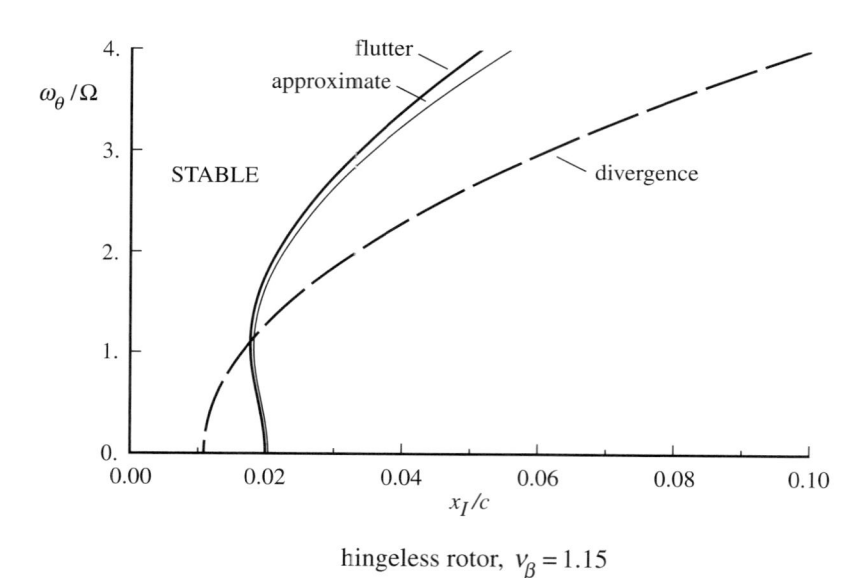

Figure 20.2. Flutter and divergence boundaries ($x_A = 0$).

Figure 20.2 is an example of the flutter and divergence boundaries for articulated ($\nu_\beta = 1$, $K_P = 0$) and hingeless ($\nu_\beta = 1.15$) rotors with uniform properties and $\gamma = 8$, $c/R = 0.0628$ ($\sigma = 0.08$ and $N = 4$) and $\widehat{I}_f = 0.0004$. The pitch axis is at the aerodynamic center, $x_A = 0$. The shed wake effects are neglected, so $C'(k_e) = 1$. Shown are the solutions of the full equations for the flutter boundary (equations 20.21 and 20.22), the approximate solution (equation 20.28), and the divergence boundary (equation 20.19). There is more influence of the flap frequency when x_A/c is non-zero.

Both flutter and divergence stability are increased by increasing the control system stiffness ω_θ or decreasing I_x by moving the blade center-of-gravity toward the leading edge. A conservative approach is to place the center-of-gravity at the aerodynamic center or just aft of it. Considering loads as well as stability, rotor blades are generally mass balanced about the quarter chord. Most blade designs require a leading edge weight to accomplish this balance. Although the control system stiffness is an important flutter parameter, calculating it is difficult because of the complicated geometry of the control system and the influence of actuator stiffness. Mechanical or frictional damping in the control system and pitch bearing can also influence flutter stability. Usually such damping is nonlinear, requiring numerical integration of the equations of motion to determine the stability. Alternatively, a nonlinear damper can be represented in a linear stability analysis by using an equivalent viscous damping, with a value such that the same amount of energy is dissipated during a cycle of motion. Brooks and Baker (1958) observed a stabilizing effect of compressibility that was attributed to the rearward shift of the aerodynamic center above the critical Mach number.

This analysis has used dimensionless parameters, so the system stiffness is represented by $(\omega_\theta/\Omega)^2$. In terms of dimensional quantities, a given rotor has a fixed value of ω_θ. The minimum allowable ω_θ/Ω at the flutter boundary then becomes a restriction on the maximum rotor speed Ω. For design of the rotor, the dimensionless parameter ω_θ/Ω is useful, but when the rotor has been built flutter places a limit on Ω. Flutter testing of rotors is conducted by increasing the rotor speed until the flutter or divergence boundary is encountered as a result of decreasing ω_θ/Ω. The best indication of flutter when testing rotors is in the control loads, which are a measure of the pitch motion.

20.1.4 Shed Wake Influence

For certain operating conditions the wake can have a significant impact on the rotor blade flutter stability. The effect of the rotor returning shed wake on the unsteady aerodynamic loads can be accounted for by using the lift deficiency function $C'(k)$, either within the spanwise integrals or at an effective radius ($k_e = \omega c/2r_e$). In chapter 10, Theodorsen's, Loewy's, and a number of approximate lift deficiency functions are derived. A lift deficiency function follows from the assumption of purely harmonic motion, which is appropriate on the stability boundary. Since C' is a complex number depending on the flutter frequency ω, the full equations for the flutter boundary (equations 20.21 and 20.22) must be solved in complex form, typically for ω and I_x given ω_θ^2. An iterative solution is required, so C' can be evaluated from ω of the last iteration. Alternatively, ω and ω_θ^2 can be found given I_x; there are either two solutions or none (Figure 20.1).

When the returning shed wake influence is included, there can be several unstable regions, instead of a single region as with quasistatic aerodynamics. Daughaday, DuWaldt, and Gates (1957) investigated pitch-flap flutter for a two-bladed rotor, varying rotor speed and blade chordwise center-of-gravity position. A stabilizing effect of the returning wake was observed for pitch mode frequency near 2/rev and was predicted using Loewy's theory. Thus if Loewy's function is used, there can be more than two solutions for ω and ω_θ^2 at a given I_x. Flutter calculations based on quasistatic aerodynamics ($C' = 1$) are usually conservative; see Miller and Ellis (1956) and Jones (1958). The quasistatic flutter boundary tends to form an envelope around the boundaries calculated, including the wake influence through the lift deficiency function. The effect of the wake is to divide the flutter instability region into several regions because of an increased stability in narrow ranges about certain critical values of ω_θ corresponding to harmonic excitation. Such a modification of the flutter boundary is of little practical significance, although the corresponding influence on the flutter frequency can be significant.

In certain operating conditions the returning wake of the rotor can also produce a single-degree-of-freedom instability, called wake-excited flutter. Wake-excited flutter has been observed on rotors operating in hover at low collective (Brooks and Baker (1958)). The returning shed wake reduces the circulatory loads that are responsible for flap damping (see section 10.7), which can lead to a pitch-flap flutter instability. The circulatory lift does not influence the pitch moment for oscillation about the quarter chord, but the blade pitch damping moments are affected by the returning shed wake if the pitch axis is off the aerodynamic center. For oscillation near certain harmonics of the rotor speed, about the leading edge or about the midchord, the damping in pitch can be negative. Thus a single-degree-of-freedom instability is possible. Wake-excited flutter usually occurs under the operating conditions for which the returning wake has the strongest influence: low collective pitch as in run-up on the ground or autorotation, low forward speed or hover, and a pitch natural frequency near a harmonic of the rotor speed.

20.1.5 Forward Flight

The aerodynamics of the rotor in forward flight introduce periodic coefficients in the equations of motion for flap and pitch of the blade. The eigenvalues of these periodic-coefficient linear differential equations can be obtained by the methods discussed in section 15.6. At high advance ratio ($\mu > 0.5$) such an analysis is essential to properly evaluate the stability, including the influence of the periodic coefficients; reverse flow must also be included in the aerodynamic model at such high speeds. At low to moderate advance ratio, a constant coefficient approximation can be sufficiently accurate. If the mean values of the coefficients in the rotating frame are used, the only effect of forward flight retained is an order μ^2 increase in M_θ and m_θ. As for the flapping dynamics, the averaging of the coefficients should therefore be performed in the non-rotating frame. The constant coefficient approximation in forward flight best models the lower frequency behavior of rotors with a large number of blades. Since the pitch natural frequency tends to be relatively high, it can be expected that the exact solution of the periodic coefficient equations is required for pitch-flap flutter more often than when only the flapping motion is involved.

Alternatively, the dynamic stability can be assessed from the results of a direct numerical integration of the equations of motion. Such an approach is also necessary if nonlinear effects are to be included in the analysis, such as those due to blade stall or compressibility. The evaluation of the stability of periodic systems from the transient motions is not an elementary matter.

20.1.6 Coupled Blades

The flutter analysis developed here has so far considered a single independent blade. Even in the shaft-fixed case, all the blades are coupled through the rotor control system. The load-path through the control system, which determines the stiffness of the restraint at the pitch bearing, depends on the pitch motion of all the blades. Thus the rigid pitch natural frequency ω_θ, which is a primary parameter of the flutter stability, cannot in general be defined for the motion of an individual blade alone. The flutter analysis must consider the entire rotor.

In the non-rotating frame the coupling of the blades through the control system appears as a different stiffness or natural frequency for each of the non-rotating equations of motion for θ_0, θ_{nc}, θ_{ns}, and $\theta_{N/2}$ (see section 16.5.4). With three or more blades the rotating control system is axisymmetric, so the coupling is due to the non-rotating control system. The primary load-paths are through the collective and cyclic pitch control systems (θ_0, θ_{1c}, and θ_{1s} equations). The higher modes involve flexibility of the pitch horn and pitch link, and bending of the swashplate (θ_{2c}, θ_{2s}, ..., $\theta_{N/2}$ equations as required). In hover, the only coupling of the non-rotating equations is between θ_{nc} and θ_{ns} (and also the corresponding β_{nc} and β_{ns} equations). The equation for the collective mode is identical to that of a single blade; hence the flutter solution can be obtained by considering an independent blade with the appropriate collective natural frequency ω_{θ_0}. Similarly, the equation of motion for $\theta_{N/2}$ is not coupled with the other pitch degrees of freedom and is the same as an independent blade with pitch frequency $\omega_{\theta_{N/2}}$. If $\omega_{\theta_{nc}} = \omega_{\theta_{ns}}$, then the coupled equations for θ_{nc} and θ_{ns} are also equivalent to a single independent blade. For the reactionless modes this is the case because of the axisymmetry of the rotating control system. The effective natural frequencies of the lateral and longitudinal cyclic control are not likely to be equal ($\omega_{\theta_{1c}} \neq \omega_{\theta_{1s}}$). If the difference in cyclic control system stiffnesses is significant, the flutter analysis must consider a four-degree-of-freedom problem for θ_{1c}, θ_{1s}, β_{1c}, and β_{1s}. The coupled motion of the θ_{1c} and θ_{1s} degrees of freedom is a progressive mode at a frequency near $\omega_\theta + 1/\text{rev}$ and a regressive mode at a frequency near $\omega_\theta - 1/\text{rev}$. The natural frequencies of the reactionless modes (θ_{2c}, θ_{2s}, ..., $\theta_{N/2}$) are generally higher than those of the collective and cyclic modes. The stiffness of the cyclic control system is usually less than that of the collective control system, in which case the critical flutter problem involves the cyclic pitch degrees of freedom (θ_{1c} and θ_{1s}).

It can also be necessary to consider aerodynamic effects that differ for each non-rotating mode of the rotor. For example, the cyclic modes involve identical motion of each blade at a given azimuth and thus are more susceptible to excitation by a disturbance at a particular point in the fixed frame, such as aerodynamic interference due to the fuselage or tail rotor. When Loewy's lift deficiency function is used to account for the returning shed wake effects, there is a separate function C' for each non-rotating mode (see section 10.7), although an equivalent single blade model is again often possible.

20.2 Flap-Lag Dynamics

Let us examine next the stability of the coupled flap and lag motion of a rotor blade. The flap or lag motion alone has positive aerodynamic damping, although it is low for the lag mode. The blade in-plane and out-of-plane motions are coupled by Coriolis and aerodynamic forces, which can produce an instability. The problem considered here has only two degrees of freedom: the fundamental flap and lag modes.

20.2.1 Flap-Lag Equations

The equations of motion for the first out-of-plane and first in-plane modes of the rotor blade were derived in section 16.4.1 (equations 16.94 and 16.97):

$$\widehat{I}_\beta(\ddot{\beta} + v_\beta^2 \beta) - \widehat{I}_{\beta\zeta} 2\beta_0 \dot{\zeta} - \chi(\widehat{I}_\zeta \widehat{K}_\zeta - \widehat{I}_\beta \widehat{K}_\beta)\theta_s \zeta = \gamma M_F \quad (20.38)$$

$$\widehat{I}_\zeta(\ddot{\zeta} + v_\zeta^2 \zeta) + \widehat{C}_\zeta \dot{\zeta} + \widehat{I}_{\beta\zeta} 2\beta_0 \dot{\beta} - \chi(\widehat{I}_\zeta \widehat{K}_\zeta - \widehat{I}_\beta \widehat{K}_\beta)\theta_s \beta = \gamma M_L \quad (20.39)$$

Here β is the flap degree of freedom with rotating natural frequency v_β, and ζ is the lag degree of freedom with natural frequency v_ζ. The degrees of freedom β and ζ are purely out-of-plane and purely in-plane, respectively. Structural coupling of the flap and lag motion is accounted for by the off-diagonal stiffness terms, from equation 16.108. The parameter χ is a measure of the distribution of the hinge stiffness inboard and outboard of the pitch bearing. For $\chi = 0$ all the flexibility is inboard (uncoupled), with axes parallel and normal to the hub plane. For $\chi = 1$ all the flexibility is outboard of the pitch bearing (fully coupled), with axes at the pitch angle θ_s. With no hinge offset, the flap and lag spring constants are related to the blade frequencies: $\widehat{K}_\beta = K_\beta/I_\beta \Omega^2 = v_\beta^2 - 1$ and $\widehat{K}_\zeta = K_\zeta/I_\zeta \Omega^2 = v_\zeta^2$. The inertial constants $\widehat{I}_\beta = \widehat{I}_\zeta = \widehat{I}_{\beta\zeta} = 1$, assuming that the mode shapes are $\eta_\beta = \eta_\zeta = r$. Mechanical or structural damping of the lag motion has been included as a viscous damping coefficient $\widehat{C}_\zeta = C_\zeta/I_b \Omega$ (dimensional C_ζ). The flap and lag motion of the blade are coupled inertially by Coriolis forces, which have been linearized about the trim coning angle β_0. Lag velocity $\dot{\zeta}$ aft produces a downward flap acceleration of the blade or, in the d'Alembert view, an upward flap moment. Upward flap velocity $\dot{\beta}$ produces a lead Coriolis moment on the blade.

The aerodynamic flap and lag moments were derived in section 16.8.3:

$$\int_e^1 \eta_\beta \frac{F_z}{ac} dr = M_F = M_\theta(-K_{P_\beta}\beta - K_{P_\zeta}\zeta) + M_{\dot{\beta}}\dot{\beta} + M_\beta \beta + M_{\dot{\zeta}}\dot{\zeta} + M_\zeta \zeta \quad (20.40)$$

$$\int_e^1 \eta_\zeta \frac{F_x}{ac} dr = M_L = Q_\theta(-K_{P_\beta}\beta - K_{P_\zeta}\zeta) + Q_{\dot{\beta}}\dot{\beta} + Q_\beta \beta + Q_{\dot{\zeta}}\dot{\zeta} + Q_\zeta \zeta \quad (20.41)$$

Since only the stability is of interest here, the pitch control input has not been included. There is, however, a pitch change due to kinematic pitch-flap and pitch-lag coupling: $\Delta\theta = -K_{P_\beta}\beta - K_{P_\zeta}\zeta$, which produces flap and lag moments. The sign convention for positive coupling is that a nose-down pitch change is produced by flap-up or lag-back deflection of the blade. For hover or vertical flight the aerodynamic coefficients are constant, and $M_\beta = M_\zeta = Q_\beta = Q_\zeta = 0$. Assuming uniform inflow, $\eta_\beta = \eta_\zeta = r$, and neglecting tip losses, the aerodynamic coefficients are

$$M_\theta = \frac{1}{8} \quad (20.42)$$

$$M_{\dot{\beta}} = -\frac{1}{8} \quad (20.43)$$

$$M_{\dot{\zeta}} = -\left(\frac{3}{2}\frac{C_T}{\sigma a} + \frac{5}{24}\lambda_{HP} + \frac{1}{80}\theta_{tw}\right) \quad (20.44)$$

$$Q_\theta = \frac{\lambda_{HP}}{6} \quad (20.45)$$

$$Q_{\dot\beta} = \frac{3}{4}\frac{C_T}{\sigma a} - \frac{7}{48}\lambda_{HP} + \frac{1}{160}\theta_{tw} \qquad (20.46)$$

$$Q_{\dot\zeta} = -\left[\lambda_{HP}\left(\frac{C_T}{\sigma a} + \frac{\lambda_{HP}}{4}\right) + \frac{c_d}{4a}\right] \qquad (20.47)$$

The hover trim induced velocity is $\lambda_{HP} = \kappa\sqrt{C_T/2}$. The aerodynamic forces are determined by a single parameter of the rotor operating state: the thrust coefficient C_T. The collective pitch of the rotor can also be used as the parameter, since the collective is related to the rotor thrust by

$$\theta_{0.75} = \frac{6C_T}{\sigma a} + \frac{3}{2}\lambda_{HP} \qquad (20.48)$$

The Coriolis forces also require the rotor coning angle, which for hover is

$$\begin{aligned}
\beta_0 &= \frac{1}{\widehat{I}_\beta \nu_\beta^2}\left(\gamma\int_0^1 \frac{1}{2}r^3\alpha\,dr + \widehat{K}_\beta\beta_p\right)\\
&= \frac{1}{\widehat{I}_\beta \nu_\beta^2}\left(\gamma\left(\frac{3}{4}\frac{C_T}{\sigma a} + \frac{1}{48}\lambda_{HP} + \frac{1}{160}\theta_{tw}\right) + \widehat{I}_\beta\widehat{K}_\beta\beta_p\right)
\end{aligned} \qquad (20.49)$$

where β_p is the precone angle. Precone is used to reduce the mean flap bending moments at the hinge spring, which are proportional to $(\beta_0 - \beta_p)$. The ideal coning angle β_{ideal} is the coning obtained with spring $K_\beta = 0$; hence flap frequency $\nu_\beta = 1$. With precone $\beta_p = \beta_{\text{ideal}}$ (which depends on C_T/σ), the mean hinge spring moment is zero; see section 6.14.

The differential equations for the rotor flap and lag motion in hover are thus

$$\begin{bmatrix}\widehat{I}_\beta & 0\\ 0 & \widehat{I}_\zeta\end{bmatrix}\begin{pmatrix}\ddot\beta\\ \ddot\zeta\end{pmatrix} + \begin{bmatrix}-\gamma M_{\dot\beta} & -\gamma M_{\dot\zeta} - 2\widehat{I}_{\beta\zeta}\beta_0\\ -\gamma Q_{\dot\beta} + 2\widehat{I}_{\beta\zeta}\beta_0 & -\gamma Q_{\dot\zeta} + \widehat{C}_\zeta\end{bmatrix}\begin{pmatrix}\dot\beta\\ \dot\zeta\end{pmatrix}$$
$$+ \begin{bmatrix}\widehat{I}_\beta\nu_\beta^2 + K_{P_\beta}\gamma M_\theta & -\chi(\widehat{I}_\zeta\widehat{K}_\zeta - \widehat{I}_\beta\widehat{K}_\beta)\theta_s + K_{P_\zeta}\gamma M_\theta\\ -\chi(\widehat{I}_\zeta\widehat{K}_\zeta - \widehat{I}_\beta\widehat{K}_\beta)\theta_s + K_{P_\beta}\gamma Q_\theta & \widehat{I}_\zeta\nu_\zeta^2 + K_{P_\zeta}\gamma Q_\theta\end{bmatrix}\begin{pmatrix}\beta\\ \zeta\end{pmatrix} = 0$$
$$(20.50)$$

and the characteristic equation is

$$\left(\widehat{I}_\beta s^2 - \gamma M_{\dot\beta}s + \widehat{I}_\beta\nu_\beta^2 + K_{P_\beta}\gamma M_\theta\right)\left(\widehat{I}_\zeta s^2 + (-\gamma Q_{\dot\zeta} + \widehat{C}_\zeta)s + \widehat{I}_\zeta\nu_\zeta^2 + K_{P_\zeta}\gamma Q_\theta\right)$$
$$- \left((-\gamma M_{\dot\zeta} - 2\widehat{I}_{\beta\zeta}\beta_0)s - \chi(\widehat{I}_\zeta\widehat{K}_\zeta - \widehat{I}_\beta\widehat{K}_\beta)\theta_s + K_{P_\zeta}\gamma M_\theta\right)$$
$$\times \left((-\gamma Q_{\dot\beta} + 2\widehat{I}_{\beta\zeta}\beta_0)s - \chi(\widehat{I}_\zeta\widehat{K}_\zeta - \widehat{I}_\beta\widehat{K}_\beta)\theta_s + K_{P_\beta}\gamma Q_\theta\right) = 0 \qquad (20.51)$$

which can be solved for the four roots of the system (a complex conjugate pair each for the flap and lag modes). The uncoupled motion has positive damping for both flap and lag, so an instability can only be encountered because of the coupling terms. Divergence is seldom a factor in flap-lag dynamics, assuming that K_{P_β} and K_{P_ζ} are not so negative that there is a net negative flap or lag spring.

When there is no pitch-lag coupling ($K_{P_\zeta} = 0$) and no structural coupling ($\chi = 0$), the only lag influence on the flap equation is the velocity term. The flap moment due to $\dot\zeta$ consists of aerodynamic and Coriolis terms:

$$\gamma M_{\dot\zeta} + 2\widehat{I}_{\beta\zeta}\beta_0 = -\gamma\left(\frac{\lambda_{\mathrm{HP}}}{6} + 2\int_0^1 \frac{1}{2}r^3\alpha\,dr\right) + 2\frac{\widehat{I}_{\beta\zeta}}{\widehat{I}_\beta v_\beta^2}\left(\gamma\int_0^1 \frac{1}{2}r^3\alpha\,dr + \widehat{I}_\beta\widehat{K}_\beta\beta_p\right)$$

$$\cong -2\gamma\frac{v_\beta^2 - 1}{v_\beta^2}\int_0^1 \frac{1}{2}r^3\alpha\,dr + 2\frac{\widehat{K}_\beta\beta_p}{v_\beta^2} \tag{20.52}$$

Thus for an articulated rotor with no flap hinge spring or offset ($v_\beta = 1$ and $K_\beta = 0$), the aerodynamic and Coriolis flap moments due to lag velocity nearly cancel, and the flap equation is decoupled from the lag motion. The flap and lag motion are stable in this case. Lag motion produced by the Coriolis forces due to flapping is important for the blade loads and vibration, but not for stability. When there is no flap hinge offset, $v_\beta^2 = 1 + \widehat{K}_\beta$. In addition, the ideal precone is $\beta_{\text{ideal}} = \gamma\int_0^1 \frac{1}{2}r^3\alpha\,dr$. So if $v_\beta > 1$ because of a hinge spring, but with ideal precone, the total flap moment due to lag velocity is still zero. With ideal precone the hinge spring does not contribute to the balance of flap moments determining the coning, and the solution for β_0 is the same as for the articulated rotor. Hence for an articulated rotor with a flap frequency near 1/rev and small pitch-flap and pitch-lag coupling, in hover or low forward speed, the flap-lag motion of the blade is expected to remain stable.

Consider next the case of zero thrust, so that all the aerodynamic coefficients except M_θ, $M_{\dot\beta}$, and $Q_{\dot\zeta}$ are zero or nearly so. Assume also zero precone (so β_0) and no structural coupling ($\theta_s = 0$). Then the only remaining coupling of the flap and lag equations is a flap moment due to ζ, acting through M_θ when there is kinematic pitch-lag coupling ($K_{P_\zeta} \neq 0$). The lag equation is decoupled from the flap motion, so the system is stable. Hence a flap-lag instability is a high thrust or high collective phenomenon. The flap-lag stability boundary gives a critical thrust level or equivalently a collective pitch or coning angle limit.

20.2.2 Articulated Rotors

For an articulated rotor the lag frequency is small, typically $v_\zeta = 0.25$ to 0.30/rev. Recall that $v_\zeta \cong \frac{3}{2}e$, where e is the lag hinge offset; see section 6.15. With no hinge springs, the structural coupling terms are zero. For this case, an approximate solution for the flap-lag stability can be obtained, which shows the influence of pitch-lag and pitch-flap coupling. When $K_{P_\zeta} \neq 0$, the flap moment produced by ζ through M_θ dominates the small flap moments due to the lag velocity $\dot\zeta$, and the latter can be neglected. All the aerodynamic lag moments due to flapping are neglected compared to the Coriolis term. The term $K_{P_\zeta}\gamma Q_\theta$ is considered to be included in the lag spring $\widehat{I}_\zeta v_\zeta^2$, and we write $\overline{C}_\zeta = \widehat{C}_\zeta - \gamma Q_{\dot\zeta}$. With these approximations, the equations of motion in Laplace form are

$$\begin{bmatrix} \widehat{I}_\beta s^2 - \gamma M_{\dot\beta}s + \widehat{I}_\beta v_\beta^2 + K_{P_\beta}\gamma M_\theta & K_{P_\zeta}\gamma M_\theta \\ 2\widehat{I}_{\beta\zeta}\beta_0 s & \widehat{I}_\zeta s^2 + \overline{C}_\zeta s + \widehat{I}_\zeta v_\zeta^2 \end{bmatrix}\begin{pmatrix} \beta \\ \zeta \end{pmatrix} = 0 \tag{20.53}$$

The equations are now coupled only because of the Coriolis lag moment and the flap moment produced by pitch-lag coupling.

Since the flap damping is high, the lag mode is most likely to go unstable. The frequency of that mode is near ν_ζ, which is small for an articulated rotor. The eigenvalue at the flutter boundary is small, and hence the flap equation can be approximated by a quasistatic balance of the flap moments due to β and ζ:

$$(\widehat{I}_\beta \nu_\beta^2 + K_{P_\beta} \gamma M_\theta)\beta + (K_{P_\zeta} \gamma M_\theta)\zeta = 0 \tag{20.54}$$

or

$$\beta = -\frac{K_{P_\zeta} \gamma M_\theta}{\widehat{I}_\beta \nu_\beta^2 + K_{P_\beta} \gamma M_\theta} \zeta \tag{20.55}$$

This is the flap motion that accompanies the lag oscillation in the flutter mode because of the pitch-lag coupling. On substituting for this flap motion in the Coriolis lag moment, the lag equation becomes

$$\left(\widehat{I}_\zeta s^2 + \left(\overline{C}_\zeta - 2\widehat{I}_{\beta\zeta}\beta_0 \frac{K_{P_\zeta} \gamma M_\theta}{\widehat{I}_\beta \nu_{\beta e}^2}\right)s + \widehat{I}_\zeta \nu_\zeta^2\right)\zeta = 0 \tag{20.56}$$

For an articulated rotor, the flap response to the moments produced by the pitch-lag coupling is in phase with the low-frequency lag motion. Hence the Coriolis lag moment due to flap velocity results in a lag damping term, which determines the stability of the lag motion.

The criterion for stability follows directly from the requirement of positive net lag damping:

$$\overline{C}_\zeta - 2\widehat{I}_{\beta\zeta}\beta_0 \frac{K_{P_\zeta} \gamma M_\theta}{\widehat{I}_\beta \nu_\beta^2 + K_{P_\beta} \gamma M_\theta} > 0 \tag{20.57}$$

Morduchow and Hinchey (1950) derived the equations of motion for rigid flap and lag of an articulated rotor in hover and found that positive pitch-lag coupling ($K_{P_\zeta} > 0$) destabilizes the lag mode. Chou (1958) investigated a lag instability of a fully articulated rotor blade due to pitch-lag coupling, which was encountered in a rotor test at high collective and low rotor speed. A lag oscillation of about 30° amplitude and 0.32/rev frequency was observed, with the flap motion at the same frequency. Upon examination, the control linkage was found to give positive pitch-lag coupling. Chou obtained a criterion for stability (equation 20.57) by considering the Coriolis damping of the lag motion arising from the flapping produced by the pitch-lag coupling. He also obtained the stability criterion directly by means of Routh's discriminant and demonstrated that for articulated rotors the result is equivalent to the approximate criterion. Blake, Burkam, and Loewy (1961) extended Chou's analysis of the articulated rotor to include all the aerodynamic terms and concluded that equation 20.57 is generally conservative.

Including the aerodynamic moment $Q_{\dot\beta}$, and noting that $2\widehat{I}_{\beta\zeta}\beta_0 - \gamma Q_{\dot\beta} \cong \widehat{I}_{\beta\zeta}\beta_0$, the stability criterion is

$$\widehat{C}_\zeta - \gamma Q_{\dot\zeta} - \widehat{I}_{\beta\zeta}\beta_0 \frac{K_{P_\zeta} \gamma M_\theta}{\widehat{I}_\beta \nu_\beta^2 + K_{P_\beta} \gamma M_\theta} > 0 \tag{20.58}$$

This expression gives the lag damping required for stability or alternatively the maximum pitch-lag coupling allowed. For a given rotor, the stability decreases as the coning angle β_0 increases with thrust. Positive pitch-lag coupling (lag back, pitch down) is destabilizing for articulated rotors.

20.2.3 Stability Boundary

An oscillatory instability occurs when a pair of complex conjugate roots crosses the imaginary axis into the right half-plane. The stability boundary is obtained by substituting $s = i\omega$ in the characteristic equation, where ω is a real and positive frequency. The real and imaginary parts of the characteristic equation are then

$$\widehat{I}_\beta \widehat{I}_\zeta (-\omega^2 + v_{\beta e}^2)(-\omega^2 + v_{\zeta e}^2) + \gamma M_{\dot\beta} \overline{C}_\zeta \omega^2 + (-\gamma M_{\dot\zeta} - 2\widehat{I}_{\beta\zeta}\beta_0)(-\gamma Q_{\dot\beta} + 2\widehat{I}_{\beta\zeta}\beta_0)\omega^2$$
$$- (K_{P_\zeta}\gamma M_\theta - \chi(\widehat{I}_\zeta \widehat{K}_\zeta - \widehat{I}_\beta \widehat{K}_\beta)\theta_s)(K_{P_\beta}\gamma Q_\theta - \chi(\widehat{I}_\zeta \widehat{K}_\zeta - \widehat{I}_\beta \widehat{K}_\beta)\theta_s) = 0 \quad (20.59)$$

and

$$-\gamma M_{\dot\beta}\widehat{I}_\zeta(-\omega^2 + v_{\zeta e}^2) + \overline{C}_\zeta \widehat{I}_\beta(-\omega^2 + v_{\beta e}^2)$$
$$- (-\gamma M_{\dot\zeta} - 2\widehat{I}_{\beta\zeta}\beta_0)(K_{P_\beta}\gamma Q_\theta - \chi(\widehat{I}_\zeta \widehat{K}_\zeta - \widehat{I}_\beta \widehat{K}_\beta)\theta_s)$$
$$- (-\gamma Q_{\dot\beta} + 2\widehat{I}_{\beta\zeta}\beta_0)(K_{P_\zeta}\gamma M_\theta - \chi(\widehat{I}_\zeta \widehat{K}_\zeta - \widehat{I}_\beta \widehat{K}_\beta)\theta_s) = 0 \quad (20.60)$$

The pitch-flap and pitch-lag coupling contribute to the effective stiffnesses: $\widehat{I}_\beta v_{\beta e}^2 = \widehat{I}_\beta v_\beta^2 + K_{P_\beta}\gamma M_\theta$ and $\widehat{I}_\zeta v_{\zeta e}^2 = \widehat{I}_\zeta v_\zeta^2 + K_{P_\zeta}\gamma Q_\theta$. The total lag damping is $\overline{C}_\zeta = \widehat{C}_\zeta - \gamma Q_{\dot\zeta}$. The imaginary part of the characteristic equation is solved for the flutter frequency:

$$(-\gamma M_{\dot\beta}\widehat{I}_\zeta + \overline{C}_\zeta \widehat{I}_\beta)\omega^2 = -\gamma M_{\dot\beta}\widehat{I}_\zeta v_{\zeta e}^2 + \overline{C}_\zeta \widehat{I}_\beta v_{\beta e}^2$$
$$+ (\gamma M_{\dot\zeta} + 2\widehat{I}_{\beta\zeta}\beta_0)K_{P_\beta}\gamma Q_\theta + (\gamma Q_{\dot\beta} - 2\widehat{I}_{\beta\zeta}\beta_0)K_{P_\zeta}\gamma M_\theta$$
$$- (\gamma M_{\dot\zeta} + \gamma Q_{\dot\beta})\chi(\widehat{I}_\zeta \widehat{K}_\zeta - \widehat{I}_\beta \widehat{K}_\beta)\theta_s = 0 \quad (20.61)$$

Substituting for ω^2 in the real part gives the equation of the stability boundary.

20.2.4 Hingeless Rotors

Consider now the stability of the flap-lag motion of a hingeless rotor blade, as modeled by purely out-of-plane and purely in-plane modes with arbitrary natural frequencies v_β and v_ζ. Assuming no pitch-flap or pitch-lag coupling ($K_{P_\beta} = K_{P_\zeta} = 0$) and no structural coupling ($\chi = 0$), the characteristic equation is

$$(\widehat{I}_\beta s^2 - \gamma M_{\dot\beta} s + \widehat{I}_\beta v_\beta^2)(\widehat{I}_\zeta s^2 + \overline{C}_\zeta s + \widehat{I}_\zeta v_\zeta^2)$$
$$- (\gamma M_{\dot\zeta} + 2\widehat{I}_{\beta\zeta}\beta_0)(\gamma Q_{\dot\beta} - 2\widehat{I}_{\beta\zeta}\beta_0) s^2 = 0 \quad (20.62)$$

again writing $\overline{C}_\zeta = \widehat{C}_\zeta - \gamma Q_{\dot\zeta}$ for the total lag damping. The imaginary part of the characteristic equation is solved for the flutter frequency:

$$\omega^2 = \frac{-\gamma M_{\dot\beta}\widehat{I}_\zeta v_\zeta^2 + \overline{C}_\zeta \widehat{I}_\beta v_\beta^2}{-\gamma M_{\dot\beta}\widehat{I}_\zeta + \overline{C}_\zeta \widehat{I}_\beta} = v_\zeta^2 + \left(\frac{\overline{C}_\zeta \widehat{I}_\beta}{-\gamma M_{\dot\beta}\widehat{I}_\zeta + \overline{C}_\zeta \widehat{I}_\beta}\right)(v_\beta^2 - v_\zeta^2) \quad (20.63)$$

If the flap damping is much higher than the lag damping, the flutter frequency is near v_ζ, implying that the lag mode is unstable. Substituting for ω^2 in the real part of the characteristic equation gives the equation of the flutter boundary:

$$(\gamma M_{\dot\zeta} + 2\widehat{I}_{\beta\zeta}\beta_0)(\gamma Q_{\dot\beta} - 2\widehat{I}_{\beta\zeta}\beta_0)$$
$$= -\gamma M_{\dot\beta}\overline{C}_\zeta\left[1 + \frac{\widehat{I}_\beta^2 \widehat{I}_\zeta^2 (v_\beta^2 - v_\zeta^2)^2}{(-\gamma M_{\dot\beta}\widehat{I}_\zeta + \overline{C}_\zeta \widehat{I}_\beta)(-\gamma M_{\dot\beta}\widehat{I}_\zeta v_\zeta^2 + \overline{C}_\zeta \widehat{I}_\beta v_\beta^2)}\right] \quad (20.64)$$

20.2 Flap-Lag Dynamics

The left-hand side is the product of the coupling terms, which for stability must be less than the product of the damping terms on the right-hand side. This equation can be considered as a criterion for the minimum lag damping required for stability, or for the maximum allowable rotor thrust, which determines the aerodynamic forces and coning angle in the coupling terms. Alternatively, equation 20.64 can be viewed as defining a stability boundary on the v_β vs. v_ζ plane.

Now in the flap equation, the aerodynamic and Coriolis moments due to $\dot\zeta$ are nearly equal in magnitude but opposite in sign, so

$$\gamma M_{\dot\zeta} + 2\widehat{I}_{\beta\zeta}\beta_0 \cong -\gamma\frac{\lambda_{HP}}{6} - 2\gamma\frac{v_\beta^2 - 1}{v_\beta^2}\frac{3}{4}\frac{C_T}{\sigma a} + 2\frac{\widehat{K}_\beta\beta_p}{v_\beta^2} \tag{20.65}$$

When $v_\beta > 1$, the aerodynamic term is larger. In the lag equation, the aerodynamic moment due to $\dot\beta$ has about one-half the magnitude of the Coriolis terms and opposite sign, so

$$\gamma Q_{\dot\beta} - 2\widehat{I}_{\beta\zeta}\beta_0 \cong -\gamma\frac{3\lambda_{HP}}{16} + \gamma\frac{v_\beta^2 - 2}{v_\beta^2}\frac{3}{4}\frac{C_T}{\sigma a} - 2\frac{\widehat{K}_\beta\beta_p}{v_\beta^2} \tag{20.66}$$

Hence

$$\left(\gamma M_{\dot\zeta} + 2\widehat{I}_{\beta\zeta}\beta_0\right)\left(\gamma Q_{\dot\beta} - 2\widehat{I}_{\beta\zeta}\beta_0\right)$$
$$\cong \frac{2(v_\beta^2-1)(2-v_\beta^2)}{v_\beta^4}\left(\gamma\frac{3}{4}\frac{C_T}{\sigma a} - \frac{\widehat{K}_\beta\beta_p}{v_\beta^2-1}\right)\left(\gamma\frac{3}{4}\frac{C_T}{\sigma a} + 2\frac{\widehat{K}_\beta\beta_p}{2-v_\beta^2}\right)$$
$$+ \gamma^2\frac{\lambda_{HP}}{8}\left(\frac{C_T}{\sigma a} + \frac{\lambda_{HP}}{4}\right) \tag{20.67}$$

The last term can be combined with $-\gamma M_{\dot\beta}\overline{C}_\zeta$ on the right-hand side of the equation to give

$$-\gamma M_{\dot\beta}\overline{C}_\zeta - \gamma^2\frac{\lambda_{HP}}{8}\left(\frac{C_T}{\sigma a} + \frac{\lambda_{HP}}{4}\right) = -\gamma M_{\dot\beta}\left(\frac{\gamma c_d}{4a} + \widehat{C}_\zeta\right) \tag{20.68}$$

The aerodynamic damping $Q_{\dot\zeta}$ in $\overline{C}_\zeta$ cancels, except for the viscous drag term. The criterion for flap-lag stability of the hovering rotor with no pitch-flap or pitch-lag coupling is thus

$$\left(\frac{6C_T}{\sigma a} - \frac{8}{\gamma}\frac{\widehat{K}_\beta\beta_p}{v_\beta^2-1}\right)\left(\frac{6C_T}{\sigma a} + \frac{16}{\gamma}\frac{\widehat{K}_\beta\beta_p}{2-v_\beta^2}\right)\left(\frac{\gamma}{8}\right)^2\frac{2(v_\beta^2-1)(2-v_\beta^2)}{v_\beta^4}$$
$$< -\gamma M_{\dot\beta}\left[\frac{\gamma c_d}{4a} + \widehat{C}_\zeta + \left(\frac{8}{\gamma}\right)^2\frac{(v_\beta^2-v_\zeta^2)^2\overline{C}_\zeta}{(1+(8\overline{C}_\zeta/\gamma))(v_\zeta^2+(8\overline{C}_\zeta/\gamma)v_\beta^2)}\right] \tag{20.69}$$

This approximation was derived by Ormiston and Hodges (1972). Since the flap damping $(-\gamma M_{\dot\beta})$ and lag damping $(\overline{C}_\zeta)$ are positive, the right-hand side of equation 20.69 is always positive. The flap-lag motion is stable if the left-hand side is zero or negative. One such case is the articulated rotor, for which $v_\beta = 1$ and the left-hand side is zero, due to the decoupling of the flap equation from the lag motion when there is no pitch-lag coupling. In general, the flap-lag motion is stable unless $1 < v_\beta^2 < 2$, or $1 < v_\beta < 1.414$, but that covers the usual range of flap frequencies for articulated and hingeless rotors. The left-hand side is positive for high enough rotor thrust or collective pitch, and the flap-lag motion is unstable at some critical C_T depending on the lag damping. The term in brackets on the right-hand side is of order $\widehat{C}_\zeta$, so the dimensionless lag damping $\widehat{C}_\zeta$ required for stability is of the order $\overline{\alpha}^2 = (6C_T/\sigma a)^2$, which is small. Hence an articulated rotor with ν_β slightly above 1/rev and a mechanical damper giving a high level of lag damping is almost certainly stable (assuming $K_{P_\zeta} = 0$). For a hingeless rotor, however, ν_β is significantly above 1/rev and the structural lag damping is small, so a flap-lag instability is possible.

Let us consider the case for which the flap-lag motion is least stable. The second term on the right-hand side of equation 20.69 has a minimum value (zero) when $\nu_\beta = \nu_\zeta$. Furthermore, the factor $2(\nu_\beta^2 - 1)(2 - \nu_\beta^2)/\nu_\beta^4$ on the left-hand side has a maximum value of $\frac{1}{4}$ at $\nu_\beta^2 = \frac{4}{3}$, or $\nu_\beta \cong 1.15$. Thus the stiff in-plane hingeless rotor with $\nu_\beta = \nu_\zeta = 1.15$ has the minimum flap-lag stability margin. For this case the stability criterion becomes

$$\left(\frac{\gamma}{8}\right)^2 \left(\frac{6C_T}{\sigma a}\right)^2 - (3\widehat{K}_\beta \beta_p)^2 < \frac{\gamma}{2}\left(\frac{\gamma c_d}{4a} + \widehat{C}_\zeta\right) \tag{20.70}$$

Recall that the ideal precone is $\beta_p \cong \gamma \frac{3}{4}\frac{C_T}{\sigma a}$, and $\widehat{K}_\beta = \nu_\beta^2 - 1 = \frac{1}{3}$ here. So the left-hand side is zero with ideal precone, and the flap-lag motion is stable. Neglecting the precone term, and writing the lag damping in terms of a structural damping coefficient g_s ($\widehat{C}_\zeta = g_s \nu_\zeta$), the stability criterion is

$$\frac{6C_T}{\sigma a} < 4\sqrt{\frac{c_d}{2a} + \frac{4g_s}{\gamma\sqrt{3}}} \tag{20.71}$$

Since the blade viscous drag damping alone gives roughly $C_T/\sigma < 0.10$, any reasonable level of structural damping should be sufficient to stabilize the flap-lag motion. Figure 20.3 shows flap-lag stability boundaries in terms of the natural frequencies of the blade flap and lag motion, for a rotor with $\gamma = 8$, $\sigma = 0.08$, and $c_d = 0.0080$. Shown are the solution of the complete equation for the stability boundary (equation 20.64) and the approximate solution (equation 20.69). At a given rotor thrust or collective, the motion is unstable inside a roughly elliptical region centered on the worst case $\nu_\beta = \nu_\zeta = 1.15$. These results are for a rotor with no precone, no pitch-flap or pitch-lag coupling, and no structural flap-lag coupling. A small amount of lag damping is sufficient to stabilize the motion.

To summarize the results for the case of no pitch-flap or pitch-lag coupling, the articulated rotor with flap frequency near 1/rev, small lag frequency, and large lag damping is likely stable. The worst case for flap-lag stability is a stiff in-plane hingeless rotor with equal rotating flap and lag frequencies: $\nu_\beta = \nu_\zeta = \sqrt{4/3}$. With small precone, little structural damping, and high thrust, a flap-lag instability is possible. The motion is stabilized by separating the flap and lag frequencies (moving away from the line $\nu_\beta = \nu_\zeta$) and by keeping the flap frequency away from the critical value $\nu_\beta = 1.15$/rev.

Structural coupling between the flap and lag motion is important, because even a small amount of out-of-plane motion in the lag mode greatly increases its aerodynamic damping and thus is very stabilizing. For rigid flap and lag motion, structural coupling is introduced in terms of the distribution of the hinge stiffness inboard and outboard of the pitch bearing. For $\chi = 0$ all the flexibility is inboard (uncoupled), with axes parallel and normal to the hub plane. To lowest order, structural coupling adds the term X^2/ω^2 to the right-hand side of equation 20.64, where

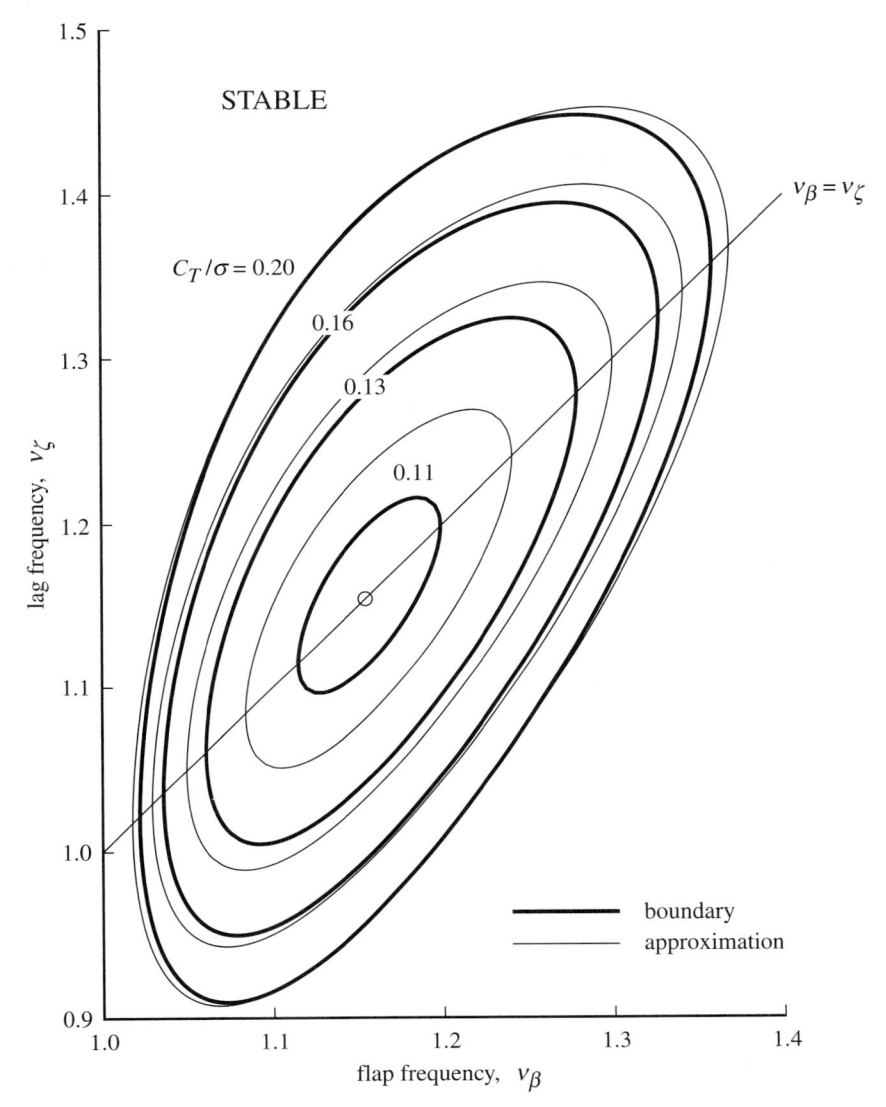

Figure 20.3. Flap-lag stability boundaries as a function of rotor thrust, no pitch-flap or pitch-lag coupling, and no structural flap-lag coupling.

$X = \chi(\widehat{I}_\zeta \widehat{K}_\zeta - \widehat{I}_\beta \widehat{K}_\beta)\theta_s = \chi(\widehat{I}_\zeta v_\zeta^2 - \widehat{I}_\beta(v_\beta^2 - 1))\theta_s$, implying an increase in stability. Figures 20.4 and 20.5 show the stabilizing effect of $\chi > 0$. See Ormiston and Hodges (1972) and Burkham and Miao (1972).

20.2.5 Pitch-Flap and Pitch-Lag Coupling

Introducing pitch-flap coupling K_{P_β}, but still no pitch-lag or structural flap-lag coupling, there is a small shift of the flutter frequency due to the lag moment:

$$\omega^2 = v_\zeta^2 + \left(\frac{\overline{C}_\zeta \widehat{I}_\beta}{-\gamma M_{\dot\beta}\widehat{I}_\zeta + \overline{C}_\zeta \widehat{I}_\beta}\right)(v_\beta^2 - v_\zeta^2) + \frac{(\gamma M_{\dot\zeta} + 2\widehat{I}_{\beta\zeta}\beta_0)K_{P_\beta}\gamma Q_\theta}{-\gamma M_{\dot\beta}\widehat{I}_\zeta + \overline{C}_\zeta \widehat{I}_\beta} \qquad (20.72)$$

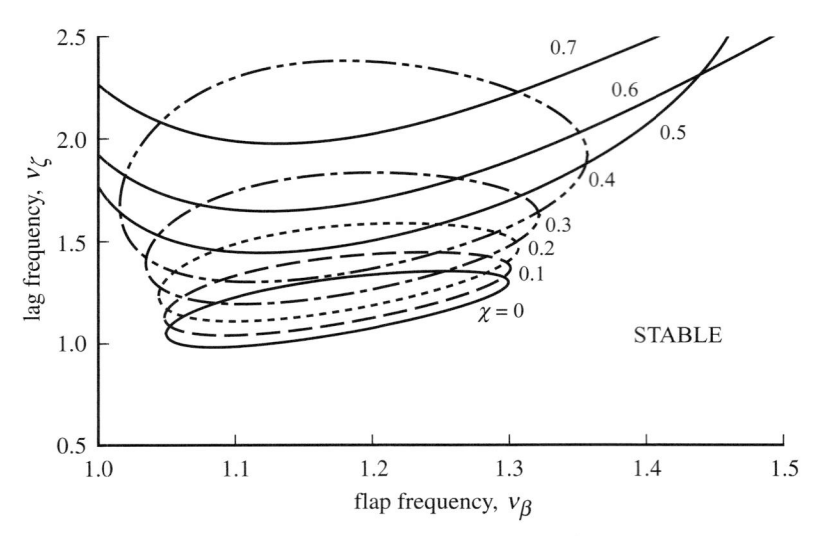

Figure 20.4. Flap-lag stability boundaries as a function of structural flap-lag coupling, no pitch-flap or pitch lag coupling, and $C_T/\sigma = 0.14$.

but the principal change is replacing the flap stiffness v_β^2 with the effective stiffness $v_{\beta e}^2 = v_\beta^2 + \frac{\gamma}{8} K_{P_\beta}$ in the characteristic equation and hence in the stability boundary. The flap-lag motion is then stable unless the effective frequency is in the range $1 < v_{\beta e}^2 < 2$. By using negative δ_3 or positive pitch-flap coupling ($K_{P_\beta} = \tan \delta_3 < 0$, so flap-up/pitch-up), the negative aerodynamic spring can lower the effective flap frequency below 1/rev, thus assuring flap-lag stability regardless of the lag frequency or thrust. Gaffey (1969) observed the favorable influence of negative δ_3 on the flap-lag stability of a proprotor operating in high inflow.

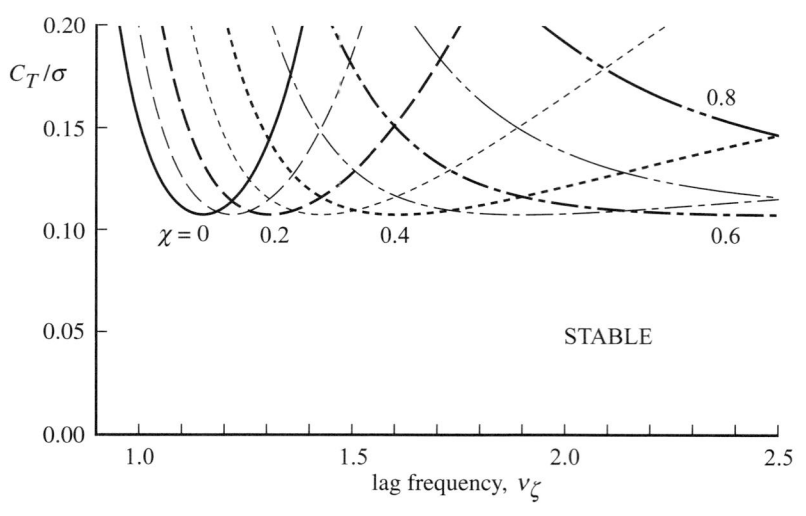

Figure 20.5. Flap-lag stability boundaries as a function of structural flap-lag coupling, no pitch-flap or pitch lag coupling, and $v_\beta = 1.15$.

20.2 Flap-Lag Dynamics

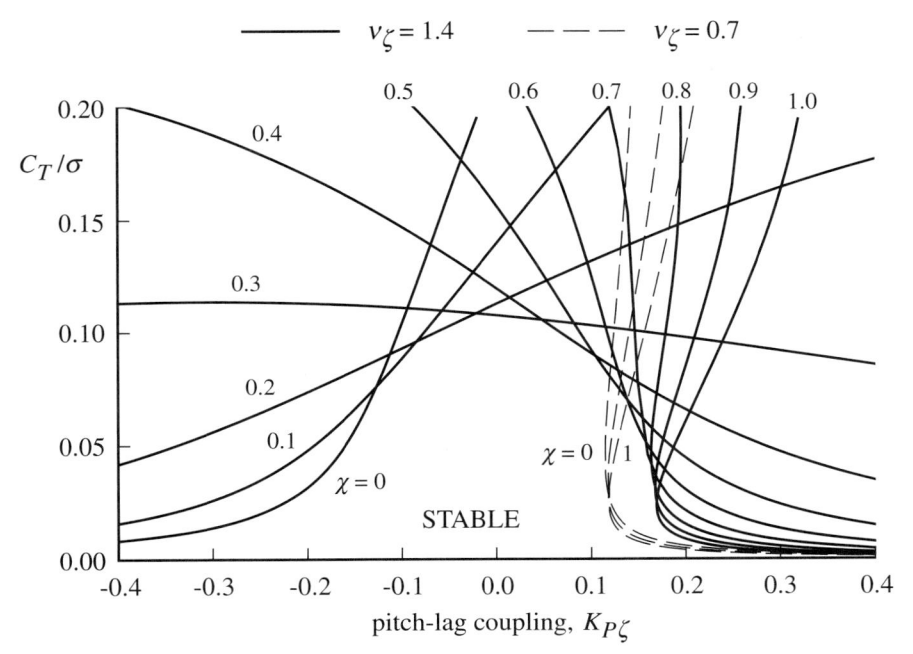

Figure 20.6. Flap-lag stability boundaries as a function of pitch-lag coupling and structural flap-lag coupling, $v_\beta = 1.15$, and $v_\zeta = 1.4$ and 0.7.

Consider a rotor with large pitch-lag coupling K_{P_ζ}. The flutter frequency is approximately

$$\omega^2 = v_{\zeta e}^2 + \left(\frac{\overline{C}_\zeta \widehat{I}_\beta}{-\gamma M_{\dot\beta} \widehat{I}_\zeta + \overline{C}_\zeta \widehat{I}_\beta} \right) (v_{\beta e}^2 - v_{\zeta e}^2) + \frac{K_{P_\zeta} \gamma M_\theta (\gamma Q_{\dot\beta} - 2\widehat{I}_{\beta\zeta} \beta_0)}{-\gamma M_{\dot\beta} \widehat{I}_\zeta + \overline{C}_\zeta \widehat{I}_\beta} \quad (20.73)$$

Substituting for ω^2 in the real part of the characteristic equation, a few terms can be omitted for large K_{P_ζ}, giving the stability boundary:

$$\widehat{I}_\beta \widehat{I}_\zeta (v_{\beta e}^2 - v_{\zeta e}^2) \left(\overline{C}_\zeta \widehat{I}_\beta (v_{\beta e}^2 - v_{\zeta e}^2) + K_{P_\zeta} \gamma M_\theta (\gamma Q_{\dot\beta} - 2\widehat{I}_{\beta\zeta} \beta_0) \right)$$
$$+ \left(\gamma M_{\dot\beta} \overline{C}_\zeta + (\gamma M_{\dot\zeta} + 2\widehat{I}_{\beta\zeta} \beta_0)(\gamma Q_{\dot\beta} - 2\widehat{I}_{\beta\zeta} \beta_0) \right) \gamma M_{\dot\beta} \widehat{I}_\zeta v_{\zeta e}^2$$
$$- (K_{P_\zeta} \gamma M_\theta - X)(K_{P_\beta} \gamma Q_\theta - X) \gamma M_{\dot\beta} \widehat{I}_\zeta = 0 \quad (20.74)$$

where $X = \chi (\widehat{I}_\zeta v_\zeta^2 - \widehat{I}_\beta (v_\beta^2 - 1)) \theta_s$. Equation 20.74 can be solved for the damping $\widehat{C}_\zeta$ required for stability. For small lag frequency and no structural coupling, it reduces to the result of section 20.2.2. Figure 20.6 shows the influence of pitch-lag coupling on the stability boundary for stiff in-plane ($v_\zeta = 1.4$) and soft in-plane ($v_\zeta = 0.7$) cases. For articulated and soft in-plane hingeless rotors, positive pitch-lag coupling (lag back, pitch down) is destabilizing, with little influence of structural coupling. For stiff in-plane hingeless rotors, negative pitch-lag coupling (lag back, pitch up) is destabilizing with no structural coupling ($\chi = 0$), whereas positive pitch-lag coupling is destabilizing with full structural coupling ($\chi = 1$).

The stabilizing influence of structural flap-lag coupling and negative pitch-lag coupling for a soft in-plane hingeless rotor was verified experimentally by Bousman,

Sharpe, and Ormiston (1976). The effects of the two couplings together was more than the sum of the individual effects, although the experimental data exhibited behavior that was more complicated than predicted by the simple theory.

20.2.6 Blade Stall

When the hovering rotor is operating at very large thrust, the blade stalls along much of the radius. For the blade sections in stall, the lift-curve slope is zero or negative, resulting in a reduction of the aerodynamic damping of the blade flap modes. Without large flap mode damping, the flap-lag motion can be unstable. Figure 20.7 shows the damping ratio of the flap and lag modes of hingeless rotors, calculated with and without stall in the airfoil tables. For these calculations there is no lag damper or blade structural damping; hence the very small lag mode damping at low thrust. At large C_T/σ (depending on the airfoil maximum lift coefficient), stall is evident in the drop of the flap mode damping. This phenomenon was observed and identified by Huber (1973) in flight tests and by Ormiston and Bousman (1975) in model rotor tests. In hover the thrust required for instability due to stall is well above what can be sustained with available power. For forward flight the phenomenon is not encountered, because stall then occurs only over part of the azimuth.

20.2.7 Elastic Blade and Flap-Lag-Torsion Stability

Although hingeless rotors have been developed based on a rigid blade analysis (Huber (1973)), an elastic blade model is required to fully capture the dynamic behavior, particularly when the important phenomena have not been already established for the design concept. The stability results from an elastic flap-lag blade model are similar to the results from the rigid blade model; see Hodges and Ormiston (1976) and Friedmann and Straub (1980). The elastic blade model includes a representation of structural flap-lag coupling. Blade rigid pitch motion (due to control system flexibility) and elastic torsion motion are important for stability and loads calculations, unless the fundamental torsion mode frequency is very high (20/rev or more). Multi-mode resonances (such coalescence of the frequencies of third flap, second lag, and first torsion modes) can lead to instabilities. If the torsion mode frequency is not too low, the principal effect of the pitch and torsion flexibility is the introduction of effective pitch-flap and pitch-lag coupling; see Hodges and Ormiston (1973). For blade flap-lag-torsion models and stability calculations, see Burkam and Miao (1972) and Hodges and Ormiston (1976, 1977).

Pitch-flap and pitch-lag coupling are key parameters controlling the blade dynamics. For an articulated rotor these couplings are determined by the geometry of the root hinges and control system, but for a hingeless rotor they depend on the bending and torsion loads acting on the blade. Thus a fully coupled flap-lag-torsion model of the blade motion is required for an accurate analysis of the aeroelastic stability of a hingeless rotor. Section 16.5.2 discusses structural pitch-flap and pitch-lag coupling. Based on a rigid blade model, an estimate of the pitch-lag coupling is derived:

$$K_{P_\zeta} = \frac{1}{\widehat{K}_\theta}(v_\beta^2 - 1 - v_\zeta^2)(\beta - \beta_p) \qquad (20.75)$$

(equation 16.145, without the droop angle). Trim elastic flap deflection relative to the precone angle produces a flap moment at the blade root. Lag motion rotates

20.2 Flap-Lag Dynamics

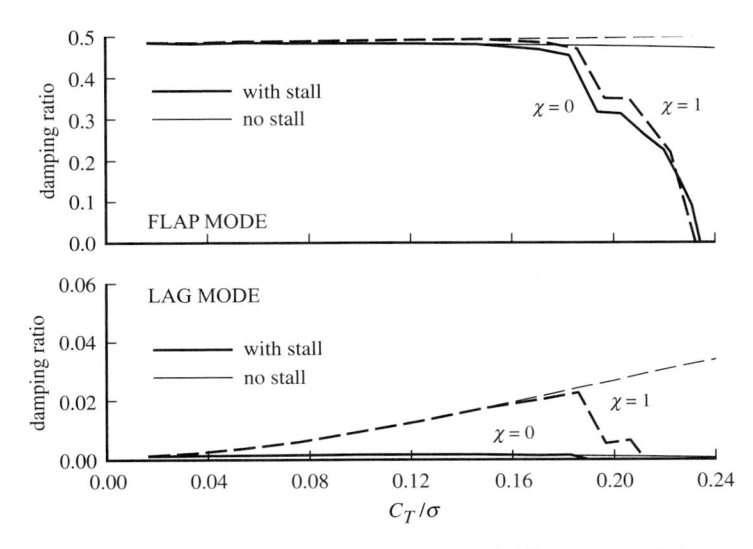

stiff in-plane hingeless rotor, $v_\beta = 1.15$ and $v_\zeta = 1.4$

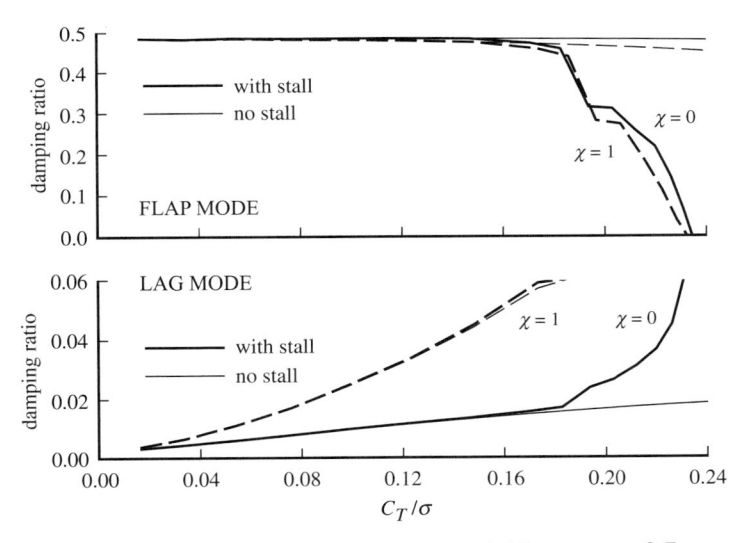

soft in-plane hingeless rotor, $v_\beta = 1.15$ and $v_\zeta = 0.7$

Figure 20.7. Influence of blade stall on flap-lag stability in hover.

this flap moment in the hub plane, so the flap moment has a component about the pitch bearing. Due to pitch flexibility, this moment produces a blade pitch change and hence pitch-lag coupling. Figure 20.8 illustrates the variation of the effective pitch-lag coupling with thrust and design parameters (precone and pitch stiffness); see Huber (1973). In particular, precone has a significant influence on the blade stability through this coupling, in addition to its action through the Coriolis forces. The rotor design approach can be to minimize the couplings (as with a matched stiffness design, $v_\beta^2 - 1 = v_\zeta^2$) or to introduce couplings favorable for stability and loads.

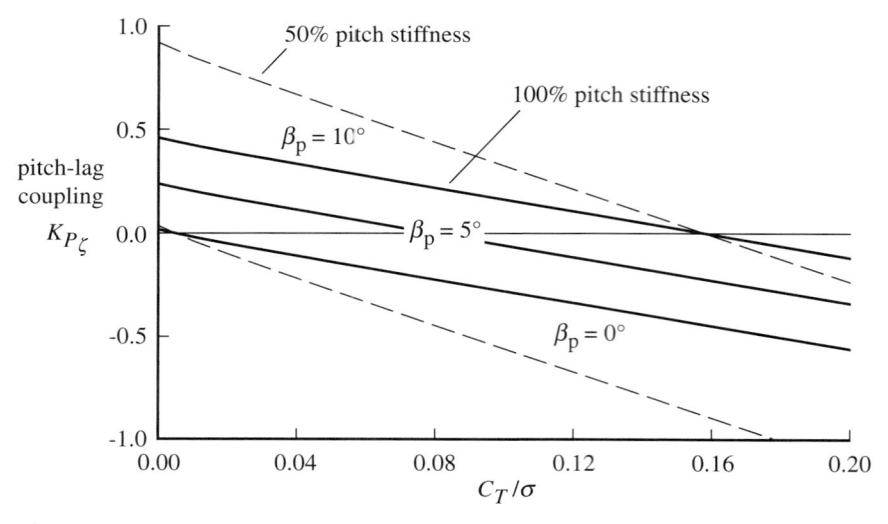

Figure 20.8. Effective pitch-lag coupling of soft in-plane hingeless rotor blade.

20.3 Ground Resonance

Ground resonance is a dynamic instability involving the blade lag motion coupled with in-plane motion of the rotor hub. This instability is characterized by a resonance of the frequency of the rotor lag motion (specifically the low-frequency lag mode in the non-rotating frame) and a natural frequency of the structure supporting the rotor. Since the lag frequency depends on the rotor rotational speed, such resonances define certain critical speed ranges for the rotor. An instability is possible at a resonance if the rotating lag frequency ν_ζ is below 1/rev, as for articulated and soft in-plane hingeless rotors. With articulated rotors, the critical mode is usually an oscillation of the helicopter on the landing gear when in contact with the ground; hence the name ground resonance. Sometimes the phenomenon can occur in flight as well, particularly with a hingeless rotor, and then it is called air resonance.

The hub in-plane motions are coupled with the cyclic lag modes ζ_{1c} and ζ_{1s}, which correspond to lateral and longitudinal shifts of the net rotor center-of-gravity from the axis of rotation. Since the low-frequency lag mode involves whirling of the rotor center-of-gravity about the shaft, ground resonance is potentially very destructive, and avoiding this instability is an important consideration in helicopter design. The basic requirement is that resonances of the support structure with the lag mode be kept out of the operating range of the helicopter. Generally, resonances above 120% normal operating speed or below 40% normal speed are acceptable. The rotor has little energy at low speed, so sometimes it is possible to run up through very low-frequency resonances without a large amplitude motion building up. For a fairly large range about the normal rotor speed range, either resonances must be avoided or sufficient system damping provided to prevent any instability.

The classical ground resonance analysis considers four degrees of freedom: longitudinal and lateral in-plane motion of the rotor hub, and the two cyclic lag degrees of freedom. The actual vibration modes of the rotor support, such as the motion of the helicopter on its landing gear, usually involve tilt of the shaft as well, but the in-plane motion of the hub is the dominant factor in ground resonance. The rotor aerodynamic forces have little influence on ground resonance, compared to the structural and inertial forces. Damping of the rotor and support comes almost entirely from mechanical dampers or structural damping. Thus the aerodynamic forces are neglected in elementary ground resonance analysis. Such a model provides a good description of the fundamental characteristics of ground resonance and even gives good numerical results, particularly for articulated rotors. With hingeless rotors a more complete model is required, including the rotor aerodynamics and flap motion and a better description of the support motion. The basic analysis of ground resonance is the work of Coleman and Feingold (1958).

20.3.1 Ground Resonance Equations

The degrees of freedom involved in ground resonance are the cyclic rotor lag modes ζ_{1c} and ζ_{1s}, which produce a shift of the net rotor center-of-gravity; and the hub longitudinal and lateral displacements, x_h and y_h. For now, only the case of a rotor with three or more blades is considered. The equations of motion for the rotor lag degrees of freedom, including the influence of the hub motion, were derived in section 16.7 (equations 16.248 and 16.249). Dropping the aerodynamic forces, but including a lag damping term, the equations of motion in the non-rotating frame are

$$I_\zeta \left(\ddot{\zeta}_{1c} + \widehat{C}_\zeta (\dot{\zeta}_{1c} + \zeta_{1s}) + 2\dot{\zeta}_{1s} + (v_\zeta^2 - 1)\zeta_{1c} \right) - S_\zeta \ddot{y}_h = 0 \quad (20.76)$$

$$I_\zeta \left(\ddot{\zeta}_{1s} + \widehat{C}_\zeta (\dot{\zeta}_{1s} - \zeta_{1c}) - 2\dot{\zeta}_{1c} + (v_\zeta^2 - 1)\zeta_{1s} \right) + S_\zeta \ddot{x}_h = 0 \quad (20.77)$$

Here v_ζ is the rotating natural frequency of the lag motion and $C_\zeta = I_\zeta \widehat{C}_\zeta$ is the lag damping (in the rotating frame, due to aerodynamic, structural, or mechanical damping). The first and second moments of the blade lag inertia are $S_\zeta = \int_0^1 \eta_\zeta m \, dr$ and $I_\zeta = \int_0^1 \eta_\zeta^2 m \, dr$. Since the hub in-plane motion is coupled by the inertial forces with ζ_{1c} and ζ_{1s} only, the other non-rotating lag degrees of freedom are not involved in ground resonance. The in-plane forces acting on the rotor hub were also derived in section 16.7 (equations 16.258 and 16.259). Retaining only the inertial terms, the rotor drag and side force are

$$H = -\frac{N}{2} S_\zeta \ddot{\zeta}_{1s} - NM_b \ddot{x}_h \quad (20.78)$$

$$Y = \frac{N}{2} S_\zeta \ddot{\zeta}_{1c} - NM_b \ddot{y}_h \quad (20.79)$$

where N is the number of blades and the blade mass is $M_b = \int_0^1 m \, dr$. The rotor support structure is represented by a mass-spring-damper system in the longitudinal and lateral directions, excited by the rotor hub forces:

$$\widetilde{M}_x \ddot{x}_h + C_x \dot{x}_h + K_x x_h = H \quad (20.80)$$

$$\widetilde{M}_y \ddot{y}_h + C_y \dot{y}_h + K_y y_h = Y \quad (20.81)$$

These equations are often a good model of the actual helicopter or wind-tunnel support dynamics if the generalized mass and damping of the appropriate vibration modes are used, as determined from the hub impedance. Substituting for the rotor hub forces gives

$$M_x \ddot{x}_h + C_x \dot{x}_h + K_x x_h + \frac{N}{2} S_\zeta \ddot{\zeta}_{1s} = 0 \quad (20.82)$$

$$M_y \ddot{y}_h + C_y \dot{y}_h + K_y y_h - \frac{N}{2} S_\zeta \ddot{\zeta}_{1c} = 0 \quad (20.83)$$

where $M_x = \widetilde{M}_x + NM_b$ and $M_y = \widetilde{M}_y + NM_b$ are the total mass, including the rotor. Write $K_z = M_x\omega_x^2$ and $C_x = M_x\widehat{C}_x$, and similarly define the natural frequency ω_y and damping $\widehat{C}_y$ of the lateral mode.

These equations are dimensionless. In particular, the lag frequency ν_ζ is per revolution. Normalized inertias (divided by the characteristic inertia I_b) are not introduced, so dimensional results are obtained by simply including factors of Ω and R. The coupled lag and support equations of motion describing the ground resonance dynamics are then

$$\begin{bmatrix} I_\zeta & 0 & -S_\zeta & 0 \\ 0 & I_\zeta & 0 & S_\zeta \\ -\frac{N}{2}S_\zeta & 0 & M_y & 0 \\ 0 & \frac{N}{2}S_\zeta & 0 & M_x \end{bmatrix} \begin{pmatrix} \ddot{\zeta}_{1c} \\ \ddot{\zeta}_{1s} \\ \ddot{y}_h \\ \ddot{x}_h \end{pmatrix} + \begin{bmatrix} I_\zeta\widehat{C}_\zeta & 2I_\zeta & 0 & 0 \\ -2I_\zeta & I_\zeta\widehat{C}_\zeta & 0 & 0 \\ 0 & 0 & M_y\widehat{C}_y & 0 \\ 0 & 0 & 0 & M_x\widehat{C}_x \end{bmatrix} \begin{pmatrix} \dot{\zeta}_{1c} \\ \dot{\zeta}_{1s} \\ \dot{y}_h \\ \dot{x}_h \end{pmatrix}$$

$$+ \begin{bmatrix} I_\zeta(\nu_\zeta^2 - 1) & I_\zeta\widehat{C}_\zeta & 0 & 0 \\ -I_\zeta\widehat{C}_\zeta & I_\zeta(\nu_\zeta^2 - 1) & 0 & 0 \\ 0 & 0 & M_y\omega_y^2 & 0 \\ 0 & 0 & 0 & M_x\omega_x^2 \end{bmatrix} \begin{pmatrix} \zeta_{1c} \\ \zeta_{1s} \\ y_h \\ x_h \end{pmatrix} = 0 \qquad (20.84)$$

Symmetry is recovered if the support equations are divided by $\frac{N}{2}$. The only coupling of the rotor and support motion is due to the inertial terms. In terms of the Laplace variables, the characteristic equation of this system is

$$I_\zeta^2 \left((s^2 + \widehat{C}_\zeta s + \nu_\zeta^2 - 1)^2 + (2s + \widehat{C}_\zeta)^2\right) M_y(s^2 + \widehat{C}_y s + \omega_y^2) M_x(s^2 + \widehat{C}_x s + \omega_x^2)$$

$$- I_\zeta (s^2 + \widehat{C}_\zeta s + \nu_\zeta^2 - 1) \left(M_y(s^2 + \widehat{C}_y s + \omega_y^2) + M_x(s^2 + \widehat{C}_x s + \omega_x^2)\right) (NS_\zeta^2/2) s^4$$

$$+ (NS_\zeta^2/2)^2 s^8 = 0 \qquad (20.85)$$

The solution of this eighth-order polynomial gives the four eigenvalues of the system (and their complex conjugates) and hence the ground resonance stability.

A divergence type instability is not possible for this system. On setting $s = 0$, the characteristic equation gives the divergence stability criterion:

$$I_\zeta^2 \left((\nu_\zeta^2 - 1)^2 + \widehat{C}_\zeta^2\right) M_y\omega_y^2 M_x\omega_x^2 > 0 \qquad (20.86)$$

which is always satisfied, assuming that either $\nu_\zeta \neq 1$ or $C_\zeta \neq 0$.

Consider the uncoupled dynamics, obtained by setting $S_\zeta = 0$. The uncoupled hub motion consists of damped oscillations with natural frequencies ω_x and ω_y. The uncoupled (shaft-fixed) lag motion is a damped oscillation with eigenvalue

$$s_R = (\widehat{C}_\zeta/2) + i\sqrt{\nu_\zeta^2 - (\widehat{C}_\zeta/2)^2} \qquad (20.87)$$

in the rotating frame. In the non-rotating frame there are two cyclic lag modes, with eigenvalues $s_{NR} = s_R \pm i$. The high-frequency lag mode ($s_{NR} = s_R + i$) corresponds to a progressive whirling motion of the rotor center-of-gravity at frequency $(\text{Im}\,s_R + 1)$/rev. The low-frequency lag mode ($s_{NR} = s_R - i$) is a whirling motion of the rotor center-of-gravity at frequency $|\text{Im}\,s_R - 1|$/rev, regressive if $\text{Im}\,s_R > 1$/rev (such as for a stiff in-plane hingeless rotor) or a progressive whirling mode if $\text{Im}\,s_R < 1$/rev, as for an articulated rotor. Thus for $S_\zeta = 0$, the characteristic equation factors into a product of the rotor and the support characteristic equations, with the solutions

$$s = (\widehat{C}_\zeta/2) + i\sqrt{\nu_\zeta^2 - (\widehat{C}_\zeta/2)^2} \pm i \qquad (20.88)$$

$$s = (\widehat{C}_x/2) + i\sqrt{\omega_x^2 - (\widehat{C}_x/2)^2} \tag{20.89}$$

$$s = (\widehat{C}_y/2) + i\sqrt{\omega_y^2 - (\widehat{C}_y/2)^2} \tag{20.90}$$

and their conjugates. The uncoupled rotor and support motion is stable, and a ground resonance instability can only be due to the inertial coupling when $S_\zeta > 0$. When there is no damping and $S_\zeta = 0$ as well, the solution is $s = \pm i\omega$ for frequencies $\omega = \nu_\zeta \pm 1$, ω_x, and ω_y.

The coupling of the rotor and support is governed by the parameter $S_\zeta = \int_0^1 \eta_\zeta m \, dr$. For an articulated rotor, S_ζ is the product of the blade mass and the radial distance of the blade center-of-gravity from the lag hinge. The measure of the coupling between the rotor and support terms in the characteristic equation is

$$\frac{N}{2} \frac{S_\zeta^2}{I_\zeta M_x} \cong \frac{3}{8} \frac{NM_b}{M_x} = \frac{3}{8} \frac{M_{\text{rotor}}}{M_{\text{support}}} \tag{20.91}$$

The mass of the support mode is usually much larger than the rotor mass, so this parameter is quite small. Although an exact analytical solution of the eighth-order ground resonance characteristic equation is not available, useful results can be obtained on the basis of the assumption that S_ζ is small, specifically $S_\zeta^2/I_\zeta M_x \ll 1$. The solution for small S_ζ gives accurate numerical results as well, for most practical cases.

The natural frequency of the lag motion, ν_ζ, determines the fundamental character of the ground resonance dynamics. The dimensionless blade lag frequency varies with the rotor speed according to $\nu_\zeta^2 = K_1/\Omega^2 + K_2$, where K_1 and K_2 are the Southwell coefficients (see section 18.1). For an articulated rotor, $K_1 = 0$ and $\nu_\zeta^2 = K_2 = \frac{3}{2}\frac{e}{1-e}$, where e is the lag hinge offset. For a hingeless rotor K_1 is not zero: the dimensional non-rotating lag frequency is $\nu_{NR} = \sqrt{K_1}$. The lag frequency at high rotor speed approaches $\nu_\zeta = \sqrt{K_2}$ (per-rev). For a soft in-plane rotor (hingeless or articulated), $K_2 < 1$; for a stiff in-plane rotor, probably $K_2 > 1$, although $\nu_\zeta > 1/\text{rev}$ can always be found at low rotor speed even if $K_2 < 1$ (assuming $K_1 \neq 0$).

For a given rotor and support, it is preferable to present the ground resonance solution in terms of dimensional frequencies, using the rotor speed Ω as a parameter. The dimensional support frequencies ω_x and ω_y are constants, and the dimensional lag frequency depends on the rotor speed according to $\nu_\zeta^2 = K_1 + K_2\Omega^2$. Hence at low speed $\nu_\zeta \cong \nu_{NR} = \sqrt{K_1}$, and at high speed ν_ζ/Ω approaches $\sqrt{K_2}/\text{rev}$. The variation of ν_ζ with Ω determines the resonances with the various natural frequencies of the support. In the present ground resonance analysis, dimensionless parameters are used.

20.3.2 No-Damping Case

Consider first the case of no damping of the lag or support motion. On setting $C_\zeta = C_x = C_y = 0$, the characteristic equation becomes

$$\left((s^2 + \nu_\zeta^2 - 1)^2 + 4s^2\right)(s^2 + \omega_y^2)(s^2 + \omega_x^2)$$

$$- (s^2 + \nu_\zeta^2 - 1)\left(M_y(s^2 + \omega_y^2) + M_x(s^2 + \omega_x^2)\right)\frac{NS_\zeta^2/2I_\zeta}{M_y M_x}s^4$$

$$+ \frac{(NS_\zeta^2/2I_\zeta)^2}{M_y M_x}s^8 = 0 \tag{20.92}$$

The eigenvalues of a polynomial with real coefficients must appear as complex conjugate pairs. If there is no damping, however, the characteristic equation is actually a polynomial in s^2. So the substitution $s = i\omega$, or $s^2 = -\omega^2$ (with ω complex), gives a polynomial in ω^2 that also has real coefficients. Complex conjugate solutions for ω correspond to solutions for s that are symmetric about the imaginary s-axis. Complex solutions of the characteristic equation with zero damping occur in groups of four roots, symmetric about both the real and imaginary axes. There is one root in each quadrant of the s-plane. Since two of these roots are in the right half-plane ($\text{Re}\, s > 0$), such complex solutions correspond to an unstable system. Moreover, there is no way that the system can be stable with all the roots in the left half-plane. The requirement of symmetry about the imaginary axis is satisfied if all the roots are on the imaginary axis, which corresponds to neutral stability. Neutral stability is the best that can be achieved when there is no damping in the rotor or support to extract energy from the system.

Therefore, for the case of no damping, there are boundaries not between stable and unstable conditions, but rather between neutrally stable and unstable conditions. Inside a neutral stability region, all the roots are on the imaginary axis. At the stability boundary, two roots meet at positive frequency and two at negative frequency, and they break off from the imaginary axis. Then inside an unstable region there are four complex roots, corresponding to the support mode in resonance and the low-frequency lag mode. Substituting $s = i\omega$ (where ω is a real number) now defines the entire neutral stability region, not just the flutter boundary. The simplest means of defining the stability boundary in this case is to solve the characteristic equation assuming $s = i\omega$. Where all eight solutions cannot be obtained with ω real, it must be an unstable condition. The uncoupled solution ($S_\zeta = 0$) is exactly $s = \pm i\omega$, with $\omega = \nu_\zeta \pm 1$, ω_x, and ω_y. Since an instability involves four roots, it requires a resonance of a support mode and a rotor mode. At such a resonance, the coupling due to S_ζ produces the instability under certain conditions.

On substituting $s = i\omega$, the characteristic equation for the case of no damping becomes

$$((\nu_\zeta^2 - 1 - \omega^2)^2 - 4\omega^2)(\omega_y^2 - \omega^2)(\omega_x^2 - \omega^2)$$
$$- (\nu_\zeta^2 - 1 - \omega^2)\left(M_y(\omega_y^2 - \omega^2) + M_x(\omega_x^2 - \omega^2)\right)\frac{NS_\zeta^2/2I_\zeta}{M_y M_x}\omega^4$$
$$+ \frac{(NS_\zeta^2/2I_\zeta)^2}{M_y M_x}\omega^8 = 0 \tag{20.93}$$

This polynomial can be solved numerically for ω^2. Alternatively, values of ω^2 can be assumed, so the characteristic equation becomes a quadratic for the dimensional frequency ($\nu_\zeta^2 - \Omega^2$). Thus the solution for ω^2 as a function of Ω or ν_ζ can be constructed. Where less than four values of ω^2 are obtained for a given Ω, the motion is unstable. The most general approach is to calculate the eigenvalues of the ground resonance equations of motion.

Although an exact solution of the characteristic equation can be obtained numerically, it is useful to have an analytical solution based on the assumption that the coupling parameter S_ζ is small. The roots for small coupling should be near the exact solution for $S_\zeta = 0$: $\omega = \nu_\zeta \pm 1$, ω_x, and ω_y. Therefore, we find a correction of order S_ζ^2 to the uncoupled roots. Considering the solution near $\omega = \omega_x$, write

20.3 Ground Resonance

the eigenvalues as $\omega^2 = \omega_x^2 + S_\zeta^2 s_1$. Then to lowest order in S_ζ^2, the characteristic equation is

$$-\left((\nu_\zeta^2 - 1 - \omega_x^2)^2 - 4\omega_x^2\right)(\omega_y^2 - \omega_x^2)s_1 - (\nu_\zeta^2 - 1 - \omega_x^2)M_y(\omega_y^2 - \omega_x^2)\frac{N/2I_\zeta}{M_y M_x}\omega_x^4 = 0 \quad (20.94)$$

which gives

$$\omega^2 = \omega_x^2 \left[1 - \frac{(\nu_\zeta^2 - 1 - \omega_x^2)\omega_x^2}{(\nu_\zeta^2 - 1 - \omega_x^2)^2 - 4\omega_x^2}\frac{NS_\zeta^2/2I_\zeta}{M_x}\right] \quad (20.95)$$

The order S_ζ^2 solution near $\omega = \omega_y$ is similar. For the solution near $\omega = \nu_\zeta \pm 1$, write $\omega^2 = (\nu_\zeta \pm 1)^2 + S_\zeta^2 s_1$. Then to lowest order in S_ζ^2, the characteristic equation is

$$s_1 2\nu_\zeta (\omega_y^2 - (\nu_\zeta \pm 1)^2)(\omega_x^2 - (\nu_\zeta \pm 1)^2)$$
$$+ (\nu_\zeta \pm 1)^5 \left(M_y(\omega_y^2 - (\nu_\zeta \pm 1)^2) + M_x(\omega_x^2 - (\nu_\zeta \pm 1)^2)\right)\frac{N/2I_\zeta}{M_y M_x} = 0 \quad (20.96)$$

or

$$\omega^2 = (\nu_\zeta \pm 1)^2 \left[1 - \frac{(\nu_\zeta \pm 1)^3}{2\nu_\zeta}\frac{M_y(\omega_y^2 - (\nu_\zeta \pm 1)^2) + M_x(\omega_x^2 - (\nu_\zeta \pm 1)^2)}{(\omega_y^2 - (\nu_\zeta \pm 1)^2)(\omega_x^2 - (\nu_\zeta \pm 1)^2)}\frac{NS_\zeta^2/2I_\zeta}{M_y M_x}\right] \quad (20.97)$$

Near a resonance of ω_x or ω_y with a rotor root $(\nu_\zeta \pm 1)$, this expansion is not valid.

In the limit $\Omega = 0$, the dimensional lag frequency approaches the non-rotating lag frequency ν_{NR}. In terms of the dimensionless frequencies, ω_x and ω_y become infinite proportional to $1/\Omega$, and $\nu_\zeta \cong \nu_{NR}$. Then the solution near $\omega = \omega_x$ becomes

$$\omega^2 = \omega_x^2 \left[1 + \frac{\omega_x^2}{\omega_x^2 - \nu_{NR}^2}\frac{NS_\zeta^2/2I_\zeta}{M_x}\right] \quad (20.98)$$

Hence the solution ω^2 is increased at low rotor speed if $\omega_x > \nu_{NR}$ and decreased if $\omega_x > \nu_{NR}$. The solution near $\omega = \nu_\zeta \pm 1$ for $\Omega = 0$ is

$$\omega^2 = \nu_{NR}^2 \left[1 - \frac{\nu_{NR}^2}{2}\frac{M_y(\omega_y^2 - \nu_{NR}^2) + M_x(\omega_x^2 - \nu_{NR}^2)}{(\omega_y^2 - \nu_{NR}^2)(\omega_x^2 - \nu_{NR}^2)}\frac{NS_\zeta^2/2I_\zeta}{M_y M_x}\right] \quad (20.99)$$

The direction of the shift in ω^2 depends on the magnitude of the non-rotating lag frequency ν_{NR} relative to the support frequencies ω_x and ω_y.

In the limit of large rotor speed, the dimensional lag frequency becomes infinite proportionally to Ω^2. In terms of the dimensionless frequencies, ω_x and ω_y approach zero as $1/\Omega$, while the lag frequency ν_ζ approaches a constant per-rev value. Then the solution near $\omega = \omega_x$ becomes

$$\omega^2 = \omega_x^2 \left[1 - \frac{\omega_x^2}{\nu_\zeta^2 - 1}\frac{NS_\zeta^2/2I_\zeta}{M_x}\right] \quad (20.100)$$

The solution ω^2 is increased if $\nu_\zeta < 1/\text{rev}$. The solution near $\omega = \nu_\zeta \pm 1$ for Ω approaching infinity is

$$\omega^2 = (\nu_\zeta \pm 1)^2 \left[1 + \frac{\nu_\zeta \pm 1}{2\nu_\zeta}(M_y + M_x)\frac{NS_\zeta^2/2I_\zeta}{M_y M_x}\right] \quad (20.101)$$

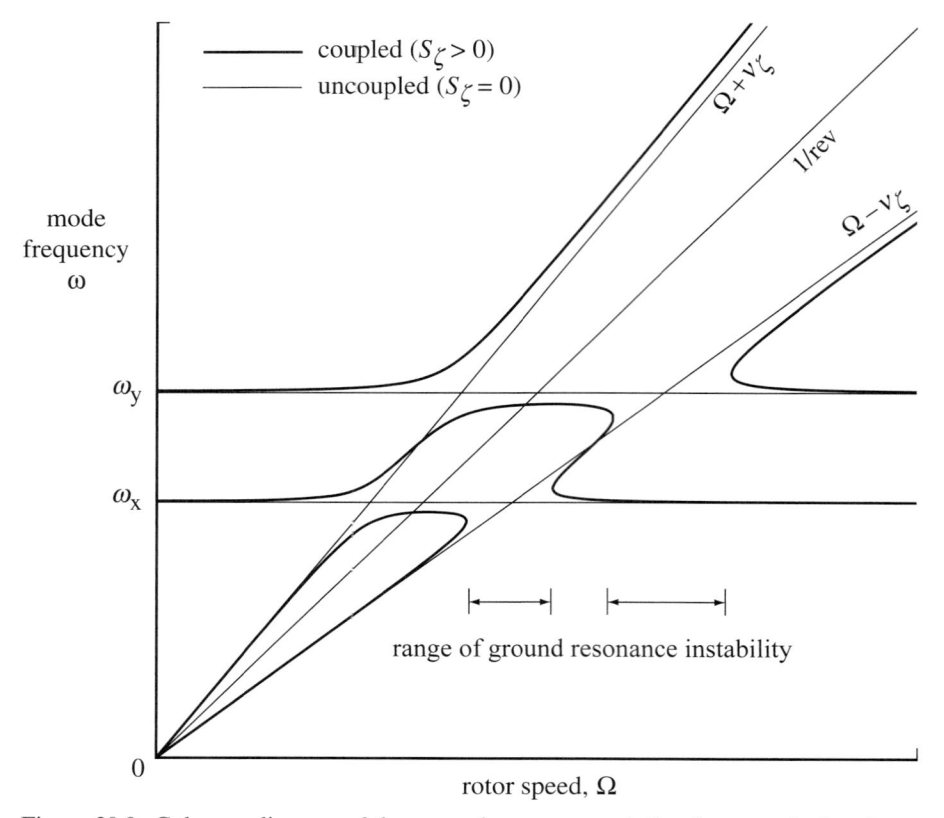

Figure 20.9. Coleman diagram of the ground resonance solution for an articulated rotor.

so ω^2 is increased for the $\omega = \nu_\zeta + 1$ solution and also for the $\omega = \nu_\zeta - 1$ solution if $\nu_\zeta > 1/\text{rev}$.

Finally consider the limit $\omega = 0$, for which the characteristic equation reduces to $(\nu_\zeta^2 - 1)^2 = 0$, which gives $\nu_\zeta - 1 = 0$. With the rotor speed as a parameter, the $\omega = 0$ solution defines where the roots intercept the Ω-axis, namely at the rotor speed for which $\nu_\zeta = 1/\text{rev}$. In the uncoupled case, the solution $\omega = \nu_\zeta - 1$ intercepts the Ω-axis at this point. The low-frequency lag mode root intercepts the Ω-axis at the same point for all values of S_ζ.

The ground resonance solution for the case of no damping can be presented graphically in a form known as a Coleman diagram, which is a plot of the dimensional frequencies ω (the roots of the characteristic equation) as a function of the rotor speed Ω. The dimensional solution for the uncoupled case ($S_\zeta = 0$) is $\omega = \omega_x, \omega_y, \Omega \pm \nu_\zeta$, plus the corresponding negative frequencies, for a total of eight roots. The negative solutions for ω are mirror images of the positive solutions and so need not be plotted. The solution for $S_\zeta > 0$ can easily be sketched using these results for the influence of small S_ζ in the limits $\Omega = 0$ and $\Omega = \infty$, plus the knowledge that for a coupled system the loci of roots never cross. The character of the ground resonance solution depends primarily on the lag frequency $\nu_\zeta^2 = K_1 + K_2 \Omega^2$ (dimensional).

Figures 20.9 to 20.11 present the Coleman diagrams for three types of rotors: an articulated rotor ($K_1 = 0$ and $K_2 < 1$), a soft in-plane hingeless rotor ($K_1 > 0$ and $K_2 < 1$), and a stiff in-plane hingeless rotor ($K_1 > 0$ and $K_2 > 1$). The solution for these cases was obtained by finding the eigenvalues of the ground resonance

20.3 Ground Resonance

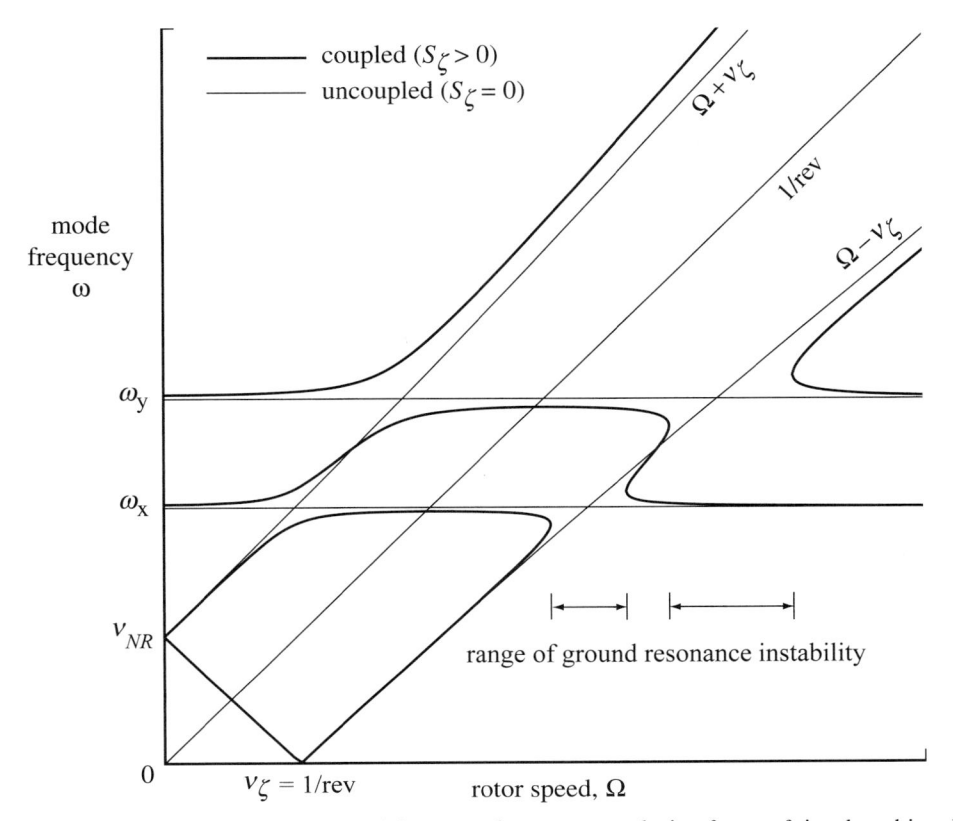

Figure 20.10. Coleman diagram of the ground resonance solution for a soft in-plane hingeless rotor.

equations, for small coupling ($S_\zeta^2/I_\zeta M_x \ll 1$) and K_1 and K_2 values chosen for exposition. The uncoupled roots for the support modes are horizontal lines at $\omega = \omega_x$ and $\omega = \omega_y$, and the uncoupled roots for the rotor are the low- and high-frequency rotor modes at $\omega = \nu_\zeta \pm \Omega$, which approach $\nu_{NR} = \sqrt{K_1}$ at low rotor speed and are asymptotic to constant per-rev values ($\sqrt{K_2} \pm 1$)/rev) at high rotor speed. Thus the lag mode frequencies are in resonance with the support mode frequencies at some rotor speed.

For $S_\zeta > 0$, the solution is displaced from the uncoupled frequencies, as indicated by the results for small coupling. If there are four positive solutions for ω at a given rotor speed, then the system is stable (neutrally stable for this case of zero damping). For the articulated and soft in-plane hingeless rotors (Figures 20.9 and 20.10), there are ranges of Ω where only two positive real solutions for ω exist, occurring around the resonances of the low-frequency lag mode ($\Omega - \nu_\zeta$) with a support mode (ω_x or ω_y). The characteristic equation has four complex solutions in these ranges, so the system is unstable. The eigenvalue solution gives a single frequency (with $\pm$ real parts) bridging the gaps. For the stiff in-plane hingeless rotor (Figure 20.11), four positive solutions for ω exist at all rotor speeds, and a ground resonance instability does not occur.

In summary, a ground resonance instability can occur at a resonance of a rotor mode and a support mode. The resonances of the high-frequency lag mode ($\omega = 1 + \nu_\zeta$) are always stable, but resonances of the low-frequency lag mode ($\omega = 1 - \nu_\zeta$)

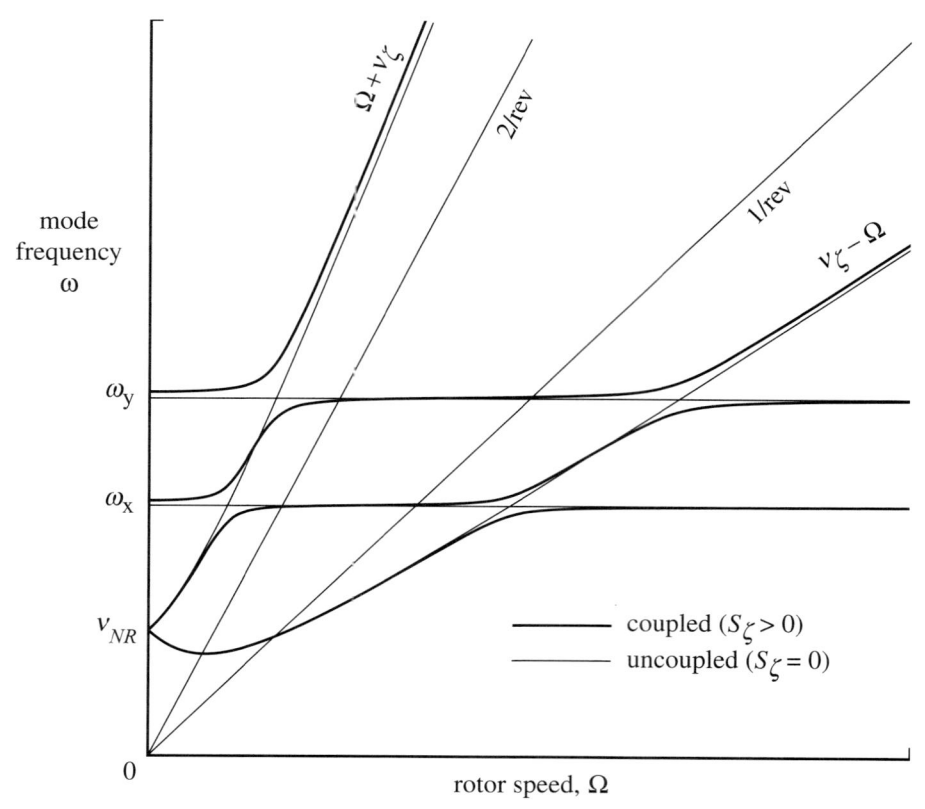

Figure 20.11. Coleman diagram of the ground resonance solution for a stiff in-plane hingeless rotor.

are unstable if the rotating natural frequency v_ζ is below 1/rev, as for articulated and soft in-plane hingeless rotors. The placement of the rotor lag frequency determines whether or not a ground resonance instability can occur.

20.3.3 Damping Required for Ground Resonance Stability

For the case with damping of the lag and support motion, the stability boundary is obtained by setting $s = i\omega$ in the characteristic equation:

$$I_\zeta^2 \left((-\omega^2 + \widehat{C}_\zeta i\omega + v_\zeta^2 - 1)^2 + (2i\omega + \widehat{C}_\zeta)^2\right)$$
$$\times M_y(-\omega^2 + \widehat{C}_y i\omega + \omega_y^2) M_x(-\omega^2 + \widehat{C}_x i\omega + \omega_x^2)$$
$$- I_\zeta (-\omega^2 + \widehat{C}_\zeta i\omega + v_\zeta^2 - 1)$$
$$\times \left(M_y(-\omega^2 + \widehat{C}_y i\omega + \omega_y^2) + M_x(-\omega^2 + \widehat{C}_x i\omega + \omega_x^2)\right) (NS_\zeta^2/2)\omega^4$$
$$+ (NS_\zeta^2/2)^2 \omega^8 = 0 \qquad (20.102)$$

It is not possible to obtain an analytical solution for the stability boundary by eliminating ω^2 from the real and imaginary parts of this equation. An inverse solution is possible in which both the real and imaginary equations are solved for ω given the rotor speed, or for the rotor speed Ω at a range of flutter frequencies. The real

20.3 Ground Resonance

solutions for ω are then plotted as a function of Ω, as in the Coleman diagram. A point where the solutions of the real and imaginary equations cross identifies a flutter frequency for which $s = i\omega$ satisfies the characteristic equation and hence defines the stability boundary. The Coleman diagrams shown in Figures 20.9 to 20.11 are the solution of the real part for the case of no damping. A more direct procedure is to solve the characteristic equation for all roots or to solve the differential equations for the eigenvalues and plot the frequency and damping ratio of the modes as a function of Ω or some other parameter.

An analytical solution for an approximate stability criterion can be obtained by again assuming S_ζ is small. An instability can occur at a resonance of the low-frequency mode with a support mode. When there is no damping and $S_\zeta > 0$, such a resonance is unstable if $\nu_\zeta < 1/\text{rev}$. Consider the damping required to stabilize this motion. Since the point of exact resonance of the uncoupled frequencies is roughly in the center of the instability region, that point is expected to be the most critical case, requiring the most damping to stabilize. Therefore we derive the stability boundary exactly at resonance, $\omega_x = 1 - \nu_\zeta$. The solution is expanded for small S_ζ^2, so $\omega^2 \cong \omega_x^2$. Since the instability is due to the inertial coupling S_ζ, the damping $(\widehat{C}_\zeta, \widehat{C}_x, \widehat{C}_y)$ at the stability boundary must also be of order S_ζ. For now $\omega_x \neq \omega_y$ is assumed. Then to lowest order in S_ζ^2, the characteristic equation for $s = i\omega$ gives the stability boundary:

$$I_\zeta^2 \left(4\nu_\zeta \widehat{C}_\zeta i\omega_x\right) M_y(\omega_y^2 - \omega_x^2) M_x(\widehat{C}_x i\omega_x) + I_\zeta 2(1-\nu_\zeta) M_y(\omega_y^2 - \omega_x^2)(NS_\zeta^2/2)\omega_x^4 \tag{20.103}$$

Hence the criterion for stability is

$$\frac{C_\zeta C_x}{\omega_x^2} > \frac{N}{4} \frac{1-\nu_\zeta}{\nu_\zeta} S_\zeta^2 \tag{20.104}$$

which was first obtained by Deutsch (1946). For stiff in-plane rotors ($\nu_\zeta > 1/\text{rev}$), the right-hand side is negative and the motion is always stable. For a soft in-plane rotor ($\nu_\zeta < 1/\text{rev}$), the product of the lag and support damping must be greater than this critical value for stability. The criterion for the lateral mode resonance at $\omega_y = 1 - \nu_\zeta$ is similar. The stability boundary for a resonance with the high-frequency lag mode, $\omega_x = 1 + \nu_\zeta$, gives the criterion

$$\frac{C_\zeta C_x}{\omega_x^2} > -\frac{N}{4} \frac{1+\nu_\zeta}{\nu_\zeta} S_\zeta^2 \tag{20.105}$$

which is always satisfied, even for zero damping.

For the case of an isotropic support ($\omega_x = \omega_y$), the characteristic equation to lowest order in S_ζ gives instead the following stability criterion:

$$\frac{C_\zeta}{\omega_x^2} \frac{C_y C_x}{C_y + C_x} > \frac{N}{4} \frac{1-\nu_\zeta}{\nu_\zeta} S_\zeta^2 \tag{20.106}$$

For isotropic support damping as well ($C_x = C_y$) this becomes

$$\frac{C_\zeta C_x}{\omega_x^2} > \frac{N}{2} \frac{1-\nu_\zeta}{\nu_\zeta} S_\zeta^2 \tag{20.107}$$

The isotropic case requires twice the damping as the anisotropic support because equal lateral and longitudinal support frequencies allow a whirling motion of the hub that couples well with the whirling motion of the low-frequency lag mode. The definition of an isotropic support requires that the frequencies ω_x and ω_y be of order

$S_\zeta^2/I_\zeta M_x$ apart, which is a small frequency difference. So in practice the isotropic case is relevant when the rotor support structure is truly axisymmetric.

The resonance of the low-frequency lag mode with a support mode is unstable if the lag frequency is below 1/rev and the product of the lag and support damping is below a critical level. The other resonances of the lag and support modes are stable even with no damping. The damping required for ground resonance stability is proportional to the inertial coupling parameter S_ζ. The damping required is also proportional to $(1 - \nu_\zeta)/\nu_\zeta$. For the small lag frequency typical of articulated rotors, a large amount of lag damping is required, so mechanical lag dampers are needed to ensure ground resonance stability. For typical soft in-plane hingeless rotors, the factor $(1 - \nu_\zeta)/\nu_\zeta$ is an order of magnitude smaller than for articulated rotors, so the blade structural damping may provide a sufficient level of C_ζ. For ground resonance stability, a high lag frequency is desired, but if ν_ζ is too close to 1/rev the blade loads and vibration are excessive. Thus even a hingeless rotor can require mechanical lag dampers for stability.

The Deutsch criterion (equation 20.104) defines a threshold for the product of damping. Both lag damping and support damping are required for stability. The parameters in equation 20.104 are dimensional, except the lag frequency ν_ζ is per-rev. The lag damping C_ζ is effective angular damping about the lag hinge (ft-lb/rad/sec or m-N/rad/sec). The support damping C_x is effective linear damping at the hub (lb/ft/sec or N/m/sec). The right-hand size of equation 20.104 is defined by the rotor: number of blades N, lag mode first moment of inertia $S_\zeta = \int_0^R \eta_\zeta m \, dr$, and lag frequency ν_ζ. The support mode is defined by the natural frequency ω_x (rad/sec) and the damping C_x. These parameters can be obtained for each vibration mode of the rotor support from the measured frequency response of the hub to excitation by in-plane forces. For uncoupled support modes, the damping is given by the peak response of displacement to force at the hub node: $C_x = 1/(\omega_x |x/F|_{\max})$.

The Deutsch criterion defines the lag damping required at the resonance of the low-frequency lag mode with ω_x, which occurs at the rotor speed $\Omega = \omega_x/(1 - \nu_\zeta)$. Then for each lateral and longitudinal support mode a critical ground resonance rotor speed is obtained, as well as the lag damping required at that Ω to stabilize the motion. By comparing the lag damping required with the damping available as a function of rotor speed, the ground resonance stability can be assessed for a given rotor and helicopter.

Ground resonance stability with articulated and soft in-plane hingeless rotors can be achieved by providing a sufficient level of damping of the rotor lag motion and of the support motion. Instabilities can also be avoided by a proper placement of the natural frequencies of the airframe to avoid resonances, but often there are too many other constraints on the structural design for this to be a practical means of handling the ground resonance problem. With a stiff in-plane rotor (for example, two-bladed teetering rotors and some hingeless rotor designs) the resonances are all stable. With articulated rotors, mechanical dampers on the landing gear and at the lag hinges are standard features of the helicopter design. The linear analysis developed here has assumed viscous damping, in which the force opposing the motion is proportional to the velocity of the motion. The actual damping of the rotor and support is almost certainly nonlinear, particularly if mechanical dampers are used. As described in section 18.6, nonlinear lag dampers can be described by an equivalent viscous damping coefficient, based on the energy dissipated during a cycle of motion. By this means the linear analysis can be applied to the real rotor.

The equivalent viscous damping depends on the frequency and amplitude of the lag motion. For example, frictional damping (restoring force proportional to the sign of $\dot{\zeta}$) gives an equivalent viscous damping coefficient proportional to $1/(\omega\zeta_{amp})$; whereas hydraulic damping (restoring force proportional to $\dot{\zeta}|\dot{\zeta}|$) gives an equivalent damping proportional to $\omega\zeta_{amp}$. The frequency of the ground resonance mode can be assumed to be near the lag frequency, $\omega \cong \nu_\zeta \Omega$ in the rotating frame, so that the rotor speed defines the frequency for the lag dampers. Then the lag damping level required for stability can be interpreted as a limitation on the lag amplitude. The damping of the support is also likely to be nonlinear, because of the complex structure of the helicopter and the presence of nonlinear elements such as oleo struts and tires. The analysis should use the lowest equivalent viscous damping that is likely to be encountered. Since calculation of the support characteristics is difficult at best, the ground resonance analysis must rely on the measured airframe frequencies and damping. The ground resonance instability is a simple phenomenon physically, and therefore with good measurements of the rotor and support damping the stability can be accurately predicted.

20.3.4 Complex Variable Representation of Motion

For the case of an axisymmetric support, complex variables can be used for a more compact analysis (see section 19.7). Coleman (1943) used complex combinations of the lag degrees of freedom and the shaft motion degrees of freedom to facilitate derivation and solution of the ground resonance equations. Let $z = \zeta_{1c} + i\zeta_{1s}$ and $h = y_h - ix_h$. Then equations 20.84 become

$$I_\zeta \left(\ddot{z} + \widehat{C}_\zeta (\dot{z} - iz) - 2i\dot{z} + (\nu_\zeta^2 - 1)z \right) - S_\zeta \ddot{h} = 0 \quad (20.108)$$

$$M_x \left(\ddot{h} + \widehat{C}_x \dot{h} + \omega_x^2 h \right) - \frac{N}{2} S_\zeta \ddot{z} = 0 \quad (20.109)$$

The characteristic equation

$$I_\zeta \left(s^2 + (\widehat{C}_\zeta - 2i)s + \nu_\zeta^2 - 1 - i\widehat{C}_\zeta \right) M_x (s^2 + \widehat{C}_x s + \omega_x^2) - \frac{N}{2} S_\zeta^2 s^4 = 0 \quad (20.110)$$

becomes for zero damping

$$I_\zeta (s^2 - 2is + \nu_\zeta^2 - 1) M_x (s^2 + \omega_x^2) - \frac{N}{2} S_\zeta^2 s^4 = 0 \quad (20.111)$$

The stability boundary for zero damping is obtained by setting $s = i\omega$:

$$I_\zeta \left(-\omega^2 + 2\Omega\omega + \Omega^2(\nu_\zeta^2 - 1) \right) M_x(-\omega^2 + \omega_x^2) - \frac{N}{2} S_\zeta^2 \omega^4 = 0 \quad (20.112)$$

which is a quadratic equation for Ω given ω. On the stability boundary at resonance, $s = i\omega_x = i(1 - \nu_\zeta)$, the characteristic equation

$$-C_\zeta i\nu_\zeta C_x i\omega_x - \frac{N}{2} S_\zeta^2 \omega_x^2 (1 - \nu_\zeta) = 0 \quad (20.113)$$

gives

$$\frac{C_\zeta C_x}{\omega_x^2} > \frac{N}{2} \frac{1 - \nu_\zeta}{\nu_\zeta} S_\zeta^2 \quad (20.114)$$

the stability criterion for a symmetric support (equation 20.107).

20.3.5 Two-Bladed Rotor

Now let us consider the ground resonance stability of a two-bladed rotor. Because the rotor inertial properties are not axisymmetric as they are when $N \geq 3$, the cyclic lag degrees of freedom are not applicable to the two-bladed rotor. Instead, the lag motion is described by the differential lag degree-of-freedom ζ_1. The equation of motion for ζ_1 was obtained in section 16.7,

$$I_\zeta \left(\ddot{\zeta}_1 + \widehat{C}_\zeta + v_\zeta^2 \zeta_1 \right) + S_\zeta \left(\ddot{x}_h \sin \psi - \ddot{y}_h \cos \psi \right) = 0 \qquad (20.115)$$

(equation 16.262), and the hub forces in sections 16.6.2 and 16.7:

$$H = -NM_b \ddot{x}_h - NS_\zeta \left((\ddot{\zeta}_1 - \zeta_1) \sin \psi + 2\dot{\zeta}_1 \cos \psi \right) \qquad (20.116)$$

$$Y = -NM_b \ddot{y}_h + NS_\zeta \left((\ddot{\zeta}_1 - \zeta_1) \cos \psi - 2\dot{\zeta}_1 \sin \psi \right) \qquad (20.117)$$

(equations 16.238, 16.239, 16.264, and 16.263). The aerodynamic forces have been dropped and a lag damper included in the equation for ζ_1. Again the hub longitudinal and lateral equations of motion are

$$\widetilde{M}_x \ddot{x}_h + C_x \dot{x}_h + K_x x_h = H \qquad (20.118)$$

$$\widetilde{M}_y \ddot{y}_h + C_y \dot{y}_h + K_y y_h = Y \qquad (20.119)$$

The mass, spring, and damping of the support modes are defined by $M_x = \widetilde{M}_x + NM_b$, $K_x = M_x \omega_x^2$, and $C_x = M_x \widehat{C}_x$; and similarly for the lateral mode. Then the equations of motion describing ground resonance of a two-bladed rotor are

$$\begin{bmatrix} I_\zeta & -S_\zeta \cos \psi & S_\zeta \sin \psi \\ -NS_\zeta \cos \psi & M_y & 0 \\ NS_\zeta \sin \psi & 0 & M_x \end{bmatrix} \begin{pmatrix} \ddot{\zeta}_1 \\ \ddot{y}_h \\ \ddot{x}_h \end{pmatrix} + \begin{bmatrix} I_\zeta \widehat{C}_\zeta & 0 & 0 \\ 2NS_\zeta \sin \psi & M_y \widehat{C}_y & 0 \\ 2NS_\zeta \cos \psi & 0 & M_x \widehat{C}_x \end{bmatrix} \begin{pmatrix} \dot{\zeta}_1 \\ \dot{y}_h \\ \dot{x}_h \end{pmatrix}$$

$$+ \begin{bmatrix} I_\zeta v_\zeta^2 & 0 & 0 \\ NS_\zeta \cos \psi & M_y \omega_y^2 & 0 \\ -NS_\zeta \sin \psi & 0 & M_x \omega_x^2 \end{bmatrix} \begin{pmatrix} \zeta_1 \\ y_h \\ x_h \end{pmatrix} = 0 \qquad (20.120)$$

These equations have periodic coefficients because of the inertial asymmetry of the rotor when $N = 2$, and the fact that the lag degree-of-freedom ζ_1 is really still in the rotating frame. The methods for analyzing the stability of such equations are discussed in section 15.6.

For the case of an isotropic support, constant coefficient differential equations can be obtained in the rotating frame. Assume $\omega_x = \omega_y$, $M_x = M_y$, and $C_x = C_y$; and define the hub deflections in the rotating frame:

$$y_r = y_h \cos \psi - x_h \sin \psi \qquad (20.121)$$

$$x_r = y_h \sin \psi + x_h \cos \psi \qquad (20.122)$$

A similar transformation of the hub in-plane forces generates the differential equations for x_r and y_r in the rotating frame. Then the equations describing the ground resonance dynamics with an isotropic support are

$$\begin{bmatrix} I_\zeta & -S_\zeta & 0 \\ -NS_\zeta & M_y & 0 \\ 0 & 0 & M_y \end{bmatrix} \begin{pmatrix} \ddot\zeta_1 \\ \ddot y_r \\ \ddot x_r \end{pmatrix} + \begin{bmatrix} I_\zeta \widehat C_\zeta & 0 & -2S_\zeta \\ 0 & M_y \widehat C_y & 2M_y \\ 2NS_\zeta & -2M_y & M_y \widehat C_y \end{bmatrix} \begin{pmatrix} \dot\zeta_1 \\ \dot y_r \\ \dot x_r \end{pmatrix}$$
$$+ \begin{bmatrix} I_\zeta v_\zeta^2 & S_\zeta & 0 \\ NS_\zeta & M_y(\omega_y^2 - 1) & M_y \widehat C_y \\ 0 & -M_y \widehat C_y & M_y(\omega_y^2 - 1) \end{bmatrix} \begin{pmatrix} \zeta_1 \\ y_r \\ x_r \end{pmatrix} = 0 \qquad (20.123)$$

which have constant coefficients. In the rotating frame there are Coriolis and centrifugal forces coupling the equations for the support motion. The characteristic equation is

$$M_y^2 \left((s^2 + \widehat C_y s + \omega_y^2 - 1)^2 + (2s + \widehat C_y)^2 \right) I_\zeta (s^2 + \widehat C_\zeta s + v_\zeta^2)$$
$$+ M_y \left((s^2 + \widehat C_y s + \omega_y^2 - 1)(4s^2 - (s^2 - 1)^2) - 4s(s^2 - 1)(2s + \widehat C_y) \right) NS_\zeta^2 = 0$$
$$(20.124)$$

For the case of no damping, the characteristic equation reduces to

$$M_y^2 \left((s^2 + \omega_y^2 - 1)^2 + 4s^2 \right) I_\zeta (s^2 + v_\zeta^2)$$
$$+ M_y \left((s^2 + \omega_y^2 - 1)(4s^2 - (s^2 - 1)^2) - 8s^2(s^2 - 1) \right) NS_\zeta^2 = 0 \quad (20.125)$$

The uncoupled solution ($S_\zeta = 0$) is $s = i\omega$, where $\omega = v_\zeta$ and $\omega = \omega_y \pm 1/\text{rev}$. Hence the support mode frequencies in the rotating frame are shifted by $\pm\Omega$ from the frequencies in the non-rotating frame, and the rotor mode is at the rotating lag natural frequency v_ζ.

As with $N \geq 3$, the solution of the characteristic equation for the case of no damping can be expanded in S_ζ about the uncoupled solution. The solution near $\omega = v_\zeta$ is

$$\omega^2 = v_\zeta^2 + \frac{(v_\zeta^2 - \omega_y^2 + 1)(4v_\zeta^2 + (v_\zeta^2 + 1)^2) - 8v_\zeta^2(v_\zeta^2 + 1)}{(v_\zeta^2 - \omega_y^2 + 1)^2 - 4v_\zeta^2} \frac{NS_\zeta^2 / I_\zeta}{M_y} \qquad (20.126)$$

and near $\omega = \omega_y \pm 1$, it is

$$\omega^2 = (\omega_y \pm 1)^2 - \frac{\omega_y^3 (\omega_y \pm 1)}{2(v_\zeta^2 - (\omega_y \pm 1)^2)} \frac{NS_\zeta^2 / I_\zeta}{M_y} \qquad (20.127)$$

With these expressions, the directions the solutions shift when $S_\zeta > 0$ can be established for the limits $\Omega = 0$ and Ω approaching infinity.

Figures 20.12 and 20.13 present Coleman diagrams for articulated (soft in-plane) and stiff in-plane two-bladed rotors. As in the case of three or more blades, a ground resonance instability appears with soft in-plane rotors ($v_\zeta < 1/\text{rev}$) at the resonance of the support and the low-frequency lag mode, which in the rotating frame means $v_\zeta = \Omega - \omega_y$.

For $N = 2$ the center of the ground resonance instability range is shifted to a rotor speed above the uncoupled resonance, in contrast to the $N \geq 3$ case, for which the instability range remains centered about the resonance. This suggests that

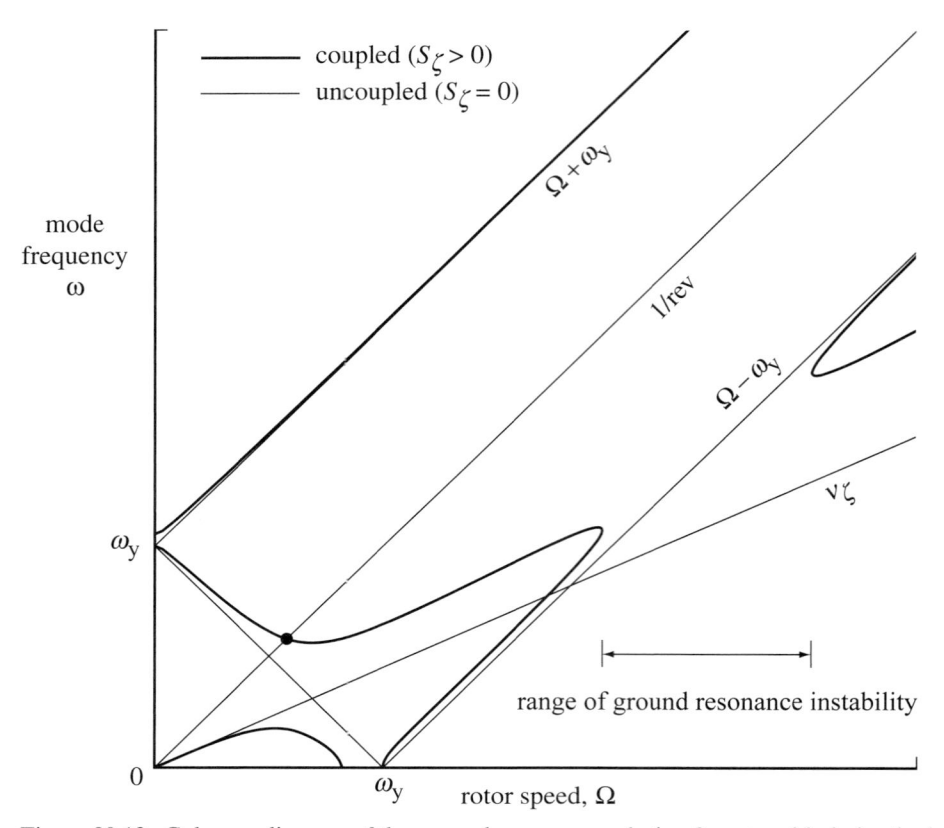

Figure 20.12. Coleman diagram of the ground resonance solution for a two-bladed articulated rotor ($\nu_\zeta < 1/\text{rev}$) on an isotropic support.

for large enough coupling the instability region might be shifted above the rotor operating range. To examine this possibility, consider the intersection of the locus with the 1/rev line, as indicated in Figure 20.12. On substituting $s^2 = -1$, for the case of no damping, the characteristic equation reduces to

$$M_y \omega_y^2 \left(M_y(\omega_y^2 - 4) I_\zeta (\nu_\zeta^2 - 1) - 8NS_\zeta^2 \right) = 0 \qquad (20.128)$$

Now since $\omega = 1/\text{rev}$ in the rotating frame corresponds to $\omega = 0$ and to $\omega = 2/\text{rev}$ in the non-rotating frame, the uncoupled solutions are $\omega_y = 0$ and $\omega_y = 2/\text{rev}$, and $\nu_\zeta = 1/\text{rev}$. The $\omega_y = 2/\text{rev}$ solution ($\omega = \omega_y - \Omega$) is of interest here. For the coupled case ($S_\zeta > 0$), this resonance occurs at the rotor speed

$$\Omega^2 = \frac{(\omega_y/2)^2}{1 - \dfrac{2}{1 - \nu_\zeta^2} \dfrac{NS_\zeta^2/I_\zeta}{M_y}} \qquad (20.129)$$

which increases with S_ζ for the soft in-plane rotor. The ground resonance instability always occurs at a rotor speed above this value, and thus it provides a conservative criterion for avoiding ground resonance. If

$$\frac{NS_\zeta^2/I_\zeta}{M_y} > \frac{1 - \nu_\zeta^2}{2} \qquad (20.130)$$

20.3 Ground Resonance

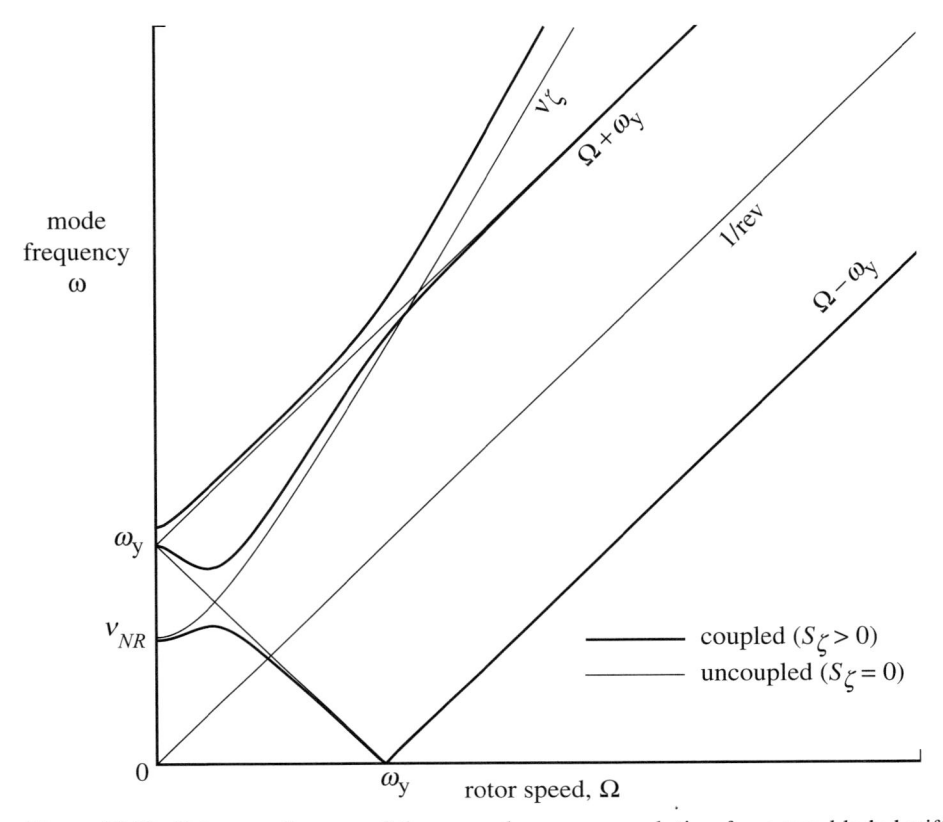

Figure 20.13. Coleman diagram of the ground resonance solution for a two-bladed stiff in-plane rotor ($v_\zeta > 1/\text{rev}$) on an isotropic support.

then both the 1/rev resonance (2/rev in the non-rotating frame) and the instability region are swept to $\Omega = \infty$. The inertial coupling required is rather large, however, even when $v_\zeta = 0.85$ or so, which is about the upper limit for soft in-plane hingeless rotors.

In Figure 20.12 there is also a region, just below $\Omega = \omega_y$, where there are only two real solutions for ω and hence the motion is unstable. This phenomenon occurs also for the stiff in-plane rotor (a very narrow range in Figure 20.13), but is not present for rotors with three or more blades. In this instability region two roots of the characteristic equation are on the real axis, with zero frequency: one positive and one negative. The frequency $\omega = 0$ in the rotating frame corresponds to $\omega = 1/\text{rev}$ in the non-rotating frame. Hence this instability is associated with the shaft critical speeds, where the rotor speed passes through the support natural frequency ω_y. On setting $s = 0$, the characteristic equation becomes

$$(\omega_y^2 - 1)\left(M_y(\omega_y^2 - 1)I_\zeta v_\zeta^2 - NS_\zeta^2\right) = 0 \qquad (20.131)$$

which has the two solutions, $\omega_y^2 = 1$ and

$$\omega_y^2 = 1 + \frac{1}{v_\zeta^2}\frac{NS_\zeta^2/I_\zeta}{M_y} \qquad (20.132)$$

These equations are valid for large S_ζ and give the two points where the coupled roots intercept the s-axis. The shaft critical instability thus occurs in the range

$$\frac{\omega_y^2}{1 + \frac{1}{v_\zeta^2}\frac{NS_\zeta^2/I_\zeta}{M_y}} < \Omega^2 < \omega_y^2 \qquad (20.133)$$

For an articulated rotor (with small v_ζ) this rotor speed range can be large even though the inertial coupling is small.

Next, consider the damping required to stabilize the ground resonance motion of a two-bladed rotor. With lag and support damping, $s = i\omega$ defines the stability boundary. As in the $N \geq 3$ case, the equation for the stability boundary is expanded in S_ζ^2 about the resonance $v_\zeta = 1 - \omega_y$. To lowest order in S_ζ^2, the characteristic equation gives the stability criterion

$$C_\zeta C_y > -\frac{\omega_y^3(\omega_y - 1)}{2v_\zeta^2} NS_\zeta^2 \qquad (20.134)$$

or since $\omega_y(\omega_y - 1)/v_\zeta = v_\zeta - 1$,

$$\frac{C_\zeta C_y}{\omega_y^2} > \frac{N}{2}\frac{1 - v_\zeta}{v_\zeta} S_\zeta^2 \qquad (20.135)$$

This is the same criterion as for a rotor with three or more blades on an isotropic support (equation 20.107). For a stiff in-plane rotor ($v_\zeta > 1/\text{rev}$) the resonance is always stable.

The damping required to stabilize the shaft critical mode (a divergence instability in the rotating frame) is obtained by substituting $s = 0$ in the characteristic equation:

$$M_y^2\left((\omega_y^2 - 1)^2 + \widehat{C}_y^2\right)I_\zeta v_\zeta^2 - M_y(\omega_y^2 - 1)NS_\zeta^2 = 0 \qquad (20.136)$$

The motion is stable if the support damping satisfies the criterion

$$C_y^2 > M_y(\omega_y^2 - 1)\left[\frac{1}{v_\zeta^2} NS_\zeta^2/I_\zeta - M_y(\omega_y^2 - 1)\right] \qquad (20.137)$$

The damping required is zero at the end points of the shaft critical speed region and has a maximum midway between them at

$$\omega_y^2 = 1 + \frac{1}{2v_\zeta^2}\frac{NS_\zeta^2/I_\zeta}{M_y} \qquad (20.138)$$

or

$$\Omega^2 = \frac{\omega_y^2}{1 + \frac{1}{2v_\zeta^2}\frac{NS_\zeta^2/I_\zeta}{M_y}} \qquad (20.139)$$

The damping required to stabilize the entire range is

$$C_y > \Omega\frac{1}{2v_\zeta^2} NS_\zeta^2/I_\zeta \cong \omega_y\frac{1}{2v_\zeta^2} NS_\zeta^2/I_\zeta \qquad (20.140)$$

This result is exact for all values of the inertial coupling S_ζ. In contrast to the ground resonance instability, only support damping is required to stabilize this motion. The level of support damping specified by this criterion is usually not very large.

Coleman and Feingold (1947) investigated the general case of a two-bladed rotor on an anisotropic support and obtained the stability by the infinite determinant method for analyzing periodic-coefficient differential equations. They found that the dynamic behavior, specifically the possible instabilities, are much the same as for the case with isotropic support. However, the periodic coefficients introduce additional resonances. A ground resonance instability can occur for soft in-plane rotors at frequencies near $\omega_y = 1 - \nu_\zeta + 2n$/rev, or at $\Omega = \omega_y/(1 - \nu_\zeta + 2n)$ where n is a positive integer. The shaft critical speed occurs at $\omega_y = 1 + 2n$/rev or at $\Omega = \omega_y/(1 + 2n)$. Thus for given rotor lag and support frequencies, additional resonances occur at rotor speeds lower than the fundamental. These resonances due to the periodic coefficients tend to occur at low Ω and have a narrower instability range than the fundamental resonance. Hence much less damping is required to stabilize the motion in these regions.

The most common two-bladed rotor design is the stiff in-plane teetering configuration. Ground resonance is not a concern since the lag frequency is above 1/rev, and only a low level of support damping is required to handle the shaft critical speed instability.

20.3.6 Air Resonance

The mechanical instability involving the coupling of the blade lag motion with the in-plane motion of a body mode can occur in flight, and then it is called air resonance. When the aircraft is not in contact with the ground, the frequency of the resonant body mode is determined by the flapping stiffness and by the damping provided by the aerodynamic forces due to flap. Although the mechanism of the instability is the same as for ground resonance, the analysis of air resonance is more complicated, requiring models of flap motion and aerodynamics. See Ormiston (1991) for further exposition.

Articulated rotors generally do not encounter air resonance. With a small lag frequency and small flap stiffness, resonances of the regressive lag mode and body modes occur at low rotor speed, well below the operating speeds required for flight. Air resonance must be considered for soft in-plane hingeless and bearingless rotors. The higher lag frequency (smaller regressive lag mode frequency) and higher flap stiffness (higher support frequencies) raise the rotor speed of resonances, possibly to near operating speeds. Moreover, hingeless rotors typically do not have much mechanical or structural damping of the lag motion. The body pitch and roll degrees of freedom produce in-plane hub motion and couple with the low-frequency flap motion; hence these are the support modes usually involved with air resonance of helicopters.

A tiltrotor with a soft in-plane rotor can encounter air resonance involving the fundamental vibration modes of the wing. In-plane motion of the rotor hub is produced by the wing chord bending motion in the helicopter configuration and by the wing vertical bending motion in the airplane configuration.

20.3.7 Dynamic Inflow

The rotor wake has a strong influence on blade flap dynamics, as discussed in section 19.5. Dynamic inflow (section 11.3) provides a global, low-frequency model for the wake-induced velocity. The simplest model has three inflow states, consisting of uniform (λ_0) and linear (λ_c, λ_s) perturbations of the wake-induced downwash at the rotor disk. Rotor aerodynamic hub moments produce perturbations of the linear inflow gradient over the disk, so the wake influence is largest with hingeless rotors. Hence the wake influence is an important factor for air resonance.

Bousman (1981) established experimentally the existence of a rotor aerodynamic state in a test of the ground resonance stability of a model rotor. The principal objective of the test was to explore the potential for pitch-flap coupling, flap-lag structural coupling, and matched flap and lag stiffness to stabilize the rotor regressive lag mode. The model was a three-bladed hingeless rotor, supported in a gimbal frame that allowed pitch and roll motion. The configurations tested included a non-matched stiffness rotor (configuration 1, $v_\zeta = 0.59$ and $v_\beta = 1.11$ at maximum rotor speed) and a matched stiffness rotor (configuration 4, $v_\zeta = 0.59$ and $v_\beta = 1.16$ at maximum rotor speed). Matched stiffness (equal non-rotating flap and lag stiffness) was obtained by increasing the stiffness of the flap flexure. Hence configuration 4 could generate larger hub moments, resulting in larger participation of inflow states. The model was tested in hover, at zero collective. The experimental method found all modes that could be excited, not just the regressive-lag and body modes.

Figures 20.14 and 20.15 show the measured modal frequency and damping for the non-matched and matched stiffness configurations. For the non-matched stiffness configuration (lower flap stiffness), the frequencies are reasonably well predicted without a dynamic inflow model, but the pitch and roll mode damping are over-predicted. Without the dynamic inflow model, the frequency and damping of the matched stiffness configuration are accurately calculated only for the regressive lag mode; the calculations are poor for all other modes. The calculated damping of the flap regressive mode is so high that the mode should not have been observable in the test.

Bousman hypothesized that the differences between measured and calculated results were attributable to the influence of inflow dynamics, which was confirmed in calculations by Johnson (1982). The frequency and damping for both matched and non-matched stiffness configurations are predicted well by including the three-state dynamic inflow model in the analysis (Figures 20.14 and 20.15). The implication is that because of the organization of the wake, the air responds as a whole to motion of the rotor, a response well described by the loading derivatives and time lag of simple dynamic inflow models. In addition to showing the large effect of the wake on coupled airframe-rotor dynamics of hingeless rotors, Bousman's test constitutes unique experimental evidence that the air about the rotor (the wake) behaves as characterized by dynamic inflow states.

20.3.8 History

Coleman and Feingold (1958) developed the analysis of ground resonance. This report republished work by Coleman (1943) on the case of three or more blades, by Feingold (1943) on the case of a two-bladed rotor with an isotropic support, and by Coleman and Feingold (1947) on the undamped two-bladed rotor with an anisotropic support. According to the foreword of Coleman and Feingold (1958),

> During the early part of World War II, some of the helicopters designed for military use were observed during ground tests to exhibit a violent oscillatory rotor instability which endangered the safety of the aircraft. This instability was at first attributed to rotor-blade

20.3 Ground Resonance

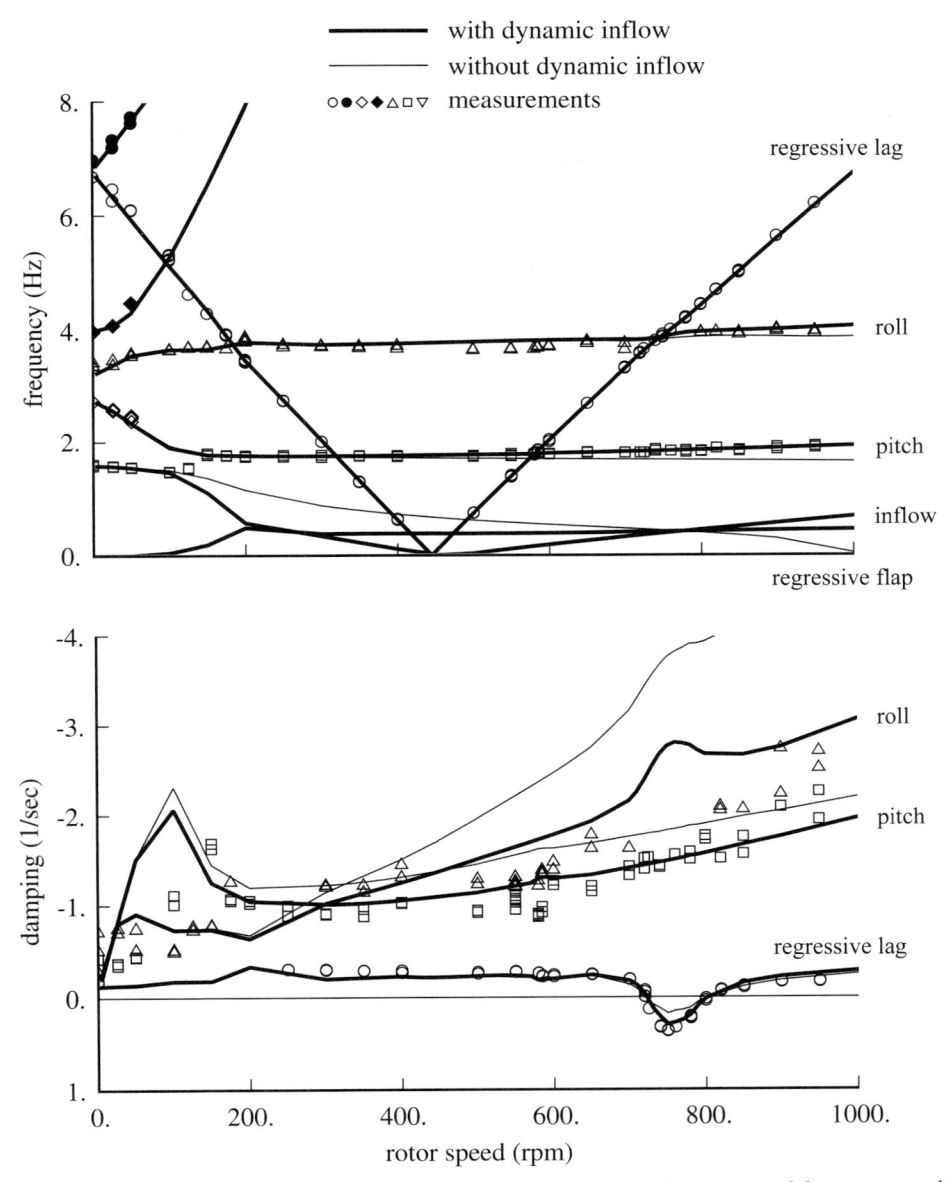

Figure 20.14. Modal frequencies and damping as a function of rotor speed for non-matched stiffness configuration, comparing model rotor measurements and calculations with and without dynamic inflow.

flutter, but a careful analysis indicated it to be caused by a hitherto unknown phenomenon in which the rotational energy of the rotor was converted into oscillatory energy of the blades. This phenomenon was usually critical when the helicopter was operating on or near the ground and, hence, was called ground resonance. An oscillatory instability of such magnitude as resulted from this phenomenon would generate forces that could quickly destroy a helicopter. The research efforts of the National Advisory Committee for Aeronautics were therefore enlisted to investigate the difficulties introduced by this phenomenon. During the interval between 1942 and 1947, a theory of the self-excited instability of hinged rotor blades was worked out by Robert P. Coleman and Arnold

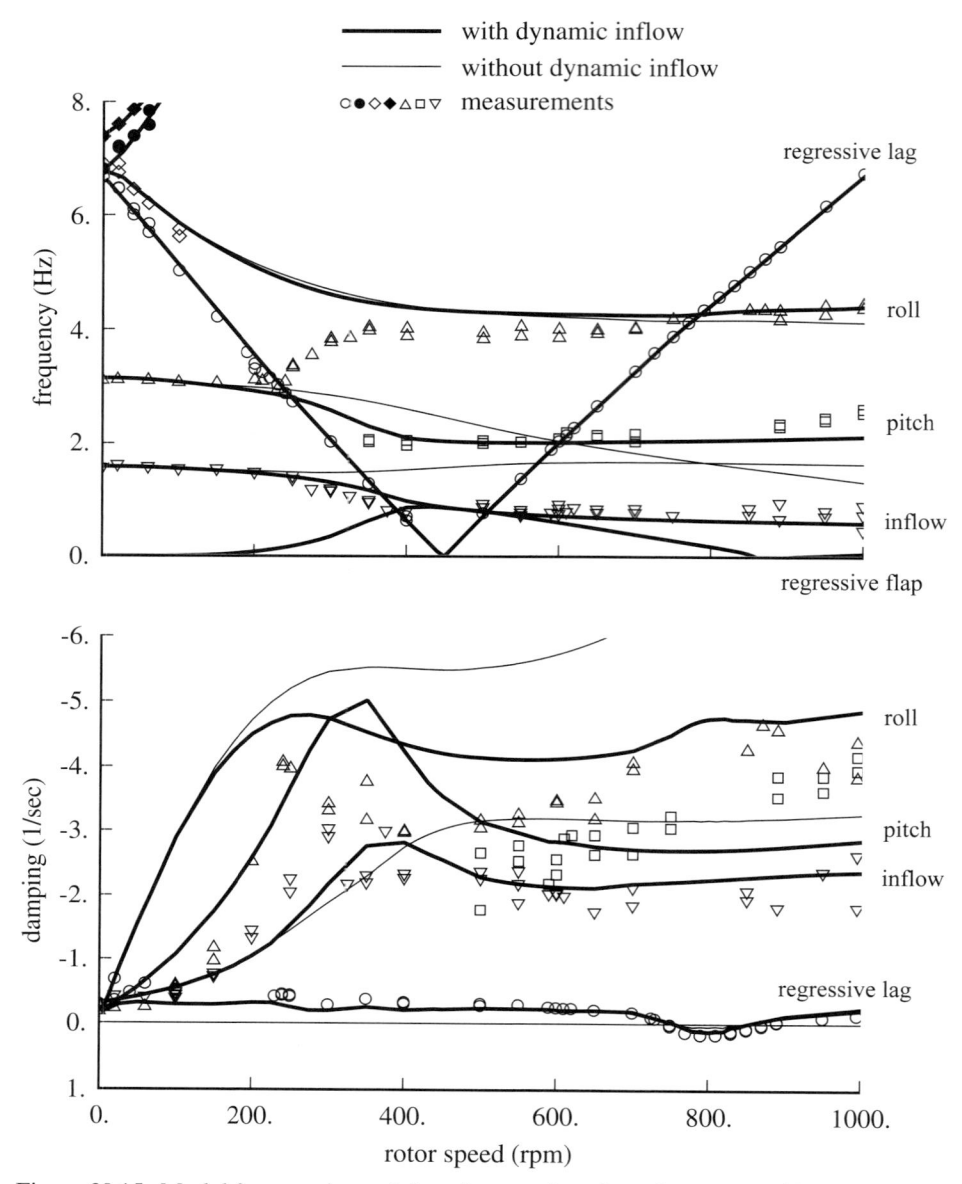

Figure 20.15. Modal frequencies and damping as a function of rotor speed for matched stiffness configuration, comparing model rotor measurements and calculations with and without dynamic inflow.

M. Feingold at the Langley Aeronautic Laboratory. This theory defined the important parameters and provided design information which made it possible to eliminate this type of instability.

Coleman's 1943 report was a corrected version of a July 1942 Advance Restricted Report of the same title. The original reports were combined into a single volume by George W. Brooks, who also contributed an appendix.

20.3 Ground Resonance

Ground resonance was a problem encountered by a number of autogyros. When the YG-1 rotor destroyed itself in the NACA Langley full-scale wind tunnel, the problem was recognized as an inherent instability of the aircraft. Gustafson (1970) mentioned the 1937 accident in the NACA wind tunnel, but noted that "the studies of [Coleman] were originally inspired by industry test experience for an advanced autogiro." Brooks (1988) described the 1941 Kellett XR-2 development:

> The accident to XR-2 proved to be important to the whole development of rotary wings. During one of the first tests of a jump take-off, ground resonance set in at high rotor speed. This built up so rapidly that the aircraft broke up before anything could be done to stop it. In less than five seconds, the rotor pylon support structure collapsed and the fuselage broke in two places, between the engine and pylon and pylon and tail. This dramatic further demonstration of a problem that had recurred repeatedly throughout the development of rotary-wing aircraft had an important effect in influencing the United States Army Air Force, the NACA and Kellett into tackling the basic problem of ground resonance. Bob Wagner of Kellett and Prewitt Coleman of NACA came up independently with mathematical solutions for the proper configuration and for damping to prevent ground resonance. This was a major step in the development of rotary-wing aircraft. Paul Stanley of the Autogiro Company of America had also arrived at mathematical and engineering solutions to the problem with the result that Pitcairn Autogiros are claimed to have largely avoided ground resonance.

There was contemporary development of ground resonance analysis by Wagner, Deutsch, and Focke (see Coleman (1943), Focke (1943), Deutsch (1946), and Wagner (1954)). Coleman's work provided the foundation for future engineering treatment of ground resonance in helicopters.

Coleman (1943) developed the theory of ground resonance stability for rotors with three or more blades, on symmetric or asymmetric support, including the effects of damping. "Of the large number of degrees of freedom of a hinged rotor, the important ones for the present problem have been found to be hinge deflection of the blades in the plane of rotation and horizontal deflections of the pylon. Other degrees of freedom, such as the flapping hinge motion of the blades, the bending or torsion of the blades, and the torsion of the drive shaft, are considered unimportant in the problem of self-excited oscillations." By using "special linear combinations of the hinge deflections" in the fixed reference frame (equivalent to multiblade coordinates), equations with constant coefficients were obtained, and only the first-harmonic coordinates of the lag motion were coupled with the pylon motion. "The physical meaning of this partial separation of variables is that...involves a motion of the common center of mass of the blades and, thus, a coupling effect with the pylon." Introducing these new coordinates was a crucial step, enabling a practical mathematical solution for ground resonance stability. Complex combinations of the pylon degrees of freedom and the lag hinge degrees of freedom reduced the number of equations from four to two. The case of no damping reduces to a single equation that is a quadratic function of the rotor speed, for an assumed value of the whirl frequency (eigenvalue). The resulting plot of lag and body mode frequencies as a function of rotor speed is the Coleman diagram, in which the absence of a solution at a rotor speed indicates instability.

20.4 Whirl Flutter

Whirl flutter is the coupled motion of the airframe and a propeller or proprotor that becomes unstable at high forward speed, with the rotor operating in axial flow. The phenomenon was encountered on propeller-driven aircraft, with a rigid propeller mounted on a pylon with pitch and yaw flexibility. For a tiltrotor, the wing elastic motions (vertical bending, chordwise bending, and torsion) couple with a flexible proprotor. The rigid body and elastic motion of the blades have a significant effect on the whirl flutter character. Demonstrating whirl flutter stability and analysis capability was a major issue in the development of the tiltrotor concept. The derivation of the whirl flutter equations and the propeller whirl flutter solution of this section follows Johnson (1974b).

20.4.1 Whirl Flutter Equations

Consider a proprotor or propeller on an aircraft in forward flight, hence operating in axial flow with the rotor z-axis forward, x-axis upward, and y-axis to the right. The aerodynamic environment is axisymmetric, with high inflow ratio: $\mu_z = V/\Omega R$, $\lambda = \lambda_i + \mu_z$. The rotor is mounted on a pylon, with a pivot a distance h on the shaft axes aft of the hub. The pivot has a pitch degree-of-freedom α_y and a yaw degree-of-freedom α_x. So the in-plane hub displacements are $x_h = h\alpha_y$ and $y_h = -h\alpha_x$. The equations of motion of the pylon are

$$\widetilde{I}_y \ddot{\alpha}_y + \widetilde{C}_y \dot{\alpha}_y + \widetilde{K}_y \alpha_y = M_y + hH \tag{20.141}$$

$$\widetilde{I}_x \ddot{\alpha}_x + \widetilde{C}_x \dot{\alpha}_x + \widetilde{K}_x \alpha_x = M_x - hY \tag{20.142}$$

where the right-hand side is the rotor hub forces and moments. The hub forces include the inertial reactions to the blade mass, which are added pylon inertia to obtain the total moments of inertia: $I_x = \widetilde{I}_x + NM_b h^2$. The pylon equations are normalized by dividing by $\frac{N}{2} I_b$:

$$\widehat{I}_y \ddot{\alpha}_y + \widehat{C}_y \dot{\alpha}_y + \widehat{K}_y \alpha_y = \gamma \frac{2C_{My}}{\sigma a} + h\gamma \frac{2C_H}{\sigma a} \tag{20.143}$$

$$\widehat{I}_x \ddot{\alpha}_x + \widehat{C}_x \dot{\alpha}_y + \widehat{K}_x \alpha_x = \gamma \frac{2C_{Mx}}{\sigma a} - h\gamma \frac{2C_Y}{\sigma a} \tag{20.144}$$

where $\widehat{I}_x = I_x/(\frac{N}{2} I_b)$, $\widehat{C}_x = \widetilde{C}_x/(\frac{N}{2} I_b \Omega)$, and $\widehat{K}_x = \widetilde{K}_x/(\frac{N}{2} I_b \Omega^2) = \widehat{I}_x \omega_x^2$. The pylon motion couples with the tip-path-plane tilt degrees of freedom of the rotor. Substituting for x_h and y_h, equation 19.26 for the flap motion becomes

$$\begin{pmatrix} \ddot{\beta}_{1c} \\ \ddot{\beta}_{1s} \end{pmatrix} + \begin{bmatrix} -\gamma M_{\dot{\beta}} & 2 \\ -2 & -\gamma M_{\dot{\beta}} \end{bmatrix} \begin{pmatrix} \dot{\beta}_{1c} \\ \dot{\beta}_{1s} \end{pmatrix} + \begin{bmatrix} v_\beta^2 - 1 & -\gamma M_{\dot{\beta}} \\ \gamma M_{\dot{\beta}} & v_\beta^2 - 1 \end{bmatrix} \begin{pmatrix} \beta_{1c} \\ \beta_{1s} \end{pmatrix}$$

$$= \begin{pmatrix} \ddot{\alpha}_y \\ -\ddot{\alpha}_x \end{pmatrix} + \begin{bmatrix} -\gamma M_{\dot{\beta}} & 2 \\ -2 & -\gamma M_{\dot{\beta}} \end{bmatrix} \begin{pmatrix} \dot{\alpha}_y \\ -\dot{\alpha}_x \end{pmatrix} + \gamma M_\mu \begin{pmatrix} -h\dot{\alpha}_x + \lambda \alpha_x \\ -h\dot{\alpha}_y + \lambda \alpha_y \end{pmatrix} \tag{20.145}$$

for a rotor with three or more blades, no pitch-flap coupling, and using $I_\beta = I_{\beta\alpha} = 1$. The in-plane velocity perturbations due to shaft tilt relative to the velocity vector

($\lambda\alpha$ terms) have been included now; see section 16.8.7. From equation 19.73, the hub force is

$$\begin{pmatrix} \dfrac{2C_H}{\sigma a} \\ \dfrac{2C_Y}{\sigma a} \end{pmatrix} = R_\beta \begin{pmatrix} \beta_{1c} \\ \beta_{1s} \end{pmatrix} + \begin{bmatrix} R_{\dot\beta} & H_{\dot\beta} \\ -H_{\dot\beta} & R_{\dot\beta} \end{bmatrix} \begin{pmatrix} \dot\beta_{1c} + \beta_{1s} - \dot\alpha_y \\ \dot\beta_{1s} - \beta_{1c} + \dot\alpha_x \end{pmatrix}$$

$$+ \begin{bmatrix} -(H_\mu + R_\mu) & R_r \\ -R_r & -(H_\mu + R_\mu) \end{bmatrix} \begin{pmatrix} h\dot\alpha_y - \lambda\alpha_y \\ -h\dot\alpha_x + \lambda\alpha_x \end{pmatrix} \quad (20.146)$$

The flapping equations can be used to express the hub moment (equation 19.75) in terms of the degrees of freedom:

$$\gamma \begin{pmatrix} -\dfrac{2C_{My}}{\sigma a} \\ \dfrac{2C_{Mx}}{\sigma a} \end{pmatrix} = (\nu_\beta^2 - 1) \begin{pmatrix} \beta_{1c} \\ \beta_{1s} \end{pmatrix}$$

$$= -\begin{pmatrix} \ddot\beta_{1c} \\ \ddot\beta_{1s} \end{pmatrix} - \begin{bmatrix} -\gamma M_{\dot\beta} & 2 \\ -2 & -\gamma M_{\dot\beta} \end{bmatrix} \begin{pmatrix} \dot\beta_{1c} \\ \dot\beta_{1s} \end{pmatrix} - \begin{bmatrix} 0 & -\gamma M_{\dot\beta} \\ \gamma M_{\dot\beta} & 0 \end{bmatrix} \begin{pmatrix} \beta_{1c} \\ \beta_{1s} \end{pmatrix}$$

$$+ \begin{pmatrix} \ddot\alpha_y \\ -\ddot\alpha_x \end{pmatrix} + \begin{bmatrix} -\gamma M_{\dot\beta} & 2 \\ -2 & -\gamma M_{\dot\beta} \end{bmatrix} \begin{pmatrix} \dot\alpha_y \\ -\dot\alpha_x \end{pmatrix} + \gamma M_\mu \begin{pmatrix} -h\dot\alpha_x + \lambda\alpha_x \\ -h\dot\alpha_y + \lambda\alpha_y \end{pmatrix} \quad (20.147)$$

The equations of motion for the tip-path-plane tilt and pylon angular motion are thus

$$\begin{pmatrix} \ddot\beta_{1c} \\ \ddot\beta_{1s} \end{pmatrix} + \begin{bmatrix} -1 & 0 \\ 0 & 1 \end{bmatrix} \begin{pmatrix} \ddot\alpha_y \\ \ddot\alpha_x \end{pmatrix}$$

$$+ \begin{bmatrix} -\gamma M_{\dot\beta} & 2 \\ -2 & -\gamma M_{\dot\beta} \end{bmatrix} \begin{pmatrix} \dot\beta_{1c} \\ \dot\beta_{1s} \end{pmatrix} + \begin{bmatrix} \gamma M_{\dot\beta} & 2 + h\gamma M_\mu \\ 2 + h\gamma M_\mu & -\gamma M_{\dot\beta} \end{bmatrix} \begin{pmatrix} \dot\alpha_y \\ \dot\alpha_x \end{pmatrix}$$

$$+ \begin{bmatrix} \nu_\beta^2 - 1 & -\gamma M_{\dot\beta} \\ \gamma M_{\dot\beta} & \nu_\beta^2 - 1 \end{bmatrix} \begin{pmatrix} \beta_{1c} \\ \beta_{1s} \end{pmatrix} + \begin{bmatrix} 0 & -\gamma\lambda M_\mu \\ -\gamma\lambda M_\mu & 0 \end{bmatrix} \begin{pmatrix} \alpha_y \\ \alpha_x \end{pmatrix} = 0 \quad (20.148)$$

and

$$\begin{bmatrix} -1 & 0 \\ 0 & 1 \end{bmatrix} \begin{pmatrix} \ddot\beta_{1c} \\ \ddot\beta_{1s} \end{pmatrix} + \begin{bmatrix} \widehat{I}_y + 1 & 0 \\ 0 & \widehat{I}_x + 1 \end{bmatrix} \begin{pmatrix} \ddot\alpha_y \\ \ddot\alpha_x \end{pmatrix}$$

$$+ \begin{bmatrix} \gamma M_{\dot\beta} - h\gamma R_{\dot\beta} & -2 - h\gamma H_{\dot\beta} \\ -2 - h\gamma H_{\dot\beta} & -(\gamma M_{\dot\beta} - h\gamma R_{\dot\beta}) \end{bmatrix} \begin{pmatrix} \dot\beta_{1c} \\ \dot\beta_{1s} \end{pmatrix}$$

$$+ \begin{bmatrix} \widehat{C}_y - \gamma M_{\dot\beta} + h\gamma R_{\dot\beta} + h^2\gamma(H_\mu + R_\mu) & -(2 + h\gamma M_\mu + h\gamma H_{\dot\beta} - h^2\gamma R_r) \\ 2 + h\gamma M_\mu + h\gamma H_{\dot\beta} - h^2\gamma R_r & \widehat{C}_x - \gamma M_{\dot\beta} + h\gamma R_{\dot\beta} + h^2\gamma(H_\mu + R_\mu) \end{bmatrix} \begin{pmatrix} \dot\alpha_y \\ \dot\alpha_x \end{pmatrix}$$

$$+ \begin{bmatrix} h\gamma(H_{\dot\beta} - R_\beta) & \gamma M_{\dot\beta} - h\gamma R_{\dot\beta} \\ \gamma M_{\dot\beta} - h\gamma R_{\dot\beta} & -h\gamma(H_{\dot\beta} - R_\beta) \end{bmatrix} \begin{pmatrix} \beta_{1c} \\ \beta_{1s} \end{pmatrix}$$

$$+ \begin{bmatrix} \widehat{K}_y - h\gamma\lambda(H_\mu + R_\mu) & \gamma\lambda M_\mu - h\gamma\lambda R_r \\ -(\gamma\lambda M_\mu - h\gamma\lambda R_r) & \widehat{K}_x - h\gamma\lambda(H_\mu + R_\mu) \end{bmatrix} \begin{pmatrix} \alpha_y \\ \alpha_x \end{pmatrix} = 0 \quad (20.149)$$

The aerodynamic coefficients for high inflow are given in section 16.8.9.

20.4.2 Propeller Whirl Flutter

Consider a propeller with three or more blades on a soft-mounted pylon. The pylon flexibility might be low because of structural damage. Assuming the blade frequencies are well above the pylon natural frequencies, the blade equations can be dropped and the equations for propeller whirl flutter are

$$\begin{bmatrix} I_y & 0 \\ 0 & I_x \end{bmatrix} \begin{pmatrix} \ddot{\alpha}_y \\ \ddot{\alpha}_x \end{pmatrix} + \begin{bmatrix} C_y & -D \\ D & C_x \end{bmatrix} \begin{pmatrix} \dot{\alpha}_y \\ \dot{\alpha}_x \end{pmatrix} + \begin{bmatrix} K_y & L \\ -L & K_x \end{bmatrix} \begin{pmatrix} \alpha_y \\ \alpha_x \end{pmatrix} = 0 \quad (20.150)$$

where $I_x = \widehat{I}_x + 1$ (the sum of pylon and rotor moments of inertia), $C_x = \widehat{C}_x + C_a$, $K_x = \widehat{K}_x - K_a$, and

$$C_a = -\gamma M_{\dot{\beta}} + h\gamma R_{\dot{\beta}} + h^2\gamma(H_\mu + R_\mu)$$

$$\cong \gamma \int_0^1 \frac{r^4}{2U} dr + h^2\gamma \int_0^1 \frac{\lambda^2}{2U} dr \cong \frac{\gamma}{8} \cos\phi_e + h^2\frac{\gamma}{2}\lambda \sin\phi_e \quad (20.151)$$

$$D = 2 + h\gamma M_\mu + h\gamma H_{\dot{\beta}} - h^2\gamma R_r$$

$$\cong 2 + h\gamma \int_0^1 \frac{\lambda r^2}{2U} dr - h\gamma \int_0^1 \frac{\lambda r^2}{2U} dr \cong 2 \quad (20.152)$$

$$K_a = h\gamma\lambda(H_\mu + R_\mu) \cong h\gamma \int_0^1 \frac{\lambda^3}{2U} dr \cong h\frac{\gamma}{2}\lambda^2 \sin\phi_e \quad (20.153)$$

$$L = \gamma\lambda M_\mu - h\gamma\lambda R_r \cong \gamma \int_0^1 \frac{\lambda^2 r^2}{2U} dr \cong \frac{\gamma}{4}\lambda^2 \cos\phi_e \quad (20.154)$$

using the high-inflow results of section 16.8.9. For the last approximations, the inflow angle is evaluated at an effective radius r_e, so $\phi_e = \tan^{-1}\lambda/r_e$. The sum

$$M_\mu + H_{\dot{\beta}} = \int_0^1 r(F_{zT} + F_{xP}) dr = \int_0^1 rU\left(3\frac{c_\ell}{2a} + M\frac{c_{\ell M}}{2a} - \frac{c_{d\alpha}}{2a}\right) dr \quad (20.155)$$

is always smaller than the Coriolis coupling in D. The aerodynamic damping C_a is positive, and the term from H_μ increases with speed. The aerodynamic spring $-K_a$ is negative. Both K_a and the aerodynamic coupling L increase with speed.

The characteristic equation for propeller whirl flutter is

$$(I_y s^2 + C_y s + K_y)(I_x s^2 + C_x s + K_x) + (-Ds + L)^2 = 0 \quad (20.156)$$

Without aerodynamics this reduces to

$$(I_y s^2 + \widehat{C}_y s + \widehat{K}_y)(I_x s^2 + \widehat{C}_x s + \widehat{K}_x) + 4s^2 = 0 \quad (20.157)$$

which always has stable roots, even with the Coriolis coupling of the pitch and yaw motion. Hence whirl flutter is caused by the aerodynamic forces on the rotor. If the spring about one axis is very large (and the other large enough to preclude divergence), the remaining degree of freedom is stable. Whirl flutter instability requires both pitch and yaw motion of the pylon. The sum of $\dot{\alpha}_y$ times the α_y equation and $\dot{\alpha}_x$ times the α_x equation gives an energy balance for the whirl flutter motion:

$$\frac{d}{d\psi}\left(\frac{1}{2}I_y\dot{\alpha}_y^2 + \frac{1}{2}I_x\dot{\alpha}_x^2 + \frac{1}{2}K_y\alpha_y^2 + \frac{1}{2}K_x\alpha_x^2\right) = -\left(C_y\dot{\alpha}_y^2 + C_x\dot{\alpha}_x^2 + L(\alpha_x\dot{\alpha}_y - \alpha_y\dot{\alpha}_x)\right) \quad (20.158)$$

20.4 Whirl Flutter

which was derived by Young and Lytwyn (1967). The left-hand side is the time rate of change of the sum of the kinetic and potential energies of the system. If the right-hand side is negative, the energy is decreasing with time, and the system is stable. The damping terms (C_y and C_x) are always extracting energy from the system. The Coriolis coupling (D) does not enter the energy balance. An instability is only possible if $L(\alpha_x \dot\alpha_y - \alpha_y \dot\alpha_x)$ is sufficiently large and negative. The coefficient L is positive, so an instability has $\alpha_x \dot\alpha_y - \alpha_y \dot\alpha_x < 0$. Writing $\alpha_x = a\cos\theta$ and $\alpha_y = a\sin\theta$, this gives $a^2 \dot\theta < 0$. So an instability occurs with the shaft whirling in a direction opposite the rotor rotation, which is called a backward whirl mode. Whirl flutter is caused by the aerodynamic coupling spring $L \cong \gamma\lambda M_\mu \cong \frac{\gamma}{4}\lambda^2 \cos\phi_e$. Pitch α_y tilts the rotor relative to the axial flow λ, producing an in-plane velocity $\lambda\alpha_y$, which gives a roll moment on the hub through the aerodynamic coefficient M_μ. L increases with forward speed. Thus whirl flutter for a rigid propeller on a pylon is a high inflow instability caused by the aerodynamic spring coupling M_μ, occurring in a backward whirl mode.

The divergence stability criterion is obtained by setting $s = 0$ in the characteristic equation, which gives $K_y K_x + L^2 > 0$ or

$$(\widehat{K}_y - h\gamma\lambda(H_\mu + R_\mu))(\widehat{K}_x - h\gamma\lambda(H_\mu + R_\mu)) + (\gamma\lambda M_\mu - h\gamma\lambda R_r)^2$$
$$\cong (\widehat{K}_y - h\gamma\lambda H_\mu)(\widehat{K}_x - h\gamma\lambda H_\mu) + (\gamma\lambda M_\mu)^2 > 0 \qquad (20.159)$$

Divergence can occur if one of the structural springs (but not both) is sufficiently smaller than the negative aerodynamic spring $K_a = h\gamma\lambda(H_\mu + R_\mu)$. The stability boundary is a hyperbola on the $\widehat{K}_y$ vs. $\widehat{K}_x$ plane. The isotropic case ($\widehat{K}_y = \widehat{K}_x$) is stable.

The flutter stability boundary is obtained by setting $s = i\omega$ (where ω is real) in the characteristic equation. The real part is

$$(K_y - I_y\omega^2)(K_x - I_x\omega^2) - (D^2 + C_y C_x)\omega^2 + L^2 = 0 \qquad (20.160)$$

The imaginary part can be solved for the flutter frequency:

$$\omega^2 = \frac{K_y C_x + K_x C_y - 2DL}{I_y C_x + I_x C_y} \qquad (20.161)$$

Substituting for ω^2 in the real part gives the equation of the stability boundary. The resulting equation is a parabola in terms of the pylon stiffnesses. Figure 20.16 illustrates the flutter and divergence stability boundaries for propeller whirl flutter. The boundaries are shown in terms of the pylon pitch and yaw frequencies ($\omega_x/\Omega = (\widehat{K}_x/I_x)^{1/2}$ and $\omega_y/\Omega = (\widehat{K}_y/I_y)^{1/2}$), for several values of the inflow ratio $\lambda = V/\Omega R$. The properties for this example are $h = 0.3$, $\gamma = 4$, $I_x = I_y = 3$, and $\widehat{C}_x = \widehat{C}_y = 0$.

The stiffness required for stability is greatest for the isotropic case (Figure 20.16). With equal inertia, stiffness, and damping of the pylon pitch and roll motions, the characteristic equation becomes

$$Is^2 + Cs + K \pm i(-Ds + L) = 0 \qquad (20.162)$$

Substituting $s = i\omega$, the imaginary part gives $\omega = \mp L/C$, and the stability boundary is

$$K - I\left(\frac{L}{C}\right)^2 - D\frac{L}{C} > 0 \qquad (20.163)$$

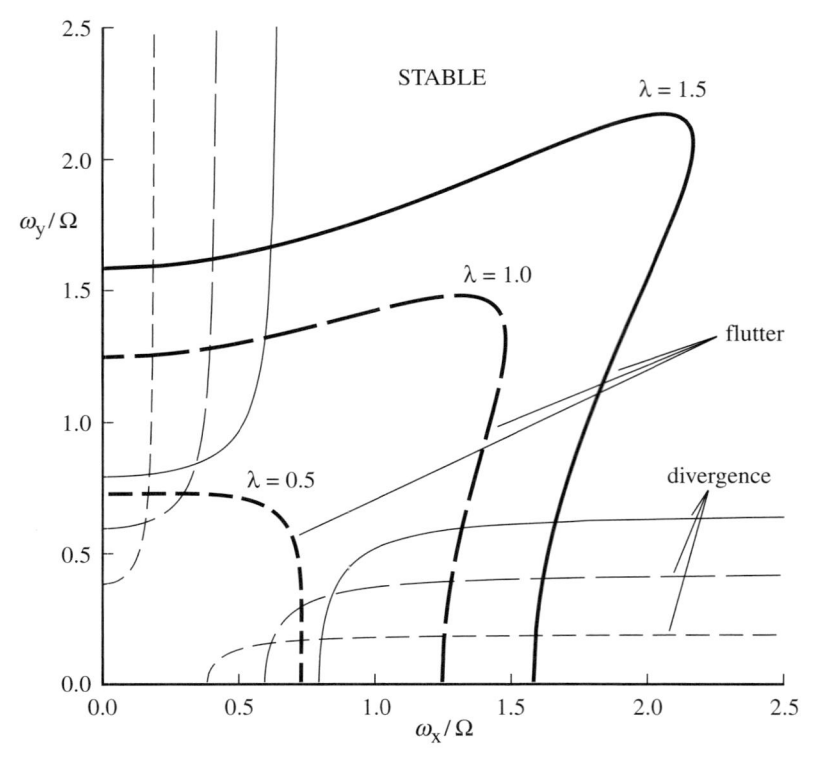

Figure 20.16. Flutter and divergence stability boundaries for propeller whirl flutter.

The required stiffness is

$$\widehat{K} > K_a + \frac{L}{\widehat{C} + C_a}\left(D + (\widehat{I}+1)\frac{L}{\widehat{C}+C_a}\right) \qquad (20.164)$$

with $K_a \cong h\gamma\lambda H_\mu$, $D \cong 2$, $C_a \cong -\gamma M_{\dot\beta} + h^2\gamma H_\mu$, and $L \cong \gamma\lambda M_\mu$. The flutter frequency can be written $\omega^2 = (L/C)^2 = (K - DL/C)/I$, as from the general equations.

20.4.3 Tiltrotor Whirl Flutter

Whirl flutter of a tiltrotor aircraft involves the wing elastic motion coupled with flexible proprotors. The tiltrotor configuration is described in section 8.2. With large proprotors, neglecting the blade motion is not possible. Indeed, most tiltrotor designs have dynamic characteristics similar to those of hingeless rotors, notably the importance of pitch-lag and flap-lag coupling. The blade dynamic design must accommodate large ranges of rotor speed and collective pitch. The attachment of the tiltrotor nacelle is relatively stiff, so the wing provides the flexibility of the rotor support. The flutter modes are thus the fundamental wing motions (vertical bending, chordwise bending, and torsion), both symmetric and anti-symmetric for the complete aircraft. The rotor rotational speed perturbations influence the stability, so the drive train dynamics must also be included in the model. Tiltrotor whirl flutter fundamentally involves a large number of degrees of freedom: aircraft rigid-body motion, wing bending, torsion modes; rotor flap motion; lowest frequency blade lag and torsion degrees of freedom; rotor rotational speed; and inter-connect shaft elastic deflection.

Usually symmetric flight conditions are considered, so symmetric and anti-symmetric subsystems can be analyzed separately for stability.

The aerodynamic analysis is simpler for the proprotor (high inflow, axial flight) than for the helicopter rotor (low inflow, edgewise flight). Axial flight implies a symmetric aerodynamic environment and hence constant coefficient equations of motion. In high inflow, both the in-plane and the out-of-plane blade motion produce a first-order change in the blade angle-of-attack; hence, through the lift-curve slope a first-order change in lift, which has both in-plane and out-of-plane components. So the lift-curve slope terms dominate the aerodynamic forces (see section 16.8.9 and Johnson (1974a)), and the coefficients depend mainly on the inflow ratio $\lambda \cong V/\Omega R$. In contrast, for the rotor with low inflow, in-plane motion produces lift and drag perturbations due to the dynamic pressure change and tilts the mean lift and drag forces; therefore the in-plane forces or forces due to the in-plane motion are small and depend on the blade trim loading. Lift changes due to angle-of-attack perturbations, normally responsible for the high aerodynamic damping of the rotor flap motion, in the proprotor also produce a high aerodynamic damping of the blade in-plane motion.

The blade flap motion leads to a 90° phase shift in the aerodynamic loads produced by rotor shaft motion. Consequently the mechanisms of tiltrotor whirl flutter are fundamentally different from those of propeller whirl flutter. Hall (1966) identified the principal mechanism of tiltrotor whirl flutter as the in-plane hub force due to shaft angular velocity. From equation 19.80, the moment on a point a distance h aft of the hub is

$$\Delta \begin{pmatrix} \dfrac{2C_{My}}{\sigma a} \\ \dfrac{2C_{Mx}}{\sigma a} \end{pmatrix} = -h \left(H_{\dot{\beta}} - R_{\beta} \right) \frac{16}{\gamma} \begin{pmatrix} \dot{\alpha}_y \\ \dot{\alpha}_x \end{pmatrix} \qquad (20.165)$$

The factor $16/\gamma$ arises from the flapping required to produce a moment on the rotor disk, such that the rotor disk follows the shaft rate. The factor $H_{\dot{\beta}} - R_{\beta}$ is the hub force due to tip-path-plane tilt. For low inflow, $H_{\dot{\beta}} - R_{\beta} = (2C_T/\sigma a) + \widehat{H}_{\dot{\beta}} = (2C_T/\sigma a) - (\lambda_{\text{HP}}/4)$. As discussed in section 16.8.6, the hub force due to tip-path-plane tilt has several sources. The slope of the blade due to flap deflection tilts the blade lift radially, producing an in-plane component of the thrust ($R_{\beta} = -C_T/\sigma a$). The rotating-frame flap velocity due to tip-path-plane tilt changes the blade angle-of-attack, which tilts the blade lift chordwise and thereby produces an in-plane component of the thrust ($C_T/\sigma a$ term in $H_{\dot{\beta}}$). The flap velocity changes the blade lift magnitude as well, and since the lift has an in-plane component due to the trim induced velocity, in-plane hub forces are produced by these lift magnitude changes ($\widehat{H}_{\dot{\beta}} = -\lambda_{\text{HP}}/4$). For low inflow, $\widehat{H}_{\dot{\beta}}$ reduces the thrust vector tilt with the tip-path plane, but the net contribution of the rotor to helicopter pitch and roll damping is still positive. For high inflow (section 16.8.9), the inflow term dominates:

$$H_{\dot{\beta}} \cong -\int_0^1 \frac{\lambda r^2}{2U} dr \cong -\frac{\lambda}{4} \cos \phi_e \qquad (20.166)$$

Thus the moments in equation 20.165 are destabilizing at high inflow. Large λ means that the trim section lift direction is tilted far from the z-axis. Hence lift changes due to the rotor motion have a large in-plane component. The stability of a multi-degree-of-freedom system is complex and cannot be explained by a single destabilizing term. There are stabilizing aerodynamic loads, which also increase with inflow ratio, and

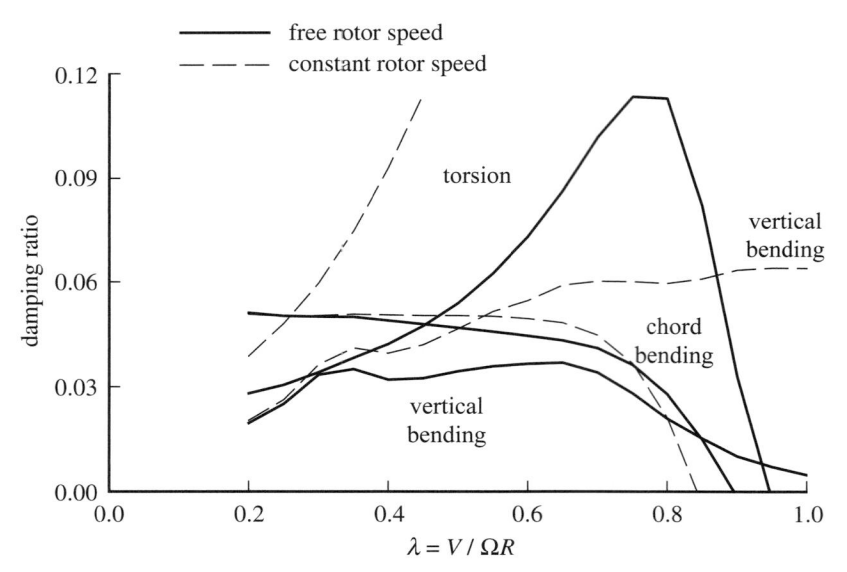

Figure 20.17. Wing mode damping of tiltrotor on a cantilever wing; influence of rotor rotational-speed degree of freedom.

equation 20.165 is a quasistatic result. Gaffey, Yen, and Kvaternik (1969) discussed the variation of the rotor loads with frequency.

At high enough forward speed, the coupled wing and proprotor motion becomes unstable. Figure 20.17 illustrates the behavior of the damping of the fundamental wing bending and torsion modes. These calculations are based on a model of the XV-15 proprotor on a cantilever wing. The rotor has a gimballed, stiff in-plane hub configuration. The rotor is trimmed to provide thrust corresponding to a constant D/q of the aircraft. The free rotor speed case is the reference for the following figures.

The rotor rotational-speed degree of freedom has a major influence on the whirl flutter stability. Vertical bending of the wing is accompanied by a roll motion of the rotor shaft. If the rotor rotational speed is fixed relative to the pylon, this roll motion is transmitted to the rotor, and the high aerodynamic damping of the rotor greatly stabilizes the wing mode (Figure 20.17). If the rotor rotational-speed degree of freedom is free relative to the pylon (in the analysis the rotor mean thrust and torque can be maintained), this source of damping is absent, and the instability occurs at significantly lower speed. Typically, the engine inertia, engine damping, and rotor-speed governor offer little restraint of the rotational-speed degree of freedom in the symmetric motions of a tilting proprotor aircraft. In the anti-symmetric motions, the interconnect shaft constrains the rotor speed, introducing a differential speed mode with a natural frequency of the same order as the wing modes. Thus the two cases of Figure 20.17 characterize the behavior of the tiltrotor symmetric and anti-symmetric motions.

Early analyses of tiltrotor whirl flutter considered only the rotor flap motion, but the results are generally very optimistic (Figure 20.18). For a stiff in-plane rotor, the frequency of the regressive lag mode varies significantly with collective pitch (hence with forward speed), crossing the wing mode frequencies. Thus the lag motion can couple strongly with the whirl flutter modes. With a soft in-plane rotor, air resonance

20.4 Whirl Flutter

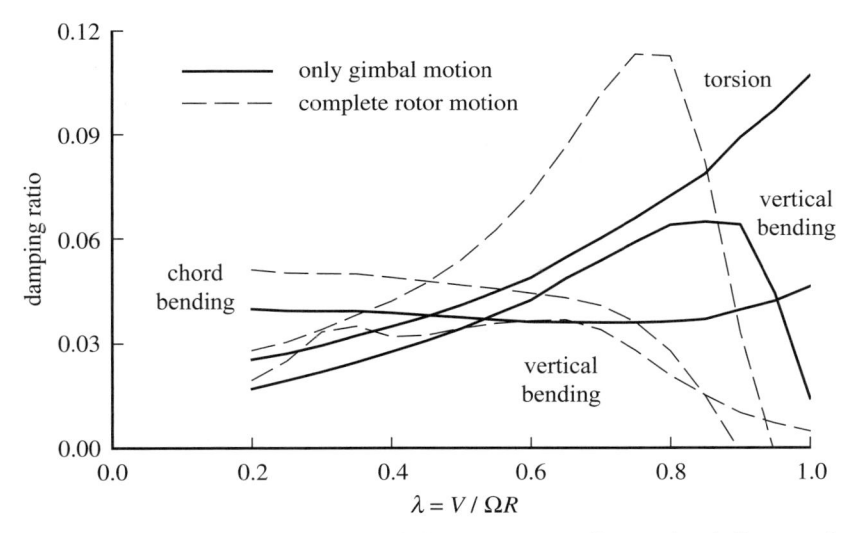

Figure 20.18. Wing mode damping of tiltrotor on a cantilever wing; influence of rotor degrees of freedom.

is possible at low flight speeds, particularly involving the wing vertical-bending mode. At operating flight speeds, the air resonance is stabilized by the rotor aerodynamic lag damping in high inflow and the wing aerodynamic damping.

For hingeless proprotors (including the gimballed, stiff in-plane design), blade pitch motion has a significant influence on whirl flutter through the introduction of effective pitch-lag coupling. The blade precone is normally selected for hover, so in propeller configuration the precone is too large, and there is a downward elastic coning deflection of the blade. With no droop and small thrust, the effective pitch-lag coupling is negative and proportional to the precone (equation 20.75). Negative pitch-lag coupling has a destabilizing influence on the whirl flutter. Figures 20.19 and 20.20 show the stabilizing influence of reduced precone or increased control system

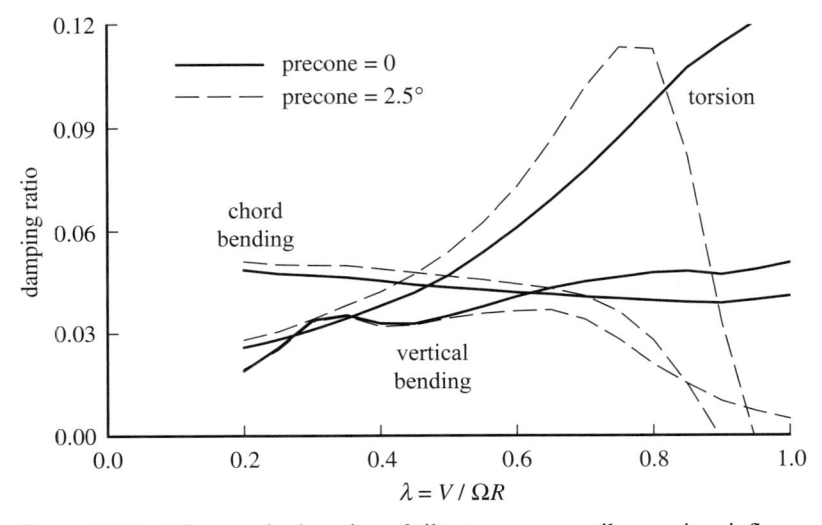

Figure 20.19. Wing mode damping of tiltrotor on a cantilever wing; influence of precone on pitch-lag coupling.

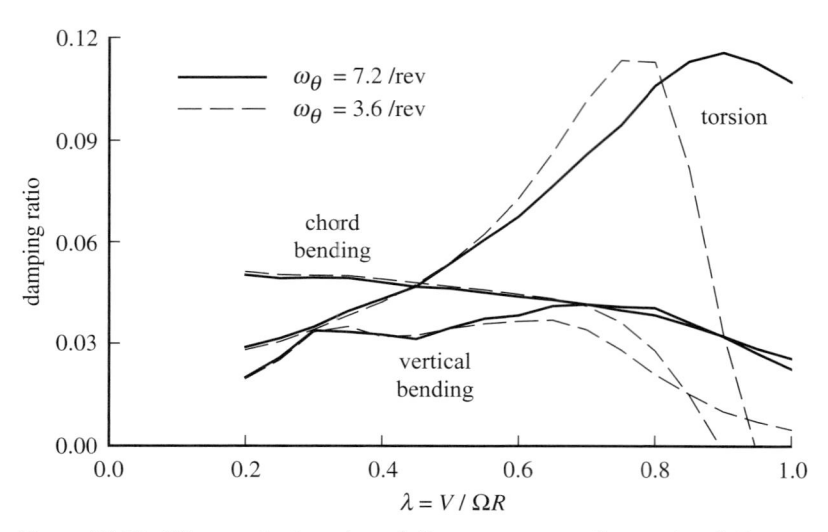

Figure 20.20. Wing mode damping of tiltrotor on a cantilever wing; influence of pitch/torsion stiffness on pitch-lag coupling.

stiffness through the reduction in magnitude of pitch-lag coupling. Blade droop has a similar effect, while not increasing hover coning loads as does reduced precone (since droop becomes aft blade sweep at the low collective pitch angles of hover).

The gimballed, stiff-plane proprotor typically uses negative pitch-flap coupling to stabilized the blade flap-lag motion at high inflow (Gaffey (1969)). Figure 20.21 shows that negative δ_3 is destabilizing for the whirl flutter modes, but without it the rotor exhibits a flap-lag instability at even lower speed.

Figure 20.22 shows the influence of operating altitude on the whirl flutter stability. With the decreased air density at higher altitude, the stability boundary occurs at higher speed, which is consistent with the maximum speed characteristics of the aircraft. This example has a helical tip Mach number of 0.75 at $\lambda = 1$ and sea level

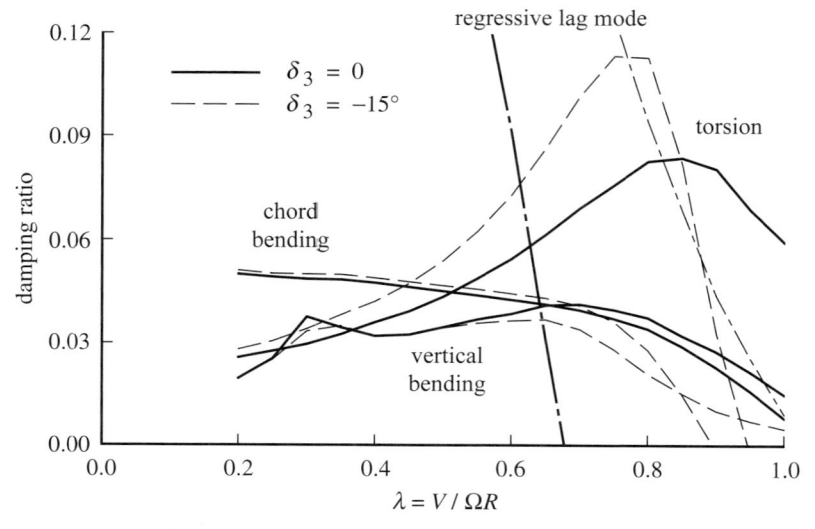

Figure 20.21. Wing mode and rotor lag mode damping of tiltrotor on a cantilever wing; influence of pitch-flap coupling.

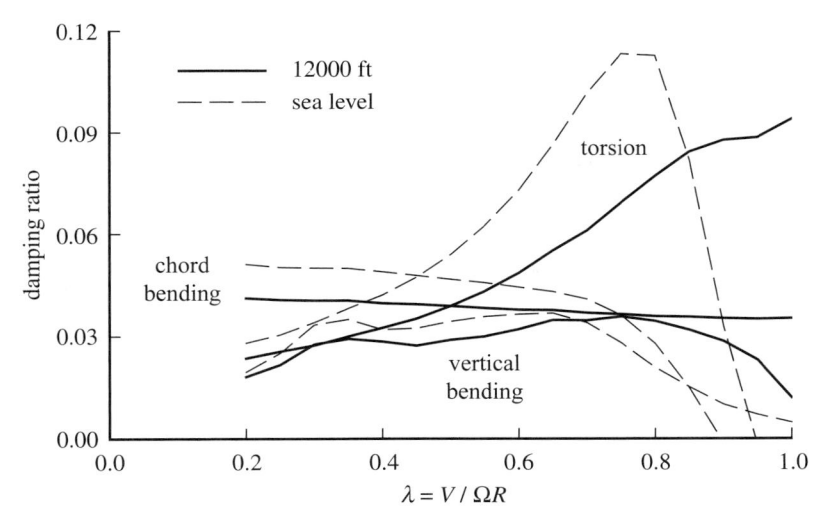

Figure 20.22. Wing mode damping of tiltrotor on a cantilever wing; influence of operating condition.

conditions, which increases at the higher altitude. With increasing Mach number, the blade lift-curve slope first increases, which increases the aerodynamic forces involved in whirl flutter and so has an unfavorable influence on the stability. After lift divergence (at a Mach number of around 0.7 to 0.8, well below drag divergence), the lift-curve slope decreases. If the blade section Mach number is above the lift divergence Mach number over a large fraction of the blade tip, the reduction in aerodynamic forces significantly increases the stability. This phenomenon becomes particularly important as the speed of sound decreases at higher altitude, allowing the effects of reduced density to dominate the stability change.

20.5 REFERENCES

Blake, B.B., Burkam, J.E., and Loewy, R.G. "Recent Studies of the Pitch-Lag Instabilities of Articulated Rotors." Journal of the American Helicopter Society, 6:3 (July 1961).

Bousman, W.G. "An Experimental Investigation of the Effects of Aeroelastic Couplings on Aeromechanical Stability of a Hingeless Rotor Helicopter." Journal of the American Helicopter Society, 26:1 (January 1981).

Bousman, W.G., Sharpe, D.L., and Ormiston, R.A. "An Experimental Study of Techniques for Increasing the Lead-Lag Damping of Soft Inplane Hingeless Rotors." American Helicopter Society 32nd Annual National V/STOL Forum, Washington, DC, May 1976.

Brooks, G.W., and Baker, J.E. "An Experimental Investigation of the Effect of Various Parameters Including Tip Mach Number on the Flutter of Some Model Helicopter Rotor Blades." NACA TN 4005, September 1958.

Brooks, P.W. *Cierva Autogiros*. Washington, DC: Smithsonian Institution Press, 1988.

Burkam, J.E., and Miao, W.-L. "Exploration of Aeroelastic Stability Boundaries with a Soft-in-Plane Hingeless-Rotor Model." Journal of the American Helicopter Society, 17:4 (October 1972).

Chou, P.C. "Pitch-Lag Instability of Helicopter Rotors." Journal of the American Helicopter Society, 3:3 (July 1958).

Coleman, R.P. "Theory of Self-Excited Mechanical Oscillations of Hinged Rotor Blades." NACA ARR 3G29, July 1943.

Coleman, R.P., and Feingold, A.M. "Theory of Ground Vibrations of a Two-Blade Helicopter Rotor on Anisotropic Flexible Supports." NACA TN 1184, January 1947.

Coleman, R.P., and Feingold, A.M. "Theory of Self-Excited Mechanical Oscillations of Helicopter Rotors with Hinged Blades." NACA Report 1351, 1958.

Daughaday, H., DuWaldt, F., and Gates, C. "Investigation of Helicopter Blade Flutter and Load Amplification Problems." Journal of the American Helicopter Society, 2:3 (July 1957).

Deutsch, M.L. "Ground Vibrations of Helicopters." Journal of the Aeronautical Sciences, 13:5 (May 1946).

Feingold, A.M. "Theory of Mechanical Oscillations of Rotors With Two Hinged Blades." NACA ARR 3I13, September 1943.

Focke, H. "Fortschritte des Hubschraubers", Schriften der Deutschen Akademie der Luftfahrt Forschung (Vortrag gehalten auf der 7, Wissenschaftssitzung der Ordentlichen Mitglieder am 1 Oktober 1943), No. 1070/43g, October 1943; "Improvements of the Helicopter," Publications of the German Academy of Aeronautical Research (lecture held at 7th scientific meeting of the regular members on October 1, 1943).

Friedmann, P.P., and Straub, F. "Application of the Finite Element Method to Rotary-Wing Aeroelasticity." Journal of the American Helicopter Society, 25:1 (January 1980).

Gaffey, T.M. "The Effect of Positive Pitch-Flap Coupling (Negative δ_3) on Rotor Blade Motion Stability and Flapping." Journal of the American Helicopter Society, 14:2 (April 1969).

Gaffey, T.M., Yen, J.G., and Kvaternik, R.G. "Analysis and Model Tests of the Proprotor Dynamics of a Tilt-Proprotor VTOL Aircraft." U.S. Air Force V/STOL Technology and Planning Conference, Las Vegas, Nevada, September 1969.

Gustafson, F.B. "History of NACA/NASA Rotating-Wing Aircraft Research, 1915–1970." Vertiflite, 16:7 (July 1970), 16:10 (October 1970).

Hall, W.E., Jr. "Prop-Rotor Stability at High Advance Ratios." Journal of the American Helicopter Society, 11:2 (April 1966).

Hodges, D.H., and Ormiston, R.A. "Stability of Elastic Bending and Torsion of Uniform Cantilevered Rotor Blades in Hover." AIAA Paper No. 73-405, March 1973.

Hodges, D.H., and Ormiston, R.A. "Stability of Elastic Bending and torsion of Uniform Cantilever Rotor Blades in Hover with Variable Structural Coupling." NASA TN D-8192, April 1976.

Hodges, D.H., and Ormiston, R.A. "Stability of Hingeless Rotor Blades in Hover with Pitch-Link Flexibility." AIAA Journal, 15:4 (April 1977).

Huber, H.B. "Effect of Torsion-Flap-Lag Coupling on Hingeless Rotor Stability." American Helicopter Society 29th Annual National Forum, Washington, DC, May 1973.

Johnson, W. "Theory and Comparison with Tests of Two Full-Scale Proprotors." NASA SP-352, February 1974a.

Johnson, W. "Dynamics of Tilting Proprotor Aircraft in Cruise Flight." NASA TN D-7677, May 1974b.

Johnson, W. "Influence of Unsteady Aerodynamics on Hingeless Rotor Ground Resonance." Journal of Aircraft, 19:8 (August 1982).

Jones, J.P. "The Influence of the Wake on the Flutter and Vibration of Rotor Blades." The Aeronautical Quarterly, 9:Part 3 (August 1958).

Miller, R.H., and Ellis, C.W. "Helicopter Blade Vibration and Flutter." Journal of the American Helicopter Society, 1:3 (July 1956).

Morduchow, M., and Hinchey, F.G. "Theoretical Analysis of Oscillations in Hovering of Helicopter Blades with Inclined and Offset Flapping and Lagging Hinge Axes." NACA TN 2226, December 1950.

Ormiston, R.A. "Rotor-Fuselage Dynamics of Helicopter Air and Ground Resonance." Journal of the American Helicopter Society, 36:2 (April 1991).

Ormiston, R.A., and Bousman, W.G. "A Study of Stall-Induced Flap-Lag Instability of Hingeless Rotors." Journal of the American Helicopter Society, 20:1 (January 1975).

Ormiston, R.A., and Hodges, D.H. "Linear Flap-Lag Dynamics of Hingeless Helicopter Rotor Blades in Hover." Journal of the American Helicopter Society, *17*:2 (April 1972).

Wagner, R.A. "Vibrations Handbook for Helicopters." Wright Air Development Center, 1954 (Volume 7 of *Rotary Wing Aircraft Handbooks and History*, Edited by E.K. Liberatore).

Young, M.I., and Lytwyn, R.T. "The Influence of Blade Flapping Restraint on the Dynamic Stability of Low Disk Loading Propeller-Rotors." Journal of the American Helicopter Society, *12*:4 (October 1967).

21 Flight Dynamics

Handling qualities are defined as "those qualities or characteristics of an aircraft that govern the ease and precision with which a pilot is able to perform the tasks required in support of an aircraft role" (Cooper and Harper (1969)). Generally the terms "flying qualities" and "handling qualities" are interchangeable, although the titles of specifications more often refer to flying qualities. Handling qualities involve the aircraft, the pilot, the tasks, and the environment (Padfield (1998)). Most of this chapter deals only with the aircraft flight dynamics or stability and control characteristics: the equations and fundamental behavior of the rotorcraft rigid-body motion. Simplifications and approximations are made to focus on the fundamental behavior of the aircraft. A more rigorous approach is needed to obtain models sufficient for rotorcraft flight control system design. Padfield (2007) covers rotorcraft flight dynamics and handling qualities in depth.

21.1 Control

Rotorcraft control requires the ability to produce moments and forces on the vehicle to establish equilibrium and thereby hold the aircraft in a desired trim state, and to produce accelerations and thereby change the aircraft velocity, position, and orientation. Like airplane control, rotorcraft control is accomplished primarily by producing moments about all three aircraft axes: pitch, roll, and yaw. The helicopter has in addition direct control over the vertical force on the aircraft, corresponding to its VTOL capability. This additional control variable is part of the versatility of the helicopter, but also makes the piloting task more difficult. The control task is eased by the use of a rotor speed governor to automatically manage the power.

Direct control over moments on the aircraft is satisfactory for trajectory control in forward flight. In hover and at low speed, direct control over the forces would be more desirable, to obtain direct command of the helicopter velocity and displacement. Such control is available only for the vertical force. The lateral and longitudinal velocities of the helicopter in hover must be controlled using pitch and roll moments about the aircraft center-of-gravity, which is a more difficult task. The pilot directly commands a change in pitch or roll attitude that then produces a longitudinal or lateral force and finally the desired velocity of the helicopter. There is considerable coupling of the forces and moments produced by the helicopter controls, so that any control application to produce a particular moment requires some compensating control inputs on the other axes as well. Moreover, without an automatic stability augmentation system, the helicopter is not dynamically or statically stable, particularly in hover. Consequently, the pilot is required to provide the feedback control to stabilize the vehicle, an operation that demands constant attention. The use of an automatic control system to augment the helicopter stability and control characteristics is desirable, and for some applications it is essential.

The rotor is almost universally used to control the helicopter. In forward flight, fixed aerodynamic surfaces such as a horizontal stabilizer and elevator can be used as well. The rotor controls consist of cyclic and collective pitch. A collective pitch change gives a change in the mean blade angle-of-attack, which changes the thrust magnitude. Cyclic pitch control gives a 1/rev pitch motion in the rotating frame, which tilts the tip-path plane. The thrust vector tilts with the tip-path plane, producing a moment about the helicopter center-of-gravity below the rotor hub. With an offset-hinge articulated rotor or a hingeless rotor, the tip-path-plane tilt also produces a moment at the rotor hub. Thus command of the rotor collective and cyclic pitch gives efficient control over the magnitude and direction of the rotor thrust vector. The 1/rev pitch change of the blades required for cyclic control is obtained using a swashplate mechanism of some kind (section 6.1.2).

The pilot's controls for the helicopter consist of a cyclic stick for control of longitudinal and lateral moments, a collective stick for control of the vertical force, foot pedals for control of the yaw moment, and a throttle for control of the rotor speed and torque (power management). These controls are similar in function to those of the airplane, with the addition of the collective stick, which is used for direct height control in hover and low-speed flight. Maintaining the rotor speed at the proper value is important. Since the rotor power required varies with both thrust and forward speed, the throttle must be coordinated with the collective and cyclic stick motions, which is a pilot task on small helicopters. All turbine-powered rotorcraft have a speed governor on the engine to automatically handle the power management. The cyclic stick controls the longitudinal and lateral motions in hover, but the helicopter is characterized by considerable coupling between the controls. The manner in which the pilot's cyclic and collective control sticks are connected to the cyclic and collective pitch of each rotor depends on the aircraft configuration; see section 8.2 and Table 8.1.

The connection of the pilot's controls to the rotor can be by a direct mechanical linkage (at least for small helicopters), or actuators can be used to produce or augment the rotor control inputs commanded by displacements of the pilot's sticks. In systems with a direct mechanical linkage between the pilot's controls and the rotor, blade feathering moments are transmitted through the control system to the pilot's sticks. The collective stick force comes from the mean blade pitch moment, and the cyclic stick forces come from the 1/rev pitch moments. The proper behavior of these control forces is important for good handling qualities. The general requirements for good control forces are low friction, low vibration, and logical control force transients. A moderate but increasing force gradient opposing any stick motion is desirable. In a helicopter with a mechanical control system, the feathering moments and hence the stick forces are sensitive to the blade dynamics and geometry. Moreover, the vibratory blade pitch moments as well as the steady loads are transmitted through the control system (section 18.3). For low steady and vibratory control forces, the proper airfoil section must be chosen, and the appropriate chordwise offsets of the aerodynamic center and center-of-gravity must be used. Because stick forces increase with the helicopter size, larger helicopters use actuators between the pilot's sticks and the rotor controls. With irreversible actuators, the blade pitch moments are not transmitted back to the pilot. The control forces must be provided by the actuators or by an automatic force-feel system in this case.

The analysis developed in Chapter 6 gives the principal characteristics of the control required to trim the helicopter. The conditions for force and moment equilibrium of the helicopter determine the orientation of the rotor hub plane and tip-path plane for a given flight state. The flapping relative to the shaft is therefore determined as a function of the helicopter center-of-gravity position and aircraft speed. The cyclic and collective control required to obtain this rotor thrust and tip-path-plane orientation was derived in section 6.5 for an articulated rotor and in section 6.14 for general flap frequency. The basic behavior of the rotor control is contained in these equations. The rotor collective pitch varies directly with the helicopter gross weight. The collective pitch varies significantly with flight speed also, mainly as a result of the variation of the inflow ratio with speed. The tip-path plane tilts rearward and to the advancing side relative to the control plane, roughly in proportion to the advance ratio. Thus a forward tilt of the control plane is required to maintain the tip-path-plane orientation as speed increases, and therefore a forward displacement of the cyclic stick is needed, which is the desired control motion for increased speed. However, a lateral cyclic stick displacement is also required as speed increases to counter the lateral flapping. The lateral flapping is sensitive to nonuniform inflow, which generally increases this undesirable coupling, particularly at low speeds.

21.2 Aircraft Motion

Now let us examine helicopter flying qualities in terms of the dynamic stability and response to control of the aircraft rigid-body motions. The equations of motion for the six rigid-body degrees of freedom are obtained from equilibrium about the center-of-gravity of forces and moments on the entire aircraft:

$$M\left(\dot{v}^{BI/B} + \widetilde{\omega}^{BI/B} v^{BI/B}\right) = F^B + F^B_{\text{grav}} \tag{21.1}$$

$$I^B \dot{\omega}^{BI/B} + \widetilde{\omega}^{BI/B} I^B \omega^{BI/B} = M^B \tag{21.2}$$

where v and ω are the linear and angular velocities, B designates the aircraft body axes, and I the inertial frame. See section 2.4 for the notation; $\widetilde{\omega}$ is the cross-product matrix from the angular velocity ω. The origin of the body axes is at the aircraft center-of-gravity. F^B and M^B are the total force and moment acting on the aircraft, about the origin of B, in B axes. The gravitational force $F^B_{\text{grav}} = MC^{BI}g^I$ is calculated from the acceleration produced by gravity (constant in inertial axes). The aircraft mass is M, and the symmetric moment of inertia matrix is

$$I^B = \begin{bmatrix} I_x & -I_{xy} & -I_{xz} \\ -I_{yx} & I_y & -I_{yz} \\ -I_{zx} & -I_{zy} & I_z \end{bmatrix} \tag{21.3}$$

where $I_x = \int (y^2 + z^2) dm$, $I_{xz} = \int (xz) dm$, and so on (in body axes, hence constant). The mass and moments of inertia include the rotor mass.

Aircraft convention for flight dynamics has the z-axis downward, x-axis forward, and y-axis to the right. For airplane stability and control analysis, aligning the x-axis with the trim velocity vector (stability axes) is common, but that approach is not appropriate for rotorcraft, since the analysis must cover hover and low-speed flight as well as high-speed cruise. Extension of the equations to an elastic airframe is simplest if center-of-mass mean axes are used for the rigid-body motion. Then the elastic free vibration modes of the aircraft are not coupled inertially or structurally with the rigid-body motion, although the origin of the B axes (center-of-mass) is not associated with a physical point on the airframe. The elastic and rigid-body degrees of freedom are coupled only by the applied loads from aerodynamic surfaces and rotors.

Aircraft convention uses a body-axes velocity and Euler angle representation of the rigid-body motion. The equations of motion are written in terms of the velocities $v^{BI/B}$ and $\omega^{BI/B}$, which must be related to displacement degrees of freedom $\dot{\xi}$ and angular degrees of freedom $\dot{\eta}$. The parameters $\dot{\xi}$ are the velocity measured in B axes, $\dot{\xi} = v^{BI/B} = C^{BI}\dot{x}^{BI/I} = (\dot{x}_F\ \dot{y}_F\ \dot{z}_F)^T$. The inertial acceleration is $\dot{v} + \tilde{\omega}v = \ddot{\xi} + \tilde{\omega}\dot{\xi}$. The aircraft position

$$x^{BI/I} = \int^t C^{IB}\dot{\xi}\, dt \qquad (21.4)$$

is not obtained directly from the degrees of freedom ξ, but rather from the integral of $\dot{\xi}$, which is path dependent. Because the velocity variables are measured relative to the body axes, the aircraft position depends on the angular motion (the rotation matrix C^{IB}) as well as on the velocities. For most aircraft flight behavior, the position relative to the inertial frame origin is not important. For a rotorcraft operating near the ground, the position is required for the wake model and ground effect. In such cases, the displacement degrees of freedom can be measured relative to the inertial axes, an unconventional description that gives the position in space directly.

The angular motion is described by the Tait-Bryan convention: first yaw ψ_F about the z-axis, then pitch θ_F about the y-axis, then roll ϕ_F about the x-axis. The rotation degrees of freedom are $\eta = (\phi_F\ \theta_F\ \psi_F)^T$; the first angle in η is the last rotation. The rotation matrix, angular velocity, and angular acceleration are

$$C^{BI} = X_{\phi_F} Y_{\theta_F} Z_{\psi_F} \qquad (21.5)$$

$$\omega^{BI/B} = R\dot{\eta} \qquad (21.6)$$

$$\dot{\omega}^{BI/B} = R\ddot{\eta} + \dot{R}\dot{\eta} \qquad (21.7)$$

$$R = \begin{bmatrix} 1 & 0 & -\sin\theta_F \\ 0 & \cos\phi_F & \sin\phi_F\cos\theta_F \\ 0 & -\sin\phi_F & \cos\phi_F\cos\theta_F \end{bmatrix} \qquad (21.8)$$

$$\dot{R}\dot{\eta} = \begin{pmatrix} \dot{\phi}_F \\ \dot{\theta}_F\cos\phi_F \\ -\dot{\theta}_F\sin\phi_F \end{pmatrix}^{\sim} \begin{pmatrix} \dot{\phi}_F + \dot{\psi}_F\sin\theta_F \\ -\dot{\psi}_F\sin\phi_F\cos\theta_F \\ -\dot{\psi}_F\cos\phi_F\cos\theta_F \end{pmatrix} \qquad (21.9)$$

The gravity vector is $g^B = C^{BI}g^I = g(-\sin\theta_F\ \sin\phi_F\cos\theta_F\ \cos\phi_F\cos\theta_F)^T$ in body axes. Superscripts B and I can be omitted now, since only the aircraft motion is considered here.

The rigid-body equations of motion are next linearized about the trim flight state: steady flight at velocity V and climb angle θ_c (negative for descent), zero angular velocity (no turns), and symmetric flight with zero trim yaw and roll angles. The trim aircraft velocity is

$$v_0 = \begin{pmatrix} V\cos(\theta_{F0} - \theta_c) \\ 0 \\ V\sin(\theta_{F0} - \theta_c) \end{pmatrix} = \begin{pmatrix} V_x \\ 0 \\ -V_z \end{pmatrix} \qquad (21.10)$$

Now $v = (\dot{x}_F\ \dot{y}_F\ \dot{z}_F)^T$ and $\omega = R(\dot{\phi}_F\ \dot{\theta}_F\ \dot{\psi}_F)^T$ are the perturbation motion relative to trim. The linearized equations of motion are

$$M(\dot{v} - \widetilde{v}_0 \omega) = F + F_{\text{grav}} \tag{21.11}$$

$$I\dot{\omega} = M \tag{21.12}$$

or

$$M(\ddot{x}_F - V_z \dot{\theta}_F) = F_x - Mg\cos\theta_{F0}\theta_F \tag{21.13}$$

$$M(\ddot{y}_F + V_x \cos\theta_{F0} \dot{\psi}_F + V_z(\dot{\phi}_F - \sin\theta_{F0}\dot{\psi}_F)) = F_y + Mg\cos\theta_{F0}\phi_F \tag{21.14}$$

$$M(\ddot{z}_F - V_x \dot{\theta}_F) = F_z - Mg\sin\theta_{F0}\theta_F \tag{21.15}$$

$$I \begin{pmatrix} \ddot{\phi}_F - \sin\theta_{F0}\ddot{\psi}_F \\ \ddot{\theta}_F \\ \cos\theta_{F0}\ddot{\psi}_F \end{pmatrix} = \begin{pmatrix} M_x \\ M_y \\ M_z \end{pmatrix} \tag{21.16}$$

In terms of rotor parameters, $V_x/\Omega R = \mu$ and $V_z/\Omega R = \mu_z$. The trim state can be more generally defined. The aircraft can be trimmed to zero sideslip angle instead of zero roll angle. Yaw angle has no influence on the flight dynamics, so turning flight (non-zero yaw rate) is a steady-state trim condition.

In state variable form, the aircraft degrees of freedom are the body axis velocity $v^{BI/B} = (u\ v\ w)^T$, angular velocity $\omega^{BI/B} = (p\ q\ r)^T$, and the pitch and roll angles $(\theta_F\ \phi_F)^T$. The state equations are force and moment equilibrium (equations 21.11 and 21.12) plus the kinematic relation $\omega^{BI/B} = R(\dot{\phi}_F\ \dot{\theta}_F\ \dot{\psi}_F)$. For the simplified aircraft geometry considered here, and when the emphasis is on the rotor response, it is convenient to work instead with the linear $(x_F\ y_F\ z_F)$ and angular $(\phi_F\ \theta_F\ \psi_F)$ displacement variables. See Padfield (2007) for the complete equations.

For an airplane with lateral symmetry (symmetric about the x–z plane), the equations separate into longitudinal and lateral-directional sets. Symmetry implies $I_{xy} = I_{yz} = 0$ and that the forces and moments depend only on the appropriate aircraft motion. A symmetric trim flight state has already been assumed. The longitudinal equations of motion are

$$M(\ddot{x}_F - V_z \dot{\theta}_F) = F_x - Mg\cos\theta_{F0}\theta_F \tag{21.17}$$

$$M(\ddot{z}_F - V_x \dot{\theta}_F) = F_z - Mg\sin\theta_{F0}\theta_F \tag{21.18}$$

$$I_y \ddot{\theta}_F = M_y \tag{21.19}$$

and the lateral-directional equations of motion are

$$M(\ddot{y}_F + V_x \cos\theta_{F0} \dot{\psi}_F + V_z(\dot{\phi}_F - \sin\theta_{F0}\dot{\psi}_F)) = F_y + Mg\cos\theta_{F0}\phi_F \tag{21.20}$$

$$I_x(\ddot{\phi}_F - \sin\theta_{F0}\ddot{\psi}_F) - I_{xz}(\cos\theta_{F0}\ddot{\psi}_F) = M_x \tag{21.21}$$

$$I_z(\cos\theta_{F0}\ddot{\psi}_F) - I_{xz}(\ddot{\phi}_F - \sin\theta_{F0}\ddot{\psi}_F) = M_z \tag{21.22}$$

It is unlikely that the axes are aligned with the inertial principal axes; hence $I_{xz} \neq 0$. Often the inertial terms in the roll and yaw equations are diagonalized, the transformation resulting in stability derivatives that are combinations of roll and yaw moments. For rotorcraft, I_{xz} can be large enough that the influence on yaw and roll rate derivatives is significant. The present analysis is not rigorous enough to merit such a complication, so instead the coupling by I_{xz} is ignored.

A helicopter has inherent asymmetry due to the direction of rotation of the main rotor. So rotorcraft do not have lateral symmetry, except for configurations such as the side-by-side helicopter or the tiltrotor. Nonetheless, the helicopter flight dynamics are first examined assuming that the longitudinal and lateral-directional motions are separable. The model is then simple enough for analytical work, since only half the equations are considered for each set. Although the model does describe the principal behavior of the helicopter, the coupling of longitudinal and lateral motions is also considered. The main rotor in hover does have symmetry about the shaft axis, which implies that the vertical-directional motions separate from the remaining longitudinal-lateral motions.

The applied forces and moments are linearized relative to the perturbed aircraft motion, as well as to the pilot's controls and gust velocities. Dividing the load by the corresponding inertia gives the stability derivative form:

$$\begin{pmatrix} F_x/M \\ M_y/I_y \\ F_z/M \end{pmatrix} = \begin{bmatrix} X_u & X_q & X_w \\ M_u & M_q & M_w \\ Z_u & Z_q & Z_w \end{bmatrix} \begin{pmatrix} \dot{x}_F \\ \dot{\theta}_F \\ \dot{z}_F \end{pmatrix}$$
$$+ \begin{bmatrix} X_\theta & X_{\theta_0} & X_u & X_w \\ M_\theta & M_{\theta_0} & M_u & M_w \\ Z_\theta & Z_{\theta_0} & Z_u & Z_w \end{bmatrix} \begin{pmatrix} \theta_s \\ \theta_0 \\ u_G \\ w_G \end{pmatrix} \qquad (21.23)$$

$$\begin{pmatrix} F_y/M \\ M_x/I_x \\ M_z/I_z \end{pmatrix} = \begin{bmatrix} Y_v & Y_p & Y_r \\ L_v & L_p & L_r \\ N_v & N_p & N_r \end{bmatrix} \begin{pmatrix} \dot{y}_F \\ \dot{\phi}_F \\ \dot{\psi}_F \end{pmatrix} + \begin{bmatrix} Y_\theta & Y_{\theta_p} & Y_v \\ L_\theta & L_{\theta_p} & Y_v \\ N_\theta & N_{\theta_p} & N_v \end{bmatrix} \begin{pmatrix} \theta_c \\ \theta_p \\ v_G \end{pmatrix} \qquad (21.24)$$

The notation uses X, Y, and Z for the longitudinal, lateral, and vertical force derivatives and L, M, and N for the roll, pitch, and yaw moment derivatives. The derivatives with respect to the linear velocity are designated by subscripts u, v, and w and, with respect to angular velocity, by subscripts p, q, and r (angular velocity in body axes, approximated here by Euler angle rates). The pilot's controls are collective θ_0, lateral cyclic θ_c, longitudinal cyclic θ_s, and pedal θ_p. The connection of the pilot's controls to the rotor controls depends on the rotorcraft configuration (section 8.2). The control derivatives are given here in terms of the amplitude of the rotor collective or cyclic pitch, ignoring control system rigging and the actual units of pilot stick deflection. For simplicity, the axes of the gust velocity components (u_G, v_G, w_G) are the aircraft body axes.

The equations in stability derivative form have units of acceleration (ft/sec^2 or m/sec^2) and angular acceleration (rad/sec^2). The dimensionless versions are scaled with rotor radius R and rotor speed Ω. The notation used reflects the aeromechanics focus of this work.

21.3 Motion and Loads

A complete analysis of rotorcraft flight dynamics requires adding to the aircraft rigid-body motion the equations for the blade degrees of freedom, particularly the rotor blade flap motion. Because of the high flap damping, the flapping transients die out in less than one revolution of the rotor; see section 19.3. Hence the rotor flap motion responds on a much shorter time scale than the inputs from the pilot, from shaft motion due to the helicopter rigid-body degrees of freedom, or from aerodynamic gusts. So for flight dynamics considering only the low-frequency response of the rotor, neglecting the transient flapping dynamics is sufficient. Then the rotors contribute to the stability derivatives, without adding degrees of freedom to the model. It is assumed that the rotor speed is constant.

In the frequency domain, there is usually sufficient separation between the flight dynamics motion and the motion of rotor modes to justify the stability derivative model. However, there are exceptions. In particular, a high-gain flight control system can be active at frequencies of the rotor domain. An airframe structural mode might couple with the flight control system and rotor motion. The low-damped lag mode of a hingeless or bearingless rotor can influence the flight dynamics.

Hohenemser (1939) originated the use of the quasistatic rotor response in investigations of helicopter flying qualities, on the basis of the very low frequency of the rigid-body motions compared to the rotor rotational speed. Miller (1948) compared the roots obtained for the helicopter longitudinal dynamics, using the complete rotor dynamics and the low-frequency response. The question was further examined by Kaufman and Peress (1956), and by Hohenemser and Yin (1974a, 1974b) with an emphasis on hingeless rotors. These investigations considered the poles, frequency response, and step response in hover and forward flight, up to high advance ratio, including feedback control. The general conclusion is that the quasistatic approximation for the rotor dynamics is quite a good model for the rotor in a flight dynamics analysis.

The influence of the rotor wake on the flight dynamics is included in terms of quasistatic lift deficiency functions derived using dynamic inflow. The time lags of the dynamic inflow model (section 11.3) are neglected here to avoid increasing the order of the dynamic systems examined. However, the eigenvalues of the dynamic inflow states can be small enough to significantly influence the flight dynamics; see Padfield (2007).

To use the rotor low-frequency response from Chapter 19 to obtain the stability derivatives, the rotor variables must be related to the aircraft motion and loads. Figure 16.8 shows the conventions for rotor hub loads, and Figure 16.13 shows the hub motion. The rotor is assumed to rotate counter-clockwise. For the purpose of exposition in this chapter, the geometry is simplified. Figure 21.1 shows the geometry for the single main rotor and tail rotor configuration. The description of the aircraft motion has the origin at the center-of-gravity, with the z-axis down and the x-axis forward. The center-of-gravity is on the shaft axis, a distance h below the main rotor hub. The aircraft axes are parallel to the rotor hub-plane axes, with directions of the x-axis and z-axis reversed. The tail rotor is on the x-axis, a distance ℓ_{tr} aft of the center-of-gravity. For this simple geometry, the linear and angular displacements of the main rotor hub are

$$\begin{pmatrix} x_h \\ y_h \\ z_h \end{pmatrix}_{rotor} = \begin{pmatrix} -x_F + h\theta_F \\ y_F + h\phi_F \\ -z_F \end{pmatrix} \qquad (21.25)$$

$$\begin{pmatrix} \alpha_x \\ \alpha_y \\ \alpha_z \end{pmatrix}_{rotor} = \begin{pmatrix} -\phi_F \\ \theta_F \\ -\psi_F \end{pmatrix} \qquad (21.26)$$

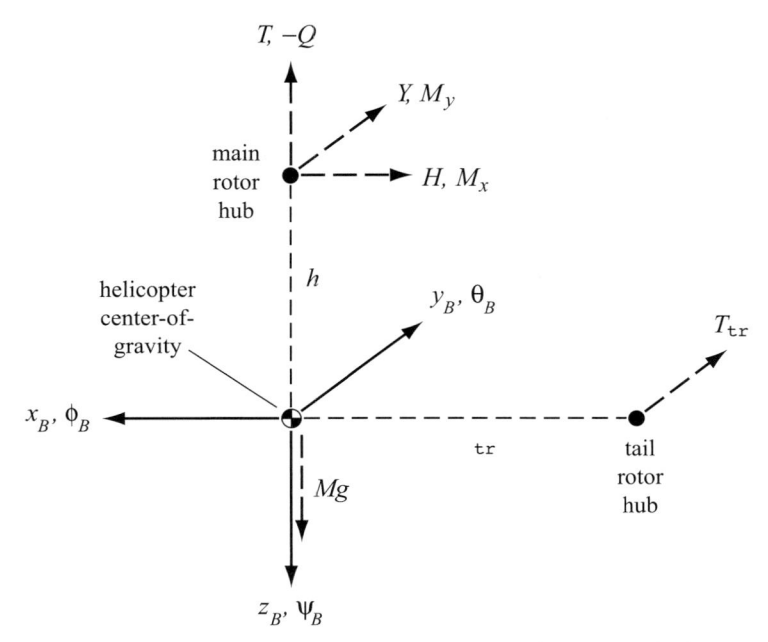

Figure 21.1. Simplified geometry of the single main rotor and tail rotor configuration.

and the rotor loads acting on the airframe are

$$\begin{pmatrix} F_x \\ F_y \\ F_z \end{pmatrix} = \begin{pmatrix} -H \\ Y \\ -T \end{pmatrix}_{rotor} \tag{21.27}$$

$$\begin{pmatrix} M_x \\ M_y \\ M_z \end{pmatrix} = \begin{pmatrix} -M_x + hY \\ M_y + hH \\ Q - \ell_{tr}T_{tr} \end{pmatrix}_{rotor} \tag{21.28}$$

In forward flight there are also aerodynamic loads acting on the fuselage and tail.

The tandem helicopter configuration has a front rotor that is a distance ℓ_f forward of the aircraft center-of-gravity and a rear rotor that is a distance ℓ_r aft of the aircraft center-of-gravity. The rotor hub heights are h_f and h_r. For this simple geometry, the linear and angular displacements of the main rotor hubs are

$$\begin{pmatrix} x_h \\ y_h \\ z_h \end{pmatrix}_{front} = \begin{pmatrix} -x_F + h_f \theta_F \\ y_F + h_f \phi_F + \ell_f \psi_F \\ -z_F + \ell_f \theta_F \end{pmatrix} \tag{21.29}$$

$$\begin{pmatrix} x_h \\ y_h \\ z_h \end{pmatrix}_{rear} = \begin{pmatrix} -x_F + h_r \theta_F \\ y_F + h_r \phi_F - \ell_r \psi_F \\ -z_F - \ell_r \theta_F \end{pmatrix} \tag{21.30}$$

$$\begin{pmatrix} \alpha_x \\ \alpha_y \\ \alpha_z \end{pmatrix}_{front} = \begin{pmatrix} \alpha_x \\ \alpha_y \\ \alpha_z \end{pmatrix}_{rear} = \begin{pmatrix} -\phi_F \\ \theta_F \\ -\psi_F \end{pmatrix} \tag{21.31}$$

and the rotor loads acting on the airframe are

$$\begin{pmatrix} F_x \\ F_y \\ F_z \end{pmatrix} = \begin{pmatrix} -H \\ Y \\ -T \end{pmatrix}_{\text{front}} + \begin{pmatrix} -H \\ Y \\ -T \end{pmatrix}_{\text{rear}} \qquad (21.32)$$

$$\begin{pmatrix} M_x \\ M_y \\ M_z \end{pmatrix} = \begin{pmatrix} -M_x + h_f Y \\ M_y + h_f H + \ell_f T \\ Q - \ell_f Y \end{pmatrix}_{\text{front}} + \begin{pmatrix} -M_x + h_r Y \\ M_y + h_r H - \ell_r T \\ Q - \ell_r Y \end{pmatrix}_{\text{rear}} \qquad (21.33)$$

The two rotors are assumed to have the same shaft inclination, but differential tilt of the shafts relative to the airframe has an important influence on tandem helicopter flight dynamics.

The side-by-side helicopter configuration has right and left main rotors, a distance $\pm \ell$ from the aircraft center-line. The rotor hub height is h above the center-of-gravity. The tiltrotor has similar geometry in hover and helicopter mode flight, the wing-tip-mounted rotors tilting forward to act as propellers in cruise flight. These aircraft configurations have true lateral symmetry (symmetric about the x–z plane). For the side-by-side configuration, the linear and angular displacements of the main rotor hubs are

$$\begin{pmatrix} x_h \\ y_h \\ z_h \end{pmatrix}_{\text{rotor}} = \begin{pmatrix} -x_F + h\theta_F \pm \ell \psi_F \\ y_F + h\phi_F \\ -z_F \mp \ell \phi_F \end{pmatrix} \qquad (21.34)$$

$$\begin{pmatrix} \alpha_x \\ \alpha_y \\ \alpha_z \end{pmatrix}_{\text{rotor}} = \begin{pmatrix} -\phi_F \\ \theta_F \\ -\psi_F \end{pmatrix} \qquad (21.35)$$

and the rotor loads acting on the airframe are

$$\begin{pmatrix} F_x \\ F_y \\ F_z \end{pmatrix} = \begin{pmatrix} -H \\ Y \\ -T \end{pmatrix}_{\text{right}} + \begin{pmatrix} -H \\ Y \\ -T \end{pmatrix}_{\text{left}} \qquad (21.36)$$

$$\begin{pmatrix} M_x \\ M_y \\ M_z \end{pmatrix} = \begin{pmatrix} -M_x + hY - \ell T \\ M_y + hH \\ Q + \ell H \end{pmatrix}_{\text{right}} + \begin{pmatrix} -M_x + hY + \ell T \\ M_y + hH \\ Q - \ell H \end{pmatrix}_{\text{left}} \qquad (21.37)$$

For this simplified geometry, the two rotors are assumed to have the same shaft inclination.

The results for the low-frequency rotor response are dimensionless (based on ρ, Ω, and R) and in rotor coefficient form. The stability derivatives (load divided by inertia) are then

$$\frac{\text{Force}}{M} = \frac{1}{\widehat{M}} \gamma \frac{2 C_F}{\sigma a} \qquad (21.38)$$

$$\frac{\text{Moment}}{I} = \frac{1}{\widehat{I}} \gamma \frac{2 C_M}{\sigma a} \qquad (21.39)$$

where $\widehat{M} = M/(\frac{1}{2} N I_b)$ and $\widehat{I} = I/(\frac{1}{2} N I_b) = \widehat{M} k^2$ are the normalized inertias of the airframe, N is the number of blades, and I_b is the characteristic inertia of the blade. The airframe radius of gyration is k. The Lock number appears in this transformation as the ratio of aerodynamic and inertial forces, but the blade inertia I_b is not a good choice to represent inertial forces in the flight dynamics equations. Thus we use here the factor

$$G = \frac{\gamma}{\widehat{M}} = \frac{\rho a N c R^2}{2M} = \frac{a\rho A_b R}{2M} \tag{21.40}$$

G is a dimensionless parameter representing the ratio of the rotor aerodynamic forces to the aircraft inertial forces. Since vertical force trim gives $Mg = T$ (dimensional), the normalized mass is related to the rotor trim blade loading: $\widehat{M}g = \gamma(2C_T/\sigma a)_{\text{trim}}$, and $G = a(g/\Omega^2 R^2)/(2(C_T/\sigma)_{\text{trim}})$. The dimensionless gravitational constant g is divided by $\Omega^2 R$, so increases proportionally to R for fixed tip speed. Then for fixed design blade loading C_T/σ and tip speed ΩR, the normalized mass $\widehat{M}$ decreases with rotor size, proportionally to $1/R$; G increases with size, proportionally to R.

Dimensionless quantities are used in the following analyses. Dimensional stability derivatives and poles are recovered by introducing factors of Ω and R as required.

21.4 Hover Flight Dynamics

The flying qualities of the helicopter have different characteristics in hover and in forward flight. The hovering analysis is simpler, because of the axisymmetry of the rotor aerodynamics in vertical flight.

Consider the rigid-body motions of a single main rotor and tail rotor helicopter in hover. The aircraft is assumed to have complete axisymmetry, so the vertical and the longitudinal-lateral dynamics are completely separated. Such separation is a basic feature of the rotor in hover and generally holds for the hover flying qualities, even though the entire helicopter is not truly axisymmetric.

Because the effects of the tail rotor are neglected except on the helicopter yaw motion, the yaw dynamics are decoupled from the other degrees of freedom. The direction of rotor rotation discriminates between the left and right sides of the rotor, so the helicopter does not have a lateral symmetry plane. However, to begin the analysis, it is assumed that the longitudinal and lateral dynamics are also decoupled. The hover equations of motion in Laplace form are thus

$$(s - Z_w)\dot{z}_F = Z_{\theta_0}\theta_0 + Z_w w_G \tag{21.41}$$

$$(s - X_u)\dot{x}_F + (-X_q s + g)\theta_F = X_\theta \theta_s + X_u u_G \tag{21.42}$$

$$-M_u \dot{x}_F + (s^2 - M_q s)\theta_F = M_\theta \theta_s + M_u u_G \tag{21.43}$$

$$(s - Y_v)\dot{y}_F + (-Y_p s - g)\phi_F = Y_\theta \theta_c + Y_v v_G \tag{21.44}$$

$$-L_v \dot{y}_F + (s^2 - L_p s)\phi_F = L_\theta \theta_c + L_v v_G \tag{21.45}$$

$$(s - N_r)\dot{\psi}_F = N_{\theta_p}\theta_p + N_v(\dot{y}_F + v_G) \tag{21.46}$$

It has been assumed that the trim pitch angle θ_{F0} is small.

21.4.1 Rotor Forces and Moments

The low-frequency rotor response including the effects of the flap motion are derived in Chapter 19. The hub reactions due to shaft motion, pitch control, and aerodynamic gusts are derived in sections 19.4 and 19.5. Without pitch-cone coupling, the low-frequency thrust response (equation 19.71) in terms of the aircraft motion and control is

$$\frac{C_T}{\sigma a} = C'T_\theta \theta_0 - \frac{1}{2} C'T_\lambda (\dot{z}_F + w_G) \qquad (21.47)$$

The aerodynamic coefficients are $T_\theta = 1/6$ and $T_\lambda = -1/4$ (section 16.8.8). The effect of the axial velocity on the inflow gives the factor of $\frac{1}{2}$ for the T_λ term. The hover lift deficiency function C' accounts for the reduction of the aerodynamic forces by the wake; see section 19.4. The inertial reactions of the rotor have been accounted for by including the rotor mass in the helicopter mass. The thrust perturbations due to collective pitch and vertical velocity of the helicopter are produced by direct changes in the blade angle-of-attack. Since vertical velocity $\dot{z}_F$ increases the angle-of-attack, T_λ is negative. Then

$$Z = -\frac{2\gamma}{\widehat{M}} \frac{C_T}{\sigma a} = -2G \frac{C_T}{\sigma a} \qquad (21.48)$$

gives the stability derivatives. The tail rotor thrust gives a yaw moment; hence the stability derivatives,

$$\left(\frac{C_T}{\sigma a}\right)_{tr} = C'_{tr} T_\theta \theta_p - \frac{1}{2} C'_{tr} T_\lambda \frac{(\Omega R)_{mr}}{(\Omega R)_{tr}} (\dot{y}_F + v_G - \ell_{tr} \dot{\psi}_F) \qquad (21.49)$$

$$N = -\ell_{tr} 2G \frac{(A_b (\Omega R)^2)_{tr}}{(A_b (\Omega R)^2)_{mr}} \left(\frac{C_T}{\sigma a}\right)_{tr} \qquad (21.50)$$

accounting for the definition of C_T/σ. The tail rotor collective pitch is the aircraft pedal control, and the tail rotor axial velocity is produced by side velocity and gust and by the yaw rate.

In terms of the aircraft motion and control, the low-frequency response of the rotor tip-path-plane tilt (equation 19.103) is

$$\begin{pmatrix} \beta_{1c} \\ \beta_{1s} \end{pmatrix} = \frac{1}{1 + N_e^2} \begin{bmatrix} 1 & N_e \\ -N_e & 1 \end{bmatrix} \left\{ \begin{pmatrix} -\theta_s \\ \theta_c \end{pmatrix} + 8M_\mu \begin{pmatrix} -\dot{x}_F + h\dot{\theta}_F - u_G \\ \dot{y}_F + h\dot{\phi}_F + v_G \end{pmatrix} \right.$$
$$\left. + \begin{bmatrix} \frac{16}{\gamma_e} & -(1 - K_R) \\ 1 - K_R & \frac{16}{\gamma_e} \end{bmatrix} \begin{pmatrix} \dot{\theta}_F \\ \dot{\phi}_F \end{pmatrix} \right\} \qquad (21.51)$$

including the hub in-plane velocity terms, but without pitch-flap coupling. The influence of the wake enters through the effective Lock number $\gamma_e = C\gamma$, stiffness $N_e = N_\star/C = (v_\beta^2 - 1)/\frac{\gamma_e}{8}$ (with C the lift deficiency function), and coupling due to wake curvature, K_R; see section 19.5. The aerodynamic coefficient is

$$M_\mu = \frac{2C_T}{\sigma a} + \frac{\lambda_{HP}}{4} \qquad (21.52)$$

(section 16.8.8); $M_\theta = -M_{\dot{\mu}} = 1/8$ has already been used. The parameter $N_\star$ defines the lateral-longitudinal coupling of the rotor response due to a flap frequency $v_\beta > 1$ (and in general to pitch-flap coupling as well). This coupling produces a decrease in the magnitude of the tip-path-plane response and a phase shift by $\Delta\psi = \tan^{-1} N_\star$, as discussed in section 19.3. Longitudinal cyclic θ_s, longitudinal hub velocity ($\dot{x}_F - h\dot{\theta}_F + u_G$), and shaft rolling velocity $\dot{\phi}_F$ produce lateral aerodynamic moments on the rotor disk. The rotor flap motion responds with maximum amplitude ($90 - \tan^{-1} N_\star$) degrees after the maximum excitation, and hence with longitudinal tip-path-plane tilt β_{1c}. The resulting flapping velocity in the rotating frame then produces a lateral aerodynamic moment on the disk. The tip-path plane tilts until the aerodynamic flap moments are in equilibrium. Since this new equilibrium position is reached quickly, the static response can be used for the low-frequency rotor dynamics. The lateral tip-path-plane tilt response has a similar origin. The remaining term in the flap response is the lag of the tip-path-plane tilt required to precess the rotor disk to follow the shaft angular velocity. A pitching rate $\dot{\theta}_F$ requires a roll moment on the rotor disk for the tip-path plane to follow the shaft; this roll moment is provided by the longitudinal flapping β_{1c}. Similarly the tip-path plane follows a shaft rolling velocity $\dot{\phi}_F$ with a steady lateral tilt relative to the shaft. The moment stability derivatives are then

$$\begin{pmatrix} M \\ L \end{pmatrix} = -\frac{\nu_\beta^2 - 1}{\widehat{I}} \begin{pmatrix} \beta_{1c} \\ \beta_{1s} \end{pmatrix} + \frac{h}{k^2} \begin{pmatrix} -X \\ Y \end{pmatrix} = -\frac{G}{k^2} \frac{\nu_\beta^2 - 1}{\gamma} \begin{pmatrix} \beta_{1c} \\ \beta_{1s} \end{pmatrix} + \frac{h}{k^2} \begin{pmatrix} -X \\ Y \end{pmatrix} \quad (21.53)$$

where X and Y are the corresponding force derivatives.

The low-frequency response of the rotor hub in-plane forces (equation 19.102, with the hub velocity terms but no pitch-flap coupling) is

$$\begin{pmatrix} \frac{2C_H}{\sigma a} \\ \frac{2C_Y}{\sigma a} \end{pmatrix} = \left\{ -\frac{2C_T}{\sigma a} + \begin{bmatrix} (1-C)\frac{C_T}{\sigma a} - C\widehat{H}_{\dot{\beta}} & CR_{\dot{\beta}} \\ -CR_{\dot{\beta}} & (1-C)\frac{C_T}{\sigma a} - C\widehat{H}_{\dot{\beta}} \end{bmatrix} \right\} \begin{pmatrix} \beta_{1c} \\ \beta_{1s} \end{pmatrix}$$

$$- \begin{bmatrix} (1-C)\frac{C_T}{\sigma a} - C\widehat{H}_{\dot{\beta}} & CR_{\dot{\beta}} \\ -CR_{\dot{\beta}} & (1-C)\frac{C_T}{\sigma a} - C\widehat{H}_{\dot{\beta}} \end{bmatrix} \begin{pmatrix} -\theta_s \\ \theta_c \end{pmatrix}$$

$$+ (1-C)8M_\mu \begin{bmatrix} -H_{\dot{\beta}} & R_{\dot{\beta}} \\ -R_{\dot{\beta}} & -H_{\dot{\beta}} \end{bmatrix} \begin{pmatrix} -\dot{x}_F + h\dot{\theta}_F - u_G \\ \dot{y}_F + h\dot{\phi}_F + v_G \end{pmatrix}$$

$$+ \begin{bmatrix} -(H_\mu + R_\mu) & R_r \\ -R_r & -(H_\mu + R_\mu) \end{bmatrix} \begin{pmatrix} -\dot{x}_F + h\dot{\theta}_F - u_G \\ \dot{y}_F + h\dot{\phi}_F + v_G \end{pmatrix}$$

$$- C \begin{bmatrix} -H_{\dot{\beta}} & R_{\dot{\beta}} \\ -R_{\dot{\beta}} & -H_{\dot{\beta}} \end{bmatrix} (1 - K_R) \begin{pmatrix} -\dot{\phi}_F \\ \dot{\theta}_F \end{pmatrix} \quad (21.54)$$

The derivation of this result has already used $M_\theta = -M_{\dot{\beta}} = 1/8$, $R_\beta = -C_T/\sigma a$, $H_\theta = -\widehat{H}_{\dot{\beta}}$, and $R_\theta = -R_{\dot{\beta}}$. The aerodynamic coefficients are given in section 16.8.8; in particular,

$$H_{\dot{\beta}} = \frac{C_T}{\sigma a} + \widehat{H}_{\dot{\beta}} = \frac{C_T}{\sigma a} - \frac{\lambda_{HP}}{4} \quad (21.55)$$

The force stability derivatives are then

$$\begin{pmatrix} -X \\ Y \end{pmatrix} = \frac{\gamma}{\widehat{M}} \begin{pmatrix} \frac{2C_H}{\sigma a} \\ \frac{2C_Y}{\sigma a} \end{pmatrix} = G \begin{pmatrix} \frac{2C_H}{\sigma a} \\ \frac{2C_Y}{\sigma a} \end{pmatrix} \quad (21.56)$$

The principal hub force is the in-plane component of the thrust vector when it is tilted with the tip-path plane. Combining the hub force and the hub moment, the total moment about the helicopter center-of-gravity is

$$\begin{pmatrix} -\dfrac{2C_{My}}{\sigma a} \\ \dfrac{2C_{Mx}}{\sigma a} \end{pmatrix}_{CG} = \left[\dfrac{\nu_\beta^2 - 1}{\gamma} + h\dfrac{2C_T}{\sigma a} \right] \begin{pmatrix} \beta_{1c} \\ \beta_{1s} \end{pmatrix} - h \begin{pmatrix} \dfrac{2C_H}{\sigma a} \\ \dfrac{2C_Y}{\sigma a} \end{pmatrix}_{TPP} \qquad (21.57)$$

where the last term is due to the in-plane hub forces relative to the tip-path plane. The pitch and roll moments have a greater role in the helicopter flight dynamics than the in-plane hub forces. For an articulated rotor the moments and forces are always directly proportional. The hub forces and moments are determined primarily by the rotor flap response. Longitudinal and lateral control plane tilt produce, respectively, pitch and roll moments about the center-of-gravity, which the pilot can use to control the aircraft. Longitudinal hub velocity $\dot{x}_F$ produces an in-plane force H opposing the motion and a corresponding pitch moment responsible for the speed stability of the helicopter. Similarly, there is a side force Y due to lateral velocity $\dot{y}_F$ and a corresponding roll moment similar to the dihedral effect of an airplane wing. By the same mechanism, rotor hub reactions due to longitudinal and lateral gusts are produced. Pitch rate $\dot{\theta}_F$ of the helicopter produces a pitch moment M_y due to the lag of the tip-path plane required to precess the rotor; similarly, roll rate $\dot{\phi}_F$ produces a roll moment M_x. Since these moments oppose the helicopter motion, the rotor provides damping of the helicopter angular motion. When $\nu_\beta > 1$, as is the case with an offset flap hinge or a hingeless rotor, the moments are increased as a result of the hub moment capability, and the coupling of the helicopter lateral and longitudinal motion is changed.

21.4.2 Hover Stability Derivatives

Substituting the rotor flap response in the rotor force and moment expressions gives the stability derivatives. The rotor aero coefficients are from section 16.8.8. The vertical force derivatives are

$$Z_{\theta_0} = -G2C'T_\theta = -GC'\dfrac{1}{3} \qquad (21.58)$$

$$Z_w = GC'T_\lambda = -GC'\dfrac{1}{4} \qquad (21.59)$$

where the lift deficiency function is

$$C' = \dfrac{1}{1 + \sigma a/16\lambda_i} \qquad (21.60)$$

and Z_w includes a factor of $\tfrac{1}{2}$ from the wake influence. The scaling factor is $G = (a\rho A_b R)/2M = g/(2C_T/\sigma a)$. The tail rotor produces the directional moment derivatives for the single main rotor and tail rotor configuration:

$$N_{\theta_p} = -\dfrac{G}{k_z^2}\ell_{tr}\dfrac{(A_b(\Omega R)^2)_{tr}}{(A_b(\Omega R)^2)_{mr}}2C'_{tr}T_\theta = -\dfrac{\ell_{tr}}{k_z^2}GC'_{tr}\dfrac{1}{3}\dfrac{(A_b(\Omega R)^2)_{tr}}{(A_b(\Omega R)^2)_{mr}} \qquad (21.61)$$

$$N_v = -\dfrac{G}{k_z^2}\ell_{tr}\dfrac{(A_b(\Omega R)^2)_{tr}}{(A_b(\Omega R)^2)_{mr}}C'_{tr}T_\lambda\dfrac{(\Omega R)_{mr}}{(\Omega R)_{tr}} = \dfrac{\ell_{tr}}{k_z^2}GC'_{tr}\dfrac{1}{4}\dfrac{(A_b(\Omega R))_{tr}}{(A_b(\Omega R))_{mr}} \qquad (21.62)$$

$$N_r = -\ell_{tr}N_v = -\dfrac{\ell_{tr}^2}{k_z^2}GC'_{tr}\dfrac{1}{4}\dfrac{(A_b(\Omega R))_{tr}}{(A_b(\Omega R))_{mr}} \qquad (21.63)$$

21.4 Hover Flight Dynamics

Moment derivatives are produced by the hub moments (if $v_\beta > 1$) and the hub forces acting a distance h above the center-of-gravity. The pitch moment derivatives are

$$M_\theta = \frac{G}{k_y^2}\left(\frac{v_\beta^2-1}{\gamma}\right)\frac{1}{1+N_e^2} - \frac{h}{k_y^2}X_\theta \tag{21.64}$$

$$M_{\theta_c} = -\frac{G}{k_y^2}\left(\frac{v_\beta^2-1}{\gamma}\right)\frac{1}{1+N_e^2}N_e - \frac{h}{k_y^2}X_{\theta_c} \tag{21.65}$$

$$M_u = \frac{G}{k_y^2}\left(\frac{v_\beta^2-1}{\gamma}\right)\frac{1}{1+N_e^2}8M_\mu - \frac{h}{k_y^2}X_u \tag{21.66}$$

$$M_v = -\frac{G}{k_y^2}\left(\frac{v_\beta^2-1}{\gamma}\right)\frac{1}{1+N_e^2}N_e 8M_\mu - \frac{h}{k_y^2}X_v \tag{21.67}$$

$$M_q = -\frac{G}{k_y^2}\left(\frac{v_\beta^2-1}{\gamma}\right)\frac{1}{1+N_e^2}\left(\frac{16}{\gamma_e} + 8M_\mu h + N_e(1-K_R)\right) - \frac{h}{k_y^2}X_q \tag{21.68}$$

$$M_p = \frac{G}{k_y^2}\left(\frac{v_\beta^2-1}{\gamma}\right)\frac{1}{1+N_e^2}\left((1-K_R) - N_e\left(\frac{16}{\gamma_e} + 8M_\mu h\right)\right) - \frac{h}{k_y^2}X_p \tag{21.69}$$

including the pitch moment derivatives that couple with lateral motion and control (M_{θ_c}, M_v, and M_p). The effective Lock number $\gamma_e = C\gamma$ and stiffness $N_e = N_\star/C = (v_\beta^2-1)/\frac{\gamma_e}{8}$ (there is no pitch-flap coupling) are obtained from the lift deficiency function

$$C = \frac{1}{1+\sigma a/8\lambda_i} \tag{21.70}$$

The $(v_\beta^2-1)/\gamma = N_e C/8$ terms are from the hub spring, whereas N_e comes from the flap response. The corresponding roll moment derivatives are $L_\theta = -M_\theta$, $L_{\theta_s} = M_{\theta_c}$, $L_v = -M_u$, $L_u = M_v$, $L_p = M_q$, and $L_q = -M_p$, with k_y replaced by k_x. The drag force derivatives are

$$X_\theta = -G\left(\frac{2C_T}{\sigma a} + CR_{\dot\beta}N_e + \left((1-C)\frac{C_T}{\sigma a} - C\widehat{H}_{\dot\beta}\right)N_e^2\right)\frac{1}{1+N_e^2} \tag{21.71}$$

$$X_{\theta_c} = G\left(\left((1+C)\frac{C_T}{\sigma a} + C\widehat{H}_{\dot\beta}\right)N_e + CR_{\dot\beta}N_e^2\right)\frac{1}{1+N_e^2} \tag{21.72}$$

$$X_u = -G\left(\left[\left(2+(1-C)N_e^2\right)\frac{C_T}{\sigma a} + \left(1+(1-C)N_e^2\right)\widehat{H}_{\dot\beta} + CR_{\dot\beta}N_e\right]8M_\mu\frac{1}{1+N_e^2}\right.$$
$$\left. + (H_\mu + R_\mu)\right) \tag{21.73}$$

$$X_v = G\left(\left[(1+C)\frac{C_T}{\sigma a} + C\widehat{H}_{\dot\beta}\right]N_e - \left(1+(1-C)N_e^2\right)R_{\dot\beta}\right]8M_\mu\frac{1}{1+N_e^2} - R_r\right) \tag{21.74}$$

$$X_q = G\left(\left[(1+C)\frac{C_T}{\sigma a} + C\widehat{H}_{\dot\beta}\right) + CR_{\dot\beta}N_e\right]\left(\frac{16}{\gamma_e} + N_e(1-K_R)\right)\frac{1}{1+N_e^2} - hX_u\right) \tag{21.75}$$

$$X_p = G\left(\left[\left((1+C)\frac{C_T}{\sigma a} + C\widehat{H}_{\dot\beta}\right)N_e - CR_{\dot\beta}\right]\left(\frac{16}{\gamma_e} + N_e(1-K_R)\right)\frac{1}{1+N_e^2}\right.$$
$$\left. - \frac{C_T}{\sigma a}(1-K_R)\right) + hX_v \tag{21.76}$$

The corresponding side force derivatives are $Y_\theta = X_\theta$, $Y_{\theta_s} = -X_{\theta_c}$, $Y_v = X_u$, $Y_u = -X_v$, $Y_p = -X_q$, and $Y_q = X_p$.

The longitudinal and lateral derivatives are much simplified for teetering rotors and articulated rotors with no hinge offset ($\nu_\beta = 1$):

$$X_\theta = -G\left(\frac{2C_T}{\sigma a}\right) = -g \tag{21.77}$$

$$X_{\theta_c} = 0 \tag{21.78}$$

$$X_u = -G\left(\left(\frac{2C_T}{\sigma a} + \widehat{H}_{\dot\beta}\right)8M_\mu + (H_\mu + R_\mu)\right)$$
$$= -g8M_\mu - G\left(\widehat{H}_{\dot\beta}8M_\mu + (H_\mu + R_\mu)\right) \tag{21.79}$$

$$X_v = -G\left(R_{\dot\beta}8M_\mu + R_r\right) \tag{21.80}$$

$$X_q = G\left((1+C)\frac{C_T}{\sigma a} + C\widehat{H}_{\dot\beta}\right)\left(\frac{16}{\gamma_e}\right) - hX_u$$
$$= g\frac{16}{\gamma}\left(\frac{1+C}{2C} + \frac{\widehat{H}_{\dot\beta}}{2C_T/\sigma a}\right) - hX_u \tag{21.81}$$

$$X_p = -G\left(\frac{C_T}{\sigma a}(1-K_R) + CR_{\dot\beta}\left(\frac{16}{\gamma_e}\right)\right) + hX_v$$
$$= -g\frac{1-K_R}{2} - G\frac{16}{\gamma}R_{\dot\beta} + hX_v \tag{21.82}$$

with the substitution $G = g/(2C_T/\sigma a)$. The moments are entirely due to the hub forces acting a distance h above the center-of-gravity, so the pitch moment derivatives are proportional to the drag force derivatives: $M = -(h/k_y^2)X$. Since $\widehat{H}_{\dot\beta}M_\mu/M_{\dot\beta}H_\mu \cong 1$, the speed stability is given primarily by the first term, $X_u \cong -g8M_\mu$. For the articulated rotor, the hub forces are primarily due to the tilt of the thrust vector with the tip-path plane, and hence they are proportional to $2C_T/\sigma a$. The tip-path-plane tilt due to cyclic pitch is given by $-M_\theta/M_{\dot\beta} = 1$, the tilt due to helicopter speed perturbations is given by $-M_\mu/M_{\dot\beta} = 8M_\mu$, and the tilt due to the helicopter angular velocity is given by $16/\gamma$. For pitch motions of the rotor the thrust vector does not remain perpendicular to the tip-path plane; rather, the thrust vector lags the tip-path plane by a significant amount. The $2C_T/\sigma a$ term in X_q is the thrust tilt with the tip-path plane, and the $\widehat{H}_{\dot\beta}$ term is the additional direct hub force. Since $\widehat{H}_{\dot\beta}$ is negative, it reduces the rotor pitch damping by the factor

$$f = \frac{(1+C)C_T/\sigma a + C\widehat{H}_{\dot\beta}}{2C_T/\sigma a} = \frac{1+C}{2} - C\frac{\lambda_{HP}}{8C_T/\sigma a} \tag{21.83}$$

including the wake influence (equation 19.110). See section 19.5 and Figure 19.12.

The major factor introduced by $\nu_\beta > 1$ (offset hinges or hingeless rotor) is the hub moment produced by tip-path-plane tilt, which increases the capability of the rotor to produce moments about the helicopter center-of-gravity. There is also increased coupling of the lateral and longitudinal motions. The pitch and roll moment derivatives are increased roughly by the factor

$$\frac{M_{\nu>1}}{M_{\nu=1}} \cong 1 + \frac{(\nu_\beta^2 - 1)/\gamma}{h2C_T/\sigma a} \qquad (21.84)$$

The force derivatives vary little with the flap frequency, but the moments dominate the longitudinal and lateral dynamics. The moment derivatives can be roughly doubled by using flap hinge offset. For a typical hingeless rotor the control derivatives (M_θ and L_θ) and speed stability (M_u and L_v) are increased by a factor of three or four compared to the articulated ($\nu_\beta = 1$) case. The pitch and roll damping (M_q and L_p) are increased even more, because the $\widehat{H}_\beta$ term reduces the damping produced by the thrust vector tilt but does not influence the hub moment contribution.

21.4.3 Vertical Dynamics

Vertical force equilibrium gives the equation of motion for the helicopter vertical velocity response to collective and vertical gust:

$$(s - Z_w)\dot{z}_F = Z_{\theta_0}\theta_0 + Z_w w_G \qquad (21.85)$$

with the stability derivatives in section 21.4.2. The vertical dynamics are described by a first-order differential equation for $\dot{z}_F$, with time constant

$$\tau_z = -\frac{1}{Z_w} = \frac{4}{GC'} = \frac{8C_T/\sigma a}{gC'} \qquad (21.86)$$

The dimensional time to half-amplitude is

$$t_{\frac{1}{2}} = 0.693 \Omega R \frac{8C_T/\sigma a}{gC'} \qquad (21.87)$$

or typically $t_{\frac{1}{2}} \cong 28 C_T/\sigma$, which is about 2 sec. Since the vertical time constant is proportional to the rotor tip speed and blade loading, it is about the same for all helicopters and specifically does not vary directly with aircraft size. For a given helicopter, the time constant is proportional to the gross weight.

The equation for vertical dynamics has a single pole at $s = Z_w$ and no zeros. The pole is proportional to size (through g), and typically the non-dimensional value is $s = -0.005$ to -0.02. The small value justifies the use of the rotor low-frequency response. The static (long time) response to control is $\dot{z}_F/\theta_0 = -Z_{\theta_0}/Z_w = -\frac{4}{3}$ or, dimensionally, $\dot{z}_F/\theta_0 = -\frac{4}{3}\Omega R$ (collective pitch in radians). The static response is determined by the balance of the rotor aerodynamic forces and hence does not depend on G or the lift deficiency function C'. The inflow perturbation due to the vertical velocity of the rotor reduces the vertical damping and hence increases $\dot{z}_F/\theta_0$ by a factor of 2, since the larger mass flow through the rotor in climb decreases the induced velocity (see section 16.8.8). The collective pitch change required to produce a small steady-state climb rate, $\Delta\theta \cong \frac{3}{4}\lambda_c$ (equation 4.31), corresponds to the control sensitivity $\dot{z}_F/\theta_0 = -\frac{4}{3}$. The short time response is

$$\ddot{z}_F = Z_{\theta_0}\theta_0 + Z_w w_G \qquad (21.88)$$

or, dimensionally, with the vertical acceleration in g's, is

$$\frac{\ddot{z}_F}{g} = -\frac{C'}{6C_T/\sigma a}\left(\theta_0 + \frac{3}{4}\frac{w_G}{\Omega R}\right) \qquad (21.89)$$

The response to collective pitch is about 0.15 g/degree.

The rotor rotational speed and ground effect are also important factors in the vertical dynamics. If the throttle is fixed so that the rotor speed can vary during the vertical motions, the vertical damping and control capability of the helicopter are reduced. The additional degree of freedom (Ω) also introduces an overshoot since the vertical dynamics are then second order. So the helicopter height control is more difficult if the rotor speed is not fixed. An automatic governor to control the rotor speed eases the piloting task. There is no spring on vertical displacements of the helicopter, so for height control the pilot must provide the feedback of height to collective. Near the ground, ground effect provides a vertical spring term, due to the rotor induced velocity changes with height. This vertical spring force helps the tasks of precision height control in hover and the helicopter flare in landing.

21.4.4 Directional Dynamics

Yaw moment equilibrium gives the equation of motion for the helicopter yaw rate response to pedal and sideward velocity:

$$(s - N_r)\dot{\psi}_F = N_{\theta_p}\theta_p + N_v(\dot{y}_F + v_G) \qquad (21.90)$$

The stability derivatives are given in section 21.4.2 for the single main rotor and tail rotor configuration, considering only the contributions of the tail rotor thrust. The control variable is the tail rotor collective pitch, $\theta_p = \theta_{0tr}$. The axial damping of the tail rotor (T_λ) gives thrust perturbations due to yaw rate, sideward velocity, and lateral gusts. Here it is assumed that the lateral dynamics do not depend on the yaw motion, so $\dot{y}_F$ affects the directional dynamics as an input. The yaw dynamics are described by a first-order differential equation with time constant

$$\tau_r = -\frac{1}{Z_r} = \frac{k_z^2}{\ell_{tr}^2}\frac{4}{GC'_{tr}}\frac{(A_b(\Omega R)^2)_{mr}}{(A_b(\Omega R)^2)_{tr}} \qquad (21.91)$$

Usually the tail length is slightly larger than the main rotor radius, the yaw moment of inertia gives about $k_z^2 = 0.1$, and the ratio of blade areas $A_{b_{tr}}/A_{b_{mr}} = 0.04$ to 0.10. The time to half-amplitude is typically about 3 sec. As for the vertical dynamics, the yaw time constant is proportional to the rotor tip speed and blade loading and does not vary directly with aircraft size.

The steady-state response to directional control is

$$\frac{\dot{\psi}_F}{\theta_p} = -\frac{N_{\theta_p}}{N_r} = -\frac{4}{3\ell_{tr}}\frac{(\Omega R)_{tr}}{(\Omega R)_{mr}} \qquad (21.92)$$

or dimensionally

$$\frac{\dot{\psi}_F}{\theta_p} = -\frac{4}{3\ell_{tr}}(\Omega R)_{tr} \qquad (21.93)$$

The yaw rate commanded by the directional control tends to decrease with the helicopter size, through ℓ_{tr}. After a small first-order lag, the tail rotor collective produces a high steady-state yaw rate even for large helicopters. Tail rotor pitch-cone coupling decreases all of the directional stability derivatives by the factor $v_\beta^2/v_{\beta e}^2 = 1/(1 + K_P\gamma/8v_\beta^2)$ (see section 19.4). So positive pitch-cone coupling increases the yaw time constant, reduces the short time response, and reduces the response to lateral gusts. The steady-state yaw response to control is not influenced by pitch-cone coupling. As for the vertical motion, the tail rotor induced velocity perturbations due to its axial velocity have reduced the yaw damping and increased the control sensitivity.

Sideward velocity of the helicopter produces a change in tail rotor thrust and hence a yaw of the helicopter. The steady-state response is $\dot\psi_F/\dot y_F = -N_v/N_r = 1/\ell_{tr}$. Thus the lateral and yaw motions of the helicopter in hover are coupled. Sideward velocity is produced by lateral cyclic control, but to maintain the helicopter heading a pedal control input is required as well. The tail rotor control required to prevent yaw during the lateral motions of the helicopter is $\theta_p/\dot y_F = -N_v/N_{\theta_p} = \frac{3}{4}/(\Omega R)_{tr}$ (dimensional). Neglecting the influence of the tail rotor thrust on the lateral dynamics, the poles of the yaw and lateral dynamics are still uncoupled, but a coordination of the pedal control with lateral cyclic is required. In forward flight this yaw moment due to sideslip is still present and provides the directional stability of the helicopter.

Assuming that the rotor speed is constant, the main rotor rotational damping torque increases the helicopter yaw damping. The thrust changes in response to vertical motion and inputs also produce torque changes that couple the vertical and yaw control, and coordination of the pedal control with collective stick inputs is needed to maintain the helicopter heading during vertical motions.

Tail rotor design and operation are complicated; see section 5.6.5. When the yaw rate or translational velocity is large, the tail rotor can be operating in the vortex ring state The tail rotor regularly operates in an adverse aerodynamic environment due to the wake of the main rotor, fuselage, and vertical tail. The directional control and yaw damping are influenced by these factors. In general, however, the tail rotor is a powerful and efficient design solution for the torque balance, directional stability, and control of single main rotor helicopters.

21.4.5 Longitudinal Dynamics

The dynamics of a helicopter in hover separate into vertical and lateral-longitudinal motions. It is assumed for now that the lateral and longitudinal dynamics can also be analyzed separately. Neglecting the influence of vertical and lateral motions, the longitudinal dynamics consist of just two degrees of freedom: pitch θ_F (positive nose upward) and longitudinal velocity $\dot x_F$ (positive forward). The inputs considered are longitudinal cyclic θ_s and longitudinal aerodynamic gust velocity u_G. Equilibrium of longitudinal forces and pitch moments give the differential equations for the longitudinal motions of the helicopter:

$$\begin{bmatrix} s - X_u & -X_q s + g \\ -M_u & s^2 - M_q s \end{bmatrix} \begin{pmatrix} \dot x_F \\ \theta_F \end{pmatrix} = \begin{pmatrix} X_\theta \\ M_\theta \end{pmatrix} \theta_s + \begin{pmatrix} X_u \\ M_u \end{pmatrix} u_G \qquad (21.94)$$

The velocity and pitch perturbations produce longitudinal forces and pitch moments through the rotor response. With the body-axes velocity representation of the aircraft motion, pitch angle produces a component of the aircraft weight in the x-direction.

The characteristic equation is

$$\Delta = s^3 - (X_u + M_q)s^2 + (X_u M_q - X_q M_u)s + gM_u = 0 \quad (21.95)$$

The three solutions of this polynomial are the poles of the helicopter longitudinal dynamics. The equations of motion invert to

$$\begin{pmatrix} \dot{x}_F \\ \theta_F \end{pmatrix} = \frac{1}{\Delta} \begin{pmatrix} X_\theta s^2 + (X_q M_\theta - X_\theta M_q)s - M_\theta g \\ M_\theta s + (X_\theta M_u - X_u M_\theta) \end{pmatrix} \theta_s$$

$$+ \frac{1}{\Delta} \begin{pmatrix} X_u s^2 + (X_q M_u - X_u M_q)s - M_u g \\ M_u s \end{pmatrix} u_G \quad (21.96)$$

The longitudinal velocity response to cyclic and gusts has two zeros, whereas the pitch response has one zero.

Consider first the case of an articulated rotor with no flap hinge offset, so the flap frequency $\nu_\beta = 1$. The expressions for the stability derivatives are simplified (section 21.4.2). The moments are entirely due to the hub forces acting a distance h above the center-of-gravity, so the pitch moment derivatives are proportional to the drag force derivatives: $M = -(h/k_y^2)X$. The response solution becomes

$$\begin{pmatrix} \dot{x}_F \\ \theta_F \end{pmatrix} = \frac{1}{\Delta} \begin{pmatrix} X_\theta s^2 - M_\theta g \\ M_\theta s \end{pmatrix} \theta_s + \frac{1}{\Delta} \begin{pmatrix} X_u s^2 - M_u g \\ M_u s \end{pmatrix} u_G$$

$$= \frac{1}{\Delta} (X_\theta \theta_s + X_u u_G) \begin{pmatrix} s^2 + gh/k_y^2 \\ -(h/k_y^2)s \end{pmatrix} \quad (21.97)$$

The pitch response has a single zero at the origin, $s = 0$. The longitudinal velocity response has two zeros on the imaginary axis at $s = \pm i(gh/k_y^2)^{1/2}$. The magnitude of these zeros is usually several times that of the poles, so they influence the response only for large gain.

For $\nu_\beta = 1$, the characteristic equation reduces to

$$\Delta = s^3 - (X_u + M_q)s^2 - gM_u = s^3 - M_q s^2 + M_u\big((k_y^2/h)s^2 + g\big) = 0 \quad (21.98)$$

The rotor gives the helicopter positive damping for pitch ($M_q < 0$), positive damping for longitudinal velocity ($X_u < 0$), and positive speed stability ($M_u > 0$). Since X_u is small compared to M_q, the dynamics are dominated by the pitch derivatives M_q and M_u. Figure 21.2 illustrates the variation of the hover longitudinal roots with M_u and M_q. For $M_u = 0$, the solution is $s = 0, 0$, and M_q. As M_u increases, the pitch damping root increases in magnitude, while the two roots at the origin become an unstable oscillatory mode. For moderate M_u the oscillatory roots approach asymptotes at $\pm 60°$ from the real axis. For large M_u, these roots approach $s = \pm i(-gM_u/X_u)^{1/2} = \pm i(gh/k_y^2)^{1/2}$. For $M_q = 0$, the solution (neglecting X_u) is

$$s = -(gM_u)^{1/3}, \quad (gM_u)^{1/3}\frac{1}{2}(1 + i\sqrt{3}), \quad (gM_u)^{1/3}\frac{1}{2}(1 - i\sqrt{3}) \quad (21.99)$$

As M_q increases, the real root increases in magnitude, and the frequency of the oscillatory mode decreases while it becomes less unstable.

The uncoupled pitch and longitudinal motions of the hovering helicopter both have positive damping. The instability of the longitudinal dynamics for hover must

21.4 Hover Flight Dynamics

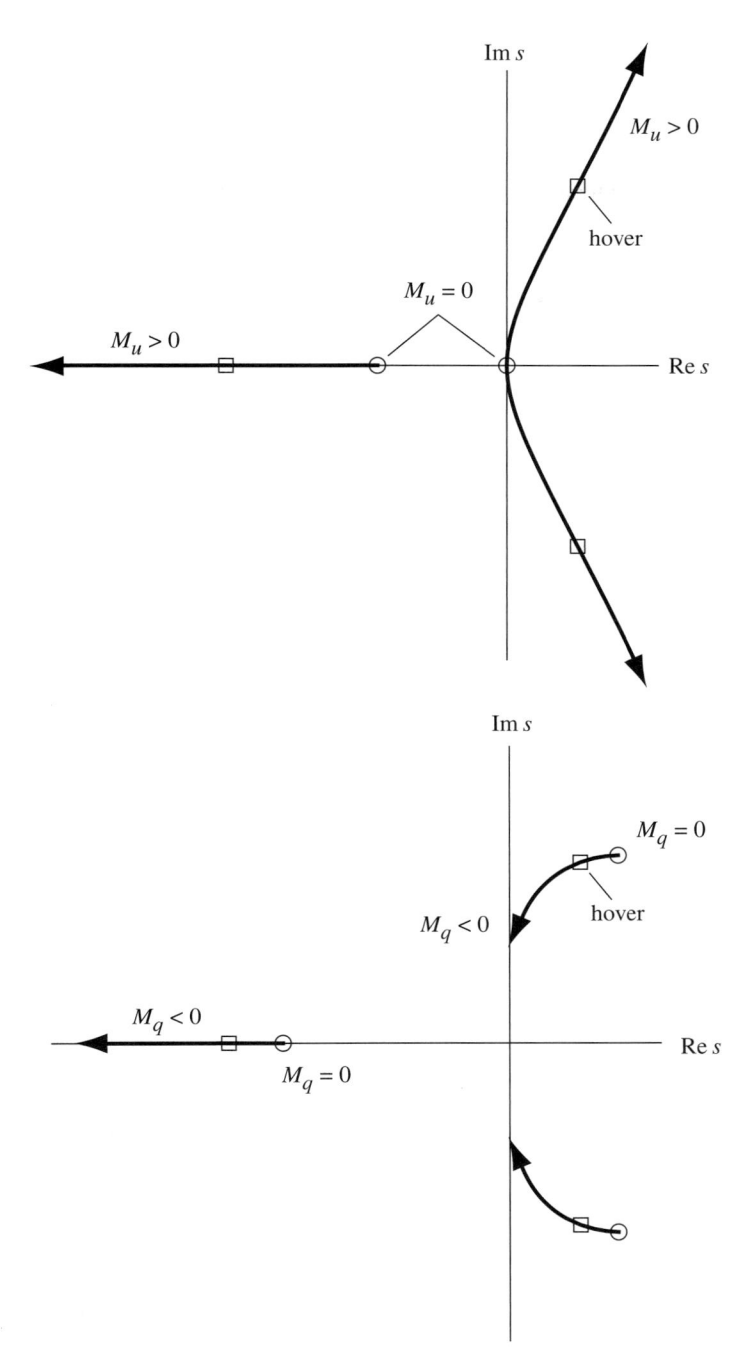

Figure 21.2. Influence of speed stability ($M_u > 0$) and pitch damping ($M_q < 0$) on the helicopter longitudinal roots.

therefore be a result of the coupling of the motion by the pitch moment due to longitudinal velocity (the speed stability M_u) and the longitudinal component of the gravitational force due to pitch. The approximate characteristic equation

$$s^3 - M_q s^2 + gM_u = 0 \qquad (21.100)$$

is equivalent to the following differential equations:

$$\begin{bmatrix} s & g \\ -M_u & s^2 - M_q s \end{bmatrix} \begin{pmatrix} \dot{x}_F \\ \theta_F \end{pmatrix} = 0 \qquad (21.101)$$

These equations contain the essence of the longitudinal dynamics in hover, with the following interpretation. Consider an oscillatory motion at frequency ω. Because of the rotor speed stability M_u, longitudinal velocity $\dot{x}_F$ produces a nose-up pitch moment. This moment is balanced by the pitch inertia, so the pitch response is 180° out-of-phase with the longitudinal velocity: $\theta_F = -(M_u/\omega^2)\dot{x}_F$, and forward velocity produces a nose-down pitch motion. The pitch motion rotates the body axes relative to the vertical, so there is a forward component of the gravitational force. Thus in the longitudinal velocity equation there is a force $g\theta_F = -(g/\omega^2)M_u \dot{x}_F$, which is negative damping, and therefore the coupled motion is unstable. The pitch moments are also reacted by the pitch damping M_q, which strongly influences the phase and frequency of the coupled motion.

For a more general view of the role of the speed stability, consider the divergence and flutter stability given by the characteristic equation 21.95. The requirement for static or divergence stability is that the constant term in the characteristic equation be positive, which is satisfied since $M_u > 0$. The requirement for dynamic or flutter stability can be obtained by applying Routh's criterion. The coefficients of the characteristic equation are positive, so all the roots are in the left-hand plane (stable) if

$$gM_u + (X_u + M_q)(X_u M_q - X_q M_u) < 0 \qquad (21.102)$$

The second term is zero for an articulated rotor, so the criterion reduces to $M_u < 0$, and the motion is unstable. The speed stability thus has a dominant role in determining the stability of the helicopter dynamics in hover. Because the criteria for static and dynamic stability are conflicting, the helicopter motion is unstable regardless of the sign or magnitude of M_u (see Figure 21.2).

Table 21.1 presents a numerical example of the helicopter dynamics in hover for a representative aircraft: single main rotor and tail rotor configuration, Lock number $\gamma = 8$, solidity $\sigma = 0.08$, rotor height above center-of-gravity $h = 0.3$, tail rotor area times tail length $A_{b_{tr}} \ell_{tr}/A_{b_{mr}} R = 0.08$, moments of inertia $k_x^2 = 0.02$, and $k_y^2 = k_z^2 = 0.1$ (where $k^2 = I/MR^2$ is the dimensionless radius of gyration), tip speed $\Omega R = 650$ ft/sec, and radius $R = 20$ ft. The operating condition is $C_T/\sigma = 0.08$, so $g/\Omega^2 R = 0.00152$ and $G = g/(2C_T/\sigma a) = 0.0543$. Table 21.1 has results with the wake effects, through the lift deficiency functions $C = 0.53$ and $C' = 0.70$, and the wake curvature $K_R = 1.5$, and without the wake effects ($C = C' = 1$ and $K_R = 0$). Table 21.1 considers the flap frequencies $\nu_\beta = 1/\text{rev}$ and $\nu_\beta = 1.1/\text{rev}$. Figure 21.3 shows the influence of ν_β on the poles.

The helicopter longitudinal dynamics are characterized by three modes: a negative real root (stable convergence) principally due to the pitch damping and a complex conjugate pair in the right half-plane (a mildly unstable oscillation) due to the coupling of the pitch and longitudinal velocity by the speed stability. Flap-hinge offset of an articulated rotor does not radically alter the character of the helicopter flight dynamics, although there is an important quantitative improvement in the handling qualities due to the hub moment capability. For a hingeless rotor the flap frequency is large enough to have a major impact on the dynamics. All three of these

21.4 Hover Flight Dynamics

Table 21.1. *Example of hovering helicopter dynamics*

	No wake			No wake
	Vertical		Directional	
s	−0.0093	−0.0136	−0.0072	−0.0119
$t_{\frac{1}{2}}$	2.3	1.6	2.9	1.8
flap frequency: $\nu_\beta = 1$/rev			$\nu_\beta = 1.10$/rev	
longitudinal pitch mode				
s	−0.018	−0.017	−0.057	−0.043
$t_{\frac{1}{2}}$	1.2	1.3	0.4	0.5
$\dot{x}_F/\theta_F$	0.089	0.098	0.030	0.038
oscillatory mode				
s	0.0046±i0.012	0.0057±i0.012	0.0014±i0.013	0.0027±i0.015
ω	0.40	0.41	0.42	0.50
ζ	−0.35	−0.42	−0.10	−0.17
T	15.9	15.5	14.8	12.5
t_2	4.6	3.7	15.8	7.9
$\dot{x}_F/\theta_F$	0.114 (112°)	0.108 (116°)	0.115 (98°)	0.096 (101°)
$z(\dot{x}_F/\theta_s)$	±i0.068	±i0.068	−0.0115±i0.129	−0.0080±i0.131
$z(\theta_F/\theta_s)$	0	0	−0.0001	−0.0001
$z(\dot{x}_F/u_G)$	±i0.068	±i0.068	−0.0138±i0.118	−0.0095±i0.121
lateral roll mode				
s	−0.049	−0.035	−0.268	−0.188
$t_{\frac{1}{2}}$	0.4	0.6	0.1	0.1
$\dot{y}_F/\phi_F$	−0.034	−0.045	−0.008	−0.010
oscillatory mode				
s	0.0032±i0.017	0.0062±i0.020	0.0002±i0.014	0.0006±i0.017
ω	0.57	0.65	0.44	0.55
ζ	−0.18	−0.30	−0.01	−0.03
T	11.1	9.7	14.3	11.5
t_2	6.7	3.4	130.5	37.2
$\dot{y}_F/\phi_F$	0.085 (−79°)	0.071 (−72°)	0.112 (−88°)	0.090 (−87°)
$z(\dot{y}_F/\theta_c)$	±i0.151	±i0.151	−0.058±i0.283	−0.040±i0.292
$z(\phi_F/\theta_c)$	0	0	−0.0001	−0.0001
$z(\dot{y}_F/v_G)$	±i0.151	±i0.151	−0.069±i0.258	−0.048±i0.268

Notes: root s (dimensionless), zero z (dimensionless), time to half-amplitude $t_{\frac{1}{2}}$ (sec), frequency ω (rad/sec), damping ratio ζ, period T (sec), time to double-amplitude t_2 (sec); eigenvector $\dot{x}_F/\theta_F$ or $\dot{y}_F/\phi_F$ (dimensionless; magnitude and phase); no wake column: $C = C' = 1$ and $K_R = 0$.

roots are small compared to the rotor speed, justifying the use of the low-frequency rotor response.

Figure 21.4 shows the time to half-amplitude of the pitch or short period mode, as a function of flap frequency ($\nu_\beta = 1.00, 1.05, 1.10$, and 1.15) and aircraft size (rotor radius $R = 10, 20, 30$, and 40 ft). Typically for $\nu_\beta = 1$, the pitch mode has a time to half-amplitude of $t_{\frac{1}{2}} = 1$ to 2 sec. This real root is roughly the same as the pole of the vertical motion. With offset hinges, the pitch mode time to half-amplitude is less than 0.5 sec. The high pitch damping of the hingeless rotor greatly increases the

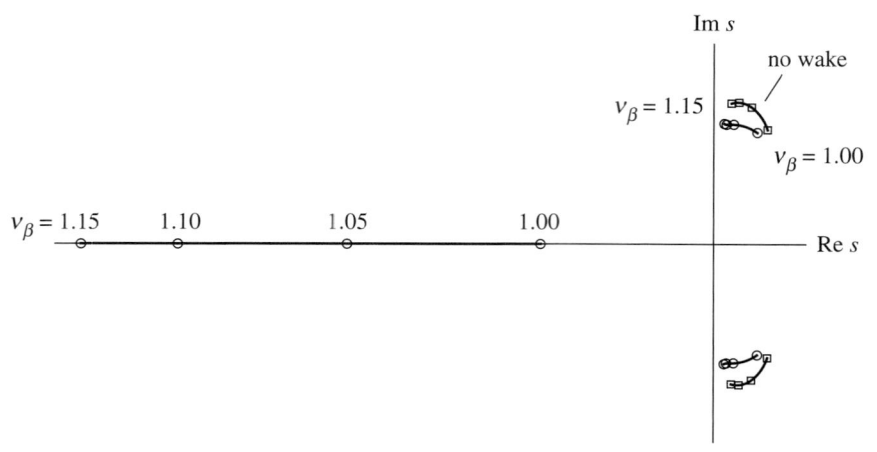

Figure 21.3. Influence of flap frequency on poles of longitudinal dynamics of hovering helicopter: $v_\beta = 1$ to 1.15/rev.

magnitude of the real root, so typically $t_{\frac{1}{2}} = 0.2$ to 0.5 sec. The dimensional pole of the pitch mode is relatively independent of the aircraft size.

Figure 21.5 shows the variation of the longitudinal oscillatory or long period mode with flap frequency and aircraft size: frequency ω (rad/sec), damping ratio ζ (positive for stable), period T (sec), and time to double-amplitude t_2 (sec, negative for stable). The period is typically $T = 10$ to 20 sec (a frequency of 0.05 to 0.10 Hz, or 0.3 to 0.6 rad/sec). The period is not much influenced by flap frequency and scales with size roughly proportionally to $\sqrt{R}$. For $v_\beta = 1$, the time to double-amplitude is typically $t_2 = 3$ to 7 sec. The high pitch damping of the hingeless rotor counters the increased speed stability, so the time to double-amplitude is increased substantially, to $t_2 = 10$ to 20 sec (hingeless rotors have not often been designed for large helicopters). The dimensionless real part of the oscillatory pole is roughly independent of size, so the time to double-amplitude is roughly proportional to R. The instability is milder for a large helicopter, because both the period and the time to double-amplitude increase with size.

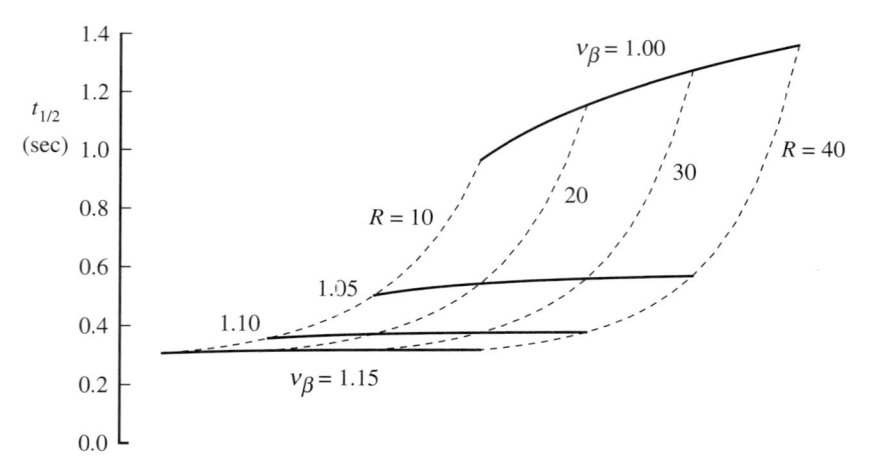

Figure 21.4. Influence of flap frequency and rotor size on the time to half-amplitude of the pitch mode: $v_\beta = 1$ to 1.15/rev and radius $R = 10$ to 40 ft.

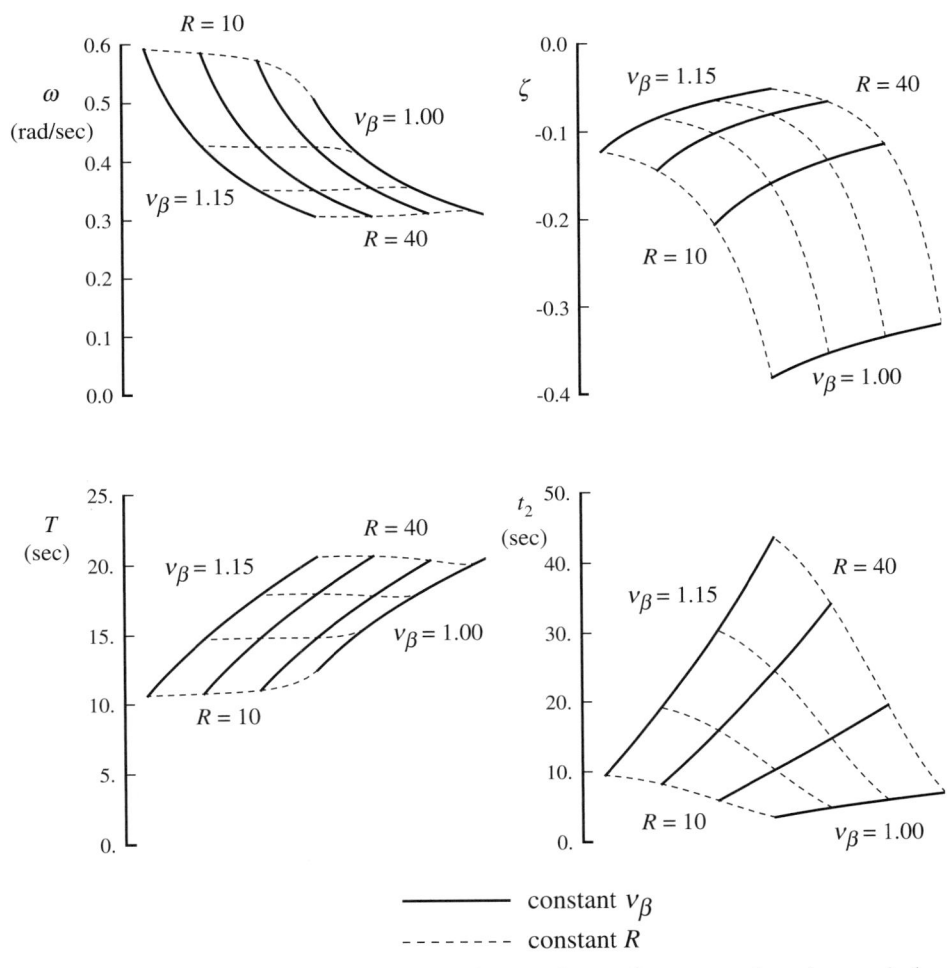

Figure 21.5. Influence of flap frequency and rotor size on frequency, damping, period, and time to double-amplitude of the longitudinal oscillatory mode: $v_\beta = 1$ to 1.15/rev and radius $R = 10$ to 40 ft.

The variation of the longitudinal roots in Figures 21.4 and 21.5 are for constant tip speed and C_T/σ. For fixed radius and tip speed, the pitch mode time to half-amplitude increases with C_T/σ, except for $v_\beta = 1$. The oscillatory mode period decreases with C_T/σ and the damping ratio becomes more negative, so the instability is worse at high blade loading. For fixed radius and blade loading, the pitch mode time to half-amplitude increases with tip speed. The oscillatory mode period increases with tip speed, and the damping ratio becomes more negative.

21.4.6 Response to Control and Loop Closures

The steady-state response to control ($s = 0$) is

$$\begin{pmatrix} \dot{x}_F \\ \theta_F \end{pmatrix} = \begin{pmatrix} -M_\theta/M_u \\ (X_\theta M_u - X_u M_\theta)/gM_u \end{pmatrix} \theta_s \cong \begin{pmatrix} -1/8M_\mu \\ 0 \end{pmatrix} \theta_s \qquad (21.103)$$

The last approximation is quite good in general and is exact for an articulated rotor with no hinge offset. Cyclic pitch control produces longitudinal velocity $\dot{x}_F$ but no attitude change in the steady-state perturbation from hover. Longitudinal cyclic produces a pitch moment $M_\theta \theta_s$, and equilibrium is achieved when the moment due to the speed stability is sufficient to cancel the control moment. At this equilibrium state there is no net force or moment on the helicopter (since $X_\theta M_u - X_u M_\theta \cong 0$) and hence no pitch attitude change. The zero steady-state pitch response to cyclic implies neutral static stability of the hovering helicopter relative to pitch attitude changes. The gradient of the cyclic control with the steady-state velocity perturbation $(\theta_s/\dot{x}_F)$ is a measure of the rotor speed stability.

Because the longitudinal motion in hover is unstable, a finite steady-state response is achieved only if the pilot or an automatic control system intervenes to ensure that the transients die out. This solution for the steady-state response is thus best interpreted as the gradient of the control to trim the helicopter for small velocity and pitch changes from hover.

The short time response to cyclic control and longitudinal gusts ($s \to \infty$) is

$$\begin{pmatrix} \ddot{x}_F \\ \ddot{\theta}_F \end{pmatrix} = \begin{pmatrix} X_\theta \\ M_\theta \end{pmatrix} \theta_s + \begin{pmatrix} X_u \\ M_u \end{pmatrix} u_G \qquad (21.104)$$

For an articulated rotor the longitudinal acceleration response to control is $(\ddot{x}_F/g)/\theta_s = X_\theta/g = -1$ g/rad, which is quite small. This result is independent of any parameters of the rotor; the response varies little with the flap frequency also. The primary response to longitudinal cyclic is the pitch acceleration. For an articulated rotor $\ddot{\theta}_F/\theta_s = M_\theta = hk/k_y^2$, and for a hingeless rotor the pitch acceleration is three or four times larger. Similarly, the longitudinal acceleration response to gusts is small, whereas the pitch acceleration response is large, especially with a hingeless rotor.

Consider a short period approximation for the hover longitudinal dynamics. Since control inputs at first produce primarily pitch motion of the helicopter, the longitudinal velocity can be neglected for a short time analysis. If $\dot{x}_F$ is neglected, the equation of motion for pitch becomes

$$(s - M_q)\dot{\theta}_F = M_\theta \theta_s + M_u u_G \qquad (21.105)$$

So initially cyclic pitch commands the helicopter pitch rate, with a first-order lag given by the pole $s = M_q$. This pole is an approximation to the pitch root of the longitudinal dynamics, but the approximation is not very good for articulated rotors, where the speed stability increases the magnitude of the root significantly. The initial response is here

$$\ddot{\theta}_F = M_\theta \theta_s + M_u u_G \qquad (21.106)$$

which is the same as obtained with the complete equations. The steady-state limit of this approximation is the pitch rate $\dot{\theta}_F/\theta_s = -M_\theta/M_q$. For still larger times the longitudinal velocity motion enters the dynamics, and the complete equations must be considered.

To achieve stable flight, the longitudinal dynamics of the hovering helicopter require feedback control, either from the pilot or from an automatic control system (perhaps a mechanical system, often using a gyro). The longitudinal velocity and pitch attitude must be sensed and, after appropriate compensation, fed back to the longitudinal cyclic pitch.

21.4 Hover Flight Dynamics

$\theta_s = K x_F$

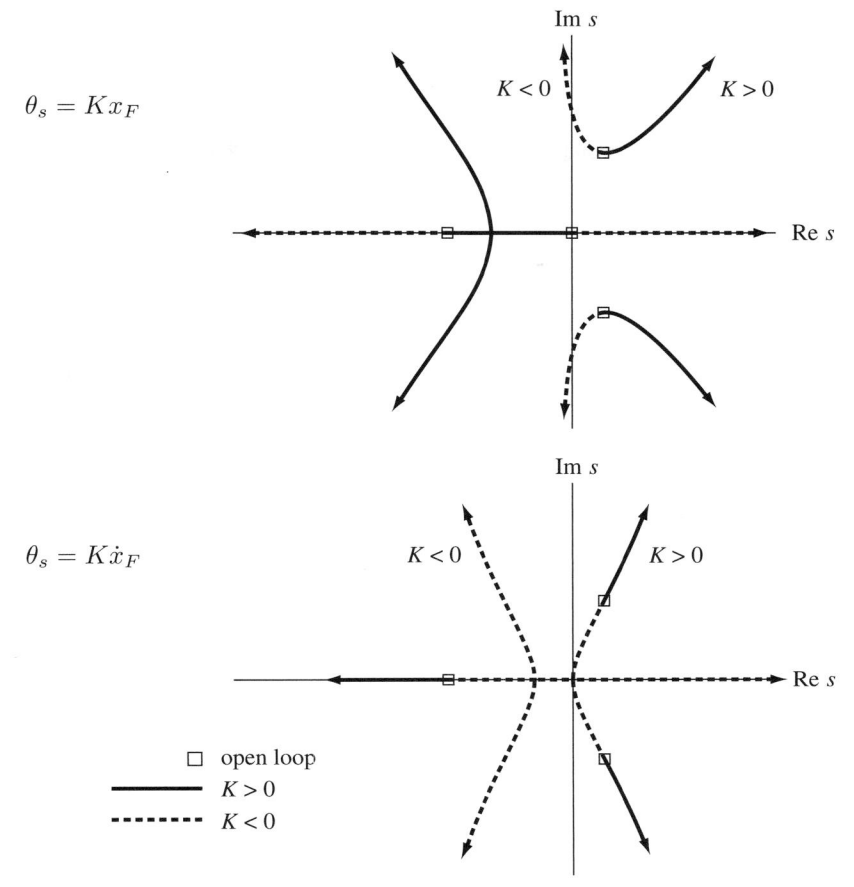

$\theta_s = K \dot{x}_F$

□ open loop
—— $K > 0$
------ $K < 0$

Figure 21.6. Longitudinal velocity feedback to cyclic for an articulated rotor helicopter in hover.

Figure 21.6 shows the root loci for feedback of the helicopter longitudinal displacement or longitudinal velocity. Since the open loop zeros of the response are large compared to the poles, they do not influence the loci except for high gains. None of these networks based on measuring x_F or $\dot{x}_F$ is satisfactory. Negative feedback ($K > 0$) of x_F or $\dot{x}_F$ destabilizes the oscillatory mode, whereas positive feedback produces a static instability. Longitudinal velocity feedback $\theta_s = K\dot{x}_F$ is equivalent to changing the speed stability, so these results are consistent with Figure 21.2.

Figure 21.7 shows the root loci for feedback of the helicopter pitch attitude, pitch rate, or a combination of attitude and rate. The pitch response has a single open loop zero at the origin. Pitch attitude feedback can stabilize the oscillatory mode with positive gain, although it is limited to rather low damping for an articulated rotor. Attitude feedback, however, decreases the damping of the real root, which is undesirable. Pitch rate feedback both increases the real root damping and stabilizes the oscillatory mode. The period and time to double-amplitude of the oscillatory root are both increased, but the mode remains unstable. Pitch rate feedback is equivalent to increasing the rotor pitch damping derivatives X_q and M_q (Figure 21.2). These results suggest the use of a combination of pitch attitude feedback to stabilize the oscillation and pitch rate feedback to keep the pitch damping high. Introducing

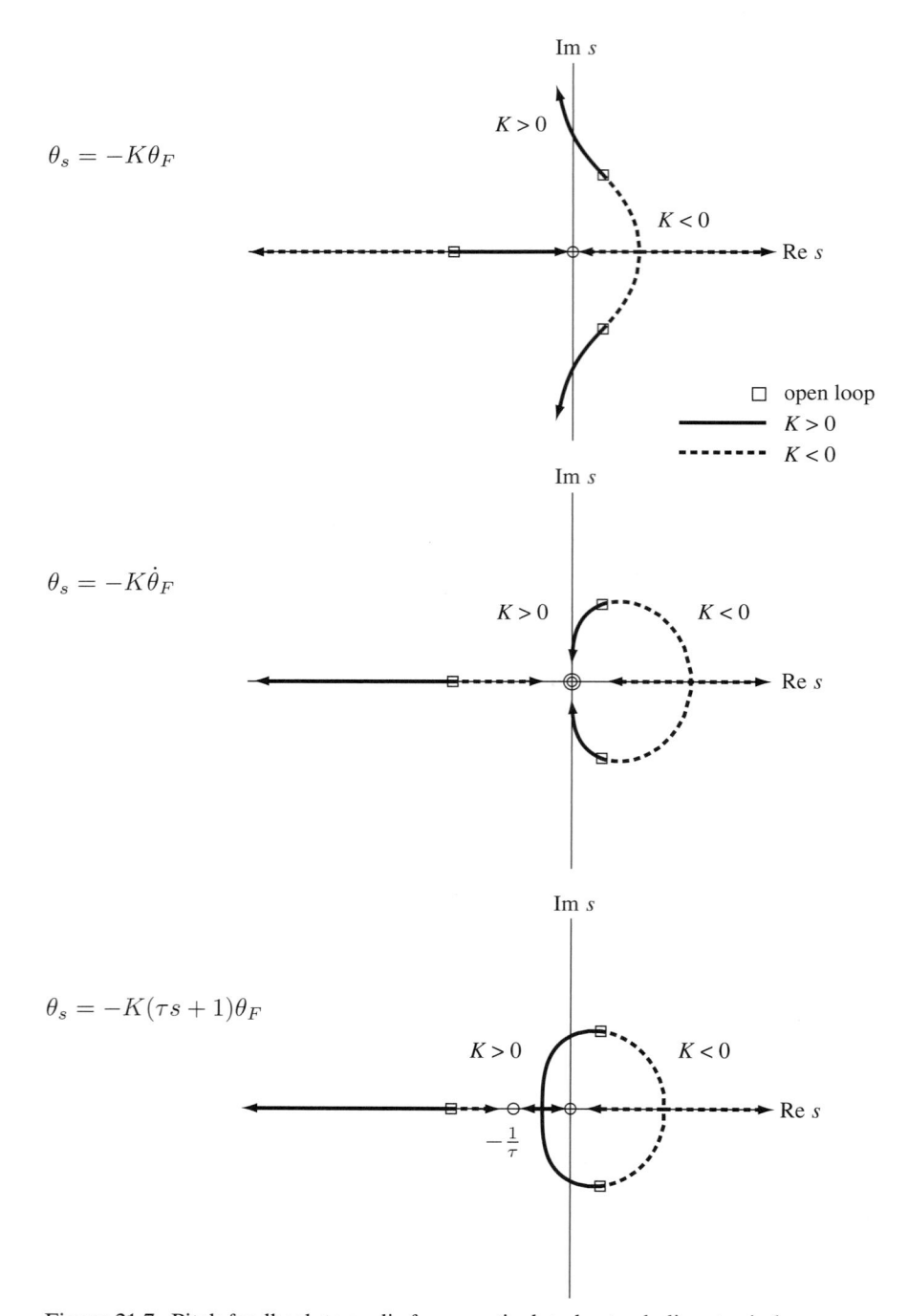

Figure 21.7. Pitch feedback to cyclic for an articulated rotor helicopter in hover.

some attitude feedback into the rate network pulls one of the zeros off the origin into the left half-plane, bringing the oscillatory branches of the loci with it, while maintaining the real root damping increase of the rate feedback. The lead τ must be large enough for the zero to remain close to the origin. If the zero gets to the left of the open loop pole, the network operates more like attitude feedback, and the real root damping decreases. Typically a lead of $\tau = 2$ to 4 sec is required for articulated rotor helicopters. Thus rate plus attitude feedback of pitch (or attitude feedback with lead) produces the desired handling qualities of high pitch damping and a stable oscillatory mode. Such feedback is not the most desirable for the pilot, however. The lead required is a bit higher than can be easily handled, so the hovering task is tiring for the pilot and the task requires training. Thus an automatic stability augmentation system is frequently desirable.

Figure 21.8 shows the root loci for feedback of the helicopter pitch attitude or rate with some lag in the network. Mechanical systems in particular are likely to introduce such a lag, with typically $\ell = 1$ sec. So there is an additional pole somewhat to the left of the real root of the hover dynamics. Generally the flying qualities deteriorate with lag in the system. With a large enough lag the attitude feedback no longer stabilizes the oscillatory mode, and for rate feedback the lag introduces a limit on the increase in the real root damping. As long as the lag is significantly smaller than the time constant of the aircraft pitch root, the root loci are not greatly influenced. In particular, the lagged rate plus attitude feedback remains satisfactory as long as the lead character of the network dominates (the lag pole must be to the left of the lead zero and preferably also to the left of the pitch root). Lagged rate feedback $(\ell s + 1)\theta_s = -K\dot{\theta}_F$ is of interest since there are mechanical systems that produce such control (see section 21.9). This system is generally similar to pure rate feedback. Although rate or lagged rate feedback does not produce a stable system, it definitely improves the helicopter flight dynamics.

In summary, the longitudinal dynamics of a hovering helicopter are described by a stable real root due to the pitch damping and a mildly unstable oscillatory root due to the speed stability. The instability is not too objectionable, since the period and time to double-amplitude are long enough for the motion to be controllable by the pilot. However, the control characteristics of the helicopter require the pilot to adopt a fairly complex compensation scheme in order to successfully stabilize this mode. The pilot has good control over the angular acceleration of the helicopter, but the direct control of translation is poor. The lack of significant direct command of the helicopter velocity (equation 21.104) makes more difficult the tasks of precisely controlling the longitudinal or lateral position in hover and providing the feedback required to stabilize the oscillatory modes. The control sensitivity (the pitch and roll rate commanded by cyclic) is high in hover. The combination of high sensitivity and only indirect control of translational velocity is conducive to pilot-induced oscillations and increases the difficulty of the control task. The pilot must work with the direct control of the helicopter attitude and thus is required to anticipate the velocity response that results, providing a significant lead in the pitch feedback to achieve stability.

The handling qualities are improved with offset flap hinges or with a hingeless rotor because of the increased pitch damping and control capability, which reduce the instability of the oscillatory mode. Increased control capability makes the task of controlling the helicopter easier. However, the gust sensitivity is also increased with a hingeless rotor. For a helicopter with an offset-hinge articulated rotor or a hingeless rotor, the root loci of feedback are similar to those in Figures 21.6 to 21.8, but the quantitative differences in the poles have important influences on the gain, lead, and lag requirements of the compensation. Thus rate plus attitude feedback is again required for satisfactory dynamics, but the increased damping and control capability mean that less lead and a lower gain are required, which eases the piloting task.

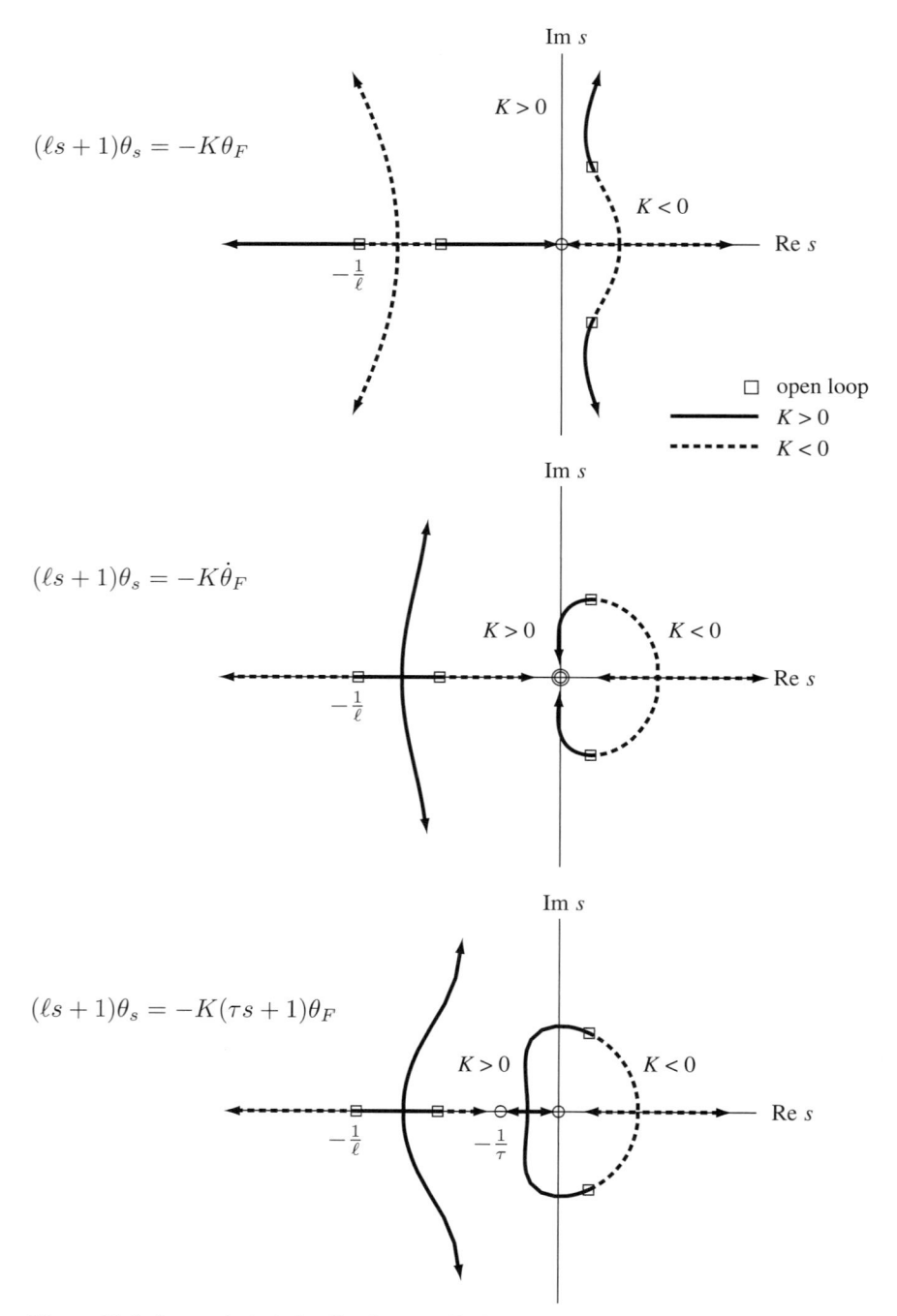

Figure 21.8. Lagged pitch feedback to cyclic for an articulated rotor helicopter in hover.

21.4.7 Lateral Dynamics

Next consider the lateral dynamics of the hovering helicopter, still assuming a separation of the longitudinal and lateral motions. The degrees of freedom involved are the lateral velocity $\dot{y}_F$ (positive right) and roll angle ϕ_F (positive right). The lateral cyclic control θ_c and lateral gust velocity v_G are also included. Equilibrium of lateral

21.4 Hover Flight Dynamics

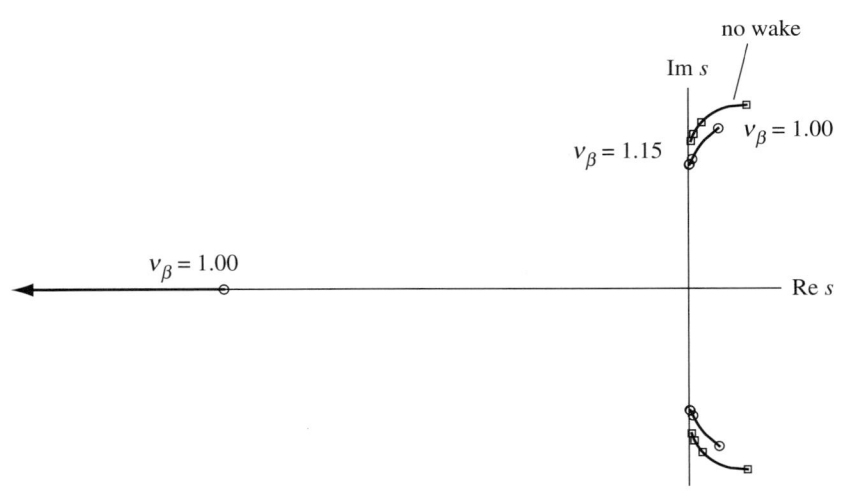

Figure 21.9. Influence of flap frequency on poles of lateral dynamics of hovering helicopter: $v_\beta = 1$ to 1.15/rev.

forces and roll moments give the differential equations for the lateral motions of the helicopter:

$$\begin{bmatrix} s - Y_v & -Y_p s - g \\ -L_v & s^2 - L_p s \end{bmatrix} \begin{pmatrix} \dot{y}_F \\ \phi_F \end{pmatrix} = \begin{pmatrix} Y_\theta \\ L_\theta \end{pmatrix} \theta_c + \begin{pmatrix} Y_v \\ L_v \end{pmatrix} v_G \qquad (21.107)$$

The rotor in hover is axisymmetric. The only physical difference between the longitudinal and lateral dynamics of the hovering helicopter is that the roll moment of inertia I_x is much smaller than the pitch moment of inertia I_y. Hence the lateral stability derivatives are equal to the corresponding longitudinal derivatives, except that in the moment derivatives k_y^2 must be replaced by k_x^2; see section 21.4.2. There are sign changes due to the reversal of the orientation of the angular motion relative to the linear motion when transforming from the longitudinal dynamics to the lateral. The effect of the smaller roll inertia is to increase the magnitude of the roll stability derivatives relative to the pitch derivatives.

The lateral dynamics are described by a real convergence mode due to the roll damping L_p, and an unstable oscillatory mode due to the rotor dihedral effect or speed stability L_v. Table 21.1 presents a numerical example of the helicopter dynamics in hover, including the lateral dynamics. The ratio of moment of inertia is $I_x/I_y = 0.2$. Figure 21.9 shows the influence of v_β on the poles.

Figure 21.10 shows the time to half-amplitude of the roll mode, as a function of flap frequency ($v_\beta = 1.00, 1.05, 1.10,$ and 1.15) and aircraft size (rotor radius $R = 10$, 20, 30, and 40 ft). Typically for an articulated rotor the roll mode has a time to half-amplitude of $t_{\frac{1}{2}} = 0.4$ to 0.5 sec; hinge offset gives $t_{\frac{1}{2}} = 0.1$ to 0.2 sec. The roll response is much quicker than the pitch response, due to the smaller roll inertia. The high roll damping of the hingeless rotor greatly increases the magnitude of the real root, so typically $t_{\frac{1}{2}} < 0.1$ sec. The dimensional pole of the roll mode is relatively independent of the aircraft size.

Figure 21.11 shows the variation of the lateral oscillatory mode with flap frequency and aircraft size: frequency ω (rad/sec), damping ratio ζ (positive for stable), period T (sec), and time to double-amplitude t_2 (sec, negative for stable). The period is typically $T = 10$ to 20 sec (a frequency of 0.05 to 0.10 Hz, or 0.3 to 0.6 rad/sec).

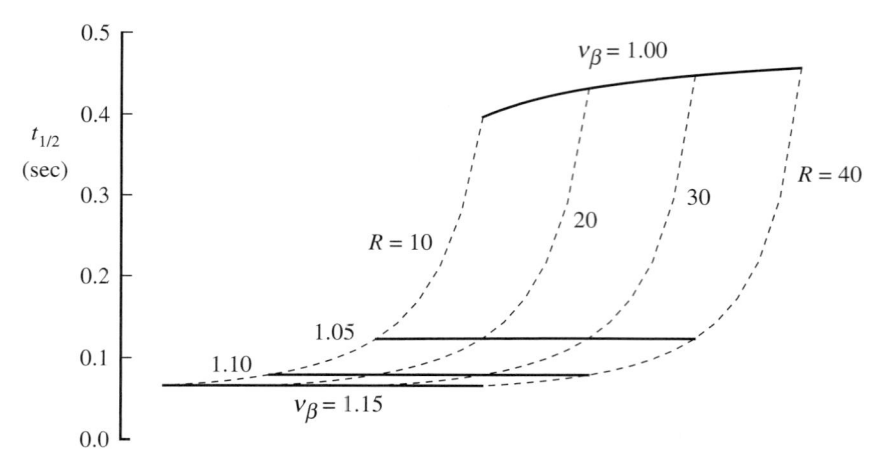

Figure 21.10. Influence of flap frequency and rotor size on the time to half-amplitude of the roll mode: $\nu_\beta = 1$ to 1.15/rev and radius $R = 10$ to 40 ft.

Because of the large roll damping with hinge offset or hingeless blades, the time to double-amplitude is much larger than for the longitudinal mode.

To stabilize the lateral motion, rate plus attitude feedback of the helicopter roll to lateral cyclic is required. Although the rotor dynamics are axisymmetric in hover, the aircraft is not. Also, the pilot's perception of motion and ability to apply appropriate control are different for longitudinal and lateral dynamics. Controlling roll motion can be more or less difficult than controlling pitch motion, depending on the pilot's task.

A short period approximation for the lateral dynamics consists of just the roll motion:

$$(s - L_p)\dot{\phi}_F = L_\theta \theta_c \tag{21.108}$$

The single pole of this equation is a good approximation for the actual roll mode, even for articulated rotors, because of the smaller inertia than for the longitudinal motion. The steady-state response is $\dot{\phi}_F/\theta_c = -L_\theta/L_p$, so lateral cyclic commands a roll rate with a small first-order lag.

21.4.8 Coupled Longitudinal and Lateral Dynamics

Consider now the coupled longitudinal and lateral motions of the single main rotor helicopter in hover. The longitudinal and lateral dynamics are in fact strongly coupled by the rotor forces. Including this coupling, the equations of motion are

$$\begin{bmatrix} s - X_u & -X_q s + g & -X_v & -X_p s \\ -M_u & s^2 - M_q s & -M_v & -M_p s \\ -Y_u & -Y_q s & s - Y_v & -Y_p s - g \\ -L_u & -L_q s & -L_v & s^2 - L_p s \end{bmatrix} \begin{pmatrix} \dot{x}_F \\ \theta_F \\ \dot{y}_F \\ \phi_F \end{pmatrix}$$

$$= \begin{bmatrix} X_\theta & X_u & X_{\theta_c} & X_v \\ M_\theta & M_u & M_{\theta_c} & M_v \\ Y_{\theta_s} & Y_u & Y_\theta & Y_v \\ L_{\theta_s} & L_u & L_\theta & L_v \end{bmatrix} \begin{pmatrix} \theta_s \\ u_G \\ \theta_c \\ v_G \end{pmatrix} \tag{21.109}$$

21.4 Hover Flight Dynamics

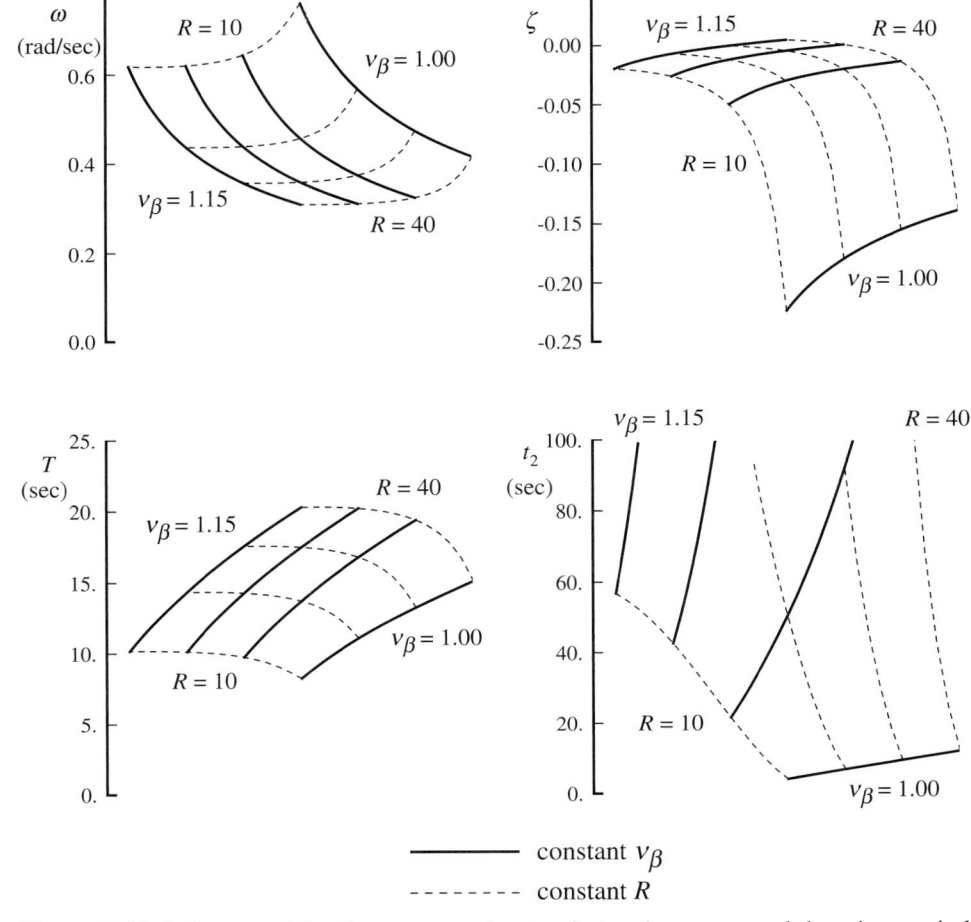

Figure 21.11. Influence of flap frequency and rotor size on frequency and damping, period, and time to double-amplitude of the lateral oscillatory mode: $\nu_\beta = 1$ to 1.15/rev and radius $R = 10$ to 40 ft.

For an articulated rotor ($\nu_\beta = 1$), the coupling derivatives are

$$X_{\theta_c} = 0 \tag{21.110}$$

$$X_v = -G\left(R_{\dot\beta} 8 M_\mu + R_r\right) \tag{21.111}$$

$$X_p = -G\left(\frac{C_T}{\sigma a}(1 - K_R) + CR_{\dot\beta}\left(\frac{16}{\gamma_e}\right)\right) + hX_v$$

$$= -g\frac{1 - K_R}{2} - G\frac{16}{\gamma} R_{\dot\beta} + hX_v \tag{21.112}$$

with the substitution $G = g/(2C_T/\sigma a)$. The side force derivatives equal the corresponding drag force derivatives, and the moment derivatives are h/k^2 times the force derivatives; see section 21.4.2. The coupling consists principally of the moment derivatives due to angular velocity, M_p and L_q. A roll rate of the rotor produces a lateral gradient of velocity over the rotor disk; hence a roll aerodynamic moment. The rotor responds with longitudinal tip-path-plane tilt, until the roll moment due

Table 21.2. *Example of hovering helicopter coupled dynamics,* $\nu_\beta = 1/\text{rev}$

	Uncoupled	Coupled	Coupled, $K_R = 0$
longitudinal			
pitch mode			
s	−0.018	−0.019	−0.020
$t_{\frac{1}{2}}$	1.2	1.1	1.0
$\dot{x}_F/\theta_F$	0.089	0.087	0.081
$\dot{y}_F/\theta_F$		0.020	0.043
ϕ_F/θ_F		−0.24	−0.56
oscillatory mode			
s	0.0046±i0.012	0.0049±i0.012	0.0053±i0.011
ω	0.40	0.39	0.37
ζ	−0.35	−0.38	−0.42
T	15.9	16.2	17.1
t_2	4.6	4.3	4.0
$\dot{x}_F/\theta_F$	0.114 (112°)	0.115 (114°)	0.119 (116°)
$\dot{y}_F/\theta_F$		0.042 (−132°)	0.068 (−137°)
ϕ_F/θ_F		0.36 (−65°)	0.56 (−72°)
lateral			
roll mode			
s	−0.049	−0.048	−0.044
$t_{\frac{1}{2}}$	0.4	0.4	0.5
$\dot{y}_F/\phi_F$	−0.034	−0.035	−0.038
$\dot{x}_F/\phi_F$		−0.005	−0.010
θ_F/ϕ_F		−0.095	−0.212
oscillatory mode			
s	0.0032±i0.017	0.0024±i0.018	0.0008±i0.019
ω	0.57	0.58	0.61
ζ	−0.18	−0.14	−0.04
T	11.1	10.9	10.4
t_2	6.7	8.8	28.4
$\dot{y}_F/\phi_F$	0.085 (−79°)	0.084 (−82°)	0.080 (−88°)
$\dot{x}_F/\phi_F$		0.020 (19°)	0.031 (6°)
θ_F/ϕ_F		0.26 (−80°)	0.41 (−87°)

Notes: root s (dimensionless), time to half-amplitude $t_{\frac{1}{2}}$ (sec), frequency ω (rad/sec), damping ratio ζ, period T (sec), time to double-amplitude t_2 (sec); eigenvector $\dot{x}_F/\theta_F$, $\dot{y}_F/\theta_F$, ϕ_F/θ_F; or $\dot{y}_F/\phi_F$, $\dot{x}_F/\phi_F$, θ_F/ϕ_F (dimensionless; magnitude and phase).

to flapping equals the roll moment due to roll rate (the angle-of-attack change due to roll rate is cancelled by that due to the flap velocity): $\beta_{1c} = -\dot{\phi}_F$. Including the effect of wake curvature on the induced velocity, the flapping is $\beta_{1c} = -(1 - K_R)\dot{\phi}_F$ (equation 21.51). The flap deflection tilts the thrust of the blade, producing an in-plane force $2C_H/\sigma a = R_\beta \beta_{1c} = (C_T/\sigma a)(1 - K_R)\dot{\phi}_F$ (equation 21.54; the $H_{\dot{\beta}}$ terms cancel). The result is $M_p = (gh/k_y^2)(1 - K_R)/2$. With $K_R = 1.5$ for hover, the wake curvature effect changes the sign of this coupling.

Tables 21.2 and 21.3 compare the coupled and uncoupled solutions for the hovering helicopter longitudinal-lateral dynamics, for flap frequencies of $\nu_\beta = 1/\text{rev}$ and $\nu_\beta = 1.1/\text{rev}$, respectively. The aircraft considered is the same as for Table 21.1.

21.4 Hover Flight Dynamics

Table 21.3. *Example of hovering helicopter coupled dynamics,* $v_\beta = 1.10$/rev

	Uncoupled	Coupled	Coupled, $K_R = 0$
longitudinal			
pitch mode			
s	−0.057	−0.072	−0.065
$t_{\frac{1}{2}}$	0.4	0.3	0.3
$\dot{x}_F/\theta_F$	0.030	0.024	0.027
$\dot{y}_F/\theta_F$		−0.015	−0.003
ϕ_F/θ_F		0.64	0.081
oscillatory mode			
s	$0.0014 \pm i0.013$	$0.0014 \pm i0.013$	$0.0023 \pm i0.012$
ω	0.42	0.41	0.41
ζ	−0.10	−0.11	−0.18
T	14.8	15.4	15.5
t_2	15.8	15.1	9.5
$\dot{x}_F/\theta_F$	0.115 (98°)	0.119 (98°)	0.120 (102°)
$\dot{y}_F/\theta_F$		0.072 (−9°)	0.095 (−157°)
ϕ_F/θ_F		0.61 (73°)	0.79 (−79°)
lateral			
roll mode			
s	−0.268	−0.252	−0.311
$t_{\frac{1}{2}}$	0.1	0.1	0.1
$\dot{y}_F/\phi_F$	−0.008	−0.009	−0.008
$\dot{x}_F/\phi_F$		0.001	−0.001
θ_F/ϕ_F		0.129	0.014
oscillatory mode			
s	$0.0002 \pm i0.014$	$-0.0004 \pm i0.014$	$-0.0011 \pm i0.013$
ω	0.44	0.44	0.41
ζ	−0.01	0.03	0.09
T	14.3	14.3	15.2
t_2	130	$t_{\frac{1}{2}} = 57$	$t_{\frac{1}{2}} = 19.2$
$\dot{y}_F/\phi_F$	0.112 (−88°)	0.113 (−90°)	0.118 (−93°)
$\dot{x}_F/\phi_F$		0.061 (179°)	0.086 (−7°)
θ_F/ϕ_F		0.54 (90°)	0.74 (−93°)

Notes: root s (dimensionless), time to half-amplitude $t_{\frac{1}{2}}$ (sec), frequency ω (rad/sec), damping ratio ζ, period T (sec), time to double-amplitude t_2 (sec); eigenvector $\dot{x}_F/\theta_F$, $\dot{y}_F/\theta_F$, ϕ_F/θ_F; or $\dot{y}_F/\phi_F$, $\dot{x}_F/\phi_F$, θ_F/ϕ_F (dimensionless; magnitude and phase).

The coupling tends to destabilize the longitudinal oscillation and stabilize the lateral oscillation for these cases. The pitch and roll real roots are given fairly well by the uncoupled equations, particularly for the hingeless rotor. The uncoupled equations correctly give the basic characteristics and most of the quantitative results of the coupled dynamics. The eigenvectors show that even when the roots are not influenced much, the coupling still introduces considerable roll motion in the longitudinal dynamics and pitch motion in the lateral dynamics. This example includes the wake effects, and coupled results are also shown without the wake curvature influence ($K_R = 0$). For the hingeless rotor, $K_R = 1.5$ changes the sign of ϕ_F/θ_F in the eigenvector.

21.5 Forward Flight

Next let us examine the dynamics of the helicopter in forward flight. Forward speed introduces new forces acting on the helicopter: centrifugal forces due to the rotation of the trim velocity vector by the angular velocity of the body axes, aerodynamic forces on the fuselage and tail, and major rotor forces that are proportional to the advance ratio. As a result, the handling qualities in forward flight are very different from those in hover. In forward flight, the vertical and lateral-longitudinal dynamics are coupled by both the rotor forces and the body accelerations. It is again assumed that the longitudinal dynamics (longitudinal velocity, pitch attitude, and vertical velocity) and the lateral dynamics (lateral velocity, roll attitude, and yaw rate) can be analyzed separately. Such an analysis provides a reasonable description of the helicopter flight dynamics, although in fact all six degrees of freedom are coupled.

The simplified geometry of Figure 21.1 is still used, as for the hover analysis. So the coordinate axes are aligned with the rotor shaft and hub plane, and the helicopter center-of-gravity is directly below the rotor hub. The forces and moments about the helicopter center-of-gravity are then obtained from the hub reactions simply by a translation along the z-axis, with no rotations. For numerical work an arbitrary reference axis system in the body can be used, and in general the center-of-gravity is offset from the shaft axis. The trim Euler angles are assumed to be small, and only level flight is considered so there is no climb velocity. The inertial cross-coupling of roll and yaw is neglected ($I_{xz} = 0$).

Forward flight influences the inertial forces by introducing the centrifugal acceleration required to turn the trim velocity vector when the body axes are rotated by the angular velocity. Principally there is a vertical acceleration due to the pitch rate and a lateral acceleration due to the yaw rate. These forces couple the vertical and lateral-longitudinal dynamics. The forces and moments acting on the helicopter are produced by the main rotor, tail rotor, fuselage and tail aerodynamics, and gravity. The axis system has been chosen specifically to simplify the role of the main rotor forces and moments. The only tail rotor contribution considered is the yaw moment produced by its thrust. The aerodynamic forces on the fuselage and tail are not considered in detail here.

The linear acceleration relative to inertial space is important to the pilot and passengers. In terms of the body-axis degrees of freedom, the inertial acceleration is $a = \dot{v} - \tilde{v}\omega$, or

$$a_x = -\ddot{x}_F + \mu_z \dot{\theta}_F \tag{21.113}$$

$$a_y = \ddot{y}_F + \mu_z \dot{\phi}_F + (\mu \cos\theta_{F0} - \mu_z \sin\theta_{F0})\dot{\psi}_F \tag{21.114}$$

$$a_z = -\ddot{z}_F + \mu\dot{\theta}_F \tag{21.115}$$

with a_x positive aft and a_z positive up. In forward flight, a pitch rate produces a vertical acceleration a_z, which is the centripetal acceleration required to turn the velocity vector. Similarly, yaw rate produces a lateral inertial acceleration. As in hover, the short time response to longitudinal cyclic is primarily a pitch rate. Thus in forward flight longitudinal cyclic controls the helicopter vertical acceleration, giving the pilot control over the flight path trajectory. The collective stick controls vertical flight path at low speed (and even beyond minimum power speed), but is used mainly for lift trim in high-speed flight. The helicopter roll axis and velocity vector coincide, so there is little inertial acceleration due to rolling of the body axes, and lateral cyclic still commands roll rate.

21.5.1 Forward Flight Stability Derivatives

As for hover, the rotor forces and moments acting on the helicopter in forward flight are obtained from the low-frequency response, and therefore the rotor dynamics do not add degrees of freedom to the system. The quasistatic rotor hub reactions, including the effects of the flap motion, are given in section 19.3. Forward flight introduces order μ^2 changes to the rotor stability derivatives present in hover, so there is no radical change in those derivatives for typical helicopter speeds (up to about $\mu = 0.5$). In forward flight there are also derivatives of order μ that couple the vertical and lateral-longitudinal dynamics of the rotor. The most important of these new stability derivatives is the pitching moment due to angle-of-attack perturbations of the helicopter.

As described in section 19.3, lateral aerodynamic moment due to vertical velocity of the helicopter produces a longitudinal tip-path-plane tilt of

$$\Delta \beta_{1c} = \frac{M_\lambda^{1s}}{M_{\dot\beta}} (\dot z_h - w_G) = 2\mu(\dot z_h - w_G) \tag{21.116}$$

(equation 19.63). In terms of the aircraft velocity, the flapping is $\Delta\beta_{1c} = -2\mu(\dot z_F + w_G)$. Downward velocity of the helicopter ($\dot z_F > 0$) increases the angle-of-attack of the blades. In forward flight the resulting lift increase is largest on the advancing side and smallest on the retreating side, so there is a lateral moment of the rotor disk (toward the retreating side) that is proportional to μ. The rotor responds 90° later with rearward tip-path-plane tilt until the lateral moment due to the flapping velocity is sufficient to establish equilibrium again. This flap deflection tilts the thrust vector and with offset hinges or hingeless blades produces a hub moment. Ignoring the amplitude and phase shift due to $\nu_\beta > 1$ (ignoring N_e), the hub reactions produce the rotor contributions to the angle-of-attack stability derivatives:

$$\Delta X_w \cong G \frac{2C_T}{\sigma a} \frac{\partial \beta_{1c}}{\partial \dot z_F} = -G \frac{2C_T}{\sigma a} 2\mu \tag{21.117}$$

$$\Delta M_w \cong -\frac{G}{k_y^2}\left(\frac{\nu_\beta^2 - 1}{\gamma} + h\frac{2C_T}{\sigma a}\right)\frac{\partial \beta_{1c}}{\partial \dot z_F} = \frac{G}{k_y^2}\left(\frac{\nu_\beta^2 - 1}{\gamma} + h\frac{2C_T}{\sigma a}\right)2\mu \tag{21.118}$$

where $G = a\rho A_b R/2M = g/(2C_T/\sigma a)$. So downward vertical velocity results in a nose-up pitch moment on the helicopter. For hingeless rotors there is also the direct hub moment term, which greatly increases the pitch moment. In forward flight, the helicopter angle-of-attack perturbation is $\dot z_F/\mu$. An angle-of-attack increase then produces a pitch-up moment on the helicopter, which would tend to increase the angle-of-attack further. Hence the rotor is the source of an angle-of-attack instability of the helicopter in forward flight.

The derivation of the aerodynamic forces acting on the helicopter fuselage and tail can be obtained from an airplane stability and control text. All of the airframe contributions to the stability derivatives are proportional to the forward speed and therefore are zero in hover. The fuselage drag in forward flight produces damping forces X_u and Z_u. The fuselage trim aerodynamic pitch moment produces a contribution (often unstable) to the speed stability M_u. The helicopter fuselage also produces unstable moments due to angle-of-attack changes: a pitch moment M_w and a yaw moment N_v. The horizontal and vertical tails give the remaining contributions to the stability derivatives, assuming that the helicopter does not have a wing. The horizontal tail produces a pitch moment due to angle-of-attack M_w, which contributes to the static angle-of-attack stability, countering the destabilizing contribution of the main rotor. The horizontal tail also gives a pitch moment due to the helicopter pitch rate M_q that adds to the rotor pitch damping, as well as corresponding contributions to the vertical force derivatives Z_w and Z_q, since the tail lift is involved. Finally, the horizontal tail contributes to the helicopter speed stability M_u and produces forces X_w and X_u. Since these last three derivatives are proportional to the tail lift coefficient, they depend on how the horizontal tail is used to trim the helicopter. The vertical tail contributions to the lateral derivatives are similar to the horizontal tail contributions to the longitudinal derivatives. The major effects of the vertical tail are the yaw moments due to lateral velocity N_v and the yaw rate N_r, which add to the directional stability and yaw damping of the helicopter. There are also the corresponding side force derivatives and derivatives due to the vertical tail lift coefficient. The vertical tail is often given a non-zero incidence angle so it develops a yaw moment in forward flight to counter the main rotor torque. The aerodynamic environment of the helicopter fuselage and tail is very complex, which makes estimation of the stability derivatives difficult. The fuselage is usually aerodynamically blunt and rough, and the tail surfaces must operate in the wake of the main rotor, tail rotor, and fuselage. Using experimental data for the aerodynamic characteristics of the helicopter airframe is therefore often essential.

The rotor angle-of-attack instability has an adverse effect on the forward flight handling qualities, and a horizontal tail is used to reduce this instability. The pitch moment produced by the horizontal tail lift due to the vertical velocity is $M_{y\text{tail}} = -\frac{1}{2}\rho V S_t \ell_t a_t \dot{z}_F$, where S_t is the horizontal tail area, ℓ_t the tail length, and a_t the tail lift-curve slope (including the effects of the wing or rotor wake at the tail). Thus the tail contribution to the angle-of-attack derivative is

$$\Delta M_w = \frac{\partial M_y/\partial \dot{z}_F}{I} = -\frac{\rho \mu S_t \ell_t a_t}{2Mk_y^2} = -\frac{a_t \rho S_t \ell_t}{2M} \frac{\mu}{k_y^2} \qquad (21.119)$$

Since both the tail and rotor moments are proportional to μ, their relative contributions to the angle-of-attack stability are independent of speed. From this expression, the tail size $(S_t \ell_t)$ needed to offset the instability can be estimated. The dependence on speed is actually more complex, because of the wake interference effects on the tail and the rotor terms of higher order in μ.

21.5.2 Longitudinal Dynamics

Consider the longitudinal dynamics of a helicopter in forward flight, consisting of three degrees of freedom: longitudinal velocity $\dot{x}_F$, pitch attitude θ_F, and vertical velocity $\dot{z}_F$. The controls are collective and longitudinal cyclic (θ_0 and θ_s), and excitation by vertical and longitudinal gust velocities is included. The equations of motion are

$$\begin{bmatrix} s - X_u & -X_q s + g & -X_w \\ -M_u & s^2 - M_q s & -M_w \\ -Z_u & -Z_q s - \mu s & s - Z_w \end{bmatrix} \begin{pmatrix} \dot{x}_F \\ \theta_F \\ \dot{z}_F \end{pmatrix} = \begin{bmatrix} X_\theta & X_{\theta_0} & X_u & X_w \\ M_\theta & M_{\theta_0} & M_u & M_w \\ Z_\theta & Z_{\theta_0} & Z_u & Z_w \end{bmatrix} \begin{pmatrix} \theta_s \\ \theta_0 \\ u_G \\ w_G \end{pmatrix}$$
$$(21.120)$$

21.5 Forward Flight

The stability derivatives now have contributions from the rotors and the aerodynamic forces on the fuselage and tail. Forward flight introduces the vertical acceleration due to pitch rate of the body axes, as well as rotor stability derivatives coupling the vertical and pitch motions. The most important new term is the angle-of-attack derivative M_w, which has unstable contributions ($M_w > 0$) from the rotor and fuselage and a stable contribution from the horizontal tail. The derivatives Z_q and Z_u are small ($Z_q = 0$ for an articulated rotor), and the force X_w is much less important than the pitch moment M_w. The characteristic equation is

$$(s - Z_w)\left(s^3 - (X_u + M_q)s^2 + (X_u M_q - X_q M_u)s + gM_u\right)$$
$$- M_w\left((\mu + Z_q)s^2 + (Z_u X_q - Z_q X_u - \mu X_u)s - gZ_u\right)$$
$$- X_w\left(Z_u s^2 - (Z_u M_q - Z_q M_u - \mu M_u)s\right) = 0 \quad (21.121)$$

The first term is the characteristic equation for hover (for $M_w = X_w = 0$), which decouples into the vertical and longitudinal motions.

The influence of forward flight is most easily examined for the case of an articulated rotor with no flap hinge offset, for which all the drag force and pitch moment derivatives are related by $M = -(h/k_y^2)X$. The characteristic equation then reduces to

$$(s - Z_w)\left(s^3 - (X_u + M_q)s^2 + gM_u\right) - M_w\left((\mu + Z_q - (k_y^2/h)Z_u)s^2 - gZ_u\right) = 0 \quad (21.122)$$

For the articulated rotor, $Z_q = 0$ and Z_u is small, so the last term is nearly just $-M_w \mu s^2$. For an offset-hinge articulated rotor or a hingeless rotor, the term $-M_w \mu s^2$ still dominates the influence of forward flight on the characteristic equation. So the characteristic equation is approximately

$$(s - Z_w)\left(s^3 - (X_u + M_q)s^2 + (X_u M_q - X_q M_u)s + gM_u\right) - M_w \mu s^2 = 0 \quad (21.123)$$

The influence of forward flight on the longitudinal dynamics of the helicopter thus consists primarily of the following forces: the pitch moment due to vertical velocity, the vertical acceleration due to pitch rate, and the longitudinal inertia of the helicopter. Their product is $-M_w \mu s^2$, which increases as the square of the speed (since M_w is proportional to μ).

Figure 21.12 illustrates the influence of forward flight on the longitudinal dynamics of a single main rotor helicopter. The parameters of Table 21.1 are used: Lock number $\gamma = 8$, solidity $\sigma = 0.08$, rotor height above center-of-gravity $h = 0.3$, pitch moment of inertia $k_y^2 = 0.1$, and blade loading $C_T/\sigma = 0.08$. The rotor has a 5% hinge offset, so $\nu_\beta = 1.039$. The horizontal tail size is $a_t S_t \ell_t / a A_b R = 0.08$. For the purposes of exposition, M_w was evaluated using equations 21.118 and 21.119, and the hover values were used for the other derivatives.

The influence of forward flight is shown in Figure 21.12 for a net angle-of-attack instability due to the rotor ($M_w > 0$) and with a large enough horizontal tail for angle-of-attack stability ($M_w < 0$). The hover poles are the real roots of the vertical and pitch modes, plus the long period, mildly unstable oscillatory mode. At high speed, two of the poles approach the origin (because of the s^2 factor in $-M_w \mu s^2$). With the main rotor alone (no tail), the angle-of-attack instability in forward flight reduces the damping of the smaller real root (usually the vertical mode) and increases the damping of the larger real root. The influence of forward speed on the oscillatory

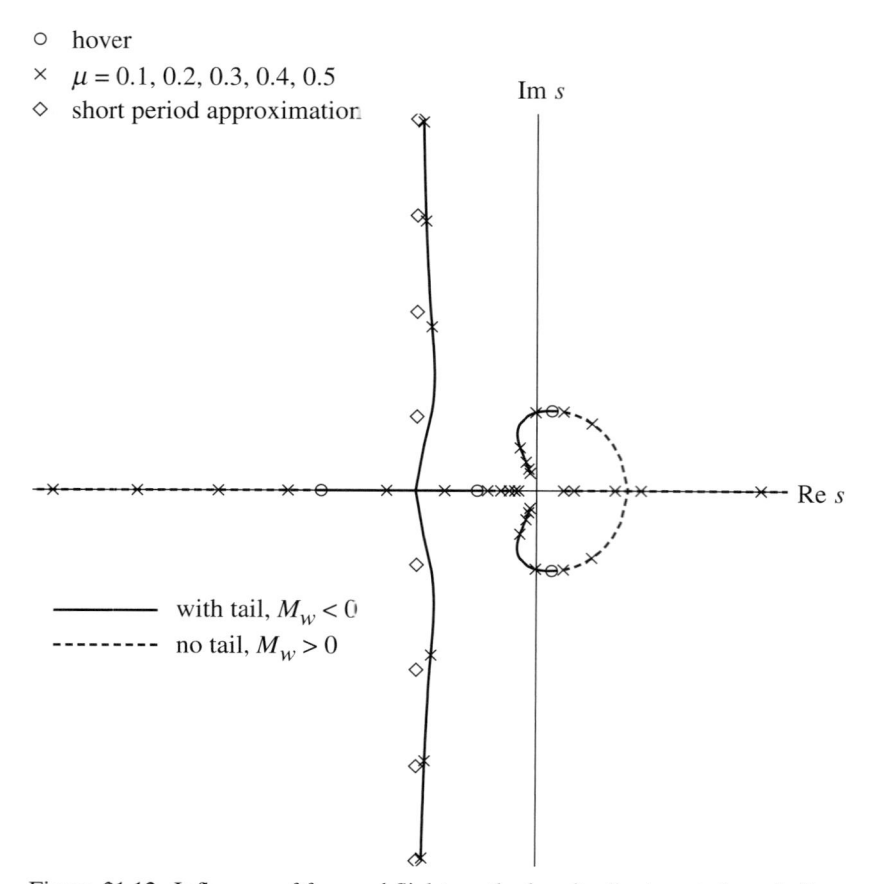

Figure 21.12. Influence of forward flight on the longitudinal roots for a helicopter without a horizontal tail ($M_w > 0$) and with a horizontal tail ($M_w < 0$).

mode is to increase the period and decrease the time to double-amplitude (reduced damping). Hence the forward flight dynamics of a helicopter without a horizontal tail are characterized by two real roots and an unstable oscillatory mode, with a degradation of the handling qualities due to the angle-of-attack instability. With a hingeless rotor M_w can be large enough at very high speed to replace the oscillatory mode with two positive real roots, one with an unacceptably small time to double-amplitude.

The helicopter can have net static stability with respect to angle-of-attack by using a large enough horizontal tail. In that case, forward speed transforms the pitch and vertical roots of hover into an oscillatory mode with a short period and high damping. The long period hover mode is stabilized and usually moved into the left half-plane, with the period increased somewhat. The forward flight longitudinal dynamics of a helicopter with a horizontal tail are thus characterized by a short period mode due to the damping of vertical and pitch motion and a long period mode stabilized by the static stability with respect to angle-of-attack. A horizontal tail large enough to produce a high level of static stability is not always practical, particularly with hingeless rotors. Moreover, the tail effectiveness is reduced at low speeds by interference with the rotor and fuselage wakes. The improvement of the handling qualities is so significant, however, that most single main rotor helicopter designs have a horizontal tail.

The behavior of the roots with increasing speed is often more complicated than predicted using the present simple model (Figure 21.12). With a horizontal tail, the oscillatory mode can be stable with a large time to half-amplitude and a period about twice the hover value. The short period mode period is around 4 to 6 sec, and time to half-amplitude is about 1 sec. If the tail is not large enough or effective enough, the oscillatory mode is unstable, with a time to double-amplitude that can be small and a period about equal the hover value. All the stability derivatives vary with speed, the influence of flap frequency is strong, and the tail effectiveness depends on the aerodynamic interference. So making generalizations about the rotorcraft longitudinal flight dynamics in forward flight is difficult.

Nonuniform inflow can be an important factor in the forward flight dynamics, producing significant changes in the stability derivatives. For example, the speed stability derivative is particularly sensitive to longitudinal variations of the inflow. In the present analysis the rotor speed is assumed to be constant. For helicopters in autorotation, in partial-power descent, or without a tight governor there can be significant rotor speed perturbations, which have a major influence on the flight dynamics.

21.5.3 Short Period Approximation

Consider a short period approximation for the helicopter longitudinal dynamics. The initial response to control and gusts is primarily vertical and pitch acceleration, with little longitudinal acceleration. The control over the longitudinal motion is indirect, so a significant $\dot{x}_F$ response takes a while to develop. Therefore, as an approximation for short times, the longitudinal velocity degree of freedom is neglected. The equations of motion reduce to

$$\begin{bmatrix} s - M_q & -M_w \\ -\mu & s - Z_w \end{bmatrix} \begin{pmatrix} \dot{\theta}_F \\ \dot{z}_F \end{pmatrix} = \begin{bmatrix} M_\theta & M_{\theta_0} & M_u & M_w \\ Z_\theta & Z_{\theta_0} & Z_u & Z_w \end{bmatrix} \begin{pmatrix} \theta_s \\ \theta_0 \\ u_G \\ w_G \end{pmatrix} \quad (21.124)$$

neglecting also the small vertical force due to $\dot{\theta}_F$, produced by the Z_q derivative. These equations retain the principal coupling of forward flight: the angle-of-attack derivative M_w and the vertical acceleration due to pitch rate. Since the gravitational spring term on the pitch motion appears in the longitudinal force equation, the short period approximation gives a second-order system for the two degrees of freedom $\dot{\theta}_F$ and $\dot{z}_F$. The equations of motion invert to

$$\begin{pmatrix} \dot{\theta}_F \\ \dot{z}_F \end{pmatrix} = \frac{1}{\Delta} \begin{bmatrix} s - Z_w & M_w \\ \mu & s - M_q \end{bmatrix} \begin{bmatrix} M_\theta & M_{\theta_0} & M_u & M_w \\ Z_\theta & Z_{\theta_0} & Z_u & Z_w \end{bmatrix} \begin{pmatrix} \theta_s \\ \theta_0 \\ u_G \\ w_G \end{pmatrix} \quad (21.125)$$

with the characteristic equation

$$\Delta = (s - M_q)(s - Z_w) - M_w \mu = 0 \quad (21.126)$$

In hover the pitch and vertical motions decouple, and the two solutions of the characteristic equation are $s = Z_w$ and $s = M_q$. The first solution is exactly the hover vertical pole. The second solution is the pole of the short period approximation for the hover longitudinal dynamics, which is an approximation for the pitch root in hover. Figure 21.12 shows the short period approximation for the poles of the longitudinal dynamics, for the $M_w < 0$ case. The approximation is good for $\mu > 0.2$ or so. The short period natural frequency and damping are thus

$$\omega_n^2 \cong M_q Z_w - M_w \mu \tag{21.127}$$

$$2\zeta \omega_n \cong -(M_q + Z_w) \tag{21.128}$$

giving the frequency $\omega = \omega_n \sqrt{1-\zeta^2}$ and time constant $\tau = 1/\zeta \omega_n$. The characteristic equation of the short period approximation can be written $\Delta = (s - s_z)(s - s_\theta)$, where s_z and s_θ are the two roots, a complex conjugate pair in forward flight if $M_w < 0$. In the limit of very small time (s approaching infinity), the short period approximation gives the same vertical and pitch acceleration response as the complete model.

The principal concern in the short period longitudinal dynamics is the normal acceleration response of the helicopter. In terms of the body-axis degrees of freedom, the vertical acceleration in inertial space is $a_z = -\ddot{z}_F + \mu \dot{\theta}_F$. The pitch rate is the main source of normal acceleration in forward flight. The response of $a_z = -s\dot{z}_F + \mu \dot{\theta}_F$ to longitudinal cyclic, as given by the short period approximation, is

$$\frac{a_z}{\theta_s} = \frac{1}{\Delta}\left(-sZ_\theta(s - M_q) + \mu(Z_\theta M_w - M_\theta Z_w)\right)$$

$$\cong -Z_\theta + \frac{-\mu M_\theta Z_w}{(s - s_z)(s - s_\theta)} \tag{21.129}$$

The initial response is $a_z/\theta_s = -Z_\theta$. This is a small vertical acceleration produced immediately after the control application by the thrust increment due to cyclic. The response of the pitch rate is zero initially, but as it builds up, the pitch rate contributes to the normal acceleration. The steady-state response (of the short period approximation) is

$$\frac{a_z}{\theta_s} = -Z_\theta - \frac{\mu M_\theta Z_w}{s_z s_\theta} = -Z_\theta - \frac{\mu M_\theta Z_w}{M_q Z_w - M_w \mu} \tag{21.130}$$

The second term is the acceleration due to the steady-state pitch rate response to cyclic.

Figure 21.13 illustrates the helicopter normal acceleration in response to a step increase of longitudinal cyclic in forward flight. By using the low-frequency rotor response, the lag in the development of the rotor forces has been neglected. Thus immediately after the application of the cyclic control there is a thrust increment that produces a small vertical acceleration. The pitch angular velocity of the helicopter is initially zero, but it builds up to the steady-state values given by the short period approximation, with a second-order transient defined by the two short period poles. This body-axis pitch rate produces the major portion of the normal acceleration in forward flight. For times beyond the validity of the short period approximation, the long period mode enters the response. A low-frequency, slowly decaying (or growing) oscillation of the normal acceleration develops (Figure 21.13). The pilot must then take the corrective action required to bring the helicopter back to the trim state.

The longitudinal cyclic command of the normal acceleration of the helicopter, and hence of its flight path trajectory, is an important aspect of the handling qualities.

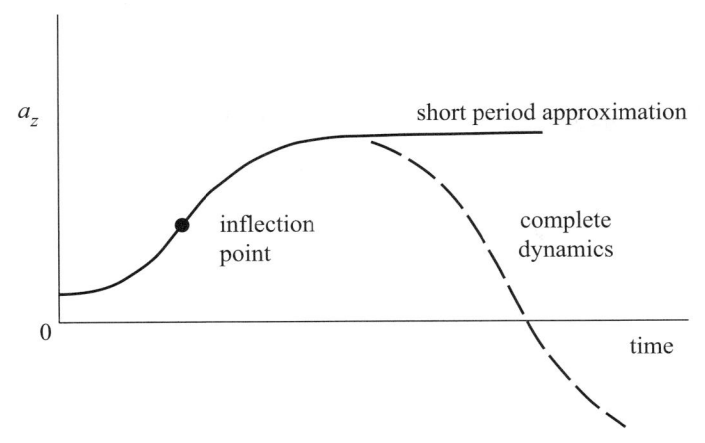

Figure 21.13. Normal acceleration response to a step increase of longitudinal cyclic.

The helicopter pitch control sensitivity is high, and a reasonable normal acceleration is achieved eventually. However, there is a delay after the application of the cyclic control (which initially produces only the small vertical acceleration Z_θ) until the pitch rate develops enough to give that normal acceleration. Thus the helicopter is characterized by a lag before the maximum normal acceleration in response to longitudinal cyclic is achieved, which can make control difficult if the lag is too long. A simple requirement of the helicopter handling qualities in forward flight is that the maximum acceleration be achieved within a certain time after the application of the cyclic control. This requirement has been quantified by using the inflection point on the trace of a_z, as a function of time (Figure 21.13). The smaller the inflection point time t_I, the sooner the maximum acceleration is achieved. Moreover, the existence of an inflection point means that the normal acceleration response is not divergent. Beyond t_I the normal acceleration curve is concave downward. The specification was that the time history of the normal acceleration due to a step input of longitudinal cyclic should be concave downward within a time t_I. The concave downward requirement was intended to ensure acceptable maneuverability characteristics of the helicopter. The problem this specification addressed is the delay in development of the normal acceleration after control application, which must not be too long for satisfactory handling qualities. Long superseded as a handling quality specification, the concave downward requirement still serves to characterize the helicopter short period response.

The stability derivatives involved in the short period response are Z_w, M_q, and M_w. The vertical damping is essentially fixed by the basic rotor design parameters. So the initial response $a_z/\theta_s \cong -\mu M_\theta Z_w/s^2$ is fixed. The inflection time can be reduced by decreasing the steady-state response $a_z/\theta_s = -\mu M_\theta Z_w/(s_z s_\theta)$; hence by increasing $s_z s_\theta = M_q Z_w - M_w \mu$. Increasing the pitch damping improves the normal acceleration response, as does increasing the static stability with respect to angle-of-attack. The inverse of the normal acceleration response is the longitudinal cyclic stick displacement per g of normal acceleration. The steady-state response of the short period approximation gives

$$\frac{\theta_s}{a_z/g} = \frac{g s_z s_\theta}{-\mu M_\theta Z_w} = \frac{g(M_q Z_w - M_w \mu)}{-\mu M_\theta Z_w} \qquad (21.131)$$

So a minimum criterion for $s_z s_\theta$ corresponds to a minimum stick gradient. Considering either the concave downward requirement or the stick gradient with load factor, the helicopter maneuver capability is an important factor in sizing the horizontal tail. The angle-of-attack stability due to the tail is an effective means of achieving the required normal acceleration response. Increasing the pitch damping by using rate feedback or a hingeless rotor also improves the short period response by reducing the control sensitivity. A gyro stabilizer bar (section 21.9) is a mechanical means to introduce lagged feedback of pitch rate in order to increase the pitch and roll damping.

21.5.4 Lateral-Directional Dynamics

Consider the lateral-directional dynamics of a helicopter in forward flight, consisting of three degrees of freedom: lateral velocity $\dot{y}_F$, roll attitude ϕ_F, and yaw rate $\dot{\psi}_F$. The controls are lateral cyclic and pedal (θ_c and θ_p), and excitation by lateral gust velocity is included. The equations of motion are

$$\begin{bmatrix} s - Y_v & -Y_p s - g & \mu \\ -L_v & s^2 - L_p s & 0 \\ -N_v & 0 & s - N_r \end{bmatrix} \begin{pmatrix} \dot{y}_F \\ \phi_F \\ \dot{\psi}_F \end{pmatrix} = \begin{bmatrix} Y_\theta & 0 & Y_v \\ L_\theta & 0 & L_v \\ 0 & N_{\theta_p} & N_v \end{bmatrix} \begin{pmatrix} \theta_c \\ \theta_p \\ v_G \end{pmatrix} \quad (21.132)$$

for the single main rotor and tail rotor configuration. In forward flight there is a lateral acceleration due to the yawing velocity of the body axes. The main rotor side force and roll moment due to yaw rate (Y_r and L_r) are small and are neglected. The main rotor torque perturbations are neglected compared to the tail rotor thrust contributions to the yaw moments, so N_θ and N_p are omitted. In this approximation, the only stability derivatives retained are those present in hover as well as in forward flight. The coupling of the yaw and lateral dynamics is due to the lateral acceleration produced by the yaw rate and the directional stability N_v (yaw moment produced by lateral velocity). The directional stability is due to the change in tail rotor thrust during lateral velocity of the helicopter. N_v does not influence the hover roots, but rather is responsible for a yaw response during lateral motions of the hovering helicopter. In forward flight the rotor is not axisymmetric, so the side force and roll moment derivatives do not equal the corresponding longitudinal stability derivatives.

The characteristic equation of the helicopter lateral dynamics in forward flight is

$$(s - N_r)\left(s^3 - (Y_v + L_p)s^2 + (Y_v L_p - Y_p L_v)s - gL_v\right) + \mu N_v(s - L_p)s = 0$$
$$(21.133)$$

The first term is the product of the characteristic equations for uncoupled yaw and lateral dynamics (sideward velocity and roll). The second term is due to the coupling in forward flight by the lateral acceleration and directional stability. Figure 21.14 shows the variation of the lateral-directional roots with advance ratio. The parameters of Table 21.1 are used to construct this example. The hover roots are the uncoupled yaw root, the roll root (much larger, requiring change of scale in the figure), and the unstable lateral oscillation. The directional stability is always positive ($N_v > 0$). Since the tail rotor gives a high level of directional stability, forward speed quickly stabilizes the oscillatory mode. For high speed, the two real roots approach $s = L_p$ (the roll mode) and $s = 0$ (the spiral mode). The other two roots approach the vertical asymptote, and the inertial coupling in forward flight transforms the hover

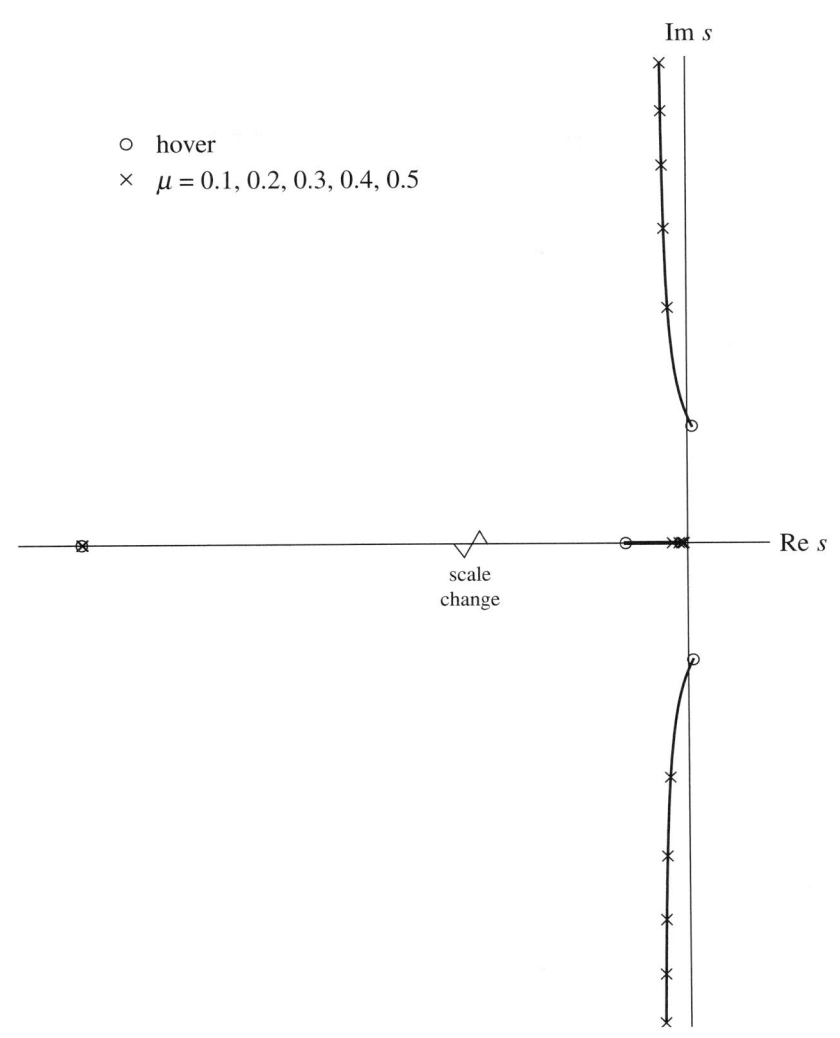

Figure 21.14. Influence of forward flight on the helicopter lateral roots (real scale doubled to show the roll root).

oscillatory mode into a stable, short period oscillation. This vertical asymptote is a good estimate of the damping of the short period oscillation. From the rules for constructing root loci, the asymptote is at

$$\text{Re}s = \frac{1}{2}\left((Y_v + L_p + N_r) - L_p\right) = \frac{1}{2}(Y_v + N_r) \cong \frac{1}{2}N_r \qquad (21.134)$$

which is just half the hover yaw root, if the yaw damping N_r does not vary too much with speed.

In forward flight, the roll root approaches $s = L_p$, the value given by the uncoupled damping. This roll root typically has a time to half-amplitude of about 0.5 sec with articulated rotors and is much smaller with hinge offset or hingeless blades. The derivative L_p does vary somewhat with speed, so although the forward flight root has nearly the value of the uncoupled roll damping, that value varies with advance ratio. The other real root (the yaw mode in hover) approaches the origin, indicating that the roll response becomes a rate change rather than an attitude change. Forward speed transforms the unstable, long period oscillation of hover into a well-damped short period mode (Dutch roll). The time to half-amplitude of this mode is about the same as the hover yaw root (N_r does vary with speed), and the period is around 2 to 4 sec. Sideslip to the right produces a yaw to the right through the directional stability. In forward flight this yaw motion implies a lateral centrifugal acceleration on the helicopter or equivalently an inertial force acting to the left. In contrast with hover, therefore, the lateral velocity of the helicopter produces a force opposing the motion, and the oscillatory mode is stable in forward flight. The directional stability N_v produced by the tail rotor is large, so the behavior of the lateral roots with speed is described well by the present simple model (Figure 21.14).

As a short period approximation to the lateral dynamics in forward flight, neglect the lateral velocity, since it builds up much more slowly than the roll or yaw motion. Then, since roll and yaw are not coupled in the present simplified model, the equations of motion for the short period approximation reduce to simply the uncoupled roll response: $(s - L_p)\dot{\phi}_F = L_\theta \theta_c$. In forward flight this is a good approximation for the roll dynamics, because the directional stability tends to decouple the lateral velocity from the roll motion. Hence lateral cyclic commands the roll rate, with a small first-order time lag due to the inertia. The steady-state response of the short period approximation gives $\dot{\phi}_F/\theta_c = -L_\theta/L_p$.

This elementary analysis of the helicopter lateral-directional dynamics is sufficient to show the basic influence of forward speed and the tail rotor on the motion. A better analysis must also include the details of the fuselage inertial and aerodynamic forces, the vertical tail aerodynamics, and the tail rotor position and direction of rotation. The mutual aerodynamic interference between the fuselage, the vertical tail, the tail rotor, and the main rotor can greatly influence the flight dynamics and hence is an important consideration in the aircraft design.

21.6 Static Stability

Static stability is defined as a tendency for a system to return toward the equilibrium position when disturbed, which implies a force or moment opposing slow perturbations from equilibrium. Static stability is related to divergence of a system. The divergence stability boundary is defined by the criterion that one pole of the system be at the origin, so divergence stability is assured if the constant term of the characteristic equation is positive. In contrast, dynamic stability means that all disturbances from equilibrium die out eventually, which requires that all poles of the system be in the left half-plane. Static stability can also be related to the steady-state response of the system to control. The presence of a force or moment opposing perturbations from equilibrium implies that, to change the equilibrium trim state, a force or moment must be applied to the aircraft by means of a control deflection. The amount of the control required (the control gradient) is a measure of the force or moment produced by the perturbation from trim and hence of the static stability. The sign of the control gradient indicates whether the system is statically stable or unstable. For simple systems all these definitions of static stability are equivalent and have elementary interpretations. For complex systems, the definition and interpretation of static stability are more difficult. For the helicopter, more than one stability derivative is involved even in most static effects, so relating stick gradients, static stability, and the dynamic stability characteristics is difficult.

The helicopter in hover has neutral static stability with respect to pitch or roll perturbations, since no moments to oppose the motion are generated directly by such perturbations. The hovering helicopter does have positive static stability with respect to longitudinal or lateral velocity perturbations, because of the speed stability derivatives M_u and L_v. Analogous behavior is found in airplane lateral dynamics, where static stability with respect to lateral velocity perturbations (sideslip) is produced by the wing dihedral effect, but there is neutral static stability with respect to roll angle perturbations.

The rotor in forward flight is statically unstable with respect to angle-of-attack. The fuselage and tail contribute significantly to the overall helicopter static stability with respect to angle-of-attack and speed in forward flight.

The important control gradients for the helicopter are the longitudinal cyclic displacements required to change speed and to change the normal load factor. For static stability with respect to speed, a forward displacement of the longitudinal cyclic stick is required to increase speed. This stick gradient is primarily related to the speed stability derivative M_u. Generally, as the forward speed of the helicopter increases the rotor tip-path plane flaps back, and a forward tilt of the control plane is needed to maintain the aircraft trim. At low forward speeds, some helicopters exhibit an unstable stick gradient with velocity. For acceptable maneuver characteristics, an aft displacement of the longitudinal cyclic stick is required to increase the helicopter vertical load factor in forward flight. This control gradient is related to the angle-of-attack and pitch damping derivatives M_w and M_q. A minimum gradient or a maximum control sensitivity is required for acceptable maneuver characteristics.

21.7 Twin Main Rotor Configurations

There are major differences between the flight dynamics of helicopters with two main rotors and those with the single main rotor and tail rotor configuration. The most common twin main rotor configuration is the tandem rotor helicopter, in which the main rotors have a typical longitudinal separation of $1.65R$ between the shafts; hence 35% overlap of the rotor disks. The tandem helicopter in hover has longitudinal symmetry (about the y–z plane), if it is possible to ignore such differences as the vertical rotor separation (the rear rotor is elevated above the front rotor to avoid the wake of the latter), the inertial and aerodynamic effects of the rear rotor pylon, and the offset of the helicopter center-of-gravity from midway between the rotors. Consequently, the tandem helicopter dynamics separate into symmetric (roll, side velocity, and vertical velocity) and anti-symmetric motions (pitch, longitudinal velocity, and yaw), at least to a better approximation than the decoupling of the lateral and longitudinal motions of a single main rotor helicopter. The vertical dynamics of the tandem helicopter are identical to those of the single main rotor helicopter. The lateral dynamics are equivalent to the truly uncoupled lateral dynamics of a single main rotor helicopter, but with quantitative differences because the fuselage of a tandem helicopter usually has a higher roll inertia. The longitudinal and yaw dynamics of the tandem helicopter involve new phenomena. The pitch control is by differential collective, and an additional source of pitch damping is the differential thrust of the main rotors due to the rotor axial velocity during pitch motions. Yaw control of the tandem rotor helicopter is obtained by differential lateral cyclic; the yaw damping is provided by the drag damping forces of the rotors.

With coaxial, contra-rotating main rotors the helicopter behaves as if it had a single main rotor with truly decoupled longitudinal and lateral dynamics. The yaw control and damping are obtained from the main rotor torque, instead of from a tail rotor.

The side-by-side helicopter configuration has true lateral symmetry, so there is a separation of the symmetric and anti-symmetric motions in both hover and forward flight. In hover the dynamics are basically the same as for the tandem rotor helicopter except for the interchange of the pitch and roll axes. Hence the longitudinal and vertical dynamics (the symmetric motions) are similar to the dynamics of a single main rotor helicopter. The lateral and yaw dynamics of the side-by-side configuration are similar to the longitudinal and yaw dynamics of the tandem rotor configuration. The interchange of the pitch and roll axes has a major impact on the flight dynamics, since different requirements are placed on lateral and longitudinal behavior.

21.7.1 Tandem Helicopter

The tandem helicopter geometry is simplified, as described in section 21.3. The front rotor is a distance ℓ_f forward of the aircraft center-of-gravity, and the rear rotor is a distance ℓ_r aft of the aircraft center-of-gravity, so the separation is $\ell = \ell_f + \ell_r$. The rotor hub heights are h_f and h_r. Write D_f, D_r for the single rotor derivatives (section 21.4.2) evaluated for the front and rear rotors, respectively. The trim thrust and inflow of the two rotors are likely different, because of center-of-gravity position and interference; hence the rotor aerodynamic coefficients and lift deficiency function are different. The front and rear control derivatives can include different gains. The two rotors are assumed to have the same radius, blade area, and tip speed. The parameter G is based on single rotor properties still and on total aircraft inertia. Hence the tandem rotor derivatives are the sum of front and rear rotor derivatives (not the average), and G is half that of single rotor: $G = a(g/\Omega R^2)/(4(C_T/\sigma)_{\text{trim}})$. From the transformation of loads in section 21.3, the vertical stability derivatives are

$$Z_{\theta_0} = Z_{f\theta_0} + Z_{r\theta_0} \tag{21.135}$$

$$Z_{\theta_s} = Z_{f\theta_0} - Z_{r\theta_0} \tag{21.136}$$

$$Z_w = Z_{fw} + Z_{rw} \tag{21.137}$$

$$Z_q = -\ell_f Z_{fw} + \ell_r Z_{rw} \tag{21.138}$$

If there are different collective gains for pitch control, then Z_{θ_s} is non-zero. If the rotor has pitch-cone coupling or the center-of-gravity is not midway between the rotors, then Z_q is non-zero. The longitudinal stability derivatives are

$$X_\theta = 0 \tag{21.139}$$

$$X_u = X_{fu} + X_{ru} \tag{21.140}$$

$$X_q = X_{fq} + X_{rq} \tag{21.141}$$

$$M_\theta = -\frac{1}{k_y^2}\left(\ell_f Z_{f\theta_0} + \ell_r Z_{r\theta_0}\right) \tag{21.142}$$

$$M_{\theta_0} = -\frac{1}{k_y^2}\left(\ell_f Z_{f\theta_0} - \ell_r Z_{r\theta_0}\right) \tag{21.143}$$

21.7 Twin Main Rotor Configurations

$$M_u = M_{fu} + M_{ru} \tag{21.144}$$

$$M_q = M_{fq} + M_{rq} + \frac{1}{k_y^2}\left(\ell_f^2 Z_{fw} + \ell_r^2 Z_{rw}\right) \tag{21.145}$$

$$M_w = -\frac{1}{k_y^2}\left(\ell_f Z_{fw} - \ell_r Z_{rw}\right) \tag{21.146}$$

$$M_{\Delta w} = -\frac{1}{k_y^2}\left(\ell_f Z_{fw} + \ell_r Z_{rw}\right) \tag{21.147}$$

The total rotor collective is $\theta_0 \pm \Delta\theta_0$. The total vertical gust at each rotor is $w_G \pm \Delta w_G$. Longitudinal cyclic is not used, so $X_\theta = 0$ and M_θ comes from the thrust due to differential collective. The derivatives M_w and M_{θ_0} couple the longitudinal and vertical motion. The lateral stability derivatives (Y_θ, Y_v, Y_p, L_θ, L_v, L_p) are just the sum of the front and rear rotor contributions. The directional stability derivatives are

$$N_{\theta_p} = \frac{1}{k_z^2}\left(\ell_f Y_{f\theta} + \ell_r Y_{r\theta}\right) \tag{21.148}$$

$$N_r = \frac{1}{k_z^2}\left(\ell_f^2 Y_{fv} + \ell_r^2 Y_{rv}\right) \tag{21.149}$$

$$N_v = \frac{1}{k_z^2}\left(\ell_f Y_{fv} - \ell_r Y_{rv}\right) \tag{21.150}$$

$$N_{\Delta v} = \frac{1}{k_z^2}\left(\ell_f Y_{fv} + \ell_r Y_{rv}\right) \tag{21.151}$$

where the pedal control is differential lateral cyclic and the lateral gust velocity is $v_G \pm \Delta v_G$ at the two rotors. The moments of inertia of the tandem helicopter are roughly twice those of a single main rotor helicopter, giving approximately $k_x^2 = 0.04$ and $k_y^2 = k_z^2 = 0.20$.

Consider the longitudinal dynamics of a tandem rotor helicopter in hover. Complete longitudinal symmetry is assumed so that the symmetric and anti-symmetric motions decouple, and it is assumed that the yaw and longitudinal motions can also be analyzed separately. The longitudinal degrees of freedom are then the pitch angle θ_F and longitudinal velocity $\dot{x}_F$; excitation is by longitudinal gust u_G. Pitch control is by differential main rotor collective $\Delta\theta_0$ ($\Delta\theta_0$ at the front rotor and $-\Delta\theta_0$ at the rear rotor), and a differential vertical gust velocity Δw_G is also included.

Compared with the single main rotor case, the derivatives X_u, M_u, and X_q are unchanged, although the speed stability M_u is numerically smaller because of the large pitch inertia. The pitch control by differential collective produces a pure pitch moment, so $X_\theta = 0$. The control derivative M_θ for the tandem helicopter typically has a value around three times that possible with the longitudinal cyclic control of a single articulated main rotor with no flap hinge offset; it is comparable to the control capability possible with a single hingeless rotor. The pitch damping M_q is dominated by the differential thrust term because of the large rotor thrust change produced by an axial velocity perturbation. The pitch damping derivative M_q of a tandem helicopter is typically four times that of a single main rotor helicopter with no flap-hinge offset or about twice that possible with a single offset-hinge articulated rotor, and moreover M_q of a tandem helicopter varies little with the rotor loading. Thus the longitudinal dynamics of a tandem helicopter are characterized by high control capability and pitch damping. The longitudinal handling qualities in hover should be noticeably better than those of a single main rotor helicopter of comparable size.

The equations of motion for pitch and longitudinal velocity are

$$\begin{bmatrix} s - X_u & -X_q s + g \\ -M_u & s^2 - M_q s \end{bmatrix} \begin{pmatrix} \dot{x}_F \\ \theta_F \end{pmatrix} = \begin{bmatrix} 0 & X_u & 0 \\ M_\theta & M_u & M_{\Delta w} \end{bmatrix} \begin{pmatrix} \theta_s \\ u_G \\ \Delta w_G \end{pmatrix} \quad (21.152)$$

With articulated rotors ($\nu_\beta = 1$; tandem rotors usually have low flap-hinge offset), $M_u = -(h/k_y^2)X_u$. The pitch damping is written $M_q = -(h/k_y^2)X_q + \Delta M_q$, where the rotor differential thrust due to pitch rate produces

$$\Delta M_q = \frac{1}{k_y^2}(\ell_f^2 Z_{fw} + \ell_r^2 Z_{rw}) \cong -\frac{\ell^2}{4k_y^2} 2GC'\frac{1}{4} = -\frac{\ell^2}{16k_y^2}\frac{C'}{2C_T/\sigma a} \quad (21.153)$$

Then the characteristic equation for the longitudinal dynamics of a tandem helicopter is

$$\Delta = s^3 - (X_u + M_q)s^2 + X_u \Delta M_q s + g M_u = 0 \quad (21.154)$$

and the response to control and gust is

$$\begin{pmatrix} \dot{x}_F \\ \theta_F \end{pmatrix} = \frac{1}{\Delta}\begin{pmatrix} X_q s - g \\ s - X_u \end{pmatrix}(M_\theta \theta_s + M_{\Delta w}\Delta w_G) + \frac{1}{\Delta}\begin{pmatrix} s^2 - \Delta M_q s + gh/k_y^2 \\ -(h/k_y^2)s \end{pmatrix} X_u u_G \quad (21.155)$$

Since $X_u + M_q \cong \Delta M_q$, the characteristic equation can be written as

$$\Delta = s^3 + gM_u - \Delta M_q(s - X_u)s = s^3 - \Delta M_q s^2 + M_u(-(k_y^2/h)\Delta M_q s + g) = 0 \quad (21.156)$$

With $\Delta M_q = 0$, the three poles are the cubic roots of $-gM_u$, as for the single main rotor case. For $M_q < 0$, the oscillatory roots do not approach the origin (Figure 21.2), but instead approach $s = X_u$ and $s = 0$. Since X_u is negative, the roots are pulled into the left half-plane, and the oscillatory mode is stabilized by the magnitude and mechanism of the tandem rotor pitch damping. With $M_u = 0$, the three poles are $s = 0, 0$, and ΔM_q. Because of the ΔM_q term, as M_u increases the oscillatory roots approach a vertical asymptote (unlike Figure 21.2) at

$$\text{Re}s = \frac{1}{2}\left(\Delta M_q - \frac{gh/k_y^2}{\Delta M_q}\right) \quad (21.157)$$

The mode can be mildly stable if this asymptote is in the left half-plane. By Routh's criterion the oscillatory mode is stable if

$$-(X_u + M_q)X_u\Delta M_q - gM_u = -X_u\left((X_u + M_q)\Delta M_q - gh/k_y^2\right)$$

$$\cong -X_u\left((\Delta M_q)^2 - gh/k_y^2\right) > 0 \quad (21.158)$$

or $|\Delta M_q| > (gh/k_y^2)^{1/2}$. Using equation 21.153, this is a limit on blade loading:

$$\frac{2C_T}{\sigma a} < \frac{\ell^2 C'}{16}\sqrt{\frac{g}{hk_y^2}} \quad (21.159)$$

The requirement is typically $C_T/\sigma < 0.03$, increasing with size as $R^{1/2}$.

In summary, the longitudinal dynamics of a tandem helicopter in hover are described by a stable real root due to the large pitch damping and by a mildly unstable low-frequency oscillatory mode (for the case of identical tilt of the main rotor shafts). The pole-zero configuration of the hover dynamics of the tandem helicopter is basically the same as for the single rotor helicopter, so the root loci for the various loop closures are similar. The higher pitch damping and control capability of the tandem helicopter ease the control tasks somewhat. The speed stability M_u depends on the relative tilt of the rotor shafts. By tilting the shafts outward, the speed stability can be reduced, because of the in-plane component of the rotor thrust damping.

For an articulated rotor ($\nu_\beta = 1$), the directional stability derivatives are

$$N_{\theta_p} \cong \frac{\ell}{2k_z^2} 2Y_\theta = -\frac{g\ell}{2k_z^2} \qquad (21.160)$$

$$N_r \cong \frac{\ell^2}{4k_z^2} 2Y_v = -\frac{g\ell^2}{4k_z^2} 8M_\mu \qquad (21.161)$$

The yaw damping is due to the rotor side forces produced by the lateral velocity through the rotor drag damping. Yaw control is by the rotor side forces due to differential lateral cyclic. The directional stability N_v is small. Compared to the yaw control provided by a tail rotor, the tandem helicopter tends to have a smaller control derivative N_{θ_p} because of the large yaw inertia. Also, with articulated rotors the yaw control capability of a tandem helicopter is proportional to the rotor loading. The yaw damping moment of a tandem helicopter is typically half that possible with a tail rotor, and it depends on the rotor loading. The damping derivative N_r is further reduced by the larger yaw inertia. Thus the yaw time to half-amplitude is typically much larger than with a tail rotor. There is some coupling between the yaw and longitudinal dynamics of the tandem helicopter. For example, differential collective produces a net torque increment on the helicopter, so the pedal control must be coordinated with the longitudinal cyclic stick to maintain the heading during pitch maneuvers. Generally, the yaw handling qualities of a tandem helicopter are not as good as those with a tail rotor, because of the lower damping and slower response.

In hover the longitudinal handling qualities of the tandem helicopter are somewhat better than those of the single main rotor configuration because of the higher pitch damping and control capability; the lateral handling qualities are somewhat worse because of the lower yaw damping and higher yaw and roll inertias. In forward flight the tandem helicopter has a large angle-of-attack instability due to the main rotors (and the fuselage), but a large horizontal tail is not practical. Thus there is a degradation of the longitudinal handling qualities in forward flight with the angle-of-attack instability producing an unstable oscillation or even a real divergence. The tandem helicopter does not have much directional stability even in hover, although some can be obtained with the center-of-gravity forward of the midpoint between the rotors. There is a large unstable contribution to N_v from the fuselage in forward flight, and the rear rotor pylon is not very effective as a vertical tail. Hence a directional instability is likely, and the lateral dynamics retain the unstable long period oscillation in forward flight.

In forward flight there are a number of unfavorable effects on the flight dynamics of tandem helicopters that arise from the aerodynamic interference between the two rotors, specifically the influence of the front rotor downwash on the rear rotor thrust. The helicopter instability with respect to angle-of-attack is increased by the aerodynamic interference. An angle-of-attack increase (hence greater downward vertical velocity of the helicopter) increases the thrust of the rotors and therefore also their induced velocities. The increased downwash of the front rotor wake at the rear rotor produces a decrease of the rear rotor thrust and hence a net nose-up pitch moment on the helicopter. Since the rear rotor is closer to stall than the front rotor, the rear rotor thrust increases are further limited, giving a larger angle-of-attack instability at high loading. The use of pitch-flap coupling on the front rotor improves the angle-of-attack stability by reducing the lift-curve slope of the front rotor relative to the rear rotor. The angle-of-attack instability is also reduced with a forward center-of-gravity position or a reduction in thrust coefficient.

The tandem helicopter can have an instability with respect to speed in forward flight. Each rotor has speed stability as usual, but the change of the rear rotor thrust with speed due to the wake of the front rotor produces an unstable moment. A speed increase reduces the induced downwash of the front rotor and hence reduces the downwash of the front rotor wake at the rear rotor. The resulting increase of the rear rotor thrust produces a nose-down pitch moment, which is a speed instability. Since this speed instability due to the rotor thrust perturbation is large, the tandem helicopter can easily have a net speed instability. The rear rotor is closer to stall because of the downwash of the front rotor, and therefore the speed instability is reduced at high loadings. The speed stability can be improved by using the longitudinal dihedral of the shafts or swashplate, so the tip-path planes are tilted toward each other (the aft rotor shaft tilted more forward). The thrust perturbations due to the axial components of the helicopter longitudinal velocity perturbation produce a nose-up moment, increasing the speed stability. The effectiveness of such dihedral is reduced by the higher collective required to trim the rear rotor in forward flight when its shaft is tilted forward. Also, the amount of tip-path plane tilt is limited by interference between the rotors and fuselage.

21.7.2 Side-by-Side Helicopter or Tiltrotor

The side-by-side helicopter configuration has right and left main rotors, a distance $\pm \ell$ from the aircraft center-line. The rotor hub height is h above the center-of-gravity. The tiltrotor has similar geometry in hover and helicopter mode flight, the wing-tip-mounted rotors tilting forward to act as propellers in cruise flight. These aircraft configurations have true lateral symmetry (symmetric about the x–z plane). From the transformation of loads in section 21.3, the vertical stability derivatives are

$$Z_{\theta_0} = Z_{r\theta_0} + Z_{l\theta_0} \qquad (21.162)$$

$$Z_w = Z_{rw} + Z_{lw} \qquad (21.163)$$

The longitudinal stability derivatives are just the sum of the left and right rotor contributions. The directional stability derivatives are

$$N_{\theta_p} = \frac{\ell}{k_z^2}(X_{r\theta} + X_{l\theta}) \qquad (21.164)$$

$$N_r = \frac{\ell^2}{k_z^2}(X_{ru} + X_{lu}) \qquad (21.165)$$

$$N_v = 0 \qquad (21.166)$$

where the pedal control is differential longitudinal cyclic. The lateral stability derivatives are

$$Y_\theta = 0 \tag{21.167}$$

$$Y_v = Y_{rv} + Y_{lv} \tag{21.168}$$

$$Y_p = Y_{rp} + Y_{lp} \tag{21.169}$$

$$L_\theta = -\frac{\ell}{k_x^2}(Z_{r\theta_0} + Z_{l\theta_0}) \tag{21.170}$$

$$L_v = L_{rv} + L_{lv} \tag{21.171}$$

$$L_p = L_{rp} + L_{lp} + \frac{\ell^2}{k_x^2}(Z_{rw} + Z_{lw}) \tag{21.172}$$

Lateral cyclic is not used, so $Y_\theta = 0$ and L_θ comes from the thrust due to differential collective.

The moments of inertia of the side-by-side configuration are much larger than those of a single main rotor helicopter. For a tiltrotor, the large mass of the nacelles on the wing tips gives typically $k_x^2 = 0.6$, $k_y^2 = 0.25$, and $k_z^2 = 0.8$.

21.8 Hingeless Rotor Helicopters

The capability of the hingeless rotor to transmit large hub moments to the helicopter has a major impact on its handling qualities. The articulated rotor in contrast can achieve only a limited hub moment with offset hinges, roughly comparable to the moment about the center-of-gravity due to the rotor thrust tilt. The hingeless rotor gives the helicopter a high control capability compared to the articulated rotor, and the damping in pitch and roll are increased by an even larger factor. The high damping also means an increased gust sensitivity, so a high-speed hingeless rotor helicopter often requires some sort of automatic control system for gust alleviation. The lateral-longitudinal coupling of the control response is also increased substantially, but can be handled satisfactorily by proper phasing of the swashplate. The increased lateral-longitudinal coupling of the transient motion and response to external disturbances remains, however. The angle-of-attack instability of the hingeless rotor in forward flight is much larger than that of an articulated rotor and requires a larger horizontal tail volume or an automatic control system to prevent a degradation of the handling qualities. The hingeless rotor is able to maintain its control capability and damping at low load factor, in contrast to the articulated rotor, which produces moments on the helicopter primarily by tilting the thrust vector.

It is often necessary to include in the analysis the blade lag and torsion degrees of freedom, as well as the flap motion, to accurately predict the flight dynamics of a hingeless rotor helicopter. The inertial and structural couplings involved in the hingeless rotor blade dynamics can have a major impact on the flight dynamics. The use of the low-frequency rotor response is generally an acceptable approximation even with a hingeless rotor.

21.9 Control Gyros and Stability Augmentation

Helicopter handling qualities can be improved by using an automatic control system. For certain operations, such as instrument flight rules (IFR) flight, a stability and control augmentation system is essential. The use of such systems naturally increases the cost and complexity of the helicopter. A gyroscope is a basic element of a helicopter automatic control system. Since the rotor can be considered a gyro, a control gyro can be used to sense the same inertial forces acting on the rotor. Such a control system can be entirely mechanical, or the gyro can just be the sensor with the control inputs provided by electro-hydraulic servos.

Consider a gyro gimballed to the rotor shaft and rotating at speed Ω_G, which for mechanical systems is usually the same as the rotor rotational speed Ω. The undisturbed gyro plane is parallel to the rotor hub plane. The gyro tilt relative to the shaft is described in the non-rotating frame by the pitch angle β_{Gc} and the roll angle β_{Gs} (positive for tilt forward and to the left, respectively). The gyro responds to the rotor shaft pitch and roll motions (θ_F and ϕ_F). The gyro is assumed to be axisymmetric, consisting of three or more identical, equally spaced radial elements. Let I_G be the gyro pitch and roll inertia. Damping C_R in the rotating frame and damping C_F in the non-rotating frame are included, but act only on the motion of the gyro relative to the shaft (so the damping is mechanical, not aerodynamic). The equations of motion for the gyro degrees of freedom follow from equilibrium of pitch and roll moments. By analogy with the dynamics of the flapping rotor, the gyro equations of motion are

$$\begin{bmatrix} I_G s^2 + (C_R + C_F)s & 2\Omega_G I_G s + C_R \Omega_G \\ -(2\Omega_G I_G s + C_R \Omega_G) & I_G s^2 + (C_R + C_F)s \end{bmatrix} \begin{pmatrix} \beta_{Gc} \\ \beta_{Gs} \end{pmatrix}$$

$$= \begin{bmatrix} I_G s & 2\Omega_G I_G \\ -2\Omega_G I_G & I_G s \end{bmatrix} \begin{pmatrix} \dot{\theta}_F \\ \dot{\phi}_F \end{pmatrix} + \begin{pmatrix} -M_{Gy} \\ M_{Gx} \end{pmatrix} \quad (21.173)$$

where M_{Gy} and M_{Gx} are pitch and roll moments acting on the gyro. Compare with equation 19.26 for the rotor tip-path-plane response. There is no structural spring, so the gyro natural frequency is $\nu_G = \Omega_G$. C_R is similar to rotating-frame flap damping $M_{\dot{\beta}}$, but acts only on gyro tilt relative to the shaft.

Typically the gyro is connected to the rotor cyclic pitch in such a way that the control plane tilt is proportional to the gyro tilt:

$$\begin{pmatrix} -\theta_s \\ \theta_c \end{pmatrix} = K_G \begin{bmatrix} \cos \Delta\psi_G & \sin \Delta\psi_G \\ -\sin \Delta\psi_G & \cos \Delta\psi_G \end{bmatrix} \begin{pmatrix} \beta_{Gc} \\ \beta_{Gs} \end{pmatrix} \quad (21.174)$$

where K_G is the gain and $\Delta\psi_G$ is the azimuthal phasing. There can be a direct mechanical link from the gyro to the blade pitch horns that produces this control input. In that case the blade feathering moments are also transmitted to the gyro, which at least increases its effective inertia and damping. Alternatively, actuators can provide the swashplate inputs proportional to the gyro tilt, with the appropriate gain and compensation network.

With C_R non-zero, the gyro response to helicopter angular velocity to order s is

$$\begin{pmatrix} \beta_{Gc} \\ \beta_{Gs} \end{pmatrix} \cong \frac{2I_G/C_R}{4I_G s + C_R} \begin{bmatrix} 2I_G s + C_R & \frac{1}{\Omega_G}\left(\frac{1}{2}C_R + C_F\right)s \\ -\frac{1}{\Omega_G}\left(\frac{1}{2}C_R + C_F\right)s & 2I_G s + C_R \end{bmatrix} \begin{pmatrix} \dot{\theta}_F \\ \dot{\phi}_F \end{pmatrix}$$

$$\cong \frac{2I_G/C_R}{4(I_G/C_R)s + 1} \begin{pmatrix} \dot{\theta}_F \\ \dot{\phi}_F \end{pmatrix} \quad (21.175)$$

21.9 Control Gyros and Stability Augmentation

where the helicopter angular acceleration terms are neglected in the last expression. Hence with control plane tilt proportional to the gyro tilt, this control system provides lagged rate feedback of the helicopter pitch and roll, which greatly improves the helicopter handling qualities. This response results from the damping in the rotating frame, C_R. A mechanical system in the rotating frame must provide the same feedback for both the pitch and roll axes. If there is no damping in the rotating frame ($C_R = 0$), the low-frequency response of the gyro to shaft pitch and roll is

$$\begin{pmatrix} \beta_{Gc} \\ \beta_{Gs} \end{pmatrix} \cong \frac{I_G}{2I_G C_F s + 4I_G^2 \Omega_G^2 + C_F^2} \begin{bmatrix} C_F s + 4\Omega_G^2 I_G & -(2I_G \Omega_G s - 2\Omega_G C_F) \\ 2I_G \Omega_G s - 2\Omega_G C_F & C_F s + 4\Omega_G^2 I_G \end{bmatrix} \begin{pmatrix} \theta_F \\ \phi_F \end{pmatrix} \tag{21.176}$$

If there is no damping in either the rotating or non-rotating frames, the response is exactly

$$\begin{pmatrix} \beta_{Gc} \\ \beta_{Gs} \end{pmatrix} = \begin{pmatrix} \theta_F \\ \phi_F \end{pmatrix} \tag{21.177}$$

In this case the gyro remains fixed in space and thus provides pure attitude sensing. The rate feedback obtained with damping in the rotating frame is most beneficial; see section 21.4.6.

The low-frequency response of the gyro to applied moments is

$$\begin{pmatrix} \beta_{Gc} \\ \beta_{Gs} \end{pmatrix} \cong \frac{-1/C_R \Omega_G}{4I_G s + C_R} \begin{bmatrix} 2I_G s + C_R & \frac{1}{\Omega_G}(C_R + C_F) s \\ -\frac{1}{\Omega_G}(C_R + C_F) s & 2I_G s + C_R \end{bmatrix} \begin{pmatrix} M_{Gx} \\ M_{Gy} \end{pmatrix}$$

$$\cong -\frac{-1/C_R \Omega_G}{4(I_G/C_R)s + 1} \begin{pmatrix} M_{Gx} \\ M_{Gy} \end{pmatrix} \tag{21.178}$$

So the gyro tilt senses the applied moments if there is damping in the rotating frame. If there is no damping in the rotating frame ($C_R = 0$), the response is

$$\begin{pmatrix} \dot{\beta}_{Gc} \\ \dot{\beta}_{Gs} \end{pmatrix} \cong \frac{-1}{2I_G C_F s + 4I_G^2 \Omega_G^2 + C_F^2} \begin{bmatrix} 2I_G \Omega_G & C_F \\ -C_F & 2I_G \Omega_G \end{bmatrix} \begin{pmatrix} M_{Gx} \\ M_{Gy} \end{pmatrix} \tag{21.179}$$

With no damping at all, the response is

$$\begin{pmatrix} \dot{\beta}_{Gc} \\ \dot{\beta}_{Gs} \end{pmatrix} = \frac{-1}{I_G(s^2 + 4\Omega_G^2)} \begin{bmatrix} 2\Omega_G & s \\ -s & 2\Omega_G \end{bmatrix} \begin{pmatrix} M_{Gx} \\ M_{Gy} \end{pmatrix}$$

$$\cong -\frac{1}{2I_G \Omega_G} \begin{pmatrix} M_{Gx} \\ M_{Gy} \end{pmatrix} \tag{21.180}$$

so the gyro senses the integral of the moments and responds with an angular velocity proportional to the applied moment after a 90° azimuthal lag.

Consider now a hub moment feedback control system utilizing a gyro. Such a system can be used with a hingeless rotor to alleviate the large hub moments produced by gusts. The hub moment of a hingeless rotor is

$$\begin{pmatrix} -\frac{2C_{My}}{\sigma a} \\ \frac{2C_{Mx}}{\sigma a} \end{pmatrix} = \frac{v_\beta^2 - 1}{\gamma} \begin{pmatrix} \beta_{1c} \\ \beta_{1s} \end{pmatrix} \tag{21.181}$$

where ν_β is the rotor blade flap frequency. Hub moment feedback then is equivalent to tip-path-plane tilt feedback, which can be used with articulated rotors as well. The hub moment or tip-path-plane tilt is sensed and transmitted to the gyro by some means, producing moments on the gyro:

$$\begin{pmatrix} M_{Gx} \\ M_{Gy} \end{pmatrix} = K_\beta \begin{bmatrix} \cos \Delta\psi_\beta & -\sin \Delta\psi_\beta \\ \sin \Delta\psi_\beta & \cos \Delta\psi_\beta \end{bmatrix} \begin{pmatrix} \beta_{1c} \\ \beta_{1s} \end{pmatrix} \qquad (21.182)$$

where K_β is the feedback gain and $\Delta\psi_\beta$ is the azimuthal phase. Then a control plane tilt is produced proportional to the gyro tilt as a response to this moment. Thus a control can be applied to the rotor to cancel the hub moment due to external disturbances. This hub moment feedback acts as well to reduce the hub moments that constitute the control capability and damping of the rotor. The gyro can sense the helicopter angular velocity and hence replace the pitch and roll damping, and the pilot can control the helicopter with such a system by applying moments directly to the gyro. The performance of a tip-path-plane tilt feedback system can be analyzed using the low-frequency response of the gyro and rotor. The analysis for the hingeless rotor is complicated by the azimuthal phase shifts, but the fundamental behavior of such a control system can be established by considering an articulated rotor. The low-frequency response of a hovering articulated rotor ($\nu_\beta = 1$) is

$$\begin{pmatrix} \beta_{1c} \\ \beta_{1s} \end{pmatrix} = \begin{pmatrix} -\theta_s \\ \theta_c \end{pmatrix} + 8M_\mu \begin{pmatrix} \dot{x}_h - u_G \\ \dot{y}_h + v_G \end{pmatrix} + \frac{16}{\gamma} \begin{pmatrix} \dot{\theta}_F \\ \dot{\phi}_F \end{pmatrix} + \begin{pmatrix} -\dot{\phi}_F \\ \dot{\theta}_F \end{pmatrix} \qquad (21.183)$$

(equation 19.103). The low-frequency gyro response with damping in the rotating frame is

$$\begin{pmatrix} \beta_{Gc} \\ \beta_{Gs} \end{pmatrix} = -\frac{1}{C_R \Omega_G} \begin{pmatrix} M_{Gx} \\ M_{Gy} \end{pmatrix} + \frac{2I_G}{C_R} \begin{pmatrix} \dot{\theta}_F \\ \dot{\phi}_F \end{pmatrix} \qquad (21.184)$$

For an articulated rotor, the control plane tilt should be proportional to the gyro tilt with no phase shift:

$$\begin{pmatrix} -\theta_s \\ \theta_c \end{pmatrix} = \begin{pmatrix} -\theta_s \\ \theta_c \end{pmatrix}_{\text{con}} + K_G \begin{pmatrix} \beta_{Gc} \\ \beta_{Gs} \end{pmatrix} \qquad (21.185)$$

Direct cyclic input from the pilot has been included, although such control is washed out by the feedback. Finally, the moments applied to the gyro are proportional to the rotor tip-path-plane tilt, again with no phase shift required:

$$\begin{pmatrix} M_{Gx} \\ M_{Gy} \end{pmatrix} = \begin{pmatrix} M_{Gx} \\ M_{Gy} \end{pmatrix}_{\text{con}} + K_\beta \begin{pmatrix} \beta_{1c} \\ \beta_{1s} \end{pmatrix} \qquad (21.186)$$

The applied moments from the pilot, by which the rotor can be controlled, have also been included. The complete control law for this system (using the gyro low-frequency response) is thus

$$\begin{pmatrix} -\theta_s \\ \theta_c \end{pmatrix} = \begin{pmatrix} -\theta_s \\ \theta_c \end{pmatrix}_{\text{con}} + K_G \left\{ \frac{2I_G}{C_R} \begin{pmatrix} \dot{\theta}_F \\ \dot{\phi}_F \end{pmatrix} - \frac{1}{C_R \Omega_G} \left[\begin{pmatrix} M_{Gx} \\ M_{Gy} \end{pmatrix}_{\text{con}} + K_\beta \begin{pmatrix} \beta_{1c} \\ \beta_{1s} \end{pmatrix} \right] \right\}$$
(21.187)

The total gain on the tip-path-plane tilt is $K = K_G K_\beta / C_R \Omega_G$. The moments on the gyro should be kept low, so a high gain should be obtained not from K_β but from K_G/C_R. Hence a low but finite damping in the rotating frame and a high ratio of the swashplate tilt to gyro tilt are desired. On substituting for this control law, the rotor flap response becomes

$$\begin{pmatrix} \beta_{1c} \\ \beta_{1s} \end{pmatrix} = \frac{1}{1+K} \left\{ \begin{pmatrix} -\theta_s \\ \theta_c \end{pmatrix}_{\text{con}} + 8M_\mu \begin{pmatrix} \dot{x}_h - u_G \\ \dot{y}_h + v_G \end{pmatrix} + \frac{16}{\gamma} \begin{pmatrix} \dot{\theta}_F \\ \dot{\phi}_F \end{pmatrix} + \begin{pmatrix} -\dot{\phi}_F \\ \dot{\theta}_F \end{pmatrix} \right\}$$
$$- \frac{K}{1+K} \frac{1}{K_\beta} \begin{pmatrix} M_{Gx} \\ M_{Gy} \end{pmatrix}_{\text{con}} + \frac{K}{1+K} \frac{2I_G \Omega_G}{K_\beta} \begin{pmatrix} \dot{\theta}_F \\ \dot{\phi}_F \end{pmatrix} \qquad (21.188)$$

For large gain the response reduces to

$$\begin{pmatrix} \beta_{1c} \\ \beta_{1s} \end{pmatrix} = -\frac{1}{K_\beta} \begin{pmatrix} M_{Gx} \\ M_{Gy} \end{pmatrix}_{\text{con}} + \frac{2I_G \Omega_G}{K_\beta} \begin{pmatrix} \dot{\theta}_F \\ \dot{\phi}_F \end{pmatrix} \qquad (21.189)$$

A hub moment feedback control system therefore reduces the direct rotor response to control, shaft motion, and gusts. The gust alleviation and the reduction of the speed stability are beneficial. In forward flight the rotor angle-of-attack instability is also reduced, which substantially improves the longitudinal flying qualities. The response to direct cyclic pitch inputs is also reduced, but the pilot can control the rotor by applying moments to the gyro instead. The hub moment feedback reduces not only the rotor angular damping but also the response to the shaft angular velocity that couples the lateral and longitudinal motions. With damping in the rotating system, the gyro supplies feedback of the pitch and roll rate to replace the rotor damping of the helicopter. The performance of the hub moment feedback system is similar with a hingeless rotor. The rotor response to external disturbances is reduced, as are the direct rotor forces due to the helicopter motion (including the speed stability and angle-of-attack instability), but angular damping is provided to replace the reduced damping from the rotor. The principal additional consideration with hingeless rotors is the selection of the azimuthal phase angles in the feedback loop, such that the helicopter longitudinal and lateral dynamics and control response are not coupled. With high gain, it may not be sufficient to consider only the quasistatic performance of the system based on the rotor and gyro low-frequency response in assessing the control system behavior. Moreover, this feedback control system is often unstable at high gain, which is an important factor in its design.

The Bell stabilizer bar, developed for two-bladed teetering rotors, is a two-arm gyro mounted on the rotor hub at right angles to the rotor blades. Both the rotor and the gyro dynamics are actually described by periodic-coefficient differential equations, but the low-frequency response is identical to that described earlier (see section 19.8). The gyro arms are linked to the pitch horns, where their input is mechanically mixed with the pilot's control input from the swashplate. There is mechanical damping in the rotating frame, between the gyro and the rotor shaft. Hence this stabilizer bar provides lagged rate feedback of the helicopter pitch and roll. Such a system is simple mechanically. The same feedback is provided for both pitch and roll, which is not desirable since the inertia is smaller in roll than in pitch. Miller (1950) showed that such a stabilizer bar is equivalent to lagged rate feedback for low frequency. Sissingh (1967) discussed this system and compared it with others, including hub moment feedback systems.

The Hiller control rotor, also developed for two-bladed rotors, is a two-arm gyro with a small airfoil on each arm. The airfoils provide aerodynamic damping in the rotating frame, so the gyro gives pitch and roll rate feedback. The gyro tilt produces cyclic pitch of the main rotor, and the pilot controls the rotor by means of cyclic pitch of the airfoils on the gyro arms, thus producing a moment on the gyro. With this configuration the cyclic stick control forces are less sensitive to the main rotor conditions, and the mechanical gyro damper is eliminated. Stuart (1948) described this control rotor and its influence on the handling qualities of the helicopter. Miller (1950) showed that such a system provides lagged rate feedback of pitch and roll, as well as feedback of longitudinal velocity and main rotor flapping due to the aerodynamic forces on the gyro.

The Lockheed gyro stabilizer is a hub moment feedback system developed for three-bladed and four-bladed hingeless rotors. The gyro arms are linked to the blade pitch horns, so gyro tilt provides the cyclic pitch. The pilot controls the helicopter by applying moments to the gyro. One version of this gyro stabilizer used feathering moment feedback. The blades were swept forward of the pitch axis, so that the flap moment had a component about the pitch axis. The feathering moment was transmitted through the pitch links to the gyro, providing hub moment feedback. The gyro damping was due to the flap damping of the main rotor. A later design used direct hub moment feedback. Arms on the blades detected the flap motion, which was then transmitted to springs that applied a moment to the gyro. The gyro had mechanical damping. The gyro response was sensed and then fed to the swashplate by hydraulic actuators, so the blade feathering moments did not influence the gyro at all. Sissingh (1967) analyzed the quasistatic performance of the Lockheed hub moment feedback system. Johnson and Hohenemser (1970) analyzed thrust and tip-path-plane-tilt integral feedback systems, particularly their influence on the rotor flap and lag dynamics, including the high-gain stability. In a review of this paper, Sissingh (1970) analyzed the high-gain stability of hub moment feedback systems.

The blade elastic pitch motion produced by inertial and aerodynamic feathering moments when the center-of-gravity is offset from the aerodynamic center can be used to provide stability augmentation for the helicopter flight dynamics. A forward shift of the blade center-of-gravity increases the helicopter pitch damping. A pitch angular velocity $\dot{\theta}_F$ of the helicopter and rotor combined with the blade velocity Ωr gives a Coriolis force on the blade, downward on the advancing side and upward on the retreating side. This Coriolis force acts at the blade center-of-gravity to produce a blade pitch moment. The response of a torsionally flexible blade with center-of-gravity forward of the pitch axis is thus a 1/rev variation equivalent to longitudinal cyclic, $\theta_s < 0$ for $\dot{\theta}_F > 0$, which implies increased pitch damping. An aft shift of the aerodynamic center of the blade also increases the helicopter pitch damping. With a pitch rate $\dot{\theta}_F$, the rotor flaps forward to provide a lateral moment on the disk (toward the retreating side), which precesses the rotor to follow the shaft. The lift forces producing this moment act at the aerodynamic center of the blade, producing a blade feathering moment also. When the aerodynamic center is aft of the pitch axis, longitudinal cyclic $\theta_s < 0$ is produced, which increases the pitch damping. Miller (1948) analyzed the rigid flap and rigid pitch dynamics of an articulated rotor, obtaining the low-frequency response of the blade pitch to the helicopter motions when the center-of-gravity and aerodynamic center are offset from the pitch axis (by x_I and x_A, respectively). He found that the elastic torsion response of the blade provides pitch and roll rate feedback proportional nearly to $(x_A - x_I)$. For increased damping the center-of-gravity should be forward of the aerodynamic center ($x_A > x_I$, which is favorable for flutter and divergence stability also). If $x_A \neq 0$, the blade pitch also responds to the longitudinal and lateral velocity of the helicopter ($\dot{x}_F$ and $\dot{y}_F$), and hence the speed stability is influenced. When $x_A = 0$, the longitudinal feedback reduces to

$$\theta_s = \frac{3x_I/R}{K_\theta/I_b\Omega^2} \dot{\theta}_F \qquad (21.190)$$

where K_θ is the control system stiffness. Damping in the non-rotating cyclic control system introduces a lag in the feedback; damping in the rotating system gives a lag and also couples the longitudinal and lateral feedback. Hence chordwise offset of the blade center-of-gravity from the aerodynamic center provides lagged rate feedback of the helicopter pitch and roll motion. A high gain requires torsionally flexible blades or a large center-of-gravity shift forward, which means a large leading edge balance weight. The influence of the blade torsional moments on the control loads and stick forces must also be considered. Simons and Modha (2007) described an influence of pitch and roll rate on the blade torsion motion. The normal Coriolis acceleration from the product of the rotor disk angular velocity ($-\dot{\alpha}_y \cos \psi_m + \dot{\alpha}_x \sin \psi_m$, about the blade flap axis) and the radial velocity (Ωx) gives a pitch moment in equation 16.136

$$\widehat{I}_f(\ddot{\theta} + (\omega^2 + 1)\theta) = 2\widehat{I}_f(-\dot{\alpha}_y \cos \psi_m + \dot{\alpha}_x \sin \psi_m) + \ldots \qquad (21.191)$$

The 1/rev pitch response is

$$\Delta\theta = \frac{2}{\omega^2}(-\dot{\alpha}_y \cos \psi_m + \dot{\alpha}_x \sin \psi_m) \qquad (21.192)$$

where $\omega^2 = K_\theta/I_f\Omega^2$ is the non-rotating pitch frequency (structural spring). Substituting this cyclic pitch in the flap response (equation 19.56), the cross response to pitch and roll rate is reduced by the factor $(1 - 2/\omega^2)$. The use of the blade torsion dynamics to provide stability augmentation for the flight dynamics was discussed further by Miller (1950), McIntyre (1962), and Reichert and Huber (1971).

21.10 Flying Qualities Specifications

The helicopter user or purchaser, or the appropriate regulatory agency, must determine what characteristics are required for acceptable flying qualities of the aircraft and establish quantitative measurements of the desired characteristics. The specification and evaluation of flying qualities are concerned with many properties of the helicopter, among them control displacement and force gradients; static stability; dynamic stability, particularly the long period roots; transient response characteristics, especially the short period behavior; control power, damping, and control sensitivity; and control coupling. The requirements vary greatly with the use intended for the vehicle. Stricter specifications are continually developed as more is learned about measuring helicopter handling qualities and designing helicopters to achieve the desired characteristics.

The requirements of specifications "are intended to assure that no limitations on flight safety or on the capability to perform intended missions will result from deficiencies in flying qualities" (ADS-33 (2000)). Notable flying qualities specifications for helicopters include MIL-H-8501 (1952), MIL-H-8501A (1961), AGARD R-577 (1970), MIL-F-83300 (1970), FAR Airworthiness Standards Parts 27 and 29, and ADS-33E-PRF (2000). MIL-H-8501A and ADS-33F-PRF are summarized here, because they are widely used and highly detailed specifications.

NACA research on helicopter flying qualities led to the first formal specification, MIL-H-8501. Although remaining a U.S. Army requirement until 1995, MIL-H-8501A was based on the flying characteristics of the first generation of production helicopters and reflected both the aircraft and the operations of its era. Despite being derived from NACA research, there was no data base or background information to support the requirements (Padfield (2012)). The dynamic requirements in MIL-H-8501A are based on fully attended operation and reasonably good visibility, for which moderate instabilities are acceptable as long as the frequencies are sufficiently low. With increased capabilities of the aircraft and flight control systems, a new handling qualities specification for rotorcraft was needed. In 1974 the U.S. Army and NASA started an effort to develop a handling qualities data base and design criteria for a new specification, and in 1982 the specification development began. The Aeronautical Design Standard for rotorcraft handling qualities (ADS-33A) was released in 1987. The fifth revision, ADS-33E-PRF, was released in 2000. There has been no modern development of civilian specifications for rotorcraft flying qualities comparable to ADS-33.

21.10.1 MIL-H-8501A

The specification MIL-H-8501A (1961) defines the flying and ground handling qualities required for military helicopters. The requirements cover simple time-domain parameters (control stick force and position gradients with speed), frequency and damping of oscillatory modes, normal acceleration response to step input (the concave-downward requirement), and angular displacements in response to control steps (as function of helicopter weight). There is some distinction between day visual flight rules and night instrument flight rules requirements, but flight in low-visibility conditions is not considered.

For static stability, MIL-H-8501A specifies the minimum and maximum initial force gradient of the longitudinal and lateral sticks and requires that the gradient always be positive. The longitudinal stick should have a stable force and position gradient with respect to speed. At low speed (transition) a moderate degree of instability is permitted for visual flight rules (VFR) operations, but is not desirable. A stable gradient of the pedal and lateral cyclic stick with sideslip angle is required, and positive directional stability and effective dihedral (lateral speed stability) are required in forward flight. For IFR operations, the directional and lateral controls must have stable force and displacement gradients. The transient control forces, control force coupling, control margin, and other factors are also addressed.

The dynamic stability characteristics in forward flight are specified in terms of the period and damping of the long period modes. Figure 21.15 summarizes the requirements for VFR and IFR operations. So there is no excessive delay in the development of the helicopter angular velocity in response to control, the specification requires that the roll, pitch, and yaw acceleration be in the proper direction within 0.2 sec after the control displacement.

To ensure acceptable maneuver stability characteristics (normal acceleration in forward flight, pitch rate in hover and at low speeds) the concave downward requirement is used: the time history of the normal acceleration and angular velocity of the helicopter should be concave downward within 2 sec after a step displacement of the longitudinal stick. Preferably, the normal acceleration should be concave

21.10 Flying Qualities Specifications

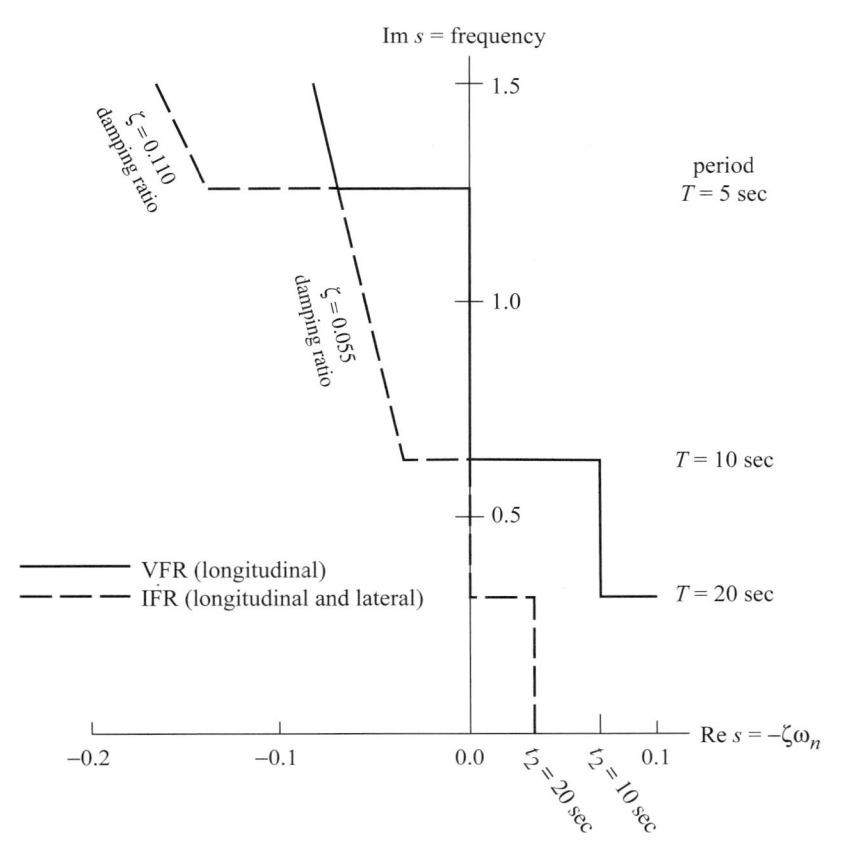

Figure 21.15. MIL-H-8501A specification of helicopter dynamic stability (long period modes) in forward flight.

downward throughout the maneuver (until the maximum acceleration), and the angular velocity should be concave downward after 0.2 sec.

To ensure that the pilot has a reasonable time for corrective action following perturbations from the trim attitude, the following requirement is used: within 10 sec after a longitudinal cyclic pulse lasting at least 0.5 sec (to simulate a disturbance), the normal acceleration should not increase by more than 0.25 g, and during the subsequent nose-down motion the normal acceleration should not decrease more than 0.25 g below the trim value.

The minimum helicopter control power in hover is specified by requiring that the attitude change be at least α_{min} one second after a 1-inch step control displacement from trim (0.5 sec for roll), where the value of α_{min} depends on the axis (pitch, roll, or yaw) and the helicopter gross weight. A similar requirement is given for the response to maximum control displacement. The directional control sensitivity for hover, as well as the roll control sensitivity at all speeds, should not be so high that the pilot tends to over-control the helicopter. In any case, the roll sensitivity must be less than 20 deg/sec/in, and the yaw sensitivity must be low enough that the yaw angle change after one second is less than 50 deg/in. For satisfactory initial control response characteristics, a minimum level of pitch, roll, and yaw damping

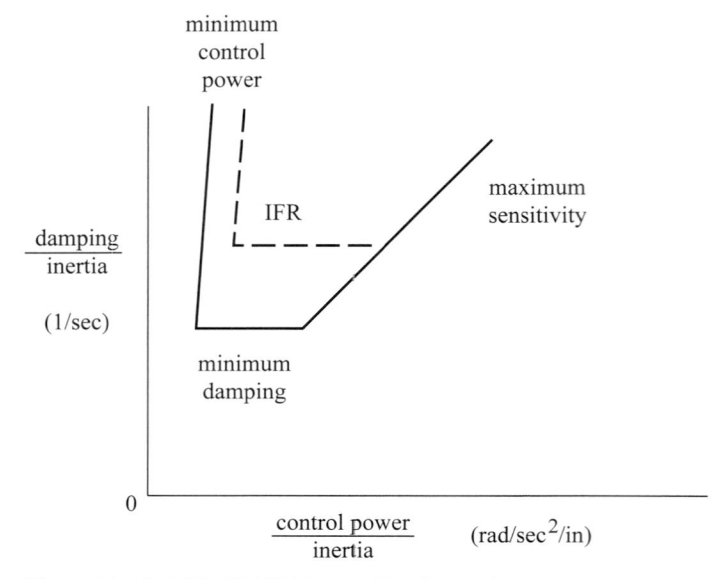

Figure 21.16. MIL-H-8501A specification of hover control power, damping, and control sensitivity.

that depends on the moment of inertia about the corresponding axis is required. The minimum control power and damping required in pitch and roll are increased for IFR operations. The specifications of hover control power, damping, and control sensitivity can be expressed in terms of requirements for the ratio of damping to inertia and the ratio of control power to inertia (with units of (ft-lb-sec)/(slug-ft^2) = 1/sec and (ft-lb/in)/(slug-ft^2) = rad/sec^2/in, respectively). Figure 21.16 illustrates the requirements for the derivatives. The ratio of damping to inertia is the inverse of the time constant, the ratio of control power to inertia is the initial angular acceleration, and the ratio of control power to damping is the control sensitivity. If the control power is too low and damping too high, the control inputs are large and the response is sluggish. If the damping is too low and the control sensitivity too high, the response is not precise and there is a tendency to over-control. If both derivatives are low, the response is inadequate, and large control inputs are needed. Harmony is desired, requiring a good match of damping and control sensitivity.

MIL-H-8501A also provides some specifications for the ride quality (maximum vibration) at the crew and passenger stations and for the control stick vibrations; see section 18.4.

Gustafson, Amer, Haig, and Reeder (1949) introduced the concave downward requirement, based on flight test investigations of helicopter longitudinal handling qualities. They concluded that the most important consideration is prevention of a prolonged stick-fixed divergence of the normal acceleration in response to cyclic control. In addition, a continuous development of the normal acceleration is desirable, rather than a pause in the development during the first second of the maneuver. So they developed a requirement to preclude the divergent tendencies of the normal acceleration: the time history of normal acceleration should be concave downward within 2 sec after a sudden displacement of the longitudinal control stick. To reduce the difficulty of anticipating the results of a control deflection, the time history of the normal acceleration should preferably be concave downward from the beginning of the maneuver to the maximum acceleration after a step displacement of the longitudinal stick, or at least the slope of the normal acceleration should be positive throughout the maneuver. To ensure that an oscillation rather than a sudden divergence occurs in response to disturbances, after a pulse displacement of the longitudinal stick the normal acceleration should remain between 1.25 g and 0.75 g. Amer (1951) found that these maneuver requirements were also applicable to tandem helicopters, although a stricter criterion might be needed because of the possible instability with respect to speed. Reeder and Whitten (1952) investigated the influence of the helicopter pitch damping on the longitudinal maneuver characteristics by using a stabilizer bar to provide increased damping due to lagged rate feedback. Correlating the pilot opinion of the handling qualities with the concave downward requirement, they concluded that the requirement is applicable for variations of the pitch damping as well, although the original development was concerned primarily with the influence of the helicopter angle-of-attack instability.

Crimi, Reeder, and Whitten (1953) conducted instrument flight trials of a helicopter. They concluded that the flying qualities requirements based on VFR operations were adequate for instrument flight at forward speeds above minimum power, although close and constant attention to the instruments was necessary. At low speeds and during precision maneuvers, lateral directional problems were encountered, making instrument flight possible only for very short periods. Amer and Tapscott (1954) developed flying qualities criteria for the lateral dynamics of both single main rotor and tandem helicopters in forward flight. They found that pedal-fixed directional stability is required. For a reasonably damped stick-fixed oscillation, if the period is less than 10 sec, the time to half-amplitude should be less than 2 cycles (giving a damping ratio greater than 5.5% critical) and there should be no residual oscillation. Gustafson and Tapscott (1958) described the stability characteristics of single main rotor and tandem helicopters, including criteria for derivatives and the influence of design parameters.

Salmirs and Tapscott (1959) conducted a flight test investigation of the influence of damping and control power on the helicopter handling qualities. A definite improvement of the handling qualities (based on pilot opinion) was found with increased damping. They presented charts summarizing the results in terms of the ratio of damping to inertia and of control power to inertia (analogous to Figure 21.16) and defining the requirements for good IFR, acceptable IFR, acceptable VFR, and unacceptable characteristics. They also gave the results in terms of the damping required as a function of the helicopter attitude change 1 sec after a 1-inch control displacement and in terms of the roll damping required as a function of the roll control sensitivity.

Kelly and Garren (1968) investigated the influence of pitch damping, control power, speed stability, and angle-of-attack stability on helicopter longitudinal handling qualities. The optimum values of these parameters were found to be largely independent of changes in the other parameters and of the operating conditions considered. A minimum level of pitch damping was established, although the improvement in handling qualities continued as the damping was increased further. They concluded that the helicopter should have neutral or slightly positive stability with respect to angle-of-attack. A minimum control power and an optimum speed stability were also established.

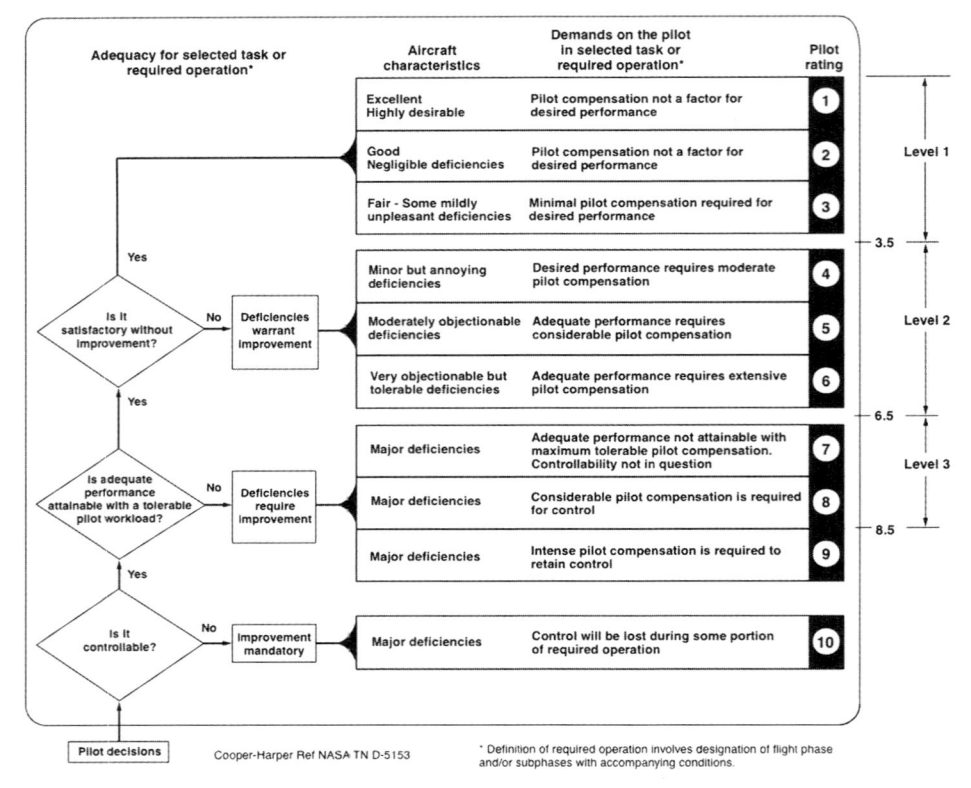

Figure 21.17. Cooper-Harper handling qualities rating (HQR) scale and handling quality levels.

21.10.2 Handling Qualities Rating

Cooper and Harper (1969) developed a handling qualities rating (HQR) scale that allows a pilot to translate an assessment of the adequacy of an aircraft for a selected task or required operation to a standard categorization of the aircraft characteristics and the demands on the pilot (Figure 21.17). An HQR is associated not with the aircraft alone, but depends on the task and required performance. The workload demands are described in terms of pilot compensation needed. The primary questions in the decision tree further allow characterization of the level of handling qualities:

> Level 1: The aircraft is satisfactory without improvement (HQR = 1 to 3). The aircraft characteristics can range from excellent (pilot compensation not a factor in desired performance) to fair (some mildly unpleasant deficiencies, minimal pilot compensation required).
> Level 2: The aircraft has deficiencies that warrant improvement, resulting in adequate performance with a tolerable pilot workload (HQR = 4 to 6). The deficiencies can range from minor but annoying to very objectionable but tolerable, so moderate to extensive pilot compensation is required.
> Level 3: The aircraft is controllable, but has deficiencies that require improvement (HQR = 7 to 8). These major deficiencies require maximum tolerable to considerable pilot compensation.

HQR = 9: The aircraft has major deficiencies, such that intense pilot compensation is required to maintain control.

HQR = 10: The aircraft is not controllable, and improvement is mandatory.

Level 1 encompasses good flying qualities, enabling the pilot to achieve the desired level of performance with acceptable workload, so minimal control compensation is required (Padfield (1998)). Level 1 implies flight with margins in pilot workload and mission capability. Level 2 means flying qualities with tolerable deficiencies, such that the pilot can achieve adequate performance, but only with a high workload. Level 3 means the aircraft cannot perform the required missions, because of excessive workload or insufficient effectiveness.

The specifications require level 1 handling qualities within the operational flight envelope. Many rotorcraft have level 2 handling qualities for specific tasks.

21.10.3 Bandwidth Requirements

The bandwidth of the aircraft response to control is a primary factor (hence a good basis for a criterion) in precise closed-loop tracking tasks, which involve high-frequency, small attitude changes. McRuer and Krendel (1974) presented the crossover model of human pilot behavior: for a large range of aircraft dynamic behavior, the pilot adapts a control strategy such that the open-loop (pilot times aircraft) transfer function takes the form

$$Y_p Y_a \cong \frac{\omega_c e^{-\tau_e s}}{s} \qquad (21.193)$$

where ω_c is called the crossover frequency, since the amplitude is unity at $\omega = \omega_c$. Thus at frequencies around the input bandwidth, the pilot strategy is to create a range of -20 dB/decade slope for the amplitude ratio and adjust the loop gain so the unit-amplitude crossover frequency is near the higher edge of this region, while maintaining an adequate stability margin (Heffley (1979)).

The crossover model is an approximate quantitative description of the human-machine system for single-loop compensation. Increasing ω_c implies increasing gain, but the closed-loop stability boundary ($-180°$ phase shift open loop) is at $\omega_c = \pi/2\tau_e$; the phase margin is $90 - \tau_e \Omega_c$. The bandwidth criterion is an application of the crossover model concept. The open-loop bandwidth is defined as the crossover frequency for a simple, pure gain pilot with 45° phase margin or 6 dB gain margin (Blanken, Bivens, and Whalley (1985)).

The bandwidth is obtained from the frequency response of the aircraft angular attitude to cockpit controls; see Figure 21.18 and Hoh (1988). This frequency response is for the aircraft with all augmentation loops closed; hence, wiht all elements of the flight control system active, but without the pilot in the loop. From the phase, the frequency for neutral stability is found, ω_{180}. Then the phase bandwidth is the frequency for 45° of phase margin, and the gain bandwidth is the frequency for 6 dB of gain margin. These margins permit the pilot to increase the gain or phase lag of the compensation without causing an instability. The bandwidth frequency is then defined as the lesser of the two frequencies for rate response, ω_{BW} = minimum of $\omega_{BW_{gain}}$ and $\omega_{BW_{phase}}$, or equal to the phase bandwidth for attitude-type response, $\omega_{BW} = \omega_{BW_{phase}}$.

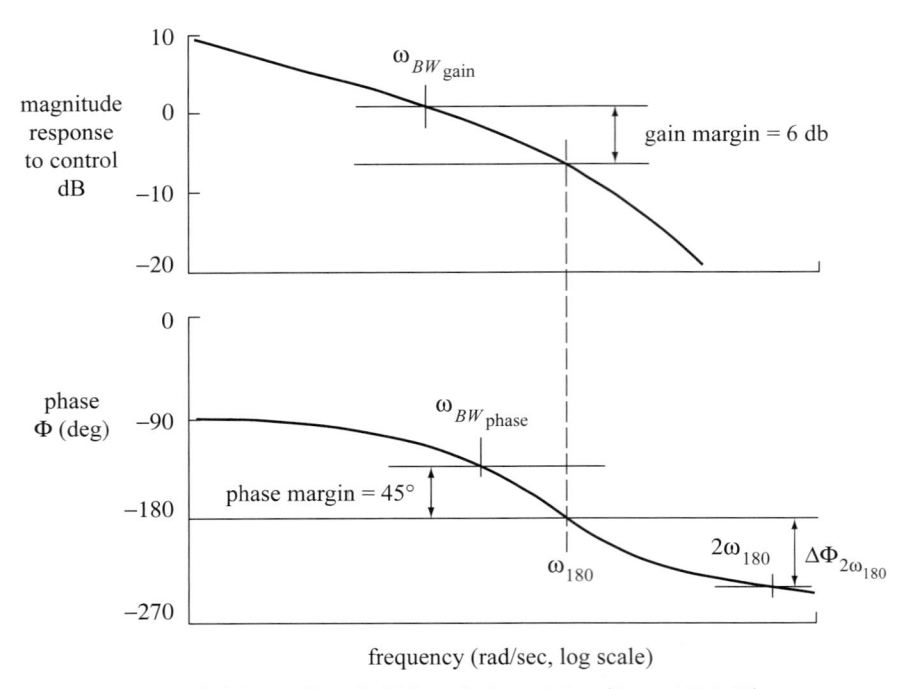

Figure 21.18. Definitions of bandwidth and phase delay (from ADS-33).

Bandwidth is a quantitative measure of the pilot's capability to make rapid and precise control input to minimize errors and thereby improve closed-loop tracking performance; these tasks can dominate the evaluation of handling qualities. High bandwidth reflects faster and more predictable aircraft response to controls. A low-phase bandwidth means the aircraft response is sluggish. A low-gain bandwidth means the aircraft is prone to pilot-induced oscillation, since small changes in pilot gain result in a rapid reduction in phase margin. Thus low values of bandwidth indicate a need for pilot lead equalization to achieve the required mission performance, and excessive requirements for lead equalization result in degraded handling quality ratings. See Blanken, Bivens, and Whalley (1985), and Hoh (1988).

Pilots and hence handling quality ratings are also sensitive to the shape of the phase curve at frequencies beyond the bandwidth frequency. This shape is characterized by the phase delay:

$$\tau_p = \frac{\Delta \Phi_{2\omega_{180}}}{57.3(2\omega_{180})} \qquad (21.194)$$

see Figure 21.18 and Hoh (1988). The phase delay is a measure of the steepness of the phase after $-180°$. A large value of phase delay means a small frequency margin between normal tracking at $45°$ of phase margin and the instability, implying that the aircraft is prone to pilot-induced oscillation. Although quantified as a time constant, τ_p is not a time domain parameter. Also, τ_p is not a system transport delay, but rather a measure of the shape of the phase curve around the instability frequency.

21.10 Flying Qualities Specifications

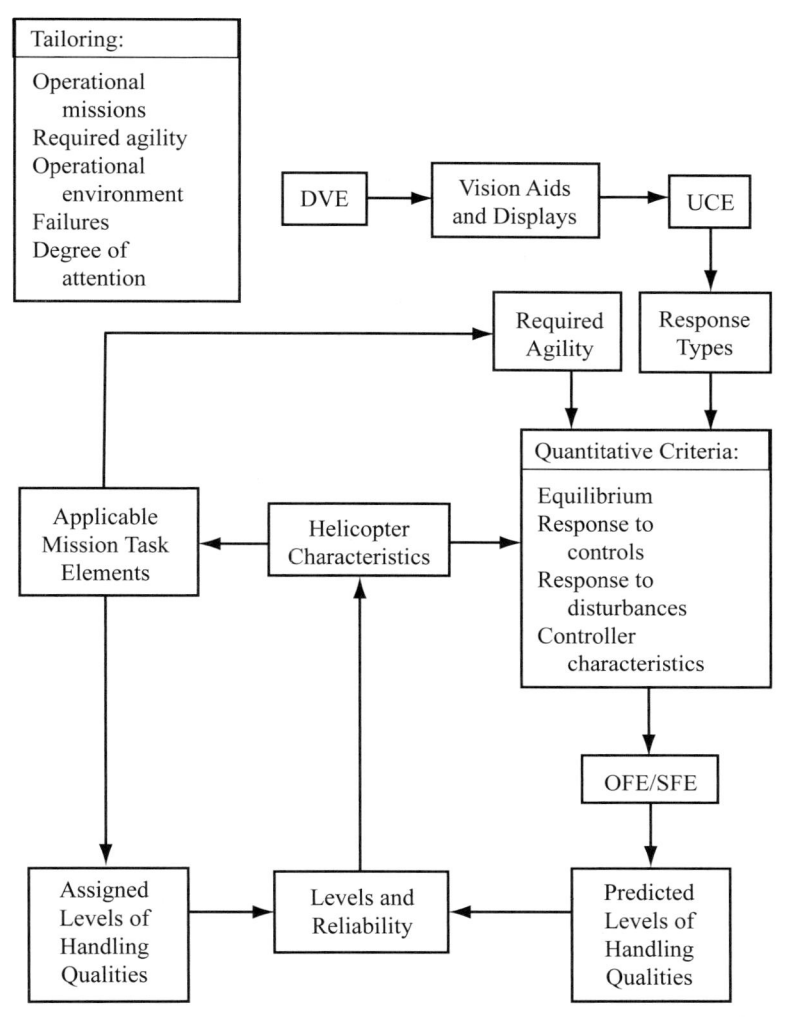

Figure 21.19. Structure of ADS-33; from Blanken, Hoh, and Mitchell (2007).

21.10.4 ADS-33

The Aeronautical Design Standard for rotorcraft handling qualities (ADS-33E-PRF) is based on a number of key concepts: specific flight test maneuvers, increased stabilization in reduced cue environment, frequency domain criteria, and the use of handling quality ratings. Frequency domain methods were new to rotorcraft specifications, but the response criteria are much simpler in the frequency domain, producing a much clearer separation of good and bad configurations. The discussion of ADS-33 in this section is based on Key and Hoh (1987), Hoh (1988), Key (1988), Padfield (1998, 2012), and Blanken, Hoh, and Mitchell (2007).

The structure of ADS-33 is outlined in Figure 21.19. ADS-33 is a mission-oriented specification, defining the response characteristics needed to accomplish specific tasks in a defined cue environment, with an acceptable workload. First the user must define the missions to be performed and the operational environment. Then the requirements are tailored to this rotorcraft, in terms of specific mission task elements (MTEs), degraded visual environment (DVE), and the operational flight envelope (OFE). From the degraded visual environment and available vision aids and displaces, the usable cue environment (UCE) is defined, and from the UCE the required aircraft response type follows for each task. From the required agility and response types, the specification provides quantitative criteria (equilibrium, response to controls, response to disturbances, controller characteristics) for the predicted levels of handling qualities. Predicted levels are obtained from measured characteristics of the rotorcraft, initially by analysis from basic aerodynamic and flight control characteristics.

For the mission task elements, assigned levels of handling qualities are obtained by test pilots using the Cooper-Harper handling quality rating (HQR) scale to assess workload and task performance, initially in piloted simulation tests. Both predicted and assigned levels are ultimately obtained from flight tests.

The MTEs are a set of flight demonstration maneuvers that have been tailored with specific requirements on desired and adequate performance for the particular class of rotorcraft. The tasks are constructed specifically to permit accurate evaluate of the handling qualities. Performance criteria are given for both good visual environment (GVE) and degraded visual environment (DVE). ADS-33 has a suite of MTEs intended to cover scout and attack, utility, and cargo aircraft. Expanding the application of ADS-33 to other aircraft classes requires first the development of appropriate MTEs.

The aircraft must be capable of performing its missions within the operational flight envelope; hence level 1 flying qualities must be achieved within the OFE. Within the service flight envelope (SFE), defined by flight boundaries due to aircraft limits rather than mission requirement, the aircraft must achieve level 2 flying qualities.

The UCE concept was developed to help specify the level of control augmentation required when the pilot can no longer make aggressive and precise maneuvers due to inadequacies in visual cueing. The UCE encompasses all visual cues, inside and outside the cockpit, including displays. The UCE is determined for a MTE/DVE combination from a subjective evaluation of the pilot's ability to accomplish aggressive and precise maneuvers. The task-tailored requirements specify the appropriate response type for each MTE and UCE. This response type can be achieved aerodynamically or by means of a stability and command augmentation system (SCAS). Additional stabilization is an effective way to compensate for missing visual cues and situations of divided attention.

Rate response type is preferred for MTEs involving fully attended operations in a good visual environment (GVE). In particular, rate response type is needed for high-agility MTEs. Most helicopters meet the requirement for rate response type. Attitude-command-attitude-hold (ACAH) is required for MTEs needing a high level of stabilization, such as operations in poor visibility or limited displays. More aggressive maneuvers with worse UCE can require even more stabilization, including translational rate command with position hold.

Dynamic response criteria define the flying qualities of the aircraft in terms of response to control and disturbances, with specifications for level 1, level 2, and level 3 (Figures 21.20 to 21.22). Two flight regimes are considered: low speed and hover (to 45 knots) and forward flight. The dynamic response criteria distinguish the amplitude and frequency of the maneuver: small, medium, and large amplitude for high and midfrequency ranges. The criteria for moderate-large motions are described in terms of agility metrics (such as response quickness and control power) and for

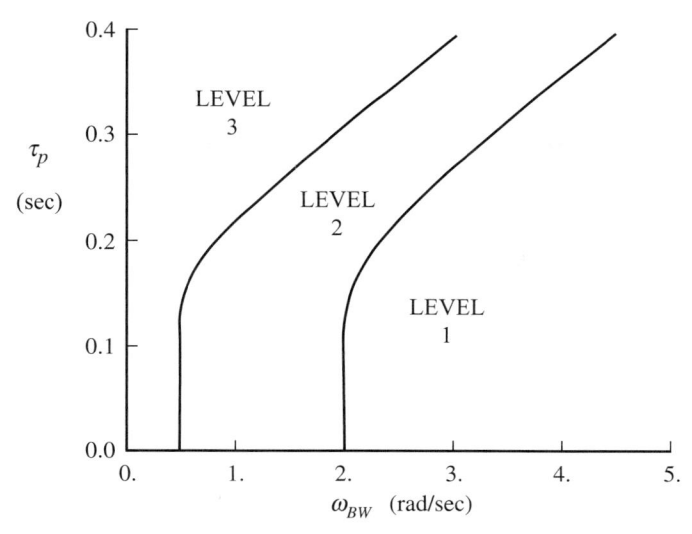

Figure 21.20. Representative ADS-33 requirement for small-amplitude attitude changes.

small-amplitude motions in terms of stability margins; see Padfield (2012). The response bandwidth is the primary factor for precise closed-loop tracking, which involves small attitude changes. For pursuit tracking, larger attitude changes and angular rates are used, and the pilot needs the ability to achieve rapid, but less precise, changes in attitude. Very large-amplitude maneuvers are usually accomplished open-loop, with large angular rates desired.

Small, high-frequency attitude changes primarily involve closed-loop tracking and can be characterized by the bandwidth frequency. Bandwidth is a measure of the maximum closed-loop frequency that the pilot can achieve without threatening stability. The bandwidth is determined from the phase or gain margin of the open-loop frequency response of the attitude to pilot control input. The criterion is a minimum bandwidth as a function of phase delay (Figure 21.20). A larger bandwidth is required for large phase delay, which is a measure of how quickly the frequency response phase increases.

The lower frequency response, between steady state and the bandwidth, is most affected by divided attention. Full attention is required if the midterm response is unstable. The midterm response is specified in terms of the damping and frequency of the flight modes (Figure 21.21). For level 1 handling qualities, divided attention MTEs must have a damping ratio of at least 0.35, with no divergences or lightly damped oscillations. This requirement eliminates many unaugmented helicopters. Full attention MTEs can have some midterm divergences and low-damped oscillations. The use of attitude hold response type constitutes compliance with the requirement, as long as oscillatory modes have a damping ratio of at least 0.35.

For moderate-amplitude maneuvers, the pilot is concerned with the ability to achieve rapid but less precise changes in attitude. The response requirement is in terms of quickness: the ratio of the peak attitude rate to the attitude change ($q_{\text{peak}}/\Delta\theta$ for pitch) achieved during a rapid attitude-change maneuver; see Figure 21.22. At low amplitude this quantity can be related to the bandwidth, so the criterion effectively allows decreased bandwidth with increasingly large maneuvers. For large amplitude the boundaries are interpreted as measures of agility.

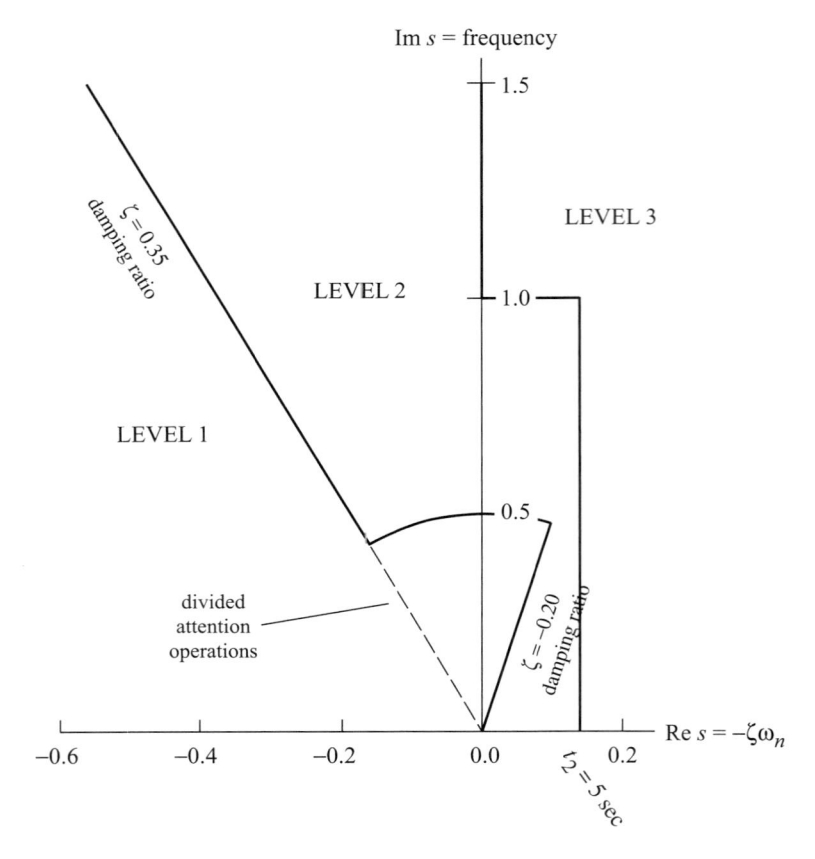

Figure 21.21. ADS-33 limitations on low frequency pitch and roll oscillations.

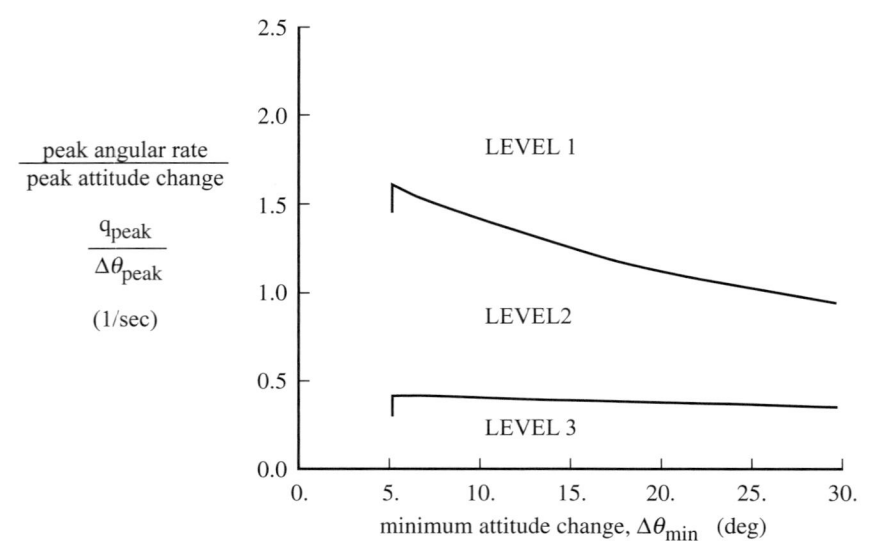

Figure 21.22. Representative ADS-33 requirement for moderate-amplitude attitude changes.

For large-amplitude attitude changes, the requirement is a measure of control power, specified as lower limits on the maximum steady angular rate that can be achieved with full control deflection. Three levels of aggressiveness are defined, for various MTEs.

21.11 REFERENCES

Advisory Group for Aerospace Research and Development. "V/STOL Handling, Criteria and Discussion." AGARD R-577-70, December 1970.
Amer, K.B. "Some Flying-Qualities Studies of a Tandem Helicopter." NACA RM L51H20a, October 1951.
Amer, K.B., and Tapscott, R.J. "Studies of the Lateral-Directional Flying Qualities of a Tandem Helicopter in Forward Flight." NACA Report 1207, 1954.
Blanken, C.L., Bivens, C.C., and Whalley, M.S. "An Investigation of the Use of Bandwidth Criteria for Rotorcraft Handling-Qualities Specifications." American Helicopter Society International Conference on Rotorcraft Basic Research, Research Triangle Park, NC, February 1985.
Blanken, C.L., Hoh, R.H., and Mitchell, D.G. "Test Guide for ADS-33E-PRF." American Helicopter Society 63rd Annual Forum, Virginia Beach, VA, May 2007.
Cooper, G.E., and Harper, Robert P., Jr. "The Use of Pilot Rating in the Evaluation of Aircraft Handling Qualities." NASA TN D-5153, April 1969.
Crim, A.D., Reeder, J.P., and Whitten, J.B. "Initial Results of Instrument-Flying Trials Conducted in a Single-Rotor Helicopter." NACA Report 1137, 1953.
Gustafson, F.B., Amer, K.B., Haig, C.R., and Reeder, J.P. "Longitudinal Flying Qualities of Several Single-Rotor Helicopters in Forward Flight." NACA TN 1983, November 1949.
Gustafson, F.B., and Tapscott, R.J. "Methods for Obtaining Desired Helicopter Stability Characteristics and Procedures for Stability Predictions." NACA Report 1350, 1958.
Heffley, R.K. "A Compilation and Analysis of Helicopter Handling Qualities Data." NASA CR 3145, August 1979.
Hoh, R.H. "Dynamic Requirements in the New Handling Qualities Specification for U.S. Military Rotorcraft." Royal Aeronautical Society International Conference on Helicopter Handling Qualities and Control, London, UK, November 1988.
Hohenemser, K. "Dynamic Stability of a Helicopter with Hinged Rotor Blades." NACA TM 907, September 1939.
Hohenemser, K.H., and Yin, S.K. "On the Use of First Order Rotor Dynamics in Multiblade Coordinates." American Helicopter Society 30th Annual National Forum, Washington, DC, May 1974.
Hohenemser, K.H., and Yin, S.K. "Methods Studies Toward Simplified Rotor-Body Dynamics." NASA CR 137570, June 1974.
Johnson, R.L., and Hohenemser, K.H. "On the Dynamics of Lifting Rotors with Thrust or Tilting Moment Feedback Controls." Journal of the American Helicopter Society, *15*:1 (January 1970).
Kaufman, L., and Peress, K. "A Review of Methods for Predicting Helicopter Longitudinal Response." Journal of the Aeronautical Sciences, *23*:3 (March 1956).
Kelly, J.R., and Garren, J.F., Jr. "Study of the Optimum Values of Several Parameters Affecting Longitudinal Handling Qualities of VTOL Aircraft." NASA TN D-4624, July 1968.
Key, D.L. "A New Handling Qualities Specification for U.S. Military Rotorcraft." Royal Aeronautical Society International Conference on Helicopter Handling Qualities and Control, London, UK, November 1988.
Key, D.L., and Hoh, R.H. "New Handling-Qualities Requirements and How They Can Be Met." American Helicopter Society 43rd Annual Forum, St. Louis, MO, May 1987.
McIntyre, H.H. "Longitudinal Dynamic Stability of a Helicopter with Torsionally Flexible Blades and Servo-Flap Control." American Helicopter Society 18th Annual National Forum, Washington, DC, May 1962.

McRuer, D.T., and Krendel, E.S. "Mathematical Models of Human Pilot Behavior." AGARD AG-188, January 1974

Miller, R.H. "Helicopter Control and Stability in Hovering Flight." Journal of the Aeronautical Sciences, 15:8 (August 1948).

Miller, R.H. "A Method for Improving the Inherent Stability and Control Characteristics of Helicopters." Journal of the Aeronautical Sciences, 17:6 (June 1950).

Padfield, G.D. "Controlling Tension Between Performance and Safety in Helicopter Operations. A Perspective on Flying Qualities." Twenty-Fourth European Rotorcraft Forum, Marseilles, France, September 1998.

Padfield, G.D. *Helicopter Flight Dynamics*. Second Edition. Oxford: Blackwell Science Ltd, 2007.

Padfield, G.D. "Rotorcraft Handling Qualities Engineering. Managing the Tension Between Safety and Performance." American Helicopter Society 68th Annual Forum, Fort Worth, TX, May 2012.

Reeder, J.P., and Whitten, J.B. "Some Effects of Varying the Damping in Pitch and Roll on the Flying Qualities of a Small Single-Rotor Helicopter." NACA TN 2459, January 1952.

Reichert, G., and Huber, H. "Influence of Elastic Coupling Effects on the Handling Qualities of a Hingeless Rotor Helicopter." AGARD CP 121, September 1971.

Salmirs, S., and Tapscott, R.J. "The Effects of Various Combinations of Damping and Control Power on Helicopter Handling Qualities During Both Instrument and Visual Flight." NASA TN D-58, October 1959.

Simons, I.A., and Modha, A.N. "Gyroscopic Feathering Moments and the "Bell Stabilizer Bar" on Helicopter Rotors." Journal of the American Helicopter Society, 52:1 (January 2007).

Sissingh, G.J. "Response Characteristics of the Gyro-Controlled Lockheed Rotor System." Journal of the American Helicopter Society, 12:4 (October 1967).

Sissingh, G.J. "Review and Discussion of 'On the Dynamics of Lifting Rotors with Thrust or Tilting Moment Feedback Controls.'" Journal of the American Helicopter Society, 15:1 (January 1970).

Stuart, J., III. "The Helicopter Control Rotor." Aeronautical Engineering Review, 7:8 (August 1948).

United States Army. "Handling Qualities Requirements for Military Rotorcraft." Aeronautical Design Standard Performance Specification, ADS-33E-PRF, February 2000.

United States Government. "Requirements for Helicopter Flying Qualities." Military Specification MIL-H-8501, November 1952.

United States Government. "General Requirements for Helicopter Flying and Ground Handling Qualities." Military Specification MIL-H-8501A, September 1961.

United States Government. "Flying Qualities of Piloted V/STOL Aircraft." Military Specification MIL-F-83300, December 1970.

United States Government. "Airworthiness Standards: Normal Category Rotorcraft." CFR Title 14, Part 27, 2011.

United States Government. "Airworthiness Standards: Transport Category Rotorcraft." CFR Title 14, Part 29, 2011.

22 Comprehensive Analysis

The digital computer programs that calculate the aeromechanical behavior of rotorcraft are called comprehensive analyses. Comprehensive analyses bring together the most advanced models of the geometry, structure, dynamics, and aerodynamics available in rotary-wing technology, subject to the requirements for accuracy and the constraints of economy. These computer programs calculate rotorcraft performance and trim, blade motion and airloading, structural loads, vibration, noise, aeroelastic stability, and flight dynamics. The multidisciplinary nature of rotorcraft problems means that similar models are required for all of these jobs. A comprehensive analysis performs these calculations with a consistent, balanced, yet high level of technology. Because the tasks require a similar level of technology and similar models, they are best performed with a single tool. The development of computer programs for rotorcraft started with the alternative approach of developing multiple codes separately for individual disciplines, such as performance, dynamics, and handling qualities. Often the range of application of a particular analysis was restricted, perhaps to improve efficiency, but more often for historical reasons. Such experience with early codes provided solid evidence of the resulting inefficient use of development and application resources and inevitable disparities in treatment of the various problems.

There are several implications of the word "comprehensive" in rotorcraft aeromechanics, all encompassed by the ideal analysis. Comprehensive refers to the need for a single tool to perform all computations, for all operating conditions and all rotorcraft configurations, at all stages of the design process. The technology is comprehensive, covering all disciplines with a high technology level. The models are comprehensive, covering a wide range of problems, rotorcraft configurations, and rotor types and dealing with the entire aircraft. The analysis is readily adapted to new configurations and new designs. The software is comprehensive, with the flexibility to adapt or extend the codes to new problems and new models. The software is reliable and accurate, yet efficient and economical, characteristics achieved through correlation and verification. The software is built with good programming practices and extensive documentation, ensuring ease of test and maintenance. Helicopter problems are inherently complex and multidisciplinary, so helicopter analyses are always driven toward consideration of these "comprehensive" issues.

Design and development of rotorcraft require the capability to calculate rotor performance and maneuver loads. To provide such calculations, a comprehensive analysis has a rotor wake model, accounts for drag and stall and compressibility of the rotor aerodynamics, includes nonlinear dynamics and elasticity of the rotor blades and airframe, and models the entire aircraft. The entire aircraft in flight is analyzed, although often the code treats just the rotor. Calculating vibration, aeroelastic stability, and flight dynamics within the comprehensive analysis is best, but can be accomplished with separate codes. The aeromechanics of a rotor alone in a steady operating condition are certainly complicated, but the capability to analyze multiple rotors and maneuvers is very important.

The tasks of a comprehensive analysis consist of trim, transient, and flutter. The trim task considers steady-state, unaccelerated flight conditions, solving for the equilibrium (periodic) motion of the rotors and airframe. Steady-state conditions include level flight, climb, descent, and turns, as well as constrained operation, such as a rotor in a wind tunnel. From the solution for the periodic motion, the aircraft performance, mean and vibratory structural loads, vibration, and noise can be evaluated. The rotor motion can be calculated for a given control setting, but usually the operating state is specified in terms of speed, gross weight, and other parameters, not in terms of the control positions. So an inverse problem is solved: the control positions and aircraft orientation are found for a specified operating condition, typically by a Newton-Raphson iteration as the outer loop of the analysis. The controls and other variables are adjusted to achieve equilibrium of the net forces and moments on the aircraft or to drive quantities such as thrust and tip-path-plane tilt to target values. The periodic motion can be represented by Fourier series, the coefficients obtained by a harmonic balance or time-finite-element method; or the equations can be integrated in time, perhaps with periodic shooting to get the periodic motion. Regardless of the procedure adopted to solve the equations, the system is nonlinear as well as periodic for most rotor problems, and an iterative method is required, perhaps with multiple nested loops. Aerodynamic loads and wakes are not fundamentally represented by differential equations, although state space models (such as dynamic inflow) are useful. The integral equations of wake models introduce additional iteration and convergence issues. The fundamental assumption of the trim solution, that the motion is periodic, is not true for the most common helicopter configuration of a main rotor and tail rotor. Handling the main rotor and tail rotor configuration, for which the rotors have different periods, requires either a very long period or an approximate solution, neglecting interactions between the rotors at the wrong period, both through the airframe and through the wakes.

The transient task numerically integrates the equations of motion in time, starting from the trim solution. Prescribed control or gust inputs produce a non-equilibrium flight path or rotor motion. With an inverse solution method or an appropriate autopilot model, a specified maneuver can be analyzed.

The flutter task obtains linear differential equations for the system by numerically or analytically perturbing the equations of motion about the trim solution. Quasistatic reduction of the equations can be used, eliminating high-frequency modes of motion or even all rotor degrees of freedom, the latter resulting in a stability derivative model of the aircraft for flight dynamics investigations. Eigen-analysis of the linearized equations gives the system stability (from the eigenvalues), as well as frequency response and time-history response. For many rotor problems the linearized equations have periodic coefficients and must be analyzed using Floquet theory. Time-invariant equations obtained by averaging the coefficients (after introducing multiblade coordinates) are often sufficiently accurate and permit more extensive and more informative analysis. The aeroelastic stability can also be evaluated from the response obtained in a numerical integration of the equations of motion (the transient task). The disadvantages of this approach are that more computation is required and obtaining quantitative measures of the stability from the time history of the motion can be difficult. Unless the stability problem is inherently nonlinear, analysis of the linearized equations is more productive.

Many numerical analyses of rotor behavior using the computer have been developed, and their application in helicopter design, testing, and evaluation is now routine. These analyses have greatly expanded the knowledge of rotor behavior and improved the ability to predict it. Yet even with the most advanced models there remain deficiencies in the predictive capability for rotary wings, which can be traced to the magnitude of the problem of calculating rotor behavior and to a fundamental lack of understanding in a number of areas of rotor aerodynamics and dynamics.

Current comprehensive analyses are based on beam models of the blade structure, lifting-line models of the blade aerodynamics, and vortex wake models. The structural dynamics are modeled using finite elements and multibody dynamics, giving exact geometry and kinematics with the capability to represent arbitrary designs. The comprehensive analysis can be coupled with computational fluid dynamic (CFD) codes for better airload calculations. The code input allows description of arbitrary geometry and configurations.

A modern comprehensive analysis can analyze arbitrary configurations and rotor types, including novel and advanced designs, with arbitrary geometry, covering whatever the designer can invent. The system configuration is defined and changed by input to the analysis, so changing the code is not necessary as long as the required physics are available. The mathematical model allows nonlinearities (structural, aerodynamic, and kinematics) and arbitrary large motion, including rigid-body motions and large rotations of components relative to each other. Hence the code can model the true geometry of a rotorcraft, including multiple load-paths (such as a swashplate and control system, lag dampers, tension-torsion straps, and bearingless rotors) and unique configurations of the blade and hub. Separating the specification of the configuration, the aeromechanical model, and the solution procedure is essential for expandability of the analysis, for otherwise the smallest change involves the entire analysis and growth becomes increasingly harder as each new feature is added. This approach also leads naturally to more general and more rigorous models.

Comprehensive analyses have their origins in the programs developed as soon as digital computers first became available to engineers in the 1960s, because rotor design was severely limited by the simplifying assumptions required for practical analyses up to that time. Figure 22.1 identifies some major comprehensive analyses, with the developer and approximately the time the code was introduced. For a history of rotorcraft comprehensive analyses, see Johnson (2012).

From the beginning, the focus of these comprehensive analyses was on flight dynamics and structural loads, requiring models of the aerodynamics, dynamics, and structure of the entire aircraft. The codes of the 1970s are considered the first generation of helicopter comprehensive analyses. These powerful and useful tools exhibited a number of common limitations. Codes were restricted in the range of helicopter and rotor configurations and the range of analysis. New configurations required developing new equations of motion. Software development processes were poor, documentation was absent, and upgrades and maintenance were difficult. Recently available technology was utilized neither uniformly nor well.

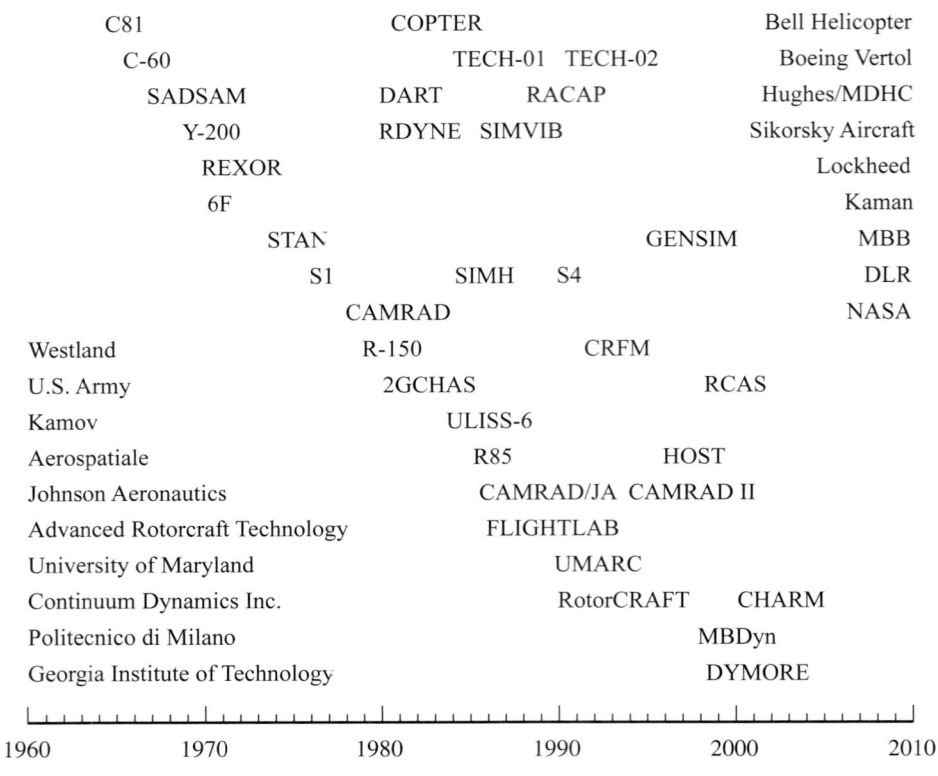

	C81		COPTER		Bell Helicopter
	C-60		TECH-01 TECH-02		Boeing Vertol
	SADSAM		DART RACAP		Hughes/MDHC
	Y-200		RDYNE SIMVIB		Sikorsky Aircraft
	REXOR				Lockheed
	6F				Kaman
		STAN		GENSIM	MBB
		S1	SIMH S4		DLR
		CAMRAD			NASA
Westland		R-150		CRFM	
U.S. Army		2GCHAS			RCAS
Kamov		ULISS-6			
Aerospatiale		R85		HOST	
Johnson Aeronautics		CAMRAD/JA CAMRAD II			
Advanced Rotorcraft Technology		FLIGHTLAB			
University of Maryland		UMARC			
Continuum Dynamics Inc.		RotorCRAFT		CHARM	
Politecnico di Milano		MBDyn			
Georgia Institute of Technology		DYMORE			

1960   1970   1980   1990   2000   2010

Figure 22.1. Comprehensive analyses.

Major new codes were developed beginning in the 1980s, often in direct response to the recognition of these limitations. The emphasis was on flexibility and versatility, through theory using assembly of primitive substructures, modular and structured software architecture, and modern software development methodology. In many cases the approach was to start by restructuring a first-generation code, with a subsequent reduction in cost and development time, simplified validation, and early achievement of the required capabilities.

Finite elements are needed to model the complexity of rotor structures, and finite element models were developed for rotor blade analysis in the early 1980s. Multibody dynamics technology is needed to model the mechanisms found in rotors. Finite element and multibody dynamics modeling capability, including input-driven definition of the geometry, was fully integrated into comprehensive analyses in the 1990s.

Significant attention was given in the 1990s and 2000s to the rotor aerodynamics, including wakes and wake geometry. Methods for coupling CFD codes with rotorcraft comprehensive analyses, with the latter handling all structural dynamic response and aircraft trim and trajectory calculations, matured in the 2000s. Versions of comprehensive analyses that can be executed in real time have been demonstrated. Current activities recognize there is still the need for reliable and efficient calculation of structural loads and vibration in the extremes of the aircraft operating capability.

The next generation of rotorcraft comprehensive analyses will be driven and enabled by the tremendous capabilities of high-performance computing, particularly modular and scalable software executed on multiple cores (Johnson and Datta

(2008)). To effectively use multiple cores, without bottlenecks at serial calculations, all elements of the analysis must be parallel and scalable. The code should scale down as well, allowing execution on smaller machines, at least with the appropriate choice of model options.

All of the tasks of a comprehensive analysis benefit from high-fidelity aeromechanics models for more accurate results. A hierarchy of models is needed, since lower fidelity models are usually faster and smaller and, within limits of their development, can sometimes be more accurate and more widely applicable. Moreover, efficient execution of the code and understanding the results demand use of the simplest possible models for each problem. All approximate or empirical models should have a path to first principles, as enabled in the future by hardware and algorithms.

Navier-Stokes solutions are required to calculate performance and handle drag, stall, and wake formation. A limitation continues to be turbulence modeling. Still, physical approximations can be useful. Euler solutions bring reduced computation, but require drag estimation from lower fidelity models, such as lifting-line theory and airfoil tables. Wake or potential flow models can be coupled with near-body Navier-Stokes solutions for cases when wake capture does not work or is too costly. The aerodynamic solution domain may have to encompass the acoustic far field as an alternative to using quadrupoles in the acoustics analysis.

Modeling blades, hub, and controls requires solutions of the exact structural dynamic equations with correct geometry, although still with small strain; multibody dynamics, for exact kinematics and joints of mechanisms; and three-dimensional structural models, including anisotropic and composite materials. Three-dimensional structural models are needed for correct modeling of coupling and load-paths and the non-beam-like parts of the system. They are needed for ends, short beams, open sections, transitions, and joints, because making beam models fit all parts of a rotor blade is always a problem.

As the interface with the design and analysis environment, a geometry engine is essential for productivity. The geometry engine delivers the descriptions required for the aerodynamic and structural analysis, including discretization, and must also provide communication and standards needed for design and optimization.

The flutter task, encompassing stability calculation and control design, requires state models and order reduction for both aerodynamics and structures. Nonlinear models are linearized about the trim solution, so the flutter and trim tasks must be accomplished in the same code.

High-performance computing offers the opportunity for tremendous expansion of rotorcraft analysis and design capability. Experience with current codes makes clear what the requirements are for the next generation of comprehensive analyses. As usual, rotorcraft calculations demand the widest possible integration of disciplines, a fact that makes comprehensive analyses challenging and keeps their development interesting.

22.1 REFERENCES

Johnson, W. "A History of Rotorcraft Comprehensive Analyses." NASA TP 2012-216012, April 2012.

Johnson, W., and Datta, A. "Requirements for Next Generation Comprehensive Analysis of Rotorcraft." American Helicopter Society Specialists' Conference on Aeromechanics, San Francisco, CA, January 2008.

Index

actuator disk, 40, 414
   axial flow, 430
   ellipsoidal coordinates, 424
   forward flight, 134, 430
   potential theory, 424
   vortex theory, 74, 414
advance ratio, 30, 154, 163
air resonance, 827
aircraft
   drag, 11, 291
   structural dynamics, 721
airfoil, 294
   criteria, 294
   drag, 298
airloads
   blade-vortex interaction, 536
   examples, 303, 336
   stall, 450
anti-torque, 281
articulated rotor, 5, 6, 156, 271
autogyro, 7, 279
   performance, 253
autorotation, 7, 98, 101, 285
   forward flight, 128
   index, 289
   parachute, 7, 106
   stall, 107
   vertical flight, 105

beam, 671
   finite elements, 686
   history, 703
   linear, 672
   nonlinear, 680
   structural loads, 697
beam equations, 676, 677, 697, 702
beam frequency, 680
bearingless rotor, 6, 274
bifilar, 727
blade element theory, 52, 308
   forward flight, 165

   history, 52
   hover, 54
blade element-momentum theory, 59, 81, 87
   induced velocity, 59
   tandem, 116
blade loading, 11, 31, 152, 283
   stall, 448
blade-vortex interaction, 345, 403, 535
   airloads, 536
   unsteady aerodynamics, 407
boundary layer, 485

climb
   forward flight, 128
   induced power, 43
   induced velocity, 43, 94
   momentum theory, 42
   power, 261
   vertical flight, 109
coaxial
   configuration, 278
   hover, 112
   momentum theory, 112, 138
collective, 159
complex variable representation
   flap, 783
   ground resonance, 821
compound helicopter
   configuration, 280
   performance, 252, 253
comprehensive analyses, 915
   history, 917
   next generation, 918
compressibility, 193, 283
   computational aerodynamics, 474
   unsteady aerodynamics, 390, 407
computational aerodynamics
   boundary element, 471
   CFD/CSD coupling, 483
   compressibility, 474
   history, 480
   hover, 482

computational aerodynamics (cont.)
   Navier-Stokes, 481
   potential theory, 462
   small disturbance, 477
   transonic, 477
configuration, 1, 6, 276
   coaxial, 278
   compound helicopter, 280
   lift-offset rotor, 280
   reaction drive, 278, 280
   side-by-side, 277
   single main rotor, 7, 276
   synchropter, 278
   tandem, 7, 277
   tiltrotor, 280
   twin main rotor, 7, 277
coning, 157, 181, 196
constant coefficient approximation, 557
   flap equation, 647, 755
control, 278, 844
coupling
   extension-torsion, 694
   flap-lag, 602
   Oehmichen, 782
   pitch-flap, 215, 610, 782, 805
   pitch-lag, 610, 807
cyclic, 159

delta-3, 215, 218
descent
   forward flight, 128
   induced velocity, 94
   power, 261
design trends, 8, 14
differential
   momentum theory, 77
dimensions, 27
disk loading, 1, 9, 31
download, 117
drag
   aircraft, 11, 291
   airfoil, 298
drag force, 168, 171, 250
ducted fan
   momentum theory, 141
dynamic inflow, 432, 663
   flap response, 773
   ground resonance, 827
   history, 436
   hub loads, 775
dynamic stall, 443
   Boeing, 458
   empirical models, 457
   Leishman-Beddoes, 458
   ONERA, 459

equations of motion
   derivation, 584
equivalence of flap and feather, 162
examples
   airloads, 200, 303, 336

   angle-of-attack, 199, 336
   flight dynamics, 864, 873, 876
   forward flight, 196, 336
   hover, 68, 87
   induced velocity, 336
   lateral flapping, 338
   nonuniform inflow, 336
   twist, 200, 336

fan-in-fin, 282
figure of merit, 8, 44, 59, 245, 266
finite elements
   beam, 686
   history, 705
flap, 5, 29, 153, 156, 157, 178, 590, 749
   complex variable representation, 783
   free wake geometry, 338
   frequency, 217
   high speed, 200, 752
   hinge offset, 208
   hinge spring, 201
   response phase, 206, 211, 217
   reverse flow, 189
   second harmonic, 186
   stability, 752
   structural loads, 597
   two-bladed rotor, 784
   weight moment, 192
flap bending, 592
flap eigenvalues
   forward flight, 752
   hover, 750
   non-rotating frame, 760
flap equation, 180, 591, 749
   aerodynamics, 639
   bending, 594
   constant coefficient approximation, 647, 755
   gimballed rotor, 214
   hinge offset, 210
   hinge spring, 205
   non-rotating frame, 596, 644, 759
   teetering rotor, 215
   with lag, 600
   with pitch, 607
   with torsion, 615
flap frequency, 179, 275, 591, 711
   bending, 594
   gimballed rotor, 214
   hinge offset, 210
   hinge spring, 205
   hingeless rotor, 212
   pitch-flap coupling, 217
   teetering rotor, 215
flap response
   dynamic inflow, 773
   hover, 758, 761
   lift deficiency function, 774
   low frequency, 766
flap-lag, 599
   stability, 797
   stall, 808

flap-lag bending, 604
   equations, 606
flap-lag equations, 600, 601, 798
   aerodynamics, 642, 660
   high inflow aerodynamics, 664
   non-rotating frame, 630
   shaft motion, 629, 656
flap-lag-torsion equations, 702
flight dynamics, 844
   control gyro, 895
   directional, 860
   examples, 864, 873, 876
   forward flight, 878
   hingeless rotor, 895
   hover, 853
   lateral, 873
   lateral-directional, 886
   longitudinal, 861, 880
   longitudinal-lateral, 874
   loop closure, 867
   short period, 883
   side-by-side, 894
   stability derivatives, 856, 879
   static stability, 888
   tandem, 890
   tiltrotor, 894
   twin rotor, 889
   vertical flight, 859
flight dynamics equations, 846, 849
   side-by-side, 852
   single main rotor, 850
   tandem, 851
flight dynamics specifications, 901
Floquet theory, 568
flow states, 96
flying qualities, 844
flying qualities specifications, 901
   ADS-33, 909
   bandwidth, 907
   MIL-H-8501A, 902
forward flight, 7, 152
   actuator disk, 134
   autorotation, 128
   blade element theory, 165
   climb, 128
   descent, 128
   examples, 196, 336
   flight dynamics, 878
   ground effect, 144
   ground vortex, 145
   induced power, 125, 127, 131, 176
   induced velocity, 125, 136
   momentum theory, 123, 137
   noise, 512, 514
   power, 173, 246, 259
   profile power, 176, 230
   span loading, 131
   twin rotor, 137
   vortex theory, 133, 423
   wake, 123
   wake geometry, 133, 327

Fourier series, 155, 545
frequency
   beam, 680
   blade passage, 493, 720
   fan plot, 711
   flap, 179, 205, 210, 212, 214, 215, 217, 591, 711
   flap bending, 594
   lag, 221, 601, 711
   lag bending, 604
   modal, 710
   pitch, 609
   Sturm-Liouville theory, 582
   torsion, 616

gimballed rotor, 213, 274
   flap equation, 214
   flap frequency, 214
ground effect
   forward flight, 144
   hover, 119
ground resonance, 156, 810
   complex variable representation, 821
   damping requirement, 818
   dynamic inflow, 827
   equations, 811
   history, 828
   no damping, 813
   two-bladed rotor, 822
ground vortex
   forward flight, 145

handling qualities, 844
   rating, 906
harmonic analysis, 548
height-velocity diagram, 286
helicopter
   performance, 251
high inflow aerodynamics
   flap-lag equations, 664
   hub loads, 664
high speed
   flap, 200
   flap stability, 752
   thrust, 200
higher harmonic control, 730
   effectiveness, 740
   frequency domain, 736
   identification, 732
   time domain, 737
hingeless rotor, 5, 6, 157, 212, 274
   flap frequency, 212
   flight dynamics, 895
history
   aeromechanics, 237
   autogyro, 19
   beam, 703
   blade element theory, 52
   comprehensive analyses, 917
   computational aerodynamics, 480
   dynamic inflow, 436
   finite elements, 705

history *(cont.)*
   free wake geometry, 361
   ground resonance, 828
   helicopter invention, 14
   multiblade coordinates, 560
   nonuniform inflow, 359
   vortex theory, 72
hover, 7, 39
   blade element theory, 54
   coaxial, 112
   computational aerodynamics, 482
   examples, 68, 87
   flap response, 758, 761
   flight dynamics, 853
   free wake geometry, 87
   ground effect, 119
   induced power, 42, 58
   induced velocity, 41, 57, 69
   momentum theory, 40
   noise, 505
   nonuniform inflow, 83
   performance, 60, 68, 87
   power, 43, 245, 254
   profile power, 58
   thrust, 57
   twin rotor, 111
   unsteady aerodynamics, 391
   vortex theory, 71
   wake, 73
   wake geometry, 84, 327
hub, 3
hub loads, 619
   aerodynamics, 649, 660
   dynamic inflow, 775
   high inflow aerodynamics, 664
   lift deficiency function, 771, 778
   low frequency, 769, 853
   non-rotating frame, 624, 653
   shaft motion, 630
   two-bladed rotor, 627, 654, 659
hub moment, 169
   hinge offset, 212
   hinge spring, 207

ideal rotor, 66, 68, 87
individual blade control, 730
   effectiveness, 740
induced power
   climb, 43
   factor, 51, 58, 63, 64, 68, 127, 132, 176, 269
   forward flight, 125, 127, 131, 176
   free wake geometry, 329
   hover, 42, 58
   nonuniform inflow, 63
induced velocity
   blade element-momentum theory, 59
   climb, 43, 94
   descent, 94
   forward flight, 125, 136
   hover, 41, 57, 69
   linear, 184

   turbulent wake state, 101
   vortex ring state, 101
   vortex theory, 76
inflow ratio, 30, 163
interference, 146
   hub-tail, 148
   rotor-airframe, 146
   rotor-tail, 147
   tail rotor, 149
   twin rotor, 111, 137

lag, 5, 29, 156, 158, 220
lag bending, 603
lag damper, 156, 740
lag equation, 221
   bending, 603
   with flap, 601
lag frequency, 221, 601, 711
   bending, 604
lift deficiency function
   flap response, 774
   hub loads, 771, 778
   Loewy, 394
   Theodorsen, 372, 378
   time-varying velocity, 385
   vortex theory, 422
lift force, 250
lift-offset rotor
   configuration, 280
lift-to-drag ratio, 11, 267
lifting-line theory, 307
   rotary wing, 311
   second order, 309, 312
   unsteady aerodynamics, 376
lifting-surface theory, 466
linear equations, 564
   constant, 566
   eigenvalues, 570
   Floquet theory, 568
   periodic, 568, 753
Lock number, 28, 179, 275
low frequency
   flap response, 766
   hub loads, 769, 853

mean lift coefficient, 64
momentum theory, 39, 45, 92
   climb, 42
   coaxial, 112, 138
   differential, 48, 59, 77
   ducted fan, 141
   forward flight, 123, 137
   hover, 40
   side-by-side, 140
   swirl, 48
   tandem, 115, 138, 140
   vertical flight, 93
multiblade coordinates, 549
   degrees of freedom, 549
   equations of motion, 553
   history, 560

reactionless mode, 557
two-bladed rotor, 557

no-feathering plane, 160, 162
noise, 493
   blade-vortex interaction, 535
   blade-vortex interacton, 495
   broadband, 493, 499, 532
   certification, 540
   Doppler shift, 516
   far field, 527
   Farassat, 526
   Ffowcs Williams-Hawkings, 519
   forward flight, 512, 514
   high speed impulsive, 495, 536
   hover, 505
   impulsive, 495, 535
   Kirchhoff, 522
   Lighthill, 518
   quadrupole, 526
   rotational, 494, 502
   sound pressure level, 498
   spectrum, 496
   thickness, 516
non-rotating frame
   flap equation, 596, 644, 759
   flap-lag equations, 630
   hub loads, 624, 653
   pitch equation, 618
nonuniform inflow, 186, 318
   dual-peak loading, 322
   examples, 336
   history, 359
   hover, 83
   induced power, 63
normal working state, 96

optimum rotor, 67, 68, 87
optimum windmill, 111

performance, 243
   autogyro, 253
   ceiling, 263
   charts, 253
   climb and descent, 261
   compound helicopter, 252, 253
   drag/lift, 248
   endurance, 264
   energy balance, 244
   force balance, 243
   forward flight, 259
   helicopter, 251
   hover, 60, 68, 87, 254
   metrics, 266
   power/thrust, 250
   range, 264
   referred, 266
   rotorcraft, 251
   scale with solidity, 61
   single main rotor, 252
   speed, 262

   tail rotor, 252
   taper, 87
   tiltrotor, 252
   twist, 87
pitch, 6, 29, 159
pitch equation
   non-rotating frame, 618
   with bending, 617
   with flap, 609
pitch frequency, 609
   torsion, 616
pitch-flap, 606
   aerodynamics, 666
   elastic, 613
   stability, 788
   wake, 795
pitch-flap coupling
   flap frequency, 217
   tail rotor, 218
pitch-flap equations, 607, 609, 788
   aerodynamics, 667
   elastic, 615, 617
power
   climb, 176, 261
   descent, 261
   forward flight, 173, 246, 259
   hover, 43, 245, 254
   induced, 1, 8, 42, 43, 58, 63, 92, 98, 125, 127, 131, 176, 186
   parasite, 176
   profile, 58, 176, 230
   vertical flight, 245
   wake geometry, 329
power loading, 9, 31
power-off landing, 7, 285, 288
precone, 205
profile power, 230
   forward flight, 176, 230
   hover, 58
   mean drag coefficient, 58, 232, 269

radial drag, 226
reverse flow, 154, 188
   flap, 189
   unsteady aerodynamics, 380
Reynolds number, 194
root cutout, 64, 195
rotorcraft
   performance, 251

section aerodynamics, 54, 165, 633
   autorotation, 106
shaft motion, 628
   aerodynamics, 655
   two-bladed rotor, 632
side force, 168, 171
side-by-side
   configuration, 277
   flight dynamics, 894
   flight dynamics equations, 852
   momentum theory, 140

single main rotor
    configuration, 7, 276
    flight dynamics equations, 850
    performance, 252
solidity, 11, 28, 56
    equivalent, 65
solution methods, 573
speed
    limitations, 283
    maximum, 262
stability
    flap-lag, 797
    pitch-flap, 788
    whirl flutter, 832
stall, 283, 442
    airloads, 450
    autorotation, 107
    blade loading, 448
    control loads, 442, 450
    criteria, 451
    flap-lag stability, 808
    limit, 448, 454
    rotor, 448
    static delay, 487
structural dynamics
    aircraft, 721
structural loads, 715
    beam, 697
    flap, 597
sum of harmonics, 547
swashplate, 159

tail design, 147
tail rotor, 149, 219, 276, 281
    fan-in-fin, 282
    performance, 252
    pitch-flap coupling, 218
takeoff, 290
tandem
    blade element-momentum theory, 116
    configuration, 277
    flight dynamics, 890
    flight dynamics equations, 851
    momentum theory, 115, 138, 140
taper, 87
teetering rotor, 6, 214, 271
    flap equation, 215
    flap frequency, 215
thrust, 168, 171, 195, 196, 209
    high speed, 200
    hover, 57
tiltrotor
    configuration, 280
    flight dynamics, 894
    performance, 252
    whirl flutter, 832, 836
tip loading
    vortex theory, 79
tip loss, 61, 81, 195
tip shape, 90

tip vortex, 306
    core, 339
tip-path plane, 158, 162
    tilt, 181, 196
track and balance, 721
trailing-edge flap, 740
trim, 176, 223
turbulent wake state, 98, 101
    induced velocity, 101
twin rotor
    flight dynamics, 889
    forward flight, 137
    hover, 111
    interference, 111, 137
twist, 87
    ideal, 66
two-bladed rotor
    flap, 784
    ground resonance, 822
    hub loads, 627, 654, 659
    shaft motion, 632

unsteady aerodynamics, 366
    airfoil, 366
    blade-vortex interaction, 407
    compressibility, 390, 407
    hover, 391
    Leishman-Beddoes, 390
    lifting-line theory, 376
    ONERA, 390
    reverse flow, 380
    rotary wing, 387
    Sears, 375
    Theodorsen, 372
    time-varying velocity, 382
    trailing-edge flap, 381
    vortex theory, 419

vertical drag, 117
vertical flight
    autorotation, 105
    climb, 109
    flight dynamics, 859
    induced power, 98
    momentum theory, 93
    power, 245
vibration, 717
    bifilar absorber, 727
    isolation, 725
    pendulum absorber, 726
    reduction, 723
    requirements, 722
vortex elements
    arc, 356
    line, 351
    sheet, 353
vortex ring state, 96, 101, 102
    induced velocity, 101
vortex theory, 77
    actuator disk, 74
    axial flow, 414

forward flight, 133, 423
history, 72
hover, 71
induced velocity, 76
lift deficiency function, 422
tip loading, 79
unsteady, 419

wake, 303
consolidation, 324
curved elements, 325
discretized, 319
dual-peak loading, 322, 324
forward flight, 123
hover, 73
hover visualization, 306
optimum, 78
shed vorticity, 303, 376
tip vortex, 306

trailed vorticity, 303
vortex core, 339
vortex elements, 351
wake geometry, 326
forward flight, 133, 327
free, 87, 328, 330, 331
history, 361
hover, 84, 327
power, 329
prescribed, 84, 330, 331
rigid, 330
stability, 335
whirl flutter, 832
equations, 832
propeller, 834
tiltrotor, 836
windmill brake state, 98

yawed flow, 226